T0137220

Studies in Systems, Decision and Control

Volume 142

Series editor

Janusz Kacprzyk, Polish Academy of Sciences, Warsaw, Poland
e-mail: kacprzyk@ibspan.waw.pl

The series "Studies in Systems, Decision and Control" (SSDC) covers both new developments and advances, as well as the state of the art, in the various areas of broadly perceived systems, decision making and control- quickly, up to date and with a high quality. The intent is to cover the theory, applications, and perspectives on the state of the art and future developments relevant to systems, decision making, control, complex processes and related areas, as embedded in the fields of engineering, computer science, physics, economics, social and life sciences, as well as the paradigms and methodologies behind them. The series contains monographs, textbooks, lecture notes and edited volumes in systems, decision making and control spanning the areas of Cyber-Physical Systems, Autonomous Systems, Sensor Networks, Control Systems, Energy Systems, Automotive Systems, Biological Systems, Vehicular Networking and Connected Vehicles, Aerospace Systems, Automation, Manufacturing, Smart Grids, Nonlinear Systems, Power Systems, Robotics, Social Systems, Economic Systems and other. Of particular value to both the contributors and the readership are the short publication timeframe and the world-wide distribution and exposure which enable both a wide and rapid dissemination of research output.

More information about this series at http://www.springer.com/series/13304

Eduardo Gil · Eva Gil · Juan Gil
María Ángeles Gil
Editors

The Mathematics
of the Uncertain

A Tribute to Pedro Gil

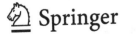 Springer

Editors
Eduardo Gil
Liberbank
Oviedo
Spain

Eva Gil
Viajes El Corte Inglés
Oviedo
Spain

Juan Gil
Funl Media
Oxford
UK

María Ángeles Gil
Departamento de Estadística e Investigación
  Operativa y Didáctica de la Matemática
Facultad de Ciencias
Universidad de Oviedo
Oviedo
Spain

ISSN 2198-4182   ISSN 2198-4190   (electronic)
Studies in Systems, Decision and Control
ISBN 978-3-030-08869-9   ISBN 978-3-319-73848-2   (eBook)
https://doi.org/10.1007/978-3-319-73848-2

Printed on acid-free paper

This Springer imprint is published by Springer Nature
The registered company is Springer International Publishing AG
The registered company address is: Gewerbestrasse 11, 6330 Cham, Switzerland

Pedro Gil in the homage paid to him by the University of Oviedo on the occasion
of his retirement (November 2010)

To Pilar,
Pedro Gil's beloved wife and inspiration

A Pilar,
querida compañera de vida e inspiración de
Pedro

Foreword

Good teacher, good scholar, and all round good person, Pedro Gil is, in all facets of his life, a role model to uphold and follow. Therefore, it is an honour for me to write the foreword to this magnificent work that I can only describe as essential given the quality of person to whom it pays tribute. It is also right and necessary for me to write it, and it is so for a number of reasons. Significantly, because I have the good fortune to be the rector of the University of Oviedo I am thus able to endorse it in the name of an institution in which Pedro worked for 34 prolific years. Moreover, it is emotional for me because I am also that student who was lucky enough to have him as a teacher at a crucial time in his career, when I had everything still to decide and, thanks to teachers like him, ended up taking the right decisions.

I could personally bear testimony to Pedro's extraordinary worth shortly after starting my university studies. During the 1976–1977 academic year, I was a second-year student at the University of Oviedo earning a bachelor's degree in Chemistry. I had at the time the feeling, that I was steadily becoming certain about, that my vocation was in the discipline of physical chemistry, a field closely related to Mathematics, the area of knowledge to which Pedro dedicated his academic career. He was at the time a young assistant professor who had just arrived from the Complutense University of Madrid, along with his wife and daughter. I am certain that it was not by chance that he joined this institution but rather thanks to some formula, one of those so extraordinary that seem to be random but that in fact can be resolved with dedication and talent. This institution will never show enough gratitude for having had Pedro as one of its most outstanding members.

My year, the one that greeted him, had a particularly large number of students, given that it coexisted with the last units of the previous chemistry syllabus in the Faculty of Sciences. Fearful as we were of that second year, reputedly the most demanding of the degree, we met a teacher much younger than any other we had had until then. He managed from the first day an unusual complicity with us and he was, even though always serious, friendly, and approachable, and all the while capable of rendering intelligible a subject as dense and obscure as Mathematics. To say that among his students he was known as Father Pedro or Saint Pedro gives a good idea of the appreciation that the students had of him and the great love and respect with which he is remembered.

Pedro Gil was that teacher who, scrupulously respectful, waited in the hall to start the class, leaving us five or ten minutes for delegated faculty members to report on the corresponding actions. Years later, sharing different forums on university management, he would also be one of the most conciliatory and reasonable interlocutors I have ever met.

Good friend, resolute scholar, loyal to the institution, and also to his mentors and role models whom he always acknowledged paying them a just tribute, Pedro was a man of wholesome character who made a significant contribution to the progress of this university that misses him so much. He is in Pilar, his life partner; in his children Eva, Juan, and Eduardo, and in his brothers and sisters. Her sister, María Ángeles, who is committed to keeping the memory of Pedro alive and vindicating the intelligent scholar who, with the mind of a statistician, anticipated many years ago the growing interest in the development of methods for the mathematical treatment of uncertainty. That is why his inaugural lecture of the 1996–1997 academic year at the University of Oviedo bore the title that has now also been adopted for this book in his memory, in my opinion, with great success.

This work, which includes contributions from many of Pedro's fellow mathematicians, disciples, and friends, is a good test of how uncertainty can come from different sources and how it can be modelled and studied with different tools and from different mathematical fields. But above all else is the realization that, as he asked in that magnificent speech, those of us who follow him should do so by learning from the past while looking to the future, firmly supported by the great legacy that he left us and that he was so humble about. And for this, as he asked, we must play with intelligence the part that corresponds to the university and that is none other than to prepare the men and women who must bear to society the fruit it needs for its development. Fruit that measure as certainties; seeds that Pedro sowed with humility and greatness.

Buen docente, buen investigador y hombre bueno, Pedro Gil es, en todas las facetas de su vida, un modelo a reivindicar y seguir. Por eso es un honor para mí realizar la introducción a esta magnífica obra que me atrevo a calificar de imprescindible, al tenor de la persona a quien rinde tributo y porque es justa y es necesaria. Lo es por muchos motivos. El primero, porque tengo la fortuna de firmarlo como rector de la Universidad de Oviedo, institución en la que Pedro trabajó durante treinta y cuatro fructíferos años. Pero lo hago además emocionado como estudiante que disfrutó la suerte de recibir su magisterio en un momento crucial de mi carrera, cuando todo estaba por decidir y que, gracias a encontrar docentes como él, terminó por decidirse de la mejor manera.

Personalmente, pude comprobar la extraordinaria valía de Pedro al poco de comenzar mis estudios universitarios. En el curso académico 1976–1977, yo era un estudiante de segundo curso de la Licenciatura de Química de la Universidad de Oviedo y mi intuición iba asentándose como la certeza de que mi vocación me encaminaba hacia la Química Física, un campo con muchas conexiones con las Matemáticas a las que Pedro dedicó su gran conocimiento. Él por entonces era un joven profesor agregado que acababa de llegar de la Universidad Complutense de Madrid, junto con su mujer y su hija mayor, gracias a que alguna fórmula, de esas tan milagrosas que parecen azar pero que sólo se resuelven con dedicación y talento, le puso en el camino de esta institución, que nunca mostrará el suficiente agradecimiento por haber contado con Pedro como uno de sus miembros más destacados.

La promoción a la que pertenezco y que le recibió a su llegada, era especialmente numerosa ya que coincidía un cambio de plan de estudios que hizo que conviviésemos con las últimas unidades del anterior Plan de Química en la Facultad de Ciencias. Temerosos como estábamos de que llegase el segundo curso, con fama de ser el más exigente de la licenciatura, nos encontramos con un profesor mucho más joven que los que habíamos tenido hasta entonces, el cual estableció desde el primer día una complicidad insólita con nosotros y que, desde la seriedad, era afable, cercano y conseguía que una asignatura en principio densa y oscura como son las Matemáticas resultase comprensible. Decir que para muchos de sus alumnos y alumnas era el Padre Pedro o San Pedro da buena idea de la percepción que el estudiantado tenía de él y el gran cariño y respeto con los que se le recuerda.

Pedro Gil fue el profesor que, escrupulosamente respetuoso, esperaba en el pasillo a comenzar la clase dejándonos cinco o diez minutos para que las personas delegadas de facultad informasen de las acciones reivindicativas que correspondiesen. Años más tarde, compartiendo foros diversos sobre gestión universitaria, sería también uno de los interlocutores más conciliadores y razonables que jamás he conocido. Buen amigo, investigador resuelto, leal a la institución y a sus referentes a quienes siempre reconocía rindiéndoles un justo tributo, fue Pedro un hombre de personalidad completa que hizo mucho por el progreso de esta Universidad que tanto le echa en falta. Está en Pilar, su compañera de vida; en sus hijos Eva, Juan y Eduardo, y en sus hermanos. Su hermana, María Ángeles, quien constantemente se encarga de mantener vivo el recuerdo de Pedro y de reivindicar la memoria del sabio inteligente que, con la visión propia del estadístico, anticipó hace muchos años el interés creciente que tendría el desarrollo de métodos para el tratamiento matemático de la incertidumbre. Por eso su conferencia para la inauguración del curso 1996-1997 en la Universidad de Oviedo llevó el título que ahora se ha adoptado también para este libro en su memoria, en mi opinión, con gran acierto.

Esta obra, que incluye contribuciones de muchos matemáticos compañeros, discípulos y amigos de Pedro, es una buena prueba de cómo la incertidumbre puede provenir de distintas fuentes y cómo puede modelarse y estudiarse con diferentes herramientas y desde diferentes campos matemáticos. Pero sobre todo es la constatación de que, como él pidió en aquel magnífico discurso, quienes le seguimos lo hacemos mirando al pasado a la vez que dirigimos nuestras expectativas hacia el horizonte futuro, apoyados firmemente en el gran legado que nos dejó y que él siempre valoró con modestia. Y para ello, como pedía, debemos jugar con inteligencia la baza que le corresponde a la universidad y que no es otra que preparar a los hombres y mujeres que deben rendir a la sociedad los frutos que necesita para su desarrollo. Frutos que medren como certezas; semillas que Pedro sembró con humildad y grandeza.

Oviedo, Spain
November 2017

Santiago García Granda
The Rector of the University of Oviedo

Preface

Sucede que recuerdo
tu boina gris,
tu voz que ahora descansa
junto a los barcos,
tu forma de calmar
las tempestades,
la calidez de tu mano,
tu oído siempre atento.
Y ahora hay una puerta
cerrada
por la que no cabe mi voz.
Y demasiadas preguntas
huérfanas de toda lógica.
Me faltan los números
y tú te difuminas.
Sucede que no llegas.

Eva Gil Sanmamed, November 2017

This scientific book is in response to a need we felt to pay tribute to our beloved biological and scientific father Pedro Gil. And for this endeavour we wished this tribute to be paid along with many of the colleagues and friends Pedro met in his career.

In the first part of the preface we are not attempting to highlight his professional achievements, which will be described to a certain detail at the end of the book, but to offer a rather personal view of Pedro.

When each of us met Pedro for the very first time, we did not realize about him being a reputed academician. It might be precisely the first remarkable fact that can be noticed about Pedro: his reputation came without being noticed. Just as if he had no interest to underline it, he treated his circle as equals, never imposing but suggesting his opinions when he was asked.

As an outstanding mathematician, **reason** was above feelings. Beside his hidden smile, his bright eyes granted the serenity, the kind and sincere manners that are essential to find a solution. But prestige did came by; and, as a matter of fact, it usually comes for those who are able to connect their way of being, their dreams and defeats, their personal brilliance to their working fields. And let us state that, from what the authors know about Pedro's personal life, we can assure Pedro was a genius.

Pedro put his family and friends before the rest of their "lives". He was a smart person, smart in many ways, smart enough to always give a thought on what he was told before showing an answer, before making a judgment.

The second remarkable virtue of Pedro's nature might be his **willpower**. Indeed, it made him overtake several obstacles on his personal life. Always helped by his personal circle, from which his beloved wife Pilar played a preeminent role, he managed not only to recover but to teach us a colossal truth: it can be done.

Last, but not least, we cannot forget about what brought us here, the most remarkable fact about his life and scientific work: his **love for teaching**. You may have noticed that a verb seems to connect all previous paragraphs: teaching. Ahead of a researcher, and he was a remarkable one, Pedro Gil was a teacher, a professor, a maestro who enjoyed teaching above all, helping to develop the future's minds.

Let us emphasize that it is not only about students where his legacy ended, but we would like to include some of our feelings:

"My vision on my father is obviously biased. I have spent my whole life as a learner, from the best teacher in life I could have ever found. I cannot attach a paper to this book as a tribute, there are quite some other "children" that will do so as you can notice from the book's size and the expected crowded auditorium when it is presented. My homage, dad, will be trying to bring to my life your genius about all you taught me."

"Pedro was my supervisor, my boss, and my guidance in many respects, both personal and professional. When Pedro arrived to Asturias, as a twenty eight-year-old "senior" professor to head the Department of Mathematics of the Faculty of Sciences of the University of Oviedo, I joined the germ of his professional team. Since then, he has guided all my professional steps, trying to instil in me (actually in all his disciples) what he considered to be the best way to proceed: to love teaching and to take care of all students, and not only about the most brilliant ones; to love researching, since it is crucial for a university career, and to encourage and support all the young members of the Department in their careers.

And I have witnessed the first row of how he built with much effort and personal commitment the current team of the Department of Statistics, OR and Math Teaching at the University of Oviedo. In addition to the already pointed out Pedro's skills, as leader of that team one should highlight his generosity in giving sound advices and encouraging us to become leaders of research groups, his strong support to all the initiatives we undertake, and him having shared all the achievements from his own efforts with us."

To shortly summarize Pedro's scientific path, one can look at his scientific genealogy skeleton from above, which has been built on the basis of the information gathered in the Mathematics Genealogy Project (http://www.genealogy.ams.org/).

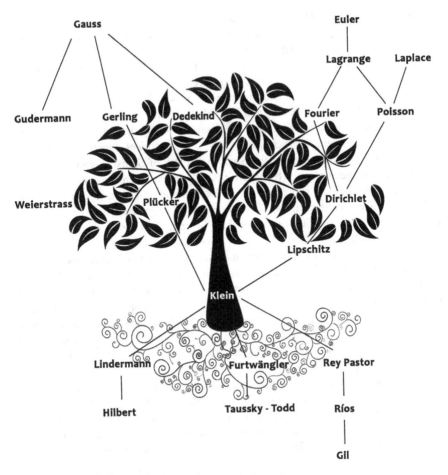

Pedro Gil's genealogy skeleton from above (*design* Eva Gil)

We consider there is no need for comments to be added about his magnificent "scientific pedigree".

On the other hand, also on the basis of the information from this genealogy project, updated as of late November 2017, Pedro Gil had 20 Ph.D. students and 81 Ph.D. linear descendants. More than half descendants are working at the University of Oviedo.

It should be remarked that when Pedro arrived to Oviedo, he started with a quite small team where members have neither research nor teaching expertise. Consequently, he should not simply lead that team, but guide it in both aspects. And this complex task required him to invest an immense capacity and a very hard work. And he did and, in spite of his youth, he succeeded in this endeavour, to get such a big scientific offspring.

As a natural consequence from all this, Pedro has unequivocally had a strong influence on many people. This influence has especially affected colleagues and students either in choosing the B.Sc. in Math, or in following an academical career, or in leading groups, and so on. In fact, one of Pedro's most recognized virtues is the one associated with his counselling, irrespectively of the framework, the problem, and the person who is asking for the advice. Pedro was always ready to provide with "*his wise advices*", and his colleagues have had the opportunity to enjoy his presence even after retiring. He enjoyed visiting them at the Department, and also joining them for activities such as Department lunches or dinners, attending all the Conferences of the Spanish Society of Statistics and OR he had presided (always accompanied by his wife Pilar, who has been considered to a great extent as a "member and supporter" of the Department), and even sharing teaching with one of his scientific children in a course on *Mathematics for everyday life*, a course he delivered for elder people of the University of Oviedo just a few weeks prior to his death.

Even regarding his closest biological relatives, Pedro's influence was clear: at the "horizontal" level, one brother and one sister have got their B.Sc. in Mathematics after Pedro getting his own, and the same happened with two of Pedro's brothers in law; at the "vertical" level, his two sons are mathematicians; and at the "oblique" level, a niece and a nephew are mathematicians too. Most of them were also specialized in Statistics + Operations Research. Pedro's enthusiasm and eagerness for his work were certainly contagious, and most of us could not draw away from them.

Since Pedro left us, we are happy and proud to confirm a deterministic fact that he fortunately knew: he has been (actually, he is) a very beloved person. And we all are permanently missing him. To illustrate this assertion along with his scientific influence we can consider a small sample of sentences from the contributions in this book, although readers are invited to have a look at the whole "population":

"*...Pedro was always a reference both in a human perspective as well as in the scientific field. His capacity for team building taking into account every one beside him and his generosity makes him a good example of what a scientist should be...*",

"*...Fleing from imposition and taxation, Pedro has taught his many disciples the need to devote a significant part of their working time to conduct forefront research. This vision and policy, which can seem to be obvious nowadays, was especially laudable at the time he created the embryo of his current university department*",

"*...he was always very approachable and helpful towards his friends and colleagues, as well as being well grounded in everyday reality. It was enough to simply call him any time you needed his valuable help. Pedro, what do you say if we...? Pedro, what do you think about...? ... he always got a kind look, close attention,... and a quick, sensitive answer, full of wisdom and affection...*",

"*... you passed away when your presence was more necessary than ever...*",

"*...in your absence we will find ourselves often wondering what quiet advice would you give us at some given situation, just as if we were sitting in front of you under the lost look of the punch-drunk boxer of the picture hanging on the wall of your office...*".

This book is a collection of contributions authored by colleagues, students/descendants, and friends of Pedro who have either developed methods to solve mathematical problems involving uncertainty or have applied such methods to real-life cases. Since the modelling and the management of uncertainty are more and more challenging and appealing topics, and Mathematics offer suitable tools in this respect, nowadays there is an increasing interest on studies like those gathered in the book.

As Pedro highlighted in the inaugural lecture of the academic year 1996–1997 in the University of Oviedo (**Part I** of this book), there are three well-known branches of the "Mathematics of the Uncertain", namely

- the Mathematics of Chance (Probability and Statistics),
- the Mathematics of Communication (Information Theory), and
- the Mathematics of Imprecision (Fuzzy Sets Theory and others).

These three branches often intertwine, since different sources of uncertainty can coexist, and they are not exhaustive. That is, they do not constitute a classical partition of the Mathematics of the Uncertain.

For these reasons, papers in the book have been classified into four classes, one per well-known branches (**Parts II–IV** in the book), and an additional one (**Part V** in the book) including papers in other mathematical fields which are concerned, to a greater or lesser extent, with uncertainty. Furthermore, the assessment of papers to **Parts II–V**, and especially to **Parts II–IV**, has often been "infected" with uncertainty, because some of the papers could be probably included in two different ones.

The book ends (**Part VI**) with some biographical sketches of Pedro Gil in connection with his professional career, both at the Complutense University of Madrid and the University of Oviedo, and one of his most outstanding contributions: his crucial and essential role in launching the Bachelor of Mathematics in the University of Oviedo.

To end this edited multiauthors book, we should deeply thank to all those contributing it. We know well how much affection for Pedro is involved in all the papers in it, as well as in all the members of his Department. We must express our most special gratitude to Asun Lubiano and Antonia Salas, two of Pedro's linear scientific descendants, for their extraordinarily meticulous proofreading of the whole book: one more proof of their enormous affection for Pedro.

Oviedo, Spain   Eduardo Gil
Oviedo, Spain   Eva Gil
Oxford, UK   Juan Gil
Oviedo, Spain   María Ángeles Gil
November 2017

Contents

Part I
Pedro Gil's Vision on the Mathematics of the Uncertain

The Mathematics of the Uncertain

Pedro Gil

> *Mathematics is the language in which God has written the universe*
>
> (Galileo Galilei)

Abstract This paper corresponds to the English translation of a shortened version of the Inaugural Lecture of the academical year 1996–1997 in the University of Oviedo, and published as Pedro Gil Álvarez (1996) Las matemáticas de lo incierto. Rev Astur Econ 7:203–219. It explains to a certain extent Pedro Gil's view on the mathematical tools and approaches to model, handle and deal with uncertainty.

When I received the gratifying invitation to give this inaugural lecture, I recalled that many years ago (in 1913, to be precise), and in this same University, it had been the father, or godfather, of Spanish and Latin American mathematicians of the twentieth century, Julio Rey Pastor, who had found himself in such an endeavour. So off I hurried to our beloved library and there I found the speech written by the master.

I must admit that my first temptation after reading his speech was to discreetly withdraw from the mission, arguing some personal problem, so as to avoid comparisons, even if I myself were to be the most critical with myself.

My next idea was to read the actual speech written by D. Julio; after all, it essentially covers the history of Spanish Mathematics of the sixteenth century, and his previous and subsequent comments on our beloved University are, unfortunately, quite current.

I finally decided to look to the future, and not dwell on the past, opening myself up to the judgement of the public, whom I suppose somewhat terrified at the prospect

---

The Inaugural Lecture of the academical year 1996–1997 in the University of Oviedo, Spain, was delivered by Pedro Gil, Full Professor of Statistics and Operations Research.

P. Gil (✉)
Departamento de Estadística, I.O. y D.M., Universidad de Oviedo, Oviedo, Spain
e-mail: magil@uniovi.es

3

of being subjected to a tediously boring lecture of what has been called the 'queen of sciences', although I very much doubt that such laudatory term would be accepted by many students or even by some professors of our Institution.

Due to their historical development, Mathematics is indeed the most accomplished building of science; but it is also true that many believe it is too complex for ordinary mortals to navigate through, that only a selected few can achieve it. The origin of such ideas rests both with the magical nature that has always been associated with numbers and geometric figures, and with the large number of students failing to pass the subject (a failure not necessarily always attributable to students) during their educational years.

Thus I started looking for a topic to present before you today, "But (again, quoting Rey Pastor) to present a topic within this discipline, with all its apparatus and, for the uninitiated, almost eerie symbolic notations, and to take advantage of your helpless position, when the law compels you to hear me, presenting one of my insignificant mathematical works, would have been an unheard-of case of cruelty, that will surely reach the limits of your tolerance."

However, my limited knowledge prevents me from attempting to present any other subject than the one that has been the center of my scientific life; I lack the versatility to face other possibilities, as many of my colleagues, who I envy, are able to do.

I have therefore decided to talk about several mathematical topics; I will however try to explain them, as much as possible, using concepts rather than formulas.

It is my desire to try, as much as possible, to avoid the use of cryptic language, which we are so used to hear from the so-called 'specialists' (see, for instance, the explanations given by some economists, doctors, etc. in the media), so that the objectives of Mathematics - in general - and Statistics and other sciences assimilated - in particular - are not only a sect's belief system, but may also be understood, to a greater or lesser extent, by those who, in line with this lecture, we could refer to as the 'average citizens'.

The general topic chosen is the mathematical study of uncertainty. Uncertainty is inherent in nature; these days nobody questions that the movement of particles, the distribution of genes and chromosomes, and the behavior of the individuals themselves need theories based on the study of uncertainty rather than on deterministic laws.

What criteria do you follow to make decisions in situations of uncertainty? How to generalize particular data to discover new phenomena or create new theories? What is, as Professor Rao points out, the process involved in these tasks: an art, a technology or a science?

There were no attempts to answer these questions until the beginning of this century, by trying to quantify uncertainty. In the last fifty years, even though it cannot be said that the success achieved is complete, it is certainly true that the results obtained have produced a revolution in all spheres of knowledge, our habits of thought have changed and this has made possible remarkable discoveries that our prejudices about determinism had previously prevented.

Now, with the new century at hand, we want to give a simple, but not easy, vision of these issues that have surreptitiously sneaked into our daily lives.

The aim is thus clear; when the lecture finishes, it will be your turn to judge whether or not it has been achieved.

I have called the lecture "The Mathematics of the Uncertain" and I have broken it into three small paragraphs and an epilogue. Let us begin.

1   The Mathematics of Chance: Probability Calculus and Statistics

Chance is perhaps the pseudonym of God when he does not want to sign

(Théophile Gautier)

What began in the seventeenth century as an amusement for idle people fond of gambling - recall the Chevalier de Méré who fostered the acclaimed correspondence between Pascal and Fermat - or as a curiosity for politicians with an interest in arithmetics, take the Graunt's or Halley's tables as an example, has now become the most powerful weapon of human sciences and nature.

For some unscrupulous individuals it has even become a double-edged sword that has very often made possible the 'pseudo-demonstrations' deceit that is used to convince us simultaneously of the veracity of both an statement and its opposite, to the point of putting, for example, our hearts in our mouths with the serious threat of contracting very serious illnesses by the consumption of humble carrots (on the other hand critical in order to get a wonderful eagle eyesight).

1.1   The Use of Statistics...

But, curiosities aside, let us focus on how statistical science works.

I will not attempt to provide a rigorous definition; there is no shortage of witty ones from which to choose from, such as the well-known one that defines Statistics as "the science that teaches us that if John has eaten a full chicken and Peter has not eaten anything, they have eaten half a chicken each," or the one that jokes about "the ideal state of our body temperature is your head in the oven and your feet in the freezer", both of them agreeing in their criticism towards the abuse of the mean value. And there are many more serious ones among which it would be very difficult to choose.

I will nevertheless mention the definition given by one of the great mathematicians of our century, Profesor Fréchet, who defines Statistics as "a science that deals with solving problems that, with mathematical rigor, have no solution". In my opinion, in this sentence, the greatness and limitations of this our science are simultaneously explained: its greatness as it acts as a support for other sciences to which other branches of Mathematics cannot help; and the limitations that are inherent to all statistical results, found in the very nature of statistical science, when considered as the study of situations in which there is random intervention.

But, while giving a definition of Statistics is not an easy task, it is however easy to point out its three essential areas, which we will now pass in brief review:

Descriptive Statistics that deals with the presentation of available information in an orderly manner; this area of Statistics was already in existence when the word statistic still referred only to the works of 'state', relative to censuses, births or deaths.

Probability Calculus, which is the mathematical model (and, as such, utopian) to represent the aforementioned data; it constitutes the most refined area in terms of rigor, and has allowed Statistics to be considered as a part of Mathematics for some sixty years (much more recent than the, until recently, called 'Modern Mathematics!'). At that time, in Spain, Statistics was only part of the Faculty of Law's syllabus, perhaps the only one that does not include them nowadays.

And finally, Inference that is nourished by the two previous areas and that is, strictly speaking, the one deserving the name of proper statistical science. Its purpose is to find out, sometimes almost guess, something (be it much or little) about those laws of chance that influence a phenomenon and either are only partly known or totally unknown.

All its conclusions are subject to limitations of a probabilistic nature: nothing is affirmed or denied with total certainty, only known to occur with a certain probability. In other words, we can assure, with some certainty, that things are going to happen as predicted, but there is always the risk (sometimes very high, if the right precautions are not taken) for the opposite to happen or, at least, for it not to happen exactly as predicted.

As indicated, the main purpose of Statistics is to conduct surveys, typically on a characteristic of interest in the population under study.

If there was the possibility of observing such a characteristic in all 'individuals' (note that the notation used indicates the origins of Statistics as a state science), the investigation would be meaningless, as it would obtain indisputable results; this is the case of censuses or a count of the votes cast.

If, on the other hand, as is so often the case, studying the whole population is not possible due to its economic or time cost, or because it is a destructive process, etc., the researcher must be satisfied with the study of a 'sample', which is a part of the population and, through it, infer the population values. For example, it would do no good to crash all the vehicles produced to find out their resistance to frontal and lateral impacts; there would be no cars left to sell.

In principle, it is reasonable to think that the larger the sample size (i.e. the number of individuals that compose it), the more accurate the survey will be; and this can be proven even if, at times, an excessive increase in the sample size does not lead to a substantial increase in the certainty obtained, but only to the corresponding expenditure, as in some of our most well-known and prestigious official pollsters. The optimal sample size in some sense must, therefore, be sought.

It is imperative for the researcher to keep the mathematical rigor of the theoretical results that he obtains and, if in some step approximations were made, these must be explicitly added to the list of limitations of the conclusion.

Finally, professional ethics should be observed when presenting the conclusions, clearly pointing out, as has already been mentioned, that none of them constitutes in itself an absolute truth but rather that it has - at best - a high probability of being true.

It would be childish trying to give an example of each of the sciences that have benefited from the development of statistical methods; I explicitly decline to do so in favor of the time available to criticize its incorrect uses.

1.2   ...and the Abuse of Statistics

And that's all the guidelines for the correct use of Statistics that I am going to give today; unfortunately the 'abuses' committed in his name are often of great consequence for the recipients of information, that is, ourselves. Who has not heard or read the sentence "there are lies, damned lies, and statistics"? Why are there books with titles like 'Lying with statistics'? Why is the same data used by someone to convince us of something and then by someone else to justify the contrary?

The answer to these questions can often be found in the lack of statistical culture of those who receive the information; I am a zealous supporter of making such a culture compulsory in our educational system because, in the words of my teacher and mentor, Professor Sixto Ríos, "these days the basic statistical concepts must be an indispensable part of the educated man's mental equipment", this way the chances of deception would be lower.

Sometimes, frequently, the cause is the lack of expertise and know-how of those responsible for the development of statistical studies; not in vain is Statistics one of the specialties with the highest levels of professional encroachment.

Finally, it is no secret that in this slippery slope of approximations and lack of absolute certainties, it is indeed easy to manipulate (even legitimately, without deceit) the results. A forefather of these manipulations was Mendel himself, as it has been proven that, quite possibly, he falsified the results obtained in his experiments with peas, to give greater strength to his famous laws.

Let us look at other examples:

If we were to assert that "after traveling 100 km at 75 km/h, and then another 100 km at 125 km/h, the average speed was 100 km/h." anyone here today would be, if he or she does not stop to think about it twice, willing to admit it as true; a simple calculation shows, however, that in such a case the average speed does not reach 94 km/h. What happened? Quite simply that we have been misled by the use of an average.

Bernard Shaw himself used to say rather ironically that "the carrying of umbrellas enlarges the chest, prolongs life, and confers comparative immunity from disease; for the Statistics show that the classes which use these articles are bigger, healthier, and live longer than the class which never dreams of possessing such things". It does not take much statistical knowledge to understand that the cause of this difference was not the umbrella (a luxury article at the time) but the quality of life of those who had it.

Similar affirmations to Shaw's mockery are common currency nowadays; the briefest glance at the press will show us the 'statistical demonstrations' of how harmful the consumption of this or that food is to our health (or how marvelous another of a similar composition is), or of how unhealthy wine is to the heart (or how excellent it is to drink it at meals to raise the spirits), or even of how bad the smoke of our cigarette is to the one that passes by our side (or the beneficial influence of cigarettes to fight dementia); and so we could carry on almost indefinitely.

What secret lies in these contradictions? Can we blame the media (so in fashion these days) for distorting and politicising the actual conclusions? The answer to that is, of course, no. The fallacy in the 'demonstration' lies, as has already been pointed out, in two essential factors: the informant may conceal the limitations of the results (the original informant, not the journalist) and, even if this did not happen, we would need a broader statistical culture to be able to properly evaluate the information received.

1.3 Polls and More

And the opinion polls? The one resounding failure of Statistics that impacts our daily life is the damn polls. Some of you may think that I am slyly swinging the lead, tiptoeing around a subject that raises controversy each time an election is called (something that, on the other hand, is far too frequent since we learned how to do it and our elders could remember how it was done). The answer is no; it is not my intention to do so, and, therefore, I will try to shed some light on some aspects of the polls that are not always well understood, although, even if I am still trying to avoid them, I will have to make use of some formulas to explain these aspects.

Why pollsters' results do not match each other? What criteria do the poll designers use to decide the sample size? How should the results obtained be objectively interpreted?

We will try to deal only with the essential elements assuming, in any case, that the right sampling procedures have been followed.

The essential aim is to find the percentage of a certain population that is inclined towards one of the options available to it; we will refer to this percentage as p. This value p is the population parameter that is unknown (in fact it will not be known until an exhaustive poll is carried out, that is, until the day following the elections in the case of a poll about voting for different parties) and that needs to be estimated through the sample; the value of the percentage corresponding to the responses of the surveyed individuals is chosen for this purpose; we will denote this percentage as the value p^*.

It is possible (with the mathematical resources available in Probability Calculus) to give an interval in which the unknown percentage p is found a percentage of times as high as you may want. Now everything seems under control, except for one small detail: what error are we willing to accept? Or, to put it in another way: what amplitude do we want the interval to have?

But there is still another factor under our control, the size of the sample - which must be decided before proceeding to the survey -; this size depends on the error that is considered to be acceptable. In general, such an error is half the length of the constructed interval (a small percentage), that is, the distance from p^* (which is the interval midpoint) to each of the extremes of the interval; if we represent it by E, in most common conditions, the required sample size equals

$$N = \frac{10000}{E^2}.$$

It can be proven that, for values $E = 2\%$, $E = 1\%$, and $E = 0.5\%$, the 'magic' numbers of the sample sizes in the polls and 'macro polls' appear: 2,500, 10,000 and 40,000 respectively. Thus, for instance, an estimated value $p^* = 30\%$ allows us to say that the population value p is within the intervals (28, 32), (29, 31) or (29.5, 30.5) (or, in its more common notation $30 \pm 2\%$, $30 \pm 1\%$, $30 \pm 0.5\%$) with a 'certainty' of 95.5%, which means (warning!) that if we were able to build many of these intervals with many different samples of the same size, the true value of p would be found in 95.5% of them.

And, as in any statistical procedure, recall the limitations of what has been achieved; whatever the margin of error E we are willing to admit, 4.5% of the intervals we could construct would not contain the true value of p. Is this 4.5% small enough or is it too large? The answer is specific to each person that commissions a study, it depends on the risks that he or she is willing to take. I personally would not risk commissioning a study that one in twenty times (approximately) would fail (what if mine is the one that fails?). Note that in case of failure, not only the value of p estimated by p^* is incorrect but also any value within the interval.

Of course, in no event does it follow from the former exposition that the results of the polls may be extrapolated to the composition of the parliament, partly because of the margins of error (of the 2,500 surveys required for a 2% error, to our beloved Asturias, considered as the standard region, will correspond 50, a value that, assuming that all the sampling and other processes are correct, corresponds to an error of almost 15%), and in part to our electoral law: in this sense only a count similar to that made with the first ballot papers at selected polling stations in each constituency (which is indeed the first sampling of the real population, that of the votes cast) is valid.

In addition, in each poll, there is a non-negligible percentage of 'don't know/no opinion' answers, whose classification is not easy because it relies on many individual factors (not just on loyalty of vote). And, even though there are indirect estimation techniques, they do not seem to be to the liking of our pollsters.

The aforementioned 'abuses' are also applicable to television and radio audience surveys, thanks to which we have the programs we have; the limitations are identical (or even worse, for until very recently only 2,000 households were sampled, and these were not even chosen at random because they had given their consent to have the gadget connected). Thankfully, and this is applicable to polls, the sampling procedures help improve the results (albeit sometimes unintentionally) by companies engaged in this type of research. No offense meant, but I would like to point out that those responsible for the research are not statisticians, but professionals from other fields such as Sociology, Political Science, Economics, etc.

Let's leave the polls aside to point out some other fallacies that we would typically find in Statistics: the constant generation of random results leads to the appearance of any possible outcome; thus, the initial surprise at a combination like 1-2-3-4-5-6 appearing in a lottery draw should not lead us to believe that such a sequence is less likely to occur than any other.

It is a situation similar to that of the monkey that hitting keys at random on a typewriter will almost surely type the complete works of Shakespeare in a very long but finite amount of time (it is estimated that the probability of reproducing Hamlet with its 27,000 letters and spaces is approximately one divided by an amount that has 41,600 zeros, which gives an idea of the magnitude of the time that we would need to be waiting to obtain the desired result).

We can also point out, within a different kind of fallacies, the well-known one of the player who considers that his probability of success increases with the number of previous failures; regarding this one Polya mentions the anecdote of the doctor who gives encouragement to his patient as follows: "You have a very serious disease; of ten people who have got this disease only one survives. But you are lucky. You will survive because you have come to me. I have already had nine patients who all died of it".

As a counterpart to the above, one more example: for a long time the so-called theory of accumulation was common belief, according to it, short runs of events occur unexpectedly frequently; as our proverbs say, "troubles seldom come singly" or the more graphical "it never rains, it pours".

And so on and so forth ad infinitum. The origin of these fallacies may be found in the continued (and mistaken) belief that phenomena are always deterministic, an opinion held by a scientist as remarkable as Einstein himself. In any case, the brilliant physicist accepted Bose's theory of the random behavior of particles, perhaps convinced by the so-called law of large numbers that introduces order into disorder.

2 The Mathematics of Communication: Information Theory

They saw what seemed to be tongues of fire
that separated and came to rest on each of them.
All of them were filled with the Holy Spirit
and began to speak in other languages as the Spirit enabled them.

(The Acts of the Apostles, 2:1–4)

Information is power. It is said that it will be the great power of the next century and, judging by the current situation, such statement will surely become a fact.

But, can you measure the amount of information? Can you give value to the information contained in a story in the news, or to that in a photo, or to the information that is transmitted over the telephone line or the internet? Is it true that, as the proverb tells us, a picture is worth a thousand words?

The answer was given simultaneously by two of the great scientists of our century: Norbert Wiener and Claude Shannon. It is not the first time that something similar occurs in the history of Mathematics (I recall from my childhood readings the race between Newton and Leibnitz to be the first to patent differential calculus), but there are small nuances that differentiate the position of both researchers: Wiener deals with the information he receives once the experience is over; that is, he measures that information a posteriori; Shannon, on the other hand, observes the a priori situation and measures the information he hopes to obtain from it; that is the reason why the entropy of the latter is the mean value of the information of the former.

There are many other information concepts that were developed earlier than these ones, and others that have been developed since the 1950s. And all have things in common. Chance? Or is it something else? As Philip Jourdain would say, it can refer to the same Mathematical Concept (with capital letters) that is expressed through different mathematical developments (with lower case).

When Hartley in 1928 suggests the logarithm of the number of results as a measure of uncertainty in the face of an experience, he is not aware of the fact that he is assuming that the results are equiprobable, and in order to avoid conflicts with the situations in which such equiprobability is not true, argues that probabilities have a lot to do with psychological motivations (perhaps he was already thinking of a subjective probability!) and that the problem must therefore be studied by psychologists and not by mathematicians or engineers.

He is wrong to adopt this strategy and avoid the problem.

But he succeeds in finding the way forward: information can be defined as the measure of the decrease of uncertainty, and therefore we can start by analysing the latter.

And it is thus that from Hartley's idea, the Information Theory was developed by Shannon, who defines the entropy for any probabilistic system as a measure of the associated uncertainty (this way linking uncertainty with probability), inspired by the quantity of the same name defined in Thermodynamics by Boltzmann. It is the well-known formula:

$$H = -\sum_i p_i \log p_i.$$

Shannon also defines the information between two systems as the average decrease of uncertainty that occurs in one of them by the knowledge of the other's result. From these initial concepts on to information channels, their capacity, the sources of information, the study of noise, etc., it is a dizzying research involving many scientists.

All this starts from a simple schema that it is well-known today to all the communication sciences: the system of source, encoder, transmitter, receiver, decoder, and destination. All this on the idea of the element of surprise: the less probable an event is, the more information it provides when it takes place (the notorious 'man bites dog' of journalism). And, in addition, in order to transmit messages whose content is considered only in terms of their syntactic value, i.e. as meaningless signifiers. And all this working wonderfully and allowing the phone company Bell (the company that employed Shannon) to achieve its goal: to offer its customers a greater reliability in the transmission of the conversations, ensuring that noises cause minimal interference.

All sorted then? No, it had only just begun; at least the next two problems were still unsolved.

2.1 Coding and Cryptology

How could we reproduce in the destination the original messages when, as it travels through the channel, a distortion occurs? How could we share storage and transmission resources while respecting the privacy of information?

It is true that it would always be possible to set a system of repetition of messages between sender and receiver and run it as many times as necessary until the absence of error could be guaranteed; or the message could be encoded in such a particular way that not even the source knew its content (as in certain exams that all of us teachers have had sometimes the pleasure to mark). But it is no less true that we could neither afford the phone bills nor the effort of decoding of doing such a thing.

It was Shannon himself who first proved the existence of a coding–decoding system that ensures that the probability of error is as small as one wants, but it is a demonstration of the kind to which mathematicians are so fond of as, in Bertrand Russell's words, it is part of "the subject in which we never know what we are talking about, nor whether what we are saying is true".

It has not been possible to build a real system validating Shannon's theorem but, in the effort, a powerful theory of codification, of fundamentally algebraic nature, has been developed and it has provided satisfactory solutions to the initial problem; all the digital systems to record, preserve and transmit the information we use today (computers, compact discs, latest generation mobile phones, etc.) are a good example of the power of the tool obtained. And all this using only the digits 0 and 1!

Regarding the second, and very current, question, the answer is positive: the importance of data protection is ever growing to prevent anyone from using the data fraudulently or more simply, for purposes other than those intended. These days it is ever more important that data are protected from hackers that can bring us to a small but respectable ruin (with the passwords of our bank accounts) or make us descend into a global tension of unpredictable consequences (by compromising the defense systems of the great powers).

And thus a new science called cryptology, with a branch of encryption (cryptography) and a branch of decryption (cryptanalysis) emerges as a natural extension of the theory of codification, and its origins are as old as mankind: let us recall the writing about the strip rolled to the cane or the 'scytale' of the Lacedaemonians or even Caesar's cipher by shifting by a constant magnitude the letters of the alphabet.

2.2 Chicken and Egg: Information or Probability?

Let's move on to the next point: is there information without probability? Which of the two is the primary concept? How to take into account the semantic value of the information transmitted? The axiomatic Information Theory, created by Professors Kampé de Fériet and Forte around 30 years ago, provides an answer to these questions.

In the practical situations in which the theory developed by Wiener and Shannon has been successfully applied, the probabilities used originate from frequencies corresponding to the repetition of an experience a large enough number of times.

It is certainly valid to use the expressions of uncertainty and information for a priori probabilities, but such probability values cannot be determined if it is not in function of all the information available on how a result is produced; for example, if it is known that the center of gravity is not the same as the geometric center of the die, then the same probability should not be assigned to the six faces.

It is however often the case that situations in which there is undoubtedly information do not allow for the consideration of probability; a classic example of this type is the figure of Cardinal Roncalli, whose election as Pope John XXIII by means of a single conclave would prevent any frequentist consideration. In addition, in a poll prior to such election, the opinion of a taxi driver, for example, should not be given the same value as that of a high-ranking Vatican official, and, yet, before the election there was 'uncertainty' and, once the result was known, 'information'.

Another example: when trying to approximate the value of pi, the successive bounding intervals that may be obtained ($3.1 < \pi < 3.2$, $3.14 < \pi < 3.15$, etc.) increase the information received, which could be measured, aside from some scaling factor, by the time required to achieve the desired precision. This perfectly valid measure also differs from the framework of Shannon's theory and it makes no sense to speak of the probability that the π number is between certain values.

And yet another situation in which the 'subjectivity' of the value of information is made clear. Take the example of a lottery that involves the random selection of numbers from 1 to 999999 where the winning number is, by way of illustration, the number 123456. According to the probabilistic model, the information received is the logarithm of the number of possible outcomes. However, when reading the result, three different readers may receive very different information: reader A, who does not play in that draw, receives null information; reader B who owns a non winning number receives information that may be calculated, for example, by means of the expense incurred in purchasing it; finally a third reader C, who has a winning number, receives a very large amount of information; the probability of 123456 winning is, evidently, the same in all cases.

As pointed out above, the entropy and, consequently, the amount of information associated, do not take into account the semantic value of the results, but only their syntactic value, regardless of the 'importance' of the message to be transmitted; this has led certain authors to consider measures in which, even though it is taken into consideration, probability alone is not considered enough. The messages to be transmitted may have the same value for operators or for encoding and decoding machines, but not for the human sender and receiver that, ultimately, are affected by those messages.

In any case, there is communication between sender and receiver, embodied in an explicit or implicit proposition (in its classical logic sense). We will assign to this proposition a measure of the amount of information it transmits.

With the sets of objects and properties we can construct the family $\mathscr{P}$ of the elementary propositions formed by the initials, their respective negations and all the compound propositions that can be obtained from the elementary ones via the classical operations of the distributive logic, constructing a network of propositions $\mathscr{F}$, which, together with $\mathscr{P}$, provides the appropriate system to define an information measure.

In its set theory formulation, the axiomatic Information Theory starts from a measurable space such as that which allows for the construction of probability; but there are certain aspects of interest that require us to find some way to examine this initial system.

The last element, which is crucial to construct the axiomatic Information Theory, is the concept of independence, whose analysis we will carry out by means of propositions:

Two propositions, P and Q, are logically independent when each one does not imply or exclude the other. But this condition may not suffice; for instance, the proposition 'X is a smoker' does not imply or exclude the proposition 'X has lung cancer' and vice versa: however, who wouldn't balk at considering both propositions as independent? If nothing else, the first allows for making predictions about the second and therefore provides information about it. Thus, as far as information is concerned, we must distinguish two levels of independence:

(a) The logical or syntactic level, which is evidenced by the condition of independence shown, based on the usual logical operations.
(b) A second level, semantic, more demanding than the previous one, which judges the independence of propositions by their meaning and that goes beyond the scope of the information system we deal with.

These three elements form a measurable information system. Its study opens new horizons for the study of models, more general than probabilistic ones, in which either there is no probability or at least not only probability. Perhaps, as it has been pointed out by some scholars of the subject, this is the connecting science able to provide a backbone to the currently widely dispersed scientific activities of our time.

3 The Mathematics of Imprecision: Fuzzy Sets

Both precision and certainty are false ideals.
They are impossible to attain, and therefore [...]
one should never try to be more precise
than the problem situation demands...

(Karl Popper)

The development of Mathematics applied to human problems reached a new milestone thirty years ago when Professor Zadeh published the work Fuzzy Sets. It is been thirty years and, as Dubois and Prade point out, "Professor Zadeh's appeal to look for non-conventional techniques has been misunderstood and interpreted as an example of permissive thinking that tries to escape the rigor of Mathematics"; as an actual fact, there are many scientists that do not want to believe in the advantages of the model in order to adapt to the study of the phenomena that surround us, why?

First, because it is a most daring act to confront the all-too-powerful 'modern Mathematics'. The basis of set theory is that each element either belongs or does not belong to any given set. Likewise, the classical logic guarantees that a proposition is true or false. Yes or no, there is no maybe, as the saying goes.

As another example, in Mathematics a function is either continuous or discontinuous; it can not be continuous to a certain degree. Similarly, a matrix is symmetric or not; it can not be somewhat symmetrical, more or less symmetrical, or symmetrical to some degree. Similarly, a paper published in a mathematical journal is expected to contain accurate definitions, axioms, and theorems. An article would typically not be considered acceptable for publication if its conclusions were established like affirmations that were not unequivocally certain.

In clear contrast to the idealized world of pure Mathematics, our perception of the real world is full of concepts that have no sharply defined boundaries, such as tall, fat, many, most, slowly, old, familiar, relevant, much older than, kind, etc.

And this is precisely the crux of the matter in the words of Professor Zadeh himself: "The key elements in human thinking are not numbers, but labels of fuzzy sets, that is, classes of objects in which the transition from membership to nonmembership is gradual rather than abrupt (...). It is clear that 'the class of all real numbers much greater than one,' or 'the class of beautiful women' (surprising example in a man as gallant as Professor Zadeh for whom all women are beautiful), or 'the class of tall men' do not constitute sets in the usual mathematical sense of the term (...) although they play an important role in human thought, particularly in the fields of form recognition, communication of information and abstraction".

Once the problem was stated, what was the solution? It is often the case that the great scientific advances are of a near-impossible comprehension for a layman in the matter (and even for many not proficient enough). But, surprisingly, not in this case; using again Professor Zadeh's own words: "A fuzzy set in a referential is characterized by a membership function which associates with each element a real number between zero and one, its membership degree. Thus, the closer the value of the function at a specific point is to one, the higher the degree of membership of the element to the set will be. When the set in question is a classical set its membership function can take only two values, one or zero, depending on whether the element considered belongs to the set or not".

So easy? Well, yes, the problem was solved: there are men that are clearly tall (degree of belonging one or next to one), others not so much (degree of membership intermediate, between zero and one) and others that are clearly not tall (degree of membership zero or close to zero).

In addition, the advantage of the formulation in terms of sets over a logical formulation, is in the immense field provided to the new theory by the already-established Mathematics. As an example, in the area of the Fuzzy Set Theory it is natural to refer to the existence of fuzzy numbers (so useful to give inaccurate assessments; in exam marks, for instance, what is the boundary between A and B grades?); or to fuzzy functions (very useful to issue certain orders, such as 'the heating system should be set high if the day is cold', or would it be better to say that if the outside thermometer shows 2 degrees above zero, the thermostat should be set at 73 degrees, if it shows 3 it should be set at 71, etc., also specifying that the temperatures are in celsius?), etc.

Some highly regarded researchers have made fierce criticisms of fuzzy sets. Consider for instance the comments made by Professor Kalman, pioneer of Systems Theory, in 1972: "No doubt Professor Zadeh's enthusiasm for fuzziness has been reinforced by the prevailing climate in the U.S., one of unprecedented permissiveness. Fuzzification, is a kind of scientific permissiveness; it tends to result in socially appealing slogans unaccompanied by the discipline of hard scientific work and patient observation. Let me say quite categorically that there is no such thing as a fuzzy scientific concept, in my opinion".

On the other hand, on top of the criticism of mathematicians that we could refer to as 'pure', is the attack of statisticians, particularly Bayesians, who have tried to consider the new working tool in situations of uncertainty as a particular case of their own methods. It seemed as if they had armed themselves to the teeth to defend their stronghold, that of decisions in an uncertain environment, within which they considered themselves the legitimate and unique residents, showing most of the time a great ignorance of the model. There are many who have tried to explain the degrees of propriety, properly normalized, as probabilities, ignoring that in many situations of a fuzzy nature it makes no sense to talk about probability: there is no probability involved in John Doe being high, or today being a good day; that is, there are situations that depend on the vagueness of the concepts used (high, good) and not on chance.

As for the 'confrontation' with statistical methods, I shall tell you a little story that happened to me not many years ago. I had been invited to form part of a doctoral thesis committee in which a model of medical diagnosis supported by probabilistic calculations was presented. During the defense, fuzzy models were rejected with a number of somewhat unconvincing arguments. Imagine my surprise when I realized that the input variables presented took 'values' that met the fuzzy criteria such as 'much', 'little', 'fever', 'high cholesterol', etc.!

The confrontation is meaningless and, as the title of the lecture given by Zadeh last year in this University smartly pointed out, "the theory of probability and fuzzy logic are complementary, rather than competitive". Moreover, as it has already been pointed out with regard to other mathematical aspects, the concept of probability and the usual statistical techniques may also be blurred.

Today, as if it were a mythological episode, creatures have almost devoured their master: it is not difficult to find those who ask for 'Fuzzy sets' in the same language as they use for the 'Pythagorean theorem', the 'Hilbert spaces' or the 'Gaussian distribution'.

3.1 Fuzzy Technology and...

These days applications of fuzzy logic are too current to ignore; a new technology, mainly developed in Japan, is born. Fuzzy drivers are simple and robust; but, perhaps more importantly, fuzzy control allows for the execution of tasks, such as parking a car, which do not lend themselves to resolution by conventional methods: the focus drivers of Sanyo or Panasonic camcorders, the Nissan ABS brake controller, the Mitsubishi air conditioner, the Sony writing and the Hitachi speech recognition systems, the subway control system in the town of Sendai, microwave controllers, dryers, lifts, and a long etcetera that provides better and cheaper features to the users, ultimately collaborating to achieve a better quality of life and greater freedom of the human being. Fuzzy is well seen by users, is a guarantee of good operation; fuzzy is, in marketing terms, a 'selling point'.

Nonetheless, too many scientists remain unconvinced that fuzzy logic has something important to offer. The aforementioned regard for what is quantitative and precise, and the disdain for the qualitative and imprecise, is too deep-rooted in society to let it go without resistance.

Let us briefly discuss some recent details about fuzziness: Zadeh uses the term granulation to refer to the process of forming fuzzy classes of objects, which are grouped by similarity. When the number of different classes to be handled is too high, such classes should be grouped together to form granules. This is what we do to determine, for example, the colors: the range of greens is so wide that the green color becomes a cluster, of course not at all sharp, of the different wavelengths that correspond to different shades of green.

The need for granulation is thus due to the limited ability of people to store details. From this point of view, fuzziness and granulation are consequences of complexity, and play a key role in the tolerance of imprecision to achieve efficiency, robustness and low cost in the final product.

One important implication of this observation, as Zadeh pointed out in his investiture as doctor honoris causa by our University, is that with the rapid growth in complexity of the information processing tasks that the computers are asked to perform, we are reaching a point at which computers will have to be designed to be able to process fuzzy information.

In fact, it is the ability to manipulate fuzzy concepts that distinguishes human intelligence from the intelligence of current generation computers. Without that ability, we can not build machines able to pick up non-stereotypical stories, translate well from one natural language to another, or do many other jobs that humans can do with ease because of their ability to granulate and manipulate the remaining fuzzy concepts.

All this is leading to a situation that scandalizes the more orthodox scientists: we are providing means to obtain results that can be denominated 'computation with words'. This will sometimes be used as a substitute to the usual computation using numbers, not in vain the variables that fuzzy logic deals with are called linguistic variables, and not in vain this system is the one used by humanity from immemorial time to calculate and reason when the information available is not accurate enough to justify the use of numbers. This takes advantage of the tolerance of imprecision to achieve efficiency and better relationship with reality. And, furthermore, provides a basis for the development of programming languages that could approach the natural languages in appearance and in capacity for expression.

Neil Postman says in Technopolis that more Mathematics, more science and more computers will not solve the problem of hunger, loneliness, etc. and that the computerized society is dangerously moving towards a global and authoritarian society.

I do not share this pessimism: I think perhaps in a few years the computation with words will become a methodology in its own right, whose final model is the human mind. Maybe this will get the machines to become a little more 'human'.

4 Epilogue

I started this lesson by referring to Professor Rey Pastor and his inaugural lecture of the 1913–1914 academic year and, despite the years that have gone by, it has not been possible for organizations of all kinds to take advantage of professionals of the areas of Statistics, Operations Research and Applied Mathematics in general, which, with the rigor characterizing the so-called exact sciences, would give guidance to management for the appropriate decision making in the presence of uncertainty, polishing the information so that it is presented adequately to the eye of the managers.

Other countries and also, within Spain itself, other more dynamic regions have already understood this, and today authentic armies of mathematicians crowd some departments of the world's largest banks or powerful multinational corporations.

What is going to happen in Asturias? Going back again to the words of Professor Julio Rey we could say that "in Mathematics it is not Asturias (Professor Rey said Spain) a modern nation; but it is neither a decadent nation nor an inept nation. It is simply a backward nation, which has not yet been incorporated into modern civilization, but that has retained enough energy and enthusiasm to bridge the gap caused by years of isolation and disorientation".

Today, fortunately, from the recreated Faculty of Sciences, we have been providing the Asturian society for the last two years with classes of mathematicians, with a preparation we are proud of (in which the practical aspect of theories is considered fundamental without neglecting the formal one) and capable of providing the society with the resources it needs for its development.

They are ready. It is now the society that should give them the opportunity to deliver: in teaching, in private organizations, in public administrations... The University has played its part: the preparation. It has been a good and intelligent move. Asturian society has to play theirs, employment in even a more smart way, so that the future of Mathematics in Asturias is not uncertain.

Part II
The Mathematics of Chance: Probability, Statistics and Applications

Using Mathematica to Calculate Shortest Confidence Intervals

Ramón Ardanuy

Abstract We discuss how to use Mathematica software to determine shortest confidence intervals for bell-shaped continuous probability distributions. This is done both for situations corresponding to classical statistics (variances and standard deviations of a normal distribution) as well as Bayesian statistics (posterior distributions corresponding to betas, gammas and inverse-gammas).

1 Preliminaries

In the evaluation of confidence intervals of a given parameter θ drawn from a sample $\mathbf{X} = (X_1, \ldots X_n)$ with size n, a standard procedure in classical statistics is using a certain random variable $T = T(\mathbf{X}, \theta)$, the pivot function, whose distribution is known, see Bartoszynski and Niewiadomska-Bugaj [1] and Rohatgi [7]; by contrast, Bayesian statistics works with the posterior distribution of θ, $\xi_n(\theta|x)$, see De Groot [3]. In both cases, if γ denotes the confidence degree and $h(t)$ the density of the distribution, the problem is obtaining an interval (a, b) such that

$$\int_a^b h(t)dt = \gamma.$$

To this end we fix two values $\alpha_1, \alpha_2 \geq 0$ satisfying $\alpha_1 + \alpha_2 = 1 - \gamma$ and

$$\int_{-\infty}^a h(t)dt = \alpha_1, \int_b^\infty h(t)dt = \alpha_2.$$

Namely, the limits of the confidence interval a and b are taken to be the quantiles α_1 and $1 - \alpha_2$ of the probability distribution (see Fig. 1); these values a and b are those that are then used to derive the confidence interval for the parameter θ.

R. Ardanuy (✉)
Departamento de Estadística, Universidad de Salamanca, Salamanca, Spain
e-mail: raa@usal.es

© Springer International Publishing AG 2018
E. Gil et al. (eds.), *The Mathematics of the Uncertain*, Studies in Systems,
Decision and Control 142, https://doi.org/10.1007/978-3-319-73848-2_2

23

Fig. 1 Calculation of values
a and b

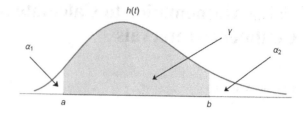

For convenient use of statistical tables, in practice it is convenient to take $\alpha_1 = \alpha_2 = (1-\gamma)/2$. When the density h is symmetric and bell-shaped the corresponding confidence interval is of minimum length, however, when these assumptions are waived the problem arises as to how to choose a, b to insure that the relevant confidence interval has expected minimum length, see Guenter [4], Kendall and Stuart [6], Wilks [9, 10], Zacks [12], etc. The associated numerical problem is tedious but in some cases it can be solved using Mathematica software.

2 Confidence Interval for the Variance and Standard Deviation of a Normal

Let $\mathbf{X} = (X_1, \dots X_n)$ be a sample with size n drawn from a normal population $X_i \stackrel{iid}{=} \mathcal{N}(\mu, \sigma^2)$ with sample mean $\bar{X}$. Depending on whether the "theoretical mean" μ is or not known the following estimators for the variance are used

$$\hat{\sigma}^2 = \frac{1}{n}\sum_{i=1}^{n}(X_i - \mu)^2 \qquad \text{or} \qquad \hat{\sigma}^2 = \frac{1}{n-1}\sum_{i=1}^{n}(X_i - \bar{X})^2. \tag{1}$$

In both cases they define an unbiased estimator with $g = n$ or $g = n-1$ degrees of freedom. To evaluate the confidence interval for σ^2, and also for σ, one uses the pivot function

$$T = \frac{g\hat{\sigma}^2}{\sigma^2} \sim \chi_g^2 \tag{2}$$

which has a chi-squared distribution with g degrees of freedom. Requiring that the unknowns a, b satisfy $\mathbb{P}(a < \chi_g^2 < b) = \gamma$, confidence intervals for the variance and standard deviation follow from

$$\gamma = \mathbb{P}\left(a < \frac{g\hat{\sigma}^2}{\sigma^2} < b\right) = \mathbb{P}\left(\frac{g\hat{\sigma}^2}{b} < \sigma^2 < \frac{g\hat{\sigma}^2}{a}\right) = \mathbb{P}\left(\sqrt{\frac{g\hat{\sigma}^2}{b}} < \sigma < \sqrt{\frac{g\hat{\sigma}^2}{a}}\right).$$

As commented, in practice one takes $\alpha_1 = \alpha_2 = (1-\gamma)/2$ that give the standard quantiles

$$a_0 = \chi_{g;(1-\gamma)/2}^2, \quad b_0 = \chi_{g;(1+\gamma)/2}^2. \tag{3}$$

However to obtain shortest confidence intervals we minimize the lengths

$$L = g\hat{\sigma}^2 \left(\frac{1}{a} - \frac{1}{b} \right) \quad \text{or} \quad L = \sqrt{g\hat{\sigma}^2} \left(\frac{1}{\sqrt{a}} - \frac{1}{\sqrt{b}} \right), \qquad (4)$$

subject to

$$\mathbb{P}\left(a < \chi_g^2 < b \right) = \gamma. \qquad (5)$$

Denoting as $H_g(t)$ and $h_g(t)$ the corresponding cdf and pdf of the χ_g^2 then (4) and (5) are equivalent to

$$\text{minimize:} \qquad \frac{1}{a} - \frac{1}{b} \quad \text{or} \quad \frac{1}{\sqrt{a}} - \frac{1}{\sqrt{b}}$$
$$\text{subject to:} \qquad H_g(b) - H_g(a) = \gamma.$$

To this end we resort to Lagrange multipliers method. In the case of the variance we differentiate the function

$$\varphi(a, b, \lambda) = \frac{1}{a} - \frac{1}{b} - \lambda \left(H_g(b) - H_g(a) - \gamma \right)$$

and obtain

$$0 = \frac{\partial \varphi(a, b, \lambda)}{\partial a} = -\frac{1}{a^2} + \lambda h_g(a) \implies a^2 h_g(a) = \frac{1}{\lambda}$$

$$0 = \frac{\partial \varphi(a, b, \lambda)}{\partial b} = \frac{1}{b^2} - \lambda h_g(b) \implies b^2 h_g(b) = \frac{1}{\lambda}$$

$$0 = \frac{\partial \varphi(a, b, \lambda)}{\partial \lambda} = - \left(H_g(b) - H_g(a) - \gamma \right)$$

which in turn reduces to a system for two unknowns a, b

$$H_g(b) - H_g(a) = \gamma \quad a^2 h_g(a) = b^2 h_g(b). \qquad (6)$$

Tables with the results for the unknowns a, b can be found in Tate and Klett [8], and also in Bartoszynski and Niewiadomska-Bugaj [1].

The connection between the length of the shortest interval associated with the pivot function (2) that follows solving (6) and that obtained from (3) is

$$r = \left(\frac{1}{a} - \frac{1}{b} \right) \Big/ \left(\frac{1}{a_0} - \frac{1}{b_0} \right).$$

For the standard deviation we must consider the function

$$\varphi(a, b, \lambda) = \frac{1}{\sqrt{a}} - \frac{1}{\sqrt{b}} - \lambda \left(H_g(b) - H_g(a) - \gamma \right).$$

By differentiation we obtain a system for two unknowns a, b

$$H_g(b) - H_g(a) = \gamma \quad a^{3/2} h_g(a) = b^{3/2} h_g(b). \tag{7}$$

Crisman [2] give tables with the values of a and b. In this case the length of the shortest interval that follows solving (7) and that obtained from (3) are related as

$$r = \left(\frac{1}{\sqrt{a}} - \frac{1}{\sqrt{b}} \right) \bigg/ \left(\frac{1}{\sqrt{a_0}} - \frac{1}{\sqrt{b_0}} \right).$$

Systems (6) and (7) are conveniently solved using Mathematica software (see Wolfram [11] for the corresponding programming language), in which the data to be introduced are the confidence level gamma and the degrees of freedom g, that in the example solved are $gamma=0.95$, $g=10$.

```
(* FOR THE C.I. OF THE VARIANCE AND STANDARD DEVIATION *)
Clear[gamma, g, distribution, a0, b0, t, h, H, w, a, b, r];
(* DATA *)
gamma = 0.95;    (* CONFIDENCE COEFFICIENT *)
g = 10;          (* DEGREES OF FREEDOM *)
(* COMPUTATIONS *)
Print["Data: Conf. coeff. gamma = ", gamma, " d.f. g = ", g];
distribution := ChiSquareDistribution[g];
a0 = Quantile[distribution, (1 - gamma)/2];
b0 = Quantile[distribution, (1 + gamma)/2];
Print["Usual solution:      a0 = ", a0, "    b0 = ", b0];
h[t_] := PDF[distribution, t];
H[t_] := CDF[distribution, t];
(* RESULTS FOR THE VARIANCE *)
w = FindRoot[{H[b] - H[a] == gamma,
    a^2*h[a] == b^2*h[b]}, {{a, a0}, {b, b0}}];
{a, b} = {a, b} /. w;
r = (1/a - 1/b)/(1/a0 - 1/b0);
Print["Shortest solution for sigma^2:  a = ", a, " b = ", b,
   "    r = ", r];
(* RESULTS FOR THE STANDARD DEVIATION *)
w = FindRoot[{H[b] - H[a] == gamma,
    a^(3/2)*h[a] == b^(3/2)*h[b]}, {{a, a0}, {b, b0}}];
{a, b} = {a, b} /. w;
r = (1/Sqrt[a] - 1/Sqrt[b])/(1/Sqrt[a0] - 1/Sqrt[b0]);
```

```
Print["Shortest solution for sigma:    a = ", a, " b = ", b,
 "    r = ", r];
```

```
Data:    Conf. coeff. gamma = 0.95        d.f. g = 10
Usual solution:                a0 = 3.24697  b0 = 20.4832
Shortest solution for sigma^2:  a = 3.88553   b = 27.2662   r = 0.851563
Shortest solution for sigma:    a = 3.77287   b = 24.2303   r = 0.933155
```

Thus for the variance we obtain $r = 0.851563$ which means that the length of the shortest interval associated with the pivot function (2) is 85.16% compared to the one usually used. Corresponding to the standard deviation one obtains a ratio of 93.32%.

3 Confidence Interval for Ratios of Variances and Standard Deviations of Independent Normal Distributions

In classical statistics one currently uses the pivot function

$$T = \frac{\hat{\sigma}_1^2/\hat{\sigma}_2^2}{\sigma_1^2/\sigma_2^2} \equiv F_{g_1,g_2}. \tag{8}$$

Here $\hat{\sigma}_1^2, \hat{\sigma}_2^2$ are independent estimators of the group variances similar to those given by (1) depending on whether or not the corresponding means are known. Hence T is a ratio of two independent appropriately scaled chi-square variables with parameters g_1, g_2 and so it has a Fisher–Snedecor distribution F_{g_1,g_2}.

Confidence intervals for the variance ratio σ_1^2/σ_2^2 and standard deviation ratio σ_1/σ_2 are then

$$\gamma = \mathbb{P}\left(a < \frac{\hat{\sigma}_1^2/\hat{\sigma}_2^2}{\sigma_1^2/\sigma_2^2} < b\right) = \mathbb{P}\left(\frac{\hat{\sigma}_1^2/\hat{\sigma}_2^2}{b} < \frac{\sigma_1^2}{\sigma_2^2} < \frac{\hat{\sigma}_1^2/\hat{\sigma}_2^2}{a}\right)$$

$$= \mathbb{P}\left(\frac{\hat{\sigma}_1/\hat{\sigma}_2}{\sqrt{b}} < \frac{\sigma_1}{\sigma_2} < \frac{\hat{\sigma}_1/\hat{\sigma}_2}{\sqrt{a}}\right).$$

In practice in the construction of the confidence interval it is useful to choose parameters as $\alpha_1 = \alpha_2 = (1 - \gamma)/2$ so as to guarantee equality of probabilities at the end-points. To this end we must choose the quantiles

$$a_0 = F_{g_1,g_2;(1-\gamma)/2} \quad b_0 = F_{g_1,g_2;(1+\gamma)/2}. \tag{9}$$

Here we are interested in minimizing the respective lengths

$$L = \frac{\hat{\sigma}_1^2}{\hat{\sigma}_2^2} \left(\frac{1}{a} - \frac{1}{b} \right) \qquad \text{or} \qquad L = \frac{\hat{\sigma}_1}{\hat{\sigma}_2} \left(\frac{1}{\sqrt{a}} - \frac{1}{\sqrt{b}} \right), \tag{10}$$

under the condition

$$\mathbb{P}\left(a < F_{g_1,g_2} < b \right) = \gamma. \tag{11}$$

Denoting as H_{g_1,g_2} and h_{g_1,g_2} the cumulative distribution and density functions of F_{g_1,g_2} Eqs. (10) and (11) read

$$\begin{aligned} \text{minimize:} &\qquad \frac{1}{a} - \frac{1}{b} \quad \text{or} \quad \frac{1}{\sqrt{a}} - \frac{1}{\sqrt{b}} \\ \text{subject to:} &\qquad H_{g_1,g_2}(b) - H_{g_1,g_2}(a) = \gamma. \end{aligned}$$

To this end we use Lagrange multipliers method, and obtain, in the case of the ratio of variances the equations for the unknowns a, b

$$H_{g_1,g_2}(b) - H_{g_1,g_2}(a) = \gamma, \quad b^2 h_{g_1,g_2}(b) = a^2 h_{g_1,g_2}(a). \tag{12}$$

John [5] gives tables to solve this problem.

The lengths of the shortest intervals associated to the pivot function (8) obtained via (12) and standard obtained from (9) are related as

$$r = \left(\frac{1}{a} - \frac{1}{b} \right) \Big/ \left(\frac{1}{a_0} - \frac{1}{b_0} \right).$$

In the case of the ratio of standard deviations we obtain the system of equations

$$H_{g_1,g_2}(b) - H_{g_1,g_2}(a) = \gamma, \quad a^{3/2} h_{g_1,g_2}(a) = b^{3/2} h_{g_1,g_2}(b). \tag{13}$$

The connection between the shortest intervals associated to the pivot function (8) obtained via (13) and (9) is

$$r = \left(\frac{1}{\sqrt{a}} - \frac{1}{\sqrt{b}} \right) \Big/ \left(\frac{1}{\sqrt{a_0}} - \frac{1}{\sqrt{b_0}} \right).$$

Systems (12) and (13) can be solved numerically using the Mathematica code given below. The relevant parameters are the confidence level gamma along with the degrees of freedom g_1, g_2. In the example we take gamma equal to 0.90, $g_1 = 5$ and $g_2 = 10$.

```
(* FOR THE C.I. OF THE RATIOS OF VARIANCES AND
STD. DEVIATIONS *)
Clear[gamma,g1,g2,distribution,a0,b0,t,h,H,w,a,b,r];
(* DATA *)
gamma = 0.90;    (* CONFIDENCE COEFFICIENT *)
g1 = 5;          (* df OF NUMERATOR *)
g2 = 10;         (* df OF DENOMINATOR *)
(* COMPUTATIONS *)
Print["Data:  Conf. coeff. gamma = ", gamma,
"d.f. Num. g1 = ", g1," d.f. Den. g2 = ", g2];
distribution := FRatioDistribution[g1, g2];
a0 = Quantile[distribution, (1 - gamma)/2];
b0 = Quantile[distribution, (1 + gamma)/2];
Print["Usual solution:                       a0 = ", a0,
   "  b0 = ", b0];
h[t_] := PDF[distribution, t];
H[t_] := CDF[distribution, t];
(* RESULTS FOR THE RATIO OF VARIANCES *)
w = FindRoot[{H[b] - H[a] == gamma,
    a^2*h[a] == b^2*h[b]}, {{a, a0}, {b, b0}}];
{a, b} = {a, b} /. w;
r = (1/a - 1/b)/(1/a0 - 1/b0);
Print["Shortest solution for sigma1^2/sigma2^2:
a = ", a, "  b = ", b, "  r = ", r];
(* RESULTS FOR THE RATIO OF STANDARDS DEVIATIONS *)
w = FindRoot[{H[b] - H[a] == gamma,
    a^(3/2)*h[a] == b^(3/2)*h[b]}, {{a, a0}, {b, b0}}];
{a, b} = {a, b} /. w;
r = (1/Sqrt[a] - 1/Sqrt[b])/(1/Sqrt[a0] - 1/Sqrt[b0]);
Print["Shortest solution for sigma1/sigma2:
a = ", a, "  b = ", b, "  r = ", r];
```

```
Data:  Conf. coeff. gamma = 0.9   d.f. Num. g1 = 5   d.f. Den. g2 = 10
Usual solution:                    a0 = 0.21119   b0 = 3.32583
Shortest solution for sigma1^2/sigma2^2:   a = 0.300861   b = 9.55443
                                           r = 0.725947
Shortest solution for sigma1/sigma2:    a = 0.284173   b = 5.43858
                                        r = 0.889052
```

Note that for the variances ratio one obtains $r = 0.725947$; this means that the length of the shortest interval associated to the pivot function (8) is 72.59 % times to that usually used; by contrast it is 88.91% times for the standard deviation ratio.

4 Bayesian Confidence Intervals

Using a Bayesian statistics confidence intervals are evaluated via the posterior distribution of parameter θ. One needs to determine end-points $a < b$ such that

$$\mathbb{P}\left(a < \theta < b \middle| X = \mathbf{x}\right) = \gamma \tag{14}$$

To this end we fix two values $\alpha_1, \alpha_2 \geq 0$ and $\alpha_1 + \alpha_2 = 1 - \gamma$. and define a and b satisfying

$$\mathbb{P}\left(\theta \leq a \middle| X = \mathbf{x}\right) = \alpha_1, \mathbb{P}\left(\theta \geq b \middle| X = \mathbf{x}\right) = \alpha_2.$$

Under the additional assumption that $\xi_\eta(\theta|x)$ has a density $h(\theta)$ then a is the quantile α_1 while b is the quantile $1 - \alpha_2$ of the posterior probability distribution (see Fig. 1). The confidence interval follows from these values. In applications it is convenient to take $\alpha_1 = \alpha_2 = (1 - \gamma)/2$ since then statistical tables provide the values of a, b for given confidence levels. When the density h is symmetric and bell-shaped the corresponding confidence interval is of minimum length. If this is not the case one needs to decide how to choose a, b to guarantee that the relevant confidence interval has expected minimum length subject to condition (14), namely

$$\text{minimize } L = b - a$$

$$\text{subject to } H(b) - H(a) = \gamma.$$

Lagrange multiplier method yields then system

$$H(b) - H(a) = \gamma, \quad h(b) = h(a). \tag{15}$$

The connection between the shortest length associated to (15) and standard interval $a_0 < \theta < b_0$ corresponding to equal probabilities $\alpha_1 = \alpha_2 = (1 - \gamma)/2$ is

$$r = \frac{b - a}{b_0 - a_0}.$$

For a beta distribution $\beta(u, v)$, which usually appears when estimating a proportion θ, and whose density functions is:

$$h(\theta) = \frac{1}{B(u, v)}\theta^{u-1}(1 - \theta)^{v-1} \propto \theta^{u-1}(1 - \theta)^{v-1}, \quad 0 < \theta < 1$$

conditions $u > 1, v > 1$ guarantee a bell-shaped distribution. To implement the previous ideas we have used the Mathematica code given below, corresponding to gamma equal to 0.95 and distribution `BetaDistribution[3, 8]`.

```
(* BAYESIAN CONFIDENCE INTERVAL *)
Clear[gamma, a, a0, b, b0, distribution, h, H, theta, w, r];
(* DATA *)
gamma = 0.95;   (* CONFIDENCE COEFFICIENT *)
distribution := BetaDistribution[3, 8];
(* COMPUTATIONS *)
a0 = Quantile[distribution, (1 - gamma)/2];
b0 = Quantile[distribution, (1 + gamma)/2];
h[theta_] := PDF[distribution, theta];
H[theta_] := CDF[distribution, theta];
w = FindRoot[{H[b] - H[a] == gamma, h[a] == h[b]},
{{a, a0}, {b, b0}}];
{a, b} = {a, b} /. w;
r = (b - a)/(b0 - a0);
(* RESULTS *)
Print["Data:    Conf. coeff. gamma = ", gamma,
"Distribution = ",  distribution];
Print["Usual C.I.:      a0 = ", a0, "       b0 = ", b0];
Print["Shortest C.I.:    a = ", a, "        b = ", b];
Print["Ratio of C.I.'s lengths:      r = ", r];
```

```
Data:    Conf. coeff. gamma = 0.95   Distribution = BetaDistribution[3,8]
Usual C.I.:      a0 = 0.0667395     b0 = 0.556095
Shortest C.I.:    a = 0.0464466     b = 0.522428
Ratio of C.I.'s lengths:      r = 0.97267
```

The length ratio that follows is $r = 0.97267$ which means that the length of the shortest interval is 97.27% times the one usually used.

To estimate the mean of a Poisson variable, the Gamma distribution $\Gamma(p, \alpha)$, p, $\alpha > 0$ with density

$$h(\theta) = \frac{1}{\alpha^p \Gamma(p)} e^{-\theta/\alpha} \theta^{p-1} \propto e^{-\theta/\alpha} \theta^{p-1} \quad \theta > 0$$

plays an important role. To guarantee a bell-shaped form it is convenient to choose the parameter $p > 1$. In this case we use the previous code with given data for the confidence level gamma, posterior distribution, specific values for $p, \alpha > 0$ and the sentence distribution:= GammaDistribution $[p, \alpha]$. In the concrete case gamma equal to 0.90, $p = 3$, $\alpha = 8$ the following results are obtained:

```
Data:    Conf. coeff. gamma = 0.9   Distribution = GammaDistribution[3,8]
Usual C.I.:      a0 = 6.54153     b0 = 50.3663
Shortest C.I.:    a = 3.53062     b = 43.8334
Ratio of C.I.'s lengths:      r = 0.919634
```

For this example the length ratio is $r = 0.919634$ which means that the length of the shortest interval is 91.96% times the one usually used.

Corresponding to the problem of estimating the mean of an exponential variable or the variance of a normal the inverse-Gamma distribution $I\Gamma(p, \alpha)$, $p, \alpha > 0$ usually comes about. The relevant pdf has the form

$$h(\theta) = \frac{\alpha^p}{\Gamma(p)} e^{-\alpha/\theta} \theta^{-1-p} \propto e^{-\alpha/\theta} \theta^{-1-p} \quad \theta > 0.$$

In this case we use again the previous code with the corresponding degree of confidence level gamma, posterior distribution, specific values for $p, \alpha > 0$ and the sentence distribution:= InverseGammaDistribution $[p, \alpha]$. In the concrete case gamma equal to 0.95, $p = 4$, $\alpha = 10$ the following results are obtained:

```
Data: Conf. coeff. gamma = 0.95
      Distribution = InverseGammaDistribution[4,10]
Usual C.I.:       a0 = 1.14061      b0 = 9.17545
Shortest C.I.:    a = 0.802739      b = 7.39994
Ratio of C.I.'s lengths:           r = 0.821075
```

For this example the length ratio is $r = 0.821075$ which means that the length of the shortest interval is 82.11% times the one usually used.

Finally, the same or similar type of Mathematica program can be used to determine shortest confidence intervals for other continuous bell-shaped posterior distributions.

References

1. Bartoszynski R, Niewiadomska-Bugaj M (1996) Probability and statistical inference. Wiley, New York
2. Crisman R (1975) Shortest confidence interval for the standard deviation of a normal distribution. J Undergr Math 7:57–62
3. De Groot MH (2004) Optimal statistical decisions. Wiley, Hoboken
4. Guenter WC (1969) Shortest confidence intervals. Am Stat 23(1):22–25
5. John S (1975) Tables for comparing two normal variances or two Gamma means. J Am Stat Assoc 70:344–347
6. Kendall MG, Stuart A (1979) The advanced theory of statistics, vol 2, 4th edn. Charles Griffin & Co Limited, London
7. Rohatgi VK (1976) An introduction to probability theory and mathematicas statistics. Wiley, New York
8. Tate RF, Klett GW (1959) Optimum confidence intervals for the variance of a normal distribution. J Am Stat Assoc 54:674–682
9. Wilks SS (1938) Shortest average confidence intervals from large samples. Ann Math Stat 9(3):166–175
10. Wilks SS (1962) Mathematical statistics. Wiley, New York
11. Wolfram S (2017) An elementary introduction to the wolfram language, 2nd edn. Wolfram Media, Champaign
12. Zacks S (1971) The theory of statistical inference. Wiley, New York

An Optimal Transportation Approach for Assessing Almost Stochastic Order

Eustasio del Barrio, Juan A. Cuesta-Albertos and Carlos Matrán

Abstract When stochastic dominance does not hold, we can improve agreement to stochastic order by suitably trimming both distributions. In this work we consider the L_2–Wasserstein distance, $\mathscr{W}_2$, to stochastic order of these trimmed versions. Our characterization for that distance naturally leads to consider a $\mathscr{W}_2$-based index of disagreement with stochastic order, $\varepsilon_{\mathscr{W}_2}(F, G)$. We provide asymptotic results allowing to test $H_0 : \varepsilon_{\mathscr{W}_2}(F, G) \geq \varepsilon_0$ versus $H_a : \varepsilon_{\mathscr{W}_2}(F, G) < \varepsilon_0$, that, under rejection, would give statistical guarantee of almost stochastic dominance. We include a simulation study showing a good performance of the index under the normal model.

1 Introduction

Let P, Q be probability distributions on the real line with distribution functions (d.f.'s in the sequel) F, G, respectively. Stochastic dominance of Q over P, denoted $P \leq_{st} Q$, is defined in terms of the d.f.'s by $F(x) \geq G(x)$ for every $x \in \mathbb{R}$ (throughout we will also use the alternative notation $F \leq_{st} G$). The meaning of this relation is that random outcomes produced by the second law tend to be larger than those produced by the first one. We gain a better understanding of this stochastic order by considering a quantile representation. For a d.f. F, the quantile function associated to F, that we will denote by F^{-1}, is defined by

$$F^{-1}(t) = \inf\{x : t \leq F(x)\}, \ t \in (0, 1).$$

E. del Barrio (✉) · C. Matrán
Departamento de Estadística e Investigación Operativa and IMUVA,
Universidad de Valladolid, 47002 Valladolid, Spain
e-mail: tasio@eio.uva.es

C. Matrán
e-mail: matran@eio.uva.es

J. A. Cuesta-Albertos
Departamento de Matemáticas, Estadística y Computación,
Universidad de Cantabria, 39005 Cantabria, Spain
e-mail: cuestaj@unican.es

© Springer International Publishing AG 2018
E. Gil et al. (eds.), *The Mathematics of the Uncertain*, Studies in Systems,
Decision and Control 142, https://doi.org/10.1007/978-3-319-73848-2_3

The following well-known statements (see e.g. [14]) are equivalent to $F \leq_{st} G$:

(a) There exist random variables X, Y defined on some probability space (Ω, σ, μ), with respective laws P and Q ($\mathcal{L}(X) = P, \mathcal{L}(Y) = Q$), satisfying $\mu(X \leq Y) = 1$.

(b) $F^{-1}(t) \leq G^{-1}(t)$ for every $t \in (0, 1)$.

Quantile functions (also called 'monotone rearrangements' in other contexts) are characterized by $F^{-1}(t) \leq x$ if and only if $t \leq F(x)$. Therefore it is straightforward that, when considered as random variables defined on the unit interval with the Lebesgue measure, they satisfy $\mathcal{L}(F^{-1}) = P, \mathcal{L}(G^{-1}) = Q$. This representation quickly shows that (a) and (b) are equivalent and, more importantly in the present setting, allows us to relate characteristics and measure agreement or disagreements with the stochastic order.

From the previous considerations it becomes clear that guaranteeing stochastic dominance, $F \leq_{st} G$, should be the goal when comparing, for instance, treatments. However, the rejection of $F \nleq_{st} G$, on the basis of two data samples is an ill posed statistical problem: As showed in [7] and noted in [12, 15], or [4], the 'non-data test', namely the test which rejects with probability α, regardless the data, is uniformly most powerful for testing the nonparametric hypotheses $H_0 : F \nleq_{st} G$ versus $H_a : F \leq_{st} G$. This fact motivates recent research looking for suitable indices measuring approximate versions of stochastic dominance. Here, suitability of an index must be understood in terms of computability and interpretability, but also in terms of statistical performance. Usually, as already suggested in a general context in [13], such measures of nearness involve the use of some kind of distance to the null. This will also be the approach here, with the choice of the L_2-Wasserstein distance between probabilities. For P, Q in the set $\mathcal{F}_2(\mathbb{R}^d)$ of Borel probabilities on $\mathbb{R}^d$ with finite second order moments, this distance is defined as

$$\mathcal{W}_2(P, Q) := \min \sqrt{\int \int \|x - y\|^2 d\nu(x, y)}, \nu \in \mathcal{F}_2(\mathbb{R}^d \times \mathbb{R}^d) \text{ with marginals } P, Q.$$

In the univariate case, $\mathcal{W}_2$ equals the L_2-distance between quantile functions, namely,

$$\mathcal{W}_2(P, Q) = \sqrt{\left(\int_0^1 |F^{-1}(t) - G^{-1}(t)|^2 dt \right)}. \tag{1}$$

Statistical applications based on optimal transportation, and particularly on the L_2 version of the Wasserstein distance, are receiving considerable attention in recent times (see e.g. [8–10, 17] or [5]). We should also mention our papers [1, 2], dealing with similarity of distributions (as a relaxation of homogeneity) through this distance, and also [4] (and [3]) which introduced an index of disagreement from stochastic dominance based on the idea of similarity. The key to this index is the existence, for a given (small enough) π, of mixture decompositions

$$F = (1 - \pi)\tilde{F} + \pi H_F$$
$$G = (1 - \pi)\tilde{G} + \pi H_G,$$
for some d.f.'s $\tilde{F}, H_F, \tilde{G}, H_G$ with $\tilde{F} \leq_{st} \tilde{G}$. (2)

If model (2) holds then stochastic order holds after removing contaminating π-fractions from each population. The minimum π compatible with (2), denoted by $\pi(F, G)$, can be taken as a measure of deviation from stochastic order, see [4] for details. The analysis in [4] is based on the connection between contamination models and trimmed probabilities. We recall that an α-trimming of a probability, P, is any other probability, say $\tilde{P}$, such that

$$\tilde{P}(A) = \int_A \tau \, dP \quad \text{for every event } A$$

for some function τ taking values in $\left[0, \frac{1}{1-\alpha}\right]$. Like the trimming methods, commonly used in Robust Statistics, consisting of removing disturbing observations, the function τ allows to discard or downplay the influence of some regions on the sample space. On the real line, writing $\mathscr{R}_\alpha(F)$ for the set of trimmings of F, it turns out (see [4]) that

$$F = (1 - \alpha)\tilde{F} + \alpha H_F \text{ for some d.f.'s } \tilde{F}, H_F \text{ if and only if } \tilde{F} \in \mathscr{R}_\alpha(F). \quad (3)$$

The contaminated stochastic order model (2) can also be recast in terms of trimmings. If we denote

$$\mathscr{F}_{st} := \{(H_1, H_2) \in \mathscr{F}_2 \times \mathscr{F}_2 : \quad H_1 \leq_{st} H_2\},$$

then, for $F, G \in \mathscr{F}_2$, (2) holds if and only if

$$(\mathscr{R}_\pi(F) \times \mathscr{R}_\pi(G)) \cap \mathscr{F}_{st} \neq \emptyset \quad (4)$$

or, equivalently (this follows from compactness of $\mathscr{R}_\pi(F) \times \mathscr{R}_\pi(G)$ with respect to d_2; we omit details), if and only if

$$d_2(\mathscr{R}_\pi(F) \times \mathscr{R}_\pi(G), \mathscr{F}_{st}) = 0, \quad (5)$$

where d_2 denotes the metric on the set $\mathscr{F}_2 \times \mathscr{F}_2$ given by

$$d_2((F_1, F_2), (G_1, G_2)) = \sqrt{\mathscr{W}_2^2(F_1, G_1) + \mathscr{W}_2^2(F_2, G_2)}$$

and, for $A, B \subset \mathscr{F}_2 \times \mathscr{F}_2, d_2(A, B) = \inf_{a \in A, b \in B} d_2(a, b)$.

For fixed π, $d_2(\mathscr{R}_\pi(F) \times \mathscr{R}_\pi(G), \mathscr{F}_{st})$ can be used as a measure of deviation from the contaminated stochastic order model (2). In this work we obtain a simple explicit characterization of this measure (see Theorem 2.1 below) that can be used for statistical purposes. Later, we use this characterization to introduce a new index,

$\varepsilon_{\mathcal{W}_2}$, see (9), to evaluate disagreement with respect to the (non-contaminated) stochastic order. We also provide asymptotic theory (Theorem 2.2) about the behavior of this index, that allows addressing the goal of statistical assessment of $\varepsilon_{\mathcal{W}_2}$-almost stochastic dominance. This index has some similarity with that proposed in [15] for which, in contrast, asymptotics are not yet available.

The remaining sections of this work are organized as follows. Section 2 presents the announced results, introduces the new index $\varepsilon_{\mathcal{W}_2}$ and discusses its application in the statistical assessment of almost stochastic order. This includes an illustration of the meaning of the index in the case of normal distributions and a small simulation study. Finally, the more technical proof of Theorem 2.2 is given in an appendix.

2 Main Results

A fact that eases the use of trimming in the stochastic dominance setting is that the set $\mathscr{R}_\alpha(F)$ has a minimum and a maximum for the stochastic order. Both can be easily characterized as follows (see Proposition 2.3 in [4]).

Proposition 2.1 *Consider a d.f. F and $\pi \in [0, 1)$. Define the d.f.'s*

$$F^\pi(x) = \max\left(\tfrac{1}{1-\pi}(F(x) - \pi), 0\right) \quad \text{and} \quad F_\pi(x) = \min\left(\tfrac{1}{1-\pi}F(x), 1\right).$$

Then $F^\pi, F_\pi \in \mathscr{R}_\pi(F)$ and any other $\tilde{F} \in \mathscr{R}_\pi(F)$ satisfies $F_\pi \leq_{st} \tilde{F} \leq_{st} F^\pi$.

Keeping the notation and recalling the characterization of the stochastic order in terms of quantile functions, a simple computation shows the relations

$$(F_\pi)^{-1}(t) = F^{-1}((1 - \pi)t), \quad (F^\pi)^{-1}(t) = F^{-1}(\pi + (1 - \pi)t), \qquad (6)$$

so we can restate this proposition in the following new way.

Proposition 2.2 *If $\tilde{F} \in \mathscr{R}_\pi(F)$, then its quantile function satisfies*

$$F^{-1}((1 - \pi)t) \leq \tilde{F}^{-1}(t) \leq F^{-1}(\pi + (1 - \pi)t), \quad 0 < t < 1. \qquad (7)$$

We use (7) to prove our next result, the announced characterization for $d_2(\mathscr{R}_\pi(F) \times \mathscr{R}_\pi(G), \mathscr{F}_{st})$, a quantity that measures deviation from the contaminated stochastic order model (2). We keep the notation in (6) and define

$$(L_\pi)^{-1}(t) = \begin{cases} (F_\pi)^{-1}(t) & \text{if} (F_\pi)^{-1}(t) \leq (G^\pi)^{-1}(t) \\ \tfrac{1}{2}((F_\pi)^{-1}(t) + (G^\pi)^{-1}(t)) & \text{if} (F_\pi)^{-1}(t) > (G^\pi)^{-1}(t) \end{cases}$$

$$(U_\pi)^{-1}(t) = \begin{cases} (G^\pi)^{-1}(t) & \text{if} (F_\pi)^{-1}(t) \leq (G^\pi)^{-1}(t) \\ \tfrac{1}{2}((F_\pi)^{-1}(t) + (G^\pi)^{-1}(t)) & \text{if} (F_\pi)^{-1}(t) > (G^\pi)^{-1}(t). \end{cases}$$

Theorem 2.1 *With the above notation, if $F, G \in \mathscr{F}_2$, then L_π^{-1}, U_π^{-1} are the quantile functions of a pair $(L_\pi, U_\pi) \in \mathscr{F}_{st}$. Furthermore, if $x_+ := \max(x, 0)$,*

$$d_2(\mathscr{R}_\pi(F) \times \mathscr{R}_\pi(G), \mathscr{F}_{st}) = d_2((F_\pi, G^\pi), (L_\pi, U_\pi))$$

$$= \sqrt{\frac{1}{2} \int_0^1 (F^{-1}((1 - \pi)t) - G^{-1}(\pi + (1 - \pi)t))_+^2 \, dt}.$$

Proof To see that $(L_\pi)^{-1}$ is a quantile function we note that

$$(L_\pi)^{-1}(t) = \min\left((F_\pi)^{-1}(t), \tfrac{1}{2}\left((F_\pi)^{-1}(t) + (G^\pi)^{-1}(t)\right)\right).$$

This shows that $(L_\pi)^{-1}$ is nondecreasing and left continuous, hence a quantile function. That $L_\pi \in \mathscr{F}_2$ follows from the elementary bounds

$$-\left(\left|(F_\pi)^{-1}(t)\right| + \left|(G^\pi)^{-1}(t)\right|\right) \leq (L_\pi)^{-1}(t) \leq (F_\pi)^{-1}(t).$$

A similar argument works for U_π. Obviously $L_\pi \leq_{st} U_\pi$ and, therefore, $(L_\pi, U_\pi) \in \mathscr{F}_{st}$. Now, for any $(U_1, U_2) \in \mathscr{R}_\pi(F) \times \mathscr{R}_\pi(G)$ and $(V_1, V_2) \in \mathscr{F}_{st}$ we have $U_1^{-1}(t) \geq (F_\pi)^{-1}(t), U_2^{-1}(t) \leq (G^\pi)^{-1}(t), V_1^{-1}(t) \leq V_2^{-1}(t)$. We define $A_\pi := \{t \in (0, 1) : (F_\pi)^{-1}(t) > (G^\pi)^{-1}(t)\}$. Then

$$d_2((U_1, U_2), (V_1, V_2)) = \int_0^1 ((U_1^{-1}(t) - V_1^{-1}(t))^2 + (U_2^{-1}(t) - V_2^{-1}(t))^2) dt$$

$$\geq \int_{A_\pi} ((U_1^{-1}(t) - V_1^{-1}(t))^2 + (U_2^{-1}(t) - V_2^{-1}(t))^2) dt$$

$$\geq \int_{A_\pi} (((F_\pi)^{-1}(t) - (L_\pi)^{-1}(t))^2 + ((G^\pi)^{-1}(t) - (U_\pi)^{-1}(t))^2) dt$$

$$= \int_0^1 (((F_\pi)^{-1}(t) - (L_\pi)^{-1}(t))^2 + ((G^\pi)^{-1}(t) - (U_\pi)^{-1}(t))^2) dt$$

$$= d_2((F_\pi, G^\pi), (L_\pi, U_\pi)),$$

where the second inequality is just the trivial fact that if $f > g$, then the minimum value $\min_{a,b,c,d}(a - b)^2 + (c - d)^2$, for $a \geq f, c \leq g, b \leq d$ is just attained at $a = f, c = g, b = d = \frac{f+g}{2}$. To complete the proof we note that

$$d_2((F_\pi, G^\pi), (L_\pi, U_\pi)) = \frac{1}{2} \int_{A_\pi} ((F_\pi)^{-1}(t) - (G^\pi)^{-1}(t))^2 dt$$

$$= \frac{1}{2} \int_0^1 (F^{-1}((1 - \pi)t) - G^{-1}(\pi + (1 - \pi)t))_+^2 \, dt. \qquad \square$$

Particularizing for $\pi = 0$, we have the following corollary of Theorem 2.1:

Corollary 2.1 *Under the assumptions in Theorem 2.1, the quantile functions $L_0^{-1} = \inf\{F^{-1}, (F^{-1} + G^{-1})/2\}$ and $U_0^{-1} = \sup\{G^{-1}, (F^{-1} + G^{-1})/2\}$ verify:*

$$d_2((F, G), \mathscr{F}_{st}) = d_2((F, G), (L_0, U_0)) = \sqrt{\frac{1}{2} \int_0^1 (F^{-1}(t) - G^{-1}(t))_+^2 dt}. \quad (8)$$

Also note that, avoiding the factor $1/2$, the square of the distance in (8) is just the part of $\mathscr{W}_2^2(F, G)$ due to the violation of stochastic dominance. Therefore, for distinct d.f.'s F, G the quotient

$$\varepsilon_{\mathscr{W}_2}(F, G) := \frac{\int (F^{-1}(t) - G^{-1}(t))_+^2 dt}{\mathscr{W}_2^2(F, G)} \quad (9)$$

can be considered a normalized index of such violation. It satisfies $0 \le \varepsilon_{\mathscr{W}_2}(F, G) \le 1$, with the extreme values 0 and 1 corresponding, respectively, to perfect stochastic dominance of G over F and vice-versa. When $\varepsilon_{\mathscr{W}_2}(F, G) < \varepsilon$ we would say that F is almost stochastically dominated, at level ε, by G (w.r.t. $\varepsilon_{\mathscr{W}_2}$). We notice that [15], following a very different motivation, introduced a related index consisting in the quotient $\int (G(x) - F(x))_+ dx / \int |G(x) - F(x)| dx$.

The next result gives the mathematical background to carry statistical analyses based on the index $\varepsilon_{\mathscr{W}_2}(F, G)$. Its proof is given in the appendix.

Theorem 2.2 *Let F, G be distinct d.f.'s in $\mathscr{F}_2$ and assume $n, m \to \infty$ with $\frac{n}{n+m} \to \lambda \in (0, 1)$. If F_n and G_m are the sample d.f.'s based on independent samples of F and G with sizes n and m respectively, then $\varepsilon_{\mathscr{W}_2}(F_n, G_m) \to \varepsilon_{\mathscr{W}_2}(F, G)$ a.s. If, additionally, F and G have bounded convex supports, then*

$$\sqrt{\frac{mn}{m+n}} \left(\varepsilon_{\mathscr{W}_2}(F_n, G_m) - \varepsilon_{\mathscr{W}_2}(F, G) \right) \to_w N(0, \sigma_\lambda^2(F, G)), \quad (10)$$

where, if $u_\pm(x) = \int_0^x 2(s - G^{-1}(F(s)))_\pm ds$, and X, Y are r.v.'s such that $\mathscr{L}(X) = F$ and $\mathscr{L}(Y) = G$, then

$$\sigma_\lambda^2(F, G) = \frac{1}{\mathscr{W}_2^8(F, G)} [(1 - \lambda) Var(u_-(X)) + \lambda Var(u_+(Y))].$$

A critical analysis of the problem of assessing improvement in a treatment comparison from the perspective of stochastic dominance appears in [6]. It is argued there that under, say, normality assumptions, improvement with the new treatment is often assessed using a one sided test for the mean, while the really interesting test would be that of $F \not\le_{st} G$ versus $F \le_{st} G$. Since, as argued in the Introduction, this is not a feasible statistical task, [6] emphasizes on the alternative, feasible goal, of testing that slightly relaxed versions of stochastic dominance hold. In the present setting, such

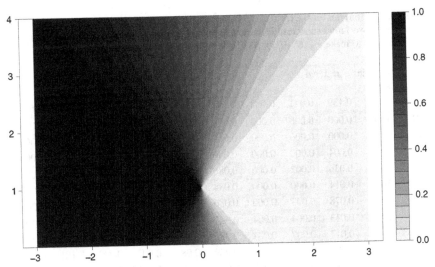

Fig. 1 Contour-plot of $\varepsilon_{\mathscr{W}_2}(N(0, 1), N(\mu, \sigma^2))$ as in (9) for different values of μ (X-axis) and σ (Y-axis)

a strategy leads to testing, at a given confidence level, $H_0 : \varepsilon_{\mathscr{W}_2}(F, G) \geq \varepsilon_0$ versus $H_a : \varepsilon_{\mathscr{W}_2}(F, G) < \varepsilon_0$, where ε_0 is a small enough prefixed amount of disagreement with the stochastic order.

Following the scheme in [4, 6], from the asymptotic normality obtained in Theorem 2.2 we propose to reject H_0 if

$$\sqrt{\frac{nm}{n+m}}(\varepsilon_{\mathscr{W}_2}(F_n, G_m) - \varepsilon_0) < \hat{\sigma}_{n,m}\Phi^{-1}(\alpha), \qquad (11)$$

where $\hat{\sigma}_{n,m}$ is an estimator of $\sigma_\lambda(F, G)$ (for example a bootstrap estimator). This rejection rule provides a consistent test of asymptotic level α. Also,

$$\hat{U} := \varepsilon_{\mathscr{W}_2}(F_n, G_m) - \sqrt{\frac{n+m}{nm}}\hat{\sigma}_{n,m}\Phi^{-1}(\alpha) \qquad (12)$$

provides an upper confidence bound for $\varepsilon_{\mathscr{W}_2}(F, G)$ with asymptotic level $1 - \alpha$.

Let us take now a closer look at the $\varepsilon_{\mathscr{W}_2}$ index for distributions in a location-scale family. For simplicity, we focus on normal laws. It is an elementary fact that $\varepsilon_{\mathscr{W}_2}$ is invariant to changes in location and scale and we can, consequently, restrict ourselves to the analysis of the values of $\varepsilon_{\mathscr{W}_2}(N(0, 1), N(\mu, \sigma^2))$, $\mu \in \mathbb{R}, \sigma > 0$. Moreover, it is easy to see that $\varepsilon_{\mathscr{W}_2}$ is constant when (μ, σ) moves along directed rays from $(0, 1)$. This fact is showed in Fig. 1. We see that $\mu > 0$ corresponds to $\varepsilon_{\mathscr{W}_2}(N(0, 1), N(\mu, \sigma^2)) < \frac{1}{2}$, with $\varepsilon_{\mathscr{W}_2}(N(0, 1), N(\mu, 1)) = 0$, but the index can be made arbitrarily close to $\frac{1}{2}$ by taking σ large enough.

Table 1 Rejection rates for $\varepsilon_{\mathcal{W}_2}(N(0, 1), N(\mu, \sigma^2)) \geq \varepsilon_0$ at level $\alpha = .05$ along 1,000 simulations. Upper (resp. lower) rows show results for nonparametric (resp. parametric) comparisons. For each σ, μ is chosen to make $\varepsilon_{\mathcal{W}_2}(N(0, 1), N(\mu, \sigma^2)) = 0.01, 0.05$ and 0.10 (first, second and third columns, resp.)

ε_0	Sample size	$\sigma = 1.1\ \mu$			$\sigma = 1.5\ \mu$			$\sigma = 2\ \mu$		
		0.139	0.091	0.068	0.697	0.455	0.341	1.395	0.909	0.683
0.01	100	0.000	0.000	0.000	0.053	0.007	0.000	0.180	0.009	0.004
		0.000	0.000	0.000	0.062	0.006	0.000	0.112	0.003	0.000
	1000	0.004	0.000	0.000	0.086	0.000	0.000	0.116	0.000	0.000
		0.036	0.002	0.000	0.086	0.000	0.000	0.086	0.000	0.000
	5000	0.014	0.000	0.000	0.084	0.000	0.000	0.077	0.000	0.000
		0.078	0.003	0.000	0.086	0.000	0.000	0.060	0.000	0.000
0.05	100	0.013	0.004	0.004	0.321	0.060	0.019	0.677	0.138	0.028
		0.017	0.007	0.004	0.382	0.064	0.027	0.690	0.086	0.017
	1000	0.101	0.017	0.004	0.929	0.088	0.003	0.999	0.101	0.000
		0.219	0.041	0.015	0.982	0.087	0.002	1.000	0.085	0.000
	5000	0.488	0.056	0.009	1.000	0.067	0.000	1.000	0.070	0.000
		0.704	0.099	0.009	1.000	0.069	0.000	1.000	0.057	0.000
0.10	100	0.034	0.017	0.006	0.608	0.210	0.092	0.930	0.402	0.148
		0.040	0.022	0.009	0.658	0.205	0.073	0.941	0.364	0.109
	1000	0.267	0.082	0.020	1.000	0.545	0.076	1.000	0.861	0.096
		0.431	0.132	0.047	1.000	0.642	0.076	1.000	0.928	0.084
	5000	0.867	0.246	0.058	1.000	0.970	0.056	1.000	1.000	0.078
		0.960	0.356	0.087	1.000	0.994	0.058	1.000	1.000	0.069

Finally, we present in Table 1 some simulations showing the performance of the proposed nonparametric procedure. We see the observed rejection rates for the test (11), when $F = N(0, 1)$ and $G = N(\mu, \sigma^2)$ for several choices of μ, σ. We show also the rejection rates based on a natural competitor, the parametric maximum likelihood estimator $\hat{\varepsilon}_{\mathcal{W}_2} := \varepsilon_{\mathcal{W}_2}\left(F_{N(\bar{X}_n, S_X^2)}, F_{N(\bar{Y}_m, S_Y^2)}\right)$. This estimator is, of course, highly nonrobust and useless in practice without the a priori knowledge that F and G are normal, but we use it here as a benchmark. We see a reasonable amount of agreement of the rejection frequencies to the nominal level of the test, even if it is slightly liberal for σ close to one and small ε_0, but the nonparametric procedure does not perform worse than the parametric benchmark. We also see that it is possible to get statistical evidence that almost stochastic order does hold. For instance, for $\mu = .697$, $\sigma = 1.5$ (true $\varepsilon_{\mathcal{W}_2} = 0.01$) sizes $n = m = 1000$ suffice to conclude that $\varepsilon_{\mathcal{W}_2} < 0.05$ with probability close to 0.93.

3 Appendix

We prove here central limit theorems for the index $\varepsilon_{\mathcal{W}_2}$ in (9). We will assume that $U_1, \ldots, U_n, V_1, \ldots, V_m$ are i.i.d. r.v.'s, uniformly distributed on $(0, 1)$. We consider independent samples i.i.d. $X_1, \ldots, X_n$ and $Y_1, \ldots, Y_m$ such that $\mathcal{L}(X_i) = F$ and $\mathcal{L}(Y_j) = G$. We note that, without loss of generality, we can assume that the X_i and Y_j are generated from the U_i and the V_j through $X_i = F^{-1}(U_i)$, $Y_j = G^{-1}(V_j)$. We write F_n, G_m, $H_{n,1}$ and $H_{m,2}$ for the empirical d.f.'s on the X_i, the Y_j, U_i and the V_j, respectively. Note that, in particular, $F_n^{-1}(t) = F^{-1}(H_{n,1}^{-1}(t))$, $G_m^{-1}(t) = G^{-1}(H_{m,2})$. Finally, $\alpha_{n,1}$ and $\alpha_{m,2}$ will denote the empirical processes associated to the U_i and the Y_j, namely, $\alpha_{n,1}(t) = \sqrt{n}(H_{n,1}(t) - t), 0 \leq t \leq 1$, and similarly for $\alpha_{m,2}$ and we will write $\alpha_{n,1}(h)$ instead of $\int_0^1 h(t) d\alpha_{n,1}(t)$.

We introduce the statistics $S_n = \int_0^1 (F_n^{-1} - G^{-1})^2$, $S_n^{\pm} = \int_0^1 (F_n^{-1} - G^{-1})_{\pm}^2$, and write S, S^+, S^- for the corresponding population counterparts. Note that, to ensure that S is finite, it is enough that $F, G \in \mathcal{F}_2$. However, to simplify the arguments our proof will require bounded supports. We set

$$T_n = \sqrt{n}(S_n - S), \quad T_n^+ = \sqrt{n}(S_n^+ - S^+), \quad T_n^- = \sqrt{n}(S_n^- - S^-),$$

$c(x) = 2x, c_+(x) = 2x_+, c_-(x) = 2x_-$ and define

$$v(t) = \int_0^{F^{-1}(t)} c(s - G^{-1}(F(s))) ds \tag{13}$$

and similarly v_+ and v_- replacing c with c_+ and c_-, respectively. Observe that $v = v_+ - v_-$. With this notation we have the following result.

Theorem 3.1 *If F and G have bounded support and G^{-1} is continuous on $(0, 1)$ then*

$$T_n = \alpha_{n,1}(v) + o_P(1), \quad T_n^+ = \alpha_{n,1}(v_+) + o_P(1), \quad T_n^- = \alpha_{n,1}(v_-) + o_P(1).$$

Proof We assume that $|F^{-1}(t)| \leq M$, $|G^{-1}(t)| \leq M$ for all $t \in (0, 1)$ and some $M > 0$. The continuity and boundedness assumption on G^{-1} allows us to assume that G^{-1} is a continuous function on $[0,1]$, hence, uniformly continuous and its modulus of continuity,

$$\omega(\delta) = \sup_{|t_1 - t_2| \leq \delta} |G^{-1}(t_1) - G^{-1}(t_2)|,$$

satisfies $\omega(\delta) \to 0$ as $\delta \to 0$. It is convenient at this point to note that T_n is a function of the U_i and also of F and we stress this fact writing $T_n(F)$ instead of T_n in this proof, and the same for T_n^+ and T_n^-. Similarly, we set $\tilde{T}_n(F) = \alpha_{n,1}(v)$, $\tilde{T}_n^+(F) = \alpha_{n,1}(v_+)$, $\tilde{T}_n^-(F) = \alpha_{n,1}(v_-)$. We claim now that

$$E|T_n(F) - \tilde{T}_n(F)|^2 \leq 16M^2 E\left(\|\alpha_{n,1}\|^2 \omega^2\left(\frac{\|\alpha_{n,1}\|}{\sqrt{n}}\right)\right), \tag{14}$$

where $\|\alpha_{n,1}\| = \sup_{0 \le t \le 1} |\alpha_{n,1}(t)|$. To check this, let us assume first that F is finitely supported, say on $-M \le x_1 < \ldots < x_k \le M$ with $F(x_j) = s_j$, $j = 1, \ldots, k$. This means that $F^{-1}(t) = x_i$ if $s_{i-1} < t \le s_i$ (we set $s_0 = 0$ for convenience) and we have

$$\int_0^1 (F^{-1} - G^{-1})^2 = \sum_{i=1}^k \int_{s_{i-1}}^{s_i} (x_i - G^{-1}(t))^2 dt = \int_0^1 (x_k - G^{-1}(t))^2 dt$$
$$- \sum_{i=1}^{k-1} \int_0^{s_i} \left[(x_{i+1} - G^{-1}(t))^2 - (x_i - G^{-1}(t))^2 \right] dt$$
$$= \int_0^1 (x_k - G^{-1}(t))^2 dt - \sum_{i=1}^{k-1} \int_0^{s_i} \left[\int_{x_i}^{x_{i+1}} c(s - G^{-1}(t)) ds \right] dt.$$

A similar expression holds for $\int_0^1 (F_n^{-1} - G^{-1})^2$ replacing s_i with $H_{n,1}(s_i)$ and we see that

$$T_n(F) = -\sqrt{n} \sum_{i=1}^{k-1} \int_{s_i}^{H_{n,1}(s_i)} \left(\int_{x_i}^{x_{i+1}} c(s - G^{-1}(t)) ds \right) dt.$$

We can argue analogously to check that

$$\tilde{T}_n(F) = -\sum_{i=1}^{k-1} \alpha_{n,1}(s_i) \left(\int_{x_i}^{x_{i+1}} c(s - G^{-1}(s_i)) ds \right)$$
$$= -\sqrt{n} \sum_{i=1}^{k-1} \int_{s_i}^{H_{n,1}(s_i)} \left(\int_{x_i}^{x_{i+1}} c(s - G^{-1}(s_i)) ds \right) dt.$$

Hence, we see that

$$|T_n(F) - \tilde{T}_n(F)| \le 2 \sum_{i=1}^{k-1} |\alpha_{n,1}(s_i)| (x_{i+1} - x_i) \omega\left(\frac{\|\alpha_{n,1}\|}{\sqrt{n}}\right)$$
$$\le 2\|\alpha_n\| (x_k - x_1) \omega\left(\frac{\|\alpha_{n,1}\|}{\sqrt{n}}\right) \le 4M \|\alpha_n\| \omega\left(\frac{\|\alpha_{n,1}\|}{\sqrt{n}}\right)$$

and (14) follows. For general F take finitely supported F_m such that $\hat{F}_m \to_w F$, $\hat{F}_m$ supported in $[-M, M]$. Then, for fixed n, $E|T_n(\hat{F}_m) - T_n(F)|^2 \to 0$ and $E|\tilde{T}_n(\hat{F}_m) - \tilde{T}_n(F)|^2 \to 0$ as $m \to \infty$. As a consequence, we conclude that (14) holds also in this case.

Now, by the Dvoretzky–Kiefer–Wolfowitz inequality (see [16]) we have $P(\|\alpha_{n,1}\| > t) \le 2e^{-2t^2}$, $t > 0$. This entails that $\|\alpha_{n,1}\|^2$ is uniformly integrable and also that $\omega\left(\|\alpha_{n,1}\| n^{-1/2}\right)$ vanishes in probability. Since, on the other hand, $\omega^2\left(\|\alpha_{n,1}\| n^{-1/2}\right) \|\alpha_{n,1}\|^2 \le M^2 \|\alpha_{n,1}\|^2$ we conclude that

$$E\left(\|\alpha_{n,1}\|^2 \omega^2\left(\frac{\|\alpha_{n,1}\|}{\sqrt{n}}\right)\right) \to 0 \tag{15}$$

as $n \to \infty$ and this proves the first claim in the Theorem. For the others, we can argue as above to see that (14) also holds if we replace $T_n(F)$ and $\tilde{T}_n(F)$ with the corresponding pairs $T_n^+(F)$ and $\tilde{T}_n^+(F)$ or $T_n^-(F)$ and $\tilde{T}_n^-(F)$. This completes the proof.

$\square$

From Theorem 3.1 we obtain the following CLT.

Corollary 3.1 *If F and G have bounded support and G^{-1} is continuous, then*

$$\sqrt{n}(\varepsilon_{\mathcal{W}_2}(F_n, G) - \varepsilon_{\mathcal{W}_2}(F, G)) \to_w N(0, \sigma^2)$$

with $\sigma^2 = \frac{Var(v_-(U))}{\mathcal{W}_2^8(F,G)}$, v_- as in (13) and U a uniform r.v. on $(0, 1)$.

Proof Observe that $\sqrt{n}(\varepsilon_{\mathcal{W}_2}(F_n, G) - \varepsilon_{\mathcal{W}_2}(F, G)) = \sqrt{n}(\frac{S_n^+}{S_n} - \frac{S^+}{S}) = \frac{1}{SS_n}$ $(T_n^+ - T_n)$. From Theorem 3.1, $(T_n^+ - T_n) = \alpha_{n,1}(v_+ - v) + o_P(1) = -\alpha_{n,1}(v_-) + o_P(1)$, while $S_n \to S$ a.s. $\qquad\square$

Remark 3.1 For the two-sample analogue of Corollary 3.1 it is important to observe that the conclusion of Theorem 3.1 holds if we replace T_n by $\hat{T}_{n,m} := \sqrt{n}(\int_0^1(F_n^{-1} - G_m^{-1})^2 - \int_0^1(F^{-1} - G_m^{-1})^2)$ and $m \to \infty$. In fact, in the finitely supported case, keeping the notation in the proof of Theorem 3.1, we have

$$\hat{T}_{n,m} = -\sqrt{n}\sum_{i=1}^{k-1}\int_{S_i}^{H_{n,1}(s_i)}\left(\int_{x_i}^{x_{i+1}} c(s - G_m^{-1}(t))ds\right)dt,$$

from which we see that

$$|\hat{T}_{n,m} - T_n| \leq 2\sqrt{n}\sum_{i=1}^{k-1}\left|\int_{S_i}^{H_{n,1}(s_i)}\left(\int_{x_i}^{x_{i+1}} |G_m^{-1}(t) - G^{-1}(t)|ds\right)dt\right|$$

$$\leq 4M\|\alpha_{n,1}\|\sup_{0\leq t\leq 1}\|G^{-1}(H_{m,2}^{-1}(t)) - G^{-1}(t)\| \to 0$$

in probability, since G^{-1} is continuous and

$$\sup_{t\in(0,1)}|H_{m,2}^{-1}(t) - t| = \sup_{x\in(0,1)}|H_{m,2}(x) - x| \to 0$$

in probability. Similar statements are true for T_n^+ and T_n^-. $\qquad\square$

Proof (Proof of Theorem 2.2.) Convergence in the L_2−Wasserstein distance sense is characterized through weak convergence plus convergence of second order moments. Therefore the a.s. consistency $\varepsilon_{\mathcal{W}_2}(F_n, G_m) \to_{a.s.} \varepsilon_{\mathcal{W}_2}(F, G)$ essentially follows from the strong law of large numbers (see [11] for details and more general results). For the asymptotic law, we write

$$\left(\varepsilon_{\mathcal{W}_2}(F_n, G_m) - \varepsilon_{\mathcal{W}_2}(F, G)\right)$$
$$= (\varepsilon_{\mathcal{W}_2}(F_n, G_m) - \varepsilon_{\mathcal{W}_2}(F, G_m)) + (\varepsilon_{\mathcal{W}_2}(F, G_m) - \varepsilon_{\mathcal{W}_2}(F, G)).$$

By Theorem 3.1 and Remark 3.1 arguing as in the proof of Corollary 3.1 we see that $\sqrt{n}(\varepsilon_{\mathcal{W}_2}(F_n, G_m) - \varepsilon_{\mathcal{W}_2}(F, G_m)) \to_w N\left(0, \frac{Var(v_-(U))}{\mathcal{W}_2^8(F,G)}\right)$. A minor modification of the proof of Corollary 3.1 yields that $\sqrt{m}(\varepsilon_{\mathscr{W}_2}(F, G_m) - \varepsilon_{\mathscr{W}_2}(F, G)) \to_w$ $N\left(0, \frac{Var(\nu_+(U'))}{\mathscr{W}_2^8(F,G)}\right)$ with ν_+ as in (13) and U' a U(0,1) law, and also that $\sqrt{n}(\varepsilon_{\mathscr{W}_2}(F_n, G_m) - \varepsilon_{\mathscr{W}_2}(F, G_m))$ and $\sqrt{m}(\varepsilon_{\mathscr{W}_2}(F, G_m) - \varepsilon_{\mathscr{W}_2}(F, G))$ are asymptotically independent. The result follows. $\qquad\square$

Acknowledgements The authors want to acknowledge the support by the Spanish Ministerio de Economía y Competitividad y fondos FEDER, Grants MTM2014-56235-C2-1-P and MTM2014-56235-C2-2.

References

1. Álvarez-Esteban PC, del Barrio E, Cuesta-Albertos JA, Matrán C (2011) Uniqueness and approximate computation of optimal incomplete transportation plans. Ann Inst Henri Poincaré - Probab et Stat 47:358–375
2. Álvarez-Esteban PC, del Barrio E, Cuesta-Albertos JA, Matrán C (2012) Similarity of samples and trimming. Bernoulli 18:606–634
3. Álvarez-Esteban PC, del Barrio E, Cuesta-Albertos JA, Matrán C (2014). A contamination model for approximate stochastic order: extended version. arXiv.org/abs/1412.1920
4. Álvarez-Esteban PC, del Barrio E, Cuesta-Albertos JA, Matrán C (2016) A contamination model for the stochastic order. Test 25:751–774
5. Álvarez-Esteban PC, del Barrio E, Cuesta-Albertos JA, Matrán C (2017). Wide Consensus aggregation in the Wasserstein Space. Application to location-scatter families. arXiv.org/abs/1511.05350
6. Álvarez-Esteban PC, del Barrio E, Cuesta-Albertos JA, Matrán C (2017) Models for the assessment of treatment improvement: the ideal and the feasible. arXiv.org/abs/1612.01291
7. Berger RL (1988) A nonparametric, intersection-union test for stochastic order. In: Gupta SS, Berger JO (eds) Statistical decision theory and related topics IV, vol 2. Springer, New York
8. Boissard E, Le Gouic T, Loubes JM (2015) Distribution's template estimate with wasserstein metrics. Bernoulli 21:740–759
9. Carlier G, Chernozhukov V, Galichon A (2016) Vector quantile regression: an optimal transport approach. Ann Stat 44:1165–1192
10. Chernozhukov V, Galichon A, Hallin M, Henry M (2014) Monge–Kantorovich depth, quantiles, ranks, and signs. Ann Stat 45(1):223–256
11. Cuesta JA, Matrán C (1992) A review on strong convergence of weighted sums of random elements based on wasserstein metrics. J Stat Plan Inference 30:359–370
12. Davidson R, Duclos JY (2007) Testing for restricted stochastic dominance. Working paper, department of economics, McGill University
13. Hodges JL, Lehmann EL (1954) Testing the approximate validity of statistical hypotheses. J R Stat Soc Ser B 16:261–268
14. Lehmann EL (1955) Ordered families of distributions. Ann Math Stat 26:399–419
15. Leshno M, Levy H (2002) Preferred by all and preferred by most decision makers: almost stochastic dominance. Manag Sci 48:1074–1085
16. Massart P (1990) The tight constant in the Dvoretsky–Kiefer–Wolfovitz inequality. Ann Probab 18:1269–1283
17. Rippl T, Munk A, Sturm A (2016) Limit laws of the empirical wasserstein distance: gaussian distributions. J Multivar Anal 151:90–109

An Alternative to the Variation Coefficient

Carlo Bertoluzza, Rosa Casals, Gloria Naval and Antonia Salas

Abstract The aim of this paper is to introduce an invariant by translation coefficient different from the variation one (widely used in literature but not fulfilling that property) that allows us to study whether the mean is a good representation of the distribution or not. The value of this new coefficient for a normally distributed random variable is obtained in order to establish a criterion, similar to the one used in the symmetry or kurtosis coefficients, to decide the grade of representation of the mean.

1 Introduction

Why Defining an Alternative Coefficient?

Variation coefficient is widely used in literature (see for instance [2–4]) in order to obtain a grade of representation of the mean for different distributions, since it is a relative dispersion measure providing the number of times that the standard deviation is contained in the mean of the corresponding distribution. However, this coefficient is not invariant by translations which, in our opinion, it is a quite significant issue. That is why we consider necessary to define a coefficient that evaluates the spread or distance of the values of the distribution with respect to a central one (allowing us to measure the grade of representation of this central value as a numerical summary of that distribution) that is invariant by changes in the origin of the distribution.

To Pedro: our advisor, preceptor but, above all, our friend.

C. Bertoluzza
Dipartimento di Informatica e Sistemistica, Universitá di Pavia, Pavia, Italy

R. Casals · G. Naval · A. Salas (✉)
Dpto. Estadística e I.O. y D.M., Universidad de Oviedo, Oviedo, Spain
e-mail: antonia@uniovi.es

R. Casals
e-mail: rmcasals@uniovi.es

G. Naval
e-mail: glorianaval@uniovi.es

© Springer International Publishing AG 2018
E. Gil et al. (eds.), *The Mathematics of the Uncertain*, Studies in Systems, Decision and Control 142, https://doi.org/10.1007/978-3-319-73848-2_4

45

Table 1 Data in Example 1.1

X	1.7	1.85	1.72	1.65	1.73	1.58	1.65	1.69	1.6	1.67
Y	170	185	172	165	173	158	165	169	160	167
Z	70	85	72	65	73	58	65	69	60	67

In order to show what we mean, let us introduce a numerical example.

Example 1.1 Let us consider the height of a group of 10 individuals given in three different ways: X *(height in meters)*; $Y = 100X$ *(height in centimeters)*; $Z = Y - 100$ *(centimeters above one meter)*. The observed values, expressed in the three alternative ways, are shown in Table 1.

Being the same data should provide us with exactly the same goodness of representation of the mean as a numerical summary of this distribution, regardless of the units (X vs Y) or the referential origin (X vs Z) considered to describe them. The variation coefficient for variable X is

$$\bar{x} = 1.684 \quad S_x^2 = 0.005164 \quad \Rightarrow \quad VC(X) = \frac{\sqrt{0.005164}}{1.684} = 0.042672789$$

and for variable Y is

$$\bar{y} = 168.4 \quad S_y^2 = 51.64 \quad \Rightarrow \quad VC(Y) = \frac{\sqrt{51.64}}{168.4} = 0.042672789$$

which means that a scale change does not affect it. However

$$\bar{z} = 68.4 \quad S_z^2 = 51.64 \quad \Rightarrow \quad VC(Z) = \frac{\sqrt{51.64}}{68.4} = 0.105059908$$

so translations do affect the value of this coefficient. This representation of the mean of Z can be understood as a worse one than that of X since the variation coefficient is greater for Z than for X. This coefficient can undergo dramatic changes for different expressions of the same set of values. Let us consider a new variable $T = Y - 170$ that provides the centimeters above/below 170. Despite the sign, the value of the variation coefficient is much greater for T than for Y

$$\bar{t} = -1.6 \quad S_t^2 = 51.64 \quad \Rightarrow \quad VC(T) = \frac{\sqrt{51.64}}{-1.6} = -4.49131106.$$

This means, with the interpretation broadly given to this coefficient in literature, that the grade of representation of -1.6 as a numerical summary of the average difference between the height (in centimeters) of those individuals and 170 is significantly smaller than that of 168.4 cm as the average height of the same group of individuals. The same holds with 68.4 cm exceeding the meter as a representation of these data (T versus Z). This makes no sense for us, since the information given by these distributions is intrinsically the same, although the values differ from one variable to another.

Example 1.2 Let us consider a normally distributed random variable $X \sim N(\mu, \sigma)$. Assume that we are working with a given value of the variance (say 1 for sake of simplicity). The bell-shape density function that characterizes the distribution is exactly the same, independently of the location of the mean. Hence, a constant value for a coefficient giving the goodness of representation of the mean was to be expected. However, as $VC(X) = \frac{SD(X)}{E(X)}$, that is, $VC(X) = \frac{\sigma}{\mu}$, this coefficient increases as the mean decreases. That is the same as saying about the mean that the closer to zero, the less representative of the distribution.

2 Definition of a New Coefficient

Is it possible to express how spread out the values of a distribution are with respect to a central value, and, consequently, how good this central value is as a representation of the distribution, in such a way that translation changes do not affect that goodness of representation (as it happens with the variation coefficient)?

We may agree that moving the values to another location keeping static the 'distances' between them and their relative positions between each other should not affect the representation of the mean as a numerical summary of the data, since the dispersion of the data is exactly the same in both locations, being the value of the referential point (mean) the only noticeable modification. 'Moving the values to another location' has modified the mean, while 'keeping static their relative positions between each other' does not change the variance of the distribution. So, as the variance averages the squared distances of the data with respect to their mean, why not introducing some invariant by translations measure in the denominator of the 'variation coefficient' so that the value it takes is exactly the same wherever the data are located? Doing so, the corresponding value of the ratio will show how good the mean is to summarize the values of a distribution just paying attention to the relative positions within the data but not to the actual location with respect to a given origin.

Definition 2.1 For any variable X, the ***representation coefficient of*** X is the ratio between its standard deviation and its range, that is

$$RC(X) = \frac{D(X)}{R(X)} = \frac{D(X)}{\max(X) - \min(X)}.$$

This coefficient is well defined for any nondegenerate variable X and takes nonnegative values smaller than 1. It is quite easy to see that the closer to one, the less representative the mean.

A More General Representation Coefficient

Nevertheless, outliers can strongly affect the value of this coefficient, so it may be better to exclude that part of the distribution on calculating a grade of the representation of the mean. We intend to present a coefficient which is not so sensitive to the presence of extreme values of the distribution. There are different ways of detecting outliers but, regardless of the method, any value that is really far from the rest of the observations is said to be an outlier. For instance, $\mu \pm 2.7\sigma$ are the outliers cutoffs given in [1] when working on a normal distribution.

For any $r \in (0, 1)$ let $P_{100r}(X)$ denote the $100r$-th percentile of X, that is, $r = P(X \le P_{100r}(X))$.

Definition 2.2 The **$100r\%$ *trimmed representation coefficient of*** X is defined as

$$RC_{100r}(X) = \frac{D(X)}{P_{100(1-\frac{r}{2})}(X) - P_{\frac{100r}{2}}(X)} \ \forall r \in [0, 1],$$

where $P_0(X) = \min(X)$ and $P_{100}(X) = \max(X)$ (so this coefficient is a generalization of the latter one).

Remark 2.1 This coefficient is generally well defined. Once again there are some exceptions, since the denominator is zero for any value of r when X is degenerate. It is also zero for not so large values of r when X does not take too many different values (then its 'extreme' percentiles coincide, that is, it is 'almost sure' degenerate; median and mode will coincide in that case and the mean is rarely going to be chosen as a representation of such a variable).

Remark 2.2 In general, any of these trimmed representation coefficients compare the value of a dispersion measure with respect to the mean with a dispersion measure which does not refer to any particular central value and that does not take into account outliers (to the extent the experimenter wants to). Then, it can be used to measure how good the mean is as a numerical summary of the distribution.

Remark 2.3 Although it can be defined for any value of $r \in [0, 1]$, it is not sensitive to eliminate too many 'outliers'; so, it makes not much sense to calculate it for $r > 0.5$. For instance, the $100r\%$ trimmed representation coefficient equals the $100(1-r)\%$ one in absolute value.

Property 1 (Invariance by translations). *The trimmed representation coefficient is invariant by translations, that is, $RC_{100r}(X + k) = RC_{100r}(X)$ for all $r \in [0, 1]$, $k \in \mathbb{R}$.*

Proof Trivial since $D(X + k) = D(X) \ \forall k$ and $P(X + k \le y) = P(X \le y - k) \ \forall y \Rightarrow P_{100r}(X + k) = k + P_{100r}(X)$. Hence $P_{100(1-\frac{r}{2})}(X + k) - P_{\frac{100r}{2}}(X + k) = P_{100(1-\frac{r}{2})}(X) - P_{\frac{100r}{2}}(X)$. $\qquad\square$

Property 2 (Absolute invariance by scale). *The trimmed representation coefficient is invariant by change of scale in absolute value, that is, $|RC_{100r}(kX)| = |RC_{100r}(X)| \ \forall r \in [0, 1]$ and $\forall k \in \mathbb{R}$.*

Proof Trivial since $D(kX) = |k| D(X) \; \forall k$. On the other hand, $P(kX \le y) = P(X \le \frac{y}{k}) \; \forall y \; \forall k > 0 \Rightarrow P_{100r}(kX) = k P_{100r}(X)$ and $P(kX \le y) = P(X \ge \frac{y}{k}) \forall y \forall k < 0 \Rightarrow P_{100r}(kX) = k P_{100(1-r)}(X)$. Hence $|P_{100(1-\frac{r}{2})}(kX) - P_{\frac{100r}{2}}(kX)| = |k| \, |P_{100(1-\frac{r}{2})}(X) - P_{\frac{100r}{2}}(X)|$. $\qquad \square$

Example 2.1 (*Example* 1.1 *continued*) For the data of the height used above, one can obtain

$$RC_{10}(X) = \frac{\sqrt{0.005164}}{1.85 - 1.58} = 0.266151766, \quad RC_{10}(Y) = \frac{\sqrt{51.64}}{185 - 158} = 0.266151766,$$

$$RC_{10}(Z) = \frac{\sqrt{51.64}}{85 - 58} = 0.266151766, \quad RC_{10}(T) = \frac{\sqrt{51.64}}{15 - (-12)} = 0.266151766.$$

Analogously, $RC_{50}(X) = \frac{\sqrt{0.005164}}{1.72 - 1.65} = RC_{50}(Y) = RC_{50}(Z) = RC_{50}(T)$. Invariance holds for any r.

3 The Trimmed Representation Coefficient for a Normally Distributed Random Variable

Invariance of the Trimmed Coefficient with Respect to the Parameters

Let $X \sim N(0, 1)$ and let $P_{100r}(X)$ be the $100r$-th percentile of X. Then for $Y \sim N(\mu, \sigma)$

$$r = P(Y \le P_{100r}(Y)) = P\left(\frac{Y - \mu}{\sigma} \le \frac{P_{100r}(Y) - \mu}{\sigma} \right)$$

$$\Rightarrow P_{100r}(Y) = \mu + \sigma P_{100r}(X)$$

which means that the $100r\%$ trimmed representation coefficient for a normal distribution does not depend on its parameters.

$$RC_{100r}(N(\mu, \sigma)) = \frac{\sigma}{(\mu + \sigma P_{100(1-\frac{r}{2})}(X)) - (\mu + \sigma P_{\frac{100r}{2}}(X))}$$

$$= \frac{1}{P_{100(1-\frac{r}{2})}(X) - P_{\frac{100r}{2}}(X)} = RC_{100r}(N(0, 1)).$$

Value of the Trimmed Representation Coefficient for Some Usual $100r\%$

Let us use programm R for obtaining the corresponding percentiles of the normal distribution. As we have just proved, we can reduce our calculations to the standard normal.

Table 2 Values of the trimmed representation coefficient for the standard normal distribution

$100r\%$	1%	2%	2.5%	5%	10%	20%	50%
RC_{100r}	0.142857	0.214929	0.2230746	0.2551067	0.3039783	0.3901519	0.7413008

The 5th and 95th percentiles of $N(0, 1)$ are

```
> qnorm(c(.95,.05), mean=0, sd=1, lower.tail=TRUE)
[1]  1.644854 -1.644854
```

so the 10% trimmed representation coefficient of $N(0, 1)$ takes value

```
> 1.644854+1.644854
[1] 3.289708
> 1/3.289708
[1] 0.3039783
```

Proceeding in an analogous way for other percentages, we can obtain the values shown in Table 2.

As it can be seen, this coefficient takes quite similar values for $r \in [0.02, 0.05]$ which are the most usual percentages for outliers cutoffs in literature.

4 Interpretation of the Trimmed Representation Coefficient

The value of this coefficient for the normal distribution can be used as a reference to determine whether the mean is good enough or not as a representation of the distribution, analogously as, for example, the kurtosis coefficient $\gamma_2(Y) = \frac{\mu_4(Y)}{[D(Y)]^4} - 3$ greater than 0 means that the distribution is sharper than the bell-shaped one, with $\mu_4(Y)$ being the 4th central moment of Y.

Property 3 *The closer to zero the (trimmed) representation coefficient of X, the better its mean as a representation of the distribution of X.*

Definition 4.1 The mean of a variable X is said to be more representative than the mean of the normal distribution for a $100r\%$ if $RC_{100r}(X) \leq RC_{100r}(N(0, 1))$.

As we have just obtained, the trimmed representation coefficient of a normal distribution increases as the order (r) increases. So, the fewer values included in the interval of reference of the denominator, the larger the coefficient.

Whenever $RC_{100r}(X) > 0.7413008$ it can be said that the mean of X is not representative for the distribution since the central 50% of the observations of a distribution is the minimal set of values to be considered on studying their representation by means of a central value.

Example 4.1 (Example 1.1 continued) As we have just obtained, $RC_{10}(X) = 0.266151766 < 0.3039783 = RC_{10}(N(0, 1))$, so, the mean of these data is a better representation of the distribution than that of the Gauss one. On the other side, if one pay attention just to the data included in the box of a box-whiskers diagram, one can conclude that those data are more disperse than those from a normal distribution since $RC_{50}(X) = 1.026585384 > 0.7413008 = RC_{50}(N(0, 1))$.

5 Trimmed Representation Coefficient for Other Theoretical Distributions

Exponential Distribution

Let us consider $f(x) = \lambda e^{-\lambda x}$ $x > 0$ the density function of an exponential random variable X.

Property 4 *The trimmed representation coefficient of an exponentially distributed random variable does not depend on the value of its parameter.*

Proof On calculating percentiles for an exponentially distributed random variable X of parameter λ, one obtain

$$r = P(X < P_{100r}) = \int_0^{P_{100r}} \lambda e^{-\lambda x} dx = 1 - e^{-\lambda P_{100r}} \Rightarrow P_{100r} = -\frac{ln(1 - r)}{\lambda}.$$

Hence

$$RC_{100r}(Exp(\lambda)) = \frac{\frac{1}{\lambda}}{\frac{1}{\lambda}\ln(1 - \frac{r}{2}) - \frac{1}{\lambda}\ln(\frac{r}{2})} = \frac{1}{\ln(1 - \frac{r}{2}) - \ln(\frac{r}{2})}. \qquad \square$$

As we have proved that the value of the parameter is not essential for calculating the trimmed representation coefficient, for the sake of simplicity we will use R and calculate the coefficient for an exponentially distributed random variable of rate equal to one.

```
> qexp(c(.95,.05,.75,.25), rate=1, lower.tail=TRUE)
[1] 2.99573227 0.05129329 1.38629436 0.28768207
> 1/(2.99573227-0.05129329)
[1] 0.3396233
> 1/(1.38629436-0.28768207)
[1] 0.9102392
> qexp(c(0.9,0.1,0.975,0.025), rate=1, lower.tail=TRUE)
[1] 2.30258509 0.10536052 3.68887945 0.02531781
> 2.30258509-0.10536052
[1] 2.197225
> 1/2.197225
```

```
[1] 0.4551195
> 3.68887945-0.02531781
[1] 3.663562
> 1/3.663562
[1] 0.2729584
> qexp(c(0.99,0.01,0.995,0.005), rate=1, lower.tail=TRUE)
[1] 4.605170186 0.010050336 5.298317367 0.005012542
> 4.605170186-0.010050336
[1] 4.59512
> 1/4.59512
[1] 0.2176222
> 5.298317367-0.005012542
[1] 5.293305
> 1/5.293305
[1] 0.1889179
```

The values of the trimmed representation coefficient for an exponential distribution are shown in Table 3. So, we can conclude that the mean of an exponential distribution is slightly less representative of its distribution than that of the normal one.

Uniform Distribution

Let us consider $X \sim \mathcal{U}(a, b)$, that is $f(x) = \frac{1}{b-a}$ if $a < x < b$.

Property 5 *The trimmed representation coefficient of a uniformly distributed random variable does not depend on the value of its parameters.*

Proof On calculating percentiles for a uniformly distributed random variable X on the interval (a, b), one obtain

$$r = P(X < P_{100r}) = \int_a^{P_{100r}} \frac{1}{b-a} dx = \frac{P_{100r} - a}{b - a} \Rightarrow P_{100r} = a + r(b - a).$$

Hence,

$$RC_{100r}(U(a, b)) = \frac{\sqrt{\frac{(b-a)^2}{12}}}{a + (1 - \frac{r}{2})(b - a) - a - \frac{r}{2}(b - a)} = \frac{1}{(1 - r)\sqrt{12}}. \quad \square$$

Table 3 Values of the trimmed representation coefficient for the exponential distribution with mean one

100r%	1%	2%	5%	10%	20%	50%
RC_{100r}	0.1889179	0.2176222	0.2729584	0.3396233	0.4551195	0.9102392

Table 4 Values of the trimmed representation coefficient for the (continuous) uniform distribution

$100r\%$	1%	2%	5%	10%	20%	50%
RC_{100r}	0.29159105	0.29456646	0.30386856	0.32075015	0.36084392	0.57735027

Table 5 Values of the trimmed representation coefficient for some discrete distributions

$100r\%$	1%	2%	5%	10%	20%
$RC_{100r}(\mathscr{P}(1))$	0.25	0.25	$\frac{1}{3}$	$\frac{1}{3}$	0.5
$RC_{100r}(\mathscr{B}(10, 0.5))$	0.19764235	0.19764235	0.26352314	0.395284471	0.79056942

As we have just obtained that the trimmed representation coefficient of a uniformly distributed random variable depends on the order, we can obtain for any interval (a, b), the values given in Table 4.

So, whenever at most 10% of the extreme observations of a distribution are excluded, the mean is said to be a better representation of the data of a normal distribution than of a uniform one.

Some Well-known Discrete Distributions

Finally, let us use program R in order to obtain some percentiles from a Poisson of mean 1 and also from a Binomial with 10 trials and $p = 0.5$. The obtained values are shown below, and summarized in Table 5.

```
> qpois(c(0.25,.75), lambda=1, lower.tail=TRUE)
[1] 0 2
> qpois(c(0.1,.9), lambda=1, lower.tail=TRUE)
[1] 0 2
> qpois(c(0.05,.95), lambda=1, lower.tail=TRUE)
[1] 0 3
> qpois(c(0.025,.975), lambda=1, lower.tail=TRUE)
[1] 0 3
> qpois(c(0.01,.99), lambda=1, lower.tail=TRUE)
[1] 0 4
> qpois(c(0.005,.995), lambda=1, lower.tail=TRUE)
[1] 0 4
> qbinom(c(.005,.995), size=10, prob=0.5, lower.tail=TRUE)
[1] 1 9
> qbinom(c(.01,.99), size=10, prob=0.5, lower.tail=TRUE)
[1] 1 9
> qbinom(c(.025,.975), size=10, prob=0.5, lower.tail=TRUE)
[1] 2 8
> qbinom(c(.05,.95), size=10, prob=0.5, lower.tail=TRUE)
[1] 2 8
> qbinom(c(.1,.9), size=10, prob=0.5, lower.tail=TRUE)
```

```
[1] 3 7
> qbinom(c(.25,.75), size=10, prob=0.5, lower.tail=TRUE)
[1] 4 6
```

All these values are larger than the corresponding ones for the normal distribution, so these variables are worse represented by their means than the normal one.

Acknowledgements The research in this paper has been partially supported by the Spanish Ministry of Economía and Competitividad Grant MTM2015-63971-P and the Principality of Asturias/FEDER Grant GRUPIN14-101. Their financial support is gratefully acknowledged.

References

1. Jobson JD (1991) Applied multivariate data analysis, vol 1. Regression and experimental design. Springer, New York
2. McPherson G (1990) Statistics in scientific investigation. Its basis, application and interpretation. Springer, New York
3. Peña D (2002) Análisis de datos multivariantes. McGrawHill/Interamericana de España, Madrid
4. Rodríguez Muñiz LJ, Tomeo-Perucha V (2011) Métodos Estadísticos para Ingeniería. Garceta Grupo Editorial, Madrid

Nonparametric Mean Estimation for Big-But-Biased Data

Ricardo Cao and Laura Borrajo

Abstract Crawford (The hidden biases in big data, Harvard Business Review, Cambridge, 2013, [2]) has recently warned about the risks of the sentence *with enough data, the numbers speak for themselves*. Some of the problems coming from ignoring sampling bias in big data statistical analysis have been recently reported by Cao (Inferencia estadística con datos de gran volumen, La Gaceta de la RSME 18:393–417, 2015, [1]). The problem of nonparametric statistical inference in big data under the presence of sampling bias is considered in this work. The mean estimation problem is studied in this setup, in a nonparametric framework, when the biasing weight function is known (unrealistic) as well as for unknown weight functions (realistic). Two different scenarios are considered to remedy the problem of ignoring the weight function: (*i*) having a small sized simple random sample of the real population and (*ii*) having observed a sample from a doubly biased distribution. In both cases the problem is related to nonparametric density estimation. A simulated dataset is used to illustrate the performance of the nonparametric methods proposed in this work.

1 Introduction

With the peak of the Big Data and related to the false assumption that *with enough data, the numbers speak for themselves*, appears what Crawford [2] calls *data fundamentalism*, the notion that massive data sets always reflect objective and absolute

R. Cao (✉) · L. Borrajo
Research Group MODES, Department of Mathematics, CITIC,
A Coruña, Spain
e-mail: ricardo.cao@udc.es

L. Borrajo
e-mail: laura.borrajo@udc.es

R. Cao
ITMATI, Universidade da Coruña, Campus de Elviña,
15071 A Coruña, Spain

© Springer International Publishing AG 2018
E. Gil et al. (eds.), *The Mathematics of the Uncertain*, Studies in Systems,
Decision and Control 142, https://doi.org/10.1007/978-3-319-73848-2_5

truth. However, like any other human creation, data and data sets are not totally objective. Occasionally, a large sample is not completely representative of the population, but it is biased: Big-but-biased Data (BBBD = B3D). Hidden biases often come from the data collection procedure. Sampling methods in which the data or individuals in the sample are self-select, often incur in sampling bias.

A good example cited by Crawford [2] is the data collected in the city of Boston through the StreetBump smartphone app, created with the objective of solving the problem with potholes in this city, where 20,000 of them are patched every year. This app passively detects bumps by recording the accelerometers of the phone and GPS data while driving, instantly reporting them to the traffic department of the city. Thus, the city could plan their repair and the management of resources in the most efficient possible way. However, one of the problems observed was that some segments of the population, such as people in lower income groups, have a low rate of smartphone use, a rate that is even lower in the older residents, where smartphone penetration is as low as 16%. Therefore, these data provide a big but very biased sample of the population of potholes in the city, with the consequent impact on the underestimation of the number of potholes in certain neighborhoods and the deficient management of resources.

Another example cited by Crawford [2] is the database of more than 20 million tweets generated by Hurricane Sandy between October 27 and November 1, 2012. A combined analysis of Twitter and Foursquare data produced some expected findings, such as an increase in grocery shopping the night before the storm, and other more surprising, such as an increase in nightlife the day after the hurricane. However, these data don't represent an unbiased sample of the population. It is well known that the greatest number of tweets about Sandy came from Manhattan, due to the high level of smartphone owners and Twitter use in New York. Only a few messages were originated in the most affected areas by the catastrophe, not only because of the lower penetration of the smartphone market in those areas, but also because of the fact that power outages caused many problems with internet access and many devices run out of battery in the hours after the storm.

In other examples, such as those cited in Hargittai [3], survey data show that people do not select into the use of sites randomly; instead, use is biased in certain ways yielding samples that limit the generalizability of findings.

In Sect. 2 the proposed bias context in this work is presented. In Sect. 3 the problem of nonparametric mean estimation is considered, when the weight function is known (unrealistic) and unknown (realistic). Two different scenarios are considered to solve the unknown weight function problem: (i) to have a simple random sample of a small size of the real population and (ii) to observe a sample from a twice biased distribution. In both cases, mean estimators that avoid the sampling bias are proposed. These estimators are related to nonparametric density estimation. Section 4 illustrates the behavior of the proposed estimators with simulated data.

2 The Samples and Their Sizes

In this context, let us consider a population with cumulative distribution function F (density f) and consider a simple random sample

$$\mathbf{X} = (X_1, \ldots, X_n)$$

of size n from this population. Assume for a while that we are not able to observe this sample but we observe, instead, another sample,

$$\mathbf{Y} = (Y_1, \ldots, Y_N)$$

of a much larger sample size ($N >> n$) from a biased distribution G (density g). We assume that, for some weight function $w(x) \geq 0$,

$$g(x) = w(x)f(x) .$$

3 Mean Estimation in B3D

To deal with the bias problem in Big Data, we decided to start with a simple problem: the estimation of the mean $\mu = \int x \, dF(x)$ in a nonparametric context. If the sample $\mathbf{X}$ is observed, then it would be enough to use its sample mean, $\overline{X}$, which is an unbiased estimator of μ. Therefore, the mean squared error of this estimator is its variance $Var(\overline{X}) = \frac{\sigma^2}{n}$. In that case, the Central Limit Theorem allows us to make inference about μ.

However, in the proposed B3D context, when we are unable to observe $\mathbf{X}$, the question may arise whether it is feasible to estimate the mean. Indeed, assuming that the underlying distribution, F, is continuous, with density f, the mean of the big and biased sample, $\overline{Y}$, is not a consistent estimator of $\mu = \int xf(x) \, dx$, but of

$$\mu_g = \int xg(x) \, dx = \int xw(x)f(x) \, dx .$$

On the other hand, since $w(x) = \frac{g(x)}{f(x)}$, it is easy to check that the mean of the ratio $\frac{Y}{w(Y)}$ is equal to the population mean:

$$E\left(\frac{Y}{w(Y)}\right) = \int \frac{y}{w(y)} g(y) \, dy = \int \frac{y}{g(y)/f(y)} g(y) \, dy$$
$$= \int yf(y) \, dy = \mu.$$

Therefore, if we know the weight function w (unrealistic case), we can use the previous expectation to motivate an estimator for μ:

$$\hat{\mu}^{BBBS,w} = \frac{1}{N} \sum_{i=1}^{N} \frac{Y_i}{w(Y_i)}. \tag{1}$$

This estimator is the sample mean of the population $Z = Y/w(Y)$, so under suitable conditions we immediately obtain its bias, variance and asymptotic distribution:

$$E\left(\hat{\mu}^{BBBS,w}\right) = \mu,$$

$$Var\left(\hat{\mu}^{BBBS,w}\right) = \frac{\sigma_Z^2}{N},$$

$$\frac{\sqrt{N}\left(\hat{\mu}^{BBBS,w} - \mu\right)}{\sigma_Z} \to N(0, 1),$$

where $\sigma_Z^2 = \int y^2 f(y)^2 g(y)^{-1} dy - \mu^2 = \int x^2 f(x) w(x)^{-1} dx - \mu^2$.

However, the most realistic situation is that the weight function w is unknown. Having this in mind, we propose another estimator that can be used when w is unknown.

Taking into account that the expectation of the inverse of the weight function w is 1,

$$E\left(\frac{1}{w(Y)}\right) = \int \frac{1}{w(y)} g(y) \, dy = \int f(y) \, dy = 1,$$

we can motivate another estimator simply considering

$$\frac{E\left(\frac{Y}{w(Y)}\right)}{E\left(\frac{1}{w(Y)}\right)} = \mu.$$

If we know the weight function w, we can use the previous expectations to motivate an estimator for the mean. This is just replacing the theoretical expectations by the empirical means:

$$\hat{\mu}^{BBBS_1,w} = \frac{\dfrac{1}{N} \sum_{i=1}^{N} \dfrac{Y_i}{w(Y_i)}}{\dfrac{1}{N} \sum_{i=1}^{N} \dfrac{1}{w(Y_i)}}. \tag{2}$$

If the weight function w is unknown, we must estimate it. As the biased sample is not enough to do it, we propose different scenarios with some additional information:

1. To have a simple random sample of small size, n, $(X_1, \ldots, X_n)$, of the true population
2. To replicate the mechanism by which the biased sample was obtained to have a sample $(Z_1, \ldots, Z_n)$ from a twice biased distribution M (density m)
3. To enrich the biased sample with information of the frequencies (v_i, for $i = 1, \ldots, N$) of each datum of the observed sample Y

Scenarios 1 and 2 are considered in this work.

Lloyd and Jones [4] analyzed the nonparametric kernel density estimator for biased data with unknown weight function, but in a different context: close to Scenario 2, using two biased samples and being $w(x)$ the probability of acceptance.

Scenario 1. Let us consider a population with cumulative distribution function F (density f). Let us assume that we observe the Big Data sample,

$$\mathbf{Y} = (Y_1, \ldots, Y_N) \,,$$

of size N from the biased distribution G (density g), but we also observe a simple random sample,

$$\mathbf{X} = (X_1, \ldots, X_n) \,,$$

of a much smaller sample size ($n << N$) of that population. We assume that

$$g(x) = w(x) f(x)$$

for some weight function w, such that $w(x) \geq 0, \forall x$.

The Parzen–Rosenblatt kernel density estimator (KDE) can be used to estimate $f(x)$ and $g(x)$:

$$\hat{f}_h(x) = \frac{1}{n} \sum_{i=1}^{n} K_h(x - X_i) \,,$$

$$\hat{g}_b(x) = \frac{1}{N} \sum_{i=1}^{N} K_b(x - Y_i) \,,$$

where $K_h(u) = \frac{1}{h} K\left(\frac{u}{h}\right)$, being K a kernel function and h and b two bandwidths. The weight function w can be easily estimated as the ratio of both estimated densities:

$$\hat{w}_{h,b}(x) = \frac{\hat{g}_b(x)}{\hat{f}_h(x)} \,.$$

Similar to the known w case, replacing respectively in (1) and (2) the weight function by its estimated version, $\hat{w}_{h,b}$, we propose these two new estimators for the mean in Scenario 1, that depend on the bandwidths h and b:

$$\hat{\mu}^{BBBS,\hat{w}_{h,b}} = \frac{1}{N}\sum_{i=1}^{N}\frac{Y_i}{\hat{w}_{h,b}(Y_i)} = \frac{1}{N}\sum_{i=1}^{N}\frac{Y_i\hat{f}_h(Y_i)}{\hat{g}_b(Y_i)},$$

$$\hat{\mu}^{BBBS_1,\hat{w}_{h,b}} = \frac{\dfrac{1}{N}\sum_{i=1}^{N}\dfrac{Y_i}{\hat{w}_{h,b}(Y_i)}}{\dfrac{1}{N}\sum_{i=1}^{N}\dfrac{1}{\hat{w}_{h,b}(Y_i)}} = \frac{\dfrac{1}{N}\sum_{i=1}^{N}Y_i\dfrac{\hat{f}_h(Y_i)}{\hat{g}_b(Y_i)}}{\dfrac{1}{N}\sum_{i=1}^{N}\dfrac{\hat{f}_h(Y_i)}{\hat{g}_b(Y_i)}}. \tag{3}$$

Scenario 2. Let us consider a population with cumulative distribution function F (density f). Assume for a while that we are not able to observe a simple random sample of the population, but we observe, instead, another sample

$$\mathbf{Y} = (Y_1, \dots, Y_N)$$

of size N from a biased distribution G (density g). We also observe a sample

$$\mathbf{Z} = (Z_1, \dots, Z_n)$$

of size $n << N$ from a twice biased distribution M (density m).

Table 1 describes the process to follow in this scenario, where it should be noted that the function w_0 must be the same in both cases.

We assume that density g corresponds to density f weighted by a given weight function w and density m corresponds to the product of density g by a weight function w_2:

$$g(x) = f(x)w(x)$$
$$m(x) = g(x)w_2(x),$$

being $w(x), w_2(x) \geq 0, \forall x$.

Table 1 Process to obtain the twice biased sample in scenario 2

Step	Population density	Biasing weight	Biased density	Normalized biasing weight
1	$f(x)$	$w_0(x)$	$g(x) \propto w_0(x)f(x)$	$w(x) = \dfrac{w_0(x)}{\int w_0(y)f(y)dy}$
2	$g(x)$	$w_0(x)$	$m(x) \propto w_0(x)g(x)$	$w_2(x) = \dfrac{w_0(x)}{\int w_0(y)g(y)dy}$

The relation between both weight functions is given by $w_2(x) = \rho w(x)$, where

$$\rho = \frac{\int f(y) w_0(y) dy}{\int g(y) w_0(y) dy}.$$

Concerning mean estimation in this situation, since $w_2(x) = \frac{m(x)}{g(x)}$ and $\frac{m(x)}{g(x)} = \rho \frac{g(x)}{f(x)}$, it is easy to check that the expectation of $\frac{Y}{w_2(Y)}$ is the mean divided by the constant ρ:

$$E\left(\frac{Y}{w_2(Y)}\right) = \int \frac{y}{w_2(y)} g(y) \, dy = \int \frac{y}{m(y)/g(y)} g(y) \, dy$$

$$= \int \frac{y}{\rho g(y)/f(y)} g(y) \, dy = \frac{1}{\rho} \int y f(y) \, dy = \frac{1}{\rho} \mu,$$

and the expectation of the inverse of the weight function w_2 is the inverse of that constant ρ:

$$E\left(\frac{1}{w_2(Y)}\right) = \int \frac{1}{w_2(y)} g(y) \, dy = \frac{1}{\rho} \int f(y) \, dy = \frac{1}{\rho}.$$

Consequently, the ratio of these expectations is exactly the mean

$$\frac{E\left(\dfrac{Y}{w_2(Y)}\right)}{E\left(\dfrac{1}{w_2(Y)}\right)} = \mu.$$

Therefore, if we know the weight function w_2, we can use the previous expectations to motivate an estimator for the mean

$$\hat{\mu}^{BBBS_2,w_2} = \frac{\dfrac{1}{N}\sum_{i=1}^{N}\dfrac{Y_i}{w_2(Y_i)}}{\dfrac{1}{N}\sum_{i=1}^{N}\dfrac{1}{w_2(Y_i)}}. \tag{4}$$

Following the same procedure as for Scenario 1, we use the Parzen–Rosenblatt KDE to estimate $g(x)$ and $m(x)$:

$$\hat{g}_h(x) = \frac{1}{N} \sum_{i=1}^{N} K_h(x - Y_i),$$

$$\hat{m}_b(x) = \frac{1}{n} \sum_{i=1}^{n} K_b(x - Z_i),$$

easily estimating the weight function w_2 as the ratio of the estimated densities

$$\hat{w}_{2h,b}(x) = \frac{\hat{m}_b(x)}{\hat{g}_h(x)}.$$

Plugging in (4) the estimated version of the weight function w_2, $\hat{w}_{2h,b}$, a new estimator for the mean in Scenario 2 is defined, depending on both bandwidths h and b:

$$\hat{\mu}^{BBBS_2, \hat{w}_{2h,b}} = \frac{\dfrac{1}{N} \sum_{i=1}^{N} \dfrac{Y_i}{\hat{w}_{2h,b}(Y_i)}}{\dfrac{1}{N} \sum_{i=1}^{N} \dfrac{1}{\hat{w}_{2h,b}(Y_i)}} = \frac{\dfrac{1}{N} \sum_{i=1}^{N} Y_i \dfrac{\hat{g}_h(Y_i)}{\hat{m}_b(Y_i)}}{\dfrac{1}{N} \sum_{i=1}^{N} \dfrac{\hat{g}_h(Y_i)}{\hat{m}_b(Y_i)}}. \tag{5}$$

4   Case Study with Simulated Data

This section shows the behavior of the proposed estimators with simulated data.

Scenario 1. Let us consider a population with uniform distribution and density

$$f(x) = \begin{cases} \frac{1}{2} & \text{if } x \in [0, 2] \\ 0 & \text{if } x \notin [0, 2] \end{cases}$$

and the following class of weight functions

$$w(x) = \begin{cases} \varepsilon^k & \text{if } x \in [0, \varepsilon) \\ x^k & \text{if } x \in [\varepsilon, 2] \\ 0 & \text{if } x \notin [0, 2] \end{cases}$$

leading to the following expression for the biased density

$$g(x) = \begin{cases} \dfrac{k+1}{k\varepsilon^{k+1} + 2^{k+1}} \varepsilon^k & \text{if } x \in [0, \varepsilon) \\ \dfrac{k+1}{k\varepsilon^{k+1} + 2^{k+1}} x^k & \text{if } x \in [\varepsilon, 2] \\ 0 & \text{if } x \notin [0, 2] \end{cases}$$

Fig. 1 Densities f (solid line) and g (dashed line) involved in the case study with simulated data in Scenario 1 considering $k = 1$ and $\varepsilon = 1.5$ for the biasing function, w (dotted line)

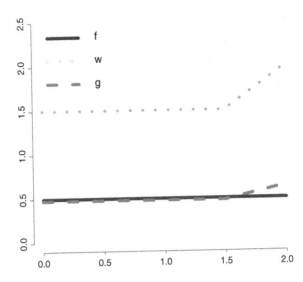

Table 2 Mean estimation with $\hat{\mu}^{BBBS_1,\hat{w}_{h,b}}$ in Scenario 1 for different values of h and b ($\overline{X} = 0.9921699$, $\overline{Y} = 1.032857$, $\hat{\mu}^{BBBS_1,w} = 0.999512$, $\mu = 1$)

$h \setminus b$	0.005	0.05	0.1	0.5	1
0.005	**0.9925275**	0.9871555	0.9843994	**1.000753**	1.014442
0.05	**0.9970936**	0.9938152	0.9916681	**1.006271**	1.018525
0.1	**1.0000137**	0.9981007	0.9970839	1.010459	1.021351
0.5	**0.9994619**	0.9995382	0.9999970	1.011548	1.019937
1	**0.9988319**	0.9990898	0.9998400	1.013350	1.022221

In addition to the simple random sample of the population, we also observe a Big Data sample from the biased distribution G (see Fig. 1).

Table 2 shows the results obtained in the mean estimation with $\hat{\mu}^{BBBS_1,\hat{w}_{h,b}}$, being $n = 10^3$ and $N = 10^6$ the sizes of the samples involved. For certain combinations of bandwidths b and h (numbers in bold), the proposed estimator (3) improves the estimation performed using the simple random sample and the Big Data sample. In this case, it is observed that the estimation given by both samples is significantly close to the real value, which is logical since the Big Data sample is only slightly biased, as can be seen in Fig. 1.

Scenario 2. Let us consider again the proposed model in Scenario 1. Let us suppose that it is not possible to observe a simple random sample of the population, but we observe a large sample from the biased distribution G (density g) and another sample of smaller sample size from the twice biased distribution M, with density

Fig. 2 Densities f (solid line), g (dashed line) and m (dashed-dotted line) involved in the case study with simulated data in Scenario 2 considering $k = 1$ and $\varepsilon = 0.2$ for the biasing function, w (dotted line)

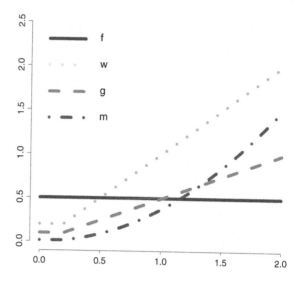

Table 3 Mean estimation with $\hat{\mu}^{BBBS_2,\hat{w}_{2h,b}}$ in Scenario 2 for different values of h and b ($\overline{Y} = 1.320286$, $\overline{Z} = 1.489173$, $\hat{\mu}^{BBBS_2,w} = 0.9987603$, $\mu = 1$)

$h \setminus b$	0.005	0.05	0.1	0.5	1
0.005	0.2832286	**1.0600380**	**1.0861935**	1.358966	1.463947
0.05	0.2872110	**1.0532916**	**1.0787246**	1.345344	1.450534
0.1	0.2897269	**1.0437671**	**1.0682164**	1.329955	1.436189
0.5	0.2856396	**0.9007445**	**0.9206018**	1.205823	1.343689
1	0.2807338	**0.7556730**	**0.7716217**	1.100725	1.283538

$$
m(x) = \begin{cases} \dfrac{2k+1}{2k\varepsilon^{2k+1} + 2^{2k+1}}\varepsilon^{2k} & \text{if } x \in [0, \varepsilon) \\ \dfrac{2k+1}{2k\varepsilon^{2k+1} + 2^{2k+1}}x^{2k} & \text{if } x \in [\varepsilon, 2] \\ 0 & \text{if } x \notin [0, 2] \end{cases}
$$

Figure 2 shows the situation analyzed in Scenario 2 for the particular choice of $k = 1$ and $\varepsilon = 0.2$. Table 3 illustrates the behavior of the estimator (5), considering $n = 10^3$ and $N = 10^6$, the sizes of the samples involved. In this case, the estimation given by the Big Data sample is not as close to the real value as in the previous scenario, since the bias is greater (see Fig. 2). Again, the proposed estimator improves the estimation performed by both samples for some particular combinations of the chosen bandwidths b and h (numbers in bold).

Acknowledgements This research has been supported by MINECO Grants MTM2014-52876-R and MTM2017-82724-R and by the Xunta de Galicia (Grupos de Referencia Competitiva ED431C-2016-015 and Centro Singular de Investigación de Galicia ED431G/01), all of them through the European Regional Development Fund (ERDF). The second author's research was sponsored by the Xunta de Galicia predoctoral grant (with reference ED481A-2016/367) for the universities of the Galician University System, public research organizations in Galicia and other entities of the Galician R&D&I System, whose funding comes from the European Social Fund (ESF) in 80% and in the remaining 20% from the General Secretary of Universities, belonging to the Ministry of Culture, Education and University Management of the Xunta de Galicia.

References

1. Cao R (2015) Inferencia estadística con datos de gran volumen. La Gaceta de la RSME 18:393–417
2. Crawford K (2013) The hidden biases in big data. Harvard Business Review, Cambridge. https://hbr.org/2013/04/the-hidden-biases-in-big-data
3. Hargittai E (2015) Is bigger always better? potential biases of big data derived from social network sites. Ann Am Acad Pol Soc Sci 659:63–76
4. Lloyd CJ, Jones MC (2000) Nonparametric density estimation from biased data with unknown biasing function. J Am Stat Assoc 95:865–876

Band Depths Based on Multiple Time Instances

Ignacio Cascos and Ilya Molchanov

Abstract Bands of vector-valued functions $f : T \mapsto \mathbb{R}^d$ are defined by considering convex hulls generated by their values concatenated at m different values of the argument. The obtained m-bands are families of functions, ranging from the conventional band in case the time points are individually considered (for $m = 1$) to the convex hull in the functional space if the number m of simultaneously considered time points becomes large enough to fill the whole time domain. These bands give rise to a depth concept that is new both for real-valued and vector-valued functions.

1 Introduction

The statistical concept of *depth* is well known for random vectors in the Euclidean space. It describes the relative position of x from $\mathbb{R}^d$ with respect to a probability distribution on $\mathbb{R}^d$ or with respect to a sample $x_1, \ldots, x_n \in \mathbb{R}^d$ from it. Given a centrally symmetric distribution (for an appropriate notion of symmetry), the point of central symmetry is the *deepest* point (center of the distribution), while the depth of outward points is low. The concept of depth has been used in the context of trimming multivariate data, to derive depth-based estimators (e.g. depth-weighted L-estimators or ranks based on the center-outward ordering induced by the depth), to assess robustness of statistical procedures, and for classification purposes, to name a few areas, see [2, 9, 17] for extensive surveys and further references.

Often, the relative position of a point x with respect to a sample is defined with respect to the convex hull of the sample or a part of the sample. For instance, the classical concept of the simplicial depth appears as the fraction of $(d + 1)$-tuples

I. Cascos (✉)
Department of Statistics, Universidad Carlos III de Madrid, Av. Universidad 30,
28911 Leganés (Madrid), Spain
e-mail: ignacio.cascos@uc3m.es

I. Molchanov
Department of Mathematical Statistics and Actuarial Science, University of Berne,
Alpeneggstr. 22, 3012 Berne, Switzerland
e-mail: ilya.molchanov@stat.unibe.ch

© Springer International Publishing AG 2018
E. Gil et al. (eds.), *The Mathematics of the Uncertain*, Studies in Systems,
Decision and Control 142, https://doi.org/10.1007/978-3-319-73848-2_6

67

of sampled points whose convex hull contains x, see [10]. Its population version is given by the probability that x is contained in the convex hull of $(d + 1)$ i.i.d. copies of the random vector.

In high-dimensional spaces the curse of dimensionality comes into play and the convex hull of a finite set of sampled points forms a rather "thin" set and so it is very unlikely to expect that many points belong to it. Even the convex hull of the whole sample becomes rather small if the space dimension d is much larger than the sample size n. The situation is even worse for infinite-dimensional spaces that are typical in functional data analysis. In view of this, a direct generalisation of the simplicial depth and convex hull depth concepts leads to the situation where most points in the space have depth zero, see also [8], who discuss problems inherent with the half-space depth in infinite-dimensional spaces, most importantly zero depth and the lack of consistency, see also [15].

One possible way to overcome such difficulties is to consider the depth for the collection of function values at any given time argument value t and then integrate (maybe weightedly) over the argument space. This idea goes back to [6] and has been further studied in [4, 14].

Another approach is based on considering the position of a function relative to the band generated by functions from the sample. The band generated by real-valued functions is defined as the interval-valued function determined by the pointwise minimum and maximum of the functions from the sample. The corresponding band depth has been studied in [11, 12]. In the multivariate case the band becomes a set-valued function that at each point equals the convex hull of the values of functions from the sample, see [13]. Another multivariate generalisation of the band depth in [7] is based on taking convex combinations of band depths associated to each component. Yet another multivariate functional depth concept was studied in [4] by integrating the half-space depth over the time domain, see also [3]. It is argued in [7] that the multivariate setting makes it possible to incorporate other functional data parameters, such as derivatives, into the sample. It is also possible to combine a function with its smoothed version, possibly with different bandwidths.

In this paper we suggest a new concept of multivariate functional depth based on taking convex hulls of the functions' values at $m \geq 1$ time points combined to build a new higher-dimensional vector. In a sense, this concept pulls together values of the function at different points and so naturally incorporates the time dependency effects, and so better reflects the shape of curves. Two examples at which these m-band depths are used are presented.

The constructions described in Sect. 3 remind very much the conventional simplicial band depth, where the main point is to check if a point belongs to the convex hull of a subsample. The underlying convex hull in the functional space is replaced by the band, as in [11]. It is shown that the introduced band depth satisfies the main properties described in [4, 13]. The theoretical computation of the m-band depth is usually unfeasible, since it requires computing the probabilities that a point belongs to a convex hull of random points. Still, its empirical variant is consistent and rather easy to compute.

2 Regions Formed by Samples in Functional Spaces

m-Bands

Let $\mathbb{E}$ be a linear space of functions $f : T \mapsto \mathbb{R}^d$ whose argument t belongs to a rather general topological space T. For example, $\mathbb{E}$ may be the family of continuous functions on an interval T or a collection of d-vectors if T is a finite set.

Consider functions $f_1, \ldots, f_j \in \mathbb{E}$. The convex hull $\text{conv}(f_1, \ldots, f_j)$ of these functions is the family of functions $f \in \mathbb{E}$ that can be represented as

$$f(t) = \sum_{i=1}^{j} \lambda_i f_i(t), \quad t \in T ,$$

for some non-negative constants $\lambda_1, \ldots, \lambda_j$ that sum up to one.

If the coefficients $\lambda_1, \ldots, \lambda_j$ are allowed to be arbitrary functions of t, we arrive at the family of functions $f \in \mathbb{E}$ such that, for all $t \in T$, the value $f(t)$ belongs to the convex hull of $f_1(t), \ldots, f_j(t)$. Following [11, 13] for univariate (resp. multivariate) functions, the set of such functions is called the *band* generated by $f_1, \ldots, f_j$ and is denoted by $\text{band}(f_1, \ldots, f_j)$. It is obvious that

$$\text{conv}(f_1, \ldots, f_j) \subset \text{band}(f_1, \ldots, f_j).$$

If $d = 1$ (as in [11]), then $\text{band}(f_1, \ldots, f_j)$ consists of all functions f such that

$$\min_{i=1,\ldots,j} f_i(t) \le f(t) \le \max_{i=1,\ldots,j} f_i(t), \quad t \in T . \tag{1}$$

In order to obtain a set of functions with interior points, one should avoid the case when the convex hull of $f_1(t), \ldots, f_j(t)$ is of a lower dimension than d at some t. In particular, for this j should be greater than d.

We define nested families of functions that lie between the band and the convex hull generated by the sample.

Definition 2.1 The m-band, $\text{band}_m(f_1, \ldots, f_j)$, generated by $f_1, \ldots, f_j \in \mathbb{E}$ is the family of functions $f \in \mathbb{E}$ such that, for all $t_1, \ldots, t_m \in T$, the vector $(f(t_1), \ldots, f(t_m))$ belongs to the convex hull of $\{(f_i(t_1), \ldots, f_i(t_m)), i = 1, \ldots, j\}$, i.e.

$$(f(t_1), \ldots, f(t_m)) = \sum_{i=1}^{j} \lambda_i (f_i(t_1), \ldots, f_i(t_m)) \tag{2}$$

for non-negative real numbers $\lambda_1, \ldots, \lambda_j$ that sum up to one and may depend on $(t_1, \ldots, t_m)$.

Example 2.1 (Special cases) If $T = \{t\}$ is a singleton, the functions become vectors in $\mathbb{R}^d$ and the m-band is their convex hull for all $m \ge 1$.

If T is a finite set of cardinality k and $d = 1$, then the functions $f_1, \ldots, f_j$ of $t \in T$ can be viewed as vectors $x_i = (x_{i1}, \ldots, x_{ik}) \in \mathbb{R}^k$, $i = 1, \ldots, j$. The 1-band is the smallest hyperrectangle that contains $x_1, \ldots, x_j$, which is given by $\times [a_l, b_l]$ for $a_l = \min(x_{il}, i = 1, \ldots, j)$ and $b_l = \max(x_{il}, i = 1, \ldots, j)$ for $l = 1, \ldots, k$. The 2-band is obtained as the largest set such that its projections on each 2-dimensional coordinate plane equals the projection of the convex hull of $x_1, \ldots, x_k$. The k-band coincides with the convex hull of $x_1, \ldots, x_j$.

If $m = 1$ and $d = 1$, then we recover the band introduced in [11] and given by (1), so that $\text{band}(f_1, \ldots, f_j) = \text{band}_1(f_1, \ldots, f_j)$.

If $f \in \text{band}_m(f_1, \ldots, f_j)$, then each convex combination of the values for $f_1, \ldots, f_j$ and f can be written as a convex combination of the values of $f_1, \ldots, f_j$ and so

$$\text{band}_m(f_1, \ldots, f_j) = \text{band}_m(f_1, \ldots, f_j, f).$$

The m-band is additive with respect to the Minkowski (elementwise) addition. In particular,

$$\text{band}_m(g + f_1, \ldots, g + f_j) = g + \text{band}_m(f_1, \ldots, f_j) \tag{3}$$

for all $g \in \mathbb{E}$. The m-band is equivariant with respect to linear transformations, that is,

$$\text{band}_m(Af_1, \ldots, Af_j) = \{Af : f \in \text{band}_m(f_1, \ldots, f_j)\} \tag{4}$$

for all $A : T \mapsto \mathbb{R}^{d \times d}$ with $A(t)$ nonsingular for all $t \in T$. If all functions generating an m-band are affected by the same phase variation, the phase of the m-band is affected as shown below,

$$\text{band}_m(f_1 \circ h, \ldots, f_j \circ h) = \{f \circ h : f \in \text{band}_m(f_1, \ldots, f_j)\} \tag{5}$$

for any bijection $h : T \mapsto T$. If $d = 1$ and $\mathbb{E}$ consists of continuously differentiable functions on $T = \mathbb{R}$, then $f \in \text{band}_m(f_1, \ldots, f_j)$ yields that f' belongs to $\text{band}_{m-1}(f_1', \ldots, f_j')$. This can be extended for higher derivatives.

It is obvious that $\text{band}_m(f_1, \ldots, f_j)$ is a convex subset of $\mathbb{E}$; since the points $t_1, \ldots, t_m$ in Definition 2.1 are not necessarily distinct, it decreases if m grows. The following result shows that the m-band turns into the convex hull for large m.

Proposition 2.1 *Assume that all functions from $\mathbb{E}$ are jointly separable, that is there exists a countable set $Q \subset T$ such that, for all $f \in \mathbb{E}$ and $t \in T$, $f(t)$ is the limit of $f(t_n)$ for $t_n \in Q$ and $t_n \to t$. Then, for each $f_1, \ldots, f_j \in \mathbb{E}$,*

$$\text{band}_m(f_1, \ldots, f_j) \downarrow \text{conv}(f_1, \ldots, f_j) \quad as \ m \to \infty.$$

Proof Consider an increasing family T_n of finite subsets of T such that $T_n \uparrow Q$ and a certain function $f \in \mathbb{E}$. If m_n is the cardinality of T_n, and f belongs to the

m_n-band of $f_1, \ldots, f_j$, then the values $(f(t), t \in T_n)$ equal a convex combination of $(f_i(t), t \in T_n), i = 1, \ldots, j$, with coefficients λ_{ni}. By passing to a subsequence, assume that $\lambda_{ni} \to \lambda_i$ as $n \to \infty$ for all $i = 1, \ldots, j$. Using the nesting property of T_n, we obtain that

$$f(t) = \sum \lambda_i f_i(t), \quad t \in T_n.$$

Now it suffices to let $n \to \infty$ and appeal to the separability of f. □

Moreover, under a rather weak assumption, the m-band coincides with the convex hull for sufficiently large m. A set of points in the d-dimensional Euclidean space is said to be in general position if no $(d-1)$-dimensional hyperplane contains more than d points. In particular, if the set contains at most $d+1$ points, they will be in general position if and only if they are all extreme points of their convex hull, equivalently, any point from their convex hull is obtained as their unique convex combination.

Proposition 2.2 If $j \leq d(m-1) + 1$ and there exists $t_1, \ldots, t_{m-1} \in T$ such that the vectors $(f_i(t_1), \ldots, f_i(t_{m-1})) \in \mathbb{R}^{d(m-1)}$, $i = 1, \ldots, j$, are in general position, then

$$\mathrm{band}_m(f_1, \ldots, f_j) = \mathrm{conv}(f_1, \ldots, f_j).$$

Proof Let $f \in \mathrm{band}_m(f_1, \ldots, f_j)$. In view of (2), $(f(t_1), \ldots, f(t_{m-1}))$ equals a convex combination of $(f_i(t_1), \ldots, f_i(t_{m-1})), i = 1, \ldots, j$, which is unique by the general position condition. By considering an arbitrary $t_m \in T$, we see that $f(t_m)$ is obtained by the same convex combination, so that f is a convex combination of functions $f_1, \ldots, f_j$. □

In particular, if $d = 1$, then the 2-band of two functions coincides with their convex hull. It suffices to note that if f_1 and f_2 are not equal, then $f_1(t_1)$ and $f_2(t_1)$ are different for some t_1 and so are in general position. The same holds for any dimension $d \geq 2$.

Example 2.2 (Linear and affine functions) Let $f_1, \ldots, f_j$ be constant functions. Then their 1-band is the collection of functions lying between the maximum and minimum values of $f_1, \ldots, f_j$. The 2-band consists of constant functions only and coincides with the convex hull.

Together with (3), this implies that the 2-band generated by functions $f_i(t) = a(t) + b_i, i = 1, \ldots, j$, is the set of functions $a(t) + b$ for b from the convex hull of $b_1, \ldots, b_j$.

If $f_i(t) = a_i t + b_i, i = 1, \ldots, j$, are affine functions of $t \in \mathbb{R}$, then their 3-band consists of affine functions only and also equals the convex hull. Indeed,

$$(f(t_1), f(t_2), f(t_3)) = \sum \lambda_i (a_i(t_1, t_2, t_3) + b_i(1, 1, 1))$$

yields that

$$\frac{f(t_3) - f(t_1)}{f(t_2) - f(t_1)} = \frac{t_3 - t_1}{t_2 - t_1}.$$

Therefore each f from band$_3(f_1, \ldots, f_j)$ is an affine function.

Example 2.3 (*Monotone functions*) Let $d = 1$ and let $f_1, \ldots, f_j$ be non-decreasing (respectively non-increasing) functions. Then their 2-band is a collection of non-decreasing (resp. non-increasing) functions.

If all functions $f_1, \ldots, f_j$ are convex (resp. concave), then their 3-band is a collection of convex (resp. concave) functions.

Remark 2.1 The definition of the m-band can be easily extended for subsets F of a general topological linear space $\mathbb{E}$. Consider a certain family of continuous linear functionals u_t, $t \in T$. An element $x \in \mathbb{E}$ is said to belong to the m-band of F if, for each $t_1, \ldots, t_m \in T$, the vector $(u_{t_1}(x), \ldots, u_{t_m}(x))$ belongs to the convex hull of $\{(u_{t_1}(y), \ldots, u_{t_m}(y)) : y \in F\}$. Then Definition 2.1 corresponds to the case of $\mathbb{E}$ being a functional space and $u_t(f) = f(t)$ for $t \in T$.

While the conventional closed convex hull arises as the intersection of all closed half-spaces that contain a given set, its m-band variant arises from the intersection of half-spaces determined by the chosen functionals u_t for $t \in T$.

Space Reduction and Time Share

The m-band reduces to a 1-band by defining functions on the product space T^m.

Proposition 2.3 *For each j, the m-band* band$_m(f_1, \ldots, f_j)$ *coincides with band* $(f_1^{(m)}, \ldots, f_j^{(m)})$, *where $f_i^{(m)} : T^m \mapsto (\mathbb{R}^d)^m$ is defined as*

$$f^{(m)}(t_1, \ldots, t_m) = (f(t_1), \ldots, f(t_m)).$$

Proof It suffices to note that $f^{(m)}(t_1, \ldots, t_m)$ belongs to the convex hull of $f_i^{(m)}(t_1, \ldots, t_m)$, $i = 1, \ldots, j$, if and only if $(f(t_1), \ldots, f(t_m))$ belongs to the convex hull of $(f_i(t_1), \ldots, f_i(t_m))$, $i = 1, \ldots, j$. $\qquad \square$

In the framework of Proposition 2.3, it is possible to introduce further bands (called *space-reduced*) by restricting the functions $f_i^{(m)}$ to a subset S of T^m. For instance, the 1-band generated by functions $f_1^{(2)}, \ldots, f_j^{(2)}$ for the arguments $(t_1, t_2) \in \mathbb{R}^2$ such that $|t_1 - t_2| = h$ describes the joint behaviour of the values of functions separated by the lag h. If $m = 1$, then the space reduction is equivalent to restricting the parameter space, which can be useful, e.g. for discretisation purposes.

It is possible to quantify the closedness of f to the band by determining the proportion of the m-tuple of time values from T^m when the values of f belong to the band. Define the m-band time-share as

$$\mathrm{TS}_m(f; f_1, \ldots, f_j)$$
$$= \{(t_1, \ldots, t_m) \in T^m : f^{(m)}(t_1, \ldots, t_m) \in \mathrm{conv}(\{f_i^{(m)}(t_1, \ldots, t_m)\}_{i=1}^j).$$

If the functions take values in $\mathbb{R}$, then $\mathrm{TS}_1(f; f_1, \ldots, f_j)$ turns into the modified band depth defined in [11, Sect. 5]. If f belongs to the m-band of $f_1, \ldots, f_j$, then $\mathrm{TS}_m(f; f_1, \ldots, f_j) = T^m$, while if f belongs to the 1-band of $f_1, \ldots, f_j$, then $\{(t, \ldots, t) : t \in T\} \subset \mathrm{TS}_m(f; f_1, \ldots, f_j)$. It is also straightforward to incorporate the space reduction by replacing T^m with a subset S.

3 Simplicial-Type Band Depths

Band Depth

In the following, we consider the event that a function f belongs to a band generated by i.i.d. random functions $\xi_1, \ldots, \xi_j$ with the common distribution P. The m-band depth of the function f with respect to P is defined by

$$\mathrm{bd}_m^{(j)}(f; P) = \mathbf{P}\{f \in \mathrm{band}_m(\xi_1, \ldots, \xi_j)\}$$
$$= \mathbf{P}\{(f(t_1), \ldots, f(t_m)) \in \mathrm{conv}(\{(\xi_i(t_1), \ldots, \xi_i(t_m))\}_{i=1}^j) \, \forall t_1, \ldots, t_m \in T\}. \tag{6}$$

If m increases, then the m-band narrows, and so the m-band depth decreases.

We recall that when $d = 1$ the 1-band coincides with the band introduced in [11]. Nevertheless the band depth defined in [11] is the sum of $\mathrm{bd}_m^{(j)}(f; P)$ with j ranging from 2 to a fixed value J. The same construction can be applied to our m-bands.

The m-band depth of f is influenced by the choice of j, and it increases with j. Unlike the finite-dimensional setting, where j is typically chosen as the dimension of the space plus one [10], there is no canonical choice of j for the functional spaces. In order to ensure that the m-band generated by $\xi_1, \ldots, \xi_j$ differs from the convex hull, it is essential to choose j sufficiently large, and in any case at least $d(m-1) + 2$, see Proposition 2.2. Furthermore, we must impose stronger conditions on j to avoid the zero-depth problem.

Proposition 3.1 *If $j \leq dm$ and the joint distribution of the marginals of P at some fixed m time points is absolutely continuous, then $\mathrm{bd}_m^{(j)}(\cdot; P) = 0$.*

Proof If $j \leq dm$ and $\{(\xi_i(t_1), \ldots, \xi_i(t_m))\}_{i=1}^j$ are independent and absolutely continuous in $\mathbb{R}^{dm}$, the probability that any fixed $x \in \mathbb{R}^{dm}$ lies in their convex hull is zero. □

A theoretical calculation of the m-band depth given by (6) is not feasible in most cases. In applications, it can be replaced by its empirical variant defined in exactly the same way as in [11] for the 1-band case. Let $f_1, \ldots, f_n$ be a sample from P. Fix any $j \in \{dm + 1, \ldots, n\}$ and define

$$\text{bd}_m^{(j)}(f; f_1, \ldots, f_n) = \binom{n}{j}^{-1} \sum_{1 \le i_1 < \cdots < i_j \le n} \mathbf{1}_{f \in \text{band}_m(f_{i_1}, \ldots, f_{i_j})},$$

so that $\text{bd}_m^{(j)}(f; f_1, \ldots, f_n)$ is the proportion of j-tuples from $f_1, \ldots, f_n$ such that f lies in the m-band generated by the j-tuple. The choice of j affects the results. It is computationally advantageous to keep j small, while it is also possible to sum up the depths over a range of the values for j, as in [11].

Time-Share Depth

Assume now that T is equipped with a probability measure μ, for example, the normalised Lebesgue measure in case T is a bounded subset of the Euclidean space or the normalised counting measure if T is discrete. Extend μ to the product measure $\mu^{(m)}$ on T^m. Define the time-share depth by

$$\text{td}_m^{(j)}(f; P) = \mathbf{E}\mu^{(m)}(\text{TS}_m(f; \xi_1, \ldots, \xi_j)).$$

If T is a subset of the Euclidean space, Fubini's Theorem yields that the time-share depth is the average of the probability that $(f(t_1), \ldots, f(t_m))$ lies in the convex hull of j points in $\mathbb{R}^{dm}$,

$$\text{td}_m^{(j)}(f; P) = \int \mathbf{P}\{(f(t_1), \ldots, f(t_m)) \in \text{conv}(\{(\xi_i(t_1), \ldots, \xi_i(t_m))\}_{i=1}^{j})\}$$

$$d\mu^{(m)}(t_1, \ldots, t_m). \quad (7)$$

For any $j \in \{dm + 1, \ldots, n\}$, the empirical time-share depth is given by

$$\text{td}_m^{(j)}(f; f_1, \ldots, f_n) = \binom{n}{j}^{-1} \sum_{1 \le i_1 < \cdots < i_j \le n} \mu^{(m)}(\text{TS}_m(f; f_{i_1}, \ldots, f_{i_j})).$$

Example 3.1 (*Univariate case*) Assume that T is a singleton. Then necessarily $m = 1$, the function f is represented by a point x in $\mathbb{R}^d$, and the band depth of x for $j = d + 1$ coincides with the simplicial depth, see [10].

Example 3.2 Let $\xi(t) = a(t) + X$, $t \in T$, where X is a random variable. Then $\text{band}(\xi_1, \ldots, \xi_j)$ for i.i.d. $\xi_i(t) = a(t) + X_i$, $i = 1, \ldots, j$, is the set of functions bounded above by $a(t) + \max X_i$ and below by $a(t) + \min X_i$. Then

$$\text{bd}_1^{(j)}(a; P) = 1 - \mathbf{P}\{X > 0\}^j - \mathbf{P}\{X < 0\}^j.$$

By Example 2.2, $\text{band}_2(\xi_1, \ldots, \xi_j)$ consists of functions $a(t) + b$ for the constant $b \in [\min X_i, \max X_i]$. Only such functions may have a positive 2-band depth.

Example 3.3 Let now $\xi(t) = a(t) + X$, where $a : T \to \mathbb{R}^d$ and X is an absolutely continuous random vector in $\mathbb{R}^d$ which is angularly symmetric about the origin. Then

$$\mathrm{bd}_1^{(j)}(a; P) = 1 - 2^{1-j} \sum_{i=0}^{d-1} \binom{j-1}{i} \tag{8}$$

being the probability that the origin belongs to the convex hull of $X_1, \ldots, X_j$, see
[16].

Properties of the Band Depths

Theorem 3.1 *For any $j \geq dm + 1$ we have:*

1. affine invariance. $\mathrm{bd}_m^{(j)}(Af + g; P_{A,g}) = \mathrm{bd}_m^{(j)}(f; P)$ *and* $\mathrm{td}_m^{(j)}(Af + g; P_{A,g}) = \mathrm{td}_m^{(j)}(f; P)$ *for all $g \in \mathbb{E}$ and $A : T \mapsto \mathbb{R}^{d \times d}$ with $A(t)$ nonsingular for $t \in T$, where $P_{A,g}(F) = P(A^{-1}(F - g))$ for any measurable subset F of $\mathbb{E}$.*
2. phase invariance. $\mathrm{bd}_m^{(j)}(f \circ h; P^h) = \mathrm{bd}_m^{(j)}(f; P)$ *for any one-to-one transformation $h : T \mapsto T$, where $P^h(F) = P(F \circ h^{-1})$ for any measurable subset F of $\mathbb{E}$ when h^{-1} is the inverse mapping of h.*
3. vanishing at infinity. $\mathrm{bd}_m^{(j)}(f; P) \to 0$ *if the supremum of $\|f\|$ over T converges to infinity, and $\mathrm{td}_m^{(j)}(f; P) \to 0$ if the infimum of $\|f\|$ over T converges to infinity.*

The affine invariance of both depths follows from the affine invariance of the m-bands, see (3), (4), while the phase-invariance of the band depth follows from (5).

In practice, the functions are going to be evaluated over a finite set of time points, thus $T = \{t_1, \ldots, t_k\}$ and probability P is a distribution on $(\mathbb{R}^d)^k$. Furthermore, the sample of functions $f_1, \ldots, f_n$ to be used to determine an empirical m-band depth should have size at least $n \geq j \geq dm + 1$.

Theorem 3.2 *If P is absolutely continuous, for any $n \geq j \geq dm + 1$ we have:*

4. maximality at the center. *If P is angularly symmetric about the point $(f(t_1), \ldots, f(t_k))$, function f will be the deepest with regard to the time-share depth, and $\mathrm{td}_m^{(j)}(f; P) = 1 - 2^{1-j} \sum_{i=0}^{dm-1} \binom{j-1}{i}$.*
5. consistency. *Band depth $\sup_{f \in \mathbb{E}} |\mathrm{bd}_m^{(j)}(f; f_1, \ldots, f_n) - \mathrm{bd}_m^{(j)}(f; P)| \to 0$ a.s. and time-share depth $\sup_{f \in \mathbb{E}} |\mathrm{td}_m^{(j)}(f; f_1, \ldots, f_n) - \mathrm{td}_m^{(j)}(f; P)| \to 0$ a.s.*

The properties of the time-share depth rely on Formula (7) that makes possible to write it as an average of the probability that a point lies in the convex hull of independent copies of a random vector. The maximality at center follows from the main result in [16] which determines the probability inside the integral in (7), see (8), while the consistency can be proved in a similar way to [13, Theorem 3] extending the uniform consistency of the empirical simplicial depth [5, Theorem 1] to the one of the probability that a point lies in the convex hull of a fixed number of independent copies of a random vector. Such an extension, which relies on probabilities of intersections of open half-spaces, can be adapted to prove the consistency of the empirical m-band depth.

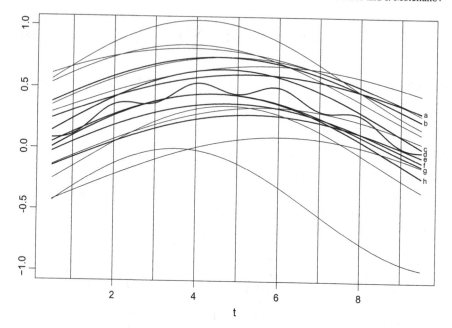

Fig. 1 17 curves evaluated at $\{1, 2, \ldots, 9\}$. The eight deepest curves are thicker than the others and each of them is assigned a letter from a to h. Five deepest curves for $\mathrm{bd}_1^{(4)}$ (in order): d,c,f,g,a, for $\mathrm{td}_1^{(4)}$: d,c,g,a,f, for $\mathrm{bd}_2^{(4)}$: g,b,f,e,a, and for $\mathrm{td}_2^{(4)}$: g,f,c,d,h

4   Data Examples

Simulated Data

Figure 1 shows 17 curves which are evaluated at $T = \{1, 2, \ldots, 9\}$. Among the 17 curves, there is a clear shape outlier (marked as d) that lies *deep* within the bunch of curves. Such an outlier will not be detected by the outliergram from [1] due to its high depth value with regard to both of the 1-band depth and half-region depth (see [12]). Nevertheless, its anomalous shape is detected by any m-band depth with $m \geq 2$.

It is remarkable that curve d, which is the deepest curve with respect to the usual band depth and modified band depth ($\mathrm{bd}_1^{(4)}$ and $\mathrm{td}_1^{(4)}$) is among the less deep curves for the 2-band depth ($\mathrm{bd}_2^{(4)}$) and is only the fourth deepest curve for its time-share depth ($\mathrm{td}_2^{(4)}$). The reason for this last fact is that if we restrict to either of the sets of time points $\{1, 3, 5, 7, 9\}$ or $\{2, 4, 6, 8\}$, curve d is not a shape outlier with respect to them.

Real Data

The nominal Gross Domestic Product per capita of the 28 countries of the European Union (2004–2013) was obtained from the EUROSTAT web-site and is represented

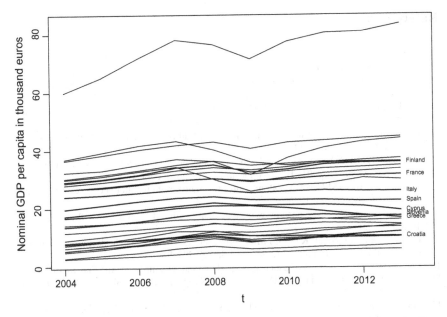

Fig. 2 Evolution of the nominal GDP per capita between 2004 and 2013 at the EU countries. Five deepest curves for $bd_1^{(5)}$ (in order): Cyprus, Spain, Italy, Greece, and Slovenia, for $bd_2^{(5)}$: Spain, Slovenia, France, Croatia, and Finland, and for $bd_2^{(5)}$ space-reduced with $S = \{(t_1, t_2) : |t_1 - t_2| = 1\}$: Croatia, Slovenia, Spain, Finland, and France

in Fig. 2. The missing observation that corresponds to Greece, 2013 was replaced by the value obtained from the FOCUSECONOMICS web-site.

The deepest curve with regard to the band depth $(bd_1^{(5)})$ is the one of Cyprus. Interestingly, Cyprus suffered the 2012–13 Cypriot financial crisis at the end of the considered period and its GDP per capita experienced a decay in 2013 in comparison with its 2012 figure much greater than the one of any other of the EU countries. Also the Greek curve is among the five deepest ones for $bd_1^{(5)}$ despite being the only country with a constant decrement in the second half of the considered time period. If we consider 2-bands, that take into account the shape of the curves, these two curves are not any more considered representative of the evolution of the GDP per capita in the EU.

Acknowledgements The authors would like to thank the Editors for the opportunity to contribute in this tribute to Pedro Gil, to whom we do sincerely appreciate. We will always remember Pedro when we order a "café con hielo".

Most of this work was carried over while IM was supported by the Chair of Excellence Programme of the University Carlos III and the Santander Bank. At that time both authors benefited from discussions with Professor Juan Romo. IM is grateful to the Department of Statistics of the University Carlos III in Madrid for the hospitality.

References

1. Arribas-Gil A, Romo J (2014) Shape outlier detection and visualization for functional data: the outliergram. Biostatistics 15:603–619
2. Cascos I (2010) Data depth: multivariate statistics and geometry. In: Kendall WS, Molchanov I (eds) New perspectives in stochastic geometry. Oxford University Press, Oxford
3. Chakraborty A, Chaudhuri P (2014) On data depth in infinite dimensional spaces. Ann Inst Stat Math 66(2):303–324
4. Claeskens G, Hubert M, Slaets L, Vakili K (2014) Multivariate functional halfspace depth. J Am Stat Assoc 109(505):411–423
5. Dümbgen L (1992) Limit theorems for the simplicial depth. Stat Prob Lett 14:119–128
6. Fraiman R, Muniz G (2001) Trimmed means for functional data. Test 10:419–414
7. Ieva F, Paganoni AM (2013) Depth measures for multivariate functional data. Commun Stat-Theor Methods 42:1265–1276
8. Kuelbs J, Zinn J (2013) Concerns with functional depth. ALEA Lat Am J Prob Math Stat 10(2):831–855
9. Liu R, Parelis JM, Singh K (1999) Multivariate analysis by data depth: descriptive statistics, graphs and inference (with discussion). Ann Stat 27:783–858
10. Liu RY (1990) On a notion of data depth based on random simplices. Ann Stat 18:405–414
11. López-Pintado S, Romo J (2009) On the concept of depth for functional data. J Am Stat Assoc 104:718–734
12. López-Pintado S, Romo J (2011) A half-region depth for functional data. Comput Stat Data Anal 55:1679–1695
13. López-Pintado S, Sun Y, Lin JK, Genton MG (2014) Simplicial band depth for multivariate functional data. Adv Data Anal Classif 8:321–338
14. Nagy S (2017) Integrated depth for measurable functions and sets. Stat Prob Lett 123:165–170
15. Nieto-Reyes A, Battey H (2016) A topologically valid definition of depth for functional data. Stat Sci 31(1):61–79
16. Wendel JG (1962) A problem in geometric probability. Math Scand 11:109–111
17. Zuo Y, Serfling R (2000) General notions of statistical depth function. Ann Stat 28:461–482

On the Combination of Depth-Based Ranks

Ignacio Cascos and Ignacio Montes

Abstract The depth of a multivariate observation assesses its degree of centrality with respect to a probability distribution, and thus it can be interpreted as a measurement of the fit of the observation wrt the distribution. If such depth is transformed into a (depth-based) rank, then we obtain a kind of p-value of a goodness-of-fit test run on a single observation. For a sample of observations, the goal is to combine their ranks in order to decide whether they were taken from some prescribed distribution. From the meta-analysis literature, it is well known that there does not exist a combination procedure for such p-values (or ranks) that outperforms the remaining ones in all possible scenarios. Here we explore several combination procedures of the depth-based ranks and analyse their behaviour in the detection of some given shifts from a prescribed distribution.

1 Introduction

In multivariate Statistics, a depth function assesses the degree of centrality of an observation with respect to a probability distribution or a data cloud, see [2, 8, 12, 18, 19]. Based on the ordering provided by such a depth function, it is possible to build a rank for multivariate observations defined as the proportion of observations that are at most as central as the given one. A central observation has thus a rank close to 1, while the rank of a peripheral observation is close to 0.

Based on such depth-based ranks, Liu [9] proposed three control charts for multivariate observations. Specifically, she proposed an individuals chart that monitors the rank of each individual observation, a chart for rational samples of a given size, and

I. Cascos (✉)
Department of Statistics, Universidad Carlos III de Madrid, Av. Universidad 30,
28911 Leganés (Madrid), Spain
e-mail: ignacio.cascos@uc3m.es

I. Montes
Department of Statistics and O.R., University of Oviedo,
C/ Federico García Lorca 18, 33007 Oviedo, Spain
e-mail: imontes@uniovi.es

© Springer International Publishing AG 2018
E. Gil et al. (eds.), *The Mathematics of the Uncertain*, Studies in Systems,
Decision and Control 142, https://doi.org/10.1007/978-3-319-73848-2_7

a chart with memory that combines all observations until the current one. These two last charts combine the information of the individual observations by averaging their respective (depth-based) ranks. The idea is simple, since peripheral observations have small ranks, an alarm must be risen at the individuals chart whenever a very small rank is detected. When several observations are considered, the alarm is risen when their average rank is small. We will show here that other combination procedures different from averaging ranks provide better results at some given scenarios.

In the Statistical Process Control literature, there is a considerably large number of proposals of nonparametric univariate control charts that monitor on-going processes either in location, scale, or in location and scale, see [5, 13]. Notice that the (usual) rank of a univariate observation, defined as the cdf evaluated at it, locates the observation throughout the range of the random variable, while the depth-based ranks that we consider here only establish how central an observation is wrt a distribution. In this sense, if each point of a control chart is to be interpreted as the statistic of a goodness-of-fit test with H_0 establishing that the distribution of the process has not departed from some prescribed one, and H_1 that some shift has altered the location parameter (and the scale parameter might have also increased), then the test built out of a (classical) univariate rank is two-sided, while the one built out of a depth-based rank is one-sided. The fact that the test is one-sided, means that the control chart has only one Control Limit, and it is also relevant that the departure of an observation from the center cannot be compensated by some other observation, simply because when we use a depth function, we miss the information of the direction of the departure. A possible alternative to the use of a depth-based rank that would exploit the information of the direction of the departure is to estimate a parameter over each sample and then evaluate the depth over such estimates of the parameter either with a parameter depth notion as in [4] or with a classical depth evaluated over artificial samples of parameters as in [11].

Consider now a depth-based rank evaluated over a single observation. When the process distribution has not departed from the original one, and under fairly weak assumptions, the distribution of such rank is uniform in the unit interval and an alarm is to be risen if it is very small, so essentially the rank can be considered as the p-value of a goodness-of-fit test. If we have a rational sample of a given size, the combination of ranks is equivalent to the combination of p-values, so we face a meta-analysis problem, see [1, 14], with p-values coming from one-sided tests.

In Sect. 2 we introduce several classical notions of depth, together with the depth-based rank and some classical control charts built from it. In Sect. 3 some alternative combination methods of ranks are presented, and the performance of the control charts built from them is presented in Sect. 4. Finally, an application of the proposed procedure is discussed in Sect. 5.

2 Depths, Depth-Based Ranks, and Control Charts

The standard charts for monitoring a production process are the individuals X-chart (for individual observations), the $\overline{X}$-chart (for rational samples of a given fixed size), and the CUSUM-chart (for samples of an increasing size, and thus with memory). These charts are very simple and efficient in the univariate framework under some mild conditions. However, their multivariate generalizations are quite sensitive to departures from the distributional assumptions.

In order to avoid these problems, Liu [9] proposed alternative charts based on data depth. Given a distribution P in the k-dimensional Euclidean space, the *simplicial depth* of $y \in \mathbb{R}^k$ with respect to P is defined in Liu [8] as

$$\mathrm{SD}_P(y) = \Pr\{y \in \mathrm{co}\{Y_1, \ldots, Y_{k+1}\}\},$$

where $Y_1, \ldots, Y_{k+1}$ are $k+1$ independent random variables with distribution P and co stands for the convex hull. The empirical simplicial depth built out of a sample $Y_1, \ldots, Y_m$ is defined as the U-statistic

$$\mathrm{SD}_m(y) = \binom{m}{k+1}^{-1} \sum_{1 \le i_1 < \ldots < i_{k+1} \le m} I(y \in \mathrm{co}\{Y_{i_1}, \ldots, Y_{i_{k+1}}\}).$$

Other alternative notions of data depth are Tukey's *half-space depth* (see [16, 18])

$$\mathrm{HD}_P(y) = \inf\{P(H) \mid H \text{ closed half-space with } y \in H\},$$

and Koshevoy and Mosler's [7] *zonoid depth*

$$\mathrm{ZD}_P(y) = \sup\left\{\alpha \in (0, 1] \mid y = \int xg(x)\,\mathrm{d}P(x),\right.$$
$$\left. \text{with } g : \mathbb{R}^k \mapsto [0, \alpha^{-1}] \text{ measurable}, \int g(x)\,\mathrm{d}P(x) = 1\right\}.$$

The empirical half-space and zonoid depths are obtained after substituting P by an empirical distribution P_m and denoted by $\mathrm{HD}_m(y)$ and $\mathrm{ZD}_m(y)$.

Based on any of the previously introduced depth notions, denoted by D_P, we can define the rank of an observation with respect to P as

$$r_P(y) = \Pr\{\mathrm{D}_P(Y) \le \mathrm{D}_P(y) \mid Y \sim P\}.$$

When the available information about P appears in terms of a sample $Y_1, \ldots, Y_m$, the empirical rank is given by

$$r_m(y) = \#\{Y_j \mid \mathrm{D}_m(Y_j) \le \mathrm{D}_m(y), \ j = 1, \ldots, m\}/m,$$

where # stands for the cardinality of a set. Both r_P and r_m measure how central is a point y with respect to a distribution (either a population P or an empirical distribution P_m), in the sense that the smaller the rank of y is, the more peripheral y is with regard to the distribution.

For any of the previous depth notions, if $X \sim P$ absolutely continuous, then $r_P(X)$ follows a uniform distribution in the unit interval, see Liu and Singh [10]. Further, since the empirical depths are uniformly consistent estimators of the population depths, then each empirical rank $r_m(X)$ weakly converges to a uniform distribution in the unit interval. The uniform consistency of the three empirical depths introduced above for absolutely continuous distributions can be found at [8] (simplicial depth), [6] (half-space depth), and [3] (zonoid depth).

Given a sample of observations $X_1, \ldots, X_n$, which we will denote by $X^{(n)}$, Liu and Singh [10] define a quality index to quantify the quality of $X^{(n)}$ as a random sample from P,

$$Q_P(X^{(n)}) = \frac{1}{n} \sum_{i=1}^{n} r_P(X_i), \tag{1}$$

as usual, the quality with respect to the empirical probability P_m is obtained after substituting the rank with respect to P by the empirical rank and will be denoted $Q_m(X^{(n)})$. We turn now our attention to the specific control charts proposed by Liu [9].

Q-Chart

The Q-chart is a non-parametric and multivariate generalization of Shewart's $\overline{X}$-chart. Rational subgroups of size n are subsequently considered, $X_1^{(n)}, X_2^{(n)}, \ldots$ and their quality indices $Q_P(X_1^{(n)}), Q_P(X_2^{(n)}), \ldots$ are plotted in a time chart. The unique (lower) Control Limit is set at the α-quantile of a sum of n independent uniform random variables in order to obtain a control chart with false alarm rate (significance level at the goodness-of-fit test) α.

S-Chart

The S-chart is a non-parametric and multivariate counterpart of the CUSUM-chart. Given the new observations $X_1, \ldots, X_n$, it monitors the cumulative sum of their ranks, that is, for $j = 1, \ldots, n$

$$S_j = \sum_{i=1}^{j} \left(r_P(X_i) - \frac{1}{2} \right) = j \left(Q_P(X^{(j)}) - \frac{1}{2} \right).$$

The process is considered to be out-of-control when the cumulative sum represented by S_j is too small. Under the assumption that j is large enough in order to apply the CLT, the Control Limit is given by $-(z_\alpha \sqrt{j/12})$. If instead of P, the historical information about the process appears as a sample of size m, the Control Limit is corrected due to the variability of such sample to adopt the expression $-(z_\alpha \sqrt{(j + j^2/m)/12})$.

A modification of the S-chart appears in the form of the S^*-chart, which monitors the statistic $S_j^* = S_j/\sqrt{j/12}$. The Control Limit for the S_j^*-chart is constant at $-z_\alpha$.

In Fig. 3 left, Q-, S- and S^*-charts are represented.

3 The Proposal

The combination of several ranks is the same problem as the combination of several p-values in a meta-analysis procedure. In Eq. (1) those ranks were averaged in order for an alarm to be risen whenever such average was too small. One reasonable property that any method of combination must satisfy is admissibility (following the jargon proposed at [1]). This property says that if an alarm is risen at a sample of ranks $r_1, \ldots, r_n$, it would also be risen at any other sample $r_1^*, \ldots, r_n^*$ with $r_i^* \leq r_i$ for each i.

We take advantage of the uniformity of the ranks and apply the inverse transform for some distribution models for which the distribution of the sum of independent random variables is well-established. That is, each individual rank is to be transformed and then, the random variables obtained will be added. The distribution models we consider in our weighting scheme are right-skewed and supported on the positive half-line in order to be sensitive to peripheral observations when they are applied to the counter-rank $(1 - r)$.

Since we apply an increasing transformation to each counter-rank (resp. rank), alarms are risen for large (resp. small) aggregated results. If F denotes the cdf of a continuous distribution, we can aggregate the transformations of the rank in terms of the quantile function F^{-1}:

$$Q_P^F(X^{(n)}) = \sum_{i=1}^{n} F^{-1}(1 - r_P(X_i)), \tag{2}$$

and rise an alarm for large values of the Q statistic. Alternatively, it is possible to use the counter-ranks, whose Q index will be denoted by Q^-, and rise an alarm for small values of the statistic.

When F stands for an Exp(1) distribution or a Beta(0.5, 1) distribution, we obtain Q indices whose distribution is known when $X \sim P$,

$$Q_P^{\text{Exp}}(X^{(n)}) = -\sum_{i=1}^{n} \log(r_P(X_i)) \sim \text{Gamma}(n, 1), \tag{3}$$

$$Q_P^{\text{Beta}}(X^{(n)}) = \sum_{i=1}^{n} (1 - r_P(X_i))^2. \tag{4}$$

Expression (3) can be alternatively obtained after taking the logarithm of the product of the ranks and corresponds to Fisher's (meta-analysis) method, which usually appears multiplied times 2 in order for its distribution to be χ^2_{2n}. The distribution of (3) is the one of the sum of squares of uniform random variables in the unit interval, and a tractable explicit expression for it, when n is not too large, is given at [17]. For large values of n, we can apply the CLT.

Other possible choices for F in (2) are the uniform distribution in the unit interval and the chi-squared distribution χ^2_1. In the first case, the index Q amounts to the sum of the counter-ranks and we obtain $n - nQ_P(X^{(n)})$, which is a decreasing transformation of (1) since the control limit here is an upper one, while in the second the Q index would follow a χ^2_n distribution.

Remark 3.1 The aggregation method used at (2) is the addition, which is related with the product in (3), but it is also possible to consider the maximum, minimum, or any given intermediate observation.

Q-Chart

The Q-chart is based on subgroups of size n denoted $X^{(n)}_1, X^{(n)}_2, \ldots$ For transformations based on the quantile function, the aggregations $Q^F_P(X^{(n)}_1), Q^F_P(X^{(n)}_2), \ldots$ are plotted in a time chart, together with a unique upper Control Limit, which can be exactly computed. In case the counter-ranks are used, the unique Control Limit is a lower one.

S-Chart

Given the observations $X_1, X_2, \ldots$ consider the cumulative sum of the transformed ranks and subtract the mean of distribution F from it, $\mu(F)$, as many times as observations are available

$$S^F_j = Q^F_P(X^{(j)}) - j\mu(F), \text{ for } j = 1, 2, \ldots$$

where again $X^{(j)}$ denotes the first j observations. If each X_i follows distribution P, then each S^F_j is a random variable centred at 0, whose distribution can be approximated to a normal by the CLT, and thus the upper Control Limit is established at $z_\alpha \sigma(F)\sqrt{j}$.

4 Comparison Results

As stated before, there is no weighting of the ranks that outperforms the remaining ones in the detection of all possible shifts in a distribution. It is actually the distribution of the rank of the observation of a shifted process (known to be uniform in case there is no shift) what determines which is the best weighting for the detection of each individual shift. In Fig. 1 below we have represented the density mass function of

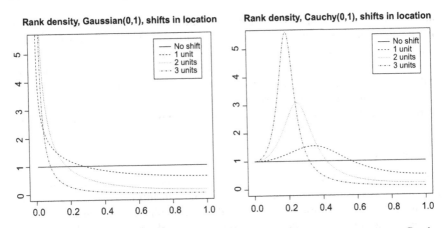

Fig. 1 Density mass functions of the (depth-based) ranks of a standard Gaussian and a Cauchy distribution after a shift in the location parameter

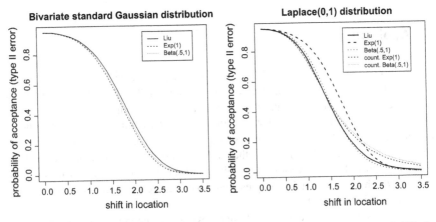

Fig. 2 Operating Characteristic curves of several Q and Q^- charts at the detection of shifts in location for samples of size $n = 5$ and false alarm rate $\alpha = 0.05$

the (depth-based) rank of a standard Gaussian (left) and a Cauchy distribution (right) unshifted and after several possible shifts in the location parameter.

If the location parameter of a Gaussian distribution is shifted, the distribution of the rank is close to the one of a Beta$(1, a)$ distribution with $a > 1$, as conjectured in [15, Sect. 5.2], but if the reference distribution has heavy tails, the situation is quite different, as can be observed in Fig. 1 right.

In order to compare the transformation of the ranks introduced in the previous section, at Fig. 2 we have represented the Operating Characteristic curves obtained for the detection of shifts in location, on samples of size $n = 5$, with $\alpha = 0.05$, and two different distribution models of the process. The lines correspond to the probability of not detecting a shift, which is always 0.95 if there is no shift in location (at $x = 0$)

and should be as low as possible for any real shift. On the left (bivariate Gaussian distribution), the best transformation is the logarithmic one, then the square, and finally the pure averaging of the ranks. On the right we considered the Laplace (double exponential) distribution since it has heavier tails. Here it turns out that if the shift in location is small, the use of the counter-ranks in order to produce index Q^- together with the logarithmic or square transformation is a good option. For large shifts, the square transformation seems to be the best option, while overall the pure average of the ranks is the best option.

Proposed Procedure

For some reference distribution or historical dataset, fix a sample size, a false alarm rate, and a target shift. The target shift is the minimal shift in whatever parameters of the reference distribution that should be detected. With these reference values, we can apply the shift to the historical dataset and resample from it. Finally, we select the transformation that detects such a shift with the largest probability.

5 Application

The data used in this section is borrowed from [9], where 580 observations were simulated from a bivariate standard normal distribution. The first 500 observations were used as the historical dataset and all depths were computed with respect to them. For the second group of 80 observations, the first 40 of them were kept without modifications and the last 40 were first multiplied by a scale factor of 2 and then vector $(2, 2)$ was added to them, so they suffered a shift in location and scale. The second group of 80 observations was split in 20 samples of size $n = 4$ in order to obtain a Q-chart. The first 10 samples were taken before the shift, while the last 10 were obtained after the shift.

The first row of Fig. 3 contains Q-charts, while the second and third contain S- and S^*-charts. On the left column the pure average of the ranks was considered (as in [9]), while the logarithmic and square transforms were applied on the middle and right columns.

In the first row of Fig. 3 and for $\alpha = 0.025$, the three Q-charts have the same behaviour (fail to detect the shift at sample number 18). If we take $\alpha = 0.1$ instead, the last two charts detect that shift, but the three of them classify as suspicious sample number 7. As for the S-charts, the chart on the left does not detect the shift until the 49th observation (the shift already occurred at the 41st), while the other two charts already detect it at the 47th.

As observed in Fig. 2 left, the logarithmic and square transformation detect shifts in a bivariate Gaussian distribution better than the pure average.

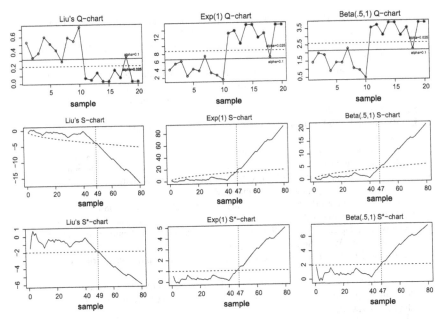

Fig. 3 Q- and S-charts as described by Liu [9] and other two combination schemes

Acknowledgements The authors would like to thank the Editors for sharing them the opportunity to contribute to this tribute to Pedro Gil. We are lucky to have known Pedro in both of his sides, the professional one, as teacher and Head of Department, and, thanks to a close relationship with his family, the personal one. He has constituted a model to follow for his students and coworkers. In his absence we will find ourselves often wondering what quiet advice would he give us at some given situation, just as if we were sitting in front of him under the lost look of the punch-drunk boxer of the painting hanging on the wall of his office.

This work was started while Ignacio Montes was with the Department of Statistics of the Universidad Carlos III de Madrid. We acknowledge the financial support by projects ECO2015-66593 and TIN2014-59543-P.

References

1. Birnbaum A (1954) Combining independent tests of significance. J Am Stat Assoc 49:559–574
2. Cascos I (2010) Data depth: multivariate statistics and geometry. In: Kendall WS, Molchanov I (eds) New perspectives in stochastic geometry. Oxford University Press, Oxford
3. Cascos I, López-Díaz M (2016) On the uniform consistency of the zonoid depth. J Multiv Anal 143:394–397
4. Cascos I, López-Díaz M (2018) Control charts based on parameter depths. Appl Math Model 53:487–509
5. Chowdhury S, Mukherjee A, Chakraborti S (2014) A new distribution-free control chart for joint monitoring of unknown location and scale parameters of continuous distributions. Qual Reliab Engin Int 30:191–204

6. Donoho D, Gasko M (1992) Breakdown properties of location estimates based on halfspace depth and projected outlyingness. Ann Stat 20:1803–1827
7. Koshevoy G, Mosler K (1997) Zonoid trimming for multivariate distributions. Ann Stat 25:1998–2017
8. Liu RY (1990) On a notion of data depth based on random simplices. Ann Stat 18:405–414
9. Liu RY (1995) Control charts for multivariate processes. J Am Stat Assoc 90:1380–1387
10. Liu RY, Singh K (1993) A quality index based on data depth and multivariate rank test. J Am Stat Assoc 88:252–260
11. Liu RY, Singh K (1997) Notions of limiting p values based on data depth and bootstrap. J Am Stat Assoc 92:266–277
12. Liu RY, Parelius J, Singh K (1999) Multivariate analysis by data depth: descriptive statistics, graphics and inference. With discussion and a rejoinder by Liu and Singh. Ann Stat 27:783–858
13. Mukherjee A, Chakraborti S (2012) A distribution-free control chart for the joint monitoring of location and scale. Qual Reliab Engin Int 28:335–352
14. Owen AB (2009) Karl Pearson's meta-analysis revisited. Ann Stat 37:3867–3892
15. Porzio GC, Ragozini G (2007) Multivariate control charts from a data mining perspective. In: Liao TW, Triantaphyllou E (eds) Recent advances in data mining of enterprise data: algorithms and applications. World Scientific, Singapore
16. Rousseeuw PJ, Ruts I (1999) The depth function of a population distribution. Metrika 49:213–244
17. Tibken B (1997) Solution to the volume of the intersection of a cube and a ball in N-space (Problem 96-19*, SIAM Rev 38(4):669–669). In: Rousseau CC, Rühr OG (eds) Problems and solutions. SIAM Rev 39(4):761–789
18. Tukey JW (1975) Mathematics and the picturing of data. In: James RD (ed) Proceedings of the international congress of mathematicians, vol 2. Canadian Mathematical Congress, USA
19. Zuo Y, Serfling R (2000) General notions of statistical depth function. Ann Stat 28:461–482

A Bayesian Network Model for the Probabilistic Safety Assessment of Roads

Enrique Castillo, Zacarías Grande and Elena Mora

Abstract A Bayesian network model for probabilistic safety analysis of roads and highways is introduced. After indicating how the list of variables and the conditional probability tables of the Bayesian network model are built, based on a video of the road, a short discussion about how maximum likelihood and Bayesian network methods can be applied to estimate the model parameters using standard methods. Next, a partitioning technique is suggested to convert the non-linear problem of computing marginal and conditional probabilities after evidence into a problem whose complexity becomes linear in the number of variables. Finally an example of application is used to illustrate the proposed methodology and some conclusions are drawn.

1 Introduction

This is our contribution to honor and remember Prof. Pedro Gil to whom this volume is dedicated (for some works related to this paper see [10, 11, 16, 17]). We decided to write a paper on applications of Probability and Statistics to demonstrate how these areas of knowledge can help Society in solving important problems as Pedro dedicated his life to help many students and colleagues.

Probabilistic safety analysis (PSA) is a well known and the best technique to assess the safety level of a system, used for nuclear power plants and extended recently to the case of railway lines (see for example, [3–5] or [12]). In this work we deal with the case of roads, where a huge number of accidents occur. This is one of the actual main concerns of our society. In our analysis, we restrict to Bayesian networks (BN), due to its incredible power to reproduce multidimensional random variables (see [2]). Though superior to other existing methods, such as fault or event trees, only very recently BNs have been applied to the analysis of traffic infrastructures.

Until very recently, most of the proposed BN models have been used: (a) to predict how frequent different traffic incident types occur, such as [1, 9, 14, 19, 21]

E. Castillo (✉) · Z. Grande · E. Mora
Department of Applied Mathematics and Computational Sciences,
University of Cantabria, Santander, Spain
e-mail: castie@unican.es

© Springer International Publishing AG 2018
E. Gil et al. (eds.), *The Mathematics of the Uncertain*, Studies in Systems,
Decision and Control 142, https://doi.org/10.1007/978-3-319-73848-2_8

(b) to classify traffic incidents according to their injury severity, such as [8], (c) to analyze and prevent traffic incidents, such as [15], (d) to perform a transportation safety assessment, such as [20], or (e) to perform a safety analysis of expressways, such as [18]. Most existing publications work with global safety assessments but do not provide means to identify the risks associated with different locations of the road segments. In addition, these methods use data to learn not only the qualitative structure (structural learning), using the K2 or related algorithms, such as [8, 15] or, but to quantify the conditional probability tables (parametric learning) using Bayesian parameter estimation methods, the EM (expectation maximization)-algorithm or other statistical methods, such as [1] or [14]. Contrary, [20] used a group of experts for designing the BN. The data used were global data, that is, mainly valid only to estimate global safety indices.

In this work we deal with the probabilistic safety assessment of road and highways and propose a Bayesian network model, which is described below.

2 Building the Bayesian Network

To build a Bayesian network we need to define its variables first and later its two components: (a) the acyclic graph, that is, the qualitative structure of the model and (b) the conditional probability tables, that is, the quantitative information.

To identify the main variables, see [13], we record a video of the road and identify all possible items causing incidents (traffic light signals, stop signals, intersections, curves, roundabouts, tunnels, acceleration and deceleration lines, etc.). Next, we identify the list of variables used in our mode, which are shown in Table 1 together with its definitions and their possible values.

Next, we obtain the acyclic graph by identifying direct dependencies between variables. To clarify ideas, in Fig. 1 the part of the acyclic graph corresponding to the items of a fraction of a road and highways, the corresponding kilometer points (KP) and cumulated probabilities of incidents are shown.

Finally, the conditional probability tables are defined using closed formulas, as illustrated in Fig. 2, where the particular case of the probability of an incident due to an error at a traffic signal is used to illustrate how closed form formulas are obtained.

Once the Bayesian network model has been built, the marginal probabilities of the incident type nodes can be calculated and the ENSI[1] values for each location calculated.

Figure 3 shows the main elements and the scanner plot associated with one section of the CA-132, showing the cumulated curve of ENSI values. With the help of these scanner plots the riskiest locations and sections can be easily located.

---

[1]ENSI refers to the expected number of equivalent severe incidents, where 6.4 medium incidents and 230 light incidents are considered equivalent as one severe incident. The relative mean costs of these incidents have been used to determine these factors.

Table 1 Notation: list of variables with their possible definitions and values

Variable	Definition	Values
D: Driver's attention	This variable represents the driver's attention	Distracted, Attentive, Alert
T: Driver fatigue	Measures driver fatigue	A positive value that increases with driving time
Sd: Driver decision on speed	Represents the action of the driver in cases where he must adjust the speed	Correct, Error I, Error II
Dri: Driver Type	This variable reflects the quality of the driver	Professional, Experienced, Standard, Bad
It: Traffic intensity	This variable measures traffic intensity	Light, Medium or Dense
Vis: Visibility	This variable measures the visibility existing at the point considered	Good, Average and Poor
Vt: Type of vehicle	Refers to the type of vehicle	Heavy Vehicle, Automobile and Moto
S: Speed	Is the circulation speed at the point considered	Set of positive values
V: Vehicle failure	Consider the possibility of a vehicle failure	No fault, Minor fault, Medium fault or Serious fault
E: Failure in the environment	It represents the possibility of unwanted events, such as obstructions of the road by stones, trees or other materials, animals as well as defects of the road, clearings, embankments, etc.	No fault, Minor fault, Medium fault or Serious fault
P: Pavement condition	Represents the state of the pavement	Good condition, Mild failure, Medium failure or Serious fault
Co: Collision	Represents the possibilities of collision with other vehicles that circulate on the road in the same or opposite direction	No Collision, Mild Collision, Medium Collision, or Severe Collision
W: Weather	Represents the type of climate	Fair, Rain/Snow, Wind, Fog, Snow/Ice
SS: Signal status	Represents the status of the signal	Green, Yellow, Red
AS: Decision of the driver at a traffic light	Represent the decision of the driver at a traffic light	Correct, Error I, Error II
DS: Decision of the driver in a signal	Represents the decision of the driver in a signal, such as Stop, yields or speed limit signals	Correct, Error
TF: Technical failure	It represents the possible failure of a vehicle, signal, etc.	Yes, No

(continued)

Table 1 (continued)

Variable	Definition	Values
CF: Crossover Frequency	Represents the frequency of crossing between two elements of brakes such as vehicles and trains or pedestrians and vehicles at a specific point on the road, such as pedestrian crossings or level crossings	Yes, No
I: Incident	Represents possible incidents that may occur at a particular location on the road or in a no-signal range	Incident, Mild Incident, Medium Incident, and Serious Incident

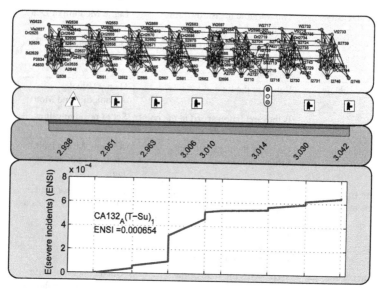

Fig. 1 Part of the acyclic graph corresponding to the indicated items, kilometer points (KP) and cumulated probabilities of incidents

3 Learning the Bayesian Network Model

In this section we address the problem of parametric learning of the Bayesian network (see [7] for more details).

Let $\{X_1, X_2, \ldots, X_s\}$ be the set of variables (nodes) of the Bayesian network, which are denoted by capital letters. Consider a sample of m vehicles circulating by the road and let $\{x_{1i}, x_{2i}, \ldots, x_{si}\}; i = 1, 2, \ldots, m$ be the sample values, where we have used lower case letters to refer to the particular values of the corresponding variables in the sample.

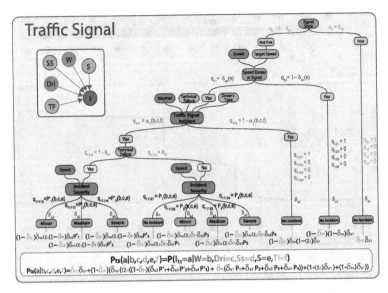

Fig. 2 Illustration of the closed formula for the light signal failure conditional probability table

Let $\theta^k_{X_k;\mathscr{P}^k} = P(X_k|\mathscr{P}^k)$ be the conditional probabilities of the node X_k given their parents $\mathscr{P}^k$, which are considered the parameters of the Bayesian network to be estimated. For the parameters to be a valid conditional probability they must satisfy the constraints:

$$\theta^k_{t_k;\mathscr{P}^k} = 1 - \sum_{a=1}^{t_k-1} \theta^k_{a;\mathscr{P}^k}; \quad \forall \mathscr{P}^k, \tag{1}$$

where t_k is the number of possible values of node X_k.

The log-likelihood of the sample becomes:

$$\log L(\mathbf{x};\boldsymbol{\theta}) = \sum_{\mathscr{P}^k} \sum_{k=1}^{s} \sum_{x_k=1}^{t_k} \left(\sum_{i=1}^{m} n^k_{x_{ki};\mathscr{P}^k_i} \right) \log \theta^k_{x_k;\mathscr{P}^k}$$

$$= \sum_{k=1}^{s} \sum_{\mathscr{P}^k} \left(\sum_{x_k=1}^{t_k} \left(\sum_{i=1}^{m} n^k_{x_{ki};\mathscr{P}^k_i} \right) \log \theta^k_{x_k;\mathscr{P}^k} \right), \tag{2}$$

where x_{ki} and $\mathscr{P}^k_i$ are the values of the node X_k and its parents in sample data i, and $n^k_{x_{ki};\mathscr{P}^k_i}$ is the number of observed vehicles in the sample such that $X_k = x_k$ and the parent values of X_k in the sample are $\mathscr{P}^k$.

Equation (2) reveals that the maximization of the log-likelihood function is equivalent to the maximization of the summands corresponding to the different conditional probabilities of the nodes one by one and separately. More precisely, we need to maximize

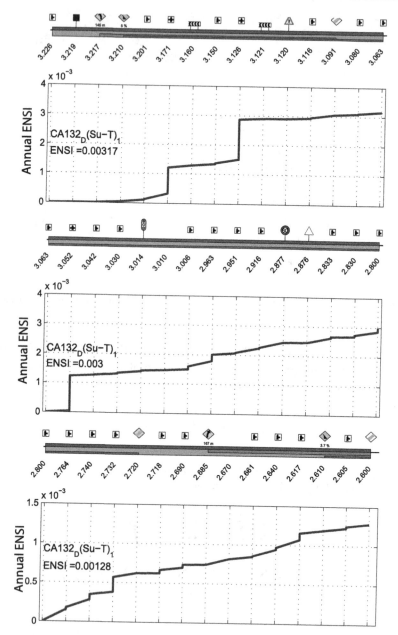

Fig. 3 Scanner of one section of the CA-132, showing the cumulated curve of ENSI values. Note how the riskiest points are emphasized

$$\left(\sum_{i=1}^{m} n_{x_{ki};\mathscr{P}_i^k}^k\right)\log\theta_{x_k;\mathscr{P}^k}^k; \quad \forall\mathscr{P}^k, \quad \forall k = 1, 2, \ldots, n, \tag{3}$$

which implies a very important reduction in complexity and CPU time.

In addition, given that the parameters must satisfy the constraints in (1), the maximization of the expressions in (3) leads to

$$\hat{\theta}_{x_k;\mathscr{P}^k}^k = \frac{\sum_{i=1}^{m} n_{x_{ki};\mathscr{P}_i^k}^k}{\sum_{x_k=1}^{t_k}\sum_{i=1}^{m} n_{x_{ki};\mathscr{P}_i^k}^k}; \quad \forall\mathscr{P}^k, \quad \forall k = 1, 2, \ldots, n; \ x_k = 1, 2, \ldots, t_k - 1, \tag{4}$$

which are the well known classical estimates, that is, the sample proportions.

If we consider Bayesian estimates associated with the Dirichlet conjugate distributions, we get

$$\hat{\theta}_{x_k;\mathscr{P}^k}^k = \frac{\sum_{i=1}^{m} n_{x_{ki};\mathscr{P}^k-i}^k + n_{x_k;\mathscr{P}^k}^{k0}}{\sum_{x_k=1}^{t_k}\sum_{i=1}^{m} n_{x_{ki};\mathscr{P}_i^k}^k + N_0}; \quad \forall\mathscr{P}^k, \quad \forall k = 1, 2, \ldots, s; \ x_k = 1, 2, \ldots, t_k - 1,$$

$$\tag{5}$$

where $n_{x_k;\mathscr{P}^k}^{k0}$ are the prior parameters and $N_0 = \sum_{x_k=1}^{t_k} n_{x_k;\mathscr{P}^k}^{k0}$.

If there are no observable parents, Formula (5) must be replaced by the following formula

$$\hat{\theta}_{x_k;\mathscr{P}^k,\hat{\mathscr{P}}^k}^k = \frac{\sum_{i=1}^{m} n_{x_{ki};\mathscr{P}_i^k}^k P(\bar{\mathscr{P}}^k) + n_{x_k;\mathscr{P}^k}^{k0}}{\sum_{\bar{\mathscr{P}}^k}\sum_{x_k=1}^{t_k}\sum_{i=1}^{m} n_{x_{ki};\mathscr{P}^k}^k P(\bar{\mathscr{P}}_i^k) + N_0};$$

$$\forall\mathscr{P}^k, \quad \forall k = 1, 2, \ldots, s; \ x_k = 1, 2, \ldots, t_k - 1 \tag{6}$$

where now $\mathscr{P}^k$ and $\bar{\mathscr{P}}^k$ refer to the subsets of parents which are observed and unobserved, respectively, and $P(\bar{\mathscr{P}}^k)$ is the joint probability of the unobserved parents of node X_k, which can be easily obtained from the X_k-parents clique.

It is clear from expression (6) that the effect of the prior information $n_{x_k;\mathscr{P}^k}^{k0}$ and N_0 on the parameter estimates $\hat{\theta}_{x_k;\mathscr{P}^k,\hat{\mathscr{P}}^k}^k$ of the observable nodes becomes negligible when the sample size is large. However, this can take place only for very large sample sizes if the true values of the parameter is very small.

4 Network Partition

Since Bayesian networks associated with real cases imply a very high number of variables (many thousands), it is necessary to reduce the complexity of the calculations; otherwise, the memory and the CPU requirements will exhaust the computer capacity. To solve this problem the Bayesian network is partitioned into a series of, as small as possible, subnetworks, such that the results of the computations is not modified (see [6]).

Figure 4 shows an example, in which the acyclic graph of the Bayesian network of a segment of a road is given (upper plot). It has been partitioned into three sub-Bayesian networks, denoted A, B and C, that can be identified by their different background colors.

The partitions are selected based on the conditional independence property shown in Fig. 4, where it can be seen that the nodes in set C are independent of the nodes of set A given the nodes in B, because any path from set A to set C passes by set B in the moral graph of the set $A \cup B \cup C$ and its ancestors (see [2]). This means that the variables in B contain all the information the variables in A have on the variables in C. Consequently, variables in A are not needed to get information on the variables in C given the variables in B.

The selected partition is not arbitrary at all. The key property for a partition to be valid is to contain a set of separators (subsets of nodes) such that the conditional

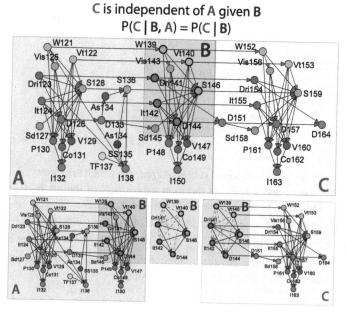

Fig. 4 Illustration of how a Bayesian network can be partitioned into a sequence of Bayesian subnetworks to obtain the marginal probabilities (forward process)

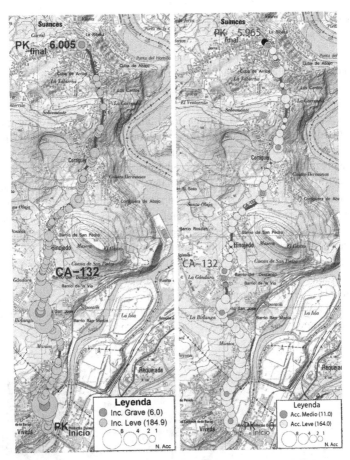

Fig. 5 On the left is the graphical representation of the accident prediction for the CA-132 road provided by the Bayesian network model, which shows the frequencies of light incidents (medium gray), severe (light gray) and severe (dark gray) by means of circles with a diameter proportional to the square root of the frequency of the incidents at the given points. Also shown on the right side is the observed accident rate so that its similarity can be observed

probability of the posterior nodes becomes independent on the previous nodes given the separator subset. Consequently, the separator subset and the partitions have been selected to satisfy this condition.

This partition procedure leads to a computation time linear in the length of the road or highway. Consequently, the CPU times are reduced substantially. This means that some small cases requiring some hours of CPU, could be calculated in a few minutes. This time reduction is even more impressive for larger networks.

5 Example of Application

In this example we consider the case of the autonomic road CA-132 between Viveda and Suances, in Cantabria (Spain), with a length of 5.95 km, which is shown in Fig. 5, where on its left side is the graphical representation of the accident prediction provided by our Bayesian network model, which shows the frequencies of light incidents (medium gray), severe (light gray) and severe (dark gray) by means of circles with a diameter proportional to the square root of the frequency of the incidents at the given points. Also shown on the right side is the observed accident rate so that similarities between predicted and observed scenarios can be observed.

Figure 6 illustrates the riskiest location in this road, which correspond to an intersection indicated by the blue circle in the upper right picture, directly obtained from Google maps. The problem is due to the lack of visibility produced by a house and a wall and the presence of a close bus stop in the upper curve, which impedes car and bus drivers to see the slow vehicles incorporating and leaving the CA-132 road at this intersection.

The left picture shows that this location is at the beginning of the road (PK 0.380) and the lowest right plot warns us about a high frequency of possible incidents, pointing to the intersection location by means of a sudden jump in the cumulated incident frequency (ENSI) curve.

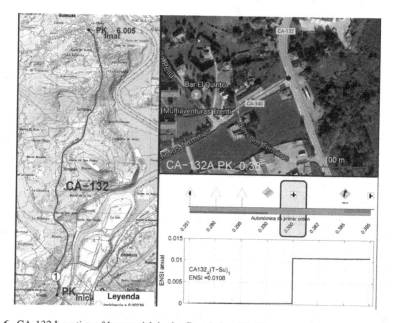

Fig. 6 CA-132 Location of largest risk in the Cantabrian CA-132 autonomic road

Table 2 Disaggregated ENSI: variable value combinations that contribute more to the total $ENSIindividual$ value of node I322-380Int $(T P_1)$

n	Weather	Vehicle type	Driver's attention	Speed (km/h)	ENSI	%
1	Bad	Car	Alert	24	5.32e-10	16.38
2	Very bad	Car	Alert	24	3.93e-10	12.11
3	Humid	Car	Alert	24	3.46e-10	10.67
4	Very bad	Car	Alert	15	2.27e-10	7.01
5	Good	Bus	Alert	24	2.24e-10	6.90
6	Humid	Bus	Alert	24	1.84e-10	5.66
7	Good	Car	Alert	48	1.7e-10	5.23
8	Humid	Car	Alert	39	1.44e-10	4.43
9	Humid	Car	Alert	48	1.29e-10	3.98
10	Good	Car	Alert	39	1.11e-10	3.42

Table 2 provides the most frequent combinations of variable values leading to incidents at this intersection and shows the disaggregated ENSI values associated with incidents at this location.

It is interesting to see that most incidents at this intersection take place with cars, attentive drivers, not good weather and low speeds. This means that slow vehicles combined with weather inclementness are the main cause of incidents. Buses appear to be in the list of causes of these incidents, but on a secondary place.

Note also that when these incidents occur with good weather the speeds are higher than when the weather is bad.

Based on Table 2, safety can be improved by as cheap solution, such as reducing the speed limit in the area or by means of a more expensive solution, as constructing a small roundabout. Running again the computer program will indicate the ENSI value reductions and whether or not the tested solutions are satisfactory enough to solve the problem.

6   Conclusions

Based on the previous study, we can conclude that Bayesian network models allow to:

1. Perform a road scanner.
2. Identify the most dangerous points of a road.
3. Determine the most frequent circumstances that produce the incidents.
4. Define the causes of occurrences or hypothetical incidents.
5. Help in making corrective decisions.
6. Optimize resources for improving safety and maintenance.
7. Predict accident concentration zones, before they occur.

References

1. Alizadeh SS, Mortazavi SB, Sepehri MM (2014) Prediction of vehicle traffic accidents using Bayesian networks. Sci J Pure Appl Sci 3(6):356–362
2. Castillo E, Gutiérrez JM, Hadi AS (1997) Expert systems and probabilistic network models. Springer, New York
3. Castillo E, Grande Z, Calviño A (2016) Bayesian networks based probabilistic risk analysis for railway lines. Comput Aided Civil Infrastruct Eng 31:681–700
4. Castillo E, Grande Z, Calviño A (2016) A Markovian–Bayesian network for risk analysis of high speed and conventional railway lines integrating human errors. Comput Aided Civil Infrastruct Eng 31:193–218
5. Castillo E, Grande Z, Calviño A, Nogal M, O'Connor A (2016) Probabilistic safety analysis of high speed railway lines including human errors. In: Pombo J (ed) Proceedings of the 3rd international conference on railway technology: research, development and maintenance. Civil-Comp Press, Stirlingshire
6. Castillo E, Grande Z, Mora E (2017) Complexity reduction and sensitivity analysis in road probabilistic safety assessment bayesian network models. Comput Aided Civil Infrastruct Eng 32(7):546–561
7. Castillo E, Grande Z, Mora E, Xu X, Lo HK (2017) Proactive, backward analysis and learning in road probabilistic Bayesian network models. Comput Aided Civil Infrastruct Eng 32(10):820–835
8. de Oña J, Mujalli RO, Calvo FJ (2011) Analysis of traffic accident injury severity on spanish rural highways using Bayesian networks. Accid Anal Prev 43(1):402–411
9. Deublein M, Schubert M, Adey BT, de Soto García B (2015) A Bayesian network model to predict accidents on Swiss highways. Infrastruct Asset Manag 2(4):145–158
10. Gil MA, Gil P, Pardo L (2012) Historical evolution of statistics in Spain. Bol Estad Investig Oper 28(1):8–23
11. Gil P (1975) Measures of uncertainty and information in statistical decision problems. Rev Real Acad Ci Exact Fs Natur (RACSAM) 69(3):549–610
12. Grande Z, Castillo E, Nogal M, O'Connor AJ (2016) Probabilistic safety analysis of high speed and conventional lines using Bayesian networks. In: Colomer-Ferrándiz JV, Insa-Franco R, Ruiz-Sánchez T (eds) Libro Actas XII Cong Ingen Transp. Universitat Politécnica de Valéncia, Valencia
13. Grande Z, Castillo E, Mora E (2017) Highway and road probabilistic safety assessment based on Bayesian network model. Comput Aided Civil Infrastruct Eng 32(5):379–396
14. Krol A (2014) Application of the Bayesian network to identify the correlations in the circumstances of the road accidents for selected streets in Katowice. Logistyka 4:2947–2957
15. Lin F, Jiang Y, Xu ZX, Dai L (2011) The analysis and prevent in traffic accidents based on Bayesian network. Adv Eng Forum 1:21–25
16. Miranda E, Couso I, Gil P (2003) Extreme points of credal sets generated by 2-alternating capacities. Int J Approx Reason 33(1):95–115
17. Miranda P, Combarro EF, Gil P (2006) Extreme points of some families of non-additive measures. Eur J Oper Res 3:1865–1884
18. Wang L, Lu H-P, Zheng Y, Qian Z (2014) Safety analysis for expressway based on Bayesian network: a case study in China. Comput Model New Technol 18(12C):438–444
19. Xu H, Zhang H, Zong F (2010) Bayesian network-based road traffic accident causality analysis. In: 2010 WASE International Conference on Information Engineering (ICIE), Vol 4. IEEE Computer Society, Los Alamitos
20. Zhang K, Shi P (2015) Transportation security assessment method for a mountainous freeway using a Bayesian network. In: Peng Q, Wang KCP, Liu X, Chen B (eds) Proceedings of the 5th international conference on transportation engineering, vol 368. ASCE Library, Reston
21. Zong F, Xu H, Zhang H (2013) Prediction for traffic accident severity: comparing the Bayesian network and regression models. Math Prob Eng 2013:ID475194

The Spiking Problem in the Context of the Isotonic Regression

Ana Colubi, J. Santos Domínguez-Menchero and Gil González-Rodríguez

Abstract The usual estimators of the regression under isotonicity are known to present the so-called *spiking problem*, that is, they are very sensitive at the tails. Three design-based strategies in order to alleviate this effect are discussed. The proposed strategies will provide uniform consistency on the (closed and bounded) working interval. Firstly, the usual isotonic regression with a suitable number of observations at the edges of the interval is considered. Secondly, a reallocation of part of the edge observations at some artificial adjacent points is suggested. Finally, a strategy based on constraining the isotonic regression to take values within some horizontal bands is investigated. Simulation studies illustrate the performance of the proposed estimators in practice.

1 Introduction

The estimation of the regression function when it is assumed to be isotonic has been considered for long (e.g., [3, 15]). The best-known estimator for this problem is the so-called isotonic regression, that is, the isotonic function that better fits the data with respect to a weighted empirical L_2 distance.

This estimator is known to be uniformly consistent in a closed and compact interval strictly contained in the design interval, under general conditions and fixed design (see [4, 13]). However, it is very sensitive at the tails, i.e., it presents a spiking problem. To guarantee uniform consistency can be specially valuable in situations like those in [5, 6, 16, 19]. This problem, and some ways to solve it, has been deeply studied in [7] and the results are summarized and discussed here. Namely, three specific alternatives are considered, and the results of a comparative simulation

A. Colubi (✉) · J. S. Domínguez-Menchero · G. González-Rodríguez
Dpto. de Estadística e I.O. y D.M., Universidad de Oviedo, 33007 Oviedo, Spain
e-mail: colubi@uniovi.es

J. S. Domínguez-Menchero
e-mail: jsdm@euniovi.es

G. González-Rodríguez
e-mail: gil@uniovi.es

© Springer International Publishing AG 2018
E. Gil et al. (eds.), *The Mathematics of the Uncertain*, Studies in Systems, Decision and Control 142, https://doi.org/10.1007/978-3-319-73848-2_9

101

study are shown. Firstly, we suggest to suitable choose the number of observations at the first and last design point. A second approach is developed by reallocating part of the observations considered at the first and last design point at some artificial adjacent points. Finally, a third method constrains the isotonic regression to take values within some horizontal band. All the three methods show a good performance and improve the one of the methods in [1, 17, 19].

The rest of the paper is organized as follows. The notation and the three suggested strategies to estimate the regression under isotonicity are presented in Sect. 2. Then, the simulation results are provided in Sect. 3. Finally, Sect. 4 concludes the paper with a brief discussion and a sincere acknowledgment.

2 The New Procedures

Let $Y(x) = m(x) + \varepsilon(x)$ be a regression model with $x \in A = [a, b] \subseteq \mathbb{R}$, where the errors $\varepsilon(x)$ have 0 mean and the regression function m is continuous and isotonic.

We assume a fixed design $\{x_{1,n}, \ldots, x_{n,n}\} \subset A$ with $x_{i,n} < x_{j,n}$ ($1 \leq i < j \leq n$, $n \in \mathbb{N}$) and $r_n(i)$ independent observations $Y^1(x_{i,n}), \ldots, Y^{r_n(i)}(x_{i,n})$ on each design point $x_{i,n}$ ($1 \leq i \leq n, n \in \mathbb{N}$). Thus, $\{Y^j(x_{i,n})\}$ is a triangular array of row-wise independent random variables.

Let $\mathscr{T}_n^I$ be the set of real isotonic functions defined on $\{x_{1,n}, \ldots, x_{n,n}\}$. For any non-negative weighting function $w : A \to \mathbb{R}^+$ the argmin of

$$\sum_{i=1}^{n} \sum_{j=1}^{r_n(i)} w(x_{i,n})(Y^j(x_{i,n}) - f(x_{i,n}))^2$$

on $\mathscr{T}_n^I$ is the well-known isotonic regression estimator $\widehat{m}_I$. The isotonic regression estimator can be computed by means of the algorithm PAVA (see [2]). An implementation in R can be found in [9]. This estimator has been widely analyzed in the literature, and it has been used as well as a base for hypothesis testing (see, for instance, [4–6, 8, 11, 12, 14, 16, 18, 19]).

The estimator as above-defined is just determined on the fixed design points. However, it can be defined on the whole A by means of any isotonic extension. We will denote by $\widehat{m}_I^*$ any isotonic extension of $\widehat{m}_I$ on A verifying that $\widehat{m}_I^*(x) = \widehat{m}_I(x_{1,n})$ if $x \leq x_{1,n}$ and $\widehat{m}_I^*(x) = \widehat{m}_I(x_{n,n})$ if $x \geq x_{n,n}$.

Let $F : [0, \infty) \to [0, 1]$ be the function given by $F(y) = \sup_{x \in A} P(|\varepsilon(x)| > y)$, for all $y \in [0, \infty)$, and assume that $\lim_{y \to \infty} F(y) = 0$ and $\int_0^\infty y |dF(y)| < \infty$. In addition, let assume that the fixed design setting verifies that for all $x \in (a, b]$, $\min\{\{x_{1,n}, \ldots, x_{n,n}\} \cap (x, b]\}$ tends to x as $n \to \infty$ and, analogously, for all $x \in [a, b)$, $\max\{\{x_{1,n}, \ldots, x_{n,n}\} \cap [a, x)\}$ tends to x as $n \to \infty$. Moreover, assume that $\limsup_{n \to \infty} N_n(A)/N_n(J) < \infty$ for all $J \subseteq A$ with

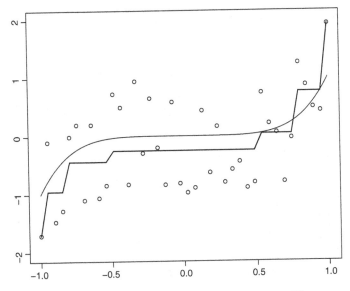

Fig. 1 Simulated example with $n = 40$ and $r_{40}(i) = 1$ for all $i \in \{1, \ldots, 40\}$

$$N_n(J) = \sum_{\{i : x_{i,n} \in J\}} r_n(i).$$

Under such conditions, [13] proved the uniform consistency of $\widehat{m}_I^*$ (with constant weighting function) on any closed and bounded subset $B \subsetneq A$.

To illustrate the spiking problem, in Fig. 1 we have chosen $A = [-1, 1], w(x) = 1$, $m(x) = x^5$, $\varepsilon(x)$ distributed as a $U(-1, 1)$ for all $x \in A$ and equally spaced design points. We see that if $r_n(i) = 1$ for all $i \in \{1, \ldots, n\}$, then the estimates are not suitable at the tails. However, with the same total sample sizes, the behaviour improves if more observations per design point at the edges are considered (see Fig. 2). This idea suggested the strategies in [7].

Three different approaches are proposed in order to estimate the regression function under isotonicity uniformly consistently in the working interval A. More general weighting functions can be considered in such a way that w is bounded and bounded away from 0.

The first approach, $P1$, is to consider a fixed design so that the number of observations at the ending design points verify:

$$\limsup_{n \to \infty} N_n(A) / \min(r_n(1), r_n(n)) < \infty.$$

The second procedure, $P2$, is based on a reallocation of the observations at the ending points by considering some artificial adjacent design points. For each $n \in \mathbb{N}$, let $k_n^a, k_n^b \in \mathbb{N}$ and consider k_n^a independent observations $Z_a^1, \ldots Z_a^{k_n^a}$ distributed as

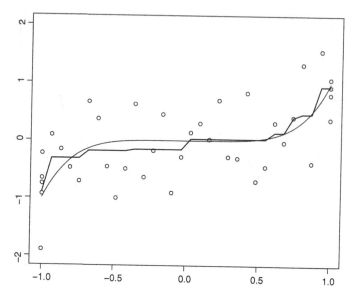

Fig. 2 Simulated example with $n = 32$, $r_{32}(1) = 5$, $r_{32}(32) = 5$ and $r_{32}(i) = 1$ for all $i \in \{2, \ldots, 31\}$

$Y(a)$ and k_n^b independent observations $Z_b^1, \ldots Z_b^{k_n^b}$ distributed as $Y(b)$. Let $\widehat{m}_2$ be the isotonic regression based on the design points

$$B_n = \cup_{j=1}^{k_n^a} \{a - j/k_n^a\} \cup \{x_{1,n}, \ldots, x_{n,n}\} \cup \cup_{j=1}^{k_n^b} \{b + j/k_n^b\},$$

associating the observation Z_a^j with the point $a - j/k_n^a$ for each $j = 1, \ldots, k_n^a$, the observation Z_b^j with the point $b + j/k_n^a$ for each $j = 1, \ldots, k_n^b$, and the rest of the observations with the original points. Thus, $\widehat{m}_2$ is the argmin of

$$\sum_{i=1}^{n} \sum_{j=1}^{r_n(i)} w(x_{i,n}) \left(Y^j(x_{i,n}) - f(x_{i,n}) \right)^2$$

$$+ \sum_{j=1}^{k_n^a} w(a) \left(Z_a^j - f(a - j/k_n^a) \right)^2 + \sum_{j=1}^{k_n^b} w(b) \left(Z_b^j - f(b + j/k_n^b) \right)^2$$

on the set of isotonic functions defined on B_n. Since the estimator $\widehat{m}_2$ is just an isotonic regression, it can also be computed by means of PAVA. In this setting, we assume that

$$\limsup_{n \to \infty} \frac{\max(k_n^a, k_n^b)}{N_n(A)} < \infty; \ \limsup_{n \to \infty} \frac{N_n(A)}{\min(k_n^a, k_n^b)} < \infty.$$

Lastly, we introduce the design $P3$ with the idea of reducing the number of observations at the ending design points with respect to the previous designs. In this sense, an optimal condition could be to simply assume that

$$\lim_{n \to \infty} \min(r_n(1), r_n(n)) = \infty.$$

For any $i \in \{1, \ldots, n\}$ let

$$\overline{Y(x_{i,n})} = \frac{1}{r_n(i)} \sum_{j=1}^{r_n(i)} Y^j(x_{i,n}),$$

$$\overline{Y} = \frac{1}{\sum_{i=1}^{n} r_n(i)} \sum_{i=1}^{n} \sum_{j=1}^{r_n(i)} Y^j(x_{i,n}),$$

and let $\widehat{m}_3$ be the isotonic regression estimator restricted to take values in

$$[\min(\overline{Y(x_{1,n})}, \overline{Y}), \max(\overline{Y}, \overline{Y(x_{n,n})})].$$

That is, $\widehat{m}_3$ is the argmin of

$$\sum_{i=1}^{n} \sum_{j=1}^{r_n(i)} w(x_{i,n}) \left(Y^j(x_{i,n}) - f(x_{i,n}) \right)^2$$

on the set of isotonic functions defined on $\mathscr{T}_n^I$ and taking values restricted to $[\min(\overline{Y(x_{1,n})}, \overline{Y}), \max(\overline{Y}, \overline{Y(x_{n,n})})]$. It can be easily computed by using the results in [10].

3 Simulations

Some simulations are shown in order to illustrate the performance of the three proposed procedures ($P1$, $P2$, $P3$), leading to estimators of the regression function that are uniformly consistent on the whole working interval A. Empirical comparisons to the classical estimator, using a single observation per design point ($P0$), as well as to the median isotonic regression ($L1$) of [17], the M-estimator (M) of [1] and the penalized estimator (W) of [19], are also summarized.

In all cases A has been chosen as $[-1, 1]$, $w(x) = 1$ for all $x \in A$ and the regression function is defined as $m(x) = x^p$ for values of p in $\{0, 1, 5\}$ in order to cover different situations with different increasing rate. The errors are i.i.d. following different distributions, namely $\mathcal{N}(0, 1)$, $U(0, 1)$, and Student's t with 3 degrees of freedom, respectively denoted by $D1$, $D2$ and $D3$. $D3$ is used in [1] to illustrate the robustness of the M-estimator.

The same overall sample size N is considered for all the procedures, and the design points are chosen as $x_{i,n} = 2(i-1)/(n-1) - 1$ for any $i \in \{1, \ldots, n\}$ and any $n \in \mathbb{N}$.

For the procedures $P0$, $L1$ and W, we have selected $n = N$ and $r_n(i) = 1$ for all $i \in \{1, \ldots, N\}$, whereas for $P1$ and $P3$, $r_n(1) = r_n(n) = \lceil N/10 \rceil + 1$ with $n = N - 2\lceil N/10 \rceil$ and $r_n(i) = 1$ for all $i \in \{2, \ldots, n-1\}$. Finally, for $P2$, $k_n^a = k_n^b = \lceil N/10 \rceil$ with $n = N - 2(\lceil N/10 \rceil)$ and $r_n(i) = 1$ for all $i \in \{1, \ldots, n\}$. With this setup, the same number of observations at the boundary of A is selected for $P1$, $P2$ and $P3$. Finally, we select $w(x) = 1$ for all $x \in A$.

As the errors are i.i.d., the estimator W is implemented by replacing $Y^1(x_{1,N})$ and $Y^1(x_{N,N})$ by $Y^1(x_{1,N}) + r\sqrt{N}$ and $Y^1(x_{N,N}) - r\sqrt{N}$ respectively, by taking $r = 0.15\,\hat{\sigma}_n$, where $\hat{\sigma}_n$ is the standard deviation of the residuals $Y^1(x_{i,N}) - \hat{m}(x_{i,N})$, $i \in \{1, \ldots, N\}$.

The estimation of the regression function has been extended to the whole interval A by linear interpolation between design points (remaining constant in the tails) in all cases. As the aim is to illustrate the uniform consistency (and the behaviour at the tails of A), we will focus on the supremum norm error (SNE) for comparative purposes. Specifically, if m^* stands for any of the isotonic estimators, the SNE is approximated as follows

$$SNE(m^*) = \sup_{i=1,\ldots,G} |m^*(z_i) - m(z_i)|,$$

being $z_i = (i-1)/(G-1)$ and $G = 1000$. The distribution of SNE is approximated by Monte Carlo based on 10000 random samples (except for the procedure M, for which only 200 iterations have been used due to its high computational cost) by using $N = 20, 100$ and the different error distributions above-mentioned. It should be noted that the procedure for calculating M is very slow due to the $max - min$ formulae and this fact has limited the number of simulations. Tables 1 and 2 summarize the simulation results.

According to the results, $P0$, M and $L1$ clearly have the worst performance. In almost all cases, $P1$, $P2$ and $P3$ are the best ones, as expected. When the regression is constant, $P2$ and $P3$ stand out. This also happens when $p = 1.5$ for the Student's t distributions. W is close to $P1$ in some cases.

4 Discussion and Acknowledgement

The spiking problem is present not only in classical isotonic regression procedures, but also in L_1-based methods, and M-estimators explicitly developed for robustness. The strategies presented in this paper are convenient to alleviate it.

Table 1 Expected value of SNE. $N = 20$

$p = 0$	D1	D2	D3	$p = 1$	D1	D2	D3	$p = 5$	D1	D2	D3
M	0.888	0.538	1.344	M	1.088	0.683	1.432	M	1.079	0.687	1.429
$P0$	0.894	0.519	1.363	$P0$	1.040	0.648	1.561	$P0$	1.077	0.684	1.592
$L1$	0.925	0.563	1.302	$L1$	1.074	0.700	1.477	$L1$	1.098	0.711	1.518
W	0.602	0.335	0.974	W	0.876	0.568	1.309	W	0.925	0.636	1.344
$P1$	0.557	0.321	0.882	$P1$	0.829	0.561	1.192	$P1$	0.841	0.579	1.211
$P2$	0.468	0.274	0.730	$P2$	0.799	0.555	1.102	$P2$	0.830	0.598	1.124
$P3$	0.456	0.265	0.701	$P3$	0.827	0.562	1.126	$P3$	0.831	0.575	1.128

Table 2 Expected value of SNE. $N = 100$

$p = 0$	D1	D2	D3	$p = 1$	D1	D2	D3	$p = 5$	D1	D2	D3
M	0.926	0.505	1.134	M	0.997	0.586	1.319	M	1.026	0.632	1.398
$P0$	0.889	0.518	1.389	$P0$	0.946	0.574	1.493	$P0$	0.993	0.610	1.506
$L1$	0.922	0.564	1.324	$L1$	0.982	0.631	1.409	$L1$	1.023	0.661	1.418
W	0.367	0.202	0.663	W	0.567	0.375	0.936	W	0.662	0.467	0.971
$P1$	0.309	0.180	0.506	$P1$	0.531	0.375	0.765	$P1$	0.600	0.442	0.819
$P2$	0.234	0.136	0.381	$P2$	0.515	0.371	0.713	$P2$	0.592	0.443	0.775
$P3$	0.230	0.133	0.372	$P3$	0.523	0.371	0.744	$P3$	0.580	0.433	0.776

A suitable increase of the number of observations at the edges can result in uniform consistency on the whole (bounded) interval. This simple procedure improves the method in [19] based on penalized loglikelihood functions, at least for a moderate number of observations.

A reallocation of the observations at the ending points by considering artificial design points improves both methods. In general, the best results are obtained by increasing the number of observations at the edges and by constraining the estimator to take values in a horizontal band. Moreover, the empirical results suggest that the uniform consistency could be established under milder conditions on the number of observations at the edges.

The three authors wish to dedicate this work to Pedro Gil, who provided them with knowledge, values, friendship and the best environment to develop their joint research at the University of Oviedo. Pedro, our sincere acknowledgement.

References

1. Álvarez EE, Yohai VJ (2012) M-estimators for isotonic regression. J Stat Plan Inference 142:2351–2368
2. Ayer M, Brunk HD, Ewing WT, Reid WT, Silverman E (1955) An empirical distribution function for sampling with incomplete information. Ann Math Stat 26:641–647
3. Barlow RE, Bartholomew D, Bremner JM, Brunk HD (1972) Statistical inference under order restrictions: the theory and applications of isotonic regression. Wiley, New York
4. Brunk HD (1970) Estimation of isotonic regression. In: Brunk HD, Puri ML (eds) Nonparametric techniques in statistical inference. Cambridge University Press, London
5. Colubi A, Domínguez-Menchero JS, González-Rodríguez G (2006) Testing constancy for isotonic regressions. Scand J Stat 33:463–475
6. Colubi A, Domínguez-Menchero JS, González-Rodríguez G (2007) A test for constancy of isotonic regressions using the L_2-norm. Stat Sin 17:713–724
7. Colubi A, Domínguez-Menchero JS, González-Rodríguez G (2017) New designs to consistently estimate the isotonic regression. Submitted
8. Cuesta-Albertos JA, Domínguez-Menchero JS, Matrán C (1995) Consistency of L_p-best monotone approximations. J Stat Plan Inference 47:295–318
9. de Leeuw J, Hornik K, Mair P (1973) Isotone optimization in R: pool-adjacent-violators algorithm (PAVA) and active set methods. J Stat Softw 32:1–24
10. Domínguez-Menchero JS, González-Rodríguez G (2007) Analyzing an extension of the isotonic regression problem. Metrika 66:19–30
11. Domínguez-Menchero JS, López-Palomo MJ (1997) On the estimation of monotone uniform approximations. Stat Probab Lett 35:355–362
12. Domínguez-Menchero JS, González-Rodríguez G, López-Palomo MJ (2005) An L_2 point of view in testing monotone regression. Nonpar Stat 17:135–153
13. Hanson DL, Pledger G, Wright FT (1973) On consistency in monotonic regression. Ann Stat 1:401–421
14. Mukerjee H (1988) Monotone nonparametric regression. Ann Stat 16:741–775
15. Robertson T, Wright FT, Dykstra RL (1988) Order restricted statistical inference. Wiley, New York
16. Sampson AR, Singh H, Whitaker LR (2009) Simultaneous confidence bands for isotonic functions. J Stat Plan Inference 139:828–842
17. Wang Y, Huang J (2002) Limiting distribution for monotone median regression. J Stat Plan Inference 107:281–287
18. Wright FT (1981) The asymptotic behavior of monotone regression estimates. Ann Stat 9:443–448
19. Wu WB, Woodroofe M, Mentz G (2001) Isotonic regression: another look at the changepoint problem. Biometrika 88:793–804

Estimation of the Inter-occurrence Time Between Events from Incomplete Data. Analysis of Periods of Unemployment in Spain

José Antonio Cristóbal, José Tomás Alcalá and Pilar Olave

Abstract This work analyzes the stochastic variable representing the waiting times between two consecutive events of a stationary renewal process, such as periods of unemployment of different individuals in a specific population. This data has been obtained through cross-sectional sampling: a specific point in time t is chosen, and a random sample of the individuals who are unemployed at the time t is extracted. For each individual we are interested in the unemployment period including t. However, in practice these values are not observable, and the only fact we can ascertain is the time from the inclusion of the individual in the previous unemployment period to the sampling time t. Our data also includes the corresponding values of a certain set of covariates (for example, the time spent on employment training courses or the age of the individual). Using non parametric techniques, an estimation of the conditional mean has been obtained, given the sex and age groups of the individuals. This is a more natural approach than other methods based on the estimation of the hazard rate, thus avoiding pre-established forms for the inclusion of covariates.

1 Introduction. Objectives and Methodology

The main objective of this work is to analyze, by way of nonparametric statistical techniques, the random variable representing the duration of unemployment of an individual in the population described below. However, the methodology is directly applicable to other situations where the variable of interest is the time between two consecutive events of a point process that satisfies certain rather general properties (stationary renewal processes). Specifically, the analyzed population consists of people residing in their main households in Spain who have been the subject of a labor

J. A. Cristóbal (✉) · J. T. Alcalá · P. Olave
University of Zaragoza, Zaragoza, Spain
e-mail: cristo@unizar.es

J. T. Alcalá
e-mail: jtalcala@unizar.es

P. Olave
e-mail: polave@unizar.es

© Springer International Publishing AG 2018
E. Gil et al. (eds.), *The Mathematics of the Uncertain*, Studies in Systems, Decision and Control 142, https://doi.org/10.1007/978-3-319-73848-2_10

force survey known as EPA (Encuesta de Población Activa in Spanish, see [10]) conducted by the INE (the Spanish National Statistics Institute). The data corresponding to the first quarter of 2014, published in anonymized files, forms the database used for this study. We have taken measurements relating to people aged between 16 and 70 years old who have had some previous work experience but were unemployed at the time of sampling. They fulfill the condition that they have been actively searching for a job in the previous 4 weeks and are available for a new job in the subsequent 15 days. The number of months from the termination of their last employment until the time of sampling has been measured, that is, the time of recurrence for these individuals. First, it is worth noting that there is substantial scientific literature concerning such an important issue for our society as the number of individuals who are unemployed at a given time or their ratios to the total population, representing a significant effort to try to understand the evolution of unemployment and its determinants. Nevertheless, in this paper we focus on an analysis of the variable representing the duration of the unemployment of individuals of this population, and on their behavior in relation to different values of certain socio-economic variables. Some works dealing with the study of the length of time in unemployment in Spain in relation to the probability of leaving that state are, among others, [12], who start from a set of data obtained under length biased sampling.

It is important to note that in this study the information is taken under a cross-sectional sample. There is a 100% censorship rate, and we can only observe for any individual the unemployment duration at the time of sampling, not the total duration. But despite this drawback, we have the advantage of a much lower cost in data collection and, more importantly, we can make inferences from more recent times and obtain a valuable snapshot of the variable of interest (duration of unemployment) at a time very close to the current situation.

There are several papers in the literature that approach this problem of dealing with the analysis of a duration from recurrence times, although most of them use parametric techniques (see, for example [1] and references therein). In [15], an overview of some parametric and semi-parametric approaches in the literature to model the survival function under a cross-sectional sample is presented, but it does not address the non-parametric approach. There are other works analyzing such durations nonparametrically, but they include a longitudinal sampling over a certain period, for example [16] who use an estimation of the hazard ratio proposed in [3] to obtain nonparametric estimators of conditional quantile functions for the duration of unemployment in Germany at the end of the last century.

However, we have preferred to obtain a nonparametric estimation of the conditional distribution function rather than only doing so with the survival function. In [6] the regression function of the duration variable against a covariate was estimated, but imposing on the latter a mixed character with nontrivial discrete and continuous components to convert the problem into an equivalent two-sample problem.

Another aspect to consider in retrospective studies when data such as unemployment durations are analyzed is the presence of the rounding effect or heaping effect. In such a case, the distribution of the frequencies presents some well-defined peaks (sometimes called "pipes" or "telescopes") at some specific points corresponding to multiples of a certain value. Before using the raw data, this effect needs to be corrected redistributing these frequencies with some methodology, assigning the excess of frequencies at particular points to the values that they have originated.

The paper is organized as follows. Section 2 introduces a technique to correct the heaping effect in the anonymized files from the EPA, corresponding to first quarter of 2014. Section 3 describes a methodology for estimating the mean of the (unobservable) distribution of the total unemployment duration of the population. This applies to distributions conditioned by different age groups and the results are compared for both sexes. Based on this estimation of the conditional mean, we obtain in Sect. 4 a nonparametric estimator for all conditional density functions. From these density estimators we calculate some quantiles for different sex and age groups, and we compare some central and other extreme quantiles.

2 Redistribution of the Observations. Correction of the Heaping Effect

As stated in the previous section, when data are collected from a retrospective survey, people answering questions about the occurrence time of a certain event are often affected by a well-known type of memory error described in the literature as the heaping or rounding effect. In [14], frequently cited as seminal, the authors study the consequences of this heaping effect on the estimation of the parameters of a certain model for the duration of unemployment, specifically addressing the case of young people in Italy. A Similar approach was made in the analysis of unemployment data from the German Socio-Economic Panel in [11]. Since then a number of works have attempted to jointly estimate the parameters of a temporal model and the parameters involved in the definition of the heaping effect (see, e.g. [2] and references cited therein, among others), though most involve longitudinal sampling. Finally, we note that the heaping effect is a special case of "coarsening" or "lack of precision" in the data (see [8] for a general theory of coarsening and [9] for some applications in the context of studies of time-to-events). In the present case, in the EPA the individuals are asked how many months have passed since they left their last job. When this number of months is high, the response tends to accumulate in multiples of 12 resulting in a higher value than the real one, erroneously assigning frequencies to these points (we will call these singular points) when in fact they correspond to values of other points in their neighborhood. Thus, the observed T variable (in this case expressed in months) takes values at points $1, 2, \ldots, N$ and is originated by another latent variable T^L, where both are related by the following scheme:

- If $T = 12k$, for some natural number k, then:

$$
\begin{cases}
T^L = 12k, & \text{with probability } \alpha_k \\
T^L = 12k + i, & \text{with probability } (1 - \alpha_k)/11, \\
& (i = \pm 1, \pm 2, \pm 3, \pm 4, \pm 5, +6)
\end{cases} \tag{1}
$$

- In other cases, $T^L = T$

Therefore, if we write $p_j = \Pr(T = j)$ and $p_j^L = \Pr(T^L = j)$, we have that $p_{12k}^L = \alpha_k p_{12k}$ and $p_{12k+i}^L = p_{12k+i} + \frac{1-\alpha_k}{11} p_{12k}$.

Now, if we set the value of p_{12k}^L by linear interpolation between the two adjacent values $p_{12k}^L = (p_{12k-1}^L + p_{12k+1}^L)/2$, finally it is obtained that:

$$\alpha_k = \frac{p_{12k} + 11(p_{12k-1} + p_{12k+1})/2}{12 p_{12k}}. \tag{2}$$

This expression (2) allows us to estimate the parameters α_k introduced to model the heaping effect, starting from a condition of smoothness for the probabilities of the latent variable in the neighborhood of the singular points. If we denote the observed frequencies of the T variable in the i-month by f_i, the natural estimator of α_k is obtained by changing in (2) each p_i by the corresponding f_i value and then $\hat{\alpha}_k = \frac{f_{12k}+11(f_{12k-1}+f_{12k+1})/2}{12 f_{12k}}$. Bearing in mind expression (2), the frequencies of the latent variable would be estimated by the amounts given by

$$q_{12k} = \frac{f_{12k} + 11(f_{12k-1} + f_{12k+1})/2}{12},$$

$$q_{12k+i} = f_{12k+i} + \frac{f_{12k} - (f_{12k-1} + f_{12k+1})/2}{12}. \tag{3}$$

In our problem, starting from anonymized data files of the EPA for the first quarter of 2014 (and after a weighting built from the elevation factors of the survey), we have split the data by sex, as well as by the 9 age groups as follows: 16–24, 25–29, 30–34, 35–39, 40–44, 45–49, 50–54, 55–59, 60–70 years. In each of the 18 conditional distributions we have corrected the heaping effect using the expression (3) to redistribute the conditional monthly frequencies of each group. Some cases of raw and corrected frequencies are represented in Fig. 1.

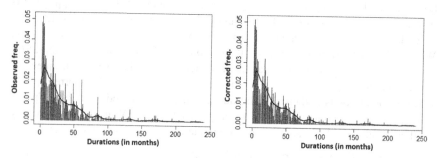

Fig. 1 Observed (left) and corrected (right) frequency groups for Women 30–35 years. A smoothed density estimator is added in both cases

3 Conditional Mean Estimation for Each Sex over the Different Age Ranges

In each of the subpopulations introduced in Sect. 2, let the random variable X be the unemployment time of an individual, which is not directly observable with our cross-sectional sampling (because no follow-up of the individual is performed). Thus, at a fixed time t_0, a random sample of individuals who are unemployed at this time (actively searching for a job and available to start a new job, as described in the Introduction) is drawn up, and the survey collects the elapsed time Y for each individual from when he or she became unemployed for the last time until the sampling instant t_0. If we call X^w the random variable measuring the time from the last starting point of a period of unemployment until the end of that unemployment period, then X^w is a length biased version of the X variable (see [5]), because the longer the time an individual is unemployed, the greater the probability that the individual is in the sample.

Note that the variable X^w itself is not observable with our sampling. What actually happens is that $Y = U \cdot X^w$, where U is a uniform variable in the unit interval, independent of X^w. Thus, the observed variable Y is obtained by the multiplicative censorship which occurs in 100% of cases (see Fig. 2). This is equivalent to saying that t_0 is chosen uniformly over the length interval X^w. Since X^w is length biased with respect to the X variable, its density function verifies:

$$f_X^w(x) = \frac{x f_X(x)}{E(X)} \tag{4}$$

where f_X is the density of the X variable, and the expectation $E(X)$ is assumed finite. From expression (4) it is straightforward that:

$$f_Y(y) = \frac{\int_y^\infty f_X(x)dx}{E(X)} = \frac{1 - F_X(y)}{E(X)} \tag{5}$$

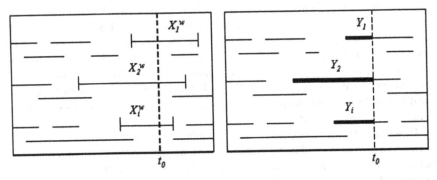

Fig. 2 Sampling scheme. The intervals correspond to periods of unemployment of the individuals

which is the known expression relating the density of the recurrence times Y with the distribution function of the original X variable. In [6], the authors studied the situation with a covariate Z, having both a discrete and an absolutely continuous part, using this mixed nature for a non-parametric estimation of the regression function with respect to such covariate. In the present context we provide below another method for obtaining a non-parametric estimator of the mean in each subpopulation defined above. If $F_Y(y)$ represents the distribution function of the observed Y variable, let us denote by $Q(p) = F_Y^{-1}(p)$, $(0 < p < 1)$ the corresponding quantile function, and let $L(p) = \frac{Q(p)}{p}$ be an auxiliary function.

Note that, because of the positivity of the recurrence times ($Y > 0$), and assuming they have a continuous distribution function F_Y, it follows that $F_Y(0) = 0$, and thus $Q(0) = 0$. Moreover, if the Q function has a right derivative $Q'_0 = Q'(0+)$ at the origin, then it is fulfilled that:

$$L_0 = \lim_{p \to 0+} \frac{Q(p)}{p} = Q'_0 = \frac{1}{F'_Y(0+)} = \frac{1}{f_Y(0+)} = \mu \tag{6}$$

bearing in mind the definition of Q and expression (5). Therefore, a way to estimate $\mu = E(X)$ is to estimate L_0, the right-sided limit at the origin of the auxiliary function L.

Note that every n^{th}−derivative of the L function is related to the derivative of order $n + 1$ of the Q function (assuming its existence). For example, it is easy to see that $L'_0 = (\mu^2/2) f_X(0+)$, whereby this derivative is always non-negative and therefore L is not decreasing at the origin. In addition, in order to cancel the derivative at the origin of the L function, it is necessary that the density function of X at the origin should also vanish (we assume that $\mu > 0$, avoiding the degenerate case in the distribution of X).

A very important case of the distribution of X leading to a derivative $L'_0 > 0$ is the exponential distribution with density function $f_X(x) = a \exp(-ax)$, where it is easy to see that the Y variable has the same exponential density function and $L(p) = -ln(1 - p)/ap$, with a right-derivative at the origin $L'_0 = 1/2a$. As an example of a notable case in which L'_0 vanishes we can mention a gamma distribution (with a shape parameter $p = 2$) where the X variable has a density $f_X(x) = a^2 x \exp(-ax)$, in which case the Y variable becomes a mixture with equal weights of an exponential distribution and a gamma distribution with a shape parameter $p = 2$. Now the L function does not have a simple expression involving elemental functions, but its derivative $L'_0 = 0$, because $f_X(0+) = 0$. Figure 3 shows graphs of the L function in the two above-mentioned cases with the scale parameter $a = 1$. Note that, despite the different forms of the distribution functions in both cases, the auxiliary function L has a very similar aspect.

Thus, considering expression (6) and the subsequent comments, our proposal is to estimate μ in each conditional distribution as the value at the origin of a function belonging to a certain Θ parametric model explained below, that minimizes the weighted distance to a set of values $Y_{(r;n)}/(r/n)$, $(r_0 \leq r \leq r_1)$.

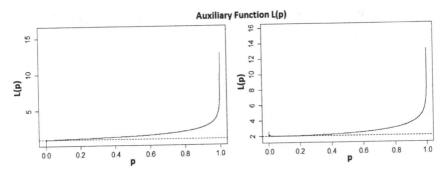

Fig. 3 Graphs of the auxiliary function $L(p)$ for a variable X having an exponential distribution with $a = 1$ (left) and a gamma distribution with $a = 1, p = 2$ (right)

Note that these ratios represent the analogous estimator of $Q(p)/p$, the ratio between the empirical quantiles (order statistics $Y_{(r;n)}$) and their order. For problems dealing with the estimation of the quantile function from its empirical analogues, there is an extensive literature in which a cut of the first order statistics is recommended to estimate the quantile function at an initial end point (out of the sample), a recommendation we have followed. Some years ago, in [4] a simulation study to analyze the optimum trim level is carried out. In our case, however, we found by some simulations that the L function is less sensitive (under certain conditions) to this trim level, so in our analysis we have not made a study of the optimization of the cut. Thus, given the parametric space:

$$\Theta = \{L(p) = -\mu \frac{\ln(1 - p^k)}{p^k}, \quad \mu > 0, \ k = 1, 2)\} \tag{7}$$

(note that in this parametric model we allow that the derivative at the origin of L is positive or zero, depending on whether the parameter k is 1 or 2, and always $L(0) = \mu$), we calculate the L^* function, the solution of the following weighted least squares problem:

$$\min_{L \in \Theta} \sum_{r=r_0}^{r_1} \left[L(r/n) - \frac{Y_{(r;n)}}{(r/n)} \right]^2 \omega_r \tag{8}$$

where ω_r is a weighting variable introduced to correct the heteroscedasticity, and defined as the inverse of the bootstrap variance of the sample quantile $Y_{(r;n)}$ (see, for example, [7]), and the endpoints r_0 and r_1 are assumed to be dependent on n in order that the consistency is achieved. Finally, the estimate of μ is given by $\hat{\mu} = L^*(0)$.

We have carried out this process with the above-mentioned 18 subpopulations and replaced the values $r_0 = 0.10, r_1 = 0.70$ in the above expression. The results are shown in Fig. 4. The graph on the left represents the estimated male unemployment

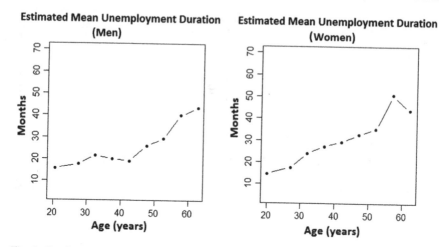

Fig. 4 Graphics of the estimated mean for durations of unemployment in different age groups for men (left) and women (right)

mean for different age groups and the graph on the right represents the estimated female unemployment mean.

4 Non-parametric Estimation of Conditional Distributions and Quantile Comparison

In this section, we first estimate the density function of the recurrence time Y in each subpopulation defined in Sect. 2. Note that, according to (5), such density should be non-increasing throughout its domain. It is known, (see [13]), that the nonparametric maximum likelihood estimator (NPMLE) for such a density function based on a sample y_i $(i = 1, \ldots, n)$ of that variable can be obtained by applying to these data the "pool adjacent violator algorithm" (PAVA), and is given by the step left continuous function:

$$\hat{f}(y) = \min_{0 \le r \le k-1} \max_{k \le s \le n} \frac{s - r}{n(y_s - y_r)}, \quad y_{(k-1)} < y \le y_k$$

$$\hat{f}(y) = 0, \quad \text{for } y > y_n. \tag{9}$$

In [17], the authors proved that although $\hat{f}(y)$ is a consistent estimator of $f(y)$ for all $0 < y < \infty$ at which f is continuous, however $\hat{f}(0+) = \hat{f}(y_1)$ is not a consistent estimator for $f(0+)$. As in our case $f(0+)$ takes the important value $1/\mu$, we can use instead the estimator proposed in [17], based on the maximization of a penalized log-likelihood function:

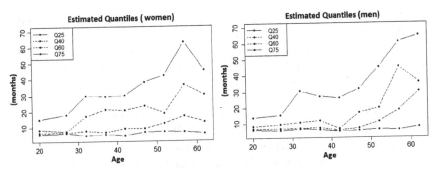

Fig. 5 Graphics of the estimators for the 0.25 and 0.75 quantiles (in solid lines) and the 0.40 and 0.60 quantiles (in dashed line) of the unemployment duration for different age groups for men (right) and women (left)

$$l_\alpha(f) = \sum_{i=1}^{n} \ln f(y_i) - n\alpha f(0+) \tag{10}$$

where f ranks among all continuous non-increasing densities and $\alpha > 0$ is a penalty parameter, which acts as a smoothing parameter. In fact, in [17], it is proved that this penalized estimator $\tilde{f}$ is consistent (also at the origin), and is the NPMLE applied to a linearly deformed set of data $\alpha + \gamma y_i$, $(i = 1, \ldots, n)$.

Given that $\tilde{f}(0+) = 1/\mu$, if we substitute here μ by the estimated value $\hat{\mu} = L^*(0)$, we obtain an equation, which allows us to calculate estimators of the two parameters α and γ. Then, using PAVA on the values of the linearly deformed sample, we build the estimator $\tilde{f}$.

Replacing now the empirical distribution of the sample y_i $(i = 1, \ldots, n)$ with that given at the beginning this section for each of the conditional distributions defined in Sect. 2, we have built an estimator of the density function of the recurrence time in each of these conditional distributions. Then, using expression (5) repeatedly, we obtain estimators of the different quantiles of the unobservable X, in each of the subpopulations.

Specifically, Fig. 5 shows the 0.25 and 0.75 quantiles (solid lines) and the 0.40 and 0.60 quantiles (dashed lines) for both the male and the female population in the corresponding age groups. We preferred to replace the graph of the median by the joint plots of the 0.4 and 0.6 quantiles, because the median may have some sudden changes that they can hinder the perception of the trend of this measure over the various age groups (note that the distributions of the unemployment duration in each subpopulation have been estimated through discrete distributions, being therefore step functions).

Note that, in the case of the male population, the 0.40 and 0.60 quantiles are not very far apart (except for the older age groups), meaning that the median will not suffer abrupt changes because of the discontinuity of the estimations. However, in the case of the female population, the distance between these two quantiles is large (at least for the intermediate age groups), which indicates that the 20% of the central probability is spread over a very wide range, so there are areas with very little probability in this segment. A slight displacement of these areas towards one side can cause a sudden change in the median value. This is the reason why we preferred to give information on the results of the estimation by the two 0.40 and 0.60 quantiles rather than the median.

In the case of the female population there is a substantial gap between the two estimated quantiles for individuals aged between 30 and 45 years which means that there are areas of low density at this period of women's working lives. This situation implies a bimodal shape in the underlying density estimation, which can be explained as a mixture of densities of two populations with clearly differentiated modes: the density with the lower modal value explains the favorable effect on employment for this age group while the other with the higher value corresponds to the group of women who have had major problems in finding employment because of the birth of their children and the early years childcare. Among men, the two estimates of the above quantiles for these age groups are much closer, thus the densities do not have a bimodal shape with clearly differentiated modes.

Acknowledgements The authors acknowledge funding from the Spanish Ministry of Economy and Competitiveness (DGICT) through project ECO2016-79392-P and the Government of Aragón, Spain (DGA, Consolidated Research Group S11).

This work is dedicated to the memory of our colleague and friend Pedro Gil, great teacher and great person, with whom we shared that time in Santiago.

References

1. Allison PD (1985) Survival analysis of backward recurrence times. J Am Stat Assoc 80:315–332
2. Bar HY, Lillard DR (2012) Accounting for heaping in retrospectively reported event data. A mixture-model approach. Stat Med 31(27):3347–3365
3. Beran R (1981) Nonparametric regression with randomly censored survival data. Technical Report, University of California, Berkeley
4. Boos DD (1984) Using extreme value theory to estimate large percentiles. Technometrics 26(1):33–39
5. Cristóbal JA, Alcalá JT (2001) An overview of nonparametric contributions to the problem of functional estimation from biased data. Test 10(2):309–332
6. Cristóbal JA, Alcalá JT, Ojeda JL (2007) Nonparametric estimation of a regression function from backward recurrence times in a cross-sectional sampling. Lifetime Data Anal 13:273–293
7. Efron B (1982) The Jackknife, the bootstrap and other resampling plans. SIAM, Philadelphia
8. Heitjan DF, Rubin DB (1991) Ignorability and coarse data. Ann Stat 19:2244–2253
9. Holt D, McDonald JW, Skinner CJ (1991) The effect of measurements error on event history analysis. In: Biemer PP, Groves RM, Lyberg LE, Mathiowetz NA, Sudman S (eds) Measurement errors in surveys. Wiley, New York
10. INE (2014) Encuesta de Población Activa. Instituto Nacional de Estadística, Primer Trim 2014, Madrid
11. Kraus F, Steiner V (1995) Modeling heaping effects in unemployment duration models - with an application to retrospective event data in the German socio-economic panel. Centre for

European Economic Research-ZEW Discussion Paper No 95-09. https://www.econstor.eu/bitstream/10419/29385/1/257629254.pdf

12. Olave P, Andrés EM, Alcalá JT (2008) Studying the relationship between unemployment periods in Spain: a nonparametric approach. Appl Econ Lett 15:683–687
13. Prakasa Rao BLS (1983) Nonparametric function estimation. Wiley, New York
14. Torelli N, Trivellato U (1993) Modeling inaccuracies in job-search duration data. J Econom 59(1–2):187–211
15. van Es B, Klaassen CAJ, Oudshoorn K (2000) Survival analysis under cross-sectional sampling: length bias and multiplicative censoring. J Stat Plan Inference 91:295–312
16. Wichert L, Wilke RA (2008) Simple non-parametric estimators for unemployment duration analysis. Appl Stat 57:117–126
17. Woodroofe M, Sun J (1993) A penalized maximum likelihood estimate of $f(0+)$ when f is non-increasing. Stat Sin 3:501–515

Modelling Dynamic Hotel Pricing
with Isotonic Functions

J. Santos Domínguez-Menchero and Emilio Torres-Manzanera

Abstract Hotel room rates and available rooms via online channels vary from day to day. From the customer's point of view, when should a tourist book a room? In order to detect the optimal time to purchase a hotel room, modelling dynamic hotel pricing behaviour patterns must be considered. Nonparametric techniques using isotonic functions are shown to be useful to estimate a reliable model that analyses the evolution of room rates. Online hotel prices were observed for eight months, collecting more than 133,000 valid records. The results of this study show that a generally recommended strategy is to make the booking at least 15 days prior to the arrival date. Other booking strategy patterns are also detailed.

1 Introduction

From a consumer point of view, dynamic pricing provides an opportunity to purchase products at different prices at different times. With electronic commerce business can be conducted anywhere, at any time, especially in the field of tourism. In such conditions, the decision of the customer to purchase airline seats or hotel rooms on the internet depends on several factors such as information quality, time, past experiences and frequency. But the most important factor influencing hotel selection is price; [5] pointed out that willingness of customers to book depends on the price presentation on the internet.

In this context, customers choose a destination and then select accommodation based on price and available rooms, using the most convenient distribution channel. Internet channels allow cost reduction in the final price, detailed information for consumers and the ability to instantly acquire the product. Specialised worldwide online hotel reservation agencies like Booking.com cover most of the top level category hotels and tourists use them surpassing other purchasing channels including travel

J. S. Domínguez-Menchero (✉) · E. Torres-Manzanera
Departamento de Estadística e I.O. y D.M., University of Oviedo, Oviedo, Spain
e-mail: jsdm@uniovi.es

E. Torres-Manzanera
e-mail: torres@uniovi.es

© Springer International Publishing AG 2018
E. Gil et al. (eds.), *The Mathematics of the Uncertain*, Studies in Systems,
Decision and Control 142, https://doi.org/10.1007/978-3-319-73848-2_11

121

agencies and telephone contacts. In particular, 46.2% of Spanish tourists and 49.0% of foreigners use these types of commercial distribution when travelling to Spain [3, 4].

Internet channels overload with information to both consumers and managers. Room rates and availability vary from day to day. What sets markets with dynamic pricing apart from others is that discounts can be offered at any time without notice, whereas in the clothing sector, for example, there are pre-determined dates for sales and consumers know that they can find lower prices on products at those times. Luck or chance is therefore a determining factor in finding lower prices in online markets for products with dynamic pricing. Thus, it is pertinent to conduct studies such as the introduced in this paper which try to reduce uncertainty and detect dynamic pricing behaviour patterns of products such as hotel rooms. This information enables consumers to optimise the waiting time before making a purchase (optimal patience) to obtain the best price, while reducing uncertainty and chance.

More specifically, the objective of this paper is to detect the optimal time to purchase a hotel room once a destination is chosen. From the customer's point of view, does time play a significant role in price and availability when booking a hotel room? When should the tourist book the room: the day before, one month before, or is there no pattern in the dynamic pricing? To answer these research questions, this paper presents a study carried out in Bilbao, one of the leading cities in Spain, during a nine-month period. It is a city with good tourism infrastructure for both holiday and business tourism. Bilbao can therefore be considered a prototype of the current state of tourism management.

The core results present an isotonic nonparametric model that analyses the evolution of room rates in the most important Internet Distribution System (IDS) channel over the course of eight months. The analysis shows that a recommended strategy in general is to wait to book a hotel room until 15 days prior to arrival. The study also collects information on what occurs during that 15-day period, thus providing useful information for consumers who wish to delay booking.

The remainder of this research note is organized as follows: Sect. 2 details data sources and the mathematical model, and Sect. 3 shows the main results.

2   Data Sources and Methodology

For the purpose of estimating the optimal time of purchase, mid- to high-end hotels in Bilbao, Spain, where observed for eight months between July 2011 and March 2012. For each of the 212 days in this period, the booking price for a double room was collected 28, 27, ..., 0 days in advance via IDS channels. The IDS channels observed were Booking.com and Activehotels.es, using the method proposed by [2]. Due to a failure in the crawling system, for two weeks in August some of the data was missing. Ultimately, the database contained a total of 133,368 valid records.

The average price considering different lags in the booking day was estimated. In particular, if t_0 represents the target day, ranged from July 2001 to March 2012, and

$l = -28, -27, \ldots, 0$, then $p_{t_0}^l$ is taken to be the geometric mean of the prices of a double room for the day t_0 when it was booked with $|l|$ advanced booking days.

For each day t_0, we define the price variation rate as

$$\Delta_{t_0,l} = 100\left(\frac{p_{t_0}^0}{p_{t_0}^l} - 1\right), \quad l = -28, -27, \ldots, 0. \tag{1}$$

This rate represents the price variation (in percent) that we would have if we wait until the target day instead of purchasing $|l|$ days in advance.

The effect of time lag for each day is given by

$$\Delta_{t_0,l}^{t_0,l+1} = 100\left(\frac{p_{t_0}^{l+1}}{p_{t_0}^l} - 1\right), \quad l = -28, -27, \ldots, -1. \tag{2}$$

This rate measures the variation between two consecutive booking days in advance, on a percent scale. The analysis of the evolution of both types of rate as a function of the number of advance booking days will show any pattern changes in the booking price. Therefore, we can obtain the optimal time of the purchase scenario.

Figure 1 shows the price variation rate for mid- to high-end hotels in the city of Bilbao. Our interest is to obtain a model that explains the price variation rate in terms of the booking day in advance,

$$\Delta_{t_0,l} = g(l) + \varepsilon_{t_0,l}, \tag{3}$$

with $\varepsilon_{t_0,l}$ the error of the model.

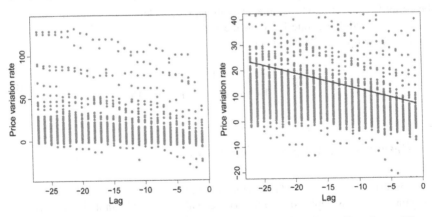

Fig. 1 Price variation rate in terms of the number of booking days in advance. Raw data and linear regression model (different scales)

In a simple linear model, the explanatory function g can be estimated by the linear regression of the observed $\Delta_{t_0,l}$. This model yields the following formula

$$\hat{g}_{linear}(l) = -0.64l + 6.16. \tag{4}$$

The linear regression model assumes a constant increase of the variation rate for each day, which is obviously not a natural assumption. In particular, the estimation of the variation rate would strictly increase indefinitely when $|l|$ increases. Later we will give more reasons that disappointing the use of the linear model. Nevertheless, the model can be taken in advance in order to show a non increasing trend between the lag and the price variation rate. Using this information, data can be modelled using isotonic regression techniques. Some problems in the field of tourism management have been solved using these techniques [7–9]. One of the main advantages in this context is that they split the time intervals where the price remains constant. This information is relevant to making an optimum purchasing decision.

Thus, we can assume that the explanatory function g in Eq. (3) is non-increasing. Let n_l be the number of observations of $\Delta_{t_0,l}$ at lag l for all target day, t_0, and I_l the mean of the observations. Then, the antitonic estimator of g is:

$$\hat{g}_{antitonic}(l) = - \max_{i \leq l} \min_{j \geq l} -\frac{\sum_{k=i}^{j} n_k I_k}{\sum_{k=i}^{j} n_k}. \tag{5}$$

This expression is known as the max-min formula and the function is defined as lineal between adjacent points. Some algorithms have been undertaken to process with additional restrictions in the field of tourism [1]. In this research the R statistics language [6] was used.

In practical applications, it is desirable to have an explicit equation which is as simple as possible. Since the isotonic regression is piecewise linear, its aspect can be simplified by eliminating small variations or constance intervals. In this case, we have considered a line between the points with $l = -15, -1$. Therefore, the simplified two-piece isotonic model is:

$$\hat{g}(l) = \begin{cases} -1.24l + 1.12 & \text{if } l \leq -15 \\ 19.70 & \text{if } l \geq -15 \end{cases}. \tag{6}$$

3   Research Findings and Conclusion

The analysis shows, in the first place, that up to 15 days prior to arrival the price tends to remain stable, as depicted in Fig. 2. This can also be seen in Table 1, which shows the price variation rates. This table also shows the rates per day. If the geometric mean with $1 \leq |l| \leq 15$ is calculated, then it is observed that for each day after the first 15 days the price goes up 1.21%. If the customer's strategy is to wait to as long as

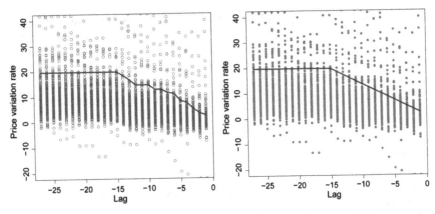

Fig. 2 Original isotonic model (left) and simplified two-piece isotonic model (right)

Table 1 Price variation rates in function of the lag (%)

Lag	Rate variation	
$|\ell|$	Target day	Per day
1	2.36	2.36
2	3.60	1.21
3	4.83	1.20
4	6.07	1.18
5	7.31	1.17
6	8.55	1.15
7	9.80	1.14
8	11.03	1.13
9	13.51	1.21
10	14.74	1.10
11	15.98	1.09
12	16.07	1.08
13	17.22	1.07
14	18.46	1.06
15	19.70	1.05
≥ 16	19.70	0.00

possible to book without loosing money, the best strategy would be to book exactly 15 days in advance.

The results clearly reject other models such as the linear model, which, although simpler, are not capable of extracting the information obtained with the isotonic model presented here. More specifically, as mentioned earlier, a linear regression assumes a constant decrease in price as consumers move away from the target day, which is absurd. Even the linear regression is used to decide about a short period of time it is useless. Suppose it is 30 days prior to arrival. The linear regression for the Bilbao statistics would mean a 0.75% price increase for every day consumers delay the booking, which is at considerable variance with the actual trend obtained from the isotonic model. Most notably, up to 15 days prior to arrival it predicts an 8.27% increase as compared to the 0.0% in our model. Linear regression would call for a sooner-the-better strategy, which, as described earlier, is not the most advantageous.

References

1. Domínguez-Menchero JS, González-Rodríguez G (2007) Analyzing an extension of the isotonic regression problem. Metrika 66:19–30
2. Gerrikagoitia J, Alzua A, Ibarguren I, Roman I, Espinosa N (2011) UNWTO Algarve forum - tourism and science: bridging theory and practice. World Tourism Organization, Algarve
3. Instituto de Estudios Turísticos (2011) Movimientos turísticos de los españoles (FAMILITUR). Informe anual, (2010) Ministerio de Industria. Turismo y Comercio, Madrid
4. Instituto de Estudios Turísticos (2011) Movimientos turísticos en fronteras (FRONTUR). Informe anual, (2010) Ministerio de Industria. Turismo y Comercio, Madrid
5. Noone BM, Mattila AS (2009) Hotel revenue management and the Internet: the effect of price presentation strategies on customers willingness to book. Int J Hosp Manag 28(2):272–279
6. R Development Core Team (2011) R: a language and environment for statistical computing. R Foundation for Statistical Computing, Vienna
7. Torres E, Domínguez-Menchero JS (2006) The impact of second homes on local taxes. Fisc Stud 27(2):231–250
8. Torres-Manzanera E, Domínguez-Menchero JS, Valdés L, Aza R (2005) Optimal purchase timing in the airline market. J Air Transp Manag 11(6):363–367
9. Valdés L, Torres E, Domínguez-Menchero JS (2007) A model to study the economic impact of collective accommodation in a region. Tour Manag 28(1):152–161

On Multistage Stochastic Mixed 0–1 Optimization with Time-Consistent Stochastic Dominance Risk Averse Management

Laureano F. Escudero and Juan F. Monge

Abstract A new time-consistent risk averse measure is considered, so-called Expected Conditional Stochastic Dominance (ECSD), for multistage stochastic mixed 0–1 optimization, where first- and second-order stochastic dominance risk averse functionals are taken into account. As a result of the ECSD modeling, its problem solving is much more difficult than the Risk Neutral counterpart, so, it is unrealistic to solve the problem up to optimality by plain use of MIP solvers. Instead of it, decomposition algorithms of some type should be used. Computational results are reported for instances of a well-known real-life problem, where a decomposition matheuristic algorithm is tested in its efficiency and computing effort, having the plain use of a MIP solver as a benchmark for computational purposes.

1 Introduction

Stochastic optimization is currently one of the most robust tools for decision making. It is broadly used in real-world applications in a wide range of problems from different areas in energy, finance, production, distribution, supply chain management, etc. The continuous optimization problems under uncertainty have been studied in [4] for risk neutral (RN) problems, and [16] for risk averse measures, among others. A survey on exact and inexact decomposition algorithms is performed in [9]. It is well known that an optimization (say, minimization) mixed 0–1 problem under uncertainty with a finite number of possible supporting scenarios has a mixed 0–1 Deterministic Equivalent Model (DEM). Traditionally, special attention has been given to optimizing the DEM by minimizing the objective function expected value

L. F. Escudero (✉)
Area Estadística e Investigación Operativa, Universidad Rey Juan Carlos,
28933 Móstoles (Madrid), Spain
e-mail: laureano.escudero@urjc.es

J. F. Monge
Centro de Investigación Operativa, Universidad Miguel Hernández,
03202 Elche (Alicante), Spain
e-mail: monge@umh.es

© Springer International Publishing AG 2018
E. Gil et al. (eds.), *The Mathematics of the Uncertain*, Studies in Systems,
Decision and Control 142, https://doi.org/10.1007/978-3-319-73848-2_12

in the scenarios, subject to the satisfaction of all the problem constraints in the defined scenarios, i.e., the so-called risk neutral (RN) approach. Currently, we are able to solve huge DEMs by using different types of decomposition approaches, e.g., see in [1] our last improvement of the Branch-and-Fix Coordination (BFC) methodology. However, the optimization of the RN model has the inconvenience of providing a solution that ignores the potential variability of the objective function value (say, cost) in the scenarios and, so, it does not avoid the potential high negative impact of the proposed solution in the low-probability high-cost scenarios to occur.

However, there are some risk averse approaches that, additionally, deal with risk management, see a good survey in [2], among others. As we know, the first risk averse measure, so-called Chance Constraint (CC) functional, was introduced in the seminal work [5], where the problem's feasible set is restricted to satisfying an upper bound on the probability (in our case, the expected fraction of scenarios) of having shortfall on satisfying each of the modeler-driven constraints. It is also so-called first-order stochastic dominance (FSD) that for a two-stage setting was introduced in [12], where a Lagrangean-based decomposition algorithm was used for problem solving. Another of the first risk averse measures, so-called Integrated Chance Constraint (ICC) functional, was introduced in [14] and expanded in [15]. The ICC type 1 is also so-called second-order stochastic dominance (SSD) that for a two-stage setting was introduced in [11], where a Lagrangean-based decomposition algorithm was used for problem solving. The multistage risk averse time-inconsistent measure based on the SSD functional was introduced in [6] for a set of profiles related to given thresholds on a multifunction setting (including the objective function). As an extension of the two-stage SD mixed-integer linear recourse, FSD and SSD measures were jointly used in a multistage mixed 0–1 setting, so-called Time Stochastic Dominance (TSD), see our works [7, 9], where the decomposition algorithms use scenario clustering in the BFC scheme in the first work and a Lagrangean-based scheme in the other one.

The risk averse functional subject of this work, so-called Expected Conditional Stochastic Dominance (ECSD), is a time-consistent risk averse functional, since it bounds the scenario probability of having shortfall on satisfying different rhs for given constraints, and it also bounds the expected shortfall, both for the scenario groups with a one-to-one correspondence with nodes in the scenario tree related to a chosen set of stages in the time horizon. Roughly, a risk averse measure is time-consistent if the solution at any node in the scenario tree does not depend on the scenarios that cannot occur at that node. Obviously, RN is a time-consistent measure. Some other time-consistent measures, mainly the Expected Conditional Value-at-Risk, have been studied in [3, 13], among others.

The rest of the paper is organized as follows. For completeness and notation presentation, Sect. 2 deals with the main concepts of risk neutral-based mathematical optimization under uncertainty. Section 3 deals with the risk averse measure ECSD. Section 4 reports some results of our computational experiment. Section 5 concludes.

2 Risk Neutral Measure in Multistage Mixed 0–1 Stochastic Problems

The DEM, compact representation, of the multistage mixed 0–1 model for minimizing the objective function expected value in the scenarios (i.e., a RN approach) can be expressed

$$z_{RN} = \min \sum_{g \in \mathcal{G}} w_g (a_g x_g + b_g y_g)$$

$$\text{s.t.} \sum_{q \in \mathscr{A}_g} (A_q^g x_q + B_q^g y_q) = h_g \qquad \forall g \in \mathcal{G} \tag{1}$$

$$x_g \in \{0, 1\}^{nx(g)}, \quad y_g \in \mathbb{R}^{ny(g)} \qquad \forall g \in \mathcal{G},$$

where $\mathcal{G}$ is the set of nodes in the scenario tree, w_g is the weight computed for node $g \in \mathcal{G}$, such that $w_g = \sum_{\omega \in \Omega_g} w_\omega$, where w_ω is the modeler-driven probability of scenario ω, for $\omega \in \Omega$, where Ω is the finite set of scenarios considered in the supporting tree; x_g and y_g are the $nx(g)$ and $ny(g)$-dimensioned vectors of the 0–1 variables and the continuous variables, resp., for node $g \in \mathcal{G}$; a_g and b_g are the vectors of the objective function coefficients for x_g and y_g, resp.; $\mathscr{A}_g$ is the set included by the same node g and its ancestors in the scenario tree with nonzero coefficients in the constraints of node $g \in \mathcal{G}$; A_q^g and B_q^g are the constraint matrices; and h_g is the rhs for node $g \in \mathcal{G}$. All vectors and matrices are with the adequate dimensions. Notice that the non-anticipativity constraints (NAC) are implicitly satisfied.

Additionally, let the following notation to be used throughout this work. $\mathcal{T}$ is the set of stages $\{1, 2, \ldots, T\}$ in the time horizon with $T = |\mathcal{T}|$; $\mathcal{G}_t \subseteq \mathcal{G}$ is the set of nodes in stage $t \in \mathcal{T}$, where $|\mathcal{G}_1| = 1$. Let $\Omega_g \subset \Omega$ be the set of scenarios in group g, where it has a one-to-one correspondence with node g in the scenario tree, for $g \in \mathcal{G}$; for easing notation, let $\omega \equiv g$ for $g \in \mathcal{G}_T$ and, then, $\omega \in \Omega_g$, where $|\Omega_g| = 1$; $t(g)$ is the stage to which node g belongs to, such that $g \in \mathcal{G}_{t(g)}$; $\mathscr{A}_g$ is the set of ancestor nodes in the scenario tree to node g (including itself), for $g \in \mathcal{G}$ (such that $\mathscr{A}_g \subseteq \tilde{\mathscr{A}}_g$); and $\tilde{\mathscr{S}}_g$ is the set included by the same node g and its successors in the scenario tree, for $g \in \mathcal{G} \setminus \mathcal{G}_T$.

3 ECSD Risk Averse Measure in Multistage Mixed 0–1 Stochastic Problems

The RN model (1) aims to minimize the objective function expected value alone subject to the constraint system in the model. As stated above the main criticism that can be made to this very popular strategy is that it ignores the variability of the objective function value in the scenarios and, in particular, the "right" tail of the non-wanted scenarios. The Expected Conditional Stochastic Dominance (ECSD) risk averse measure also minimizes the objective function expected value but, besides the set of RN constraints, a set of profiles is considered for a set of functions (including the objective one) in a given scenario subset. Each profile is included by a threshold on the function value for any scenario that belong to a given group (to be defined below) up to the end of the time horizon, a bound target on the fraction of those scenarios having excess on reaching it (the so-called first-order ECSD), a bound target on the expected excess (the so-called second-order ECSD) and a bound target on the excess for any of those scenarios. Observe that the type of risk reduction functional that is considered in this work also allows other functions besides the objective one, such as environmental functions in industrial strategic investment, mitigation effects of natural disasters, and preservation of cultural, monumental and strategic assets, among others; all of them can be accommodated in the framework presented below.

So, let the additional notation, $\mathscr{F}$ is the set of indexes of the risk reduction oriented functions to consider, such that a_g^f and b_g^f are the vectors of the coefficients of the variables in x_g and y_g for $g \in \mathscr{G}$ in the function indexed with f, for $f \in \mathscr{F}$. Let us consider that $f = 1$ is the index for the objective function and, so, $a_g^1 \equiv a_g$ and $b_g^1 \equiv b_g$. Let also $\tilde{\mathscr{T}}^f \subseteq \mathscr{T} \setminus \{T\}$ be the modeler-driven stage set for function $f \in \mathscr{F}$, such that the risk reduction measure ECSD is performed for each group of scenarios Ω_g, for $g \in \mathscr{G}_t$, $t \in \tilde{\mathscr{T}}^f$. Let also $\mathscr{P}_g^f$ denote the set of indexes of the related profiles for the function indexed with $f \in \mathscr{F}$ and the scenarios in set Ω_g, for $g \in \mathscr{G}_t$, $t \in \tilde{\mathscr{T}}^f$, such that the profile indexed with p, for $p \in \mathscr{P}_g^f$, is included by the 4-tuple $(\phi^p, \tilde{e}^p, \overline{e}^p, \overline{v}^p)$, where,

ϕ^p, threshold for the value of function f up to the last stage T in the time horizon in scenario ω, for $\omega \in \Omega_g$.

$\tilde{e}^p$, bound *target* on the excess of any of those scenarios on reaching the threshold.

$\overline{e}^p$, bound *target* on the expected excess.

$\overline{v}^p$, bound *target* on the expected fraction of those scenarios with excess.

Let the following additional variables for scenario ω and profile p, for $\omega \in \Omega_g$, $p \in \mathscr{P}_g^f$, $g \in \mathscr{G}_t$, $t \in \tilde{\mathscr{T}}^f$, $f \in \mathscr{F}$:

e_ω^p, continuous variable that takes the excess of the value of function f on reaching threshold ϕ^p, for scenario ω.

v_ω^p, 0–1 variable that takes the value 1 if the value of function f has an excess on reaching threshold ϕ^p and otherwise, 0, for scenario ω.

The ECSD model can be expressed

$$z_{ECSD} = \min \sum_{g \in \mathscr{G}} w_g (a_g^1 x_g + b_g^1 y_g)$$

$$+ \sum_{f \in \mathscr{F}} \sum_{t \in \tilde{\mathscr{T}}^f} \sum_{g \in \mathscr{G}_t} \sum_{p \in \mathscr{P}_g^f} (M_{\epsilon_{\tilde{e}^p}} \epsilon_{\tilde{e}^p} + M_{\epsilon_{\overline{e}^p}} \epsilon_{\overline{e}^p} + M_{\epsilon_{\overline{v}^p}} \epsilon_{\overline{v}^p})$$

$$\text{s.t.} \sum_{q \in \mathscr{A}_g} (A_q^g x_q + B_q^g y_q) = h_g \qquad\qquad \forall g \in \mathscr{G}$$

$$x_g \in \{0,1\}^{nx(g)}, \quad y_g \in \mathbb{R}^{ny(g)} \qquad\qquad \forall g \in \mathscr{G}$$

$$\sum_{g \in \mathscr{A}_\omega} (a_q^f x_q + b_q^f y_q) - e_\omega^p \leq \phi^p \qquad\qquad \forall \omega \in \Omega_g, \ p \in \mathscr{P}_g^f, \ g \in \mathscr{G}_t, \ t \in \tilde{\mathscr{T}}^f, \ f \in \mathscr{F}$$

$$0 \leq e_\omega^p \leq \tilde{e}^p v_\omega^p + \epsilon_{\tilde{e}^p},$$

$$v_\omega^p \in \{0, 1\}, \; \epsilon_{\bar{e}^p} \in \mathbb{R}^+ \qquad\qquad \forall \omega \in \Omega_g, \; p \in \mathcal{P}_g^f, \; g \in \mathcal{G}_t, \; t \in \tilde{\mathcal{T}}^f, \; f \in \mathcal{F}$$

$$\sum_{\omega \in \Omega_g} w_\omega e_\omega^p \le \bar{e}^p + \epsilon_{\bar{e}^p}, \; \epsilon_{\bar{e}^p} \in \mathbb{R}^+ \; \forall p \in \mathcal{P}_g^f, \; g \in \mathcal{G}_t, \; t \in \tilde{\mathcal{T}}^f, \; f \in \mathcal{F}$$

$$\sum_{\omega \in \Omega_g} w_\omega v_\omega^p \le \bar{v}^p + \epsilon_{\bar{v}^p}, \; \epsilon_{\bar{v}^p} \in \mathbb{R}^+ \; \forall p \in \mathcal{P}_g^f, \; g \in \mathcal{G}_t, \; t \in \tilde{\mathcal{T}}^f, \; f \in \mathcal{F},$$

$$(2)$$

where the last terms in the objective function prevent the potential infeasibility of the risk averse constraint system, such that the big M-parameters are the related penalization and the ϵ-variables are the slack ones to avoid the infeasibility.

The concept of *expected conditional risk avere measure* (ECRM) is introduced in [13], and its time-consistency is defined and proved. We show elsewhere [10] that ECSD is a member of the family of ECRMs and, therefore, it is time consistent. Let us assume that the decisions in a given problem have been made up to node g, for $g \in \bigcup_{\{t \in \mathcal{T} : t < T\}} \mathcal{G}_t$, according to the solution obtained in the original model (2) solved at stage $t = 1$. Now, let the submodel solved at node g, such that it is supported by the subtree rooted with node g whose successor nodes are in set $\tilde{S}_g$. Then, the rationale behind a time-consistent risk averse measure is that the solution value to be obtained in the submodel solved at stage $t(g)$ for node g should have the same value as in the original model solved at stage $t = 1$.

4   Some Computational Experience

Let the tactical supply chain planning (TSCP) problem presented in our work [10] to be the pilot case where to consider the performance of the RN and ECSD risk measures. Its deterministic version is based on a real-life case in the assembly sector. TSCP has a broad applicability, specifically, in sectors such as car, computer and domestic appliances manufacturing, among others. It is the case in which a company with multiple raw material suppliers, plants, products, tiers of production in the bill of material (BoM) and markets needs to satisfy a product demand vector over a given time horizon. The goal is to determine a raw material supplying plan and a master production, inventory and distribution planning that best makes use of the available resources and their capacity extension acquisitions in the whole supply chain for each period of a given time horizon. The resources' best use consists of minimizing the raw material supplying commitment cost, the production and inventory costs in the plants, and the product backlog and demand lost penalization along the time horizon. The raw material supplying commitment cost is frequently modeled by a piecewise linear, concave and nondecreasing function of the total volume to commit for the whole time horizon. Typical types of constraints (some of them related to either-or decisions) are as follows: Balance equations of end-products and components, conditional lower and upper bounds for raw material supplying and product release, resource consumption bounds and capacity extension

Table 1 STSCP Models' dimensions

ID	Scenario tree						RN model			ECSD model		
	Structure	T	t^*	$|\mathscr{G}_{t^*}|$	$|\mathscr{G}|$	$|\Omega|$	nc	$n01$	m	nc	$n01$	m
L1	$2^4 2^4$	8	5	16	255	128	57064	18126	90364	57064	18126	90424
L2	$2^4 2^5$	9	5	16	511	256	113768	36302	180604	113768	36302	180664
L3	$2^4 2^4$	8	5	16	255	128	72424	18131	99324	72424	18131	99384
L4	$2^4 2^5$	9	5	16	511	256	144488	36307	198524	144488	36307	198584

acquisitions, and balance equations of lost demand and backlogging, among others. There are different types of resources at different levels for groups of consecutive periods (so-called stages) along the time horizon. The cost of the resources' capacity extension acquisition is expressed as a piecewise discrete and nondecreasing function. Another important feature of the problem is that the burden of raw material stocking is frequently transferred to the suppliers.

The experiment was conducted on a PC with a 2.5 GHz dual-core Intel Core i5 processor, 8 Gb of RAM and the operating system was OS X 10.9, where the MIP solver to use is CPLEX v12.5 and its optimality tolerance is set up to 0.001. The decomposition algorithm to use is our matheuristic SDP-ECSD [10].

The problem's dimensions of the testbed under consideration are up to 9 stages, 20 end-products, 30 market centers each, 20 subassemblies, 40 raw materials, and 25 types of resources. Table 1 presents the structure of the scenario tree for each instance as well as the dimensions of the two stochastic formulations. The set of stages $\mathscr{T}$ has been split in two parts for problem solving. The first column of the table is the identifier of the instance, and the second one gives the predefined structure $A_1^{B_1} A_2^{B_2}$ of the scenario tree, where A_i denotes the number of children that each node has in each stage in part i, and B_i is the number of its stages, for $i = 1, 2$. The period subset $\tilde{\mathscr{T}}$ is singleton and $t^* \in \tilde{\mathscr{T}}$, where t^* is the period defining the groups of scenarios for cost risk reduction in the ECSD measure. The headings of the columns for the dimensions of the models are as follows: nc, number of continuous variables; $n01$, number of 0–1 variables; and m, number of constraints.

The results of solving RN model (1) are shown in Table 2. The first column refers again to the identifier of the instance. The following three columns reports the CPLEX

Table 2 RN model (1) solved with CPLEX and the SDP-ECSD matheuristic

ID	CPLEX			SDP-ECSD		
	$\bar{z}_{CPX}$	t_{CPX}	$OG\%$	$\bar{z}_{RN}$	t_{RN}	$GG\%$
L1	358305	7200	0.62	362350	1409	1.13
L2	212922	7200	1.14	215236	3884	1.09
L3	218343	7200	0.72	221997	5032	1.67
L4	398303	7200	0.28	403833	8055	1.39

Table 3 ECSD model (2) solved with the SDP-ECSD matheuristic

ID	$\bar{z}_{ECSD}$	t_{ECSD}	$dev_{ECSD}\%$
L1	375127	21065	3.53
L2	215544	25032	0.14
L3	253152	22103	14.03
L4	431792	19403	6.92

results, where $\bar{z}_{CPX}$ is the RN cost value, t_{CPX} is the elapsed time (in seconds) to obtain it, and $OG\%$ is its optimality gap (in %). The optimization of the instances reaches the allowed time (2h) without proving the 0.1%-optimality of the solution. Another block of columns in Table 2 reports the SDP-ECSD results. The headings are as follows: $\bar{z}_{RN}$ and t_{RN}, RN solution value and related elapsed time (in seconds); and $GG\%$, goodness gap, i.e., the deviation of the solution value obtained by the matheuristic from the value obtained by CPLEX, expressed as $GG\% = (\bar{z}_{RN} - \bar{z}_{CPX})/\bar{z}_{CPX}\%$. We can observe that, generally, the elapsed time that is required by SDP-ECSD is very small. On the other hand, the goodness gap of its RN value versus the one provided by CPLEX is very small as well; notice that it goes form 1.09 to 1.39%.

The violations of the $\bar{e}_{RN}^p$ and $\bar{v}_{RN}^p$ ECSD bounds for $p = 1, 2$ by the RN solution are up to 165 and 700%, respectively. The details are not shown but they are available from the authors under request, see also [10].

The results of solving ECSD model (2) by matheuristic SDP-ECSD are shown in Table 3. Some headings are as follows: $\bar{z}_{ECSD}$ and t_{ECSD}, ESCD incumbent value and related elapsed time (in seconds); and $dev_{ECSD}\%$, deviation of the ESCD cost from the RN one (see Table 2), expressed as $dev_{ECSD}\% = (\bar{z}_{ECSD} - \bar{z}_{RN})/\bar{z}_{RN}\%$.

Notice that the elapsed time required by the matheuristic for solving the ECSD model (2) is much greater than the time required for solving the RN model (1). It confirms the common knowledge, namely, the stochastic dominance strategy are computationally much harder than the RN one (requiring an elapsed time that is one order of magnitude higher than the time required for obtaining the RN solution). It is due to the cross scenario constraints for satisfying the risk reduction measure. Notice that, probably, CPLEX could not even solve the ECSD model, since it could not do it for the RN one. On the other hand, the deviation of the ECSD cost (due to the satisfaction of the cost risk reduction constraint system) could even reach the increment of the 14.04% of the RN cost (where, as notice above, the violations of the two types of risk reduction bounds are up to 165 and 700%).

5 Conclusions

Frequently there are problems with high variability in the functions to consider (beside the one to minimize). So, a risk reduction functional is required for avoiding low-probability high-negative function values incurred by the solutions obtained from the minimization of the objective function expected (cost) value. In this work we have considered the time-consistent multi-function first- and second-order stochastic dominance functional ECSD for risk management to control. It allows to personalize the type of risk reduction profiles for groups of scenarios. That high risk management could increase the cost function value while satisfying the risk reduction constraint system. Any way, a specialization of decomposition algorithms is required for problem solving in a affordable computing effort. We have used our SDP-ECSD decomposition algorithm for problem solving in a pilot case from the tactical supply chain field; the solution's quality is good enough and the computing effort is very affordable.

References

1. Aldasoro U, Escudero LF, Merino M, Pérez G (2017) A parallel branch-and-fix coordination based matheuristic algorithm for solving large sized multistage stochastic mixed 0–1 problems. Eur J Oper Res 258:590–606
2. Alonso-Ayuso A, Carvallo F, Escudero LF, Guignard M, Pi J, Puranmalka R, Weintraub A (2014) On the optimization of copper extraction in mining under uncertainty in copper prices. Eur J Oper Res 233:711–726
3. Alonso-Ayuso A, Escudero LF, Guignard M, Weintraub A (2017) Risk management for forestry planning under uncertainty in demand and prices. Eur J Oper Res. http://doi.org/10.1016/j.ejor.2017.12.22
4. Birge JR, Louveaux FV (2011) Introduction to stochastic programming, 2nd edn. Springer, New York
5. Charnes A, Cooper WW, Symonds SH (1958) Cost horizons and certainty equivalents: an approach to stochastic programming of heating oil. Manag Sci 4:235–263
6. Dentcheva D, Ruszczynski A (2003) Optimization with stochastic dominance constraints. SIAM J Optim 14:548–566
7. Escudero LF, Garín A, Merino M, Pérez G (2016) On time stochastic dominance induced by mixed integer-linear recourse in multistage stochastic programs. Eur J Oper Res 249:164–176
8. Escudero LF, Garín A, Monge JF, Unzueta A (2018) On multistage stochastic mixed 0-1 bilinear optimization based on endogenous uncertainty and time consistent stochastic dominance risk management. Submitted
9. Escudero LF, Garín A, Unzueta A (2017) Cluster Lagrangean decomposition for risk averse in multistage stochastic optimization. Comput Oper Res 85:154–171
10. Escudero LF, Monge JF, Romero-Morales D (2017) On time-consistent stochastic dominance risk averse measure for tactical supply chain planning under uncertainty. Comput Oper Res. http://doi.org/10.1016/j.ejor.2017.07.11
11. Gollmer R, Gotzes U, Schultz R (2011) A note on second-order stochastic dominance constraints induced by mixed-integer linear recourse. Math Program Ser B 126:179–190
12. Gollmer R, Neise F, Schultz R (2008) Stochastic programs with first-order stochastic dominance constraints induced by mixed-integer linear recourse. SIAM J Optim 19:552–571
13. Homem-de-Mello T, Pagnoncelli BK (2016) Risk aversion in multistage stochastic programming: a modeling and algorithmic perspective. Eur J Oper Res 249:188–199
14. Klein Haneveld WK (1986) Duality in stochastic linear and dynamic programming. Lecture notes in economics and mathematical systems, vol 274. Springer, Berlin
15. Klein Haneveld WK, van der Vlerk MH (2006) Integrated chance constraints: reduced forms and an algorithm. Comput Manag Sci 3:245–269
16. Pflug GCh, Römisch W (2007) Modeling, measuring and managing risk. World Scientific, Singapore

Bivariate Copula Additive Models for Location, Scale and Shape with Applications in Biomedicine

Jenifer Espasandín-Domínguez, Carmen Carollo-Limeres,
Luis Coladas-Uría, Carmen Cadarso-Suárez, Oscar Lado-Baleato
and Francisco Gude

Abstract In many biomedical applications it is worthwhile to model not only the effect that covariates have on the mean but also on other parameters of the response distribution such as variance. Moreover, it is sometimes necessary to study the association between two or more variables and how such associations may depend on certain factors or covariates. Different models of flexible regression have recently been proposed in statistical literature but in this work we will focus on the study of Copula Additive Models for Location, Scale and Shape since this novel approach permits to model the dependence of two variables through copula functions and where covariates are also modelled in a flexible manner. Lastly, the benefits of using these models with real biomedical data will be illustrated.

1 Introduction

Regression models are used to represent the dependence of a response variable of interest as a function of a set of predictor variables (known as covariates). Specifically, in classical regression models, it is common to study the mean of the response variable as a function of the values of the explanatory variables. However, focusing solely on the estimation of means can lead to errors when modelling data pertaining to complex structures. Thus, in some applications is it important not only to explain the effect of covariates as a function of the mean of the response, but also know the complete distribution of the response. It is also sometimes necessary to model multivariate

J. Espasandín-Domínguez (✉) · C. Carollo-Limeres · C. Cadarso-Suárez
Group of Biostatistics, Center for Research in Molecular Medicine and Chronic Diseases
(CiMUS), University of Santiago de Compostela, La Coruña, Spain
e-mail: jenifer.espasandin@usc.es

L. Coladas-Uría · O. Lado-Baleato
Department of Statistics, Mathematical Analysis, and Optimization, University of Santiago
de Compostela, La Coruña, Spain

F. Gude
Clinical Epidemiology Unit, Hospital Clínico Universitario de Santiago de Compostela,
La Coruña, Spain

© Springer International Publishing AG 2018
E. Gil et al. (eds.), *The Mathematics of the Uncertain*, Studies in Systems,
Decision and Control 142, https://doi.org/10.1007/978-3-319-73848-2_13

135

responses as well as to determine the relationship between them. In most published regression studies for multivariate responses, a specific distribution is assumed for the response variable for no apparent reason and there are few contributions using non-parametric predictors. In recent years, different regression methodologies for bivariate responses based on copula functions have been developed in statistical literature. A major advantage of the copula approach is that the marginal distributions may also come from different non-standard families (see, Marra and Radice [7]). However, most of the existing multivariate distributions are simple extensions of the univariate distributions and often have the restrictive properties that all of the marginal distributions are of the same type (e.g., by construction, all marginal distributions of a normal multivariate are normal).

Dependence modelling using copula functions has become very popular in recent years as a multivariate modelling tool in many fields where multivariate dependence is of interest and standard multivariate normality is in question. This methodology has now proven to be particularly useful in the field of medicine. In this manuscript we propose the use of bivariate copula regression models to simultaneously study two measures of quality of life obtained through the SF-36 questionnaire (Alonso et al. [2]) in order to determine the covariables that influence them and their interactions. More specifically, we will focus on the use of Bivariate Copula Additive Models for Location, Scale and Shape (CGAMLSS models; Marra and Radice [7]). This novel approach extends the use of GAMLSS (Rigby and Stasinopoulos [10]) to situations in which each parameter of a multivariate response is modelled simultaneously on some conditional covariates using different copula functions. Furthermore, this type of regression model enables the modelling of all distributional parameters using additive predictors that allow for several types of covariate effects (such as non-linear effects of continuous covariates, random effects or interactions).

Alternative approaches to Marra and Radice [7] have been formulated in the statistical literature in both the frequentist and the bayesian frameworks. However, to the best of our knowledge, they only cover parts of the flexibility of those mentioned above either because they only allow for the consideration of non-linear or normal marginal effects (e.g. Sabeti et al. [11]) or because they fail to consider additive predictors (e.g. Acar et al. [1]). In the frequentist domain, attention must also be drawn to Vector Generalized Additive Models (VGAM, Yee and Wild [20]). VGAMs allow one to model each parameter of a bivariate non-standard response as a function of flexible covariate effects. However, the main drawback is that smoothing parameters for non linear effects are not selected automatically and copula model specifications are limited. To the best of our knowledge, the only competitor in the frequentist domain to the methodology introduced by Marra and Radice [7] is a technique introduced by Vatter and Chavez-Demoulin [14]. The main difference is that the method employed by these authors is based on a two-stage technique where the parameters of the marginal distributions and the copula function are estimated separately, whereas the CGAMLSS method is based on the simultaneous estimation of all the model's parameters. Marra and Radice [7] show that CGAMLSS outperforms the two stage approach (via simulation studies).

In the Bayesian framework, Klein and Kneib [6] recently introduced structured additive conditional copula regression models. However, in this study we propose the use of CGAMLSS since it provides greater flexibility and allows for a great variety of copula functions as well as different non-standard marginal distributions for the responses.

The rest of the chapter is organized as follows: Sect. 2 provides an introduction to Bivariate Copula Additive Models for Location, Scale and Shape, including some details of on the estimation and model selection. In Sect. 3 we will show a practical application to real biomedical data. Lastly, Sect. 5 provides a summary and some comments regarding on directions of future research.

2 Bivariate Copula Additive Models for Location, Scale and Shape

Given the nature of the data we present in Sect. 3, in this paper we will study copula-based models for a pair of continuous random variables, Y_1 and Y_2. We denote the generic covariate vector by z_i. The joint cumulative distribution function (cdf) of Y_1 and Y_2 can be expressed in terms of the marginal cdfs of Y_1 and Y_2 and a copula function C that binds them together (Marra and Radice [7]):

$$F(y_1, y_2|\vartheta) = C(F_1(y_1|\mu_1, \sigma_1, \nu_1), F_2(y_2|\mu_2, \sigma_2, \nu_2); \theta)$$

where $\vartheta = (\mu_1, \sigma_1, \nu_1, \mu_2, \sigma_2, \nu_2, \theta)^T$, $F_1(y_1|\mu_1, \sigma_1, \nu_1)$ and $F_2(y_2|\mu_2, \sigma_2, \nu_2)$ are the marginal cdfs of Y_1 and Y_2 taking values in $(0, 1)$, μ_m, σ_m, ν_m, for $m = 1, 2$ are the marginal distribution parameters. In this case, we have considered three parameter distributions: μ_m, σ_m, ν_m representing the location, the scale and shape, respectively. However, the computational framework can be extended to parametric distributions with more than three parameters. $C(\cdot, \cdot)$ is a uniquely defined two-place copula function which does not depend on the marginals, and θ is an association copula parameter measuring the dependence between the two random variables (Sklar [12]). Some of the classic copulae are shown in Fig. 1.

The Gaussian and Frank copulas permit modelling of positive and negative dependence. Gaussian and Frank show weaker tail dependences and Frank exhibits a slightly stronger dependence in the middle of the distribution. Clayton is asymmetric with a strong lower tail dependence but a weaker upper tail dependence. Just the opposite is true for the Gumbel and Joe copulas. There are also rotated versions of the Clayton, Gumbel and Joe Copulae.

2.1 Model Formulation

We assume a fully parametric specification for the distribution of the bivariate response vector where potentially all parameters of the bivariate response distrib-

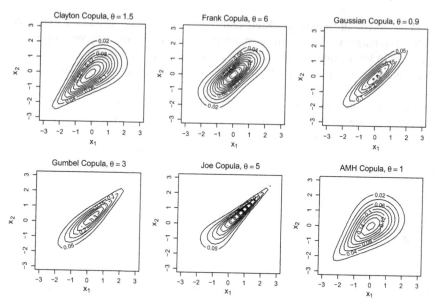

Fig. 1 Contour plots of some classic copula functions with standard normal margins for simulated data. The copula parameter is denoted by θ

ution can be related to regression predictors formed from covariates collected in the vector z_i (as binary, categorical, continuous or spatial variables) via additive predictors η and known monotonic link functions which ensure that the restrictions on the parameter spaces are maintained (see Marra and Radice [7] for available link functions).

$$\eta_i = \beta_0 + \sum_{k=1}^{K} f_k(\mathbf{z_{ki}}), i = 1, \dots, n, \tag{1}$$

where β_0 is an overall intercept, and the function f_k represents the different covariate effects. K functions f are chosen according to the type of covariate considered ($\mathbf{z_{ki}}$). For instance, to model the possible non-linear effects of the continuous covariates, different spline definitions with penalizations are available such as penalized low-ranking thin plate splines (Wood [17]), or roP-splines as proposed by Eilers and Marx [4] and other possible smoothers. From a statistical point of view, an important feature of this type of model is the possibility of modelling the effects not only of the continuous covariates but also spatial and random effects in a flexible, unified manner as well as allowing for complex interactions between different types of variables, e.g., factor-curve or surface interactions. It also allows for the modelling of spatio-temporal trends.

2.2 Estimation of the Model

The maximum likelihood function of a bivariate copula regression model is given by (Marra and Radice [7]):

$$l(\delta) = \sum_{i=1}^{n} log \{C(F_{1i}(y_{1i}|\mu_{1i}, \sigma_{1i}, v_{1i}), F_2(y_{2i}|\mu_{2i}, \sigma_{2i}, v_{2i}); \theta_i)\}$$

$$+ n \sum_{i=1}^{n} \sum_{m=1}^{2} log \{f_m(y_{mi} \mid \mu_{mi}, \sigma_{mi}, v_{mi}))\} \tag{2}$$

where $\delta = (\boldsymbol{\beta}_{\mu1}^T, \boldsymbol{\beta}_{\mu2}^T, \boldsymbol{\beta}_{\sigma1}^T, \boldsymbol{\beta}_{\sigma2}^T, \boldsymbol{\beta}_{v1}^T, \boldsymbol{\beta}_{v2}^T, \boldsymbol{\beta}_{\theta}^T)$, $C(\cdot, \cdot, \theta)$ is the density function of a copula function and $f_m(y_m|\mu_m, \sigma_m, v_m)$ the density of the marginals.

Inference in CGAMLSS models is based on penalised maximum likelihood estimation, achieved via a Trust Region algorithm (Conn et al. [3]) with integrated automatic multiple smoothing parameter selection (Marra and Radice [7]).

Due to the flexible structure of the predictors, the use of a classical (non-penalized) optimization algorithm would excessively smooth the effects and therefore would not reflect the true trends found in the data. Instead of maximizing the function of log-likelihood $l(\delta)$ the penalized log-likelihood function is maximised $l_p(\delta)$ such that:

$$l_p(\delta) = l(\delta) - \frac{1}{2}\delta^T S\delta \tag{3}$$

where $S = diag(\boldsymbol{D}_{\mu_1}\boldsymbol{\lambda}_{\mu_1}^T, \boldsymbol{D}_{\mu_2}\boldsymbol{\lambda}_{\mu_2}^T, \boldsymbol{D}_{\sigma_1}\boldsymbol{\lambda}_{\sigma_1}^T, \boldsymbol{D}_{\sigma_2}\boldsymbol{\lambda}_{\sigma_2}^T, \boldsymbol{D}_{v_1}\boldsymbol{\lambda}_{v_1}^T, \boldsymbol{D}_{v_2}\boldsymbol{\lambda}_{v_2}^T, \boldsymbol{D}_{\theta}\boldsymbol{\lambda}_{\theta}^T)$, with each λ defined as $(\lambda_1, \ldots, \lambda_K)^T$, it is the parameter that controls the degree of smoothing of non-linear effects.

To maximize expression (3) with an automatic selection of the smoothing parameter, λ, Marra and Radice [7] propose the estimation of the parametric vector δ by means of a Trust Region algorithm. In each iteration of this algorithm, a quadratic approximation of the target function in a radius defined around the previous point is minimized and the smoothing parameter λ is automatically selected. The algorithm consists of the following steps:

1. For a fixed λ:
   $\min_p \breve{l}_p \stackrel{def}{=} -\{l_p(\delta^{[a]}) + \boldsymbol{p}^T \boldsymbol{g}_p^{[a]} + \frac{1}{2}\boldsymbol{p}^T \boldsymbol{H}_p^{[a]} \boldsymbol{p}\}$ so that $\|\boldsymbol{p}\| \leq \Delta^{[a]}$,
   where $[a]$ is the iteration index, $\boldsymbol{g}_p^{[a]}$ and $\boldsymbol{H}_p^{[a]}$ are respectively the gradient vector and Hessian matrix of the log-likelihood function, penalized by the smoothing matrix S, $\| \cdot \|$ is the Euclidean distance $\Delta^{[a]}$ the radius taken by the Trust Region algorithm in each iteration.

2. Having estimated $\delta^{[a]} = \arg \min_p \breve{l}_p(\delta^{[a+1]}) + \delta^{[a]}$, λ is selected by solving an expression equivalent to the Un-Biased Risk Estimator (Wood [18]).

3. Having estimated $\lambda[a]$, step 1 is repeated until the change in the estimate of vector δ does not improve the log-likelihood function given by $\frac{|l(\delta^{[a+1]})-l(\delta[a])|}{0.1+|l(\delta^{[a+1]})|} < 1e^{-07}$.

3 Application to Biomedical Data

In this work, bivariate copula additive regression models will be applied to a biomedical database derived from a cross-sectional study performed in the municipality of A-Estrada in Galicia. An outline of the study (AEGIS, A-Estrada Glycation and Inflammation Study) is available at www.clinicaltrials.gov, code No. NCT01796184. It was aimed to investigate glycation and inflammation processes and their association with lifestyles and common diseases. In this manuscript, we will focus on health-related quality of life (HRQOL) as measured with the Short-Form 36 (SF-36) questionnaire (Ware and Sherbourne [16], Alonso et al. [2]).

The survey addresses limitations in physical functions and role activities due to health problems, bodily pain, general health perceptions, vitality (energy and fatigue), limitations in role activity, mental health and social limitations.

In this study we will jointly focus on vitality and physical function. Vitality is defined by the score obtained in 4 items with 6 possible answers per item, low scores being associated with a continuous feeling of fatigue, anguish and depression while the high scores denote feelings of happiness, serenity and calm. This scale has been validated as a measure of the fatigue associated with different diseases (Neuberger [9]). The physical function is defined by responses to 10 items with 3 options per item, a low score being indicative of a limitation in carrying out daily tasks, including getting dressed and bathing, while high scores correspond to patients with no physical limitations in their daily life and who are able to engage in vigorous physical activity (Ware et al. [15]).

The relationship between physical function and vitality has been previously studied (Wu et al. [19], Sturgeon et al. [13]), focusing on the study of fatigue in different diseases. These authors found positive relationship between both scales in patients with liver cirrhosis and chronic pain, respectively. In this study we are interested in investigating the relationship between physical function and vitality after controlling by age, sex and obesity (BMI, Body Mass Index), and in addition, the effect of these covariates on the relationship.

The data used come from the AEGIS project. At the beginning of the study, the municipality had an adult population of 18474 individuals, 3500 of whom were selected by random sampling stratified by age. After applying the exclusion criteria (death, change of address and impossibility of informed consent), a total of 1,516 people participated between 2012 and 2015 and were given the Spanish version of the SF-36 questionnaire (Alonso et al. [2]). The following variables were considered to be covariates in order to determine which factors influence vitality and physical condition and the relationship between them: gender, age (in years) and body mass index (BMI). Vitality and physical function were considered the bivariate response: (Vt, Pf).

3.1 Model Building

The aim of this section is to construct a bivariate model to study the relationship between vitality and physical activity. Proper response distribution selection, a suitable copula function and relevant covariates in the additive predictors are difficulties that researchers need to solve when formulating these types of models. Many model selection approaches have been proposed in statistical literature. According to Marra and Radice [7], the Akaike and Bayesian Information Criterion (AIC/BIC) and normalized quantile residuals were used to choose a suitable copula function and response distributions in CGAMLSS models. In addition to the statistical criteria mentioned above, it is essential to have expert knowledge from physicians or previous biomedical studies.

As mentioned above, this type of model can handle a variety of complex distributions of the response variable. We first need to select the distribution of the marginal variables Vt and Pt. Following the AIC and normalized quantile residuals, we choose the normal and Gumbel distributions for vitality and physical function respectively. Concerning the choice of copula, we started off with the normal and then, based on the (negative or positive) sign of the dependence, we only used copula functions consistent with the sign. In this case, using a Gaussian copula, the overall Kendall's $\hat{\tau}$ and $\hat{\theta}$ obtained was positive and significant ($\hat{\theta} \in (0.39, 0.54)$ and $\hat{\tau} \in (0.26, 0.36)$). This result suggests the use of another copula function to study positive dependencies such as Frank, AMH, FGM, Clayton, Joe, Gumbel and survival copulas which are copulas rotated by 180°. In this study the AIC showed that the Gumbel copula provided the best and most parsimonious fit (see Table 1).

Residuals can also be used to check the performance of a particular model (Klein and Kneib [6]). If the estimated model is close to the true model, the quantile residuals approximately follow a typical standard distribution, even if the model distribution itself is not standard. In practical terms, residuals can be assessed graphically as

Table 1 Comparison of AIC and BIC values under different copula assumptions

Copula Function	AIC	BIC
Gumbel	6445.62	6702.88
Gaussian	6458.86	6678.75
AMH	6478.33	6688.49
Frank	6487.06	6731.59
Survival gumbel	6491.46	6714.87
Clayton	6516.61	6736.25
Joe	6531.74	6801.11
Survival joe	6551.73	6788.67
FGM	6571.68	6785.52
Survival clayton	6880.60	7102.29

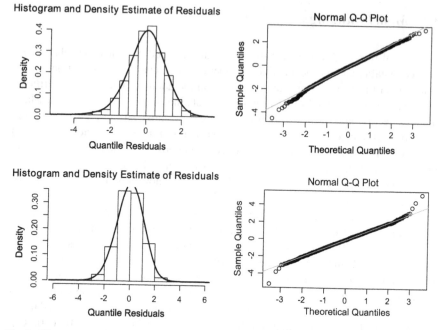

Fig. 2 Histograms and normal Quantile-quantile plots of normalized quantile residuals for vitality (top) and physical condition (bottom) for the selected model: the closer the residuals to the bisecting line, the better the fit to the data

quantile-quantile-plots. Figure 2 shows quantile-quantile plots for the selected model using a gumbel copula.

Thus, the CGAMLSS described in Sect. 2 was used with a Gumbel copula and a standard distribution response for vitality and a Gumbel distribution for physical condition with two covariate-dependent parameters (corresponding to the mean and the scale parameter σ^2) using. This model is expressed as follows:

$$\begin{cases} \eta_i^{\mu_1} = \beta_{0i}^{\mu_1} + gender\beta_i^{\mu_1} + gender * f_i^{\mu_1}(age) + f_i^{\mu_1}(BMI) \\ \eta_i^{\sigma_1^2} = \beta_{0i}^{\sigma_1^2} + gender\beta_i^{\mu_1} + f_i^{\sigma_1^2}(age), \\ \eta_i^{\mu_2} = \beta_{0i}^{\mu_2} + gender\beta_i^{\mu_2} + f_i^{\mu_2}(age) + f_i^{\mu_2}(BMI), \\ \eta_i^{\sigma_2} = \beta_{0i}^{\sigma_2} + gender\beta_i^{\sigma_2} + f_i^{\sigma_2}(age), \\ \eta_i^{\theta} = \beta_{0i}^{\theta} + f_i^{\theta}(age). \end{cases} \quad (4)$$

Eliminating the parameter index for the sake of simplicity, the predictors (η_i) are an additive composition of an intercept β_0 representing the overall level of the predictor, linear effects $gender$, and functions $f_i(z)$ reflecting the non-linear effects of continuous covariates z (age, BMI). The first and third equations of the above Formula (4) refer to the μ_1 and μ_2 parameter of *vitality* and *physical condition*. The second and fourth equations refer to the σ_1^2 and σ_2^2 parameters and the last one to the parameter association, θ.

In (4), f smooth functions of age and BMI are estimated using penalized low rank thin plate splines (Wood [17]) with default second order penalties (the option set by default in the GJRM R Package). By way of comparison, other smoothers were used to estimate the non-linear effect of continuous covariates but the results obtained were the same.

Statistical analyses were performed using GJRM R Package (Marra and Radice [7]). All results are summarised in Figs. 3, 4, 5 and 6.

Vitality diminishes in women throughout their lifetime while in men it diminishes up to age 60 and then increases coinciding with retirement age. This late rise in vitality may be due to the substitution of work with leisure activity (Mein et al. [8]). The highest levels of vitality for this population are found in BMI mean values ranging from 25–30 kg/m^2 with lower values at the extremes (see Fig. 3). In the other hand, the physical function remains stable until age 40 and thereafter decreases progressively

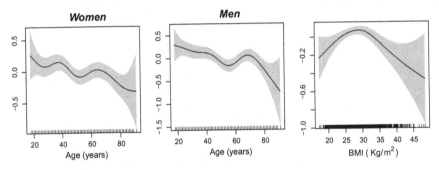

Fig. 3 Estimated smooth effects of age by gender and BMI on the mean of vitality and associated 95% point-wise intervals obtained when fitting a Gumbel copula model with normal and Gumbel margins

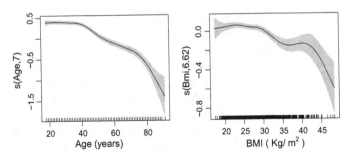

Fig. 4 Estimated smooth effects of age and BMI on the mean of physical condition and associated 95% point-wise intervals obtained when fitting a Gumbel copula model with normal and Gumbel margins

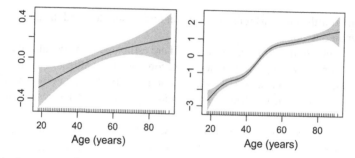

Fig. 5 Estimated smooth effects of age on the variance of vitality (left) and physical condition (right) with associated 95% point-wise intervals obtained when fitting a Gumbel copula model with normal and Gumbel margins

Fig. 6 Smooth function estimate and associated 95% point-wise confidence intervals for the association between physical condition and vitality condition

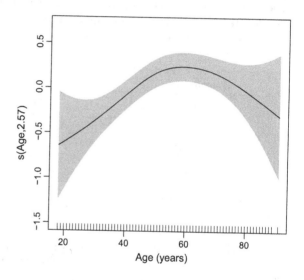

with age. BMI remains stable up to values of 30kg/m^2 (threshold value defining obesity) and then decreases progressively (see Fig. 4).

The variability of both variables, vitality and physical function, increases with age. Regarding Fig. 5, it is worth noting that the decrease in the mean of the two components as age increases may be rather abrupt depending on the individual. As shown in the Fig. 6, the ratio between physical function and vitality increases linearly up to age 60 and then levels off. The increasing association between the two variables with age could mean that the loss impairment of physical functions is not associated with the same degree of loss of vitality in young people and the elderly.

4 Discussion

In this study we have shown the usefulness of bivariate copula additive models (CGAMLSS) using real biomedical data. CGAMLSS allows for the easy inclusion of potentially any parametric continuous marginal distribution and copula function. This approach allowed us to take covariate effects into account, including the smooth estimation of continuous variables, random effects, possible spatial trends or categorical variables, among others. In particular, the CGAMLSS used in this study proved to be very useful in the joint modelling of vitality and physical condition. The use of this approach revealed hitherto unreported effects regarding the mean and variance and the relationship between these two variables. In this work we have simultaneously studied two response variables but it would be interesting to extend this joint analysis considering more than two responses. For example, simultaneously studying not only quality of life and physical condition but also other variables such as mental state which is likewise reflected in the SF-36 questionnaire. In this connection, trivariate probit models are also available in the GJRM package (Filippou et al. [5]) although approximations have not yet been developed for continuous responses.

Another benefit of this type of model is that it allows for a broad family of response variables. In this study we have focused on two continuous response variables. However, this package also fits bivariate regression models with binary responses (where link functions are not restricted to only probit) or bivariate models with binary/discrete/continuous margins in the presence of associated responses/endogeneity. This could become very useful in the field of medicine. In fact, the authors of this manuscript are currently investigating the concordance of the different diagnostic criteria of diabetes. Two of these criteria, established by the American Diabetes Association (2014), are fasting glucose levels ($\geqslant 126$ mg/dL) and glycosylated haemoglobin ($A1C \geqslant 6.5$). On this basis we are working on the development of a bivariate binary model that allows us to investigate whether there is concordance between these two diagnostic criteria and determine if these threshold levels are the most appropriate. There are also other types of diagnostic criteria defined by the ADA such as the measure of plasma glucose at 2 h (≥ 200 mg/dL) in an oral glucose tolerance test. From a statistical point of view, we will work on the development of regression models for trivariate copulas with a view to simultaneously studying the effectiveness of these three criteria commonly used in clinical practice.

References

1. Acar EF, Craiu VR, Yao F (2013) Statistical testing of covariate effects in conditional copula models. Electron J Stat 7:2822–2850
2. Alonso J, Prieto L, Antó JM (1995) La versión española del SF-36 Health Survey (Cuestionario de Salud SF-36): Un instrumento para la medida de los resultados clínicos. Med Clín 104:771–776
3. Conn AR, Gould NI, Toint PL (2000) Trust region methods. MOS-SIAM series on optimization. SAIM, Philadelphia

4. Eilers PH, Marx BD (1996) Flexible smoothing with B-splines and penalties. Stat Sci 11:89–121
5. Filippou P, Marra G, Radice R (2017) Penalized likelihood estimation of a trivariate additive probit model. Biostatistics 18(3):569–585
6. Klein N, Kneib T (2016) Simultaneous inference in structured additive conditional copula regression models: a unifying Bayesian approach. Stat Comput 26:841–860
7. Marra G, Radice R (2017) Bivariate copula additive models for location, scale and shape. Comput Stat Data Anal 112:99–113
8. Mein G, Martikainen P, Hemingway H, Stansfeld S, Marmot M (2003) Is retirement good or bad for mental and physical health functioning? whitehall II longitudinal study of civil servants. J Epidemiol Commun Health 57:46–49
9. Neuberger GB (2003) Measures of fatigue: the fatigue questionnaire, fatigue severity scale, multidimensional assessment of fatigue scale, and short form 36 vitality (Energy Fatigue) subscale of the short form health survey. Arthr Care Res 49(S5):S175–S183
10. Rigby A, Stasinopoulos DM (2005) Generalized additive models for location, scale and shape (with discussion). Appl Stat 54:507–554
11. Sabeti A, Wei M, Craiu RV (2014) Additive models for conditional copulas. Stat 3(1):300–312
12. Sklar A (1973) Random variables, joint distributions, and copulas. Kybernetica 9:449–460
13. Sturgeon JA, Darnall BD, Kao MCJ, Mackey SC (2015) Physical and psychological correlates of fatigue and physical function: a Stanford-NIH open source pain registry study. J Pain 16:291–298
14. Vatter T, Chavez-Demoulin V (2015) Generalized additive models for conditional dependence structures. J Multivar Anal 141:147–167
15. Ware JE, Kosinski M, Dewey JE, Gandek B (2000) SF-36 health survey: manual and interpretation guide. Quality Metric, Boston
16. Ware JE Jr, Sherbourne CD (1992) The MOS 36-item short-form health survey (SF-36): I. Conceptual framework and item selection. Med Care 30(6):473–483
17. Wood SN (2003) Thin plate regression splines. J R Stat Soc Ser B 65:95–114
18. Wood SN (2006) Generalized additive models: an introduction with R. Chapman and Hall/CRC, London
19. Wu LJ, Wu MS, Lien GS, Chen FC, Tsai JC (2012) Fatigue and physical activity levels in patients with liver cirrhosis. J Clin Nurs 21:129–138
20. Yee TW, Wild CJ (1996) Vector generalized additive models. J R Stat Soc Ser B 58:481–493

Domain Mean Estimators Assisted by Nested Error Regression Models

María Dolores Esteban, Domingo Morales and María del Mar Rueda

Abstract This paper introduces estimators of domain means assisted by nested error regression models. The new estimators are modifications of empirical best linear unbiased predictors that takes into account the sampling weights. They are obtained by summing up the model-based predicted values adjusted by a weighted sum residuals. The paper studies the sampling-design properties of the introduced estimators by means of simulation experiments. The simulation results show that the new estimators present a good balance between sampling bias and mean squared error.

1 Introduction

Estimation of domain (subpopulation) indicators is an important objective in most surveys, especially in large surveys conducted by national statistical agencies. These agencies employ design-based estimation procedures whenever possible, that is, whenever the domain sample sizes are large enough for having precise estimates. If domain sample sizes are small (small areas), then insufficient precision is more likely to happen. The small area estimation (SAE) theory deals with this kind of estimation settings. See the book of Rao and Molina [8] for an introduction to SAE.

The design-based inference approach constructs estimators with good sampling distribution properties. In particular, domain direct estimators are calculated by using only the available domain data. Because of the lack of precision, these estimators might not be appropriate for estimating parameters of small areas. This is why survey practitioners often apply model-based or model-assisted small area estimators.

M. D. Esteban · D. Morales (✉)
Universidad Miguel Hernández de Elche, 03202 Elche, Spain
e-mail: d.morales@umh.es

M. D. Esteban
e-mail: md.esteban@umh.es

M. del Mar Rueda
Universidad de Granada, 18010 Granada, Spain
e-mail: mrueda@ugr.es

© Springer International Publishing AG 2018
E. Gil et al. (eds.), *The Mathematics of the Uncertain*, Studies in Systems,
Decision and Control 142, https://doi.org/10.1007/978-3-319-73848-2_14

The unit-level model-based approach (prediction theory) relies on superpopulation models, which assume that the target variable census vector is a realization of a random vector with a distribution function that incorporates the auxiliary information. Cassel et al. [2] and Valliant et al. [12] described the survey sampling prediction theory and showed that it gives a general framework for statistical inference on the characteristics of finite populations. Well-known estimators of population totals or means encountered in the classical theory, as expansion, ratio, regression, and other estimators, can be obtained as predictors under some special model in the general prediction theory.

The unit-level model-based approach is commonly used in SAE. The basic SAE unit-level model is the nested error regression (NER) model. Battese et al. [1] applied this model to the prediction of United States county crop areas using survey and satellite data. Since then, the empirical best linear unbiased predictors (EBLUP) of domains means based on the NER model are being widely applied. The model-based estimators, when the assumed model is correct, tend to be better than other estimators. However, when the assumed model is incorrect, the model-based estimators are biased and they can do worse than even the naïve estimators.

Särndal et al. [11] presented the model-assisted approach to inference in finite populations, where the superpopulation model is not the basis of the inferences. The model-assisted methodology considers the properties under the design-based distribution, but employs the model to motivate the choice of estimators. Under this approach, the generalized regression (GREG) estimators play a fundamental role. GREG estimation was introduced for domain estimation in Särndal [9], Hidiroglou and Särndal [5], and Särndal and Hidiroglou [10] and were developed further (including computational tools) in Estevao et al. [3]. More recently, Lehtonen and Veijanen [6, 7] discussed GREG estimators of means and proportions and presented empirical studies based on simulation experiments. This paper uses this approach for introducing estimators of small area means. The new estimators are obtained by summing up model-based predicted values and adjusting by design-based weighted sum of residuals. Thus, the model and the sampling design are used in the definition of the estimators.

The article is arranged as follows. Section 2 discusses some aspects of model-based domain estimation. It describes the EBLUP of a mean under a NER superpopulation model. Section 3 introduces the new model-assisted counterparts of the considered EBLUP and EBP estimators. Section 4 reports a Monte Carlo simulation experiment. Section 5 gives some conclusions.

2 Model-Based Estimation

Let U be a population of size N partitioned into D domains or areas $U_1, \ldots, U_D$ of sizes $N_1, \ldots, N_D$. The model-based approach relies on superpopulation models, which assume that the census vector

$$\boldsymbol{y} = \operatorname*{col}_{1 \le d \le D} (\boldsymbol{y}_d), \quad \boldsymbol{y}_d = \operatorname*{col}_{1 \le j \le N_D} (y_{dj}),$$

is a realization of a random vector following a superpopulation model that incorporates the auxiliary information $\boldsymbol{x}_{dj} = (x_{dj1}, \ldots, x_{djp})$, $j \in U_d$, $d = 1, \ldots, D$.

By using the column and mean operator, we define

$$\boldsymbol{X} = \operatorname*{col}_{1 \le d \le D} (\boldsymbol{X}_d), \quad \boldsymbol{X}_d = \operatorname*{col}_{1 \le j \le N_d} (\boldsymbol{x}_{dj}), \quad \overline{\boldsymbol{X}}_d = \frac{1}{N_d} \sum_{j=1}^{N_d} \boldsymbol{x}_{dj}, \quad \boldsymbol{\beta} = \operatorname*{col}_{1 \le k \le p} (\beta_k).$$

The NER superpopulation model is

$$y_{dj} = \boldsymbol{x}_{dj}\boldsymbol{\beta} + u_d + e_{dj}, \quad d = 1, \ldots, D, \ j = 1, \ldots, N_d, \tag{1}$$

where the random effects $\{u_d\}$ and the errors $\{e_{dj}\}$ are mutually independent with $u_d \sim N(0, \sigma_u^2)$ and $e_{dj} \sim N(0, \sigma_e^2)$. Let us define $\boldsymbol{e}_d = \operatorname*{col}_{1 \le j \le N_D} (e_{dj})$. Then, the model (1) can be written as

$$\boldsymbol{y}_d = \boldsymbol{X}_d\boldsymbol{\beta} + u_d + \boldsymbol{e}_d, \quad d = 1, \ldots, D.$$

The vectors $\boldsymbol{y}_d$ are independent with $\boldsymbol{y}_d \sim N(\boldsymbol{\mu}_d, \boldsymbol{V}_d)$, $\boldsymbol{\mu}_d = \boldsymbol{X}_d\boldsymbol{\beta}$ and $\boldsymbol{V}_d = \sigma_u^2 \boldsymbol{1}_{N_d} \boldsymbol{1}'_{N_d} + \sigma_e^2 \boldsymbol{I}_{N_d}$, where $\boldsymbol{1}_K = \operatorname*{col}_{1 \le j \le K} (1)$ and $\boldsymbol{I}_K = \operatorname*{diag}_{1 \le j \le K} (1)$ are the 1-column vector and the identity matrix of sizes K and $K \times K$ respectively.

In practice, inference is based on a subset (sample) $s = \cup_{d=1}^{D} s_d$ of the finite population U. Let $\boldsymbol{y}_{ds}$ be the sub-vector of $\boldsymbol{y}_d$ corresponding to sample elements and $\boldsymbol{y}_{dr}$ the sub-vector of $\boldsymbol{y}_d$ corresponding to the out-of-sample elements. Without lack of generality, we can write $\boldsymbol{y}_d = (\boldsymbol{y}'_{ds}, \boldsymbol{y}'_{dr})'$. Define also $\boldsymbol{X}_{ds}$ and $\boldsymbol{V}_{ds}$ as the corresponding decompositions of $\boldsymbol{X}_d$ and $\boldsymbol{V}_d$. The sample vector $\boldsymbol{y}_{ds}$ follows the corresponding submodel of (1), i.e.

$$y_{dj} = \boldsymbol{x}_{dj}\boldsymbol{\beta} + u_d + e_{dj}, \quad d = 1, \ldots, D, \ j = 1, \ldots, n_d, \tag{2}$$

where we change N and N_d by the sample counterparts n and n_d respectively. When $\sigma_e^2 > 0$ and $\sigma_u^2 > 0$ are known, the best linear unbiased estimator (BLUE) of $\boldsymbol{\beta}$ and the best linear unbiased predictor (BLUP) of u_d, $d = 1, \ldots, D$, are

$$\tilde{\boldsymbol{\beta}} = \left(\sum_{d=1}^{D} \boldsymbol{X}'_{ds} \boldsymbol{V}_{ds}^{-1} \boldsymbol{X}_{ds} \right)^{-1} \sum_{d=1}^{D} \boldsymbol{X}'_{ds} \boldsymbol{V}_{ds}^{-1} \boldsymbol{y}_{ds}, \quad \tilde{u}_d = \sigma_u^2 \boldsymbol{1}'_{n_d} \boldsymbol{V}_{ds}^{-1} \left(\boldsymbol{y}_{ds} - \boldsymbol{X}_{ds} \tilde{\boldsymbol{\beta}} \right). \tag{3}$$

Replacing σ_e^2 and σ_u^2 by estimators $\hat{\sigma}_e^2$ and $\hat{\sigma}_u^2$ in (3), the empirical BLUE (EBLUE) of $\boldsymbol{\beta}$ and the EBLUP of u_d, $d = 1, \ldots, D$, are

$$\hat{\beta} = \left(\sum_{d=1}^{D} X'_{ds} \hat{V}_{ds}^{-1} X_{ds} \right)^{-1} \sum_{d=1}^{D} X'_{ds} \hat{V}_{ds}^{-1} y_{ds}, \quad \hat{u}_d = \hat{\sigma}_u^2 \mathbf{1}'_{n_d} \hat{V}_{ds}^{-1} \left(y_{ds} - X_{ds} \hat{\beta} \right), \quad (4)$$

where $\hat{V}_d = \hat{\sigma}_u^2 \mathbf{1}_{n_d} \mathbf{1}'_{n_d} + \hat{\sigma}_e^2 I_{n_d}$. The distribution of y_{dr}, given the sample data y_s, is

$$y_{dr}|y_s \sim y_{dr}|y_{ds} \sim N(\mu_{dr|s}, V_{dr|s}), \tag{5}$$

where

$$\mu_{dr|s} = X_{dr}\beta + \sigma_u^2 \mathbf{1}_{N_d-n_d} \mathbf{1}'_{n_d} V_{ds}^{-1}(y_{ds} - X_{ds}\beta),$$

$$V_{dr|s} = \sigma_u^2(1 - \gamma_d)\mathbf{1}_{N_d-n_d} \mathbf{1}'_{N_d-n_d} + \sigma_e^2 I_{N_d-n_d}, \quad \gamma_d = \frac{n_d \sigma_u^2}{n_d \sigma_u^2 + \sigma_e^2}.$$

Under the conditioned distribution (5), the predicted values are

$$\hat{y}_{ds}^{eb} = y_{ds}, \quad \hat{y}_{dr}^{eb} = \hat{\mu}_{dr|s} = X_{dr}\hat{\beta} + \hat{\sigma}_u^2 \mathbf{1}_{N_d-n_d} \mathbf{1}'_{n_d} \hat{V}_{ds}^{-1}(y_{ds} - X_{ds}\hat{\beta}),$$

or equivalently $\hat{y}_{dj}^{eb} = y_{dj}$ if $j \in s_d$ and $\hat{y}_{dj}^{eb} = x_{dj}\hat{\beta} + \hat{u}_d$ if $j \in r_d = U_d - s_d$. The EBLUP of $\overline{Y}_d$ is

$$\hat{\overline{Y}}_d^{eb} = \frac{1}{N_d} \sum_{j=1}^{N_d} \hat{y}_{dj}^{eb2} = \frac{1}{N_d} \sum_{j \in s_d} y_{dj} + \frac{1}{N_d} \sum_{j \in r_d} \{x_{dj}\hat{\beta} + \hat{u}_d\}$$

$$= (1 - f_d)\left[\overline{X}_d \hat{\beta} + \hat{u}_d\right] + f_d\left[\hat{\overline{Y}}_d + (\overline{X}_d - \hat{\overline{X}}_d)\hat{\beta}\right],$$

where $\hat{\overline{Y}}_d = \frac{1}{n_d} \sum_{j \in s_d} y_{dj}$, $\hat{\overline{X}}_d = \frac{1}{n_d} \sum_{j \in s_d} x_{dj}$, $f_d = \frac{n_d}{N_d}$.

3　Model-Assisted Estimation

The design-based approach relies on the sampling distribution. Let $s \subset U$ be a random sample of fixed size n drawn according to a specified sampling design $\pi(s)$ and let $s_1, \ldots, s_D$ be the corresponding domain subsamples of sizes $n_1, \ldots, n_D$. The first-order inclusion probability is the probability of obtaining the unit j of domain d, while sampling from the population according to the sampling design. It is $\pi_{dj} = \sum_{j \in s_d} \pi(s)$.

Hájek [4] proposed the ratio estimators of means

$$\hat{\overline{Y}}_d^{dir} = \frac{1}{\hat{N}_d^{dir}} \sum_{j \in s_d} \pi_{dj}^{-1} y_{dj}, \quad \hat{\overline{X}}_d^{dir} = \frac{1}{\hat{N}_d^{dir}} \sum_{j \in s_d} \pi_{dj}^{-1} x_{dj}, \quad \hat{N}_d^{dir} = \sum_{j \in s_d} \pi_{dj}^{-1},$$

which are direct estimators as they are calculated by employing only the corresponding domain data.

We follow Cassel et al. [2] and introduce a design-unbiased estimator of the domain mean $\overline{Y}_d$ assisted by model (1). The estimator is obtained by summing up the predicted values $x_{dj}\hat{\beta} + \hat{u}_d$ and adjusting by residuals, i.e.

$$
\hat{\overline{Y}}_d^{ma} = \frac{1}{N_d}\left\{ \sum_{j \in U_d}(x_{dj}\hat{\beta} + \hat{u}_d) + \sum_{j \in s_d} \pi_{dj}^{-1}(y_{dj} - x_{dj}\hat{\beta} - \hat{u}_d) \right\}
$$

$$
= \overline{X}_d\hat{\beta} + \hat{u}_d + \frac{\hat{N}_d^{dir}}{N_d}\left(\hat{\overline{Y}}_d^{dir} - \hat{\overline{X}}_d^{dir}\hat{\beta} - \hat{u}_d \right).
$$

4 Simulations

This section presents a design-based simulation experiment, where stratified random samples are drawn from a fixed artificial population. The target parameter is the domain mean. The simulation is designed to compare the model-assisted estimator $\hat{\overline{Y}}_d^{ma}$ with its model-based counterpart $\hat{\overline{Y}}_d^{eb}$.

The artificial population is constructed with size $N = 43900$ and it is divided in 5 strata of sizes $N_1 = 8820$, $N_2 = 9120$, $N_3 = 9100$, $N_4 = 8760$ and $N_5 = 8100$ respectively. Each strata is partitioned in domains. The numbers of domains within each strata are $D_1 = 18$, $D_2 = 16$, $D_3 = 14$, $D_4 = 12$ and $D_5 = 10$, with $D = D_1 + \cdots + D_5 = 70$. Auxiliary variables are generated from normal distributions $x_0 \sim N(10, 1)$ and $x_1 \sim N(\mu_t, 1)$, with means $\mu_t = 8, 9, 10, 11, 12$ in strata $t = 1, \ldots, 5$ respectively.

The y-variables, y_0, y_1 and y_{01}, are drawn from the linear mixed models

$$
M_k : y_{k,dj} = \beta_0 + \beta_1 x_{k,dj} + u_d + e_{dj}, \quad k = 0, 1, \ d = 1, \ldots, D, \ j = 1, \ldots, N_d,
$$

$$
M_{01} : y_{01,dj} = \beta_0 + \beta_1 x_{0,dj} + \beta_2 x_{1,dj} + u_d + e_{dj}, \quad d = 1, \ldots, D, \ j = 1, \ldots, N_d,
$$

where $\beta_0 = 0$, $\beta_1 = 2$, $\beta_2 = 2$, the random effects are i.i.d. $u_d \sim N(0, \sigma_u^2)$ with $\sigma_u^2 = 1$, the random errors are i.i.d. $e_{dj} \sim N(0, \sigma_e^2)$ with $\sigma_e^2 = 1$, and they are all independent. For each variable $y \in \{y_0, y_1, y_{01}\}$, the target parameters are the domain means $\overline{Y}_d = \frac{1}{N_d}\sum_{j=1}^{N_d} y_{dj}$, $d = 1, \ldots, D$, and their global means, $\overline{Y} = \frac{1}{N}\sum_{d=1}^{D}\sum_{j=1}^{N_d} y_{dj}$, are $\overline{Y} = 19.932$, $\overline{Y} = 19.851$ and $\overline{Y} = 39.804$ respectively.

The simulation is carried out under a stratified sampling design with optimal Neyman allocation (S). The global sample size is $n = 3050$. As $D = 70$, there will be around $n/D = 43.57$ sampled units per domain in average. For each strata t, $t = 1, 2, 3, 4, 5$, the sampling weights depend on the stratum variance of the target variable. Table 1 presents the sampling weights. As the sampling weights depend on the target variable y, the S design is informative.

Table 1 Sampling weights by strata

Variable	1	2	3	4	5
y_0	14.459	13.653	14.399	14.723	14.862
y_1	14.180	13.797	14.399	14.624	15.112
y_{01}	14.365	13.924	14.422	14.797	14.516

Table 2 ABIAS (left) and MSE (right)

$M(y, x)$	DIR	EB	MA	DIR	EB	MA
$M(y_0, x_0)$	0.0086	0.0187	0.0041	0.1160	0.0227	0.0231
$M(y_1, x_1)$	0.0088	0.0193	0.0040	0.1168	0.0226	0.0231
$M(y_{01}, x_0)$	0.0113	0.0372	0.0087	0.2134	0.1167	0.1181
$M(y_{01}, x_1)$	0.0113	0.0785	0.0094	0.2134	0.1041	0.1161
$M(y_1, x_0)$	0.0088	0.0368	0.0090	0.1168	0.1152	0.1167
$M(y_0, x_1)$	0.0086	0.0833	0.0089	0.1160	0.1044	0.1151

The simulation calculates empirical biases and MSEs of small area estimators of means. This simulation shows that model-based and model-assisted estimators are more robust from model misspecification when the sampling design is less informative. We calculate the Hájek (DIR) and the EB and MA estimators of domain means based or assisted by the NER model $M(y, x)$ with y and x as target and auxiliary variables respectively.

Table 2 presents the empirical average absolute biases, ABIAS, and the empirical average mean squared errors MSE. The table is divided in three parts. The first part concerns the results for the correct models $M(y_0 \mid x_0)$ and $M(y_1 \mid x_1)$. The second and the third part give the simulation results of the incomplete models $M(y_{01} \mid x_0)$ and $M(y_{01} \mid x_1)$ and the incorrect models $M(y_1 \mid x_0)$ and $M(y_0 \mid x_1)$ respectively.

When the correct or incomplete models are employed for constructing the EB and MA estimators, the EB estimator has the greatest biases and the lowest MSEs. The MA estimator has the lowest biases and similar MSEs to the EB estimator. The direct estimator has the greatest MSEs. See Figs. 1 and 2.

When the incorrect models are employed for constructing the EB and MA estimators the EB estimator has the greatest biases. The MA and direct estimators have similar biases. Concerning MSE, all the estimators behaves similarly. See also Fig. 3.

Figures 1, 2 and 3 present the boxplots of $BIAS_d$ and MSE_d, $d = 1, \ldots, D$, for the target parameters $\overline{Y}_{0,d}$ and $\overline{Y}_{01,d}$, when the EB and MA estimators rely on the correct model $M(y_0 \mid x_0)$, the incomplete model $M(y_{01} \mid x_0)$ and the incorrect model $M(y_0 \mid x_1)$ respectively.

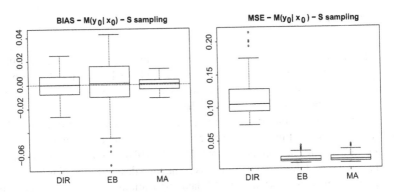

Fig. 1 Boxplots of $BIAS_d$ (left) and MSE_d (right), $d = 1, \ldots, D$, for $M(y_0 \mid x_0)$

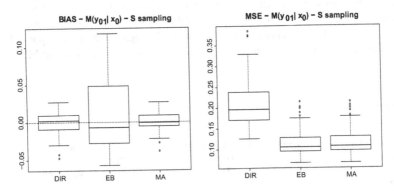

Fig. 2 Boxplots of $BIAS_d$ (left) and MSE_d (right), $d = 1, \ldots, D$, for $M(y_{01} \mid x_0)$

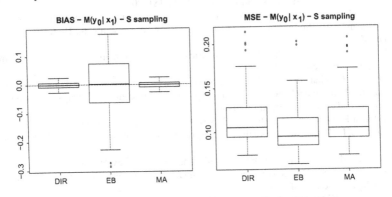

Fig. 3 Boxplots of $BIAS_d$ (left) and MSE_d (right), $d = 1, \ldots, D$, for $M(y_0 \mid x_1)$

5 Conclusions

This paper introduces model-assisted estimators of domain means. The new estimators are assisted by the NER models and they are unbiased under the design-based distribution. The simulation experiment shows that they are more robust against deviations from model specifications than the EBLUPs.

The Hájek (DIR) estimator uses only the considered domain information. It is basically an unbiased estimator with respect to the sampling design distribution but with a big variance in small area problems. The EBLUP does not contains calibrated sampling weights and therefore it could has some high design-based bias because of non response. This is a drawback in real data applications. The MA estimators try to collect the good properties of DIR and EBLUP estimators. They are constructed from the NER model and therefore they introduce the auxiliary information in the estimation process. They can employ the calibrated weights in a design-based bias correction term. The correction term gives protection against the non response bias, so that the new estimator hast a good balance between sampling bias and MSE.

References

1. Battese GE, Harter RM, Fuller WA (1988) An error-components model for prediction of county crop areas using survey and satellite data. J Am Stat Assoc 83:28–36
2. Cassel C, Särndal C, Wretman J (1977) Foundations of inference in survey sampling. Wiley, New York
3. Estevao VM, Hidiroglou MA, Särndal CE (1995) Methodological principles for a generalized estimation system at statistics Canada. J Off Stat 11:181–204
4. Hájek J (1971) Comment on an essay on the logical foundations of survey sampling, part one. In: Godambe VP, Sprott DA (eds) The foundations of survey sampling. Holt, Rinehart, and Winston, Toronto
5. Hidiroglou MA, Särndal CE (1985) An empirical study of some regression estimators for small domains. Surv Method 11:65–77
6. Lehtonen R, Veijanen A (2009) Design-based methods of estimation for domains and small areas. In: Pfeffermann D, Rao CR (eds) Sample surveys: inference and analysis. Handbook of statistics, vol 29B. Elsevier, Amsterdam
7. Lehtonen R, Veijanen A (2016) Model-assisted method for small area estimation of poverty indicators. In: Pratesi M (ed) Analysis of poverty data by small area estimation. Wiley, Chichester
8. Rao JNK, Molina I (2015) Small area estimation, 2nd edn. Wiley, Hoboken
9. Särndal CE (1984) Design-consistent versus model-dependent estimation for small domains. J Am Stat Assoc 79:624–631
10. Särndal CE, Hidiroglou MA (1989) Small domain estimation: a conditional analysis. J Am Stat Assoc 84:266–275
11. Särndal CE, Swensson B, Wretman J (1992) Model assisted survey sampling. Springer, New York
12. Valliant R, Dorfman AH, Royall RR (1999) Finite population sampling and inference: a prediction approach. Wiley, New York

Robust Approaches for Fuzzy Clusterwise Regression Based on Trimming and Constraints

Luis Angel García-Escudero, Alfonso Gordaliza,
Francesca Greselin and Agustín Mayo-Iscar

Abstract Three different approaches for robust fuzzy clusterwise regression are reviewed. They are all based on the simultaneous application of trimming and constraints. The first one follows from the joint modeling of the response and explanatory variables through a normal component fitted in each cluster. The second one assumes normally distributed error terms conditional on the explanatory variables while the third approach is an extension of the Cluster Weighted Model. A fixed proportion of "most outlying" observations are trimmed. The use of appropriate constraints turns these problem into mathematically well-defined ones and, additionally, serves to avoid the detection of non-interesting or "spurious" linear clusters. The third proposal is specially appealing because it is able to protect us against outliers in the explanatory variables which may act as "bad leverage" points. Feasible and practical algorithms are outlined. Their performances, in terms of robustness, are illustrated in some simple simulated examples.

1 Introduction

The detection of clusters around linear subspaces, instead of just around points or centroids, is often needed in Cluster Analysis. This problem is meaningful, not only because clusters are frequently arranged this way, but also because sometimes it is

L. A. García-Escudero · A. Gordaliza · A. Mayo-Iscar
Departamento de Estadística e Investigación Operativa and IMUVA,
Universidad de Valladolid, Valladolid, Spain
e-mail: lagarcia@eio.uva.es

A. Gordaliza
e-mail: alfonsog@eio.uva.es

A. Mayo-Iscar
e-mail: agustinm@eio.uva.es

F. Greselin (✉)
Department of Statistics and Quantitative Methods, Milano-Bicocca University,
Milano, Italy
e-mail: francesca.greselin@unimib.it

© Springer International Publishing AG 2018
E. Gil et al. (eds.), *The Mathematics of the Uncertain*, Studies in Systems,
Decision and Control 142, https://doi.org/10.1007/978-3-319-73848-2_15

155

interesting to discover different relations between a response variable and some other explanatory variables within each cluster. These problems are commonly known as clusterwise linear regression or switching regression models. Some others seminal references are [4, 19, 21, 26]. All those "hard" or 0–1 clustering procedures partition the data into G completely disjoint clusters. Alternatively, fuzzy clustering methods provide nonnegative membership values of observations to clusters and overlapping clusters are so generated [2, 17, 25]. This fuzzy approach can be certainly useful in clusterwise regression applications. There already exist many proposals addressing fuzzy clustering around linear subspaces. For instance, [17] provides an adaptation of the fuzzy c-means in [2] by minimizing a weighted sum of distances of each observation from the estimated regression line and where these weights depend on the fuzzy membership values. See also [18, 29] and references therein.

Robustness is also a desirable property for (fuzzy) clustering techniques due to the well-know harmful effect that (even a small fraction) outlying observations may have in them. Several methods have been recently proposed to improve clustering techniques robustness performance. For instances, many proposals can be found in [6, 10, 22] (hard) and in [1, 3] (fuzzy).

In this work, we are going to review three recent approaches for robust fuzzy clusterwise regression derived from considering a maximum likelihood approach with trimming and constraints. These methods can be seen as extensions of that introduced in [7]. Trimming is probably the simpler way to achieve robustness, being also very easy to understand. Particularly, we consider an impartial trimming approach, where the adjective "impartial" means that the data set itself tell us which are the observations that should be trimmed, as in [9]. When an observation with index i is detected as an outlier, we set membership values $u_{ig} = 0$ for every $g = 1, \ldots, G$. This is in contrast with [29] which sets $u_{ig} = 1/G$ for outlying observations. A fuzzy Classification Maximum Likelihood approach is applied in the three considered approaches. The maximization of fuzzified likelihoods is not a new idea in fuzzy clustering [15, 24, 27, 28]. It is important to fix some type of constraint on the scatter parameters, because, otherwise, that maximization is a mathematically ill-defined problem. Therefore, appropriate constraints on the scatter parameters must be added. These constraints are also useful to avoid the detection of non-interesting ("spurious") local maxima.

In the three reviewed methods, the third one is particularly appealing because it simultaneously protects us against "vertical outliers" and even "bad leverage" points. This approach, recently introduced in [13], is a trimmed and fuzzified version of the Cluster Weighted Model (CWM) in [14].

2 Three Different Approaches

Let $\widetilde{\mathbf{X}} = (\mathbf{X}', Y)'$ be a random vector in $\mathbb{R}^d \times \mathbb{R}$, where the first d components $\mathbf{X}$ are the values taken by the explanatory variables or covariates, and Y is the value taken by a response variable. Let us assume that $\{\widetilde{\mathbf{x}}_i\}_{i=1}^n = \{(\mathbf{x}_i', y_i)'\}_{i=1}^n$ is a random sample of size n, drawn from $\widetilde{\mathbf{X}}$. We use the notation $\phi_d(\cdot; \mathbf{m}, \mathbf{S})$ for the density of the d-variate Gaussian distribution with mean vector $\mathbf{m}$ and covariance matrix $\mathbf{S}$ and $\{\lambda_l(\mathbf{S})\}_{l=1}^d$ are the set of eigenvalues of the $d \times d$ matrix $\mathbf{S}$.

2.1 FTCLUST-Based Approach

The simplest approach follows from the application of the FTCLUST methodology introduced in [7] in dimension $d + 1$. We propose maximizing

$$\sum_{i=1}^n \sum_{g=1}^G u_{ig}^m \log \left(\pi_g \phi_{d+1}(\widetilde{\mathbf{x}}_i; \widetilde{\boldsymbol{\mu}}_g, \widetilde{\boldsymbol{\Sigma}}_g)\right), \tag{1}$$

where the membership values $\mathbf{u}_{ig} \in [0, 1]$ are required to satisfy

$$\sum_{g=1}^G u_{ig} = 1 \text{ if } i \in \mathscr{I} \text{ and } \sum_{g=1}^G u_{ig} = 0 \text{ if } i \notin \mathscr{I}, \tag{2}$$

for a subset $\mathscr{I} \subset \{1, 2, \ldots, n\}$ with $\#\mathscr{I} = [n(1-\alpha)]$. The parameter $\alpha \in [0, 1)$ is the fixed trimming level and $m \geq 1$ is the fuzzifier parameter. Note that the observations with indexes outside $\mathscr{I}$ do not contribute to the summation in (1). The target function in (1) is unbounded as we can easily see just by taking $|\widetilde{\boldsymbol{\Sigma}}_g| \to 0$. Thus, as done in [7], we introduce an additional constraint when maximizing (1) that forces the set of the eigenvalues of the scatter matrices to satisfy

$$\lambda_{l_1}(\widetilde{\boldsymbol{\Sigma}}_{g_1}) \leq c\lambda_{l_2}(\widetilde{\boldsymbol{\Sigma}}_{g_2}) \quad \text{for } 1 \leq l_1 \neq l_2 \leq d+1 \text{ and } 1 \leq g_1 \neq g_2 \leq G. \tag{3}$$

This type of constraints are an extension of those in [9, 20]. The use of constraints on the scatter parameters goes back to Hathaway's seminal work [16].

Let $\widetilde{\boldsymbol{\mu}}_g$ and $\widetilde{\boldsymbol{\Sigma}}_g$ be the vectors and matrices obtained from the previous constrained optimization problem with

$$\widetilde{\boldsymbol{\mu}}_{\mathbf{g}} = \begin{pmatrix} \mu_1^g \\ \mu_2^g \end{pmatrix} \text{ and } \widetilde{\boldsymbol{\Sigma}}_g = \left(\begin{array}{c|c} \Sigma_{11}^g & \Sigma_{12}^g \\ \hline \Sigma_{21}^g & \Sigma^g \end{array}\right). \tag{4}$$

From these (optimal) vectors and matrices, we obtain G linear structures as

$$y = \mu_2^g + \Sigma_{21}^g (\Sigma_{11}^g)^{-1}(\mathbf{x} - \mu_1^g) \text{ for } g = 1, \ldots, G. \tag{5}$$

The constant $1 \leq c < \infty$ guarantees that the constrained maximization of (1) is a mathematically well-defined problem and serves to avoid the detection of "spurious" local maximizers. Some weights π_g are also included in (8). They are useful when the number of clusters is misspecified, because they can be set close to 0 whenever

G is larger than needed [7, 9]. More details on the weights will be given later. We are thus considering a fuzzy classification EM-type approach as in [28].

The problem with that approach is that linear clusters are generally, by definition, elongated clusters. Therefore, eigenvalues close to 0 on the $\widetilde{\boldsymbol{\Sigma}}_g$ matrices often appear in most of clusterwise regression problems. This fact implies that large c values for the eigenvalues ratio constraint are required. Unfortunately, those large c values do not protect us correctly against "spurious" local maximizers. Moreover, the FTCLUST's good robustness properties are lost with such large c values.

2.2 Robust Fuzzy Clusterwise Regression

A different approach, which directly takes into account the underlying linear relations within each group is reviewed in this section. In clusterwise regression, it is frequently assumed that the conditional relation between Y given $\mathbf{X} = \mathbf{x}$ in the gth group can be written as $Y = \mathbf{b}_g'\mathbf{x} + b_g^0 + \varepsilon_g$ with $\varepsilon_g \sim N_1(0, \sigma_g^2)$. In that case, a robust fuzzy clusterwise regression approach can be derived through the maximization of

$$\sum_{i=1}^{n} \sum_{g=1}^{G} u_{ig}^m \log\left(\pi_g \phi_1\left(y_i; \mathbf{b}_g'\mathbf{x}_i + b_g^0, \sigma_g^2\right)\right), \tag{6}$$

where the u_{ig} membership values and the π_g weights satisfy the same requirements as in Sect. 2.1; vector $\mathbf{b}_g$ and the constant b_g^0 are, respectively, the regression slope and the intercept for the gth cluster. Again, constraints on the residual variances can be set as

$$\sigma_{g_1}^2 \leq c_\varepsilon \sigma_{g_2}^2 \qquad \text{for every } 1 \leq g_1 \neq g_2 \leq G, \tag{7}$$

for a fixed $1 \leq c_\varepsilon < \infty$ constant. These constraints again convert the maximization of (6) into a mathematically well-defined problem (see what happens when $\sigma_g^2 \to 0$). This approach has been introduced in [11].

We have applied this methodology, for a simulated data set, with $\alpha = 0$ and $c_\varepsilon = 10$ in Fig. 1a and with $\alpha = 0.05$ and $c_\varepsilon = 10$ in Fig. 1b. The simulated data set includes a small 5% fraction of background scattered noise. As seen in Fig. 1a, the detected linear structures when $\alpha = 0$ are not the correct ones and many misclassified observations are found.

This approach provides improved robustness performance by applying trimming that certainly protect us against "vertical outliers" (outliers only in y). However, as will be seen in Sect. 2.3, it does not provide great protection against "leverage points" (outliers in $\mathbf{x}$). It is well known that leverage points can be extremely harmful in Regression Analysis. Additional protection, in that case, can be obtained by applying a "second trimming" stage as described in [10], which can be straightforwardly adapted to the fuzzy clustering framework.

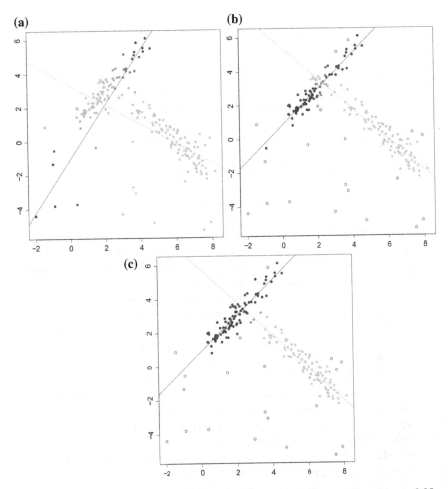

Fig. 1 **a** Fuzzy clusterwise regression with $\alpha = 0$. **b** Fuzzy clusterwise regression with $\alpha = 0.05$. **c** Fuzzy robust CWM with $\alpha = 0.05$. Fuzzy membership values are represented by using a mixture of red and green colors (grayscale converted). Trimmed observations are represented by empty black circles

2.3 Robust Fuzzy Cluster-Weighted Model (CWM)

Finally, a third approach is obtained throughout the "fuzzification" and "robustifica-tion" of the Cluster Weighted Model (CWM in the sequel) introduced in [14]. This approach has been recently proposed in [13] as a fuzzification of the "hard" robust CWM in [12]. We just focus on the the linear CWM with Gaussian components where the conditional relationship between Y given $\mathbf{X} = \mathbf{x}$ in the gth group is expressed by $Y = \mathbf{b}'_g \mathbf{x} + b^0_g + \varepsilon_g$ with $\varepsilon_g \sim N_1(0, \sigma^2_g)$ but we also assume that $\mathbf{X} \sim N_d(\boldsymbol{\mu}_g, \boldsymbol{\Sigma}_g)$.

Under these assumptions, we now consider the maximization of

$$\sum_{i=1}^{n}\sum_{g=1}^{G} u_{ig}^m \log \left(\pi_g \phi_1(y; \mathbf{b}_g' \mathbf{x}_i + b_g^0, \sigma_g^2) \phi_d(\mathbf{x}_i; \boldsymbol{\mu}_g, \boldsymbol{\Sigma}_g)\right), \tag{8}$$

with the same notation as in the statements of the two previous problems. We have that (8) is unbounded, and consequently, we introduce two further constraints as done in [12]. The first one has to do with the eigenvalues of the $\boldsymbol{\Sigma}_g$ matrices throughout

$$\lambda_{l_1}(\boldsymbol{\Sigma}_{g_1}) \leq c_X \lambda_{l_2}(\boldsymbol{\Sigma}_{g_2}) \quad \text{for every } 1 \leq l_1 \neq l_2 \leq d \text{ and } 1 \leq g_1 \neq g_2 \leq G. \tag{9}$$

A second constraint is added on the regression error terms as

$$\sigma_{g_1}^2 \leq c_\varepsilon \sigma_{g_2}^2 \quad \text{for every } 1 \leq g_1 \neq g_2 \leq G. \tag{10}$$

Notice that the two (not necessarily equal) constants $1 \leq c_X < \infty$ and $1 \leq c_\varepsilon < \infty$ serve to avoid "spurious" solutions whenever they assume moderate values. Moreover, a very flexible methodology is obtained because of the asymmetric treatment given to the marginal and conditional distributions.

Figure 1c shows the results of applying the fuzzy robust CWM with $\alpha = 0.05$ and $c_X = c_\varepsilon = 10$ for the same simulated data set as above. The methodology in Sect. 2.2 was perfectly able to recover the two underlying linear structures (recall Fig. 1b). However, the cluster assignments are not so satisfactory because some observations which clearly belong to the cluster in the left have higher membership values to the cluster in the right. This issue is due to the fact that they are very close to the regression line fitted by using mainly the observations in the cluster in the right when this line is being elongated. On the other hand, the fuzzy robust CWM take advantage of the information conveyed in the marginal distribution of $\mathbf{X}$ and so it is able to obtain more sensible membership values.

A second interesting feature of this fuzzy robust CWM is that it addresses the previously commented problems with "bad leverage" points in a very natural way because these observations take anomalous values on the explanatory variables. Therefore, their contribution to (8) is not very large and they are trimmed. For instance, we see in Fig. 2a how a 5% fraction of concentrated observations ($y \simeq 4$) are acting as bad leverage points when using the fuzzy clusterwise regression even though we had chosen a trimming level $\alpha = 0.05$ (equal to true contamination level) for it. The robust fuzzy CWM, with the same trimming level, successfully trim bad leverage observations.

3 Algorithms and Tuning Parameters

In this section, we briefly outline the proposed algorithms to implement the previously reviewed approaches. Note that the target function in all of them can be written as

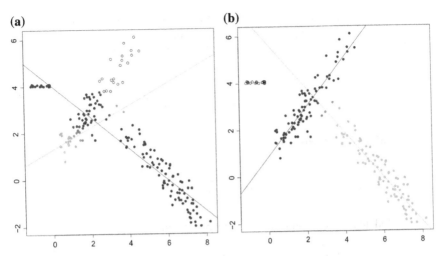

Fig. 2 **a** Robust fuzzy clusterwise regression with $\alpha = 0.05$ and $c_\varepsilon = 10$ for a data set with a 5% fraction of concentrated noise ($y \simeq 4$). **b** Robust fuzzy CWM with $\alpha = 0.05$ and $c_X = c_\varepsilon = 10$ for the same data set

$$\sum_{i=1}^{n} \sum_{g=1}^{G} u_{ig}^{m} \log(\pi_g \varphi(\widetilde{\mathbf{x}}_i; \theta_g)), \tag{11}$$

where the function $\varphi(\cdot)$ and the set of parameters θ_g depend on the specific method.

1. *Initialization*: Initial random values are assigned to parameters θ_g and π_g. This is achieved by drawing small random subsamples from the original data set, and using them to estimate initial parameters values.
2. *Iterative steps*: Repeat the following steps until convergence or reaching a maximum number of iterations:

   2.1. *Membership values*: If $\max_{g=1,\ldots,G} \pi_g \varphi(\widetilde{\mathbf{x}}_i; \theta) \geq 1$, then

$$u_{ig} = I\{\pi_g \varphi(\widetilde{\mathbf{x}}_i; \theta_g) = \max_{q=1,\ldots,k} \pi_q \varphi(\widetilde{\mathbf{x}}_i; \theta_q)\}, \tag{12}$$

   where $I\{\cdot\}$ is the 0–1 indicator function. If $\max_{g=1,\ldots,G} \pi_g \varphi(\widetilde{\mathbf{x}}_i; \theta_g) < 1$, we set

$$u_{ig} = \left(\sum_{q=1}^{G} \left(\frac{\log(\pi_g \varphi(\widetilde{\mathbf{x}}_i; \theta_g))}{\log(\pi_q \varphi(\widetilde{\mathbf{x}}_i; \theta_q))} \right)^{\frac{1}{m-1}} \right)^{-1}. \tag{13}$$

   2.2. *Trimmed observations*: Compute

$$r_i = \sum_{g=1}^{G} u_{ig}^{m} \log(p_g \varphi_g(\mathbf{x}_i, y_i; \theta)) \tag{14}$$

and sort them as $r_{(1)} \leq r_{(2)} \leq \ldots \leq r_{(n)}$. Set membership values $u_{ig} = 0$, $g = 1, \ldots, G$, for all the indexes i such that $r_i < r_{([n\alpha])}$.

2.3. *Update parameters*: Use previous u_{ig} to update weights as

$$\pi_g = \sum_{i=1}^{n} u_{ig}^m \Big/ \sum_{i=1}^{n} \sum_{g=1}^{G} u_{ig}^m, \qquad (15)$$

and $\boldsymbol{\mu}_g$ (analogously, $\widetilde{\boldsymbol{\mu}}_g$) as

$$\boldsymbol{\mu}_g = \frac{\sum_{i=1}^{n} u_{ig}^m \mathbf{x}_i}{\sum_{i=1}^{n} u_{ig}^m}. \qquad (16)$$

Update intercepts and slope vectors by computing

$$\mathbf{b}_g = \left(\frac{\sum_{i=1}^{n} u_{ig}^m \mathbf{x}_i \mathbf{x}_i'}{\sum_{i=1}^{n} u_{ig}^m} - \frac{\sum_{i=1}^{n} u_{ig}^m \mathbf{x}_i}{\sum_{i=1}^{n} u_{ig}^m} \cdot \frac{\sum_{i=1}^{n} u_{ig}^m \mathbf{x}_i'}{\sum_{i=1}^{n} u_{ig}^m} \right)^{-1}$$
$$\cdot \left(\frac{\sum_{i=1}^{n} u_{ig}^m y_i \mathbf{x}_i}{\sum_{i=1}^{n} u_{ig}^m} - \frac{\sum_{i=1}^{n} u_{ig}^m y_i}{\sum_{i=1}^{n} u_{ig}^m} \cdot \frac{\sum_{i=1}^{n} u_{ig}^m \mathbf{x}_i}{\sum_{i=1}^{n} u_{ig}^m} \right),$$

and

$$b_g^0 = \frac{\sum_{i=1}^{n} u_{ig}^m y_i}{\sum_{i=1}^{n} u_{ig}^m} - \mathbf{b}_g' \frac{\sum_{i=1}^{n} u_{ig}^m \mathbf{x}_i}{\sum_{i=1}^{n} u_{ig}^m}. \qquad (17)$$

All previous formulae are typical in fuzzy clustering. The most difficult part is how to update the constrained scatter parameters. To update σ_g^2 and Σ_g^2, we start from the weighted sample covariance matrices

$$T_g = \frac{\sum_{i=1}^{n} u_{ig}^m (\mathbf{x}_i - \boldsymbol{\mu}_g)(\mathbf{x}_i - \boldsymbol{\mu}_g)'}{\sum_{i=1}^{n} u_{ig}^m}, \qquad (18)$$

and the weighted residual variances

$$d_g^2 = \frac{\sum_{i=1}^{n} u_{ig}^m (y_i - b_g^0 - \mathbf{x}_i' \mathbf{b}_g)^2}{\sum_{i=1}^{n} u_{ig}^m}. \qquad (19)$$

Then, to update Σ_g (analogously, $\widetilde{\Sigma}_g$), the singular-value decomposition $T_g = U_g' E_g U_g$ is considered, with U_g being an orthogonal matrix and $E_g = \text{diag}(e_{g1}, e_{g2}, \ldots, e_{gd})$ a diagonal matrix. As done in [7, 8], these eigenvalues must be optimally truncated. The optimal truncation value is obtained by minimizing a real valued function. Analogously, in case that the d_j^2 error residual variances do not satisfy the required constraint, the d_j^2 must be optimally truncated too [11].

3. *Return* the set of parameters θ_g yielding the highest value of (11) obtained after all the random initializations and iterative steps.

Note that trimming is done through "concentration steps" [23] and imposing the required constraint on the scatter parameters is an important ingredient of this algorithm.

As can be seen, several parameters have to be chosen when applying the proposed methods in real data problems. We observed that the estimated parameters θ_g do not necessarily depend on all the tuning parameters, in a critical way. For instance, usually a trimming level slightly greater than the one needed to remove contamination is not problematic. However, monitoring the sizes of the sorted r_i values in (14) is useful to set sensible α values. Regarding the constraints on the scatter parameters, our suggestion is not choosing excessively high values for both c_X and c_ε (at least in the approaches described in Sects. 2.2 and 2.3). The choice of the fuzzifier parameter m depends on the desired degree of fuzziness in the clustering solution. Unfortunately, as happens with other likelihood-based fuzzy clustering approaches, the effect of m is affected by the scale of the measured variables (see [7, 11]). The joint monitoring of the proportions of "hard assignments" and "relative entropies" ($\sum_{g=1}^{G} \sum_{i=1}^{n} u_{ig} \log u_{ig} / [n(1 - \alpha)] \log(G)$) provide useful heuristical tools aimed at addressing this issue.

References

1. Banerjee A, Davé RN (2012) Robust clustering. WIREs Data Min Know Discov 2:29–59
2. Bezdek JC (1981) Pattern recognition with fuzzy objective function algoritms. Plenum Press, New York
3. Davé RN, Krishnapuram R (1997) Robust clustering methods: a unified view. IEEE Trans Fuzzy Syst 5:270–293
4. DeSarbo WS, Cron WL (1988) A maximum likelihood methodology for clusterwise linear regression. J Classif 5:249–282
5. Dotto F, Farcomeni A, García-Escudero LA, Mayo-Iscar A (2017) A fuzzy approach to robust clusterwise regression. Adv Data Anal Classif 11(4):691–710
6. Farcomeni A, Greco L (2015) Robust methods for data reduction. Chapman & Hall/CRC, London
7. Fritz H, García-Escudero LA, Mayo-Iscar A (2013) Robust constrained fuzzy clustering. Inf Sci 245:38–52
8. Fritz H, García-Escudero LA, Mayo-Iscar A (2013) A fast algorithm for robust constrained clustering. Comput Stat Data Anal 61:124–136
9. García-Escudero LA, Gordaliza A, Matrán C, Mayo-Iscar A (2008) A general trimming approach to robust cluster analysis. Ann Stat 36:1324–1345
10. García-Escudero LA, Gordaliza A, Matrán C, Mayo-Iscar A (2010) A review of robust clustering methods. Adv Data Anal Classif 4:89–109
11. García-Escudero LA, Gordaliza A, San Martín R, Mayo-Iscar A (2010) Robust clusterwise linear regresssion through trimming. Comput Stat Data Anal 54:3057–3069

12. García-Escudero LA, Gordaliza A, Greselin F, Ingrassia S, Mayo-Iscar A (2016) The joint role of trimming and constraints in robust estimation for mixtures of gaussian factor analyzers. Comput Stat Data Anal 99:131–147
13. García-Escudero LA, Greselin F, Mayo-Iscar A (2017) Robust fuzzy cluster weighted modeling. Submitted
14. Gershenfeld N, Schoner B, Metois E (1999) Cluster-weighted modelling for time-series analysis. Nature 397:329–332
15. Gustafson EE, Kessel WC (1979) Fuzzy clustering with a fuzzy covariance matrix. In: Proceedings of 1978 IEEE conference on decision and control including the 17th symposium on adaptive processes. IEEE, Piscataway
16. Hathaway R (1985) A constrained formulation of maximum-likelihood estimation for normal mixture distributions. Ann Stat 13:795–800
17. Hathaway RJ, Bezdek JC (1993) Switching regression models and fuzzy clustering. IEEE Trans Fuzzy Syst 1:195–204
18. Honda K, Ohyama T, Ichihashi H, Hotsu A (2008) FCM-type switching regression with alternating least square method. In: Proceedings of 2008 IEEE international conference on Fuzzy systems (IEEE World congress on computational intelligence). IEEE, Piscataway
19. Hosmer DW Jr (1974) Maximum likelihood estimates of the parameters of a mixture of two regression lines. Commun Stat 3:995–1006
20. Ingrassia S, Rocci R (2007) Constrained monotone EM algorithms for finite mixture of multivariate gaussians. Comput Stat Data Anal 51:5339–5351
21. Lenstra AK, Lenstra JK, Rinnoy Kan AHG, Wansbeek TJ (1982) Two lines least squares. Ann Discr Math 16:201–211
22. Ritter G (2015) Robust cluster analysis and variable selection. Monographs on statistics and applied probability. Chapman & Hall/CRC, London
23. Rousseeuw PJ, Van Driessen K (1999) A fast algorithm for the minimum covariance determinant estimator. Technometrics 41:212–223
24. Rousseeuw PJ, Kaufman L, Trauwaert E (1996) Fuzzy clustering using scatter matrices. Comp Stat Data Anal 23:135–151
25. Ruspini E (1969) A new approach to clustering. Inf Control 15:22–32
26. Späth H (1982) A fast algorithm for clusterwise regression. Computing 29:175–181
27. Trauwaert E, Kaufman L, Rousseeuw PJ (1991) Fuzzy clustering algorithms based on the maximum likelihood principle. Fuzzy Sets Syst 42:213–227
28. Yang M-S (1993) On a class of fuzzy classification maximum likelihood procedures. Fuzzy Sets Syst 57:365–375
29. Wu WL, Yang MS, Hsieh NJ (2009) Alternative fuzzy switching regression. In: Proceedings of international multiconference of engineers and computer scientists, IMECS 2009, vol I. IAENG, Hong Kong

Robust Morphometric Analysis Based on Landmarks

Alfonso García-Pérez

Abstract Procrustes Analysis is a Morphometric method based on Configurations of Landmarks that estimates the superimposition parameters by least-squares; for this reason, the procedure is very sensitive to outliers. There are classical results, based on the normality of the observations, to test whether there are significant differences between individuals. In this paper we determine a Von Mises plus Saddlepoint approximation for the tail probability (p-value) of this test for the Procrustes Statistic, when the observations come from a model close to the normal.

1 Introduction

This paper is about a robust classification problem of n individuals based on their shapes, i.e., using their geometric information. The usual (classical or robust) methods based on Multivariate Analysis can not extract all the geometric information from the individuals. For this reason, in recent years, morphometrics methods based on Configurations of landmarks have been developed. A landmark is a peculiar point whose position is common in all the individuals to classify. For instance, when we classify skulls, the landmarks could be the center of the supraorbital arch, the chin, etc.; or, if we classify projectile points found in an archaeological site, the landmarks could be the ends of the points.

In all the cases, the mathematical (geometric) information that we obtain from the individuals is the k coordinates of their p landmarks, $l_i = (c_{i1}, \ldots, c_{ik})$, $i = 1, \ldots, p$.

The matrix of landmarks coordinates is called a Configuration. For each individual with p landmarks of dimension k (where k is equal to 2 or 3) we have a collection of landmark coordinates expressed in $p \times k$ matrix as

A. García-Pérez (✉)
Departamento de Estadística I. O. y C. N., Universidad Nacional de Educación
a Distancia (UNED), Paseo Senda del Rey 9, 28040 Madrid, Spain
e-mail: agar-per@ccia.uned.es

© Springer International Publishing AG 2018
E. Gil et al. (eds.), *The Mathematics of the Uncertain*, Studies in Systems,
Decision and Control 142, https://doi.org/10.1007/978-3-319-73848-2_16

$$M = \begin{pmatrix} c_{11} \cdots c_{1k} \\ \cdots \cdots \cdots \\ c_{p1} \cdots c_{pk} \end{pmatrix}$$

There are many morphometric methods; see for instance [1] or [3]. In this paper we consider Superimposition Methods; namely, Procrustes Analysis, obtaining the Procrustes coordinates and adapting the Configurations to a common (local) reference system, matching them at the common center. For these reasons, a Local Coordinate Reference System is needed and a Geographical Information System very useful.

A common graphical representation of a Configuration is a scatter plot of its landmarks coordinates. Joining the resulting points with segments we obtaining a polygon where the landmarks coordinates define the vertices of the polygon.

Because we use the shape of the individuals in their classification and shape is a property of an object that is invariant under scaling, rotation and translation (otherwise, for instance, an object and itself with double size could be classified into two different groups), in order to classify them with a Procrustes Analysis, we have first to remove the effect of Size (scale), Location (translation) and Orientation (rotation) to standardize them and match them in a common center in order to make them comparable.

This means that we have to estimate by least-squares the superimposition parameters α, β and Γ (scale, translation and rotation) in order to minimize the full Procrustes distance d_F between Configurations M_1 and M_2, i.e.,

$$\min d_F(M_1, M_2) = \min \|M_2 - \alpha M_1 \Gamma - 1_p \beta'\|$$

$$= \sqrt{trace[(M_2 - \alpha M_1 \Gamma - 1_p \beta')'(M_2 - \alpha M_1 \Gamma - 1_p \beta')]}$$

where α is a scalar representing the Size, β is a vector of k values corresponding to a Location parameter formed by the centroid coordinates, 1_p is a column vector of dimension $p \times 1$ and Γ a $k \times k$ square rotation matrix.

The idea that we pursue with this transformation is to match both Configurations, i.e., a superimposition of M_1 onto M_2.

It is possible to use a Classical Morphometric Analysis from a descriptive point of view. This is briefly exposed, together with its robustification by replacing the classical estimators with robust ones, in [6].

2 Classical Morphometric Analysis from an Inferential Point of View

Instead of considering a descriptive morphometric analysis it is more interesting to test if there are significant differences between two Configurations. From a classical point of view, we have the following result in [8, 11]: If X_1 and X_2 are two scaled and centered Configurations of dimension $p \times k$, the *Residual Distance* between Configurations X_1 and X_2 is defined as

$$||X_2 - X_1||^2 = trace\left[(X_2 - X_1)'(X_2 - X_1)\right].$$

As saw before, the $k \times k$ square rotation matrix Γ is determined such that the Procrustes distance between these two Configurations X_1 and X_2 (i.e., between landmarks) is minimal

$$\min_{\Gamma} ||X_2 - X_1\Gamma||^2 = \min_{\Gamma} trace\left[(X_2 - X_1\Gamma)'(X_2 - X_1\Gamma)\right].$$

This minimum obtained after matching (i.e., after translation, rotation and scaling) is called the *Procrustes statistic*:

$$G(X_1, X_2) = \min_{\Gamma} ||X_2 - X_1\Gamma||^2.$$

Under the null hypothesis H_0 that there is no systematic differences between Configurations X_1 and X_2, i.e., they belong to the same group, or more precisely, that if η is a constant, they are of the form

$$X_2 = X_1 + \eta\,e$$

where the $p \times k$ landmarks coordinates of Configuration e are univariate i.i.d. $N(0, 1)$, then

$$G(X_1, X_2) \approx \eta^2\,\chi_g^2$$

i.e., $G_s(X_1, X_2) = G(X_1, X_2)/\eta^2 \approx \chi_g^2$, where $g = kp - k(k+1)/2 - 1$. Hence, we can compute tail probabilities (p-values) for testing H_0. It must be $p > (k+1)/2 + 1/k$ and obviously an integer.

3 Robust Morphometric Analysis from an Inferential Point of View

The standard normality of the landmarks coordinates is a very hard assumption. For this reason we shall use robust methods for testing H_0 assuming that the $p \times k$ landmarks coordinates of e follow, not a standard normal distribution but a contaminated normal model:

$$\frac{X_2 - X_1}{\eta} \rightsquigarrow (1 - \varepsilon)N(0, 1) + \varepsilon N(0, \nu).$$

In this section we are going to compute the tail probabilities (p-values), assuming this contaminated model, using a VOM+SAD approximation.

We use this scale contaminated normal mixture model because the Configurations are matched at the common centroid that is the new origin and equal to 0, being the contamination in the scale the natural source of contamination in the observations.

3.1 Von Mises Approximations for the p-Value of the Procrustes Statistic

In order to test the null hypothesis H_0 that there is no systematic differences between the standardized Configurations X_1 and X_2, using the Procrustes statistic $G_s(X_1, X_2)$ that follows a χ_g^2 distribution under a normal model, we have the following result.

Proposition 3.1 *Let $G_s(X_1, X_2)$ be the Procrustes statistic, that follows a χ_g^2 distribution when the underlying model is a normal distribution, $\Phi_{\mu,\sigma}$. If the previous null hypothesis H_0 holds, the von Mises (VOM) approximation for the functional tail probability (if F is close to the normal $\Phi_{\mu,\sigma}$) is*

$$P_F\{G_s(X_1, X_2) > t\} \simeq g \int_{-\infty}^{\infty} P\{\chi_{g-1}^2 > t - (\tfrac{x-\mu}{\sigma})^2\} dF(x) - (g - 1)P\{\chi_g^2 > t\}.$$

Proof The von Mises (VOM) approximation for the functional tail probability is (if F is close to the normal $\Phi_{\mu,\sigma}$)

$$p_g^F = P_F\{G_s(X_1, X_2) > t\} \simeq p_g^\Phi + \int \text{TAIF}(x; t; \chi_g^2, \Phi_{\mu,\sigma}) dF(x) \qquad (1)$$

where TAIF is the Tail Area Influence Function defined in [4].

Replacing the normal model by the contaminated normal model $\Phi^\varepsilon = (1 - \varepsilon) \Phi_{\mu,\sigma} + \varepsilon \delta_x$ and computing the derivative at $\varepsilon = 0$ we obtain that

$$\text{TAIF}(x; t; \chi_g^2, \Phi_{\mu,\sigma}) = \frac{\partial}{\partial \varepsilon} P_{\Phi^\varepsilon}\{G_s(X_1, X_2) > t\}\Big|_{\varepsilon=0}$$
$$= g P\{\chi_{g-1}^2 > t - (x - \mu)^2/\sigma^2\} - g P\{\chi_g^2 > t\}$$

integrating now, we obtain the result. $\qquad \square$

Considering a scale contaminated normal (SCN) model

$$(1 - \varepsilon)N(0, 1) + \varepsilon N(0, v)$$

Table 1 *Exact* and approximate *p*-values with $g = 3$	t	"Exact"	Approximate
	6	$0'149$	$0'148$
	8	$0'077$	$0'076$
	10	$0'042$	$0'042$
	12	$0'024$	$0'025$
	14	$0'016$	$0'016$
	16	$0'011$	$0'011$
	18	$0'007$	$0'008$

the VOM approximation is

$$p_g^F \simeq (1 - g\,\varepsilon)P\{\chi_g^2 > t\} + g\,\varepsilon \int_{-\infty}^{\infty} P\{\chi_{g-1}^2 > t - x^2\}\,d\Phi_{0,\nu}(x).$$

In Table 1 appear, [10], the *Exact* values (obtained through a simulation of 100.000 samples) and the VOM approximations when $\varepsilon = 0'05$, $\nu = 2$ and $g = 3$.

To obtain the previous numerical results we had to deal with numerical integration. Sometimes, we would like to have analytic expressions of p_g^F to value the effect of contamination ε, etc. For this reason, and for controlling the relative error of the approximation, in the next section we shall compute the Saddlepoint approximation for the *p*-value of the Procrustes Statistic.

3.2 Saddlepoint Approximations for the p-Value of the Procrustes Statistic

Using Lugannani and Rice formula, [9], for the sample mean of g independent square normal variables, we obtain the VOM+SAD approximation given in the next result.

Proposition 3.2 *Let $G_s(X_1, X_2)$ be the Procrustes statistic, that follows a χ_g^2 distribution when the underlying model is a normal distribution, $\Phi_{\mu,\sigma}$. If the null hypothesis H_0 holds, the saddlepoint approximation of the von Mises expansion, VOM+SAD approximation, for the functional tail probability (if F is close to the normal $\Phi_{\mu,\sigma}$) is*

$$P_F\{G_s(X_1, X_2) > t\} \simeq P\{\chi_g^2 > t\} - B + B \int_{-\infty}^{\infty} \frac{\sqrt{g}}{\sqrt{t}}\, e^{\frac{(t-g)(x-\mu)^2}{2t\sigma^2}}\, dF(x) \quad (2)$$

where $B = \dfrac{g\sqrt{g}}{\sqrt{\pi}\,(t-g)}\, e^{-(t-g-g\cdot\log(t/g))/2}$.

Proof If $G_s(X_1, X_2)$ follows a χ_g^2 distribution, and $Y_1, \ldots, Y_g$ are g independent gamma distributions $\gamma(1/2, 1/2)$ with moment generating function M and cumulant generating function $K = \log M$, it is, following [2, 5, 9] or [7],

$$P_\Phi \left\{ \frac{G_s(X_1, X_2)}{g} > t \right\} = P \left\{ \frac{1}{g} \sum_{i=1}^{g} Y_i > t \right\}$$

$$= 1 - \Phi_s(w) + \phi_s(w) \left\{ \frac{1}{r} - \frac{1}{w} + O(g^{-3/2}) \right\} \quad (3)$$

where Φ_s and ϕ_s are the cumulative distribution and density functions of the standard normal distribution.

If K is the cumulant generating function, that is the functional of $\Phi_{\mu,\sigma}$,

$$K(\theta) = \log \int_{-\infty}^{\infty} e^{\theta (u-\mu)^2/\sigma^2} \, d\Phi_{\mu,\sigma}(u)$$

and z_0 is the (functional) saddlepoint, i.e., it is the solution of the equation $K'(z_0) = t$, the functionals that appear in (3) are

$$w = sign(z_0) \sqrt{2g \cdot (z_0 t - K(z_0))} = \sqrt{g} \, sign(z_0) \sqrt{2 (z_0 t - K(z_0))} := \sqrt{g} \, w_1$$

$$r = z_0 \sqrt{g \cdot K''(z_0)} = \sqrt{g} \, z_0 \sqrt{K''(z_0)} := \sqrt{g} \, r_1.$$

As we saw before, the VOM approximation for the tail probability depends on the TAIF. To obtain the TAIF of $G_s(X_1, X_2)/g$ at $\Phi_{\mu,\sigma}$ we have to replace the model $\Phi_{\mu,\sigma}$ by the contaminated model $\Phi^\varepsilon = (1 - \varepsilon)\Phi_{\mu,\sigma} + \varepsilon \, \delta_x$ in all the functionals in the right side of (3) that depend on $\Phi_{\mu,\sigma}$, and then to obtain the derivative at $\varepsilon = 0$; this process is represented with a dot over the functional. Since $\phi_s'(w) = -\phi_s(w) \, w$ and $\phi_s(w) \leq 1$, we obtain that

$$\mathrm{TAIF} \left(x; t; \frac{G_s(X_1, X_2)}{g}, \Phi_{\mu,\sigma} \right) = \frac{\partial}{\partial \varepsilon} P_{\Phi^\varepsilon} \left\{ \frac{G_s(X_1, X_2)}{g} > t \right\} \Bigg|_{\varepsilon=0}$$

$$= -\phi_s(w) \, \dot{w} + \phi_s'(w) \, \dot{w} \left\{ \frac{1}{r} - \frac{1}{w} + O(g^{-3/2}) \right\} + \phi_s(w) \left\{ -\frac{\dot{r}}{r^2} + \frac{\dot{w}}{w^2} + O(g^{-3/2}) \right\}$$

$$= \phi_s(w) \left[-\frac{w \, \dot{w}}{r} - \frac{\dot{r}}{r^2} + \frac{\dot{w}}{w^2} \right] + O(g^{-1})$$

$$= \phi_s(w) \left[-\frac{\sqrt{g}\, w_1 \sqrt{g}\, \dot{w}_1}{\sqrt{g}\, r_1} - \frac{\sqrt{g}\, \dot{r}_1}{g\, r_1^2} + \frac{\sqrt{g}\, \dot{w}_1}{g\, w_1^2} \right] + O(g^{-1})$$

$$= \frac{\phi_s(w)}{r_1} \left[-\sqrt{g} \cdot w_1 \, \dot{w}_1 \right] + O(g^{-1/2})$$

because the functionals w_1, $\dot{w}_1$, r_1 and $\dot{r}_1$ do not depend on g. Since

$$\dot{w}_1 = sign(z_0) \frac{2(\dot{z}_0\, t - \dot{K}\,(z_0))}{2\sqrt{2(z_0\, t - K(z_0))}} = \frac{\dot{z}_0\, t - \dot{K}\,(z_0))}{w_1}$$

it will be

$$\mathrm{TAIF} \left(x; t; \frac{G_s(X_1, X_2)}{g}, \Phi_{\mu,\sigma} \right) = \frac{\phi_s(w)}{r_1} \sqrt{g} \left[\dot{K}\,(z_0) - \dot{z}_0\, t \right] + O(g^{-1/2}). \quad (4)$$

Hence, we have to compute the influence functions $\dot{K}\,(z_0)$ and $\dot{z}_0$. To do this, because

$$K'(\theta) = \frac{\displaystyle\int_{-\infty}^{\infty} e^{\theta\,(u-\mu)^2/\sigma^2} \left(\frac{u-\mu}{\sigma} \right)^2 d\Phi_{\mu,\sigma}(u)}{\displaystyle\int_{-\infty}^{\infty} e^{\theta\,(u-\mu)^2/\sigma^2} \, d\Phi_{\mu,\sigma}(u)}$$

from the saddlepoint equation, $K'(z_0) = t$, we obtain

$$\int_{-\infty}^{\infty} e^{z_0\,(u-\mu)^2/\sigma^2} \left[\left(\frac{u-\mu}{\sigma} \right)^2 - t \right] d\Phi_{\mu,\sigma}(u) = 0.$$

Replacing again the model by the contaminated model $\Phi^\varepsilon = (1-\varepsilon)\, \Phi_{\mu,\sigma} + \varepsilon\, \delta_x$ before obtaining the derivative at $\varepsilon = 0$, and making the change of variable $(u - \mu)/\sigma = y$, we obtain

$$\dot{z}_0 \left[\int_{-\infty}^{\infty} e^{z_0\, y^2} y^4 \, d\Phi_s(y) - t \int_{-\infty}^{\infty} e^{z_0\, y^2} y^2 \, d\Phi_s(y) \right] + e^{z_0\,(x-\mu)^2/\sigma^2} \left[\left(\frac{x-\mu}{\sigma} \right)^2 - t \right] = 0$$

i.e.,

$$\dot{z}_0 = \frac{1}{2}\, t^{-5/2} e^{\frac{(t-1)(x-\mu)^2}{2t\sigma^2}} \left[t - \left(\frac{x-\mu}{\sigma} \right)^2 \right].$$

In a similar way, we obtain that

$$\dot{K}(z_0) = \frac{3}{2}t^{-1/2}e^{z_0(x-\mu)^2/\sigma^2} - \frac{1}{2}t^{-3/2}e^{z_0(x-\mu)^2/\sigma^2}\left(\frac{x-\mu}{\sigma}\right)^2 - 1.$$

Also it is

$$r_1 = z_0\sqrt{K''(z_0)} = \frac{t-1}{\sqrt{2}} \quad \text{and} \quad \phi_s(w) = \frac{1}{\sqrt{2\pi}}e^{-g\cdot(t-1-\log t)/2}.$$

Therefore, from (4), it will be

$$\text{TAIF}\left(x;t;\frac{G_s(X_1,X_2)}{g},\Phi_{\mu,\sigma}\right) = A\left(\frac{1}{\sqrt{t}}e^{\frac{(t-1)(x-\mu)^2}{2t\sigma^2}} - 1\right) + O(g^{-1/2})$$

where

$$A = \frac{\sqrt{g}}{\sqrt{\pi}(t-1)}e^{-g\cdot(t-1-\log t)/2}.$$

From (1), we obtain now the VOM+SAD approximation for the p-value of the test statistic $G_s(X_1,X_2)/g$,

$$P_F\left\{\frac{G_s(X_1,X_2)}{g} > t\right\} \simeq P\{\chi_g^2 > gt\} - A + A\int_{-\infty}^{\infty}\frac{1}{\sqrt{t}}e^{\frac{(t-1)(x-\mu)^2}{2t\sigma^2}}\,dF(x)$$

and from this, we obtain the approximation (2) for the test statistic $G_s(X_1,X_2)$. $\square$

If F is the location contaminated normal mixture (LCN),

$$F = (1-\varepsilon)N(0,1) + \varepsilon N(\theta,1)$$

the VOM+SAD approximation is

$$P_F\{G_s(X_1,X_2) > t\} \simeq P\{\chi_g^2 > t\} + \varepsilon B\left[e^{-(1-t/g)\theta^2/2} - 1\right].$$

In Table 2 appear the *Exact* values (obtained through simulation of 100.000 samples), the VOM and the VOM+SAD approximations when $\varepsilon = 0'01$, $\theta = 1$ and $g = 5$.

Corollary 3.1 *To test the null hypothesis H_0 that there is no systematic differences between the standardized Configurations X_1 and X_2 with p landmarks of dimension k (i.e., X_1 and X_2 belong to the same classification group) using the Procrustes statistic $G_s(X_1,X_2)$ and assuming that the error difference between Configurations*

Table 2 Exact and approximate p-values with $g = 5$

t	"Exact"	VOM appr.	VOM+SAD appr.
9	$0'1125$	$0'1129$	$0'1136$
11	$0'0538$	$0'0539$	$0'0545$
13	$0'0251$	$0'0249$	$0'0253$
15	$0'0114$	$0'0112$	$0'0115$
17	$0'0050$	$0'0049$	$0'0051$
19	$0'0022$	$0'0022$	$0'0023$

$$\frac{X_2 - X_1}{\eta}$$

follows a scale contamination normal model $(1 - \varepsilon)N(0, 1) + \varepsilon N(0, v)$, *the VOM+SAD approximation for the tail probability (p-value) is*

$$P\{G_s(X_1, X_2) > t\} \approx P\{\chi_g^2 > t\} + \varepsilon \frac{g^{3/2}}{\sqrt{\pi}(t - g)} \left[\frac{\sqrt{g}}{\sqrt{t - v^2(t - g)}} - 1 \right]$$

$$\cdot \exp\left\{ -\frac{1}{2}\left(t - g - g \cdot \log \frac{t}{g} \right) \right\} \tag{5}$$

where $g = kp - k(k + 1)/2 - 1$. It must be $p > (k + 1)/2 + 1/k$ and obviously an integer.

Then, if $k = 2$, it is $g = 2p - 4$ and $p > 2$. And if $k = 3$, it is $g = 3p - 7$ and $p \geq 3$.

There are some applications of this approximation in [6]. There we test if there are significance differences between dots of *Notch tips and bay leaves*, of Solutrense period, that were found in caves of Asturias (Spain). We do this analysis using a photo of the "Museo Arqueológico de Asturias" (Oviedo), including this photo in QGIS as a raster layer.

4 Conclusions

Classical Morphometric Analysis based on Landmarks is not robust because it is based on sample means and least-squares estimation using a Normal distribution as model.

In this paper we consider a Contaminated Normal Model to make robust inferences. Namely, for this mixture model we obtain an von Mises approximation of the

p-value of a test for the null hypothesis of no significance differences between two individuals based on their shapes.

We also obtain a very accurate saddlepoint approximation of this von Mises approximation.

Acknowledgements This work is partially supported by Grant MTM2015-67057-P from Ministerio de Economía, Industria y Competitividad (Spain).

References

1. Claude J (2008) Morphometrics with R. Springer, New York
2. Daniels HE (1983) Saddlepoint approximations for estimating equations. Biometrika 70:89–96
3. Dryden IL, Mardia KV (2016) Statistical shape analysis (with applications in R). Wiley, Chichester
4. Field CA, Ronchetti E (1985) A tail area influence function and its application to testing. Commun Stat 4:19–41
5. García-Pérez A (2006) Chi-square tests under models close to the normal distribution. Metrika 63:343–354
6. García-Pérez A, Cabrero-Ortega MY (2017) Robust morphometric analysis based on landmarks. Applications. arXiv:1703.04642
7. Jensen JL (1995) Saddlepoint approximations. Clarendon Press, Oxford
8. Langron SP, Collins AJ (1985) Perturbation theory for procrustes analysis. J R Stat Soc Ser B 47:277–284
9. Lugannani R, Rice S (1980) Saddle point approximation for the distribution of the sum of independent random variables. Adv Appl Probab 12:475–490
10. R Development Core Team (2016) R: A language and environment for statistical computing. R Foundation for Statistical Computing, Viena. http://www.R-project.org
11. Sibson R (1979) Studies in the robustness of multidimensional scaling: perturbational analysis of classical scaling. J R Stat Soc Ser B Stat Methodol 41:217–229

Smoothing-Based Tests with Directional Random Variables

Eduardo García-Portugués, Rosa M. Crujeiras
and Wenceslao González-Manteiga

Abstract Testing procedures for assessing specific parametric model forms, or for checking the plausibility of simplifying assumptions, play a central role in the mathematical treatment of the uncertain. No *certain* answers are obtained by testing methods, but at least the *uncertainty* of these answers is properly quantified. This is the case for tests designed on the two most general data generating mechanisms in practice: distribution/density and regression models. Testing proposals are usually formulated on the Euclidean space, but important challenges arise in non-Euclidean settings, such as when directional variables (i.e., random vectors on the hypersphere) are involved. This work reviews some of the smoothing-based testing procedures for density and regression models that comprise directional variables. The asymptotic distributions of the revised proposals are presented, jointly with some numerical illustrations justifying the need of employing resampling mechanisms for effective test calibration.

E. García-Portugués (✉)
Department of Statistics, Carlos III University of Madrid, Avenida de la Universidad 30,
28911 Leganes, Spain
e-mail: edgarcia@est-econ.uc3m.es

E. García-Portugués
UC3M-BS Institute of Financial Big Data, Carlos III University of Madrid, Calle Madrid 135,
28903 Getafe, Spain

R. M. Crujeiras · W. González-Manteiga
Department of Statistics, Mathematical Analysis and Optimization,
Universidade de Santiago de Compostela, Rúa Lope Gomez de Marzoa s/n,
15782 Santiago de Compostela, Spain
e-mail: rosa.crujeiras@usc.es

W. González-Manteiga
e-mail: wenceslao.gonzalez@usc.es

© Springer International Publishing AG 2018
E. Gil et al. (eds.), *The Mathematics of the Uncertain*, Studies in Systems,
Decision and Control 142, https://doi.org/10.1007/978-3-319-73848-2_17

175

1 On Goodness-of-Fit Tests and Smoothing

In the early years of the 20th century, K. Pearson and colleagues initiate the development of testing methods for assessing the goodness-of-fit of a certain parametric model. Pearson[19] presents his celebrated χ^2 test as a criterion to check if a given *system of deviations* from a theoretical distribution could be supposed to come from random sampling, but it is not until a couple of years later when Elderton [5] coins the term goodness-of-fit *of theory to observation*. Also at the beginning of last century, Pearson [20] introduces the first ideas for goodness-of-fit tests in regression models. With no theoretical support from probability theory (which was developed almost at the same time, and therefore, its impact on statistics was noticed some years later), these works set the basis for the construction of testing procedures with the aim of assessing a certain parametric null hypothesis for density/distribution (see [2, 4], as two influential papers) and regression models (see [13] for a complete review on goodness-of-fit tests in this setting).

This work focus on a certain class of tests that makes use of nonparametric (smooth) estimators of the target function, that is, the density or the regression functions. First, consider the problem of testing a certain parametric density model

$$H_0 : f \in \mathscr{F}_\Theta \quad \text{versus} \quad H_1 : f \notin \mathscr{F}_\Theta, \tag{1}$$

with $\mathscr{F}_\Theta = \{f_\theta : \theta \in \Theta\}$ a parametric density family. From a smoothing-based perspective, a pilot estimator $\hat{f}$ constructed from $X_1, \ldots, X_n$, a sample from the random variable (rv) X, will be confronted with a parametric estimator by the use of a certain discrepancy measure. Bickel and Rosenblatt [2] consider the classical Kernel Density Estimator (KDE) $\hat{f}_g(x) = \frac{1}{ng} \sum_{i=1}^{n} K\left(\frac{x-X_i}{g}\right)$, with kernel K and bandwidth g, to be compared with a parametric estimator $f_{\hat{\theta}}$ under the null through an L^2-distance. In general, test statistics for (1) can be built as $T_n = d(\hat{f}, f_{\hat{\theta}})$, being d a discrepancy measure between both estimators.

The ideas of goodness-of-fit tests for density curves have been naturally extended in the nineties of the last century to regression models. Consider, as a reference, a regression model $Y = m(X) + \varepsilon$, where the goal is to test

$$H_0 : m \in \mathscr{M}_\Theta \quad \text{versus} \quad H_1 : m \notin \mathscr{M}_\Theta \tag{2}$$

in an omnibus way from a sample $\{(X_i, Y_i)\}_{i=1}^n$ of (X, Y). Here $m(x) = \mathbb{E}[Y|X = x]$ is the regression function of Y over X, and ε is a random error such that $\mathbb{E}[\varepsilon|X] = 0$. A pilot estimator $\hat{m}(x) = \sum_{i=1}^{n} W_{n,i}(x)Y_i$ can be constructed using nonparametric weights, such as the Nadaraya-Watson weights given by $W_{n,i}(x) = K\left(\frac{x-X_i}{g}\right) / \sum_{j=1}^{n} K\left(\frac{x-X_j}{g}\right)$. Other possible weights, such as the ones from local linear estimation, k-nearest neighbours, or splines, can be also considered. Using this kind of pilot estimators, tests statistics can be built (similarly to the density case) as $T_n = d\left(\hat{m}, m_{\hat{\theta}}\right)$.

In the presence of directional random variables, and considering the previous smoothing ideas, similar tests can be developed.

2 Goodness-of-Fit Tests with Directional Data

The statistical analysis of *directional data*, this is, elements in the q-sphere $\Omega_q = \{\mathbf{x} \in \mathbb{R}^{q+1} : \mathbf{x}'\mathbf{x} = 1\}$, is notably different from the analysis of *linear* (Euclidean) data. In particular, no canonical ordering exists in Ω_q, which makes rank-based inference ill-defined. We refer to the book of Mardia and Jupp [18] for a comprehensive treatment of statistical inference with directional data, and for a collection of applications. Some smooth estimators for density and regression in this context are briefly revised below. These estimators are used as pilots for the testing proposals introduced in the subsequent sections.

2.1 Smooth Estimation of Density and Regression

Let $\mathbf{X}_1, \ldots, \mathbf{X}_n$ denote a sample from the directional rv $\mathbf{X}$ with density f. Hall et al. [14] and Bai et al. [1][1] introduce a KDE for directional data, which is defined as follows:

$$\hat{f}_h(\mathbf{x}) = \frac{1}{n} \sum_{i=1}^{n} L_h(\mathbf{x}, \mathbf{X}_i), \quad L_h(\mathbf{x}, \mathbf{X}_i) = \frac{c_{h,q}(L)}{n} L\left(\frac{1 - \mathbf{x}'\mathbf{X}_i}{h^2}\right), \quad (3)$$

with $L : \mathbb{R}_0^+ \to \mathbb{R}_0^+$ being the kernel, $h > 0$ the bandwidth parameter, and

$$c_{h,q}(L)^{-1} = \lambda_{h,q}(L)h^q, \quad \lambda_{h,q}(L) = \omega_{q-1} \int_0^{2h^{-2}} L(r) r^{\frac{q}{2}-1} (2 - rh^2)^{\frac{q}{2}-1} \, dr,$$

with $\lim_{h \to 0} \lambda_{h,q}(L) = \lambda_q(L) = 2^{\frac{q}{2}-1} \omega_{q-1} \int_0^\infty L(r) r^{\frac{q}{2}-1} \, dr$. ω_q denotes both the area of Ω_q, $\omega_q = 2\pi^{\frac{q+1}{2}} / \Gamma\left(\frac{q+1}{2}\right)$, and the Lebesgue measure in Ω_q. For the consistency of (3), it is required that $h = h_n \to 0$ when $n \to \infty$ at a rate slower than $nh^q \to \infty$.

A directional rv usually appears related to another linear or directional rv, being cylindrical and toroidal data the most common situations in practice. In these scenarios, the modelling approach can be focused on the estimation of the joint density or the regression function. From the first perspective, in order to estimate the density of a directional-linear rv $(\mathbf{X}, Y)$ in $\Omega_q \times \mathbb{R}$, García-Portugués et al. [9] propose a KDE

---

[1]Hall et al. [14]'s (1.3) is equivalent to Bai et al. [1]'s (1.3), but the latter employs a notation with a more direct connection with the usual KDE.

adapted to this setting:

$$\hat{f}_{h,g}(\mathbf{x}, y) = \frac{1}{n} \sum_{i=1}^{n} LK_{h,g} ((\mathbf{x}, y)(\mathbf{X}_i, Y_i)), \tag{4}$$

where $LK_{h,g} ((\mathbf{x}, y), (\mathbf{X}_i, Y_i)) = L_h (\mathbf{x}, \mathbf{X}_i) \times \frac{1}{g} K \left(\frac{y - Y_i}{g} \right)$ is a directional-linear product kernel, and h, g are two bandwidth sequences such that, for the consistency of (4), $h, g \to 0$ and $nh^q g \to \infty$.

In a toroidal scenario, a directional–directional KDE for the density of a rv $(\mathbf{X}_1, \mathbf{X}_2)$ in $\Omega_{q_1} \times \Omega_{q_2}$ can be derived adapting (4):

$$\hat{f}_{h_1, h_2}(\mathbf{x}_1, \mathbf{x}_2) = \frac{1}{n} \sum_{i=1}^{n} LL_{h_1, h_2} ((\mathbf{x}_1, \mathbf{x}_2), (\mathbf{X}_{1i}, \mathbf{X}_{2i})), \tag{5}$$

with $LL_{h_1, h_2} ((\mathbf{x}_1, \mathbf{x}_2), (\mathbf{X}_{1i}, \mathbf{X}_{2i})) = L_{h_1} (\mathbf{x}_1, \mathbf{X}_{1i}) \times L_{h_1} (\mathbf{x}_2, \mathbf{X}_{2i})$, with $h_1, h_2 \to 0$ and $nh_1^{q_1} h_2^{q_2} \to \infty$ required for consistency.

Considering now a regression setting with scalar response and directional covariate, let $\{(\mathbf{X}_i, Y_i)\}_{i=1}^{n}$ be a sample from the regression model $Y = m(\mathbf{X}) + \varepsilon$, where $m(\mathbf{x}) = \mathbb{E}[Y|\mathbf{X} = \mathbf{x}] : \Omega_q \to \mathbb{R}$ is the regression function of Y over $\mathbf{X}$, and ε is a random error such that $\mathbb{E}[\varepsilon|\mathbf{X}] = 0$. A nonparametric estimator for m, following the *local linear* ideas (see [6]), can be constructed as follows. Consider a Taylor expansion in a vicinity of $\mathbf{X}_i$:

$$m(\mathbf{X}_i) \approx m(\mathbf{x}) + \nabla m(\mathbf{x})'(\mathbf{I}_{q+1} - \mathbf{x}\mathbf{x}')(\mathbf{X}_i - \mathbf{x}) = \beta_0 + \boldsymbol{\beta}_1' \mathbf{B}_\mathbf{x}'(\mathbf{X}_i - \mathbf{x}), \tag{6}$$

where $\mathbf{B}_\mathbf{x}' \mathbf{B}_\mathbf{x} = \mathbf{I}_q$, $\mathbf{B}_\mathbf{x} \mathbf{B}_\mathbf{x}' = \mathbf{I}_{q+1} - \mathbf{x}\mathbf{x}'$, and $\mathbf{I}_q$ is the identity matrix of dimension q. From the extension of m to $\mathbf{x} \in \mathbb{R}^{q+1} \backslash \{\mathbf{0}\}$ by $m(\mathbf{x}/ ||\mathbf{x}||)$, since $\nabla m(\mathbf{x})'\mathbf{x} = 0$, the central expression in (6) follows. This motivates the weighted least squares problem

$$\min_{(\beta_0, \boldsymbol{\beta}_1) \in \mathbb{R}^{q+1}} \sum_{i=1}^{n} \left(Y_i - \beta_0 - \delta_{p,1} \boldsymbol{\beta}_1' \mathbf{B}_\mathbf{x}'(\mathbf{X}_i - \mathbf{x}) \right)^2 L_h(\mathbf{x}, \mathbf{X}_i), \tag{7}$$

where $\delta_{r,s}$ is Kronecker delta, used to control both the local constant ($p = 0$) and local linear ($p = 1$) fits. The estimate $\hat{\beta}_0$ solving (7) provides a local linear estimator for m:

$$\hat{m}_{h,p}(\mathbf{x}) = \sum_{i=1}^{n} W_{n,i}^p(\mathbf{x})Y_i, \quad W_{n,i}^p(\mathbf{x}) = \mathbf{e}_1' \left(\mathscr{X}_{\mathbf{x},p}' \mathscr{W}_\mathbf{x} \mathscr{X}_{\mathbf{x},p} \right)^{-1} \mathscr{X}_{\mathbf{x},p}' \mathscr{W}_\mathbf{x} \mathbf{e}_i, \tag{8}$$

where $\mathbf{Y} = (Y_1 \ldots, Y_n)'$, $\mathscr{W}_\mathbf{x} = \text{diag}(L_h(\mathbf{x}, \mathbf{X}_1), \ldots, L_h(\mathbf{x}, \mathbf{X}_n))$, $\mathbf{e}_i$ is the i-th unit canonical vector, and $\mathscr{X}_{\mathbf{x},1}$ is the $n \times (q + 1)$ matrix with the i-th row given by

$(1, (\mathbf{X}_i - \mathbf{x})'\mathbf{B}_\mathbf{x})$ (if $p = 0$, $\mathcal{X}_{\mathbf{x},0} = (1, \ldots, 1)'$). For the consistency of (8), $h \to 0$ and $nh^q \to \infty$ are required.

2.2 Density-Based Tests

Testing (1) allows to check whether there are significant evidences against assuming the density has a given parametric nature, f_{θ_0}, with parameter θ_0 either specified (simple hypothesis) or unspecified (composite hypothesis). In the spirit of Fan [7]'s test, Boente et al. [3] propose the next test statistic for addressing (1):

$$T_{n,1} = \int_{\Omega_q} \left(\hat{f}_h(\mathbf{x}) - L_h f_{\hat{\theta}}(\mathbf{x})\right)^2 \omega_q(d\mathbf{x}),$$

where $L_h f_{\theta_0}(\mathbf{x}) = \int_{\Omega_q} L_h(\mathbf{x}, \mathbf{y}) f_{\theta_0}(\mathbf{y}) \omega_q(d\mathbf{y}) = \mathbb{E}_{f_{\theta_0}}\left[\hat{f}_h(\mathbf{x})\right]$ is the expectation of (3) under f_{θ_0}. This term is included in order to match the asymptotic biases of the nonparametric and parametric estimators.

The asymptotic distribution of $T_{n,1}$ is settled on Zhao and Wu [22]'s central limit theorem for the integrated squared error of (3), $I_n = \int_{\Omega_q} (\hat{f}(\mathbf{x}) - f(\mathbf{x}))^2 \omega_q(d\mathbf{x})$. The result is given under three different rates for $h \to 0$. The relevant one for $T_{n,1}$ is $nh^{q+4} \to 0$, when the integrated variance dominates the integrated bias (not dominant under H_0), and is given next:

$$nh^{\frac{q}{2}}(I_n - \mathbb{E}[I_n]) \xrightarrow{d} \mathcal{N}\left(0, 2v_d^2 R(f)\right),$$

with $R(f) = \int_{\Omega_q} f(\mathbf{x})^2 \omega_q(d\mathbf{x})$ (the functional $R(\cdot)$ denotes the integration of the squared argument on its domain of definition) and

$$v_d^2 = \gamma_q \lambda_q(L)^{-4} \int_0^\infty r^{\frac{q}{2}-1} \left\{\int_0^\infty \rho^{\frac{q}{2}-1} L(\rho)\varphi_q(r, \rho)\, d\rho\right\}^2 dr,$$

$$\varphi_q(r, \rho) = \begin{cases} L\left(r + \rho - 2(r\rho)^{\frac{1}{2}}\right) + L\left(r + \rho + 2(r\rho)^{\frac{1}{2}}\right), & q = 1, \\ \int_{-1}^1 \left(1 - \theta^2\right)^{\frac{q-3}{2}} L\left(r + \rho - 2\theta(r\rho)^{\frac{1}{2}}\right) d\theta, & q \geq 2, \end{cases}$$

$$\gamma_q = \begin{cases} 2^{-\frac{1}{2}}, & q = 1, \\ \omega_{q-1}\omega_{q-2}^2 2^{\frac{3q}{2}-3}, & q \geq 2. \end{cases}$$

Under certain regularity conditions on f_{θ_0} and L (A1–A3 in [3]), if $\hat{\theta} - \theta_0 = \mathcal{O}_\mathbb{P}(n^{-\frac{1}{2}})$ under H_0, then

$$nh^{\frac{q}{2}} \left(T_{n,1} - \frac{\lambda_q(L^2)\lambda_q(L)^{-2}}{nh^q} \right) \xrightarrow{d} \mathcal{N} \left(0, 2v_d^2 R(f_{\theta_0}) \right).$$

Hence, asymptotically, the test rejects H_0 at level α whenever $T_{n,1} > t_{\alpha;n,q,\theta_0} = (nh^q)^{-1}\lambda_q(L^2)\lambda_q(L)^{-2} + h^{\frac{q}{2}} v_d \sqrt{2R(f_{\theta_0})} z_\alpha$. Under local Pitman alternatives of the kind $H_{1P} : f = f_{\theta_0} + (nh^{\frac{q}{2}})^{\frac{1}{2}} \Delta$ ($\Delta = 0$ gives H_0), where $\Delta : \Omega_q \to \mathbb{R}$ is such that $\int_{\Omega_q} \Delta(\mathbf{x}) \omega_q(d\mathbf{x}) = 0$, and if $\hat{\theta} - \theta_0 = \mathcal{O}_{\mathbb{P}}(n^{-\frac{1}{2}})$ under H_{1P}, the test rejects if $T_{n,1} > t_{\alpha;n,q,\theta_0} - R(\Delta)$. Hence, the larger the L^2-norm of Δ, the larger the power.

If f is a directional-linear density, testing (1) can be done using

$$T_{n,2} = \int_{\Omega_q \times \mathbb{R}} \left(\hat{f}_{h,g}(\mathbf{x}, y) - LK_{h,g} f_{\hat{\theta}}(\mathbf{x}, y) \right)^2 dy \, \omega_q(d\mathbf{x}),$$

where $LK_{h,g} f_{\theta_0}(\mathbf{x}, y) = \int_{\Omega_q \times \mathbb{R}} LK_{h,g}((\mathbf{x}, y), (\mathbf{z}, t)) f_{\theta_0}(\mathbf{z}, t) \, dt \, \omega_q(d\mathbf{z})$ is the expected value of $\hat{f}_{h,g}(\mathbf{x}, y)$ under H_0. Under regularity assumptions for the density and kernels (A1, A2 and A5 in [11]), and $\hat{\theta} - \theta_0 = \mathcal{O}_{\mathbb{P}}(n^{-\frac{1}{2}})$ under $H_{1P} :$ $f = f_{\theta_0} + (nh^{\frac{q}{2}})^{\frac{1}{2}} \Delta$ ($\Delta : \Omega_q \times \mathbb{R} \to \mathbb{R}$ is such that $\int_{\Omega_q \times \mathbb{R}} \Delta(\mathbf{x}, y) \, dy \, \omega_q(d\mathbf{x}) = 0$), the limit law of $T_{n,2}$ under H_{1P} is

$$n(h^q g)^{\frac{1}{2}} \left(T_{n,2} - \frac{\lambda_q(L^2)\lambda_q(L)^{-2} R(K)}{nh^q g} \right) \xrightarrow{d} \mathcal{N} \left(R(\Delta), 2v_d^2 v_l^2 R(f_{\theta_0}) \right), \quad (9)$$

where $v_l^2 = \int_{\mathbb{R}} \left\{ \int_{\mathbb{R}} K(u) K(u+v) \, du \right\}^2 dv$. v_d^2 and v_l^2 are the variance components associated to the smoothing and, for the Gaussian and von Mises kernels, their expressions are remarkably simple: $v_l^2 = (8\pi)^{-\frac{1}{2}}$ and $v_d^2 = (8\pi)^{-\frac{q}{2}}$.

Estimator (4) allows also to check the independence between the rv's $\mathbf{X}$ and Y in an omnibus way, for arbitrary dimensions. This degree of generality contrasts with the available tests for assessing the independence between directional and linear variables, mostly focused on the circular case and on the examination of association coefficients (e.g. [8, 16, 17]). Independence can be tested à la Rosenblatt [21] by considering the problem

$$H_0 : f_{\mathbf{X},Y} = f_{\mathbf{X}} f_Y \quad \text{versus} \quad H_1 : f_{\mathbf{X},Y} \neq f_{\mathbf{X}} f_Y, \quad (10)$$

where $f_{\mathbf{X},Y}$ is the joint directional-linear density, and $f_{\mathbf{X}}$ and f_Y are the marginals. To that aim, [10] propose the statistic

$$T_{n,3} = \int_{\Omega_q \times \mathbb{R}} \left(\hat{f}_{h,g}(\mathbf{x}, y) - \hat{f}_h(\mathbf{x}) \hat{f}_g(y) \right)^2 dy \, \omega_q(d\mathbf{x}).$$

Under the same conditions on the density and kernels required for (9), and with the additional bandwidths' bond $h^q g^{-1} \to c$, $0 < c < \infty$, the asymptotic distribution of $T_{n,2}$ under independence is

$$n(h^q g)^{\frac{1}{2}} \left(T_{n,3} - A_n \right) \xrightarrow{d} \mathcal{N} \left(0, 2v_d^2 v_l^2 R(f_X) R(f_Y) \right), \tag{11}$$

where $A_n = \frac{\lambda_q(L^2)\lambda_q(L)^{-2}R(K)}{nh^q g} - \frac{\lambda_q(L^2)\lambda_q(L)^{-2}R(f_Y)}{nh^q} - \frac{R(K)R(f_X)}{ng}$. Note that (11) is similar to (9), plus two extra bias terms given by the marginal KDEs.

$T_{n,2}$ and $T_{n,3}$ can be modified to work with a directional-directional rv by using the KDE in (5). The statistics for (1) and (10) are now:

$$T_{n,4} = \int_{\Omega_{q_1} \times \Omega_{q_2}} \left(\hat{f}_{h_1,h_2}(\mathbf{x}_1, \mathbf{x}_2) - L K_{h_1,h_2} f_{\hat{\theta}}(\mathbf{x}_1, \mathbf{x}_2) \right)^2 \omega_{q_2}(d\mathbf{x}_2)\, \omega_{q_1}(d\mathbf{x}_1),$$

$$T_{n,5} = \int_{\Omega_{q_1} \times \Omega_{q_2}} \left(\hat{f}_{h_1,h_2}(\mathbf{x}_1, \mathbf{x}_2) - \hat{f}_{h_1}(\mathbf{x}_1) \hat{f}_{h_2}(\mathbf{x}_2) \right)^2 \omega_{q_2}(d\mathbf{x}_2)\, \omega_{q_1}(d\mathbf{x}_1),$$

respectively. Under the directional-directional analogues of the assumptions required for (9) and (11), the asymptotic rejection rule of $T_{n,4}$ is $T_{n,4} > (nh_1^{q_1} h_2^{q_2})^{-1}$ $\lambda_{q_1}(L^2)\lambda_{q_2}(L^2)(\lambda_{q_1}(L)\lambda_{q_2}(L)^{-2} + (h_1^{q_1} h_2^{q_2})^{\frac{1}{2}} v_{d_1} v_{d_2} \sqrt{2R(f_{\theta_0})} z_\alpha$ and, under independence, $n(h_1^{q_1} h_2^{q_2})^{\frac{1}{2}} \left(T_{n,5} - B_n \right) \xrightarrow{d} \mathcal{N}\left(0, 2v_{d_1}^2 v_{d_2}^2 R(f_{X_1}) \quad R(f_{X_2}) \right)$, with $B_n = \frac{\lambda_{q_1}(L^2)\lambda_{q_1}(L)^{-2}\lambda_{q_2}(L^2)\lambda_{q_2}(L)^{-2}}{nh_1^{q_1} h_2^{q_2}} - \frac{\lambda_{q_1}(L^2)\lambda_{q_1}(L)^{-2}R(f_{X_2})}{nh_1^{q_1}} - \frac{\lambda_{q_2}(L^2)\lambda_{q_2}(L)^{-2}R(f_{X_1})}{nh_2^{q_2}}$.

2.3 Regression-Based Tests

The testing of (2) (i.e., the assessment of whether m has a parametric structure m_{θ_0}, with θ_0 either specified or unspecified) is rooted on the nonparametric estimator for m introduced in (8). In a similar way to Härdle and Mammen [15] in the linear setting, problem (2) may be approached with the test statistic

$$T_{n,6} = \int_{\Omega_q} \left(\hat{m}_{h,p}(\mathbf{x}) - \mathscr{L}_{h,p} m_{\hat{\theta}}(\mathbf{x}) \right)^2 \hat{f}_h(\mathbf{x}) w(\mathbf{x})\, \omega_q(d\mathbf{x}),$$

where $\mathscr{L}_{h,p} m_{\theta_0}(\mathbf{x}) = \sum_{i=1}^n W_{n,i}^p(\mathbf{x}) m_{\theta_0}(\mathbf{X}_i)$ is the smoothing of m_{θ_0}, included to reduce the asymptotic bias (see [15]), and $w : \Omega_q \to \mathbb{R}_0^+$ is an optional weight function. The inclusion of $\hat{f}_h$ has the benefits of avoiding the presence of the density of $\mathbf{X}$ in the asymptotic bias and variance, and of mitigating the effects of the squared difference in sparse areas of $\mathbf{X}$.

Under H_0, $\hat{\theta} - \theta_0 = \mathcal{O}_{\mathbb{P}}(n^{-\frac{1}{2}})$, and certain regularity conditions (A1–A3 and A5 in [12]), the limit distribution of $T_{n,6}$ is

$$nh^{\frac{q}{2}} \left(T_{n,6} - \frac{\lambda_q(L^2)\lambda_q(L)^{-2}}{nh^q} \int_{\Omega_q} \sigma_{\theta_0}^2(\mathbf{x}) w(\mathbf{x})\, \omega_q(d\mathbf{x}) \right) \xrightarrow{d} \mathcal{N}\left(0, 2v_d^2 R\left(\sigma_{\theta_0}^2 w\right) \right),$$

where $\sigma_{\theta_0}^2(\mathbf{x}) = \mathbb{E}\left[(Y - m_{\theta_0}(\mathbf{X}))^2 | \mathbf{X} = \mathbf{x} \right]$, this is, $\mathbb{V}\mathrm{ar}\left[Y | \mathbf{X} = \mathbf{x} \right]$ under H_0.

3 Convergence Towards the Asymptotic Distribution

Unfortunately, the asymptotic distributions of the test statistics $T_{n,k}$, $k = 1, \ldots, 6$ are almost useless in practise. In addition to the unknown quantities present in the asymptotic distributions, the convergences toward the limits are slow and depend on the bandwidth sequences. This forces the consideration of resampling mechanisms for calibrating the distributions of the statistics under the null: parametric bootstraps in $T_{n,1}$, $T_{n,2}$, and $T_{n,4}$ [3, 11]; a *wild* bootstrap for $T_{n,6}$ [12]; and a permutation approach for $T_{n,3}$ and $T_{n,5}$ [10]. The purpose of this section is to illustrate, as an example, the convergence to the asymptotic distribution of the statistics $T_{n,3}$ and $T_{n,6}$ via insightful numerical experiments.

First, for $T_{n,3}$ we considered a circular-linear framework ($q = 1$), with a von Mises density with mean $\boldsymbol{\mu} = (0, 1)$ and concentration $\kappa = 1$ for the circular variable, and a $\mathcal{N}(0, 1)$ for the linear density. We also took von Mises and normal kernels. These choices gave $R(f_{\mathbf{X}}) = (2\pi)^{-1}\mathscr{I}_0(2)\mathscr{I}_0(1)^{-2}$ ($\mathscr{I}_0$ stands for the modified Bessel function of first kind and order 0), $R(f_Y) = \left(2\pi^{\frac{1}{2}}\right)^{-1}$, $v_d^2 = v_l^2 = (8\pi)^{-\frac{1}{2}}$, and $R(K) = \lambda_1(L^2)\lambda_1(L)^{-2} = \left(2\pi^{\frac{1}{2}}\right)^{-1}$. We simulated $M = 500$ samples of size $n = 5^k \times 10^l$, $k = 0, 1$, $l = 1, \ldots, 6$ under independence, obtaining $\left\{n\left(\frac{h_n g_n}{2v_d^2 v_l^2 R(f_{\mathbf{X}})R(f_Y)}\right)^{\frac{1}{2}}\left(T_{n,3}^j - A_n\right)\right\}_{j=1}^M$. We took $h_n = g_n = 2n^{-\frac{1}{3}}$ as a compromise between fast convergence and avoiding numerical instabilities. Figure 1 shows several density estimates for the sample of standardized statistics, jointly with the p-values of the Kolmogorov–Smirnov (K–S) test for $\mathcal{N}(0, 1)$, and of the Shapiro–Wilk (S–W) test for normality. Both tests are significant up to a very large sample size (close to $n = 5 \times 10^5$ data), which is apparent from the visual disagreement between the finite sample and asymptotic distributions for $n = 10^3$.

Second, for $T_{n,6}$, the regression model $Y = 1 + \varepsilon$ is considered, with $\varepsilon \sim \mathcal{N}\left(0, \frac{1}{4}\right)$, and $\mathbf{X}$ uniformly distributed on the circle. The composite hypothesis is $H_0 : m \equiv c$,

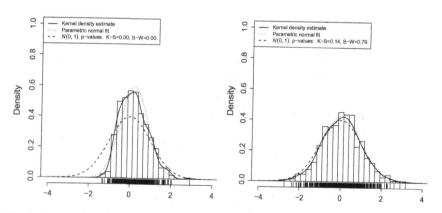

Fig. 1 Asymptotic and empirical distributions for the standardized statistic $T_{n,3}$, for sample sizes $n = 10^3$ (left) and $n = 5 \times 10^5$ (right)

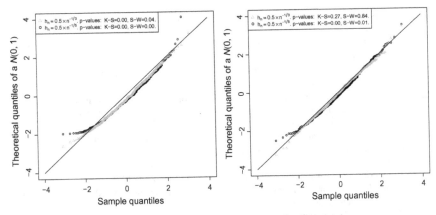

Fig. 2 QQ-plot comparing the sample quantiles of $\left\{nh^{\frac{1}{2}}\left(\frac{128}{\pi}\right)^{\frac{1}{4}}\left(T_n^j - \frac{\sqrt{\pi}}{4}nh\right)\right\}_{j=1}^{M}$ with the ones of the asymptotic distribution, for $n = 10^3$ (left) and $n = 5 \times 10^5$ (right)

for $c \in \mathbb{R}$ unknown. H_0 is checked using the local constant estimator with von Mises kernel and $w \equiv 1$. Figure 2 shows the QQ-plots computed from the sample $\left\{nh^{\frac{1}{2}}\left(\frac{128}{\pi}\right)^{\frac{1}{4}}\left(T_n^j - \frac{\sqrt{\pi}}{4}nh\right)\right\}_{j=1}^{M}$, for the bandwidth sequences $h_n = \frac{n^{-r}}{2}, r = \frac{1}{3}, \frac{1}{5}$, which were chosen in order to illustrate their impact in the convergence to the asymptotic distribution. Specifically, it can be seen that the effect of undersmoothing boosts the convergence since the bias is mitigated. Again, up to large sample sizes, the degree of disagreement between the finite sample and the asymptotic distributions is quite evident.

Acknowledgements The authors acknowledge the support of project MTM2016-76969-P from the Spanish State Research Agency (AEI), Spanish Ministry of Economy, Industry and Competitiveness, and European Regional Development Fund (ERDF). We also thank Eduardo Gil, Eva Gil, Juan J. Gil, and María Angeles Gil for inviting us to contribute to this volume, in memory of Pedro.

References

1. Bai ZD, Rao CR, Zhao LC (1988) Kernel estimators of density function of directional data. J Multivar Anal 27(1):24–39
2. Bickel PJ, Rosenblatt M (1973) On some global measures of the deviations of density function estimates. Ann Stat 1(6):1071–1095
3. Boente G, Rodríguez D, González-Manteiga W (2014) Goodness-of-fit test for directional data. Scand J Stat 41(1):259–275
4. Durbin J (1973) Weak convergence of the sample distribution function when parameters are estimated. Ann Stat 1:279–290
5. Elderton WP (1902) Tables for testing the goodness of fit of theory to observation. Biometrika 1(2):155–163
6. Fan J, Gijbels I (1996) Local polynomial modelling and its applications, vol 66. Monographs on statistics and applied probability. Chapman and Hall, London

7. Fan Y (1994) Testing the goodness of fit of a parametric density function by kernel method. Econom Theory 10(2):316–356
8. Fisher NI, Lee AJ (1981) Nonparametric measures of angular-linear association. Biometrika 68(3):629–636
9. García-Portugués E, Crujeiras RM, González-Manteiga W (2013) Kernel density estimation for directional-linear data. J Multivar Anal 121:152–175
10. García-Portugués E, Barros AMG, Crujeiras RM, González-Manteiga W, Pereira J (2014) A test for directional-linear independence, with applications to wildfire orientation and size. Stoch Environ Res Risk Assess 28(5):1261–1275
11. García-Portugués E, Crujeiras RM, González-Manteiga W (2015) Central limit theorems for directional and linear data with applications. Stat Sin 25:1207–1229
12. García-Portugués E, Van Keilegom I, Crujeiras R, González-Manteiga W (2016) Testing parametric models in linear-directional regression. Scand J Stat 43(4):1178–1191
13. González-Manteiga W, Crujeiras RM (2013) An updated review of goodness-of-fit tests for regression models. Test 22(3):361–411
14. Hall P, Watson GS, Cabrera J (1987) Kernel density estimation with spherical data. Biometrika 74(4):751–762
15. Härdle W, Mammen E (1993) Comparing nonparametric versus parametric regression fits. Ann Stat 21(4):1926–1947
16. Liddell IG, Ord JK (1978) Linear-circular correlation coefficients: some further results. Biometrika 65(2):448–450
17. Mardia KV (1976) Linear-circular correlation coefficients and rhythmometry. Biometrika 63(2):403–405
18. Mardia KV, Jupp PE (2000) Directional statistics, 2nd edn. Wiley series in probability and statistics. Wiley, Chichester
19. Pearson K (1900) On the criterion that a given system of deviations from the probable in the case of a correlated system of variables is such that it can be reasonably supposed to have arisen from random sampling. The London, Edinburgh, and Dublin Philos Mag and J Sci Ser-5 50(302):157–175
20. Pearson K (1916) On the application of "goodness of fit" tables to test regression curves and theoretical curves used to describe observational or experimental data. Biometrika 11(3): 239–261
21. Rosenblatt M (1975) A quadratic measure of deviation of two-dimensional density estimates and a test of independence. Ann Stat 3(1):1–14
22. Zhao L, Wu C (2001) Central limit theorem for integrated square error of kernel estimators of spherical density. Sci China Ser A 44(4):474–483

Asymptotic Behavior of Fractional Blending Systems

F. Javier Girón and M. Lina Martínez

Abstract After a brief description of the dynamic systems of ageing wines and spirits known as "fractional blending systems" or "criaderas and solera" systems, a general mathematical model is presented to determine, on the one hand, the distribution of the age of liquids of all the scales of the system and, on the other hand, the mean or average age of the liquids as the system is run. A theorem on the existence of an asymptotic equilibrium distribution of the "fractional blending systems" is given. This result refers to the existence of a unique asymptotic distribution of the ages which turns out to be a generalization of the Pascal distribution. This, in turn, implies the existence and uniqueness of an equilibrium mean or average of the ages.

1 Introduction

It is a common perception of winemakers that brandies and/or wines aged using static versus dynamic (sometimes called fractional blending) ageing systems differ in many aspects. The most common of these perceptions is the fact that brandies and wines aged using a fractional blending system are more homogeneous than those obtained using a statical ageing system. This is due, in part, to the mixing of liquids

This paper is dedicated to the memory of a great friend of mine, the late Pedro Gil, a brilliant statistician but an even better and extraordinary human being. We met at the Department of Statistics of the Universidad Complutense de Madrid, in the seventies, where we soon became good friends and shared many scientific and vital experiences. In spite of living in different cities we always maintained an everlasting friendship.

F. J. Girón (✉)
Real Academia de Ciencias Exactas, Físicas y Naturales, C/ Valverde 22, 28004 Madrid, Spain
e-mail: fj_giron@uma.es

M. L. Martínez
Faculty of Sciences, Department of Statistics, Universidad de Málaga, Campus de Teatinos s/n, 29071 Málaga, Spain
e-mail: mlmartinez@uma.es

© Springer International Publishing AG 2018
E. Gil et al. (eds.), *The Mathematics of the Uncertain*, Studies in Systems, Decision and Control 142, https://doi.org/10.1007/978-3-319-73848-2_18

185

of different ages which occurs in fractional blending systems, and constitute the final product that goes to the market, i.e. the one obtained from the *solera*.

In Sect. 2, we briefly describe the main features and ways of operation of the dynamical systems. A general mathematical model for describing the behavior of the *criaderas and solera dynamic systems*, sometimes called *fractional blending systems*, is presented in Sect. 3. The same model applies to sherry wines and brandies, and it allows for the computation of the distribution of the age of the wines and/or brandies at any period of the system for all criaderas and the solera —which constitute the setup of the dynamical system—, and, as a byproduct, the computation of the mean or average age. This section includes a general theorem that proves the asymptotic behavior of the dynamical systems. In a previous paper (see Girón [1]), a simpler model for the computation of the average age, which included the demonstration of the existence of a limiting average age for all elements of the system, was given.

2 Description of the Dynamical Systems of Ageing

The general process of maturing wines and spirits, known as "static" or by "vintages or añadas", is described as follows: an oak cask is replenished with the liquid that remains without manipulation during the established period of maturation. In some occasions, before the end of the maturation period, the liquid may be transferred to another cask.

On the other hand, the process known as "dynamic", or "by criaderas and solera system", is a "successive fractional blending system". Fractional blending system means a mixing process consisting of partial withdrawal with the corresponding replenishment. This process is genuinely Spanish and is unique (see Martínez De la Ossa et al. [3], and Quirós and Carrascal [4]) and has been practiced for over 200 years (Quirós and Carrascal [4]).

It consists of a stock of casks, divided into "escalas", each of them formed by an approximate number of butts. The first of these "escalas" is the "solera", which contains the oldest wine or brandy, being the next older "escala" the first "criadera" and so on (second "criadera", third "criadera", etc.) A proportion of the "solera" is drawn off (operation called "saca") for bottling and selling in the market and the "solera" is replenished ("rocío") with the same quantity of liquid from the first "criadera". The latter is replenished with the second "criadera" and so on up to the last "criadera", which is filled with just new liquid. This system gives uniformity to the product (see Fig. 1).

The number of "escalas" or steps usually is 3–4 for manzanillas de Sanlucar, 5–6 or even 9 for fino wines, and up to 16–20 for brandies. The "saca" of the "solera" is normally done 3–4 times per year; it may be done less frequently but it is uncommon to do it more frequently.

Fig. 1 Scheme of a dynamical system of *criaderas y solera* illustrating the operations of withdrawing off (*sacas*) and filling up (*rocíos*), which are known under the name of running up the scales (*correr las escalas*)

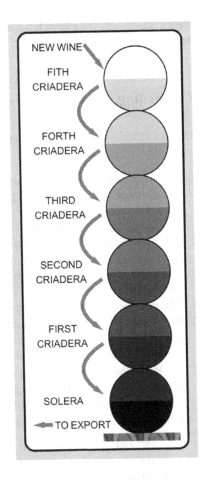

3 A General Mathematical Model to Determine the Age Distribution in a System of Criaderas and Soleras

3.1 Introducing the Mathematical Model

In this section, a mathematical model for describing the behavior of the dynamic systems based on criaderas and soleras is presented, which shows the evolution of the distribution of the age of the wines and/or brandies. The main advantage of this model is that it provides a rigorous proof of the fact that dynamical systems become *stationary*, i.e. the distribution of the age of the criaderas and solera achieve a limit distribution which does not depend on the initial distribution of ages of the liquids in the system. As a byproduct, the *average or mean age* of the final product obtained from the soleras -which is a mixture of several ages, due to the dynamic nature of these systems-, also reaches a limit. This last fact was well known to the winemakers from empirical experience, but no rigorous proof of it has been given yet.

3.2 Setting up and Development of the Model

The dynamical systems depends -and, consequently, differ- on the following basic parameters:

1. The frequency of drawing off: whether it is annual, every six months, quarterly, etc., denoted by t, which, depending on the frequency of the drawing off, usually takes the values $t = 1, 1/2, 1/3, 1/4$, and $1/6$. Its inverse $n_a = 1/t$ denotes the annual number of drawing offs.
2. The proportion of liquid withdrawn each time, denoted by p and measured in a $(0,1)$ scale. The remaining proportion in the barrel is denoted as $q = 1 - p$.
3. The number of scales used -barrels or "escalas", called "criaderas", and the last one "solera"-, is denoted by k.
4. The number of periods the system is run, denoted by n.
5. The vector $\mathbf{F}^{(0)} = (F_0^{(0)}, F_1^{(0)}, \ldots, F_k^{(0)})'$, which denotes the distribution of the age of the content of each barrel at the start -initial conditions of the system-, where $F_0^{(0)}$ denotes the "sobretablas distribution" of the liquid to fill up the first barrel, i.e. the last criadera (see Fig. 1); $F_1^{(0)}$ the corresponding distribution of the filling up of the second barrel, and so on up to the "solera barrel". This operation, as explained before, is known as the *rocío*. In the case of brandies, the *rocío* filling is a zero age liquid at the first stage.

After the end of the first period t, the distribution of the age of the barrels in the system $\mathbf{F}^{(1)} = (F_0^{(1)}, F_1^{(1)}, \ldots, F_k^{(1)})$ is the following mixture of distributions, as exemplified by Fig. 1.

To this mixture we have to add t years. As we are working with distribution functions, the addition of a constant t has to be entered in the formulas as the convolution of the mixture with a distribution degenerated at t.

This is accomplished through the convolution -represented by the operator $*$- of the distribution with a Dirac delta, δ_t, concentrated at t.

$$
\begin{aligned}
F_0^{(1)} &= F_0^{(0)} & &= F_0^{(0)} \\
F_1^{(1)} &= q(F_1^{(0)} * \delta_t) + p(F_0^{(0)} * \delta_t) & &= (q F_1^{(0)} + p F_0^{(0)}) * \delta_t \\
F_2^{(1)} &= q(F_2^{(0)} * \delta_t) + p(F_1^{(0)} * \delta_t) & &= (q F_2^{(0)} + p F_1^{(0)}) * \delta_t \qquad (3.1) \\
&\cdots\cdots\cdots\cdots\cdots\cdots & &\cdots\cdots\cdots\cdots\cdots \\
F_k^{(1)} &= q(F_k^{(0)} * \delta_t) + p(F_{k-1}^{(0)} * \delta_t) & &= (q F_k^{(0)} + p F_{k-1}^{(0)}) * \delta_t.
\end{aligned}
$$

$$\Downarrow \qquad\qquad\qquad \Downarrow$$

$$\text{saca} \qquad\qquad \text{rocío}$$

The first equation of Formulae (3.1) means that the age of the *rocío* liquid has always the same distribution: typically, but not necessarily, a Dirac delta at 0, i.e. 0 age liquid or, equivalently, $F_0^{(0)} = F_0^{(1)} = \cdots = F_0^{(n)} = \delta_0$.

In general, after n periods, the age of the liquid in each barrel, is obtained through the recurrence formulas

$$
\begin{aligned}
F_0^{(n)} &= F_0^{(n-1)} \\
F_1^{(n)} &= (q F_1^{(n-1)} + p F_0^{(n-1)}) * \delta_t \\
F_2^{(n)} &= (q F_2^{(n-1)} + p F_1^{(n-1)}) * \delta_t
\end{aligned}
\tag{3.2}
$$

$$\cdots\cdots\cdots\cdots\cdots\cdots\cdots\cdots\cdots$$

$$
F_k^{(n)} = (q F_k^{(n-1)} + p F_{k-1}^{(n-1)}) * \delta_t.
$$

As the distributions involved are discrete ones, the computation of the distribution of $F_j^{(n)}$ for any $j = 1, \ldots, k$, and $n = 1, 2, \ldots$ is a very complex one due to the presence of a convolution at every period.

One way to solve these recurrence equations is through the use of characteristic functions. The characteristic function of a mixture of distributions is the same mixture of the corresponding characteristic functions of the terms in the mixture, and the characteristic function of a convolution of independent random variables, as is our case, is the product of the corresponding characteristic functions. Further, the use of characteristic function greatly simplifies the analysis of the limiting behavior of the recurrence Eq. (3.2) when n tends to infinity.

Therefore, if we denote by $\varphi_j^{(n)}(s)$ the characteristic function of $F_j^{(n)}$, Eq. (3.2) can be written as

$$
\begin{aligned}
\varphi_0^{(n)}(s) &= \varphi_0^{(n-1)}(s) \\
\varphi_1^{(n)}(s) &= e^{its} (q \varphi_1^{(n-1)}(s) + p \varphi_0^{(n-1)}(s)) \\
\varphi_2^{(n)}(s) &= e^{its} (q \varphi_2^{(n-1)}(s) + p \varphi_1^{(n-1)}(s))
\end{aligned}
\tag{3.3}
$$

$$\cdots\cdots\cdots\cdots\cdots\cdots\cdots\cdots\cdots\cdots\cdots$$

$$
\varphi_k^{(n)}(s) = e^{its} (q \varphi_k^{(n-1)}(s) + p \varphi_{k-1}^{(n-1)}(s)).
$$

Defining, for all n, the vector of characteristic functions

$$
\boldsymbol{\varphi}^{(n)}(s) = (\varphi_0^{(n)}(s), \varphi_1^{(n)}(s), \ldots, \varphi_k^{(n)}(s))',
\tag{3.4}
$$

Equation (3.3) can be written in matrix form as:

$$\varphi^{(n)}(s) = \begin{pmatrix} 1 & 0 & 0 & \cdots & 0 & 0 & 0 \\ p\,e^{its} & q\,e^{its} & 0 & \cdots & 0 & 0 & 0 \\ 0 & p\,e^{its} & q\,e^{its} & \cdots & 0 & 0 & 0 \\ \multicolumn{7}{c}{\cdots\cdots\cdots\cdots\cdots\cdots\cdots\cdots\cdots} \\ 0 & 0 & 0 & \cdots & q\,e^{its} & 0 & 0 \\ 0 & 0 & 0 & \cdots & p\,e^{its} & q\,e^{its} & 0 \\ 0 & 0 & 0 & \cdots & 0 & p\,e^{its} & q\,e^{its} \end{pmatrix} \varphi^{(n-1)}(s). \tag{3.5}$$

Denoting by $\mathbf{Q}(t, p, s)$ the square $(k + 1) \times (k + 1)$ matrix in Eq. (3.5), these equations adopt the simple form

$$\varphi^{(n)}(s) = \mathbf{Q}(t, p, s)\,\varphi^{(n-1)}(s). \tag{3.6}$$

In terms of the initial conditions $\varphi^{(0)}(s)$, this last equation can be written as

$$\varphi^{(n)}(s) = \mathbf{Q}(t, p, s)^n\,\varphi^{(0)}(s), \tag{3.7}$$

where $\mathbf{Q}(t, p, s)^n$ denotes the n-th power of matrix $\mathbf{Q}(t, p, s)$.

This last formula allows for easy computations of the characteristic functions of the distributions of all criaderas and solera of the system for any period n, where $\varphi^{(0)}(s)$ is the vector of characteristic functions of the initial conditions $\mathbf{F}^{(0)}$. From these characteristic functions it can be easily derived the distributions of the age of the criaderas and soleras at any period.

3.3 Asymptotic Behavior: The Equilibrium Distribution

Note that, from Eq. (3.7), the limiting behavior of the system only depends on the limit of the matrix power $\mathbf{Q}^n$ when n increases to infinity, and the initial conditions, as the next lemma shows.

Lemma 3.1 *The limit of* $\mathbf{Q}(t, p, s)^n$ *as n goes to infinity is the following* $(k + 1) \times (k + 1)$ *matrix whose columns, except the first one, are zero vectors.*

$$\lim_{n \to \infty} \mathbf{Q}(t, p, s)^n = \begin{pmatrix} 1 & 0\,0\ldots 0 \\ \dfrac{p\,e^{its}}{1 - q\,e^{its}} & 0\,0\ldots 0 \\ \left(\dfrac{p\,e^{its}}{1 - q\,e^{its}}\right)^2 & 0\,0\ldots 0 \\ \cdots\cdots\cdots\cdots\cdots \\ \left(\dfrac{p\,e^{its}}{1 - q\,e^{its}}\right)^k & 0\,0\ldots 0 \end{pmatrix}. \tag{3.8}$$

The following theorem follows straightforward from Lemma 3.1.

Theorem 3.1 *The limit of the vector of characteristic functions* $\varphi^{(n)}(s)$, *when* n *tends to infinity, is*

$$\varphi^{(\infty)}(s) = \lim_{n \to \infty} \varphi^{(n)}(s) = \lim_{n \to \infty} \mathbf{Q}(t, p, s)^n \, \varphi^{(0)}(s)$$

$$= \varphi_0^{(0)}(s)\left(1, \frac{p \, e^{its}}{1 - q \, e^{its}}, \left(\frac{p \, e^{its}}{1 - q \, e^{its}}\right)^2, \dots, \left(\frac{p \, e^{its}}{1 - q \, e^{its}}\right)^k\right)'. \qquad (3.9)$$

Further, the vector of characteristic functions $\varphi^{(\infty)}(s)$, *is the unique solution of the equation*

$$\psi = \mathbf{Q}(t, p, s)\, \psi.$$

This asymptotic result deserves several comments.

From Eq. (3.9), the limiting distribution of the age of barrel j, or equivalently of the *criadera* $k - j$, for $j = 1, 2, \dots, k$, has the following characteristic function:

$$\varphi_j^{(\infty)}(s) = \varphi_0^{(0)}(s) \cdot \left(\frac{p \, e^{its}}{1 - q \, e^{its}}\right)^j.$$

For $t = 1$, the second factor of the last equation corresponds to the characteristic function of a Pascal distribution (see Johnson et al. [2]) with parameters j and p, Pas(j, p), for any $j = 1, 2, \dots, k$. If $t \neq 1$, the characteristic function corresponds to the distribution of t times a Pascal distribution, which we denote as a generalized Pascal distribution of parameters t, j and p, denoted by Pas(t, j, p).

A Pascal distribution of parameters j (a positive integer) and $p \in (0, 1)$ (a probability), is a discrete distribution -related to waiting times in independent Bernoulli trials with the same probability p-, which takes values $x = j, j + 1, j + 2, \dots$ with probability mass

$$\Pr(x \mid j, p) = \binom{x - 1}{j - 1} p^j q^{x-j},$$

and 0 for values $x = 1, \dots, j - 1$.

If we denote by $A(j, p)$ a random variable with a Pascal distribution of parameters j and p, then the distribution of the random variable $A(t, j, p) = t \, A(j, p)$ is said to follow a *generalized Pascal distribution* of parameters t, j and p, Pas(t, j, p). Its probability mass for values of $z = jt, (j + 1)t(j + 2)t, \dots$, is given by

$$\Pr(A(t, j, p) = z) = \binom{\frac{z}{t} - 1}{j - 1} p^j q^{\frac{z}{t}-j},$$

and 0 otherwise.

If the distribution of the age of the *sobretablas* liquid is $F_0^{(0)}$, then the asymptotic distribution of the age $F_j^{(\infty)}$ at stages $j = 1, \dots, k$ are the convolution of $F_0^{(0)}$ with a generalized Pascal distribution Pas(t, j, p).

Corollary 3.1 follows immediately from Theorem 3.1 and by the Lévy-Cramér theorem.

Corollary 3.1 *For all $j = 1, \ldots, k$, the sequence of distribution functions $F_j^{(n)}$ converges in distribution to the equilibrium distribution $F_j^{(\infty)} = F_0^{(0)} * \mathrm{Pas}(t, j, p)$.*

Another important consequence of Theorem 3.1 refers to the increasing of "the age" of wines or spirits, from the last criadera to the solera. It is taken as a matter of fact among wine makers; on the other hand, it is obviously intuitive that the oldest wines are in the solera of the dynamical systems. But "the age" of any scale is not a single number but it is a distribution function.

There is a natural partial ordering of one-dimensional distribution functions called *stochastic dominance of first order* (see Shaked and Shanthikumar [5]) which is defined as follows:

Let F and G be two distribution functions. We say that F *stochastically dominates* G, and will be denoted by $G \precsim_{sd} F$ or equivalently $F \succsim_{sd} G$, if $F(x) \leq G(x)$ for all $x \in \mathbb{R}$.

Next theorem proves that the distribution of age of the solera stochastically dominates that of the first criadera and, in turn, this one dominates that of the second, and so on; further, this fact holds true whatever the initial distribution of the "sobretablas" $F_0^{(0)}$ is.

Theorem 3.2 *For any distribution of the age of the "sobretablas", $F_0^{(0)}$, the following stochastic dominance orders holds*

$$F_1^{(\infty)} \precsim_{sd} F_2^{(\infty)} \precsim_{sd} \cdots F_j^{(\infty)} \precsim_{sd} \cdots \precsim_{sd} F_k^{(\infty)}.$$

Proof First, we use the fact that Pascal distributions are stochastically ordered according to the integer parameter j for all t and p fixed, as follows:

$$\mathrm{Pas}(t, 1, p) \precsim_{sd} \mathrm{Pas}(t, 2, p) \precsim_{sd} \cdots \precsim_{sd} \mathrm{Pas}(t, j, p) \precsim_{sd} \cdots \precsim_{sd} \mathrm{Pas}(t, k, p).$$

Second, we note that stochastic ordering is preserved when convolved with any arbitrary distribution. If we take this distribution to be $F_0^{(0)}$, we obtain:

$$F_0^{(0)} * \mathrm{Pas}(t, 1, p) \precsim_{sd} \cdots \precsim_{sd} F_0^{(0)} * \mathrm{Pas}(t, j, p) \precsim_{sd} \cdots \precsim_{sd} F_0^{(0)} * \mathrm{Pas}(t, k, p).$$

and the theorem follows suit as $F_j^{(\infty)} = F_0^{(0)} * \mathrm{Pas}(t, j, p)$, for $j = 1, 2, \ldots, k$. $\square$

3.4 On the Average Age of the Dynamical Systems

The computation of the mean or average age of dynamical systems, and its asymptotic behavior, is of vital importance to winemakers. Some results can be derived from Theorem 3.1 and Corollary 3.1.

In fact, the convergence in distribution of the sequence of distribution of age in all scales $\{F_j^{(n)}\}_n$ to the equilibrium distribution $F_j^{(\infty)}$, for all $j = 1, \ldots, k$, implies the convergence of the sequence of means. Thus, if we denote by E_0 the mean of the *sobretablas* distribution, and we take into account that the mean of a generalized Pascal distribution is $j\,t/p$, then the limit of the mean age of the $k - j$-th "criadera" is $E_0 + j\,t/p$.

If we take expectations in Eq. (3.2), we can state, without proof, a more precise result that further establishes the existence of an equilibrium vector for the mean age.

If we denote by $E_j^{(n)}$ the mean of the distribution $F_j^{(n)}$, and by $\mathbf{E}^{(n)} = (E_0, E_1^{(n)}, \ldots,$ $E_j^{(n)}, \ldots, E_k^{(n)})$ at period n, the vector of the mean ages of the sobretablas, the criaderas, and the solera, we have:

Theorem 3.3 *The limit of the vector of the mean ages of the dynamical system, $\mathbf{E}^{(n)}$, when the number of periods n tends to infinity, is*

$$\mathbf{E}^{(\infty)} = \lim_{n \to \infty} \mathbf{E}^{(n)} = (E_0, E_0 + t/p, \ldots, E_0 + j\,t/p, \ldots, E_0 + k\,t/p)'.$$

Further, the limit vector, $\mathbf{E}^{(\infty)}$, is the unique equilibrium (or invariant) vector of the mean ages, that is, if $\mathbf{E}^{(0)} = \mathbf{E}^{(\infty)}$, then

$$\mathbf{E}^{(0)} = \mathbf{E}^{(1)} = \cdots = \mathbf{E}^{(n)} = \cdots = \mathbf{E}^{(\infty)}.$$

In particular, if $E_0 = 0$ -the most frequent value for ageing brandies and some wines-, the limit age of the "solera" is:

$$\text{equilibrium or limit age of the solera} = \frac{k\,t}{p}.$$

This last formula tells that, in order to increase the mean age of the solera liquids, the frequency of drawing offs (sacas) t should be taken as large as possible: however, in practice t should be less or equal than one year. On the other hand, k should be as large as possible, but logistic and economic reasons prevent k from being too large, except for wines and brandies of certified quality, as for instance VOS and VORS sherrys.

Diminishing p produces very old liquids but it drastically reduces the convergence rate to reach the equilibrium age. Therefore, a compromise among the parameters t, k and p should be taken when setting up a new dynamical system.

4 Discussion

A rigorous proof of the existence of an asymptotic equilibrium of the "criaderas y solera" dynamical systems is presented in the paper. The asymptotic results refer both to the distribution of the ages of liquids in the system for all scales -including the solera- as well as to the mean or average age of the liquids in the system.

Further, explicit results for the asymptotic or equilibrium distributions and means of the liquids are given (see Theorems 3.1 and 3.3 and Corollary 3.1).

These theoretical results can prove useful in practice to devise and optimize dynamical systems that approach the equilibrium in a fewer number of periods. Another possible application would be the setting up of complex hierarchical dynamical systems substituting the sobretablas distribution for that of the solera of another dynamical system.

References

1. Girón FJ (1991) Un modelo general para determinar la edad de un sistema de soleras. Trab Estadíst 6(2):3–10
2. Johnson NL, Kemp AW, Kotz S (2005) Univariate discrete distributions, 3rd edn. Wiley, New York
3. Martínez De la Ossa E, Pérez L, Caro I (1987) Variations of the major volatiles through aging of sherry. Am J Enol Vitic 38:293–297
4. Quirós JM, Carrascal V (1992) Ageing Brandy de Jerez by the Solera system. In: Cantagrel R (ed) 1er Symposium Scienifique International Cognac. Cognac
5. Shaked M, Shanthikumar JG (2007) Stochastic orders. Springer series in statistics. Springer, New York

Multiple Hypothesis Tests: A Bayesian Approach

Miguel A. Gómez-Villegas and Beatriz González-Pérez

Abstract Multiple hypothesis tests is a topic which has recently shown a major expansion, mainly due to the expansion of the methodology developed in connection with genomics. These new methods allow scientists to handle simultaneously thousands of null hypotheses. The frequentist approach to this problem consists of using different error measures in testing so that to ensure the Type I error remains below a desired level. This paper introduces a parametric Bayesian analysis to determine the hypotheses to be considered as being significant (i.e., useful) for a posterior deeper analysis. The results are to be compared with the frequentist methodology of the false discovery rate (FDR). Differences between both approaches are shown by means of simulation examples.

1 Introduction

This article is my memory to Pedro Gil Álvarez who was my professor circa 1970 at the Faculty of CC. Mathematics at the Complutense University.

I always thought Pedro was a very intelligent person. When I was in my 4th and 5th year of College he was in charge of the labs of the most diverse subjects in the field of statistics. Later, I realized that Teaching Assistants were scarce in the department. As a consequence, the professors who were at the beginning of their careers had to perform a remarkable effort to deal with students who were just at the initial level of statistics. This reinforced my feeling that he was very smart.

In the field of simultaneous inference, multiple hypothesis testing deals with the testing of more than one hypothesis at time. A single hypothesis test can be described as follows

M. A. Gómez-Villegas (✉) · B. González-Pérez
Departamento de Estadística e Investigación Operativa I. Facultad de
Ciencias Matemáticas, Instituto de Matemática Interdisciplinar (IMI),
Universidad Complutense de Madrid, 28040 Madrid, Spain
e-mail: ma.gv@mat.ucm.es

B. González-Pérez
e-mail: beatrizg@mat.ucm.es

© Springer International Publishing AG 2018
E. Gil et al. (eds.), *The Mathematics of the Uncertain*, Studies in Systems,
Decision and Control 142, https://doi.org/10.1007/978-3-319-73848-2_19

195

$$H = 0 : \theta \in \Theta_0 \quad \text{versus} \quad H = 1 : \theta \in \Theta_1 \tag{1}$$

with $\Theta_0 \cap \Theta_1 = \emptyset$. A statistic, $T(X)$, is observed and a value $T(x) = t$ is obtained.

From the frequentist point of view, the null hypothesis will be rejected if the observed value, t, is over a certain threshold. This threshold, which is arbitrary, settles a certain rejection region, Γ, in such a way that if $t \in \Gamma$ then $H = 0$ is rejected while if $t \notin \Gamma$, then $H = 0$ is accepted. The rejection region defines the Type I error, that is to reject the null hypothesis when it is true, $\theta \in \Theta_0$ but $t \in \Gamma$.

When testing a single hypothesis as (1), an acceptable maximum Type I error probability is specified and the conclusions are obtained based on a statistic which meets this specification. Then, the maximum Type I error probability is fixed at a certain level, which is known as significance level, α

$$\sup_{\theta \in \Theta_0} Pr(T \in \Gamma_\alpha | \theta) = Pr(T \in \Gamma_\alpha | H = 0) = \alpha \tag{2}$$

and a frequentist measure of the evidence against the null hypothesis is the p–value, defined as the minimum false positive rate at which an observed statistic can be called significant,

$$p - value(t) = \sup_{\theta | H = 0} Pr(T \in \Gamma_t | H = 0) \tag{3}$$

where Γ_t is the critical region for $T = t$. Alternatively, the probability that, the statistic is as or more extreme than the observed one, t, under the null hypothesis,

$$p - value(t) = Pr(|T(X)| \geq |t| | H = 0)$$

can be used as test statistics (see Lehmann and Romano [8, p. 63]).

But when many hypothesis are tested, to fix an individual Type I error probability for each one may have consequences if the set of hypothesis are evaluated as a whole. A review about multiple hypothesis testing can be seen in Shaffer [10].

The question is, basically, whether the probability of a false positive increases with the number of tests. For example, if the significance level is fixed at 0.05 for each test and a set of 100 tests are evaluated, the expected number of false positives is 5. Then, 5 hypothesis will be rejected simply by chance. The level 0.05 has been widely used in the literature since Fisher proposed it, and its intense use has produced basically correct scientific inferences. Otherwise it would not have remained as a reference level so long. But the 0.05 level was applied to a single hypothesis, not to a great number of simultaneous hypothesis; this is the reason for introducing measures of evidence that take into account all the hypothesis which are tested simultaneously.

Multiple hypothesis testing has been widely used in the past in different fields as Shaffer [10] pointed out. Recently, the field of genomics, and in particular the DNA microarray experiments where thousands of hypothesis can be tested simultaneously, have influenced the revitalization of the procedures of the multiple hypothesis tests, see Dudoit et al. [4] In this context, consider m hypothesis tests

Table 1 Multiple hypothesis

	Number accepted	Number rejected	
$H = 0$	U	V	m_0
$H = 1$	T	S	m_1
	W	R	m

$$H_i = 0 \quad \text{versus} \quad H_i = 1, \quad i = 1, \ldots, m \qquad (4)$$

and we want to test the m null hypothesis simultaneously. Benjamini and Hochberg [2] propose Table 1 to summarize the problem.

In Sect. 2 we have summarized some of the frequentist procedures to test m null hypothesis simultaneously, as in (4). In Sect. 3 we have proposed two Bayesian approaches for testing (4) and we compare, for simulated data, the results obtained for these Bayesian methods with the frequentist results. Section 3 includes a hierarchical model. Finally, Sect. 4 contains some conclusions and comments.

2 Frequentist Procedures

2.1 Type I Error Rates

Shaffer [10] picked out the generalizations of the Type I error described above to the multiple testing problems: the family wise error rate, FWER, the per–comparison error rate PCER, the per-family error rate, PFER, and the false discovery rate, FDR. This last one was introduced by Benjamini and Hochberg [2].

In any case, the procedure to carry out a multiple hypothesis test consists of controlling a particular Type I error rate at a certain level α and producing a list of R rejected hypothesis. If the level α is fixed to control the Type I error rate only when all the null hypothesis are true, $m_0 = m$, one speaks of weak control, whereas strong control referrers to the control of the Type I error rate under any possible combination of the true and false null hypothesis. See Shaffer [10] and Dudoit et al. [4] for more details about this setting.

2.2 p-Values

As defined above, the p-value $p_i(t_i)$ for a single hypothesis H_i can be viewed as the level of the test at which the hypothesis H_i would be rejected, given the value of a statistic $T_i = t_i$. The smaller the p-value, $p_i(t_i)$, the stronger the evidence against the null H_i. With a fixed significance level, α, rejecting H_i when $p_i \leq \alpha$ assumes that the Type I error rate is controlled at level α.

The concept of the p-value can be extended to the multiple testing problem under the concept of the adjusted p-value, see Dudoit et al. [4]. Given a test procedure, for example a FDR procedure (see Benjamini and Hochberg [2]) defined as in (6), the adjusted p-value for a single hypothesis H_i is defined as

$$\tilde{p}_i = \inf\{\alpha \in [0, 1] : H_i \text{ is rejected at nominal FDR} = \alpha\} \qquad (5)$$

in words, the nominal level of the entire test procedure at which the hypothesis H_i would be rejected, given the values of all the test statistics.

Westfall and Young [13] estimated the adjusted p-values by resampling methods. A recent discussion about p-values can be seen in Wasserstein and Lazar [12].

2.3 The False Discovery Rate

Benjamini and Hochberg [2] introduced this concept, less conservative than the others, to control the expected proportion of Type I errors among the rejected hypothesis. So, the FDR is defined as, from Table 1,

$$FDR = E\left[\frac{V}{R}\right] Pr(R > 0). \qquad (6)$$

If $R = 0$, then $FDR = 0$. Benjamini and Hochberg [2] derived a procedure for strong control of the FDR for independent test statistics. This procedure for control of the FDR at level α can be resumed as follows:

- The observed p-values are computed and ordered: $p_{(1)} \leq p_{(2)} \leq \cdots \leq p_{(m)}$.
- Compute $R_{BH} = \max\{i : p_{(i)} \leq \alpha(i/m)\}$.
- Reject null hypothesis corresponding to $p_{(1)}, \ldots, p_{(R_{BH})}$. If R_{BH} does not exist, no hypothesis is rejected.

Benjamini and Yekutieli [3] showed that the procedure above controls the FDR at level α, under certain conditions of dependency.

The adjusted p-values corresponding to this control are

$$\tilde{p}_{(i)} = \min_{j=i,\ldots,m} \left\{\min\left(\frac{m}{j} p_{(j)}, 1\right)\right\}. \qquad (7)$$

In a microarray setting, Dudoit et al. [4] proposed the FDR controlling procedures as alternatives to other approaches. They argue that in this context one may be willing to bear a few false positives as long as their number is small in comparison to the number of rejected hypothesis.

3 A Bayesian Approach

Consider the problem of multiple hypothesis testing given in (4). As in Table 1, we denote by m_1 the unknown number of hypotheses where $H_i = 1, i = 1, \ldots, m$. In this section we propose two different Bayesian methods for testing (4).

Bayesian inference on multiple hypothesis has been widely studied. We are studying it in the same way as the approach proposed by Waller and Duncan [11], Hobert [7], Barbieri and Berger [1] and Scott and Berger [9].

Our objective is to give a Bayesian estimator for m_1, for instance, the mean or the mode of the posterior density of m_1. We denote θ_i by the prior probability of the null $H_i = 0, \theta_i = P(H_i = 0)$, and consequently $1 - \theta_i = P(H_i = 1), i = 1, \ldots, m$, and supposing that the m hypothesis are independent, then $H_i|\theta_i \sim Bernoulli(1 - \theta_i)$ and $m_1 = \sum_{i=1}^{m} H_i$. An initial Bayesian approach is to assume that all θ_i are equal to θ. Then, $m_1|\theta \sim Binomial(m, 1 - \theta)$ and we can estimate m_1 with $\hat{m}_1 = m(1 - \hat{\theta})$, where $\hat{\theta}$ is an estimator of θ (a location parameter of the posterior density of θ).

For each hypothesis H_i, a vector $T_i = (X_{i1}, \ldots, X_{in})$ is observed. Suppose that for all i, under $H_i = 0$ the density is $f(t|0)$, while under $H_i = 1$ the density is $f(t|1)$. Thus, the observations T_i (assume i.i.d. random variables) come from a mixture of both densities:

$$f(t_i|\theta) = f(t_i|H_i = 0)Pr(H_i = 0|\theta) + f(t_i|H_i = 1)Pr(H_i = 1|\theta) \quad (8)$$
$$= \theta f(t_i|0) + (1 - \theta)f(t_i|1)$$

and the likelihood can be written as

$$f(t_1, t_2, \ldots, t_m|\theta) = \prod_{i=1}^{m} f(t_i|\theta) = \prod_{i=1}^{m} [\theta f(t_i|0) + (1 - \theta)f(t_i|1)], \quad (9)$$

where $t_i = (x_{i1}, \ldots, x_{in})$.

The prior distribution for the parameter θ can be thought of as a beta distribution, $Beta(a, b)$, because of its versatility to model a density over the interval $[0, 1]$. Then,

$$\pi(\theta|a, b) = \frac{\Gamma(a + b)}{\Gamma(a)\Gamma(b)}\theta^{a-1}(1 - \theta)^{b-1}, \quad 0 \le \theta \le 1 \quad (10)$$

and the posterior density of θ, given $t_1, \ldots, t_m, a, b$, is given by

$$\pi(\theta|t_1, \ldots, t_m, a, b) = \frac{\pi(\theta|a, b) \prod_{i=1}^{m} [\theta f(t_i|0) + (1 - \theta)f(t_i|1)]}{\int_0^1 \pi(\theta|a, b) \prod_{i=1}^{m} [\theta f(t_i|0) + (1 - \theta)f(t_i|1)] d\theta} \quad (11)$$

where, in the first step, a and b are known and previously fixed.

Then, we compute and order the posterior probabilities

$$P(H_i = 0|T_i = t_i) = \frac{f(t_i|0)\theta}{f(t_i|0)\theta + f(t_i|1)(1 - \theta)}, \quad i = 1, \ldots, m. \quad (12)$$

These probabilities will be estimated using $\widehat{\theta}$, the estimated value of θ obtained through (11), then

$$\widehat{P}(H_i = 0|T_i = t_i) = \frac{f(t_i|0)\widehat{\theta}}{f(t_i|0)\widehat{\theta} + f(t_i|1)(1 - \widehat{\theta})}, \quad i = 1, \ldots, m. \quad (13)$$

Really, the Bayesian method involves computing $P(H_i = 0|T_1 = t_1, \ldots, T_m = t_m)$, but the approximation proposed by (12) and (13) simplifies the simulation's task a lot.

Finally, we will use $\widehat{\theta}$ and the estimated posterior probabilities in multiple hypothesis testing using the two following methods:

Method 1: We estimate the percentage of hypothesis where $H_i = 0$ by using the mean $\widehat{\theta}$ of the posterior density (11) as an estimator of θ. In this case, $\widehat{m}_1 = m(1 - \widehat{\theta})$. Then, the $\widehat{m}_1$ hypotheses with the lowest estimated posterior probabilities, see (13), will be rejected (will be declared interesting).

Method 2: We use the following Bayesian decision procedure

- Given $\widehat{\theta}$, the observed posterior probabilities are computed from (13) and ordered.
- Compute $\widehat{i} = \max\{i : \widehat{P}\left(H_{(i)} = 0|T_{(i)} = t_{(i)}\right) \leq 0.5\}$. Thresholds other than 0.5 can be used.
- Reject the null hypothesis corresponding to $t_{(1)}, \ldots, t_{(\widehat{i})}$. If $\widehat{i}$ does not exist, no hypothesis is rejected.

Then, the hypotheses with the estimated posterior probabilities lower than 0.5 will be rejected.

From the Bayesian point of view, Method 2 is formally more correct than Method 1. But, experimentally, when simulations with m_1 known are run, Method 1 adjusts the results better when the hypothesis are close and the sample size, n, is small which is usually the case in these kinds of problems. Whereas if the hypothesis are not close both methods provide similar results.

The next example, used by the authors previously cited, shows how the methodology is applied to an example with a normal model.

Example 3.1 If under $H_i = 0$ the model is $N(0, 1)$ and under $H_i = 1$ is $N(1, 1)$, and n observations are taken for all $i = 1, \ldots, m$, then

$$f(t_i|\theta) = \theta \prod_{j=1}^{n} f(x_{ij}|0) + (1 - \theta) \prod_{j=1}^{n} f(x_{ij}|1)$$

$$= (2\pi)^{-n/2} e^{-(1/2)\sum_{j=1}^{n} x_{ij}^2} (\theta + (1 - \theta)e^{n(\bar{x}_i - 1/2)}).$$

Then the joint distribution of the nm observations is

$$f(t_1, \ldots, t_m|\theta) = \prod_{i=1}^{m} f(t_i|\theta) \tag{14}$$

$$= (2\pi)^{-nm/2} e^{-(1/2)\sum_{i=1}^{m}\sum_{j=1}^{n} x_{ij}^2} \prod_{i=1}^{m}(\theta + (1-\theta)e^{n(\bar{x}_i - 1/2)}).$$

And the posterior density of θ given $t_1, \ldots, t_m, a, b$, is

$$\pi(\theta|t_1, \ldots, t_m, a, b) \propto \pi(\theta|a, b) f(t_1, \ldots, t_m|\theta)$$

$$\propto \theta^{a-1}(1-\theta)^{b-1} \prod_{i=1}^{m}\left(\theta + (1-\theta)e^{n(\bar{x}_i - 1/2)}\right).$$

whereas the posterior probability of the null is estimated by

$$\hat{P}(H_i = 0|T_i = t_i) = \left(1 + \frac{1 - \widehat{\theta}}{\widehat{\theta}} e^{n(\bar{x}_i - 1/2)}\right)^{-1} \tag{15}$$

for $i = 1, \ldots, m$ hypothesis, with $\widehat{\theta} = E[\theta|t_1, \ldots, t_m, a, b]$.

We use Montecarlo integration to estimate $\widehat{\theta}$. For this, we simulate a random sample $\theta_1, \ldots, \theta_k$ from the prior distribution $Beta(a, b)$. Then, we estimate $\widehat{\theta}$ as:

$$\widehat{\theta} = E[\theta|t_1, \ldots, t_m, a, b] = \frac{\int_0^1 \theta f(t_1, \ldots, t_m|\theta)\pi(\theta|a, b)d\theta}{\int_0^1 f(t_1, \ldots, t_m|\theta)\pi(\theta|a, b)d\theta}$$

$$\approx \frac{\sum_{l=1}^{k} \theta_l f(t_1, \ldots, t_m|\theta_l)}{\sum_{l=1}^{k} f(t_1, \ldots, t_m|\theta_l)}.$$

Usually, in a multiple hypothesis setting, the point is to identify a small proportion of interesting cases that will be investigated in detail. Then, the number of accepted hypothesis would be greater than 90% (see Efron [5]). Because of this, it seems appropriate to consider a $Beta(a, 1)$ density as the prior distribution for θ, since this prior gives a high probability to small intervals of θ close to 1. Moreover, this prior includes a wide list of densities, the noninformative $Beta(1, 1)$ density among others even though we propose $a \geq 9$ to be coherent with the initial assumption that no more than 10% of null hypothesis would be declared interesting.

In this case, the posterior density is given by

$$\pi(\theta|t_1, \ldots, t_m, a) \propto \theta^{a-1} \prod_{i=1}^{m}\left(\theta + (1-\theta)e^{n(\bar{x}_i - 1/2)}\right). \tag{16}$$

Table 2 Results with Method 1 with a prior density $\theta \sim Beta(a, 1)$ (for different values of a) and for simulated data from (14) with $\theta = 0.9$, different values of m and $n = 5$ observations per (each m) hypothesis

$m = 500$	$a = 1$	$a = 7$	$a = 11$	$a = 25$	$a = 50$
$\widehat{\theta}$	0.8901	0.8929	0.8942	0.8996	0.9091
$\widehat{m}_1(m_1 = 45)$	55	54	53	50	45
$prob_1$	0.6067	0.6126	0.6080	0.6097	0.6080
% Type I error	5.055	4.835	4.615	4.396	3.516
% Type II error	28.9	28.9	28.9	33.3	35.6
$m = 1000$	$a = 1$	$a = 7$	$a = 11$	$a = 25$	$a = 50$
$\widehat{\theta}$	0.8769	0.8786	0.8798	0.8826	0.8884
$\widehat{m}_1(m_1 = 107)$	123	121	120	117	112
$prob_1$	0.6635	0.6663	0.6632	0.6387	0.6273
% Type I error	4.927	4.703	4.591	4.367	3.807
% Type II error	26.168	26.168	26.168	27.103	27.103
$m = 5000$	$a = 1$	$a = 7$	$a = 11$	$a = 25$	$a = 50$
$\widehat{\theta}$	0.8984	0.8985	0.8983	0.8991	0.8999
$\widehat{m}_1(m_1 = 517)$	508	508	508	504	500
$prob_1$	0.6648	0.6649	0.6646	0.6637	0.6591
% Type I error	3.747	3.747	3.747	3.681	3.636
% Type II error	34.236	34.236	34.236	34.429	34.816
$m = 10000$	$a = 1$	$a = 7$	$a = 11$	$a = 25$	$a = 50$
$\widehat{\theta}$	0.9057	0.9057	0.9053	0.9057	0.9062
$\widehat{m}_1(m_1 = 944)$	943	943	947	943	938
$prob_1$	0.6721	0.6722	0.6735	0.6721	0.6715
% Type I error	3.379	3.379	3.412	3.379	3.335
% Type II error	32.521	32.521	32.415	32.521	32.627

A simulation from a mixture of a $N(0, 1)$ (90%) and a $N(1, 1)$ (10%) is carried out for $m = 500$, 1000, 5000 and 10000 hypothesis, with $n = 5$ observations of each hypothesis. A $Beta(a, 1)$ prior density for θ is taken where $a = 1, 7, 11, 25, 50$.

First, we use Method 1 and calculate $\widehat{\theta}, \widehat{m}_1 = m(1 - \widehat{\theta})$, – the number of rejected hypothesis with method 1–,

$$prob_1 = \widehat{P}\left(H_{(\widehat{m}_1)} = 0 | T_{(\widehat{m}_1)} = t_{(\widehat{m}_1)}\right)$$

– the highest posterior probability rejecting null hypothesis with Method 1–, and the percentage of Type I and Type II errors. The results are shown in Table 2.

Table 3 Results with Method 2 with a prior density $\theta \sim Beta(a, 1)$ (for different values of a) for the same data sets of Table 2

$m = 500$	$a = 1$	$a = 7$	$a = 11$	$a = 25$	$a = 50$
$\widehat{\theta}$	0.8901	0.8929	0.8942	0.8996	0.9091
$\widehat{i}\,(m_1 = 45)$	36	36	35	34	33
% Type I error	2.418	2.418	2.198	1.978	1.758
% Type II error	44.4	44.4	44.4	44.4	44.4
$m = 1000$	$a = 1$	$a = 7$	$a = 11$	$a = 25$	$a = 50$
$\widehat{\theta}$	0.8769	0.8786	0.8798	0.8826	0.8884
$\widehat{i}\,(m_1 = 107)$	96	95	95	94	90
% Type I error	2.352	2.352	2.352	2.352	2.239
% Type II error	29.906	30.841	30.841	31.776	34.579
$m = 5000$	$a = 1$	$a = 7$	$a = 11$	$a = 25$	$a = 50$
$\widehat{\theta}$	0.8984	0.8985	0.8983	0.8991	0.8999
$\widehat{i}\,(m_1 = 517)$	348	348	349	347	345
% Type I error	1.851	1.851	1.851	1.851	1.807
% Type II error	48.743	48.743	48.549	48.936	48.936
$m = 10000$	$a = 1$	$a = 7$	$a = 11$	$a = 25$	$a = 50$
$\widehat{\theta}$	0.9057	0.9057	0.9053	0.9057	0.9062
$\widehat{i}\,(m_1 = 944)$	654	654	656	654	652
% Type I error	1.557	1.557	1.568	1.557	1.546
% Type II error	45.657	45.657	45.551	45.657	45.763

Then, with the same data sets and $\widehat{\theta}$ from Method 1 used to compute the posterior probabilities (13), we use Method 2 and calculate $\widehat{i}$, the number of hypothesis rejected with Method 2, and the percentage of Type I and Type II errors. The results are shown in Table 3.

Observe that these two testing Bayesian procedures are robust with respect to the value of a, in the sense that a does not strongly influence the results. This new issue is good because the known Bayesian methods for testing (4) depends strongly on the parameters.

In order to compare the two proposed Bayesian methods with the FDR procedure of Benjamini and Hochberg [2], Table 4 shows, for the same data sets and $a = 11$ (an intermediate value of a), the number of null hypothesis rejected (R_{BH}, $\widehat{m}_1$, $\widehat{i}$, respectively), and the percentage of Type I and Type II errors.

Note that, with our Bayesian methods, simulations show that the number of rejected null hypotheses is more adjusted to the true than the frequentist method of Benjamini and Hochberg [2] is. For comparisons, see Table 4. In this sense the

Table 4 Results with the procedure of Benjamini and Hochberg [2] and Method 1 and Method 2 with $a = 11$, for the same data sets of Tables 2 and 3

		$\alpha = 0.05$	$\alpha = 0.05$	$\alpha = 0.05$	$a = 11$	$a = 11$	$a = 11$
		BH Meth.	BH Meth.	BH Meth.	Meth. 1	Meth. 1	Meth. 1
m	m_1	R_{BH}	% Type I	% Type II	$\hat{m}_1$	% Type I	% Type II
500	45	9	0	80	53	4.62	28.90
1000	107	36	0.22	68.22	120	4.59	26.17
5000	517	106	0.11	80.46	508	3.75	34.24
10000	944	187	0.03	80.51	947	3.41	32.42
		$\alpha = 0.1$	$\alpha = 0.1$	$\alpha = 0.1$	$a = 11$	$a = 11$	$a = 11$
		BH Meth.	BH Meth.	BH Meth.	Meth. 2	Meth. 2	Meth. 2
m	m_1	R_{BH}	% Type I	% Type II	$\hat{i}$	% Type I	% Type II
500	45	18	0.44	64.40	35	2.20	44.40
1000	107	59	0.56	49.53	95	2.35	30.84
5000	517	161	0.29	71.37	349	1.85	48.55
10000	944	307	0.23	69.70	656	1.57	45.55

procedure of Benjamini and Hochberg [2] is more conservative than each of the Bayesian methods.

One of the problems in multiple hypothesis testing with frequentist procedures is the fact that only a small number of interesting hypothesis are detected. In fact, if we want that the frequentist method achieves similar results to Method 2, simulations show that a value of $\alpha > 0.2$ is needed, but this value is not admissible.

Moreover, with our procedures the percentages of Type I errors are admissible – it does not exceed 5%–, and Method 2 is more conservative than Method 1, and the percentages of Type II errors are less than the same percentages with the frequentist procedure.

4   A Simple Hierarchical Model

Really, the parameters of the prior distribution are usually unknown and then a simple hierarchical model must be used. If we want to take a non informative prior about (a, b), Gelman et al. [6] suggests, in a different context, the use of

$$\pi(a, b) \propto (a + b)^{-5/2}. \tag{17}$$

In Sect. 2 we justified the choice of a $Beta(a, 1)$ prior to model our initial opinion about θ. Then, we propose to use $\pi(a) \propto (a + 1)^{-5/2}$. Furthermore, if we suppose (see Sect. 2) that under $H_i = 0$ the model is $N(0, 1)$ and under $H_i = 1$ is $N(1, 1)$, then the posterior density for (θ, a) when nm observations are taken is

$$\pi(\theta, a|t_1, \ldots, t_m) \propto \pi(\theta|a)\pi(a)f(t_1, \ldots, t_m|\theta)$$

$$\propto \theta^{a-1}(a+1)^{-5/2}\prod_{i=1}^{m}\left(\theta + (1-\theta)e^{n(\bar{x}_i - 1/2)}\right).$$

We use Montecarlo integration to estimate $\widehat{\theta}$. Then, we simulate a random sample $a_1, \ldots, a_h$ from $\pi(a) \propto (a+1)^{-5/2}$ and for each a_l, a sample $\theta_1^l, \ldots, \theta_k^l$ from the prior distribution $\pi(\theta|a_l) \propto \theta^{a_l - 1}$, for $l = 1, \ldots, h$, is obtained. Finally, we estimate $\widehat{\theta}$ as:

$$\widehat{\theta} = E[\pi(\theta|t_1, \ldots, t_m, a, b)] = \frac{\int_0^\infty \int_0^1 \theta f(t_1, \ldots, t_m|\theta)\pi(\theta|a)\pi(a)d\theta da}{\int_0^\infty \int_0^1 f(t_1, \ldots, t_m|\theta)\pi(\theta|a)\pi(a)d\theta da}$$

$$\approx \frac{\sum_{l=1}^{h}\sum_{i=1}^{k}\theta_i^l f(t_1, \ldots, t_m|\theta_i^l)}{\sum_{l=1}^{k}\sum_{i=1}^{k} f(t_1, \ldots, t_m|\theta_i^l)}.$$

The same data set as in Example 3.1 is used, where a simulation from a mixture of a $N(0, 1)$ (90%) and a $N(1, 1)$ (10%) was carried out for $m = 500, 1000, 5000$ and 10000 hypothesis, with $n = 5$ observations of each hypothesis.

First, we use Method 1 to calculate $\widehat{\theta}$, $\widehat{m}_1$, the highest posterior probability of rejecting the null hypothesis and the percentage of Type I and Type II errors. The results are shown in Table 5.

Finally, we use Method 2 with the same data set, and taking $\widehat{\theta}$ from Method 1, to calculate the number of hypothesis rejected and the percentage of Type I and Type II errors. The results are shown in Table 6.

Table 5 Results using Method 1 for the hierarchical case

	$m = 500$	$m = 1000$	$m = 5000$	$m = 10000$
$\widehat{\theta}$ ($\theta = 0.9$)	0.8910	0.8781	0.8982	0.9055
$\widehat{m}_1$	55	122	509	945
$prob_1$	0.6089	0.6659	0.6655	0.6721
% Type I error	5.01	4.82	3.77	3.40
% Type II error	28.89	26.17	34.24	32.52

Table 6 Results using Method 2 for the hierarchical case

	$m = 500$	$m = 1000$	$m = 5000$	$m = 10000$
$\widehat{\theta}$ ($\theta = 0.9$)	0.8910	0.8781	0.8982	0.9055
$\widehat{m}_1$	36	95	351	655
% Type I error	2.42	2.35	1.87	1.56
% Type II error	44.44	30.84	48.36	45.55

In Sect. 3 it was shown that the results are not significantly affected by changes in the parameter a of the beta prior distribution for θ. This is the reason that, for the hierarchical case, we obtain similar results to those obtained in the previous section, because in this case the different possible values of a are replaced by the mean of its prior distribution.

5 Conclusions

The proposed methodology in this paper involves to provide a Bayesian estimator for θ (the percentage of true null hypothesis in (4)), for instance, the mean or the mode of the posterior density of θ. Posterior probabilities of $H_i = 0, i = 1, \ldots, m$, are calculated and estimated by using a prior density $Beta(a, 1)$ for θ. Based on this estimator of θ, we propose two different Bayesian approaches to test (4).

Simulations show that these two Bayesian procedures are robust with respect to the value of a, in the sense that the parameter a does not strongly influence the results. This new issue is good because the known Bayesian methods for testing (4) depends strongly on the parameters.

It is well known that detecting a small number of interesting hypothesis is one of the problems in multiple hypothesis testing with frequentist approaches. In this sense, another important conclusion is that our Bayesian methods are less conservative than the procedure of Benjamini and Hochberg [2], because it allows us to reject a higher number of null hypothesis to test (4). In fact, with each of our Bayesian methods, methods 1 and 2, computations show that the number of rejected null hypotheses is more adjusted to the true than the frequentist is.

Moreover, the analyzed examples show that with our procedures the percentages of Type I errors are admissible (they do not exceed 5%), Method 2 being more conservative than Method 1, and the percentages of Type II errors are less than the same percentages with the frequentist procedure.

Acknowledgements Research supported from MEC, UCM, CM-UCM.

References

1. Barbieri M, Berger J (2004) Optimal predictive model selection. Ann Stat 32:870–897
2. Benjamini Y, Hochberg Y (1995) Controlling the false discovery rate: a practical and powerful approach to multiple testing. J R Stat Soc Ser B 57:289–300
3. Benjamini Y, Yekutieli D (2001) The control of the false dicovery rate in multiple testing under dependency. Ann Stat 29:1165–1188
4. Dudoit S, Shaffer JP, Boldrick JC (2003) Multiple hypothesis testing in microarray experiments. Stat Sci 18(1):71–103
5. Efron B (2004) Large-scale silmultaneous hypothesis testing: the choice of the null hypothesis. J Am Stat Assoc 99(465):96–103

6. Gelman A, Carlin JB, Stern HS, Rubin DB (2004) Bayesian data analysis. Chapman and Hall/CRC, London
7. Hobert J (2000) Hierarhical models: a current computational perspective. J Am Stat Assoc 95:1312–1316
8. Lehmann EL, Romano JP (2005) Testing statistical hypotheses, 3rd edn. Wiley, New York
9. Scott JG, Berger JO (2006) An exploration of aspects of Bayesian multiple testing. J Stat Plan Infer 136:2144–2162
10. Shaffer JP (1995) Multiple hypothesis testing: a review. Ann Rev Psychol 46:561–584
11. Waller RA, Duncan DB (1969) A Bayes rule for the symmetric multiple comparison problem. J Am Stat Assoc 64:1484–1503
12. Wasserstein RL, Lazar NA (2016) The ASA's statement on p-values: context, process, and purpose. Am Stat 70(2):129–133
13. Westfall PH, Young SS (1993) Resampling-based multiple testing: examples and methods for p-value adjustment. Wiley, New York

δ-Records Observations in Models with Random Trend

Raúl Gouet, Miguel Lafuente, F. Javier López and Gerardo Sanz

Abstract In this paper we prove a Law of Large Numbers for the number of δ-records in a sequence of random variables with an underlying trend. Our results generalizes results appeared in the literature for the i.i.d. case and for records in models with random trend. Two examples to illustrate the application of our results are included.

1 Introduction

Greek and Roman believed that memory resided in the human heart. As a consequence, the action of keeping in mind extraordinary events of all kind, derived in the coining of the word recordari, directly composed by *re-* meaning restore, and *-cor* from *cordis* meaning heart in English. Thus, it is in an unsurprising way that the modern term record is very present in quotidian life as well as in a wide range of specialized domains like meteorology, economics, finance, seismology and sports, and as a direct consequence of this, in mathematical research.

This work is dedicated to the memory of Pedro Gil, who his students, colleagues and friends will miss. For all those who were lucky enough to know him, his professional career and his excellent human quality leave a difficult space to fill.

R. Gouet
Departamento Ingeniería Matemática y Centro de Modelamiento Matemático
(UMI 2807, CNRS), Universidad de Chile, Santiago, Chile
e-mail: rgouet@dim.uchile.cl

M. Lafuente (✉) · F. Javier López · G. Sanz
Facultad de Ciencias, Departamento de Métodos Estadísticos and BIFI,
Universidad de Zaragoza, Zaragoza, Spain
e-mail: miguellb@unizar.es

F. Javier López
e-mail: javier.lopez@unizar.es

G. Sanz
e-mail: gerardo.sanz@unizar.es

Mathematically, we consider and observation to be a record if it is the greatest observed until it took place. Around this definition, Record Theory and Extreme Value Theory have been quite well studied for decades, specially in the framework of the analysis of independent, identically distributed (i.i.d.) observations. Interesting introductory books about this topic are [1, 2, 20]. Nevertheless, these disciplines had to evolve in many different aspects in an attempt to cover other phenomena. In particular, we are going to consider two generalizations of very distinct nature regarding the study of usual records over i.i.d. observations.

The first one concerns the mathematical definition of record. In the last years, some record-related concepts have appeared in the literature, such as near-records (see [3, 17, 22]), geometric records (see [9, 14]) or records with confirmation [21]. One of these concepts, with interesting applications in statistical inference, is the notion of δ-records [12], which will be the object of study of the present paper.

Definition 1.1 Let $\{X_n\}_{n\in\mathbb{N}}$ be a sequence of random variables and $\delta \in \mathbb{R}$, we say that the jth entry is a δ-record if

$$X_j > \max\{X_1, \ldots, X_{j-1}\} + \delta. \tag{1}$$

By convention, the first observation X_1 is considered to be a δ-record for every δ.

As we may notice in the above definition, choosing $\delta = 0$ takes us to the usual record concept. It is straightforward to remark that $\delta > 0$ will result in a fewer number of δ-records than usual records while $\delta < 0$ leads us to the opposite feature.

In the latter years, δ-records have been studied revealing themselves as useful tools in a wide variety of settings without greatly increasing the methodological difficulties compared to the record scenario. Along this line, the structure of the process [15], distribution [18, 19], asymptotic properties [13] and elements of inference [11, 13] have been explored under i.i.d. assumptions.

On the other hand, it is common to see in some practical situations a number of records much higher than expected by the i.i.d. theory. Approximately, we should expect a number of the magnitude of $log(n)$ being n the number of observation as was first remarked by Rényi [23] in 1962. In order to try to cover this kind of phenomena, or simply, in cases where we theoretically conjecture the existence of an increasing trend underlying the observations, the Linear Drift Model (LDM), appears as an effective tool of introducing time dependence among observations. From the point of view of extremes, the LDM was introduced by Ballerini and Resnick [4] in 1985 showing a Law of Large Numbers and a Central Limit Theorem for the record rate as long as distributions are light-tailed. The LDM assumes that the observations $\{Y_n\}_{n\in\mathbb{N}}$ can be written in the following form

$$Y_n = X_n + cn, \tag{2}$$

being $c > 0$ the slope of the trend, and X_n i.i.d. random variables. In this context, we may highlight the contributions of De Haan and Verkade [7] exploring the behaviour under heavy-tailed distributions, and Borovkov studying inter-record times, Markovianity and characterizations [6]. Regarding the applications, one may read the introduction of [10] and the references therein for applications in different fields of physics, as well as [25, 26] to see how the study of records in the LDM can be helpful in the problem of the global warming climate.

Further generalisations of the LDM can be found in [5], where the i.i.d. assumption is relaxed to strict stationarity, and [16] where the proposed random trend model can be expressed as

$$Y_n = X_n + T_n, \tag{3}$$

where $(X_n)_{n \in \mathbb{N}}$ is a stationary ergodic sequence of random variables and T_n represents an stochastic drift process with ergodic stationary increments.

It is easy to check that the expression in (3) represents a flexible model that comprises the LDM, its correspondent generalisation in [5] and other models like some drifted random walks.

In this work we will provide a Law of Large Numbers for the δ-record rate in the model with random trend in the case $\delta \leq 0$. Our result represents an extension of the analogous results in [13], devoted to the i.i.d. case, and in [16], where the asymptotic record rate for a model with random trend is proved. Moreover, for the particular case $\delta = 0$, our proof is much simpler than the one given in [16].

2 Notation

We will denote by $\vee$ and $\wedge$ the maximum and minimum operator respectively over a set.

From now on, we will work under the assumptions of the following extension of the LDM introduced in (3). We consider a sequence of observations

$$Y_n = X_n + T_n, \tag{4}$$

where $T_n := \sum_{k=1}^n \tau_k$, $n \geq 1$ is the stochastic trend process and (X_n, τ_{n+1}) is a bivariate, strictly stationary and ergodic sequence and $0 < c := E[\tau_1] < \infty$ appears as the mean increment or slope of the trend. Also, we will assume $E[X_1^+] < \infty$ where we have $x^+ := x \vee 0$.

While the decomposition $Y_n = X_n' + nc$ with $X_n' := X_n + T_n - nc$ could be done, it does not eliminate the random trend including it into the residuals of LDM because X_n' is not stationary in general. Finally, we remark that (X_n) and (τ_n) are allowed to be dependent as long as they have a finite expectation.

As any stationary single-ended sequence can be extended to a double-ended one, for theoretical purposes we can consider the indices as defined in $\mathbb{Z}$ yielding the model in (4) to

$$Y_n = X_n + T_n, \tag{5}$$

where $T_n := \sum_{k=1}^n \tau_k$, for $n \geq 1$ and, $T_n := \tau_1 - \sum_{k=n}^0 \tau_k$, given $n \leq 0$.

According to Definition 1.1, the condition of Y_n being a δ-record in the sequence $\{Y_s, \ldots, Y_n\}$ for a previously chosen $\delta \in \mathbb{R}$ can be written as

$$Y_n > \bigvee_{k=s}^{n-1} Y_k + \delta. \tag{6}$$

So, we will write M_n for the maximum up to the nth observation starting at an observation indexed by 1, in contraposition with the starred M_n^* where the maximum is taken up to the nth observation in the double-ended sequence:

$$M_n := \bigvee_{i=1}^n Y_i, \qquad\qquad M_n^* := \bigvee_{i \leq n} Y_i, \tag{7}$$

where we will see later that $M_n^* < \infty$ for all $n \in \mathbb{Z}$.

In the same way, given $n \in \mathbb{N}$, the random variables $1_{n,\delta}$, will represent the occurrence of a δ-record over the single-ended sequence

$$1_{n,\delta} = \begin{cases} 1 \text{ if } Y_n > M_{n-1} + \delta, \\ 0 \text{ otherwise.} \end{cases} \tag{8}$$

while, for $n \in \mathbb{Z}$, $1_{n,\delta}^*$ will be the analogous over the double-ended one

$$1_{n,\delta}^* = \begin{cases} 1 \text{ if } Y_n > M_{n-1}^* + \delta, \\ 0 \text{ otherwise.} \end{cases} \tag{9}$$

Also, for simplicity, 1_n and 1_n^* will be written to denote the corresponding record indicators, i.e., in the case $\delta = 0$. The fact that $E[1_{n,\delta}] = P[Y_n$ is a δ-record] is forthright.

The sum of these indicators will be the total number of δ-records at time n, $N_{n,\delta}$ and $N_{n,\delta}^*$, both quantities starting at point 1 just changing the definition of δ-records

$$N_{n,\delta} = \sum_{j=1}^n 1_{n,\delta} \qquad\qquad N_{n,\delta} = \sum_{j=1}^n 1_{n,\delta}^*. \tag{10}$$

3 A Law of Large Numbers for δ-records with Random Trend

In order to prove the Law of Large Numbers, we will require the following theorems which we state for completeness.

Lemma 3.1 (Birkhoff's Ergodic Theorem) *Let $X = (X_1, X_2, \ldots)$ be a stationary (strict sense) ergodic random sequence with $E[|X_1|] < \infty$. Then,*

$$\lim \frac{1}{n} \sum_{k=1}^{n} X_k(\omega) = E(X_1) \ a.s. \ and \ in \ L_1. \tag{11}$$

Proof See [24, p. 385, Theorem 3]. □

Lemma 3.2 (Dubins–Freedman strong law) *Let $(U_n)_{n\geq 1}$ be a sequence of nonnegative and bounded random variables, adapted to the increasing family of σ-algebras $(\mathcal{G}_n)_{n\geq 0}$. Then*

$$\left\{ \sum_{n\geq 1} U_n = \infty \right\} = \left\{ \sum_{n\geq 1} E[U_n \mid \mathcal{G}_{n-1}] = \infty \right\} \ a.s. \tag{12}$$

and

$$\frac{\sum_{k=1}^{n} U_k}{\sum_{k=1}^{n} E[U_k \mid \mathcal{G}_{k-1}]} \to 1 \ on \ \left\{ \sum_{n\geq 1} E[U_n \mid \mathcal{G}_{n-1}] = \infty \right\} \ a.s. \tag{13}$$

Proof See [8]. □

Finally, we are able to prove the next Law of Large Numbers for the number of δ-records, or equivalently, for the asymptotic δ-record rate, the proof is organized in six short steps.

Theorem 3.1 $N_{n,\delta}/n \to E[1_{1,\delta}^*] = P[X_1 > \bigvee_{k\geq 1}\{X_{1-k} - \sum_{j=2-k}^{1} \tau_j\} + \delta] > 0$ *a.s. and in L_1 as $n \to \infty$ if $\delta \leq 0$.*

Proof 1. $M_n \to \infty$ and $N_{n,\delta} \to \infty$ a.s. Since M_n is an increasing sequence by construction, it either converges to a finite limit or diverges to ∞ a.s. Also, we have $\forall a \in \mathbb{R}$

$$P[M_n > a] \geq P[X_n > a - T_n] \geq P[X_n > a - nc/2, T_n \geq nc/2] \to 1, \tag{14}$$

in view of $P[X_n > a - nc/2] = P[X_0 > a - nc/2] \to 1$ and $P[T_n \geq nc/2] \to 1$, by Birkhoff's theorem. Thus, $M_n \to \infty$, and then $N_n \to \infty$. Since $1_n \leq 1_{n,\delta} \ \forall n \in \mathbb{N}$ and $\delta \leq 0$, the result is straightforward.

2. $P[\bigvee_{k\geq 1}\{X_{n-k} - \sum_{j=n+1-k}^{n} \tau_j\} \in \mathbb{R}] = 1$ and $M_n^* < \infty$ a.s. for all $n \in \mathbb{Z}$. Stationarity will guarantee the result if we prove that

$$P\left[X_{-k} > \sum_{j=-k+1}^{0} \tau_j, \ i.o. \right] = 0. \tag{15}$$

We know by Birkhoff's Theorem that $P[\sum_{j=-k+1}^{0} \tau_j \leq kc/2, \ i.o.] = 0$, thus

$$P\left[X_{-k} > \sum_{j=-k+1}^{0} \tau_j \ i.o.\right] \leq P\left[X_{-k} > kc/2 \ i.o.\right]$$

$$= P\left[X_0 > kc/2 \ i.o.\right] \text{ for } k \geq 1. \qquad (16)$$

Also, since $E[X_0^+] < \infty$ we know $\sum_{k=1}^{\infty} P[X_0 > kc/2] < \infty$ and the result holds by the Borel–Cantelli lemma.

3. $\exists \ 0 < N < \infty$ a.s. such that $M_n^* = M_n$ and $1_{n,\delta} = 1_{n,\delta}^*$ a.s. $\forall n > N$. As a consequence of the previous result, we have

$$P\left[\bigvee_{k\geq 1}\left\{X_{n-k} - \sum_{j=n+1-k}^{n} \tau_j\right\} \in \mathbb{R}\right] = 1 \ \forall n \in \mathbb{Z}, \qquad (17)$$

which together with $M_n \to \infty$ imply that $\exists \ 0 < N < \infty$ a.s. such that $1_{N,0}^* = 1$ almost surely. On the other hand, given $n \in \mathbb{N}$ we have $1_{n,0} \geq 1_{n,0}^*$ by construction, and thus $1_{N,0} = 1$ a.s. Now, we trivially have $M_n^* = M_n$ and $1_{n,\delta} = 1_{n,\delta}^*$ a.s. $\forall n > N$.

4. $E[1_{1,\delta}^*] > 0$ if $\delta \leq 0$.

Knowing that $E[1_{1,\delta}^*] \geq E[1_1^*]$ by definition, it suffices to check $E[1_1^*] > 0$. Let us assume $E[1_1^*] = P[X_1 > \bigvee_{k\geq 1}\{X_{1-k} - \sum_{j=2-k}^{1} \tau_j\}] = 0$, then

$$P\left[X_1 > \bigvee_{k\geq 1}\left\{X_{1-k} - \sum_{j=2-k}^{1} \tau_j\right\} \ \Big| \ \mathcal{F}_0\right] = 0 \ \text{ a.s.} \qquad (18)$$

and so stationarity would imply

$$P\left[X_n > \bigvee_{k\geq 1}\left\{X_{n-k} - \sum_{j=n+1-k}^{n} \tau_j\right\} \ \Big| \ \mathcal{F}_{n-1}\right] = 0 \ \text{ a.s. } \forall n \in \mathbb{N}. \qquad (19)$$

Since $\bigvee_{k\geq 1}\{X_{n-k} - \sum_{j=n+1-k}^{n} \tau_j\}$ and $\bigvee_{k=1}^{n-1}\{X_{n-k} - \sum_{j=n+1-k}^{n} \tau_j\}$ couples by the previous reasoning, then

$$\sum_{n=1}^{\infty} P\left[X_n > \bigvee_{k\geq 1}\left\{X_{n-k} - \sum_{j=n+1-k}^{n} \tau_j\right\} \ \Big| \ \mathcal{F}_{n-1}\right] < \infty \qquad (20)$$

and the Dubins–Freedman strong law imply

$$\sum_{n=1}^{\infty} P\left[X_n > \bigvee_{k \geq 1}\left\{X_{n-k} - \sum_{j=n+1-k}^{n} \tau_j\right\}\right] = \sum_{n=1}^{\infty} 1_n < \infty, \qquad (21)$$

resulting in a contradiction.

5. $N_{n,\delta}^*/n \to E[1_{1,\delta}^*] > 0$ a.s. as $n \to \infty$.
   Choosing N such that the coupling of the part three has taken place, we have

$$\sum_{k=N+1}^{\infty} 1_{n,\delta} = \sum_{k=N+1}^{\infty} 1_{n,\delta}^* \quad \text{a.s.} \qquad (22)$$

Since we have that $1_{n,\delta}^*$ is a strictly stationary and ergodic sequence by construction, we get by means of the Birkhoff's Theorem the desired result.

$$\frac{1}{n}\sum_{k=1}^{n} 1_{n,\delta}^* \to \mathbb{E}[1_{1,\delta}^*] \quad \text{a.s.} \qquad (23)$$

6. $N_{n,\delta}/n \to E[1_{1,\delta}^*] > 0$ a.s. as $n \to \infty$.
   Since we know that $\mathbb{E}[1_{1,\delta}^*] > 0$ then $\sum_{k=1}^{n} 1_{k,\delta}^* \to \infty$ a.s. as $n \to \infty$. Because of the couple we have at same time $\sum_{k=1}^{\infty} 1_{n,\delta} = \infty$ a.s. by (22). Now, for any $\{a_n\}_{n \in \mathbb{N}}$ real sequence such that $\{a_n\} \to \infty$ we have

$$\left|\frac{N_{n,\delta} - N_{n,\delta}^*}{a_n}\right| \leq \left|\frac{N}{a_n}\right| \to 0 \quad \text{a.s.} \qquad (24)$$

since N does not depend on n. Finally, we can conclude from $\left|\frac{N_{n,\delta}-N_{n,\delta}^*}{n}\right| \to$
0 a.s. and $\frac{N_{n,\delta}^*}{n} \to \mathbb{E}[1_{1,\delta}^*]$ a.s. that $\frac{N_{n,\delta}}{n} \to \mathbb{E}[1_{1,\delta}^*]$ a.s.
Finally, convergence in L_1 is forthright by the dominated convergence theorem. $\qquad \square$

In order to facilitate the comparison of our result with the case of the usual (upper) records, we show here some examples that were analysed in [16].

Example 3.1 (LDM with Gumbel residuals)
   Let $(X_n)_{n \geq 1}$ be i.i.d. random variables with standard Gumbel distribution, i.e., $F_{X_n} = \exp(-e^{-x})$ for $x \in \mathbb{R}$ and $\tau_n = c$, $n \in \mathbb{N}$ entailing $T_n = cn$ as in the LDM. From Theorem 3.1 we know

$$\lim_{n \to \infty} \frac{N_{n,\delta}}{n} = E[1_1^*]. \qquad (25)$$

On the other hand we have

$$E[1^*_{1,\delta}] = \mathbb{P}\left[X_1 > \bigvee_{k \geq 1}\{X_{1-k} - ck\} + \delta\right] = \int_{-\infty}^{\infty} \prod_{j=1}^{\infty} F(x + cj - \delta)F(dx), \quad (26)$$

together with the fact

$$F(x + cj - \delta) = \exp(-\exp(-x - cj + \delta)) = F(x)^{\exp(-cj+\delta)}. \quad (27)$$

Since $-c < 0$, we arrive to

$$E[1^*_{1,\delta}] = \int_{-\infty}^{\infty} \prod_{j=1}^{\infty} F(x + cj - \delta)F(dx)$$

$$= \int_{-\infty}^{\infty} F(x)^{\exp(\delta)\sum_{j=1}^{\infty}\exp(-cj)} F(dx) = \int_{-\infty}^{\infty} F(x)^{e^{\delta}e^{-c}/(1-e^{-c})} F(dx)$$

$$= \left[\frac{u^{e^{\delta}e^{-c}/(1-e^{-c})}}{e^{\delta}e^{-c}/(1 - e^{-c}) + 1}\right]_{u=0}^{u=1} = \frac{1 - e^{-c}}{e^{\delta}e^{-c} + 1 - e^{-c}}. \quad (28)$$

Note that imposing $\delta = 0$ we are able to recover the asymptotic record rate

$$E[1^*_1] = 1 - e^{-c}, \quad (29)$$

that coincides with the results in [4, 16].

Example 3.2 (Drifted random walk)

Consider now a random walk, $S_n = \sum_{i=1}^{n} \tau_n$, where $(\tau_n, n \geq 1)$ is a stationary ergodic sequence with $E[\tau_1] > 0$. We want to get the asymptotic δ-record rate of that random walk. For this we need to fit this situation to our general framework in (4), for which we take $X_n = 0 \; \forall n \in \mathbb{N}$ and $T_n \equiv S_n$.

Now

$$E[1^*_{1,\delta}] = P\left[X_1 > \bigvee_{k \geq 1}\left\{X_{1-k} - \sum_{j=2-k}^{1}\tau_j\right\} + \delta\right] = P\left[0 > \bigvee_{k \geq 1}\left\{-\sum_{j=2-k}^{1}\tau_j\right\} + \delta\right]$$

$$= P\left[\bigwedge_{k \geq 1}\left\{\sum_{j=2-k}^{1}\tau_j\right\} > \delta\right]. \quad (30)$$

Again, this result generalizes the analogous in [16] for $\delta = 0$.

Acknowledgements Support from Grant MTM2014-53340-P of MINECO is gratefully acknowledged. The authors are members of the research group "Modelos Estocásticos", supported by DGA and the European Social Fund.

References

1. Ahsanullah M (1995) Record statitics. Nova Science Publishers, New York
2. Arnold B, Balakrishnan N, Nagajara H (1998) Records. Wiley, New York
3. Balakrishnan N, Pakes A, Stepanov A (2005) On the number and sum of near-record observations. Adv Appl Probab 37:765–780
4. Ballerini R, Resnick S (1985) Records from improving populations. J Appl Probab 22:487–502
5. Ballerini R, Resnick S (1987) Records in the presence of a linear trend. Adv Appl Probab 19:801–828
6. Borovkov K (1999) On records and related processes for sequences with trends. J Appl Probab 36:668–681
7. De Haan L, Verkade E (1987) On extreme-value theory in the presence of a trend. J Appl Probab 24:62–76
8. Dubins L, Freedman D (1965) A sharper form of the Borel–Cantelli lemma and the strong law. Ann Stat 36:800–807
9. Eliazar I (2005) On geometric record times. Phys A 348:181–198
10. Franke J, Wergen G, Krug J (2010) Records and sequences of records from random variables with a linear trend. J Stat Mech P10013
11. Gouet R, López J, Maldonado L, Sanz G (2014) Statistical inference for the geometric distribution based on δ-records. Comput Stat Data Anal 78:21–32
12. Gouet R, López J, Sanz G (2007) Asymptotic normality for the counting process of weak records and δ-records in discrete models. Bernoulli 13:754–781
13. Gouet R, López J, Sanz G (2012) On δ-record observations: asymptotic rates for the counting process and elements of maximum likelihood estimation. Test 21:188–214
14. Gouet R, López J, Sanz G (2012) On geometric records: rate of appearance and magnitude. J Stat Mech P01005
15. Gouet R, López J, Sanz G (2015) On the point process of near-records values. Test 24:302–321
16. Gouet R, López J, Sanz G (2015) Records from stationary observations subject to a random trend. Adv Appl Probab 47:1175–1189
17. Hashorva E (2003) On the number of near-maximum insurance claim under dependence. Insur Math Econ 32:37–49
18. López-Blázquez F, Salamanca-Miño B (2013) Distribution theory of δ-record values: case δ ≤ 0. Test 22:715–738
19. López-Blázquez F, Salamanca-Miño B (2015) Distribution theory of δ-record values: case δ ≥ 0. Test 24:558–582
20. Nevzorov V (2001) Records: mathematical theory. American Mathematical Society, Providence
21. Nevzorov V, Stepanov A (2014) Records with confirmation. Stat Probab Lett 95:39–47
22. Pakes A (2007) Limit theorems for numbers of near-records. Extremes 10:207–224
23. Rényi A (1962) Théorie des éléments saillants d'une suite d'observations. Ann Scient l'Univ Clermont-Ferrand 2(8):7–13
24. Shiryayev AN (1984) Probability. Springer, New York
25. Wergen G, Hense A, Krug J (2012) Record occurrence and record values in daily and monthly temperatures. Clim Dyn 42:1275–1289
26. Wergen G, Krug J (2010) Record-breaking temperatures reveal a warming climate. Europhys Lett 92(3):30008

Recent Developments and Advances in Joint Modelling of Longitudinal and Survival Data

Ipek Guler, Christel Faes, Francisco Gude and Carmen Cadarso-Suárez

Abstract In many biomedical studies, patient are followed-up repeatedly during the research study and different types of outcomes are collected such as longitudinal biomarkers and a time-to-event information. Commonly, it is of interest to study the association between the longitudinal biomarkers and the time-to-event. This chapter gives an overview of joint models for a single longitudinal and survival data with its extensions to multivariate longitudinal and time-to-event models.

1 Introduction

There exist various methods to study the association between a longitudinal outcome and the time-to-event process in the literature. Focusing on a case which has a single longitudinal biomarker and a survival data, the earliest methods are the extended Cox model (Anderson and Gill [1]) and a two-stage approach method (Self and Pawitan [21]). Although these methods have advantages in terms of fast computing, they also have several limitations. The extended Cox model assumes that the covariates are external and not related to the failure mechanism (Kalbfleisch and Prentice [14]); also, this model does not take into account the measurement error of the longitudinal process. In the two-stage approach no survival information is used for the longitudinal process such that informative drop-out is not accounted for. If the main interest is on the association between the longitudinal and survival data, joint models are required to feature this correlation (Tsiatis and Davidian [23], Wulfsohn and Tsiatis [24]).

I. Guler (✉) · C. Cadarso-Suárez
Center for Research in Molecular Medicine and Chronic Diseases (CiMUS),
University of Santiago de Compostela, 15782 Santiago de
Compostela, A Coruña, Spain
e-mail: ipek.guler@usc.es

C. Faes
I-Biostat, Hasselt University, Hasselt, Diepenbeek, Belgium

F. Gude
Clinical Epidemiology Unit. Hospital Clinico, Santiago de Compostela, Spain

© Springer International Publishing AG 2018
E. Gil et al. (eds.), *The Mathematics of the Uncertain*, Studies in Systems,
Decision and Control 142, https://doi.org/10.1007/978-3-319-73848-2_21

219

Previous research on joint modelling is concentrated on the association of a single longitudinal outcome and survival process. There are a wide number of clinical applications and developed statistical software (Crowther et al. [6], Guo and Carlin [10], Philipson et al. [16], Rizopoulos [19], Zhang et al. [26]).

Recently, there are already several extensions in joint modeling approaches such as the use of flexible longitudinal profiles using multiplicative random effects (Ding and Wang [8]), alternatives to the common parametric assumptions for the random effects distribution (Brown et al. [3]), and handling multiple failure times (Elashoff et al. [9]). Nice overviews of this field are given by Tsiatis and Davidian [23], and Yu et al. [25]. The more recent work is done for multiple longitudinal biomarkers with time-to-event data which are focused mainly on the bayesian framework (Rizopoulos and Ghosh [20], Tang et al. [22], among others).

In this chapter we will give an overview of joint models for longitudinal and survival data within the context of shared random effects framework and recent extensions with a real data illustration on Orthotopic Liver Trasplantation.

2   Joint Modelling of Longitudinal and Survival Data

Joint models have gained increasing attention over the last two decades, especially in biomedical investigations (Wulfsohn and Tsiatis [24], Henderson et al. [12], Rizopoulos [18]).

The joint models are based on a joint likelihood calculation of longitudinal and time-to-event data within different frameworks to calculate the conditional distributions. For instance, the shared random effects framework is based on the simultaneous estimation of both longitudinal and time-to-event through an incorporation of shared random effects which underlines the conditional distributions (Wulfsohn and Tsiatis [24]).

There are different factorizations of this joint distribution which generate various modeling strategies. For a general idea, let the Y be the longitudinal process, T the survival processes, and U a latent random effect. Then, JMLS can be grouped into the following modeling classes:

Selection Models: In these models a latent random effect, U, underlines only the longitudinal process Y, and the calculation of joint likelihood consists of a factorization into the conditional distribution of the longitudinal process given the random effect on the one hand and the conditional distribution of the survival process given the longitudinal outcome on the other hand. In this type of model the focus is only on the time-to-event process, thus can be used for survival analysis with endogenous variables

$$f(Y, T, U) = f(U)f(Y \mid U)f(T \mid Y).$$

Pattern-Mixture Models: These models are similar to the selection models, but factorization is reversed. In this setting, the factorization of the joint likelihood is conducted into the conditional distribution of the longitudinal outcome given survival process on the one hand, and conditional distribution of the survival outcome given the random effect on the other hand. This type of models can be used for the longitudinal studies with a drop-out process generated by non-ignorable mechanism

$$f(Y, T, U) = f(U)f(T \mid U)f(Y \mid T).$$

Shared Random Effect Models: In these models the latent random effect underlines both longitudinal and survival process

$$f(Y, T, U) = f(U)f(Y \mid U)f(T \mid U).$$

In JLMS, standard methods make use of two submodels, in order to specify the full joint likelihood. The longitudinal process is modeled by a linear mixed model as follows

$$Y_{tij} = \beta_0 + \beta_1 t_{ij} + U_{0i} + U_{1i}t_{ij} + \varepsilon_{ij} \tag{1}$$

where Y_{ij} is response variable measured on subject $i = 1, ..., n$ at time point t_{ij}, with $j = 1, ..., m_i$. The β_0, β_1 represent the coefficients of the fixed effects, (i.e. the intercept and the time effect respectively), and U_{0i}, U_{1i} are the random intercept and random slope effects respectively. Here we assume

$$\begin{pmatrix} U_{0i} \\ U_{1i} \end{pmatrix} \sim N \left(\begin{pmatrix} 0 \\ 0 \end{pmatrix}, \Sigma \right)$$

with

$$\Sigma = \begin{pmatrix} \sigma_{u_0}^2 & \sigma_{u_0}\sigma_{u_1}\rho_{12} \\ \sigma_{u_0}\sigma_{u_1}\rho_{12} & \sigma_{u_1}^2 \end{pmatrix}.$$

In the Σ expression, $\sigma_{u_0}^2$ and $\sigma_{u_1}^2$ are the variances of the random effects and ρ_{12} represents the correlation between them.

The survival process is usually modeled by using the following Cox proportional hazard model (Cox [5])

$$\lambda(t) = \lambda_0(t) \exp(\beta X + \alpha\omega_i(t)), \tag{2}$$

where $\lambda_0(t)$ is the unspecified baseline risk function, X is a matrix of fixed effects including the baseline covariates (such as age, gender, etc.), $\omega_i(t)$ is a function reflecting the association structure between the longitudinal and survival data including the same random effects U and α is the coefficient of this association. Rizopoulos [18], use the true value of the longitudinal biomarker at time t as the association function, $\omega_i(t)$, in the survival model. In this case, the association structure is defined via:

$$\omega_i(t) = \beta_0 + \beta_1 t_{ij} + U_{0i} + U_{1i} t_{ij}.$$

The survival sub-model becomes,

$$\lambda(t) = \lambda_0(t) \exp(\beta X + \alpha(\beta_0 + \beta_1 t_{ij} + U_{0i} + U_{1i} t_{ij}))$$

in which α represent the association between the longitudinal biomarker and the risk for death at time t taking into account the true value of the longitudinal biomarker both with fixed and random effects predictions.

3 Illustrative Example

3.1 Liver Transplantation Data

This dataset includes information from a follow-up study of 642 patients undergoing liver transplantation at a single, tertiary care transplant hospital from July 1994 to July 2011. Patients were followed up until 12 July 2012. Patient survival was defined as the period of time between transplantation and end of follow-up or death. Median follow-up was 5.6 years (range: [0.1, 17.5]). The sample had a survival rate of 85% at one year and 65% at ten years. Pre-transplant variables were recorded, including age, gender, body mass index (BMI), indication for OLT, Model for End-stage Liver Disease (MELD) score, haemoglobin, haematocrit, platelet count, prothrombin time, serum total bilirubin levels, serum creatinine levels, fasting blood glucose and prior diagnosis of diabetes.

The study sought to investigate the ability of postoperative glucose profiles to predict mortality in patients who underwent OLT, differentiating between those with and those without pre-existing diabetes mellitus. OLT is the established treatment for end-stage liver disease and acute fulminant hepatic failure. Advances in both medical management and surgical techniques have led to an increase in the number of long-term survivors. However, alterations in glucose metabolism are common among patients undergoing surgery, and are associated with increased risk of mortality and morbidity. These abnormalities, particularly hyperglycaemia, are also common in critically ill patients, even those without a diagnosis of diabetes. Thus, insulin is administered via continuous perfusion to maintain glycaemia figures between 120 and180 mg/dL, The postoperative glucose profiles for individuals with and without previous diabetes, can be seen in Fig. 1.

Fig. 1 Subject specific trajectories (grey) and overall curves (black) of Glucose levels for the diabetic and non-diabetic patients

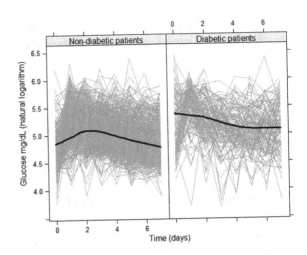

3.2 Application Study

In this section we illustrate three different longitudinal sub-model and the final survival model after a variable selection using Akaike Information Criterian (Akaike [2]) using the illustrative example Liver Transplantation Data.

The three longitudinal sub-model consist of three different posibilities introducing only the random intercept, random intercept and slope and in case of necessary a spline smoothing model. The glucose measurements are transformed using a logarithmic transformation as a response variable in the longitudinal model. As we observe in Fig. 1, the overall trajectories of Glucose measurements follow non-linear trends over time for diabetic and non-diabetic patients. Thus, we use the spline model for our application study.

(a) *Intercept Model*

$$\log(Glucose)_{i,diab} = \beta_0 + \beta_1 time_i + \beta_2 vhc_i + \beta_3 meld_i + U_{0i} + \varepsilon(t_{ij}),$$
$$\log(Glucose)_{i,nodiab} = \beta_0 + \beta_1 time_i + \beta_2 age_i + \beta_3 carc_i + \beta_4 meld_i$$
$$+ \beta_5 bmi_i + \beta_6 T H_i + U_{0i} + \varepsilon(t_{ij}),$$

where *time* is the time that repeated measurements are taken and U_{0i} is the random intercept effect for each patient.

(b) *Slope Model*

$$\log(Glucose)_{i,diab} = \beta_0 + \beta_1 time_i + \beta_2 vhc_i + \beta_3 meld_i + U_{0i} +$$
$$+ U_{1i} t_{ij} + \varepsilon_i(t_{ij}),$$
$$\log(Glucose)_{i,nodiab} = \beta_0 + \beta_1 time_i + \beta_2 age_i + \beta_3 carc_i + \beta_4 meld_i$$
$$+ \beta_5 bmi_i + \beta_6 T H_i + U_{0i} + U_{1i} t_{ij} + \varepsilon_i(t_{ij}),$$

in this model we additionally have $U_{1i}(t_{ij})$ which represents the random slope effect of the different Glucose trajectories of each patient.

(c) *Spline Model*

$$\log(Glucose)_{i,diab} = (\beta_0 + b_{i0}) + (\beta_1 + b_{i1})B_n(t, d1) + (\beta_2 + b_{12})B_n(t, d_2)$$
$$+ (\beta_3 + b_{i3})B_n(t, d_3) + \beta_4 vhc_i + \beta_5 meld_i + \varepsilon_i(t),$$
$$\log(Glucose)_{i,nodiab} = (\beta_0 + b_{i0}) + (\beta_1 + b_{i1})B_n(t, d1) + (\beta_2 + b_{12})B_n(t, d_2)$$
$$+ (\beta_3 + b_{i3})B_n(t, d_3) + \beta_4 age_i + \beta_5 carc_i$$
$$+ \beta_6 meld_i + \beta_7 bmi_i + \beta_8 TH_i + \varepsilon_i(t),$$

where $\{B_n(t, d_k); k = 1, 2, 3\}$ denotes a B-spline basis matrix for a natural cubic spline (de Boor [7]).

To study the survival process, significant covariates were selected by using a backward stepwise procedure. The final separate models for diabetic and non-diabetic patients are the following:

$$h_{i,diab}(t) = h_0(t)\exp(\lambda_1 vhc_i + \lambda_2 meld_i + \alpha \log(Glucose)_i(t)),$$
$$h_{i,nodiab}(t) = h_0(t)\exp(\lambda_1 age_i + \lambda_2 carc_i + \lambda_3 meld_i$$
$$+ \lambda_4 bmi_i + \lambda_5 TH_i + \alpha \log(Glucose)_i(t)),$$

where $h_0(t)$ is the baseline risk function, t is the time-to-event and $\log(Glucose)$ is the true (unobserved) value of the longitudinal outcome.

3.3 Results

The results indicate that final models for diabetic and non-diabetic patients take a relative risk model with Weibull baseline risk function and with a spline longitudinal sub-model. In Table 1 we synthesized all the information of both joint models. We observed that non-diabetic patients with higher Glucose level have a worse survival through the coefficients of association 0.0032 (95% CI: 0.002–0.004). However, for diabetic patients the association between the Glucose levels and survival process is not statistically significant, 0.0002 (95% CI: −0.0033–0.0004).

Among non-diabetic patients who underwent liver transplantation, glucose levels rose until reaching a peak on h 24–36 and then declined to their former levels over the course of the following days (all within the context of insulin being administered via continuous perfusion). This behaviour could reflect glycaemic response to stress ("stress hyperglycaemia", see Fig. 1), typically attributed to insulin resistance caused by endogenous and exogenous catecholamines and glucocorticoids. Blood glucose profiles were observed to be statistically associated with long-term mortality among patients without diabetes ($p < 0.01$). Due to having non-linear trends for longitudinal biomarker, the interpretation of the coefficient of association (α) becomes compromised. Thus, the overall glucose profiles are shown in Fig. 1.

Table 1 Fitted values of the final model for the joint model approaches for diabetic (on the left) and non-diabetic (on the right) patients

Joint Models (JM) - Diabetic patients

		Coef	Std. error
Longitudinal process	Intercept (β_0)	5.3261	0.0591
	β_1	-0.4126	0.0487
	β_2	-0.1523	0.0760
	β_3	-0.2286	0.0376
	β_4	-0.0696	0.0508
	β_5	-0.0006	0.0039
Survival process	Glucose	0.0002	0.0020
	vhc	0.8412	0.3349
	Meld	0.0464	0.0300
Loglikelihood	-869.1573		

Joint Models (JM) - Non-diabetic patients

		Coef	Std. error
Longitudinal process	Intercept (β_0)	4.4489	0.0602
	β_1	-0.1849	0.0183
	β_2	0.6881	0.0332
	β_3	-0.3461	0.0136
	β_4	0.0032	0.0007
	β_5	-0.0130	0.0195
	β_6	0.0026	0.0012
	β_7	0.0030	0.0021
	β_8	0.0053	0.0009
Survival process	Glucose	0.0032	0.0010
	Age	0.0293	0.0079
	carc	0.6604	0.1899
	Meld	0.0755	0.0116
	bmi	-0.0517	0.0224
	TH	0.0216	0.0073
Loglikelihood	-2659.553		

4   Extensions on Joint Modelling Framework

This section presents some recent extensions to standart joint modelling framework. We focus on the developments which are gained remarkable attention recently by the statisticians and clinicians.

4.1 Dynamic Predictions

In biomedical follow-up studies the objective is often to identify prognostic factors that can be used to guide clinical management of patients. Beside the estimation of the risk for death, the predictions for individual patients is also important for the this type of studies. Due to current trends in medical practice towards to personalized medicine joint modeling approaches present new features such as dynamic predictions (Rizopoulos and Ghosh [20]). In this concept of modeling, the survival predictions has a dynamic nature, that is the predictions at time t are dynamically changed by a new longitudinal measure taken at time t.

In particular, for a specific patient and at a specific time point during follow-up, all available information is used (including both baseline information and accumulated biomarker levels) to produce predictions of survival probabilities. This information gives a better understanding of the disease dynamics and provides optimal decision at that specific time point t.

Dynamic predictions also allow to update the prediction when we have new information recorded for the patient. Thus, the conditional probability is of primary interest, described as,

$$\pi_i(u/t) = P(T_i^* \geq u/T_i^* > t, Y_i(t), \omega_i, D_n),$$

where u is the followed-up time $(u > t)$, D_n denotes the sample on which joint model was fitted and ω_i is the baseline covariates. Rizopoulos and Ghosh [20] uses a Bayesian formulation of the problem and Monte Carlo estimates of $\pi_i(u/t)$.

Figure 2 shows the dynamic predictions of our application study presented in Sect. 3, for particular cases: a diabetic patient (subject 51) and a non-diabetic patient (subject 40), with their longitudinal observations to observe the effect of the longitudinal outcome to the survival probability of these patients.

4.2 Advances in Multivariate Longitudinal and Survival

Many biomedical study collects several longitudinal biomarkers and time-to-event data during the follow-up time and the interest is to study the relationship between the longitudinal biomarkers with their effect on time-to-event. Recently there exist several extensions to study such association in the literature. A recent overview is made by Hickey et al. [11]. Those extensions are mainly focused on the Bayesian estimation techniques (Ibrahim et al. [13], Rizopoulos and Ghosh [20], Brown et al. [3], Chi and Ibrahim [4], Proust-Lima et al. [17], Liu and Li [15], Tang et al. [22]). On a frequentist approach setting, the likelihood calculation of joint models are getting compromised.

The joint models for multivariate longitudinal and survival data consist of two sub-models: (i) Multivariate longitudinal sub-model (ii) Survival sub-model. In the

Fig. 2 Dynamic predictions for the final model for a diabetic patient (subject 51) and a non-diabetic patient (subject 40)

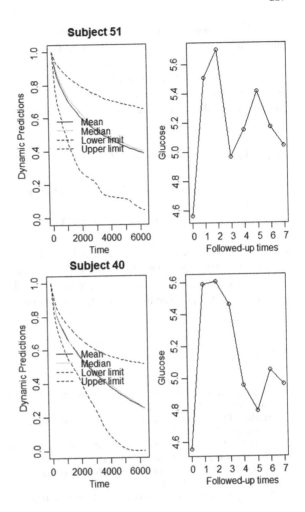

following the sub-models and computational issues on the likelihood calculation of joint models for multivariate longitudinal and survival data are presented.

4.2.1 Multivariate Longitudinal Sub-model

The longitudinal sub-model can be fitted using linear mixed model as presented in Sect. 3.2 for each longitudinal biomarker k. Thus the multivariate longitudinal sub-model is presented as follows,

$$Y_{ik}(t_{ij}) = \beta_k X_{ik}(t_{ij}) + u_{ik} Z_{ik}(t_{ij}) + \varepsilon_{ijk} \tag{3}$$

where X_{ik}, u_{ik} and ε_{ijk} are recorded for each patient at time t for each longitudinal biomarker k. Using a multivariate normal distribution assumption for the error terms and random effects u_{ik}, we can assume different correlation structures. The correlation matrix of the random effects u_{ik} captures the correlation between the longitudinal biomarkers. This correlation matrix, in a shared random effects framework, gets dimensionally higher when the number of random effects is large.

4.2.2 Time-to-Event Sub-model

Time-to-event sub-model is usually fitted by a Cox proportional hazard model as the following,

$$\lambda_i(t) = \lambda_0(t) \exp(\beta X_i + \alpha_k \omega_{ik}(t)) \tag{4}$$

where $\lambda_0(t)$ is the unspecified baseline risk function, X is a matrix of fixed effects including the baseline covariates (such as age, gender, etc.), $\omega_{ik}(t)$ is a function reflecting the association structure between the longitudinal biomarker k and survival data including the same random effects U and α_k is the coefficient of this association.

References

1. Andersen PK, Gill RD (1982) Cox's regression model for counting processes: a large sample study. Ann Stat 10(4):1100–1120
2. Akaike H (1974) A new look at the statistical model identification. IEEE Trans Autom Control 19(6):716–723
3. Brown ER, Ibrahim JG, Degruttola V (2005) A flexible B-spline model for multiple longitudinal biomarkers and survival. Biometrics 61(1):64–73
4. Chi YY, Ibrahim JG (2006) Joint models for multivariate longitudinal and multivariate survival data. Biometrics 62:432–445
5. Cox D (1972) Regression models and life-tables (with discussion). J R Stat Soc Ser B 34(2):187–220
6. Crowther MJ, Abrams KR, Lambert PC (2013) Joint modeling of longitudinal and survival data. Stata J 13:165–184
7. de Boor C (1978) A practical guide to splines. Series applied mathematical sciences, vol 27. Springer, New York
8. Ding J, Wang JL (2008) Modeling longitudinal data with nonparametric multiplicative random effects jointly with survival data. Biometrics 64(2):546–556
9. Elashoff R, Li G, Li N (2008) A joint model for longitudinal measurements and survival data in the presence of multiple failure types. Biometrics 64(3):762–771
10. Guo X, Carlin BP (2004) Separate and joint modeling of longitudinal and event time data using standard computer packages. Am Stat 58:16–24
11. Hickey GL, Philipson P, Jorgensen A, Kolamunnage-Dona R (2016) Joint modelling of time-to-event and multivariate longitudinal outcomes: Recent developments and issues. BMC Med Res Method 16(1):1–15
12. Henderson R, Diggle PJ, Dobson A (2000) A joint modelling of longitudinal measurements and event time data. Biostatistics 1:465–480
13. Ibrahim JG, Chen M-H, Sinha D (2004) Bayesian methods for joint modeling of longitudinal and survival data with applications to cancer vaccine trials. Stat Sin 14:863–883

14. Kalbfleisch JD, Prentice RL (2002) The Statistical analysis of failure time data, 2nd edn. Wiley, Hoboken

15. Liu F, Li Q (2016) A Bayesian model for joint analysis of multivariate repeated measures and time to event data in crossover trials. Stat Methods Med Res 25(5):2180–2192

16. Philipson P, Sousa I, Diggle PJ, Williamson PR, Kolamunnage-Dona R, Henderson R (2012) Package joineR: joint modelling of repeated measurements and time-to-event data. R Foundation for Statistical Computing, Austria. https://CRAN.R-project.org/package=joineR

17. Proust-Lima C, Joly P, Dartigues J-F, Jacqmin-Gadda H (2009) Joint modelling of multivariate longitudinal outcomes and a time-to-event: a nonlinear latent class approach. Comput Stat Data Anal 53:1142–1154

18. Rizopoulos D (2012) Joint models for longitudinal and time-to-event data, with applications in R. Chapman and Hall/CRC, Boca Raton

19. Rizopoulos D (2014) The R package JMbayes for fitting joint models for longitudinal and time-to-event data using MCMC. arXiv:1404

20. Rizopoulos D, Ghosh P (2011) A bayesian semiparametric multivariate joint model for multiple longitudinal outcomes and a time-to-event. Stat Med 30:1366–1380

21. Self S, Pawitan Y (1992) Modeling a marker of disease progression and onset of disease. In: Jewell N, Dietz K, Farewell V (eds) AIDS epidemiology: methodological issues. Birkhäuser, Boston

22. Tang N, Tang A, Pan D (2014) Semiparametric bayesian joint models of multivariate longitudinal and survival data. Comput Stat Data Anal 77:113–129

23. Tsiatis AA, Davidian M (2004) Joint modeling of longitudinal and time-to-event data: an overview. Stat Sin 14:809–834

24. Wulfsohn MS, Tsiatis AA (1997) A joint model for survival and longitudinal data measured with error. Biometrics 53:330–339

25. Yu M, Taylor J, Sandler H (2008) Individualized prediction in prostate cancer studies using a joint longitudinal-survival-cure model. J Am Stat Assoc 103:178–187

26. Zhang D, Chen M-H, Ibrahim JG, Boye ME, Shen W (2009) JMFit: a SAS macro for joint models of longitudinal and survival data. J Stat Softw 30:1–3

Long-Term Survival After Abdominal Aortic Aneurysm Repair

José Manuel Llaneza, Norberto Corral, Sara Busto, Amer Zanabili
and Manuel Alonso

Abstract In this paper, the 10-year survival rate for elective abdominal aortic aneurysm repair in the Hospital Universitario Central de Asturias is statistically analyzed. It is a retrospective, observational and descriptive study of demographic and clinical variables on patients undergoing either open or endovascular abdominal aortic aneurysm repair with respect to the variable *exitus*. The study comprises 256 patients who were treated for abdominal aortic aneurysm repair along the period fro January 2003 to December 2006. Groups of patients did not show a homogeneous behaviour, since diabetes, old age, cardiac and renal pathologies had a negative effect on their control and monitoring. The average 10-year survival rate amounts 37.9%, cancer being the major cause of mortality (lung cancer being the most frequent one) and cardiovascular diseases were the second major cause of patients death.

1 Introduction

Abdominal aortic aneurysm (AAA) is a vascular disease with a prevalence greater than 4% on the population over 65 years old. This prevalence compels healthcare systems to spend numerous resources to the early diagnosis, the monitoring and the surgical repair of the AAA when required. This is especially critical since AAAs tend naturally to grow, what puts pressure on nearby structures and may eventually rupture with catastrophic consequences. Nowadays, it is well-known that preventing aneurysm rupture by repairing it leads to excellent postoperative survival rates. This type of surgery is prophylactically recommended, so to avoid the aneurysm sac rupture, what would substantially increase the perioperative mortality.

J. M. Llaneza (✉) · S. Busto · A. Zanabili · M. Alonso
Servicio de Cirugía Vascular, Hospital Universitario
Central de Asturias, Oviedo, Asturias, Spain
e-mail: lanezacoto@gmail.com

N. Corral
Departamento de Estadística, I.O. y D.M., Universidad de Oviedo,
Oviedo, Asturias, Spain
e-mail: norbert@uniovi.es

© Springer International Publishing AG 2018
E. Gil et al. (eds.), *The Mathematics of the Uncertain*, Studies in Systems,
Decision and Control 142, https://doi.org/10.1007/978-3-319-73848-2_22

Along the last years the anaesthesia and resuscitation techniques have improved. As a consequence, the elective surgery outcomes, both those associated with open AAA repair (OAR) and those associated with endovascular one (EVAR), have also improved. Nevertheless, there are different concomitant risk factors in connection with AAA development so that, in several cases in which the AAA is treated, the survival cannot be reliably predicted.

In the literature on the topic, one can find many monitoring studies allowing us to identify short-term mortality factors for these patients, but long-term ones are very scarce. This paper aims to examine the 10-year survival rate of patients who have been treated with elective AAA in the Hospital Universitario Central de Asturias (HUCA), as well as to statistically analyze both the OAR and the EVAR results.

2 Materials and Methods

This was a retrospective, observational and descriptive study of the patients who underwent elective surgical repair of AAA in HUCA from January 2003 to December 2006.

As the election criterion in HUCA, patients undergoing AAA repair had a maximum diameter of AAA over 5.5 cm, an enlargement over 5 mm along the last six months or an enlargement over 1 cm along the last year. These patients were treated either with OAR or with EVAR, depending on the anatomical characteristics of the AAA, the age or the risk factors for surgery. The study in this paper does not include patients who had an emergency surgery.

The monitoring of these patients was addressed through outpatients periodical appointments of the HUCA Vascular Unit. For OAR patients, the first appointment was scheduled six months after surgery, and later by annual check ups, which included clinical examination and a doppler study and medical image test (usually eco-doppler) every five years thereafter, or whenever it is required due to suspecting the patient having complications. EVAR patients underwent a computerized axial tomography (CAT) with contrast (angio-CT) during the first postoperative month, and six months later. In case the EVAR patient was free from endoleak, the monitoring is annual and consisted of an eco-doppler in our vascular lab and a simple abdominal X-ray. In case the EVAR patient showed either (type I and III) high-pressure endoleak or type II endoleak with aneurysm sac enlargement, angio-CT with contrast was considered to treat him/her. Once the problem was solved, monitoring consisted of ecographic controls.

Demographic and clinical analyzed variables were vascular risk factors, lung, cardiac and renal associated pathologies as well as risk scales collected in a registry database of all the AAA in the HUCA. To fill in the information about variables *exitus* (death) and *exitus date*, the information was drawn from the clinical history of HUCA, the digitalized clinical histories of the public hospitals in the Principality of Asturias, and the mortality records in the Principality.

Causes of *exitus* were classified into 4 groups, namely, AAA-related death, cardiovascular disease (ICD-10), cancer, and other diseases.

The statistical analysis was carried out by the Department of Statistics, OR and TM in the Faculty of Sciences of the University of Oviedo by using (open access) R packages. Comparisons for qualitative variables were based on chi-square tests. Comparisons for quantitative variables were based either on Student *t* tests or on Kruskall–Wallis non parametric test. Survival analyses were based on the Kaplan–Meier test. Finally, the inferential analysis about the influence of some of the variables on the mortality considered Cox regression model.

3 Results

From January 2003 to December 2006, 258 patients underwent elective AAA surgery in HUCA, either by OAR or EVAR. Along these four years the surgical trend experienced a drastic change (see Fig. 1). More concretely, whereas in the first year 57 out of 74 patients underwent OAR and 17 underwent EVAR, in the fourth year 17 out of 56 patients underwent OAR and 39 underwent EVAR. This change can be easily seen in the graphical display in Fig. 1. Actually, this trend to consider EVAR more frequently than OAR has been consolidated in the Vascular Surgery Unit of HUCA along the years (currently, around 4 out of 5 elective AAA patients undergo EVAR).

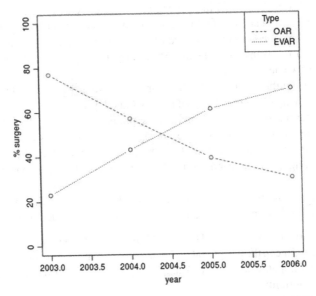

Fig. 1 Evolution of the numbers of elective patients undergoing OAR and EVAR in the period 2003–2006 in HUCA

Table 1 Univariate descriptive and inferential comparative analyses between OAR and EVAR groups for different demographic and clinical variables

Variable	OAR (134)	EVAR (122)	p-value
Age	69.25 ± 7.68	75.75 ± 7.08	0.000
Octogenerians	7 (5.2%)	42 (33.9%)	0.000
Smoking	125 (93%)	104 (83.9%)	0.017
Dyslipidemia	71 (53%)	45 (36.3%)	0.007
Cardiopathy	52 (38.8%)	68 (54.8%)	0.010
ASA scale	2.54	2.73	0.046
TIA	74.01	69.63	0.020
Glasgow score	74.23 ± 10.96	82.56 ± 9.98	0.000

If we constrain to the four-year period 2003–2006, 134 of the 258 treated patients (i.e., 51.94%) underwent OAR and 124 (i.e., 48.06%) underwent EVAR. In the second group 2 patients were discarded from the study: one of them was lost to clinical follow-up, and for the other one the endovascular stenting failed so that we cannot consider him as a suitable surgery candidate because of having comorbidity. Concerning the 30-day and in-hospital mortality there were 6 in the OAR group (4.5%) and 5 in the EVAR one (4.1%).

In conducting the univariate analysis the results were gathered in Table 1. One can easily observe that OAR and EVAR lead to significantly different results (i.e., p-values are lower than 0.05) in connection with age, percentage of octogenarian patients, major associated risk factors, cardiopathies, and surgical risk scales (where ASA score corresponds to the American Society of Anaesthesiologists scale, TIA is the acronym for Transient Ischemic Attack, and Glasgow score means the well-known Glasgow aneurysm score).

Due to the significant different behaviour of the two groups, separate (instead of comparative) analyses were performed for OAR and EVAR groups in order to examine the long-term evolution of patients and the predictive mortality factors. It would be further emphasized that there were not significant differences associated either with the aneurysm size itself or with the sex of the patient (in fact, the sample only includes 8 women, so one cannot reliably draw statistical conclusions concerning sex).

By means of a multivariate logistic regression, for which only the small p-values (lower than 0.05) have been gathered in Table 2, one can conclude that

- for OAR patients, the age, the Diabetes Mellitus, the chronic kidney failure, the level of creatinine in the peripheral blood, the heart failure, the atrial fibrillation, the ischemic cardiopathy, and the ASA and Glasgow scores, showed to be predictive factors for mortality;
- for EVAR patients, only the atrial fibrillation, the heart failure and the level of creatinine in the peripheral blood showed to be predictive factors for mortality.

Table 2 Predictive factors of mortality for OAR and EVAR groups

OAR (134)		EVAR (122)	
Predictive factor	p-value	Predictive factor	p-value
Age	< 0.001	Atrial fibrillation	< 0.001
ASA score	0.025	Heart failure	0.030
Diabetes mellitus	0.023	Creatinine	0.030
Ischemic cardiopathy	0.016		
Atrial fibrillation	0.004		
Heart failure	0.015		
Kidney disease	< 0.001		
Creatinine	0.008		
Glasgow score	< 0.001		

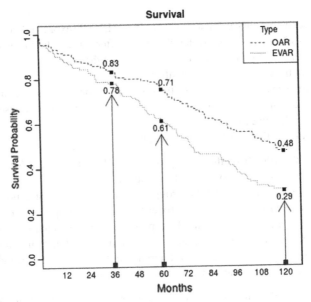

Fig. 2 Long-term (10-year) survival evolution for OAR and EVAR patients in HUCA

On the other hand, the mortality has been directly analyzed. Early mortality was slightly greater for OAR than for EVAR. In OAR group there were 6 *exitus*, what means 4.69% of perioperative mortality, since death occur within 30 days after surgery. In EVAR group there were 5 *exitus*, what means 4.10% of perioperative mortality. However, the situation is reversed afterwards, and this seems to be due to the high mean age and presence of risk factors of patients undergoing EVAR. In fact, many of these patients were included in the EVAR group, because of them being quite old and not fulfilling criteria for OAR, which involved a high surgical risk.

The long-term mortality can be examined on the basis of Fig. 2.

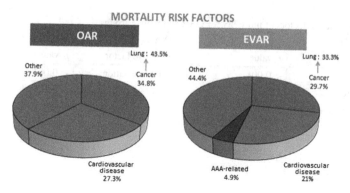

Fig. 3 Long-term (10-year) distribution of the mortality risks in the subsample of non-perioperative *exitus* for OAR and EVAR in HUCA

It can be seen how curves for survival OAR and EVAR (expressed in proportions) distance themselves over the considered 10 years after surgery. In this way, along the 36 months after surgery the distance was up to 5%, whereas this distance amounted around 20% at the end of the 10 years after surgery.

If mortality is analyzed for each of the 4 groups associated with the main causes of *exitus*, conclusions differ from the AAA-related death and other diseases to cancer and cardiovascular diseases.

In connection with the AAA-related death:

- there were no *exitus* from the OAR group after 10 years, since AAA was rightly excluded and there were no infectious complications;
- there were 4 *exitus* from the EVAR group after 10 years, what corresponds to 4.9% of the global *exitus* in EVAR group (over the 81 non-perioperative *exitus*); three of the deaths were associated with the rupture of the aneurysmal sac (one of them due to a type II endoleak which did not respond to the treatment of either selective embolizations of the sac, and two of them due to type I endoleaks), and the other one was associated with an endoprosthetic infection.

Regarding the other diseases group, Fig. 3 shows that mortality is more frequent for EVAR than for OAR group. It seems reasonable since EVAR group included a higher comorbidity and more octogenarian patients at the moment of the prosthesis placement.

Nevertheless, the general mortality risk causes are similar for OAR and EVAR patients, tumor pathology and especially lung cancer prevailing over cardiovascular diseases with close proportions in both groups. This trend was preserved along the follow-up as one can see in the 5-year information gathered in Table 3.

Table 3 5-year and 10-year percentages of mortality causes (over non-perioperative *exitus*)

	OAR		EVAR	
	5-year percentage (%)	10-year percentage (%)	5-year percentage (%)	10-year percentage (%)
Cancer	40.8	34.8	30	29.7
Cardiovascular disease	33.3	27.3	25	21
Other	25.9	37.9	45	44.4
AAA-related	0	0	0	4.9

4 Discussion

The early survival of the elective AAA repaired patients has dramatically improved along the last 50 years. The emergence and rapid therapeutic implementation of EVAR has been crucial for such an improvement. This technique allowed vascular specialists to incorporate a relevant cohort of old patients with high comorbidity to undergo AAA repair, in such a way that when studies with age adjustments are performed the improved life expectancy is not lost. These convenient expectations can be found, for instance, in [12] where 70% 5-year survival of the elective OAR is shown to be very similar to our own experience showing 74.6% 5-year survival.

However, in spite of the technical improvements, mortality in AAA patients is still high, and in some cases it doubles that of the general population. The main cause for this is a generalized arteriosclerosis with a high rate for stroke, heart attack, and major lower limb amputations due to ischaemia, as reported by classical studies like the one in [3].

The influence of the arteriosclerosis risk factors is not revealed only in the long-term mortality, but it is well-known that the 30-day mortality is not negligible when we refer to AAA surgery in hospitals conducting a small to moderate number of these surgeries. In all these cases the cardiopathy-based mortality is highly present, as it can be seen in [4, 11, 13] where 57.9% of perioperative mortality after AAA repair is associated with cardiac pathology. Data in this paper are quite similar to those in the last citations, with an important presence of vascular pathology in patients dying, just the second cause for mortality after cancer pathologies. Anyway, ischemic pathology was only influential in the OAR group, whereas arrhythmia by atrial fibrillation and heart failure were always clearly influential in mortality for both groups.

The aneurysm size was considered in the literature to be related to mortality in both the pre-surgery and the perioperative period and to the long-term survival (see, for instance, [2, 14]). One of the long-term mortality theories is the potential direct relationship between the aneurysm size and the arteriosclerosis intensity. Nevertheless, this theory has never been confirmed. Furthermore, one should be aware that aneurysm diameter also depends on other factors like age, gender, smoking, high blood pressure and high cholesterol level, among others. All this makes that the aneurysm size takes often a rather misleading value, so that it has never been independently analyzed as a risk factor (see [1, 11]).

In connection with sex, survival studies indicate in general that women have a higher life expectancy than men. However, in case of an underlying aneurysm pathology, mortality of AAA repaired women is equal or greater than mortality of AAA repaired men (see [6, 7, 10]). This leads to conclude that the advantage in the general population is reduced so that there is no substantial influence of gender. Our results agree with such an idea of the sex not influencing mortality by AAA repair, although since the number of women in our study is quite small, conclusions are not actually reliable.

Patients suffering chronic obstructive pulmonary disease (COPD) have shown worse survival (see [7, 10, 15]), especially when they are associated with domiciliary oxygen. In our study COPD was not identified as a risk factor for the long-term mortality, but statistical conclusions are limited in this respect since only a few of our cases were associated with such a permanent oxygen need.

The study in [10] also considers the chronic kidney failure as a factor of bad prognosis. This meta-analysis examines 16 studies, and authors pointed out the limitations caused because of these studies considering different meanings and interpretations of the variable kidney failure. In the study in this paper, kidney failure has only behaved as a risk factor in the OAR group, albeit the level of creatinine in blood has behaved as a mortality risk factor in OAR and EVAR groups.

Other major risk factors, as high blood pressure and Diabetes Mellitus have been also analyzed. The first one was not influential in our study, although in [10] it has been highlighted as a risk factor but by considering different diagnoses criteria and treatments; when the left ventricular hypertrophy is added, heterogeneity disappears and the negative influence of the high blood pressure is clear (see [5, 8]). The variable Diabetes Mellitus is not unequivocally defined, in connection with both its diagnosis and its treatment and potential complications, although it seems evident that the long-term mortality tends to increase (see [10]). In case of HUCA, the Diabetes Mellitus has only behaved as a risk factor in the OUR group.

The study in this paper does not consider either socio-economic or ethnic variables which could affect survival in a negative way (see [9]), since our populations seems to be rather homogeneous in these respects.

5 Conclusions

The global survival of our data was 37.9%, the percentages of survival being noticeably greater for the OAR, although EVAR patients were definitely older and having a more important associated comorbidity. The major cause of *exitus* was cancer, especially lung one, for both groups of surgery. The second cause was cardiovascular diseases, and they materialize in a rather constant way during the follow-up.

Limitations of our study are evident since it concerns a few patients and a short monitoring. Furthermore, the change prompted by the Unit therapeutic choice was clear. Anyway, to continue this type of analyses will be decisive in supporting and suggesting future surgical corrections and the choice of the treatment.

Acknowledgements The research in this paper has been partially supported by the Principality of Asturias/FEDER Grant GRUPIN14-101 and the Spanish Grant MTM2015-63971-P from the Ministerio de Economía y Competitividad. Its financial support is gratefully acknowledged.

Dedication

I was fortunate to meet Pedro at the middle 80s of last century. My need of help in order to do my PhD dissertation, as it happens to most of the doctors on dealing with the design of a study and the statistical analysis of the data, as well as the departure of my friend María Ángeles to her American adventure, made me to end up, quite defenseless, at the Department of Statistics in the University of Oviedo, where I met wonderful people and will never be grateful enough to Norberto, Teresa and Teófilo for their warm support.

Despite Pedro's kind nature, the first reach out was difficult and things did not go well. At that time, he would be just fed up with people coming with a pile of data and the unique purpose of seeing what could be drawn from them, but without actual interest on a scientific collaboration. Nevertheless, my intention was to plan a well-designed work before beginning with the clinical examinations but I couldn't calmly explain him my goals. At that initial period he did not look upon me favorably, and I used to literally keep clear of him, even keeping me hidden, sometimes, in an office so we could not meet each other. Several months later, the relationship dramatically changed, and from that moment I met one of the most positively influential persons in my life. I had the opportunity of talking with him in the faculty, during the years of preparation of my dissertation, in those long and friendly late-evening get-togethers, when we had finished our work and met, generally at Norberto's office, before coming back home. Pedro used to recap the day, concerning about everyone's issues, my dissertation included although it was far from the daily duties of the Department, turning later to speak about any other subject, none of us paying attention to the hour until we were there. I learned a lot from those conversations, not of Mathematics, but about coping with problems and living life.

Once I defended my dissertation, although I was in contact with the group of Statistics, I did not see Pedro so often. Unfortunately, his health issues appeared some years later. During his disease, he had a desirable composure at each conversation that we had during his visits to external consultations at the old Hospital. Tobacco could not be part of his life any more due to his throat and circulatory issues. Knowing Pedro, it seemed quite impossible to give him up from cigarettes. So I cruelly let him know the truth while he stared at me as if he were saying "Ask me for anything, but do not bother me with tobacco". Surprisingly he quitted smoking almost at the first attempt. When the first examinations had concluded, I remind telling him, with my usual tact, that he was much older internally than what was written on his identity card. Then his mathematical mind appeared and we tried to calculate the age of his arteries in comparison with the age at his official documents. We were always joking around that difference and trying to adjust it with every new examination.

Another issue was to achieve that Pedro could walk with ease by Oviedo's hills. The peripheral circulatory disease causes the so-called sporadic failure that obliges those suffering it to stop every so often. Being able of walking farther depends on the training. As he always said, with snide humor, worse than bearing with his pains on walking was putting up with his colleagues' comments who pulled his legs when he stopped and encouraged him to go on. His willpower prevailed here, too, and he finally achieved to do an almost normal life. Apart from what is strictly professional, we hardly needed anything to bring up-to-date; a glance and a couple of sentences were enough most of the times due to our mutual understanding.

Along the years I treated Pedro, he never boasted about honors or merits. I knew about them through his mates at the department and, above all, at that surprise jubilee meeting where all his achievements were revised. There I confirmed his humility and how much everybody cared about him.

Shortly after his heart warn for the first time. He had just arrived from Paris and one could realize how happy he was with his children's welfare. He was aware of his condition, never having long face, but always looking forward. We argued a lot about how his badly damaged heart was going to be fixed, but none of us were in the loop on that subject. After several days of arguments, sitting at the hospital bed, he told me that he had realized that none of us was a cardiology expert so we should better let the doctors at the haemodynamics service work. During his stay at the hospital we spoke at length about many subjects apart from his disease. I slipped away to Pedro's room whenever I could, even at strange hours due to a long-lasting surgery. For me, it was meeting a friend from outside the hospital in order to keep ourselves up to date. We spoke openly about everything. Once again, it was a continuous lesson, because time and experience had made a good work in his mind. He was not only great at Mathematics but also knew why it was worth living; at least it seems to me that he did. He had a wonderful point of view; in the majority of the matters, his reasoning was so quiet, clear and organized that everything looked easier.

I will always remember your love for the well-done work and your smile on seeing me, even in those periods when things were not going well.

Thanks a lot, Buddy!

Tuve la suerte de conocer a Pedro en mitad de los 80 del siglo pasado. La necesidad de ayuda para hacer la tesis doctoral, como nos ocurre a la mayoría de los médicos cuando se aborda el diseño de un estudio y el tratamiento estadístico de los datos, unida a la marcha de mi amiga María Ángeles a su aventura americana, hizo que aterrizase, un poco desvalido, en el Departamento de Estadística de la Universidad de Oviedo, donde conocí a gente maravillosa y nunca agradeceré lo suficiente a Norberto, Teresa y Teófilo el apoyo que me brindaron.

A pesar del carácter amable de Pedro' la primera toma de contacto fue difícil y la cosa no pintó bien. Por aquella época ya debía de estar un poco harto de que apareciese por allí gente con un montón de datos para ver qué se podía sacar de ellos, pero sin estar realmente interesados en una colaboración científica. Mi intención, sin embargo, era plantear un trabajo bien diseñado y luego comenzar con las exploraciones clínicas. No pude llegar a explicarle con tranquilidad mis objetivos, así que en esa primera época no me miraba con 'buenos ojos' y yo procuraba evitarlo, literalmente, e incluso alguna vez me tuvieron 'escondido' en un despacho para no encontrarnos. Pasados los primeros meses, la relación cambió radicalmente, y a partir de ahí conocí al que creo fue una de las personas que influyó de forma más positiva en mi vida. Tuve la oportunidad de conversar con él en la facultad, durante los años que duró la preparación de la tesis, en aquellas largas y amigables tertulias a última hora de la tarde, cuando el trabajo finalizaba y nos reuníamos, generalmente en el despacho de Norberto, antes de irnos a casa. Pedro hacía un repaso de cómo había ido el día, interesándose por las cosas de cada uno, incluso de mi tesis, que tan lejos quedaba del quehacer diario del Departamento y posteriormente hablábamos de cualquiera tema, sin importarnos la hora a ninguno de los que allí estábamos. De esas conversaciones aprendí muchas cosas, no de matemáticas, sino de cómo encarar los problemas y a caminar por la vida.

Una vez leída la tesis, aunque seguí en relación con la gente de Estadística, no veía a Pedro con tanta frecuencia. Pasado el tiempo, por desgracia, aparecieron sus problemas de salud. Durante su enfermedad tuvo una entereza envidiable en las conversaciones que manteníamos en sus visitas a la consulta externa del viejo Hospital. Sus problemas de garganta y circulatorios hacían que el tabaco no pudiese estar presente de ninguna de las maneras en su vida. Conociendo a Pedro parecía una tarea imposible conseguir separarlo de los cigarrillos, así que le informé crudamente de la realidad mientras me miraba como diciendo "pídeme cualquier otra cosa, pero deja de fastidiarme con el tabaco". Increíblemente consiguió dejar de fumar prácticamente a la primera. Una vez hechas las primeras exploraciones, recuerdo que le comenté, con mi delicadeza habitual, que por dentro era mucho más viejo que lo que decía su carnet de identidad. Entonces comenzó a aflorar su mente matemática e intentábamos calcular la edad de sus arterias en comparación con la edad que figuraba en el DNI. Siempre bromeábamos con esa diferencia e intentábamos ajustarla con las exploraciones que se iba haciendo.

La segunda parte del problema era conseguir que Pedro pudiese caminar con soltura por las cuestas de Oviedo. La enfermedad circulatoria periférica provoca lo que se llama claudicación intermitente y obliga, a quien la padece, a tener que pararse cada cierta distancia. Conseguir caminar más metros depende del entrenamiento que se haga. Como siempre me comentaba, de forma socarrona, los dolores de caminar no eran lo peor; lo malo era aguantar los comentarios de los compañeros de la facultad que le tomaban el pelo por las veces que tenía que pararse y le animaban a continuar; también aquí se impuso su fuerza de voluntad y terminó consiguiendo hacer una vida prácticamente normal. Aparte de lo puramente profesional, poco necesitábamos para ponernos al día, ya que la relación de complicidad fluía claramente entre nosotros, un par de miradas y cuatro frases bastaban para entendernos en la mayoría de las cosas.

En todos los años que traté con Pedro, nunca alardeó de honores ni méritos. Si los conocí fue a través de sus compañeros de departamento y, sobre todo, en aquella reunión sorpresa por su jubilación donde se repasaron sus logros; en ella pude confirmar su humildad y sobre todo comprobar lo que la gente le quería.

Desde esa reunión pasó poco tiempo hasta que su corazón dio el primer aviso. Venía de París y se le notaba muy feliz con lo bien que les iba a sus hijos. El fue consciente de su situación, siempre sin una mala cara y con ganas de mirar hacia delante. Discutíamos mucho de cómo le iban a arreglar su maltrecho corazón, aunque los dos estábamos bastante pez en esas cosas. Tras varios días de conversaciones sentados en la cama del hospital, me comentó que se había dado cuenta de que realmente ninguno de los dos teníamos ni idea de cardiología y era mejor dejar trabajar a los médicos de la unidad de hemodinámica. Durante su ingreso charlamos largamente de temas al margen de su enfermedad y siempre que tenía un hueco me escapaba hasta la habitación de Pedro para visitarle, a veces en horas un poco raras porque había tocado quirófano largo, pero para mí era como reunirme con un amigo fuera del hospital para comentar cómo nos había ido el día. Hablamos abiertamente de todo y nuevamente fue una enseñanza continua, pues el paso del tiempo y la experiencia hicieron un buen trabajo en su cabeza, que no sólo sabía mucho de Matemáticas sino que tenía una idea muy clara de por qué merece la pena vivir, o al menos eso me parecía a mí. Tenía una perspectiva maravillosa de las cosas y en la mayoría de los temas que tratábamos su razonamiento era tranquilo, claro y organizado haciendo que todo pareciese siempre más sencillo.

Te recordaré siempre por tu aprecio al trabajo bien hecho y la sonrisa que me dedicabas al verme, incluso en los momentos en que las cosas no iban bien.

¡Muchas gracias, Amigo!

José Manuel Llaneza, November 2017

References

1. Bahia SS, Holt PJE, Jackson D, Patterson BO, Hinchliffe RJ, Thompson MM, Karthilke-salingam A (2015) Systematic review and meta-analysis of long-term survival after elective infrarenal abdominal aortic aneurysm repair 1969–2011: 5 year survival remains poor despite advances in medical care and treatment strategies. Eur J Vasc Endovasc Surg 50:320–330
2. Brady AR, Fowkes FG, Thompson SG, Powell JT (2001) Aortic aneurysm diameter and risk of cardiovascular mortality. Arterioscler Thromb Vasc Biol 21:1203–1207
3. Brown OW, Hollier LH, Pairlorelo PC, Kazmier FJ, McCready RA (1981) Abdominal aortic aneurysm and coronary artery disease. Arch Surg 116:1484–1488
4. Conrad MF, Crawford RS, Pedraza JD, Brewster DC, LaMuraglia M (2007) Long-term durability of open abdominal aortic aneurysm repair. J Vasc Surg 46(4):669–675
5. Galinaes EL, Reynolds S, Dombrovskiy VY, Vogel TR (2015) The impact of preoperative statin therapy on open and endovascular abdominal aortic aneurysm repair outcomes. Vascular 23(4):344–349
6. Gloviczki P, Huang Y, Oderich GS, Duncan AA, Kaira M, Fleming MD (2015) Clinical presentation, comorbidities, and age but not female gender predict survival after endovascular repair of abdominal aortic aneurysm. J Vasc Surg 61(4):853–861
7. Grootenboer N, Myriam Hunink MG, Hendriks JM, van Sambeek MRHM, Buth J (2013) Sex differences in 30-day and 5-year outcomes after endovascular repair of abdominal aortic aneurysms in the EUROSTAR study. J Vasc Surg 58(1):42–49
8. Hertzer NR, Mascha EJ (2005) A personal experience with factors influencing survival after elective open repair of ifrarenal aortic aneurysms. J Vasc Surg 42(5):898–905
9. Khashram M, Pitama S, William JA, Jones GT, Roake JA (2017) Survival disparity following abdominal aortic aneurysm repair highlights inequality in ethnic and socio-economic status. Eur J Vasc Endovasc Surg 54(6):689–696
10. Khashram M, William JA, Hider PN, Jones GT, Roake JA (2016) Systematic review and meta-analysis of factors influencing survival following abdominal aneurysm repair. Eur J Endovasc Surg 51:203–215
11. Kertai MD, Boersma E, Westerhout CM, van Domburg R, Klein J, Bax JJ, Van Urk H, Poldermans D (2004) Association between long-term statin use and mortality after sucessful abdominal aortic aneurysm surgery. Am J Med 116(2):96–103
12. Mani K, Björck M, Lundkvist J, Wanhainen A (2009) Improved long-term survival after abdominal aortic aneurysm repair. Circulation 120:201–211
13. Norman PE, Semmens JB, Lawrence-Brown MM, Holman CD (1998) Long-term relative survival after surgery for abdominal aortic aneurysm in Western Australia: population based study. BMJ 317:852–856

14. N Engl J Med (2002) Long-term outcomes of inmediate repair compared with surveillance of small abdominal aortic aneurysm. 346:1445–1452
15. Von Elm E, Atman D, Egger M, Pocock SJ, Gotzsche PC, Vandenbroucke JP (2007) Strengthening the reporting of observational studies in epidemiology (STROBE) statement: guidelines for reporting observational studies. Br Med J 335(7624):806–808

Some Comments on Stochastic Orders and Posets

María Concepción López-Díaz and Miguel López-Díaz

Abstract In this paper, we review some relations between partially ordered sets and stochastic orders. We focus our attention on analyzing if the property of being order-isomorphic is transferred from partially ordered sets to the stochastic orders generated by such partially ordered sets.

1 Introduction

The aim of this manuscript is to summarize some relations between partially ordered sets and stochastic orders defined by means of such ordered sets.

The theory of ordered sets plays an important role in many mathematical fields, as algebra, analysis, graph theory, combinatorics (see for instance [1, 10, 23, 28], etc.). It is applied in different areas as computer science, coding theory, cryptography, (see for instance [6, 12, 22], etc.), as well as being by itself a remarkable area of interest for researchers.

Consider $(\mathscr{X}, \preceq_{\mathscr{X}})$ a *poset (partially ordered set)*, that is, $\mathscr{X}$ is a set and $\preceq_{\mathscr{X}}$ is a binary relation defined on $\mathscr{X}$ which satisfies the reflexivity, transitivity and antisymmetric properties. Some of the most useful concepts related to ordered sets are the following.

A subset $U \subset \mathscr{X}$ is said to be an *upper set* if given $x_1, x_2 \in \mathscr{X}$ with $x_1 \in U$ and $x_1 \preceq_{\mathscr{X}} x_2$, then $x_2 \in U$.

An *upper quadrant set* is a subset of $\mathscr{X}$ of the form $Q_x^{\preceq_{\mathscr{X}}} = \{z \in \mathscr{X} \mid x \preceq_{\mathscr{X}} z\}$, with $x \in \mathscr{X}$. We will denote by $\mathscr{Q}^{\preceq_{\mathscr{X}}}$ the class of upper quadrant sets determined by the partial order $\preceq_{\mathscr{X}}$ on $\mathscr{X}$. Note that any upper quadrant set is an upper set.

M. C. López-Díaz
Departamento de Matemáticas, Universidad de Oviedo, Oviedo, Spain
e-mail: cld@uniovi.es

M. López-Díaz (✉)
Departamento de Estadística e I.O. y D.M., Universidad de Oviedo, Oviedo, Spain
e-mail: mld@uniovi.es

© Springer International Publishing AG 2018
E. Gil et al. (eds.), *The Mathematics of the Uncertain*, Studies in Systems, Decision and Control 142, https://doi.org/10.1007/978-3-319-73848-2_23

Let $x_1, x_2 \in \mathscr{X}$. We will say that x_2 *covers* x_1 if $x_1 \preceq_{\mathscr{X}} x_2$, $x_1 \neq x_2$ and there is not $x_3 \in \mathscr{X}$, $x_1 \neq x_3 \neq x_2$, with $x_1 \preceq_{\mathscr{X}} x_3 \preceq_{\mathscr{X}} x_2$.

In the case where $\mathscr{X}$ is a finite set, the *Hasse diagram* of the poset $(\mathscr{X}, \preceq_{\mathscr{X}})$ is a directed graph with vertices set $\mathscr{X}$ and an edge from x to y if y covers x.

To construct a Hasse diagram we will draw the points of $\mathscr{X}$ in the plane such that if $x_1 \preceq_{\mathscr{X}} x_2$, the point for x_2 has a larger y-(vertical) coordinate than the point for x_1.

Hasse diagrams are very useful representations of ordered sets which describe the order relations between elements in a graphical way.

Mappings from an ordered set to another ordered set which preserve the order relations, play a remarkable role in order theory. These mappings are said to be order-preserving mappings.

Let $(\mathscr{X}, \preceq_{\mathscr{X}})$ and $(\mathscr{Y}, \preceq_{\mathscr{Y}})$ be posets. A mapping $\phi : \mathscr{X} \to \mathscr{Y}$ is said to be *order-preserving* if for any $x_1, x_2 \in \mathscr{X}$ with $x_1 \preceq_{\mathscr{X}} x_2$, we have that $\phi(x_1) \preceq_{\mathscr{Y}} \phi(x_2)$. In particular, if $(\mathscr{X}, \preceq_{\mathscr{X}})$ is a poset, a mapping $f : \mathscr{X} \to \mathbb{R}$ is said to be $\preceq_{\mathscr{X}}$-*preserving* if for any $x_1, x_2 \in \mathscr{X}$ with $x_1 \preceq_{\mathscr{X}} x_2$, we have that $f(x_1) \leq f(x_2)$. Note that the class of mappings which are $\preceq_{\mathscr{X}}$-preserving is the class of order-preserving mappings when we consider the posets $(\mathscr{X}, \preceq_{\mathscr{X}})$ and $(\mathbb{R}, \leq)$, $\leq$ being the usual order on the real line.

A very interesting class of order-preserving mappings is the class of bijective order-preserving maps. Let $(\mathscr{X}, \preceq_{\mathscr{X}})$ and $(\mathscr{Y}, \preceq_{\mathscr{Y}})$ be posets. A mapping $\phi : \mathscr{X} \to \mathscr{Y}$ is said to be an *order-isomorphism* if:

(i) ϕ is order-preserving,
(ii) there exists $\phi^{-1} : \mathscr{Y} \to \mathscr{X}$ inverse of ϕ,
(iii) ϕ^{-1} is order-preserving.

Clearly a mapping $\phi : \mathscr{X} \to \mathscr{Y}$ is an order-isomorphism if and only if

(i) ϕ is bijective,
(ii) for all $x_1, x_2 \in \mathscr{X}$, it holds that $x_1 \preceq_{\mathscr{X}} x_2$ if and only if $\phi(x_1) \preceq_{\mathscr{Y}} \phi(x_2)$.

Two posets $(\mathscr{X}, \preceq_{\mathscr{X}})$ and $(\mathscr{Y}, \preceq_{\mathscr{Y}})$ are said to be *order-isomorphic* if there exists an order-isomorphism $\phi : \mathscr{X} \to \mathscr{Y}$.

Note that two order-isomorphic sets are indistinguishable for the ordered sets theory, because they have the same order structure.

The reader is referred, for instance, to [14, 26, 27] for an introduction to the theory of ordered sets.

In probabilistic and statistic frameworks, one of the main aims is to compare random magnitudes in accordance with an appropriate criterion. Ordered sets of probabilities have a great importance in this context. The study of such ordered sets involves the analysis of theoretical and applied problems. Pre-orders on sets of probabilities are called stochastic orders (see for instance [3, 25, 29]). Stochastic orders have been successfully applied in areas like medicine, ecology, veterinary science, biology, economics, quality control theory, shape analysis, communications and others (see for instance [2, 5, 7–9, 16, 18–21, 25, 30], etc.).

A stochastic order is said to be integral if there exists a class of real measurable mappings satisfying that $P \stackrel{\sim}{\preceq} Q$ when

$$\int_{\mathscr{X}} f \, dP \le \int_{\mathscr{X}} f \, dQ$$

for all f in such a class, such that the above integrals exist, P and Q being probabilities. That class of mappings is said to be a generator of the order. We should note that there could be different generators of the same stochastic order (the reader is referred to [24] and Chap. 2 of [25] for integral stochastic orderings).

It is possible to define a stochastic order on the class of probabilities associated with a set endowed with a partial order by means of the class of preserving mappings as follows (see, for instance, [15, 25]).

Let $(\mathscr{X}, \preceq_{\mathscr{X}})$ be a poset. Consider a σ-algebra $\mathscr{A}$ on $\mathscr{X}$. Let $\mathscr{F}^{\mathscr{X}}$ stand for the class of real measurable $\preceq_{\mathscr{X}}$-preserving mappings. Let $\mathscr{P}_{\mathscr{X}}$ denote the set of probabilities associated with the measurable space $(\mathscr{X}, \mathscr{A})$. Define a pre-order $(\preceq_{\mathscr{X}_g})$ in that class by: let $P_1, P_2 \in \mathscr{P}_{\mathscr{X}}$, then

$$P_1 \preceq_{\mathscr{X}_g} P_2 \quad \text{when} \quad \int_{\mathscr{X}} f \, dP_1 \le \int_{\mathscr{X}} f \, dP_2$$

for all $f \in \mathscr{F}^{\mathscr{X}}$ for which both integrals exist.

Conditions for an integral stochastic order to be generated by a poset are studied in [17].

In some frameworks is quite useful to pay attention to the property of being order-isomorphic for stochastic orders generated by partially ordered sets, when such partially ordered sets are order-isomorphic.

2 On Order-Isomorphisms of Stochastic Orders Generated by Posets

In this section we consider the question proposed in Sect. 1 on the property of being order-isomorphic for posets and stochastic orders generated by such posets.

If two posets are order-isomorphic, then the stochastic orders generated by those posets are also order-isomorphic. It is interesting to remark that the converse is not true in general. We describe particular conditions under which the converse holds. These results are mainly included in [19].

Let us consider $(\mathscr{X}, \preceq_{\mathscr{X}})$ a poset. We will denote by $\mathscr{B}_{\mathscr{X}}$ the σ-algebra generated by the class of upper quadrant sets, that is, $\mathscr{B}_{\mathscr{X}} = \sigma(\mathscr{Q}^{\preceq_{\mathscr{X}}})$.

The usual Borel σ-algebra on $\mathbb{R}$ will be denoted by $\mathscr{B}$.

The symbol $\mathscr{F}^{\mathscr{X}}$ will represent the set of mappings $f : \mathscr{X} \to \mathbb{R}$ which are measurable with respect to $\mathscr{B}_{\mathscr{X}}$ and $\mathscr{B}$, and $\preceq_{\mathscr{X}}$-preserving.

On the other hand, $\mathscr{P}_{\mathscr{X}}$ will stand for the set of probability measures on the measurable space $(\mathscr{X}, \mathscr{B}_{\mathscr{X}})$. Moreover, $\mathscr{P}_{\mathscr{X}}{}^0$ will denote the subset of $\mathscr{P}_{\mathscr{X}}$ composed by degenerated probabilities, that is,

$$\mathscr{P}_{\mathscr{X}}{}^0 = \{P_x \in \mathscr{P}_{\mathscr{X}} \mid x \in \mathscr{X}, P_x(B) = 1 \text{ if } x \in B, P_x(B) = 0 \text{ otherwise}, B \in \mathscr{B}_{\mathscr{X}}\}.$$

The class $\mathscr{F}^{\mathscr{X}}$ of all measurable $\preceq_{\mathscr{X}}$-preserving mappings generates a stochastic order on $\mathscr{P}_{\mathscr{X}}$, denoted by $\preceq_{\mathscr{X}g}$, as follows: if $P_1, P_2 \in \mathscr{P}_{\mathscr{X}}$, then

$$P_1 \preceq_{\mathscr{X}g} P_2 \quad \text{when} \quad \int_{\mathscr{X}} f dP_1 \leq \int_{\mathscr{X}} f dP_2$$

for all $f \in \mathscr{F}^{\mathscr{X}}$ for which both integrals exist.

Note that $(\mathscr{P}_{\mathscr{X}}, \preceq_{\mathscr{X}g})$ is a poset. Reflexivity and transitivity are obvious. Now if $P_1 \preceq_{\mathscr{X}g} P_2$ and $P_2 \preceq_{\mathscr{X}g} P_1$, since $I_U \in \mathscr{F}^{\mathscr{X}}$, I_U being the indicator function of U, then $P_1(U) = P_2(U)$ for all $U \in \mathscr{A}^{\preceq}$, where $\mathscr{A}^{\preceq}$ is the class of all finite intersections of upper quadrant sets. Since $\sigma(\mathscr{A}^{\preceq}) = \sigma(\mathscr{Q}^{\preceq})$ and $\mathscr{A}^{\preceq}$ is a π-system, then $P_1 = P_2$ (see, for instance, [4, p. 42]), that is, $\preceq_{\mathscr{X}g}$ satisfies the antisymmetric property.

The following result can be found in [13]. We should note that the finiteness of $\mathscr{X}$ is essential in that result.

Let $(\mathscr{X}, \preceq_{\mathscr{X}})$ be a poset with $\mathscr{X}$ finite. Let $P_1, P_2 \in \mathscr{P}_{\mathscr{X}}$, then $P_1 \preceq_{\mathscr{X}g} P_2$ if and only if $P_1(U) \leq P_2(U)$ for any U upper set.

If $\mathscr{X}$ is not finite, an upper set does not necessarily belong to $\mathscr{B}_{\mathscr{X}}$, and so the above result does not hold.

Now, let us consider order-isomorphisms. Let $(\mathscr{X}, \preceq_{\mathscr{X}})$ and $(\mathscr{Y}, \preceq_{\mathscr{Y}})$ be posets and let $\phi : \mathscr{X} \to \mathscr{Y}$ be an order-isomorphism. It is not hard to prove that ϕ is measurable with respect to the σ-algebras $\mathscr{B}_{\mathscr{X}}$ and $\mathscr{B}_{\mathscr{Y}}$.

In relation to the property of being order-isomorphic, the following result is obtained in [19].

Let $(\mathscr{X}, \preceq_{\mathscr{X}})$ and $(\mathscr{Y}, \preceq_{\mathscr{Y}})$ be order-isomorphic posets. Then the posets $(\mathscr{P}_{\mathscr{X}}, \preceq_{\mathscr{X}g})$ and $(\mathscr{P}_{\mathscr{Y}}, \preceq_{\mathscr{Y}g})$ are order-isomorphic.

We should remark that an order-isomorphism ∇ between $(\mathscr{P}_{\mathscr{X}}, \preceq_{\mathscr{X}g})$ and $(\mathscr{P}_{\mathscr{Y}}, \preceq_{\mathscr{Y}g})$ could be defined by means of an order-isomorphism ϕ between $(\mathscr{X}, \preceq_{\mathscr{X}})$ and $(\mathscr{Y}, \preceq_{\mathscr{Y}})$. This fact is quite useful in many applications.

Example 2.1 (*from* [19]) Consider the posets $(\mathscr{X}, \preceq_{\mathscr{X}})$ and $(\mathscr{Y}, \preceq_{\mathscr{Y}})$ whose Hasse diagrams (taken from [27]) are given in Fig. 1.

Let $\phi : \mathscr{X} \to \mathscr{Y}$ be the mapping such that $\phi(x_1) = y_1$, $\phi(x_2) = y_2$, $\phi(x_3) = y_4$, $\phi(x_4) = y_3$, $\phi(x_5) = y_5$, $\phi(x_6) = y_7$, $\phi(x_7) = y_8$, $\phi(x_8) = y_6$, $\phi(x_9) = y_{10}$, $\phi(x_{10}) = y_9$ and $\phi(x_{11}) = y_{11}$. Then, ϕ is an order-isomorphism and the posets $(\mathscr{X}, \preceq_{\mathscr{X}})$ and $(\mathscr{Y}, \preceq_{\mathscr{Y}})$ are order-isomorphic. As a consequence, we conclude that the posets $(\mathscr{P}_{\mathscr{X}}, \preceq_{\mathscr{X}g})$ and $(\mathscr{P}_{\mathscr{Y}}, \preceq_{\mathscr{Y}g})$ are order-isomorphic.

The converse of the result above is not true in general, as the following example shows (see [19]).

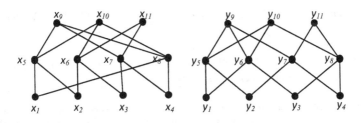

Fig. 1 Hasse diagrams in Example 2.1

Fig. 2 Hasse diagrams in Example 2.2

Example 2.2 Consider the sets $\mathscr{X} = \{x_1, x_2, x_3\}$ and $\mathscr{Y} = \{y_1, y_2\}$, and the posets $(\mathscr{X}, \preceq_{\mathscr{X}})$ and $(\mathscr{Y}, \preceq_{\mathscr{Y}})$ with the Hasse diagrams given in Fig. 2. That posets are not order-isomorphic since $|\mathscr{X}| \neq |\mathscr{Y}|$.

Note that $\mathscr{X}$ and $\mathscr{Y}$ are finite. Then, by means of the result in [13] given above, if $P_1, P_2 \in \mathscr{P}_{\mathscr{X}}$ with $P_1 \neq P_2$, we have that neither $P_1 \preceq_{\mathscr{X}_g} P_2$ nor $P_2 \preceq_{\mathscr{X}_g} P_1$ are satisfied, and the same happens with the probabilities of $\mathscr{P}_{\mathscr{Y}}$. As a consequence, if we define a bijection between $\mathscr{P}_{\mathscr{X}}$ and $\mathscr{P}_{\mathscr{Y}}$, it will be an order-isomorphism. It is not hard to prove that there exists a bijection between both sets (see for instance [11]), and so $(\mathscr{P}_{\mathscr{X}}, \preceq_{\mathscr{X}_g})$ and $(\mathscr{P}_{\mathscr{Y}}, \preceq_{\mathscr{Y}_g})$ are order-isomorphic.

Now, the question is to find conditions under which an order-isomorphism between $(\mathscr{P}_{\mathscr{X}}, \preceq_{\mathscr{X}_g})$ and $(\mathscr{P}_{\mathscr{Y}}, \preceq_{\mathscr{Y}_g})$ implies the existence of an order-isomorphism between $(\mathscr{X}, \preceq_{\mathscr{X}})$ and $(\mathscr{Y}, \preceq_{\mathscr{Y}})$. Some of such conditions were proved in [19].

The following result holds.

Let $(\mathscr{X}, \preceq_{\mathscr{X}})$ and $(\mathscr{Y}, \preceq_{\mathscr{Y}})$ be posets, consider the posets $(\mathscr{P}_{\mathscr{X}}, \preceq_{\mathscr{X}_g})$ and $(\mathscr{P}_{\mathscr{Y}}, \preceq_{\mathscr{Y}_g})$. Let $\nabla : \mathscr{P}_{\mathscr{X}} \to \mathscr{P}_{\mathscr{Y}}$ be an order-isomorphism such that $\nabla(\mathscr{P}_{\mathscr{X}}^0) = \mathscr{P}_{\mathscr{Y}}^0$, then $(\mathscr{X}, \preceq_{\mathscr{X}})$ and $(\mathscr{Y}, \preceq_{\mathscr{Y}})$ are order-isomorphic.

Another condition is provided by the following concept in [19], which in conjunction with an order-isomorphism between $(\mathscr{P}_{\mathscr{X}}, \preceq_{\mathscr{X}_g})$ and $(\mathscr{P}_{\mathscr{Y}}, \preceq_{\mathscr{Y}_g})$, lead to an order-isomorphism between $(\mathscr{X}, \preceq_{\mathscr{X}})$ and $(\mathscr{Y}, \preceq_{\mathscr{Y}})$. Let $(\mathscr{X}, \preceq_{\mathscr{X}})$ be a poset and $I \subset \mathscr{X}$. A mapping $\Upsilon : I \to \mathscr{X}$ is said to conserve $\preceq_{\mathscr{X}}$ in each point separately, when given $x_1, x_2 \in I$, if one of the following relations holds: $x_1 \preceq_{\mathscr{X}} x_2$, $\Upsilon(x_1) \preceq_{\mathscr{X}} x_2, x_1 \preceq_{\mathscr{X}} \Upsilon(x_2)$, then the three relations are satisfied simultaneously.

Given $(\mathscr{X}, \preceq_{\mathscr{X}})$ a poset and S a subset of $\mathscr{X}$, we will denote by $\overline{S}$ the complement of S in $\mathscr{X}$, that is, $\overline{S} = \mathscr{X} \setminus S$.

The following result gives a new condition involving mappings which conserve orders in each point separately (see [19]).

Let $(\mathscr{X}, \preceq_{\mathscr{X}})$ and $(\mathscr{Y}, \preceq_{\mathscr{Y}})$ be posets, let us consider the posets $(\mathscr{P}_{\mathscr{X}}, \preceq_{\mathscr{X}_g})$ and $(\mathscr{P}_{\mathscr{Y}}, \preceq_{\mathscr{Y}_g})$. Let $\nabla : \mathscr{P}_{\mathscr{X}} \to \mathscr{P}_{\mathscr{Y}}$ be an order-isomorphism. Let us define the sets $L = \nabla^{-1}(\mathscr{P}_{\mathscr{Y}}^0) \cap \overline{\mathscr{P}_{\mathscr{X}}^0}$ and $M = \nabla(\mathscr{P}_{\mathscr{X}}^0) \cap \overline{\mathscr{P}_{\mathscr{Y}}^0}$. If L and M are nonempty sets and if there exists an order-isomorphism $\Omega : L \to M$ such that $\Omega^{-1} \circ \nabla : \nabla^{-1}(M) \to \mathcal{P}_{\mathcal{X}}$ and $\nabla^{-1} \circ \Omega : L \to \mathcal{P}_{\mathcal{X}}$ conserve $\preceq \mathcal{X}_g$ in each point separately, then the posets $(\mathcal{X}, \preceq_{\mathcal{X}})$ and $(\mathcal{Y}, \preceq_{\mathcal{Y}})$ are order-isomorphic.

The proof can be found in [19], together with an interesting application to the analysis of chemical components of seaweeds.

Acknowledgements The authors are indebted to the Spanish Ministry of Science and Innovation and Principado de Asturias since this research is financed by Grants MTM2011-22993, MTM2013-45588-C3-1-P, MTM2015-63971-P, FC-15-GRUPIN14-101 and FC-15-GRUPIN14-142.

References

1. Arhippainen J, Kauppi J, Mattas J (2017) Order structure, multipliers, and Gelfand representation of vector-valued function algebras. Banach J Math Anal 11:207–222
2. Ayala G, López-Díaz MC, López-Díaz M, Martínez-Costa L (2012) Studying hypertension in ocular fundus images using Hausdorff dispersion ordering. Math Med Biol 2:131–143
3. Belzunce F, Martínez-Riquelme C, Mulero J (2016) An introduction to stochastic orders. Elsevier/Academic Press, Amsterdam
4. Billingsley P (1995) Probability and measure. Wiley, New York
5. Blaszczyszyn B, Yogeshwaran D (2009) Directionally convex ordering of random measures, short noise fields, and some applications to wireless communications. Adv Appl Prob 41:623–646
6. Brualdi RA, Graves JS, Lawrence KM (1995) Codes with a poset metric. Discret Math 147:57–72
7. Carleos C, López-Díaz M (2010) An indexed dispersion criterion for testing the sex-biased dispersal of lek mating behavior of capercaillies. Environ Ecol Stat 17:283–301
8. Carleos C, López-Díaz MC, López-Díaz M (2010) A stochastic order of shape variability with an application to cell nuclei involved in mastitis. J Math Imaging Vis 38:95–107
9. Carleos C, López-Díaz MC, López-Díaz M (2014) Ranking star-shaped valued mappings with respect to shape variability. J Math Imaging Vis 48:1–12
10. Cirulis J (2017) The diamond partial order for strong rickart rings. Linear Multilinear Algebra 65:192–203
11. Dugundji J (1966) Topology. Allyn and Bacon, Boston
12. Garg VK (2015) Introduction to lattice theory with computer science applications. Wiley, Hoboken
13. Giovagnoli A, Wynn HP (2008) Stochastic orderings for discrete random variables. Stat Prob Lett 78:2827–2835
14. Harzheim E (2005) Ordered sets. Advances in mathematics, vol 7. Springer, New York
15. Kamae T, Krengel U, O'Brien GL (1977) Stochastic inequalities on partially ordered spaces. Ann Probab 5:899–912
16. Kijima M, Ohnishi M (1999) Stochastic orders and their applications in financial optimization. Math Meth Oper Res 50:351–372
17. López-Díaz C, López-Díaz M (2012) When is an integral stochastic order generated by a poset? J Inequalities Appl 2012: Art.265
18. López-Díaz M (2011) A test for the bidirectional stochastic order with an application to quality control theory. Appl Math Comput 217:7762–7771
19. López-Díaz MC, López-Díaz M (2013) On order-isomorphisms of stochastic orders generated by partially ordered sets with applications to the analysis of chemical components of seaweeds. MATCH Commun Math Comput Chem 69:463–486

20. López-Díaz MC, López-Díaz M (2013) A note on the family of extremality stochastic orders. Insur Math Econ 53:230–236
21. López-Díaz MC, López-Díaz M (2017) Martínez-Fernández S (2017) A stochastic comparison of customer classifiers with an application to customer attrition in commercial banking. Scand Actuar J 7:606–627
22. Lu X, Wen Q, Wang L, Du J (2016) A lattice-based signcryption scheme without trapdoors. J Electr Inf Technol 38:2287–2293
23. McConville T (2017) Lattice structure of GridTamari orders. J Comb Theory A 148:27–56
24. Müller A (1997) Stochastic orders generated by integrals: a unified study. Adv Appl Probab 29:414–428
25. Müller A, Stoyan D (2002) Comparison methods for stochastic models and risks. Wiley, Chichester
26. Neggers J, Kim HS (1998) Basic posets. World Scientific Publishing, Singapore
27. Schröder BSW (2003) Ordered sets. An introduction. Birkhäuser, Boston
28. Sebastian Vadakkenveettil B, Unnikrishnan A, Balakrishnan K, Padinjare Pisharath Balakrishna R (2017) Morphological filtering on hypergraphs. Discret Appl Math 216:307–320
29. Shaked M, Shanthikumar JG (2007) Stochastic orders. Springer, New York
30. Tepedelenlioglu C, Rajan A, Zhang Y (2011) Applications of stochastic ordering to wireless communications. IEEE Trans Wirel Commun 10:4249–4257

Optimal Experimental Design for Model Selection: A Partial Review

Jesús López–Fidalgo and Chiara Tommasi

Abstract Model selection is a core topic in modern Statistics. This is a review of what has been researched on optimal experimental design for model selection. The aim is to find good designs for increasing the test power for discriminating between rival models. This topic has a special impact nowadays in the area of experimental design.

1 Introduction

Model selection is nowadays one of the hot topics in Statistics and finding optimal experimental designs for that purpose will save time, money and risk to the researchers. A joint solution to the problem of identifying the maximum information both for discriminating between rival models and for fitting the best of them is very much desirable. Obtaining a good model is crucial for prediction in a world of presence of massive data. Optimal Experimental Design (OED) theory and practice has a lot to say about that. The reason for that belief is that OED is very much concerned with the information behind the data. Traditionally this has been used for designing informative experiments in order to save data. This is still very important if we think, e.g. in experimentation with animals or humans. In the contemporary scenario with great quantities of observations, frequently with low quality data, we need tools to clean the data and discriminate which is the best model to extract the information from it.

OED searches both for better estimates and predictions as well as for optimizing the cost of experimenting. The research on OED has experienced a noteworthy

J. López–Fidalgo (✉)
ICS, Unidad de Estadística, Biblioteca de Humanidades,
Universidad de Navarra, 31080 Pamplona, Spain
e-mail: fidalgo@unav.es

C. Tommasi
Department of Economics, Business and Statistics, University of Milano,
Via Conservatorio 7, 20122 Milano, Italy
e-mail: chiara.tommasi@unimi.it

© Springer International Publishing AG 2018
E. Gil et al. (eds.), *The Mathematics of the Uncertain*, Studies in Systems,
Decision and Control 142, https://doi.org/10.1007/978-3-319-73848-2_24

increase in the last few years. Main beneficiaries of the theory of optimal design are the areas of engineering, well-being in society and very much in health with clinical trials and Phase I/II experimentation.

We dare to say there are two important research directions in OED at this moment. On the one hand the most traditional interest in outlining correctly complex, but real, situations. On the other hand there is special interest nowadays in developing efficient algorithms for finding optimal designs, which is a non-standard task in this area where the aim is a probability measure in a non-Euclidean space. Both directions must go together, complementing each other, in the research.

Whenever a variable is under the control of the experimenter OED has a lot to say. The main idea is to search for the most informative experimental conditions with the optimal number of replicates at each of these experimental conditions. The sample size has to be fixed in advance and this is usually a difficult task to be solved, frequently needing number theory. A wonderful idea was introduced by Kiefer [21] extending the concept of experimental design to any probability measure on a compact design space. Then the so called General Equivalence Theorem (GET) [22, 40] provided a fantastic tool for checking whether a particular design is optimal or not. Additionally this gives the way of computing either closed–formed optimal designs or developing numerical algorithms for their computation. What one considers optimal here depends very much on the aim, say either estimating the parameters, predicting results or discriminating between models; as well as on the model itself.

Most of the optimality criteria focus on the inverse of the Fisher Information Matrix (FIM). For linear models this is proportional to the covariance matrix of the estimates of the parameters. The usual inferences and predictions are based on this matrix, so some appropriate function of it has to be optimized. For a linear model this matrix depends only on the design and thus the optimization challenge is just to look for the best design according to the corresponding criterion. If the model is nonlinear the inverse of the FIM is asymptotically proportional to the covariance matrix, but the FIM now depends on the unknown, and no yet, estimated parameters at the time of planning the experimentation. There are different ways to approach this issue,

(i) Locally optimal designs [15], assuming some nominal values of the parameters where the inverse of the FIM is locally approximated to the covariance function.

(ii) Adaptive sequential designs, where the next design point takes into account the observed data from previous experiments. In this case the design is a stochastic process. This approach is rather popular, especially in medicine and pharmacological studies with clinical trials.

(iii) Bayesian optimal designs [36], assuming a prior distribution on the parameters and a joint utility function including both objectives at the same time, estimating the parameters of the model and finding the optimal design for that.

Another important difficulty in OED arises for correlated observations, e.g. over time or space. The world of industry is a champion on the used of experimental designs. Computational aspects of OED have always been an important issue. Note the contribution of Fisher [20], which gives to the experimental design theory the current statistical approach.

Although the OED appears in 1918 with an extensive article by Kristine Smith [32], it will not be developed at some extent until the 50s. The volume by Cox in 1958 [16] contains a short and descriptive introduction to the basic ideas of the theory. Pioneering monographs were written by Fedorov [19] and Silvey [31]. A number of monographs have arrived later and can not be mentioned here for space reasons.

The most significant advances in theory and practice of Optimal Design of Experiments can be followed through the volumes of conference proceedings mODa (Model-Oriented Data Analysis and Design) apart from other relevant collections coming from workshops and conferences. The family of algorithms traditionally used in the calculation of optimal designs are not, in a general sense, the best possible. Although there is an important effort in this aspect it remains as an important challenge for the theory.

2 Different Approaches to the Problem

One of the main criticisms to OED is that a design has to be found for a particular model and the model has to be guessed without having the data yet. Honestly speaking we should say there are not solutions to this issue, which in fact is still present even with the data at hand. George Box used to say that "Models, of course, are never true, but fortunately it is only necessary that they be useful", e.g. [13]. On the other hand it is quite common that many models come from the experience, retrospective data or intuitions of the practitioner. Sometimes they are analytically derived, e.g. as a solution of differential equations, as happens with the majority of pharmacokinetics models [25, 29, 30]. But, this issue is more severe in Experimental Design. In a variety of situations before having the data two or more models may be potential candidates. Once the data are collected a model has to be chosen after a model selection procedure, then in a second step it has to be fitted. Thus, the optimal design in this case faces two different objectives. On the one hand, it has to be good for discriminating between models. This means to organize the experiment in such a way some distances between the fitted models from the data are as large as possible in order to make clear the differences between them [7, 11, 12, 14, 23, 36, 37]. Appropriate distances need to be chosen according to the existing statistical tests for discriminating [17]. On the other hand the design has to be good for inferences with the chosen model, either for estimating the parameters, estimating some functions of them or making predictions. Even for linear model this is not trivial at all.

There are different approaches to this issue with some controversial and therefore some unclear features of them. Optimal designs from different reasonable perspectives may be rather different even for simple models. There is the need of investigating this issue providing clear directions to model selection. One of the first attempts to tackle the discimination paradigm from an OED point of view consisted in embedding two (or more) rival models in a more general model and designing to estimate the additional parameters [5, 10]. This is the so called D_s-optimal design, which makes much sense from an intuitive point of view. But it is not clear that this criterion increases the power of the usual discrimination tests. As a matter of fact [24] have proved that T- and D_s-optimality coincide only in the case that the optimal values of the parameters in the T-criterion are the nominal values for the D_s-criterion, which is something very artificial and may be far from reality. Otherwise, the designs could be rather different. There is nothing definitively proved about its relationship with the test power of the likelihood ratio test or any other discrimination test. In [28] three different approaches were considered for discriminating between models: (*i*) augmenting a given design in an optimal way, (*ii*) evaluating a mixture of the various criteria, and (*iii*) optimizing an objective subject to achieving a prescribed efficiency level for the others.

Another intuitive idea is maximizing the distance between two models assuming one of them is considered as the "true" model, which is the model in the alternative hypothesis (T–optimality) [11, 12, 23]. It happens that this focuses on maximizing the non–centrality parameter of the likelihood ratio test statistic, which is a function of the test power. The traditional F discrimination test is a particular case for nested linear models. This criterion was extended to generalized linear models (GLM) [27] as well as heteroscedasticity and multiple response [38]. As a definitive extension of T–optimality [23] gave a criterion based on the Kullback-Leibler distance, KL–optimality, which accounts for the likelihood ratio test power, which is also related to the AIC. In particular, assuming just two rival models with pdf's

$$f_i(y, x, \theta_i), \qquad i = 1, 2,$$

and assuming f_t is the "true" model for either $t = 1$ or $t = 2$, the Kullback–Leibler distance between them is

$$\mathcal{I}[f_t(y, x, \theta_t), f_i(y, x, \theta_i)] = \int f_t(y, x, \theta_t) \log\left[\frac{f_t(y, x, \theta_t)}{f_i(y, x, \theta_i)}\right] dy,$$

where $i \neq t$, y is the vector of responses, θ_i and θ_t are the parameters in the two models and x is the vector of experimental conditions at which the response y is observed.

Then, KL–optimality is defined by the following objective function,

$$I_{i,t}(\xi) = \min_{\theta_i \in \Omega_i} \int_{\mathscr{X}} \mathcal{I}[f_t(y, x, \theta_t), f_i(y, x, \theta_i)] \, \xi(dx).$$

The relationship with the celebrated AIC criterion is illustrated as follows. Let $L_i(\theta_i)$ and $L_t(\theta_t)$ be the log-likelihoods of each model. It is assumed that θ_t^* is known. Then

$$\min_{\theta_i} \mathcal{I}[f_t(y, x, \theta_t^*), f_i(y, x, \theta_i)] = E_t[L_t(\theta_t^*)] - \max_{\theta_i} E_t[L_i(\theta_i)],$$

where E_t stands for the expectation according to distribution given by f_t. The expected Akaike criterion for model i is

$$E_t[AIC_i] = 2\{m_i - E_t[L_i(\hat{\theta}_i)]\},$$

where m_i is the number of parameters of model i and $\hat{\theta}_i$ is the Maximum Likelihood Estimator (MLE) of θ_i, $\hat{\theta}_i = \arg\max_{\theta_i} L_i(\theta_i)$.

Thus, if the minimum can be exchanged with the expectation then KL-optimality also minimizes the AIC of model i.

Another way to see this relationship is using the so called relative AIC of model i with respect to model t,

$$\exp\left[\frac{AIC_t - AIC_i}{2}\right] = \exp\left[\log\frac{L_i(\hat{\theta}_i)}{L_t(\theta_t^*)} - m_i\right],$$

where the log-likelihood ratio appears in the right hand side.

López Fidalgo et al. [23] have proved also that T-optimality and all the mentioned existing extensions are particular cases of KL-optimality. This means the discrimination can be considered among non-Normal models or even for correlated observations [2, 9]. The criterion has been generalized in different ways for more than two rival models, essentially assuming convex combinations of the efficiencies for several models [34]. Following this idea [37] considered a max–min criterion and provided a couple of suggesting examples with different probability distributions or different mean in GLMs. Compound criteria with D-optimality have been used to search for good designs also for fitting the model [35]. The sequential [26], Bayesian [36] and Copula [1] perspectives have also been taken into account. The computational issue still needs a lot of work [18].

Summarizing all this, Model Selection is a major topic in contemporary Statistics and the OED perspective can provide a significant improvement to this problem. Finding a joint solution to the problem of identifying the maximum of information, both for discriminating between rival models and for fitting the best them, is still a challenging topic.

3 Hot Areas for Further Research

In what follows some ideas for potential further work are presented.

3.1 Robustness of KL-Optimal Designs from Different Points of View

It would be interesting to check which features in the rival models seem more sensitive for a successful discrimination using the KL-criterion. In particular, it would be useful to check how this criterion compared to others is able to detect differences between rival models in the probability distribution, the mean function, the variance structure or the dimension of the model, e.g. more variables in the mean function. Analytic results would be very much welcomed, but simulations studies will be there in any case. For generalized linear models the link function must be added to this list. Atkinson [6] worked out all the details for making inference on an extended model that includes several submodels as GLMs.

3.2   Other Divergence Measures

Once a deep knowledge of what the Kullback-Leibler divergence does, better solutions may be looked for. López Fidalgo et al. [23] proved that KL-optimality accounts for maximizing the power of the likelihood ratio test, but there are other statistical tests frequently used in practice [17, 33, 39]. The test power in these cases is likely to be connected with other divergence measures [8]. In particular, there is the need to check for the meaning of the D_s–optimality of the differing parameters from nested models. At first sight it seems that for nested models D_s- and KL-optimality should be the same. This is true just in the very particular case of linear models on the parameters when the nested model differs from the root model in just one parameter. In this case the KL-optimal design is the D_1-optimal design for estimating that parameter. For other models the optimal designs may be very different with very low relative efficiencies with respect to each other [24]. For other discrimination tests appropriate criteria must be found, taking into account other information divergence measures, such as those derived from the Rényi entropy of f-divergences.

3.3   Bayesian Paradigm

As mentioned above most of the optimality criteria focus on the inverse of the information matrix. If the model is nonlinear the FIM depends on the parameters of the model. KL-optimality, and so T-optimality, does not focus on this matrix, but directly on the probability measure (the design). Nevertheless, even for linear models nominal values of the parameters of the "true" model are needed. One way for approaching this issue is the use of the Bayesian paradigm assuming a prior distribution on the parameters and a joint utility function including both objectives at the same time, estimating the parameters of the model and finding the optimal design for that. This theory has been considered in the literature as an important and necessary approach. The Kullback-Leibler divergence jointly with the Shannon information is used, in a different way than for discrimination, to develop a Bayesian D-optimality criterion. Tommasi and López–Fidalgo [36] introduced the Bayesian approach for discriminating between two rival models. Following-up this idea a systematization of the Bayesian theory for discriminating between models jointly with fitting purposes would be welcomed. Additionally, this will avoid the annoying assumption of the "true model".

Utility functions focused on discrimination must be considered. This is very much related again to looking for different divergence measures. Some but not all of the classic optimality criteria have a utility based Bayesian version. Mixing utility functions may help to describe several simultaneous goals. Additionally, [37] considered a max-min criterion for more than two rival models and gave a relationship with a Bayesian criterion assuming a prior distribution of the weights of each model. This particular point may be explored using mathematical programming techniques applied here. This would be very useful for computational purposes. Max–min criteria are not easy to deal with because of the lack of differentiability. The GET is still applicable but an annoying auxiliary probability measure has to be found.

Summarizing these ideas, the Bayesian paradigm applied here brings mainly three results: (i) a convincing way of dealing with the unknown parameters, (ii) better justified criteria not supported on the artificial assumption of a "true" model and (iii) efficient computational techniques for more than two rival models.

3.4 Correlated Observations

The Big Data world is a scenario of correlations, which needs to be considered from different perspectives. In contrast with other criteria for discrimination, KL-optimality is still valid in this situation. Campos-Barreiro and López–Fidalgo [9] proved that a standard generalization of T-optimality to correlated observations can be done just when the covariance matrix of the observations is assumed completely known. Designs for models with a partially unknown covariance structure have been widely studied (see, e.g., [4]). Most of this work has been done for D-optimality. The approximation of the covariance matrix by the inverse of the FIM holds in this situation under some assumptions. If these conditions are not satisfied we have usually performed simulations with the designs obtained in the last steps of the algorithm in order to check the monotonicity between the determinants of the FIM and the covariance matrix. KL-optimality is not based on the FIM. Therefore, there is no problem with the mentioned approximation.

A lot of work can be done on the area of optimal designs for discrimination in presence of correlated observations. In particular, the usual time series models require a discrimination process to select the best model, e.g. the best values of p, d and q in an ARIMA(p, d, q) model. This is an area where things are not so simple from an experimental design point of view. Amo-Salas et al. [3] considered a rather simple time series model where the implicit covariance structure is worked out from the model. This can be done analytically in very simple models, but it needs some new results to be able to find a proper criterion both for discriminating and for estimating the models.

Amo-Salas et al. [4] provided theoretical results for a function to be definite positive. These results are then used to generate potential covariance structures from

Bernstein polynomials. These results provide a powerful tool for spatio-temporal modeling, which is not trivial since the covariance structure needs to be such that the generated covariance matrix is nondefinite negative. Another interesting point would be a deep study of the covariance structure behind the usual time series models in order to look for appropriate optimal designs.

4 Computing Optimal Designs

While iterative procedures are very much needed for OED in general, they are specially needed for finding optimal information for discriminating between models. For KL-optimality some classical algorithms have been adapted, but much more work has to be done here. López–Fidalgo et al. [23] provided a general algorithm based on the directional derivative,

(i)   For a given design ξ_s let

$$\theta_{i,s} = \arg\min_{\theta_i \in \Omega_i} \int \mathscr{I}(f_t, f_i, x, \theta_i)\xi_s(dx)$$
$$x_s = \arg\max_{x \in \chi} \mathscr{I}(f_t, f_i, x, \theta_{i,s}).$$

(ii)   For a chosen α_s with $0 \leq \alpha_s \leq 1$ let

$$\xi_{s+1} = (1 - \alpha_s)\xi_s + \alpha_s \xi_{x_s},$$

where ξ_{x_s} is a design with measure concentrated at the single point x_s. Typical conditions for the sequences $\{\alpha_s\}$ are

$$\lim_{s \to \infty} \alpha_s = 0, \quad \sum_{s=0}^{\infty} \alpha_s = \infty, \quad \sum_{s=0}^{\infty} \alpha_s^2 < \infty.$$

This algorithm becomes slow after a while and needs to be combined with a finer algorithm in the last part of the procedure. Tommasi et al. [37] provided another algorithm, this time for a max-min criterion considering more than two rival models. Convenient algorithms need to be adapted to this criterion and then their performance need to be evaluated.

Other algorithms to be adapted include exact methods such as multiplicative, interior point method, active set method, sequential quadratic programming, Nelder Mead and metaheuristic algorithms such as particle swarm optimization, simulated annealing, genetic algorithm, and hybridizations of these methods.

A friendly software is critical for the actual application of these ideas. Nowadays there are web application frameworks for most of the commercial and non commercial

Mathematical an Statistical software offering a product available for anyone without having that particular software in his or her computer.

5 Discussion

Although OED for discrimination between models is quite popular at this moment there is not a definitive systematization of the features and objectives in this field converging to the joint concern of discriminating and fitting the model. As described above, there are some controversial and unsolved issues around all that has been done so far. Model Selection is essentially demanded for the current need of a scientific and correct massive data treatment. KL-optimality is currently the most general justified criterion. Nevertheless, other criteria have to be developed according to some other statistical tests not based on the likelihood ratio test [17]. This includes a deep study of D_s-optimality as well as trying additional divergence measures by justifying and comparing them.

Checking the robustness of KL-optimality for different aspects is an important task in order to have a clear idea of their strengths and limitations. A helpful Bayesian approach can be made mainly in two ways. On the one hand for dealing with the unknown parameters at the time of planning an experiment. On the other hand for the case of more than two models, which introduces a rather more complex approach with a multicriteria perspective. The later is very much related to massive data consideration. Then in a parallel way there is the case of correlated data in the so called Spatio–Temporal Statistics, which nowadays is present everywhere with great quantities of possible data collection. Implementing algorithms and software for obtaining optimal designs is very much desirable in the whole area of OED.

Model selection and so designing experiments for that purpose is very much demanded today in our world of data analytics in the gates of the so called Industry 4.0.

Acknowledgements The authors have been sponsored by Spanish Research Agency and Fondos FEDER MTM2013-47879-C2-1-P and MTM2016-80539-C2-R1. Prof. López–Fidalgo wants to appreciate the invitation to write a paper for this book. Although there was not direct collaboration in the topic with Pedro, he was always a reference both in a human perspective as well as in the scientific field. His capacity for team building taking into account every one beside him and his generosity makes him a good example of what a scientist should be.

References

1. Aletti G, May C, Tommasi CH (2016) Best estimation of functional linear models. J Multivar Anal 151:54–68
2. Amo-Salas M, López-Fidalgo J, López-Ríos VI (2012) Optimal designs for two nested pharmacokinetic models with correlated observations. Commun Stat-Simul C 41(7):944–963

3. Amo-Salas M, López-Fidalgo J, Pedregal DJ (2015) Experimental designs for autoregressive models applied to industrial maintenance. Reliab Eng Syst Saf 133:87–94
4. Amo-Salas M, López-Fidalgo J, Porcu E (2013) Optimal designs for some stochastic processes whose covariance is a function of the mean. Test 22(1):159–181
5. Atkinson AC (1972) Planning experiments to detect inadequate regression models. Biometrika 59:275–293
6. Atkinson AC (1995) Multivariate transformations, regression diagnostics and seemingly unrelated regression. In: Kitsos CP, Müller WG (eds) MODA 4: advances in model-oriented data analysis. Physica, Heidelberg
7. Atkinson AC (2008) DT-optimum designs for model discrimination and parameter estimation. J Stat Plan Inference 138:56–64
8. Basseville M (2013) Divergence measures for statistical data processing-an annotated bibliography. Signal Process 93:621–633
9. Campos-Barreiro S, López-Fidalgo J (2016) KL-optimal experimental design for discriminating between two growth models applied to a beef farm. Math Biosci Eng 13(1):67–82
10. Atkinson AC, Cox DR (1974) Planning experiments for discriminating between models (With discussion by Wynn HP, Titterington DM, Laycock PJ, Lindley DV, Hill DH, Herzberg AM, Tukey PA, O'Hagan A, Fedorov VV, Dickey J, Kiefer J, Smith CAB). J R Stat Soc Ser B 36:321–348
11. Atkinson AC, Fedorov VV (1975) The design of experiments for discriminating between two rival models. Biometrika 62:57–70
12. Atkinson AC, Fedorov VV (1975) Optimal design: experiments for discriminating between several models. Biometrika 62:289–303
13. Box GEP (1979) Some problems of statistics and everyday life. J Am Stat Assoc 74:1–4
14. Braess D, Dette H (2013) Optimal discriminating designs for several competing regression models. Ann Stat 1(2):897–922
15. Chernoff H (2000) Sequential analysis and optimal design. SIAM Society for Industrial and Applied Mathematics, Philadelphia
16. Cox DR (1958) Planning of experiments. Wiley, New York
17. Cox DR (1962) Further results on tests of separate families of hypotheses. J R Stat Soc Ser B 24(2):406–424
18. Deldossi L, Osmetti SA, Tommasi Ch (2016) PKL-Optimality criterion in copula models for efficacy-toxicity response. In: Müller Ch, Kunert J, Atkinson AC (eds) MODA 11: advances in model-oriented data analysis. Physica, Heidelberg
19. Fedorov VV (1972) Theory of optimal experiments. Academic Press, New York
20. Fisher RA (1960) The design of experiments, 7th edn. Oliver and Boyd, Edinburgh
21. Kieffer J (1959) Optimum experimental designs. J R Stat Soc Ser B 21:272–319
22. Kieffer J, Wolfowitz J (1960) The equivalence of two extremum problems. Can J Math 12:363–366
23. López-Fidalgo J, Tommasi Ch, Trandafir C (2007) An optimal experimental design criterion for discriminating between non-normal models. J R Stat Soc Ser B 69:231–242
24. López-Fidalgo J, Tommasi Ch, Trandafir C (2007) Optimal designs for discriminating between some extensions of the Michaelis-Menten model. J Stat Plan Inference 138:3797–3804
25. López-Fidalgo J, Wong WK (2002) Design for the Michaelis-Menten model. J Theor Biol 215:1–11
26. May C, Tommasi Ch (2014) Model selection and parameter estimation in non-linear nested models. Stat Sin 24:63–82
27. Ponce de León A, Atkinson A (1992) The design of experiments to discriminate between two rival generalized linear models. In: Fahrmeir L, Francis B, Gilchrist R, Tutz G (eds) Advances in GLM and statistical modelling. Lecture notes in statistics. Springer, New York
28. Pukelsheim F, Rosenberger JL (1993) Experimental designs for model discrimination. J Am Stat Assoc 88(442):642–649
29. Rodríguez Aragón LJ, López-Fidalgo J (2007) T-, D- and c-optimum designs for BET and GAB adsorption isotherms. Chemom Intell Lab Syst 89(1):36–44

30. Sánchez G, López-Fidalgo J (2003) Mathematical techniques for solving analytically large compartmental systems. Health Phys 85(2):184–193
31. Silvey SD (1980) Optimal design. Chapman and Hall, London
32. Smith K (1918) On the standard deviations of adjusted and interpolates values of an observed polynomial functions and its constants and the guidance they give towards a proper choice of the distribution of observations. Biometrika 12:1–85
33. Stewart WE, Shon Y, Box GEP (1998) Discrimination and goodness of fit of multiresponse mechanistic models. AIChE J 44(6):1404–1412
34. Tommasi Ch (2007) Optimal designs for discriminating among several non-normal models. In: López-Fidalgo J, Rodríguez-Díaz JM, Torsney B (eds) Advances in model-oriented design and analysis moda 8, series contributions to statistics, vol 8. Physica, Heidelberg
35. Tommasi Ch (2009) Optimal designs for both model discrimination and parameter estimation. J Stat Plan Inference 139:4123–4132
36. Tommasi Ch, López-Fidalgo J (2010) Bayesian optimum designs for discriminating between models with any distribution. Comput Stat Data Anal 54:143–150
37. Tommasi Ch, Martín-Martín R, López-Fidalgo J (2016) Max-min optimal discriminating designs for several statistical models. Comput Stat Data Anal 26(6):1163–1172
38. Ucinski D, Bogacka B (2004) T-Optimum designs for multiresponse dynamic heteroscedastic models. In: Di Bucchianico A, Läuter H, Wynn HP (eds) MODA 7 - advances in model-oriented design and analysis, series contributions to statistics. Physica, Heidelberg
39. van de Wal M, de Jager B (2001) A review of methods for input/output selection. Automatica 37:487–510
40. Whittle P (1973) Some general points in the theory of optimal experimental design. J R Stat Soc Ser B 1:123–130

A Statistical Analysis of the Treatment of Type 2 Diabetes in the Presence of Chronic Kidney Disease in Patients Hospitalized for Heart Failure

Juan Ignacio López-Gil and María Asunción Lubiano

Abstract Many patients suffering chronic kidney disease are associated with a type 2 diabetes mellitus. Different therapies for the treatment of type 2 diabetes have been considered. This paper aims to check whether these therapies can be affected by the presence of the kidney disease. The study was conducted on a sample of patients who were hospitalized for heart failure in CAULE (Complejo Asistencial Universitario de León).

1 Introduction

There is an extensive recent literature involving jointly the type 2 diabetes mellitus (T2DM), the chronic kidney disease (CKD) and the heart failure (HF) as well as their medical/pharmacological treatment (see, for instance, [1–5, 7–9, 12, 13]).

The study in this paper is constrained to a sample from a subpopulation of type 2 DM patients. This subpopulation refers to the type 2 DM patients who were hospitalized for heart failure. The sample corresponded to that of 248 patients who were admitted in CAULE along a certain recent period.

The statistical analysis has aimed to get conclusions on the dependence of the treatment received by type 2 DM patients who were hospitalized for heart failure (e.g., diet, insulin, metformin, sulfonylureas, etc. or the combination of some of them) and the presence/absence of CKD. Data management and statistical analysis were

---

Pedro Gil was our uncle (López-Gil) and 'scientific grandfather' (Lubiano), and we have known each other thanks to him. Pedro was the first person helping us when we both join the University of Oviedo to start the BSc in Medicine (López-Gil) and the Ph.D. in Math and teaching asssistance in Statistics (Lubiano). This first paper we are coauthoring should be dedicated to Pedro's memory.

J. I. López-Gil
Servicio de Atención Primaria, Centro de Salud José Aguado, Leon, Spain

M. A. Lubiano (✉)
Departamento de Estadística, I.O. y D.M., Universidad de Oviedo, Oviedo, Spain
e-mail: lubiano@uniovi.es

Table 1 Numbers of hospitalized for heart failure patients in CAULE receiving single or combined treatments for type 2 DM, and classified in accordance with the presence/absence of CKD and with the degree of CKD according to the classification with KDIGO 2012 [10] based on GFR categories (n-G2 = normal-Stage/Degree 2, Gi = Stage/Degree i)

Treatment	CKD		CKD degrees			
	YES	NO	n-G2	G3	G4	G5
Diet	9	26	26	6	3	0
Glibenclamide	1	0	0	1	0	0
Glicazide	0	2	2	0	0	0
Glimepiride	2	4	4	1	0	1
Insulin	44	32	32	31	13	0
Insulin + glicazide	0	1	1	0	0	0
Insulin + linagliptin	3	2	2	2	1	0
Insulin + metformin	5	6	6	3	2	0
Insulin + metformin + lixisenatide	0	1	1	0	0	0
Insulin + metformin + repaglinide	0	1	1	0	0	0
Insulin + repaglinide	1	0	0	1	0	0
Insulin + saxagliptin	0	1	1	0	0	0
Insulin + sitagliptin	1	1	1	1	0	0
Insulin + sitagliptin + metformin	3	1	1	2	1	0
Insulin + vildagliptin	3	0	0	2	0	1
Insulin + vildagliptin + metformin	0	3	3	0	0	0
Linagliptin	6	2	2	5	1	0
Linagliptin + metformin	1	0	0	0	1	0
Linagliptin + metformin + repaglinide	1	0	0	1	0	0
Linagliptin + repaglinide	1	0	0	1	0	0
Liraglutide + metformin	0	1	1	0	0	0
Metformin	4	34	34	2	2	0
Metformin + glicazide	1	0	0	1	0	0
Metformin + glimepiride	0	1	1	0	0	0
Repaglinide	3	0	0	0	2	1
Repaglinide + linagliptin + metformin	1	0	0	0	1	0
Sitagliptin	3	1	1	2	1	0
Sitagliptin + glimepiride	1	0	0	1	0	0
Sitagliptin + metformin	1	8	8	0	1	0
Sitagliptin + metformin + glibenclamide	0	1	1	0	0	0
Sitagliptin + metformin + glicazide	0	1	1	0	0	0
Sitagliptin + repaglinide	1	0	0	0	1	0
Vildagliptin	4	2	2	3	1	0
Vildagliptin + glicazide	0	1	1	0	0	0
Vildagliptin + glimepiride	0	1	1	0	0	0
Vildagliptin + metformin	4	10	10	3	1	0

carried out using SPSS software (IBM SPSS Statistics version 24). The analysis has been mainly based on testing from contingency tables built from data in Table 1.

Of 248 type 2 DM hospitalized for heart failure patients, 104 suffered CKD, whereas 144 did not.

2 Analyzing the Dependence of Receiving Each Type 2 DM Treatment and the Presence of CKD

This section aims to discuss whether there is a statistical dependence between receiving each of the single type 2 DM treatment and the presence of CKD. The discussion for each single treatment has been carried out on the basis of a 2×2 contingency table independence test. More concretely, whenever the Pearson chi-square test is reliable because of the cells counts being large enough it has been applied; on the contrary, we have considered the exact Fisher procedure.

As an example of such contingency tables analysis, assume the selected treatment is metformin. Then, the associated contigency table is the one given in Table 2.

The value of the Pearson chi-square statistic is equal to 19.176, whence the p-value is 0.000012 and hence there is a significant dependence between the metformin treatment and the presence of CKD.

Separate conclusions for each of the treatments can be found gathered in Table 3.

The obtained p-values for testing independence in the considered 2×2 contingency tables have been quite conclusive. In this way, for each of the involved treatments these p-values have been either much greater than 0.1 or much lower than 0.01 (but for the diet which is slightly lower than 0.05). Consequently, we can conclude that the diet treatment of type 2 DM hospitalized for heart failure patients seems to be significantly dependent on the presence of CKD, linagliptin and repaglinide treatments are quite significantly dependent and insulin and metformin treatments are very significantly dependent.

In cases the presence of CKD influences the diabetic treatment, we can consider a deeper analysis which is to be presented in the following section.

Table 2 2×2 contingency table of type 2 DM hospitalized for heart failure patients in CAULE where the two involved variables are receiving or not metformin treatment and suffering or not CKD

	Suffering from CKD	Not suffering from CKD
With metformin treatment	21	68
Without metformin treatment	83	76

Table 3 Summary of statistical conclusions from 2 × 2 contingency table of type 2 DM hospitalized for heart failure patients in CAULE where the two involved variables are receiving or not the given treatment and suffering or not CKD

Treatment	Suffering from CKD	Not suffering from CKD	Differences
Diet	9	26	**significant** ($p < 0.05$)
Glibenclamide	1	1	Not significant ($p = 1$)
Glicazide	1	5	Not significant ($p \gg 0.1$)
Glimepiride	3	6	Not significant ($p \gg 0.1$)
Insulin	60	49	**significant** ($p \ll 0.01$)
Linagliptin	13	4	**Significant** ($p < 0.01$)
Liraglutide	0	1	Not significant ($p = 1$)
Lixisenatide	0	1	Not significant ($p = 1$)
Metformin	21	68	**Significant** ($p \ll 0.01$)
Repaglinide	8	1	**Significant** ($p < 0.01$)
Saxagliptin	0	1	Not significant ($p = 1$)
Sitagliptin	10	13	Not significant ($p \gg 0.1$)
Vildagliptin	11	17	Not significant ($p \gg 0.1$)

3 Analyzing the Dependence of Receiving Each Type 2 DM Treatment and the Degree of CKD

For each of the five treatments for which differences have been shown to be significant one can develop independence contingency table tests concerning the degrees of CKD [10].

However, because of the required expected (theoretical) frequencies conditions for applying contingency tests, instead of having 2 × 4 tables, we have reduced them by removing either G5 or both n-G2 and G5. The conclusions have been gathered in next subsections.

3.1 Diet Treatment Dependence on the CKD Degree

In case of diet (for which the influence of the CKD was slightly significant), the Pearson chi-square test p-value equals 0.131 (Table 4).

Table 4 2 × 3 contingency table of type 2 DM hospitalized for heart failure patients in CAULE where the two involved variables are following or not a diet and degrees of CKD

Treatment	CKD degrees		
	n-G2	G3	G4
Diet	26	6	3
Not diet	118	63	29

Table 5 2 × 3 contingency table of type 2 DM hospitalized for heart failure patients in CAULE where the two involved variables are following or not insulin treatment and degrees of CKD

Treatment	CKD degrees		
	n-G2	G3	G4
Insulin	49	42	17
Not insulin	95	27	15

Close p-values are obtained for the likelihood ratio and the linear-by-linear association tests. Consequently, there are not statistical evidences for the CKD degree influencing to prescribe or not diet.

3.2 Insulin Treatment Dependence on the CKD Degree

In case of insulin (for which the influence of the CKD was strongly significant), the Pearson chi-square test p-value equals 0.001 (Table 5).

Close p-values are obtained for the likelihood ratio and the linear-by-linear association tests. Consequently, as for the presence of CKD, the CKD degree significantly affects the use of insulin therapy.

3.3 Linagliptin Treatment Dependence on the CKD Degree

In case of linagliptin (for which the influence of the CKD was quite significant), the exact Fisher test p-value equals 1, but one should take into account that only degrees G3 and G4 could be considered for the expected frequencies being right (Table 6).

Close asymptotic p-values are obtained for the Pearson chi-square, the chi-square with Yates correction for continuity, the likelihood ratio, and the linear-by-linear association tests. Consequently, although the influence of the CKD was quite significant on prescribing the linagliptin treatment, there is no statistical evidence for the CKD degree influencing it.

Table 6 2 × 2 contingency table of type 2 DM hospitalized for heart failure patients in CAULE where the two involved variables are following or not linagliptin treatment and degrees of CKD

Treatment	CKD degrees	
	G3	G4
Linagliptin	9	4
Not linagliptin	60	28

Table 7 2 × 3 contingency table of type 2 DM hospitalized for heart failure patients in CAULE where the two involved variables are following or not metformin treatment and degrees of CKD

Treatment	CKD degrees		
	n-G2	G3	G4
Metformin	68	12	9
Not metformin	76	57	23

3.4 Metformin Dependence on the CKD Degree

In case of metformin (for which the influence of the CKD was strongly significant), the Pearson chi-square test p-value is lower than 0.0001 (Table 7).

Close p-values are obtained for the likelihood ratio and the linear-by-linear association tests. Consequently, as for the presence of CKD, the CKD degree significantly affects the use of metformin therapy.

3.5 Repaglinide Treatment Dependence on the CKD Degree

Finally, in case of repaglinide (for which the influence of the CKD was quite significant), the exact Fisher test p-value equals 0.204, and only degrees G3 and G4 could be considered for the expected frequencies accomplishing the required conditions (Table 8).

Close asymptotic p-values are obtained for the Pearson chi-square, the chi-square with Yates correction for continuity, the likelihood ratio, and the linear-by-linear

Table 8 2 × 2 contingency table of type 2 DM hospitalized for heart failure patients in CAULE where the two involved variables are following or not repaglinide treatment and degrees of CKD

Treatment	CKD degrees	
	G3	G4
Repaglinide	3	4
Not repaglinide	66	28

association tests. Consequently, although the influence of the CKD was quite significant on prescribing the ripaglidine treatment, there is no statistical evidence for the CKD degree influencing it.

4 Concluding Remarks

From the performed contingency tables-based statistical analysis it is quite clear that CKD presence and degrees significantly influence the farmacological treatment of type 2 DM hospitalized for heart failure in case of insulin and metmorfin. For the rest of treatments, differences have not been shown to be significant.

By looking at data in Table 1, it is clear that the most usual farmacological treatment for CDK patients is insulin, whereas metformin is the most commonly employed for non-CKD. Consequently, taking into account the current Clinical Practice Guidelines (see, e.g., [6, 11]), the adequate treatment of type 2 DM in patients with CKD requires a thorough knowledge of their pharmacokinecits by all professionals involved in the treatments and, first of all, a good coordination between primary care physicians and specialists to provide a multifaceted care program to reduce progression of disease.

Acknowledgements The research in this paper has been partially supported by the Principality of Asturias/FEDER Grant GRUPIN14-101. Its financial support is gratefully acknowledged.

References

1. Cefalu WT, Buse JB, Del Prato S, Home PhD, LeRoith D, Nauck MA, Raz I, Rosenstock J, Riddle MC (2014) Beyond Metformin: safety considerations in the decision-making process for pelecting a second medication for type 2 diabetes management. Diabetes Care 37:2647–2659
2. Cernea S (2016) Heart failure and chronic kidney disease in type 2 diabetes. J Interdisc Med 1(3):252–258
3. Chin MP, Wrolstad D, Bakris GL, Chertow GM, De Zeeuw D, Goldsberry A, Linde PG, McCullough PA, McMurray JJ, Wittes J, Meyer CJ (2014) Risk factors for heart failure in patients with type 2 diabetes mellitus and stage 4 chronic kidney disease treated with bardoxolone methyl. J Cardiac Fail 20(12):953–958
4. Davie M, Chatterjee S, Khunti K (2016) The treatment of type 2 diabetes in the presence of renal impairment: What we should know about newer therapies. Clin Pharm: Adv and Appl 8:61–81
5. Filippatos G, Anker SD, Böhm M, Gheorghiade M, Køber L, Krum H, Maggioni AP, Ponikowski P, Voors AA, Zannad F, Kim S-Y, Nowack C, Palombo G, Kolkhof P, Kimmeskamp-Kirschbaum N, Pieper A, Pitt B (2016) A randomized controlled study of finerenone vs. eplerenone in patients with worsening chronic heart failure and diabetes mellitus and/or chronic kidney disease. Eur Heart J 37(27):2105–2114
6. Gómez-Huelgas R, Martínez-Castelao A, Artola S, Górriz JL, Menéndez E (2014) en nombre del Grupo de Trabajo para el Documento de Consenso Tratamiento de la diabetes tipo 2 en el paciente con enfermedad renal crónica. Med Clin (Barc) 142(2):85.e1–85.e10

7. Hahr AJ, Molitch ME (2015) Management of diabetes mellitus in patients with chronic kidney disease. Clin Diab Endocr 1:2
8. Hung Y-C, Lin C-C, Huang W-L, Chang M-P, Chen C-C (2016) Sitagliptin and risk of heart failure hospitalization in patients with type 2 diabetes on dialysis: A population-based cohort study. Scientific Reports 6:Art. #30499
9. Karagiannis T, Bekiari E, Boura P, Tsapas A (2016) Cardiovascular risk with DPP-4 inhibitors: latest evidence and clinical implications. Ther Adv Drug Saf 7(2):36–38
10. KDIGO (2013) KDIGO 2012 clinical practice guideline for the evaluation and management of chronic kidney disease. Chapter 1: definition and classification of CKD. Kidney Int Suppl 3(1):19–62
11. Marín-Peñalver JJ, Martín-Timón I, Sevillano-Collantes C, del Cañizo-Gómez FJ (2016) Update on the treatment of type 2 diabetes mellitus. World J Diabetes 7(17):354–395
12. Pálsson R, Patel UD (2014) Cardiovascular complications of diabetic kidney disease. Adv Chronic Kidney Dis 21(3):273–280
13. Patel PA, Liang L, Khazanie P, Hammill BG, Fonarow GC, Yancy CW, Bhatt DL, Curtis LH, Hernández AF (2016) Antihyperglycemic medication use among medicare beneficiaries with heart failure, diabetes mellitus, and chronic kidney disease. Circ Heart Fail 9(7):pii e002638

A Diagnostic Test Approach for Multitesting Problems

Pablo Martínez-Camblor, Sonia Pérez-Fernández and Norberto Corral

Abstract In the last decades, multiple-testing problems have received much attention. Many different methods have been proposed in order to deal with this relevant issue. Most of them are focused on controlling some weak version of the Type I error such as the False Discovery Rate. Type II error is frequently forgotten. In this work, the multitesting problem is treated from a diagnostic test approach in which the p-values play the role of the studied predictive marker. In this context, the receiver operating characteristic, ROC, curve is estimated. Several Monte Carlo simulations help for a better understanding of the problem. Finally, a real dataset studying the relationship between atosomal CpG sites and characteristic of hepatocellular carcinoma is considered.

1 Introduction

Modern science frequently produces data on thousands of different features. Probably, the -omics technologies (genomics, transcriptomics, proteomics, etc.) stand for the most relevant examples although other fields like astrophysics, brain imaging or spatial epidemiology have also increased substantially the size of the collected data. Conventionally, statistical analyses of those data often include a huge number of hypotheses to be tested at the same time. In this context, standard statistical concepts, like the p-value, lose their original probabilistic interpretation. Notice that, for any fixed nominal level, the probability of spurious effects, or false positives,

P. Martínez-Camblor (✉)
The Dartmouth Institute for Health Policy and Clinical Practice, Geisel School
of Medicine at Dartmouth, Hanover, NH, USA
e-mail: Pablo.Martinez.Camblor@Dartmouth.edu

S. Pérez-Fernández · N. Corral
Department of Statistics and Operational Research and Mathematics Didactics,
University of Oviedo, Oviedo, Spain
e-mail: perezsonia@uniovi.es

N. Corral
e-mail: norbert@uniovi.es

© Springer International Publishing AG 2018
E. Gil et al. (eds.), *The Mathematics of the Uncertain*, Studies in Systems,
Decision and Control 142, https://doi.org/10.1007/978-3-319-73848-2_26

Table 1 Number of mistakes committed when N null hypotheses are simultaneously tested

	Reject	Not Reject	Total
Null true	$r - R$	U	N_0
Alternative true	R	$u - U$	$N - N_0 (= N_1)$
Total	r	$N - r (= u)$	N

greatly increases when a massive number of null hypotheses are simultaneously tested. Classical multiple comparison procedures focus on controlling the probability of committing any Type I error, i.e., to control the family wise error rate (FWER). Unfortunately, this objective is too ambitious in this context and more liberal criteria must be used (see Farcomeni [8] for a recent and extensive review of modern multiple hypothesis testing methods).

In the multiple testing context, conventionally N (null) hypotheses are contrasted simultaneously from adequate tests. The classical Table 1 depicts schematically the possible practical situations. Of course, in practice, only the total number of hypotheses, N, and the total number of rejections, r, are really known.

Notice that controlling the FWER is equivalent to control the probability $\mathscr{P}\{r - R > 0\}$. Seeger [16] introduced and later Benjamini and Hochberg [1] revised and popularized the false discovery rate (FDR), defined as the expected proportion of spurious effects, i.e., FDR$= E[(r - R)/(r \vee 1)]$ $(a \vee b = b$ if $a \leq b)$. The FDR is a frequentist well established definition for the multiple hypothesis testing error and, undoubtedly, the most used procedure.

Several generalizations for the FWER and the FDR criteria and different procedures to implement them have been proposed (see, for instance, Sarkar [15] and references therein). Most recent works deal with the problem of controlling the tail probability of false positives. Genovese and Wasserman [11] proposed to control the tail probability of the false discovery proportion (FDP) by the so labeled FDX (false discovery exceedance), i.e., for a fixed α, to control $\mathscr{P}\{1 - R/r > \alpha\}$. In addition, we highlight the sequential goodness of fit (SGoF) strategy, proposed by Carvajal-Rodríguez et al. [2] and deeply explored by de Uña-Álvarez [5]. The SGoF method rejects an amount of null hypotheses equal to the difference between the observed and the expected amounts of p-values below a given threshold under the assumption that all nulls are true (we denote by $H_0 = \cap_{i=1}^{N} H_{0,i}$, in bold, this hypothesis). The outcome of most of those methods is a cutoff point; p-values below this threshold are declared significant while p-values above it are declared non-significant.

Considering each test as a sampling unit and its p-value as a marker of the null hypothesis credibility, the multiple hypothesis testing problem has a clear connection to the classification theory (this point of view was briefly explored by Storey [18]). In this paper, assuming that p-values follow a mixture distribution (see Sect. 2), multitesting problem is dealt with from a diagnostic test approach; in particular, the well-known receiver-operating characteristic (ROC) curve is derived. This curve does not provide a cut-off point but graphical information about the real diagnostic capacity of the studied marker, in this case, the p-value. One of the main goals of this work is pointing out the limitation of the process. Notice that, depending on the power of the study, the p-value is not always a good *diagnostic marker*. And even when it is a good diagnostic marker, the classification could be difficult depending on the prevalence of untrue nulls. In Sect. 3, some particularities of the ROC curve when it is applied on multitesting problems are considered. Section 4 is devoted to the ROC curve estimation. The performance of the proposed method is empirically studied via Monte Carlo simulations. In Sect. 5, a real data problem is analyzed. Finally, in Sect. 6, we present our main conclusions.

2 The Mixture Model

When it is assumed that there exist N independent hypotheses ($H_{0,i}$, $1 \leq i \leq N$) which are going to be tested from adequate tests and that F_0 and F_1 are the cumulative distribution functions (CDFs) for the p-values when the null is true and untrue, respectively, then the CDF of the p-values will be the mixture distribution

$$\mathbb{G}_N(t) = \pi_0 \cdot F_0(t) + (1 - \pi_0) \cdot F_1(t),$$

where π_0 ($= N_0/N$) is the true null proportion. In this case, for each $t \in [0, 1]$, the function $\mathbb{G}_N(t)$ represents the probability that the p-value associated with a randomly selected hypothesis will be less or equal to t. However, it should be noted that this model assumes that all true nulls follow the same distribution (F_0), which can be plausible, but also that all untrue nulls follow the same distribution (F_1), which is a quite more unrealistic proviso. Without this assumption, the probability that a p-value from an untrue hypothesis will be less or equal to t depends on the particular hypothesis from which this p-value has been drawn. Although depending on the particular experiment studied a random effects or hierarchical model can be more appropriate (see Efron et al. [6]), we adopt the most common situation in which the set of hypotheses to be tested are previously fixed, therefore $F_1 = N_1^{-1} \cdot \sum_{i \in J_1} F_{1,i}$, where $J_1 \subseteq \{1, \ldots, N\}$ stands for the set of indices in which the null is untrue. For each $t \in [0, 1]$, $\mathbb{G}_N(t)$ stands for the average probability that the p-value associated with a randomly selected hypothesis will be less or equal to t. This interpretation is still valid in the presence of dependency structures.

Notice that, on the usual assumption that, under the null, the CDF of each individual p-value is $t \cdot \mathbb{I}_{[0,1]}(t)$ ($\mathbb{I}_A$ stands for the standard indicator function on the set A), i.e., each individual p-value follows a uniform distribution within $[0, 1]$, it is derived

$$\mathbb{G}_N(t) = \pi_0 \cdot t \cdot \mathbb{I}_{[0,1]}(t) + (1 - \pi_0) \cdot F_1(t). \tag{1}$$

Remark 2.1 Although it is reasonable to assume that, under the null, each single p-value is uniformly distributed on $[0, 1]$, and this proviso is true when the null is simple and the distribution of the test statistic is continuous and known, the true

p-value distribution is only stochastically dominated by the uniform if its distribution is discrete or the p-value is estimated by a resampling method (see, for instance, Farcomeni [8]).

From the mixture model, the traditional BH procedure for controlling the FDR at level α proposed by Benjamini and Hochberg [1] can be seen as a plug-in method for estimating the threshold (Genovese and Wasserman [10]),

$$T_{\text{BH}}(\alpha) = \sup\{t \in [0, 1] : \mathbb{G}_N(t) \geq t/\alpha\}, \tag{2}$$

and, assuming independence among tests, the SGoF method (taking $\gamma = \alpha$) tries to estimate the threshold (de Uña-Álvarez [5]) as

$$T_{\text{SGoF}}(\alpha) = \mathbb{G}_N^{-1}\left(\mathbb{G}_N(\alpha) - \alpha - z_\alpha \cdot \sqrt{\alpha \cdot (1 - \alpha)/N} + N^{-1}\right), \tag{3}$$

where $\mathbb{G}_N^{-1}(t) = \inf\{s : \mathbb{G}_N(s) \geq t\}$ and $z_\alpha = \Phi^{-1}(1 - \alpha)$ (Φ stands for the standard normal CDF).

3 Receiver Operating Characteristic Curve in the Multitesting Problem

The ROC curve is a popular graphical method frequently used in order to study the diagnostic capacity of continuous markers. It represents in a plot the true-positive rate (TPR) against the false-positive one (FPR) of all thresholds of the marker. Both practical and theoretical aspects of the ROC curve have been extensively studied in the specialized literature (see Zhou et al. [19] for a recent review). Assuming that smaller values of the marker indicate larger confidence that a given subject is positive, let χ and ξ be two continuous random variables representing the marker values for the negative and positive subjects, respectively. Therefore, for a fixed point $t \in [0, 1]$, the ROC curve is defined as follows,

$$\mathscr{R}(t) = F_\xi(F_\chi(t)) = \mathscr{P}\{\xi \leq F_\chi(t)\} = \mathscr{P}\{F_\chi^{-1}(\xi) \leq t\} = F_{F_\chi^{-1}(\xi)}(t), \tag{4}$$

where F_χ and F_ξ denote the CDFs for the variables χ and ξ, respectively. In the current context, p-values play the role of marker values and the true and false nulls are the negative and positive subjects, respectively. Assuming the mixture model (1), the problem is simplified; in this case, the ROC curve stands for the CDF for the untrue nulls, i.e., $\mathscr{R}(t) = F_1(t)$ ($t \in [0, 1]$). Notice that the sensitivity (TPR) must be interpreted as the *average* probability that an untrue null hypothesis will be correctly classified as untrue, i.e., the average probability of rejecting an untrue null hypothesis (power). Figure 1 depicts the real ROC curve for $F_1(t) = t^a \cdot \mathbb{I}_{[0,1]}(t)$ with a such that the average sensitivity is 0.8 when the specificity (1-FPR) is 0.95. In order to

Fig. 1 ROC curve under the mixture model for $F_1(t) = t^a \cdot \mathbb{I}_{[0,1]}(t)$ with $a = log(0.8)/log(0.05)$. Remember that, in this context, $\mathscr{R}(t) = F_1(t)$. Notice that the x-axis is not in linear scale

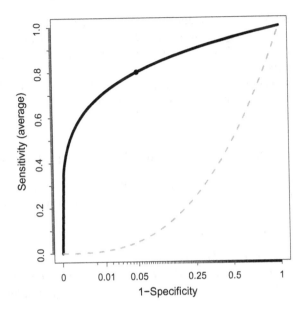

give more relevance to high specificities, the scale of the x-axis has been modified. Remember that ROC curve is equal to F_1 and the main diagonal (dotted gray line, since the scale of x-axis is modified) is equal to F_0.

Due to the fact that the ROC curve does not depend on the prevalence of the studied characteristic, $1 - \pi_0$, this graphical method provides valuable information about the real capacity of the marker (p-value) to identify the subjects (tests) as positives (rejecting) or negatives (no rejecting). Conventionally, the aforementioned thresholds strongly depend on this prevalence; particularly, for the above situation, $T_{BH}(0.1)$ is 0.0202, 0.0160 and 0.0076 for $\pi_0 = 0.75$, 0.80 and 0.90, respectively; with these cut-off points, the obtained powers would be 0.749, 0.736 and 0.697.

The SGoF method also depends on the number of tests, N; when $N = 10,000$, $T_{SGoF}(0.1)$ takes the values 0.0086, 0.0074 and 0.0043 for $\pi_0 = 0.75$, 0.80 and 0.90, leading to powers of 0.703, 0.696 and 0.668. The Youden index, Y, is achieved at point 0.06 (Youden index is often used in diagnostic tests in order to obtain an optimal threshold; it is defined as $Y = \max_{t \in \mathbb{R}}\{TPR(t) - FPR(t)\}$; at this point, the obtained FPR is obviously 0.060 and the average power 0.811 (= TPR); i.e., by using this cut-off point, in average 6% of the true nulls would be declared false and around 19% of the untrue nulls would not be rejected.

Remark 3.1 As it is well-known, the area under the ROC curve (AUC) is one of the most commonly used global index of diagnostic accuracy (Faraggi and Reiser [7]). It ranges between $1/2$, when the marker does not contribute to a correct classification, and 1, if the marker can classify perfectly all subjects. The AUC has a direct probabilistic interpretation: in particular, it is the probability that the value of the marker in a randomly chosen negative subject will be higher than the value of the marker in a randomly chosen positive subject. In the current context it can be read as the average of the power when the Type I error varies between 0 and 1. However, large Type I errors are not interesting in practice; hence, limiting the range of the Type I error and considering the partial area under the curve, pAUC (Ma et al. [13]), seems to be a more adequate measure. In this case, the AUC is 0.931 and the pAUC between 0 and 0.05 is 0.0372 (0.745 in a 0 − 1 scale).

4 Receiver Operating Characteristic Curve Estimation

In practice, we observe N p-values, $\{p_1, \ldots, p_N\}$, corresponding to N nulls to be tested $\{H_{0,1}, \ldots, H_{0,N}\}$. Assuming independence between the p-values, the Liapunov's Central Limit Theorem (see, for instance, Ibarrola et al. [12]) implies the following result:

Theorem 4.1 *Let* $\{p_1, \cdots, p_N\}$ *be an independent random sample where for each* $i \in 1, \ldots, N$, p_i *was drawn from* $F_{*,i}$. *For each* $t \in [0, 1]$, *let be* $\mathbb{G}_N(t) = (1/N) \cdot \sum_{i=1}^{N} F_{*,i}(t)$ *and* $\hat{\mathbb{G}}_N(t) = (1/N) \cdot \sum_{i=1}^{N} \mathbb{I}_{(-\infty,t]}(p_i)$, *then*

$$N \cdot \frac{\hat{\mathbb{G}}_N(t) - \mathbb{G}_N(t)}{\sqrt{\sum_{i=1}^{N} F_{*,i}(t) \cdot (1 - F_{*,i}(t))}} \xrightarrow{\mathscr{L}}_N \mathscr{N}(0, 1). \tag{5}$$

This Central Limit Theorem for non identically distributed variables implies that, if N is sufficiently large, the empirical CDF estimator provides a good approximation of the real distribution function. In Genovese and Wasserman [10] this convergence is deeply considered from a stochastic process approach.

On the other hand, the π_0 estimation has been previously considered in the specialized literature. For instance, Dalmasso et al. [4] took advantage of the logarithmic function properties and defined the family of estimators

$$\hat{\pi}_{0,k} = \frac{(1/N) \sum_{i=1}^{N} [-\log(1 - p_i)]^k}{k!}, \text{ with } k \in \mathbb{N}. \tag{6}$$

Once π_0 is estimated, from the mixture model and result (5), the estimation of the ROC curve is direct. However, assuming that for any fixed nominal level the probability of rejecting an untrue null hypothesis is higher than the probability of rejecting a true null hypothesis, and taking into account that $\mathscr{R}$ is a non-decreasing function, then

$$\hat{\mathscr{R}}(t) = \max\left\{ \sup_{s \in [0,t]} \{\hat{F}_1(s)\}, \, t \cdot \mathbb{I}_{[0,1]}(t) \right\}, \tag{7}$$

where $\hat{F}_1(s) = \min\{(\hat{\mathbb{G}}_N(s) - \hat{\pi}_{0,2} \cdot s \cdot \mathbb{I}_{[0,1]}(s)) \cdot (1 - \hat{\pi}_{0,2})^{-1}, 1\}$, is a more appropriate estimator for the ROC curve. Of course, the estimator is still valid by using other π_0 approximations.

4.1 Simulation Study

The behavior of the proposed estimator is empirically studied via Monte Carlo simulations. Two different strategies were considered: in the first one, the p-values were directly drawn from different theoretical mixture models; in the second one, the whole problem is simulated, i.e., one sample is drawn and the corresponding p-values are computed from an adequate test where the null hypothesis is $\mu = 0$.

In the first scenario we run independent random samples of size N from the distribution $\mathbb{G}_N(t) = \pi_0 \cdot t \cdot \mathbb{I}_{[0,1]}(t) + (1 - \pi_0) \cdot F_1(t)$ where $F_1 = (1/N_1) \sum_{i=1}^{N_1} t^{a_i} \cdot \mathbb{I}_{[0,1]}(t)$ with $a_i = L + (i/N_1) \cdot U$ and two parameters (L and U) selected in order to obtain different power averages; particularly, at level 0.05, power averages ($\bar{\beta}_{0.05}$) of 0.6 and 0.8 were studied.

In the second considered scenario the whole problem is simulated. We consider situations where the nulls to be tested are $H_{0,i} : \mu_i = 0$, where μ_i stands for the expected value of the ith population ($1 \leq i \leq N$). Under the null, we drawn independent samples (with size $n = 25$) from a standard normal distribution while, under the alternative, the samples were drawn from a $\mathcal{N}(\mu^*, 1)$ distribution where μ^* was taken such that the power average was the desired one ($\bar{\beta}_{0.05} = 0.6$ and 0.8 were considered). Then, the p-values were computed by using the Student t-test (parametric).

Figure 2, left, depicts the approximate shape of the involved curves in the first scenario; notice that the real one depends on the number of untrue nulls (N_1). At right, the shape of the respective ROC curves in the second scenario is displayed.

Table 2 shows the observed results for both scenarios on 1,000 Monte Carlo iterations. Particularly, we report mean $\pm$ standard deviation for the absolute difference between the real and the estimated proportion of true nulls (π_0), as well as the

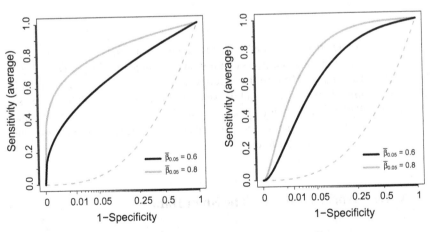

Fig. 2 Left, ROC curve under the mixture model ($F_1(t) = (1/N_1) \cdot \sum_{i=1}^{N_1} t^{a_i} \cdot \mathbb{I}_{[0,1]}(t)$ with $a_i = L + (i/N_1) \cdot U$ and $L = 0.05$) when p-values are obtained from the first described scenario. Right, ROC curve when p-values are obtained from the second described scenario

Table 2 Mean ± standard deviation for the absolute difference between π_0 and $\hat{\pi}_{0,2}$; the integrated absolute difference between the ROC curve estimation and its target; the errors $E[\hat{T}_{\mathrm{BH}}(0.05)]$ and $E[\hat{T}_{\mathrm{SGoG}}(0.05)]$ obtained from 1, 000 and 10, 000 Monte Carlo iterations for both described scenarios

N	π_0	$\bar{\beta}_{0.05}$	$\|\hat{\pi}_{0,2} - \pi_0\|$	$\int \|\hat{\mathscr{R}} - \mathscr{R}\|$	$E[\hat{T}_{\mathrm{BH}}(0.05)]$	$E[\hat{T}_{\mathrm{SGoF}}(0.05)]$
Scenario I						
1,000	0.95	0.60	0.056 ± 0.04	0.178 ± 0.09	0.190 ± 0.14	1.871 ± 2.21
		0.80	0.054 ± 0.04	0.182 ± 0.14	0.103 ± 0.07	4.287 ± 7.61
1,000	0.85	0.60	0.059 ± 0.05	0.109 ± 0.07	0.093 ± 0.07	0.350 ± 0.26
		0.80	0.052 ± 0.04	0.074 ± 0.06	0.052 ± 0.04	0.546 ± 0.44
10,000	0.95	0.60	0.019 ± 0.01	0.109 ± 0.06	0.056 ± 0.04	0.257 ± 0.19
		0.80	0.018 ± 0.01	0.075 ± 0.06	0.032 ± 0.02	0.381 ± 0.29
10,000	0.85	0.60	0.032 ± 0.02	0.088 ± 0.04	0.031 ± 0.02	0.091 ± 0.07
		0.80	0.020 ± 0.02	0.045 ± 0.02	0.016 ± 0.01	0.130 ± 0.10
Scenario II						
1,000	0.95	0.60	0.054 ± 0.04	0.168 ± 0.11	2.404 ± 2.50	0.520 ± 0.40
		0.80	0.056 ± 0.04	0.189 ± 0.15	0.446 ± 0.33	0.450 ± 0.35
1,000	0.85	0.60	0.055 ± 0.04	0.082 ± 0.06	0.455 ± 0.33	0.158 ± 0.12
		0.80	0.055 ± 0.04	0.069 ± 0.07	0.137 ± 0.11	0.164 ± 0.12
10,000	0.95	0.60	0.018 ± 0.01	0.080 ± 0.05	0.865 ± 0.70	0.114 ± 0.08
		0.80	0.017 ± 0.01	0.063 ± 0.06	0.150 ± 0.11	0.112 ± 0.08
10,000	0.85	0.60	0.020 ± 0.02	0.044 ± 0.02	0.159 ± 0.12	0.044 ± 0.03
		0.80	0.015 ± 0.01	0.027 ± 0.02	0.043 ± 0.03	0.047 ± 0.03

integrated absolute error committed by the ROC curve estimator proposed in (7). In addition, information about the errors committed (measured by $E[\hat{T}] = |\hat{T} - T|/T$) for $T_{\mathrm{BH}}(0.05)$ and $T_{\mathrm{SGoF}}(0.05)$ are also reported. For $N = 1, 000$, the observed results are disappointing; both the mean and the standard deviation of the ROC curve estimates are really large. However they decrease for $N = 10, 000$. It is worth to notice that the observed results in the second scenario are a bit better than the previous ones but, in any case, really similar to them.

5 A Real-World Example: The Shen Data

The Shen data contains information of 62 Taiwanese cases of hepatocellular carcinoma (HCC) on which tumor and adjacent non-tumor tissues were analyzed using Illumina methylation arrays (Illumina, Inc., San Diego, CA) that screen 26,538

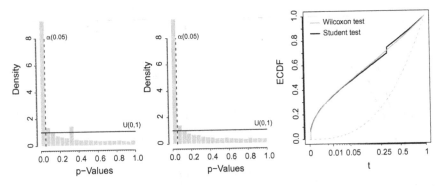

Fig. 3 Histogram for the p-values obtained from the t (left) and Wilcoxon (middle) tests, as well as $\hat{\mathbb{G}}_N(\cdot)$ for the t (black) and Wilcoxon (gray) tests

Table 3 Values of $\hat{\pi}_{0,k}$ for different k-values

	$\hat{\pi}_{0,1}$	$\hat{\pi}_{0,2}$	$\hat{\pi}_{0,3}$	$\hat{\pi}_{0,4}$
t-test	0.404	0.384	0.450	0.590
Wilcoxon test	0.424	0.427	0.541	0.791

autosomal CpG sites. The reader is referred to Shen et al. [17] for a complete information about the original study.

The data are publicly available at the Gene Expression Omnibus (GEO) page, with access number GSE37988 (https://www.ncbi.nlm.nih.gov/geo/query/acc.cgi?acc=GSE37988).

Probes corresponding to the X and Y chromosomes were removed from the dataset in order to eliminate the X-inactivation effects. We considered the raw data (without any previous quality control).

Due to the fact that each CpG is measured on the same subjects, the parametric Student t-test or the non-parametric Wilcoxon test for paired samples can be used in order to check the null of equality between, let us abuse, distributions in tumor and non-tumor tissues.

The total number of CpG sites with a p-value less than 0.05 (usual nominal level) was 12,394 (46.7%) using the t-test, and 12,592 (47.4%) using the Wilcoxon test.

Figure 3 shows the p-value histograms and $\mathbb{G}_N$ function estimates for the Student (black) and the Wilcoxon (gray) test.

Table 3 shows estimations of π_0 using different k-values of the estimator defined in (6). Results are influenced by the p-values close to 1, specially for largest k.

The BH method, $\hat{T}_{BH}(0.05)$, declares significant the 10,451 smallest p-values (cut-off point of 0.019690) for the t-test and the 10,585 smallest ones when the Wilcoxon test is used (cut-off point of 0.019943).

At these points, the estimated average sensibilities (using $\hat{\pi}_{0,2}$) were 0.627 and 0.681, for the t and the Wilcoxon test, respectively.

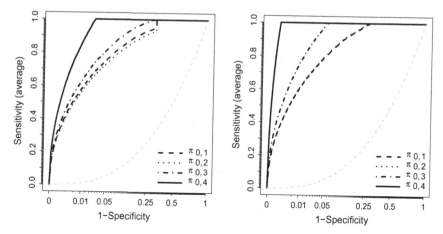

Fig. 4 ROC curve estimations for different $\hat{\pi}_{0,k}$ ($k \in \{1, 2, 3, 4\}$) for t-test (left) and Wilcoxon test (right)

Figure 4 depicts the estimated ROC curves for the Student (left) and the Wilcoxon (right) test for the different π_0 estimates considered.

Obviously, the results depend on the $\hat{\pi}_{0,k}$ value; however, it should be recalled that the x-axis scale was altered and the real difference is not as large as it seems.

6   Discussion

Currently, a number of researchers, mainly bioinformaticians, must often deal with multitesting problems. As a consequence, the development of adequate statistical tools in order to handle and control the involved Type I and Type II errors is a really hot topic in the specialized literature. Although some authors as Genovese and Wasserman [9] or Storey [18] have already considered the false non-discovery rate, most of the works are focused on trying to control, in some way, the false positive rate. However, the first quantity is crucial in order to know what the real capacity of detecting true effects is. Notice that, even when we know the exact number of false nulls, the p-value could not be an appropriate measure for separating those from the true nulls.

Actually, this work does not propose real practical solutions but it pays attention to an usually forgotten aspect of the multitesting problem. It intends to ponder the technical limitations which this complex issue provokes. Remember that, in most cases, we only have one sample of p-values drawn from an N-dimensional vector. By assuming independence among tests, one can perform some kind of correct inference; furthermore, the independence assumption is reasonable in a number of practical situations (see Clarke and Hall [3]), but it may be a source of serious mistakes and misleading conclusions. In the presence of arbitrary correlation structures, the limitations are clear. The main problem lies in the variability of the observed proportion of rejections under the null; while in the independent case, in the usual practical problems, this number is really close to the fixed nominal level, under dependent structures it can vary a lot (see Martínez-Camblor [14]) and both the π_0 and the $\mathcal{R}(\cdot)$ estimates are strongly dependent on that observed value.

The simulation studies show the limitation capacity to perform a correct estimation of the ROC curve in the multitesting context. In addition, the presence of p-values close to the extremes (zero and one) provokes precision problems in the obtained estimates; unfortunately, when the number of tests is large (most frequent case), this problem is not unusual. The accuracy problem, frequently ignored, gains relevance when the selected cut-off point strongly depends on the fifth or sixth decimal position.

In this report, we explore the interpretation of the sensibility and specificity in the multitesting problem by assuming the mixture model and, in this context, a ROC curve estimator is proposed. The explored methods allow us to give an estimation of the sensitivity average for each particular problem. Even taking into account the limitations of the procedures, in each particular problem, this quantity can help to have a better understanding of what the actual capacity of p-values to distinguish false from true null hypotheses is.

Acknowledgements The authors acknowledge support by the Grants MTM2014-55966-P and MTM2015-63971-P from the Spanish Ministerio de Economía y Competitividad and by Severo Ochoa Grant BP16118 and FC-15-GRUPIN14-101 from the Principality of Asturias.

References

1. Benjamini Y, Hochberg Y (1995) Controlling the false discovery rate: a practical and powerful approach to multiple testing. J R Stat Soc Ser B 57(1):289–300
2. Carvajal-Rodríguez A, de Uña-Álvarez J, Rolán-Álvarez E (2009) A new multitest correction (SGoF) that increases its statistical power when increasing the number of tests. BMC Bioinf 10:209
3. Clarke S, Hall P (2009) Robustness of multiple testing procedures against dependence. Ann Stat 37(1):332–358
4. Dalmasso C, Broët P, Moreau T (2005) A simple procedure for estimating the false discovery rate. Bioinformatics 21(5):660–668
5. de Uña-Álvarez J (2011) On the statistical properties of SGoF multitesting method. Statist Appl Genet Molec Biol 10(1):Art18
6. Efron B, Tibshirani R, Storey JD, Tusher V (2001) Empirical Bayes analysis of a microarray experiment. J Am Stat Assoc 96(456):1151–1160
7. Faraggi D, Reiser B (2002) Estimation of the area under the ROC curve. Stat Med 21(20):3093–3106
8. Farcomeni A (2008) A review of modern multiple hypothesis testing, with particular attention to the false discovery proportion. Stat Method Med Res 17(4):347–388
9. Genovese CR, Wasserman L (2002) Operating characteristics and extensions of the false discovery rate procedure. J R Stat Soc Ser B 64(3):499–517
10. Genovese CR, Wasserman L (2004) A stochastic process approach to false discovery control. Ann Stat 32(3):1035–1061
11. Genovese CR, Wasserman L (2006) Exceedance control of the false discovery proportion. J Am Stat Assoc 101(476):1408–1417

12. Ibarrola P, Pardo L, Quesada V (1997) Teoría de la Probabilidad. Síntesis, Madrid
13. Ma H, Bandos AI, Rockette HE, Gur D (2013) On use of partial area under the ROC curve for evaluation of diagnostic performance. Stat Med 32(20):3449–3458
14. Martínez-Camblor P (2014) On correlated z-values distribution in hypothesis testing. Comp Stat Data Anal 79:30–43
15. Sarkar SK (2007) Stepup procedures controlling generalized FWER and generalized FDR. Ann Stat 35(6):2405–2420
16. Seeger P (1968) A note on a method for the analysis of significances en masse. Technometrics 10(3):586–593
17. Shen J, Wang S, Zhang YJ, Kappil M, Wu HC, Kibriya MG, Wang Q, Jasmine F, Ahsan H, Lee PH, Yu MW, Chen CJ, Santella RM (2012) Genome-wide DNA methylation profiles in hepatocellular carcinoma. Hepatology 55(6):1799–1808
18. Storey JD (2003) The positive false discovery rate: a Bayesian interpretation and the q-value. Ann Stat 31(6):2013–2035
19. Zhou X-H, Obuchowski NA, McClish DK (2002) Statistical methods in diagnostic medicine. Wiley, New York

The Length-Time Bias in Tourism Surveys

José Manuel Menéndez-Estébanez and José Manuel González-Sariego

Abstract The length-time bias arising in many tourist surveys, is often not taken into account in stating the most common estimates. In this paper, by means of the application to tourism research on the Principality of Asturias, one can see that a model, suggested to provide us with better estimates of means and proportions under longitudinal bias conditions, shows very important differences with respect to the methods based on the usual estimators of these parameters.

1 Introduction

Nowadays, in any area in the world having a certain tourist impact there are public and private centers trying to get appropriate tools for statistical synthesis so that they can provide these centers with the widest possible information in order to establish policies and programs in connection with tourism.

This information is often exclusively given by national statistical services. Consequently, the available information from them is very general, and there is usually a need of information regarding small areas.

In the case of Spain, all the autonomous communities have either public or private bureaus, complementing the official statistics regularly provided by the Government of the nation. This complementary endeavor is carried out by handling statistical indicators of a more regional character. Particularly, in the Principality of Asturias, the Asturias Tourist Information System (SITA)[1] has a permanent and systematic mechanism for collecting, managing, organizing and disseminating information on the supply, the demand or the macroeconomic magnitudes of the Asturian tourism, among others.

---

[1]http://www.sita.org.

J. M. Menéndez-Estébanez (✉) · J. M. González-Sariego
Departamento de Estadística, I.O. y D.M., Universidad de Oviedo,
33007 Oviedo, Spain
e-mail: jmme@uniovi.es

J. M. González-Sariego
e-mail: sariego@uniovi.es

E. Gil et al. (eds.), *The Mathematics of the Uncertain*, Studies in Systems, Decision and Control 142, https://doi.org/10.1007/978-3-319-73848-2_27

285

This more detailed knowledge of a particular tourist area can be achieved in many different ways. But in case one wishes to know to a deeper extent the behavior of the visitors of the internal (i.e., domestic and inbound) tourism, individual surveys should be conducted.

For this reason, in almost all the autonomous communities tourists are frequently contacted to conduct surveys. In general, surveys are rather different because of either the orientation of the questionnaire or the considered sampling method. Unfortunately, many of the surveys are very similar because of the lack of procedures to minimize estimating errors. In this respect, it is very common that data are not weighted, that errors different from those involved in the sampling are not analyzed, bias associated with length-time is not taken into account, and so on.

These deficiencies are crucial in estimating a series of parameters that are fundamental in the tourism setting. These parameters are "the occupancy rate", "the total number of nights spent", "the average length of stay", "the proportion of visitors using private tourist accommodation", "the proportion of same-day visitors", "the average daily spending of visitors", etc.

In the following sections the problems associated with length-time bias are to be seen and a basic application of the statistical model used in Asturias to alleviate such a bias is shown.

2 Length-Time Bias

The length-time bias, also referred to as longitudinal bias, could be described as a distortion of the results due to the fact that the probabilities of population units to belong to the sample are not necessarily equal. This is due to differences in the time these units remain in the statistical framework.

There are just a few studies on this topic, and they are usually related to Medicine. For instance, Pelikan and Moskowitz [1] argue that, by taking specimens on specific dates to detect breast cancers, the results are skewed because of the greater probability in locating long-term cases in contrast to situations associated with a greater severity and a lower time of development. This bias induces a more benign interpretation of the cancer problem than the real one, if no statistical correction factors are introduced. Yoshimoto and Tanaka [6] consider similar assertions, although referring to the detection of intracranial diseases.

In the tourism context the problem is analogous to these one can find in the literature. Samples of tourists are usually chosen on specific days, which makes it more likely to interview tourists who have been in the region for many days than those who will stay for only one day. In fact, for many of the last tourists the probability of belonging to the sample will equal zero (it is enough that they stay in the area in days different from those considered for interviewing).

A consequence from such a situation is that tourists spending more days of stay in the region are over-represented in the sample. This affects all the estimates that have been mentioned in the previous section. For instance, the number of nights spent in

Fig. 1 Daily number of visitors to a region

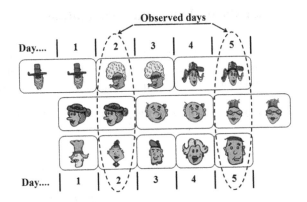

the region or the average length of stay will be higher than they actually are. On the contrary, the proportion of same-day visitors who visited the area in the considered period will be lower than the actual one. In Example 2.1, with a very elementary structure and under very ideal conditions, these dysfunctions can be clearly seen.

Example 2.1 (Estimation of the average length of stay and the proportion of same-day visitors)

A region receives two visitors every day. One of them stays only 1 day (same-day visitor) and the other stays 2 days, with 1 night being spent in the only hotel in the region. In other words, every day the region has three tourists, 2 of them being accommodated in the hotel and spending a total of 2 days of stay, and 1 same-day visitor who will only spend that given day in the region. And this scheme happens for any day of observation (Fig. 1).

Suppose now that we record these data for k days chosen at random in such a way that consecutive days are not chosen, in order to prevent a visitor from appearing repeated in the sample.

What value would be assessed to the average visitor's stay? At the first glance, the answer could be 5/3, since 2 people spent 2 days and 1 person spents only 1. However, this is not the right answer, since the average length of stay intended as the ratio between the total number of stays ($3k$) and the number of visitors ($2k$) equals 3/2.

And even more: what is the proportion of same-day visitors who visit the region during these k days? If every day in the sampling we count 2 people staying in the hotel and 1 same-day visitor, our immediate response will be 1/3. However, during this period, the region has been visited by as many same-day visitors as people have been hosted in the hotel, so the correct answer would be 1/2.

From Example 2.1, it can be deduced that estimates of the average length of stay and the proportions as the indicated one, should not be based on their natural (analogue) estimators, "sample mean" and "sample proportion", but it is necessary to support them on other statistical models that help in correcting biases.

3   Statistical Model and Formulation

Torres et al. [3] have developed a model allowing to compute means and proportions and incorporating the effect associated with the length-time bias. As particular cases of the model, the two following corollaries can be obtained.

Corollary 3.1 *The average length stay of the interviewees along some days is given by*

$$as = \frac{1}{\displaystyle\sum_{k=0}^{d} \frac{f_k}{k+1}}, \tag{1}$$

where as = "average length of stay" (in number of days of the stay), k = "number of nigths spent", d = "maximum number of nights spent", and f_k = "relative frequency of the number of nigths spent".

Corollary 3.2 *For any categorical variable, the proportion of interviewees between two categories A and B can be expressed as follows:*

$$\frac{I_A}{I_B} = \frac{as_A \cdot n_A}{as_B \cdot n_B}, \tag{2}$$

where I_A = "number of interviewees in category A", I_B = "number of interviewees in category B", as_A = "average length of stay in Category A", as_B = "average length of stay in Category B", n_A = "number of visitors in Category A", and n_B = "number of visitors in Category B".

From Eq. (2) it is immediately possible to deduce the proportion of visitors from category A with respect to the total of visitors of both categories (see Eq. (3)).

$$p_A = \frac{n_A}{n_A + n_B} = \frac{as_B \cdot I_A}{as_B \cdot I_A + as_A \cdot I_B}. \tag{3}$$

The above results are only valid under the hypotheses that the number of tourists entering the region every day is constant, and the distribution of the number of nights spent is constant over the days.

These hypotheses may seem to be very restrictive. However, we should not forget that these are the ones that are usually assumed for this type of samplings when homogeneous periods are analyzed with respect to the tourist behavior in a region.

Thus, for instance, when we determine the average length of stay for the summer season we will calculate it from the average lengths of each of the months. Otherwise, it would be necessary to distinguish within each month the peculiarities of each week and even those of each day (if it is Monday, if it is holiday, if it is first day of the month, and so on). But, in practice the latter is impossible to be considered, then a similar behavior is usually assumed for every day of the month, for the number of new visitors and for the distribution of the number of nights spent.

One can easily check that by applying Eq. (1) to Example 2.1 we get the actual output for the average stay length. Indeed, since one third of the visitors do not spend a night and the others spent one, we have that

$$f_0 = 1/3, \quad f_1 = 2/3, \quad k = 0, 1$$

and, hence,

$$as = \frac{1}{\dfrac{1/3}{1} + \dfrac{2/3}{2}} = 1.5, \tag{4}$$

whence the average stay length equals 1.5 days.

If Eq. (3) is applied to Example 2.1 we can verify that the actual result for the proportion of same-day visitors (Category A) can also be obtained, since

$$I_A = 1, \quad I_B = 2, \quad as_A = 1, \quad as_B = 2,$$

whence

$$p_A = \frac{2 \cdot 1}{2 \cdot 1 + 1 \cdot 2} = 0.5, \tag{5}$$

so the proportion of same-day visitors is 0.5, that is, there are as many same-day visitors as non-same-day visitors.

4 Application to the Principality of Asturias

Along the year and by means of individual interviews, SITA collects diverse information on the visitors of the Principality of Asturias.

It is a very complex process, since one has to take into account all types of visitors (tourists and same-day visitors), all types of accommodation establishments (hotel, rural tourism, camping and private tourist accommodation) of any area of Asturias (western, center and east) and in any period of the year. The detailed sampling methodology, which has been proposed after several years of trials and being yearly updated to improve its statistical reliability, can be found in Valdés et al. [5] and Torres et al. [4].

In the survey of tourist demand corresponding to 2016, 4,700 interviews were conducted, them being proportionally distributed along the 12 months of the year (according to the Asturias Tourist Information System [2]). In the technical datasheet (see Table 1) some relevant features from this study have been descriptively summarized.

Table 1 Technical datasheet summarizing the survey conducted on visitors to the Principality of Asturias in 2016 (Source: SITA)

FEATURES	SURVEY
UNIVERSE	over 18 years tourists in Asturias
GEOGRAPHICAL SCOPE	Principality of Asturias
INFORMATION GATHERING PROCESS	structured individual questionnaire questionaire's languages: Spanish, English, French, German
POPULATION SIZE	technically infinity
SAMPLE SIZE	4700 (1368 in accommodation establishments)
CONFIDENCE LEVEL	95% ($Z = 1.96$, $p = q = 0.5$)
SAMPLING PROCEDURE	1) in accommodation establishments: quota sampling, with quotas depending on the period/season, accommodation type, day of the week, geographical area,... 2) otherwise: after stratification by either period/season, or accommodation type, or day of the week,... the interviewer chooses at random a sample of visitors from especially attractive venues
FIELD SURVEY DATE	January to December 2016

STATISTICAL ERRORS TABLE		$N(^*)$	N	SAMPLING ERROR
COLLECTIVE ACCOMMODATION		2,257,173	2,558	±1.94%
Collective Accommodation Splits				
AREA	Western	251,140	299	±5.66%
	Central	1,321,130	1,523	±2.51%
	Eastern	684,903	736	±3.61%
PERIOD	1st Quarter	435,376	513	±4.32%
	2nd Quarter	1,197,460	1,278	±2.74%
	3rd Quarter	624,337	767	±3.54%
TRAVEL REASONS	Holidays	1,644,894	2,093	±2.14%
	Work	491,052	286	±5.79%
PRIVATE TOURIST ACCOMMODATION		2,794,675	731	±3.62%
SAME-DAY VISITORS		2,138,782	1,411	±2.61%
(*) estimated				

August has been the month with more conducted interviews, since it is the month having a greater tourist demand in Asturias. More concretely, 946 interviews have been conducted. In the two following subsections Corollaries 1 and 2 are applied on the data collected by SITA in the Principality of Asturias along August 2016.

Table 2 Distribution of nigths spent in hotels (Source: SITA)

# Nights	Percentage
1	6.31
2	15.09
3	13.91
4	17.23
5	10.79
6	10.00
7	13.85
8	2.70
9	2.53
10	3.00
11	0.26
12	1.82
14	0.57
15	1.37
21	0.57
Total	100.00

4.1 Determining the Average Stay Length

The primary information to determine the average stay length corresponds to the distribution of the number of nights visitors have spent. Data for tourists who are hosted in hotels have been collected in Table 2.

The sample mean of nigths spent in hotels, without considering the length-time bias, equals 4.98 nights, that is, the average length of stay of visitors in the hotel sector is 5.98 days.

If this computation is made in accordance with Eq. (1), so that the length-time bias is taken into account, the average stay length equals 4.72 days.

The same computations can be performed for tourists hosted in either collective or private tourist establishments. In the first case the average stay is equal to 9.19 days if the longitudinal bias is not considered, and to 8.76 days otherwise. In the second case, these means equal 13.06 days and 12.43 days, respectively.

4.2 Computing the Proportion of Visitors Hosted in Private Tourist Establishments

A very relevant information in analyzing the tourism is the one associated with data about the number of people who use private tourist accommodation. For this purpose, it is necessary to estimate the percentage of tourists spending nights in Asturias in this type of establishment.

The primary information available is the number of interviews conducted in each accommodation modality (Table 3).

If the sample proportion is used to estimate the proportion of tourists visiting Asturias and spending nights in private tourist establishments, the estimate is 0.2721.

In case the length-time bias is taken into account, by using Eq. (3) the estimate is given by

$$\frac{8.76 \times 203}{8.76 \times 203 + 12.43 \times 543} = 0.2085.$$

5   Implications from the Length-Time Bias

Involving the longitudinal bias in the usual estimates concerning the tourism framework entails outstanding effects in computing the macroeconomic touristic figures.

Regarding the accommodation establishments in the Principality of Asturias, tourism revenues for August 2016 can be determined by multiplying the average stay length by factors in Table 4.

Outputs depending on whether or not the longitudinal bias is taken into account, are shown in Table 5. On the basis of these outputs one can conclude that involving length-time bias in estimating revenues leads to an overestimation of income equal to 30.05 million Euros. And this only refers to August. For the purpose of developing the Tourism Satellite Account, the annual revenues should be considered.

Table 3 Number of interviews according to the type of accommodation (Source: SITA)

Accommodation type	# Conducted interviews	Percentage
Collective	543	72.79
Private	203	27.21
Total	746	100.00

Table 4 Average daily expenditure and number of tourists in hotels in August 2016 (Source: SITA)

Average daily expenditure (in Euros)	92.61
# Tourists	257,502

Table 5 Tourism revenue in hotels in August 2016 (Source: compiled by authors)

Involving longitudinal bias	Revenues (Million of Euros)
NO	142.61
YES	112.56

Table 6 Number of tourists hosted in private accommodation in August 2016 (Source: compiled by authors)

Involving longitudinal bias	# Tourists with private accommodation
NO	102,728
YES	72,392

On the other hand, according to the National Institute of Statistics (INE),[2] there have been a total of 274,810 tourists who have been hosted in collective establishments during August 2016. The computation of the number of tourists who spent the night in private tourist accommodation is based on the percentage of visitors hosted in such establishments, as shown in Sect. 4.2. Depending on whether or not the longitudinal bias is considered, the estimates amount 72,392 or 102,728, respectively (see Table 6).

As an immediate implication from the results in the last table, if length-time bias is not taken into account, there is an overestimation of 30,336 tourists in August 2016. As for other parameters, this leads to a very optimistic estimation of annual revenues in the region (in the detailed case the ones coming from the tourism sector of private tourist accommodation).

6 Conclusions

Involving the length-time bias in the estimates of parameters associated with tourist surveys leads to significant discrepancies w.r.t. using classical estimation methods. The estimation on the basis of the sample mean makes sense when the simple random sampling is considered. Nevertheless this is not the usual way to proceed, since the selection of a visitor is usually assumed to directly depend on his/her days of touristic stay.

The suggested corrections in the model applied in this paper can be very convenient in elaborating the main macroeconomic tourism figures. This is due to the fact that many key variables such as the number of tourists, the average duration of their stays, the average expenditure, the percentages of tourism types, and so on, are importantly affected.

The literature on case studies involving the length-time bias is rather limited. In the field of tourism, this concept seems to be quite unknown. It would be therefore essential to disseminate it among agencies working with tourist surveys, as well as to open discussions on other models of estimation, such as the one applied in this work, to enhance the quality of tourism statistical surveys and their results.

---

[2]http://www.ine.es.

References

1. Pelikan S, Moskowitz M (1993) Effects of lead time, length bias, and false-negative assurance on screening for breast cancer. Cancer 71:1998–2005
2. SITA: Sistema de Información Turística de Asturias (2017) El Turismo en Asturias en 2016. Oviedo: Consejería de Empleo, Industria y Turismo del Principado de Asturias and Universidad de Oviedo
3. Torres E, Menéndez JM, Domínguez JS (2001) Modelo descriptivo para el cálculo de medias en estadísticas turísticas a partir de observaciones exhaustivas durante un período. In: Proceeding XXVI Congreso Nacional de Estadística e Investigación Operativa, Universidad de Jaén
4. Torres E, Sustacha I, Menéndez J, Valdés L (2002) A solution to problems and disadvantages in statistical operations of surveys of visitors at accommodation establishments and at popular visitor places. In: Proceeding 6th international forum on tourism statistics, Hungarian Central Statistical Office, Budapest
5. Valdés L, de la Ballina J, Aza R, Torres E, Menéndez J, Domínguez JS, del Valle E (2001) A methodology to measure tourism expenditure and total tourism production at the regional level. In: Lennon JJ (ed) Tourism statistics international perspectives and current issues. Continuum, London
6. Yoshimoto Y, Tanaka Y (2008) Biological heterogeneity and length-biased sampling in asymptomatic neurosurgical patients. Br J Neurosurg 22:368–372

Detecting Change-Points in the Time Series of Surfaces Occupied by Pre-defined NDVI Categories in Continental Spain from 1981 to 2015

Ana F. Militino, M. Dolores Ugarte and Unai Pérez-Goya

Abstract The free access to satellite images since more than 40 years ago has provoked a rapid increase of multitemporal derived information of remote sensing data that should be summarized and analyzed for future inferences. In particular, the study of trends and trend changes is of crucial interest in many studies of phenology, climatology, agriculture, hydrology, geology or many other environmental disciplines. Overall, the normalized difference vegetation index (NDVI), as a satellite derived variable, plays a crucial role because of its usefulness for vegetation and landscape characterization, land use and land cover mapping, environmental monitoring, climate change or crop prediction models. Since the eighties, it can be retrieved all over the world from different satellites. In this work we propose to analyze its temporal evolution, looking for breakpoints or change-points in trends of the surfaces occupied by four NDVI classifications made in Spain from 1981 to 2015. The results show a decrease of bare soils and semi-bare soils starting in the middle nineties or before, and a slight increase of middle-vegetation and high-vegetation soils starting in 1990 and 2000 respectively.

1 Introduction

Since Compton Tucker [32] showed in 1979 that the Normalized Difference Vegetation Index (NDVI), generated from NOAA Advanced Very High Resolution Radiometer (AVHRR) could be used to map land cover and monitor vegetation

A. F. Militino (✉) · M. D. Ugarte · U. Pérez-Goya
Department of Statistics and O.R., Public University of Navarre, 31006 Pamplona, Spain
e-mail: militino@unavarra.es

U. Pérez-Goya
e-mail: unai.perez@unavarra.es

A. F. Militino · M. D. Ugarte
InaMat (Institute for Advanced Materials), Public University of Navarre,
31006 Pamplona, Spain
e-mail: lola@unavarra.es

© Springer International Publishing AG 2018
E. Gil et al. (eds.), *The Mathematics of the Uncertain*, Studies in Systems,
Decision and Control 142, https://doi.org/10.1007/978-3-319-73848-2_28

295

changes and desertification at continental and global scales, numerous research projects and studies have been carried out. Many of them through the analysis of NDVI3g time series. This is the third generation of images from the AVHRR sensors, retrieved from the framework of the Global Inventory Monitoring and Modeling System (GIMMS).

GIMMS NDVI3g data have been widely used during the last decades for studying large scale changing trends along years, mainly over continental or semi-continental regions. Its actual resolution of 8 Km at the Equator or 1/12 degrees is an attractive feature for monitoring changes of vegetation at any scale.

These images are not raw images, but bi-weekly composite images by means of the Maximum Value Compositing (MVC) procedure [12]. This technique suppresses clouds, atmospheric and radiometric effects and reduces the directional reflectance and off-nadir viewing effects. The result is a smaller number of output images with regard to the original ones, but with better quality and where the spatial and temporal stochastic dependence is still present.

The Normalized Difference Vegetation Index (NDVI) reflects vegetation growth and it is closely related to the amount of photosynthetically absorbed active radiation as indicated in [33]. It is calculated using the radiometric information obtained for the red (R) and near-infrared (NIR) wavelengths of the electromagnetic spectrum as $NDVI = ((NIR) - R)/((NIR) + R)$. See [27] for more details. Although numerical limits of NDVI can vary for the vegetation classification, it is widely accepted that negative NDVI values correspond to water or snow, NDVI values close to zero correspond to bare soils, sparse vegetation is approximately between 0.2–0.5, middle vegetation has thresholds between 0.5 and 0.7 and dense vegetation such as that found in forests or crops at their peak growth stage presents NDVI values between 0.7 and 1.0.

Mann–Kendall non-paramteric test is one of the most broadly used methods for parametric changes in time series of NDVI pixels. See [14, 21], or [30] as examples of the use of this test. When plotting significant changes, a patchy map can be obtained because every pixel is analysed separately. Figure 1 shows in gray-coloured pixels the Mann–Kendall statistic corresponding to significant changes in trend when this test is applied to the continental Spain from July 1981 to December 2015.

Mann–Kendall test only assumes a time dependence within the same pixel across years, but it does not encompass the spatial dependence among neighbour pixels. Therefore, close locations can present different trend changes, something questionable in real situations. Some improvements of this test have been also provided [23]. For example, [24] introduces the contextual Mann–Kendall approach that removes serial correlation through a prewhitening process. To detect spatio-temporal change-points in the NDVI trend is not a trivial task because of the different scales and dependencies between space and time, therefore we propose to aggregate pixels between pre-defined thresholds of NDVI values for estimating the occupied land cover area. Specifically, a total of 4 categories are obtained.For everyone of these

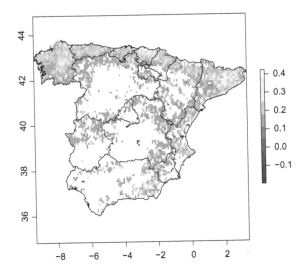

Fig. 1 Mann–Kendall test applied to NDVI3g data. Coloured pixels correspond to significant changes in trends obtained from July 1981 to December 2015

categories we have a time series of areas where different change-point methods will be applied to detect breakdown points in means and variances.

The work is divided in four sections. Section 2 explains the source of remote sensing data and the way of downloading. In Sect. 3 the different methods used to detect trend changes are briefly explained. Section 4 describes the results. The paper finishes with some conclusions.

2 Data

Remote sensing data were captured from the GIMMS NDVI3g images during the period July 1981–December 2015.

More details of GIMMS NDVI3g can be found in [25]. It has been largely used along recent years, for example in [1, 20] or [37]. The data have flags accounting for additional pixel-by-pixel information about its quality. These flags can vary between 1 and 7, where 1 or 2 indicates good quality, numbers between 3 and 6 indicate different kinds of processing, and 7 indicates missing data. GIMMS NDVI3g data are bi-weekly composite NDVI data set and it has shown to be more accurate than the GIMMS NDVI predecessors for monitoring vegetation activity and phenological change [36]. GIMMS NDVI3g data can be downloaded from http://ecocast.arc.nasa.gov/data/pub/gimms/3g.v1/. For this study, we have downloaded 828 bi-weekly images, but to preserve space Figs. 2 and 3 provide the monthly averages of NDVI3g in continental Spain for the first and second semesters respectively from 2011 to 2015.

These images have been cropped, projected and plotted in the free statistical software R [26]. In particular, library gimms [9] has been used for downloading the

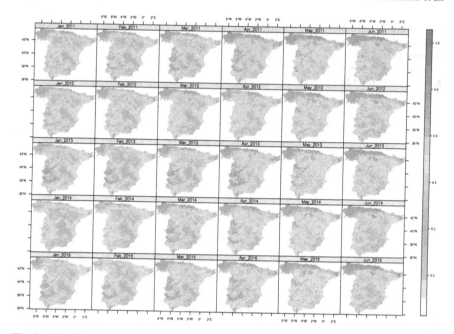

Fig. 2 GIMMS NDVI3g monthly averaged data corresponding to the six first months from 2011 to 2015

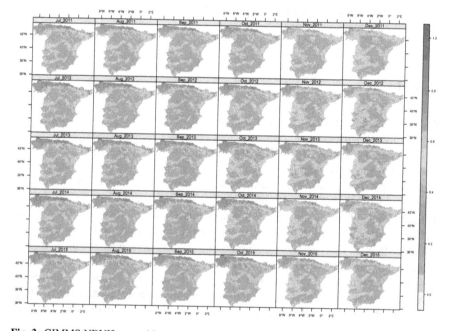

Fig. 3 GIMMS NDVI3g monthly averaged data corresponding to the six last months from 2011 to 2015

images and importing in R, yet it can also be done with `raster` library [11]. In green colors NDVI3g values closer to 1 are depicted and in brown colors the values closer to 0. In these maps, the pattern of high vegetation in the North and the Central West part of the country is predominant while middle vegetation is concentrated mainly in the watercourse of two important rivers: Guadiana and Guadalquivir and some mountain ranges. In the second semester low vegetation is predominant in the central part of Spain. This seasonality must be removed before applying change-point detection techniques.

3 Change-Point Methods

Change-points methods refer to the inference of a change in distribution for a set of observations. An excellent reference for these procedure is given in [7]. They arose in the 1950's from the process of quality control, and yet there were developed for independent and identically distributed random variables [8], the expansion to the time-ordered observations [2] was immediate. However, the application of these methods to remote sensing data is still very rare. In this work we compare four specific R packages for solving the change-point detection problem in time series of land cover areas in Spain from 1981 to 2015.

Change-Point Package: Segmented Neighborhood, Binary Segmentation and PELT

The `changepoint` package [19] contains three methods for multiple change-point detection in addition to a variety of test statistics. The change can be either in mean and/or variance settings with a similar argument structure. The implemented methods are: Segmented neighborhood, binary segmentation and PELT. See [17] for details. Binary segmentation [10, 28, 29] first applies a single change-point test statistic to the entire data. If a change-point is identified, the data is split into two at the change-point location. The single change-point procedure is repeated on the two new data sets, before and after the change. If change-points are identified in either of the new data sets, they are split further. This process continues until no change-points are found in any parts of the data. The splitting is based on likelihood ratio-tests similar to those used in cluster analysis. The segment neighborhood algorithm was proposed by [4, 5]. The algorithm minimizes a penalized expression of cost using a dynamic programming technique to obtain the optimal segmentation for $m + 1$ change-points reusing the information that was calculated for m change-points. The PELT algorithm [18] is similar to the segment neighborhood algorithm in that it provides an exact segmentation. It is computationally more efficient, due to its useof dynamic programming and pruning. The test statistics are likelihood ratio tests that can be applied to different families of distributions [7].

ecp Package: Divisive and Agglomerative Algorithms

The ecp package [13] contains two algorithms: divisive and agglomerative. These algorithms come from hierarchical cluster analysis and detect changes within the marginal distributions. They do not make any assumption regarding the nature of the change in distribution or any distribution assumptions beyond the existence of the αth absolute moment, for some $\alpha \in (0, 2)$. The agglomerative algorithm estimates change point locations through an optimal segmentation. Both approaches are able to detect any type of distributional change within the data. The divisive method provides consistent estimates of both the number and location of change points, under standard regularity assumptions. These methods also deal with the nonparametric multiple change point analysis of multivariate data. Regardless of the dimension, the nonparametric estimation can be done for both the number of change points and the positions at which they occur. These procedures have been widely used in financial modeling [31], and bioinformatics [22] to identify genes that are associated with specific cancers and other diseases or to detect credit card fraud [6].

bfast Package: Breaks For Additive Seasonal and Trend

The more specific R programm to manage with change-point detection in time series of satellite images is BFAST [34, 35]. BFAST is the acronym of "Breaks For Additive Seasonal and Trend" that integrates the decomposition of time series into trend, seasonal, and remainder components with methods for detecting change within time series. It iteratively estimates the time and number of changes characterizing the change by its magnitude and direction and using harmonic seasonal model requiring few observations.

strucchange Package: Generalized Fluctuation and F Tests

The strucchange package [40] contains Generalized fluctuation and F test for structural change in linear regression models. Here, the null hypothesis of "no structural change" is tested against the alternative that the coefficient vector varies over time for certain patterns of deviation from the null hypothesis. Significance can be also assessed through various tests. See [38, 39] for details.

4   Results

For everyone of the 828 images, NDVI3g values are assigned to 4 categories: bare soils (ndvi1) for values between 0 and 0.2, sparse vegetation (ndvi2) for values greater than 0.2 and less or equal than 0.5, middle vegetation (ndvi3) for values greater than 0.5 and less or equal than 0.7 and dense vegetation (ndvi4) for values greater than 0.7. Table 1 shows the average occupied area (in %) for the four NDVI3g classifications estimated in the continental Spain and in 15 regions. The classified areas have been

Table 1 Average percentage of the area occupied by the 4 pre-defined NDVI3g categories estimated in Spain and in 15 regions from 1981 to 2015

	num	ndvi1	ndvi2	ndvi3	ndvi4
Andalucía (An)	1	4.87	76.27	18.04	0.82
Aragón (Ar)	2	3.70	78.78	17.20	0.32
Cantabria (Ca)	3	0.19	7.90	47.68	44.23
Castilla-La Mancha (Cm)	4	5.66	81.62	12.59	0.13
Castilla y León (Cl)	5	1.60	64.57	32.00	1.83
Cataluña (Ct)	6	1.41	48.80	44.13	5.65
Comunidad de Madrid (Ma)	7	2.78	75.95	21.21	0.06
Comunidad Foral de Navarra (Na)	8	1.32	49.53	32.57	16.58
Comunidad Valenciana (Va)	9	1.76	83.04	15.20	0.00
Extremadura (Ex)	10	1.16	61.99	35.02	1.82
Galicia (Ga)	11	0.01	3.37	55.06	41.57
La Rioja (Ri)	12	0.28	62.04	33.18	4.49
País Vasco (Pv)	13	0.05	13.95	46.11	39.89
Principado de Asturias (As)	14	0.29	5.76	44.52	49.42
Región de Murcia (Mu)	15	9.67	90.15	0.18	0.00
Spain	16	3.01	64.34	26.29	6.36

calculated summing the number of pixels in these categories and multiplying by 65.95 km^2, the mean surface by pixel. In the whole territory the average percentage of bare soils is estimated in 3%, the average percentage of sparse vegetation is 64%, the average percentage of middle vegetation is 26% and 6% is the average percentage of dense vegetation. The 100% corresponds to the 504.537 km^2 of continental Spain.

Table 2 gives the two last figures of the change-point years detected by the four methods in Spain and by regions. Columns cp1, cp2, cp3 and cp4 show the year of the first detected change-point by changepoint package in ndvi1, ndvi2, ndvi3 and ndvi4 categories respectively. Columns str1, str2, str3 and str4 show the year of the first detected change-point by strucchange method in the same four categories. Similarly, columns bf1, bf2, bf3 and bf4 for bfast package. Columns prun1, prun2, prun3 and prun4 do the same with ecp package and finally, prun shows the year of the detected change-point in the overall NDVI3g. Empty places correspond to an absence of change-point.

In Spain, cp and str methods exactly coincide detecting the year of change-point in ndvi1 (year 1996) and ndvi3 (year 2000) categories. They do not coincide in ndvi2 (years 1993 and 1986) and they roughly coincide in ndvi4 (year 1989–1990), the upper category. Likely, both methods are the best candidates to explain the performance of the Spanish land cover change between July 1981 and December 2015. Figures 4 and 5 plot the detected change-points over the seasonally adjusted

Table 2 Years of change-points detected in the overall NDVI3g, and in the four pre-defined categories with `changepoint` (cp), `structchange` (str), `bfast`(bf) and `ecp`(prun) methods calculated in continental Spain and in 15 regions

regions		ndvi1				ndvi2				ndvi3				ndvi4				ndvi
		cp1	str1	bf1	prun1	cp2	str2	bf2	prun2	cp3	str3	bf3	prun3	cp4	str4	bf4	prun4	prun
(An)	1	96	96	91	86	12	10	92	13	0	0	91	13	4	1		13	86
(Ar)	2	97	86	91	86	7	7	2	13	96	96	5	10	6	10	2	13	86
(Ca)	3	87	87		13	88	88	99	86	90	90		13	90	90	98	86	13
(Cm)	4	96	96		86	10	10		13	0	96		85	4	3		13	86
(Cl)	5	94	96		86	93	86		86	93	86		13	96	97		13	86
(Ct)	6	89	88	93	86	89	89	98	86	89	89	92	86	88	88	98	12	13
(Ma)	7	97	96		86	86	86		13	86	86	92	13	87			13	86
(Na)	8	89	96	96	90	91	89	99	13	86	87		13	91	94		13	90
(Va)	9	0	0	91	1	7	7	91	12	6	6	91	12	91			13	1
Ex	10	95	95		13	86	86	4	13	86	96		13	96	1		12	13
Ga	11	99	86		13	89	89	94	86	90	90	99	85	90	90	99	13	13
Ri	12	97	87		13	89	88		13	88	88		13	97	97		13	13
Pv	13	87	87	92	86	89	89	91	13	89	94		13	89	89	97	86	86
As	14	91	86		13	89	88	91	86	89	89	94	13	89	89	97	13	13
Mu	15	86	86	91	86	86	86	91	86	7	7	91	13	15			13	86
Spain	16	96	96		86	93	86		86	0	0	91	13	90	89	99	12	86

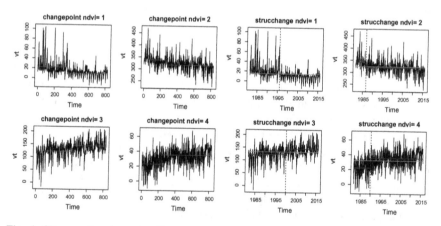

Fig. 4 Change-points in seasonally adjusted trends of GIMMS NDVI3g data obtained with `changepoint` and `strcchange` packages in the 4 pre-defined categories from 1981 to 2015

trends in the four pre-defined categories in Spain from 1981 to 2015 with the 4 methods. The proximity between `changepoint` and `strucchange` methods is clear in all categories except in ndvi2 where `changepoint` method is more exigent and conservative for detecting change-points. Both methods show a decreasing trend in the lowest categories (ndvi1, ndvi2), but an increased trend in the upper categories (ndvi3 and ndvi4). Package `bfast` does not detect any change-point neither in ndvi1 nor in ndvi2 but it detects changes in ndvi3 and ndvi4. The `prun` method estimates

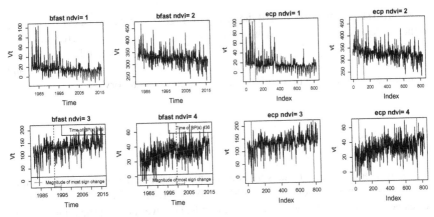

Fig. 5 Change-points in seasonally adjusted trends of GIMMS NDVI3g data obtained with `bfast` and `ecp` packages in the 4 pre-defined categories from 1981 to 2015

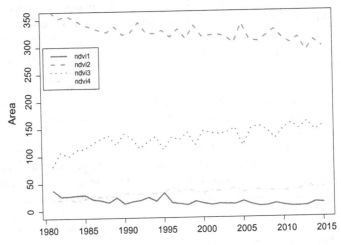

Fig. 6 Yearly averages of areas corresponding to the 4 pre-defined categories (ndvi1, ndvi2, ndvi3 and ndvi4) in Spain from 1981 to 2015

change-points in the beginning or in the last years, so it seems to be very sensible to small changes. Unfortunately, in the majority of regions, the year of the detected change-point do not coincide neither in methods nor in the categories, although the bigger the regions, the better the approximation.

Figure 6 shows the evolution of the yearly average area in hectares, corresponding to the 4 categories in Spain from the same studied period. Clearly, a decreasing trend in bare and semi-arid soil is observed, corresponding to ndvi1 and ndvi2 categories, an important increase of middle vegetated soil corresponding to ndvi3 category, and a small increase of trend in ndvi4, the dense vegetation category.

5   Conclusions

Nowadays, satellite remote sensing is a common instrument for detecting changes in land cover surfaces over time. GIMMS NDVI3g provides a world wide long time series very useful for analysing temporal trends, however and yet the quality of the series have been improved with regard to the old NDVI series, it is known that there is no concordance with other NDVI images coming from alternative sources, as MODIS TERRA or MODIS AQUA. See [3, 16] where a detailed comparison has been made in Central Europe from 2000 to 2013. Moreover, as long as we down-scaling the spatial resolution, more inaccurate estimations we obtain. From this perspective, we can say that the change-points trends found in Spanish regions are only approximate for small regions such as Navarra, Asturias, Murcia, País Vasco o Aragón. The main advantage of using GIMMS NDVI3g is that is the longest NDVI series of images with 34 years, already pre-processed, easily accessible, and from all over the world. As long as we can retrieve longest series of high spatial resolution satellite images, these results could change.

The variety of methods found in the literature for detecting change-points in ordered observations is large, and they do not necessarily provide the same points. At this regard, we consider that matching results at different categories tip the balance in the changepoint and strucchange favor. Unfortunately, this methodology cannot determine the locations where these changes have been produced, because we loose the spatial location as long as we aggregate different pixels between the same thresholds. For this aim we need not only to develop a specific spatio-temporal methodology but also a larger spatial resolution of time series of images. For example, Sentinel 2 A can provide the required spatial resolution, however, the history of these images is still too short for being reliable in the time series analysis.

There is an inherent difficulty in checking the performance of this result, because there are not previous studies similar to this one. Perhaps, this step can only be done looking for vegetation changes previously documented. The most relevant is [15] where the authors investigate the NDVI changes in trends happened in Iberian peninsula between 1981 and 2001, using GIMMS NDVI3g data with a pixel by pixel approach. The interpretation of global trends in the peninsula is limited, although the results show a slight desertification in Iberian Mountains, but is dated more than 15 years ago.

Based on the results given in changepoint and strucchange methods we can finally conclude that the detected change-points in Spain show a decrease of bare soils and semi-bare soils starting in the middle nineties or a bit before, and a slight increase of middle-vegetation and high-vegetation soils starting in 1990 and 2000 respectively. Further research is needs to confirm the results found in this pilot study.

Final Note

This work is a tribute from the Department of Statistics and Operations Research of the Public University of Navarre to our dear colleague Pedro Gil Álvarez, for his remarkable contribution to the area of Statistics and Probability, his generosity with his mates and his unforgettable and keen sense of humor.

Acknowledgements This research was supported by the Spanish Ministry of Economy, Industry, and Competitiveness (project MTM2017-82553-R) jointly financed with the European Regional Development Fund (FEDER), the Government of Navarre (PI015-2016 and PI043-2017 projects) and the Fundación CAN-Obra Social Caixa 2016.

References

1. Ahmed M, Else B, Eklundh L, Ard J, Seaquist J (2017) Dynamic response of ndvi to soil moisture variations during different hydrological regimes in the sahel region. Int J Remote Sens 38(19):5408–5429
2. Antoch J, Hušková M, Prášková Z (1997) Effect of dependence on statistics for determination of change. J Stat Plan Inference 60(2):291–310
3. Atzberger C, Klisch A, Mattiuzzi M, Vuolo F (2013) Phenological metrics derived over the european continent from ndvi3g data and modis time series. Remote Sens 6(1):257–284
4. Auger IE, Lawrence CE (1989) Algorithms for the optimal identification of segment neighborhoods. Bull Math Biol 51(1):39–54
5. Bai J, Perron P (2003) Critical values for multiple structural change tests. Econ J 6(1):72–78
6. Bolton RJ, Hand DJ (2002) Statistical fraud detection: a review. Stat Sci 17(3):235–249
7. Chen J, Gupta AK (2011) Parametric statistical change point analysis: with applications to genetics, medicine, and finance. Springer, Heidelberg
8. Csörgö M, Horváth L (1997) Limit theorems in change-point analysis, vol 18. Wiley, New York
9. Detsch F (2016) Gimms: download and process GIMMS NDVI3g data. https://CRAN.R-project.org/package=gimms
10. Edwards AW, Cavalli-Sforza LL (1965) A method for cluster analysis. Biometrics 21(2):362–375
11. Hijmans RJ (2015) Raster: geographic data analysis and modeling. https://CRAN.R-project.org/package=raster
12. Holben BN (1986) Characteristics of maximum-value composite images from temporal avhrr data. Int J Remote Sens 7(11):1417–1434
13. James NA, Matteson DS (2014) ecp: an R package for nonparametric multiple change point analysis of multivariate data. J Stat Softw 62(7):1–25
14. de Jong R, de Bruin S, de Wit A, Schaepman ME, Dent DL (2011) Analysis of monotonic greening and browning trends from global ndvi time-series. Remote Sens Environ 115(2):692–702
15. Julien Y, Sobrino JA, Mattar C, Ruescas AB, Jiménez-Muñoz JC, Sòria G, Hidalgo V, Atitar M, Franch B, Cuenca J (2011) Temporal analysis of normalized difference vegetation index (ndvi) and land surface temperature (lst) parameters to detect changes in the iberian land cover between 1981 and 2001. Int J Remote Sens 32(7):2057–2068

16. Kern A, Marjanović H, Barcza Z (2016) Evaluation of the quality of ndvi3g dataset against collection 6 modis ndvi in central Europe between 2000 and 2013. Remote Sens 8(11):955

17. Killick R, Eckley IA (2014) Changepoint: an R package for changepoint analysis. J Stat Softw 58(3):1–19

18. Killick R, Fearnhead P, Eckley IA (2012) Optimal detection of changepoints with a linear computational cost. J Am Stat Assoc 107(500):1590–1598

19. Killick R, Haynes K, Eckley IA (2016) Changepoint: an R package for changepoint analysis. https://CRAN.R-project.org/package=changepoint

20. Li H, Wang C, Zhang L, Li X, Zang S (2017) Satellite monitoring of boreal forest phenology and its climatic responses in Eurasia. Int J Remote Sens 38(19):5446–5463

21. Li Z, Huffman T, McConkey B, Townley-Smith L (2013) Monitoring and modeling spatial and temporal patterns of grassland dynamics using time-series modis ndvi with climate and stocking data. Remote Sens Environ 138:232–244

22. Matteson DS, James NA (2014) A nonparametric approach for multiple change point analysis of multivariate data. J Am Stat Assoc 109(505):334–345

23. Militino AF, Ugarte MD, Pérez-Goya U (2017) Stochastic spatio-temporal models for analysing ndvi distribution of gimms ndvi3g images. Remote Sens 9(1):76

24. Neeti N, Eastman JR (2011) A contextual Mann–Kendall approach for the assessment of trend significance in image time series. Trans GIS 15(5):599–611

25. Pinzon JE, Tucker CJ (2014) A non-stationary 1981–2012 avhrr ndvi3g time series. Remote Sens 6(8):6929–6960

26. R Core Team (2017) R: a language and environment for statistical computing. R foundation for statistical computing, Vienna, Austria. https://www.R-project.org/

27. Rouse J Jr, Haas R, Schell J, Deering D (1974) Monitoring vegetation systems in the great plains with erts. NASA spec publ 351:309

28. Scott AJ, Knott M (1974) A cluster analysis method for grouping means in the analysis of variance. Biometrics 30(3):507–512

29. Sen A, Srivastava MS (1975) On tests for detecting change in mean. Ann Stat 3(1):98–108

30. Sobrino JA, Julien Y, Morales L (2011) Changes in vegetation spring dates in the second half of the twentieth century. Int J Remote Sens 32(18):5247–5265

31. Talih M, Hengartner N (2005) Structural learning with time-varying components: tracking the cross-section of financial time series. J R Stat Soc Ser B 67(3):321–341

32. Tucker CJ (1979) Red and photographic infrared linear combinations for monitoring vegetation. Remote Sens Environ 8(2):127–150

33. Tucker CJ, Pinzon JE, Brown ME, Slayback DA, Pak EW, Mahoney R, Vermote EF, El Saleous N (2005) An extended avhrr 8-km ndvi dataset compatible with modis and spot vegetation ndvi data. Int J Remote Sens 26(20):4485–4498

34. Verbesselt J, Hyndman R, Newnham G, Culvenor D (2010a) Detecting trend and seasonal changes in satellite image time series. Remote Sens Environ 114(1):106–115

35. Verbesselt J, Hyndman R, Zeileis A, Culvenor D (2010b) Phenological change detection while accounting for abrupt and gradual trends in satellite image time series. Remote Sens Environ 114(12):2970–2980

36. Wang J, Dong J, Liu J, Huang M, Li G, Running SW, Smith WK, Harris W, Saigusa N, Kondo H, Liu Y, Hirano T, Xiao X (2014) Comparison of gross primary productivity derived from gimms ndvi3g, gimms, and modis in southeast Asia. Remote Sens 6(3):2108–2133

37. Yuan X, Li L, Chen X, Shi H (2015) Effects of precipitation intensity and temperature on ndvi-based grass change over northern china during the period from 1982 to 2011. Remote Sens 7(8):10164–10183

38. Zeileis A (2006) Implementing a class of structural change tests: an econometric computing approach. Comput Stat Data Anal 50:2987–3008

39. Zeileis A, Kleiber C, Krämer W, Hornik K (2003) Testing and dating of structural changes in practice. Comput Stat Data Anal 44:109–123
40. Zeileis A, Leisch F, Hornik K, Kleiber C (2002) Strucchange: an R package for testing for structural change in linear regression models. J Stat Softw 7(2):1–38

Choice Functions and Rejection Sets

Enrique Miranda, Arthur Van Camp and Gert de Cooman

Abstract We establish an equivalent representation of coherent choice functions in terms of a family of rejection sets, and investigate how each of the coherence axioms translates into this framework. In addition, we show that this family allows to simplify the verification of coherence in a number of particular cases.

1 Introduction

Coherent choice functions constitute an uncertainty model that is more general than sets of desirable gambles, while still preserving some of their nice properties, such as being able to deal effectively with sets of probability zero when conditioning. One of their drawbacks is the technical difficulty of verifying the coherence axioms. In this paper, we try to remedy this situation somewhat by providing an equivalent representation of choice functions in terms of those option sets that allow a subject to reject the zero gamble, which may be interpreted as those option sets that he should consider preferable to the status quo. As we shall see, this representation, in addition to capturing more intuitively the ideas underlying coherence, also helps to simplify the verification of coherence in a number of particular cases.

This paper is organized as follows: in Sect. 2, we recall the basic aspects of coherent choice functions that we shall need in the rest of the paper. Our representation in terms of rejection sets is established in Sect. 3, where we also discuss two additional properties that seem of interest for choice functions. In Sect. 4, we look in more detail at a number of particular cases: choice functions on binary spaces

E. Miranda (✉)
Department of Statistics and Operations Research Federico García Lorca, University of Oviedo, 18 33007 Oviedo, Spain
e-mail: mirandaenrique@uniovi.es

A. Van Camp · G. de Cooman
IDLab Technologiepark, Ghent University, 914 9052 Zwijnaarde (Gent), Belgium
e-mail: arthur.vancamp@ugent.be

G. de Cooman
email: gert.decooman@ugent.be

© Springer International Publishing AG 2018
E. Gil et al. (eds.), *The Mathematics of the Uncertain*, Studies in Systems, Decision and Control 142, https://doi.org/10.1007/978-3-319-73848-2_29

(that is, when the experiment on which the outcomes of the options depend on can only take two values) and those defined by means of coherent sets of desirable gambles. The paper concludes with some additional remarks in Sect. 5.

2 Coherent Choice Functions

Let Ω be a possibility space. A *gamble* on Ω is a bounded map $f : \Omega \to \mathbb{R}$. We denote by $\mathscr{L}$ the set of all gambles on Ω. Gambles will also be called *options*. For any two gambles f and g, we denote $f \le g$ if $f(\omega) \le g(\omega)$ for every ω in Ω, and we collect all the gambles f for which $f \le 0$ in $\mathscr{L}_{\le 0}$. We let $f < g$ if $f \le g$ and $f \ne g$, and collect all the gambles f for which $f < 0$ in $\mathscr{L}_{<0}$, and the gambles f for which $f > 0$ in $\mathscr{L}_{>0}$.

Choice functions are defined on finite collections of gambles. We collect all those collections in the set $\mathscr{Q}$.

Definition 2.1 A *choice function* C on a possibility space Ω is a map

$$C : \mathscr{Q} \to \mathscr{Q} \cup \{\emptyset\} : A \mapsto C(A) \text{ such that } C(A) \subseteq A.$$

We collect all the choice functions on Ω in $\mathscr{C}(\Omega)$, often denoted as $\mathscr{C}$ when the possibility space is clear from the context.

The idea underlying this simple definition is that a choice function C selects the set $C(A)$ of 'best' options in the *option set* A. Our definition resembles the one commonly used in the literature [1, 6, 8], except perhaps for an also not entirely unusual restriction to *finite* option sets [2, 5, 7].

Equivalently to a choice function C, we may consider its associated *rejection function* R, defined by $R(A) := A \setminus C(A)$ for all A in $\mathscr{Q}$. It returns the options $R(A)$ that are rejected -not selected- by C.

We focus here on a special class of choice functions, which we call *coherent*.

Definition 2.2 We call a choice function C on Ω *coherent* if for all A, A_1 and A_2 in $\mathscr{Q}$, all f and g in $\mathscr{L}$, and all λ in $\mathbb{R}_{>0}$[1]:

$C_1.$ $C(A) \ne \emptyset$;
$C_2.$ if $f < g$ then $\{g\} = C(\{f, g\})$;
$C_3.$ a. if $C(A_2) \subseteq A_2 \setminus A_1$ and $A_1 \subseteq A_2 \subseteq A$ then $C(A) \subseteq A \setminus A_1$;
      b. if $C(A_2) \subseteq A_1$ and $A \subseteq A_2 \setminus A_1$ then $C(A_2 \setminus A) \subseteq A_1$;
$C_4.$ a. if $A_1 \subseteq C(A_2)$ then $\lambda A_1 \subseteq C(\lambda A_2)$;
      b. if $A_1 \subseteq C(A_2)$ then $A_1 + \{f\} \subseteq C(A_2 + \{f\})$.

These axioms are a subset of the ones introduced in [6], duly translated from horse lotteries to gambles. We have omitted two of the coherence axioms from [6]: one is

---

[1]By $\mathbb{R}_{>0}$ we mean all the (strictly) positive real numbers.

the *Archimedean* axiom, because it is not fully compatible with the idea of deriving choice functions from coherent sets of desirable gambles [9], which is one of the goals in our approach. The other one, which we shall consider later on, is the so-called *convexity* axiom. Although this axiom leads to a number of useful properties, and in particular to a connection with lexicographic probability systems [10], we have refrained from including it in the list of coherence axioms because it is not satisfied by some interesting choice functions.

Equivalent formulations of these axioms, better suited for our subsequent proofs, are the following:

$$(C_3a) \Leftrightarrow \left(\forall A, A' \in \mathcal{Q}, \forall f \in A\right) \left(f \in R(A), A \subseteq A'\right) \Rightarrow f \in R\left(A'\right),$$
$$(C_3b) \Leftrightarrow (\forall A \in \mathcal{Q}, \forall f \in A) \{0, f\} \subseteq R(A) \Rightarrow 0 \in R(A \setminus \{f\}),$$
$$(C_3a) \Leftrightarrow (\forall A \in \mathcal{Q}, \forall \lambda > 0) \, R(\lambda A) = \lambda R(A),$$
$$(C_3b) \Leftrightarrow (\forall A \in \mathcal{Q}, \forall f \in \mathcal{L}) \, R(A + f) = R(A) + f.$$

3 A Representation in Terms of Rejection Sets

Next we give an equivalent representation of choice functions in terms of *rejection sets*. For any $f \in \mathcal{L}$ and any natural number i, we define

$$\mathbb{K}_f^i := \{A : f \in R(A), |A| = i\} \text{ and } \mathbb{K}_f := \cup_{i \in \mathbb{N}} \mathbb{K}_f^i. \tag{1}$$

We are going to characterize coherent choice functions in terms of these rejection sets. Our first result shows that we can restrict our attention to the case $f = 0$:

Proposition 3.1 *Let C be a choice function and consider the family of option sets* $\{\mathbb{K}_f : f \in \Omega\}$ *it induces by means of Eq. (1). Then*

$$C \text{ satisfies Axiom } C_4b \Leftrightarrow (\forall f \in \mathcal{L}) \, \mathbb{K}_0 + f = \mathbb{K}_f.$$

Proof For necessity, consider an option set A that includes 0. Then the option set $A + f$ includes f, and since by C_4b it holds that $R(A + f) = R(A) + f$, we conclude that $A \in \mathbb{K}_0$ if and only if $A + f \in \mathbb{K}_f$.

Conversely, for sufficiency, consider an option set A and a gamble f. Take any $g \in R(A)$, then $A \in \mathbb{K}_g$, whence by assumption $A - g \in \mathbb{K}_0$ and as a consequence $A - g + (f + g) = A + f \in \mathbb{K}_{f+g}$. Then indeed $g + f \in R(A + f)$, whence Axiom C_4b holds. $\qquad\square$

Taking this result into account, in what follows we shall restrict our attention to rejection sets $\mathbb{K}$ for which $\mathbb{K}_0 + f = \mathbb{K}_f$ for every f in $\mathcal{L}$. We can then simplify the notation above to

$$K^i := \mathbb{K}_0^i = \{A : 0 \in R(A), |A| = i\} \text{ and } K := \mathbb{K}_0 = \{A : 0 \in R(A)\}, \tag{2}$$

respectively, and denote $\mathcal{Q}_0$ the family of option sets that include the zero gamble. Our next result provides a characterisation of the different coherent axioms in terms of these sets:

Proposition 3.2 *Let C be a choice function satisfying Axiom C_4b, and consider the sets K^i, K defined in Eq. (2).*

(a) *C satisfies Axiom C_1 if and only if $(\forall A \in \mathcal{Q}_0)(\exists f \in A)\, A - f \notin K$.*

(b) *C satisfies Axiom C_2 if and only if $(\forall f \in \mathcal{L}_{>0})\{f, 0\} \subseteq K^2$.*

(c) *C satisfies Axiom C_3a if and only if $\left(\forall A \in K, \forall A' \in \mathcal{Q}_0\right)(A \subseteq A' \Rightarrow A' \in K)$.*

(d) *C satisfies Axiom C_3b if and only if $(\forall A \in K, \forall f \in A)$ $(A - f \in K \Rightarrow A \setminus \{f\} \in K)$.*

(e) *C satisfies Axiom C_4a if and only if $(\forall A \in \mathcal{Q}_0, \forall \lambda > 0)(A \in K \Leftrightarrow \lambda A \in K)$.*

Proof (a) Taking Axiom C_4b into account, Axiom C_1 holds if and only if $C(A) \neq \emptyset$ for every $A \in \mathcal{Q}_0$. This in turn is equivalent to $(\exists f \in A)\, f \in C(A)$, which by C_4b is equivalent to $0 \in C(A - f)$ or, in other words, to $A - f \notin K$.

(b) Under Axiom C_4b, Axiom C_2 is equivalent to $(\forall f \in \mathcal{L}_{>0})\{f\} = C(\{0, f\})$, or, in other words, to $(\forall f \in \mathcal{L}_{>0})\{f, 0\} \subseteq K^2$.

(c) For necessity, consider any A in K and any A' in $\mathcal{Q}_0$ such that $A' \supseteq A$. Because $A \in K, 0 \in R(A)$, whence, by Axiom C_3a, $0 \in R\left(A'\right)$. Then indeed $A' \in K$. Conversely, for sufficiency, consider any A and A' in $\mathcal{Q}_0$ such that $A \subseteq A'$, and any f in $R(A)$. Then by Axiom C_4b, $0 \in R(A - f)$, so $A - f \in K$, whence also $A' - f \in K$, because $A' - f \supseteq A - f$. Then $0 \in R\left(A' - f\right)$, and applying again C_4b, indeed $f \in R(A')$.

(d) For necessity, consider any A in K and f in A such that $A - f \in K$. Then $0 \in R(A - f)$, whence $f \in R(A)$, by Axiom C_4b. Applying Axiom C_3b, we deduce that $0 \in R(A \setminus \{f\})$, whence indeed $A \setminus \{f\} \in K$.
Conversely, for sufficiency, consider any A in $\mathcal{Q}$ and f in A such that $\{0, f\} \subseteq R(A)$. Then $A \in K$ and by Axiom C_4b, $f \in R(A)$ implies that $0 \in R(A - f)$, so $A - f \in K$. Then $A \setminus \{f\} \in K$, or, in other words, indeed $0 \in R(A \setminus \{f\})$.

(e) It suffices to note that under Axiom C_4b, Axiom C_4a is equivalent to $0 \in C(A) \Leftrightarrow 0 \in C(\lambda A)$ for every $\lambda > 0$ and every $A \in \mathcal{Q}_0$. $\qquad\square$

An immediate consequence is:

Corollary 3.1 *A choice function C is coherent if and only if it satisfies Axiom C_4b and the rejection set K it induces by Eq. (2) is increasing, scale invariant, includes $\{f, 0\}$ for every $f \in \mathcal{L}_{>0}$ and it satisfies the following two properties:*

- *$(\forall A \in \mathcal{Q}_0)(\exists f \in A)\, A - f \notin K$.*
- *$(\forall A \in K, \forall f \in A)(A - f \in K \Rightarrow A \setminus \{f\} \in K)$.*

Next we consider a couple of additional consistency axioms that were deemed interesting by [10]. The first one is the *convexity* axiom, which is given by:

C_5 if $A \subseteq A_1 \subseteq CH(A)$ then $C(A) \subseteq C(A_1)$, for all A and A_1 in $\mathcal{Q}$, where $C(A) := \left\{ \sum_{i=1}^{n} \alpha_i f_i : n \in \mathbb{N}, f_i \in A, \alpha_i \geq 0, \sum_{i=1}^{n} \alpha_i = 1 \right\}$ is the *convex hull* of A.

In terms of rejection sets, it is characterized by the following proposition:

Proposition 3.3 *Let C be a choice function satisfying Axiom C_4b. Then C satisfies Axiom C_5 if and only if $(\forall A_1 \in K, \forall A \in \mathscr{Q}_0)(A \subseteq A_1 \subseteq CH(A) \Rightarrow A \in K)$.*

Proof For necessity, application of Axiom C_5 tells us that, whenever $A \subseteq A_1 \subseteq CH(A)$ hold, $0 \in R(A_1)$ implies that $0 \in R(A)$, or, in other words, $A_1 \in K$ implies that $A \in K$.

Conversely, for sufficiency, consider two option sets A and A_1 such that $A \subseteq A_1 \subseteq CH(A)$, and let us show that $C(A) \subseteq C(A_1)$. Assume *ex absurdo* that there is some $f \in A$ such that $f \in R(A_1)$ and $f \in C(A)$. Then since $A - f \subseteq A_1 - f \subseteq CH(A - f)$, we can apply axiom C_4b and assume that, without loss of generality, $f = 0$. But then we obtain that $A_1 \in K$ while $A \notin K$, a contradiction. $\qquad\square$

A weaker property that is also useful is the so-called *separate homogeneity*, which means that for all n in $\mathbb{N}$, all $f_1, f_2, \ldots, f_n$ in $\mathscr{L}$ and all $\mu_1, \mu_2, \ldots \mu_n$ in $\mathbb{R}_{>0}$:

$$0 \in C(\{0, f_1, f_2, \ldots, f_n\}) \Leftrightarrow 0 \in C(\{0, \mu_1 f_1, \mu_2 f_2, \ldots, \mu_n f_n\}). \qquad (3)$$

This property follows from axioms C_3a, C_4a, C_5 [10, Proposition 1]. Moreover, and unlike C_5 that is linked to *lexicographic* choice functions, separate homogeneity is compatible with maximality as a decision rule, and therefore better suited for connecting choice functions with desirability. Furthermore, separate homogeneity is strictly weaker: there are classes of interesting coherent choice functions that satisfy Eq. (3) but not Axiom C_5. In terms of the rejection sets, it is trivial to prove that it can be expressed in the following manner:

Proposition 3.4 *Let C be a choice function satisfying Axiom C_4b. It satisfies separate homogeneity if and only if for all n in $\mathbb{N}$, all $f_1, f_2, \ldots, f_n$ in $\mathscr{L}$ and all $\mu_1, \mu_2, \ldots \mu_n$ in $\mathbb{R}_{>0}$, $\{0, f_1, f_2, \ldots, f_n\} \in K \Leftrightarrow \{0, \mu_1 f_1, \mu_2 f_2, \ldots, \mu_n f_n\} \in K$.*

4 Particular Cases

In this section, we consider a number of particular cases of choice functions for which the representation in terms of rejection sets simplifies somewhat.

4.1 Coherent Choice Functions Defined via Maximality

We begin by considering choice functions defined via Walley's notion of maximality [9]. A set of gambles $\mathscr{D}$ is called *coherent* when it is a convex cone that includes all non-negative gambles and does not include the zero gamble. We refer to [3, 4, 11] for a study of the notion of desirability and its variants. In particular, any coherent set of desirable gambles can be used to define a coherent choice function, by means of the formula

$$C_{\mathscr{D}}(A) := \{ f \in A : (\forall g \in A)\, g - f \notin \mathscr{D} \}. \tag{4}$$

Unlike general choice functions, the ones defined in the manner above are uniquely determined by binary comparisons. Thus, it is not surprising that for them the representation in terms of rejection sets takes a simpler form:

Proposition 4.1 *Let $\mathscr{D}$ be a coherent set of gambles and let $C_{\mathscr{D}}$ be the coherent choice function it induces by Eq. (4). Then $K = \big\{ A \in \mathscr{Q}_0 : (\exists A_1 \in K^2)\, A_1 \subseteq A \big\}$ and $K^2 = \{\{0, f\} : f \in \mathscr{D}\}$.*

Proof Consider an option set A in K. By Eq. (4), $0 \in R_{\mathscr{D}}(A)$ if and only if $A \cap \mathscr{D} \neq \emptyset$. If $|A| = 2$, then $A = \{0, f\}$ for some f in $\mathscr{D}$, and as a consequence $K^2 \supseteq \{\{0, f\} : f \in \mathscr{D}\}$. Conversely, consider any $A' \in K^2$. Then $A' = \{0, g\}$ for some g in $\mathscr{L}$. But since $0 \in R_{\mathscr{D}}(A')$, we have $g \in \mathscr{D}$, so $K^2 \subseteq \{\{0, f\} : f \in \mathscr{D}\}$, proving that indeed $K^2 = \{\{0, f\} : f \in \mathscr{D}\}$. If, on the other hand, $|A| \geq 3$, then $A \supseteq \{0, f\}$ for some f in $\mathscr{D}$. But then $0 \in R_{\mathscr{D}}(\{0, f\})$, so $A \supseteq A'$ for some $A' \in K^2$, and therefore indeed $K = \big\{ A \in \mathscr{Q}_0 : (\exists A_1 \in K^2)\, A_1 \subseteq A \big\}$. □

4.2 Coherent Choice Functions on Binary Spaces

Next, we consider coherent choice functions defined on binary spaces. It turns out that, under separate homogeneity, they are determined by rejection sets of cardinality two or three:

Proposition 4.2 *Let C be a coherent choice function on $\Omega = \{a, b\}$. If C satisfies Eq. (3), then*

$$K = \big\{ A \in \mathscr{Q}_0 : (\exists A_1 \in K^2 \cup K^3)\, A_1 \subseteq A \big\}.$$

Proof Let us prove that for every A in K there exists a A_1 in $K^2 \cup K^3$ for which $A_1 \subseteq A$.

Consider thus A in K. By Axiom C_2, we find that $A \cap \mathscr{L}_{<0} \subseteq R(A \cap \mathscr{L}_{\leq 0})$, so Axiom C_3a implies that then $A \cap \mathscr{L}_{<0} \subseteq R(A)$. Since $A \in K$ and therefore also $0 \in R(A)$, by Axiom C_3b we find that then $0 \in (A \cap \mathscr{L}_{<0}^c)$, so we can assume without loss of generality that $A \cap \mathscr{L}_{<0} = \emptyset$. There are two possibilities.

If $A \cap \mathscr{L}_{>0} \neq \emptyset$, then for any f in $A \cap \mathscr{L}_{>0}$ it follows from Axiom C_2 that $0 \in R(\{0, f\})$, whence the set $\{0, f\} \subseteq A$ belongs to K^2. So we find indeed that $A_1 := \{0, f\}$ in K^2 for which $A_1 \subseteq A$.

If $A \cap \mathscr{L}_{>0} = \emptyset$, then we can denote $A = \{f_1, \ldots, f_n, g_1, \ldots, g_m\}$ for some $n \geq 0$ and $m \geq 0$ but $\max\{m, n\} \geq 1$, where f_i belongs to the second quadrant (i.e., $f_i(a) < 0 < f_i(b)$) for every i in $\{1, \ldots, n\}$ and g_j belongs to the fourth quadrant (i.e., $g_j(a) > 0 > g_j(b)$) for every j in $\{1, \ldots, m\}$. Let $\lambda_i := \frac{-1}{f_i(a)}$ and $\mu_j := \frac{1}{g_j(a)}$ for every i in $\{1, \ldots, n\}$ and j in $\{1, \ldots, m\}$. Then, applying Eq. (3),

$$0 \in R(\{0, \lambda_1 f_1, \ldots, \lambda_n f_n, \mu_1 g_1, \ldots, \mu_m g_m\}).$$

Infer that $\lambda_i f_i(a) = -1$ for every $i \in \{1, \ldots, n\}$. Letting $i^* := \arg\max\{\lambda_i f_i(b) : i \in \{1, \ldots, n\}\}$, we infer that

$$\lambda_k f_k(b) < \lambda_{i^*} f_{i^*}(b) \Rightarrow \lambda_k f_k \in R(\{\lambda_k f_k, \lambda_{i^*} f_{i^*}\}) \Rightarrow \lambda_k f_k \in R(A),$$

where last implication follows from Axiom C_3a. Similarly, $\mu_j g_j(a) = 1$ for every $j \in \{1, \ldots, m\}$, and letting $j^* := \arg\max\{\mu_j g_j(b) : j \in \{1, \ldots, m\}\}$, we infer that

$$\mu_j g_j(b) < \mu_{j^*} g_{j^*}(b) \Rightarrow \mu_j g_j \in R(\{\mu_j g_j, \mu_{j^*} g_{j^*}\}) \Rightarrow \mu_j g_j \in R(A),$$

where again last implication follows from Axiom C_3a. If we now apply C_3b, we deduce that $0 \in R\left(\{0, \lambda_{i^*} f_{i^*}, \mu_{j^*} g_{j^*}\}\right)$, whence $0 \in R\left(\{0, f_{i^*}, g_{j^*}\}\right)$, applying Eq. (3). Thus, there is a subset of A with cardinality three that also belongs to K. $\square$

A key property in the proof of Proposition 4.2 is that separate homogeneity, together with Axiom C_2, allows to assume without loss of generality that an option set A that includes the zero gamble has at most one gamble f in the second quadrant (for which $f(a) < 0 < f(b)$) and one g in the fourth quadrant (for which $g(a) > 0 > g(b)$). Let us show that this does not necessarily happen without separate homogeneity:

Example 4.1 Consider $\Omega = \{a, b\}$ and let $\mathscr{D}$ be the coherent set of gambles $\mathscr{D} := \{f \in \mathscr{L} : f(a) < 0 < f(b) \text{ and } f(a) + f(b) > 0\} \cup \mathscr{L}_{>0}$. Let C be the choice function determined by the rejection function

$$0 \in R(A) \Leftrightarrow A \cap \mathscr{D} \neq \emptyset \text{ or } (\exists \lambda_1 > \lambda_2 > 0) \{(-\lambda_1, \lambda_1), (-\lambda_2, \lambda_2)\} \subseteq A \quad (5)$$

for all A in $\mathscr{Q}_0$. We extend the domain of R to $\mathscr{Q}$ by letting $f \in R(A) \Leftrightarrow 0 \in R(A - f)$ for all A in $\mathscr{Q}$ and f in A. Remark already that $(-\lambda, \lambda)$ lies on the border of $\mathscr{D}$ for every $\lambda > 0$: indeed, for every g in $\mathscr{D}$ we have that $(-\lambda, \lambda) + g \in \mathscr{D}$.

Let us show that C is a coherent choice function. Taking into account the last part of the definition, we see that C_4b holds, and we can restrict our attention to option sets in $\mathscr{Q}_0$. We show that C satisfies Axioms C_2, C_3a, C_3b, C_4a, and C_1, in this order.

For Axiom C_2, consider any f in $\mathscr{L}_{>0}$. Then $f \in \mathscr{D}$, so indeed $0 \in R(\{0, f\})$.

For Axiom C_3a, consider any A and A' in $\mathscr{Q}_0$ such that $A \subseteq A'$, and any f in $R(A)$. Using Axiom C_4b, then $0 \in R(A - f)$, whence $(A - f) \cap \mathscr{D} \neq \emptyset$ or $\{(-\lambda_1, \lambda_1), (-\lambda_2, \lambda_2)\} \subseteq A - f$ for some $\lambda_1 > \lambda_2 > 0$. But $A' - f \supseteq A - f$, so also $\left(A' - f\right) \cap \mathscr{D} \neq \emptyset$ or $\{(-\lambda_1, \lambda_1), (-\lambda_2, \lambda_2)\} \subseteq A' - f$, and therefore $0 \in R\left(A' - f\right)$, whence, again by Axiom C_4b, indeed $0 \in R\left(A'\right)$.

For Axiom C_3b, consider any A in $\mathscr{Q}_0$ and any f in A such that $\{0, f\} \subseteq R(A)$. We need to prove that then $0 \in R(A \setminus \{f\})$. Since $f \in R(A)$, then

(i) $\mathscr{D} \cap (A - f) \neq \emptyset$, or (ii) $\{(-\lambda_1, \lambda_1), (-\lambda_2, \lambda_2)\} \subseteq A - f$ for some $\lambda_1 > \lambda_2 > 0$.

Furthermore, since $0 \in R(A)$, then $\mathscr{D} \cap A \neq \emptyset$, or $\{(-\lambda_1, \lambda_1), (-\lambda_2, \lambda_2)\} \subseteq A$ for some $\lambda_1 > \lambda_2 > 0$. If $f \notin \mathscr{D} \cap A$ and $f \notin \{(-\lambda_1, \lambda_1), (-\lambda_2, \lambda_2)\}$, then also $\mathscr{D} \cap A \setminus \{f\} \neq \emptyset$ or $\{(-\lambda_1, \lambda_1), (-\lambda_2, \lambda_2)\} \subseteq A \setminus \{f\}$, whence $0 \in R(A \setminus \{f\})$. So assume that (a) $f \in \mathscr{D}$ or (b) $f = (-\lambda, \lambda)$ for some $\lambda > 0$.

If (a) $f \in \mathscr{D}$, then (i) or (ii) must be the case. If (i) occurs, then there is some gamble g in $(\mathscr{D} + f) \cap A$, whence $g - f \in \mathscr{D}$ for some g in A. But since $f \in \mathscr{D}$, also $g = f + g - f \in \mathscr{D}$, and therefore $0 \in R(\{0, g\})$, whence by Axiom C_3a indeed $0 \in R(A \setminus \{f\})$. If (ii) occurs, then there are $\lambda_1 > \lambda_2 > 0$ such that $f + (-\lambda_1, \lambda_1), f + (-\lambda_2, \lambda_2) \in A$, whence, since $f \in \mathscr{D}$, by construction also $f + (-\lambda_1, \lambda_1) \in \mathscr{D}$. Therefore $0 \in R(\{0, f + (-\lambda_1, \lambda_1)\})$, whence by Axiom C_3a, also $0 \in R(A \setminus \{f\})$.

If (b) $f = (-\lambda, \lambda)$ for some $\lambda > 0$, then, similarly, (i) or (ii) must be the case. If (i) occurs, then there is some g in A such that $g - f \in \mathscr{D}$. Therefore by construction also $g = f + g - f \in \mathscr{D}$, whence $0 \in R(\{0, g\})$, and then by Axiom C_3a, also $0 \in R(A \setminus \{f\})$. If (ii) occurs, then there is some $\lambda_1 > 0$ and $\lambda_2 > 0$ for which $\{f + (-\lambda_1, \lambda_1), f + (-\lambda_2, \lambda_2)\} = \{(-\lambda - \lambda_1, \lambda + \lambda_1), f + (-\lambda - \lambda_2, \lambda + \lambda_2)\} \subseteq A$. Letting $\lambda_1' := \lambda + \lambda_1$ and $\lambda_2' := \lambda + \lambda_2$, we find that $\{(-\lambda_1', \lambda_1'), f + (-\lambda_2', \lambda_2')\} \subseteq A \setminus \{f\}$, whence $0 \in R(A \setminus \{f\})$.

Axiom C_4a follows from Eq. (5), taking into account that $\mathscr{D}$ is a cone.

Finally, for Axiom C_1, assume *ex absurdo* that $C(A) = \emptyset$ for some A in $\mathscr{Q}_0$. Then $A = R(A)$ whence, by Axiom C_3b, $0 \in R(\{0\})$. But $0 \notin \mathscr{D}$ and $(-\lambda, \lambda) \notin \{0\}$ for every $\lambda > 0$, so $0 \notin R(\{0\})$, a contradiction.

On the other hand, it follows by Eq. (5) that, given the option set $A = \{0, (-1, 1), (-2, 2)\}$, we obtain $C(A) = \{(-1, 1), (-2, 2)\}$. However, the same equation implies that $0 \in C(\{0, (-1, 1)\})$. This shows that C does not satisfy separate homogeneity, and also that we cannot reduce the intersection with the second quadrant to only one gamble.

On the other hand, Proposition 4.2 also depends crucially on the assumption that $|\Omega| = 2$, as our next example shows:

Example 4.2 Consider a ternary space Ω, some n in $\mathbb{N}$, and let f_k be the gamble given by $f_k := (-1, \frac{k}{n}, -\frac{k^2}{n^2})$, for all k in $\{1, \ldots, n\}$. Let us show that for each k we can find a probability measure whose expectation operator P_k (called *linear prevision* in Walley's terminology) satisfies $P_k(f_k) > 0 > P_k(f_j)$ for every $j \in \{1, \ldots, n\} \setminus \{k\}$.

To find such expectation operators, let P be the expectation operator associated with the mass function $(0, \frac{2k}{n+2k}, \frac{n}{n+2k})$. Then $P(f_k - f_j) = \frac{k-j}{n(n+2k)}(2k - (k + j))$, whence $P(f_k - f_j) > 0$ if $k \neq j$. Moreover, $P(f_k) = \frac{k^2}{n(n+2k)} > 0$.

If we now consider any $\lambda \in (0, 1)$ and define P_k as the expectation operator associated with the mass function $(\lambda, (1 - \lambda)\frac{2k}{n+2k}, (1 - \lambda)\frac{n}{n+2k})$, we obtain

$P(f_k - f_j) = (1 - \lambda)P(f_k - f_j) > 0$ whenever $k \neq j$. Moreover,

$$P_k(f_k) = -\lambda + (1 - \lambda)P(f_k) > 0 \Leftrightarrow \lambda < \frac{P(f_k)}{1 + P(f_k)},$$

and similarly

$$P_k(f_j) = -\lambda + (1 - \lambda)P(f_j) < 0 \Leftrightarrow \lambda > \frac{P(f_j)}{1 + P(f_j)}.$$

Since, for every $j \in \{1, \ldots, n\} \setminus \{k\}$, $\frac{P(f_j)}{1+P(f_j)} < \frac{P(f_k)}{1+P(f_k)}$ because $P(f_j) < P(f_k)$, we let

$$\lambda \in \left(\max_{j \in \{1,\ldots,n\} \setminus \{k\}} \frac{P(f_j)}{1 + P(f_j)}, \frac{P(f_k)}{1 + P(f_k)} \right),$$

and for this λ we obtain $P_k(f_k) > 0 > P_k(f_j)$ for every $j \neq k$.

Now, let $\mathscr{D}_k$ be the coherent set of gambles given by $\mathscr{D}_k := \{f \in \mathscr{L} : P_k(f) > 0\}$, and let $C_{\mathscr{D}_k}$ be the coherent choice function it induces by Eq. (4). Then the choice function C given by $C(A) := \bigcup_{k=1}^{n} C_{\mathscr{D}_k}(A)$ is also coherent [9, Proposition 3], and it can be checked to satisfy separate homogeneity because all $C_{\mathscr{D}_k}$ do. If we now consider the option set $A = \{0, f_1, \ldots, f_n\}$, we get that $C_{\mathscr{D}_k}(A) = \{f_k\}$ for every k, since $P_k(f_k) > 0 > P_k(f_j)$ implies that $f_k, f_k - f_j \in \mathscr{D}_k$ for every j. As a consequence, we obtain $C(A) = \{f_1, \ldots, f_n\}$, whence $A \in K$. However, for every k it holds that $C_{\mathscr{D}_k}(A \setminus \{f_k\}) = \{0\}$, using again that $P_k(f_j) < 0$ for every $j \neq k$, and therefore $C(A \setminus \{f_k\}) = A \setminus \{f_k\}$. Thus, A has no proper subset that also belongs to the rejection class K.

5 Conclusions

It is a consequence of coherence that a choice function is uniquely determined by those option sets that allow us to reject the zero gamble, i.e., those that are considered preferable to the status quo. In this paper, we have investigated the structure of these sets and shown that the coherence axioms can be expressed more intuitively in terms of these sets. In addition, we have shown that all the necessary information is given by option sets of cardinality two when the choice function is defined via maximality, and with cardinality two or three in most (but not all) cases of interest when the possibility space is binary. Moreover, we have shown that this last result does not extend to larger possibility spaces; thus, determining an analogous representation for arbitrary spaces would be the main open problem for the future.

On Pedro Gil

I was lucky to meet Pedro Gil in many different roles: first as a teacher, then as head of department, later as a supervisor, and finally, and all throughout, as a friend.

He helped me in my first research work, on robust statistics for my final project during my BsC; and we taught together in a course on *Mathematics for everyday life* until a few weeks prior to his death.

He was brilliant as a professor, generous as a researcher, and charismatic as a leader. Few bring together unanimity in the manner he did; and still, or perhaps because of it, he was always unassuming: he would treat you in the same manner irrespective of your position, and would make you feel that, whatever your problems, someone cared and would try to help. Because above all his many qualities, he had one that few possess: he was a good man.

He has not left a void because he is still very much present.

Enrique Miranda, April 2017

Acknowledgements This paper was written during a stay from Arthur van Camp at the University of Oviedo, funded by Banco Santander via Campus de Excelencia Internacional. We would like to acknowledge this funding, as well as that of project TIN2014-59543-P of the Spanish Ministerio de Economía y Competitividad. Gert de Cooman's research was partly funded through project number 3G012512 of the Research Foundation Flanders (FWO).

References

1. Aizerman M (1985) New problems in the general choice theory. Soc Cho Welf 2(4):235–282
2. He J (2012) A generalized unification theorem for choice theoretic foundations: Avoiding the necessity of pairs and triplets. Economics discussion paper 2012–23, Kiel Institute for the World Economy. https://ssrn.com/abstract=2056939
3. Miranda E, Zaffalon M (2010) Notes on desirability and coherent lower previsions. Ann Math Artif Intell 60(3–4):251–309
4. Quaeghebeur E (2014) Desirability. In: Augustin T, Coolen F, de Cooman G, Troffaes M (eds) Introduction to imprecise probabilities, Chap. 1. Wiley, Chichester
5. Schwartz T (1972) Rationality and the myth of the maximum. Noûs 6(2):97–117
6. Seidenfeld T, Schervish M, Kadane J (2010) Coherent choice functions under uncertainty. Synthese 172(1):157–176
7. Sen A (1971) Choice functions and revealed preference. Rev Econ Stud 38(3):307–317
8. Sen A (1977) Social choice theory: a re-examination. Economics 45:53–89
9. Van Camp A, de Cooman G, Miranda E, Quaeghebeur E (2015) Modelling indifference with choice functions. In: Augustin T, Doria S, Miranda E, Quaeghebeur E (eds) ISIPTA '15: proceedings of the 9th international symposium on imprecise probability: theories and application. Aracne Editrice, Pescara
10. Van Camp A, Miranda E, de Cooman G (2016) Lexicographic choice functions without Archimedeanicity. In: Ferraro M, Giordani P, Vantaggi B, Gagolewski M, Gil MA, Grzegorzewski P, Hryniewicz O (eds) Soft methods for data science. Series advances in intelligent systems and computing vol. 456. Springer, Cham
11. Walley P (1991) Statistical reasoning with imprecise probabilities. Chapman and Hall, London

Mathematical Modeling in Biological Populations with Reproduction in a Non-predictable Environment

Manuel Molina, Manuel Mota and Alfonso Ramos

Abstract In order to mathematically model the demographic dynamics of biological populations with sexual reproduction, we consider the more realistic situation where the reproductive process occurs in a non-predictable environment. We also assume that both biological processes, mating and reproduction, are influenced by the number of couples in the population. In this framework, a class of discrete-time two-sex branching models has been introduced in (A class of two-sex branching processes with reproduction phase in a random environment. Stochastics 88:147–161) [10]. In this work, we continue the research about such a class of stochastic models, investigating the time to extinction and some applications.

1 Introduction

Branching models have been especially developed to describe biological phenomena, playing a major role in studies on population dynamics, see e.g., [7] or [8]. We focus here the interest on the development of branching models to describe the dynamics of biological populations with sexual reproduction. This research line was initiated in [3] where the bisexual Galton-Watson model was introduced. In this model, the population of the species under consideration consists of two disjoint

---

This work is my modest contribution to this volume edited as a tribute to Pedro Gil. He was an excellent professional and a great person. His wise advices were very useful for me (Manuel Molina).

M. Molina (✉) · M. Mota
Faculty of Sciences, Department of Mathematics, University of Extremadura, 06006 Badajoz, Spain
e-mail: mmolina@unex.es

M. Mota
e-mail: mota@unex.es

A. Ramos
Faculty of Veterinary, Department of Mathematics, University of Extremadura, 10003 Cáceres, Spain
e-mail: aramos@unex.es

319

types of individuals: females and males, and two biological processes are carried out: mating and reproduction. First, in the mating process the couples female-male are formed. Then, in the reproduction process, such couples produce new female and male descendants.

In [3], considering two specific mating strategies, conditions for the extinction of populations with dynamics mathematically described through the bisexual Galton-Watson model were established. By using more general mating strategies, several contributions about the probabilistic evolution of such populations have been derived, see e.g., [1, 2, 4], or [6]. From these works, the interest in this issue increased and new classes of two-sex (bisexual) branching models have been introduced and studied, see for details [11] and the references cited therein.

Significant efforts have been made to develop two-sex branching models based on the assumption that the couples which take part in the reproduction process (progenitor couples) behave over time identically with respect to the reproduction, see e.g., [9, 13], or [12]. In such models, the number of progenitor couples is determined in a predictable environment. However, in many biological species, due to several random factors, e.g., weather conditions, food supply, fertility parameters, predators or human activities, the reproduction occurs in a non-predictable environment. Stochastic models to describe the dynamics of such species have not been sufficiently developed.

In order to contribute some solution to this issue, a new class of two-sex branching models has been introduced in [10]. In each generation, the number of progenitor couples is randomly determined. Moreover, the couples reproduce according to a probability law which changes over time depending on the number of progenitor couples in the population.

The motivation behind the present work is to continue the research about such a class of models. In Sect. 2, the probability model is formally described and intuitively interpreted. Section 3 is devoted to presenting the main results. Assuming that the extinction of the population occurs, the probability distribution about the time to extinction is investigated. The class of models studied is then applied to describe the phenomena concerning to populate or re-populate habitats with biological species. The concluding remarks and some open questions for research are included in Sect. 4.

2 Two-Sex Model

Let us consider biological populations with the following basic characteristics: mating and reproduction can be affected by the current number of couples in the population; in each generation, the number of progenitor couples and the probability law governing the reproductive process are randomly determined; each progenitor couple, independently of the others, produces new female and male descendants and then disappear. In [10], the demographic dynamics of such populations has been mathematically described through the stochastic sequence $\{(F_n, M_n)\}_{n=1}^{\infty}$, F_n and M_n denoting, respectively, the number of females and males at time (generation) n, defined as follows:

$$(F_{n+1}, M_{n+1}) := \sum_{i=1}^{\phi_{n,Z_n}} (f_{n,i}^{(\phi_{n,Z_n})}, m_{n,i}^{(\phi_{n,Z_n})}), \quad Z_{n+1} := L_{Z_n}(F_{n+1}, M_{n+1}), \qquad (1)$$

where $n \in \mathbb{N}$ (non-negative integers) and the empty sum is assumed to be $(0, 0)$. The variable Z_{n+1} represents the number of couples formed in the population at time $n + 1$. Initially, we assume a positive number k_0 of couples. Given that $Z_n = k$, $\{\phi_{n,k}\}_{k=0}^{\infty}$ is a sequence of random variables taking values in $\mathbb{N}$. For each k, the variables $\phi_{n,k}$, $n \in \mathbb{N}$, are assumed to be independent and identically distributed. The role of $\phi_{n,k}$ is to determine the number of progenitor couples in the nth generation. Note that immigration/emigration of couples in the population is allowed. When $\phi_{n,k} > k$ then $\phi_{n,k} - k$ immigrant couples come to the population and they take part in the reproduction process. When $\phi_{n,k} < k$ then $k - \phi_{n,k}$ couples emigrate from the population. It is assumed that $P(\phi_{n,0} = 0) = 1$ and, for $k \in \mathbb{N}_+$ (positive integers), $P(\phi_{n,k} = 0) < 1$.

If $\phi_{n,k} = j$ then, irrespectively of n, $(f_{n,i}^{(j)}, m_{n,i}^{(j)})$, $i = 1, \ldots, j$, are assumed to be independent and identically distributed random vectors, $(f_{n,i}^{(j)}, m_{n,i}^{(j)})$ representing the numbers of females and males descending from the ith progenitor couple in the nth generation. Its probability law, denoted by $\{p_{f,m}^{(j)}, f, m \in \mathbb{N}\}$, is referred to as the offspring probability distribution when j progenitor couples take part in the reproductive process:

$$p_{f,m}^{(j)} := P(f_{n,1}^{(j)} = f, m_{n,1}^{(j)} = m), \quad f, m \in \mathbb{N}.$$

Clearly, $p_{0,0}^{(0)} := P(f_{0,1}^{(0)} = 0, m_{0,1}^{(0)} = 0) = 1$.

$\{L_k\}_{k=0}^{\infty}$ is a sequence of functions defined on $\mathbb{N}^2$ and taking values in $\mathbb{N}$. They are referred to as mating functions. The role of L_k is to determine the number of couples formed in the mating process. It is assumed to be non-decreasing in each argument and such that:

$$L_k(0, m) = L_k(f, 0) = 0, \quad f, m \in \mathbb{N}.$$
$$L_k(f_1 + f_2, m_1 + m_2) \geq L_k(f_1, m_1) + L_k(f_2, m_2), \quad f_i, m_i \in \mathbb{N}, \quad i = 1, 2.$$

Last assumption (superadditivity) expresses the fact that $f_1 + f_2$ females and $m_1 + m_2$ males coexisting together will form a number of couples that is at least as great as the total number of couples formed by f_1 females and m_1 males and f_2 females and m_2 males, living separately.

Note that $\{(F_n, M_n)\}_{n=1}^{\infty}$, defined in (1), is a discrete-time two-sex model representing the demographic dynamics of females and males in the population in an environment which changes, stochastically in time, influenced by the number of couples in each generation. In fact, if at time n, k couples have been formed then, the offspring probability distribution and the mating function governing the reproduction and mating processes are $\{p_{f,m}^{(\phi_{n,k})}, f, m \in \mathbb{N}\}$ and L_k, respectively. This class of models includes, as particular cases, the two-sex branching models investigated in [3, 12, 13, 15]. In addition to its theoretical interest, it can be used to describe the evolution of biological species characterized by a single reproductive episode before death (semelparous species), see for details [5].

3 Results

This section is devoted to stating some results about the extinction/survival of biological populations with dynamics mathematically described through the class of two-sex models (1). It is organized in three subsections. First, the necessary preliminary definitions and results are given. Then, assuming that the extinction of the population occurs, the probability distribution about the time to extinction is investigated. Finally, an application to the phenomena concerning to populate or re-populate habitats with biological species is presented.

3.1 Preliminary Results

From (1), if for some $n \in \mathbb{N}$, $\phi_{n,Z_n} = 0$ then $(F_{n+m}, M_{n+m}) = (0, 0)$ and $Z_{n+m} = 0$, $m \in \mathbb{N}_+$, which means the extinction of the population. Also, if for some $n \in \mathbb{N}_+$, $Z_n = 0$ then, using that $P(\phi_{n,0} = 0) = 1$, the extinction occurs. By simplicity, we will consider that the population is extinct when, in some generation, no couples are formed in the mating process.

Definition 3.1 Let

$$Q_{k_0} := P(\lim_{n \nearrow \infty} Z_n = 0 | Z_0 = k_0), \ k_0 \in \mathbb{N}_+$$

be the probability that the extinction occurs in a population starting with k_0 couples.

The results stated in this subsection have been proved in [10].

Proposition 3.1 *Assume that, for each $k \in \mathbb{N}_+$, one of the following conditions is satisfied:*

1. $P(\phi_{1,k} = 0) > 0$.
2. *For some $j \in \mathbb{N}_+$, $\max\{P(f_{0,1}^{(j)} = 0), P(m_{0,1}^{(j)} = 0)\} P(\phi_{1,k} = j) > 0$.*

Then:

$$P(\lim_{n \nearrow \infty} Z_n = \infty | Z_0 = k_0) = 1 - Q_{k_0}, \ k_0 \in \mathbb{N}_+. \tag{2}$$

We henceforth assume (2) holds. In order to obtain sufficient conditions for the extinction/survival of the population, the following working assumptions about the sequences $\{L_k\}_{k=0}^{\infty}$, $\{\phi_{1,k}\}_{k=0}^{\infty}$ and $\{(f_{0,1}^{(j)}, m_{0,1}^{(j)})\}_{j=1}^{\infty}$ are required:

($a1$) $L_k(f, m) \le L_{k+1}(f, m), k, f, m \in \mathbb{N}$.
($a2$) $\phi_{1,k_1} + \phi_{1,k_2}$ is stochastically smaller[1] than ϕ_{1,k_1+k_2}, $k_1, k_2 \in \mathbb{N}$.
($a3$) $f_{0,1}^{(j)}$ (resp. $m_{0,1}^{(j)}$) is stochastically smaller than $f_{0,1}^{(j+1)}$ (resp. $m_{0,1}^{(j+1)}$), $j \in \mathbb{N}_+$.

Assumption ($a1$) represents the usual behaviour in many biological species in which the mating is promoted as the number of couples in the population grows. Requirement ($a2$) means that $\phi_{1,k_1} + \phi_{1,k_2}$ is more likely to take smaller values than ϕ_{1,k_1+k_2}. Assumption ($a3$) expresses the fact that when the number of reproductive couples in the population grows then the numbers of female and male descendants take large values with higher probabilities.

Definition 3.2 Let

$$R_k := k^{-1} E[Z_{n+1}|Z_n = k], \ k \in \mathbb{N}_+$$

be the mean growth rates per couple. From (1), it is deduced that:

$$R_k = k^{-1} \sum_{j=0}^{\infty} E\left[L_k\left(\sum_{i=1}^{j} f_{1,i}^{(j)}, \sum_{i=1}^{j} m_{1,i}^{(j)}\right) \right] P(\phi_{1,k} = j).$$

Proposition 3.2 *Under assumptions ($a1$), ($a2$) and ($a3$):*

(a) $R := \lim_{k \nearrow \infty} R_k$ *exists and* $R = \sup_{k \ge 1} R_k$.
(b) *If* $R \le 1$ *then* $Q_{k_0} = 1, k_0 \in \mathbb{N}_+$.
(c) *If* $R > 1$ *and* $\sup_{k \ge 1} k^{-1} Var[Z_{n+1}|Z_n = k] < \infty$ *then there exists* $k^* \in \mathbb{N}_+$ *such that* $Q_{k_0} < 1, k_0 \ge k^*$.

3.2 Time to Extinction

Definition 3.3 Given $Z_0 = k_0 \in \mathbb{N}_+$, define

$$T_{k_0} := \sup\{k \ge 0 : Z_k > 0\}, \ k_0 \in \mathbb{N}_+.$$

This variable represents the number of generations elapsed before the possible extinction of the population occurs starting from k_0 couples.

Clearly, $P(T_{k_0} < \infty) = Q_{k_0}$. For $i \in \mathbb{N}$, let

$$\varphi_i(s) := \sum_{k=0}^{\infty} s^k P(Z_i = k), \ 0 \le s \le 1.$$

---

[1]Given the variables X and Y, we say that X is stochastically smaller than Y if, for each real number t, $P(X \le t) \ge P(Y \le t)$.

Note that $\varphi_0(s) = s^{k_0}$, $0 \leq s \leq 1$. By simplicity, we will denote:

$$\varphi_i^*(0) := \varphi_{i+1}(0) - \varphi_i(0), \ i \in \mathbb{N}.$$

Assuming that $T_{k_0} < \infty$, next result provides the probability distribution of T_{k_0} and its main moments.

Proposition 3.3 (a) $P(T_{k_0} = i | T_{k_0} < \infty) = Q_{k_0}^{-1} \varphi_i^*(0), \ i \in \mathbb{N}.$

(b) $E[T_{k_0} | T_{k_0} < \infty] = \sum_{i=1}^{\infty} (1 - Q_{k_0}^{-1} \varphi_i(0)).$

(c) $Var[T_{k_0} | T_{k_0} < \infty] = Q_{k_0}^{-1} [\sum_{i=1}^{\infty} i^2 \varphi_i^*(0) - Q_{k_0}^{-1} (\sum_{i=1}^{\infty} (Q_{k_0} - \varphi_i(0))^2].$

Proof Using that $P(T_{k_0} < \infty) = Q_{k_0}$ we deduce:

(a)

$$P(T_{k_0} = 0 | T_{k_0} < \infty) = Q_{k_0}^{-1} P(Z_1 = 0) = Q_{k_0}^{-1} \varphi_0^*(0).$$

For $i \in \mathbb{N}_+$,

$$\begin{aligned}
P(T_{k_0} = i | T_{k_0} < \infty) &= Q_{k_0}^{-1} P(T_{k_0} = i) \\
&= Q_{k_0}^{-1} (P(T_{k_0} \leq i) - P(T_{k_0} \leq i - 1)) \\
&= Q_{k_0}^{-1} (P(Z_{i+1} = 0) - P(Z_i = 0)) \\
&= Q_{k_0}^{-1} \varphi_i^*(0).
\end{aligned}$$

(b)

$$\begin{aligned}
E[T_{k_0} | T_{k_0} < \infty] &= \sum_{i=0}^{\infty} P(T_{k_0} > i | T_{k_0} < \infty) \\
&= \sum_{i=0}^{\infty} (1 - Q_{k_0}^{-1} P(T_{k_0} \leq i)) \\
&= \sum_{i=1}^{\infty} (1 - Q_{k_0}^{-1} \varphi_i(0)).
\end{aligned}$$

(c)

$$E[T_{k_0}^2 | T_{k_0} < \infty] = Q_{k_0}^{-1} \sum_{i=1}^{\infty} i^2 \varphi_i^*(0).$$

The result is deduced using that:

$$Var[T_{k_0} | T_{k_0} < \infty] = E[T_{k_0}^2 | T_{k_0} < \infty] - (E[T_{k_0} | T_{k_0} < \infty])^2. \qquad \square$$

3.3 Application

A problem of great ecological importance is to populate or re-populate habitats with biological species. Suppose that we are interested in such a problem for a biological species which has the basic characteristics indicated in Sect. 2. Thus, in a first approach, the two-sex branching model defined in (1) can be applied to describe the probabilistic evolution of the species in the corresponding habitat.

We will assume that conditions required in Proposition 3.2 hold. Also, according to this result, in order to have a positive probability for the survival of the species, we will assume that the population dynamics of the species in the habitat can be appropriately described through a model (1) where $R > 1$ and $\sup_{k \geq 1} k^{-1} Var[Z_{n+1}|Z_n = k] < \infty$.

Initially, k_0 couples of the species are introduced in the habitat. If, after some generations, the species is extinct then k_0 new couples are again introduced, and so on, until the implementation of the species in the habitat is obtained.

For each $l \in \mathbb{N}_+$, let us denote by $\{(F_{n,l}, M_{n,l})\}_{n=1}^{\infty}$ the model, defined in (1), describing the probabilistic evolution concerning the lth attempt of re-population. Let $\{Z_{n,l}\}_{n=0}^{\infty}$, with $Z_{0,l} = k_0$, be the sequence related to $\{(F_{n,l}, M_{n,l})\}_{n=1}^{\infty}$ representing the number of couples formed in the habitat in the successive generations.

Thus, we are modeling the problem to populate or re-populate the habitat with the biological species through a sequence of independent processes $\{\{(F_{n,l}, M_{n,l})\}_{n=1}^{\infty}, l \in \mathbb{N}_+\}$, each of them, with the same underlying probability model as given in (1). Consequently, all the models have the same sequences of mating functions $\{L_k\}_{k=0}^{\infty}$ and offspring probability distributions $\{p_{f,m}^{(j)}, f, m \in \mathbb{N}\}_{j=0}^{\infty}$ governing, respectively, the mating and reproduction processes.

We derive, independently of l, that:

$$P(\lim_{n \nearrow \infty} Z_{n,l} = 0 | Z_{0,l} = k_0) = Q_{k_0}.$$

Let

$$T_{k_0,l} := \sup\{k \geq 0 : Z_{k,l} > 0\}, \, l \in \mathbb{N}_+.$$

$\{T_{k_0,l}\}_{l=1}^{\infty}$ is a sequence of independent and identically distributed random variables. Its common probability law is given in Proposition 3.3. By simplicity,

$$P_{k_0,i} := P(T_{k_0,1} = i | T_{k_0,1} < \infty), \, i \in \mathbb{N}.$$

Let us denote by N_{k_0} the variable representing the number of attempts before the implementation of the species in the habitat is obtained. Notice that N_{k_0} is distributed according to the geometric probability law:

$$P(N_{k_0} = i) = Q_{k_0}^i (1 - Q_{k_0}), \, i \in \mathbb{N}. \tag{3}$$

We deduce that:

$$E[N_{k_0}] = Q_{k_0}(1 - Q_{k_0})^{-1}, \quad Var[N_{k_0}] = Q_{k_0}(1 - Q_{k_0})^{-2}. \tag{4}$$

Definition 3.4 Let

$$T_{k_0}^* := \sum_{l=1}^{N_{k_0}} T_{k_0,l} \tag{5}$$

be the total number of generations elapsed before the implementation of the biological species in the habitat occurs.

Proposition 3.4 (a) $\quad P(T_{k_0}^* = i) = Q_{k_0}^{*^{-1}}(\delta_{i,0} + \sum_{j=1}^{\infty} Q_{k_0}^j P_{k_0,i}^{(*j)}), i \in \mathbb{N}.$

(b) $\quad E[T_{k_0}^*] = Q_{k_0}^* \sum_{i=1}^{\infty}(Q_{k_0} - \varphi_i(0)).$

(c) $\quad Var[T_{k_0}^*] = Q_{k_0}^*[(1 + Q_{k_0}^*) \sum_{i=1}^{\infty} i^2 \varphi_i^*(0) - Q_{k_0}^{-1}(\sum_{i=1}^{\infty}(Q_{K_0} - \varphi_i(0)))^2].$

$\qquad Q_{k_0}^* := (1 - Q_{k_0})^{-1}, \quad \delta_{i,0} := 1 \ if \ i = 0 \ or \ 0 \ if \ i \neq 0,$

$$P_{k_0,i}^{(*j)} := \sum_{i_1+...+i_j=i} P_{k_0,i_1}...P_{k_0,i_j}.$$

Proof (a) Taking into account (3) and (5),

$$P(T_{k_0}^* = i) = \sum_{j=0}^{\infty} P(T_{k_0}^* = i|N_{k_0} = j)P(N_{k_0} = j)$$

$$= Q_{k_0}^{*^{-1}}(\delta_{i,0} + \sum_{j=1}^{\infty} Q_{k_0}^j P(\sum_{l=1}^{j} T_{k_0,l} = i))$$

$$= Q_{k_0}^{*^{-1}}(\delta_{i,0} + \sum_{j=1}^{\infty} Q_{k_0}^j P_{k_0,i}^{(*j)}), \ i \in \mathbb{N}.$$

(b) From (4), (5), and Proposition 3.3,

$$E[T_{k_0}^*] = E[N_{k_0}]E[T_{k_0,1}|T_{k_0,1} < \infty] = Q_{k_0}^* \sum_{i=1}^{\infty}(Q_{k_0} - \varphi_i(0)).$$

(c) It is derived using again (4) and (5), Proposition 3.3, and the fact that:

$$Var[T_{k_0}^*] = E[N_{k_0}]Var[T_{k_0,1}|T_{k_0,1} < \infty] + Var[N_{k_0}]E[T_{k_0,1}^2|T_{k_0,1} < \infty]. \quad \square$$

4 Conclusion

In this work, we have focused the attention on the stochastic modeling of biological populations with reproductive process in a non-predictable environment. We have continued the research about the class of discrete-time two-sex branching models introduced in [10].

By assuming the extinction of the population, we have investigated the probabilistic behaviour of the variable representing the number of generations elapsed before the extinction occurs (Proposition 3.3). We have also considered the ecological problem concerning to populate or re-repopulate habitats with biological species which reproduce sexually. By considering, as mathematical approach, such a class of two-sex models we have investigated the probabilistic evolution of the species in the habitat. In particular, we have determined the probability distribution of the variable representing the total number of generations elapsed until the implementation of the species in the habitat occurs and its main moments (Proposition 3.4).

Some open questions for research about the class of two-sex models defined in (1) are, for example: to investigate, assuming the non-extinction of the population, its limiting evolution; to study the probabilistic behaviour of the population considering that $P(\phi_{1,0} = 0) < 1$; and to develop its inferential theory.

For instance, it is important to determine appropriate estimators for the extinction probability and for the main reproductive parameters (offspring mean vectors and covariance matrices) involved in the probability model, namely:

$$(\mu_1^{(j)}, \mu_2^{(j)}); \left(\sigma_{ij}^{(j)}\right)_{i,j=1,2}, \ j \in \mathbb{N}$$

where,

$$\mu_1^{(j)} := E[f_{0,1}^{(j)}], \ \mu_2^{(j)} := E[m_{0,1}^{(j)}],$$

$$\sigma_{11}^{(j)} := Var[f_{0,1}^{(j)}], \ \sigma_{12}^{(j)} = \sigma_{21}^{(j)} := Cov[f_{0,1}^{(j)}, m_{0,1}^{(j)}], \ \sigma_{22}^{(j)} := Var[m_{0,1}^{(j)}].$$

In [14], under the more general non-parametric statistical setting, some inferential questions have been investigated. In particular, we have proposed Bayes estimators for such reproductive parameters. An application to Pacific salmon populations has been also presented.

Acknowledgements This research has been supported by the Gobierno de Extremadura, Grant GR15105, the Ministerio de Economía y Competitividad of Spain, Grant MTM2015-70522-P, and the FEDER.

References

1. Alsmeyer G, Rösler U (2002) Asexual versus promiscuous bisexual Galton-Watson processes: The extinction probability ratio. Ann Appl Probab 12:125–142
2. Bruss FT (1984) A note on extinction criteria for bisexual Galton-Watson branching processes. J Appl Probab 21:915–919
3. Daley DJ (1968) Extinction conditions for certain bisexual Galton–Watson branching processes. Z. Wahrscheinlichkeitsth 9:315–322
4. Daley DJ, Hull DM, Taylor JM (1986) Bisexual Galton-Watson branching processes with superadditive mating functions. J Appl Probab 23:585–600
5. Fleming IA (1996) Reproductive strategies of atlantic salmon: Ecology and evolution. Rev Fish Biol Fish 6:379–416
6. González M, Molina M (1996) On the limit behavior of a superadditive bisexual Galton-Watson branching process. J Appl Probab 33:960–967
7. Haccou P, Jagers P, Vatutin V (2005) Branching processes: variation, growth, and extinction of populations. Cambridge University Press, Cambridge
8. Jagers P (1975) Branching processes with biological applications. Wiley, London
9. Ma M, Xing Y (2006) The asymptotic properties of supercritical bisexual Galton-Watson branching processes with immigration of mating units. Acta Math Sci Ser B 4:603–609
10. Ma S, Molina M, Xing Y (2016) A class of two-sex branching processes with reproduction phase in a random environment. Stochastics 88:147–161
11. Molina M (2010) Two-sex branching process literature. In: González Velasco M, Puerto I, Martínez R, Molina M, Mota M, Ramos A (eds) Workshop on branching processes and their applications, vol 197. Lecture notes in statistics. Springer, Berlin
12. Molina M, Mota M, Ramos A (2002) Bisexual Galton-Watson branching process with population-size-dependent mating. J Appl Probab 39:479–490
13. Molina M, Jacob C, Ramos A (2008) Bisexual branching processes with offspring and mating depending on the number of couples in the population. Test 17:265–281
14. Molina M, Mota M, Ramos A (2016) Statistical inference in two-sex biological populations with reproduction in a random environment. Ecol Complex 30:63–69
15. Xing Y, Wang Y (2005) On the extinction of a class of population-size dependent bisexual branching processes. J Appl Probab 42:175–184

Field Substitution and Sequential Sampling Method

M. Macarena Muñoz-Conde, Joaquín Muñoz-García,
Antonio Pascual-Acosta and Rafael Pino-Mejías

16 de Marzo de 2016
Bastaría con robar
todos los relojes
para que el tiempo dejara de latir
Eva Gil Sanmamed, *"Días cotidianos"*, 2006

Abstract A solution is proposed to the problem of non-response in the implementation and analysis of a survey, correcting total non-response through a sequential process of field substitution. Substitutions in surveys are often a useful procedure when there is a framework (sample frame) of reference for all elements of the population and when substitution is selected randomly. Through this sequential procedure, the probability of an element being included in the sample is updated in accordance with the selection sequence. A range of simulations are also carried out to compare the results of the proposed method to an ideal situation of zero non-response and against two other estimators calculated from the incomplete sample. The simulations were run on populations with three different distributions.

M. M. Muñoz-Conde
Instituto de Estadística y Cartografía de Andalucía, C/Leonardo da Vinci 21,
Sevilla, Spain
e-mail: mmacarena.munoz@juntadeandalucia.es

J. Muñoz-García (✉) · A. Pascual-Acosta · R. Pino-Mejías
Departamento de Estadística e Investigación Operativa, Universidad de Sevilla,
41004 Sevilla, Spain
e-mail: joaquinm@us.es

A. Pascual-Acosta
e-mail: antoniop@us.es

R. Pino-Mejías
e-mail: rafaelp@us.es

© Springer International Publishing AG 2018
E. Gil et al. (eds.), *The Mathematics of the Uncertain*, Studies in Systems,
Decision and Control 142, https://doi.org/10.1007/978-3-319-73848-2_31

329

1   Introduction

Non-sampling errors are of great significance in the implementation and analysis of surveys; one example of such errors is the effect of non-response by interviewees. In recent years, many authors have noted a tendency among interviewees not to respond to survey questionnaires. In the interests of economy of bibliographical references, citing [5] should be sufficient. Indeed, the problem of total nonresponse is present even when the interviewee is required to respond to the survey. There are several different classifications of non-response, for example the distinction made in [6] between "total non-response", when there is no response to any part of the survey questionnaire, and "item non-response", where individual questions are not answered, and finally "partial non-response", where a response is not given to parts or sections of the questionnaire. The effect of non-response in surveys has an impact on the probability on which inferences are based, giving rise to biased estimates and increased variance: this has been noted by several authors, e.g. [3]. Because of this, studies have been carried out into diverse methods intended to correct for the effects of non-response, including weighting, imputation and statistical modelling, as described in [7]. These methods are usually applied after the completion of fieldwork and do not generally involve any significant extra costs in running the survey.

2   Field Substitution

Substitution or field substitution is a method used to correct for total non-response. Although there has been criticism of the method, it is used by many organisations that specialise in running surveys, particularly when a sample frame of reliable, accurate information is available. Field substitution consists in replacing interviewees who do not respond to the survey with other elements of the population, subpopulation, domain or set (we will generally refer to populations) with very similar characteristics to those of the non-responding individuals. The new element is often referred to as the substitute. There are several forms of field substitution, which have been described by various authors: Chapman [2] proposes two separate ways of carrying out random substitution. In the first, the substitute is specially designed to be as similar as possible to the interviewee generating the non-response in terms of the variables considered or set in the sample frame; the second is referred to as random substitution because substitutes are selected randomly from among the non-responding interviewees population or subpopulation. Vehovar [11] describes these methods in detail, drawing a distinction between two general situations, according to whether probability sampling is used. Where it is, he proposes two situations, according to whether or not the interviewer can influence the selection of substitutes. Vehovar maintains the general method proposed by Chapman. Lynn [8] extends Vehovars [11] proposal, classifying substitution under a structure composed of three dimensions: firstly, who decides whether to make a substitution (the survey manager or the interviewer); secondly, who selects the substitute (again, whether this is the survey manager or the interviewer); and thirdly, how the substitute is selected, with three possible options (simple random selection, stratified random selection and non-random selection). Within this structure, then, there would logically appear to be twelve possibilities for substitution design; from a practical standpoint, these are reduced to eight, as assuming that the interviewer was making the decision, substitutes would not be selected randomly. Smith [10] proposes considering five elements to describe the main features of substitution methods: the possibility of implementing substitution in some phase of multiphase sampling; whether substitution is permissive or controlled; whether the substitution procedure is randomised; whether substitutes are similar in terms of variables included in the sample frame; and for household surveys, whether the substitute is selected from within the same household.

3   Sequential Field Substitution

The proposed method for the substitution of elements of the population shall be the sequential and random selection of substitutes. The approach will be one of simple random sampling without replacement, which does not in principle imply any restriction, as this is the generally applied sampling method for the final selection of survey respondents. Selection of population elements will therefore be consecutive and random. Where "total non-response is absent", the decision will be made to include the element in the sample, and if "total non-response is present", they will be excluded from the sample. When this is the case, they will be substituted by continuing with random selection until a substitute is found. This procedure is followed until the desired survey sample size is reached. This results in a process whereby the first order inclusion probability is updated as they are included in the sample (or not). This update process follows the method proposed in [1], a special case of a random sampling method, the splitting method introduced in [4]. Let us assume a population U (whose elements we shall assume, on account of the above, to have shared features, as per the sample frame) with N members, from which where substitutions are not necessary, we need a sample m made up of n elements, or where substitutions are made, we need a sample m_s, also made up of n elements who do not present total non-response, guaranteeing the initially planned sample size of n. The assumption is made that the elements of the population are accessible and identifiable, in order to be able to apply sequential selection. We therefore define indicator variables,

$$I_k = \begin{cases} 1 & \text{if the } k\text{th element is included in the sample} \\ 0 & \text{if the } k\text{th element is not included in the sample} \end{cases} \quad k = 1, 2, \dots, N$$

where the index k marks the order in which population elements are accessed, and the probability of being selected is updated in accordance with the number of elements included (or not included) in the sample. As we are applying simple random sampling without replacement, every element of the population initially has probability

$\pi_j^{(0)} = \frac{n}{N}$ $j = 1, 2, \ldots, N$ of being included in the sample in the initial phase, identified as the $0 - phase\ (selection)$. This probability will be updated in the sequential process theferore in the ith phase, their probability of being selected is determined by

$$\pi_j^{(i)} = P\left[I_j = 1\right] = \frac{n - \sum\limits_{k=1}^{i} I_k}{N - i} \ para\ j \geq i + 1, i + 2, \ldots, N,\ i = 1, 2, \ldots, N - 1$$

with the following relationship between the consecutive phases:

$$\pi_j^* = P\left[I_j = 1\right] = \pi_j^{(i)} = \pi_j^{(i-1)} - \left(I_i - \pi_j^{(i-1)}\right)\frac{1}{N - i}$$
$$\text{for } j \geq i + 1, i + 2, \ldots, N,\ i = 1, 2, \ldots, N - 1$$

as $\pi_j^{(i-1)}$ is constant for any $j \geq i, i + 1, \ldots, N$, the result is $\pi_j^{(i-1)} = \pi_i^{(i-1)}$, and

$$\pi_j^* = P\left[I_j = 1\right] = \pi_j^{(i)} = \pi_j^{(i-1)} - \left(I_i - \pi_i^{(i-1)}\right)\frac{1}{N - i}$$
$$\text{for } j \geq i + 1, i + 2, \ldots, N,\ i = 1, 2, \ldots, N - 1$$

therefore

$$\pi_j^* = P\left[I_j = 1\right] = \pi_j^{(i)} = \pi_j^{(0)} - \sum\limits_{k=1}^{i}\left(I_k - \pi_k^{(k-1)}\right)\frac{1}{N - k}$$
$$\text{for } j \geq i + 1, i + 2, \ldots, N,\ i = 1, 2, \ldots, N - 1$$

and for the first element observed after the $i - th$ phase, which will be the extraction element $i + 1$,

$$\pi_{i+1}^* = P\left[I_{i+1} = 1\right] = \pi_{i+1}^{(i)} = \pi_{i+1}^{(0)} - \sum\limits_{k=1}^{i}\left(I_k - \pi_k^{(k-1)}\right)\frac{1}{N - k}$$
$$\text{for } i = 1, 2, \ldots, N - 1.$$

The random variables of the previous expression

$$\varepsilon_k = I_k - \pi_k^{(k-1)} \ para\ k = 1, 2, \ldots, i$$

comply with $E\left[\varepsilon_k\right] = 0$, $var\left(\varepsilon_k\right) = var\left(I_k\right) = P\left[I_k = 1\right]\left(1 - P\left[I_k = 1\right]\right)$ and they are pairwise uncorrelated. This makes it possible to calculate the second order inclusion probabilities of selection

$$I_{i+1} - \pi_{i+1}^{(0)} = I_{i+1} - \pi_{i+1}^{(i)} - \sum_{k=1}^{i} \left(I_k - \pi_k^{(k-1)} \right) \frac{1}{N-k} = \varepsilon_{i+1} - \sum_{k=1}^{i} \varepsilon_k \frac{1}{N-k}$$

resulting in

$$\pi_{i+1,i+1+l}^* = P\left[I_{i+1} = 1, I_{i+1+l} = 1 \right] = E\left[I_{i+1} I_{i+1+l} \right]$$

$$= \frac{n^2}{N^2} + \sum_{k=1}^{i} P\left[I_k = 1 \right] (1 - P\left[I_k = 1 \right]) \left(\frac{1}{N-k} \right)^2$$

$$- \frac{1}{N-(i-1)} P\left[I_{i+1} = 1 \right] \left(1 - P\left[I_{i+1} = 1 \right] \right)$$

for $l > 0$, $i = 1, 2, \ldots, N-1$.

This makes it possible to calculate variance for the chosen estimators for the population. Therefore, if $Y = \{y_1, y_2, \ldots, y_N\}$ is the parameter associated to the population U, and if the parametric function for the population total is taken to be $T = \sum_{i=1}^{N} y_i$, the estimator of the total, based on sample m_s, will be the estimator of the ratio for the mean of the subpopulation of elements of U, that do not present total non-response, with correction for the initial population size. This is as follows

$$\widehat{T}_{m_s} = N \frac{\sum_{i \in m_s} \frac{y_i}{\pi_i^*}}{\sum_{i \in m_s} \frac{1}{\pi_i^*}} \tag{1}$$

with $\sum_{i \in m_s} \frac{1}{\pi_i^*}$ being the estimator of the size of the subpopulation of members of U who do not present total non-response. The estimator of the approximate variance of this estimator is determined by

$$\widehat{V}\left(\widehat{T}_{m_s} \right) = N^2 \left(\sum_{i \in m_s} \frac{1}{\pi_i^*} \right)^{-1} \sum_{i,j \in m_s} \frac{\pi_{ij}^* - \pi_i^* \pi_j^*}{\pi_{ij}^*} \left(\frac{y_i - \overline{y}_{m_s}}{\pi_i^*} \right) \left(\frac{y_j - \overline{y}_{m_s}}{\pi_j^*} \right)$$

with $\overline{y}_{m_s} = \sum_{i \in m_s} y_i$ (see [9, p. 182]).

4 An Empirical Study

The proposed estimator has been empirically assessed through simulations of the non-response process, using three other estimators, described below. In this way, diverse samples are considered for population U and parameter Y, as described below. Therefore, there are also different expressions for estimators of the total and their respective estimators of variance. For a sample $m \subseteq U$ of size n, with no element presenting non-response, the estimator of the total T is:

$$\widehat{T}_m = \sum_{i \in m} \frac{y_i}{\pi_i^{(0)}} \tag{2}$$

and the estimator of variance is determined by

$$\widehat{V}\left(\widehat{T}_m\right) = \frac{N-n}{n} \frac{1}{n-1} \sum_{i \in m} \left(y_i - \bar{y}_m\right)^2$$

with $\bar{y}_m = \frac{1}{n} \sum_{i \in m} y_i$.

Considered below is the sample without correction for total non-response, meaning a sample $m_r \subset m \subseteq U$ with m_r being of size $n_r < n$, with the estimators

$$\widehat{T}_{m_r} = \sum_{i \in m_r} \frac{y_i}{\pi_{ir}} \quad \text{with } \pi_{ir} = \frac{n_r}{N} \tag{3}$$

and the estimator of variance is determined by

$$\widehat{V}\left(\widehat{T}_{m_r}\right) = \frac{N-n_r}{n_r} \frac{1}{n_r-1} \sum_{i \in m_r} \left(y_i - \bar{y}_{m_r}\right)^2$$

and the other estimator being the result of considering the sequential method but for a reduced sample,

$$\widehat{T}_{m_r} = N \frac{\sum\limits_{i \in m_r} \frac{y_i}{\pi_i^*}}{\sum\limits_{i \in m_r} \frac{1}{\pi_i^*}} \tag{4}$$

where $\sum\limits_{i \in m_r} \frac{1}{\pi_i^*}$ will be the estimator of the subpopulation of elements of U that do no present total non-response.

The estimator of approximate variance is determined by

$$\widehat{V}\left(\widehat{T}_{m_r}\right) = N^2 \left(\sum_{i \in m_r} \frac{1}{\pi_i^*}\right)^{-1} \sum_{i,j \in m_r} \frac{\pi_{ij}^* - \pi_i^* \pi_j^*}{\pi_{ij}^*} \left(\frac{y_i - \bar{y}_{m_r}}{\pi_i^*}\right) \left(\frac{y_j - \bar{y}_{m_r}}{\pi_j^*}\right)$$

with $\bar{y}_{m_r} = \sum\limits_{i \in m_r} y_i$.

In order to empirically compare the performance of the four estimators of the total, 1,000 samples were generated, each with 400 elements. Random perturbation was applied to each sample generated, or in other words for each complete sample

Table 1 Simulations from N(2,1)

P_{NR}	(1)	(2)	(3)	(4)
0.025	113.81	100.01	127.43	143.28
0.050	169.79	157.99	179.07	349.73
0.100	143.55	129.40	174.04	177.26

Table 2 Simulations from Bernoulli

P_{NR}	P_{Bern}	(1)	(2)	(3)	(4)
0.025	0.3	110.41	141.67	143.73	134.22
0.025	0.5	96.32	128.33	164.62	153.04
0.050	0.3	135.44	126.67	149.26	259.63
0.050	0.5	118.99	120.00	183.86	127.93
0.100	0.3	169.00	136.67	160.17	189.71
0.100	0.5	113.85	100.00	117.43	130.28

Table 3 Simulations from exponential

P_{NR}	(1)	(2)	(3)	(4)
0.025	283.81	185.04	275.52	343.28
0.050	179.89	169.03	179.29	199.73
0.100	1043.55	947.68	1050.70	1123.24

m, assuming a determinate value for the probability of non-response, P_{NR}, thereby obtaining an incomplete sample m_r. Based on m_r, the completed sample m_s, with 400 elements, was built by calculating all four estimators based on the samples obtained in this way. The population is made up of 50,000 members, with the measurement variable distributed according to the law N(2,1). Table 1 shows the average distance between each estimator and the population total for three values of probability of non-response and for all four estimators. It can be observed that the proposed estimator (1) generally offers values that are closer to the total than estimators (3) and (4), only being surpassed by the estimator that corresponds to completely observed samples (2).

A similar study was carried out on a Bernoulli population, with two possible values (0.3 and 0.5) for parameter p of the law. Once more, the same three values as in Table 1 were considered for the probability of non-response. The results of this are shown in Table 2. Of the six configurations, four show the better performance of method (1) compared to estimators (3) and (4), with method (3) surpassing estimator (1) in two cases. Table 2 confirms that estimator (4) performs the worst.

Finally, Table 3 shows the results of an analogous study with an exponential law with rate $= 1/2$.

While estimator (3) surpasses (1) in the first row, the conclusion generally supports that of the preceding tables. For this reason, it can be concluded that the proposed estimator (1) is a potential alternative to estimator (3), with behaviour that is generally not inferior that of the latter technique. Estimator (4) is clearly inferior to the other estimators.

References

1. Bondesson L, Thorburn D (2008) A list of sequential sampling method suitable for real time sampling. Scand J Stat 35:466–483
2. Chapman DW (1983) The impact of substitution on survey estimates. In: Madow WG, Olkin I (eds) Incomplete data in sample surveys, vol 3. Academic Press, New York
3. Cochran WG (1977) Sampling techniques, 3rd edn. Wiley, New York
4. Deville JC, Tillé Y (1998) Unequal probability sampling without replacement through a split-ting method. Biometrika 85:89–101
5. Dillman DA, Eltinge JL, Groves RM, Little RJA (2002) Survey nonresponse in design data collection, and analysis. In: Dillman DA, Eltinge JL, Little RJA (eds) Survey nonresponse. Wiley, New York
6. Kalton G, Kasprzyk D (1986) The treatment of missing survey data. Surv Methodol 12:1–16
7. Little RJA, Rubin DB (2002) Statistical analysis with missing data, 2nd edn. Wiley, Hoboken
8. Lynn P (2004) The use of substitution in surveys. Surv Stat 49:14–16
9. Särndal E, Swensson B, Wretman J (1992) Model assisted survey sampling. Springer, New York
10. Smith TW (2007) Note on the use of substitution in surveys. Report in substitution, International Social Survey Programme (ISSP)
11. Vehovar V (2003) Field Substitution redefined. Surv Stat 48:35–37

Working with (too) Few Samples

Angela L. Riffo-Campos, Francisco Montes and Guillermo Ayala

Abstract This paper is concerned with gene set differential expression analysis. We compare the transcriptomic behaviour of each gene set between different experimental conditions.The gene set is previously defined. It has been used a gene set collection downloaded from Gene Ontology.A randomization test is proposed and compared with other previous procedures using a RNA-seq experiment of colorectal cancer (CRC).

1 Introduction

The research in molecular biology and related areas, especially in Biomedicine, have been changed abruptly in the last two decades. This was due to the incorporation of new technologies, in particular to the next generation sequencing (NGS) and bioinformatics tools, giving way to the "omics" data era (see Berger et al. [4]). The omics data, are those obtained from all genetic material of an organism (Genomics), from all RNA present in a cell type (Transcriptomics), from the total number of proteins in a cell type (Proteomics), among others [30]. Thus, in a short time, a lot of biological information was generated. The data is stored in specialized centres, as the National Center for Biotechnology Information (NCBI), the European Molecular Biology Laboratory (EMBL) or the DNA Data Bank of Japan (DDBJ). The information is available for the scientific community for testing new approaches and new

A. L. Riffo-Campos · F. Montes · G. Ayala (✉)
Departamento de Estadística e Investigación Operativa, Universidad de Valencia,
Avda. Vicent Andrés Estellés, 1, 46100 Burjasot, Spain
e-mail: guillermo.ayala@uv.es

A. L. Riffo-Campos
e-mail: angela.riffo@uv.es

F. Montes
e-mail: francisco.montes@uv.es

© Springer International Publishing AG 2018
E. Gil et al. (eds.), *The Mathematics of the Uncertain*, Studies in Systems,
Decision and Control 142, https://doi.org/10.1007/978-3-319-73848-2_32

bioinformatics tools. In addition, big consortia have been formed to collect, analyze and share the biological data. Some examples are:

- The haplotype map of the human genome (HapMap: http://hapmap.ncbi.nlm.nih. gov/), to create a catalog of common genetic variants that occur in human beings [26].
- The Encyclopedia of DNA Elements (ENCODE: http://genome.ucsc.edu/ ENCODE/) project, whose objective was the identification of all functional elements in the human genome sequence [25].
- The 1000 genomes project (http://www.1000genomes.org/), which aims at obtaining a deep characterization of human genetic variations that have a frequency of at least 1% in the population studies [24].
- The Cancer Genome Atlas (TCGA: http://cancergenome.nih.gov/), which is a comprehensive and coordinate effort to accelerate the understanding of the molecular basis of cancer, making available a large amount of patient data, including omics data [28].

Nevertheless, outside the large international consortia, a big number of individuals (samples) to analyze is uncommon. This is due to first, to increase the number of individuals per experiment, entails an exponential expenditure of time, space and resources in obtaining the samples for analysis. Then, because the sequencing technologies, although increasingly accessible, are expensive. In addition, in the case of humans and experimentation models in mammals, it is added that the use of the biological material and animals in investigation is highly legislated, which makes difficult the obtaining of a great number of samples. As a result, and to exemplify, of the 1,252,392 dataset found in the Gene Expression Omnibus (GEO) repository, when searching for "human", only 16,881 have a sample size greater than 20 per experiment (https://www.ncbi.nlm.nih.gov/gds/?term=human, 06/2017). Thus, more than 98% of the dataset have 20 or less samples and usually it is about comparisons between 2 or 3 conditions.

We will focus on the transcriptomics data. When the molecular bases for the causes and development of a disease are studied, they are trying to identify genes whose function has been altered with respect to a healthy individual, studying between others, the transcriptome in both conditions. This data can be obtained mainly by two types of technology: the microarray technologies, that was developed in the 1990s and the massively parallel sequencing of RNA (RNAseq) in form of cDNA, developed in 2006 [16]. Although, there are also complementary platforms to the study of the transcriptome as the Tiling Arrays, GRO-seq and others. In addition, third-generation sequencing platforms are already a fact, but these data are currently scarce. The datasets generated from microarray technologies are usually stored in the GEO repository (https://www.ncbi.nlm.nih.gov/geo/). While, the sequencing data are usually stored in the Sequence Read Archive (SRA, https://www.ncbi.nlm.nih. gov/sra/), European Nucleotide Archive (ENA, http://www.ebi.ac.uk/ena) and also in GEO. The microarray technology allows a quantitative measurement of the difference of expression of known genes, when two or more conditions are compared, generating continuous data as result of this comparison [8]. With the RNAseq is possible to compare transcript levels in different conditions and also to know the sequence of the transcribed (cDNA), the data generated by sequencing are counts [32]. This was an important step in the large-scale study of alternative splicing, which allowed the discovering of new isoforms and their quantification.

The alternative splicing is the biological process by which the same gene can give rise to many transcripts with different biological functions [17, 31]. So, a gene can be considered as a small group of isoforms. In many cases, change in the expression of one isoform can be the cause of an alteration in cellular functioning and this change may not be evident when studying differences in the transcriptome at the gene level [35]. Therefore, the study of the transcriptome at the isoform level has become more important in the last years.

The processing of raw data from a RNA-seq starts aligning the reads (after quality control), usually short cDNA sequences, with the reference genome or transcriptome and then the number of read that belongs to a gene or isoform is counted, using some bioinformatics tools like Bowtie [13], Tophat–Cufflinks [29] or STAR [7]. After that, it is analyzed if the differences in the number of reads are statistically significant (differential expression) by region (gene/isoforms) between two or more conditions, using R packages like edgeR [20], DEGseq [33] or CummeRbund [29], for more details see Conesa et al. [6]. To finally perform a functional analysis of groups of differentially expressed genes. This is usually done through gene set enrichment analysis, by gene ontology (GO) or KEGG Orthology (KO) categories. This allows to focus the study to those genes that participate in biological mechanisms associated to a pathology, discarding those genes differentially expressed by reasons external to it.

2 The Problem and a (too) Simple Approach

For a given gene i and sample j, the random variable X_{ij} is the random expression of the gene i in the sample j. This random expression could be observed from an array or a RNA-seq experiment. The random expression matrix is

$$\mathbb{X} = [X_{ij}]_{i=1,\dots,N; j=1,\dots,n},$$

where N is the total number of genes and n is the number of samples. As usual the observed expression matrix would be

$$\mathbb{x} = [x_{ij}]_{i=1,\dots,N; j=1,\dots,n}.$$

We will have some covariable associated to the samples. The vector $y = (y_1, \dots, y_n)'$ can be numeric (observation times of the samples or an experimental factor indicating a treatment).

A gene set will be a subset of $G = \{1, \dots, N\}$. Note that the rows $u_1, \dots, u_N$ are not a random sample because the expressions corresponding to different genes are not independent. However, the columns $v_1, \ldots, v_n$ is a random sample. The number of rows (features) N is much greater than n, $N \gg n$. In fact, N is several thousands and n no more than one or two hundreds.

Let us consider the two conditions setup with sample sizes n_1 and n_2. In this paper, we assume that n_1 and n_2 are very small. These conditions corresponds to different phenotypes or different experimental conditions.

The basic problem is to quantify the association between the gene expression and the phenotype. For a given gene, let U_j with $j = 1, \ldots, n_1$ the i.i.d. random expressions under the first condition. The corresponding random sample under the second condition will be U'_j with $j = 1, \ldots, n_2$. We test the null hypothesis $H_0 : \mu = \mu'$ versus $H_1 : \mu \neq \mu'$ where μ and μ' are the means of any U_j and U'_j respectively. It is usual to test the null hypothesis of a common mean (assuming a common variance) by means the t-test. For so small sample sizes the estimation of the common variance is not efficient and perhaps is not a good choice. The observed p-value is not reliable. It is a common practice to use a randomization test. Now we work given the data i.e. given the observed u_j and u'_j. The first group is chosen randomly from the total number of samples, $n = n_1 + n_2$ and the second condition corresponds to non-chosen samples. The number of possible assignments is $\binom{n}{n_1}$. The different values (possibly not all different) will be $t_1, \ldots, t_B$. We assume that t_1 corresponds to the real conditions. In the randomization distribution each value has the same probability. This is called the permutation distribution. Usually, we will sample B times from the randomization distribution. This distribution is used to test if there is association between gene expression and phenotype. If a one-tail test is used then the p-value is given by

$$p = \frac{|\{i : i \geq 2, |t_i| > |t_1|\}|}{B - 1},$$

where $| \cdot |$ is the cardinality of the set. This p-value measures how extreme is t_1 with respect to the other values in the permutation distribution. The p-value or the t-statistic quantifies the marginal phenotype-expression association for the ith gene. If we consider a gene set $S = \{i_1, \ldots, i_{|S|}\}$ then $\mathbf{t}(S) = (t_{i_1}, \ldots, t_{i_{|S|}})$ are the observed statistics used to measure the expression-phenotype association within S. The bth random selection of n_1 samples will produce a new $\mathbf{t}$-vector and a subvector $\mathbf{t}_b(S)$. In short, $\mathbf{t}_1(S)$ will correspond with the original y and $\mathbf{t}_b(S)$ (with $b = 2, \ldots, B$th selection). For each $b = 1, \ldots, B$ we calculate

$$\mathbf{t}_{(b)} = \frac{1}{B - 1} \sum_{\substack{j \neq i; j=1}}^{B} \mathbf{t}_j. \tag{1}$$

We evaluate the distances between $\mathbf{t}_b$ and $\mathbf{t}_{(b)}$ for all $b = 1, \ldots, B$

$$d_b = d(\mathbf{t}_b, \mathbf{t}_{(b)}). \tag{2}$$

If there is no expression-phenotype association then all possible orderings of $(d_1, \ldots, d_B)$ have the same probability. We can use the randomization p-value

$$p = \frac{|\{i : i \geq 2, d_i > d_1\}|}{B - 1}.$$

In particular, the Manhattan and Euclidean distances will be used in the experimental results.

3 Experimental Results

We have downloaded the bioproject PRJNA218851, from SRA (SRX347940-SRX-347887) repository. This data consists of a gene expression profiling study by RNA-seq in 18 Korean patients with colorectal cancer (CRC). From each patient, the samples were normal colon (>5 cm from the tumor border), primary CRC that were histologically identified as adenocarcinoma, and liver metastasis (54 samples in total). A detailed description of the cohort can be found in [12].

The RNA-seq workflow consists of the following steps:

- Download of the raw data from SRA using SRA Toolkit [27].
- The quality control of the reads performed using FastQC [1].
- The preprocessing using FASTX-toolkit [10].
- The reads were aligned with the human genome (hg19) and counted using STAR [7].

3.1 A Preliminary Marginal Analysis

We have compared the normal colon with primary CRC and the normal colon with liver metastasis. A marginal differential expression analysis has been performed using the method edgeR with a common dispersion parameter [18–20]. The raw and adjusted (using the Benjamini–Hochberg correction [3]) p-values are really small i.e. a high number of genes seems to be differentially expressed genes and a low of false discovery rate has been chosen, $FDR = 10^{-30}$. The significant gene set contains a total of 129 differentially expressed genes when comparing normal colon with primary CRC (first significant gene set) and 603 genes when comparing normal colon with liver metastasis (second significant gene set). These results are consistent with the progressive deregulation of gene expression profiles in the cancer progression [21]. The most significant genes in the first significant gene set (from lower to higher raw p value) were

- The C-C motif chemokine ligand 25 (CCL25) gene (adjusted p-value $2.03E - 70$) that is related to cancer in 47 papers and the CCL25 protein regulate CRC progression and invasion.
- The COL10A1 gene (adjusted p-value $2.48E - 70$) encodes the alpha chain of type X collagen that is related to cancer in 27 papers and was proposed as biomarker for early detection of colon cancer [5, 22].
- The NKX2-1 gene (adjusted p-value $1.23E - 69$) encodes a transcription factor that is related to cancer in 140 papers and is involved in p53 pathway, promoting the invasiveness in colon cancer cells [11].
- The human regenerating gene 1B (REG1B)(adjusted p-value $2.93E - 66$) is found associated to cancer in 7 papers and was proposal as a novel candidate therapeutic target for CRC [15].
- The long non-coding RNA AFAP1-AS1 (adjusted p-value $4.18E - 65$) is associated to cancer in 26 papers and in CRC was related with poor prognosis and promotes tumorigenesis [14, 34], and facilitates tumor growth and promotes metastasis [9].

Other genes in the list that are located next to the genes before mentioned, as REG1A, FEZF1-AS1, MMP7, KRT17, COL11A1, CLDN18, COMP, MMP3, MMP1, INHBA and others, also have been related with CRC. However, genes like NPSR1 and LINC02418 have not been previously reported as differentially expressed in CRC.

In the result obtained in the second significant gene set, the top five genes in the list were: FGG ($3.84E - 123$), FGB ($5.81E - 122$), haptoglobin ($6.34E - 122$), ITIH1 ($1.24E - 121$), ALB ($1.04E - 121$) all related to cancer. The FGG, haptoglobin and ALB genes also were related to metastatic colorectal cancer [2, 23]. Thus, in general, our results are consistent with literature.

3.2   Three Different Gene Set Analysis

We have chosen as gene set collection the "biological process" ontology from Gene Ontology working only with sets with five or more genes. Three different gene set analysis have been used.

- **An over-representation analysis** using two different significant gene sets corresponding to the comparisons between control with cancer (first significant gene set) and control with metastasis (second significant gene set).
- **A gene set test** where the association phenotype-expression is quantified using the raw p-value of the edgeR method. We aggregate within each gene set using the mean. The self-contained null hypothesis is tested using the randomization distribution of the mean p-value. The p-value corresponds to the randomization distribution of the aggregated value.

- **Our approach** proposed in the previous section using the Manhattan and Euclidean distances. The statistic corresponds to the raw p-values of the edgeR method. The results commented later use the Manhattan distance.

From now on we will use a false discovery rate equal to 0.001.

The first method, the over-representation analysis, for the first significant gene set, finds only one GO category, the collagen catabolic process (GO:0030574), that contains the PEPD, ADAM15, COL13A1, FURIN and many others (see Colorectal Cancer Atlas and the Human Protein Atlas databases).

The second method, the GSA analysis, for the control-cancer comparison obtains 2003 GO categories. Our analysis, third method, for the control-cancer comparison result in 3527 GO significant categories.

For both GSA and our method, some of the first significant GO categories were the same:

- The ribosomal large subunit export from nucleus (GO:0000055), include the XPO1, RRS1 and NMD3 genes that are involved in CRC (see Colorectal Cancer Atlas and the Human Protein Atlas databases).
- The G2/M transition of mitotic cell cycle (GO:0000086) that includes the ENSA, LIN54, PPP2R1A and MAPRE1 genes involved in CRC.
- The activation of MAPKK activity (GO:0000186), include the CSF2RA, SPTB and TAOK3 genes involved in CRC.
- The mRNA splicing, via spliceosome (GO:0000398), include the HNRNPA1, BUD31, POLR2F, NUDT21, CD2BP2, METTL3 and others genes involved in CRC,

and others GO categories.

When we compare control versus metastasis a total of 69 significant GO categories has been obtained by over-representation. The second method has found 2805 significant gene sets and 4065 has been found by our method.

Finally, we have compared the adjusted p-value obtained using the three different procedures. In particular, we have added a modified version of our method using the Euclidean distance i.e. we have the three procedures just used plus the third one with the Euclidean distance. Figure 1a displays the cumulative distribution functions for the four procedures when we compare control versus cancer. The different metrics for our procedure produces very similar results. Our method provides clearly lesser p-values with respect the first and second method. The worst procedure is the first one. A similar comment applies to Fig. 1b. If we compare both plots it is clear that the p-values are clearly smaller for control-cancer comparison with respect to control-metastasis comparison. Let us remember that metastasis is later evolution of the disease.

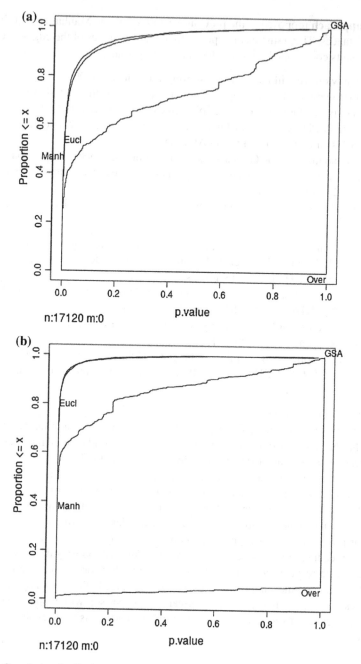

Fig. 1 **a** Cumulative distribution functions for *p*-values obtained comparing control versus cancer. The label **Over** corresponds to the first method, **GSA** to the second method, **Manh** to the third method with the Manhattan distance and **Eucl** to the third method with the Euclidean distance. **b** Cumulative distribution functions for *p*-values obtained comparing control versus metastasis. See text for details

Acknowledgements This paper has been partially supported by Grant DPI2017-87333-R.

References

1. Andrews S (2010–2016) FastQC a quality control tool for high throughput sequence data. http://www.bioinformatics.babraham.ac.uk/projects/fastqc/
2. Artac M, Uysal M, Karaagac M, Korkmaz L, Er Z, Guler T, Boruban MC, Bozcuk H (2017) Prognostic impact of neutrophil/lymphocyte ratio, platelet count, CRP, and albumin levels in metastatic colorectal cancer patients treated with FOLFIRI-Bevacizumab. J Gastrointest Cancer 48(2):176–180
3. Benjamini Y, Hochberg Y (1995) Controlling the false discovery rate: a practical and powerful approach to multiple testing. J R Stat Soc Ser B 57(1):289–300
4. Berger B, Peng J, Singh M (2013) Computational solutions for omics data. Nat Rev Genet 14(14):333–346
5. Chen H-J, Edwards R, Tucci S, Bu P-C, Milsom J, Lee S, Edelmann W, Gümüs ZH, Shen X, Lipkin S (2012) Chemokine 25induced signaling suppresses colon cancer invasion and metastasis. J Clin Investig 122(9):3184–3196
6. Conesa A, Madrigal P, Tarazona S, Gómez-Cabrero D, Cervera A, McPherson A, Szcześniak MW, Gaffney DJ, Elo LL, Zhang X, Mortazavi A (2016) A survey of best practices for RNA-seq data analysis. Genome Biol 17:13
7. Dobin A, Davis CA, Schlesinger F, Drenkow J, Zaleski C, Jha S, Batut P, Chaisson M, Gingeras TR (2013) STAR: ultrafast universal RNA-seq aligner. Bioinformatics 29(1):15–21
8. Govindarajan R, Duraiyan J, Kaliyappan K, Palanisamy M (2012) Microarray and its applications. J Pharm Biol Sci 4(Suppl 2):S310–S312
9. Han X, Wang L, Ning Y, Li S, Wang Z (2016) Long non-coding RNA AFAP1-AS1 facilitates tumor growth and promotes metastasis in colorectal cancer. Biol Res 49(1):36
10. Hannon Lab. FASTX toolkit. http://hannonlab.cshl.edu/fastx_toolkit/
11. He T-Y, Tsai L-H, Huang C-C, Chou M-C, Lee H (2014) LKB1 loss at transcriptional level promotes tumor malignancy and poor patient outcomes in colorectal cancer. Ann Surg Oncol 21(4):703–710
12. Kim S-K, Kim S-Y, Kim J-H, Roh SA, Cho D-H, Kim YS, Kim JC (2014) A nineteen gene-based risk score classifier predicts prognosis of colorectal cancer patients. Mol Oncol 8(8):1653–1666
13. Langmead B (2010) Aligning short sequencing reads with Bowtie. Curr Protoc Bioinformatics Chap. 11: Unit 11.7
14. Li Q, Dai Y, Wang F, Hou S (2016) Differentially expressed long non-coding RNAs and the prognostic potential in colorectal cancer. Neoplasma 63:977–983
15. Liu Z, Zhang Y, Xie J, Li C, Wang X, Shen J, Zhang Y, Wang S, Cheng N (2015) Regenerating gene 1B silencing inhibits colon cancer cell HCT116 proliferation and invasion. Int J Biol Markers 30(2):e217–e225
16. Mantione KJ, Kream RM, Kuzelova H, Ptacek R, Raboch J, Samuel JM, Stefano GB (2014) Comparing bioinformatic gene expression profiling methods: microarray and RNA-seq. Med Sci Monitor Basic Res 20:138–141
17. Matlin AJ, Clark F, Smith CWJ (2005) Understanding alternative splicing: towards a cellular code. Nature Rev Mol Cell Biol 6(5):386–398
18. Robinson MD, Smyth GK (2007) Moderated statistical tests for assessing differences in tag abundance. Bioinformatics 23(21):2881–2887
19. Robinson MD, Smyth GK (2008) Small-sample estimation of negative binomial dispersion, with applications to sage data. Biostatistics 9(2):321–332
20. Robinson MD, McCarthy DJ, Smyth GK (2007) edgeR: a bioconductor package for differential expression analysis of digital gene expression data. Bioinformatics 26(1):139–140

21. Sadikovic B, Al-Romaih K, Squire JA, Zielenska M (2008) Cause and consequences of genetic and epigenetic alterations in human cancer. Current Genomics 9(6):394–408

22. Solé X, Crous-Bou M, Cordero D, Olivares D, Guinó E, Sanz-Pamplona R, Rodríguez-Moranta F, Sanjuan X, de Oca J, Salazar R, Moreno V (2014) Discovery and validation of new potential biomarkers for early detection of colon cancer. PLoS ONE 9(9):e106748

23. Sun L, Hu S, Yu L, Guo C, Sun L, Yang Z, Qi J, Ran Y (2016) Serum haptoglobin as a novel molecular biomarker predicting colorectal cancer hepatic metastasis. Int J Cancer 138(11):2724–2731

24. The (1000) Genomes Project Consortium (2012) An integrated map of genetic variation from 1,092 human genomes. Nature 491(7422):56–65

25. The ENCODE Project Consortium (2012) An integrated encyclopedia of DNA elements in the human genome. Nature 489(57):91–100

26. The International HapMap Consortium (2007) A second generation human haplotype map of over 3.1 million SNPs. Nature 449(7164):851–861

27. The ncbi sequence read archive (SRA). https://submit.ncbi.nlm.nih.gov/subs/sra/

28. Tomczak K, Czerwińska P, Wiznerowicz M (2015) The Cancer Genome Atlas (TCGA): An immeasurable source of knowledge. Contemp Oncol 19(1A):A68–A77

29. Trapnell C, Roberts A, Goff L, Pertea G, Kim D, Kelley DR, Pimentel H, Salzberg SL, Rinn JL, Pachter L (2012) Differential gene and transcript expression analysis of RNA-seq experiments with TopHat and Cufflinks. Nat Protoc 7(3):562–578

30. Tseng G, Ghosh D, Zhou XJ (2015) Integrating Omics Data, 1st edn. Cambridge Univ Press, Cambridge

31. Wang ET, Sandberg R, Luo S, Khrebtukova I, Zhang L, Mayr C, Kingsmore SF, Schroth GP, Burge CB (2008) Alternative isoform regulation in human tissue transcriptomes. Nature 456(7221):470–476

32. Wang Z, Gerstein M, Snyder M (2009) RNA-Seq: A revolutionary tool for transcriptomics. Nat Rev Genet 10(1):57–63

33. Wang L, Feng Z, Wang X, Wang X (2010) Zhang X (2010) Degseq: an R package for identifying differentially expressed genes from rna-seq data. Bioinformatics 26(1):136

34. Wang F, Ni H, Sun F, Li M, Chen L (2016) Overexpression of lncRNA AFAP1-AS1 correlates with poor prognosis and promotes tumorigenesis in colorectal cancer. Biomed & Pharmacoth 81:152–159

35. Zhang Z-F, Pal S, Bi Y, Tchou J, Davuluri RV (2013) Isoform level expression profiles provide better cancer signatures than gene level expression profiles. Genome Med 5(4):33

Estimation of the Owen Value Based on Sampling

Alejandro Saavedra-Nieves, Ignacio García-Jurado and M. Gloria Fiestras-Janeiro

Abstract In this paper we introduce a procedure based on sampling to estimate the Owen value of a cooperative game. It is an adaptation of an analogous procedure for the estimation of the Shapley value, and it is specially useful when dealing with games having large sets of players. We provide some results in order to choose a sample size guaranteeing a bound for the absolute error with a given probability, and illustrate our procedure with an example taken from the game theoretical literature.

1 Introduction

Game theory is the mathematical theory of interactive decision problems. When the agents involved in one of such problems can make commitments, they usually commit to cooperate, i.e. to behave in a way that results in a social optimum. These problems are treated by cooperative game theory. According to [4], "cooperative game theory deals with coalitions and allocations, and considers groups of players willing to allocate the joint benefits derived from their cooperation". Cooperative game theory has become a useful mathematical tool in theoretical economics, political science and in other disciplines. Reference [3] is a survey with some applications of cooperative game theory.

An important solution concept in cooperative game theory is the Owen value, introduced in [7]. The Owen value proposes for every cooperative game an allocation of the joint benefits generated by the players when there exists an a priori structure that conditions their cooperative possibilities. It is a variation of the Shapley value

A. Saavedra-Nieves · M. G. Fiestras-Janeiro
Department of Statistics and Operations Research, Universidade de Vigo, 36200 Vigo, Spain
e-mail: asaavedra@uvigo.es

M. G. Fiestras-Janeiro
e-mail: fiestras@uvigo.es

I. García-Jurado (✉)
Department of Mathematics, Universidade da Coruña, 15071 A Coruña, Spain
e-mail: ignacio.garcia.jurado@udc.es

© Springer International Publishing AG 2018
E. Gil et al. (eds.), *The Mathematics of the Uncertain*, Studies in Systems, Decision and Control 142, https://doi.org/10.1007/978-3-319-73848-2_33

347

(see [8]) that has many applications in political science, logistics and cost allocation (see, for instance, [2]).

The main drawback concerning the Shapley value and the Owen value is computational, since their complexities increase exponentially with the number of players. There are many papers dealing with this issue from several perspectives. Recently, a polynomial calculation procedure of the Shapley value based on sampling was introduced in [1]. In this paper, we adapt it for the estimation of the Owen value, and provide both theoretical and experimental results for our procedure.

2   The Owen Value

In this section we give a brief introduction to cooperative games and the Owen value.

A transferable utility cooperative game (or, simply, a TU-game) is a pair (N, v), where N is the finite set of players and $v : 2^N \longrightarrow \mathbb{R}$ is a map satisfying $v(\emptyset) = 0$.

We denote by G^N the class of TU-games with set of players N. A coalition is a subset of players $T \subseteq N$ with t members. The coalition N is named the grand coalition. For each coalition $T \subseteq N$, $v(T)$ indicates the benefits that T generates when its members cooperate.[1]

The definition and analysis of rules for allocating the benefits generated by the cooperation of the players in the grand coalition is a central topic of cooperative game theory. Possibly, the most important of such rules is the Shapley value introduced in [8]. The Shapley value is defined, for every $i \in N$ and every $(N, v) \in G^N$, as

$$Sh_i(N, v) = \sum_{T \subseteq N \setminus \{i\}} \frac{t! \, (n - t - 1)!}{n!} (v(T \cup \{i\}) - v(T)).$$

An alternative formulation of the Shapley value can be given in terms of permutations. We denote by $\Pi(N)$ the set of permutations of the players in N. For each $\sigma \in \Pi(N)$, the set of predecessors of agent $i \in N$ according to σ is denoted by P_i^σ and defined as

$$P_i^\sigma = \{j \in N : \sigma(j) < \sigma(i)\}.$$

The Shapley value can be written as

$$Sh_i(N, v) = \frac{1}{|\Pi(N)|} \sum_{\sigma \in \Pi(N)} (v(P_i^\sigma \cup \{i\}) - v(P_i^\sigma))$$

for every $i \in N$ and every $(N, v) \in G^N$.

---

[1]Alternatively, a TU-game can map each coalition to the cost that its members support when cooperating. In this case, the game is said to be a cost game and is denoted by (N, c).

Now we introduce the model of TU-games with a priori unions and the Owen value.

Definition 2.1 A TU-game with a priori unions is a triplet (N, v, P) where $(N, v) \in G^N$ and $P = \{P^1, \ldots, P^m\}$ is a partition of N. P is interpreted as a coalition structure that restricts the cooperation among the players in N.

We denote by U^N the class of TU-games with a priori unions and set of players N. The Owen value is the extension of the Shapley value for TU-games with a priori unions. Take $(N, v, P) \in U^N$ and $i \in N$. Denote by P_i the union to which i belongs. The Owen value $O_i(N, v, P)$ is defined as

$$\sum_{Q \subseteq P \setminus \{P_i\}} \sum_{T \subseteq P_i \setminus \{i\}} \frac{t!(p_i - t - 1)!q!(m - q - 1)!}{p_i!m!} (v(\bigcup_{P^a \in Q} P^a \cup T \cup \{i\}) - v(\bigcup_{P^a \in Q} P^a \cup T)).$$

The Owen value can also be formulated in terms of permutations. A permutation $\sigma \in \Pi(N)$ is said to be compatible with a coalition structure P if the elements of each class of P are not separated by σ. Formally, if $\Pi_P(N)$ denotes the set of all permutations of N that are compatible with P, then $\sigma \in \Pi_P(N)$ if and only if for all $i, j, k \in N$ it holds that

$$P_i = P_j \text{ and } \sigma(i) < \sigma(k) < \sigma(j) \Rightarrow P_k = P_i = P_j.$$

Now, the Owen value of (N, v, P) can be rewritten, for every $i \in N$, as

$$O_i(N, v, P) = \frac{1}{|\Pi_P(N)|} \sum_{\sigma \in \Pi_P(N)} (v(P_i^\sigma \cup \{i\}) - v(P_i^\sigma)). \tag{1}$$

Notice that when $P = \{N\}$ or $P = \{\{i\}, i \in N\}$, $\Pi_P(N)$ is equal to $\Pi(N)$ and the Shapley value and Owen value coincide. Hence, the Owen value is an extension of the Shapley value. Since its introduction in [7], it has generated a large literature and has become an important rule for cooperative games.

3 A Sampling Procedure to Estimate the Owen Value

The computation of the Owen value for a particular TU-game with a priori unions can be a hard task from the computational point of view, because its complexity increases exponentially with the number of players. In this section, we describe and analyze a procedure to estimate the Owen value based on sampling. It is an extension of an analogous procedure to estimate the Shapley value introduced in [1].

We want to estimate the Owen value of a TU-game with a priori unions $(N, v, P) \in U^N$, with $P = \{P^1, \ldots, P^m\}$. The sampling procedure we propose is described below:

- The population of the sampling procedure is the set of all permutations of N compatible with P, i.e. $\Pi_P(N)$.
- The vector of parameters to be estimated is $O = (O_i)_{i \in N}$, where O_i denotes $O_i(N, v, P)$ for all $i \in N$.
- The characteristics to be studied in each sampling unit $\sigma \in \Pi_P(N)$ are the marginal contributions of the players according to σ, i.e. the vector $(x(\sigma)_i)_{i \in N}$, where $x(\sigma)_i = v(P_i^\sigma \cup \{i\}) - v(P_i^\sigma)$ for all $i \in N$.
- The sampling procedure takes each permutation $\sigma \in \Pi_P(N)$ with the same probability. To that aim, it chooses at random a permutation of the elements of each P^k ($k \in \{1, \dots, m\}$) and then it chooses at random a permutation of the elements of $\{1, \dots, m\}$. Combining this collection of $m + 1$ permutations, $\sigma \in \Pi_P(N)$ is obtained.
- The estimation of O is the mean of the marginal contributions vectors over the sample S, i.e. $\hat{O} = (\hat{O}_i)_{i \in N}$, where $\hat{O}_i = \frac{1}{s} \sum_{\sigma \in S} x(\sigma)_i$ for each $i \in N$ (s denoting the sample size).

Let us study now some properties of our sampling procedure from a statistical perspective. Fix $i \in N$. Clearly each estimator $\hat{O}_i$ is unbiased since

$$E(\hat{O}_i) = E\left(\frac{1}{s} \sum_{\sigma \in S} x(\sigma)_i\right) = E(x(\sigma)_i) = O_i.$$

Besides,

$$\text{Var}(\hat{O}_i) = \text{Var}\left(\frac{1}{s} \sum_{\sigma \in S} x(\sigma)_i\right) = \frac{\text{Var}(x(\sigma)_i)}{s}.$$

Hence, taking into account that

$$MSE(\hat{O}_i) = E(\hat{O}_i - O_i)^2 = (E(\hat{O}_i) - O_i)^2 + \text{Var}(\hat{O}_i),$$

it is clear that $MSE(\hat{O}_i)$ converges to zero when s tends to infinity.

Now we state and prove some results that can be helpful when choosing the sample size. This collection of results is a generalization of an analogous collection for the estimator of the Shapley value provided in [6].

Proposition 3.1 *Take* $\varepsilon > 0$, $\alpha \in (0, 1)$. *Then,*

$$s \geq \frac{\text{Var}(x(\sigma)_i)}{\alpha \varepsilon^2} \Rightarrow P(|\hat{O}_i - O_i| \geq \varepsilon) \leq \alpha.$$

Proof Clearly,

$$P(|\hat{O}_i - O_i| \geq \varepsilon) = P(|\hat{O}_i - E(\hat{O}_i)| \geq \varepsilon) = P\left(|\hat{O}_i - E(\hat{O}_i)| \geq \frac{\varepsilon \sqrt{\text{Var}(\hat{O}_i)}}{\sqrt{\text{Var}(\hat{O}_i)}}\right).$$

Applying Chebyshev's inequality, it holds that

$$P\left(|\hat{O}_i - E(\hat{O}_i)| \geq \frac{\varepsilon\sqrt{Var(\hat{O}_i)}}{\sqrt{Var(\hat{O}_i)}}\right) \leq \frac{Var(\hat{O}_i)}{\varepsilon^2} = \frac{Var(x(\sigma)_i)}{s\varepsilon^2} \leq \alpha. \qquad \square$$

Proposition 3.2 *Take $\varepsilon > 0$, $\alpha \in (0, 1)$ and denote*

$$r_i = \max_{\sigma, \sigma' \in \Pi_P(N)} (x(\sigma)_i - x(\sigma')_i).$$

Then,

$$s \geq \frac{\ln(2/\alpha)r_i^2}{2\varepsilon^2} \Rightarrow P(|\hat{O}_i - O_i| \geq \varepsilon) \leq \alpha.$$

Proof Clearly,

$$P(|\hat{O}_i - O_i| \geq \varepsilon) = P(|\hat{O}_i - E(\hat{O}_i)| \geq \varepsilon) = P(|\sum_{\sigma \in S} x(\sigma)_i - E(\sum_{\sigma \in S} x(\sigma)_i)| \geq \varepsilon s).$$

Applying Hoeffding's inequality,[2] it holds that

$$P(|\sum_{\sigma \in S} x(\sigma)_i - E(\sum_{\sigma \in S} x(\sigma)_i)| \geq \varepsilon s) \leq 2\exp(\frac{-2\varepsilon^2 s}{r_i^2}) \leq \alpha. \qquad \square$$

Propositions 3.1 and 3.2 can be used to choose the sample size when estimating the Owen value of a game with a priori unions. Proposition 3.2 refers to the range of the collection of marginal contributions, about which one usually have information. For instance, if (N, v) is a convex or a concave game[3] then it is clear that

$$r_i = |v(N) - v(N \setminus \{i\}) - v(\{i\})|.$$

Proposition 3.1 refers to the variance of the collection of marginal contributions, which is usually unknown. However, using the bound of the variance provided by the Popoviciu's inequality,[4] the following corollary follows immediately.

---

[2]*Hoeffding's inequality*: Let $\sum_{j=1}^{r} X_j$ be the sum of r independent random variables such that $a_j \leq X_j \leq b_j$ for all $j \in \{1, \ldots, r\}$. Then $P(|\sum_{j=1}^{r} X_j - E(\sum_{j=1}^{r} X_j)| \geq t) \leq 2\exp(\frac{-2t^2}{\sum_{j=1}^{r}(b_j - a_j)^2})$.

[3](N, v) is a convex game when for every $i \in N$ and every $K, T \subseteq N \setminus \{i\}$ with $K \subset T$, it holds that $v(K \cup \{i\}) - v(K) \leq v(T \cup \{i\}) - v(T)$. (N, v) is a concave game when $(N, -v)$ is convex.

[4]*Popoviciu's inequality on variances*: Let M and m be an upper and a lower bound on the values of a bounded random variable X with variance $Var(X)$. Then, $Var(X) \leq \frac{1}{4}(M - m)^2$.

Corollary 3.1 *Take $\varepsilon > 0$, $\alpha \in (0, 1)$. Then,*

$$s \geq \frac{r_i^2}{4\alpha\varepsilon^2} \;\Rightarrow\; P(|\hat{O}_i - O_i| \geq \varepsilon) \leq \alpha.$$

We can summarize the results and comments above in the following corollary.

Corollary 3.2 *(a) Take $\varepsilon > 0$, $\alpha \in (0, 1)$. Then,*

$$s \geq \min\left\{\frac{1}{4\alpha\varepsilon^2}, \frac{\ln(2/\alpha)}{2\varepsilon^2}\right\} r_i^2 \;\Rightarrow\; P(|\hat{O}_i - O_i| \geq \varepsilon) \leq \alpha.$$

(b) Take $\varepsilon > 0$, $\alpha \in (0, 1)$ and assume that (N, v) is a convex or a concave game. Then,

$$s \geq \min\left\{\frac{1}{4\alpha\varepsilon^2}, \frac{\ln(2/\alpha)}{2\varepsilon^2}\right\} \left(v(N) - v(N \setminus \{i\}) - v(\{i\})\right)^2$$

implies that $P(|\hat{O}_i - O_i| \geq \varepsilon) \leq \alpha$.

Notice that, typically, we deal with small values of α. In [6, p. 46] it is proved that if $\alpha \leq 0.23$ then

$$\min\left\{\frac{1}{4\alpha\varepsilon^2}, \frac{\ln(2/\alpha)}{2\varepsilon^2}\right\} = \frac{\ln(2/\alpha)}{2\varepsilon^2}.$$

4 An Example

In this section we illustrate the performance of our sampling procedure in an example taken from the game theoretical literature. The game in the example has many players so that its Owen value cannot be calculated using the general formula in (1); however, it can be calculated using a special formula. We compare the Owen value calculated with the Owen value estimated using the sampling procedure.

Our example is a cost game with a priori unions (N, c, P) studied in [9]. It is a special type of cost game, a so-called airport game, that can be used to define the fee for the planes operating in an airport. We now briefly explain what is an airport game (more details can be found in [5] or [4]). Suppose that $\mathscr{T}$ is the set of types of planes operating in an airport in a particular period. Denote by N_τ the set of movements that are made by planes of type $\tau \in \mathscr{T}$ and by N the set of all movements, i.e. $N = \cup_{\tau \in \mathscr{T}} N_\tau$. Let c_τ be the cost of a runway that is suitable for planes of type τ in the considered period. Without loss of generality, we assume that

$$c_1 \leq c_2 \leq \ldots \leq c_{|\mathscr{T}|}.$$

Now, for every $T \subseteq N$, $c(T)$ is defined as the cost of a runway that can be used by all the movements in T, i.e.

$$c(T) := \max\{c_\tau : T \cap N_\tau \neq \emptyset\}.$$

Defining a fee for each movement can be seen as dividing $c(N)$ among the movements. For doing this, [9] proposes to take into account that the planes are operated by airlines and, then, to model this cost allocation problem as a cost game with a priori unions (N, c, P), where P is the partition induced in N by the airlines to which the planes belong. Moreover, [9] proposes to use the Owen value to solve this cost allocation problem and provides a formula for the Owen value in this special case that makes it computable for real airport games (that typically involve large numbers of players).

Table 1 depicts the elements characterizing the cost game with a priori unions describing the movements in the first three months of 1993 at Lavacolla, the airport of Santiago de Compostela, Spain. It shows the partition of the movements in airlines, the numbers of movements, the types of planes with runway costs, and includes the calculated Owen value using the formula provided in [9]. Observe that the corresponding cost game with a priori unions has 1258 players and, hence, the Owen value of this game is a vector in $\mathbb{R}^{1258}$; however, the allocations proposed by the Owen value for all movements of the same type of plane belonging to the same airline are identical and then the Owen value of this game can be described giving its 25 possibly different components (one for each type in each airline).

Let us estimate the Owen value using our sampling procedure. What is a bearable error in this example? Notice that the runway costs are given in thousands of pesetas[5]; we consider that a bound for the absolute error of 6 thousands pesetas (about 36 euros) can be tolerated. It is a well-known fact that airport games are concave and thus, according to Corollary 3.2, the minimum sample sizes to guarantee that the absolute error is smaller than or equal to 6 with probability at least $1 - \alpha$ can be calculated; they are those in Table 2. The required sample sizes only depend on the types of planes, because r_i only depends on the type of i.

We take $\alpha = 0.1$ and estimate the Owen value using a sample of size $1.1 \cdot 10^8$. The estimated Owen value is a vector in $\mathbb{R}^{1258}$. Figure 1 depicts the absolute errors for all players and Table 3 shows a summary of those errors (the maximum and minimum error for each type of player) as well as the theoretical maximum errors with $\alpha \leq 0.1$ according to Corollary 3.2.

In view of Fig. 1 and Table 3, we conclude that our estimation is satisfactory in this example. Comparing the theoretical and observed absolute errors in Table 3 we can also conjecture that the absolute errors when using our procedure are significantly smaller than the absolute errors guaranteed by Corollary 3.2.

In fact we have performed a small simulation study to check this conjecture. We have fixed one player of Type 7, in particular Player 1258 that corresponds to one operation of a plane DC-10 of Viasa. We have estimated 1000 times the Owen value using our procedure with sample size 10^5. Table 4 displays the theoretical maximum

---

[5]The peseta was the official currency in Spain in 1993. One peseta is about 0.006 euros.

Table 1 Airlines using Lavacolla in the first three months of 1993 and the owen value

Airline	Movements	Types	Costs	Owen value
Air Europa	36	B-757(3)	32496	8.110
	172	B-737(5)	39494	11.183
Aviaco	12	DC-9(4)	34265	151.093
Britannia	6	B-737(5)	39494	369.224
British Airways	2	B-757(3)	32496	843.378
Condor Flugdienst	2	B-757(3)	32496	843.378
Caledonian Airways	2	B-757(3)	32496	843.378
Eurobelgian Airlines	2	B-737(5)	39494	1107.673
Futura	32	B-737(5)	39494	69.230
Gestair Executive Set	2	CESSNA(1)	8120	176.522
Iberia	452	DC-9(4)	34265	2.037
	438	B-727(6)	44850	9.070
Air Charter	2	B-737(5)	39494	1107.673
Corse Air	4	B-737(5)	39494	553.836
Air UK Leisure	2	B-737(5)	39494	1107.673
Ibertrans	2	CESSNA(1)	8120	176.522
LTE	36	B-757(3)	32496	46.854
Mac Aviation	6	LEARJET-25(2)	15134	120.367
Monarch Airlines Ltd	2	B-737(5)	39494	1107.673
Sobelair	6	B-737(5)	39494	369.224
Trabajos Aéreos	2	CESSNA(1)	8120	176.522
Tea Basel LTD	2	B-737(5)	39494	1107.673
Oleohidráulica Balear SA	4	CESSNA(1)	8120	88.261
Viasa	30	DC-10(7)	50000	334.778
Spanair	2	B-737(5)	39494	1107.673

absolute error for three values of α (according to Corollary 3.2), as well as the minimum, maximum and mean observed absolute errors in these 1000 estimations. Clearly, the observed errors are much smaller than the theoretical errors, that may let us think that the bound given by Corollary 3.2 is rather conservative.

Table 2 Sampling sizes for $\varepsilon = 6$ in our example

	$\alpha = 0.1$	$\alpha = 0.05$	$\alpha = 0.01$
Type 1	$2.74336 \cdot 10^6$	$3.37811 \cdot 10^6$	$4.85196 \cdot 10^6$
Type 2	$9.52967 \cdot 10^6$	$1.17346 \cdot 10^7$	$1.68544 \cdot 10^7$
Type 3	$4.39370 \cdot 10^7$	$5.41031 \cdot 10^7$	$7.77079 \cdot 10^7$
Type 4	$4.88508 \cdot 10^7$	$6.01539 \cdot 10^7$	$8.63987 \cdot 10^7$
Type 5	$6.48982 \cdot 10^7$	$7.99143 \cdot 10^7$	$1.14780 \cdot 10^8$
Type 6	$8.36942 \cdot 10^7$	$1.03059 \cdot 10^8$	$1.48023 \cdot 10^8$
Type 7	$1.04018 \cdot 10^8$	$1.28086 \cdot 10^8$	$1.83969 \cdot 10^8$

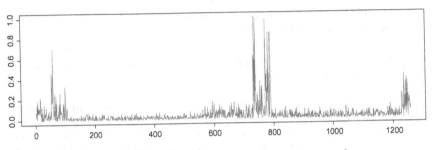

Fig. 1 Absolute error for each player in the airport problem with $s = 1.1 \cdot 10^8$

Table 3 Observed and theoretical errors for the different types of planes with $s = 1.1 \cdot 10^8$

	Type 1	Type 2	Type 3	Type 4	Type 5	Type 6	Type 7
Theoretical	0.94754	1.76601	3.79202	3.99844	4.60862	5.23363	5.83459
Maximum	0.19236	0.22385	0.70153	0.33207	1.00818	0.14056	0.42972
Minimum	0.01010	0.07045	0.00048	0.000098	0.00031	0.00017	0.01038

Table 4 Observed and theoretical errors with 1000 estimations

	$\alpha = 0.1$	$\alpha = 0.05$	$\alpha = 0.01$
Theoretical error	193.51	214.73	257.35
	Minimum	Maximum	Mean
Absolute error	0.004	28.71	6.53

5 Concluding Remarks

In this paper we have provided a procedure based on sampling to estimate the Owen value of a cooperative game that is specially useful when dealing with games having large sets of players. We have theoretical results that show that the procedure is accurate and that allow us to choose the sample size guaranteeing the a priori established requirements on the errors. We have illustrated the behavior of the

Table 5 Sampling sizes for games with $r_i = 1$ for all $i \in N$

α	$\varepsilon = 0.1$	$\varepsilon = 0.05$	$\varepsilon = 0.01$	$\varepsilon = 0.005$	$\varepsilon = 0.001$	$\varepsilon = 0.0005$
0.15	130	519	12952	51806	1295134	5180534
0.1	150	600	14979	59915	1497866	5991465
0.05	185	738	18445	73778	1844440	7377759
0.01	265	1060	26492	105967	2649159	10596635

procedure in a real example taken from the game theoretical literature. The example we consider requires the use of large samples, but this is not always the case. For instance, Table 5 displays the minimum sample sizes to guarantee that the absolute error when estimating the Owen value with our procedure is smaller than or equal to ε with probability at least $1 - \alpha$ for any game (N, v) with $r_i = 1$ for all $i \in N$; notice that this condition is satisfied, for example, by simple games that form a class of games for which the Owen value is specially relevant.

Acknowledgements This work has been supported by MINECO grants MTM2014-53395-C3-1-P, MTM2014-53395-C3-2-P, MTM2014-53395-C3-3-P, and by the Xunta de Galicia through the ERDF (Grupos de Referencia Competitiva ED431C-2016-015 and ED431C-2016-040, and Centro Singular de Investigación de Galicia ED431G/01). Authors acknowledge J. Costa and P. Saavedra-Nieves for their comments on an earlier version of this paper.
Ignacio García-Jurado and M. Gloria Fiestras-Janeiro would like to sincerely thank Pedro Gil for his affectionate help and welcome at various moments in their careers. In particular, Pedro Gil was the organizer of the first conference that they attended, held in Gijón, Spain, in 1985.

References

1. Castro J, Gómez D, Tejada J (2009) Polynomial calculation of the Shapley value based on sampling. Comput Oper Res 36:1726–1730
2. Costa J (2016) A polynomial expression for the Owen value in the maintenance cost game. Optimization 65:797–809
3. Fiestras-Janeiro MG, García-Jurado I, Mosquera MA (2011) Cooperative games and cost allocation problems. Top 19:1–22
4. González-Díaz J, García-Jurado I, Fiestras-Janeiro MG (2010) An introductory course on mathematical game theory. American Mathematical Society, Providence
5. Littlechild SC, Owen G (1973) A simple expression for the Shapley value in a special case. Manag Sci 20:370–372
6. Maleki S (2015) Addressing the computational issues of the Shapley value with applications in the smart grid. Ph.D. thesis, Faculty of Physical Sciences and Engineering Electronics and Computer Science, Southampton University, Southampton
7. Owen G (1977) Values of games with a priori unions. In: Henn R, Moeschlin O (eds) Mathematical economics and game theory. Springer, Berlin
8. Shapley LS (1953) A value for n-person games. In: Kuhn HW, Tucker AW (eds) Contributions to the theory of games II. Princeton University Press, Princeton
9. Vázquez-Brage M, van den Nouweland A, García-Jurado I (1997) Owen's coalitional value and aircraft landing fees. Math Soc Sci 34:273–286

The β-Gradient for Testing Probability Profiles

Miquel Salicrú, Juan José Barreiro and María Isabel González-Framil

Abstract With multinomial response (h,ϕ)-diversity has been widely used as an alternative to variance when characterizing the dispersion. In this context the β-gradient has been related to both the Pseudo-F and to the variation rate. In this work, asymptotic distribution of the β-gradient has been obtained and, as a result, the following inferential applications have been considered: contrast between probability profiles (hypothesis test) and reliability of the β-gradient estimation (confidence intervals). In order to illustrate the applicability of this result, the variation rate has been used to follow the time course of Barret's esophagus.

1 Introduction

Using the variance as a dispersion measure and decomposing it into a sample of observations, Fisher [5] introduced the analysis of variance to determine the significance of factors potentially affecting the variable response. Considering the introduced approach for a reduced number of fixed factors, a variety of models have been described in the literature. These models adapt well to the specific characteristics of many experimental situations. The generalization of the response (distribution), the factors (typology and structure), the relationship between observations (independence or correlation) and the broad computational implementation (parametric and non-parametric) have made the design of experiments a usual tool for multiple

M. Salicrú (✉)
Genetics, Microbiology and Statistics Dep, Universitat de Barcelona, Av Diagonal 643, 08028 Barcelona, Spain
e-mail: msalicru@ub.edu

J. J. Barreiro
Departamento de Física Matemática y de Fluidos, Carretera de Vigo, Universidad Nacional de Educación a Distancia, Torres do Pino s/n, 32001 Ourense, Spain
e-mail: jbarreiro@ourense.uned.es

M. I. González-Framil
Departamento de Medicina, Universidad de Oviedo, C/ Julián Clavería s/n, 33006 Oviedo, Spain
e-mail: framil@uniovi.es

© Springer International Publishing AG 2018
E. Gil et al. (eds.), *The Mathematics of the Uncertain*, Studies in Systems, Decision and Control 142, https://doi.org/10.1007/978-3-319-73848-2_34

357

fields of knowledge (engineering, biotechnology, agronomy, biology, medicine, psychology,...). In this context, Rao [17–19] raises the idea of considering other dispersal measures as an alternative to the variance. In particular, he explores the use and decomposition of the diversity/entropy and, with the aim of characterizing the assemblage of populations (ecology of populations), he focuses in both Gini's and quadratic entropy. The interest of these approximations is due to: (a) the differential characterisation of variability in different fields of knowledge; (b) the reestablishment of the variance as a measure of dispersion when considering normal distribution responses; and (c) the special adaptation of these measures to categorical responses with multinomial distribution (probability profiles).

Being their first uses in thermodynamics and telecommunications (Clausius [2], Shannon [22]), diversity measures have been used in multiple domains to characterise concepts and functionalities (Patil and Taillie [15]). Among other applications, diversity has been associated with the following: social structure of populations (Eagle et al. [4]); industrial clusters and risk (Attaran [1]); homogeneity in the uses of the territory and the structures of the facades (Li and Guo [12], Klir [10]); level of complexity of a language or dialect (Juola [9]); channel capacity when using MIMO antenna configurations (Jensen and Morris [7]); human variation (Lewontin [11], Molnar [14]); level of complexity of viral quasispecies (Gregori et al. [6]); and clinical imaging diagnosis (Zhang et al. [23]). Exponential growth of the diversity/entropy measures described in the literature led to the introduction of (h,ϕ)-diversity functional (Salicrú et al. [21], Pardo et al. [16]):

$$H_h^\phi (P) = h \left(\sum_{i=1}^{s} \phi (p_i) \right) \tag{1}$$

being h and ϕ functions twice differentiable with continuity and $P = (p_1, p_2, \ldots, p_s)$ a probability vector $(p_1 + p_2 + \cdots + p_s = 1, p_i \geq 0)$. By specifying functions h and ϕ, functional integrates most of the measures described in the literature. For example: for $h(x) = x$ and $\phi(x) = -x \cdot \ln x$ Shannon entropy is obtained; for $h(x) = (1 - \alpha)^{-1} \cdot (x - 1)$ and $\phi(x) = x^\alpha$ Havrda–Charvát family is obtained; and for $h(x) = (1 - \alpha)^{-1} \cdot \ln x$ and $\phi(x) = x^\alpha$ Renyi family is obtained. With independent samples of size n, the estimation of diversity can be expressed as follows:

$$H_h^\phi \left(\hat{P} \right) = h \left(\sum_{i=1}^{s} \phi (\hat{p}_i) \right) \tag{2}$$

being $\hat{P} = (\hat{p}_1, \hat{p}_2, \ldots, \hat{p}_s)$ the vector of estimated frequencies.

The broad use of these measures is not independent on the conceptual and methodological progress achieved in telecommunications, information theory, particle physics, economy, biomedicine and ecology. For the last of these areas, the decomposition of diversity for a one factor design made possible to explain the assemblage of populations in space and led to the introduction of new concepts (Jost [8]):

α-diversity (local diversity); β-diversity (gradient or variation rate); and γ-diversity (regional diversity). Formally, for r populations $\pi_1, \pi_2, \ldots, \pi_r$, with weight or probability a priori $\lambda_1, \lambda_2, \ldots, \lambda_r$ ($\lambda_1 + \lambda_2 + \cdots + \lambda_r = 1$) and a probability distribution characterized by the vectors $P_1, P_2, \ldots, P_r$, the diversity decomposes as follows (Rao [17, 18]):

$$H_h^\phi (P_0) = \sum_{i=1}^r \lambda_i H_h^\phi (P_i) + \left[H_h^\phi (P_0) - \sum_{i=1}^r \lambda_i H_h^\phi (P_i) \right]$$

being $H_h^\phi (P_0) = H_\gamma (P_1, P_2, \ldots, P_r)$ the total diversity, $\sum_{i=1}^r \lambda_i H_h^\phi (P_i) = H_\alpha (P_1, P_2, \ldots, P_r)$ the intrinsic diversity of populations (residual effect), $H_h^\phi (P_0) - \sum_{i=1}^r \lambda_i H_h^\phi (P_i) = J (P_1, P_2, \ldots, P_r)$ the diversity between populations (Jensen divergence), and $P_0 = \lambda_1 P_1 + \lambda_2 P_2 + \cdots + \lambda_r P_r$ the mixture of the populations. Conceptually, the total diversity and the intrinsic diversity of the populations are related to the γ-diversity and α-diversity described in ecology. With samples of sizes $n_1, n_2, \ldots, n_r$, the statistical equivalent to the F of Fisher–Snedecor (r-population test) and the β-gradient or variation rate between probability profiles (pseudo β-diversity in ecology) are written in the following form:

$$F_{exp}^* = \frac{H_h^\phi \left(\hat{P}_0\right) - \sum_{i=1}^r \lambda_i H_h^\phi \left(\hat{P}_i\right)}{\sum_{i=1}^r \lambda_i H_h^\phi \left(\hat{P}_i\right)} = \frac{H_h^\phi \left(\hat{P}_0\right)}{\sum_{i=1}^r \lambda_i H_h^\phi \left(\hat{P}_i\right)} - 1$$

and

$$\hat{H}_\beta = H_\beta \left(\hat{P}_1, \hat{P}_2, \ldots, \hat{P}_r\right) = \frac{H_h^\phi \left(\hat{P}_0\right)}{\sum_{i=1}^r \lambda_i H_h^\phi \left(\hat{P}_i\right)} = \frac{\hat{H}_\gamma}{\hat{H}_\alpha}. \tag{3}$$

This paper focuses its attention on assessing the variation rate in space or in time using (h,ϕ)-diversity. In Sect. 2, the asymptotic distribution of the estimated β-diversity has been obtained and consequently, the statistics and the decision criteria, which allow inferential application, have been determined. In Sect. 3, clinical monitoring of Barret's esophagus has been used to illustrate the obtained results. Finally, conclusions and future work are presented in Sect. 4.

2 Inference with the β-Gradient

The interest in the inference has been focused on both the realization of hypothesis tests (adherence to fixed value and comparison of β-gradients) and the construction of confidence intervals (precision in the estimation). For this reason, the distribution of $\hat{H}_\beta$ statistic has also been obtained.

2.1 Asymptotic Distribution

Taylor's first-order expansion of function $\hat{H}_\beta$ at point $(P_1, P_2, \ldots, P_r)$ provides the approximation:

$$\widehat{H}_\beta = H_\beta + \sum_{i=1}^{r} \sum_{j=1}^{s} t_{ij} \cdot (\hat{p}_{ij} - p_{ij}) + R_{2,1}$$

being

$$t_{ij} = \frac{1}{H_\alpha} h' \left(\sum_{k=1}^{s} \phi(p_{0k}) \right) \phi'(p_{0j}) \cdot \lambda_i - \frac{H_\gamma}{H_\alpha^2} h' \left(\sum_{k=1}^{s} \phi(p_{ik}) \right) \phi'(p_{ij}) \cdot \lambda_i$$

$P_i = (p_{i1}, p_{i2}, \ldots, p_{is})$, $P_0 = (p_{01}, p_{02}, \ldots, p_{0s})$, $p_{0k} = \lambda_1 p_{1k} + \lambda_2 p_{2k} + \cdots + \lambda_r p_{rk}$ and $R_{2,1}$ Lagrange's remainder of order 2 $(i = 1, \ldots, r; k = 1, \ldots, s)$.

Additionally, from considering the asymptotic distribution of the estimate of the vector of parameters,

$$\left(\hat{P}_1 - P_1, \hat{P}_2 - P_2, \ldots, \hat{P}_r - P_r \right) \xrightarrow{L} N \left(0, \sum_{P} \right)$$

we obtain:

$$\widehat{H}_\beta - H_\beta \xrightarrow{L} N \left(0, \sigma_\beta \right)$$

being $\sigma_\beta^2 = T' \sum_P T$, $T' = (t_{11}, \ldots, t_{1s}, \ldots, t_{r1}, \ldots, t_{rs})$ the partial derivative vector, $\sum_P$ the block matrix in which the positions of the diagonal are occupied by matrices $\sum_{P_i} / n_i$ and the rest by matrices 0 and $\sum_{P_i} = (\delta_{hk} p_{ih} - p_{ih} p_{ik})_{hk}$ the variance-covariance matrix corresponding to the probability vector P_i (δ_{hk} corresponds to the Kronecker delta). Consequently, the following result is demonstrated:

Theorem 2.1 *With independent samples of size* $n_1, n_2, \ldots, n_r$,

$$\widehat{H}_\beta - H_\beta \xrightarrow{L} N (0, \sigma_\beta)$$

being

$$\hat{\sigma}_\beta^2 = \left[\frac{H_\gamma}{H_\alpha^2} \right]^2 \sum_{i=1}^{r} \frac{\lambda_i^2}{n_i} \sigma_i^2 + \frac{1}{H_\alpha^2} \sum_{i=1}^{r} \frac{\lambda_i^2}{n_i} \sigma_0^2 - 2 \frac{H_\gamma}{H_\alpha^3} \sum_{i=1}^{r} \frac{\lambda_i^2}{n_i} \sigma_{0i} \qquad (4)$$

$$\hat{\sigma}_i^2 = \left[h' \left(\sum_{j=1}^{s} \phi(\hat{p}_{ij}) \right) \right]^2 \left\{ \sum_{j=1}^{s} \hat{p}_{ij} [\phi'(\hat{p}_{ij})]^2 - \left[\sum_{j=1}^{s} \hat{p}_{ij} \cdot \phi'(\hat{p}_{ij}) \right]^2 \right\}$$

$$\hat{\sigma}_0^2 = \left[h' \left(\sum_{j=1}^{s} \phi\left(\hat{p}_{0j}\right) \right) \right]^2 \left\{ \sum_{j=1}^{s} \hat{p}_{ij} \left[\phi'\left(\hat{p}_{0j}\right) \right]^2 - \left[\sum_{j=1}^{s} \hat{p}_{ij} \cdot \phi'\left(\hat{p}_{0j}\right) \right]^2 \right\}$$

with

$$\hat{\sigma}_{0i} = h' \left(\sum_{j=1}^{s} \phi\left(\hat{p}_{ij}\right) \right) \cdot h' \left(\sum_{j=1}^{s} \phi\left(\hat{p}_{0j}\right) \right)$$

$$\left\{ \sum_{j=1}^{s} \hat{p}_{ij} \left[\phi'\left(\hat{p}_{ij}\right) \right] \left[\phi'\left(\hat{p}_{0j}\right) \right] - \left[\sum_{j=1}^{s} \hat{p}_{ij} \cdot \phi'\left(\hat{p}_{ij}\right) \right] \left[\sum_{j=1}^{s} \hat{p}_{ij} \cdot \phi'\left(\hat{p}_{0j}\right) \right] \right\}$$

Remark 2.1 The asymptotic approximation of the quadratic form to a non-central chi-square distribution explains the convergence of $R_{2,1}$ to 0 (Dik and Gunst [3]). On the other hand, when weights of the populations have not been established a priori, they can be estimated as: $\hat{\lambda}_i = n_i / (n_1 + \cdots + n_r)$.

2.2 Inferential Applications

Analogous considerations to the described by Pardo et al. [16] lead to resolve the hypothesis test and to obtain the precision of the estimation:

a. One-sample hypothesis test

$$H_0 : H_\beta = H_\beta\left(0\right)$$

For this experimental situation the statistic to be used is:

$$Z_{exp} = \frac{\hat{H}_\beta - H_\beta\left(0\right)}{\hat{\sigma}_\beta} \approx N(0,1) \tag{5}$$

being $\hat{\sigma}_\beta$ the β-variance estimator. The p-value is obtained in the standard manner: $p = 2P\left[Z_{N(0,1)} > \left|Z_{exp}\right|\right]$ when $H_1 : H_\beta \neq H_\beta\left(0\right)$; $p = P\left[Z_{N(0,1)} > Z_{exp}\right]$ when $H_1 : H_\beta > H_\beta\left(0\right)$; and $p = 1 - P\left[Z_{N(0,1)} > Z_{exp}\right]$ when $H_1 : H_\beta < H_\beta\left(0\right)$. The asymptotic power of the test is obtained by taking into account the distribution of the statistic described in Theorem 2.1.

b. Two-sample hypothesis test

$$H_0 : H_\beta\left(1\right) = H_\beta\left(2\right)$$

For this experimental situation the statistic to be used is:

$$Z_{exp} = \frac{\hat{H}_\beta(1) - \hat{H}_\beta(2)}{\sqrt{\hat{\sigma}_\beta^2(1) + \hat{\sigma}_\beta^2(2)}} \approx N(0, 1).$$ (6)

The p-value and the asymptotic power of the test are obtained using the same method described above.

c. Confidence intervals. The confidence limits are, for β-gradient:

$$\hat{H}_\beta \pm z_{1-\alpha/2} \cdot \sigma_\beta$$ (7)

and for the difference of β-gradients $H_\beta(1) - H_\beta(2)$,

$$\hat{H}_\beta(1) - \hat{H}_\beta(2) \pm z_{1-\alpha/2}\sqrt{\hat{\sigma}_\beta^2(1) + \hat{\sigma}_\beta^2(2)}.$$ (8)

3 Example: Clinical Follow-Up Over Time

Barret's esophagus is a premalignant disorder in which the normal stratified squamous epithelium of the distal esophagus is replaced by intestinal metaplasia. Recent research has related genomic instability and clonal evolution to the progression of esophageal adenocarcinoma. More precisely, to clonal abundance, entropy and diversity. Introducing parallel schemes to the ones described for population dynamics (ecology and evolution), the number of loci showing differences has been used to characterize abundance, its relative distribution has been used to obtain the entropy (Shannon, Gini–Simpson, Hill numbers, ...) and, the genetic diverge and the loss of specific genes (p16, p53,...) have been used to characterise diversity (Maley et al. [13], Reid et al. [20]).

Seeking dysplastic changes, the diagnosis/follow-up is routinely carried out by endoscopies (image) and biopsies (molecular data). Image processing generally uses filters in space and frequency domains, first-order and second-order moments, and Fourier and Wavelet transform. On the other hand, data collected from biopsies allow the evaluation of the variation and the contrasting of the clonal diversity over time: a) the biopsy analysis provides the number of clones n and the clonal profile $\hat{P} = (\hat{p}_1, \hat{p}_2, \ldots, \hat{p}_s)$; b) the β-gradient characterizes the variation over time and the confidence interval provides reliability to the estimate; and c) the hypothesis test $H_0 : H_\beta = 1 + \delta$ vs $H_1 : H_\beta > 1 + \delta$ ($\delta \approx$ clinical relevance) allows determining the clinical significance.

In this context, the information provided by three follow-up biopsies from the same patient served to illustrate the results obtained in this study. More precisely, using the clonal profiles of Barrett's segment obtained by flow cytometry:

Control 1 (t_0). $n_1 = 3255$, $P_1 = (0.800, 0.100, 0.100, 0.000, 0.000, 0.000)$

Control 2 (t_1). $n_2 = 3425$, $P_2 = (0.750, 0.100, 0.100, 0.050, 0.000, 0.000)$

Control 3 (t_2). $n_3 = 3120$, $P_3 = (0.625, 0.125, 0.100, 0.075, 0.050, 0.025)$

$H_0 : H_\beta = 1.05$ vs $H_1 : H_\beta > 1.05$ $(\delta = 0.05)$ have been compared. The results after applying Shannon's entropy $(h(x) = x, \phi(x) = -x \cdot \ln x)$ are presented below:

$$\hat{H}_\beta = 1.0554, \quad \hat{\sigma}_\beta = 0.0023, \quad Z_{exp} = 2.35 \quad \text{and} \quad p-\text{value} = 0.0094.$$

Having demonstrated the significance of the clonal diversity variation, the confidence interval corresponding to the β-gradient establishes the precision in the estimate:

$$1.0554 \pm 0.0044 \quad (\alpha = 0.05).$$

The uncertainty associated with this estimate has been related to the randomness derived from the sample extraction process and the variation in sample medical analysis.

In environments where errors are so costly, decisions are based on observable facts, on the comparison and compatibility of information from different sources, and on the medical treatment choice that involves maximum benefit assuming minimum risk. For this reason, when considered in isolation, the information provided by the β-gradient has a limited value in itself. Its interest becomes clear when included in a wider framework.

4 Conclusions and Future Orientation

The asymptotic distribution of the β-gradient allows to obtain the reliability of the estimation and to verify the significance of differences in space and time. From the experimental perspective, this approximation provides an alternative to solve the design of one factor with multinomial response. The present and future challenges are oriented in several directions: (a) to favor clinical diagnosis, to identify complementary indicators, to obtain inferential results and to treat the information in relational models; (b) to ensure application robustness, to improve the estimation reliability when sample sizes are small; (c) to obtain cause-effect relationships, to extend the described approach to more factors and structures; and (d) to favor the application by implementing a software solution. To improve the reliability of the estimation we are testing the combination of analytical and computational techniques (bootstrap-t approaches with asymptotic variance in order to reduce computational cost). Without replicating experimental conditions we are testing the use of the asymptotic variance of (h,ϕ)-diversity as a substitute for the residual variance, and combined with the usual sums of squares, deducing the statistics that allow the extension to more factors and structures.

Acknowledgements The authors are especially thankful to Jaume F. Comas (Centros Científicos y Tecnológicos, Universidad de Barcelona) for his valuable suggestions in the carrying out of this study. This work was supported in part by research project 2014 SGR 464 (GRBIO) of the Departament d'Economia i Coneixement, Generalitat de Catalunya.

References

1. Attaran M (1986) Industrial diversity and economic performance in US areas. Ann Reg Sci 20(2):44–54
2. Clausius R (1867) Abhandlungen ber die mechanische Wärmetheorie. Druck und Verlag von Friedrich Viewegund Sohn, Braunschweig
3. Dik JJ, Gunst MCM (1985) The distribution of general quadratic forms in normal variables. Statist Neederl 39:14–26
4. Eagle N, Macy M, Claxton R (2010) Network diversity and economic development. Science 328(5981):1029–1031
5. Fisher RA (1925) In: Oliver and Boyd (eds) Statistical Methods for Research Workers. Edinburgh
6. Gregori J, Salicrú M, Domingo E, Sánchez A, Rodríguez-Frías F, Quer J (2014) Inference with viral quasispecies diversity indices: clonal and NGS approaches. Bioinformatics 30(8):1104–1111
7. Jensen MA, Morris ML (2005) Efficient capacity-based antenna selection for MIMO systems. IEEE Trans Veh Techol 54(1):110–116
8. Jost L (2007) Partitioning diversity into independent alpha and beta components. Ecology 88(10):2427–2439
9. Juola P (1998) Measuring linguistic complexity: the morphological tier. J Quant Ling 5(3):206–213
10. Klir G (2013) Architecture of systems problem solving. Springer Science and Business Media-Plenum Press, New York
11. Lewontin RC (1972) The apportionment of human diversity. In: Dobzhansky T, Hecht MK, Steere WC (eds) Evolutionary Biology, vol 6. Springer-Appleton Century Crofts, New York
12. Li J, Guo QS (2002) Analysis of dynamic evolvement in urban land-use composition based on Shannon entropy. Resourc Env Yangtze Basin 11(5):393–397
13. Maley CC, Galipeau PC, Finley JC, Wongsurawat VJ, Li X, Sanchez CA, Reid BJ (2006) Genetic clonal diversity predicts progression to esophageal adenocarcinoma. Nature Genet 38(4):468–473
14. Molnar S (2016) Human variation: races, types, and ethnic groups, 6th edn. Routledge, New York
15. Patil GP, Taillie C (1982) Diversity as a concept and its measurement. J Am Stat Assoc 77(379):548–561
16. Pardo L, Morales D, Salicrú M, Menéndez ML (1997) Large sample behavior of entropy measures when parameters are estimated. Commun Stat Theor Methods 26(2):483–501
17. Rao CR (1982a) Diversity and dissimilarity coefficients: a unified approach. Theor Popul Biol 21(1):24–43
18. Rao CR (1982b) Diversity: Its measurement, decomposition, apportionment and analysis. Sankhyā Ser A 44(1):1–22
19. Rao CR (2010) Quadratic entropy and analysis of diversity. Sankhyā Ser A 72(1):70–80
20. Reid BJ, Kostadinov R, Maley CC (2011) New strategies in Barrett's esophagus: integrating clonal evolutionary theory with clinical management. Clin Cancer Res 17(11):3512–3519

21. Salicrú M, Menéndez ML, Morales D, Pardo L (1993) Asymptotic distribution of (h,φ)-entropies. Commun Stat Theory Methods 22(7):2015–2031
22. Shannon CE (1948) A mathematical theory of communications. Bell Syst Tech J 27:379–423
23. Zhang Y, Dong Z, Wang S, Ji G, Yang J (2015) Preclinical diagnosis of magnetic resonance (MR) brain images via discrete wavelet packet transform with Tsallis entropy and generalized eigenvalue proximal support vector machine (GEPSVM). Entropy 17(4):1795–1813

Mixtures of Gaussians as a Proxy in Hybrid Bayesian Networks

Antonio Salmerón and Fernando Reche

Abstract In this paper we explore the use of mixtures of Gaussians as a proxy for mixtures of truncated basis functions in hybrid Bayesian networks. The idea is to use mixtures of Gaussians during the learning process, and move to mixtures of truncated basis functions for carrying out probabilistic inference. This would bridge the gap between efficient inference and learning in hybrid Bayesian networks, specially in scenarios where data comes in streams and models need to be continuously updated.

1 Introduction

Back in the 90s, Prof. Pedro Gil visited Almería and gave an excellent talk where he discussed the mathematics of uncertainty. It was an extremely interesting and motivating talk, in which he spoke about his goal of seeking what is in common between the different theories of uncertainty. Inspired by his thoughts, in our research group we kept on working deeper on probabilistic graphical models (PGMs) and more precisely on Bayesian networks (BNs) [7]. In spite of the adoption of probability theory as the tool for handling uncertainty, PGMs are general models in the sense that they focus on the underlying structure of the problem under study.

However, PGMs had some limitations that prevented them from being used in common real-world scenarios. More precisely, their use was problematic when the scenario involved discrete and continuous variables simultaneously, where the conditional Gaussian model [5] was the most developed solution, but their drawback was the limitation on the compatible graph structures, as discrete variables are not allowed to have continuous parents. Trying to overcome that limitation, and also with the aim of sidestepping the Gaussian assumption, we developed the mixture of truncated exponentials (MTE) model [6]. With a similar aim, mixtures of polynomials

A. Salmerón (✉) · F. Reche
Department of Mathematics, University of Almería, Almería, Spain
e-mail: antonio.salmeron@ual.es

F. Reche
e-mail: freche@ual.es

© Springer International Publishing AG 2018
E. Gil et al. (eds.), *The Mathematics of the Uncertain*, Studies in Systems, Decision and Control 142, https://doi.org/10.1007/978-3-319-73848-2_35

(MOPs) were developed later [9]. Both models can be regarded as particular cases of the more general mixtures of truncated basis functions (MoTBFs) [2].

MoTBFs offer a very efficient framework for inference in BNs with discrete and continuous variables [4], but still they have their own limitations. Namely, estimating MoTBFs from data is troublesome due to the fact that they do not belong to the exponential family. In current Big Data scenarios and more specifically when modeling streaming data, models inside the exponential family are quite convenient, as they provide sufficient statistics of dimension 1. It means that there is no need to store the full sample at any time, as knowing the value of each sufficient statistic is enough. When it comes to analyzing streaming data, it means that the estimation process can be carried out just taking into account the current value of the sufficient statistic and the newly arrived data item.

Mixtures of Gaussians (MoGs) do not belong to the exponential family, but they can be easily extended by including a hidden multinomial variable representing the components of the mixture. The extended model belongs to the exponential family. However, from the point of view of estimation, the advantages fade away due to the fact that there is no data for the hidden variable and therefore iterative algorithms like the EM algorithm [1] are required. There are, however, efficient estimation alternatives for MoGs able to deal with streaming data [8].

Our purpose in this paper is to explore the possibility of using MoGs as a proxy for MoTBFs. If that is possible, then we would benefit from the advantages of each one of them for estimation (learning) and inference respectively. The idea is to learn an MoG that could be quickly translated into an MoTBF that would be used for inference. For this to be successful, the translation procedure from MoGs to MoTBFs should be fast enough as to be negligible in terms of complexity with respect to the inference process.

2 Mixtures of Truncated Basis Functions

The MoTBF model is a generalization of MTEs and MOPs. The MTE model [6] is defined by the concept of MTE potential. Let $\mathbf{X}$ be a mixed n-dimensional random vector and let $\mathbf{Y} = (Y_1, \ldots, Y_d)^\mathsf{T}$ and $\mathbf{Z} = (Z_1, \ldots, Z_c)^\mathsf{T}$ be the discrete and continuous parts of $\mathbf{X}$, respectively, with $c + d = n$. We say that a function $f : \Omega_\mathbf{X} \mapsto \mathbb{R}_0^+$ is an *MTE potential* if for each fixed value $\mathbf{y} \in \Omega_\mathbf{Y}$ of the discrete variables $\mathbf{Y}$, the potential over the continuous variables $\mathbf{Z}$ is defined as:

$$f(\mathbf{z}) = a_0 + \sum_{i=1}^{m} a_i \exp\left\{\mathbf{b}_i^\mathsf{T} \mathbf{z}\right\}, \tag{1}$$

for all $\mathbf{z} \in \Omega_\mathbf{Z}$, where $a_i \in \mathbb{R}$ and $\mathbf{b}_i \in \mathbb{R}^c$, $i = 1, \ldots, m$. We also say that f is an MTE potential if there is a partition $\mathscr{I}_1, \ldots, \mathscr{I}_k$ of $\Omega_\mathbf{Z}$ into hypercubes and in each one of them, f is defined as in Eq. (1). An MTE potential is an *MTE density* if it integrates to 1.

MOPs were introduced in [9] as an alternative to the MTEs. The univariate MOP potential for a continuous variable Z is

$$f(z) = a_0 + \sum_{i=1}^{m} a_i z^i,$$

while for a multivariate continuous vector $\mathbf{Z} = (Z_1, \ldots, Z_c)^{\mathsf{T}}$ the potential takes the form

$$f(\mathbf{z}) = \prod_{j=1}^{c} \left\{ a_0^{(j)} + \sum_{i=1}^{m} a_i^{(j)} z^i \right\}.$$

The MoTBF framework is based on the abstract notion of real-valued *basis functions* $\psi(\cdot)$, which includes both polynomial and exponential functions as special cases. Let X be a continuous variable with domain $\Omega_X \subseteq \mathbb{R}$ and let $\psi_i : \mathbb{R} \to \mathbb{R}$, for $i = 0, \ldots, k$, define a collection of real basis functions. We say that a function $g_k : \Omega_X \mapsto \mathbb{R}_0^+$ is an MoTBF potential of order k wrt. $\Psi = \{\psi_0, \psi_1, \ldots, \psi_k\}$ if g_k can be written as [4]

$$g_k(x) = \sum_{i=0}^{k} \theta_i \, \psi_i(x), \tag{2}$$

where a_i are real numbers. The potential is a density if $\int_{\Omega_X} g_k(x)\, dx = 1$. Note that as opposed to the MTE and MOP definitions [6, 9], a marginal MoTBF potential does not employ interval refinement to improve its expressive power.

Example 1 By letting the basis functions correspond to polynomial functions, $\psi_i(x) = x^i$ for $i = 0, 1, \ldots$, the MoTBF model reduces to a MOP model for univariate distributions. Similarly, if we define the basis functions as $\psi_i(x) = \{1, \exp(-x), \exp(x), \exp(-2x), \exp(2x), \ldots\}$, the MoTBF model corresponds to an MTE model with the exception that the parameters in the exponential functions are fixed. Notice, however, that in both situations the model does not rely on a partitioning of the interval over which the distribution is defined (as opposed to the standard definitions of MOPs and MTEs).

2.1 Estimating Univariate MoTBFs from Data

We briefly describe the estimation procedure presented in [3]. It relies on the empirical cumulative distribution function (CDF) as a representation of the data, defined for a sample $D = \{x_1, \ldots, x_N\}$ as

$$G_N(x) = \frac{1}{N} \sum_{\ell=1}^{N} \mathbf{1}\{x_\ell \leq x\}, \quad x \in \mathbb{R}, \tag{3}$$

where $\mathbf{1}\{\cdot\}$ is the indicator function.

The method in [3] fits a potential, whose derivative is an MoTBF, to the empirical CDF using least squares. As an example, if we use polynomials as basis functions, $\Psi = \{1, x, x^2, x^3, \ldots\}$, the parameters of the CDF, denoted as $c_0, \ldots, c_k$, can be obtained solving the optimization problem

$$\underset{c_0,\ldots,c_k}{\text{minimize}} \quad \sum_{\ell=1}^{N} \left(G_N(x_\ell) - \sum_{i=0}^{k} c_i x_\ell^i \right)^2$$

$$\text{subject to} \quad \sum_{i=1}^{k} i\, c_i\, x^{i-1} \geq 0 \quad \forall x \in \Omega, \tag{4}$$

$$\sum_{i=0}^{k} c_i \alpha^i = 0 \text{ and } \sum_{i=0}^{k} c_i \beta^i = 1,$$

where the constraints ensure that the obtained parameters conform a valid CDF, and α and β are, respectively, the minimum and maximum of the data sample. More precisely, the first constraint guarantees that the corresponding density is nonnegative and the last two restrictions ensure that it integrates to 1, which is equivalent to say that the CDF is equal to 0 at the minimum of the variable and equal to 1 at the maximum. Note that the estimated function is not actually a density, but a CDF instead. An MoTBF density can be obtained just by derivation from the CDF.

3 MoTBF Approximation of MoGs

A translation procedure for efficiently finding an MoTBF approximation to any density function is described in [2]. It involves solving a convex optimization problem that minimizes an upper bound of the Kullback–Leibler divergence between the target density and the MoTBF approximation. As our goal is to embed the translation between MoGs and MoTBFs into a BN probabilistic inference process, we need a faster alternative. We will focus, from now on, on MOPs. A simple and fast alternative translation procedure is to use *moment matching*. More precisely, the MOP approximation would be obtained by solving a system of linear equations involving as many non-central moments as parameters the MOP approximation contains.

It is well known that the non-central moment of order k of an MoG is obtained as the weighted average of the non-central moments of each Gaussian component. In the case of MOPs, it is given by the following result.

Lemma 1 *Let* $f(x) = \sum_{i=0}^{m} a_i x^i$, $\alpha < x < \beta$ *be a MOP density with parameters* $a_0, \ldots, a_m$. *The non-central moment of order k of f is*

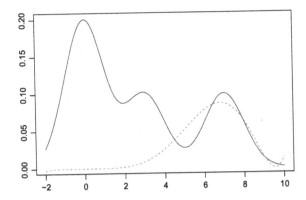

Fig. 1 A Gaussian mixture (solid line) and an MOP approximation (dashed line) obtained by moment matching

$$\mathbb{E}[X^k] = \sum_{i=0}^{m} \frac{a_i}{i+k+1}(\beta^{i+k+1} - \alpha^{i+k+1}).\tag{5}$$

Proof

$$\mathbb{E}[X^k] = \int_{\alpha}^{\beta} x^k f(x)dx = \int_{\alpha}^{\beta} \sum_{i=0}^{m} a_i x^{i+k}$$

$$= \sum_{i=0}^{m} \frac{a_i}{i+k+1} x^{i+k+1} \Big]_{\alpha}^{\beta} = \sum_{i=0}^{m} \frac{a_i}{i+k+1}(\beta^{i+k+1} - \alpha^{i+k+1}). \qquad \square$$

The problem of the moment-matching approach is that it does not guarantee that the resulting MOP is a density. In fact, there is not even guarantee that it is non negative. To no surprise, the obtained approximations are poor, as can be observed in Fig. 1.

Our proposal is to define a translation procedure based on the estimation method described in Sect. 2.1. The idea is to draw a small sample from the MoG that we want to approximate and apply the estimation method to the obtained sample, yielding a MOP density. If the sample is sufficiently small, the optimization step can be efficiently solved, as it is basically least squares on a set of sample points.

4 Evaluation

We have evaluated our proposed translation method on a series of examples implemented in R using the MoTBFs package.[1] The evaluation consisted on generating a sample of size 100 for 6 different MoG densities and then estimate a MOP density

---

[1] https://CRAN.R-project.org/package=MoTBFs.

Table 1 Parameters of the
MoGs depicted in Fig. 2

MoG	$\mathbf{w}^{\mathsf{T}}$	μ^{T}	σ^{T}
(a)	$(0.5, 0.25, 0.25)$	$(0, 3, 7)$	$(1, 1, 1)$
(b)	$(0.47, 0.06, 0.47)$	$(0, 3, 6)$	$(1, 5, 1)$
(c)	$(1/3, 1/3, 1/3)$	$(0, 3, 6)$	$(1, 1, 1)$
(d)	$(1/3, 1/3, 1/3)$	$(0, 2, 4)$	$(1, 1, 1)$
(e)	$(0.85, 0.15)$	$(0, 7)$	$(1, 1)$
(f)	$(0.4, 0.2, 0.2, 0.2)$	$(0, 3, 6, 9)$	$(1, 1, 1, 1)$

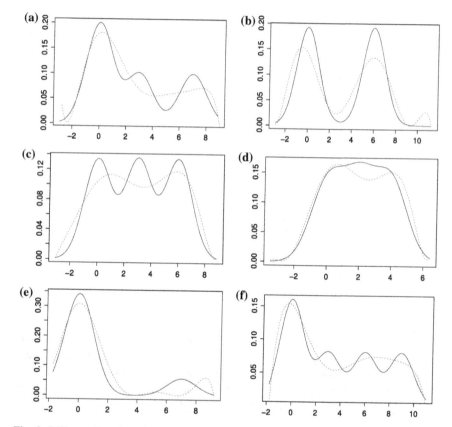

Fig. 2 Different Gaussian mixtures (solid line) and their MOP approximations (dashed line) obtained using the proposed translation procedure

with 8 parameters (i.e., degree 7) using the procedure in Sect. 2.1. The MoG densities used are described in Table 1, where $\mathbf{w}^{\mathsf{T}}$ is the vector of component weights, and μ^{T} and σ^{T} are the vector containing the means and standard deviations of each Gaussian component.

The sample MoG densities have been chosen so that they cover a wide variety of density shapes, including close to uniform (Fig. 2d), skewed (Fig. 2a, e and f) and

Table 2 Log-likelihood of the models depicted in Fig. 2

	MoG	MOP
(a)	−2.283841	−2.326817
(b)	−2.268437	−2.192675
(c)	−2.233056	−2.188837
(d)	−1.996243	−1.98867
(e)	−1.808173	−1.832089
(f)	−2.501593	−2.541394

multi-modality (Fig. 2a, c). The plots in Fig. 2 show how the MoTBF approximations are able to capture the shape of the MoG density reasonably well in all the cases.

We also carried out a comparison in terms of log-likelihood. We measured the average log-likelihood (i.e. the log-likelihood divided by the sample size) of the samples used to estimate the MOP densities, using the MOP and MoG models. The results are shown in Table 2. The results are similar and Wilcoxon's rank test does not find significant differences between both (p-value $= 0.9372$).

5 Conclusions

We have shown that MOPs can effectively approximate MoGs in a fast and accurate way. The translation is carried out by estimating a MOP density form a small artificial sample drawn from the MoG density. This paves the way to the use of MOPs (or MoTBFs in general) when learning from data streams. The solution is to use MoGs as a proxy, so that in fact what is learnt from the stream is a MoG density (which can be done with no need to store the full sample) and afterwards transform it into a MOP. Once the translations are completed, probabilistic inference can be carried out directly on MOPs.

Acknowledgements This research has been funded by the Spanish Ministry of Economy and Competitiveness, through projects TIN2013-46638-C3-1-P, TIN2016-77902-C3-3-P and by ERDF funds.

References

1. Dempster A, Larid N, Rubin D (1977) Maximum likelihood from incomplete data via the EM algorithm. J R Stat Soc Ser B 39(1):1–38
2. Langseth H, Nielsen T, Rumí R, Salmerón A (2012b) Mixtures of truncated basis functions. Int J Approx Reason 53(2):212–227
3. Langseth H, Nielsen T, Pérez-Bernabé I, Salmerón A (2014) Learning mixtures of truncated basis functions from data. Int J Approx Reason 55:940–956
4. Langseth H, Nielsen T, Rumí R, Salmerón A (2012) Inference in hybrid Bayesian networks with mixtures of truncated basis functions. In: Cano A, Gómez-Olmedo M, Nielsen T (eds)

PGM'2012: Proceedings of the 6th European workshop on probabilistic graphical models. Universidad de Granada, DECSAI

5. Lauritzen SL (1992) Propagation of probabilities, means and variances in mixed graphical association models. J Am Stat Assoc 87:1098–1108

6. Moral S, Rumí R, Salmerón A (2001) Mixtures of truncated exponentials in hybrid Bayesian networks. In: Benferhat S, Besnard P (eds) EQSCARU'2001, vol 2143. Lecture Notes in Artificial Intelligence. Springer, Berlin

7. Pearl J (1988) Probabilistic reasoning in intelligent systems. Morgan-Kaufmann, San Mateo

8. Pinto RC, Engel PM (2015) A fast incremental Gaussian mixture model. PLOS ONE 10(10):e0139, 931

9. Shenoy PP, West JC (2011) Inference in hybrid Bayesian networks using mixtures of polynomials. Int J Approx Reason 52:641–657

Sublinear Expectations: On Large Sample Behaviours, Monte Carlo Method, and Coherent Upper Previsions

Pedro Terán

Abstract Shige Peng's sublinear expectations generalize ordinary linear expectation operators. It is shown that the behaviour of sample averages of Peng i.i.d. variables may be very different from the probabilistic intuition. In particular, Peng's generalization of the Monte Carlo method is shown to be wrong. It is also observed that sublinear expectations coincide with Peter Walley's coherent upper previsions on a linear space.

1 A Tale of (Improbably) Three Pedros

It is my great pleasure to contribute to this volume honouring professor Pedro Gil. I met Pedro after participating in the local phase of the Mathematical Olympiad (which he organized) and he played a heavy role in my decision to become a Math student, a possibility I had not considered before. Some years later, I got a fellowship associated to his research project, which made it possible for me to get a doctoral degree and thus eventually led me to my current job. On that count alone, he has been one of the few most influential people in my life.

When I started as a fellow, the Department of Statistics at Oviedo had just spinned off from the Department of Mathematics. Both entities would continue to share equipment (like the copier machine) for some time but action was taken almost immediately to divide the post-graduates' room. So I found myself arriving there and being assigned no space until the works were done.

About one month later, I went to Pedro's office to inform him (in his capacity as head of the department) that I finally had a desk. 'But I have nothing to put on it anyway', I added nonchalantly. To my surprise, he smiled, emptied his pencil holder, and handed it to me. He went on to explain that it had been given to him decades earlier by Pedro Abellanas after a remarkably parallel conversation, and had been with him for all of his career. That token of appreciation I have cherished ever since.

P. Terán (✉)
Escuela Politécnica de Ingeniería. Departamento de Estadística e I.O. y D.M.,
Universidad de Oviedo, 33071 Gijón, Spain
e-mail: teranpedro@uniovi.es

© Springer International Publishing AG 2018
E. Gil et al. (eds.), *The Mathematics of the Uncertain*, Studies in Systems, Decision and Control 142, https://doi.org/10.1007/978-3-319-73848-2_36

This contribution is connected to Pedro's work on imprecise probability, a topic on which he was the advisor of Inés Couso and Enrique Miranda. It makes some observations about Shige Peng's theory of sublinear expectations. One is that a sublinear expectation is in fact the same thing as a coherent upper prevision on a linear space of gambles, in Walley's terminology. Another is that, although Peng's approach and particularly Peng's restrictive definition of independence allow one to obtain analogs of classical limit theorems, unfamiliar behaviours are possible too.

Finally, Peng's Monte Carlo method for computing sublinear expectations is wrong. That is a (rare, one should hope) instance of a counterexample to a result presented at a plenary lecture of the International Congress of Mathematicians, therefore something I'm happy to present in return for a pencil holder.

2 Introduction and Preliminaries

Shige Peng devised a purely analytical approach to stochastic analysis. While that provides a path to learning stochastic calculus without any knowledge of probability theory, we are interested here in a different side effect. Peng's technical machinery, which in essence is based on viscosity solutions of nonlinear partial differential equations, allows one to extend the reach of probability theory in such a way that the usual expectation is replaced by more general functionals which are sublinear.

Extended results include fundamental limit theorems like the law of large numbers [6] (to be discussed below), the central limit theorem, and the large deviation principle. Peng also constructed a non-additive generalization of the Brownian motion with its associated stochastic calculus. For an introduction to Peng's approach, the reader is referred to the lecture notes [6], survey [4], and 2010 ICM plenary lecture [5].

A *sublinear expectation space* is a triple $(\Omega, \mathcal{H}, \mathbb{E})$ where Ω is a set, $\mathcal{H}$ is a linear space of real functions on Ω (including the constants and closed under taking the absolute value), and $\mathbb{E} : \mathcal{H} \to \mathbb{R}$ is a mapping such that

(SE.i) $\mathbb{E}[X] \le \mathbb{E}[Y]$ if $X \le Y$,
(SE.ii) $\mathbb{E}[c] = c$ for all constant c,
(SE.iii) $\mathbb{E}[X + Y] \le \mathbb{E}[X] + \mathbb{E}[Y]$,
(SE.iv) $\mathbb{E}[cX] = c \cdot \mathbb{E}[X]$ for $c \ge 0$.

Elements of $\mathcal{H}$ will be called *random variables*. A subset $A \subset \Omega$ whose indicator function I_A is in $\mathcal{H}$ will be called an *event* (although, events are conspicuously absent from Peng's approach). Its complement will be denoted A^c.

It turns out, as a consequence of the Hahn–Banach theorem, that every sublinear expectation admits the representation

$$\mathbb{E}[X] = \sup_{E \in \mathfrak{E}} E[X]$$

for some family of linear functionals $\mathfrak{E}$ (whose restrictions to events are thus finitely additive probabilities).

Set

$$C_{l.Lip}(\mathbb{R}^n) = \{\varphi : \mathbb{R}^n \to \mathbb{R} \mid \exists C, m \geq 0 \mid \forall x, y \in \mathbb{R}^n,$$

$$|\varphi(x) - \varphi(y)| \leq C(1 + |x|^m + |y|^m)|x - y|\}.$$

In order to consider independence or identical distribution in Peng's sense, one must also assume that $X \in \mathscr{H}$ implies $\varphi(X) \in \mathscr{H}$ for each $\varphi \in C_{l.Lip}(\mathbb{R})$. A random variable X will be called *Peng independent* of a random variable Y if

$$\mathbb{E}[\varphi(X, Y)] = \mathbb{E}[\mathbb{E}[\varphi(x, Y)]_{x=X}] \text{ for all } \varphi \in C_{l.Lip}(\mathbb{R}^2),$$

where $\mathbb{E}[\varphi(x, Y)]_{x=X}$ denotes the mapping $\omega \in \Omega \mapsto \mathbb{E}[\varphi(X(\omega), Y)]$. Observe that, under this definition, 'X independent of Y' does not imply 'Y independent of X' and, in fact, these are seldom equivalent [2].

Random variables X, Y will be called *Peng identically distributed* if

$$\mathbb{E}[\varphi(X)] = \mathbb{E}[\varphi(Y)] \text{ for all } \varphi \in C_{l.Lip}(\mathbb{R}).$$

They will be called *Peng i.i.d.* if they are Peng identically distributed and X is Peng independent of Y.

A *random vector* is an element $(X_1, \ldots, X_n)$ of $\mathscr{H}^n$ such that $\varphi(X_1, \ldots, X_n) \in \mathscr{H}$ for each $\varphi \in C_{l.Lip}(\mathbb{R}^n)$. A random variable Y will be called *Peng independent* of a random vector $(X_1, \ldots, X_n)$ if

$$\mathbb{E}[\varphi(X_1, \ldots, X_n, Y)] = \mathbb{E}[\mathbb{E}[\varphi(x_1, \ldots, x_n, Y)]_{x_1=X_1, \ldots, x_n=X_n}]$$

for every $\varphi \in C_{l.Lip}(\mathbb{R}^{n+1})$. A sequence $\{X_n\}_n$ will be called *Peng independent and identically distributed* (Peng i.i.d.) if X_n is identically distributed as X_m for all n, m, and X_{n+1} is Peng independent of $(X_1, \ldots, X_n)$ for all $n \in \mathbb{N}$.

Dual to a sublinear expectation $\mathbb{E}$ is the superlinear functional $\mathscr{E}$ given by $\mathscr{E}[X] = -\mathbb{E}[-X]$. Unlike in ordinary probability theory, in general $\mathscr{E} \neq \mathbb{E}$. If $\mathbb{E}$ is linear and $\mathbb{E}[X_n] \to 0$ whenever X_n is a decreasing sequence converging to 0, then $\mathbb{E}$ is the expectation with respect to some probability measure. In that case one can check that, taking $\mathscr{H}$ to be a linear space of measurable functions, the notions above coincide with the usual ones.

Peng's PDE background and methods explain the idiosincratic choice of disregarding events and defining the distribution of a random variable only in terms of expectations of the locally Lipschitz functions in $C_{l.Lip}(\mathbb{R})$. To handle events, it is convenient to measure them with

$$\nu(A) = \mathscr{E}[I_A], \quad \overline{\nu}(A) = \mathbb{E}[I_A] = 1 - \nu(A^c).$$

The notation $\nu, \overline{\nu}$ reflects the fact that, from a probabilistic point of view, it seems to be more intuitive to use ν since $\overline{\nu}(A) = \overline{\nu}(B) = 1$ can happen for disjoint A, B; hence, a set of $\overline{\nu}$-measure 1 is not necessarily 'big' in the sense of probability theory. But a set of ν-measure 1 is 'big' because, dually, ν is the infimum of a family of finitely additive probabilities.

We will say that a sequence of random variables $\{X_n\}_n$ converges to a random variable X *almost surely* if $\nu(X_n \to X) = 1$, *in probability* if $\nu(|X_n - X| < \varepsilon) \to 1$ for each $\varepsilon > 0$, and *in law* if $\mathbb{E}[\varphi(X_n)] \to \mathbb{E}[\varphi(X)]$ for each $\varphi \in C_{l.Lip}(\mathbb{R})$.

Of these types of convergence, only convergence in law was considered by Peng. To ensure $\{X_n \to X\}$ and $\{|X_n - X| < \varepsilon\}$ are events, Peng's assumptions on $\mathscr{H}$ are not enough. That will not be a problem in this paper.

3 Counterexample to Monte Carlo Method, and Some Unexpected Behaviours

For a given sequence $\{X_n\}_n$ of random variables, we write

$$S_n = n^{-1} \sum_{i=1}^{n} X_i.$$

Peng proved the following law of large numbers for sublinear expectations.

Proposition 3.1 [6, Theorem II.3.1] *Let* $(\Omega, \mathscr{H}, \mathbb{E})$ *be a sublinear expectation space. Let* $\{X_n\}_n$ *be a sequence Peng i.i.d. as a random variable* X. *Then,*

$$\mathbb{E}[\varphi(S_n)] \to \sup_{x \in [\mathscr{E}[X], \mathbb{E}[X]]} \varphi(x)$$

for each $\varphi \in C_{l.Lip}(\mathbb{R})$.

It is interesting to note that, when $\mathbb{E}$ is the ordinary expectation against a probability measure, Peng's law is equivalent to a weak law of large numbers.

Corollary 3.1 (Weak law of large numbers) *Let* X *be an integrable random variable on a probability space* $(\Omega, \mathscr{A}, P)$. *Let* $\{X_n\}_n$ *be a sequence i.i.d. as* X. *Then*

$$S_n \to E(X)$$

in probability.

Proof It suffices to prove it in the case when X is bounded, since standard truncation techniques yield then the more general result (see e.g. the proof of [7, Theorem 7.12]). Thus we assume without loss of generality that X takes on values in a compact interval $[a, b]$.

Let f be an arbitrary real continuous bounded function, and fix $\varepsilon > 0$. By the Weierstrass approximation theorem, there exists a polynomial φ such that $\sup_{x\in[a,b]} |f(x) - \varphi(x)| < \varepsilon$. With the triangle inequality,

$$|E[f(S_n)] - f(E[X])| \leq |E[f(S_n)] - E[\varphi(S_n)]|$$

$$+ |E[\varphi(S_n)] - \varphi(E[X])| + |\varphi(E[X]) - f(E[X])|.$$

The first and third terms are bounded above by ε.

We claim now $\varphi \in C_{l.Lip}(\mathbb{R})$. Since $C_{l.Lip}(\mathbb{R})$ is a linear space, it is enough to show that the mappings $x \mapsto x^i$ $(i \geq 1)$ are in $C_{l.Lip}(\mathbb{R})$. But

$$|x^i - y^i| = |\sum_{j=0}^{i-1} x^j y^{i-1-j}| \cdot |x - y| \leq (\sum_{j=0}^{i-1} |x|^j |y|^{i-1-j})|x - y|$$

$$\leq i \max\{|x|, |y|\}^{i-1} |x - y| \leq C(1 + |x|^m + |y|^m)|x - y|$$

for the choices $C = i, m = i - 1$.

Having established that claim, and observing that the X_n are Peng i.i.d., Proposition 3.1 gives

$$E[\varphi(S_n)] \to \sup_{x\in[\mathscr{E}[X], \mathbb{E}[X]]} \varphi(x) = \varphi(E[X]),$$

proving that the second term above goes to 0. In summary,

$$\limsup_n |E[f(S_n)] - f(E[X])| \leq 2\varepsilon,$$

whence the arbitrariness of ε implies $E[f(S_n)] \to f(E[X])$. By the Portmanteau Lemma [1, Theorem 2.1], the arbitrariness of f yields $S_n \to E[X]$ in distribution, which is equivalent to convergence in probability since the limit is a constant. $\quad\square$

Based on this result, he proposed in [4, Sect. 5.2] and [5] a Monte Carlo method relying on the following formula to obtain an almost sure sample approximation to the expectation:

$$\mathbb{E}[\varphi(X)] = \limsup_n n^{-1} \sum_{i=1}^n \varphi(X_i), \tag{1}$$

for any $\varphi \in C_{l.Lip}(\mathbb{R})$ and Peng i.i.d. X_n. If one can generate a simulated sequence of X_n, that provides a way of computing the expectation of X (and, more generally, of nice functions of X) which would be quite more comfortable than the representation as a supremum of linear functionals.

We present now a sequence that satisfies Peng's law of large numbers and fails (1), showing that the latter does not follow from the former and thus cannot be used to compute the expectation.

Theorem 3.1 *There exist a sublinear expectation space* $(\Omega, \mathscr{H}, \mathbb{E})$ *and a sequence* $\{X_n\}_n \subset \mathscr{H}$ *such that*

(a) The X_n are Peng i.i.d. as some $X \in \mathscr{H}$.
(b) $\mathscr{E}[X] = 0, \mathbb{E}[X] = 1$.
(c) For every continuous $\varphi : \mathbb{R} \to \mathbb{R}$,

$$\mathbb{E}[\varphi(S_n)] \to \max_{x \in [\mathscr{E}[X], \mathbb{E}[X]]} \varphi(x).$$

(d) $S_n \to \mathscr{E}[X]$ almost surely.
(e) $v(\mathscr{E}[X] \le S_n \le \mathbb{E}[X]) = 1$.
(f) $v(S_n < \mathbb{E}[X]) = 0$.

Proof Set $\Omega = \mathbb{N} \cup \{0\}$, take $\mathscr{H}$ to be the set of all bounded (necessarily Borel measurable) functions on Ω, and define $\mathbb{E}[X] = \sup_{\omega \in \Omega} X(\omega)$ for all $X \in \mathscr{H}$. It is readily checked that $\mathbb{E}$ is a sublinear expectation.

Let $X_n(\omega)$ be the nth bit in the binary representation of ω, i.e.

$$\omega = \sum_{n=1}^{\infty} 2^{n-1} X_n(\omega)$$

with $X_n(\omega) \in \{0, 1\}$.

Proof of part (a) In this space, Peng identical distribution of two random variables X, Y amounts to the identity

$$\sup_{\omega \in \Omega} \varphi(X(\omega)) = \sup_{\omega \in \Omega} \varphi(Y(\omega)) \quad \forall \varphi \in C_{l.Lip}(\mathbb{R}).$$

Therefore, any two variables with the same range $X(\Omega) = Y(\Omega)$ are Peng identically distributed. Since the range of each X_n is $\{0, 1\}$, they are all Peng identically distributed.

To establish Peng independence, we need to show

$$\sup_{\omega \in \Omega} \varphi(X_1(\omega), \dots, X_{n+1}(\omega)) = \sup_{\omega \in \Omega} [\sup_{\omega' \in \Omega} \varphi(x_1, \dots, x_n, X_{n+1}(\omega))]_{x_i = X_i(\omega')}$$

$$= \sup_{\omega, \omega' \in \Omega} \varphi(X_1(\omega), \dots, X_n(\omega), X_{n+1}(\omega')).$$

The '$\le$' part is clear. Moreover, from the definition of the X_n, the set $\{(X_1(\omega), \dots, X_{n+1}(\omega))\}_{\omega \in \Omega}$ exhausts $\{0, 1\}^{n+1}$. Hence

$$\sup_{\omega \in \Omega} \varphi(X_1(\omega), \ldots, X_{n+1}(\omega)) = \sup_{(x_1,\ldots,x_{n+1}) \in \{0,1\}^{n+1}} \varphi(x_1, \ldots, x_{n+1})$$

$$\geq \sup_{\omega, \omega' \in \Omega} \varphi(X_1(\omega), \ldots, X_n(\omega), X_{n+1}(\omega'))$$

as well.

Proof of part (b) Setting $X = X_1$, we have

$$\mathbb{E}[X] = \sup_{\omega \in \Omega} X(\omega) = 1$$

and

$$\mathscr{E}[X] = -\mathbb{E}[-X] = \inf_{\omega \in \Omega} X(\omega) = 0.$$

Proof of part (c) As ω ranges over Ω, the $(X_1(\omega), \ldots, X_n(\omega))$ exhaust $\{0, 1\}^n$. Thus the range of S_n is $\{0, 1/n, \ldots, (n-1)/n, 1\}$. From the continuity of φ and part *(b)*,

$$\mathbb{E}[\varphi(S_n)] = \sup_{\omega \in \Omega} \varphi(S_n(\omega)) = \sup_{0 \leq k \leq n} \varphi(k/n) \to \max_{x \in [0,1]} \varphi(x) = \max_{x \in [\mathscr{E}[X], \mathbb{E}[X]]} \varphi(x).$$

Proof of part (d) In view of part *(b)*, we have to show $S_n(\omega) \to 0$ for each $\omega \in \Omega$. But that is clear, since the binary representation of each $\omega \in \Omega$ has finitely many non-zero terms.

Proof of part (e) Since X_n can only take on 0 and 1 as values, $0 \leq S_n \leq 1$ and then, from part *(b)*,

$$v(\mathscr{E}[X] \leq S_n \leq \mathbb{E}[X]) = v(0 \leq S_n \leq 1) = v(\Omega) = \mathbb{E}[1] = 1.$$

Proof of part (f) Since the first n bits of the binary representation of $2^n - 1$ are 1,

$$v(S_n < 1) = \mathscr{E}[I_{\{S_n < 1\}}] = \inf_{\omega \in \Omega} I_{\{S_n < 1\}}(\omega) \leq I_{\{S_n < 1\}}(2^n - 1) = 0$$

because $S_n(2^n - 1) = n/n = 1$. $\square$

Peng's formula (1) predicts

$$\limsup_{n} S_n = \sup_{x \in [\mathscr{E}[X], \mathbb{E}[X]]} x = \mathbb{E}[X] = 1$$

almost surely, but Theorem 3.1 *(d)* shows on the contrary that

$$S_n(\omega) \to \mathscr{E}[X] = 0 \quad \forall \omega \in \Omega.$$

Thus Peng's formula is wrong. It might have been the case that it actually followed from his law of large numbers and the latter were wrong, but Theorem 3.1 *(c)* implies that $\{X_n\}_n$ fulfils Peng's law of large numbers. Therefore the problem is that (1) does not follow from the law of large numbers.

Parts (e) and (f) suggest that the behaviour of S_n in this sublinear expectation space is quite different from the usual probability intuition. That is confirmed by the following proposition.

Proposition 3.2 *The sequence in Theorem 3.1 has the following properties:*

(a) $n^{-1} \sum_{i=1}^{n} a_i X_i \to 0$ *almost surely for any sequence* $a_n \geq 0$, *in particular* $S_n \to 0$ *almost surely.*

(b) S_n *does not converge in probability.*

(c) $S_n \to Y$ *in law, with the distribution of Y being given by* $\mathbb{E}_Y[\varphi] = \max_{x \in [0,1]} \varphi(x)$.

Proof of part (a) For each $\omega \in \Omega$, it holds that $X_n(\omega) = 0$ for $n > \log_2 \omega + 1$. Thus, since X_n takes on values 0 and 1,

$$0 \leq n^{-1} \sum_{i=1}^{n} a_n X_n(\omega) \leq n^{-1} \sum_{i \leq \log_2 \omega + 1} a_i \to 0.$$

Proof of part (b) Fix $\varepsilon \in (0, 1/4)$. Reasoning by contradiction, assume $S_n \to Y$ in probability for some random variable Y. Accordingly, $v(|S_n - Y| < \varepsilon) \to 1$. For each event A one has

$$v(A) = \mathscr{E}[I_A] = \inf_{\omega \in \Omega} I_A(\omega);$$

the right-hand side is 0 unless A contains all $\omega \in \Omega$. Therefore, for all sufficiently large n,

$$\{|S_n - Y| < \varepsilon\} = \Omega$$

and then

$$\{|S_n - S_{2n}| < 2\varepsilon\} = \Omega.$$

Whatever n may be, since the first n bits of the binary representation of $2^n - 1$ are ones and the next n bits are zeros, we have

$$S_n(2^n - 1) = \frac{n}{n} = 1, \qquad S_{2n}(2^n - 1) = \frac{n}{2n} = 1/2.$$

But $\varepsilon < 1/4$, whence $|S_n(2^n - 1) - S_{2n}(2^n - 1)| > 2\varepsilon$, a contradiction.

Proof of part (c) It follows from Theorem 3.1 *(c)* or Peng's law.   □

Thus the behaviours of S_n with respect to the three types of convergence are all different, in particular

(1) Almost sure convergence does not imply convergence in probability.
(2) The limit of the sequence of laws exists but is not the law of the almost sure limit.
(3) Almost sure convergence does not imply convergence in law.

It must be emphasized that a.s. convergence *is not* a weak notion in this context. In fact, it implies *uniform* a.s. convergence over a family of finitely additive probabilities (and, under appropriate conditions, see e.g. [3], σ-additive probabilities).

To make things even more interesting, for this sublinear expectation convergence in probability actually implies almost sure convergence; that is an instance of a more general phenomenon (see e.g. [8, Proposition 5.1]).

4 Sublinear Expectations and Coherent Upper Previsions

The purpose of this section is to observe that Peng's sublinear expectations are the same thing as the coherent upper previsions in e.g. Walley's book [10]. While Peng cites Walley, he seems to have been unaware of the equivalence.

A *coherent upper prevision* is a functional $\mathbb{E}$ on a linear space $\mathscr{H}$ of functions (called *gambles*) on a set Ω, such that, for any $n \geq 0, m \geq 1$, and $X_0, X_1, \ldots, X_n \in \mathscr{H}$,

$$\inf_{\omega \in \Omega} \Big[\sum_{k=1}^{n} (X_k(\omega) - \mathbb{E}[X_k]) - m(X_0(\omega) - \mathbb{E}[X_0]) \Big] \leq 0.$$

Proposition 4.1 *Let $\mathbb{E}$ be a functional on a linear space $\mathscr{H}$ of bounded functions (closed under taking the absolute value) on a set Ω. Then $\mathbb{E}$ is a sublinear expectation if and only if $\mathbb{E}$ is a coherent upper prevision.*

Proof Walley [10, Sect. 2.6] showed that $\mathbb{E}$ is a coherent upper prevision if and only if the following three properties are met for any $X, Y \in \mathscr{H}$:

(CUP.i) $\mathbb{E}[X] \leq \sup_{\omega \in \Omega} X(\omega)$,
(CUP.ii) $\mathbb{E}[X + Y] \leq \mathbb{E}[X] + \mathbb{E}[Y]$,
(CUP.iii) $\mathbb{E}[cX] = c \cdot \mathbb{E}[X]$ for $c \geq 0$.

Properties (CUP.ii) and (CUP.iii) are the same as properties (SE.iii) and (SE.iv) of a sublinear expectation. Thus the proof splits into two parts.

1. A sublinear expectation satisfies (CUP.i).

Let $X \in \mathscr{H}$. Then, by (SE.i) and (SE.ii),

$$\mathbb{E}[X] \leq \mathbb{E}[\sup_{\omega \in \Omega} X(\omega)] = \sup_{\omega \in \Omega} X(\omega).$$

2. A coherent upper prevision satisfies (SE.i) and (SE.ii).

(SE.i): If $X \leq Y$, then, using (CUP.ii) and (CUP.i),

$$\mathbb{E}[X] = \mathbb{E}[Y + (X - Y)] \leq \mathbb{E}[Y] + \mathbb{E}[X - Y] \leq \mathbb{E}[Y] + \sup_{\omega \in \Omega}(X(\omega) - Y(\omega)) \leq \mathbb{E}[Y].$$

(SE.ii): From (CUP.iii) and (CUP.ii),

$$0 = \mathbb{E}[0] = \mathbb{E}[c - c] \leq \mathbb{E}[c] + \mathbb{E}[-c].$$

By (CUP.i), $\mathbb{E}[c] \leq c$ and $\mathbb{E}[-c] \leq -c$. Combining the three inequalities,

$$c \leq -\mathbb{E}[-c] \leq \mathbb{E}[c] \leq c,$$

which proves $\mathbb{E}[c] = c$. $\qquad\square$

5 Concluding Remarks

The sequence $\{X_n\}_n$ in Theorem 3.1 is interesting when contemplated from the probabilistic perspective. Each X_n should have, intuitively, a Bernoulli $\mathscr{B}(1/2)$ distribution. The product space $\{0, 1\}^{\mathbb{N}}$ admits then a unique probability measure (the product measure $\mathbb{P}$) such that all those distributions are independent. But the subset of $\{0, 1\}^{\mathbb{N}}$ which corresponds to Ω, namely the 0–1 sequences with finitely many ones, is $\mathbb{P}$-null, whence a probability for Ω cannot be retrieved from $\mathbb{P}$.

Proposition 5.3 in [8] presents an example of behaviours similar to Proposition 3.2, but the independence in the sense of Peng is not satisfied there (only a rather weaker form of independence holds), and a different definition of convergence in law is used as well. Theorem 3.1 *(a)* is similar to Proposition 6.1 *(a)* in [9], but both the notions of independence and identical distribution used are different.

References

1. Billingsley P (1968) Convergence of probability measures. Wiley, New York
2. Hu M, Li X (2014) Independence under the G-expectation framework. J Theor Probab 27:1011–1020
3. Huber PJ, Strassen V (1973) Minimax tests and the Neyman-Pearson lemma for capacities. Ann Stat 1:251–263
4. Peng S (2009) Survey on normal distributions, central limit theorem, Brownian motion and the related stochastic calculus under sublinear expectations. Sci China Ser A 52:1391–1411
5. Peng S (2010) Backward stochastic differential equation, nonlinear expectation and their applications. In: Bhatia R (ed) Proc Int Congress Math Hyderabad 2010,

Vol I: Plenary lectures and ceremonies. Hindustan, New Delhi (Video available at http://videos.icm2010.in/player.php?search=&video_id=plenary_lecture&date=&highres=1&play_from_clip=8&play_to_clip=8#player)

6. Peng S (2010) Nonlinear expectations and stochastic calculus under uncertainty. arXiv:1002.4546

7. Proschan MA, Shaw PA (2016) Essentials of probability theory for statisticians. CRC, Boca Raton

8. Terán P (2014) Laws of large numbers without additivity. Trans Am Math Soc 366:5431–5451

9. Terán P (2015) On independence and the law of large numbers for upper probabilities. Unpublished manuscript

10. Walley P (1991) Statistical reasoning with imprecise probabilities. Chapman & Hall, London

A Non-iterative Estimator for Interval Sampling and Doubly Truncated Data

Jacobo de Uña-Álvarez

Abstract Interval sampling is often used in Survival Analysis and reliability studies. With interval sampling, the sampling information is restricted to the lifetimes of the individuals or units who fail between two specific dates d_0 and d_1. Thus, this sampling procedure results in randomly doubly truncated data, where the (possibly negative) left-truncation variable is the time from onset to d_0, and the right-truncation variable is the left-truncation variable plus the interval width $d_1 - d_0$. In this setting, the nonparametric maximum likelihood estimator (NPMLE) of the lifetime distribution is the Efron–Petrosian estimator, a non-explicit estimator which must be computed in an iterative way. In this paper we introduce a non-iterative, nonparametric estimator of the lifetime distribution and we investigate its performance relative to that of the NPMLE. Simulation studies and illustrative examples are provided. The main conclusion of this piece of work is that the non-iterative estimator, being much simpler, performs satisfactorily. Application of the proposed estimator for general forms of random double truncation is discussed.

1 Introduction

In Survival Analysis, reliability studies and other fields, interval sampling is often employed. With 'pure' interval sampling, the sampling information is restricted to the lifetimes of the individuals (or units) who die (or fail) between two specific dates d_0 and d_1. Thus, this sampling procedure results in randomly doubly truncated data, where the (possibly negative) left-truncation variable is the time from onset to d_0, and the right-truncation variable is the left-truncation variable plus the interval width $\tau = d_1 - d_0$. See for example Austin et al. [1] and Zhu and Wang [12]. This sort of data appear also when d_1 is a subject-specific event time of secondary interest, $d_0 = d_1 - \tau$, and the sample is restricted to lifetimes between d_0 and d_1 (Mandel et al. [5]). More generally, random double truncation refers to a setting in which

J. de Uña-Álvarez (✉)
Department of Statistics and OR, SiDOR Research Group & CINBIO,
University of Vigo, Vigo, Spain
e-mail: jacobo@uvigo.es

© Springer International Publishing AG 2018
E. Gil et al. (eds.), *The Mathematics of the Uncertain*, Studies in Systems,
Decision and Control 142, https://doi.org/10.1007/978-3-319-73848-2_37

387

the lifetime of ultimate interest X is observed only when $U \leq X \leq V$, where U and V are two other random variables (the left and right truncation times). When U (respectively V) degenerates at $-\infty$ (respectively $+\infty$), the double truncation scenario reduces to a one-sided truncation setting (Woodroofe [11]).

Estimation of the distribution function (df) F of a doubly truncated variable X has received some attention in the last decades. The nonparametric maximum likelihood estimator (NPMLE) of F was introduced by Efron and Petrosian [3] under the assumption of independence between X and (U, V). The NPMLE is a non-explicit estimator which must be computed in an iterative way. Efron and Petrosian [3] introduced two different algorithms to compute the estimator. A third iterative algorithm which simultaneously computes the NPMLE of F and that of the bivariate distribution of (U, V) was proposed by Shen [9]. These three algorithms were reviewed by Moreira et al. [7], who implemented them in the R package DTDA. One drawback of Efron–Petrosian NPMLE is precisely the nonavailability of a closed-form expression; this leads to computational issues in particular settings, where the algorithms may not converge. Another drawback is the lack of solid asymptotic theory for the NPMLE; even when some asymptotic results have been derived (Shen [9, 10]), their formal proofs are incomplete or they contain important gaps (Mandel et al. [5]). So an interesting question is that of the construction of alternative nonparametric estimators for the target F. This is the main goal of this paper.

The rest of the paper is organized as follows. In Sect. 2 a new, non-iterative nonparametric estimator of the lifetime df is introduced and discussed for several forms of double truncation. In Sect. 3 the performance of the proposed method relative to the Efron–Petrosian NPMLE is investigated through simulations. The simulated scenarios include 'pure' interval sampling, 'mixed' interval sampling (under which some extra restriction on the dates of onset is present), as well as the setting in which U and V are not necessarily linked through equation $V = U + \tau$. Section 4 provides two different real data illustrative examples. Finally, a discussion is reported in Sect. 5.

2 A Non-iterative Estimator

Let (U_i, X_i, V_i), $1 \leq i \leq n$, be the available observations, that is, a random sample with the conditional distribution of (U, X, V) given $U \leq X \leq V$. For interval sampling, one has indeed $V = U + \tau$ where $\tau = d_1 - d_0$ is the interval width. We assume that (U, V) and X are independent. Put $F(x) = P(X \leq x)$ for the lifetime cumulative df, which we assume to be continuous, and let $[a_X, b_X]$ be the support of X. In this setting, the identifiable distribution is that of X conditionally on $a_U \leq X \leq b_V$, where a_U and b_V stand respectively for the lower limit of the support of U and the upper limit of the support of V (Woodroofe [11]). Therefore, for identifiability of F, we assume $a_U \leq a_X$ and $b_X \leq b_V$.

Since closed-form estimators exist for one-sided truncation (Lynden-Bell estimator, see Woodroofe [11]), our idea to introduce a non-iterative estimator for F in our doubly truncated setting is to remove one of the truncation times. To this end, let $S^{(v)}$ denote the subsample satisfying $X_i \leq v \leq V_i$, and let $\hat{F}^{(v)}(x)$ denote the Lynden-Bell estimator computed from $S^{(v)}$. This means that the X_i's in $S^{(v)}$ are considered as left-truncated by their respective U_i's. Because of condition $X_i \leq v \leq V_i$ in $S^{(v)}$, and due to the independence between (U, V) and X, the estimator $\hat{F}^{(v)}(x)$ converges to $F^{(v)}(x) = P(X \leq x | v - \tau \leq X \leq v)$, where condition $v - \tau \leq X$ comes from the fact that the lower limit of the support of U given $V \geq v$ is precisely $v - \tau$. Now, interestingly, for $x \in [v - \tau, v]$

$$F^{(v)}(x) = \frac{F(x) - F^-(v - \tau)}{F(v) - F^-(v - \tau)}$$

where $F^-(x) = P(X < x)$. Alternatively,

$$F(v) = \frac{F(x) - F^-(v - \tau)}{F^{(v)}(x)} + F^-(v - \tau). \tag{1}$$

Then, there is some hope that the value of $F(v)$ can be recovered from the value of $F(x)$ for smaller lifetimes $x < v$, and from the Lynden-Bell estimator $\hat{F}^{(v)}(x)$, by using (1). This is indeed the case, as we will show.

Let $x_{(1)} < \ldots < x_{(n)}$ denote the ordered X_i's, and take $x = x_{(i-1)}$ and $v = x_{(i)}$ in (1). We assume that $x_{(i)} - \tau \leq x_{(i-1)}$, so (1) holds; note that this can be ensured by increasing the sample size n. Provided that $\hat{F}^{(x_{(i)})}(x_{(i-1)}) > 0$, Eq. (1) suggests

$$\hat{F}(x_{(i)}) = \frac{\hat{F}(x_{(i-1)}) - \hat{F}^-(x_{(i)} - \tau)}{\hat{F}^{(x_{(i)})}(x_{(i-1)})} + \hat{F}^-(x_{(i)} - \tau). \tag{2}$$

Equation (2) can be used to define a non-iterative estimator $\hat{F}$, which is constructed from left $(x_{(1)})$ to right $(x_{(n)})$. Explicitly, the goal is to compute $F_1, \ldots, F_n$ where $F_i = F(x_{(i)})$, $1 \leq i \leq n$. Introduce for $i = 2, \ldots, n$

$$P_i = F^-(x_{(i)} - \tau) = \sum_{j=1}^{i-1}(F_j - F_{j-1})I_{ji}, \qquad I_{ji} = I(x_{(j)} < x_{(i)} - \tau), \tag{3}$$

where $F_0 = 0$. Note that P_i in (3) just depends on $F_1, \ldots, F_{i-1}$. We now introduce formally the non-iterative estimator. Put $L_i = \hat{F}^{(x_{(i)})}(x_{(i-1)})$.

Definition 2.1 [*The non-iterative estimator*] The non-iterative estimator $\hat{F}$ is a discrete distribution with support $\{x_{(1)}, \ldots, x_{(n)}\}$ and respective cumulative masses $F_1, \ldots, F_n$ defined as follows:

(a) $F_1 = 1$;
(b) for $i = 2, \ldots, n$, $F_i = (F_{i-1} - P_i)/L_i + P_i$;
(c) normalize so $F_n = 1$, that is, redefine F_i as $F_i / \max(F_1, \ldots, F_n)$.

It is important to note that Definition 2.1 introduces indeed a non-decreasing sequence of cumulative masses $F_1 \leq \cdots \leq F_n$. To see this, write

$$F_2 - F_1 = (F_1 - P_2)/L_2 + P_2 - F_1 = (1/L_2 - 1)(F_1 - (F_1 - F_0)I_{12}) \geq 0,$$

where the inequality follows from $L_2 \leq 1$. Now, assume that $F_j - F_{j-1} \geq 0$ holds for $1 \leq j \leq i - 1$. Then,

$$F_i - F_{i-1} = (1/L_i - 1)(F_{i-1} - \sum_{j=1}^{i-1}(F_j - F_{j-1})I_{ji})$$

$$\geq (1/L_i - 1)(F_{i-1} - \sum_{j=1}^{i-1}(F_j - F_{j-1})) = 0,$$

where the inequality follows from the induction hypothesis and $L_i \leq 1$. Therefore, $F_i \geq F_{i-1}$ holds for $1 \leq i \leq n$.

According to (3), many of the P_i's can be zero, $P_i = 0$ for $2 \leq i \leq i_0$ say, where i_0/n can be close to 1 in practice. Note that $P_i = 0$ occurs at least while $x_{(i)} \leq \tau$. For such i's, the non-interactive estimator satisfies $F_i = F_{i-1}/L_i$. It can be easily seen that this simple formula defines a consistent estimator of F when $a_U^{(v)} \leq a_X^{(v)}$ for all v in the support of X, where $a_\xi^{(v)}$ is the lower bound of the support of ξ given $X \leq v \leq V$. Note that this condition is violated when e.g. $V = U + \tau$ and $P(X < v - \tau) > 0$ for some v, which will be typically the case with interval sampling.

In Definition 2.1 we have assumed $L_i > 0$, $1 \leq i \leq n$. Recall that L_i stands for the Lynden-Bell estimator $\hat{F}^{(x_{(i)})}(x_{(i-1)})$. Explicitly (see Woodroofe [11]),

$$L_i = \hat{F}^{(x_{(i)})}(x_{(i-1)}) = 1 - \prod_{j=1}^{i-1}\left\{1 - \frac{I(x_{(i)} \leq v_{[j]})}{\sum_{k=j}^{i} I(u_{[k]} \leq x_{(j)}, x_{(i)} \leq v_{[k]})}\right\},$$

where $(u_{[j]}, v_{[j]})$ denotes the (U, V)-value attached to $x_{(j)}$. From this expression it becomes clear than $L_i = 0$ may happen; in such a case, a proper modification of the Lynden-Bell estimator should be used to avoid zero denominators in Definition 2.1. In the simulation study below we use $1/n$ as a lower bound for L_i when needed.

One critical issue is that of the efficiency of the non-iterative estimator relative to the Efron–Petrosian NPMLE. In a sense, the proposed approach proceeds to estimate the target $F(x)$ at point $x = x_{(i)}$ by focusing on the information contained in the specific interval $[x_{(i)} - \tau, x_{(i)})$. Since this interval changes with $x_{(i)}$, one may expect that most of the sampling information is used in this way. Simulation results in the next Section support this guess, providing a relative efficiency above 93% for all the considered scenarios.

The non-interative estimator in Definition 2.1 was constructed by removing the right-truncation time in a sequential manner. In principle, there is no reason to think that the analogous estimator which proceeds by removing the left-truncation variable rather than the right-truncation time performs worse. This alternative estimator starts by considering a subsample of type $\tilde{S}^{(u)} = \{i : U_i \leq u \leq X_i\}$, and then makes use of the Lynden-Bell correction for right-truncation. The resulting estimator equals that in Definition 2.1 when the double truncation event $U \leq X \leq V$ is represented in the reverse way $-V \leq -X \leq -U$ and $(-V, -U)$ and $-X$ play the role of (U, V) and X respectively. For this reason, no more details on the alternative estimator will be given here; however, both the direct and the reverse estimators are considered in our simulation study below.

3 Simulation Study

In this section we investigate through simulations the finite sample performance of the non-iterative estimator introduced in Sect. 2, relative to the Efron–Petrosian NPMLE. The simulated model is as follows. The lifetime variable X is uniformly distributed on the unit interval, $X \sim U(0, 1)$. The left-truncation variable is $U \sim U(-0.25, 0.5)$, and the right-truncation time is $V = U + \tau$ with $\tau = 0.75$. This model represents a 'mixed' interval sampling situation in which only items with failures between two calendar dates d_0 and d_1 (with interval width 0.75), whereas allowed dates of onset belong to the interval $d_0 - 0.5$ and $d_0 + 0.25$. For this model we have $a_U = -0.25 \leq 0 = a_X$ and $b_X = 1 \leq 1.25 = b_V$. Therefore, the identifiability conditions for the recovery of F are fulfilled. The simulated scenario introduces an observational bias on the lifetime, in the sense that intermediate values of X are observed with probability relatively larger. Therefore, a naive application of the ordinary empirical distribution of the sample would overestimate the data frequency around 0.5, reporting a non-uniform cumulative distribution.

We have computed the bias, the standard deviation and the mean squared error (MSE) of several possible estimators along $1,000$ Monte Carlo trials of size $n = 50$, 100, 250 and 500. Specifically, the list of estimators is the following:

(I)   the Efron–Petrosian NPMLE, $\hat{F}_{EP}(x)$;
(II)   the direct estimator in Definition 2.1, $\hat{F}(x)$;
(III) the reverse estimator as defined in Sect. 2, $\hat{F}_R(x)$;
(IV)   a weighted estimator which combines the direct and the reverse estimators, $\hat{F}_w(x) = w\hat{F}(x) + (1 - w)\hat{F}_R(x), 0 < w < 1$.

Note that the estimator (I) is iterative, while the estimators (II), (III) and (IV) are non-iterative. The iterative estimator (I) was computed by using the algorithm proposed by Shen [9], which is implemented by the function shen of the R package DTDA (Moreira et al. [7]). The accuracy of each estimator was measured at the three quartiles of F, these are $x_{0.25} = 0.25$, $x_{0.5} = 0.5$ and $x_{0.75} = 0.75$. For the weighted estimator (IV), we provide the results corresponding to the 'fair' choice $w = 0.5$ as well as those of the optimal estimator which attains the minimum MSE (although in this case the optimal weight may vary depending on the quartile).

Fig. 1 Direct estimator (solid line), Efron–Petrosian NPMLE (dashed), and reverse estimator (dotted line) for a simulated trial with $n = 100$. 'Mixed' interval sampling scenario. Straight line corresponds to the true cumulative distribution

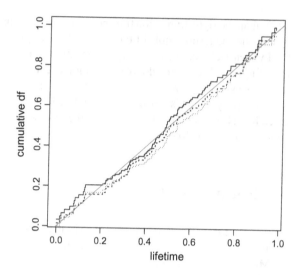

The results for the four different sample sizes n are reported separately, see Tables 1, 2, 3 and 4. Just for illustration purposes, in Fig. 1 the estimators (I), (II) and (III) are depicted for a particular trial of sample size $n = 100$; the three of them are estimating the uniform distribution quite accurately. The results in Tables 1, 2, 3 and 4 indicate that all the methods perform consistently, with a MSE which decreases with an increasing sample size. On the other hand, the bias is negligible compared to the standard deviation, so the MSE roughly equals the variance of the estimator. The direct and reverse non-iterative estimators are competitive with the NPMLE, although their MSE is somehow larger. The weighted non-iterative estimator with optimal weight is of course the best competitor, with an efficiency relative to the NPMLE ranging between 0.945 and 1.006 depending on the specific quartile and sample size. Interestingly, the optimal weight w^* which minimizes the MSE was always around $1/2$; more specifically, it ranged between 0.384 and 0.657. This explains why the estimator based on the 'fair' weight $w = 0.5$ performed relatively well, with relative efficiencies always above 0.93. Whatever the case, we have seen in the simulations that any mixture of the direct and reverse non-iterative estimators always improved both $\hat{F}(x)$ and $\hat{F}_R(x)$ in the sense of the MSE. This is further discussed below.

The following alternative scenarios were simulated too, always with $X \sim U(0, 1)$, for sample sizes $n = 250$ and $n = 500$:

(i) $X \sim U(0, 1)$, $U \sim U(-0.25, 0.5)$, $V \sim U(0.5, 1.25)$. Therefore, the marginal distributions of U and V were as above, but these two variables were generated independently. In this particular setting of independent truncation variables, the non-iterative estimator reduces to the closed-form estimator with cumulative probabilities $F_n = 1$, $F_i = F_{i+1}L_i$, $i = n - 1, \ldots 1$;

Table 1 Bias, standard deviation (SD) and mean squared error (MSE) of the several estimators along 1,000 Monte Carlo trials under 'mixed' interval sampling, case $n = 50$. Fixed weight is $w = 1/2$. Optimal weights w^* in the simulations were 0.384 ($x_{0.25}$), 0.495 ($x_{0.5}$) and 0.657 ($x_{0.75}$)

	Bias	SD	MSE	Bias	SD	MSE	Bias	SD	MSE
	$x_{0.25}$			$x_{0.5}$			$x_{0.75}$		
$\hat{F}(x)$	0.02692	0.09454	0.00966	0.01532	0.09989	0.01021	0.01881	0.08640	0.00782
$\hat{F}_R(x)$	−0.01682	0.08928	0.00825	−0.01409	0.10005	0.01021	−0.02568	0.09574	0.00983
$\hat{F}_w(x)$	0.00505	0.08641	0.00749	0.00062	0.09406	0.00885	−0.00344	0.08518	0.00727
$\hat{F}_{w^*}(x)$	−0.00003	0.08607	0.00741	0.00047	0.09406	0.00885	0.00353	0.08422	0.00711
$\hat{F}_{EP}(x)$	0.00317	0.08468	0.00718	0.00081	0.09433	0.00890	−0.00105	0.08346	0.00697

Table 2 Bias, standard deviation (SD) and mean squared error (MSE) of the several estimators along 1,000 Monte Carlo trials under 'mixed' interval sampling, case $n = 100$. Fixed weight is $w = 1/2$. Optimal weights w^* in the simulations were 0.384 ($x_{0.25}$), 0.525 ($x_{0.5}$) and 0.566 ($x_{0.75}$)

	Bias	SD	MSE	Bias	SD	MSE	Bias	SD	MSE
	$x_{0.25}$			$x_{0.5}$			$x_{0.75}$		
$\hat{F}(x)$	0.01319	0.06365	0.00423	0.00703	0.06706	0.00455	0.01081	0.05987	0.00370
$\hat{F}_R(x)$	−0.00939	0.05975	0.00366	−0.00731	0.06779	0.00465	−0.01121	0.06232	0.00401
$\hat{F}_w(x)$	0.00190	0.05775	0.00334	−0.00014	0.06345	0.00403	−0.00020	0.05716	0.00327
$\hat{F}_{w^*}(x)$	−0.00072	0.05749	0.00331	0.00022	0.06344	0.00403	0.00125	0.05706	0.00326
$\hat{F}_{EP}(x)$	0.000374	0.05590	0.00312	−0.00034	0.06308	0.00398	0.00076	0.05584	0.00312

Table 3 Bias, standard deviation (SD) and mean squared error (MSE) of the several estimators along 1,000 Monte Carlo trials under 'mixed' interval sampling, case $n = 250$. Fixed weight is $w = 1/2$. Optimal weights w^* in the simulations were 0.394 ($x_{0.25}$), 0.465 ($x_{0.5}$) and 0.576 ($x_{0.75}$)

	Bias	SD	MSE	Bias	SD	MSE	Bias	SD	MSE
	$x_{0.25}$			$x_{0.5}$			$x_{0.75}$		
$\hat{F}(x)$	0.00477	0.04220	0.00180	0.00103	0.04476	0.00200	0.00196	0.04018	0.00162
$\hat{F}_R(x)$	−0.00474	0.03994	0.00162	−0.00479	0.04389	0.00195	−0.00714	0.04124	0.00175
$\hat{F}_w(x)$	0.00001	0.03861	0.00149	−0.00188	0.04181	0.00175	−0.00259	0.03817	0.00146
$\hat{F}_{w^*}(x)$	−0.00099	0.03847	0.00148	−0.00209	0.04179	0.00175	−0.00190	0.03815	0.00146
$\hat{F}_{EP}(x)$	−0.00078	0.03746	0.00140	−0.00215	0.04141	0.00172	−0.00219	0.03725	0.00139

(ii) $X \sim U(0, 1)$, $U \sim U(-0.25, 1)$, $V = U + 0.25$. This is the situation with 'pure' interval sampling: $a_U = -\tau$ is minus the interval width ($\tau = 0.25$), and $b_U = 1$ is the upper limit of the support of X. In this case there is no observational bias because U is uniformly distributed;

(iii)   Same as scenario (ii) but with $U = (1 + \tau)\sqrt{U(0, 1)} - \tau$, which is non-uniform and results in an observational bias on X of $G(x) \equiv P(U \leq x \leq V) = \tau(\tau + 2x)/(1 - \tau)^2$.

Scenarios (i)–(iii) reported little novelty relative to results in Tables 1, 2, 3 and 4. With independent truncation times (scenario (i)), the efficiency of the (optimal) non-

Table 4 Bias, standard deviation (SD) and mean squared error (MSE) of the several estimators along 1,000 Monte Carlo trials under 'mixed' interval sampling, case $n = 500$. Fixed weight is $w = 1/2$. Optimal weights w^* in the simulations were 0.394 ($x_{0.25}$), 0.475 ($x_{0.5}$) and 0.596 ($x_{0.75}$)

	Bias	SD	MSE	Bias	SD	MSE	Bias	SD	MSE
	$x_{0.25}$			$x_{0.5}$			$x_{0.75}$		
$\hat{F}(x)$	0.00452	0.02935	0.00088	0.00238	0.03112	0.00097	0.00177	0.02801	0.00079
$\hat{F}_R(x)$	−0.00085	0.02820	0.00080	−0.00096	0.03082	0.00095	−0.00332	0.02930	0.00087
$\hat{F}_w(x)$	0.00183	0.02704	0.00073	0.00071	0.02928	0.00086	−0.00077	0.02693	0.00073
$\hat{F}_{w^*}(x)$	0.00127	0.02699	0.00073	0.00063	0.02928	0.00086	−0.00028	0.02686	0.00072
$\hat{F}_{EP}(x)$	0.00139	0.02625	0.00069	0.00051	0.02921	0.00085	−0.00036	0.02614	0.00068

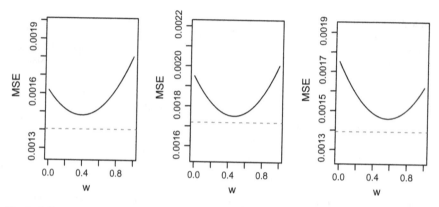

Fig. 2 MSE of the weighted non-iterative estimator $\hat{F}_w(x)$ at the three quartiles of F: $x = x_{0.25}$, $x = x_{0.5}$ and $x = x_{0.75}$ (from left to right). The dashed horizontal line corresponds to the MSE of the NPMLE. 'Mixed' interval sampling scenario, $n = 250$

iterative estimator relative to the NPMLE was always above 0.98. In this case, the best weighted estimator gave a small weight $w^* \simeq 0.2$ to the direct estimator when estimating F at the first quartile, this weight being large at the third quartile ($w^* \simeq 0.7$). This seems to be somehow the case in Tables 1, 2, 3 and 4 too, reflecting the fact that the reverse estimator performs better at $x_{0.25}$, while the direct estimator reports a smaller MSE at $x_{0.75}$. In the model with 'pure' interval sampling (scenarios (*ii*) and (*iii*)), the efficiency of the proposed estimator relative to the NPMLE was always above 0.93. For these scenarios (*ii*) and (*iii*) the weight leading to the minimum MSE ranged between 0.394 and 0.556. It is interesting to note that, for the 'pure' interval sampling scenarios and $n = 250$, $\min_{1 \le i \le n} L_i = 0$ happened for a small number of trials (2.1% at maximum), so the non-iterative estimator could not be computed. To avoid this issue we just redefined L_i as $\max(L_i, 1/n)$; the simulation results we refer to are based on such a correction.

In Fig. 2 we depict the MSE for the weighted, non-iterative estimator depending on the choice of the weight w, for the 'mixed' interval sampling scenario and $n = 250$; the dashed horizontal line corresponds to the MSE of the NPMLE. The results

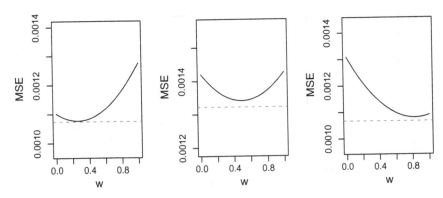

Fig. 3 MSE of the weighted non-iterative estimator $\hat{F}_w(x)$ at the three quartiles of F: $x = x_{0.25}$, $x = x_{0.5}$ and $x = x_{0.75}$ (from left to right). The dashed horizontal line corresponds to the MSE of the NPMLE. Independent truncation variables scenario (i), $n = 250$

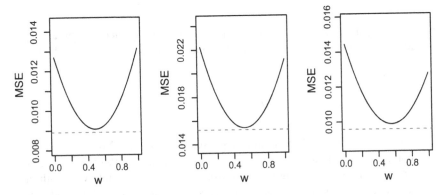

Fig. 4 MSE of the weighted non-iterative estimator $\hat{F}_w(x)$ at the three quartiles of F: $x = x_{0.25}$, $x = x_{0.5}$ and $x = x_{0.75}$ (from left to right). The dashed horizontal line corresponds to the MSE of the NPMLE. 'Pure' interval sampling scenario without observation bias (ii), $n = 250$

corresponding to the other scenarios (i)–(iii) are provided in Figs. 3, 4 and 5. Other sample sizes n provided similar plots. From Figs. 2, 3, 4 and 5 we can see that the results of the direct and reverse estimators are improved by weighting them, and that most of the times the choice $w = 0.5$ is roughly optimal, as discussed above.

4   Real Data Illustration

In this section we provide two illustrative real data analyses. Both of them correspond to the situation of 'pure' interval sampling. The results provided by the non-iterative estimator are compared to the standard Efron–Petrosian NPMLE. Since for both

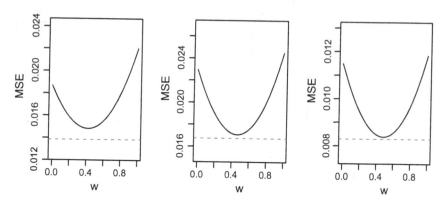

Fig. 5 MSE of the weighted non-iterative estimator $\hat{F}_w(x)$ at the three quartiles of F: $x = x_{0.25}$, $x = x_{0.5}$ and $x = x_{0.75}$ (from left to right). The dashed horizontal line corresponds to the MSE of the NPMLE. 'Pure' interval sampling scenario with observational bias (*iii*), $n = 250$

datasets ties among the X_i's occur, lifetimes were slightly modified by adding a (small) uniform random number so the strict inequalities $x_{(1)} < \cdots < x_{(n)}$ hold.

4.1 Childhood Cancer Data

This dataset refers to $n = 406$ children diagnosed from cancer in North Portugal. The data correspond to all the cases diagnosed between January 1, 1999, and December 31, 2003, so interval sampling was employed indeed, with an interval width of 5 years. The variable X of interest is age (in years) at diagnosis from childhood cancer which, by definition, is supported on the [0, 15] interval. The right-truncation time V represents the age of the individual by December 31, 2003, while $U = V - 5$ is the left-truncation time. The observed support of U is roughly $(-4.5, 14.5)$ (in years). See Moreira and de Uña-Álvarez [6] for further details.

In this case $\tau = 5$ (years) and $\max_{2 \le i \le n}(x_{(i)} - x_{(i-1)}) = 0.343$, so condition $x_{(i)} - \tau \le x_{(i-1)}, 2 \le i \le n$, holds. Condition $L_i > 0$ is satisfied too; indeed, $\min_{2 \le i \le n} L_i = 0.5$, and 398 of the L_i's are larger than 0.9. On the other hand, $P_i = 0$ for $i \le 192$, P_i being strictly positive for larger i's.

The several estimators for the cumulative df of X are depicted in Fig. 6. They all suggest a non-uniform, slightly concave df. For completeness, the naive estimator based on the ordinary empirical df of the X_i's is also displayed, being close to the estimators which correct for truncation. The reason for this is that, in this particular application, the double truncation does not introduce any observational bias on X, because the truncation times are uniformly distributed and their supports are large enough (Moreira and de Uña-Álvarez [6]). To emphasize this, in Fig. 7 the NPMLE of the biasing function $G(x) = P(U \le x \le V)$ is provided; the function in this plot is roughly constant, indicating that all the X_i's are sampled with the same probability.

Fig. 6 Direct estimator (solid line), Efron–Petrosian NPMLE (dashed), and reverse estimator (dotted line) for the childhood cancer data ($n = 406$). The naive estimator provided by the ordinary empirical df is displayed too (dotted-dashed line)

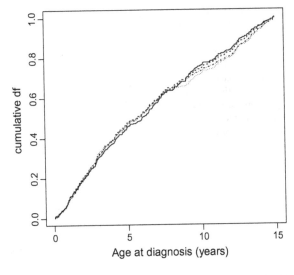

Fig. 7 Biasing function for the childhood cancer data ($n = 406$)

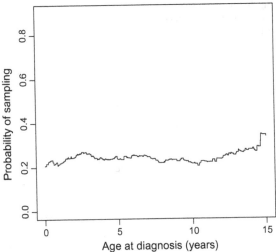

4.2 AIDS Blood Transfusion Data

Kalbfleisch and Lawless [4] reported 494 cases of transfusion-related AIDS, corresponding to cases diagnosed prior to July 1, 1986. Here, the lifetime X is the induction or incubation time (in months), which is the time elapsed from HIV infection to AIDS. Importantly, since HIV was unknown before 1982, cases developing AIDS prior to this date were not reported. Let V denote the time from HIV infection to July 1, 1986 (in months), and introduce $U = V - 54$; then, due to the interval sampling, only triplets (U, X, V) satisfying $U \leq X \leq V$ were observed (Bilker and

Fig. 8 Direct estimator
(solid line), Efron–Petrosian
NPMLE (dashed), and
reverse estimator (dotted
line) for the AIDS blood
transfusion data ($n = 298$).
The naive estimator provided
by the ordinary empirical df
is displayed too
(dotted-dashed line)

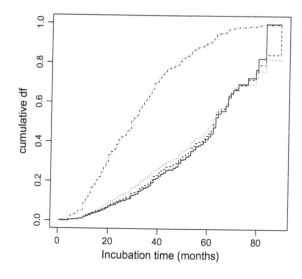

Fig. 9 Biasing function for
the AIDS blood transfusion
data ($n = 298$)

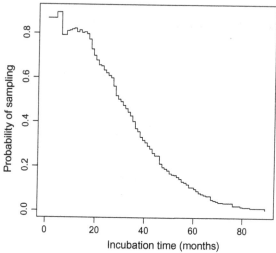

Wang [2]). We restrict our analysis to the $n = 298$ cases with consistent data, for
which the infection could be attributed to a single transfusion or short series of trans-
fusions. This dataset was recently used to illustrate regression issues under double
truncation too (Moreira et al. [8]). The observed values of X ranged from 0.5 to 89
(months), while U ranged from -48.5 to 45.5. This implies that the distribution of
X is identifiable at least on the interval $[0, 99.5]$.

For this example we have $\tau = 54$ (months) and $\max_{2 \leq i \leq n}(x_{(i)} - x_{(i-1)}) = 6$, so
again we have $x_{(i)} - \tau \leq x_{(i-1)}$, $2 \leq i \leq n$. Condition $L_i > 0$ is also fulfilled, with
$\min_{2 \leq i \leq n} L_i = 0.5$, and with 291 of the L_i-values larger than 0.9. On the other hand,
the 256 first P_i's were zero.

In Fig. 8 we depict the direct and reverse non-iterative estimators together with the Efron–Petrosian NPMLE, the three of them showing a good agreement. Unlike in the previous example, the ordinary empirical df grossly overestimates the distribution of the incubation time, due to the oversampling of relatively small values of X. The estimated observational bias can be seen in Fig. 9. The curve in Fig. 9 is roughly decreasing, which indicates that the right-truncation issue dominates in this case.

5 Discussion

In this paper a non-iterative estimator of a cumulative df from doubly truncated data has been introduced. The estimator is simpler than the Efron–Petrosian NPMLE, which has no explicit form and must be computed iteratively. We have seen through simulations that the new estimator is competitive, with a relative efficiency above 0.93 in all the simulated settings. An interesting issue is that of the selection in practice of the optimal mixture between the direct and the reverse non-iterative procedures. This point deserves more investigation. In any case, the 'fair' mixture based on the weight $w = 1/2$ exhibited a good performance.

The non-iterative estimator has been introduced in the setting of interval sampling. However, we have discussed other settings with random double truncation too. Specifically, we have seen that the non-iterative estimator reduces to a simple closed-form estimator when U and V fulfill certain support conditions relative to X. The performance of such a simple estimator has been explored through simulations too.

No formal proof of consistency has been given for the non-iterative estimator. This is currently under investigation. On its turn, although some asymptotic theory for the (iterative) NPMLE has been developed (Shen [9, 10]), the corresponding formal proofs contain some important gaps which still need to be properly attacked (Mandel et al. [5]). Therefore, the estimator proposed and investigated in this paper could be an important first step towards a more tractable theoretical framework with doubly truncated data.

Acknowledgements Work supported by the Grant MTM2014-55966-P of the Spanish Ministerio de Economía y Competitividad.

References

1. Austin MD, Simon DK, Betensky RA (2014) Computationally simple estimation and improved efficiency for special cases of double truncation. Lifetime Data Anal 20:335–354
2. Bilker WB, Wang MC (1996) A semiparametric extension of the Mann–Whitney test for randomly truncated data. Biometrics 52:10–20
3. Efron B, Petrosian V (1999) Nonparametric methods for doubly truncated data. J Am Stat Assoc 94:824–834

4. Kalbfleisch JD, Lawless JF (1989) Inference based on retrospective ascertainment: an analysis of the data on transfusion-related AIDS. J Am Stat Assoc 84:360–372

5. Mandel M, de Uña-Álvarez J, Simon DK, Betensky RA (2017) Inverse probability weighted Cox regression for doubly truncated data. Biometrics. https://doi.org/10.1111/biom.12771

6. Moreira C, de Uña-Álvarez J (2010) Bootstrapping the NPMLE for doubly truncated data. J Nonparametr Stat 22:567–583

7. Moreira C, de Uña-Álvarez J, Crujeiras R (2010) DTDA: an R package to analyze randomly truncated data. J Stat Softw 37:1–20

8. Moreira C, de Uña-Álvarez J, Meira-Machado L (2016) Nonparametric regression with doubly truncated data. Comput Stat Data Anal 93:294–307

9. Shen PS (2010) Nonparametric analysis of doubly truncated data. Ann Inst Stat Math 62:835–853

10. Shen PS (2016) Analysis of transformation models with doubly truncated data. Stat Method 30:15–30

11. Woodroofe M (1985) Estimating a distribution function with truncated data. Ann Stat 13:163–177

12. Zhu H, Wang MC (2014) Nonparametric inference on bivariate survival data with interval sampling: association estimation and testing. Biometrika 101:1–15

The Mathematics of the Uncertain: Factors Related to Survival of Spanish Firms Along Recent Crisis

Antonio Vaamonde-Liste, Manuel Meijide-Vecino,
Patricio Sánchez-Bello and Pilar Muñoz-Dueñas

Abstract Several mathematical techniques have been considered in the literature to deal with the uncertainty associated with the life of firms. In this paper, factors associated with the survival of Spanish companies have been examined by means of statistical approaches, on the basis of a large sample of 5000 Spanish companies.

1 Introduction

Firms are living beings: they are born, grow, and inevitably they die. The longest-running company in the world seems to be the japanese Hoshi Ryokan Hotel chain (founded in 578); in Spain the oldest company is probably Codorniu wine makers, active since 1551. Different mathematical and statistical models have been developed to address the uncertainty associated with the life span, and these models can be applied to the life of companies too.

There are many factors that affect the survival of companies, structural and transient factors, technical and human, avoidable or predictable factors and other unavoidable, internal to the company and external factors. Although a firm's main objective is usually to achieve economic gain, often mere survival becomes the primary goal during difficult times.

Economic crises reveal the structural weaknesses of companies -many of them planned only for happy times- and those that survive often consider a benefit, even necessary for the good order of things, the forced disappearance of uncomfortable competitors.

In this paper we study some factors associated with the survival of Spanish companies using statistical techniques such as the Kaplan–Meier method for the construction of life tables and survival curves, and the Harrington–Fleming test [7] to evaluate the significance of the factors.

A. Vaamonde-Liste (✉) · M. Meijide-Vecino · P. Sánchez-Bello · P. Muñoz-Dueñas
Departamento de Estadística e Investigación Operativa, E.U. Estudios Empresariales,
Universidad de Vigo, C/Torrecedeira 105, 36208 Vigo, Spain
e-mail: vaamonde@uvigo.es

© Springer International Publishing AG 2018
E. Gil et al. (eds.), *The Mathematics of the Uncertain*, Studies in Systems,
Decision and Control 142, https://doi.org/10.1007/978-3-319-73848-2_38

401

A sample of 5000 companies selected at random from all Spanish companies is used, with data obtained from the Ardan database, managed by the Vigo Free Trade Zone Consortium, in relation to the annual accounts deposited by the companies in the corresponding Mercantile Registry to 2014. More than 200 variables are available: balance sheets, profit and loss accounts, different ratios, year of incorporation, sector of activity, and other characteristics of companies.

2 Survival of Firms

Enterprise survival studies were initiated by Marshall [9], who uses the forest simile in which young trees are growing and displacing old trees. Schumpeter [11] relates the survival of firms with innovation, which generates competitive capacity, and did study the effect of firms size.

The first theoretical approach to business survival is due to Gibrat [5] (Gibrat's Law or Proportional Growth Act), who describes how companies are distributed by size in a sector, and states that the growth of the company is independent of its size. The models of Jovanovic [8], or Ericson and Pakes [4], among others, introduce the notions of selection and learning, creating a theory of business selection with incomplete information.

The main variable of our study is the age of the company, expressed in years. The age distribution of the Spanish companies obtained with the data of the sample is represented in the histogram in Fig. 1; most companies are close to 20 years old, although a few are more than a century old. At the time of the beginning of the crisis (2008), it is possible to distinguish a notch in the graph, around 5–6 years old, with

Fig. 1 Age of Spanish firms

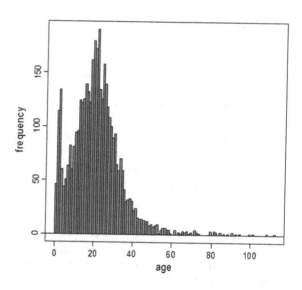

Table 1 Descriptive statistics on the age of Spanish firms

Min	1st Quartile	Median	Mean	3rd Quartile	Maximum
0.00	13.00	21.00	21.41	28.00	113.00

Table 2 Life table for Spanish firms

Time	n.risk	n.event	Survival	Std.err.	Lower 95% CI	Upper 95% CI
0	4113	7	0.998	0.000643	0.997	1.000
10	3459	114	0.969	0.002799	0.963	0.974
20	2262	67	0.946	0.003927	0.938	0.953
30	839	29	0.927	0.005225	0.917	0.937
40	269	10	0.904	0.009143	0.886	0.922
50	104	4	0.883	0.013806	0.856	0.910
60	46	4	0.833	0.027554	0.781	0.889
70	29	0	0.833	0.027554	0.781	0.889
80	20	1	0.802	0.040264	0.727	0.885

Fig. 2 Survival curve for Spanish firms

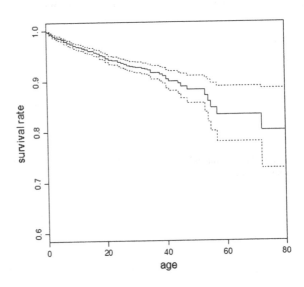

very low frequencies, a period in which many companies disappeared and very few were created.

Some descriptive statistics allow a first approximation (see Table 1).

Half of the companies are under 21 years old, a quarter less than 13 years, and another quarter more than 28 years.

We built the life table using Kaplan–Meier method, determining the probability of survival for each value of age, with a confidence interval 95% (see Table 2).

The survival curve graphically represents the survival rate (see Fig. 2).

3 Factors Affecting the Survival of Firms

The factors affecting the survival of firms are to be analyzed in this section.

3.1 Size

Size is a factor traditionally associated with survival (Dunne et al. [3], Mata and Portugal [10]). Larger companies have more advantages, compete better, and overcome difficulties more easily. Following European Union criteria, we consider "micro" enterprises those whose annual income do not reach 2 million euros, "small" up to 10, "medium" up to 50, and "large" those that exceed this figure. More than half Spanish companies are microenterprises, and almost 30% are small (see Table 3).

The survival curves for the different factor levels are displayed in Fig. 3.

The Harrington–Fleming test is performed, and outputs are gathered in Table 4.

The p-value is practically equal to zero, indicating that size is significantly related to survival. The graph shows as micro firms, whose curve is consistently below the

Table 3 Size of Spanish firms

Size	n	%
micro	2610	52.96
small	1434	29.10
medium	625	12.68
large	259	5.26

Fig. 3 Survival rate of Spanish firms by size

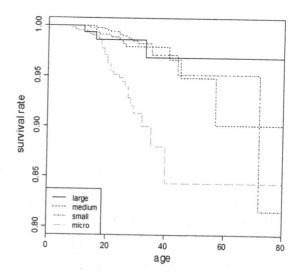

Table 4 Harrington–Fleming test details by size

Size	N	Observed	Expected	$(O - E)^2/E$	$(O - E)^2/V$
large	172	3	5.54	1.17	1.29
medium	545	11	17.11	2.18	2.93
micro	1024	42	18.07	31.67	44.36
small	1055	15	30.27	7.70	13.58
	$\chi_3^2 = 44.5, \quad p = 0.00000000117$				

Table 5 Harrington–Fleming test details by profitability

Profitability	N	Observed	Expected	$(O - E)^2/E$	$(O - E)^2/V$
low	2055	107	122	1.76	3.65
high	2057	129	114	1.88	3.65
	$\chi_1^2 = 3.7, \quad p = 0.0561$				

others, have a clearly smaller survival rate than the remaining, and there are only small differences between the other groups.

3.2 Economic Profitability

We will consider two groups of companies: with profitability below the median, and above the median. We perform the Harrington-Fleming test (see Table 5).

The p-value is over the usual 0.05 significance level, which does suggest that profitability, at least in the crisis period considered, is not a survival factor. Let's see the graph displayed in Fig. 4.

We observe how survival curves, surprisingly, intersect. Although for companies over 40 years old high profitability can ensure a significantly higher survival rate, this does not happen for the early ages, less than 30 years: apparently young companies with lower profitability survive better, which seems contradictory. However, it is well-known that in times of crisis, many companies are forced to reduce their margins by temporarily assuming low or even negative returns, in order to compete and maintain their activity during difficult times, which may explain the observed effect. Companies that are able to lower prices and take on less profit survive better.

3.3 Solvency

Solvency indicates the ability of companies to respond to payments due, is indispensable for daily dealings with suppliers and for dealing with credit institutions, and in

Fig. 4 Survival rate of
Spanish firms by profitability

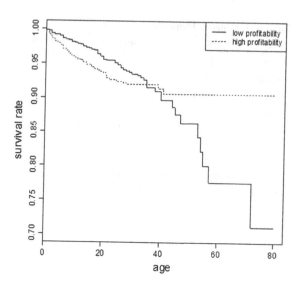

Table 6 Harrington–Fleming test details by solvency

Solvency	N	Observed	Expected	$(O - E)^2/E$	$(O - E)^2/V$
low	2079	160	106	26.9	49.7
high	2030	76	130	22.1	49.7
	$\chi_1^2 = 49.7$, $\quad p = 0.0000000000018$				

most sectors of activity is a necessary condition for the operation of the company. Their relationship with survival is clear: solvent companies survive, and those that are not, may disappear (see Görg and Spaliara [6]).

We will consider again, for simplicity, two groups of companies: with solvency below and above the median. We perform the Harrington–Fleming test (see Table 6).

The p-value is practically equal to 0, which confirms a significant relationship. Let us see the survival curves displayed in Fig. 5.

The survival rate for less solvent enterprises is consistently below that of the solvent companies.

3.4 Debt

The most indebted companies run the risk of not being able to cope with the repayment of the loans received, and disappear (Bridges and Guariglia [2]). The levels of indebtedness are different in the different industries, and a deep study must take into account these aspects. In this subsection we consider debt rates globally, in relation to the average rate.

Fig. 5 Survival rate of
Spanish firms by solvency

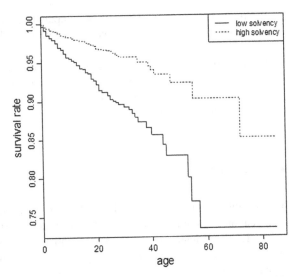

Table 7 Harrington–Fleming test details by debt

Debt	N	Observed	Expected	$(O-E)^2/E$	$(O-E)^2/V$
low	2296	105	136.7	7.34	17.5
high	1817	131	99.3	10.11	17.5
	$\chi_1^2 = 17.5,\quad p = 0.0000283$				

Fig. 6 Survival rate of
Spanish firms by debt

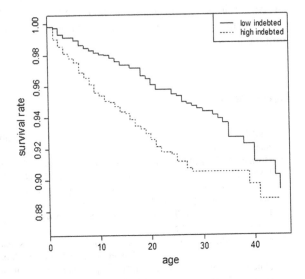

To do this we rebuild two groups of companies, according to whether their debt ratio is lower or higher than the median. We perform the Harrington–Fleming test (see Table 7).

The p-value is practically equal to 0, which indicates that the effect of this factor on survival is significant. The survival curves are displayed in Fig. 6.

The most heavily indebted companies have survival rates that are clearly lower than the less indebted ones.

4 Conclusions

It is possible to introduce rationality criteria in the uncertainty associated with the disappearance of companies, especially accentuated in times of crisis.

The Kaplan–Meier method allows estimating the survival curve of a population of firms, and helps to analyze the factors that can determine the probability of disappearance of a company.

In the case of Spanish companies, in the years immediately after the beginning of the recent crisis, it is observed that size is a determining factor, with smaller companies (micro-enterprises) having the lowest survival rates. Other factors also appear to be significantly related to corporate mortality: economic profitability, although it appears to affect young firms differently to those having a consolidated position, solvency, or indebtedness.

On some of these factors it is possible to act, so that companies improve their survival options: it is necessary to increase the size of the company to strengthen the competitive position, pay special attention to the solvency ratio, and reasonably limit the indebtedness.

References

1. Audretsch DB, Mahmood T (1994) The rate of hazard confronting new firms and plants in U.S. manufacturing. Rev Ind Organ 9(1):41–56
2. Bridges S, Guariglia A (2008) Financial constraints, global engagement and firm survival in UK: evidence from micro data. Scott J Polit Econ 55:444–464
3. Dunne T, Roberts MJ, Samuelson R (1989) The growth and failure of U.S. manufacturing plants. Q J Econ 104(4):671–698
4. Ericson R, Pakes A (1995) Markov-perfect industry dynamics: a framework for empirical work. Rev Econ Stud 62(1):53–82
5. Gibrat R (1931) Les Inégalités Économiques. Applications: Aux Inégalités des Richesses, à la Concentration des Entreprises, Aux Population des Villes, Aux Statistiques des Familles, etc., dune Loi Nouvelle: La Loi de lEffet Proportionnel. Librairie du Recueil Sirey, Paris
6. Görg H, Spaliara M-E (2014) Financial health, exports and firm survival: evidence from UK and French firms. Economica 81(323):419–444
7. Harrington DP, Fleming TR (1982) A class of rank test procedures for censored survival data. Biometrika 69(3):553–566

8. Jovanovic B (1982) Selection and the evolution of industry. Econometrica 50(3):649–670
9. Marshall A (1920) Principles of economics. MacMillan, London
10. Mata J, Portugal P (1994) Life duration of new firms. J Ind Econ 42(3):227–246
11. Schumpeter JA (1934) The theory of economic development. Harvard University Press, Cambridge

A Functional Stochastic Model for the Evolution of Spanish Stock Market

Mariano J. Valderrama, Manuel Escabias, Federico Galán
and Antonio Gil

Abstract The study of the evolution of financial series has always been a complex problem because of the nature of stock market series that usually are close to a random walk. The most usual approach has been to apply ARCH and GARCH models, as well as methods that attempt to capture stochastic volatility. In this paper we present an alternative way of approximating this problem, that consists of modeling these series by functional principal components analysis of the financial process up to a certain time frame. The study focused on the Spanish index IBEX35 over a broad period (2007–2013) and, based on continuous market trading, the sample paths were considered integrable square curves. The objective of the work is the estimation of explanatory models for the different bonds as well as the correlation between them.

1 Introduction

Functional data analysis (FDA) is a statistical methodology related to multivariate analysis and stochastic processes whose pioneer works are the ones by Deville [5], Saporta [14] and Besse and Ramsay [4], although its main development and interest for scientific comunity araised since J.O. Ramsay and B. Silverman published in 1997 the book *Functional Data Analysis* [12] as well as since the shared S-Plus, R and Matlab libraries are available for researchers around the world allowing the

M. J. Valderrama (✉) · M. Escabias
Department of Statistics and Operation Research, University of Granada,
18071 Granada, Spain
e-mail: valderra@ugr.es

M. Escabias
e-mail: escabias@ugr.es

F. Galán · A. Gil
Department of Financial Economics and Accounting, University of Granada,
18071 Granada, Spain
e-mail: fgalan_1@ugr.es

A. Gil
e-mail: amgil@ugr.es

© Springer International Publishing AG 2018
E. Gil et al. (eds.), *The Mathematics of the Uncertain*, Studies in Systems,
Decision and Control 142, https://doi.org/10.1007/978-3-319-73848-2_39

application of FDA methods and obtaining results in a simple way. The aforementioned libraries have been improved along the years, and some new ones have arisen to deal with FDA with different objectives, for example *fdakma* for K-mean alignment in FDA, *fdaMixed* in a mixed model framework, *fdatest* for interval testing, *fda.usc* that has utilities for statistical computing in FDA, *FDboost* for boosting functional regression models or *fdcov* for analysis of covariance operators, among others in R (see https://CRAN.R-project.org). Recently a web application has been launched that allows the use of some of the methods of analysis of functional data without deep knowledgement of the FDA theory to use it (http://www.statfda.com/).

The aim of FDA is to apply statistical methods when data are curves, that is mathematical functions, by using the full information of the complete curve instead of a set of discrete observations of that curve. For example Aguilera et al. [3] used FDA methods to model the occurrence of Lupus flares from curves of stress and Escabias et al. [8] to model allergic variables from weather curves.

Economic and financial series have been also modeled by FDA methods. So, Aguilera et al. [1] applied functional principal component analysis for stochastic modelling of stock-prices evolution by means of functional principal component analysis, and also Ingrassia and Costanzo [10] to classify financial time series. Moreover Aguilera et al. [2] forecasted binary longitudinal data by a functional PC-ARIMA model.

The objective of this paper is to illustrate the use of functional principal component analysis to model stock market series.

2 Some Theory on Functional Data Analysis

Values of a functional variable depend on a continuous magnitude such as time, so that a functional data set is a collection of curves $\{x_1(t), \ldots, x_n(t)\}$, with $t \in T$. Each curve can be observed at different time points of its argument t as $x_i = \left(x_i(t_0), \ldots, x_i\left(t_{m_i}\right)\right)'$ for the set of times $t_0, \ldots, t_{m_i}, i = 1, \ldots, n$ and these are not necessarily the same for each curve.

Different approaches have been taken to the study of functional data, including the nonparametric methods proposed by Ferraty and Vieu [9] and Müller [11], and the basis expansion methods used by Ramsay and Silverman [13] that we follow in the present application to reconstruct the functional form of curves in order to evaluate them at any time point t. This method assumes that the curves belong to a finite dimensional space generated a basis of functions $\{\phi_1(t), \ldots, \phi_p(t)\}$ and so they can be expressed as

$$x_i(t) = \sum_{j=1}^{p} a_{ij}\phi_j(t), \ i = 1, \ldots, n. \tag{1}$$

The functional form of the curves is determined when the basis coefficients $a_i = (a_{i1}, \ldots, a_{ip})'$ are known. These can be obtained from the discrete observations either by least squares or by interpolation (see, for example, Escabias et al. [6, 7]). In our application the least squares method is considered for functional representation.

Let $x_1(t), x_2(t), \ldots, x_n(t)$ be a set of curves all of them observed at the same time points $t_1, t_2, \ldots, t_m$. Then the available information in this situation is the matrix $X = (x_i(t_j))$. The basis coefficients of all curves are obtained by least squares approximation as $A^T = (\Phi^T \Phi)^{-1} \Phi^T X^T$ where $\Phi = (\phi_k(t_j))$ is the matrix of basic functions evaluated at sampling points.

From a set of curves $x_1(t), x_2(t), \ldots, x_n(t)$ the mean curve is defined as

$$\overline{x}(t) = \frac{1}{n} \sum_{i=1}^{n} x_i(t)$$

and the covariance surface as

$$C(s, t) = \frac{1}{n-1} \sum_{i=1}^{n} (x_i(s) - \overline{x}(s)) (x_i(t) - \overline{x}(t)).$$

Functional principal components are defined as the uncorrelated random variables obtained as linear combination of the sample curves

$$\xi_i = \int_T (x_i(t) - \overline{x}(t)) f(t) \, dt, \ i = 1, \ldots, n$$

that maximize the variance of $\xi_1, \ldots, \xi_n$. By imposing this condition, functional principal components are the solutions of a Fredholm second order integral equation

$$\int_T C(s, t) f(s) \, ds = \lambda f(t)$$

where λ denotes the variance of a functional principal component.

When curves are expressed in terms of basic functions as (1) previous equation has p solutions for λ that verify: $\lambda_1 \geq \lambda_2 \geq \cdots \geq \lambda_p$. Each λ_j generates a linear space of eigenfunctions with dimension equal to its multiplicity, so that when all them are simple the associated eigenspace have only one function $f_j(t)$ and define the funcional principal component curves:

$$\xi_{ij} = \int_T (x_i(t) - \overline{x}(t)) f_j(t) \, dt, \ j = 1, \ldots, p, \ i = 1, \ldots, n.$$

In matrix form, functional principal components are usually considered in a $n \times p$ matrix Γ. Moreover, each functional principal component cummulates a proportion of the total variability given by

$$\frac{\lambda_j}{\sum_{j=1}^{p} \lambda_j}.$$

The total variability that cummulates the functional principal components is equal to the total variability of curves.

When curves are expressed in terms of basic functions as (1), eigenfunctions are also represented in terms of the same basic functions

$$f_j(t) = \sum_{k=1}^{p} F_{jk}\phi_k(t), \quad j = 1, \ldots, p.$$

The original curves can be approximated by using a reduced set of functional principal components

$$x_i(t) = \sum_{j=1}^{q<p} \xi_{ij} f_j(t), \quad i = 1, \ldots, n,$$

and then we have an approximation of the original curves in terms of basis fuctions, that is, by knowing their basis coefficients

$$x_i(t) = \sum_{j=1}^{q<p} \xi_{ij} \sum_{k=1}^{p} F_{jk}\phi_k(t), \quad i = 1, \ldots, n.$$

3 Modeling Stock Market Data by FDA

The available data are the daily stock values of 33 companies belonging to Spanish stock market index IBEX35, including the own main index. The companies considered were the following ones:

Abengoa	BBVA	Indra
Abertis	Bolsa y Mercados	Mapfre
Acciona	Enagas	Metrovacesa
Acerinox	ENDESA	NHHoteles
ACS	FCC	OHL
Antena 3	Ferrovial	PRISA
Banco Popular	Gamesa	Red Eléctrica Española
Banco Sabadell	Gas Natural	Repsol
Banco Santander	Grifolsa	Sacyr Vallehermoso
Banesto	Iberdrola	Técnicas Reunidas
Bankinter	Inditex	Telefónica

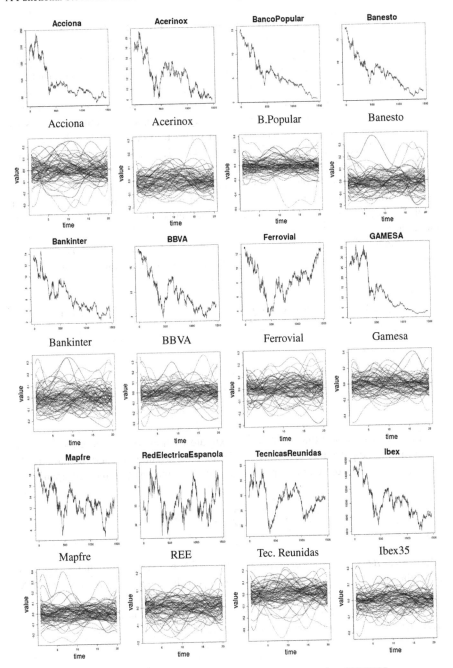

Fig. 1 Daily stock values and 10-days rates of return of some companies of IBEX35

For each company we had available 1440 daily returns since 2007 to 2013 whose observed series can be seen in Fig. 1.

The problem of modelling stock values by functional data analysis is the dependency structure that exists between nearby observations. For this reason, instead of them we have used the 10-days rates of return given for each series by $r(d) = (C(d + 10) - C(d))/C(d)$ where $C(d)$ is the correponding stock price at time $d = 1, 2, \ldots$ In addition, functional data analysis methods use curve samples as data independent, so that we have cut each series into curves of length 20: $x_{ij} = x_i(t_j) = R(20x(i - 1) + j); i = 1, 2, \ldots, j = 1, 2, \ldots, 20$. Some of these curves can be also seen in Fig. 1, where discrete observations were turned to curves by basis expansion in terms of cubic B-spline basis by least squares approximation.

After the approximation by basis expansion a functional principal component analysis was performed independently for each curve. Cumulated explained variances are summarized in Table 1.

The linear correlation between IBEX functional principal components and functional principal components of the companies involved suggested the next linear models

$$\widehat{\xi}_{i1}(\text{IBEX}) = \widehat{\beta}_{11}\widehat{\xi}_{i1}(X_1) + \widehat{\beta}_{12}\widehat{\xi}_{i1}(X_2) + \widehat{\beta}_{13}\widehat{\xi}_{i1}(X_3) + \widehat{\beta}_{14}\widehat{\xi}_{i1}(X_4),$$
$$\widehat{\xi}_{i2}(\text{IBEX}) = \widehat{\beta}_{21}\widehat{\xi}_{i2}(X_5) + \widehat{\beta}_{22}\widehat{\xi}_{i2}(X_6),$$
$$\widehat{\xi}_{i3}(\text{IBEX}) = \widehat{\beta}_{31}\widehat{\xi}_{i3}(X_7),$$
$$\widehat{\xi}_{i4}(\text{IBEX}) = \widehat{\beta}_{41}\widehat{\xi}_{i4}(X_8) + \widehat{\beta}_{42}\widehat{\xi}_{i4}(X_9),$$
$$\widehat{\xi}_{i5}(\text{IBEX}) = \widehat{\beta}_{51}\widehat{\xi}_{i5}(X_{10}) + \widehat{\beta}_{52}\widehat{\xi}_{i5}(X_{11}).$$

Taking into account that we can approximate each curve from a reduced set of functional principal components, we can formulate a prediction function of the IBEX curves from the predicted principal components obtained by the main companies

$$x_i(t) = \overline{x}(t) + \sum_{j=1}^{5} \widehat{\xi}_{ij}(\text{IBEX}) f_j(t)$$

where $\widehat{\xi}_{ij}(IBEX)$ are the predictions of the IBEX principal components obtained from the ones of the main companies, and $f_j(t)$ the eigenfunctions of the IBEX curves.

The prediction of the several fits are summarized in Table 2 and the prediction accuracy for some of the curves can be seen in Fig. 2.

Table 1 Cumulated explained variance of the functional principal components

Empresa	Comp.1	Comp.2	Comp.3	Comp.4	Comp.5	Comp.6	Comp.7	Comp.8	Comp.9	Comp.10
Abeng.	52.42	77.25	90.20	94.18	96.29	97.87	99.15	99.73	99.88	100.00
Aber.	54.05	78.67	90.80	94.77	96.65	98.14	99.10	99.62	99.85	100.00
Acc.	59.28	76.91	91.46	94.70	96.76	98.12	99.15	99.71	99.87	100.00
Acer.	50.48	78.08	91.26	94.53	96.45	98.07	99.11	99.68	99.88	100.00
ACS	53.01	77.74	89.30	93.54	96.37	97.78	98.84	99.67	99.87	100.00
Ant3	51.02	80.75	91.25	94.97	96.97	98.27	99.27	99.79	99.92	100.00
B.Pop	52.54	79.06	91.13	94.93	97.14	98.43	99.21	99.70	99.87	100.00
B.Sab.	49.40	80.72	92.44	96.15	97.63	98.70	99.27	99.78	99.93	100.00
B.San.	38.73	73.62	88.81	93.39	96.06	97.57	98.88	99.67	99.88	100.00
Ban.	47.47	80.14	90.32	94.19	96.59	98.13	99.22	99.75	99.90	100.00
Bank.	52.84	79.05	89.48	93.86	95.84	97.63	98.84	99.71	99.88	100.00
BBVA	42.90	72.30	88.57	93.62	96.02	97.51	98.88	99.67	99.87	100.00
BolMer	49.86	75.99	90.03	93.75	95.69	97.41	98.78	99.65	99.83	100.00
Enagas	52.91	75.43	89.41	93.49	95.82	97.81	99.05	99.64	99.84	100.00
END.	50.33	80.91	91.30	94.84	97.05	98.45	99.14	99.75	99.92	100.00
FCC	53.43	74.05	87.81	93.02	96.16	97.88	99.00	99.71	99.89	100.00
Ferr.	52.00	76.98	90.40	94.07	96.24	98.00	99.02	99.67	99.84	100.00
GAM.	47.35	76.04	90.26	94.18	96.72	98.09	99.14	99.73	99.88	100.00
G.Nat.	54.45	74.99	89.02	93.78	95.85	97.77	98.98	99.65	99.87	100.00
Grif.	52.97	77.91	90.82	94.90	96.72	98.30	99.29	99.71	99.88	100.00
Iber.	51.13	73.05	88.49	92.28	95.35	97.18	98.70	99.52	99.83	100.00
Indit	50.66	72.82	89.41	94.48	96.56	97.85	98.97	99.58	99.82	100.00
Indra	43.14	76.18	89.34	92.44	95.35	97.38	98.81	99.68	99.84	100.00
Mapfre	52.42	76.59	88.33	92.97	95.40	97.12	98.66	99.50	99.82	100.00
Metr.	46.18	83.36	93.75	97.60	98.43	99.04	99.50	99.85	99.94	100.00
NH	54.70	80.66	91.27	94.92	97.08	98.31	99.33	99.75	99.90	100.00
OHL	54.53	81.91	91.48	94.77	96.88	98.28	99.23	99.82	99.93	100.00
PRISA	42.64	78.04	91.89	95.22	96.92	98.16	99.21	99.75	99.91	100.00
REE	49.28	74.54	88.76	93.59	95.81	97.96	98.97	99.56	99.80	100.00
Repsol	47.05	71.46	89.37	93.02	95.77	97.32	98.72	99.64	99.87	100.00
Sacyr	53.79	76.56	91.99	95.05	96.80	97.94	99.00	99.77	99.90	100.00
Tec.Reu.	53.78	78.14	90.66	94.06	96.00	97.69	99.11	99.69	99.87	100.00
Telef.	57.21	80.01	90.14	93.17	95.79	97.41	98.72	99.61	99.86	100.00
Ibex	48.07	71.92	88.17	92.53	95.51	97.12	98.60	99.60	99.86	100.00

Table 2 Prediction analysis of the five main functional components of IBEX 35

	$\widehat{\beta}_{ij}$	Std. Error	t	p-value
$\widehat{\xi}_{i1}$ (IBEX)				
$\widehat{\xi}_{i1}(X_1 = \text{REE})$	0.26	0.08	3.35	0.00
$\widehat{\xi}_{i1}(X_2 = \text{Mapfre})$	0.29	0.04	6.77	0.00
$\widehat{\xi}_{i1}(X_3 = \text{Acerinox})$	0.15	0.05	2.77	0.01
$\widehat{\xi}_{i1}(X_4 = \text{Acciona})$	0.12	0.04	2.92	0.00
$\widehat{\xi}_{i2}$ (IBEX)				
$\widehat{\xi}_{i2}(X_5 = \text{Ferrovial})$	−0.25	0.06	−3.88	0.00
$\widehat{\xi}_{i2}(X_6 = \text{Tec.Reu})$	0.31	0.07	4.52	0.00
$\widehat{\xi}_{i3}$ (IBEX)				
$\widehat{\xi}_{i3}(X_7 = \text{BBVA})$	0.60	0.02	25.98	0.00
$\widehat{\xi}_{i4}$ (IBEX)				
$\widehat{\xi}_{i4}(X_8 = \text{Banesto})$	0.30	0.07	4.25	0.00
$\widehat{\xi}_{i4}(X_9 = \text{Bankinter})$	0.29	0.05	5.25	0.00
$\widehat{\xi}_{i5}$ (IBEX)				
$\widehat{\xi}_{i5}(X_{10} = \text{B.Popular})$	0.37	0.05	6.98	0.00
$\widehat{\xi}_{i5}(X_{11} = \text{GAMESA})$	0.16	0.05	3.38	0.00

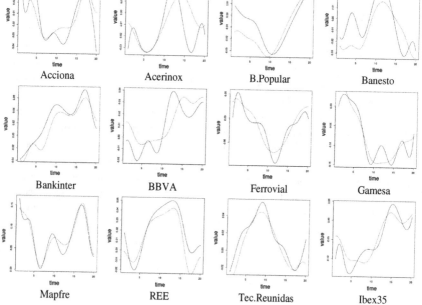

Fig. 2 Original 10-days rate of return (solid line) overlayed with the estimated one (dashed line) for some companies of IBEX35

Acknowledgements This research has been supported by project MTM2017-88708-P of Ministerio de Economía y Competitividad, Government of Spain and project P11-FQM-8068 from Consejería de Innovación, Ciencia y Empresa, Junta de Andalucía, Spain. Authors are very grateful for the invitation to participate in this volume that is a tribute to our dear master, colleague and friend Pedro Gil, whose memory will live forever.

References

1. Aguilera AM, Ocaña FA, Valderrama MJ (1999) Stochastic modelling for evolution of stock-prices by means of functional principal component analysis. Appl Stoch Mod Bus Ind 15(4):227–234
2. Aguilera AM, Escabias M, Valderrama MJ (2008) Forecasting binary longitudinal data by a functional PC-ARIMA model. Comput Stat Data Anal 52:3187–3197
3. Aguilera AM, Escabias M, Valderrama MJ (2008) Discussion of different logistic models with functional data. Application to systemic lupus erythematosus. Comput Stat Data Anal 53:151–163
4. Besse P, Ramsay JO (1986) Principal component analysis of sampled functions. Psychometrika 51(2):285–311
5. Deville JC (1974) Méthodes statistiques et numériques de l'analyse harmonique. Ann de l'INSEE 15:3–101
6. Escabias M, Aguilera AM, Valderrama MJ (2005) Modeling environmental data by functional principal component logistic regression. Environmetrics 16:95–107
7. Escabias M, Aguilera AM, Valderrama MJ (2007) Functional PLS logit regression model. Comput Stat Data Anal 51:4891–4902
8. Escabias M, Valderrama MJ, Aguilera AM, Santofimia ME, Aguilera-Morillo MC (2013) Stepwise selection of functional covariates in forecasting peak levels of olive pollen. Stoch Environ Res Risk Asses 27(2):367–376
9. Ferraty F, Vieu P (2006) Nonparametric functional data analysis. Springer, New York
10. Ingrassia S, Costanzo GD (2005) Functional principal component analysis of financial time series. In: Vichi M, Monari P, Mignani S, Montanari A (eds) New developments in classification and data analysis. Springer, Berlin
11. Müller HG (2008) Functional modeling of longitudinal data. In: Fitzmaurice G, Davidian M, Verbeke G, Molenberghs G (eds) Longitudinal data analysis. Handbooks of modern statistical methods. Chapman and Hall/CRC, New York
12. Ramsay JO, Silverman BW (1997) Functional data analysis. Springer, New York
13. Ramsay JO, Silverman BW (2005) Functional data analysis, 2nd edn. Springer, New York
14. Saporta G (1985) Data analysis for numerical and categorical individual time-series. Appl Stoch Mod Data Anal 1:109–119

Limit Behavior of Polya Urn Schemes

Ricardo Vélez and Tomás Prieto-Rumeau

In memory of our colleague and friend Pedro Gil

Abstract The long term behavior of the composition of Polya's urn schemes is analyzed by means of simple martingale arguments. The results hold even under the assumption that random numbers of balls are added to the urn depending only on the color of the ball obtained in the successive extractions.

1 Polya's Urn Schemes

Urn schemes, devised by George Polya, are a useful tool to think about probability beyond the scope of independence, too predominant in introductory courses. They help to illustrate many essential concepts such as *exchangeability*, *martingales*, *stochastic convergences* and *limit laws*.

An urn scheme considers an urn with balls of two different colors: $\mathscr{A}$mber and $\mathscr{B}$lue. Successive extractions are performed at random and the obtained ball is returned to the urn together with a certain number of additional balls of each color. Initially there is a total of t balls in the urn, α of which are amber and $\beta = t - \alpha$ are blue. The urn scheme is characterized by the parameters:

$$[\underbrace{a', \, a}_{amber} \; ; \; \underbrace{b, \, b'}_{blue}]$$

Supported by Grant MTM 2016–75497-P, from the Spanish Ministerio de Economía y Competitividad.

R. Vélez (✉) · T. Prieto-Rumeau
Department of Statistics–UNED, Senda del Rey, 9, 28040 Madrid, Spain
e-mail: rvelez@ccia.uned.es

© Springer International Publishing AG 2018
E. Gil et al. (eds.), *The Mathematics of the Uncertain*, Studies in Systems, Decision and Control 142, https://doi.org/10.1007/978-3-319-73848-2_40

421

where a' and a stand for the number of amber and blue balls respectively added when an amber ball is obtained, while b amber and b' blue balls are adjoined if a blue ball is extracted. Negative values for some of these parameters are usually admitted and to add $-x$ balls means to remove x balls of the urn. However, in this case it may happen that the balls of some color are exhausted and the extractions cannot be pursued according to the rules of the model.

The most practical assumption is that the total number of added balls is fixed, say $c > 0$; so that $a' = c - a$ and $b' = c - b$. In this case the total number of balls in the urn is not random, but independent of the successive colors previously obtained. Unless otherwise stated, this will be the considered model:

$$[c - a,\ a; b,\ c - b],\qquad c > 0.$$

We will denote

- Y_k the color of the ball obtained in the trial $k \geq 1$: $Y_k = 1$ if it is amber, $Y_k = 0$ if it is blue.
- $X_n = \sum_{k=1}^{n} Y_k$, the total number of amber balls obtained in the first n extractions.

One can allow the possibility that a and b could have random values. To this end, assume that $\{a_k\}_{k\geq 1}$ and $\{b_k\}_{k\geq 1}$ are two sequences of independent identically distributed random variables with integer values, independent also of $\{Y_k\}_{k\geq 1}$. Then if $Y_k = 1$, a_k blue balls and $c - a_k$ amber balls are added to the urn, while if $Y_k = 0$ the blue ball is returned to the urn together with b_k amber balls and $c - b_k$ blue balls. Specifically a_k and b_k would have a finite range of values (for instance $0 \leq a_k, b_k \leq c$) and $\bar{a}$ and $\bar{b}$ will denote the means of a_k and b_k respectively.

The Refs. [2–4] collect the known results about this subject. Reference [6] also contains a review of the main papers in the area.

The aim of this paper is to show that the long term frequencies of each color will always be stabilized around some value, fixed or random, and therefore it can never oscillate indefinitely. This conclusion is established in Sect. 3. As an introduction, Sect. 2 presents some well known results to be used later.

2   The Polya–Eggenberger Model

The most complete analysis of an urn scheme corresponds to the Polya–Eggenberger model for which $a = b = 0$. Explicitly, when a ball of any color is obtained, it returns to the urn only with c new balls of the same color. In these conditions, for any $(i_1, i_2, \ldots, i_n) \in \{0, 1\}^n$, it will be

$$P\{Y_1 = i_1, \ldots, Y_n = i_n\}$$
$$= \frac{\alpha\,(\alpha + c) \cdots (\alpha + (r - 1)c)\ \beta(\beta + c) \cdots (\beta + (s - 1)c)}{t\,(t + c) \cdots (t + (n - 1)c)} \tag{1}$$

where $r = \sum_{j=1}^{n} i_j$ is the number of ones and $s = n - r$ the number of zeros in $(i_1, \ldots, i_n)$. In fact, the successive factors in the denominator give the number of balls in the urn when the successive extractions are done. If i_j is the first index with value 1, at the time of extraction j, there are α amber balls in the urn; next $\alpha + c$ amber balls are in the urn when the second amber ball is obtained; and so on. Similarly, $\beta, \beta + c, \beta + 2c, \ldots$ are the number of blue balls in the urn when the successive blue balls are obtained. An obvious modification is needed for $i_1 = \cdots = i_n$, since no α (or β) factors appear if all them are zeros (or ones). Upon dividing by c^n the numerator and denominator, we get

$$P\{Y_1 = i_1, \ldots, Y_n = i_n\}$$
$$= \frac{\Gamma(\alpha/c + r)}{\Gamma(\alpha/c)} \frac{\Gamma(\beta/c + s)}{\Gamma(\beta/c)} \frac{\Gamma(t/c)}{\Gamma(t/c + n)} = \frac{\beta(\alpha/c + r, \beta/c + s)}{\beta(\alpha/c, \beta/c)}. \quad (2)$$

As these probabilities depend on $(i_1, i_2, \ldots, i_n)$ only through n and r (and $s = n - r$), any permutation of $(i_{\pi(1)}, \ldots, i_{\pi(n)})$ has the same probability or, equivalently,

$$P\{Y_{\pi(1)} = i_1, \ldots, Y_{\pi(n)} = i_n\} = P\{Y_1 = i_1, \ldots, Y_n = i_n\}$$

so that $\{Y_k\}_{k \geq 1}$ is an exchangeable sequence of random variables. We can then use the well known result:

Theorem 2.1 (de Finetti) *Let $\{Y_k\}_{k \geq 1}$ be an exchangeable sequence of random variables with values in $\{0, 1\}$. There exists a random variable U, with distribution θ on $(0, 1)$, such that, conditionally on U, $\{Y_k\}_{k \geq 1}$ are independent random variables with distribution: $P\{Y_k = 1|U\} = U$.*
Therefore, if $i_1 + \cdots + i_n = r$, it is

$$P\{Y_1 = i_1, \ldots, Y_n = i_n\} = \int_0^1 u^r (1 - u)^{n-r} \, \theta(du) \quad (3)$$

and moreover $\frac{1}{n} \sum_{k=1}^{n} Y_k \longrightarrow U$ a.s.

The last sentence states that the ratio X_n/n of amber balls obtained in the first n attempts converges, with probability 1, to a random value U that may have a different value each time the experience is performed. But long-term oscillations will never be observed in the average X_n/n nor on the proportion of amber balls in the urn:

$$p_n = \frac{\alpha + cX_n}{t + cn} \sim \frac{X_n}{n}.$$

A simple MatLab program allows to draw the graphic of Fig. 1, whose curves (n, p_n) correspond to 10 realizations of 2000 extractions from an urn, with initially

Fig. 1 Proportion (in 10 realizations of 2000 extractions) of amber balls when $a = b = 0$

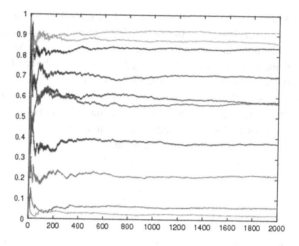

3 amber balls and 2 blue balls, to which 5 balls of the obtained color are added. This is the same kind of phenomenon observed when extracting balls from an urn with fixed composition, except for the key distinctness that the limit is not a preestablished value, but a different value of U in each replay.

Concerning the distribution θ with which the value of U is chosen, according to (3) it must be

$$\int_0^1 u^r (1 - u)^{n-r} \theta(du) = \frac{\beta(\alpha/c + r, \beta/c + s)}{\beta(\alpha/c, \beta/c)}. \tag{4}$$

But, when θ is the *beta* distribution with parameters α/c and β/c, the left member equals

$$\int_0^1 u^{\alpha/c+r-1} (1 - u)^{\beta/c+n-r-1} \frac{du}{\beta(\alpha/c, \beta/c)} = \frac{\beta(\alpha/c + r, \beta/c + n - r)}{\beta(\alpha/c, \beta/c)}$$

and the equality (3) holds. Obviously, there is no other distribution with the same property, since (4) determines, for $r = n$, all the moments of θ. Shortly, if the urn composition is chosen with distribution *beta* $(\alpha/c, \beta/c)$, by replacing only the extracted ball, the sequence $\{Y_k\}$ will have the same distribution than in the Polya–Eggenberger model.

These results are quoted without proof in [2] and were implicit in the original paper of Eggenberger and Polya. They can also be found in [5, Examples VII, 4(*a*) and 9(*a*)].

3 A Martingale Analysis

We now return to the general setting of the first section, with c balls added after each extraction k, of which a_k are blue if the extracted ball is amber and b_k are amber if the extracted ball is blue. We will analyze some kind of almost sure asymptotic behavior of the urn composition, based only on the means $\bar{a}$ and $\bar{b}$.

Let $\mathscr{F}_n = \sigma(\{Y_k, a_k, b_k\}_{1 \le k \le n})$ be the σ-field of events depending on the results of the first n moves in a stochastic Polya urn scheme. The fraction of amber balls in the urn after the first n moves:

$$p_n = \frac{\alpha + \sum_{k=1}^{n} Y_k(c - a_k) + \sum_{k=1}^{n}(1 - Y_k)b_k}{t + cn} \tag{5}$$

is $\mathscr{F}_n$-measurable. Further

$$E[p_{n+1}|\mathscr{F}_n, a_{n+1}, b_{n+1}] = \frac{p_n(t + cn) + p_n(c - a_{n+1}) + (1 - p_n)b_{n+1}}{t + c(n + 1)}$$

and

$$E[p_{n+1}|\mathscr{F}_n] = p_n + \frac{(1 - p_n)\bar{b} - p_n\bar{a}}{t + c(n + 1)}. \tag{6}$$

More precisely, p_n gives the probability of obtaining an amber ball until the stopping time $\tau = \inf\{n \ge 1 | p_n \notin [0, 1]\}$. When $\tau < \infty$ and $p_\tau > 1$, the experiment should end due to the lack of blue balls; similarly, for $\tau < \infty$ and $p_\tau < 0$, extractions are stopped due to the lack of amber balls. Obviously if $0 \le a_k, b_k \le c$, the numbers of balls of both colors are increasing and $\tau = \infty$.

Proposition 3.1 *Assume that $\bar{a} \le 0 < \bar{b}$.*

(i) As long as $p_n \in [0, 1]$, $\{p_n\}$ is a submartingale.
(ii) With probability 1, $\tau < \infty$ or $p_n \to p_\infty \in [0, 1]$ (that can be a random variable).

Moreover $p_n \xrightarrow{L^1} p_\infty$ conditional on $\{\tau = \infty\}$.

(iii) If $\bar{a} = 0$, it is $p_\infty = 1$. For $\bar{a} < 0$, $P\{\tau < \infty\} = 1$.

Proof (i) $E[p_{n+1}|\mathscr{F}_n] \ge p_n$ if and only if $p_n(\bar{a} + \bar{b}) \le \bar{b}$. And, indeed, $p_n(\bar{a} + \bar{b}) \le p_n\bar{b} \le \bar{b}$.

(ii) Conditional on $\{\tau = \infty\}$, p_n is a bounded submartingale, convergent to a random variable p_∞ almost surely and in L^1.

(iii) When $\bar{a} = 0$, $\bar{p}_n = E[p_n|\tau = \infty]$ will increase to $\bar{p}_\infty = E[p_\infty|\tau = \infty]$. Assuming that some trajectory in $\{\tau = \infty\}$ satisfies $1 - \bar{p}_\infty > \varepsilon$ for some $\varepsilon > 0$, it would verify

$$\bar{p}_{n+1} > \bar{p}_n + \frac{\bar{b}\varepsilon}{t + c(n + 1)}.$$

This gives a contradiction, because the last terms form a divergent series.

For $\bar{a} < 0$, if $\bar{b}/(\bar{a} + \bar{b}) = \delta > 1$, assume that $P\{\tau = \infty\} > 0$ and then $\bar{p}_n \leq 1$ for all the trajectories in $\{\tau = \infty\}$ and all n. But

$$\bar{p}_{n+1} \geq \bar{p}_n + (\bar{a} + \bar{b})\frac{\delta - 1}{t + c(n + 1)}$$

will give $\bar{p}_\infty = \infty$. A similar argument holds if $\bar{b}/(\bar{a} + \bar{b}) = \delta < 0$ and it is even simpler for $\bar{a} + \bar{b} = 0$.   □

Shortly, for $\bar{a} < 0 < \bar{b}$ some color exhausts in a finite time, while for $\bar{a} = 0$ the urn is asymptotically depleted of blue balls, if any color is not exhausted in a finite time. In this last case, $p_n \to 1$ and therefore:

Corollary 3.1 *When $\bar{a} = 0 < \bar{b}$, in $\{\tau = \infty\}$ it is $X_n/n \longrightarrow 1$ almost surely.*

Proof Let us consider $Z_k = Y_k - p_{k-1}$, for which $E[Z_k|\mathscr{F}_{k-1}] = 0$. Moreover, since $E[Y_k|\mathscr{F}_{k-1}] = p_{k-1}$, it is

$$E[Z_k^2] = E[Y_k] - 2E[Y_k p_{k-1}] + E[p_{k-1}^2] = E[p_{k-1}] - E[p_{k-1}^2] \leq \frac{1}{4}$$

and therefore $\sum_{k=1}^\infty k^{-2} E[Z_k^2] < \infty$. Thus, the strong law of large numbers for martingales given in [5, VII, Theorem 3] asserts that, almost surely,

$$\frac{1}{n}\sum_{k=1}^n Z_k = \frac{X_n}{n} - \frac{1}{n}\sum_{k=1}^n p_k \longrightarrow 0 \qquad \text{or} \qquad \frac{X_n}{n} \longrightarrow 1$$

since $p_n \to 1$ in $\{\tau = \infty\}$.   □

A simpler argument may be based on the redraft of (5):

$$\left(\frac{t}{n} + c\right)p_n = \frac{\alpha}{n} + c\frac{X_n}{n} - \frac{X_n}{n}\frac{1}{X_n}\sum_{k=1}^n Y_k a_k + \frac{1}{n}\sum_{k=1}^n b_k - \frac{X_n}{n}\frac{1}{X_n}\sum_{k=1}^n Y_k b_k. \quad (7)$$

The Borel–Cantelli Lemma in [1, Theorem 5.3.2] gives $\{X_n \to \infty\} = \{\sum_{n=1}^\infty p_n = \infty\}$, so that $X_n \to \infty$ almost surely in $\{\tau = \infty\}$. Thus $(1/X_n)\sum_{k=1}^n Y_k a_k \to \bar{a} = 0$ and $(1/X_n)\sum_{k=1}^n Y_k b_k \to \bar{b}$ whatever values the sequence $\{Y_k\}_{k\geq 1}$ may have. Therefore it follows $c = \bar{b} + (c - \bar{b})\lim_n X_n/n$ and $\lim_n X_n/n = 1$, under the unavoidable requirement $\bar{b} \neq c$.

Of course, for $\bar{b} \leq 0 < \bar{a}$, the possible outcomes are reversed. While $\{p_n\}$ is a supermartingale, the proportion of blue balls is a submartingale. Then, $\tau < \infty$ for $\bar{b} < 0$ and, if $\bar{b} = 0$, either $\tau < \infty$ or $p_n \to 0$.

We now turn to the case where $\bar{a}, \bar{b} \neq 0$ have the same sign or, equivalently,

$$p^\star = \frac{\bar{b}}{\bar{a} + \bar{b}} \in (0, 1).$$

Proposition 3.2 *Assume that $\bar{a}, \bar{b} > 0$.*

(i) *As long as $p_n \in [0, p^\star]$, $\{p_n\}$ is a submartingale while, when $p_n \in [p^\star, 1]$, $\{p_n\}$ is a supermartingale.*

(ii) *With probability 1, $\tau < \infty$ or $p_n \to p^\star$.*

Proof (i) If $p_n \in [0, p^\star]$ the second term in the right hand side of (6) is nonnegative; thus $\{p_n\}$ is a submartingale. On the contrary, for $p_n \in [p^\star, 1]$, the indicated term is nonpositive and $\{p_n\}$ is a supermartingale.

(ii) Within the event $\{\tau = \infty\}$ let $\bar{p} = \limsup_n p_n$ and $\underline{p} = \liminf_n p_n$ and let us consider $C = \{\tau = \infty, \underline{p} < \bar{p}\}$. The event $C \cap \{\underline{p} > p^\star\}$ has probability zero, because those trajectories are, from some n onwards, trajectories of a submartingale with two different cluster points. For the same reason, the probability of $C \cap \{\bar{p} < p^\star\}$ is also zero. Hence, $C \cap \{\underline{p} < p^\star < \bar{p}\}$ differs from C in a set of probability zero.

Now, let $C_\delta = C \cap \{\underline{p} < p^\star < p^\star + \delta < \bar{p}\}$. The trajectories in C_δ must perform an infinite number of upcrossings of the interval $(p^\star, p^\star + \delta)$ through positive steps of size

$$p_{n+1} - p_n = \begin{cases} \dfrac{c - a_{n+1} - cp_n}{t + c(n+1)} & \text{if } Y_{n+1} = 1 \\[2mm] \dfrac{b_{n+1} - cp_n}{t + c(n+1)} & \text{if } Y_{n+1} = 0 \end{cases}$$

that is less than any $\varepsilon > 0$ for n large enough. Therefore the probability of C_δ is bounded by $[(p^\star + \delta) \vee (1 - p^\star)]^{\delta/\varepsilon}$ and, this being true for any $\varepsilon > 0$, it should be $P(C_\delta) = 0$. A similar reasoning shows that $C'_\delta = C \cap \{\underline{p} < p^\star - \delta < p^\star < \bar{p}\}$ has also probability zero and consequently $\{\tau < \infty\} \cup \{\underline{p} = p^\star = \bar{p}\}$ has probability one. □

Repeating the proof of Corollary 3.1, the convergence $X_n/n \to p^\star$ follows from $p_n \to p^\star$.

In [4, Corollary 3.2] the assertion $p_n \xrightarrow{P} p^\star$ is attributed to Bagchi–Pal in 1985, when $a_k = a$ and $b_k = b$ are non random values, under the additional condition $a + b \geq c/2$. Later [4, Theorem 6.2] reproduces a similar result from Smythe in 1996, which again only asserts the convergence in probability for nonrandom parameters. For the Friedman model in 1949, in which $a = b > 0$ is a fixed non random integer, the almost sure convergence $p_n \to 1/2$ was established by Freedman in 1965. A different proof, due to Ornstein, can be found in [1, Example 5.4.5].

For $\bar{a}, \bar{b} < 0$ a result similar to Proposition 3.2 holds:

Proposition 3.3 *Assume that $\bar{a}, \bar{b} < 0$.*

(i) *As long as $p_n \in [0, p^\star]$, $\{p_n\}$ is a supermartingale while, when $p_n \in [p^\star, 1]$, $\{p_n\}$ is a submartingale.*

(ii) With probability 1, $\tau < \infty$ or $p_n \to p^*$.

Assertion *(i)* is obvious and the reasoning of *(ii)* in Proposition 3.2 needs no change to hold in this case. However the path of $\{p_n\}$ is now much more unstable since, as opposed to the previous case, there is here a trend upwards above p^* and downwards below p^*. Hence the probability of $\{\tau < \infty\}$ is greater for $-\bar{a}, -\bar{b}$ than for $\bar{a}, \bar{b} > 0$.

Finally, we may consider the state of affairs for $\bar{a} = \bar{b} = 0$. Under such circumstances (6) shows that $\{p_n\}$ is a bounded martingale while $p_n \in [0, 1]$. Therefore $p_n \to p_\infty$ in $\{\tau = \infty\}$. But so, nothing can be said about p_∞. Fortunately, one can show

Lemma 3.1 *Assume that $\bar{a} = \bar{b} = 0$.*

As long as $p_n \in [0, 1]$, (1) holds. Consequently, $\{Y_k\}_{k \geq 1}$ is an exchangeable sequence of random variables conditional on $\{\tau = \infty\}$.

Proof If $p_n \in [0, 1]$, it is

$$P\{Y_{n+1} = 1 | \mathscr{F}_n\} = p_n = \frac{\alpha + \sum_{k=1}^n Y_k(c - a_k) + \sum_{k=1}^n (1 - Y_k)b_k}{t + cn}$$

thus, since $\bar{a} = \bar{b} = 0$, we have

$$P\{Y_{n+1} = 1 | Y_1, \ldots, Y_n\} = \frac{\alpha + c \sum_{k=1}^n Y_k}{t + cn} = \frac{\alpha + cr}{t + cn}$$

in $\{X_n = r\}$, while in $\{X_n = n - s\}$,

$$P\{Y_{n+1} = 0 | Y_1, \ldots, Y_n\} = \frac{\beta + c \sum_{k=1}^n (1 - Y_k)}{t + cn} = \frac{\beta + sc}{t + cn}.$$

Hence, (1) holds and $\{Y_k\}$ is exchangeable when $\{\tau = \infty\}$.   □

According to the previous result and using again de Finetti's theorem:

Proposition 3.4 *Assume that $\bar{a} = \bar{b} = 0$.*

Conditional on $\{\tau = \infty\}$, the limit $p_\infty = \lim_n p_n$ exists almost surely and has the same beta$(\alpha/c, \beta/c)$ *distribution than in the Polya–Eggenberger model.*

4 Final Comments

Our main conclusion is that a Polya urn scheme $[c - a, a; b, c - b]$, even with random values of a and b, can never have long term oscillations. It either fails for lack of some kind of balls or, on the contrary, the proportion of amber balls converges to the fixed value $p^* = \bar{b}/(\bar{a} + \bar{b})$ or, if $\bar{a} = \bar{b} = 0$, to a random value p_∞ with distribution *beta* $(\alpha/c, \beta/c)$. The scarce role of c is quite unexpected.

Table 1 Observed values of τ in 100 simulations of the specified scheme

τ	>5000	4	5	6	7	8	9	10	12	14	15	16	19	20	21	22	23
%	42	4	5	1	4	4	3	1	3	2	3	1	2	2	1	1	3
τ		25	26	27	29	36	37	44	57	85	87	90	99	101	186	492	826
%		3	1	1	1	1	1	1	1	1	1	1	1	1	1	1	1

It is usual in the literature about Polya's urn schemes to consider only *tenable* models that *withstand the test of time...and we can perpetuate the drawing according to the given scheme on every possible stochastic path* [4, Sect. 3.1]. In our terms that means $P\{\tau = \infty\} = 1$. In the indicated reference, conditions for tenability about non random values of $a', a, b, b', \alpha, \beta$ are exposed. But such approach does not fit in the case of random parameters (unless the analysis is restricted to the case $0 \leq a_k, b_k \leq c$).

It would be desirable to complete the results here exposed with a research about the value of $P\{\tau = \infty\}$ as a function of the parameters α, β and c and the distributions of a_k, b_k. Although this issue seems unsuccessful from a theoretical point of vue, it may be analyzed by means of simulations for any fixed conditions. For instance, for the zero mean distributions $a_k = [-5, -2, 2, 5]$, $b_k = [-4, 0, 4]$, each one with equal probabilities, $\alpha = \beta = 10$ and $c = 3$, every replay of 5000 steps of the scheme gives place to a curve as in Fig. 1, although now it can overflow from [0, 1] at a time τ. In 100 replays of this scheme the observed values of τ were those gathered in Table 1.

This is a strong clue that $P\{\tau = \infty\} \simeq 0.42$. Of course this probability is very sensitive to changes in α, β and c.

References

1. Durret R (2010) Probability: theory and examples. Cambridge University Press, Cambridge
2. Johnson NL, Kotz S (1977) Urn models and their application. Wiley, New York
3. Kotz S, Balakrishnan N (1997) Advances in urn models during the past two decades. In: Balakrishnan N (ed) Advances in combinatorial methods and applications to probability and statistics. Birkhäuser, Boston
4. Mahmoud HM (2009) Polya urn models. Chapman and Hall/CRC, London
5. Feller W (1971) An introduction to probability theory and its applications, vol II. Wiley, New York
6. Flajolet P, Dumas P, Puyhaubert V (2006) Some exactly solvable models of urn process theory. In: DMTCS Proceedings, (Fourth colloquium on mathematics and computer science) AG:59–118

A Simple Step-Stress Model for Coherent Systems and Associated Inference Based on System Signatures

Xiaojun Zhu, Debanjan Mitra and Narayanaswamy Balakrishnan

Abstract Coherent systems are important structures in reliability. In this paper, we discuss the maximum likelihood estimates (MLEs) of model parameters of an n-component system with known signature having an exponential component lifetime distribution based on a simple step-stress model. We also develop confidence intervals (CIs) for the model parameters. A detailed Monte Carlo simulation study is carried out to examine the performance of the point and interval estimates. Finally, a data analysis is performed for illustrating all the inferential methods developed here.

1 Introduction

1.1 Step-Stress Experiment

The products that are tested in industrial experiments are often extremely reliable these days, and so possess large mean times to failure under normal operating conditions. Hence, very few failures (or no failures at all) are observed under conventional life-testing experiments with Type-I or Type-II censoring. This makes it very difficult to obtain adequate information about the lifetime distribution and its parameters.

To overcome this problem, reliability engineers traditionally resort to accelerated testing wherein the test units are subjected to higher stress levels than used level. The accelerated stress test may be performed using constant stress or linearly increasing stress levels. Such an accelerated testing helps to reduce the time to failure

X. Zhu
Department of Mathematical Sciences, Xi'an Jiaotong-Liverpool University,
Suzhou 215123, P.R. China

D. Mitra
Indian Institute of Management Udaipur, Udaipur 313001, Rajasthan, India

N. Balakrishnan (✉)
Department of Mathematics and Statistics, McMaster University,
Hamilton, Ontario L8S4K1, Canada
e-mail: bala@mcmaster.ca

© Springer International Publishing AG 2018
E. Gil et al. (eds.), *The Mathematics of the Uncertain*, Studies in Systems,
Decision and Control 142, https://doi.org/10.1007/978-3-319-73848-2_41

431

and makes the data collection easier and also the test to be completed rapidly. The data collected from such an experiment may then be extrapolated for estimating the underlying distribution of failure times under normal used conditions. This process, therefore, requires a model relating the level of stress to the failure distribution, such as increasing the effects of temperature, voltage, load, vibration, etc., on the lifetimes of units under test. Interested readers may refer to Nelson [14, 15] for detailed overviews on this model.

The step-stress experiment is a special case of accelerated testing which allows for different test conditions at various intermediate stages of the life-testing experiment. Suppose n identical units are placed on a life-test at an initial stress level of x_0. At prefixed times $\tau_1, \tau_2, \ldots, \tau_m$, the stress levels are changed to $x_1, x_2, \ldots, x_m$, respectively. A simple step-stress model has only two stress levels, namely, x_0 and x_1. Considerable research has been carried out on several inferential aspects associated with this model. For example, Miller and Nelson [12] and Bai et al. [2] discussed the determination of optimal time τ at which to change the stress level from x_0 to x_1.

1.2 Coherent Systems and Signatures

A reliability system usually has many components, with the failure of the system depending on the failure of one or more of its components. A reliability system is said to be a coherent system if

- it is monotone in its components (i.e., replacing a failed component by a working one cannot make the system worse);
- every component is relevant (i.e., every component influences either the functioning or the failure of the system).

Suppose a coherent system has n components whose lifetimes $X_1, \ldots, X_n$ are i.i.d. continuous random variables with distribution function $F(\cdot)$. Let the order statistics arising from these n variables be denoted by $X_{1:n} < X_{2:n} < \ldots < X_{n:n}$. Then, the system lifetime T will coincide with an order statistic $X_{i:n}$, for some $i \in \{1, 2, \ldots, n\}$. This leads to the concept of system signature.

Let P_i, $i = 1, 2, \ldots, n$, be such that $P_i = P(T = X_{i:n})$. Then, the system signature is simply the vector $\mathbf{P}$, where $\mathbf{P} = (P_1, P_2, \ldots, P_n)$. Clearly, $P_i \geq 0$ for all i, and they do not depend on the component lifetime distribution, and are such that $\sum_{i=1}^{n} P_i = 1$. The system signature $\mathbf{P}$ is a pure distribution-free measure of a system's design. The signature vector helps us to compare the performance characteristics of different systems in a complete nonparametric way without reference to the lifetime distribution of the components. Interested readers may refer to Balakrishnan et al. [5].

Xiong [18] and Balakrishnan et al. [4] have discussed inference for a simple step-stress model with Type-II censored data based on the exponential lifetime distribution. On the other hand, Balakrishnan et al. [6] studied linear inference for a coherent system with known signature based on Type-II censored data. Motivated by their research, we consider here a simple step-stress life-test on coherent systems and then develop point and interval estimation methods for the model parameters based on Type-II censored system lifetimes under exponential component lifetime distribution and known signature.

1.3 Motivation

Let us consider 30 individual component-failures obtained from a simple step-stress life-test with the change of stress done at time τ. Instead, if some systems had been formed with these components (say, 10 series or parallel or series-parallel 3-component systems) and that only such systems themselves can be tested under the step-stress life-test. Then, we will be interested in estimating the model parameters from such a system lifetime data obtained from a step-stress experiment and then use them in turn to estimate some reliability characteristics of systems and components. We may also be interested in evaluating the relative efficiency of basing the test on systems instead of on components directly. Here, we address these issues when the lifetime distribution of components is exponential and the signature of the system under consideration is known.

The rest of this paper is organized as follows. In Sect. 2, we briefly describe the basic model for a step-stress test and an n-component system. In Sect. 3, we discuss the MLEs and explain the process of setting initial values required for the iterative process by the method of Best Linear Unbiased Estimators (BLUEs). In Sect. 4, we develop interval estimation methods based on asymptotic properties of MLEs as well as by the use of bootstrap approach. A Monte Carlo simulation study is then carried out in Sect. 5 to evaluate the performance of the proposed methods of inference. The relative efficiency of basing the step-stress test on systems instead of on components directly is also evaluated here for different systems and different levels of censoring. An example is analyzed in Sect. 6 to illustrate all the inferential methods developed here. Finally, some concluding comments are made in Sect. 7.

2 Simple Step-Stress Model for Coherent Systems

Consider a coherent system with n components, and let $X_1, \ldots, X_n$ denote the lifetimes of its components. We assume that $X_1, \ldots, X_n$ are independent and identically distributed (i.i.d.) with probability density function (PDF) $f_X(\cdot)$ and cumulative distribution function (CDF) $F_X(\cdot)$. Further, let us suppose that the component lifetimes are exponentially distributed with means θ_1 and θ_2 at stress levels x_0 and x_1, respectively. At a pre-fixed time τ, the stress level is increased from x_0 to x_1. Then, the cumulative exposure model relates the life distribution of the units at one stress level to the life distribution of the units at the next stress level. The model is based on the assumption that the residual life of the experimental units depends only on the cumulative exposure the units have experienced, with no memory of how this exposure was accumulated. Thus, under the cumulative exposure model, the CDF of each component is given by (see Xiong [18] and Balakrishnan et al. [4]) as

$$F_X(t) = \begin{cases} F_1(t), & 0 < t < \tau, \\ F_2\left(\frac{\theta_2}{\theta_1}\tau + t - \tau\right), & \tau \le t < \infty, \end{cases} \tag{1}$$

where $F_1(t) = 1 - e^{-\frac{t}{\theta_1}}$ and $F_2(t) = 1 - e^{-\frac{t}{\theta_2}}$, respectively. The corresponding PDF is given by

$$f_X(t) = \begin{cases} \frac{1}{\theta_1}e^{-\frac{t}{\theta_1}}, & 0 < t < \tau, \\ \frac{1}{\theta_2}e^{-\frac{t-\tau}{\theta_2} - \frac{\tau}{\theta_1}}, & \tau \le t < \infty. \end{cases} \tag{2}$$

It is well known that the distribution of the system lifetime, say T, can be represented as a mixture of distributions of order statistics arising from the lifetimes of components in the system. Suppose the reliability system that is being tested has a known structure for which the system signature is $\mathbf{P}$. Then, the PDF and the survival function (SF) of the system lifetime T are given by Samaniego [16, 17] and Kochar et al. [10] as

$$f_T(t) = \sum_{i=1}^{n} P_i f_{i:n}(t), \tag{3}$$

$$S_T(t) = \sum_{i=1}^{n} P_i S_{i:n}(t), \tag{4}$$

where $f_{i:n}$ and $S_{i:n}$ are the PDF and SF of the ith order statistic $X_{i:n}$ given by (see Balakrishnan and Cohen [3] and Arnold et al. [1])

$$f_{i:n}(t) = \binom{n}{i} i f_X(t) \left[F_X(t)\right]^{i-1} \left[S_X(t)\right]^{n-i}, \tag{5}$$

$$S_{i:n}(t) = \sum_{j=0}^{i-1} \binom{n}{j} \left[F_X(t)\right]^j \left[S_X(t)\right]^{n-j}, \tag{6}$$

where $f_X(\cdot)$ and $F_X(\cdot) = 1 - S_X(\cdot)$ are the PDF and CDF of the component lifetime, respectively. This representation of the distribution of T, through the distributions of order statistics arising from the lifetimes of components and system signature, is referred to as Samaniego's representation.

Another representation of the distribution of T by Navarro et al. [13] is

$$f_T(t) = \sum_{i=1}^{n} a_i f_{1:i}(t) = \sum_{i=1}^{n} a_i i f_X(t) [S_X(t)]^{i-1}, \tag{7}$$

$$S_T(t) = \sum_{i=1}^{n} a_i [S_{1:i}(t)] = \sum_{i=1}^{n} a_i [S_X(t)]^i, \tag{8}$$

where $f_{1:i}(\cdot)$ and $S_{1:i}(\cdot)$ are the PDF and SF of the series system lifetimes

$$X_{1:i} = \min(X_1, \ldots, X_i), \quad i = 1, 2, \ldots, n.$$

Here, $\mathbf{a} = (a_1, \ldots, a_n)$ is called the minimal signature, with a_i's not all being non-negative and not depending on the component lifetime distribution are such that $\sum_{i=1}^{n} a_i = 1$. Note that this representation can be obtained from (3) and (4) by using binomial expansion on the term $[F_X(\cdot)]^l$.

Let m such n-component coherent systems be placed on a simple step-stress experiment with two stress levels. Suppose a Type-II right censored system life data are observed from this step-stress experiment of the form

$$T_{1:m} < T_{2:m} < \cdots < T_{r:m}.$$

If $r = m$, then a complete sample would be observed from the step-stress test. As mentioned before, let τ be the pre-fixed time at which the stress level changes from x_0 to x_1, and further let n_1 be the number of failures that occur before τ at the first stress level and n_2 be the number of failures that occur after τ at the second stress level, with $n_1 + n_2 = r$. Note that if $n_1 = r$, the test gets terminated before reaching the second stress level; otherwise, the experiment continues under the first stress level till time τ, then the stress level is increased to the second stress level, and the experiment continues till r failures are observed.

Proposition 2.1 *Suppose the component lifetime distribution belongs to the scale family. Then, if a simple step-stress life-test is conducted on a coherent system with a lifetime T, it is equivalent to conducting a simple step-stress test on each of its components with lifetimes $X_i's$, under the cumulative exposure model.*

Proof From (1), we readily find

$$F_T(t) = \begin{cases} F_{T,1}(t), & 0 < t < \tau, \\ F_{T,2}\left(\dfrac{\theta_2}{\theta_1}\tau + t - \tau\right), & \tau \le t < \infty, \end{cases}$$

$$
= \begin{cases} \sum_{i=1}^{n} P_i F_{i:n}(t), & 0 < t < \tau, \\ \sum_{i=1}^{n} P_i F_{i:n}\left(\dfrac{\theta_2}{\theta_1}\tau + t - \tau\right), & \tau \leq t < \infty, \end{cases}
$$

$$
= \begin{cases} \sum_{i=1}^{n} P_i \sum_{j=1}^{n} \binom{n}{j}\left[F_X(t)\right]^j\left[1 - F_X(t)\right]^{n-j}, & 0 < t < \tau, \\ \sum_{i=1}^{n} P_i \sum_{j=1}^{n} \binom{n}{j}\left[F_X\left(\dfrac{\theta_2}{\theta_1}\tau + t - \tau\right)\right]^j\left[1 - F_X\left(\dfrac{\theta_2}{\theta_1}\tau + t - \tau\right)\right]^{n-j}, & \tau \leq t < \infty. \end{cases}
$$

From here, the required result follows.   □

Proposition 2.2 *Under a simple step-stress life-testing experiment on a coherent system with lifetime T and exponential component lifetime distribution, the SF of T is a non-decreasing function of* θ_2.

Proof Evidently,

$$
\frac{\partial S_T(t; \theta_1, \theta_2)}{\partial \theta_2} = \begin{cases} 0, & 0 < t < \tau, \\ \frac{t-\tau}{\theta_2} f_T(t, \theta_1, \theta_2) > 0, & \tau \leq t < \infty, \end{cases}
$$

which shows the required result.   □

Now, based on the observed system failure times $t_{1:m} < t_{2:m} < \ldots < t_{r:m}$ under a simple step-stress life-test, we have the likelihood function as (see Balakrishnan and Cohen [3])

$$
L(\theta_1, \theta_2) = C \prod_{k=1}^{r} f_T(t_{k:m})[S_T(t_{r:m})]^{m-r}, \tag{9}
$$

where $C = \frac{m!}{r!(m-r)!}$.

It is clear from (9) that the likelihood function is different for the following three cases:

Case I. $n_1 = r$,
Case II. $n_1 = 0$,
Case III. $1 \leq n_1 \leq r - 1$.

Now, let us discuss these three cases separately.

Case I: $n_1 = r$

In this case, all the r failures are observed at the first stress level. The likelihood function in (9) in this case becomes

$$
L(\theta_1, \theta_2)
$$

$$\propto \prod_{k=1}^{r} \left\{ \sum_{i=1}^{n} P_i \binom{n}{i} i \frac{1}{\theta_1} e^{-\frac{t_{k:m}}{\theta_1}} \left[1 - e^{-\frac{t_{k:m}}{\theta_1}} \right]^{i-1} \left[e^{-\frac{t_{k:m}}{\theta_1}} \right]^{n-i} \right\}$$

$$\times \left\{ \sum_{i=1}^{n} P_i \sum_{j=0}^{i-1} \binom{n}{j} \left[1 - e^{-\frac{t_{r:m}}{\theta_1}} \right]^{j} \left[e^{-\frac{t_{r:m}}{\theta_1}} \right]^{n-j} \right\}^{m-r}$$

$$= \frac{1}{\theta_1^r} \prod_{k=1}^{r} \left\{ \sum_{i=1}^{n} P_i \binom{n}{i} i \sum_{l=0}^{i-1} (-1)^l \binom{i-1}{l} e^{-\frac{t_{k:m}}{\theta_1}(n-i+l+1)} \right\}$$

$$\times \left\{ \sum_{i=1}^{n} P_i \sum_{j=0}^{i-1} \binom{n}{j} \sum_{l=0}^{j} (-1)^l \binom{j}{l} e^{-\frac{t_{r:m}}{\theta_1}(n-j+l)} \right\}^{m-r}. \tag{10}$$

Case II: $n_1 = 0$

In this case, all the r failures are observed at the second stress level, and the corresponding likelihood function is given by

$$L(\theta_1, \theta_2)$$

$$\propto \prod_{k=1}^{r} \left\{ \sum_{i=1}^{n} P_i \binom{n}{i} \frac{i}{\theta_2} e^{-\frac{\frac{\theta_2}{\theta_1}\tau + t_{k:m} - \tau}{\theta_2}} \left[1 - e^{-\frac{\frac{\theta_2}{\theta_1}\tau + t_{k:m} - \tau}{\theta_2}} \right]^{i-1} \left[e^{-\frac{\frac{\theta_2}{\theta_1}\tau + t_{k:m} - \tau}{\theta_2}} \right]^{n-i} \right\}$$

$$\times \left\{ \sum_{i=1}^{n} P_i \sum_{j=0}^{i-1} \binom{n}{j} \left[1 - e^{-\frac{\frac{\theta_2}{\theta_1}\tau + t_{r:m} - \tau}{\theta_2}} \right]^{j} \left[e^{-\frac{\frac{\theta_2}{\theta_1}\tau + t_{r:m} - \tau}{\theta_2}} \right]^{n-j} \right\}^{m-r}$$

$$= \frac{1}{\theta_2^r} \prod_{k=1}^{r} \left\{ \sum_{i=1}^{n} P_i \binom{n}{i} i \sum_{l=0}^{i-1} (-1)^l \binom{i-1}{l} e^{-\frac{1}{\theta_2}\left(\frac{\theta_2}{\theta_1}\tau + t_{k:m} - \tau\right)(n-i+l+1)} \right\}$$

$$\times \left\{ \sum_{i=1}^{n} P_i \sum_{j=0}^{i-1} \binom{n}{j} \sum_{l=0}^{j} (-1)^l \binom{j}{l} e^{-\frac{1}{\theta_2}\left(\frac{\theta_2}{\theta_1}\tau + t_{r:m} - \tau\right)(n-j+l)} \right\}^{m-r}. \tag{11}$$

Case III: $1 \leq n_1 \leq r - 1$

In this case, there are n_1 failures in the first stress level and n_2 failures in the second stress level and the corresponding likelihood function is given by

$$L(\theta_1, \theta_2)$$

$$\propto \prod_{k=1}^{n_1} \left\{ \sum_{i=1}^{n} P_i \binom{n}{i} i \frac{1}{\theta_1} e^{-\frac{t_{k:m}}{\theta_1}} \left[1 - e^{-\frac{t_{k:m}}{\theta_1}} \right]^{i-1} \left[e^{-\frac{t_{k:m}}{\theta_1}} \right]^{n-i} \right\}$$

$$\times \prod_{k=n_1+1}^{r} \left\{ \sum_{i=1}^{n} P_i \binom{n}{i} \frac{i}{\theta_2} e^{-\frac{\frac{\theta_2}{\theta_1}\tau + t_{k:m} - \tau}{\theta_2}} \left[1 - e^{-\frac{\frac{\theta_2}{\theta_1}\tau + t_{k:m} - \tau}{\theta_2}} \right]^{i-1} \left[e^{-\frac{\frac{\theta_2}{\theta_1}\tau + t_{k:m} - \tau}{\theta_2}} \right]^{n-i} \right\}$$

$$\times \left\{ \sum_{i=1}^{n} P_i \sum_{j=0}^{i-1} \binom{n}{j} \left[1 - e^{-\frac{\theta_2}{\theta_1}\tau + t_{r:m} - \tau}{\theta_2}} \right]^{j} \left[e^{-\frac{\frac{\theta_2}{\theta_1}\tau + t_{r:m} - \tau}{\theta_2}} \right]^{n-j} \right\}^{m-r}$$

$$= \frac{1}{\theta_1^{n_1}} \prod_{k=1}^{n_1} \left\{ \sum_{i=1}^{n} P_i \binom{n}{i} i \sum_{l=0}^{i-1} (-1)^l \binom{i-1}{l} e^{-\frac{t_{k:m}}{\theta_1}(n-i+l+1)} \right\}$$

$$\times \frac{1}{\theta_2^{n_2}} \prod_{k=n_1+1}^{r} \left\{ \sum_{i=1}^{n} P_i \binom{n}{i} i \sum_{l=0}^{i-1} (-1)^l \binom{i-1}{l} e^{-\frac{1}{\theta_2}\left(\frac{\theta_2}{\theta_1}\tau + t_{k:m} - \tau\right)(n-i+l+1)} \right\}$$

$$\times \left\{ \sum_{i=1}^{n} P_i \sum_{j=0}^{i-1} \binom{n}{j} \sum_{l=0}^{j} (-1)^l \binom{j}{l} e^{-\frac{1}{\theta_2}\left(\frac{\theta_2}{\theta_1}\tau + t_{r:m} - \tau\right)(n-i+l)} \right\}^{m-r}. \tag{12}$$

Alternatively, the likelihood function can be also expressed in a much simpler form using the minimal signature representations in (7) and (8) as follows:

$$L(\theta_1, \theta_2)$$

$$\propto \begin{cases} \left\{ \frac{1}{\theta_1^r} \prod_{k=1}^{r} \left[\sum_{i=1}^{n} a_i i e^{-\frac{i t_{k:m}}{\theta_1}} \right] \right\} \left[\sum_{i=1}^{n} a_i e^{-\frac{i t_{r:m}}{\theta_1}} \right]^{m-r}, & n_1 = r, \\[2ex] \left\{ \frac{1}{\theta_2^r} \prod_{k=1}^{r} \left[\sum_{i=1}^{n} a_i i e^{-\frac{i}{\theta_2}\left(\frac{\theta_2}{\theta_1}\tau + t_{k:m} - \tau\right)} \right] \right\} \left[\sum_{i=1}^{n} a_i e^{-\frac{i}{\theta_2}\left(\frac{\theta_2}{\theta_1}\tau + t_{r:m} - \tau\right)} \right]^{m-r}, & n_1 = 0, \\[2ex] \left\{ \frac{1}{\theta_1^{n_1}} \prod_{k=1}^{n_1} \left[\sum_{i=1}^{n} a_i i e^{-\frac{i t_{k:m}}{\theta_1}} \right] \right\} \left\{ \frac{1}{\theta_2^{n_2}} \prod_{k=n_1+1}^{r} \left[\sum_{i=1}^{n} a_i i e^{-\frac{i}{\theta_2}\left(\frac{\theta_2}{\theta_1}\tau + t_{k:m} - \tau\right)} \right] \right\} \\[2ex] \quad \times \left[\sum_{i=1}^{n} a_i e^{-\frac{i}{\theta_2}\left(\frac{\theta_2}{\theta_1}\tau + t_{r:m} - \tau\right)} \right]^{m-r}, & 1 \le n_1 \le r-1. \end{cases} \tag{13}$$

In the following section, we will discuss the maximum likelihood estimation of the parameters θ_1 and θ_2, and we shall use the simpler form of the likelihood function in (13) for this purpose.

3 Maximum Likelihood Estimation

We readily see that in Case I, the MLE of θ_2 does not exist. For this reason, we will work under the condition $0 \leq n_1 \leq r - 1$ in what follows. The log-likelihood function, without the constant, is obtained from (13) as

$$
\ln L(\theta_1, \theta_2) = -n_1 \ln \theta_1 + \sum_{k=1}^{n_1} \ln \left(\sum_{i=1}^{n} a_i i e^{-\frac{i t_{k:m}}{\theta_1}} \right)
$$
$$
- n_2 \ln \theta_2 + \sum_{k=n_1+1}^{r} \ln \left(\sum_{i=1}^{n} a_i i e^{-\frac{i}{\theta_2} \left(\frac{\theta_2}{\theta_1} \tau + t_{k:m} - \tau \right)} \right)
$$
$$
+ (m - r) \ln \left(\sum_{i=1}^{n} a_i e^{-\frac{i}{\theta_2} \left(\frac{\theta_2}{\theta_1} \tau + t_{r:m} - \tau \right)} \right). \tag{14}
$$

Note that this equation includes Case II, i.e., $n_1 = 0$, as well. Upon taking derivatives with respect to θ_1 and θ_2, we obtain from (14) the following likelihood equations:

$$
\frac{\partial \ln L}{\partial \theta_1} = -\frac{n_1}{\theta_1} + \frac{1}{\theta_1^2} \sum_{k=1}^{n_1} \frac{t_{k:m} \sum_{i=1}^{n} a_i i^2 e^{-\frac{i t_{k:m}}{\theta_1}}}{\sum_{i=1}^{n} a_i i e^{-\frac{i t_{k:m}}{\theta_1}}}
$$
$$
+ \frac{\tau}{\theta_1^2} \sum_{k=n_1+1}^{r} \frac{\sum_{i=1}^{n} a_i i^2 e^{-\frac{i}{\theta_2} \left(\frac{\theta_2}{\theta_1} \tau + t_{k:m} - \tau \right)}}{\sum_{i=1}^{n} a_i i e^{-\frac{i}{\theta_2} \left(\frac{\theta_2}{\theta_1} \tau + t_{k:m} - \tau \right)}} + \frac{(m-r)\tau}{\theta_1^2} \frac{\sum_{i=1}^{n} a_i i e^{-\frac{i}{\theta_2} \left(\frac{\theta_2}{\theta_1} \tau + t_{r:m} - \tau \right)}}{\sum_{i=1}^{n} a_i e^{-\frac{i}{\theta_2} \left(\frac{\theta_2}{\theta_1} \tau + t_{r:m} - \tau \right)}}, \tag{15}
$$

$$
\frac{\partial \ln L}{\partial \theta_2} = -\frac{n_2}{\theta_2} + \frac{1}{\theta_2^2} \sum_{k=n_1+1}^{r} \frac{(t_{k:m} - \tau) \sum_{i=1}^{n} a_i i^2 e^{-\frac{i}{\theta_2} \left(\frac{\theta_2}{\theta_1} \tau + t_{k:m} - \tau \right)}}{\sum_{i=1}^{n} a_i i e^{-\frac{i}{\theta_2} \left(\frac{\theta_2}{\theta_1} \tau + t_{k:m} - \tau \right)}}
$$
$$
+ \frac{(m-r)(t_{r:m} - \tau)}{\theta_2^2} \frac{\sum_{i=1}^{n} a_i i e^{-\frac{i}{\theta_2} \left(\frac{\theta_2}{\theta_1} \tau + t_{r:m} - \tau \right)}}{\sum_{i=1}^{n} a_i e^{-\frac{i}{\theta_2} \left(\frac{\theta_2}{\theta_1} \tau + t_{r:m} - \tau \right)}}. \tag{16}
$$

Equating the non-linear likelihood mathematical expressions in Eqs. (15) and (16) to zero, we get the equations

$$h_1(\theta_1, \theta_2) = n_1\theta_1 - \sum_{k=1}^{n_1} \frac{t_{k:m} \sum_{i=1}^{n} a_i i^2 e^{-\frac{i t_{k:m}}{\theta_1}}}{\sum_{i=1}^{n} a_i i e^{-\frac{i t_{k:m}}{\theta_1}}} - \tau \sum_{k=n_1+1}^{r} \frac{\sum_{i=1}^{n} a_i i^2 e^{-\frac{i}{\theta_2}\left(\frac{\theta_2}{\theta_1}\tau + t_{k:m} - \tau\right)}}{\sum_{i=1}^{n} a_i i e^{-\frac{i}{\theta_2}\left(\frac{\theta_2}{\theta_1}\tau + t_{k:m} - \tau\right)}}$$

$$-(m-r)\tau \frac{\sum_{i=1}^{n} a_i i e^{-\frac{i}{\theta_2}\left(\frac{\theta_2}{\theta_1}\tau + t_{r:m} - \tau\right)}}{\sum_{i=1}^{n} a_i e^{-\frac{i}{\theta_2}\left(\frac{\theta_2}{\theta_1}\tau + t_{r:m} - \tau\right)}} = 0, \tag{17}$$

$$h_2(\theta_1, \theta_2) = n_2\theta_2 - \sum_{k=n_1+1}^{r} \frac{(t_{k:m} - \tau) \sum_{i=1}^{n} a_i i^2 e^{-\frac{i}{\theta_2}\left(\frac{\theta_2}{\theta_1}\tau + t_{k:m} - \tau\right)}}{\sum_{i=1}^{n} a_i i e^{-\frac{i}{\theta_2}\left(\frac{\theta_2}{\theta_1}\tau + t_{k:m} - \tau\right)}}$$

$$-(m-r)(t_{r:m} - \tau) \frac{\sum_{i=1}^{n} a_i i e^{-\frac{i}{\theta_2}\left(\frac{\theta_2}{\theta_1}\tau + t_{r:m} - \tau\right)}}{\sum_{i=1}^{n} a_i e^{-\frac{i}{\theta_2}\left(\frac{\theta_2}{\theta_1}\tau + t_{r:m} - \tau\right)}} = 0. \tag{18}$$

As these two equations do not have closed-form solutions, we may use the Newton-Raphson algorithm to find the solution as

$$\begin{pmatrix} \theta_1^{(i+1)} \\ \theta_2^{(i+1)} \end{pmatrix} = \begin{pmatrix} \theta_1^{(i)} \\ \theta_2^{(i)} \end{pmatrix} - \begin{pmatrix} \frac{\partial h_1(\theta_1,\theta_2)}{\partial \theta_1}, & \frac{\partial h_1(\theta_1,\theta_2)}{\partial \theta_2} \\ \frac{\partial h_2(\theta_1,\theta_2)}{\partial \theta_1}, & \frac{\partial h_2(\theta_1,\theta_2)}{\partial \theta_2} \end{pmatrix}_{(\theta_1,\theta_2)=(\theta_1^{(i)},\theta_2^{(i)})}^{-1} \begin{pmatrix} h_1(\theta_1^{(i)}, \theta_2^{(i)}) \\ h_2(\theta_1^{(i)}, \theta_2^{(i)}) \end{pmatrix}. \tag{19}$$

Alternatively, due to the form of Eqs. (17) and (18), we may find the MLEs as solutions of fixed-point equations.

3.1 Providing Initial Value

In this subsection, we discuss the issue of providing an initial value by using the BLUEs under the condition $1 \le n_1 \le r - 1$, i.e., Case III. Let

$$T_{j:m}^* = \begin{cases} \frac{T_{j:m}}{\theta_1} & \text{if } T_{j:m} < \tau, \\ \frac{T_{j:m} - \tau}{\theta_2} + \frac{\tau}{\theta_1} & \text{if } T_{j:m} \ge \tau. \end{cases} \tag{20}$$

Let us further denote $\mathbf{T}_1^* = (T_{1:m}^*, \cdots, T_{n_1:m}^*)'$ and $\mathbf{T}_2^* = (T_{n_1+1:m}^*, \cdots, T_{r:m}^*)'$. We then treat $\mathbf{T}_1^*$ as an ordinary Type-II censored sample by neglecting the fact that $T^* < \tau/\theta_1$, and do similarly for $\mathbf{T}_2^*$. Let $\boldsymbol{\mu}_1 = (\mu_{1:m}, \ldots, \mu_{n_1:m})', \boldsymbol{\mu}_2 = (\mu_{n_1+1:m}, \ldots, \mu_{r:m})', \boldsymbol{\Sigma}_1 = ((\sigma_{i,j:m}))$ for $i, j = 1, \ldots, n_1, \boldsymbol{\Sigma}_2 = ((\sigma_{i,j:m}))$ for $i, j = n_1 + 1, \ldots, r$, where $\mu_{j:m} = E(T_{j:m}^*)$ and $\sigma_{i,j:m} = cov(T_{i:m}^*, T_{j:m}^*)$. Then, according to Balakrishnan et al. [6], we have

$$\mu_{1:m} = m \sum_{\substack{j_1,\ldots,j_n \geq 0 \\ j_1+\cdots+j_n=m-1}} \binom{m-1}{j_1,\ldots,j_n} a_1^{j_1} \cdots a_n^{j_n} \sum_{i=1}^{n} \frac{a_i i}{(j_1 + 2j_2 + \cdots + nj_n + i)^2}, \quad (21)$$

$$\mu_{1:m}^{(2)} = m \sum_{\substack{j_1,\ldots,j_n \geq 0 \\ j_1+\cdots+j_n=m-1}} \binom{m-1}{j_1,\ldots,j_n} a_1^{j_1} \cdots a_n^{j_n} \sum_{i=1}^{n} \frac{2a_i i}{(j_1 + 2j_2 + \cdots + nj_n + i)^3}, \quad (22)$$

$$\mu_{s,s+1:m} = \frac{m!}{(s-1)!(m-s-1)!} \sum_{l=0}^{s-1} (-1)^l \binom{s-1}{l} \sum_{\substack{k_1,\ldots,k_n \geq 0 \\ k_1+\cdots+k_n=l}} \binom{l}{k_1,\ldots,k_n}, \quad (23)$$

where $\mu_{j:m}^{(2)} = E\left[(T_{j:m}^*)^2\right]$ and $\mu_{i,j:m} = E(T_{i:m}^* T_{j:m}^*)$. Upon using these results, along with the well-known triangle and rectangle rules for moments of order statistics (see Arnold et al. [1])

$$s\mu_{s+1:m}^{(l)} + (m-s)\mu_{s:m}^{(l)} = m\mu_{s:m-1}^{(l)}, \quad s = 1, 2, \ldots, m-1, l = 1, 2, \quad (24)$$

$$(i-1)\mu_{i,j:m} + (j-i)\mu_{i-1,j:m} + (m-j+1)\mu_{i-1,j-1:m} = m\mu_{i-1,j-1:m-1}, \quad 2 \leq i < j \leq m, \quad (25)$$

we can find all the elements of $\boldsymbol{\mu}_1$, $\boldsymbol{\mu}_2$, $\boldsymbol{\Sigma}_1$ and $\boldsymbol{\Sigma}_2$. At the first stress level, we then have the BLUE of θ_1 as

$$\tilde{\theta}_1 = \left(\frac{\boldsymbol{\mu}_1' \boldsymbol{\Sigma}_1^{-1}}{\boldsymbol{\mu}_1' \boldsymbol{\Sigma}_1^{-1} \boldsymbol{\mu}_1} \right) \mathbf{T}_1, \quad (26)$$

where $\mathbf{T}_1 = (T_{1:m}, \ldots, T_{n_1:m})$. At the second stress level, we replace θ_1 by $\tilde{\theta}_1$, and let $\tilde{\mathbf{T}}_2^* = (T_{n_1+1:m} - \tau, \ldots, T_{r:m} - \tau)$, and $\boldsymbol{\mu}_2^* = \boldsymbol{\mu}_2 - (\tau/\tilde{\theta}_1, \ldots \tau/\tilde{\theta}_1)'$. Then, we have the BLUE of θ_2 as

$$\tilde{\theta}_2 = \left(\frac{\boldsymbol{\mu}_2^{*'} \boldsymbol{\Sigma}_2^{-1}}{\boldsymbol{\mu}_2^{*'} \boldsymbol{\Sigma}_2^{-1} \boldsymbol{\mu}_2^*} \right) \tilde{\mathbf{T}}_2^*. \quad (27)$$

The values of $\tilde{\theta}_1$ and $\tilde{\theta}_2$ so obtained may be used as initial values in the Newton-Raphson iterative process in Eq. (19).

To provide the initial values in the case when $n_1 = 0$, quite naturally, one can take $\tilde{\theta}_1 = \tau$, and then use the above procedure to obtain the BLUE of θ_2, $\tilde{\theta}_2$.

4 Interval Estimation of Parameters

In this section, we describe the construction of CIs for the model parameters by parametric bootstrap approach as well as by an asymptotic approach using the observed Fisher information matrix.

4.1 Bootstrap Approach

From the given system lifetime data from the simple step-stress experiment, we compute the MLE $(\hat{\theta}_1, \hat{\theta}_2)$. Then, with $(\hat{\theta}_1, \hat{\theta}_2)$ as the values of (θ_1, θ_2), we generate an n-component system lifetime data of size m and find the corresponding MLEs $(\hat{\theta}_{1(1)}, \hat{\theta}_{2(1)}), \ldots, (\hat{\theta}_{1(B)}, \hat{\theta}_{2(B)})$ from B bootstrap samples.

Then, from this set of B estimates, we can estimate the means and variances of the estimates $\hat{\theta}_1$ and $\hat{\theta}_2$ as

$$\widehat{E(\hat{\theta}_j)} = \overline{\hat{\theta}_j} \quad \text{and} \quad \widehat{Var(\hat{\theta}_j)} = \frac{1}{B-1} \sum_{i=1}^{B} (\hat{\theta}_{j(i)} - \overline{\hat{\theta}_j})^2, \quad \text{for } j = 1, 2, \qquad (28)$$

where $\overline{\hat{\theta}_j} = \frac{1}{B} \sum_{i=1}^{B} \hat{\theta}_{j(i)}$. We can also obtain the $100(1 - \alpha)\%$ bootstrap CIs for the parameters as

$$\left(\hat{\theta}_{j(\frac{\alpha B}{2})}, \hat{\theta}_{j((1-\frac{\alpha}{2})B)} \right), \qquad \text{for } j = 1, 2, \qquad (29)$$

where $\hat{\theta}_{j(l)}$ denotes the lth smallest value of $\hat{\theta}_j$ obtained from the B bootstrap simulations; see Lehmann [11] for more details.

4.2 Asymptotic Approach Using Observed Fisher Information Matrix

Under usual regularity conditions, it can be shown that the MLEs are consistent and are asymptotically normally distributed. With (θ_1, θ_2) as the parameter vector and $(\hat{\theta}_1, \hat{\theta}_2)$ as the MLEs, as $m \to \infty$, we have

$$\sqrt{m} \begin{pmatrix} \hat{\theta}_1 - \theta_1 \\ \hat{\theta}_2 - \theta_2 \end{pmatrix} \xrightarrow{D} N_2 \left(\mathbf{0}, I_2^{-1} \right), \qquad (30)$$

where $\xrightarrow{D}$ denotes convergence in distribution and $N_2\left(\mathbf{0}, \mathbf{I}_2^{-1}\right)$ denotes the bivariate normal distribution with mean vector $\mathbf{0}$ and covariance matrix $\mathbf{I}_2^{-1}$, with $\mathbf{I}_2$ being estimated by the observed Fisher information matrix

$$I_{obs}(\theta_1, \theta_2) = \begin{pmatrix} -\frac{\partial^2 \ln L}{\partial \theta_1^2} & -\frac{\partial^2 \ln L}{\partial \theta_1 \partial \theta_2} \\ -\frac{\partial^2 \ln L}{\partial \theta_1 \partial \theta_2} & -\frac{\partial^2 \ln L}{\partial \theta_2^2} \end{pmatrix}_{\theta_1=\hat{\theta}_1, \theta_2=\hat{\theta}_2}. \tag{31}$$

5 Simulation Study

We carried out a detailed Monte Carlo simulation study for several 3 component systems, with $m = 40$ and 100 with 20% censoring (i.e., $r = 0.8m$), and by taking $\theta_1 = 1$, without loss of any generality. The values of θ_2 were chosen as 0.50 and 0.25, and τ chosen so that $F(\tau) = 0.15$ and 0.30, respectively. We then compare the inference obtained from system lifetime data with that obtained from component lifetime data for the same value of τ. For example, when $F(\tau) = 0.15$ for a parallel system, for the same τ, we found $F(\tau) = 0.53$ for component lifetimes. We computed the means and mean squared errors (MSEs) of $\hat{\theta}_1$ and $\hat{\theta}_2$ based on both system and component lifetime data obtained from the step-stress experiment. We also computed the coverage probabilities of 95% equi-tailed CIs obtained from the asymptotic approach using the observed Fisher information matrix as well as from the bootstrap approach. To evaluate the relative efficiency of basing the step-stress life-test on systems instead of on components directly, we also calculated the trace and determinant of the observed Fisher information matrix, as well as the total time on test transform (TTT) under both schemes of data observation. All these results are presented in Table 1. We observe that for system data, in all cases considered, $\hat{\theta}_1$ is positively biased, while $\hat{\theta}_2$ is negatively biased for most cases. Based on the trace and determinant of the observed Fisher information matrix, we also find that the parallel system always has the highest efficiency while the series system always has the lowest efficiency. So, as one would then expect, the parallel system always has the smallest MSEs for both $\hat{\theta}_1$ and $\hat{\theta}_2$, and a series system has the largest MSEs. Furthermore, we find that the standard errors (SEs) obtained by bootstrap are always greater than those obtained by the use of the observed Fisher information matrix, which may explain why the CI obtained by bootstrap has a larger coverage probability. It is also of interest to note that the test of parallel systems result in more information than the test of components even though the latter has a larger TTT.

444

X. Zhu et al.

Table 1 Simulated values of means and MSEs of the MLEs and coverage probabilities of 95% equi-tailed CIs from Observed Fisher Information Matrix and Bootstrap method with $\theta_1 = 1.00$ for some 3-component coherent systems

m	r	$F(\tau)$	θ_2	$E(\hat{\theta}_1)$	$MSE(\hat{\theta}_1)$	P_I	P_B	$E(\hat{\theta}_2)$	$MSE(\hat{\theta}_2)$	P_I	P_B	$tr(\mathbf{I})$	$det(\mathbf{I})$	TTT
$P = (0, 0, 1)$ (equivalently, $a = (3, -3, 1)$)														
40	32	0.15	0.50	1.0367	0.0335	0.9467	0.9744	0.4975	0.0071	0.9126	0.9332	230.1019	7169.9157	46.9008
			0.25	1.0358	0.0332	0.9464	0.9752	0.2488	0.0018	0.9130	0.9342	768.9110	28691.6979	37.8926
		0.30	0.50	1.0191	0.0183	0.9406	0.9919	0.4987	0.0106	0.9170	0.9290	194.2018	7602.2448	52.3166
			0.25	1.0182	0.0183	0.9426	0.9929	0.2498	0.0051	0.8922	0.9234	394.9594	19056.2646	46.0137
100	80	0.15	0.50	1.0178	0.0152	0.9506	0.9788	0.4988	0.0029	0.9289	0.9420	481.2647	29274.7394	117.3317
			0.25	1.0174	0.0153	0.9499	0.9798	0.2495	0.0007	0.9292	0.9454	1669.1031	117028.7801	94.7791
		0.30	0.50	1.0109	0.0087	0.9529	0.9944	0.4971	0.0041	0.9317	0.9403	401.2955	32264.1461	130.8870
			0.25	1.0105	0.0087	0.9525	0.9955	0.2487	0.0010	0.9319	0.9399	1216.3488	128943.3484	115.1122
Individual Components														
120	96	0.53	0.50	1.0048	0.0168	0.9510	0.9225	0.5003	0.0081	0.9300	0.9455	208.7292	9076.6907	79.6430
			0.25	1.0048	0.0168	0.9510	0.9225	0.2501	0.0020	0.9300	0.9455	631.2352	36306.7600	71.6418
		0.67	0.50	1.0025	0.0126	0.9449	0.9074	0.4985	0.0185	0.9219	0.9489	160.8267	6046.1850	87.9152
			0.25	1.0025	0.0126	0.9449	0.9074	0.2493	0.0046	0.9219	0.9489	390.6962	24184.7401	84.0356
300	240	0.53	0.50	1.0034	0.0064	0.9495	0.9310	0.4986	0.0007	0.9540	0.9555	498.3396	53449.9944	199.4721
			0.25	1.0034	0.0064	0.9495	0.9310	0.2493	0.0029	0.9540	0.9555	1505.4801	213799.9775	179.3757
		0.67	0.50	1.0024	0.0049	0.9485	0.9040	0.4960	0.0060	0.9385	0.9505	375.3328	33842.1317	220.1997
			0.25	1.0024	0.0049	0.9485	0.9040	0.2480	0.0015	0.9385	0.9505	889.8873	135368.5269	210.4671
$P = (0, 2/3, 1/3)$ (equivalently, $a = (1, 1, -1)$)														
40	32	0.15	0.50	1.0872	0.1149	0.9419	0.9644	0.4954	0.0085	0.9289	0.9364	159.6738	2405.0454	26.1256
			0.25	1.0879	0.1153	0.9421	0.9647	0.2477	0.0021	0.9285	0.9421	578.0724	9592.3506	19.7047
		0.30	0.50	1.0353	0.0444	0.9473	0.9850	0.4969	0.0122	0.9293	0.9398	133.2835	2914.0701	29.6104
			0.25	1.0352	0.0445	0.9482	0.9874	0.2484	0.0031	0.9286	0.9407	437.0351	11644.6047	24.9305

(continued)

Table 1 (continued)

m	r	$F(\tau)$	θ_2	$E(\hat\theta_1)$	$MSE(\hat\theta_1)$	P_I	P_B	$E(\hat\theta_2)$	$MSE(\hat\theta_2)$	P_I	P_B	$tr(\mathbf{I})$	$det(\mathbf{I})$	TTT
100	80	0.15	0.50	1.0302	0.0300	0.9553	0.9734	0.4991	0.0034	0.9477	0.9467	364.9061	13674.6879	65.4872
			0.25	1.0306	0.0300	0.9558	0.9744	0.2494	0.0008	0.9442	0.9477	1321.5172	54636.6372	49.3425
		0.30	0.50	1.0147	0.0159	0.9467	0.9839	0.4995	0.0047	0.9392	0.9482	303.2646	16350.7164	74.1975
			0.25	1.0155	0.0162	0.9455	0.9839	0.2498	0.0012	0.9375	0.9425	986.8110	65276.2409	62.4079
Individual Components														
120	96	0.30	0.50	1.0231	0.0351	0.9445	0.9455	0.4982	0.0044	0.9325	0.9365	295.6496	9615.6443	65.6081
			0.25	1.0231	0.0351	0.9445	0.9455	0.2491	0.0011	0.9325	0.9365	1065.2696	38462.5771	50.5893
		0.44	0.50	1.0085	0.0211	0.9425	0.9285	0.4998	0.0060	0.9340	0.9380	242.4652	10070.0658	74.0857
			0.25	1.0085	0.0211	0.9425	0.9285	0.2499	0.0015	0.9340	0.9380	799.8534	40280.2634	63.3059
300	240	0.30	0.50	1.0073	0.0118	0.9520	0.9430	0.4998	0.0004	0.9585	0.9590	708.7493	56144.9535	164.3009
			0.25	1.0073	0.0118	0.9520	0.9430	0.2496	0.0016	0.9585	0.9590	2557.3296	224579.8140	126.6189
		0.44	0.50	1.0038	0.0079	0.9445	0.9320	0.4996	0.0022	0.9540	0.9565	580.8794	59120.7556	185.5250
			0.25	1.0038	0.0079	0.9445	0.9320	0.2498	0.0005	0.9540	0.9565	1917.9492	236483.0224	158.4551
$\boldsymbol{P} = (1/3,\ 2/3,\ 0)$ (equivalently, $\boldsymbol{a} = (0,\ 2,\ -1)$)														
40	32	0.15	0.50	1.2004	0.5182	0.9263	0.9453	0.4965	0.0078	0.9313	0.9393	157.5542	1340.0938	13.8182
			0.25	1.2002	0.5184	0.9265	0.9446	0.2483	0.0020	0.9305	0.9351	597.2200	5354.2252	9.5798
		0.30	0.50	1.0709	0.1086	0.9320	0.9645	0.4961	0.0108	0.9295	0.9430	130.2955	1808.7314	16.0897
			0.25	1.0721	0.1089	0.9343	0.9662	0.2480	0.0027	0.9293	0.9409	462.9937	7207.8274	12.9851
100	80	0.15	0.50	1.0553	0.0817	0.9452	0.9548	0.4996	0.0031	0.9482	0.9482	360.6123	6914.5743	34.6241
			0.25	1.0562	0.0822	0.9451	0.9512	0.2499	0.0008	0.9482	0.9477	1374.5878	27597.6486	23.9855
		0.30	0.50	1.0213	0.0298	0.9518	0.9684	0.4999	0.0040	0.9498	0.9468	290.3676	9982.1165	40.2920
			0.25	1.0214	0.0298	0.9517	0.9723	0.2500	0.0010	0.9507	0.9496	1030.4460	39855.0142	32.4873
Individual Components														
120	96	0.13	0.50	1.0570	0.0993	0.9435	0.9515	0.4988	0.0032	0.9395	0.9395	353.1675	6340.0016	55.8798

(continued)

Table 1 (continued)

m	r	$F(\tau)$	θ_2	$E(\hat{\theta}_1)$	$MSE(\hat{\theta}_1)$	P_I	P_B	$E(\hat{\theta}_2)$	$MSE(\hat{\theta}_2)$	P_I	P_B	$tr(\mathbf{I})$	$det(\mathbf{I})$	TTT
300			0.25	1.0570	0.0993	0.9435	0.9515	0.2494	0.0008	0.9395	0.9395	1353.5426	25360.0063	35.9968
		0.25	0.50	1.0266	0.0430	0.9425	0.9415	0.4987	0.0039	0.9375	0.9400	310.4790	9045.3644	62.9498
			0.25	1.0266	0.0430	0.9425	0.9415	0.2493	0.0010	0.9375	0.9400	1139.8687	36181.4575	46.6019
	240	0.13	0.50	1.0141	0.0268	0.9475	0.9615	0.5001	0.0012	0.9535	0.9555	853.2532	35430.5950	139.9700
			0.25	1.0141	0.0268	0.9475	0.9615	0.2500	0.0003	0.9535	0.9555	3279.9019	141722.3801	90.1226
		0.25	0.50	1.0077	0.0141	0.9430	0.9440	0.4999	0.0014	0.9530	0.9560	747.4730	52400.1421	157.6520
			0.25	1.0077	0.0141	0.9430	0.9440	0.2499	0.0004	0.9530	0.9560	2751.1433	209600.5682	116.6455
$\boldsymbol{P} = (1, 0, 0)$ (equivalently, $\boldsymbol{a} = (0, 0, 1)$)														
40	32	0.15	0.50	1.1800	0.5077	0.9353	0.9554	0.4962	0.0098	0.9208	0.9268	128.0412	1015.0584	6.2869
			0.25	1.1787	0.5094	0.9348	0.9545	0.2482	0.0025	0.9216	0.9308	483.5945	4073.2877	4.1422
		0.30	0.50	1.0689	0.1434	0.9315	0.9705	0.4975	0.0128	0.9210	0.9310	109.9070	1324.6117	7.2876
			0.25	1.0683	0.1433	0.9313	0.9684	0.2489	0.0032	0.9208	0.9238	391.7288	5296.5761	5.6405
100	80	0.15	0.50	1.0504	0.0957	0.9358	0.9674	0.5007	0.0039	0.9438	0.9418	289.3374	4911.5227	15.8063
			0.25	1.0508	0.0953	0.9377	0.9696	0.2506	0.0010	0.9473	0.9478	1097.6901	19461.1354	10.4002
		0.30	0.50	1.0155	0.0372	0.9415	0.9765	0.5023	0.0052	0.9445	0.9455	243.3527	6863.2933	18.2988
			0.25	1.0156	0.0370	0.9426	0.9763	0.2511	0.0013	0.9441	0.9451	870.4851	27409.9294	14.1390
Individual Components														
120	96	0.05	0.50	1.1840	0.5019	0.9434	0.9549	0.4983	0.0028	0.9409	0.9394	383.0721	3440.0681	50.9716
			0.25	1.1840	0.5019	0.9434	0.9549	0.2491	0.0007	0.9409	0.9394	1503.8390	13760.2726	28.6484
		0.11	0.50	1.0735	0.1335	0.9345	0.9535	0.4986	0.0031	0.9395	0.9395	361.3883	5636.3592	54.5459
			0.25	1.0735	0.1335	0.9345	0.9535	0.2493	0.0008	0.9395	0.9395	1394.7258	22545.4368	33.9960
300	240	0.05	0.50	1.0491	0.0899	0.9470	0.9510	0.4998	0.0003	0.9500	0.9515	927.5424	17617.3989	127.7386
			0.25	1.0491	0.0899	0.9470	0.9510	0.2499	0.0011	0.9500	0.9515	3651.4276	70469.5957	71.7754
		0.11	0.50	1.0172	0.0334	0.9435	0.9575	0.5001	0.0012	0.9550	0.9550	873.1663	31154.6523	136.6355
			0.25	1.0172	0.0334	0.9435	0.9575	0.2500	0.0003	0.9550	0.9550	3379.3681	124618.6092	85.1208

6 Illustrative Example

For illustrative purpose, let us consider the Type-II censored system lifetime data in hours (with $m = 40$ and $r = 30$) obtained from a simple step-stress life-test on 3-component coherent system with signature vector $\boldsymbol{P} = (0, 2/3, 1/3)$ with the change of stress done at time $\tau = 7$ hours, presented in Table 2.

In this case, we have $n_1 = 9$ and $n_2 = r - n_1 = 30 - 9 = 21$. From Eqs. (26) and (27), we determine the initial estimates of θ_1 and θ_2 in this case to be $\tilde{\theta}_1 = 14.9468$ and $\tilde{\theta}_2 = 5.2392$. With these initial values, we solved iteratively Eq. (19) and determined the MLEs as well as their SEs, and the 95% CIs based on the asymptotic approach using observed Fisher information matrix and the bootstrap approach. These results are presented in Table 3. We notice here again that the bootstrap SEs are larger than those obtained from the observed Fisher information matrix and also the bootstrap CIs are wider than those determined from the asymptotic approach. This is in conformance with the findings from the simulation results.

Furthermore, upon using the MLEs of θ_1 and θ_2 presented in Table 3, we also estimated the 50th (namely, the median), 90 and 95th percentiles of system lifetimes at the first stress level, the second stress level and in the considered step-stress test. Note that these results can be readily obtained by the results corresponding to the choices of $\tau = \infty$, $\tau = 0$ and $\tau = 7$, respectively.

These estimates, along with their SEs and 95% CIs based on the parametric bootstrap approach, are all presented in Table 4. From these results, upon comparing

Table 2 Censored failure-time data of 3-component systems with signature vector $\boldsymbol{P} = (0, 2/3, 1/3)$

Stress level	Failure times							
First	2.0262	2.6885	2.9351	3.2039	4.0190	4.8049	5.6151	6.6172
	6.8337							
Second	7.0446	7.3139	7.6264	7.6341	7.7320	7.9985	8.1117	8.1963
	8.2014	8.4150	8.4330	9.2122	10.1894	10.7557	10.9652	11.1651
	11.2694	11.2728	11.5362	11.5508	12.4190			

Table 3 MLEs of the parameters based on censored failure-time data in Table 2, their SEs and 95% CIs based on the asymptotic approach using observed Fisher information matrix and the bootstrap approach

Parameter	MLEs	Observed fisher information matrix		Bootstrap	
		SE	Asymptotic CI	SE	Bootstrap CI
θ_1	14.9910	3.0845	(8.9455, 21.0367)	3.9656	(10.4988, 25.5718)
θ_2	5.1463	1.0718	(3.0455, 7.2472)	1.0837	(3.2666, 7.4942)

Table 4 Estimates of the percentiles of system lifetimes at the first stress level, second stress level and in the considered step-stress test, using the MLEs presented in Table 3, their SEs and 95% CIs based on the parametric bootstrap approach

τ	Parameter	MLE	SE	Bootstrap CI
$+\infty$	$q_{0.50}$	13.6229	3.6037	(9.5406, 23.2380)
	$q_{0.90}$	35.7241	9.4501	(25.0188, 60.9381)
	$q_{0.95}$	45.5769	12.0565	(31.9190, 77.7450)
0	$q_{0.50}$	4.6767	0.9848	(2.9685, 6.8103)
	$q_{0.90}$	12.2638	2.5825	(7.7845, 17.8589)
	$q_{0.95}$	15.6462	3.2947	(9.9315, 22.7845)
7	$q_{0.50}$	9.2736	0.6560	(8.0857, 10.6684)
	$q_{0.90}$	16.8607	2.0675	(13.1921, 21.3499)
	$q_{0.95}$	20.2432	2.7629	(15.3467, 26.2386)

the respective CIs, we can readily observe that the system lifetime at the first stress level is significantly better than at the second stress level (at all three percentile levels). Similarly, the lifetime of the system at the first stress level is significantly better than the system in the considered step-stress test at 90 and 95th percentiles, but not at the median level.

7 Concluding Remarks

In this paper, we have discussed inferential issues based on a Type-II censored system lifetime data on a coherent system with known signature having an exponential component lifetime distribution under a simple step-stress life-test experiment. It will naturally be of interest to extend the results developed here for other step-stress experiments such as those under time constraint (see Balakrishnan et al. [7]) and those with random stress change times (see Xiong and Milliken [19]). Moreover, the problem of determining an optimal τ in this setting will also be of great interest (see Gouno et al. [8]). Finally, it will be of practical interest to generalize the present work to the case when the component lifetime distribution is Weibull along the lines of Kateri and Balakrishnan [9]. Work on these problems is currently under progress and we hope to report these findings in a future paper.

References

1. Arnold BC, Balakrishnan N, Nagaraja HN (1992) A first course in order statistics. Wiley, New York
2. Bai DS, Kim MS, Lee SH (1989) Optimum simple step-stress accelerated life tests with censoring. IEEE Trans Reliab 38:528–532
3. Balakrishnan N, Cohen AC (1991) Order statistics and inference: estimation methods. Academic Press, Boston
4. Balakrishnan N, Kundu D, Ng HKT, Kannan N (2007) Point and interval estimation for a simple step-stress model with Type-II censoring. J Qual Technol 39:35–47
5. Balakrishnan N, Navarro J, Samaniego FJ (2012) Signature representation and preservation results for engineered systems and applications to statistical inference. In: Lisnianski A, Frenkel I (eds) Recent advances in system reliability. Springer, London
6. Balakrishnan N, Ng HKT, Navarro J (2011) Linear inference for type-II censored lifetime data of reliability systems with known signatures. IEEE Trans Reliab 60:426–440
7. Balakrishnan N, Xie Q, Kundu D (2009) Exact inference for a simple step-stress model from the exponential distribution under time constraint. Ann Inst Stat Math 61:251–274
8. Gouno E, Sen A, Balakrishnan N (2004) Optimal step-stress test under progressive Type-I censoring. IEEE Trans Reliab 53:388–393
9. Kateri M, Balakrishnan N (2008) Inference for a simple step-stress model with type-II censoring, and Weibull distribution lifetimes. IEEE Trans Reliab 57:616–626
10. Kochar S, Mukerjee H, Samaniego FJ (1999) The signature of a coherent system and its application to comparisons among systems. Nav Res Logist 46:507–523
11. Lehmann EL (1999) Elements of large sample theory. Springer, New York
12. Miller R, Nelson W (1983) Optimum simple step-stress plans for accelerated life testing. IEEE Trans Reliab 32:59–65
13. Navarro J, Ruiz JM, Sandoval CJ (2007) Properties of coherent systems with dependent components. Commun Stat-Theory Methods 36:175–191
14. Nelson W (1980) Accelerated life testing-step-stress models and data analyses. IEEE Trans Reliab 29:103–108
15. Nelson W (1990) Accelerated testing: statistical models, test plans, and data analysis. Wiley, New York
16. Samaniego FJ (1985) On closure of the IFR class under formation of coherent systems. IEEE Trans Reliab 34:69–72
17. Samaniego FJ (2007) System signatures and their applications in engineering reliability. Springer, New York
18. Xiong C (1998) Inferences on a simple step-stress model with type-II censored exponential data. IEEE Trans Reliab 47:142–146
19. Xiong C, Milliken F (1999) Step-stress life-testing with random stress-change times for exponential data. IEEE Trans Reliab 48:141–148

Part III
The Mathematics of Communication: Information Theory and Applications

Asymptotic Developments for the ϕ-Inequality Indices

María Carmen Alonso and Ignacio Martínez

Abstract This paper presents asymptotically optimal estimates of a family of divergence-inspired inequality indices in simple and stratified samplings.

1 Introduction

Inequality of a positive random variable refers to the extent to which the variable is distributed in an uneven manner among a population. In fact, the most popular inequality measures are scale invariant, and they can be viewed as a kind of indices of the relative variability/dispersion of the variable.

Inequality is especially applied in Economics, and it often concerns the difference found in various measures of income or wealth among individuals in a group, among groups in a population, or among big populations, like countries. Economic inequality is crucial for purposes of equity, equality of opportunity, and so on.

In addition to the well-known Theil index [13], several indices have been suggested in the literature aiming to well-quantifying such a feature of the distribution of a variable in an easy-to-use and well-argued way. In this respect, the additively decomposable indices (see, for instance, Bourguignon [4], Cowell [5] and Shorrocks [11]) have been shown to be interesting alternatives to Theil's index.

At the end of the eighties, and partially following Aczél and Kannapan [1], several authors (see, for instance, Gil et al. [8], Taneja et al. [12]) have commented the fact that some interesting inequality indices can be based/inspired on statistical directed divergence measures between two probability distributions. Their key idea is to replace one of the distributions in the divergence by the weighted and normalized variable values. This idea has been reconsidered in some recent studies (see, for

This work is dedicated to the memory of Pedro Gil (boss, colleague and friend). He has been deeply involved in launching the ideas and developments behind, and the research in the paper has been carried out under his guidance. We miss him a lot.

M. C. Alonso · I. Martínez (✉)
Universidad de Oviedo, Departamento de Matemáticas, Oviedo, Spain
e-mail: iml@uniovi.es

© Springer International Publishing AG 2018
E. Gil et al. (eds.), *The Mathematics of the Uncertain*, Studies in Systems, Decision and Control 142, https://doi.org/10.1007/978-3-319-73848-2_42

instance, [9, 10]), Csizár's divergence [2, 6, 7] being one of the most commonly used divergence measures.

In this paper we are going to examine the asymptotic behaviour under rather general conditions of the ϕ-inequality index (or Csizár's divergence-based inequality measure), which includes the additively decomposable inequality indices [4, 5, 8, 11] as special cases and it is defined as follows:

Definition 1.1 Let $\phi : (0, \infty) \to [0, \infty)$ be a mapping such that it is stritcly convex (or upper concave) and satisfying the following conditions/conventions:

$$\phi(1) = 0, \ \ 0 = \lim_{u \downarrow 0} \phi(u), \ \ 0 \cdot \phi(u/0) = u$$

and, the dual function $\phi^*(u) = u \cdot \phi(1/u)$, satisfying that

$$0 \cdot \phi(u/0) = u \cdot \phi^*(0), \ \ 0 \cdot \phi^*(u/0) = u \cdot \phi(0).$$

Let X be a positive random variable associated with a probability space $(\Omega, \mathcal{A}, P_\theta)$, P_θ also denoting the probability measure induced by X and θ being a real- or vectorial-valued parameter, $\theta \in \Theta$.

The (population) ϕ-inequality index associated with X is given, if it exists, by the Lebesgue-Stieltjes integral

$$\mathfrak{I}_\phi(X; \theta) = \int_{X(\Omega)} \phi\left(\frac{x}{E_\theta(X)}\right) dP_\theta(x),$$

whenever the expected value $E_\theta(X) = \int_{X(\Omega)} x \, dP_\theta(x) < \infty$.

If the family of probability measures $\{P_\theta, \ \theta \in \Theta\}$ is dominated by a σ-finite measure η over the Borel σ-field on $X(\Omega)$, and f_θ denotes the density function associated with P_θ w.r.t. η, the ϕ-inequality index associated with X is given by

$$\mathfrak{I}_\phi(X; \theta) = \int_{X(\Omega)} \phi\left(\frac{x}{E_\theta(X)}\right) f_\theta(x) \, d\eta(x).$$

2 Asymptotic Behaviour of the ϕ-Inequality Index in a Random Sampling from a Population. Application to the Simple Random Sampling from Finite Populations

Assume that $\theta = (\theta_1, \ldots, \theta_r)$. If θ is unknown and it is estimated under certain regularity conditions by means of the maximum likelihood estimator based on a simple random sample from variable X, then we are now going to study the asymptotic behaviour of the statistic corresponding to the ϕ-inequality index making use of these estimates.

The following theorem establishes the regularity conditions and main general result in this framework.

Theorem 2.1 *Let $(X_1, \ldots, X_n)$ be a simple random sample from a random variable X associated with the probability space $(\Omega, \mathcal{A}, P_\theta)$, with the parameter $\theta \in \Theta$ being unknown, $\{P_\theta, \theta \in \Theta\}$ being dominated by a σ-finite measure η and f_θ denoting the associated density function.*

Assume the following regularity conditions on P_θ and ϕ are fulfilled:

(R_1) *The true value of θ, θ_0, belongs to the interior of Θ, which is supposed to be an r-dimensional rectangle in $\mathbb{R}^r$.*

(R_2) *The set $\mathbb{X} = \{x \in \mathbb{R} : f_\theta(x) > 0\}$ does not depend on θ.*

(R_3) *For each $x \in \mathbb{X}$ and $\theta \in \Theta$ there exist the partial derivatives of orders 1, 2 and 3 w.r.t. θ_i $(i = 1 \ldots, r)$ of $\log f_\theta$, and there also exist real-valued functions g_i, h_{ij} and H_{ijk} defined on $\mathbb{X}$, with g_i and h_{ij} integrable on $\mathbb{X}$ w.r.t. η and H_{ijk} integrable on $\mathbb{X}$ w.r.t. P_θ and maybe depending on θ in a neighborhood of θ_0, such that in such a neighborhood*

$$\left| \frac{\partial \log f_\theta(x)}{\partial \theta_i} \right| \le g_i(x), \quad \left| \frac{\partial^2 \log f_\theta(x)}{\partial \theta_i \, \partial \theta_j} \right| \le h_{ij}(x),$$

$$\left| \frac{\partial^3 \log f_\theta(x)}{\partial \theta_i \, \partial \theta_j \, \partial \theta_k} \right| \le H_{ijk}(x)$$

for $i, j, k \in \{1, \ldots, r\}$.

(R_4) *For each $x \in \mathbb{X}$ and $\theta \in \Theta$ there exist the partial derivatives of orders 1, 2 and 3 w.r.t. θ_i $(i = 1 \ldots, r)$ of $\phi(x/E_\theta(X)) \cdot f_\theta$, and there also exist real-valued functions g_i^ϕ, h_{ij}^ϕ and H_{ijk}^ϕ defined on $\mathbb{X}$, with g_i^ϕ and h_{ij}^ϕ integrable on $\mathbb{X}$ w.r.t. η and H_{ijk}^ϕ integrable on $\mathbb{X}$ w.r.t. P_θ and maybe depending on θ in a neighborhood of θ_0, such that in such a neighborhood*

$$\left| \frac{\partial \phi(x/E_\theta(X)) \cdot f_\theta}{\partial \theta_i} \right| \le g_i^\phi(x), \quad \left| \frac{\partial^2 \phi(x/E_\theta(X)) \cdot f_\theta}{\partial \theta_i \, \partial \theta_j} \right| \le h_{ij}^\phi(x),$$

$$\left| \frac{\partial^3 \phi(x/E_\theta(X)) \cdot f_\theta}{\partial \theta_i \, \partial \theta_j \, \partial \theta_k} \right| \le H_{ijk}^\phi(x)$$

for $i, j, k \in \{1, \ldots, r\}$.

(R_5) *The Fisher information matrix in θ,*

$$I_X^F(\theta) = \left[\int_{\mathbb{X}} \frac{\partial \log f_\theta(x)}{\partial \theta_i} \cdot \frac{\partial \log f_\theta(x)}{\partial \theta_j} \cdot f_\theta(x) \, d\eta(x) \right]_{ij},$$

is definite and the associated quadratic form is positive definite for each $\theta \in \Theta$.

Then, if $\left\{\widehat{\boldsymbol{\theta}}_n = (\widehat{\theta}_{n1}, \ldots, \widehat{\theta}_{nr})\right\}_{n \in \mathbb{N}}$ *is a sequence of estimators of* $\boldsymbol{\theta}$ *from* $(X_1, \ldots, X_n)$ *which are solutions of the likelihood equations system*

$$
\begin{cases}
\sum_{l=1}^n \frac{\partial f_\theta(X_l)}{\partial \theta_1} = 0 \\
\quad \vdots \\
\sum_{l=1}^n \frac{\partial f_\theta(X_l)}{\partial \theta_r} = 0
\end{cases}
$$

which is a strongly consistent sequence of estimators, asymptotically normal and effi-cient (more concretely, $\widehat{\boldsymbol{\theta}}_n$ *is asymptotically distributed as an r-dimensional normal* $\mathcal{N}(\boldsymbol{\theta}_0, [I_X^F(\boldsymbol{\theta}_0)]^{-1}/n)$ *as* $n \to \infty$*), we have that*

(i) $\left\{\mathfrak{I}_\phi(X; \widehat{\boldsymbol{\theta}}_n)\right\}_{n \in \mathbb{N}}$ *is a sequence of estimators of* $\mathfrak{I}_\phi(X; \boldsymbol{\theta}_0)$ *from* $(X_1, \ldots, X_n)$ *which is strongly consistent whatever* $\boldsymbol{\theta}_0$ *may be.*

(ii) *The sequence of random variables* $\left\{\sqrt{n}[\mathfrak{I}_\phi(X; \widehat{\boldsymbol{\theta}}_n) - \mathfrak{I}_\phi(X; \boldsymbol{\theta}_0)]\right\}_{n \in \mathbb{N}}$ *con-verges in law as* $n \to \infty$ *to the one-dimensional normal distribution*

$$
\mathcal{N}\left(0, \nabla \mathfrak{I}_\phi(X; \boldsymbol{\theta}_0) [I_X^F(\boldsymbol{\theta}_0)]^{-1} \left(\nabla \mathfrak{I}_\phi(X; \boldsymbol{\theta}_0)\right)^t\right)
$$

whenever $\nabla \mathfrak{I}_\phi(X; \boldsymbol{\theta}_0) [I_X^F(\boldsymbol{\theta}_0)]^{-1} \left(\nabla \mathfrak{I}_\phi(X; \boldsymbol{\theta}_0)\right)^t > 0$, *where* ∇ *denotes the gradient vector.*

(iii) *If* $\nabla \mathfrak{I}_\phi(X; \boldsymbol{\theta}_0) [I_X^F(\boldsymbol{\theta}_0)]^{-1} \nabla \left(\mathfrak{I}_\phi(X; \boldsymbol{\theta}_0)\right)^t = 0$ *and some of the second partial derivatives of* $\mathfrak{I}_\phi(X; \boldsymbol{\theta})$ *do not vanish, then the sequence of random variables* $\left\{2n[\mathfrak{I}_\phi(X; \widehat{\boldsymbol{\theta}}_n) - \mathfrak{I}_\phi(X; \boldsymbol{\theta}_0)]\right\}_{n \in \mathbb{N}}$ *converges in law as* $n \to \infty$ *to a linear com-bination of at most r independent chi-square variables with 1 degree of freedom. Furthermore, the asymptotic distribution will be a chi-square one with q degrees of freedom if, and only if,*

$$
H(\mathfrak{I}_\phi(X; \boldsymbol{\theta}_0)) [I_X^F(\boldsymbol{\theta}_0)]^{-1} H(\mathfrak{I}_\phi(X; \boldsymbol{\theta}_0)) = H(\mathfrak{I}_\phi(X; \boldsymbol{\theta}_0)),
$$

where H is the Hessian matrix and q is the range of $H(\mathfrak{I}_\phi(X; \boldsymbol{\theta}_0))$.

On the basis of Theorem 2.1 and well-known large sample results, one can con-clude that

Corollary 2.1 *Under the situation and assumptions in Theorem 2.1, we have that the sequence of random variables*

$$
\left\{ \frac{\sqrt{n}[\mathfrak{I}_\phi(X; \widehat{\boldsymbol{\theta}}_n) - \mathfrak{I}_\phi(X; \boldsymbol{\theta}_0)]}{\sqrt{\nabla \mathfrak{I}_\phi(X; \widehat{\boldsymbol{\theta}}_n) [I_X^F(\widehat{\boldsymbol{\theta}}_n)]^{-1} \left(\nabla \mathfrak{I}_\phi(X; \widehat{\boldsymbol{\theta}}_n)\right)^t}} \right\}_{n \in \mathbb{N}}
$$

converges in law as $n \to \infty$ *to the one-dimensional standard normal distribu-tion, whenever that* $\nabla \mathfrak{I}_\phi(X; \boldsymbol{\theta}_0) [I_X^F(\boldsymbol{\theta}_0)]^{-1} \left(\nabla \mathfrak{I}_\phi(X; \boldsymbol{\theta}_0)\right)^t > 0$ *and* $\nabla \mathfrak{I}_\phi(X; \widehat{\boldsymbol{\theta}}_n) [I_X^F(\widehat{\boldsymbol{\theta}}_n)]^{-1} \left(\nabla \mathfrak{I}_\phi(X; \widehat{\boldsymbol{\theta}}_n)\right)^t > 0$.

The preceding conclusions can be easily particularized to the simple random sampling from finite populations. If we consider a population with M individuals and X is a random variable taking on R different positive values $x_1, \ldots, x_R$, and $p_t = P(X = x_t)$, $t = 1, \ldots, R$, then the population ϕ-inequality index is given by

$$\Im_\phi(X; \boldsymbol{p}) = \sum_{t=1}^{R} \phi\left(\frac{x_t}{E(X; \boldsymbol{p})}\right) \cdot p_t,$$

with $p_R = 1 - (p_1 + \ldots + p_{R-1})$ and $E(X; \boldsymbol{p}) = \sum_{t=1}^{R} x_t p_t$.

Assume that a sample of size n is chosen at random and with (or even without, in case M is large enough) replacement from the population and f_{nt} is the sample relative frequency of x_t, then the preceding results can be particularized by considering $\boldsymbol{\theta} = \boldsymbol{p} = (p_1, \ldots, p_{R-1})$ and $\widehat{\boldsymbol{\theta}}_n = \boldsymbol{f}_n = (f_{n1}, \ldots, f_{n(R-1)})$. In this way, in the simple random sampling from the final population, one can ensure that

Theorem 2.2 *In simple random sampling, if ϕ has finite derivatives of order 3, we have that*

(i) $\{\Im_\phi(X; \boldsymbol{f}_n)\}_{n \in \mathbb{N}}$ *is a sequence of estimators of $\Im_\phi(X; \boldsymbol{p}_0)$ which is strongly consistent whatever the true value $\boldsymbol{p}_0 = (p_{10}, \ldots, p_{(R-1)0})$ of $\boldsymbol{p} \in (0, 1)^{R-1}$ may be.*

(ii) *The sequence of random variables $\left\{\sqrt{n}[\Im_\phi(X; \boldsymbol{f}_n) - \Im_\phi(X; \boldsymbol{p}_0)]\right\}_{n \in \mathbb{N}}$ converges in law as $n \to \infty$ to the one-dimensional normal distribution $\mathcal{N}(0, \sigma^2(\boldsymbol{p}_0))$, where*

$$\sigma^2(\boldsymbol{p}_0) = \sum_{t=1}^{R} p_{t0} \left[\phi\left(\frac{x_t}{E(X; \boldsymbol{p}_0)}\right) - x_t \sum_{u=1}^{R} p_{u0} \frac{x_u}{(E(X; \boldsymbol{p}_0))^2} \cdot \phi'\left(\frac{x_u}{E(X; \boldsymbol{p}_0)}\right)\right]^2$$

$$- \left\{\sum_{t=1}^{R} p_{t0} \left[\phi\left(\frac{x_t}{E(X; \boldsymbol{p}_0)}\right) - x_t \sum_{u=1}^{R} p_{u0} \frac{x_u}{(E(X; \boldsymbol{p}_0))^2} \cdot \phi'\left(\frac{x_u}{E(X; \boldsymbol{p}_0)}\right)\right]\right\}^2$$

whenever $\sigma^2(\boldsymbol{p}_0) > 0$.

(iii) *If $\sigma^2(\boldsymbol{p}_0) = 0$ and for some $i, j \in \{1, \ldots, R-1\}$*

$$H_{ij} = \frac{x_i - x_R}{E(X; \boldsymbol{p}_0)} \left\{\phi'\left(\frac{x_R}{E(X; \boldsymbol{p}_0)}\right) \cdot \frac{x_R}{E(X; \boldsymbol{p}_0)} - \phi'\left(\frac{x_j}{E(X; \boldsymbol{p}_0)}\right) \cdot \frac{x_j}{E(X; \boldsymbol{p}_0)}\right.$$

$$+ \frac{x_j - x_R}{E(X; \boldsymbol{p}_0)} \sum_{t=1}^{R} p_{t0}\, \phi''\left(\frac{x_t}{E(X; \boldsymbol{p}_0)}\right) \cdot \frac{x_t^2}{(E(X; \boldsymbol{p}_0))^2}\right\}$$

$$+ \frac{x_j - x_R}{E(X; \boldsymbol{p}_0)} \left\{\phi'\left(\frac{x_R}{E(X; \boldsymbol{p}_0)}\right) \cdot \frac{x_R}{E(X; \boldsymbol{p}_0)} - 2\phi\left(\frac{x_R}{E(X; \boldsymbol{p}_0)}\right)\right.$$

$$\left.- \phi'\left(\frac{x_i}{E(X; \boldsymbol{p}_0)}\right) \cdot \frac{x_i}{E(X; \boldsymbol{p}_0)} + 2\phi\left(\frac{x_i}{E(X; \boldsymbol{p}_0)}\right)\right\} > 0,$$

then the sequence of random variables $\{2n[\Im_\phi(X;\widehat{\theta}_n) - \Im_\phi(X;\theta_0)]\}_{n\in\mathbb{N}}$ converges in law as $n \to \infty$ to a linear combination of at most $R - 1$ independent chi-square variables with 1 degree of freedom.

Consequently,

Corollary 2.2 *Under the situation and assumptions in Theorem 2.2, we have that the sequence of random variables*

$$\left\{ \frac{\sqrt{n}[\Im_\phi(X;f_n) - \Im_\phi(X;p_0)]}{\sqrt{\sigma^2(f_n)}} \right\}_{n\in\mathbb{N}}$$

converges in law as $n \to \infty$ to the one-dimensional standard normal distribution, whenever $\sigma^2(f_n) > 0$ and $\sigma^2(p_0) > 0$.

3 Asymptotic Behaviour of the ϕ-Inequality Index in a Random Sampling from Several Populations. Application to the Stratified Random Sampling from Finite Populations

Assume that the random variable X is now considered over S different populations, Ω^s, so that P_{θ^s} is the induced probability measure for X in the sth population. Let $\theta^s = (\theta_1^s, \ldots, \theta_{r_s}^s)$, and assume that $\{P_{\theta^s}, \theta^s \in \Theta^s\}$ is dominated by a σ-finite measure η^s and f_{θ^s} denoting the associated density function. If probabilities of considering each of the populations are known to be $W_1, \ldots, W_S$, the situation can be viewed as that of analyzing X over a randomized population in which X is distributed as P_{θ^s} with probability W_s with $s \in \{1, \ldots, S\}$. In accordance with this view, it seems natural to quantify the ϕ-inequality index associated with X in such a randomized population by means of the value, if it exists, given by

$$\Im_\phi(X;\theta^1, \ldots, \theta^S) = \sum_{s=1}^{S} W_s \int_{X(\Omega^s)} \phi\left(\frac{x}{E_{\theta^1,\ldots,\theta^S}(X)}\right) dP_{\theta^s}(x),$$

whenever the expected value exists

$$E_{\theta^1,\ldots,\theta^S}(X) = \sum_{s=1}^{S} W_s E(X;\theta^s) = \sum_{s=1}^{S} W_s \int_{X(\Omega^s)} x \, dP_{\theta^s}(x) < \infty.$$

The following theorem establishes the main general result in this new framework.

Theorem 3.1 *Let $(X_{11}, \ldots, X_{1n_1}), \ldots, (X_{S1}, \ldots, X_{Sn_S})$ be S independent simple random samples from random variable X associated with the probability space*

$(\Omega^1, \mathcal{A}^1, P_{\theta^1}), \ldots, (\Omega^S, \mathcal{A}^S, P_{\theta^S})$, respectively, with the parameters $\theta^s \in \Theta^s$ being unknown, $\{P_{\theta^s}, \theta^s \in \Theta^s\}$ being dominated by a σ-finite measure η^s and f_{θ^s} denoting the associated density function for each $s \in \{1, \ldots, S\}$.

Let $w_s = n_s/n$ (with $n = n_1 + \ldots + n_S$). Assume that over each of the S populations and probability spaces the regularity conditions in Theorem 2.1 are fulfilled.

Then, if for each s the sequence $\{\widehat{\boldsymbol{\theta}}^s_{n_s} = (\widehat{\theta}^s_{n_s 1}, \ldots, \widehat{\theta}^s_{n_s r_s})\}_{n_s \in \mathbb{N}}$ is a sequence of estimators of θ^s from $(X_{s1}, \ldots, X_{sn_s})$ which are solutions of the likelihood equations system

$$\begin{cases} \sum_{l=1}^{n_s} \frac{\partial f_{\theta^s}(X_{sl})}{\partial \theta^s_1} = 0 \\ \quad\vdots \\ \sum_{l=1}^{n_s} \frac{\partial f_{\theta^s}(X_{sl})}{\partial \theta^s_{r_s}} = 0 \end{cases}$$

which is a strongly consistent sequence of estimators, asymptotically normal and efficient (more concretely, $\widehat{\boldsymbol{\theta}}_{n_s}$ is asymptotically distributed as an r_s-dimensional normal $\mathcal{N}(\theta^s_0, [I^F_X(\theta^s_0)]^{-1}/n_s)$ as $n_s \to \infty$), θ^s_0 being the true value of θ^s, we have that

(i) $\{\mathfrak{I}_\phi(X; \widehat{\boldsymbol{\theta}}^1_{n_1}, \ldots, \widehat{\boldsymbol{\theta}}^S_{n_S})\}_{n \in \mathbb{N}}$ is a sequence of estimators of $\mathfrak{I}_\phi(X; \theta^1_0, \ldots, \theta^S_0)$ from $(X_{11}, \ldots, X_{1n_1}, \ldots, X_{S1}, \ldots, X_{Sn_S})$ which is strongly consistent whatever the true value $(\theta^1_0, \ldots, \theta^S_0)$ of $(\theta^1, \ldots, \theta^S)$ may be.

(ii) The sequence of random variables $\{\sqrt{n}[\mathfrak{I}_\phi(X; \widehat{\boldsymbol{\theta}}^1_{n_1}, \ldots, \widehat{\boldsymbol{\theta}}^S_{n_S}) - \mathfrak{I}_\phi(X; \theta^1_0, \ldots, \theta^S_0)]\}_{n \in \mathbb{N}}$ converges in law as $n_1 \to \infty, \ldots, n_S \to \infty$ to the one-dimensional normal distribution

$$\mathcal{N}\Big(0, \nabla \mathfrak{I}_\phi(X; \theta^1_0, \ldots, \theta^S_0) [I^F_X(\theta^1_0, \ldots, \theta^S_0)]^{-1} \big(\nabla \mathfrak{I}_\phi(X; \theta^1_0, \ldots, \theta^S_0)\big)^t\Big)$$

whenever $\nabla \mathfrak{I}_\phi(X; \theta^1_0, \ldots, \theta^S_0) [I^F_X(\theta^1_0, \ldots, \theta^S_0)]^{-1} \big(\nabla \mathfrak{I}_\phi(X; \theta^1_0, \ldots, \theta^S_0)\big)^t > 0$, with

$$\nabla \mathfrak{I}_\phi(X; \theta^1_0, \ldots, \theta^S_0) = \big(\nabla \mathfrak{I}_\phi(X; \theta^1_0), \ldots, \nabla \mathfrak{I}_\phi(X; \theta^S_0)\big),$$

$$\nabla \mathfrak{I}_\phi(X; \theta^s_0) = \left(\frac{\partial \mathfrak{I}_\phi(X; \theta^1, \ldots, \theta^S)}{\partial \theta^s_1}, \ldots, \frac{\partial \mathfrak{I}_\phi(X; \theta^1, \ldots, \theta^S)}{\partial \theta^s_{r_s}}\right)\Bigg|_{(\theta^1, \ldots, \theta^S) = (\theta^1_0, \ldots, \theta^S_0)},$$

$$[I^F_X(\theta^1_0, \ldots, \theta^S_0)]^{-1} = \begin{pmatrix} \frac{[I^F_X(\theta^1_0)]^{-1}}{w_1} & \cdots & 0 \\ \vdots & \ddots & \vdots \\ 0 & \cdots & \frac{[I^F_X(\theta^S_0)]^{-1}}{w_S} \end{pmatrix}.$$

(iii) If $\nabla \mathfrak{I}_\phi(X; \theta^1_0, \ldots, \theta^S_0) [I^F_X(\theta^1_0, \ldots, \theta^S_0)]^{-1} \big(\nabla \mathfrak{I}_\phi(X; \theta^1_0, \ldots, \theta^S_0)\big)^t = 0$ and some of the second partial derivatives of $\mathfrak{I}_\phi(X; \theta^1, \ldots, \theta^S)$ do not vanish, then the sequence of random variables $\{2n[\mathfrak{I}_\phi(X; \widehat{\boldsymbol{\theta}}^1_{n_1}, \ldots, \widehat{\boldsymbol{\theta}}^S_{n_S}) - \mathfrak{I}_\phi(X; \theta^1_0, \ldots, \theta^S_0)]\}_{n \in \mathbb{N}}$ converges in law as $n_1 \to \infty, \ldots, n_S \to \infty$ to a linear combination of at most $n_1 + \ldots + n_S$ independent chi-square variables with 1 degree of free-

dom. Furthermore, the asymptotic distribution will be a chi-square one with q degrees of freedom if, and only if,

$$H(\mathfrak{I}_\phi(X;\theta_0^1,\ldots,\theta_0^S))\,[I_X^F(\theta_0^1,\ldots,\theta_0^S)]^{-1}\,H(\mathfrak{I}_\phi(X;\theta_0^1,\ldots,\theta_0^S))$$

$$= H(\mathfrak{I}_\phi(X;\theta_0^1,\ldots,\theta_0^S)),$$

where H is the Hessian matrix and q is the range of $H(\mathfrak{I}_\phi(X;\theta_0^1,\ldots,\theta_0^S))$.

On the basis of Theorem 3.1 and well-known large sample results, one can conclude that

Corollary 3.1 *Under the situation and assumptions in Theorem 3.1, we have that the sequence of random variables*

$$\left\{ \frac{\sqrt{n}[\mathfrak{I}_\phi(X;\widehat{\theta}_{n_1}^1,\ldots,\widehat{\theta}_{n_S}^S) - \mathfrak{I}_\phi(X;\theta_0^1,\ldots,\theta_0^S)]}{\sqrt{\nabla \mathfrak{I}_\phi(X;\widehat{\theta}_{n_1}^1,\ldots,\widehat{\theta}_{n_S}^S)\,[I_X^F(\widehat{\theta}_{n_1}^1,\ldots,\widehat{\theta}_{n_S}^S)]^{-1}\big(\nabla \mathfrak{I}_\phi(X;\widehat{\theta}_{n_1}^1,\ldots,\widehat{\theta}_{n_S}^S)\big)^t}} \right\}_{n\in\mathbb{N}}$$

converges in law as $n_1 \to \infty, \ldots, n_S \to \infty$ to the one-dimensional standard normal distribution, whenever

$$\nabla \mathfrak{I}_\phi(X;\widehat{\theta}_{n_1}^1,\ldots,\widehat{\theta}_{n_S}^S)\,[I_X^F(\widehat{\theta}_{n_1}^1,\ldots,\widehat{\theta}_{n_S}^S)]^{-1}\big(\nabla \mathfrak{I}_\phi(X;\widehat{\theta}_{n_1}^1,\ldots,\widehat{\theta}_{n_S}^S)\big)^t > 0$$

and

$$\nabla \mathfrak{I}_\phi(X;\theta_0^1,\ldots,\theta_0^S)\,[I_X^F(\theta_0^1,\ldots,\theta_0^S)]^{-1}\big(\nabla \mathfrak{I}_\phi(X;\theta_0^1,\ldots,\theta_0^S)\big)^t > 0.$$

The preceding conclusions can be easily particularized to the stratified random sampling from finite populations. Consider a population with M individuals which can be partitioned into S strata in accordance with which the positive random variable X shows a rather homogeneous behaviour. Assume X takes on R different positive values $x_1,\ldots,x_R$, p_t^s denotes the probability that X takes the value x_t in the sth stratum, $t = 1,\ldots,R$, $\boldsymbol{p}^s = (p_1^s,\ldots,p_{R-1}^s)$, and M_s = number of individuals in the sth stratum of the population, $s = 1,\ldots,S$. If $W_s = M_s/M$, then the population ϕ-inequality index is given by

$$\mathfrak{I}_\phi(X;\boldsymbol{p}^1,\ldots,\boldsymbol{p}^S) = \sum_{t=1}^{R}\sum_{s=1}^{S} W_s\,\phi\left(\frac{x_t}{E(X;\boldsymbol{p}^1,\ldots,\boldsymbol{p}^S)}\right) \cdot p_t^s,$$

with $p_R^s = 1 - (p_1^s + \ldots + p_{R-1}^s)$ and $E(X;\boldsymbol{p}^1,\ldots,\boldsymbol{p}^S) = \sum_{t=1}^{R}\sum_{s=1}^{S} x_t W_s p_t^s$.

Assume that a sample of size n is chosen at random and according to a stratified random sampling with allocation $n_1,\ldots,n_S$ with (or even without, in case M_s is large enough) replacement from each stratum and independently among strata. Let f_{nt}^s be the relative frequency of x_t in the sample from the sth stratum, then the results in Theorem 3.1 can be particularized by considering $\theta^s = p^s$ and $\widehat{\theta}_n = f_n^s = (f_{n1}^s, \ldots, f_{n(R-1)}^s)$. Denote $w_s = n_s/n$ for $s \in \{1, \ldots, S\}$. In this way, in the stratified random sampling from the final population, one can ensure that

Theorem 3.2 *In the stratified random sampling, if ϕ has finite derivatives of order 3, we have that*

(i) *$\{\Im_\phi(X; f_n^1, \ldots, f_n^S)\}_{n\in\mathbb{N}}$ is a sequence of estimators of $\Im_\phi(X; p_0^1, \ldots, p_0^S)$ which is strongly consistent whatever the true value $p_0^s = (p_{10}^s, \ldots, p_{(R-1)0}^s)$ of $p^1, \ldots,$ $p^S \in (0,1)^{R-1}$ may be.*

(ii) *The sequence of random variables*

$$\left\{\sqrt{n}[\Im_\phi(X; f_n^1, \ldots, f_n^S) - \Im_\phi(X; p_0^1, \ldots, p_0^S)]\right\}_{n\in\mathbb{N}}$$

converges in law as $n_1 \to \infty, \ldots, n_S \to \infty$ to the one-dimensional normal distribution $\mathcal{N}(0, \sigma^2(p_0^1, \ldots, p_0^S))$, where

$$\sigma^2(p_0^1, \ldots, p_0^S) = \sum_{s=1}^S \frac{W_s}{w_s} \left\{ \sum_{t=1}^R p_{t0}^s \left[\phi\left(\frac{x_t}{E(X; p_0^1, \ldots, p_0^S)}\right) \right.\right.$$

$$-x_t \sum_{s^\circ=1}^S W_{s^\circ} \sum_{u=1}^R \frac{x_u}{(E(X; p_0^1, \ldots, p_0^S))^2} \cdot \phi'\left(\frac{x_u}{E(X; p_0^1, \ldots, p_0^S)}\right) \Bigg]^2$$

$$- \left\{ \sum_{t=1}^R p_{t0}^s \left[\phi\left(\frac{x_t}{E(X; p_0^1, \ldots, p_0^S)}\right) \right.\right.$$

$$\left.\left.\left. -x_t \sum_{s^\circ=1}^S W_{s^\circ} \sum_{u=1}^R p_{u0}^{s^\circ} \frac{x_u}{(E(X; p_0))^2} \cdot \phi'\left(\frac{x_u}{E(X; p_0^1, \ldots, p_0^S)}\right) \right] \right\}^2 \right\}$$

whenever $\sigma^2(p_0^1, \ldots, p_0^S) > 0$.

(iii) *If $\sigma^2(p_0^1, \ldots, p_0^S) > 0$ and for some $i, j \in \{1, \ldots, R-1\}$*

$$H_{ij}^{ss^\circ} = \frac{W_s W_{s^\circ}(x_i - x_R)}{E(X; p_0^1, \ldots, p_0^S)} \left\{ \phi'\left(\frac{x_R}{E(X; p_0^1, \ldots, p_0^S)}\right) \cdot \frac{x_R}{E(X; p_0^1, \ldots, p_0^S)} \right.$$

$$-\phi'\left(\frac{x_j}{E(X; p_0^1, \ldots, p_0^S)}\right) \cdot \frac{x_j}{E(X; p_0^1, \ldots, p_0^S)}$$

$$\left. +\frac{x_j - x_R}{E(X; p_0^1, \ldots, p_0^S)} \sum_{s^{\circ\circ}=1}^S W_{s^{\circ\circ}} \sum_{t=1}^R p_{t0}^{s^{\circ\circ}} \phi''\left(\frac{x_t}{E(X; p_0^1, \ldots, p_0^S)}\right) \cdot \frac{x_t^2}{(E(X; p_0^1, \ldots, p_0^S))^2} \right\}$$

$$+ \frac{W_S W_{s^\circ}(x_j - x_R)}{E(X; p_0^1, \ldots, p_0^S)} \left\{ \phi'\left(\frac{x_R}{E(X; p_0^1, \ldots, p_0^S)}\right) \cdot \frac{x_R}{E(X; p_0^1, \ldots, p_0^S)} \right.$$

$$-2\phi\left(\frac{x_R}{E(X; p_0^1, \ldots, p_0^S))}\right)$$

$$\left. -\phi'\left(\frac{x_i}{E(X; p_0^1, \ldots, p_0^S)}\right) \cdot \frac{x_i}{E(X; p_0^1, \ldots, p_0^S))} + 2\phi\left(\frac{x_i}{E(X; p_0^1, \ldots, p_0^S)}\right) \right\} > 0,$$

then the sequence of random variables

$$\left\{ 2n[\mathfrak{I}_\phi(X; f_n^1, \ldots, f_n^S) - \mathfrak{I}_\phi(X; p_0^1, \ldots, p_0^S)] \right\}_{n \in \mathbb{N}}$$

converges in law as $n_1 \to \infty, \ldots, n_S \to \infty$ to a linear combination of at most $S(R-1)$ independent chi-square variables with 1 degree of freedom.

Consequently,

Corollary 3.2 *Under the situation and assumptions in Theorem 3.2, we have that the sequence of random variables*

$$\left\{ \frac{\sqrt{n}[\mathfrak{I}_\phi(X; f_n^1, \ldots, f_n^S) - \mathfrak{I}_\phi(X; p_0^1, \ldots, p_0^S)]}{\sqrt{\sigma^2(f_n)}} \right\}_{n \in \mathbb{N}}$$

converges in law as $n_1 \to \infty, \ldots, n_S \to \infty$ to the one-dimensional standard normal distribution, whenever $\sigma^2(f_n^1, \ldots, f_n^S) > 0$ and $\sigma^2(p_0^1, \ldots, p_0^S) > 0$.

4 Concluding Remarks

Results in this paper could be applied in future research for inferential developments.

In Alonso et al. [3] indices in this paper were extended to deal with fuzzy-valued random elements. Some of their properties were examined, but asymptotic properties remain as an open problem to be discussed.

References

1. Aczél J, Kannappan PL (1978) A mixed theory of information. III. Inset entropies of degree β. Inf Control 39:315–322
2. Ali SM, Silvey SD (1966) A general class of coefficients of divergence of one distribution from another. J R Stat Soc Ser B 28:131–142
3. Alonso MC, Brezmes T, Lubiano MA, Bertoluzza C (2001) A generalized real-valued measure of the inequality associated with a fuzzy random variable. Int J Approx Reason 26:47–66

4. Bourguignon F (1979) Decomposable inequality measures. Econometrica 47:901–920
5. Cowell FA (1980) Generalized entropy and the measurement of distributional change. Eur Econ Rev 13:147–159
6. Csiszár I (1963) Eine informationstheoretische ungleichung und ihre Aanwendung auf den Bbeweis der ergodizität von Markhoffschen ketten. Publ Math Inst Hungar Acad Sci 8:85–108
7. Csiszár I (1967) Information-type measures of difference of probability distributions and indirect observations. Studia Sci Math Hungar 2:299–318
8. Gil MA, Pérez R, Gil P (1989) A family of measures of uncertainty involving utilities: definition, properties, applications and statistical inferences. Metrika 36:129–147
9. Magdalou B, Nock R (2011) Income distributions and decomposable divergence measures. J Econ Theory 146:2440–2454
10. Oczki J, Wędrowska E (2014) The use of Csiszár's divergence to assess dissimilarities of income distributions of EU countries. Quant Meth Econ XV(2):167–176
11. Shorrocks AF (1980) The class of additively decomposable inequality measures. Econometrica 48:613–626
12. Taneja IJ, Pardo L, Morales D, Menéndez ML (1989) On generalized information and divergence measures and their applications: a brief review. Qüestiió 13:47–73
13. Theil H (1967) Economics and information theory. North-Holland, Amsterdam

A Logistic Regression Analysis Approach for Sample Survey Data Based on Phi-Divergence Measures

Elena Castilla, Nirian Martín and Leandro Pardo

Abstract A new family of minimum distance estimators for binary logistic regression models based on ϕ-divergence measures is introduced. The so called "pseudo minimum phi-divergence estimator"(PMϕE) family is presented as an extension of "minimum phi-divergence estimator" (MϕE) for general sample survey designs and contains, as a particular case, the pseudo maximum likelihood estimator (PMLE) considered in Roberts et al. (Biometrika 74:1–12, [8]). Through a simulation study it is shown that some PMϕEs have a better behaviour, in terms of efficiency, than the PMLE.

1 Introduction

Suppose that the population of interest is partitioned into I cells or domains according to the levels of one or more factors. Let N_i ($i = 1, ..., I$) denote the ith domain size, $N = \sum_{i=1}^{I} N_i$ the population domain total and N_{i1}, the population counts, out of N_i, where the binary response (0 for failure and 1 for success) variable is equal to 1. Since N_{i1} and N_i are fixed but unknown values ($i = 1, ..., I$), $\widehat{N}_i$ denotes the survey estimator of the ith domain size N_i and $\widehat{N}_{i1}$ the corresponding estimate of the successful events N_{i1}. The ratio estimator $\widehat{p}_i = \widehat{N}_{i1}/\widehat{N}_i$, $i = 1, ..., I$, is often used to estimate the population proportion of successful events, $\pi_i = \frac{N_{i1}}{N_i}$, $i = 1, ..., I$. Standard sampling theory provides an estimator of the covariance matrix of the

---

I had the great honor to have Pedro as a friend; a deeply beloved friend. We have shared not only many scientific interests in connection with Statistical Information Theory but also, in accordance with our frequent conversations, many similarities in our youthful life experiences. The research in this paper, written with two coauthors, is dedicated in memoriam to Professor Pedro Gil (Leandro).

E. Castilla · N. Martín · L. Pardo (✉)
Complutense University of Madrid, 28040 Madrid, Spain
e-mail: lpardo@mat.ucm.es

N. Martín
Universidad Carlos III de Madrid, 28903 Madrid, Getafe, Spain
e-mail: nirian@estad.ucm.es

© Springer International Publishing AG 2018
E. Gil et al. (eds.), *The Mathematics of the Uncertain*, Studies in Systems, Decision and Control 142, https://doi.org/10.1007/978-3-319-73848-2_43

465

$\widehat{p} = (\widehat{p}_1, ..., \widehat{p}_I)^T$. Another choice is using the logistic regression

$$\pi\left(x_i^T \beta\right) = \frac{\exp\{x_i^T \beta\}}{1 + \exp\{x_i^T \beta\}} = \frac{\exp\left\{\beta_0 + \sum\limits_{s=1}^{k} \beta_j x_{ij}\right\}}{1 + \exp\left\{\beta_0 + \sum\limits_{s=1}^{k} \beta_j x_{ij}\right\}}, \quad i = 1, ..., I, \quad (1)$$

to modelize the population proportion of successful events,

$$\pi_i = \pi\left(x_i^T \beta\right) = \frac{N_{i1}}{N_i},$$

which is assumed to depend on constants $x_{ij}, j = 1, ..., k$ $(k < I)$ derived from the factor levels, summarized in a $(k+1)$-vector of known constants $x_i = (1, x_{i1}, ..., x_{ik})^T$, and also on a $(k+1)$-vector of parameters $\beta = (\beta_0, \beta_1, ..., \beta_k)^T$.

Under independent binomial sampling in each domain, it is well-known that the maximum likelihood estimator (MLE) of β, $\widehat{\beta}$, is obtained through iterative calculations from the following likelihood equations

$$X^T \text{diag}(n)\pi\left(\beta\right) = X^T \text{diag}(n)\widehat{q}, \quad (2)$$

where $X = (x_1, ..., x_I)^T$ is a full rank matrix, $\pi\left(\beta\right) = \left(\pi\left(x_1^T \beta\right), ..., \pi\left(x_I^T \beta\right)\right)^T$, $\widehat{q} = (\widehat{q}_1, ..., \widehat{q}_I)^T$ with $\widehat{q}_i = n_{i1}/n_i$, n_i being the sample size from the ith domain, $n = \sum_{i=1}^{I} n_i$ the ith sample domain total and n_{i1} the sample total of successful events the ith domain. If we consider the probability vectors

$$\widehat{p}^* = \left(\frac{n_1}{n}\widehat{q}_1, \frac{n_1}{n}\left(1 - \widehat{q}_1\right), ..., \frac{n_I}{n}\widehat{q}_I, \frac{n_I}{n}\left(1 - \widehat{q}_I\right)\right)^T$$

$$= \left(\frac{n_{11}}{n}, \frac{n_1 - n_{11}}{n}, ..., \frac{n_{I1}}{n}, \frac{n_I - n_{I1}}{n}\right)^T$$

$$= \left(\frac{n_{11}}{n}, \frac{n_{12}}{n}, ..., \frac{n_{I1}}{n}, \frac{n_{I2}}{n}\right)^T, \quad (n_{i2} = n_i - n_{i1}),$$

and

$$p^*(\beta) = \left(\frac{n_1}{n}\pi\left(x_1^T \beta\right), \frac{n_1}{n}\left(1 - \pi\left(x_1^T \beta\right)\right), ..., \frac{n_I}{n}\pi\left(x_I^T \beta\right), \frac{n_I}{n}\left(1 - \pi\left(x_I^T \beta\right)\right)\right)^T$$

the MLE of β, $\widehat{\beta}$, can be equivalently defined by

$$\widehat{\beta} = \arg\min_{\beta \in \mathbb{R}^{k+1}} d_{Kullback}\left(\widehat{p}^*, p^*\left(\beta\right)\right),$$

where $d_{Kullback}\left(\widehat{p}^*, p^*\left(\beta\right)\right)$ is the Kullback-Leibler divergence between the probability vectors $\widehat{p}^*$ and $p^*\left(\beta\right)$ defined by

$$d_{Kullback}\left(\widehat{\boldsymbol{p}}^*, \boldsymbol{p}^*(\boldsymbol{\beta})\right) = \sum_{i=1}^{I} \left[\frac{n_{i1}}{n} \log \frac{n_{i1}}{n_i \pi\left(\boldsymbol{x}_i^T \boldsymbol{\beta}\right)} + \frac{n_{i2}}{n} \log \frac{n_{i2}}{n_i\left(1 - \pi\left(\boldsymbol{x}_i^T \boldsymbol{\beta}\right)\right)}\right].$$

In Pardo et al. [5] the minimum phi-divergence estimator (MϕE) was introduced, as a natural extension of the MLE, as

$$\widehat{\boldsymbol{\beta}}_\phi = \arg \min_{\boldsymbol{\beta} \in \mathbb{R}^{k+1}} d_\phi\left(\widehat{\boldsymbol{p}}^*, \boldsymbol{p}^*(\boldsymbol{\beta})\right), \tag{3}$$

where $d_\phi\left(\widehat{\boldsymbol{p}}^*, \boldsymbol{p}^*(\boldsymbol{\beta})\right)$ is the phi-divergence measure between the probability vectors $\widehat{\boldsymbol{p}}^*$ and $\boldsymbol{p}^*(\boldsymbol{\beta})$ given by

$$d_\phi\left(\widehat{\boldsymbol{p}}^*, \boldsymbol{p}^*(\boldsymbol{\beta})\right) = \sum_{i=1}^{I} \frac{n_i}{n} \left[\pi\left(\boldsymbol{x}_i, \boldsymbol{\beta}\right) \phi\left(\frac{n_{i1}}{n_i \pi(\boldsymbol{x}_i^T \boldsymbol{\beta})}\right) (1 - \pi\left(\boldsymbol{x}_i, \boldsymbol{\beta}\right)) \phi\left(\frac{n_{i2}}{n_i\left(1 - \pi(\boldsymbol{x}_i^T \boldsymbol{\beta})\right)}\right)\right],$$

with $\phi \in \Phi^*$. By Φ^* we are denoting the class of all convex functions, $\phi(x), x > 0$, such that at $x = 1, \phi(1) = \phi'(1) = 0$, and at $x = 0, 0\phi(0/0) = 0$ and $0\phi(p/0) = p\lim_{u\to\infty} \frac{\phi(u)}{u}$. For every $\phi \in \Phi^*$, differentiable at $x = 1$, the function

$$\Psi(x) = \phi(x) - \phi'(1)(x - 1)$$

also belongs to Φ^*. Therefore, we have $d_\psi\left(\widehat{\boldsymbol{p}}^*, \boldsymbol{p}^*(\boldsymbol{\beta})\right) = d_\phi\left(\widehat{\boldsymbol{p}}^*, \boldsymbol{p}^*(\boldsymbol{\beta})\right)$ and ψ has the additional property that $\psi'(1) = 0$. Since the two divergence measures are equivalent, we can consider the set Φ^* to be equivalent to the set

$$\Phi = \Phi^* \cap \{\phi : \phi'(1) = 0\}.$$

For more details see Cressie and Pardo [3] and Pardo [6]. In what follows, we give our theoretical results for $\phi \in \Phi$, but often apply them to choices of functions in Φ^*.

An application of the MϕE in logistic regression can be seen in Pardo et al. [7]. For general sample survey designs we do not have maximum likelihood estimators due to difficulties in obtaining appropriate likelihood functions. Hence, it is a common practice to use a pseudo maximum likelihood estimator (PMLE) of $\boldsymbol{\beta}, \widehat{\boldsymbol{\beta}}_P$, obtained from (2) by replacing $n_i/n, i = 1, ..., I$, by the estimated domain relative size $w_i = \widehat{N}_i/\widehat{N}, i = 1, ..., I$, and the sample proportions $\widehat{q}_i = n_{i1}/n_i, i = 1, ..., I$, by the ratio estimate $\widehat{p}_i = \widehat{N}_{i1}/\widehat{N}_i, i = 1, ..., I$,

$$\boldsymbol{X}^T \text{diag}(\boldsymbol{w})\boldsymbol{\pi}(\boldsymbol{\beta}) = \boldsymbol{X}^T \text{diag}(\boldsymbol{w})\widehat{\boldsymbol{p}}, \tag{4}$$

where $\boldsymbol{w} = (w_1, ..., w_I)^T$.

In this paper we extend the concept of MϕE by considering the "pseudo minimum phi-divergence estimator" (PMϕE) as a natural extension of the PMLE and we solve some statistical problem for the model considered in (1). In Sect. 2 we shall introduce the PMϕE for general sample designs and we study its asymptotic behavior. A numerical example is presented in Sect. 3 and, finally, in Sect. 4 a simulation study is carried out.

2 Pseudo Minimum Phi-Divergence Estimator for General Sample Designs

For general sample designs, we should consider the kernel of the weighted loglikelihood

$$\ell_w\left(\boldsymbol{\beta}\right) = n \sum_{i=1}^{I} w_i \left[\widehat{p}_i \log \pi\left(\boldsymbol{x}_i^T \boldsymbol{\beta}\right) + (1 - \widehat{p}_i) \log(1 - \pi\left(\boldsymbol{x}_i^T \boldsymbol{\beta}\right))\right],$$

which is derived from the kernel of the likelihood for I independent binomial random variables

$$\ell\left(\boldsymbol{\beta}\right) = \sum_{i=1}^{I} n_i \left[\widehat{q}_i \log \pi\left(\boldsymbol{x}_i^T \boldsymbol{\beta}\right) + (1 - \widehat{q}_i) \log(1 - \pi\left(\boldsymbol{x}_i^T \boldsymbol{\beta}\right))\right]$$

$$= n \sum_{i=1}^{I} \frac{n_i}{n} \left[\widehat{q}_i \log \pi\left(\boldsymbol{x}_i^T \boldsymbol{\beta}\right) + (1 - \widehat{q}_i) \log(1 - \pi\left(\boldsymbol{x}_i^T \boldsymbol{\beta}\right))\right],$$

replacing $\frac{n_i}{n}$ by $w_i = \widehat{N}_i/\widehat{N}$, and $\widehat{q}_i = n_{i1}/n_i$ by $\widehat{p}_i = \widehat{N}_{i1}/\widehat{N}_i$, $i = 1, ..., I$. If we consider the two probability vectors

$$\widehat{\boldsymbol{p}}_w = (w_1 \widehat{p}_1, w_1 (1 - \widehat{p}_1), ..., w_I \widehat{p}_I, w_I (1 - \widehat{p}_I))^T$$

and

$$\boldsymbol{p}_w(\boldsymbol{\beta}) = \left(w_1 \pi\left(\boldsymbol{x}_1^T \boldsymbol{\beta}\right), w_1 \left(1 - \pi\left(\boldsymbol{x}_1^T \boldsymbol{\beta}\right)\right), ..., w_I \pi\left(\boldsymbol{x}_I^T \boldsymbol{\beta}\right), w_I \left(1 - \pi\left(\boldsymbol{x}_I^T \boldsymbol{\beta}\right)\right)\right)^T,$$

we get

$$\ell_w\left(\boldsymbol{\beta}\right) = -n d_{Kullback}\left(\widehat{\boldsymbol{p}}_w, \boldsymbol{p}_w\left(\boldsymbol{\beta}\right)\right) + k,$$

where k is a constant not depending on $\boldsymbol{\beta}$. Therefore the PMLE of $\boldsymbol{\beta}$, $\widehat{\boldsymbol{\beta}}_P$, presented in (4) can be defined as

$$\widehat{\boldsymbol{\beta}}_P = \arg \max_{\boldsymbol{\beta} \in \mathbb{R}^{k+1}} \ell_w\left(\boldsymbol{\beta}\right) = \arg \min_{\boldsymbol{\beta} \in \mathbb{R}^{k+1}} d_{Kullback}\left(\widehat{\boldsymbol{p}}_w, \boldsymbol{p}_w\left(\boldsymbol{\beta}\right)\right).$$

Based on the previous interpretation of the PMLE, in the following definition we shall present the PMϕE.

Definition 2.1 The PMϕE in a general sample design for the parameter β in the model considered in (1) is defined as

$$\widehat{\beta}_{\phi,P} = \arg \min_{\beta \in \mathbb{R}^{k+1}} d_\phi \left(\widehat{p}_w, p_w(\beta) \right),$$

where

$$d_\phi \left(\widehat{p}_w, p_w(\beta) \right) = \sum_{i=1}^{I} w_i \left[\pi(x_i, \beta) \phi \left(\frac{\widehat{p}_i}{\pi(x_i^T \beta)} \right) (1 - \pi(x_i, \beta)) \phi \left(\frac{1 - \widehat{p}_i}{1 - \pi(x_i^T \beta)} \right) \right]$$

is the phi-divergence measure between the probability vectors $\widehat{p}_w$ and $p_w(\beta)$.

The following result establishes the asymptotic distribution of the PMϕE of β, $\widehat{\beta}_{\phi,P}$.

Theorem 2.1 *Let us assume that β_0 is the true value of β and*

$$w \xrightarrow[n \to \infty]{P} W, \quad W = (W_1, ..., W_I)^T, \quad W_i = \frac{N_i}{N},$$

$$\widehat{p} \xrightarrow[n \to \infty]{P} \pi(\beta_0), \quad \sqrt{n}(\widehat{p} - \pi(\beta_0)) \xrightarrow[n \to \infty]{\mathcal{L}} \mathcal{N}(0, V).$$

Then, we have

$$\sqrt{n}(\widehat{\beta}_{\phi,P} - \beta_0) \xrightarrow[n \to \infty]{\mathcal{L}} \mathcal{N}\left(0_{k+1}, V(\beta_0)\right),$$

where

$$V(\beta_0) = (X^T \Delta X)^{-1} X^T \operatorname{diag}(W) V \operatorname{diag}(W) X (X^T \Delta X)^{-1}, \tag{5}$$

$$\Delta = \operatorname{diag}\{W_i \pi(x_i^T \beta_0)(1 - \pi(x_i^T \beta_0))\}_{i=1,...,I}.$$

Proof Based on Theorem 1 in Pardo et al. [5], we have

$$\widehat{\beta}_{\phi,P} = \beta_0 + (X^T \Delta X)^{-1} X^T \operatorname{diag}\left\{c_i^T\right\}_{i=1}^{I} \operatorname{diag}^{-1/2}\left(p_w(\beta_0)\right)(\widehat{p}_w - p_w(\beta_0))$$

$$+ o\left(\left\|\operatorname{diag}\left\{c_i^T\right\}_{i=1}^{I} \operatorname{diag}^{-1/2}\left(p_w(\beta_0)\right)(\widehat{p}_w - p_w(\beta_0))\right\| 1_{k+1}\right),$$

with

$$c_i = \left(w_i \pi(x_i^T \beta_0)(1 - \pi(x_i^T \beta_0))\right)^{1/2} \begin{pmatrix} (1 - \pi(x_i^T \beta_0))^{1/2} \\ -\pi(x_i^T \beta_0)^{1/2} \end{pmatrix}, \quad i = 1, .., I.$$

Since

$$\operatorname{diag}\left\{c_i^T\right\}_{i=1}^{I} \operatorname{diag}^{-1/2}\left(p_w(\beta_0)\right)(\widehat{p}_w - p_w(\beta_0)) = \operatorname{diag}(w)(\widehat{p} - \pi(\beta_0)),$$

$w \xrightarrow[n\to\infty]{p} W$ and $\widehat{p} \xrightarrow[n\to\infty]{p} \pi(\beta_0)$, it holds

$$\sqrt{n}(\widehat{\beta}_{\phi,P} - \beta_0) = (X^T \Delta X)^{-1} X^T diag(W) \sqrt{n}(\widehat{p} - \pi(\beta_0)) + o_p(\mathbf{1}_{k+1}).$$

From the Sluysky's theorem and taking into account $\sqrt{n}(\widehat{p} - \pi(\beta_0)) \xrightarrow[n\to\infty]{\mathscr{L}} \mathscr{N}$ $(\mathbf{0}_{k+1}, V)$, it follows the desired result. $\qquad\square$

Remark 2.1 Under independent binomial sampling in each domain, it is well-known that $V = \text{diag}\{\pi(x_i^T \beta_0)(1 - \pi(x_i^T \beta_0))\}_{i=1,...,I}\text{diag}^{-1}(W)$ and hence $V(\beta_0) = (X^T \Delta X)^{-1}$, which matches Theorem 2 in Pardo et al. [5].

Remark 2.2 The asymptotic results obtained in the current paper differ from Castilla et al. [2] in the elements tending to infinite, here the total individuals in the whole sample, n, while in the cited paper is the total number of clusters what tends to infinite.

3 A Numerical Example

In order to obtain the PMϕEs, from a practical point of view, we can give an explicit expression for ϕ. In this paper we shall focus on the Cressie-Read subfamily

$$\phi_\lambda(x) = \begin{cases} \frac{1}{\lambda(1+\lambda)}\left[x^{\lambda+1} - x - \lambda(x - 1)\right], & \lambda \in \mathbb{R} - \{-1, 0\} \\ \lim_{\upsilon\to\lambda} \frac{1}{\upsilon(1+\upsilon)}\left[x^{\upsilon+1} - x - \upsilon(x - 1)\right], & \lambda \in \{-1, 0\} \end{cases}.$$

We can observe that for $\lambda = 0$, we have

$$\phi_{\lambda=0}(x) = \lim_{\upsilon\to 0} \frac{1}{\upsilon(1+\upsilon)}\left[x^{\upsilon+1} - x - \upsilon(x - 1)\right] = x\log x - x + 1,$$

and the associated phi-divergence, coincides with the Kullback divergence, therefore the PMϕEs based on $\phi_\lambda(x)$ contains as special case the PMLE.

We shall consider the example presented in Molina and Skinner [4]. A random subsample of 50 clusters (primary sampling units) containing 1299 households was selected from the 1975 U.K. Family Expenditure Survey. These households are divided into 12 groups of sizes $n_1, ..., n_{12}$ by age of head of household (4 levels) and number of persons in the household (3 levels). The binary response is 1 if the household owns the dwelling it occupies and 0 otherwise. The number of households r_i for which the binary response is 1, together with n_i are shown in Table 1 of the cited paper.

Table 1 PMCREs for the clustered family expenditure survey data model

λ	$\widehat{\beta}_{0,\lambda,P}$	$\widehat{\beta}_{1(2),\lambda,P}$	$\widehat{\beta}_{1(3),\lambda,P}$	$\widehat{\beta}_{1(4),\lambda,P}$	$\widehat{\beta}_{2(2),\lambda,P}$	$\widehat{\beta}_{2(3),\lambda,P}$
0	−0.1585	0.4403	−0.1412	−0.4179	0.5042	0.4703
2/3	−0.1564	0.4291	−0.1436	−0.4174	0.4985	0.4735
1	−0.1574	0.4251	−0.1438	−0.4158	0.4971	0.476
2	−0.1663	0.4192	−0.1408	−0.4075	0.4974	0.4856

We denote by $\beta_{1(r)}$ the parameter associated to the level r of the factor "age of head of housholds", $r = 2, 3$ and 4 since $\beta_{1(1)} = 0$ and by $\beta_{2(s)}$ the parameter associated to the level s of the factor "number of persons in the housholds", $s = 2, 3$, since we assume $\beta_{2(1)} = 0$. The parameter vector with unknown values will be denote by

$$\beta = (\beta_0, \beta_{1(2)}, \beta_{1(3)}, \beta_{1(4)}, \beta_{2(2)}, \beta_{2(3)})^T.$$

The design matrix that we are going to consider for the example under consideration is given by

$$X = \begin{pmatrix} 1 & 1 & 1 & 1 & 1 & 1 & 1 & 1 & 1 & 1 & 1 & 1 \\ 0 & 0 & 0 & 1 & 1 & 1 & 0 & 0 & 0 & 0 & 0 & 0 \\ 0 & 0 & 0 & 0 & 0 & 0 & 1 & 1 & 1 & 0 & 0 & 0 \\ 0 & 0 & 0 & 0 & 0 & 0 & 0 & 0 & 0 & 1 & 1 & 1 \\ 0 & 1 & 0 & 0 & 1 & 0 & 0 & 1 & 0 & 0 & 1 & 0 \\ 0 & 0 & 1 & 0 & 0 & 1 & 0 & 0 & 1 & 0 & 0 & 1 \end{pmatrix}^T = (x_1, ..., x_{12})^T$$

and the logistic regression model under consideration is given by

$$\pi\left(x_i^T \beta\right) = \frac{\exp\left\{x_i^T \beta\right\}}{1 + \exp\left\{x_i^T \beta\right\}}, \quad i = 1, ..., 12,$$

equivalent to

$$\pi\left(x_i^T \beta\right) = \frac{\exp\{\beta_0 + \beta_{1(r)} + \beta_{2(s)}\}}{1 + \exp\{\beta_0 + \beta_{1(r)} + \beta_{2(s)}\}},$$

if the ith probability is associated with the rth level of the first variable ($r = 1, ..., 4$) and the sth level of the second variable ($s = 1, ..., 3$). In the following table we present the pseudo minimum Cressie-Read divergence estimators (PMCREs) of β, $\beta_{\lambda,P}$, for $\lambda \in \{0, 2/3, 1, 2\}$.

4　Simulation Study

The following simulation study has been designed by following the previous example. Since in the logistic regression model there are two factors, the first one with 4 categories and the second one with 3 categories, in total $I = 12$ domains are taken into account. Let

$$p\left(\beta\right) = \left(\tfrac{N_1}{N}\pi\left(x_1^T\beta\right), \tfrac{N_1}{N}\left(1 - \pi\left(x_1^T\beta\right)\right), ..., \tfrac{N_I}{N}\pi\left(x_I^T\beta\right), \tfrac{N_I}{N}\left(1 - \pi\left(x_I^T\beta\right)\right)\right)^T$$

be the theoretical probability vector in the logistic regression with complex sampling. The values of the components of $p\left(\beta\right)$ in which the simulation is based are given in Table 2. In total $n = 1299$ individuals are taken from the primary units of the sample, $J = 50$ clusters, of size $m_{(j)} = 26, j = 1, ..., 49, m_{(50)} = 25 \left(\sum_{j=1}^{50} m_{(j)} = n\right)$. Since the clusters are mutually independent and there is (possibly) correlation inside each cluster, we consider three possible distributions for

$$\left(n_{11(j)}, n_{1(j)} - n_{11(j)}, n_{21(j)}, n_{2(j)} - n_{21(j)}, ..., n_{12,1(j)}, n_{12(j)} - n_{12,1(j)}\right)^T$$

Table 2 Theoretical values of $p(\beta)$ in the simulation study

i	1	2	3	4	5	6	7	8	9	10	11	$I = 12$
$\frac{N_i}{N}$	$\frac{10}{1299}$	$\frac{63}{1299}$	$\frac{110}{1299}$	$\frac{14}{1299}$	$\frac{35}{1299}$	$\frac{281}{1299}$	$\frac{40}{1299}$	$\frac{110}{1299}$	$\frac{185}{1299}$	$\frac{204}{1299}$	$\frac{196}{1299}$	$\frac{51}{1299}$
$\pi(x_i^T\beta)$	$\frac{2}{10}$	$\frac{38}{63}$	$\frac{65}{110}$	$\frac{6}{14}$	$\frac{29}{35}$	$\frac{188}{281}$	$\frac{17}{40}$	$\frac{56}{110}$	$\frac{105}{185}$	$\frac{78}{204}$	$\frac{93}{196}$	$\frac{21}{51}$
$\frac{N_i}{N}\pi(x_i^T\beta)$	$\frac{2}{1299}$	$\frac{38}{1299}$	$\frac{65}{1299}$	$\frac{6}{1299}$	$\frac{29}{1299}$	$\frac{188}{1299}$	$\frac{17}{1299}$	$\frac{56}{1299}$	$\frac{105}{1299}$	$\frac{78}{1299}$	$\frac{93}{1299}$	$\frac{21}{1299}$

Table 3 Scheme of a correlated sample generation through clusters

i	j	1	2	$\cdots$	j	$\cdots$	$J = 50$	sample
1	$k = 1$	$n_{11(1)}$	$n_{11(2)}$	$\cdots$		$\cdots$	$n_{11(50)}$	n_{11}
	$k = 2$	$n_{12(1)}$	$n_{12(2)}$				$n_{12(50)}$	n_{12}
2	$k = 1$	$n_{21(1)}$	$n_{21(2)}$	$\cdots$		$\cdots$	$n_{21(50)}$	n_{21}
	$k = 2$	$n_{22(1)}$	$n_{22(2)}$				$n_{22(50)}$	n_{22}
$\vdots$		$\vdots$	$\vdots$	$\ddots$			$\vdots$	$\vdots$
i	$k = 1$	$n_{i1(1)}$	$n_{i1(2)}$		$n_{i1(j)}$		$n_{i1(50)}$	n_{i1}
	$k = 2$	$n_{i2(1)}$	$n_{i2(2)}$		$n_{i2(j)}$		$n_{i2(50)}$	n_{i2}
$\vdots$		$\vdots$	$\vdots$			$\ddots$	$\vdots$	$\vdots$
$I = 12$	$k = 1$	$n_{12,1(1)}$	$n_{12,1(2)}$	$\cdots$		$\cdots$	$n_{12,1(50)}$	$n_{12,1}$
	$k = 2$	$n_{12,2(1)}$	$n_{12,2(2)}$				$n_{12,2(50)}$	$n_{12,2}$
		$m_{(1)}$	$m_{(2)}$	$\cdots$	$m_{(j)}$	$\cdots$	$m_{(50)}$	n

Fig. 1 RMSEs for PMCREs of β with Dirichlet-multinomial (above), Random-clumped (middle) and m-inflated (below) distributions

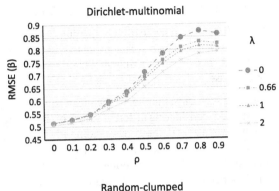

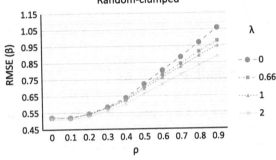

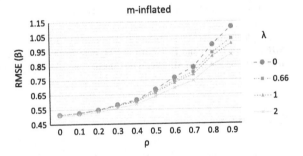

corresponding to the jth cluster (column, in Table 3), $j = 1, ..., J = 50$:

- Dirichlet-multinomial with parameters $(m_{(j)}; \rho, \boldsymbol{p}(\boldsymbol{\beta}))$, with $\rho \in \{\frac{1}{10}(i-1)\}_{i=1}^{10}$;
- Random-clumped with parameters $(m_{(j)}; \rho, \boldsymbol{p}(\boldsymbol{\beta}))$, with $\rho \in \{\frac{1}{10}(i-1)\}_{i=1}^{10}$;
- $m_{(j)}$-inflated with parameters $(m_{(j)}; \rho, \boldsymbol{p}(\boldsymbol{\beta}))$, with $\rho \in \{\frac{1}{10}(i-1)\}_{i=1}^{10}$.

For details about these distributions see Alonso et al. [1]. The values of interest for the sample are

$$n_{i1} = \sum_{j=1}^{50} n_{11(j)}, \quad i = 1, ..., I \quad \text{and} \quad n_i = \sum_{j=1}^{50} (n_{11(j)} + n_{12(j)}), \quad i = 1, ..., I.$$

Notice that the assumptions of Theorem 2.1 are held. In addition:

- If $\rho = 0$ (multinomial distribution within each cluster), then V is a diagonal matrix since the elements of $\widehat{p}$ are uncorrelated. In this case, we obtain MLEs and MϕEs.
- If $\rho > 0$, then V is not a diagonal matrix since the elements of $\widehat{p}$ are correlated. In this case, we obtain PMLEs and PMϕEs.

In these scenarios, the root of the mean square error (RMSE) for the PMCREs of β are studied, considering different values of the tuning parameter $\lambda \in \{0, 2/3, 1, 2\}$. Note that when $\lambda = 0$, the corresponding PMCRE of β is equal to the PMLE.

Results of the simulation study with 2,000 samples are shown in Fig. 1. As expected from a theoretical point of view, the RMSE increases as ρ increases. With independence to the distribution considered, estimators corresponding to $\lambda \in \{2/3, 1, 2\}$ present a better performance than the PMLE ($\lambda = 0$). This difference becomes more considerable for large values of ρ.

5 Concluding Remarks

In this paper we have considered the problem of estimating the parameters of the logistic regression model for sample survey data, introducing the family of the PMϕEs that contains as a particular case the PMLE. A simulation study is carried out in order to see that there are PMϕEs that have a better behaviour than the PMLE in relation to the mean square error.

Acknowledgements This research is partially supported by Grants MTM2015-67057-P (MINECO/FEDER) and ECO2015-66593, both from Ministerio de Economía y Competitividad and FPU16/03104 from Ministerio de Educación, Cultura y Deporte (Spain).

References

1. Alonso-Revenga JM, Martín N, Pardo L (2017) New improved estimators for overdispersion in models with clustered multinomial data and unequal cluster sizes. Stat Comput 27(1):193–217
2. Castilla E, Martín N, Pardo L (2017) Pseudo minimum phi-divergence estimator for multinomial logistic regression with complex sample design. Adv Stat Anal. https://doi.org/10.1007/s10182-017-0311-6
3. Cressie N, Pardo L (2003) Phi-divergence statistic. In: El-Shaarawi AH, Piegorsch WW (eds) Encyclopedia of environmetrics, vol 3. Wiley, New York
4. Molina EA, Skinner CJ (1992) Pseudo-likelihood and quasi-likelihood estimation for complex sampling schemes. Comput Stat Data Anal 13:395–405
5. Pardo JA, Pardo MC, Pardo L (2005) Minimum ϕ-divergence estimator in logistic regression models. Stat Pap 47:91–108
6. Pardo L (2006) Statistical inference based on divergence measures. Chapman and Hall/CRC, London
7. Pardo JA, Pardo MC, Pardo L (2006) Testing in logistic regression models based on ϕ-divergence measures. J Stat Plan Inference 136:982–1006
8. Roberts G, Rao JNK, Kumer S (1987) Logistic regression analysis of sample survey data. Biometrika 74:1–12

Some Multivariate Measures Based on Distances and Their Entropy Versions

Carles M. Cuadras and Sonia Salvo-Garrido

Abstract We study some properties of the geometric variability of a set in relation with a distance. We show that the Shannon entropy is the lower bound for the geometric variability. We prove that this quantity can be partitioned in presence of mixtures and give a version in terms of entropy. The association between two data sets is studied by means of distances for general data. A multivariate version of the intraclass correlation is given and related to the correlation ratio. We obtain an entropy version of the correlation ratio.

1 Introduction

The distances in statistics are currently used in multidimensional scaling, canonical variate analysis, principal components analysis, biplot, cluster analysis, correspondence analysis, multiple factor analysis, etc. Statistical distances are also useful in regression, discriminant analysis and multivariate analysis of variance, see [1, 2, 7]. Moreover, we can construct probability densities with distances.

Section 2 recalls classical metric scaling and the geometric variability (GV) of a finite set with respect to a given distance. When we consider a population with a probability density, it is shown that GV is bounded by the Shannon entropy. Section 3 shows how to split the GV into several parts when the observations come from $k \geq 2$ populations. An entropy version is also given. Section 4 provides a multivariate measure of association between two data sets, defined using distances. Section 5 proposes a generalization of the intraclass correlation and the correlation ratio. In both Sects. 4 and 5, we provide suitable versions in terms of entropy.

C. M. Cuadras (✉)
Universidad de Barcelona, Barcelona, Spain
e-mail: ccuadras@ub.edu

S. Salvo-Garrido
Universidad de La Frontera, Temuco, Chile
e-mail: sonia.salvo@ufrontera.cl

© Springer International Publishing AG 2018
E. Gil et al. (eds.), *The Mathematics of the Uncertain*, Studies in Systems, Decision and Control 142, https://doi.org/10.1007/978-3-319-73848-2_44

2 Geometric Variability Measuring Uncertainty

Let $\Omega = \{\omega_1, \ldots, \omega_n\}$ a set with n objects or individuals. Suppose that, following a suitable procedure, we can define, a dissimilarity or distance measure,

$$\delta : \Omega \times \Omega \to \mathbb{R}_+ \cup \{0\},$$

providing, for each pair ω_i, ω_j. the real quantity $\delta_{ij} = \delta(\omega_i, \omega_j)$ such that

$$\delta_{ij} = \delta_{ji} \geq \delta_{ii} = 0, \quad i, j \in \{1, \ldots, n\}.$$

We obtain a $n \times n$ distance matrix $\Delta = (\delta_{ij})$.

Let us suppose that Δ is a Euclidean distance matrix, i.e., we can find a configuration of points $\mathbf{x}_1, \ldots, \mathbf{x}_n \in \mathbb{R}^p$, with coordinates $\mathbf{x}_i = (x_{i1}, \ldots, x_{ip})'$, $i = 1, \ldots, n$, such that

$$\delta_{ij}^2 = (\mathbf{x}_i - \mathbf{x}_j)'(\mathbf{x}_i - \mathbf{x}_j), \quad i, j = 1, \ldots, n.$$

Thus the coordinates of the objects belonging to Ω provide a $n \times p$ matrix $\mathbf{X} = [x_{ij}]$, say, such that the Euclidean distance between each pair of rows i, j equals δ_{ij}.

In order to check that Δ is Euclidean, let $\mathbf{I}_n$ be the identity matrix, $\mathbf{1}_n$ the vector of ones and $\mathbf{H}_c = \mathbf{I}_n - n^{-1}\mathbf{1}_n\mathbf{1}_n'$ the centring matrix. Let $\mathbf{A} = -\frac{1}{2}\Delta^{(2)}$, $\mathbf{G} = \mathbf{H}_c\mathbf{A}\mathbf{H}_c$, where $\Delta^{(2)} = (\delta_{ij}^2)$. The matrix Δ is Euclidean if and only if the inner product matrix $\mathbf{G}$ is positive semidefinite. Then we can perform the spectral decomposition $\mathbf{G} = \mathbf{U}\Lambda^2\mathbf{U}$, with Λ diagonal, and obtain the matrix $\mathbf{X} = \mathbf{U}\Lambda$, which provide the coordinates. If the eigenvalues of $\mathbf{G}$ are in descending order, $\mathbf{X}$ contains the so-called principal coordinates of Ω with respect to δ.

The geometric variability (GV) of the distance δ is the average

$$V_\delta = \frac{1}{2n^2} \sum_{i,j=1}^n \delta_{ij}^2. \tag{1}$$

It can be proved that V_δ is related to the inner product matrix $\mathbf{G}$ by

$$V_\delta = \text{tr}(\mathbf{G})/n. \tag{2}$$

The GV V_δ is a measure of dispersion which can be interpreted as a measure of uncertainty. To see this, let $\mathbf{X}$ be a random vector with density $f(\mathbf{x})$, with respect to a suitable measure, and support $S \subset \mathbb{R}^p$. Let $\delta(\mathbf{x}, \mathbf{y})$ be a distance function between two observations of $\mathbf{X}$. If the expectation exists, the GV of $\mathbf{X}$ with respect to δ is defined by

$$V_\delta(\mathbf{X}) = \frac{1}{2}E_{X,X'}\left[\delta^2(\mathbf{X}, \mathbf{X}')\right],$$

where $\mathbf{X}'$ is iid as $\mathbf{X}$. Note that V_δ in (1) is a U-statistic for estimating $V_\delta(\mathbf{X})$. A variant of $V_\delta(\mathbf{X})$ is called diversity coefficient in [12].

Let us suppose that there exists a representation $\psi : S \to L$ of S in a Euclidean (or separable Hilbert) space L with inner product $\langle \cdot, \cdot \rangle$ and related norm $|| \cdot ||$, such that $\delta^2(\mathbf{x}, \mathbf{y}) = ||\psi(\mathbf{x}) - \psi(\mathbf{y})||^2$. Then it can be proved that

$$V_\delta(\mathbf{X}) = E||\psi(\mathbf{X})||^2 - ||E[\psi(\mathbf{X})]||^2, \tag{3}$$

which is formally similar to the variance, see [8].

Related to $V_\delta(\mathbf{X})$ is the proximity function from an observation $\mathbf{x}$ to the population represented by $\mathbf{X}$. It is defined by

$$\phi_\delta^2(\mathbf{x}) = E_X[\delta^2(\mathbf{x}, \mathbf{X})] - V_\delta(\mathbf{X}).$$

This proximity function and the GV performs well under affine transformations. If δ changes to $\widehat{\delta}$, where $\widetilde{\delta}^2(\mathbf{x}.\mathbf{y}) = a\delta^2(\mathbf{x}.\mathbf{y}') + b$, if $\mathbf{x} \neq \mathbf{y}$, then $V_{\widetilde{\delta}}(\mathbf{X}) = aV_\delta(\mathbf{X}) + b/2$ and $\phi_{\widehat{\delta}}^2(\mathbf{x}) = a\phi_\delta^2(\mathbf{x}) + b/2$. Thus we can consider suitable choices of a, b (if it is necessary) for generating the probability density

$$f_{\widehat{\delta}}(\mathbf{x}) = \exp\left[-\phi_{\widehat{\delta}}^2(\mathbf{x})\right]. \tag{4}$$

For example, if X has mean μ, variance σ^2, support $\mathbb{R}$ and $\delta^2(x, y) = (x - y)^2$, $x, y \in \mathbb{R}$, then $V_\delta(X) = \sigma^2$, $\phi_\delta^2(\mathbf{x}) = (x - \mu)^2$ and

$$f_{\widehat{\delta}}(x) = \exp\left[-a(x - \mu)^2 - b/2\right],$$

is the density of the normal $N(\mu, \sigma^2)$ distribution for $a = 1/(2\sigma^2), b = 2\ln(\sigma\sqrt{2\pi})$.

Therefore, we can consider a density f_δ generated by δ from (4). Let us compare f_δ to the true density f of $\mathbf{X}$. Using the Kullback-Leibler divergence, we have

$$\begin{aligned} I(f||f_\delta) &= E_X\{\ln[f(X)/f_\delta(X)]\} \\ &= -H(f) + E_X[\phi_\delta^2(\mathbf{X})] \\ &= V_\delta(\mathbf{X}) - H(f) \geq 0, \end{aligned}$$

where $H(f) = E_X[-\ln f(X)]$ is the Shannon entropy. Hence, if δ is a distance normalized to generate a density f_δ, then $V_\delta(\mathbf{X}) \geq H(f)$, and the Shannon entropy is the lower bound for the GV.

For example, if X has density f with support $\mathbb{R}$, then $H(f) \leq \ln(\sigma\sqrt{2\pi e})$, with equality if X is $N(\mu, \sigma^2)$. Thus, for any standardized distance δ generating f_δ, with GV $V_\delta(X)$, we have

$$H(f) \leq \min\{V_\delta(X), \ln(\sigma\sqrt{2\pi e})\}.$$

There is equality $V_\delta(X) = \ln(\sigma\sqrt{2\pi e})$ if the (squared) distance is $(x - y)^2/(2\sigma^2) + 2\ln(\sigma\sqrt{2\pi})$, as this distance generates the $N(\mu, \sigma^2)$ distribution.

Table 1 Distance-based generation of the normal, exponential and logistic distributions. In two cases the geometric variability GV attains the entropy, showing that the distance is suitable

Distribution and support	Squared distance	Proximity function	Generated density	Inequality $V_\delta \geq H(f)$
Any f, $\mathbb{R}$	$\pi(x-y)^2$	$\pi(x-\mu)^2$	$N(\mu, 1/2\pi)$	$\pi\sigma^2 \geq H(f)$
$N(\mu, \sigma^2)$, $\mathbb{R}$	$\pi(x-y)^2$	$\pi(x-\mu)^2$	$N(\mu, 1/2\pi)$	$\pi\sigma^2 \geq \ln(\sigma\sqrt{2\pi e})$
$N(\mu, \sigma^2)$, $\mathbb{R}$	$\frac{1}{2\sigma^2}(x-y)^2$ $+2\ln(\sigma\sqrt{2\pi})$	$\frac{1}{2\sigma^2}(x-\mu)^2$ $+\ln(\sigma\sqrt{2\pi})$	$N(\mu, \sigma^2)$	$V_\delta =$ $\ln(\sigma\sqrt{2\pi e})= H(f)$
$\alpha e^{-\alpha x}$, $\mathbb{R}_+$	$x+y$	x	e^{-x}	$\alpha^{-1} \geq 1 - \ln\alpha$
Log(s), $\mathbb{R}$	$\pi(x-y)^2$	πx^2	$N(0, 1/2\pi)$	$\pi^3 s^2/3 \geq 2 + \ln s$
Log(s), $\mathbb{R}$	$|x-y|/s +$ $2(1+\ln s)$	$x/s + \ln s +$ $2\ln(1+e^{-x/s})$	Log(s)	$V_\delta = 2 + \ln s =$ $H(f)$
$N(\mu, \Sigma)$, $\mathbb{R}^p$	$\pi\|\mathbf{x}-\mathbf{y}\|^2$	$\pi\|\mathbf{x}-\mu\|^2$	$N(\mu, \frac{1}{2\pi}I_p)$	$\pi\,\text{tr}(\Sigma) \geq$ $\frac{1}{2}\ln\left[(2\pi e)^p |\Sigma|\right]$

Clearly, $V_\delta(\mathbf{X}) = H(f)$ if $f = f_\delta$ (a.e.). This always occurs if we take the symmetric function $s(\mathbf{x}, \mathbf{y}) = -\ln f(\mathbf{x}) - \ln f(\mathbf{y})$ as "dissimilarity". Indeed, we can get $f = f_\delta$ with other distances, such as $|x-y|$ and $\sqrt{|x-y|}$. See Table 1, where Log(s) stands for the logistic density $e^{-x/s}/(1 + e^{-x/s})^2$.

To simplify the construction of f_δ, we can understand the entropy as the average of the number of bits, i.e., $H_2(f) = E_X[-\log_2 f(\mathbf{X})]$, which is proportional to $H(f)$. The basis is 2 rather than e. Thus, after considering a distance δ and computing the proximity function, we can take a suitable basis c providing a density and modify the entropy. This avoid the affine transformation of δ^2 and proves the following result, where $H_c(f) = H(f)/\ln c$.

Proposition 2.1 *Let* $\mathbf{X}$ *be a random vector with density* f *and support* S. *Let* $V_\delta(\mathbf{X})$ *and* $\phi_\delta^2(\mathbf{x})$ *be the geometric variability and the proximity function related to a distance* δ. *Suppose that* c *is a positive constant such that* $f_\delta(\mathbf{x}) = c^{-\phi_\delta^2(\mathbf{x})}$, $\mathbf{x} \in S$, *is a probability density. If* $H_c(f) = E_X[-\log_c f(\mathbf{X})]$ *is finite, then*

$$V_\delta(\mathbf{X}) \geq H_c(f),$$

with equality if $f_\delta = f$ *(a.e.).*

However, in general, a proper distance providing the true density may not exist, see [4, 8].

3 Partitioning the Uncertainty with Mixtures

Let us consider $n = n_1 + \cdots + n_k$ univariate observations of a variable. The ANOVA identity is $T = B + W$, where T is the total sum of squares. Clearly,

$$\frac{1}{n}T = \frac{1}{n}B + \sum_{i=1}^{k}\left(\frac{n_i}{n}\right)\frac{1}{n_i}W_i, \tag{5}$$

where $W_i = \sum_{h=1}^{n_i}(x_{ih} - x_{i\cdot})^2$, which satisfies $n_i^{-1}W_i = (2n_i^2)^{-1}\sum_{h,h'=1}^{n_i}(x_{ih} - x_{ih'})^2$ and similarly for $n^{-1}T$. This is a particular case of (6).

Suppose that $\mathbf{X}$ is the result of observing $\mathbf{X}_i$ with probability w_i, $i = 1, \ldots, k$. In other words, we consider the mixture $f = w_1f_1 + \cdots + w_kf_k$ concerning k densities with the same support S. Assuming that the above representation $\psi : S \to L$ exists, the GV with respect to a distance δ is given by

$$V_\delta(\mathbf{X}) = V(\mu_1, \ldots, \mu_k) + \sum_{i=1}^{k} w_i V_i, \tag{6}$$

where

$$V(\mu_1, \ldots, \mu_g) = \frac{1}{2}\sum_{i,j=1}^{k} w_i\delta^2(\mu_i, \mu_j)w_j = \sum_{i=1}^{k} w_i\delta^2(\mu_i, \mu),$$

with $\mu_i = E[\psi(\mathbf{X}_i)]$, $\delta^2(\mu_i, \mu_j) = \|\mu_i - \mu_j\|^2$, $\mu = w_1\mu_1 + \cdots + w_k\mu_k$, and $V_i = \frac{1}{2}E_{\mathbf{X}_i, \mathbf{X}_i'}[\delta^2(\mathbf{X}_i, \mathbf{X}_i')]$, where $\mathbf{X}_i, \mathbf{X}_i'$ are i.i.d. with density $f_i(\mathbf{x})$. We can interpret $V_\delta(\mathbf{X})$ as the total GV, which splits into two parts: between and within groups. This apportionment also appears in Sect. 5, see (10) and [5].

To prove (6), let us consider the above representation $\psi : S \to L$ and take $\mu = E[\psi(\mathbf{X})]$, $\mu_i = E[\psi(\mathbf{X}_i)]$. Then

$$
\begin{aligned}
\frac{1}{2}E_{\mathbf{X},\mathbf{X}'}[\delta^2(\mathbf{X}, \mathbf{X}')] &= E[\|\psi(\mathbf{X})\|^2] - \|E[\psi(\mathbf{X})]\|^2 \\
&= \sum_{i=1}^{k} w_i E[\|\psi(\mathbf{X}_i) - \mu\|^2] \\
&= \sum_{i=1}^{k} w_i E[\|\psi(\mathbf{X}_i) - \mu + \mu_i - \mu_i\|^2] \\
&= \sum_{i=1}^{k} w_i E[\|\psi(\mathbf{X}_i) - \mu_i\|^2] + \sum_{i=1}^{k} w_i\|\mu - \mu_i\|^2.
\end{aligned}
$$

Another approach for apportioning the geometric variability was considered in [12].

Next, we obtain a version of (6) in terms of entropy. As $-\ln x$ is a convex function, from the Jensen inequality,

$$E_{X_i}[-\ln f(X)] \geq E_{X_i}[-\ln f_i(X_i)].$$

Therefore

$$H(f) = \sum_{i=1}^{k} w_i E_{X_i}[-\ln f(X)]$$
$$\geq \sum_{i=1}^{k} w_i E_{X_i}[-\ln f(X_i)]$$
$$= \sum_{i=1}^{k} w_i H(f_i),$$

and the difference between $H(f)$ and $\sum_{i=1}^{k} w_i H(f_i)$ is positive. In terms of Kullback-Leibler divergence, this difference is $\sum_{i=1}^{k} w_i I(f_i\|f)$. We have the following proposition, where we may understand $H(f)$ as the "total" entropy, and similarly the other two quantities as "between" and "within" entropies, generalizing (5) and (6).

Proposition 3.1 *If $f = w_1 f_1 + \cdots + w_k f_k$ is the mixture of k densities, the following decomposition of the entropy holds:*

$$H(f) = \sum_{i=1}^{k} w_i I(f_i\|f) + \sum_{i=1}^{k} w_i H(f_i). \tag{7}$$

4 Multivariate Measure of Association

Consider two data sets observed on the same $\Omega = \{\omega_1, \ldots, \omega_n\}$. For the first set suppose that, by means of a distance function δ_x, we obtain a $n \times n$ distance matrix Δ_x. Then we find $\mathbf{A}_x = -\frac{1}{2}\Delta_x^{(2)}$, $\mathbf{G}_x = \mathbf{H}_c \mathbf{A}_x \mathbf{H}_c$ with $\mathbf{G}_x$ semidefinite positive. Thus $\mathbf{G}_x = \mathbf{U}\Lambda_x^2\mathbf{U}'$ and we get $\mathbf{X} = \mathbf{U}\Lambda_x$. the $n \times p$ matrix $\mathbf{X}$ containing the principal coordinates of Ω with respect to δ_x. For the second data set, suppose that, by means of a distance function δ_y, we similarly obtain a $n \times n$ distance matrix Δ_y and find $\mathbf{A}_y, \mathbf{G}_y$, where $\mathbf{G}_y$ is also semidefinite positive. Thus $\mathbf{G}_y = \mathbf{V}\Lambda_y^2\mathbf{V}'$ and we get $\mathbf{Y} = \mathbf{V}\Lambda_y$, the $n \times q$ matrix $\mathbf{Y}$ containing the principal coordinates of Ω with respect to δ_y.

We aim to define a coefficient of association between both data sets in terms of GV. Let us define the joint distance between two individuals ω_i, ω_j by taking principal coordinates $\mathbf{x}_i, \mathbf{y}_j$, (see [6]):

$$\delta_{xy}^2(i, j) = \delta_x^2(i, j) + \delta_y^2(i, j) - (\mathbf{x}_i - \mathbf{x}_j)\Lambda_x^{-1}\mathbf{X}'\mathbf{Y}\Lambda_y^{-1}(\mathbf{y}_i - \mathbf{y}_j)'.$$

The related inner product matrix is

$$\mathbf{G}_{xy} = \mathbf{G}_x + \mathbf{G}_y - \tfrac{1}{2}(\mathbf{G}_x^{1/2}\mathbf{G}_y^{1/2} + \mathbf{G}_y^{1/2}\mathbf{G}_x^{1/2}).$$

From $\operatorname{tr}(\mathbf{G}_{xy}) \leq \operatorname{tr}(\mathbf{G}_x) + \operatorname{tr}(\mathbf{G}_y)$ we have $V_{\delta_{xy}} \leq V_{\delta_x} + V_{\delta_y}$, with equality if $\mathbf{X}'\mathbf{Y} = \mathbf{0}$, i.e., the columns of $\mathbf{X}$ are orthogonal to the columns of $\mathbf{Y}$. Also $\delta_{xy} = \delta_x = \delta_y$ if both distances are equal, so $V_{\delta_{xy}} = V_{\delta_x} = V_{\delta_y}$. Thus a measure of association is

$$A_V = 2\frac{V_{\delta_x} + V_{\delta_y} - V_{\delta_{xy}}}{V_{\delta_x} + V_{\delta_y}},$$

which satisfies $0 \leq A_V \leq 1$. If we standardize the distances to $V_{\delta_x} = V_{\delta_y} = 1$, this measure reduces to $A_V = 2 - V_{\delta_{xy}}$. In the univariate case $p = q = 1$ we have $A_V = r$, where r is the correlation coefficient.

Now suppose that $f(\mathbf{x})$, $g(\mathbf{y})$ are the probability densities of two observations x, $\mathbf{y}$ of the random vectors $\mathbf{X}$, $\mathbf{Y}$, respectively. Let $h(\mathbf{x}, \mathbf{y})$ be the joint density, i.e., the density of an observation of $(\mathbf{X}, \mathbf{Y})$. It is well-known that $H(h) \leq H(f) + H(g)$. If there is stochastic independence $h(\mathbf{x}, \mathbf{y}) = f(\mathbf{x})g(\mathbf{y})$, then $H(h) = H(f) + H(g)$ and if $f = g$ then $H(h) = H(f) = H(g)$. This suggests a general measure of dependence similar to A_V.

Proposition 4.1 *Suppose $0 \leq \min\{H(f), H(g), H(h)\} < \infty$. A measure of association A_E in terms of entropy, which satisfies $0 \leq A_E \leq 1$, is given by*

$$A_E = 2\frac{H(f) + H(g) - H(h)}{H(f) + H(g)},$$

See [3], for other distance-based measures of multivariate association.

5 Multivariate Intraclass Correlation

Consider the random effects model

$$y_{ij} = \mu + A_i + e_{ij}, \quad i = 1, \ldots, k, \ j = 1, \ldots, n_i,$$

where A_i is a r.v. such that $E(A_i) = 0$, $\text{var}(A_i) = \sigma_A^2$, $i = 1, \ldots, k$, $E(e_{ij}) = 0$, $\text{var}(e_{ij}) = \sigma^2$. Assuming complete independence, the intraclass correlation coefficient ρ_I is defined as the ordinary correlation between two observations y_{ij}, $y_{ij'}$ belonging to the same class or group. It is readily proved that

$$\rho_I = \frac{\sigma_A^2}{\sigma_A^2 + \sigma^2}.$$

Let B and W the sums of squares between and within groups, with $k - 1$ and $n - k$ degrees of freedom, respectively. The mean squares are given by $\overline{B} = B/(k-1)$ and $\overline{W} = W/(n-k)$, where $n = n_1 + \cdots + n_k$. Following [9], ρ_I can be estimated by

$$\widehat{\rho_I} = \frac{\overline{B} - \overline{W}}{\overline{B} + (n_0 - 1)\overline{W}}, \tag{8}$$

where $n_0 = \left(n - \sum_{i=1}^{k} n_i^2/n \right) /(k-1)$. Note that $n_0 = n_1 = \cdots = n_k$ in the balanced case.

It is worth relating $\widehat{\rho}_I$ to the correlation ratio $\widehat{\eta}^2$. Since $B/W = \widehat{\eta}^2/(1 - \widehat{\eta}^2)$, a little algebra shows that

$$\widehat{\rho}_I = \frac{(n-1)\widehat{\eta}^2 - (k-1)}{(n - kn_0 + n_o - 1)\widehat{\eta}^2 + (n_0 - 1)(k-1)}. \tag{9}$$

Working with p variables the MANOVA identity is $\mathbf{T} = \mathbf{B} + \mathbf{W}$. If $\mathbf{X}$ is the full $n \times p$ data matrix obtained via principal coordinate analysis with respect to a distance δ, then $\mathbf{T} = \mathbf{X}'\mathbf{X}$ and the inner product matrix is $\mathbf{G} = \mathbf{X}\mathbf{X}'$. As $\mathrm{tr}(\mathbf{X}\mathbf{X}') = \mathrm{tr}(\mathbf{X}'\mathbf{X})$, a multivariate version of (5) holds.

Similarly, we can find the principal coordinates $\mathbf{X}_i$ in the group i. We have $\mathbf{W}_i = \mathbf{X}_i'\mathbf{X}_i$ and, since $\mathrm{tr}(\mathbf{X}_i'\mathbf{X}_i) = \mathrm{tr}(\mathbf{X}_i\mathbf{X}_i')$, from (2), we have

$$V_\delta(\text{within } i) = \frac{1}{2n_i^2} \sum_{h,h'=1}^{n_i} \delta_{h,h'}^2(i) = \frac{1}{n_i}\mathrm{tr}(\mathbf{W}_i), \tag{10}$$

where $\delta_{h,h'}(i)$ stands for the distance between two observations h, h' within the same group i. A similar expression holds for $V_\delta(\text{total})$ concerning the n observations. If we combine (2) and (5), we get the sample version of (6):

$$V_\delta(\text{total}) = V_\delta(\text{between}) + \sum_{i=1}^{k} \left(\frac{n_i}{n} \right) V_\delta(\text{within } i),$$

where $V_\delta(\text{total})$ and $V_\delta(\text{within } i)$ can be obtained directly from the given distances, whereas $V_\delta(\text{between})$ can be found by subtraction.

A multivariate generalization of the above intraclass correlation (9) is

$$\widehat{\Phi}_I = \frac{\mathrm{tr}(\mathbf{B})/(k-1) - \mathrm{tr}(\mathbf{W})/(n-k)}{\mathrm{tr}(\mathbf{B})/(k-1) + (n_0 - 1)\mathrm{tr}(\mathbf{W})/(n-k)}.$$

For general data described by distances between pairs of observations, the relation between $\mathrm{tr}(\mathbf{B})$, $\mathrm{tr}(\mathbf{W})$ and the geometric variability, suggests the general intraclass correlation given by

$$\widehat{\Phi}_I = \frac{V_\delta(\text{between})/(k-1) - \sum_{i=1}^{k} V_\delta(\text{within } i)/(n-k)}{V_\delta(\text{between})/(k-1) + (n_0 - 1)\sum_{i=1}^{k} V_\delta(\text{within } i)/(n-k)},$$

where $n_0 = \left(n - \sum_{i=1}^{k} n_i^2/n \right) /(k-1)$.

Since $\widehat{\Phi}_I$ depends on the average sample size n_0, it could not be possible to get the entropy version of this coefficient. However, $\widehat{\rho}_I$ is related to the correlation ratio $\widehat{\eta}^2 = B/(B+W)$, see (9), and for this coefficient an entropy version can be obtained.

First, a multivariate generalization using distances is given by $\widehat{\eta}^2 = V_\delta(\text{between})/V_\delta(\text{total})$. The entropy version, taking into account the mixture $f = w_1 f_1 + \cdots + w_k f_k$, see (7), is next proposed. This measure satisfies $0 \le \eta_E^2 \le 1$. Clearly, $\eta_E^2 = 0$ if $w_i = 1$ for some i and $\eta_E^2 = 1$ if $H(f_i) = 0$ for all i.

Proposition 5.1 *The correlation ratio expressed in terms of entropy is given by*

$$\eta_E^2 = \frac{\sum_{i=1}^k w_i I(f_i \| f)}{H(f)}.$$

6 Conclusions

There are many distances available in multivariate analysis. Often, the choice is heuristic and ignores the underlying probability distribution. Flury [10] claims that the correct approach in classification should be performed using probability mixtures, rather than clustering algorithms. Indeed, if the distance generates a probability density close to the true density, the choice is good. An open question is to test whether the difference between the true density and the generated density, is significant, see [11].

The decomposition of the variability in the presence of several groups, can be clarified and generalized with distances and entropies. Also, some useful measures of association, intraclass correlation and correlation ratio, can be expressed by means of distances and entropies.

Although we suppose continuous distributions, most results obtained here can be adapted to the discrete case. For example, suppose a multivariate Bernoulli distribution concerning k events $E_1, , \ldots, E_k$, with probabilities $p_1, \ldots, p_k$. If we choose the distance $\delta(E_i, E_j) = 2$ if $i \ne j$, then the geometric variability is the Gini-Simpson entropy (G-S) $1 - \sum_{i=1}^k p_i^2$. We can obtain discrete versions of the above results, for example, an inequality concerning G-S and Shannon entropies, and the apportionment of G-S in the presence of several groups.

Acknowledgements This work was supported by FONDECYT Project no. 1160249.

References

1. Anderson MJ (2001) A new method for non-parametric multivariate analysis of variance. Austral Ecol 26:32–46
2. Cuadras CM (2008) Distance-based multisample tests for multivariate data. In: Arnold BC, Balakrishnan N, Sarabia JM, Mínguez R (eds) Advances in mathematical and statistical modeling. Birkhauser, Boston

3. Cuadras CM (2011) Distance-based approach in multivariate association. In: Ingrassia S, Rocci R, Vichi M (eds) New perspectives in statistical modeling and data analysis. Springer, Berlin

4. Cuadras CM, Atkinson RA, Fortiana J (1997) Probability densities from distances and discriminant analysis. Stat Prob Lett 33:405–411

5. Cuadras CM, Cuadras D (2011) Partitioning the geometric variability in multivariate analysis and contingency tables. In: Fichet B, Piccolo D, Verde R, Vichi M (eds) Classification and multivariate analysis for complex data structures. Springer, Berlin

6. Cuadras CM, Fortiana J (1998) Visualizing categorical data with related metric scaling. In: Blasius J, Greenacre M (eds) Visualization of categorical data. Academic Press, London

7. Cuadras CM, Fortiana J (2004) Distance-based multivariate two sample tests. In: Nikulin MS, Balakrishnan N, Mesbah M, Limnios N (eds) Parametric and semiparametric models with applications to reliability, survival analysis, and quality of life. Birkhauser, Boston

8. Cuadras CM, Fortiana J, Oliva F (1997) The proximity of an individual to a population with applications in discriminant analysis. J Classif 14:117–136

9. Donner A, Wells G (1986) A comparison of confidence interval methods for the intraclass correlation coefficient. Biometrics 42:401–412

10. Flury B (1997) A first course in multivariate statistics. Springer, New York

11. Pardo L (2005) Statistical inference based on divergence measures. Chapman and Hall/CRC, London

12. Rao CR (1982) Diversity and dissimilarity coefficients: a unified approach. Theor Popul Biol 21:24–43

Cryptographic Uncertainness: Some Experiments on Finite Semifield Based Substitution Boxes

Ignacio F. Rúa and Elías F. Combarro

Abstract Substitution boxes (S-boxes) are an important part of the design of block ciphers. They provide nonlinearity and so the security of the cipher depends strongly on them. Some block ciphers use S-boxes given by lookup tables (e.g., DES) where as others use S-boxes obtained from finite field operations (e.g., AES). As a generalization of the latter, finite semifields (i.e., finite nonassociative division rings) have been suggested as algebraic structures from which S-boxes with good cryptographic properties might be obtained. In this paper we present the results of experiments on the construction of S-boxes from finite semifields of orders 256 and 64, using the left and right inverses of these rings.

1 Introduction

> [...] a new science, called *Criptology*, arises. It has a field devoted to encryption (*Cryptography*) an another one to decryption (*Cryptanalysis*). Its origins are as old as humanity: remember the writing on a strip of parchment wrapped around a staff or Lacedaemonian 'scytale'; or the Caesar cipher consisting on a constant shifting of the letters of the alphabet.

These words are part of the opening lecture of the academic year 1996–1997 delivered by Pedro Gil at University of Oviedo [7] (and mostly translated to English in Part I of this book). The lecture, which was titled "The Mathematics of the Uncertain", had a first part devoted to randomness, Probability and Statistics. The second part dealt with Information Theory and the Mathematics of communication (it even had a third and final part dedicated to fuzzy sets). It is difficult to understand modern Cryptography without a probabilistic point of view [8]. The first author to systematize this approach was Claude Shannon, the *father* of Information Theory. Appart from introducing the concepts of *entropy* and *information* in the context of communication

I. F. Rúa (✉)
Departamento de Matemáticas, Universidad de Oviedo, Oviedo, Spain
e-mail: rua@uniovi.es

E. F. Combarro
Departamento de Informática, Universidad de Oviedo, Oviedo, Spain
e-mail: efernandezca@uniovi.es

in noiseless and noisy channels [20] (just as mentioned in Pedro Gil's lecture[1]), he considered a probabilistic model of *perfect secrecy* [21]. Following this idea, semantic security (which is, from a certain point of view, the theoretical notion of a secure cryptographic system) is founded on a probabilistic setting [9].

Block ciphers (which transform a block of bits of fixed size into another block of the same size with the help of a bit-key of also fixed, perhaps different, size) are a symmetric (i.e., private) key cryptographic primitive used in many other designs (e.g., cryptosystems, message authentication codes, hash functions,...) [14]. *Substitution boxes* (called *S-boxes*) are an important part of the design of block ciphers. They provide nonlinearity to the transformation and so the security of the cipher depends strongly on them. Some block ciphers use S-boxes given by lookup tables (e.g., DES) where as others use S-boxes obtained from finite field operations (e.g., AES) [22]. As a generalization of the latter, finite semifields (i.e., finite nonassociative division rings) have been suggested as algebraic structures from which S-boxes with good cryptographic properties might be obtained [5]. This is not the first time that nonassociative structures have been considered in a cryptographic setting (just recall, for instance, [6, 10, 13, 16]).

In this paper, following the path of [5], we present the results of experiments on the construction of S-boxes from finite semifields of order 256, using the left and right inverses of these rings. We process all finite semifields of such an order and rank 4 (and not only the 28 representatives up to isotopy considered in [5, Section 5.3]), and also all finite semifields of dimension 6 over $\mathbb{F}_2$ (as this is the biggest dimension for which all finite semifields of characteristic 2 have been classified). The paper is organized as follows: in Sect. 2 basic notions of block ciphers (including properties of S-boxes) are reminded. Section 3 is devoted to finite semifields and their properties. Finally, in the last section we collect the results obtained from our computational experiments.

2 Block Ciphers and Substitution Boxes

A *block cipher* is a deterministic cipher $E : \{0, 1\}^b \times \{0, 1\}^k \to \{0, 1\}^b$ which transforms a block M of b bits of fixed size into another block C of the same size with the help of a key K of also fixed, perhaps different, size k [14]. Well-known examples of block ciphers include the previous and the current NIST standards for encryption data: DES and AES [22]. For instance, in DES $b = 64, k = 56$, whereas in AES $b = 256, k \in \{128, 196, 256\}$. These ciphers are of utmost importance because, as pointed out in [2],

---

[1]Incidentally, let us mention that we had the privilege of learning the basic aspects of Probability, Statistics and Information Theory from Pedro himself, in two courses delivered at University of Oviedo some twenty years ago.

> Block ciphers are the "work horse" of practical cryptography: not only they can be used to build a stream cipher, but they can be used to build ciphers with stronger security properties [...], as well as many other cryptographic primitives.

A common design of block ciphers is that of iterated ciphers, where a round function is used repeatedly r times to process the block of bits M using a set of round keys obtained from the master key K with the help of an auxiliary key schedule algorithm (e.g., in DES $r = 14$, in AES $r \in \{10, 12, 14\}$). In these ciphers, the ultimate transformation of the block M depends on the round function F. Traditionally, the function F can be of Feistel type (such as in DES) or a Substitution-Permutation Network (such as in AES) [22]. In either case, both use substitution boxes in the design of F.

A *substitution box* (called *S-box*) is a fixed boolean function $S : \{0, 1\}^n \to \{0, 1\}^m$, where the parameters n, m depend on the actual cipher considered (for instance, in DES $n = 6, m = 4$, in AES $n = m = 8$). S-boxes are a core part of the design of block ciphers as they provide nonlinearity to the transformation. The security of the cipher (e.g., robustness against differential or linear attacks) depends strongly on them. Some block ciphers use S-boxes given by lookup tables (e.g., DES) where as others use S-boxes obtained from finite field operations [22]. For instance, AES S-boxes identify the set $\{0, 1\}^8$ with the Galois field $\mathbb{F}_{2^8}$ of 256 elements (multiplication is taken modulo the polynomial $x^8 + x^4 + x^3 + x + 1$) and before applying an $\mathbb{F}_2$-affine transformation, the input element is changed into its multiplicative inverse in $\mathbb{F}_{2^8}$ (the zero element is replicated).

Different properties of an S-box can be introduced in order to determine its cryptographic utility, and so multiple criteria can be found in the literature (e.g., [15, 19]). In this paper we study properties #1, #3 and #4 in [5] for S-boxes of sizes 256 and 64. Namely, we identify the sets $\{0, 1\}^8$ and $\{0, 1\}^6$ with $\mathbb{F}_2^8$ and $\mathbb{F}_2^6$, and consider

1. Bijectivity: $n = m = 8$ (alt. $n = m = 6$), and the S-box must be bijective.
2. Non-linearity: the linear invariant λ_S is defined as

$$\lambda_S = \max\{ \, | -2^{n-1} + \#\{x \in \mathbb{F}_2^n \, : \, (a|x) = (b|S(x))\}| \, : \, a, b \in \mathbb{F}_2^n, b \neq 0\}$$

where $(a|x)$ denotes the usual inner product in $\mathbb{F}_2^n$, $n = 8$ (alt. $n = 6$).
3. The differential invariant δ_S is equal to

$$\delta_S = \max\{ \, \#\{x \in \mathbb{F}_2^n \, : \, S(x) \oplus S(a \oplus x) = b\} \, : \, a, b \in \mathbb{F}_2^n, a \neq 0\}$$

where $a \oplus x$ denotes bitwise addition mod 2, and $n = 8$ (alt. $n = 6$).

With respect to these properties AES S-boxes are optimal in the sense that they are bijective, have minimal non-linearity $\lambda_{AES} = 16$, and minimal differential invariant $\delta_{AES} = 4$ among non-APN functions [5]. Also, $\lambda_{\mathbb{F}_{64}} = 8$ and $\delta_{\mathbb{F}_{64}} = 4$.

3 Finite Semifields

In this section we collect definitions and facts on finite semifields [4, 11]. A finite
nonassociative ring D is called *finite semifield*, if the set of nonzero elements D^*
is closed under the product, and it has an identity element. In such a case D^* is a
multiplicative loop. That is, there exists an element $e \in D^*$ (the identity of D) such
that $ex = xe = x$, for all $x \in D$ and, for all $a, b \in D^*$, the equation $ax = b$ (resp.
$xa = b$) has a unique solution. Let us emphasize that these *left* and *right inverses*
might be different elements of the finite semifield. This is an important fact apparently
obviated in [5, 6].

Finite semifields are nonassociative finite division rings and, apart from finite
fields, *proper* finite semifields exist. The characteristic of a finite semifield D is
a prime number p, and D is a finite-dimensional algebra over $\mathbb{F}_q$ ($q = p^c$) of
dimension d, for some $c, d \in \mathbb{N}$, so that the order of D is $|D| = q^d$. Moreover,
$\mathbb{F}_q$ can be chosen to be contained in the associative-commutative center $Z(D)$ of D.
In this paper we will be interested in finite semifields of order 256, i.e., of dimension
8 over its center $Z(D) = \mathbb{F}_2$ or of rank 4 (i.e., of dimension 4 over $\mathbb{F}_4 \subseteq Z(D)$). The
finite field $\mathbb{F}_{256}$ is included in the latter case. Also, we will be interested in semifields
of order 64, i.e., 8-dimensional over $\mathbb{F}_2$. E.g., the Galois field $\mathbb{F}_{2^8}$.

Isomorphism of finite semifields is defined as usual for algebras, and the classification of finite semifields up to isomorphism can be naturally considered. Because
of the connections to finite geometries [1], the following notion must be considered.
An *isotopy* between two finite semifields D_1 and D_2 is a triple (F, G, H) of bijective
$\mathbb{F}_q$−linear maps $D_1 \to D_2$ such that $H(ab) = F(a)G(b)$, for all $a, b \in D_1$. Clearly,
any isomorphism between two semifields is an isotopy, but the converse is not necessarily true. From any finite semifield D, a projective plane $\mathscr{P}(D)$ can be constructed
[11]. Theorem 6 in [1] shows that isotopy of finite semifields is the algebraic translation of the isomorphism between the corresponding projective planes.

By [11, Theorem 5.2.1], up to six projective planes can be constructed from a
given finite semifield D using the transformations of the group S_3. Actually, S_3 acts
on the set of semifield planes of a given order producing, for each semifield D, its
Knuth orbit [11]. So, the classification of finite semifields can be reduced to the
classification of the corresponding Knuth orbits.

In the particular case of semifields of order 256 and rank 4, i.e., with center
containing $\mathbb{F}_4$, a computer-assisted classification was presented in [3]. A total amount
of 28 Knuth classes were obtained. The actual number of semifields is much bigger.
Namely, the number of isotopy classes is 51 and the number of nonisomorphic finite
semifields containing $\mathbb{F}_4$ is 75939 (these numbers were obtained with the techniques
describe in [3]). Unfortunately, a complete classification of finite semifields of order
256 has not been achieved (not even of order 128 [18]). Moreover, it is even unknown
how many of them might there exist (the number must be clearly much bigger than
those 75939 containing $\mathbb{F}_4$ in the center).

The biggest dimension for which all finite semifields of characteristic 2 have been classified is 6 [17]. There are 80 Knuth orbits of such an order containing 322 isotopy classes for a total amount of 376971 semifields.

4 Some Experiments on Finite Semifield Based S-boxes

Inspired by the S-boxes of AES, Dumas and Orfila propose in [5] the construction of S-boxes from the multiplicative structure of finite semifields. Namely, they suggest "using the inverse function" [5, Section 5.3]. As was noticed in the previous section, a distinction between left and right inverse is needed when dealing with (noncommutative) finite semifields. So, given a finite semifield D of order 256 (alt. 64) and identity e, we have considered the two following S-boxes:

$$
\begin{array}{ll}
S_r: & D \rightarrow D \\
& a \neq 0 \rightarrow b \text{ s.t. } ab = e \\
& 0 \rightarrow 0
\end{array}
\qquad
\begin{array}{ll}
S_l: & D \rightarrow D \\
& a \neq 0 \rightarrow b \text{ s.t. } ba = e \\
& 0 \rightarrow 0
\end{array}
$$

It is clear that, when D is commutative (in particular, if S is the Galois field $\mathbb{F}_{2^8}$ or $\mathbb{F}_{2^6}$), both S-boxes coincide. It is also evident that, because D is a finite semifield, the bijectivity property holds in both cases. In order to compute the linear $\lambda_{S_r}, \lambda_{S_l}$ and differential $\delta_{S_r}, \delta_{S_l}$ invariants we identify the elements of D with those of the set $\mathbb{F}_2^8$ (alt. $\mathbb{F}_2^6$). This can be straightforwardly done as the representation of finite semifields introduced in [3] is exactly that one. Moreover, in Table 2 of such a paper it is contained a complete description of all finite semifields of order 256 and rank 4, i.e., and center containing the finite field $\mathbb{F}_4$ [3, Section 4.2], up to Knuth orbit. These are the semifields also considered in [5, Section 5.3], where it is claimed that

> We thus have also tried to construct S-boxes based on all these 28 semifields up to isotopy, by using the inverse function.

It appears that the authors have only consider the 28 representatives in their construction and at most one of the two possible "inverse function". As it was said in the previous section, up to isomorphism, the actual number of finite semifields or order 256 with center containing $\mathbb{F}_4$ is much bigger. So, we have used the computational machinery described in [3, Section 3] to generate all those finite semifields. For each one of them, we have explicitly constructed the aforementioned S-boxes S_r and S_l. Since we are interested in S-boxes with "good" cryptographic properties, we have taken as a reference the invariants for the AES S-box ($\lambda_{AES} = 16, \delta_{AES} = 4$). Let us remark the following fact.

Proposition 4.1 $\lambda_{S_l} = \lambda_{S_r}$, for any finite semifield D of order 2^n.

Proof For all $x, y \in D$, we have that $y = S_l(x)$ iff $x = S_r(y)$. Therefore, for all nonzero $a, b \in \mathbb{F}_2^n$:

$$\#\{x \in \mathbb{F}_2^n : (a|x) = (b|S_l(x))\} = \#\{y \in \mathbb{F}_2^n : (b|y) = (a|S_r(y))\}.$$

On the other hand, since $(0|x) = 0$, for all $x \in \mathbb{F}_2^n$, and because the maps S_l and S_r are bijections, we have:

$$\#\{x \in \mathbb{F}_2^n : 0 = (c|S_l(x))\} = \#\{y \in \mathbb{F}_2^n : 0 = (c|S_r(y))\}$$

for all $0 \neq c \in \mathbb{F}_2^n$. Hence,

$$\lambda_{S_l} = \max\{ | -2^{n-1} + \#\{x \in \mathbb{F}_2^n : (a|x) = (b|S_l(x))\}| : a, b \in \mathbb{F}_2^n, b \neq 0\}$$

$$= \max\{ | -2^{n-1} + \#\{y \in \mathbb{F}_2^n : (b|y) = (a|S_r(x))\}| : a, b \in \mathbb{F}_2^n, a \neq 0\} = \lambda_{S_r}. \ \square$$

Our computations show that none of the generated S-boxes had a pair of invariants matching those of the finite field $\mathbb{F}_{2^8}$. So, no S-box with "good" cryptographic properties was obtained from the constructions S_r or S_l on semifields of order 256 containing $\mathbb{F}_4$ in the center. Let us mention, for the record, that the linear and differential parameters might be different for isotopic non-isomorphic finite semifields. This means that these parameters are not isotopy invariants, such as the center or nuclei sizes [12]. So, for instance, a full computation of the linear and differential parameters for finite semifields isotopic to Semifield #II of [3, Table 1], shows that we can find parameters $(\lambda_{S_r}, \delta_{S_r}) = (38, 12), (38, 14), (36, 10), (34, 10), \ldots$.

The construction of S-boxes S_r and S_l was also applied to all finite semifields of order 64. Remember that the parameters of the finite field of such an order are $\lambda_{\mathbb{F}_{64}} = 8$ and $\delta_{\mathbb{F}_{64}} = 4$. The computational results show that there are some proper semifields with $\delta_{S_l} = 4$. Namely, semifields falling in Knuth orbits #IV, V, VIII, X. Among these, only 6 proper semifields in Knuth orbit #V share the pair $(\lambda_{S_r}, \delta_{S_r}) = (8, 4)$ with the finite field $\mathbb{F}_{64}$. We have plotted in the following graph (Fig. 1) all pairs $(\lambda_{S_r}, \delta_{S_r})$ and $(\lambda_{S_l}, \delta_{S_l})$ found in our study.

We finish this short note by showing one of the S-boxes with the same parameters of the finite field S-box $S_{\mathbb{F}_{64}}$, but constructed from left inverses in a finite semifield of order 64.

Table 1 An S-Box with minimal linear and differential parameters (constructed from a proper semifield of order 64)

S_l	0	1	2	3	4	5	6	7
0	00	40	73	24	30	45	62	27
1	41	70	55	47	05	03	46	32
2	15	37	31	11	17	66	74	06
3	72	34	57	02	10	35	14	64
4	44	77	43	67	71	36	53	25
5	20	21	13	56	33	54	01	61
6	60	50	26	12	75	16	76	65
7	23	52	51	22	42	63	07	04

Fig. 1 Parameters for
semifields S-boxes S_r and S_l
of order 64

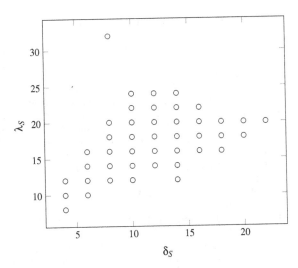

Conclusion

We have explicitly constructed S-boxes from proper finite semifields of orders 256
and 64, and computed their linear and differential parameters. The results in the
case of 256 elements are not satisfactory, since none of these S-boxes have the same
minimal invariants as those of the AES S-box. This is not surprising since only rank 4
semifields of such an order were analyzed, as this was the only subclass of semifields
of order 256 for which a complete classification has been achieved so far. On the
other hand, the case of order 64 semifields (for which a full classification is known)
is more promising. Some S-boxes have been constructed with the same parameters
of those obtained from the Galois field $\mathbb{F}_{64}$.

Acknowledgements I.F. Rúa is partially supported by MINECO-13-MTM2013-45588-C3-1-
P, and Principado de Asturias Grant GRUPIN14-142. E.F. Combarro is partially supported by
MINECO-16-TEC2015-67387-C4-3-R.

References

1. Albert AA (1960) Finite division algebras and finite planes. In: Bellman R, Hall M Jr (eds)
   Combinatorial analysis-proceedings of symposia in applied mathematics, vol 10. SIAM/AMS,
   Providence
2. Boneh D, Shoup V (2015) A graduate course in applied cryptography. https://crypto.stanford.
   edu/~dabo/cryptobook/
3. Combarro EF, Rúa IF, Ranilla J (2011) New advances in the computational exploration of
   semifields. Int J Comput Math 88(9):1990–2000
4. Cordero M, Wene GP (1999) A survey of finite semifields. Discret Math 208(209):125–137

5. Dumas JG, Orfila JB (2014) Generating S-boxes from semifields pseudo-extensions. arXiv:1411.2503

6. Figueroa R, Salzberg PM, Shiue PJ-S (1994) A family of cryptosystems based on combinatorial properties of finite geometries. In: Mullen GL, Shiue PJ-S (eds) Contemporary mathematics, vol 168. AMS, Providence

7. Gil Álvarez P (1996) Las matemáticas de lo incierto. Servicio de publicaciones, Universidad de Oviedo. http://digibuo.uniovi.es/dspace/bitstream/10651/28625/1/matematicasincierto.pdf

8. Goldreich O (2001) Foundations of cryptography. Cambridge University Press, Cambridge

9. Goldwasser S, Micali S (1984) Probabilistic encryption. J Comput Syst Sci 28(2):270–299

10. Kalka A (2012) Non-associative public-key cryptography. arXiv:1210.8270

11. Knuth DE (1965) Finite semifields and projective planes. J Algebra 2:182–217

12. Lavrauw M, Polverino O (2011) Finite semifields and Galois geometry. In: De Beule J, Storme L (eds) Current research topics in galois geometry. NOVA Academic Publishers, Hauppauge

13. Malekian E, Zakerolhosseini A (2010) A non-associative lattice-based public key cryptosystem. Secur Commun Netw 5(2):145–163

14. Menezes AJ, van Oorschot PC, Vanstone SA (1996) Handbook of applied cryptography. CRC Press, Boca Raton

15. Mister S, Adams C (1996) Practical S-box design. In: Tavares S, Meijer H (eds) SAC'96, 3rd Annual works selected areas in cryptography. Queen's University, Kingston. http://sacworkshop.org/proc/SAC_96_005.pdf

16. Rúa IF (2004) Anillos no asociativos en codificación y criptografía. Ph.D. thesis. University of Oviedo

17. Rúa IF, Combarro EF, Ranilla J (2009) Classification of semifields of order 64. J Algebra 322:4011–4029

18. Rúa IF, Combarro EF, Ranilla J (2012) Determination of division algebras with 243 elements. Finite Fields Their Appl 18:1148–1155

19. Saarinen M-JO (2012) Cryptographic analysis of all 4 × 4-bit S-boxes. In: Knudsen LR, Wu H (eds) Selected areas in cryptography, vol 7118. Lecture notes in computer science. Springer, Berlin

20. Shannon C (1948) A mathematical theory of communication. Bell Syst Tech J 27(3):379–423

21. Shannon C (1949) Communication theory of secrecy systems. Bell Syst Tech J 28(4):656–715

22. Stinson DR (2006) Cryptography: theory and practice, 3rd edn. Chapman and Hall/CRC, London

The Roll of Dices in Cryptology

María Isabel González Vasco, Santos González, Consuelo Martínez
and Adriana Suárez Corona

Abstract Probability plays a fundamental role in complexity theory, which in turn is one of the pillars of modern cryptology. However, security practitioners are not always familiar with probability theory, and thus fail to foresee the impact of (seemingly small) deviations from the theoretical description of a scheme at the implementation level. On the other hand, many cryptographic scenarios involve mutually distrusting parties, which need however to cooperate towards a joint goal. In order to attain assurance of the *good* behavior of one party, interactive validation methods (also known as *interactive proof systems*) are employed. Randomness is at the core of such methods, which most often will only provide *relative* assurance, in the sense that they will establish correctness in a probabilistic way. In this paper we will briefly discuss the role of probability theory within modern cryptology, reviewing *probabilistic proof systems* as a powerful tool towards efficient protocol design, and *provable security*, as an invaluable framework for deriving formal security proofs.

1 Introduction

As neatly put down in words by O. Goldreich [8], *"both in mathematics and in real life, proofs are meaningful only with respect to commonly agreed principles of reasoning"*.

M. I. González Vasco
Universidad Rey Juan Carlos, Madrid, Spain
e-mail: mariaisabel.vasco@urjc.es

S. González · C. Martínez (✉)
Universidad de Oviedo, Oviedo, Spain
e-mail: cmartinez@uniovi.es

S. González
e-mail: santos@uniovi.es

A. Suárez Corona
Universidad de León, León, Spain
e-mail: asuac@unileon.es

© Springer International Publishing AG 2018
E. Gil et al. (eds.), *The Mathematics of the Uncertain*, Studies in Systems, Decision and Control 142, https://doi.org/10.1007/978-3-319-73848-2_46

493

In other words, verification methods actually determine whether a certain thread of reasoning is actually a proof. While in logic such verification methods or *proof systems* are explicited by (static) axioms and rules of inference, in modern complexity theory verification methods are often interactive and have a computational nature. Cryptographic evidences are frequently constructed under this premises; its final goal is to establish the validity of a claim in a probabilistic way (i.e., not in absolute terms) and do not build on immovable principles, but rather ground on interactive, dynamic processes. *Probabilistic proof systems* are not theoretical formalisms, but rather cryptographic constructions in themselves, through which, for instance, users engaging in a protocol can verify that others have faithfully followed the prescribed steps in the protocol specification.

On the other hand, security proofs in cryptography usually try to link the robustness of a cryptographic construction to well-studied mathematical problems that are considered hard, such as factoring large numbers or computing discrete logarithms in finite fields or elliptic curves. Such reductions are in most cases probabilistic: it is evidenced that if an adversary can violate a certain security property with non-negligible probability (above his presumed trivial success rate), then one could use that adversary as an *oracle* to solve the problem that is believed to be hard. Thus, cryptographic proofs only provide security attestation which is valid as long as the computational assumption holds. For instance, if quantum computers were indeed realized various cryptographic schemes relying on the RSA, discrete logarithm and related assumptions would be broken see [18]. Nevertheless, even when proofs are not universal statements, they provide strong arguments in favor of the security of the corresponding protocols if the model is reasonable and appropriate.[1]

Ultimately we want to show how probability and algebra collaborate, contrarily to what one could initially think, to improve methods and results in cryptography, where both disciplines play an essential role. This is not the only case. Just to mention another important example let's say a few words about expanders in graph theory and refer readers to [13] for details. The existence of expanders was proved using probabilistic methods, what was not enough for applications, where explicit constructions of such graphs were needed. The first explicit construction was made by Margulis, using methods of group representations.

In this paper we will concentrate in cryptography. In the next section we will introduce probabilistic proof systems and try to give a glimpse of their fascinating role in modern cryptography, while the last section is devoted to the provable security paradigm for deriving cryptographic proofs.

2 Probabilistic Proof Systems

Cryptographic constructions should protect honest users from misbehaving participants that maliciously deviate from the prescribed protocol specifications. Prob-

---

[1]The *provable security paradigm* has been questioned by different authors, see [15].

abilistic proof systems allow users to produce evidence of their correct behavior, disclosing (most surprisingly!) often nothing else than the correctness of the underlying computation. We give a simplified introduction to this topic in this section; as our exposition will be rather informal, we refer the interested reader to [7, 8] for precise definitions and proofs.

2.1 Interactive Proof Systems

Two actors will be engaged in an interactive proof system; a *prover* (which is assumed to be computationally unbounded) and a *verifier*, bounded computationally in some sense.

Definition 2.1 An *interactive proof system (IPS)* for a set S is a two-party procedure, involving a *verifier* V and a *prover* P which interact under the following conditions:

- V executes a probabilistic polynomial time strategy,
- P executes an unbounded strategy,
- **Completeness:** for every $s \in S$, the verifier V always accepts after interacting with the prover P on common input s,
- **Soundness:** for every $x \notin S$, and every strategy of the prover P, the verifier V will reject with probability $P \geq \frac{1}{2}$ after interacting with the prover P on common input x.

Example 2.1 In Fig. 1 we can see a simple construction for an IPS for *quadratic residuosity* [1].[2] Recall that given $n \in \mathbb{Z}$, $x \in \mathbb{Z}_n^*$ is a *quadratic residue* modulo n if there is some $y \in \mathbb{Z}$ such that $x = y^2 \pmod{n}$. No efficient algorithm is known for deciding whether an integer x is actually a quadratic residue modulo n or not, unless the factorization of n in prime factors is known (see, for instance, the discussion in Chap. 2 of [14]).

It is easy to see that the above proof is complete, as indeed whenever x is really a quadratic residue, the verifier will accept with probability one. Soundness is somewhat less trivial to argue, for we need to justify that for every possible strategy of the prover, if x is not a quadratic residue modulo n, the verifier will reject the proof with probability at least $\frac{1}{2}$. Let y be the value chosen by the prover in step 1 of the interaction. Let us first assume x is not a quadratic residue, then P must decide whether to follow **step 1** from the interaction faithfully or not. Assume he does, and $b = 1$ (this happens with probability $\frac{1}{2}$, as the verifier is honest in its own interest). At this, the verifier will only accept if $z^2 = xy$, which is impossible since $(zu^{-1})^2 = z^2u^{-2} = z^2y^{-1} = x$, given that x is not a quadratic residue. Thus, the verifier rejects with probability at least $\frac{1}{2}$. However, if in **step 1** of the interaction y is

---

[2]We follow standard notation and denote by $\mathbb{Z}_n^*$ the group of units in $\{1, \ldots, n-1\}$, where product is defined modulo n. Also, as standard, throughout the paper, by "u.a.r." we mean *uniformly at random*.

Common input: $x, n \in \mathbb{Z}$
Prover's private input: w such that $w^2 = x \pmod{n}$
Interaction:

    1. P selects u u.a.r. from $\mathbb{Z}_n^*$ and sends $y = u^2 \pmod{n}$ to V
    2. V selects a bit b u.a.r. and sends it to P
    3. P sends $z = w^b u \pmod{n}$ to V.

Verification:

- If $b = 0$ and $z^2 = y \pmod{n}$, V sets Res $= 1$
- If $b = 1$ and $z^2 = xy \pmod{n}$, V sets Res $= 1$
- Otherwise, V sets Res $= 0$

Fig. 1 An IPS for quadratic residuosity

not a quadratic residue (that is, P deviates from the protocol specification here) and $b = 0$, the verifier will never accept (for the prover cannot provide z with $z^2 = y$). As a result, in either case the probability of rejection is at least $\frac{1}{2}$.

2.2 Zero Knowledge

The notion of *zero knowledge* must be understood information-theoretically; informally, a cryptographic protocol will have this property if no participant or observer may gain any information from its execution besides the prescribed output.

Goldwasser, Micali and Rackof gave in [10] a formal definition of *Zero Knowledge IPS*: it essentially states that whatever a verifier V can derive from the input $x \in S$, its prior knowledge and the interaction with P, could as well have been computed without any interaction with P. Thus, for proving that an IPS protocol is zero-knowledge, we had to argue that the *View* of V on x, defined as the concatenation of all messages sent from P to V as well as all random bits used by V during the execution of the protocol on x can actually be simulated without P. For the concrete example of Fig. 1, this boils down to proving that given x, we can construct elements *following the same distribution* as those output by the actual prover. This is indeed the case: in the first step, the simulator chooses u and further selects a bit b_1 (both u.a.r.). Now, if $b_1 = 0$, he sets $y = u^2$, while if $b_1 = 1$ he sets $y = x^{-1}u^2$. Assume further that the bit chosen by the verifier is $b = b_1$ (otherwise, simulation is aborted), in this case, the simulator sets $z = u$. It is easy to see that the simulated interaction follows the same distribution than the real interaction, for in any case the elements y and z output by the simulator/prover will be distributed u.a.r. in $\mathbb{Z}_n^*$.

Common input: Two graphs $G_1 = (V, E_1)$ and $G_0 = (V, E_0)$
Interaction:

    1. V selects for $i = 1, \ldots, m$, a bit $\alpha_i \in \{0, 1\}$ u.a.r. and constructs for each i a graph H_i, which is a random graph isomorphic to G_{α_i}.
    2. P outputs $\beta_i \in \{0, 1\}$ such that H_i is isomorphic to G_{β_i} for $i = 1, \ldots m$.

Verification:

    • V sets Res $= 1$ if and only if $\alpha_i = \beta_i$ for all $i = 1, \ldots m$, otherwise, V sets Res $= 0$.

Fig. 2 A IPS for Graph Non-Isomorphism

2.3 The Complexity Class $\mathscr{IP}$

If the error probability in the soundness condition of an interactive proof system is nonzero, it is typically called a *probabilistic proof system*.[3] There is a fundamental reason for this, as proven in Proposition 9.2. of [8]: whenever the soundness error is zero, the considered set S is in the complexity class $\mathscr{NP}$; namely, any object can be proven to belong to S by providing an evidence that can be checked in (deterministic) polynomial time.

If we denote by $\mathscr{IP}$ the class of sets having interactive proof systems, it is widely believed that $\mathscr{NP}$ is a proper subset of $\mathscr{IP}$. A nice illustrative example of the intuition behind that separation is the set S of non-isomorphic graphs, which can be proven to belong to $\mathscr{IP}$, while no deterministic (non-interactive) proof is known for that set (i.e., S is not known to be in $\mathscr{NP}$). Thus, randomness is essential in the above definition; there stems the power of probabilistic proof systems.

Example 2.2 In Fig. 2 we depict an IPS for Graph Non-Isomorphism [11]. Namely, a prover P wants to convince a verifier V that two graphs are not isomorphic. At this, given a graph G, by a *random graph isomorphic to G* we mean the graph defined by selecting uniformly at random a permutation σ on the set of vertices of G and applying it (in the natural way) to the set of edges of G.

It is easy to see that the IPS in Fig. 2 is correct and sound[4]; yet, it is not zero-knowledge; a cheating verifier may gain information from the interaction (this is the case if, for instance, he has not constructed the H_i himself from the first step, and thus ignores the bits α_i).

---

[3] Actually, the terms *interactive* and *probabilistic* are often used as synonyms in this setting.

[4] For soundness: if G_0 and G_1 were isomorphic, we take that α_i will equal 1 with probability $\frac{1}{2}$; as a result, the probability that the verifier does not reject in this case is at most $\frac{1}{2^m}$.

3 Provable Security

One of the most challenging problems in cryptography is the design of constructions that can be proven secure in a rigorous way and not only based on heuristics or on their resistance to some attacks. This last approach has often had unfortunate consequences: many proposed protocols that were designed preventing specific attacks, could however be cracked via unforeseen cryptanalytic techniques. Provable security, introduced by Goldwasser and Micali [9], tries to specify the security requirements of cryptographic constructions through a meaningful formal model capturing what "security" actually means in each concrete application scenario. Thus, a formal security model should specify two points in a well-motivated way:

- *Adversarial goal*: captures what it means to be "secure" or alternatively, when a system is considered broken.
- *Adversarial model*: captures how an adversary can interact with the users of the system, i.e., which capabilities the adversary has (computational power and available information).

Regarding adversarial capabilities, a system is *computationally secure* if only adversaries with bounded computational resources are considered. On the other hand, a system is *unconditionally secure* or *information-theoretically secure* if it is secure against adversaries with unbounded computational power [23]. While the security proofs of the first kind of systems are based on complexity theory, assuming the hardness of some computational problems, proofs of unconditional security are based on information theory.

3.1 Information Theoretic Security

Information theory goes back to Claude Shannon [17] in the late 1940s. It is one of the foundational theories behind modern computer science and has applications in many areas, including cryptography (see [16, 24]). In this section we will give an overview of the concepts of information theory needed to understand its relationship to cryptography. We follow mainly [20, 22], and heartily refer the interested reader to the inspiring note (in Spanish) [6].

Our driving example will be private key encryption. We formalize a private key encryption scheme as a tuple of algorithms (KeyGen, Enc, Dec), as follows:

- KeyGen: the key generation algorithm produces a key k, given a security parameter.
- Enc: the encryption algorithm, on input a plaintext m and a key k, outputs a ciphertext c.
- Dec: the decryption algorithm, given a ciphertext c and a key k, recovers a plaintext m.

In this setting, we also need to specify the sets of possible plaintexts $\mathscr{P}$, keys $\mathscr{K}$ and ciphertexts $\mathscr{C}$. These sets can be thought of as discrete random variables, and we will consistently thus denote by $\Pr(P = m)$, $\Pr(K = k)$ and $\Pr(C = c)$ the probability of P, K and C taking values m, k and c, respectively. Intuitively, for an encryption scheme to be considered secure we would like that a ciphertext c does not reveal any information about the corresponding plaintext m.

Definition 3.1 (*Perfect secrecy* [17]) An encryption scheme has *perfect secrecy* if

$$\Pr(P = m \mid C = c) = \Pr(P = m)$$

for all plaintext $m \in \mathscr{P}$ and all ciphertexts $c \in \mathscr{C}$.[5]

This definition implies that if an encryption scheme is perfectly secure, then $|\mathscr{K}| \geq |\mathscr{C}| \geq |\mathscr{P}|$.

Theorem 3.1 (Shannon [17]) *Let* $\Pi = (\texttt{KeyGen}, \texttt{Enc}, \texttt{Dec})$ *be an encryption scheme with* $|\mathscr{K}| = |\mathscr{C}| = |\mathscr{P}|$. *Then, it provides* perfect secrecy *if and only if:*

- *every key is used with the same probability, i.e.,* KeyGen *simulates the uniform distribution in* $\mathscr{K}$,
- *for each plaintext* $m \in \mathscr{P}$ *and each ciphertext* $c \in \mathscr{C}$ *there is a unique key* $k \in \mathscr{K}$ *such that* $\texttt{Enc}_k(m) = c$.

Definition 3.1 about perfect secrecy can be rewritten in terms of entropy and information: *entropy* is a mathematical measure of information or uncertainty about a random variable X. Similarly, *mutual information* measures how much information a random variable provides about another one, i.e. how much the knowledge of one of these variables reduces uncertainty about the other one.[6] An encryption scheme has *perfect secrecy* if the mutual information between P and C is zero. This also implies that the entropy of C must be no smaller than that of P.

3.2 Computational Security

Security definitions are normally defined through a "security game" between two players: a challenger and an adversary. Both players are probabilistic processes which communicate with each other. Thus, the game can be modeled as a probability space. The challenger generates the parameters needed in the system and may answer to queries the adversary makes to different oracles, modeling the attacker's capabilities. At the end of the game, it is determined whether the adversary wins, provided some particular event Succ occurs, which means the adversary breaks the scheme. The

---

[5]Here $\Pr(P = m \mid C = c)$ denotes conditional probability, i.e., the probability of $P = m$ once we know the ciphertext is c.

[6]A formal discussion on entropy and information can be found in [12].

proof must assess that the probability of an adversary winning the security game is sufficiently small (i.e. negligible in the security parameter considered), or more generally, that this probability differs only negligibly from a given value (that reflects his chances of trivially breaking the scheme).

We will see an example of a security model for signature schemes see [5]. First, we present the formal definition and an example of a signature scheme.

Definition 3.2 (*Signature Scheme*) A *digital signature scheme* is a triple of algorithms (KeyGen, Sign, Verify) as follows:

- KeyGen: the key generation algorithm produces a pair of keys (sk, vk), where sk is a private signing key and vk is the verification key, which is public.
- Sign: the signing algorithm, on input a plaintext m and a signing key sk, outputs a signature σ.
- Verify: the verification algorithm, given a signature σ, a verification key vk and a plaintext m, outputs either a 1, if the signature is valid, or a 0 otherwise.

Example 3.1 An example of digital signature algorithm is the RSA Full Domain Hash signature scheme [3], that is shown in Fig. 3.

The most standard security definition for signature schemes is *existential unforgeability under adaptive chosen message attack (EUF-CMA)*, that is defined below. This notion captures the intuition that the adversary should not create on its own new valid signatures, even having access to a signing oracle that may sign arbitrary messages.

Definition 3.3 (*UF-CMA security*) A signature scheme is *existentially unforgeable* against *chosen message attacks (UF-CMA)* if for any probabilistic polynomial time adversary $\mathscr{A}$ the advantage

---

Let $H : \{0,1\}^* \to \mathbb{Z}_n^*$ be a hash function

KeyGen:

- Choose two large random primes p, q.
- $n = pq$ and $\phi(n) = (p-1)(q-1)$.
- Pick $e \in \mathbb{Z}_n$ relatively prime to $\phi(n)$.
- $d = e^{-1} \pmod{\phi(n)}$.
- Return $sk = d, vk = (n, d)$.

Sign:

- Return $\sigma = (H(m))^d \pmod{n}$

Verify:

- $h = (\sigma)^e \pmod{n}$
- If $h = H(m)$, return 1. Otherwise, return 0.

Fig. 3 RSA-FDH digital signature scheme

> **Setup:** The challenger $\mathscr{C}$ runs the algorithm KeyGen and hands the public key vk to $\mathscr{A}$.
>
> **Challenge:** The adversary is allowed to ask (adaptively) queries for:
>
> - signatures on a message m. The challenger responds with $\mathrm{Sign}_{sk}(m)$.[a]
>
> **Forgery Phase:** The adversary outputs a tuple (σ^*, m^*) and wins if and only if:
>
> - $\mathrm{Ver}(m^*, \sigma^*, vk) = 1$;
> - in the challenge phase, there were no signature queries on m^*.
>
> ---
>
> [a] This is typically modeled speaking of *queries to a Sign oracle*, see the proof in the next section.

Fig. 4 UF-CMA security of a digital signature

$$\mathrm{Adv}_{\mathscr{A}}^{UF-CMA}(k) := \Pr[\mathrm{Succ}_{\mathscr{A}}^{UF-CMA}]$$

is a negligible function in the security parameter.[7] Here $\mathrm{Succ}_{\mathscr{A}}^{UF-CMA}$ denotes the event that $\mathscr{A}$ wins the game presented in Fig. 4.

Game-Hopping Proofs

A common approach in security proofs is the *game-hopping* or *sequence of games* technique discussed by Shoup [19] and Dent [4]. This way it is often easier to bound the probability of an adversarial win, especially when several unrelated assumptions are to be taken or different cryptographic primitives are involved in a cryptographic construction.

Following the game-hopping approach, a sequence of games is considered, where the first game is the original attack game and the adversary's success probability can be easily bounded for the last game. Let $\mathscr{A}$ be the adversary. For each game, the attack environment is slightly altered in such a way that we can bound the difference in $\mathscr{A}$'s success probability. Finally, keeping track of all the probability changes and the bound found for the last game, we can determine an upper bound for the success probability in the original game. We can distinguish the following types of transitions between games:

- **Transitions based on indistinguishability:** At this, the input the adversary receives is changed. Instead of receiving an element from distribution P_1, he receives an element from P_2. If the adversary is able to distinguish between two adjacent games, that gives a method for distinguishing distributions P_1 and P_2, assumed to be computationally indistinguishable.
- **Transitions based on failure events:** Here, games G_i and G_{i+1} are constructed in such a way that they proceed identically unless some "failure event" F occurs. Denote by Succ_j the event capturing an adversarial win in game G_j. If both games are defined over the same probability space then this is equivalent to saying that events $\mathrm{Succ}_i \wedge \neg F$ and $\mathrm{Succ}_{i+1} \wedge \neg F$ are the same. Applying the Difference

---

[7]Informally, a negligible function has domain in $\mathbb{N}$, range in $\mathbb{R}^+$ and goes to zero faster than the inverse of any polynomial.

Lemma [19], to prove that $|\Pr(\mathsf{Succ}_i) - \Pr(\mathsf{Succ}_{i+1})|$ is negligible, it is enough to show that $\Pr(F)$ is negligible.

- **Transitions based on large failure events**: In this kind of transition, games G_i and G_{i+1} proceed identically unless some failure event F occurs. However, the difference with the above transition is that now the probability of this failure event is large, although not overwhelming ($\Pr(\neg F)$ is not negligible). We assume the environment can detect when F occurs and events Succ_i and F are independent. The environment behaves exactly the same in the two adjacent games, unless event F occurs. In this case, the environment halts the simulation, and the adversary is considered not to have won the game. Otherwise, the adversary is considered to win in G_{i+1} if and only if it would win in G_i. Therefore, $\Pr(\mathsf{Succ}_{i+1}) = \Pr(\mathsf{Succ}_i \wedge \neg F) = \Pr(\mathsf{Succ}_i) \cdot \Pr(\neg F)$.

  Thus, $\Pr(\mathsf{Succ}_i)$ is non-negligible if and only if $\Pr(\mathsf{Succ}_{i+1})$ is non-negligible.
- **Bridging steps**: This kind of transition introduces a "bridging step", i.e. the inputs to the attacker are the same, but have been computed in different ways, which should however be completely equivalent. Therefore $\Pr(\mathsf{Succ}_i) = \Pr(\mathsf{Succ}_{i+1})$. This kind of transition is often done to prepare for one of the transitions discussed above, making the proof easier to follow.

Example 3.2 (Security of RSA-FDH signature scheme) We will show now an example of game-hopping proof to prove the security of the digital signature presented in Fig. 3. Its security relies on the following computational problem:

Definition 3.4 *(RSA problem)* Let $n = pq$ be the product of two distinct primes of length ℓ and let e be an integer relatively prime to $(p-1)(q-1)$. Given n, e and y chosen u.a.r from $\mathbb{Z}_n^*$, compute x such that $x^e = y \pmod{n}$.

If $\mathscr{A}$ is a probabilistic polynomial time adversary, we denote $\mathscr{A}$'s probability of success in solving this problem as $\mathsf{Succ}_{\mathscr{A}}^{RSA}$.

Theorem 3.2 *RSA-FDH signature scheme is EUF-CMA secure.*

Proof The proof is conducted through a sequence of games [21]. We denote by $\Pr[S_i]$ the advantage (see Definition 3.3) of the adversary when confronted with **Game i**. Moreover, we prove the security in the random oracle model [2], i.e. we consider the outputs of hash functions are truly random. The challenger will keep a list to answer $\mathscr{A}$'s queries to H.

Game 0. All the oracles are simulated as in the real protocol; thus, $\Pr[S_0]$ is exactly $\mathsf{Adv}_{\mathscr{A}}^{\mathsf{EUF\text{-}CMA}}$, as in Definition 3.3.

Game 1. In this game, we guess the index I of which query to H will be the first one for m^*. Let $\mathsf{GuessCorrect}$ be the event we guess this index correctly. This game is identical to **Game 0**, except that the execution is aborted if the event $\mathsf{GuessCorrect}$ does not occur. Let q_s be the number of queries to the $Sign$ oracle and q_h the number of queries to the random oracle H, respectively. The probability of $\mathsf{GuessCorrect}$ is upper bounded by $\frac{1}{q_s+q_h}$. Thus,

$$\Pr[S_1] \geq \Pr[S_0] \cdot \frac{1}{q_s + q_h}.$$

Game 2. This game is identical to **Game 1**, except that the simulation of the H oracle is changed. Now, the RSA problem challenge y is returned in the Ith query to H. Since y was chosen u.a.r, the distributions are identical. Thus, $\Pr[S_2] = Pr[S_1]$.

Game 3. This game behaves as the previous one, except that now, for every query x to H, an element s is chosen u.a.r. from $\mathbb{Z}_n^*$. Then, $z = s^e \pmod{n}$ is computed. The oracle will return z and an entry (x, s, z) is added to the H list. The signing oracle is changed accordingly. When asked to sign x, it looks for an entry (x, s, z) in the H list and returns s. Notice that signatures are valid and the distributions are identical. Therefore, $\Pr[S_3] = Pr[S_2]$.

Now, if the adversary returns a valid forgery (m^*, σ^*), then, m^* was the Ith query to the H list. Therefore, $H(m^*) = y$ and σ^* is a solution to the RSA problem for (n, e, y). Thus, the challenger in **Game 3** acts as an algorithm $\mathscr{B}$ which solves the RSA problem with the help of $\mathscr{A}$. Therefore, $\Pr[S_3] = \mathrm{Succ}_{\mathscr{B}}^{\mathsf{RSA}}$.

Combining all the probabilities, we get

$$\mathrm{Succ}_{\mathscr{A}}^{\mathsf{EUF-CMA}} \leq (q_s + q_h + 1)\mathrm{Succ}_{\mathscr{B}}^{\mathsf{RSA}}. \qquad \square$$

Acknowledgements This paper is affectionately dedicated to Pedro, who enthusiastically lead the first steps of so many students in the information theory pathways. Authors 2,3 and 4 have been partially supported by project MTM2013-45588-C3-1-P. Authors 2 and 3 have been partially supported by project GRUPIN 14-142, Principado de Asturias.

References

1. Barak B (2016) Lecture notes: zero knowledge proofs. http://www.boazbarak.org
2. Bellare M, Rogaway P (1993) Random oracles are practical: a paradigm for designing efficient protocols. In: Denning D, Pyle R, Ganesan R, Sandhu R, Ashby V (eds) Proceedings of 1st ACM conference on computer and communications security. ACM, New York
3. Bellare M, Rogaway P (1996) The exact security of digital signatures - how to sign with RSA and Rabin. In: Maurer U (ed) Advances in cryptology EUROCRYPT'96, vol 1070. Lecture notes in computer science. Springer, Berlin
4. Dent A (2006) A note on game-hopping proofs. IACR Cryptology ePrint Archive: Report 2006/260
5. Fiat A, Shamir A (1987) How to prove yourself: practical solutions to identification and signature problems. In: Odlyzko AM (ed) Advances in cryptology CRYPTO'86, vol 263. Lecture notes in computer science. Springer, Berlin
6. Gil P (2007) Por qué teoría de la información? Bol Soc Estad Investig Oper 23(3):8–9
7. Goldreich O (2004) The foundations of cryptography - volume 1, basic tools. Cambridge University Press, Cambridge
8. Goldreich O (2008) Computational complexity, a conceptual perspective. Cambridge Univeristy Press, Cambridge
9. Goldwasser S, Micali S (1984) Probabilistic encryption. J Comput Syst Sci 28(2):270–299

10. Goldwasser S, Micali S, Rackoff C (1985) The knowledge complexity of interactive proof-systems. In: Sedgewick R (ed) STOC'85 Proceedings of 17th annual ACM symposium on theory of computing. ACM, New York
11. Goldreich O, Micali S, Wigderson A (1991) Proofs that yield nothing but their validity or all languages in NP vave zero-knowledge proof systems. J ACM 38(3):691–729
12. Gray RM (2013) Entropy and information theory. Springer, New York
13. Jaikin A (2013) Grafos, grupos y variedades: un punto de encuentro. Gaceta Real Soc Matem Española 16(4):761–776
14. Katz J (2010) Digital signatures. Springer, New York
15. Koblitz N, Menezes A (2007) Another look at "provable security". J Cryptol 20(1):3–37
16. Maurer UM (1993) The role of information theory in cryptography. In: Farrell PG (ed) Codes and ciphers: cryptography and coding IV, proceedings of 4th IMA conference on cryptography and coding. IMA Press, Berlin
17. Shannon C (1948) A mathematical theory of communication. Bell Syst Tech J 27(3):379–423, 623–656
18. Shor P (1994) Algorithms for quantum computation: Discrete logarithms and factoring. In: SFCS'94 proceedings of the 35th annual symposium on foundations of computer science. IEEE Computer Society, Washington
19. Shoup V (2004) Sequences of games: a tool for taming complexity in security proofs. IACR cryptology ePrint archive: report 2004/332
20. Smart N (2003) Cryptography: an introduction, 3rd edn. McGraw-Hill College, New York
21. Stebila D (2014) An introduction to provable security. Lecture notes from AMSI winter school on cryptography. https://www.douglas.stebila.ca/teaching/amsi-winter-school/
22. Stinson D (1997) Cryptography: theory and practice. CRC Press, Boca Raton
23. Vernam GS (1926) Cipher printing telegraph systems for secret wire and radio telegraphic communications. J Am Inst Electron Eng 55:109–115
24. Wolf S (1998) Unconditional security in cryptography. In: Damgard I (ed) Lectures on data security, modern cryptology in theory and practice. Springer, Berlin

Tools for Detecting Leverage and Influential Observations in Generalized Linear Models

María Carmen Pardo

Abstract There has been extensive development of diagnostic measures for Generalized Linear Models fitted by the maximum likelihood method. However, there is evidence that the maximum likelihood estimator is extremely sensitive to outlying, leverage and influential observations. We propose a diagnostic measure based on minimum distance estimates to assess the effect that the estimation method has on parameter estimates. Furthermore, a new single case deletion diagnostic to detect leverage observations is developed. Finally, the paper concludes with an analysis of real data.

1 Introduction

Generalized linear models (GLM) have been introduced by Nelder and Wedderburn [6] as a unifying family of models for nonstandard cross-sectional regression analysis with non-normal responses. We focus on multinomial response models that model relationships between a multinomial response variable and a set of explanatory variables.

The usual method of fitting GLM, maximum likelihood, is extremely sensitive to outlying responses and to extreme and influential points. There is evidence that minimum distance estimates have favorable asymptotic and small sample properties in some settings. Minimum distance estimator was presented for the first time by Wolfowitz [17] and it provides a convenient method of consistently estimating unknown parameters. An extensive bibliography for minimum distance estimators is in Parr [15]. Pardo et al. [14] and Pardo JA and Pardo MC [11] defined the minimum ϕ-divergence estimation as a generalization of the maximum likelihood estimation for logistic regression and binary GLM, respectively. This estimation procedure was generalized for multinomial GLM by Pardo MC [9].

Much of the work with minimum distance estimation has focussed on hypothesis testing, but only a limited amount of work has been performed, in the area of influence

M. C. Pardo (✉)
Universidad Complutense de Madrid, Plaza de Ciencias, 3, 28040 Madrid, Spain
e-mail: mcapardo@ucm.es

© Springer International Publishing AG 2018
E. Gil et al. (eds.), *The Mathematics of the Uncertain*, Studies in Systems, Decision and Control 142, https://doi.org/10.1007/978-3-319-73848-2_47

diagnostic for this method estimation. Bedrick and Tsai [1] proposed several diagnostic tools for detecting outlying and leverage points based on the minimum power divergence estimation for binomial response models. These points can be or can not be influential points. That is to say, if removing it or adding it would markedly change the regression outputs this point is called influential. This idea was exploited by Pardo MC [8] and Pardo MC and Pardo JA [13] who show that based-maximum likelihood diagnostics for multivariate extensions of generalized linear models extend naturally to the family of the minimum ϕ-divergence estimators. Pardo MC [8] proposed a generalized hat matrix to determine how much leverage each data value can have on each fitted value to a GLM for ordinal data and a new family of residuals for detecting outliers. Pardo MC and Pardo JA [13] proposed a mean slippage model for detecting outliers.

In Sect. 2, we introduce the model, the minimum ϕ-divergence estimation method and the notation of the paper. Diagnostic Tools for detecting influential and leverage points for GLM fitted by minimum distance are presented in Sect. 3. The first one is a generalization of one existing for maximum likelihood fit and the second one is a new one based on ϕ-divergences. For illustration, the procedure is applied to a data set in Sect. 4.

2 Background and Notation for GLM

Let Y be the response variable with J possible values (labeled $1, \ldots, J$), which is observed together with m explanatory variables $x^T = (x_1, \ldots, x_m) \in \mathbb{R}^m$. Given x, Y is a multinomially distributed with probability vector $\pi^T = (\pi_1, \ldots, \pi_J)$ and $\pi_r = P(Y = r \mid x^T), r = 1, \ldots, J$.

Suppose that the x_i^T takes N different values,

$$x_i^T = (x_{i1}, \ldots, x_{im}), \quad i = 1, \ldots, N.$$

The multinomial GLM assumes that $\mu_i = \mathrm{E}\left[Y \mid x_i^T\right]$ is related to the linear predictor

$$\eta_i = Z_i^T \beta$$

by

$$\mu_i = \mathbf{h}\left(\eta_i\right) = \mathbf{h}(Z_i^T \beta), \quad i = 1, \ldots, N, \tag{1}$$

where $\mathbf{h}$ is a vectorial response function, Z_i is a $p \times (J - 1)$-design matrix obtained from x_i and β is a p-dimensional vector of unknown parameters.

Let $n(x_i)$ be the number of observations considered when the explanatory variable x^T takes the value x_i^T, in such a way that if x^T is fixed at x_i^T we have a multinomial distribution with parameters $\left(n(x_i); \pi_1(Z_i^T \beta), \ldots, \pi_{J-1}(Z_i^T \beta)\right)$.

Different models are obtained when we specify the response function and the design matrix such as cumulative models, sequential models among others (see [4]).

Suppose we observe the sample $Y_1 = y_1, \ldots, Y_N = y_N$ jointly with the explanatory variables $x_1, \ldots, x_N$, Pardo MC [9] proposed to estimate the vector $\boldsymbol{\beta}$ of unknown parameters using the minimum ϕ-divergence estimator which is obtained minimizing the ϕ-divergence measure between

$$\widehat{p} = \left(\frac{y_{11}}{n}, \ldots, \frac{y_{J1}}{n}, \frac{y_{12}}{n}, \ldots, \frac{y_{J2}}{n}, \ldots, \frac{y_{1N}}{n}, \ldots, \frac{y_{JN}}{n} \right)^T,$$

with $y_{Ji} = n(x_i) - \sum_{s=1}^{J-1} y_{si}$, $i = 1, \ldots, N$, $n = n(x_1) + \cdots + n(x_N)$ and

$$p(\boldsymbol{\beta}) = \left(\frac{n(x_1)}{n} \widetilde{\pi} \left(Z_1^T \boldsymbol{\beta} \right)^T, \ldots, \frac{n(x_N)}{n} \widetilde{\pi} \left(Z_N^T \boldsymbol{\beta} \right)^T \right)^T$$

being $\widetilde{\pi}(Z_i^T \boldsymbol{\beta})^T = \left(\pi_1(Z_i^T \boldsymbol{\beta}), \ldots, \pi_J(Z_i^T \boldsymbol{\beta}) \right)$. Therefore, the minimum ϕ-divergence estimator is defined as

$$\widehat{\boldsymbol{\beta}}_\phi = \arg \min_{\boldsymbol{\beta} \in \Theta} D_\phi \left(\widehat{p}, p(\boldsymbol{\beta}) \right) \tag{2}$$

with

$$D_\phi \left(\widehat{p}, p(\boldsymbol{\beta}) \right) = \sum_{l=1}^{J} \sum_{i=1}^{N} \pi_l(Z_i^T \boldsymbol{\beta}) \frac{n(x_i)}{n} \phi \left(\frac{y_{li}/n}{\pi_l \left(Z_i^T \boldsymbol{\beta} \right) n(x_i)/n} \right), \tag{3}$$

where $\phi \in \Phi$ and Φ is the class of all convex functions $\phi(x)$, $x > 0$, such that at $x = 1$, $\phi(1) = \phi'(1) = 0$, $\phi''(1) > 0$, and at $x = 0$, $0\phi(0/0) = 0$ and $0\phi(p/0) = p \lim_{u \to \infty} \phi(u)/u$. For more details about ϕ-divergences see Vajda [16] and Pardo L [7]. As a particular case, this family contains the maximum likelihood estimator (MLE).

An important family of ϕ-divergences in statistical problems is the power-divergence family,

$$\begin{aligned}
\phi_{(a)}(x) &= (a(a+1))^{-1} \left(x^{a+1} - x \right); \ a \neq 0, a \neq -1, \\
\phi_{(0)}(x) &= \lim_{a \to 0} \phi_{(a)}(x) = x \log x - x + 1, \\
\phi_{(-1)}(x) &= \lim_{a \to -1} \phi_{(a)}(x) = -\log x + x - 1,
\end{aligned} \tag{4}$$

which was introduced and studied by Cressie and Read [3]. Under mild regularity conditions this estimator is BAN.

3 Model Diagnostics

After selecting a preliminary model for a data set, it is of interest for an analyst to be able to find influential cases and, based on them, make decisions concerning their usefulness in a problem at hand. Observations that are badly predicted (or allocated) are termed outlies. Pardo MC [8] defined a family of residuals based on the ϕ-divergence measures to identify outlying points in multinomial GLM. As particular case, this family contains the Pearson and the Deviance residuals. Pardo JA and Pardo MC [13] proposed an outlier detection diagnostic using the influence of single cases on a family of ϕ-divergence test statistics which contains as particular case the likelihood ratio test. Although, an outlies may be an influential observation, i.e., its omission from the data set results in substantial changes to certain aspects of the fit of the model. However, an influential observation need not necessarily be an outlies. For example, when the observation distorts the form of the fitted model to such an extend that the observation itself has a small residual. Therefore, the presence of an influential observation can not necessarily be detected from a direct examination of the residuals, and so additional diagnostic techniques are required.

To analyze the influence of the observation x_i in the estimation of $\boldsymbol{\beta}$, we propose the family of influence measures

$$C_{\phi_2}^{(i)} = \frac{1}{p} \left(\widehat{\boldsymbol{\beta}}_{\phi_2} - \widehat{\boldsymbol{\beta}}_{\phi_2}^{(i)} \right)^T Cov \left(\widehat{\boldsymbol{\beta}}_{\phi_2} \right)^{-1} \left(\widehat{\boldsymbol{\beta}}_{\phi_2} - \widehat{\boldsymbol{\beta}}_{\phi_2}^{(i)} \right)$$

with

$$Cov \left(\widehat{\boldsymbol{\beta}}_{\phi_2} \right) \approx \left(\boldsymbol{Z} \widetilde{\boldsymbol{V}} \left(\boldsymbol{\beta}^0 \right) \boldsymbol{Z}^T \right)^{-1}$$

where $\boldsymbol{\beta}^0$ is the true value of the parameter $\boldsymbol{\beta}$,

$$\widetilde{\boldsymbol{V}} \left(\boldsymbol{\beta} \right) = Diag \left(\widetilde{\boldsymbol{V}}_1 \left(\boldsymbol{\beta} \right), \ldots, \widetilde{\boldsymbol{V}}_N \left(\boldsymbol{\beta} \right) \right)$$

with $\widetilde{\boldsymbol{V}}_i \left(\boldsymbol{\beta} \right) = n \boldsymbol{V}_{n,i} \left(\boldsymbol{\beta} \right)$ being

$$V_{n,i} \left(\boldsymbol{\beta} \right) = \frac{n \left(x_i \right)}{n} \frac{\partial \boldsymbol{\pi} \left(\eta_i \right)}{\partial \eta_i} \Sigma_i^{-1} \left(\boldsymbol{\beta} \right) \frac{\partial \boldsymbol{\pi} \left(\eta_i \right)}{\partial \eta_i^T} \tag{5}$$

and $\Sigma_i^{-1} \left(\boldsymbol{\beta} \right) = \left(v_{sr} \left(\boldsymbol{\beta} \right) \right)_{s,r=1,\ldots,J-1}$ with

$$v_{sr} \left(\boldsymbol{\beta} \right) = \begin{cases} \dfrac{1}{\pi_r \left(\boldsymbol{Z}_i^T \boldsymbol{\beta} \right)} + \dfrac{1}{\pi_J \left(\boldsymbol{Z}_i^T \boldsymbol{\beta} \right)} & r = s \\ \dfrac{1}{\pi_J \left(\boldsymbol{Z}_i^T \boldsymbol{\beta} \right)} & r \neq s \end{cases}.$$

The value of $C_{\phi_2}^{(i)}$ indicates the ith observation influence. For $\phi_2(x) = x \log x - x + 1$, we get the Cook's distance (Cook [2]) which is commonly used for the detection of influential observations in regression analysis. It is an overall measure of the change in the parameter estimates when one or more observations are deleted from the data set.

Pardo MC [8] defined a generalized form of the hat matrix based on minimum ϕ-divergence estimation as

$$H\left(\widehat{\boldsymbol{\beta}}_\phi\right) = \boldsymbol{V}_n\left(\widehat{\boldsymbol{\beta}}_\phi\right)^{1/2} \boldsymbol{Z}^T \boldsymbol{I}_{F,n}\left(\widehat{\boldsymbol{\beta}}_\phi\right)^{-1} \boldsymbol{Z} \boldsymbol{V}_n\left(\widehat{\boldsymbol{\beta}}_\phi\right)^{1/2}. \tag{6}$$

The square matrix $\boldsymbol{H}$ and $\boldsymbol{M} = \boldsymbol{I} - \boldsymbol{H}$ are projection block matrices, where the blocks $\boldsymbol{H}^{ij}\left(\widehat{\boldsymbol{\beta}}_\phi\right)$ and $\boldsymbol{M}^{ij}\left(\widehat{\boldsymbol{\beta}}_\phi\right)$ $(i, j = 1, \ldots, N)$, are $(J-1)$-dimensional, respectively.

Observations with relatively large values of $\widehat{h}_{ii}^{\phi_2}$, which are the diagonal elements of the hat matrix defined in (6), are distant from the others on the basis of the values of their explanatory variables; they may be influential but will not necessarily be. These observations are called leverage points and they are characterized by the fact that it greatly increases the variability of the estimates when omitted from the sample.

In the sequel, we propose a new indicator that measure the increase of variability when the explanatory variable $\boldsymbol{x}_i$ is omitted from the sample. This idea was exploited by Martin and Pardo L [5] in loglinear models and Pardo MC and Alonso [10] for generalized estimating equations methodology.

As

$$\widehat{\boldsymbol{\beta}}_{\phi_2} \simeq \mathcal{N}\left(\boldsymbol{\beta}^0, \left(\boldsymbol{Z}\boldsymbol{V}_\lambda\left(\boldsymbol{\beta}^0\right)\boldsymbol{Z}^T\right)^{-1}\right),$$

where $\boldsymbol{V}_\lambda\left(\boldsymbol{\beta}\right) = \lim_{n \to \infty} \boldsymbol{V}_n\left(\boldsymbol{\beta}\right)$, and

$$\widehat{\boldsymbol{\beta}}_{\phi_2}^{(i)} \simeq \mathcal{N}\left(\boldsymbol{\beta}^0, \left(\boldsymbol{Z}_{(i)}\boldsymbol{V}_\lambda^{(i)}\left(\boldsymbol{\beta}^0\right)\boldsymbol{Z}_{(i)}^T\right)^{-1}\right),$$

(see Remark 1 in Pardo JA and Pardo MC [12]), where the matrices $\boldsymbol{Z}_{(i)}$ and $\boldsymbol{V}_\lambda^{(i)}\left(\boldsymbol{\beta}^0\right)$ are the matrices $\boldsymbol{Z}$ and $\boldsymbol{V}_\lambda$, respectively, in which the observation $\boldsymbol{x}_i$ has been eliminated. Then to measure the variability of the estimates $\widehat{\boldsymbol{\beta}}_{\phi_2}$ and $\widehat{\boldsymbol{\beta}}_{\phi_2}^{(i)}$ is equivalent to measure the distance between $f_{\widehat{\boldsymbol{\beta}}_{\phi_2}}(\boldsymbol{x})$ and $f_{\widehat{\boldsymbol{\beta}}_{\phi_2}^{(i)}}(\boldsymbol{x})$ which are the asymptotic densities for $\widehat{\boldsymbol{\beta}}_{\phi_2}$ and $\widehat{\boldsymbol{\beta}}_{\phi_2}^{(i)}$, respectively. Therefore, we choose the ϕ-divergence measure to propose a new measure for detecting leverage points given as

$$D_\phi\left(\widehat{\boldsymbol{\beta}}_{\phi_2}, \widehat{\boldsymbol{\beta}}_{\phi_2}^{(i)}\right) = \int_{\mathbb{R}^p} f_{\widehat{\boldsymbol{\beta}}_{\phi_2}^{(i)}}(\boldsymbol{x})\,\phi\left(\frac{f_{\widehat{\boldsymbol{\beta}}_{\phi_2}}(\boldsymbol{x})}{f_{\widehat{\boldsymbol{\beta}}_{\phi_2}^{(i)}}(\boldsymbol{x})}\right) d\boldsymbol{x}.$$

In the particular case that we consider the power-divergence family introduced by Cressie and Read [3], i.e., the ϕ-divergence given in (4). We have, after several algebraic operations, that

$$\int_{\mathbb{R}^p} \frac{f_{\widehat{\boldsymbol{\beta}}_{\phi_2}}(x)^{a+1}}{f_{\widehat{\boldsymbol{\beta}}_{\phi_2}^{(i)}}(x)^a} dx = \frac{Det(I + aH^{ii}(\beta^0))^{-1/2}}{Det(M^{ii}(\beta^0))^{a/2}}$$

if $a \neq 0$ and $a \neq -1$. Then, we obtain the following expression for the new measure to detect leverage points

$$I_a^{(i)}\left(\widehat{\boldsymbol{\beta}}_{\phi_2}\right) = \frac{1}{a(a+1)}\left[\frac{Det\left(I + aH^{ii}\left(\widehat{\boldsymbol{\beta}}_{\phi_2}\right)\right)^{-1/2}}{Det\left(I - H^{ii}\left(\widehat{\boldsymbol{\beta}}_{\phi_2}\right)\right)^{a/2}} - 1\right] \quad a \neq 0, a \neq -1,$$

$$I_0^{(i)}\left(\widehat{\boldsymbol{\beta}}_{\phi_2}\right) = -\frac{1}{2}\left\{Trace\left(H^{ii}\left(\widehat{\boldsymbol{\beta}}_{\phi_2}\right)\right) + \log\left(Det\left(I - H^{ii}\left(\widehat{\boldsymbol{\beta}}_{\phi_2}\right)\right)\right)\right\},$$

$$I_{-1}^{(i)}\left(\widehat{\boldsymbol{\beta}}_{\phi_2}\right) = \frac{1}{2}\left\{Trace\left(\left(Z\widetilde{V}\left(\widehat{\boldsymbol{\beta}}_{\phi_2}\right)Z^T\right)\left(Z_{(i)}\widetilde{V}^{(i)}\left(\widehat{\boldsymbol{\beta}}_{\phi_2}\right)Z_{(i)}^T\right)^{-1} - I\right)\right.$$
$$\left. + \log\left(Det\left(M^{ii}\left(\widehat{\boldsymbol{\beta}}_{\phi_2}\right)\right)\right)\right\}.$$

Remark 3.1 An alternative way to calculate the above measures is calculating $\lambda_1, \ldots, \lambda_{J-1}$, the eigenvalues of the matrix $H^{ii}\left(\widehat{\boldsymbol{\beta}}_{\phi_2}\right)$, since

$$\frac{Det\left(I + aH^{ii}\left(\widehat{\boldsymbol{\beta}}_{\phi_2}\right)\right)^{-1/2}}{Det\left(I - H^{ii}\left(\widehat{\boldsymbol{\beta}}_{\phi_2}\right)\right)^{a/2}} = \prod_{j=1}^{J-1}\left(\frac{1 + a\lambda_j}{(1 - \lambda_j)^{-a}}\right)^{-1/2}.$$

4 Numerical Example

As an illustration of the new tools for diagnostic presented in previous section we consider data on the perspectives of students, psychology students at the University of Regensburg were asked if they expected to find adequate employment after getting their degree. The response categories were ordered with respect to their expectation. The responses were 'don't expect adequate employment' (category 1), 'not sure' (category 2), and 'immediately after the degree' (category 3). The data are given in Fahrmeir and Tutz [4]. Table 1 shows the data for different ages of the students.

To fit the cumulative logit model

$$Pr\left(Y \leq r/Age\right) = \left(1 + \exp\left(-\left(\alpha_r + \beta \log\left(Age\right)\right)\right)\right)^{-1}, \qquad r = 1, 2,$$

Table 1 Grouped data for job expectations of psychology students in Regensburg

N° Obs	Age	Y 1	2	3	$n(x_i)$
1	19	1	2	0	3
2	20	5	18	2	25
3	21	6	19	2	27
4	22	1	6	3	10
5	23	2	7	3	12
6	24	1	7	5	13
7	25	0	0	3	3
8	26	0	1	0	1
9	27	0	2	1	3
10	29	1	0	0	1
11	30	0	0	2	2
12	31	0	1	0	1
13	34	0	1	0	1

Table 2 Minimum $\phi_{(a_2)}$ divergence estimators

a_2	α_1	α_2	β
0	14.9884	18.1497	−5.4027
2/3	8.4044	11.2404	−3.2143
1	5.8553	8.526	−2.3661

Fig. 1 Index plot of $I_a^{(i)}\left(\widehat{\beta}^1\right)$ as a function of a. Shown are $a = -1/2$ solid line, $a = 0$ dash line $a = 2/3$ dot line, $a = 1$ dash dot line

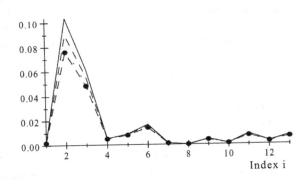

we use the minimum ϕ-divergence estimations with $\phi = \phi_{(a_2)}$ given $\phi_{(a_2)}$ in (4), $\widehat{\beta}^{a_2} \equiv \widehat{\beta}_{\phi_{(a_2)}}$, for $a_2 = 0$ (MLE), 2/3 (Cressie-Read estimator) and 1 (minimum chi-square estimator) which are shown in Table 2.

For displaying diagnostic tools, index plots are generally suggested. Figure 1 shows the measure $I_a^{(i)}\left(\widehat{\beta}^1\right)$. Index plots of $I_a^{(i)}\left(\widehat{\beta}^{a_2}\right)$ for $a_2 = 0$ and 2/3 are skipped by brevity since they are similar to Fig. 1.

Fig. 2 Index plot of $C_{\phi_{(a_2)}}^{(i)}$ as a function of a_2. Shown are $a_2 = 0$ solid line, $a_2 = 2/3$ dash line $a_2 = 1$ dot line

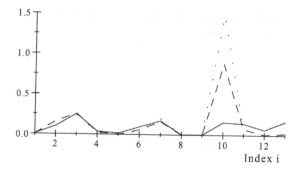

Fig. 3 $\chi^2 (2)$-probability plot of $\left(r_{i,S}^{\phi_{(a_2)}}\right)^T \left(r_{i,S}^{\phi_{(a_2)}}\right)$. Shown are $a_2 = 0$ dot, $a_2 = 2/3$ circle $a_2 = 1$ cross

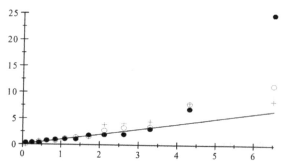

All measures identify Observations 2 and 3 as those having the highest leverage values. The three figures are very similar which means that to detect leverage points with these measures it does not matter which estimation method you use. However, there is difference among the measures obtained for different values of a. The leverage of observation 2 is higher when a decreases. Therefore, as Fahrmeir and Tutz [4] pointed out that are not leverage points since the values are primarily caused by the relatively local sample sizes, the best measure is $I_1^{(i)} \left(\widehat{\beta}^{a_2}\right)$, that is to say, to consider the Pearson divergence between $f_{\widehat{\beta}^{a_2}} (x)$ and $f_{\widehat{\beta}^{a_2(i)}} (x)$.

Next, we look for influential points drawing the index plot of the generalized Cook distance. Figure 2 shows that Observations 3, 7, 10, and 13 are the most influential. We find that Observation 10 is less influential for the estimation method corresponding to $a = 0$ (MLE).

The reason can be found in Fig. 3 where a χ^2-probability plot based on the standardized $\phi_{(a)}$-divergence residuals, $r_{i,S}^{\phi_{(a)}}$, which were defined in [8], is shown. Therefore, Observation 10 is an outlies when we estimate by MLE ($a = 0$) so it does not influence in the estimation. However, Observation 10 is not an outlies when we fit the model using the minimum $\phi_{(a_2)}$-divergence estimations with $a_2 = 2/3$ and 1 so it pintpoins as influential point.

Acknowledgements This work was partially supported by Grant MTM2013-40778-R. To the memory of Dr. Pedro Gil, an excellent statistician and great person.

References

1. Bedrick EJ, Tsai C (1993) Diagnostics for binomial response models using power divergence statistics. Comput Stat Data Anal 15:381–392
2. Cook RD (1977) Detection of influential observations in linear regression. Technometrics 19:15–18
3. Cressie NAC, Read T (1984) Multinomial goodness-of-fit tests. J R Stat Soc Ser B 46:440–464
4. Fahrmeir L, Tutz G (2001) Multivariate statistical modelling based on generalized linear models. Springer, New York
5. Martin N, Pardo L (2010) A new measure of leverage cells in multinomial loglinear models. Commun Stat-Theor Meth 39:517–530
6. Nelder JA, Wedderburn RWM (1972) Generalized linear models. J R Stat Soc Ser A 135:370–384
7. Pardo L (2006) Statistical inference based on divergence measures. Chapman & Hall, London
8. Pardo MC (2010) High Leverage points and outliers in generalized linear models for ordinal data. In: Skiadas CH (ed) Advances in data analysis, statistics for industry and technology. Birkhäuser, Boston
9. Pardo MC (2011) Testing equality restrictions in generalized linear models. Metrika 73(2):231–253
10. Pardo MC, Alonso R (2012) Influence measures based on the volume of confidence ellipsoids for GEE. Biom J 54(4):552–567
11. Pardo JA, Pardo MC (2008) Minimum phi-divergence estimator and phi-divergence statistics in generalized linear models with binary data. Methodol Comput Appl Probab 10(3):357–379
12. Pardo JA, Pardo MC (2009) Selecting and checking a binomial response generalized linear model using phi-divergences. J Stat Plan Inference 139:3038–3049
13. Pardo MC, Pardo JA (2011) Detection of outlying points in ordered polytomous. In: Pardo L, Balakrishnan N, Gil MA (eds) Modern mathematical tools and techniques in capturing complexity. Springer, Berlin
14. Pardo JA, Pardo L, Pardo MC (2006) Minimum ϕ-divergence estimator in logistic regression models. Stat Pap 47:91–108
15. Parr WC (1981) Minimum distance estimation: a bibliography. Commun Stat-Theor Meth 12:1205–1224
16. Vajda I (1989) Theory of statistical inference and information. Kluwer Academic Publishers, Dordrecht
17. Wolfowitz J (1953) Estimation by the minimum distance method. Ann Inst Stat Math 5:9–23

On Economic Applications of Information Theory

Rigoberto Pérez, Ana Jesús López, Covadonga Caso,
Mercedes Alvargonzález and María Jesús Río

Abstract In many of his writings, lectures and talks Pedro Gil emphasized the wide variety of economic applications of Information Theory. In this paper, as a tribute to him, we aim to provide some insight into this topic, mainly referred to income inequality, industrial concentration and economic forecasting.

1 Introduction

Theil's seminal book [42] set the foundations for research in the applications of Information Theory to economic topics such as income inequality, industrial concentration or aggregation in input-output analysis. Since then, information measures have provided an extremely useful toolbox for an increasing number of applications both in Economics and Econometrics.

Within this context, Pedro Gil's contributions include not only a brilliant academic background and a pioneering textbook [14] but also an outstanding leadership and a contagious enthusiasm. This paper is a tribute to him and provides an overview of some works conducted by a group of Pedro's disciples.

In the first section we shall start by setting out the information-theoretic framework for the applied analyses developed in the following sections.

R. Pérez (✉) · A. J. López · C. Caso · M. Alvargonzález · M. J. Río
Dpto. de Economía Aplicada, Universidad de Oviedo, Oviedo, Spain
e-mail: rigo@uniovi.es

A. J. López
e-mail: anaj@uniovi.es

C. Caso
e-mail: ccaso@uniovi.es

M. Alvargonzález
e-mail: malvarg@uniovi.es

M. J. Río
e-mail: mjrio@uniovi.es

© Springer International Publishing AG 2018
E. Gil et al. (eds.), *The Mathematics of the Uncertain*, Studies in Systems, Decision and Control 142, https://doi.org/10.1007/978-3-319-73848-2_48

515

2 Information Measures of Order β

Taking the well-known entropy measure proposed by Shannon [39] and the Kullback–Leibler divergence as starting point, a wide variety of generalized families of entropy and divergence measures have been proposed, in particular the β-order measures introduced by Havrda and Charvát [20].

Let X be a discrete random variable taking values $\{x_1, \ldots, x_k\}$ and $P = \{p_1, \ldots, p_k\}$ and $Q = \{q_1, \ldots, q_k\}$ two probability distributions associated with X ($p_i, q_i \geq 0$, $i = 1, \ldots, k$; $\sum_i p_i = \sum_i q_i = 1$).

Definition 2.1 The β-order entropy measure $H^\beta(P)$ associated with the probability distribution P is defined as:

$$H^\beta(P) = \frac{1}{1 - 2^{1-\beta}} \left[1 - \sum_{i=1}^{k} p_i^\beta \right] ; \ \beta > 0, \beta \neq 1. \tag{1}$$

When $\beta \to 1$ in (1), we get Shannon's entropy: $H_S(P) = -\sum_{i=1}^{k} p_i log(p_i)$.

$H^\beta(P)$ is a symmetric, continuous and concave function of its arguments p_i that achieves its maximum for the uniform distribution. It also satisfies a branching property, which is especially interesting in economic applications:

$$H^\beta(P) = H^\beta(p_1 + p_2, p_3, \ldots, p_k) + (p_1 + p_2)^\beta H^\beta \left(\frac{p_1}{p_1 + p_2}, \frac{p_2}{p_1 + p_2} \right). \tag{2}$$

Definition 2.2 The β-order divergence measure $D^\beta(Q : P)$ between the distributions Q and P is defined as:

$$D^\beta(Q : P) = \frac{1}{1 - 2^{1-\beta}} \left[\sum_{i=1}^{k} q_i \left(\frac{p_i}{q_i} \right)^{1-\beta} - 1 \right] ; \ \beta > 0, \beta \neq 1. \tag{3}$$

$D^\beta(Q : P)$ is a directed divergence measure and it is neither symmetric nor does it satisfy the triangular property. Hence, it is interpreted as a pseudo-distance that measures the gain of information resulting in the additional knowledge in Q relative to P. When $\beta \to 1$ in (3), we get the Kullback–Leibler divergence measure: $D_{KL}(Q : P) = \sum_{i=1}^{k} q_i log \left(\frac{q_i}{p_i} \right)$.

Entropy measures depend only on the probabilities of the possible values of X. In order to incorporate the nature of the values of a variable in measuring its uncertainty, we may consider a system of utilities $U = \{u_1, \ldots, u_k\}$ that quantify the quality of the values of X.

Definition 2.3 The β-measure of uncertainty involving utilities (unquietness) associated with X is given by:

$$HU^{*\beta}(P;U) = \begin{cases} \frac{1}{1-2^{1-\beta}}\left[\sum_{i=1}^{k} p_i \left(\frac{u_i}{E(u)}\right)^{1-\beta} - 1\right] & \beta > 0, \beta \neq 1 \\ -\sum_{i=1}^{k} p_i log\left(\frac{u_i}{E(u)}\right) & \beta = 1 \end{cases} \qquad (4)$$

$HU^{*\beta}$ belongs to the family of inset entropies characterized by Aczél and Kannappan [5] and can also be regarded as a directed divergence measure of order β as defined in (3). This measure was characterized by Gil et al. [17] in terms of a set of five axioms: symmetry, continuity, recursivity, relation with entropy and a normalization condition.

As population measures are usually unknown, an inferential approach to the measurement of uncertainty was proposed. In this context, our research focused on the family of *quadratic measures* ($\beta = 2$) in (1) and (4) which fulfill the essential properties of Shannon-type measures and have a more suitable behaviour in estimation problems. Pérez [32, 33] and Caso and Gil [9] derived unbiased estimators and the corresponding mean square errors for the quadratic entropy for different sampling schemes (simple and stratified random sampling, with and without replacement). Similar studies were performed for estimating income inequality (Pérez et al. [36] and Caso and Gil [10]) or analyzing the asymptotic behaviour of the β−measure of unquietness (Gil et al. [16, 17]).

3 Economic Inequality

Although the most traditional measure of income inequality is the Gini index [18], this indicator cannot be broken down in the usual sense, and therefore it is not possible to obtain the inequality of a population divided into subgroups from inequalities computed at the subgroup level. Theil [42] was the first author who proposed the use of information measures for the study of inequality, finding some conceptual and operational advantages, mainly related to the ramification property. Some other authors such as Bourguignon [8], Cowell [12] and Shorrocks [40] proposed axiomatic characterizations for the decomposable inequality measures.

According to Cowell [12] given a distribution of incomes $\{x_1, \ldots, x_k\}$ and probabilities $\{p_1, \ldots, p_k\}$ a family of measures that satisfy the principles of mean independence, population independence, symmetry, the Pigou-Dalton transfer principle and decomposability are given by the expression:

$$GE(\beta) = \frac{1}{\beta(1-\beta)}\left[\sum_{i=1}^{k} p_i \left(\frac{x_i}{E(X)}\right)^{\beta} - 1\right]; \ \beta \neq 0, \beta \neq 1.$$

For each β this measure of inequality is, except for a constant, a measure of uncertainty involving utilities (4). The parameter represents the weight given to the distances between incomes of different parts of the distribution. When β takes low values the measure is more sensitive to what occurs at the bottom of the distribution, whereas for high values of the parameter β the measures would be more sensitive to what occurs at the top.

3.1 Collective and Double-Quadratic Inequality

Since inequality measures aim to summarize the distributive imbalances in a given population it seems advisable to derive their expressions from an individual point of view, as suggested by Yitzhaki [43] and Chakravarty and Chakraborty [11] among others.

Adopting the mean value as a reference, López [22] proposes an individual indicator of inequality given by the expression: $d_i = \frac{E(X)}{x_i} - 1$ whose expected value leads to a collective inequality measure $D(X) = \sum_{i=1}^{k} d_i p_i$ which coincides, except for a constant, with the case $\beta = -1$ of the previously defined family of additively decomposable inequality measures (Pérez [33]).

Furthermore, this collective inequality appears to be connected with a poverty measure P_D derived by López [22] from relative poverty gaps g_r. More specifically, P_D can be expressed in two different ways:

$$P_D = HE(g_r) \quad \text{and} \quad P_D(X, z) = H\left(\frac{D_p + I}{1 - I}\right)$$

where z denotes the poverty line, H the head-count ratio, I the income gap ratio and D_p the collective inequality among the poor. Thus, P_D includes the "three I" referring to the incidence, intensity and inequality dimensions of poverty, as suggested by Sen [38].

On the other hand, Alvargonzález [1] derived a quadratic individual expression $d_i^2 = \left(\frac{E(X)}{x_i} - 1\right)^2$, which guarantees positive results, emphasizing the perceptions of the less-favoured people. The synthesis of the individual quadratic inequalities leads to the double quadratic measure of inequality:

$$D^+(X) = \sum_{i=1}^{k} d_i^2 p_i = \sum_{i=1}^{k} \left(\frac{E(X)}{x_i} - 1\right)^2 p_i.$$

Although this indicator does not belong to the family of decomposable inequality measures, it can be obtained from the cases $\beta = -1$ and $\beta = -2$.

The previously described collective and double quadratic inequality measures satisfy the set of commonly required properties.

Properties: Continuity, Principle of population or independence of population's size, Symmetry or impartiality, Extensibility, Maximality, Non negativity,

Normalization, Invariability due to changes of scale, Variation with changes of origin, Decomposability or Ramification, Pigou-Dalton's condition.[1]

Regarding the decomposability, given a population X of size N, with $\mu = E(X)$, divided into r subgroups with size N_j and mean incomes μ_j, the collective inequality satisfies:

$$D = D^* + \sum_{j=1}^{r} \alpha_j D_j \quad \text{with} \quad D^* = \sum_{j=1}^{r} \left(\frac{\mu}{\mu_j} - 1 \right) \frac{N_j}{N} \; ; \; \alpha_j = \frac{\mu N_j}{\mu_j N} \forall j = 1, \ldots, r,$$

while for the double quadratic inequality the additive decomposability is replaced by a wider ramification, as shown in Alvargonzález [1].

Empirical applications of these measures can be found in López [22], Pérez and López [34, 35], Alvargonzález [1] and Alvargonzález et al. [3, 4], among others.

3.2 Inequality and Economic Growth

According to the inverted-U hypothesis proposed by Kuznets [21] inequality increases in the first levels of growth and then decreases after a certain point of return. Kuznets' initial work, referring to the Lorenz curve, was extended by several authors, such as Anand and Kanbur [6]. Considering six inequality measures, these authors derive their relationship with the development level and the conditions for the fulfillment of Kuznets' hypotheses.

Following a similar procedure, Alvargonzález et al. [2] derived the functional expressions for collective and double-quadratic inequality and the necessary conditions for the existence of a turning point. In both cases the relationship between the level of inequality and economic growth is given by polynomic functions and the decomposability property allows the analysis of Kuznets conditions (initial growth and existence of turning point), expressing inequality as the sum of intersectoral and intrasectoral components.

Furthermore, the polynomic function related to the double-quadratic index provides a particularly flexible expression, allowing the possibility of existence of more than one point of return and suggesting a better capability of adaptation to different empirical realities.

---

[1]In the case of the double quadratic inequality the Pigou-Dalton's condition or progressive transfers principle holds under certain (non-restrictive) asssumptions.

Inequality-Growth relationship
Collective inequality $D = A\mu^2 + B\mu + C$, with $C = -1$
$A = \left(\frac{1}{\mu_1 - \mu_2}\right)\left(\frac{D_1 + 1}{\mu_1} - \frac{D_2 + 1}{\mu_2}\right); B = \left(\frac{1}{\mu_1 - \mu_2}\right)\left(\frac{\mu_1(D_2 + 1)}{\mu_2} - \frac{\mu_2(D_1 + 1)}{\mu_1}\right)$
Double-quadratic inequality $D = A\mu^3 + B\mu^2 + C\mu + E$, with $E = 1$
$A = \left(\frac{1}{\mu_1 - \mu_2}\right)\left(\frac{1}{\mu_1^2} - \frac{1}{\mu_2^2} + \frac{1}{\mu_1^2}D_1^+ - \frac{1}{\mu_2^2}D_2^+ + \frac{2}{\mu_1^2}D_1 - \frac{2}{\mu_2^2}D_2\right)$
$B = \left(\frac{1}{\mu_1 - \mu_2}\right)\left(-\frac{2}{\mu_1} - \frac{\mu_2}{\mu_1^2} + \frac{\mu_1}{\mu_2^2} + \frac{2}{\mu_2} - \frac{\mu_2}{\mu_1^2}D_1^+ + \frac{\mu_1}{\mu_2^2}D_2^+ \right.$ $\left. -2\frac{\mu_2}{\mu_1^2}D_1 + 2\frac{\mu_1}{\mu_2^2}D_2 - \frac{2}{\mu_1}D_1 + \frac{2}{\mu_2}D_2\right)$
$C = \left(\frac{2}{\mu_1 - \mu_2}\right)\left(\frac{\mu_2}{\mu_1} - \frac{\mu_1}{\mu_2} + \frac{\mu_2}{\mu_1}D_1 - \frac{\mu_1}{\mu_2}D_2\right)$
Empirical Evidence: Alvargonzález [1]: Cross-section estimation from World Bank; Alvargonzález [1], Alvargonzález et al. [2, 4] and López et al. [26]: Panel estimation from PWT and Deininger-Squire database; López and Cowell [23]: Panel estimation from WIID

4 Industrial Concentration

Since the nineteen thirties, industrial concentration has been considered both as a determinant of market structure and as an indicator of the level of competition, allowing us to estimate the market power. Within the suitable framework of the information measures Theil [42] proposes a concentration index derived from Shannon's measure and Río and Pérez [37] define the quadratic index, that coincides, except for a constant, with the case $\beta = 2$ in (1): $I^2(X) = 1 - \sum_{i=1}^{k} p_i^2$, where p_i represent the market shares of a sector with k firms. This index measures the level of competition between the firms of a given sector and, in the opposite sense, the sector's level of concentration.

Some authors such as Hannah and Kay [19] examine the properties that an industrial concentration measure should satisfy, proposing axiomatic characterizations. An in-depth study of Theil's index can be found in Curry and George [13] while Río and Pérez [37] exhaustively analyze the quadratic index, showing its good behaviour with regard to the characterizations.

As industrial concentration can be examined at different aggregation levels, the decomposability property plays a significant role, allowing the analysis of both global and sectoral concentration and the measurement of the influence of each industrial sector. More specifically, according to the branching property (2) the quadratic concentration for an industry with r sectors can be computed as: $I^2(X) = I_r^2 + \sum_{i=1}^{r} \alpha_i I^2(X_i)$, where we denote by I_r^2 the intersectoral concentration while $I^2(X_i)$ represents the individual concentration of sector i and α_i is the relative weight. It is important to bear in mind that, since only the entropy-based indicators satisfy the decomposability requirement, both Theil's and the quadratic concentration measures are the most adequate options.

From an empirical point of view, in order to compute the industrial concentration level we need to access information referred to the firm market shares and this task can be particularly difficult for certain sectors and small size firms. In this situation we can take profit from the suitable inferential behaviour of the quadratic index as shown in Pérez [33], Caso and Gil [9] and Río and Pérez [37] among others.

5 Economic Forecasting

The increasing amount of prospective sources and methods provides a wide variety of forecasts referring to the main economic variables, such as GDP, employment and consumer prices. In this context, Information Theory can be of great help when analyzing the combination and evaluation of forecasts, and also the definition of future scenarios.

5.1 Forecast Combination

Since the pioneering work by Bates and Granger [7] many authors have shown that combined forecasts perform quite well since they take advantage of the vast amount of existing information, reducing the level of risk.

Forecast combination techniques involve different levels of complexity and mainly focus on the estimation of the weights assigned to each individual forecast, which -among other procedures- can be performed through regression techniques, based on the relative past performance of individual forecasts.

More specifically, if we consider the alternative h-step forecasts for an economic variable (y_{t+h}) available at time t, the theory suggests the convenience of combining the individual results $\left(\hat{y}_{t+h,t}^1, \hat{y}_{t+h,t}^2 \ldots, \hat{y}_{t+h,t}^F\right)$ to obtain an aggregated prediction $\hat{y}_{t+h,t}^c$ through a vector of weights $W = (\alpha_1, \alpha_2, \ldots, \alpha_F)'$ that calibrates different degrees of experts' ability.

The efficient calculation of weights requires a number of observations larger than the number of individual forecasts. Nevertheless, sometimes it is not possible to have sufficient information to measure individual forecasting ability, either because the situation to predict is new or because new members are incorporated onto the forecasting panel. Under the described circumstances the Maximum Entropy Econometric approach provides a basis for transforming the sample information into a probability distribution that reflects our uncertainty about the individual outcomes. Moreno and López [29] apply this methodology to Shannon's entropy and the quadratic entropy by Pérez [33], leading to the expressions:

$H_S(P, W) = -P' \log P - W' \log W$ for Shannon's entropy

$H^2(P, W) = 2(1 - P'P) - P' + 2(1 - W'W)$ for quadratic entropy

whose maximization provides the optimal probability vectors $\hat{P}$ and $\hat{W}$.

Some empirical applications conducted on forecasts of Spanish GDP growth (Moreno et al. [31], Moreno and López [29]) and inflation (Moreno and López [30]) confirm that the estimated weights do not show significant differences for Shannon and quadratic uncertainty, allowing the selection of a group of forecasters with relatively high and stable weights.

5.2 Forecast Evaluation

The increasing forecasting availability and the controversial debate about the advantages of alternative forecasting procedures suggest the need for further research on evaluation metrics. Although forecasting procedures are usually compared by means of an error measure, the information-based U index was proposed by Theil [41] comparing the actual and forecasted values of a variable Y for a given period through the following expression:

$$U = \sqrt{\frac{\sum_t (\hat{g}_t - g_t)^2}{\sum_t g_t^2}} \; ; \; \hat{g}_t = \frac{\hat{y}_t - y_{t-1}}{y_{t-1}}, \; g_t = \frac{y_t - y_{t-1}}{y_{t-1}}.$$

As forecasts become more accurate, Theil's U index decreases, achieving null value when actual and forecasted rates of growth are coincident. Furthermore, an additive disaggregation of U^2 in three terms allows us to compute the relative weights of the bias, variance and covariance factors.

The uncertainty measures involving utilities (4) provide a suitable framework for the evaluation of forecasts. Thus, given a variable Y with actual values $y_1, \ldots, y_T$, and a set of forecasts $\hat{y}_1, \ldots, \hat{y}_T$ the quadratic unquietness of Y, $HU^{*2}(Y)$ and the quadratic unquietness of Y conditioned to its forecasts $HU^{*2}(Y/[\hat{Y}])$, can be respectively computed as:

$$HU^{*2}(Y) = \frac{2}{T} \sum_i \left(\frac{E(Y)}{y_i} - 1 \right), \; HU^{*2}\left(Y/[\hat{Y}] \right) = \sum_j p_{[\hat{y}_j]} \frac{E(Y)}{E_{[\hat{y}_j]}(Y)} HU^{*2}\left(Y/[\hat{y}_j] \right)$$

where $E(Y)$ denotes the expected value of the variable and $E_{[\hat{y}_j]}(Y) = \sum_t Y_t p_{Y/[\hat{y}_j]}$ is the expected value of Y conditioned to the forecasting interval $[\hat{y}_j]$. Within this framework, since the quadratic unquietness conditioned to forecasts is expected to be lower than the initial one, the difference between both results can be interpreted as the information provided by forecasts. However, with the aim of including the existing relationship between actual and forecasted values, we also take into account the linear correlation coefficient $r_{Y,\hat{y}}$, and thus in López and Pérez [24] we define the quadratic information accuracy measure (QIAM) as:

$$QIAM\left(Y, [\hat{Y}] \right) = HU^{*2}(Y) - HU^{*2}\left(Y/[\hat{Y}] \right) \left(1 - r_{Y,\hat{y}} \right)$$

whose result increases both with the reduction of unquietness related to the forecasts and with the correlation between actual and forecasted values.

Despite their similarities, Theil's U Index and QIAM differ in their interpretation, as more accurate forecasts lead to lower values of U while QIAM increases. Nevertheless, this is not a serious drawback since the comparison of different forecasting techniques is usually based on rankings.

In López and Pérez [24, 25] and López et al. [27] we provide empirical evidence based on the M3-Competition by Makridakis and Hibon [28] that includes 3003 time series and 24 forecasting techniques. The results show outstanding similarities between the rankings provided by Theil's U and QIAM, suggesting the influence of their informational content. As expected, the accuracy of the different methods depends upon the length of the forecasting horizon, thus agreeing with Makridakis and Hibon's findings from five error based accuracy measures.

However, while these authors conclude that statistically sophisticated procedures do not lead to more accurate forecasts and that combined forecasts outperform individual methods, our results differ, especially when dealing with financial time series. According to both Theil's U and QIAM the most sophisticated methods head the accuracy rankings and the combined forecast does not beat the individual methods being combined.

5.3 Forecasting Scenarios

Scenario-based forecasting allows for a wide range of possible forecasts to be generated. Although the most common practice considers best, middle and worst case scenarios, the range of possibilities can be extended including a wide variety of more and less plausible alternatives.

Assigning probabilities to the considered scenarios allows the measurement of the corresponding levels of uncertainty, which can be revised once we access to updated information. Thus, denoting by P and Q the sets of initial and final probabilities, Theil [42] proposes the use of a Shannon-based measure, given by the expression $I(Q : P) = \sum_{i=1}^{k} q_i \log \frac{q_i}{p_i}$ obtained by comparing the expected initial and final levels of uncertainty, with weights given by the final probabilities.

A similar approach can be applied to the quadratic uncertainty proposed by Pérez [33]. For each scenario the initial and final uncertainty levels are respectively given by $2(1 - p_i)$ and $2(1 - q_i)$, and the quadratic information provided by the new set of probabilities is obtained through the expression $QI(Q : P) = 2\sum_{i=1}^{k} q_i(q_i - p_i)$, which, unlike Theil's index, can be computed even for scenarios with null probability.

It can be easily proved that the quadratic information is zero if initial and final probabilities are coincident, although the opposite implication does not hold.

6 Concluding Remarks

The economic applications summarized in the previous sections illustrate our work influenced by Pedro Gil, and could be extended to a wider scope, including fields such as input-output and network analysis or the quantification of the digital divide.

We finish this paper by thanking Pedro for his inspiring work and joining his words *"I would like to express my conviction (although I am aware that now I am influenced by both my heart and my head) that Information Theory will go on succeeding in scientific and technological applications"* [15].

References

1. Alvargonzález M (2003) Medidas doble cuadráticas de información. Algunas aplicaciones económicas. PhD thesis, Universidad de Oviedo
2. Alvargonzález M, López AJ, Pérez R (2004) Growth-inequality relationship. An analytical approach and some evidence for Latin America. Appl Econ Int Dev 4(2):91–108
3. Alvargonzález M, López AJ, Pérez R (2004) The double quadratic uncertainty measures and their economic applications. In: López-Díaz M, Gil MA, Grzegorzewski P, Hryniewicz O, Lawry J (eds) Soft methodology and random information systems. Springer, Heidelberg
4. Alvargonzález M, López AJ, Pérez R (2006) Privación relativa y desigualdad. La medida doble cuadrática. Estadíst Esp 48(162):271–293
5. Aczel J, Kannappan PL (1978) A mixed theory of information III. Inset entropies of degree β. Inf Control 39:315–322
6. Anand S, Kanbur SMR (1993) The Kuznets process and the inequality-development relationship. J Dev Econ 40:25–52
7. Bates JM, Granger CWJ (1969) The combination of forecasts. Op Res Q 20(4):451–468
8. Bourguignon F (1979) Decomposable income inequality measures. Econometrica 47(4):901–920
9. Caso C, Gil MA (1988) The Gini-Simpson Index of diversity: estimation in the stratified sampling. Commun Stat-Theor Meth 17(9):2981–2995
10. Caso C, Gil MA (1989) Estimating income inequality in the stratified sampling from complete data (Parts I and II). Kybernetika 25(4):298–319
11. Chakravarty SR, Chakraborty AB (1984) Income inequality and relative deprivation. Econ Lett 14:283–287
12. Cowell FA (1980) On the structure of additive inequality measures. Rev Econ Stud 47:521–231
13. Curry B, George KD (1983) Industrial concentration: a survey. J Ind Econ XXXI(3):203–255
14. Gil P (1981) Teoría Matemática de la Información. ICE, Madrid
15. Gil P (2007) ¿Por qué la teoría de la información? BEIO 23(3):8–9
16. Gil MA, Caso C, Gil P (1989) Estudio asintótico de una clase de índices de desigualdad muestrales. Trab Estadíst 4(1):95–109
17. Gil MA, Pérez R, Gil P (1989) A family of measures of uncertainty involving utilities: definition, properties, applications and statistical inferences. Metrika 36:129–147
18. Gini C (1921) Measurement of inequality of incomes. Econ J 31:124–126
19. Hannah L, Kay JA (1977) Concentration in modern industry: theory, measurement and the UK experience. MacMillan, London
20. Harvda J, Charvát F (1967) Quantification method of classification processes. Kybernetika 3:129–147
21. Kuznets S (1955) Economic growth and income inequality. Am Econ Rev 45:1–28

22. López AJ (1991) Desigualdad de renta y pobreza: una aproximación conceptual y cuantitativa. PhD thesis, Universidad de Oviedo

23. López AJ, Cowell FA (2013) Desigualdad y crecimiento económico. ¿Círculos viciosos o virtuosos? Rev Galega Econ 22:15–36

24. López AJ, Pérez R (2012) Financial forecasting accuracy: exploring the M3-competition. In: 6th CSDA international conference of the& financial econometrics ERCIM WG on computing & statistics, London

25. López AJ, Pérez R (2017) Forecasting performance and information measures. Revisiting the M-competition. Estud Econ Apl 35(2):299–314

26. López AJ, Alvargonzález M, Pérez R (2006) Crecimiento económico y desigualdad. Nuevas extensiones del proceso de Kuznets. Estud Econ Apl 24(1):221–244

27. López AJ, Pérez R, Moreno B (2011) Forecasting performance and M-competition. Does the accuracy measure matter? In: Proceedings of the 58th world statistics congress. https://isi-web. org/index.php/publications/proceedings

28. Makridakis S, Hibon M (2000) The M-3 competition: results, conclusions and implications. Int J Forecast 16:451–476

29. Moreno B, López AJ (2007) Combining forecasts through information measures. Appl Econ Lett 14(12):899–903

30. Moreno B, López AJ (2013) Combining economic forecasts by using a maximum entropy econometric approach. Int J Forecast 32(2):124–136

31. Moreno B, López AJ, Pérez R (2007) Combinación de predicciones basada en medidas de información. Una aplicación al crecimiento económico en España. Estadíst Esp 164(49):5–32

32. Pérez R (1985) Estimación de la incertidumbre, la incertidumbre útil y la inquietud en poblaciones finitas: una aplicación a las medidas de desigualdad. Rev Real Acad Cienc Exact, Fís Nat LXXIX(4):651–654

33. Pérez R (1985) Estimación de la incertidumbre, la incertidumbre útil y la inquietud en poblaciones finitas: una aplicación a las medidas de desigualdad. PhD thesis, Universidad de Oviedo

34. Pérez R, López AJ (2001) La distribución de la renta. Una visión panorámica 1981–2001. Rev Astur Econ extra:267–286

35. Pérez R, López AJ (2004) Crecimiento y distribución de la renta. Aproximación a la desigualdad económica y social. Pap Econ Esp 20:220–233

36. Pérez R, Caso C, Gil MA (1986) Unbiased estimation of income inequality. Statist Hefte (Statist Pap) 27:227–237

37. Río MJ, Pérez R (1988) Sobre la medición de la concentración industrial. Invest Econ Supl 1988:81–88

38. Sen A (1976) Poverty: an ordinal approach to measurement. Econometrica 44(2):219–231

39. Shannon C (1948) A mathematical theory of communication. Bell Syst Tech J 27:379–423

40. Shorrocks AF (1980) The class of additively decomposable inequality measures. Econometrica 48(3):613–625

41. Theil H (1966) Applied economic forecasting. North Holland Publishing, Amsterdam

42. Theil H (1967) Economics and information theory. North Holland Publishing, Amsterdam

43. Yitzhaki S (1979) Relative deprivation and the Gini coefficient. Q J Econ 43(2):321–324

Concept Uncertainty in Adversarial Statistical Decision Theory

David Ríos Insua, Jorge González-Ortega, David Banks and Jesús Ríos

Abstract We focus on concept uncertainty which adds a new layer to the traditional risk analysis distinction between aleatory and epistemic uncertainties, when adversaries are present. The idea is illustrated with a problem in adversarial point estimation framed as a specific case of adversarial statistical decision theory.

1 Introduction

Our focus in this paper will be on problems in which two or more agents confront (as in non-cooperative game theory) and the consequences that they receive are random (as in risk analysis) and depend on the actions of all participants (as in game theory at large).

In his lecture giving title to this volume, Pedro Gil [8] dealt with several issues related to the mathematics of uncertainty. At that time, it was typical to address just two types of uncertain phenomena, see e.g. [14, 22] or [16], which could be described as follows in the class of problems we consider:

- *Aleatory uncertainty*. It refers to the randomness of the outcomes the agents receive, thus conditional on their choices.
- *Epistemic uncertainty*. It refers to the strategic choices of intelligent adversaries, as driven by unknown preferences, beliefs and capabilities.

Since that time, a number of issues have arisen prominently in several application areas. To wit: (*i*) *high-profile terrorist attacks* are demanding significant investments

D. Ríos Insua (✉) · J. González-Ortega
Institute of Mathematical Sciences, CSIC-UAM-UC3M-UCM, Madrid, Spain
e-mail: david.rios@icmat.es
e-mail: jorge.gonzalez@icmat.es

D. Banks
Department of Statistical Science, Duke University, Durham, NC, USA
e-mail: banks@stat.duke.edu

J. Ríos
Cognitive Computing Department, IBM Research Division,
Yorktown Heights, NY, USA
e-mail: banks@stat.duke.edu

© Springer International Publishing AG 2018
E. Gil et al. (eds.), *The Mathematics of the Uncertain*, Studies in Systems,
Decision and Control 142, https://doi.org/10.1007/978-3-319-73848-2_49

527

in protective responses and there is concern that not all of them are sufficiently effective; (*ii*) *key business sectors* have become more mathematically sophisticated and use this expertise to shape strategies in competitive decisions in marketing, auctions and others; and (*iii*) the on-going arms race in *cyber-security* implies that financial penalties for myopic protection are random and may be remarkably high.

These have entailed the need to consider a third type of uncertainty which will be the target of this paper:

- *Concept uncertainty*. It refers to beliefs about how opponents frame the problem in relation to questions like what kind of strategic analysis do they use, whether they are rational or how deeply do they think. Often, the solution used will determine the epistemic uncertainties that are relevant and, consequently, the aleatory uncertainties.

Most authors in this context would focus on game theoretical methods based on variants of Nash equilibria, see e.g. [7]. However, this is not satisfactory in most of the above applications since beliefs and preferences of adversaries will not be readily available, frequently violating game theoretic common knowledge assumptions, see [9]. Thus, key assumptions of the customarily proposed solution approaches would not actually hold.

In contrast, we shall turn to Adversarial Risk Analysis (ARA), described in [1, 18], an emergent paradigm for the type of problems we consider. ARA provides one-sided prescriptive support to a decision maker, maximizing her subjective expected utility, treating the adversaries' decisions as random variables. To do so, ARA supports modelling the adversaries' decision-making problems and, under assumptions about their rationality such as them being expected utility maximizers, tries to assess their probabilities and utilities. Then, ARA can predict their optimal actions. However, the uncertainty about the adversaries' probabilities and utilities is propagated into their decisions, leading to probability distributions over them. Often, such assessments lead to a hierarchy of nested decision problems, as described in [17], close to the concept of level-k thinking, see [19].

The main goals of ARA are to weaken the Nash equilibria common knowledge assumptions and provide more flexible models for opponent behaviour. ARA is explicitly Bayesian in that subjective distributions are employed to express the uncertainties of the analyst. In a comparison of methods for adversarial risk management, [12] preferred ARA precisely because it handles and apportions these separate uncertainties more explicitly. In this respect, ARA may be seen as a Fermitisation strategy, see [20], since it simplifies the assessment of complex uncertainties by considering decompositions of complex problems into simpler ones in which the assessment is easier.

We shall first stress this idea of concept uncertainty with a qualitative security example in Sect. 2. Section 3 then shows its relevance in what we call Adversarial Statistical Decision Theory (ASDT), which extends the standard Statistical Decision Theory (SDT) framework with the presence of several decision makers, focusing on point estimation. We end up with some discussion in Sect. 4.

2 A Security Example

We shall dwell on the idea of concept uncertainty through a qualitative example in relation with security. Remember that we refer to situations in which several opponents make decisions. Consider a security case with a Defender (she) and an Attacker (he) who make simultaneous decisions. We will be supporting the Defender. To fix ideas, in a context of counterterrorism, suppose the Attacker may choose to bomb a train, while the Defender could have policemen searching for it.

Typically, after the opponents both choose their actions, the outcomes and payoffs for each of them are random variables. In our example, there is a chance that the policemen will thwart the attack and a chance that the bomb will explode, killing a random number of people and causing a random amount of economic and political damage. This randomness is aleatory uncertainty and is conditional on the choices of the opponents: it does not depend upon any strategic calculation on their parts, referring to the non-strategic randomness of an outcome. The Defender should assess her beliefs about the outcome probabilities, conditional on the actions chosen by both agents. This can be addressed through traditional probabilistic risk analysis, see [2]. The Defender's beliefs should be informed by expert judgement and previous history and extracted through appropriate elicitation methods [4, 13], taking into account factors such as experts being overconfident and that previous history may be only partially relevant.

Epistemic uncertainty describes the Defender's distribution over the choice the Attacker will make, which usually integrates his preferences, beliefs and capabilities. For the bomb-on-a-train example, the Defender does not know which train the Attacker will target. His choice would depend upon factors like which train is his most valuable target (a preference), the train he thinks that has the bigger chance of being attacked (a belief) and whether or not he has a bomb (a capability). The Defender does not know these and, thus, has epistemic uncertainty. She expresses it as a distribution over all possible trains. This uncertainty is handled differently for each solution concept that the Defender thinks the Attacker might use. For example, with a Nash equilibrium concept, the Defender believes that the Attacker thinks they both know the relevant utilities and beliefs; hence, the relevant epistemic uncertainty refers to the Defender's distribution over the payoff bi-matrices that the Attacker may be using. In the case of Bayes-Nash equilibrium concept, there is additional uncertainty related to games of incomplete information [10]. Besides the distribution over the entailed payoff bi-matrices, the Defender must also express her epistemic uncertainty about the common knowledge distributions which the Attacker is assuming that they share, their common distributions on types. In principle, full description of the epistemic uncertainties is complex, even with simple solution concepts. Often, there are pragmatic approximations that may be used.

Finally, as mentioned, concept uncertainty would arise from ignorance of how one's opponent will frame the analysis. In classical game theory terms, the Defender does not know which solution concept the Attacker will use to make his decision. Concept uncertainty embraces a wide range of strategies and is an essential component in the ARA formulation: the Defender must model how the Attacker will make his decision, but there are many possible solution concepts that he might use. Some of them are:

- *Non-strategic play.* The Defender believes that the Attacker will select an action without consideration of her choice. This includes the case in which the Attacker selects actions with probabilities proportional to the perceived utility of success, [15]. It also allows for non-sentient opponents, such as a hurricane.
- *Nash or Bayes-Nash equilibrium methods.* The Defender concludes that the Attacker is assuming they both have a great deal of common knowledge.
- *Level-k thinking.* The Defender considers that the Attacker thinks k plies deep in an "I think that she thinks that I think..." kind of reasoning. Level-0 corresponds to non-strategic play.
- *Mirroring equilibrium analysis.* The Defender supposes that the Attacker is modelling her decision making in the same way that she is modelling his, and both use subjective distributions on all unknown quantities.

As a Bayesian approach, ARA enables great flexibility in tailoring the analysis to the context and in accounting for the different kinds of uncertainty that arise. Usually, the Defender does not know which solution concept the Attacker has chosen. But based on previous experience with the Attacker, and input from informants or other sources, she can place a subjective probability distribution over his possible solution concepts. She could then make the decision that maximizes her expected utility against that weighted mixture of strategies. Realistically, a full Bayesian analysis that puts positive probability on a large number of different solution concepts becomes computationally burdensome. However, in principle, the approach is simple. Each solution concept will lead (after handling the relevant epistemic and aleatory uncertainties) to a distribution over the Attacker's actions. Then, the Defender weights each distribution with her personal probability that the Attacker is using such solution concept. This generates a weighted distribution on the Attacker's action space which reflects all of the Defender's knowledge about the problem and all of her uncertainty. The approach is related to Bayesian model averaging, in which uncertainty about the model is expressed by a probability distribution over the possible models and inference is based upon a weighted average of the posterior distributions from each model, see [3, 11].

3   The Case of Adversarial Point Estimation

We study now the relevance of concept uncertainty showcasing it in the standard problem of point estimation, described in the inaugural lecture by [8]. We modify it by considering the presence of an adversary willing to deceive the relevant decision maker.

3.1 Point Estimation as a SDT Problem

We recall first the standard SDT framework for point estimation. As illustrated in the Influence Diagram (ID) in Fig. 1, we consider a decision maker D (she) who needs to make a decision $d \in \Theta$ based on an observation x which depends on a state θ taking values in the same set Θ. She obtains a loss $l_D(d, \theta) \in \Re$, which depends on the decision she makes and the state actually occurring.

To solve her decision making problem, she could describe her prior beliefs over state θ through the prior $p_D(\theta)$ and the dependence of data x on the state θ through the likelihood $p_D(x \mid \theta)$. Given such elements, she would seek the decision $d^*(x)$ that minimizes her posterior expected loss, which is

$$d^*(x) = \arg\min_d \int l_D(d, \theta) \, p_D(\theta \mid x) \, d\theta. \tag{1}$$

As an example, under the quadratic loss function $l_D(d, \theta) = (\theta - d)^2$, we easily obtain that $d^*(x) = E[\theta \mid x]$, see [6]. Note that, for optimization purposes, we may ignore the denominator $p_D(x)$ in Bayes formula and solve

$$d^*(x) = \arg\min_d \int l_D(d, \theta) \, p_D(x \mid \theta) \, p_D(\theta) \, d\theta, \tag{2}$$

which is equivalent to (1). This transformation allows us to directly involve the probabilities $p_D(\theta)$ and $p_D(x \mid \theta)$ and, in principle, avoid computing the more complex posterior $p_D(\theta \mid x)$.

We rehearse now the optimisation argument with the quadratic loss in the simplified version (2). The objective function to be minimised in this case is

$$\int \theta^2 \, p_D(x \mid \theta) \, p_D(\theta) \, d\theta + d^2 \int p_D(x \mid \theta) \, p_D(\theta) \, d\theta - 2d \int \theta \, p_D(x \mid \theta) \, p_D(\theta) \, d\theta,$$

Fig. 1 Sketch of the general SDT problem

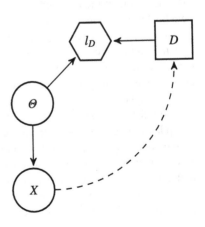

which is equivalent to minimising in d

$$d^2 p_D(x) - 2d \int \theta \, p_D(x \mid \theta) \, p_D(\theta) \, d\theta,$$

whose minimum is, as announced,

$$d^*(x) = \frac{1}{p_D(x)} \int \theta \, p_D(x \mid \theta) \, p_D(\theta) \, d\theta = \int \theta \, p_D(\theta \mid x) \, d\theta = E[\theta \mid x]. \quad (3)$$

3.2 Adversarial Point Estimation as an ASDT Problem

We extend now the SDT framework by considering a strategic adversary A (he) who makes decisions in the previous context which affect the information and/or consequences that our decision maker obtains. We may actually consider several adversarial scenarios, reflected through the Bi-Agent Influence Diagrams (BAIDs) in Fig. 2.

$\mathscr{S}1$. A somehow manages to transform state θ affecting the data process leading to a modified one $a(\theta) = \lambda$. We call this a *structural attacker*, shown in Fig. 2a.

$\mathscr{S}2$. A manages to modify the data x to be received by D, who actually observes $a(x) = y$. We call this a *data-fiddler attacker*, depicted in Fig. 2b, which reflects the typical case in adversarial machine learning, see e.g. [5] or [21].

$\mathscr{S}3$. A makes decisions a so that the losses for D and A, which we respectively designate $l_D(d, a, \theta)$ and $l_A(d, a, \theta)$, depend upon both their decisions and the relevant state. Other than that, A faces a problem structurally similar to that of D. We call this the simultaneous ASDT problem, represented in Fig. 2c.

Note that the three formulations could appear combined in certain scenarios. For example, in a cyber-security problem, A might add spam to modify its proportion (θ); alter some of the received messages (x); and, finally, in addition, undertake his own business decisions (a), which compete with those of D.

To streamline the discussion, we shall consider just the case $\mathscr{S}1$ in which A is a structural attacker capable of perturbing θ (Fig. 2a). Specifically, suppose that A just uses deterministic transformations $\lambda = a + \theta$, biasing the underlying state, with a chosen by him. We shall also assume a quadratic loss function for D. Note first that a decision maker who is not aware of the presence of the adversary would be proposing as optimal the decision in expression (2), without noticing that $p_D(x \mid \theta)$ should be updated to $p_D(x \mid \theta + a)$ as the data process is being perturbed and, therefore, systematically erring her estimation in general.

Consider now an adversary aware point estimator D. She faces a problem similar to that in Fig. 1, except for the additional uncertainty about a, as reflected in the ID in Fig. 3a, which modifies the state. Since she does not know her adversary's decision, his corresponding node (the circled A) appears as random to her. Once D makes a

Fig. 2 BAIDS for the
different scenarios in the
ASDT problem

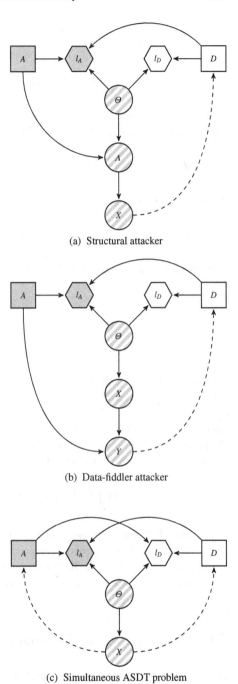

(a) Structural attacker

(b) Data-fiddler attacker

(c) Simultaneous ASDT problem

forecast $p_D(a)$ of the action a to be performed by A, the problem that she needs to solve is formulated (equivalently) as

$$d^*(x) = \arg\min_d \iiint (\theta - d)^2 \, p_D(x\,|\,\lambda)\, p_D(\lambda\,|\,\theta, a)\, p_D(\theta)\, p_D(a)\, d\lambda\, d\theta\, da. \quad (4)$$

In the same manner as the non-adversarial version in Sect. 3.1, observe that this leads to the equivalent problem of minimising in d

$$d^2\, p_D(x) - 2d \iiint \theta\, p_D(x\,|\,\lambda)\, p_D(\lambda\,|\,\theta, a)\, p_D(\theta)\, p_D(a)\, d\lambda\, d\theta\, da,$$

whose solution is

$$d^*(x) = \frac{1}{p_D(x)} \iiint \theta\, p_D(x\,|\,\lambda)\, p_D(\lambda\,|\,\theta, a)\, p_D(\theta)\, p_D(a)\, d\lambda\, d\theta\, da. \quad (5)$$

Expression (5) may then be simplified to obtain

$$d^*(x) = \frac{1}{p_D(x)} \iint \theta\, p_D(x\,|\,\lambda)\, p_D(\lambda\,|\,\theta)\, p_D(\theta)\, d\lambda\, d\theta$$

$$= \frac{1}{p_D(x)} \int \theta\, p_D(x\,|\,\theta)\, p_D(\theta)\, d\theta = \int \theta\, p_D(\theta\,|\,x)\, d\theta = E\,[\theta\,|\,x].$$

Apparently, we reach the same solution as in (3). However, these compressed expressions do not explicitly show that the probability model $p_D(\theta\,|\,x)$ is different in both cases. Therefore, it is good to use decomposed expressions such as (4) which involve probabilities that are easier to model and learn from, in line with the Fermitisation point of view mentioned in Sect. 2. In our specific case, we have $p_D(\lambda = \theta + a\,|\,\theta, a) = 1$ (and 0 otherwise). Thus, expression (4) becomes

$$d^*(x) = \arg\min_d \iint (\theta - d)^2\, p_D(x\,|\,\lambda = \theta + a)\, p_D(\theta)\, p_D(a)\, d\theta\, da. \quad (6)$$

Let us remark that a good assessment of $p_D(a)$ is therefore crucial, although it is complex because of the strategic element involved. To better assess it, D may consider A's problem reflected in Fig. 3b. To him, her decision is a chance node.

At this point we need to face concept uncertainty.

3.3 Adressing Concept Uncertainty

We consider now, as an example, two different solution concepts for the adversary and a way to aggregate them in terms of our uncertainty. A numerical example is also provided.

Fig. 3 BAID's
decomposition for a
structural attacker in the
ASDT problem

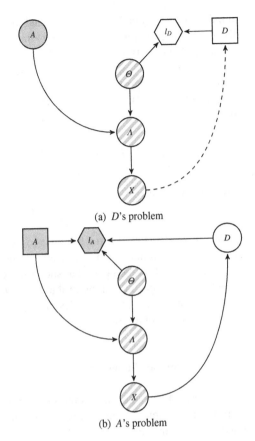

(a) *D*'s problem

(b) *A*'s problem

3.3.1 A Bayesian Adversary

Suppose first that the adversary minimises expected loss. Then, *A*'s optimal decision
would be solved through

$$a_B^* = \arg\min_a \iiint l_A(d, a, \theta) \, p_A(d \mid x) \, p_A(x \mid \lambda = \theta + a) \, p_A(\theta) \, \mathrm{d}d \, \mathrm{d}x \, \mathrm{d}\theta.$$

However, *D* does not know *A*'s probabilities and loss function. If she acknowledges her uncertainty about them through random probabilities and losses $F \sim$
$\big(L_A(d, a, \theta), P_A(d \mid x), P_A(x \mid \lambda = \theta + a), P_A(\theta)\big)$, she would solve

$$A_B^* = \arg\min_a \iiint L_A(d, a, \theta) \, P_A(d \mid x) \, P_A(x \mid \lambda = \theta + a) \, P_A(\theta) \, \mathrm{d}d \, \mathrm{d}x \, \mathrm{d}\theta$$

to find the optimal random decision A_B^*, which is a random variable whose distribution is induced by the above random probabilities and loss function. Then, *D* has

found the distribution that she would need to calculate her best decision $d^*(x)$. That distribution incorporates all of her uncertainty about A's decision making context, assuming that he is Bayesian. We shall define it through $p_D^B(a) = P(A_B^* = a)$, if we deem it to be discrete, and, similarly, if it is continuous.

In general, to approximate $p_D^B(a)$, one will typically use simulation, drawing K samples $\left(L_A^k(d, a, \theta), P_A^k(d \mid x), P_A^k(x \mid \lambda = \theta + a), P_A^k(\theta)\right), k = 1, \ldots, K$ from F, finding

$$A_{B,k}^* = \arg\min_a \iiint L_A^k(d, a, \theta) \, P_A^k(d \mid x) \, P_A^k(x \mid \lambda = \theta + a) \, P_A^k(\theta) \, dd \, dx \, d\theta$$

(7)

and approximating

$$\hat{p}_D^B(A = a) \approx \#\{A_{B,k}^* = a\}/K.$$

Within F, $P_A(d \mid x)$ is much harder to assess than the other three elements. It entails strategic thinking, since D needs to understand A's beliefs about what point estimates she will make given that she observes x. This could be the beginning of a hierarchy of decision making problems if D assumes that A considers her to be adversary aware; see [17] for a description of the potentially infinite regress in a simple class of problems.

3.3.2 A Minimax Adversary

Suppose now that rather than minimising posterior expected loss, D feels that A behaves in a minimax manner. This means that he would try to minimise his possible maximum loss through

$$a_M^* = \arg\min_a \max_{d,\theta} l_A(d, a, \theta).$$

However, as in Sect. 3.3.1, D does not know A's loss function. If she assumes some uncertainty about it through a random loss $L_A(d, a, \theta)$, she would solve

$$A_M^* = \arg\min_a \max_{d,\theta} L_A(d, a, \theta),$$

to find the optimal random decision A_M^*. Again, the distribution of A_M^* incorporates all of her uncertainty about A's decision making context, assuming this time that he is a minimax adversary. Thus, we may define it as $p_D^M(a) = P(A_M^* = a)$ in the discrete case, and, similarly, in the continuous one. To approximate $p_D^M(a)$ we would typically use simulation as in Sect. 3.3.1, drawing K samples from $L_A^k(d, a, \theta), k = 1, \ldots, K$, finding

$$A_{M,k}^* = \arg\min_a \max_{d,\theta} L_A^k(d, a, \theta).$$

This would lead to

$$\hat{p}_D^M(A = a) \approx \#\{A_{M,k}^* = a\}/K.$$

3.3.3 Aggregating Solution Concepts

The above adversary types could be extended with other solution concepts, but for the purpose of this paper two of them are enough. Now, if D believes that A minimises expected loss with weight π_B and is minimax with weight π_M, with $\pi_B + \pi_M = 1$ and $\pi_B, \pi_M \geq 0$, then the required forecast $p_D(a)$ would be based on $\pi_B \, \hat{p}_D^B(a) + \pi_M \, \hat{p}_D^M(a)$ which would be plugged in D's optimal decision making problem (6), leading D to her optimal adversary aware point estimation.

3.3.4 A Numerical Example

Consider that D believes that:

- The state follows a normal prior $\theta \sim \mathcal{N}(\mu_D, \sigma_D)$. In particular, we shall specify that $\mu_D = 0$ and $\sigma_D = 1$.
- The observations $x = (x_1, \ldots, x_n)$ will be i.i.d. around state θ as $x_i \mid \theta \sim \mathcal{N}(\theta, \rho_D)$ for $i = 1, \ldots, n$. Specifically, we shall set $\rho_D = \frac{1}{2}$.

When D is unaware of A's presence, we may compute D's optimal decision by means of expression (2), given observations $(x_1, \ldots, x_n)$, minimising in d

$$
\int (\theta - d)^2 \frac{1}{\sigma_D \sqrt{2\pi}} \exp\left(\frac{-(\theta - \mu_D)^2}{2\sigma_D^2}\right) \prod_{i=1}^{n} \frac{1}{\rho_D \sqrt{2\pi}} \exp\left(\frac{-(x_i - \theta)^2}{2\rho_D^2}\right) d\theta
$$

$$
\propto \int (\theta - d)^2 \frac{\sqrt{\rho_D^2 + n\sigma_D^2}}{\rho_D \, \sigma_D \sqrt{2\pi}} \exp\left(\frac{-(\theta - \frac{\mu_D \rho_D^2 + \sigma_D^2 \sum_{i=1}^{n} x_i}{\rho_D^2 + n\sigma_D^2})^2}{2 \frac{\rho_D^2 \sigma_D^2}{\rho_D^2 + n\sigma_D^2}}\right) d\theta,
$$

which is equivalent to minimising in d

$$
d^2 \int \frac{\sqrt{\rho_D^2 + n\sigma_D^2}}{\rho_D \, \sigma_D \sqrt{2\pi}} \exp\left(\frac{-(\theta - \frac{\mu_D \rho_D^2 + \sigma_D^2 \sum_{i=1}^{n} x_i}{\rho_D^2 + n\sigma_D^2})^2}{2 \frac{\rho_D^2 \sigma_D^2}{\rho_D^2 + n\sigma_D^2}}\right) d\theta
$$

$$
- 2d \int \theta \frac{\sqrt{\rho_D^2 + n\sigma_D^2}}{\rho_D \, \sigma_D \sqrt{2\pi}} \exp\left(\frac{-(\theta - \frac{\mu_D \rho_D^2 + \sigma_D^2 \sum_{i=1}^{n} x_i}{\rho_D^2 + n\sigma_D^2})^2}{2 \frac{\rho_D^2 \sigma_D^2}{\rho_D^2 + n\sigma_D^2}}\right) d\theta.
$$

Both integrals include the density function of the normal distribution

$$
Z \sim \mathcal{N}\left(\frac{\mu_D \rho_D^2 + \sigma_D^2 \sum_{i=1}^{n} x_i}{\rho_D^2 + n\sigma_D^2}, \frac{\rho_D \, \sigma_D}{\sqrt{\rho_D^2 + n\sigma_D^2}}\right),
$$

and we may simplify the problem to minimising in d

$$d^2 - 2d\, E[Z] = d^2 - 2d\, \frac{\mu_D\, \rho_D^2 + \sigma_D^2 \sum_{i=1}^n x_i}{\rho_D^2 + n\, \sigma_D^2},$$

whose optimal value is

$$d^*(x) = \frac{\mu_D\, \rho_D^2 + \sigma_D^2 \sum_{i=1}^n x_i}{\rho_D^2 + n\, \sigma_D^2}. \tag{8}$$

In particular, with the above specific parameters:

$$d^*(x) = \frac{4 \sum_{i=1}^n x_i}{4n + 1}.$$

Suppose now that D is aware of A's presence. Then, she needs to further model her perspective of A's probabilities and losses. Assume that D believes that:

- A thinks the state follows a normal distribution $\theta \sim \mathcal{N}(\mu_A, \sigma_A)$. Moreover, $\mu_A \sim \mathcal{U}(-1, 1)$ and $\sigma_A \sim \mathcal{U}(\frac{1}{2}, \frac{3}{2})$ reflect D's uncertainty about the actual parameters he uses.
- A's action is restricted to $a \in \{0, 1\}$, related with doing nothing ($a = 0$, with the state unperturbed as $\lambda = \theta$) or perturbing the state by adding 1 ($a = 1$, with the state becoming $\lambda = \theta + 1$).
- A takes the observations $(x_1, \ldots, x_n)$ to be i.i.d. and normal around the transformed state λ as $x_i \mid \lambda \sim \mathcal{N}(\lambda, \rho_A)$ for $i = 1, \ldots, n$ where $\rho_A \sim \mathcal{U}(\frac{1}{4}, \frac{3}{4})$ models D's uncertainty.
- A is a Bayesian adversary with probability $\pi_B = \frac{2}{3}$ and a minimax adversary with probability $\pi_M = \frac{1}{3}$.
- A's loss function incorporates a quadratic term and a penalty in relation with a, so that $L_A(d, a, \theta) = \alpha |a| - \beta (\theta - d)^2$. Note that A wins, on one hand, when D's estimate is further distant from θ, and, on the other hand, as his perturbation is smaller. We model D's uncertainty about the loss parameters with $\alpha \sim \mathcal{U}(1, \frac{3}{2})$ and $\beta \sim \mathcal{U}(1, 3)$.
- A determines that D will make her decision (point estimate of θ) uniformly around $\bar{x} = \frac{1}{n} \sum_{i=1}^n x_i$ with $d \mid x \sim \mathcal{U}(\bar{x} - \frac{1}{2}, \bar{x} + \frac{1}{2})$.

Note first that when A is a minimax adversary as in Sect. 3.3.2, his maximum loss corresponds with D correctly guessing state θ accentuated by his choice of a. Therefore, he would be making $a = 0$ and, thus:

$$\hat{p}_D^M(0) = 1, \qquad \hat{p}_D^M(1) = 0.$$

This leads to the same decision (8) as in the non-adversarial problem, as can be easily verified by replacing $p_D(a)$ with $\hat{p}_D^M(a)$ in expression (6).

We consider now that A is a Bayesian adversary as in Sect. 3.3.1. Using $K = 10^4$ iterations in the proposed simulation scheme (7), we obtain that:

$$\hat{p}_D^B(0) = 0.318, \qquad \hat{p}_D^B(1) = 0.682.$$

Thus being more likely that A will perform his perturbation. Then, making use of expression (6), we may compute D's optimal decision, given observations $(x_1, \ldots, x_n)$, by minimising in d

$$\hat{p}_D^B(0) \int (\theta - d)^2 \frac{1}{\sigma_D\sqrt{2\pi}} \exp\left(\frac{-(\theta-\mu_D)^2}{2\sigma_D^2}\right) \prod_{i=1}^n \frac{1}{\rho_D\sqrt{2\pi}} \exp\left(\frac{-(x_i-\theta)^2}{2\rho_D^2}\right) d\theta$$

$$+ \hat{p}_D^B(1) \int (\theta - d)^2 \frac{1}{\sigma_D\sqrt{2\pi}} \exp\left(\frac{-(\theta-\mu_D)^2}{2\sigma_D^2}\right) \prod_{i=1}^n \frac{1}{\rho_D\sqrt{2\pi}} \exp\left(\frac{-(x_i-\theta-1)^2}{2\rho_D^2}\right) d\theta.$$

This may be simplified in the same manner as the non-adversarial version to obtain the equivalent problem of minimising in d

$$\hat{p}_D^B(0)\,\xi(x, 0) \left(d^2 - 2d\, \frac{\mu_D\,\rho_D^2+\sigma_D^2\sum_{i=1}^n x_i}{\rho_D^2+n\,\sigma_D^2}\right)$$

$$+ \hat{p}_D^B(1)\,\xi(x, 1) \left(d^2 - 2d\, \frac{\mu_D\,\rho_D^2+\sigma_D^2\sum_{i=1}^n (x_i-1)}{\rho_D^2+n\,\sigma_D^2}\right),$$

where

$$\xi(x, a) = \exp\left(\frac{\frac{(\mu_D\,\rho_D^2+\sigma_D^2\sum_{i=1}^n (x_i-a))^2}{\rho_D^2+n\,\sigma_D^2} - \sigma_D^2\sum_{i=1}^n (x_i-a)^2}{2\,\rho_D^2\,\sigma_D^2}\right),$$

and whose minimum is

$$d^*(x) = \frac{\mu_D\,\rho_D^2}{\rho_D^2+n\,\sigma_D^2} + \frac{\hat{p}_D^B(0)\,\xi(x,0)\,\sigma_D^2\sum_{i=1}^n x_i + \hat{p}_D^B(1)\,\xi(x,1)\,\sigma_D^2\sum_{i=1}^n (x_i-1)}{(\hat{p}_D^B(0)\,\xi(x,0) + \hat{p}_D^B(1)\,\xi(x,1))\,(\rho_D^2+n\,\sigma_D^2)}.$$

In particular, making use of the specific parameters:

$$d^*(x) = \frac{4\,(0.318\,\xi(x,0)\sum_{i=1}^n x_i + 0.682\,\xi(x,1)\sum_{i=1}^n (x_i-1))}{(0.318\,\xi(x,0) + 0.682\,\xi(x,1))\,(4n+1)}$$

with

$$\xi(x, a) = \exp\left(\frac{8\,(\sum_{i=1}^n (x_i-a))^2}{4n+1} - 2\sum_{i=1}^n (x_i-a)^2\right).$$

Aggregating the different solution concepts as in Sect. 3.3.3, we may determine D's optimal decision, given observations $(x_1, \ldots, x_n)$, by minimising in d

Table 1 D's decision for each solution concept

Solution concept	Optimal solution
Non-adversarial	$\dfrac{4 \sum_{i=1}^{n} x_i}{4n + 1}$
ARA: Minimax adversary	$\dfrac{4 \sum_{i=1}^{n} x_i}{4n + 1}$
ARA: Bayesian adversary	$\dfrac{4 \left(0.318 \, \xi(x, 0) \sum_{i=1}^{n} x_i + 0.682 \, \xi(x, 1) \sum_{i=1}^{n} (x_i - 1)\right)}{(0.318 \, \xi(x, 0) + 0.682 \, \xi(x, 1)) \, (4n + 1)}$
ARA: Uncertain concept	$\dfrac{4 \left(0.545 \, \xi(x, 0) \sum_{i=1}^{n} x_i + 0.455 \, \xi(x, 1) \sum_{i=1}^{n} (x_i - 1)\right)}{(0.545 \, \xi(x, 0) + 0.455 \, \xi(x, 1)) \, (4n + 1)}$

$$(\pi_B \, \hat{p}_D^B(0) + \pi_M) \int (\theta - d)^2 \, \frac{1}{\sigma_D \sqrt{2\pi}} \exp\left(\frac{-(\theta - \mu_D)^2}{2\sigma_D^2}\right) \prod_{i=1}^{n} \frac{1}{\rho_D \sqrt{2\pi}} \exp\left(\frac{-(x_i - \theta)^2}{2\rho_D^2}\right) d\theta$$

$$+ \pi_B \, \hat{p}_D^B(1) \int (\theta - d)^2 \, \frac{1}{\sigma_D \sqrt{2\pi}} \exp\left(\frac{-(\theta - \mu_D)^2}{2\sigma_D^2}\right) \prod_{i=1}^{n} \frac{1}{\rho_D \sqrt{2\pi}} \exp\left(\frac{-(x_i - \theta - 1)^2}{2\rho_D^2}\right) d\theta,$$

whose solution is:

$$d^*(x) = \frac{\mu_D \, \rho_D^2}{\rho_D^2 + n \, \sigma_D^2}$$

$$+ \frac{(\pi_B \, \hat{p}_D^B(0) + \pi_M) \, \xi(x, 0) \, \sigma_D^2 \sum_{i=1}^{n} x_i + \pi_B \, \hat{p}_D^B(1) \, \xi(x, 1) \, \sigma_D^2 \sum_{i=1}^{n} (x_i - 1)}{((\pi_B \, \hat{p}_D^B(0) + \pi_M) \, \xi(x, 0) + \pi_B \, \hat{p}_D^B(1) \, \xi(x, 1)) \, (\rho_D^2 + n \, \sigma_D^2)}.$$

In particular, using the specific parameters:

$$d^*(x) = \frac{4 \left(0.545 \, \xi(x, 0) \sum_{i=1}^{n} x_i + 0.455 \, \xi(x, 1) \sum_{i=1}^{n} (x_i - 1)\right)}{(0.545 \, \xi(x, 0) + 0.455 \, \xi(x, 1)) \, (4n + 1)}.$$

Table 1 summarizes the different optimal solutions obtained with all the considered solution concepts.

As we mentioned in Sect. 3.2, the non-adversary aware situation would systematically err D's estimation. The ARA estimator which takes into account concept uncertainty would fully acknowledge all uncertainty present in the problem.

4 Discussion

We have illustrated the ideas of concept uncertainty in an ASDT problem under the ARA framework, emphasising point estimation. The approach is general extending beyond point estimation and may be applied to other standard statistical problems, including interval estimation, hypothesis testing and forecasting. We have dwelt only on one out of the three ASDT formulations outlined, that of structural attackers.

Moreover, we have only considered two concepts concerning Bayesian and minimax adversaries. Additional classes of adversaries include non-strategic or prospect maximising players, among others. Our model requires assessing the weights of various concepts, which would be assessed based on expert judgement. Should the situation be repeated over time, we could introduce a scheme for learning about the relevance concepts based on a Bayesian updating scheme as described in [17].

Multiple attacker cases in the ASDT problem are also of interest. Under an ARA perspective, we would support D against all A agents. This poses further introspection as we would need to differentiate attackers who behave individually from those who partially or totally coordinate. Furthermore, it could also be the case that their attacks influence somehow each other.

Acknowledgements The work of DRI is supported by the Spanish Ministry of Economy and Innovation program MTM2014-56949-C3-1-R and the AXA-ICMAT Chair on Adversarial Risk Analysis. JGO's research is financed by the Spanish Ministry of Economy and Competitiveness under FPI SO grant agreement BES-2015-072892. This work has also been partially supported by the Spanish Ministry of Economy and Competitiveness through the "Severo Ochoa" Program for Centers of Excellence in R&D (SEV-2015-0554) and project MTM2015-72907-EXP.

References

1. Banks D, Ríos J, Ríos Insua D (2015) Adversarial risk analysis. CRC Press, Boca Raton
2. Bedford T, Cooke RM (2001) Probabilistic risk analysis: foundations and methods. Cambridge University Press, Cambridge
3. Clyde M, George EI (2004) Model uncertainty. Stat Sci 19(1):81–94
4. Cooke RM (1991) Experts in uncertainty: opinion and subjective probability in science. Oxford University Press, New York
5. Dalvi N, Domingos P, Mausam, Sanghai S, Verma D (2004) Adversarial classification. In: Kohavi R, Gehrke J, DuMouchel W, Ghosh J (eds) KDD'04 proceedings of 10th ACM SIGKDD international conference on knowledge discovery and data mining. ACM Press, New York
6. French S, Ríos Insua D (2000) Statistical decision theory. Wiley, New York
7. Gibbons R (1992) Game theory for applied economists. Princeton University Press, Princeton
8. Gil P (1997) Las matemáticas de lo incierto. Inaugural year address, University of Oviedo, Oviedo. http://digibuo.uniovi.es/dspace/bitstream/10651/28625/1/matematicasincierto.pdf
9. Hargreaves-Heap S, Varoufakis Y (1995) Game theory: a critical introduction. Routledge, London
10. Harsanyi JC (1967) Games with incomplete information played by "Bayesian" players, I-III. Part I. The basic model. Manag Sci 14(3):159–182
11. Hoeting JA, Madigan D, Raftery AE, Volinsky CT (1999) Bayesian model averaging: a tutorial. Stat Sci 14(4):382–417
12. Merrick J, Parnell GS (2011) A comparative analysis of PRA and intelligent adversary methods for counterterrorism management. Risk Anal 31(9):1488–1510
13. O'Hagan A, Buck CE, Daneshkhah A, Eiser JR, Garthwaite PH, Jenkinson DJ, Oakley JE, Rakow T (2006) Uncertain judgements: eliciting experts' probabilities. Wiley, Chichester
14. Parry GW (1996) The characterization of uncertainty in probabilistic risk assessments of complex systems. Reliab Eng Syst Saf 54(2–3):119–126
15. Paté-Cornell E, Guikema S (2002) Probabilistic modeling of terrorist threats: a systems analysis approach to setting priorities among countermeasures. Mil Op Res 7(4):5–23

16. Refsgaard JC, van der Sluijs JP, Højberg AL, Vanrolleghem PA (2007) Uncertainty in the environmental modelling process - a framework and guidance. Environ Model Softw 22(11):1543–1556

17. Ríos J, Ríos Insua D (2012) Adversarial risk analysis for counterterrorism modeling. Risk Anal 32(5):894–915

18. Ríos Insua D, Ríos J, Banks D (2009) Adversarial risk analysis. J Am Stat Assoc 104(486):841–854

19. Stahl DO, Wilson PW (1995) On players' models of other players: theory and experimental evidence. Games Econ Behav 10(1):218–254

20. Tetlock P, Gardner D (2015) Superforecasting: the art and science of prediction. Random House, New York

21. Tygar JD (2011) Adversarial machine learning. IEEE Intern Comput 15(5):4–6

22. Walker WE, Harremoës P, Rotmans J, van der Sluijs JP, van Asselt MB, Janssen P, Krayer von Krauss MP (2003) Defining uncertainty: a conceptual basis for uncertainty management in model-based decision support. Integr Assess 4(1):5–17

On Some Properties of Beta-Generated Distributions

Konstantinos Zografos

Abstract This note is concerned with a broad family of univariate distributions, namely the class of the beta-generated distributions which is created on the basis of the beta distribution by incorporating on this classic model a parent distribution function F with respective density f. The class of beta-generated distributions have received a great attention in the recent literature on distribution theory. The aim of this note is to provide with some properties and characterizations of the beta-generated distributions which are already known from the theory of order statistics.

1 Introduction

Beta-generated distributions have received a considerable attention during the last fifteen years. This research activity has signalled in the paper by Eugene et al. [10], to the best of our knowledge. This paper introduced the beta-normal distribution, a model developed on the basis of the classic normal and beta distributions. Two years later Jones [12] has introduced and studied the family of beta generated distributions as a generalization of the distribution of order statistics. Based on Zografos and Balakrishnan [19], Zografos [17, 18] and the references appeared therein, for a continuous distribution function F with density f, the family of beta-generated distributions, generated by the parent model F and the parameters $\alpha, \beta > 0$, has its pdf (cf. Jones [12]).

$$g_F(x; \alpha, \beta) = \frac{1}{B(\alpha, \beta)} f(x) \{F(x)\}^{\alpha-1} \{1 - F(x)\}^{\beta-1},\tag{1}$$

It is a great honor to present this work to the memory of my beloved friend Pedro Gil, an exceptional scientist and an outstanding human personality.

K. Zografos (✉)
Department of Mathematics, University of Ioaninna, 451 10 Ioaninna, Greece
e-mail: kzograf@uoi.gr

© Springer International Publishing AG 2018
E. Gil et al. (eds.), *The Mathematics of the Uncertain*, Studies in Systems, Decision and Control 142, https://doi.org/10.1007/978-3-319-73848-2_50

for $\alpha > 0$ and $\beta > 0$, where $B(\alpha, \beta) = \int_0^1 t^{\alpha-1}(1-t)^{\beta-1}dt$ is the complete beta function. If the parameters α and β are positive integers, the beta-generated model in (1) is the distribution of the ith order statistic in a random sample of size n from distribution F, where $i = \alpha$ and $n = \alpha + \beta - 1$.

The beta-normal distribution of Eugene et al. [10] is obtained from (1) if the parent model F is that of the classic normal distribution. The family of distributions (1) can be generated by means of a transformation of a random variable Y with beta distribution, $Beta(\alpha, \beta)$, $\alpha > 0$ and $\beta > 0$. In particular, if $Y \sim Beta(\alpha, \beta)$, then the density of the random variable $X = F^{-1}(Y)$ is given by (1). This representation of X helps to generate random numbers from (1) while the case $\alpha = \beta = 1$ corresponds to the well-known quantile function representation $X = F^{-1}(U)$, where $U \sim U(0, 1)$, which is used in order to generate data from a distribution F. There is a huge literature where the family of distributions (1) is studied and particular models, obtained from (1) for special choices of the parent distribution F, are proposed as best alternatives to the existing models for modelling real data. In this direction, it is mentioned the paper by Nadarajah and Kotz [15], Akinsete et al. [1], Zografos and Balakrishnan [19], Barreto-Souza et al. [8], Paranaiba et al. [16], Zografos [18], Alexander and Sarabia [2], Cordeiro and de Castro [9], Alexander et al. [3] while alternative families of distributions have been introduced in Zografos and Balakrishnan [19], Lee et al. [14] and Alzaatreh et al. [5], among many other.

This note is motivated by the papers of Asadi et al. [6] and Baratpour et al. [7] and its main aim is to provide with some characterizations of the beta-generated distributions, defined by (1), which are obeyed for the distribution of order statistics, studied in the above mentioned two papers. More precisely, Sect. 2 follows the work by Baratpour et al. [7] and it provides with characterizations of the parent model F, in (1), by means of the Shannon entropy of this broad family of univariate distributions. Section 3 is devoted to the study of some informational properties of the family (1), by following ideas in the paper by Asadi et al. [6]. Section 4 provides with some conclusions and this note is finished with an appendix which provides the proof of a result.

2 Characterizations of Beta-Generated Distributions

The first result in this section characterizes equality of Shannon entropies of two beta-generated distributions with different parent models, by the equality of the respective parent models. This characterization of the family of beta-generated distributions is provided by means of Shannon entropy of (1). Shannon entropy is an omnipresent quantity in almost all fields of science and engineering and its explicit expression for the beta-generated distributions has been derived in Zografos and Balakrishnan [19]. In view of Corollary 1 in Zografos and Balakrishnan [19], the Shannon entropy of (1) is given by

$$\mathcal{H}_{Sh}(g_F) = \ln B(\alpha, \beta) - (\alpha - 1)[\Psi(\alpha) - \Psi(\alpha + \beta)]$$
$$-(\beta - 1)[\Psi(\beta) - \Psi(\alpha + \beta)] \qquad (2)$$
$$-E_Y[\ln f(F^{-1}(Y))],$$

where $Y \sim Beta(\alpha, \beta)$. Next proposition generalizes Theorem 2.3 in the paper by Baratpour et al. [7], for the distribution of order statistics.

Proposition 2.1 *Let random variables X and Y with respective c.d.f. and p.d.f. F, f and G, g, respectively. Denote by g_F and g_G the beta-generated distributions, obtained from (1), in the case of parent models F and G, respectively. Then,*

$$X \overset{d}{=} Y \text{ if and only if } \mathcal{H}_{Sh}(g_F) = \mathcal{H}_{Sh}(g_G), \text{ for } \alpha \geq 1, \beta \geq 1,$$

where the symbol $\overset{d}{=}$ is used to denote that the distribution of the random variable on the left coincides with the distribution of the random variable on the right.

Proof The proof follows the steps of the proof of Theorem 2.3 of Baratpour et al. [7] and it is only sketched. If $X \overset{d}{=} Y$, then $F = G$, a.e., $g_F = g_G$, a.e. and hence the respective Shannon entropies are coincide. Let now $\mathcal{H}_{Sh}(g_F) = \mathcal{H}_{Sh}(g_G)$. Based on (2), it is immediate to see that $E_Y[\ln f(F^{-1}(Y))] = E_Y[\ln g(G^{-1}(Y))]$, with $Y \sim Beta(\alpha, \beta)$. This is equivalent to

$$\int_0^1 y^{\alpha-1} \left\{ (1-y)^{\beta-1} \left[\ln f(F^{-1}(y)) - \ln g(G^{-1}(y)) \right] \right\} dy = 0.$$

If it is then applied Exercise 4 in Aliprantis and Burkinshaw (see [4, p. 90]), for $n = \alpha - 1, n \geq 0$, we obtain that

$$(1 - y)^{\beta-1} \left[\ln f(F^{-1}(y)) - \ln g(G^{-1}(y)) \right] = 0, \text{ for } y \in (0, 1),$$

or

$$\ln f(F^{-1}(y)) = \ln g(G^{-1}(y)), \text{ for } y \in (0, 1).$$

The proof is now completed by following the last part in the proof of Theorem 2.3 of Barapour et al. [7]. □

This proposition motivates the following remarks.

Remark 2.1 (*i*) If $\alpha = m$ and $\beta = n - m + 1$, then Proposition 2.1 leads to Theorem 2.3 of Barapour et al. [7]. If $\alpha = 1$ or $\alpha = n$, then previous proposition leads to Corollaries 2.3 and 2.4 of Barapour et al. [7], respectively.
(*ii*) Let X and Y be two random variables and consider independent observations $X_1, X_2, \ldots, X_{n_1}$ and $Y_1, Y_2, \ldots, Y_{n_2}$ on them. Let f and g be the respective distributions of X and Y. Based on this formulation, let's concentrate on the null hypothesis of homogeneity of the observations on X and Y, that is,

$H_0 : f$ coincides with g, or, in an equivalent form, $H_0 : X \overset{d}{=} Y$.

Taking into account Proposition 2.1, the null hypothesis is equivalent to

$$H_0 : \mathcal{H}_{Sh}(g_F) = \mathcal{H}_{Sh}(g_G).$$

A test statistic can be developed for this last hypothesis on the basis of estimators of Shannon entropies $\mathcal{H}_{Sh}(g_F)$ and $\mathcal{H}_{Sh}(g_G)$.

For the necessities of the rest of this section we consider parent density and distribution functions f_θ and F_θ, depending on a real valued parameter θ. The corresponding beta-generated distribution with parent model F_θ is then defined by

$$g_{F_\theta}^{\alpha,\beta}(x) = \frac{1}{B(\alpha, \beta)} f_\theta(x) \{F_\theta(x)\}^{\alpha-1} \{1 - F_\theta(x)\}^{\beta-1}, \ \alpha > 0, \beta > 0.$$

The hazard rate function, associated to F_θ, is defined by

$$r_\theta(x) = \frac{f_\theta(x)}{1 - F_\theta(x)}, x > 0.$$

The next proposition is motivated by Theorem 2.1 of Barapour et al. [7] and it characterizes the hazard rate function associated to the parent model $F_\theta(x)$. The proof of this proposition is given in the Appendix.

Proposition 2.2 *The hazard rate function depends only on the parameter* $\theta \in \Theta \subseteq \mathbb{R}$, *that is,* $r_\theta(x) = w(\theta)$, *for a positive real valued function* $w(\theta)$, *if and only if,* $\mathcal{H}_{Sh}(f_\theta) - \mathcal{H}_{Sh}\left(g_{F_\theta}^{1,\beta}\right) = \ln \beta$, *for* $\beta \geq 1$.

The above proposition extends Theorem 2.1 of Barapour et al. [7]. This theorem is obtained by an application of the above Proposition 2.2, for $\alpha = 1$ and $\beta = n$. Moreover, the condition that the hazard rate function depends only on the parameter θ is just the condition that the parent model F_θ is that of the exponential distribution with hazard rate function $w(\theta)$ (cf. Baratpour et al. [7, p. 50]).

A similar result to that of Proposition 2.2 can be proved on the basis of the reversed hazard rate function, associated to the parent model F_θ, which is defined by

$$\widetilde{r}_\theta(x) = \frac{f_\theta(x)}{F_\theta(x)}, x > 0.$$

Proposition 2.3 *The reversed hazard rate function depends only on the parameter* $\theta \in \Theta \subseteq \mathbb{R}$, *that is,* $\widetilde{r}_\theta(x) = v(\theta)$, *for a positive real valued function* $v(\theta)$, *if and only if,* $\mathcal{H}_{Sh}(f_\theta) - \mathcal{H}_{Sh}\left(g_{F_\theta}^{\alpha,1}\right) = \ln \alpha$, *for* $\alpha \geq 1$.

The proof of this last proposition is derived by a similar argument as that of the proof of Proposition 2.2, given in the Appendix, and Lemma 1(a) in Zografos and Balakrishnan [19]. Proposition 2.3 extends Theorem 2.2 of Barapour et al. [7].

3 On Divergence Between Beta-Generated and Parent Model

Motivated by Asadi et al. [6] and Zografos and Balakrishnan [19], the aim of this section is to derive explicit expressions for a divergence or a distance type measure between the parent model F and the beta-generated distribution g_F, with parent model F, given by (1). We will concentrate, in the sequel, to the Kullback–Leibler divergence or its symmetric version, as it was defined by Kullback (see [13, p. 6]) and it is known as Jeffreys divergence. Jeffreys divergence between the parent model F with density f and the beta-generated distribution g_F with parent F, is defined by

$$J(f, g_F) = D_{KL}(f, g_F) + D_{KL}(g_F, f),\tag{3}$$

where D_{KL} is the Kullback–Leibler divergence between the underlined densities and it is defined by

$$D_{KL}(f, g_F) = \int_{\mathbb{R}} f(x) \ln \frac{f(x)}{g_F(x;\alpha,\beta)} dx, \text{ and}$$

$$D_{KL}(g_F, f) = \int_{\mathbb{R}} g_F(x; \alpha, \beta) \ln \frac{g_F(x;\alpha,\beta)}{f(x)} dx.\tag{4}$$

Jeffreys divergence $J(f, g_F)$, in (3), obeys non-negativity and it is symmetric in its arguments, that is,

$$J(f, g_F) \geq 0 \text{, with equality if and only if } f = g_F, \text{ and}$$
$$J(f, g_F) = J(g_F, f).$$

However, it doesn't obey the triangular inequality and it isn't therefore a distance measure of the underlined densities.

In order to obtain $J(f, g_F)$ in an explicit form it is necessary to obtain $D_{KL}(f, g_F)$ and $D_{KL}(g_F, f)$, in an explicit form. To proceed in this direction, using (1) and after simple algebraic manipulations

$$D_{KL}(f, g_F) = \int_{\mathbb{R}} f(x) \ln \frac{f(x)}{g_F(x;\alpha,\beta)} dx$$

$$= \ln B(\alpha, \beta) - \int_{\mathbb{R}} f(x) \ln \left\{ [F(x)]^{\alpha-1}[1 - F(x)]^{\beta-1} \right\} dx.$$

Applying now the transformation $u = F(x)$ to the last integral

$$D_{KL}(f, g_F) = \ln B(\alpha, \beta) - (\alpha - 1) \int_0^1 \ln u\, du - (\beta - 1) \int_0^1 \ln(1 - u)du,$$

which leads to

$$D_{KL}(f, g_F) = \ln B(\alpha, \beta) + \alpha + \beta - 2, \ \alpha, \beta > 0.\tag{5}$$

In a similar manner and in view of (1) and (4)

$$D_{KL}(g_F, f) = -\ln B(\alpha, \beta) + (\alpha - 1) \int_{\mathbb{R}} g_F(x; \alpha, \beta) \ln F(x) dx$$

$$+ (\beta - 1) \int_{\mathbb{R}} g_F(x; \alpha, \beta) \ln[1 - F(x)] dx.$$

Based on Lemma 1 of Zografos and Balakrishnan [19], it is immediately obtained

$$D_{KL}(g_F, f) = -\ln B(\alpha, \beta) + (\alpha - 1)[\Psi(\alpha) - \Psi(\alpha + \beta)]$$

$$+ (\beta - 1)[\Psi(\beta) - \Psi(\alpha + \beta)]. \tag{6}$$

The explicit expressions of (5) and (6) along with that of Eq. (3) are summarized in the next proposition.

Proposition 3.1 *Kullback–Leibler and Jeffreys divergences between the parent model f and the beta-generated distribution g_F, in (1), are given by*

$$D_{KL}(f, g_F) = \ln B(\alpha, \beta) + \alpha + \beta - 2,$$

$$D_{KL}(g_F, f) = -\ln B(\alpha, \beta) + (\alpha - 1)[\Psi(\alpha) - \Psi(\alpha + \beta)]$$

$$+ (\beta - 1)[\Psi(\beta) - \Psi(\alpha + \beta)],$$

$$J(f, g_F) = (\alpha - 1)[\Psi(\alpha) - \Psi(\alpha + \beta) + 1]$$

$$+ (\beta - 1)[\Psi(\beta) - \Psi(\alpha + \beta) + 1],$$

for $\alpha, \beta > 0$.

This last proposition motivates some remarks and conclusions on the divergence between the parent model f and the beta-generated distribution g_F, of (1). It also gathers some similar results for Kullback–Leibler divergence between the parent model f and the distribution of order statistics of a random sample from f, as they have been previously derived in the work by Asadi et al. [6].

Remark 3.1 Suppose that $X_1, \ldots, X_n$ is a random sample from a population which is described by a density f or a distribution function F. In this frame, the distribution $f_{(i)}$, of the ith order statistic, $i = 1, \ldots, n$, is obtained from (1) for $\alpha = i$ and $\beta = n - i + 1$. Based on the previous proposition the respective divergences between f and $f_{(i)}$ are given by,

$$D_{KL}(f, f_{(i)}) = \ln B(i, n - i + 1) + n - 1,$$

$$D_{KL}(f_{(i)}, f) = -\ln B(i, n - i + 1) + (i - 1)[\Psi(i) - \Psi(n + 1)]$$

$$+ (n - i)[\Psi(n - i + 1) - \Psi(n + 1)],$$

$$J(f, f_{(i)}) = (i - 1)[\Psi(i) - \Psi(n + 1) + 1]$$

$$+(n - i)[\Psi(n - i + 1) - \Psi(n + 1) + 1].$$

The first two equations for $D_{KL}(f, f_{(i)})$ and $D_{KL}(f_{(i)}, f)$ are already known from Asadi et al. [6, p. 218]. We observe that all the above divergences between f and g_F or between f and $f_{(i)}$ don't depend on the parent model f. Hence, they can't discriminate between different parent models. All the above three divergences between f and $f_{(i)}$ are characterized by a monotonicity property. Based on Asadi et al. [6, p. 218], "the information discrepancy (as it is expressed by $D_{KL}(f, f_{(i)})$ or $D_{KL}(f_{(i)}, f)$ or $J(f, f_{(i)})$) between the distribution of order statistics and f decreases up to the median and then increases. Thus, amongst the order statistics, the median has the closest distribution to the data distribution.".

4 Conclusions

Some well known properties and characterizations of the distribution of order statistics are extended to the class of the beta-generated distributions and the repertory of the properties and characterizations of this broad family of univariate distributions is therefore enriched. An identification characterization has been proved in Proposition 2.1. According to this characterization two distributions coincide if and only if the Shannon entropies of the respective beta-generated distributions are coincide. This result can be exploited to develop tests of homogeneity which would be based on the empirical versions of Shannon entropies of the respective beta-generated distributions.

5 Appendix: Proof of Proposition 2.2

($\Longrightarrow$) Suppose that $r_\theta(x) = \frac{f_\theta(x)}{1 - F_\theta(x)} = w(\theta), \theta \in \Theta \subseteq \mathbb{R}$. Then,

$$\mathcal{H}_{Sh}(f_\theta) = - \int_{\mathbb{R}} f_\theta(x) \ln f_\theta(x) dx$$
$$= - \int_{\mathbb{R}} f_\theta(x) \ln [r_\theta(x)(1 - F_\theta(x))] dx$$
$$= - \ln w(\theta) - \int_{\mathbb{R}} f_\theta(x) \ln (1 - F_\theta(x)) dx,$$

or

$$\mathcal{H}_{Sh}(f_\theta) = - \ln w(\theta) - \int_{\mathbb{R}} B(1, 1) g_{F_\theta}^{1,1}(x) \ln (1 - F_\theta(x)) dx.$$

Taking into account that $B(1, 1) = 1$ and using Lemma 1(b) of Zografos and Balakrishnan [19], we get

$$\mathcal{H}_{Sh}(f_\theta) = -\ln w(\theta) - [\Psi(1) - \Psi(2)].$$

Based on

$$\Psi(x+1) - \Psi(x) = \frac{1}{x}, \tag{7}$$

(cf. Gradshteyn and Ryzhik [11, §. 365]), it is obtained that,

$$\mathcal{H}_{Sh}(f_\theta) = -\ln w(\theta) + 1. \tag{8}$$

On the other hand,

$$
\begin{aligned}
\mathcal{H}_{Sh}\left(g_{F_\theta}^{1,\beta}\right) &= -\int_{\mathbb{R}} g_{F_\theta}^{1,\beta}(x) \ln g_{F_\theta}^{1,\beta}(x) dx \\
&= -\int_{\mathbb{R}} g_{F_\theta}^{1,\beta}(x) \ln\left[\frac{1}{B(1,\beta)} f_\theta(x)(1 - F_\theta(x))^{\beta-1}\right] dx \\
&= -\int_{\mathbb{R}} g_{F_\theta}^{1,\beta}(x) \ln\left[\frac{1}{B(1,\beta)} r_\theta(x)(1 - F_\theta(x))^{\beta}\right] dx
\end{aligned} \tag{9}
$$

and taking into account that $r_\theta(x) = w(\theta)$ and $B(1, \beta) = 1/\beta$ it is obtained that

$$\mathcal{H}_{Sh}\left(g_{F_\theta}^{1,\beta}\right) = -\ln \beta - \ln w(\theta) - \beta \int_{\mathbb{R}} g_{F_\theta}^{1,\beta}(x) \ln(1 - F_\theta(x)) dx.$$

Using again Lemma 1(b) of Zografos and Balakrishnan [19] and Eq. (7) above,

$$
\begin{aligned}
\mathcal{H}_{Sh}\left(g_{F_\theta}^{1,\beta}\right) &= -\ln \beta - \ln w(\theta) - \beta[\Psi(\beta) - \Psi(\beta+1)] \\
&= -\ln \beta - \ln w(\theta) + 1.
\end{aligned} \tag{10}
$$

Equations (8) and (10) lead now to the desired result.
($\Longleftarrow$) Suppose that $\mathcal{H}_{Sh}(f_\theta) - \mathcal{H}_{Sh}(g_{F_\theta}^{1,\beta}) = \ln \beta$, for $\beta \geq 1$. Based on (9) and on the identity $B(1, \beta) = 1/\beta$,

$$\mathcal{H}_{Sh}\left(g_{F_\theta}^{1,\beta}\right) = -\ln \beta - \beta \int_{\mathbb{R}} g_{F_\theta}^{1,\beta}(x) \ln(1 - F_\theta(x)) dx - \int_{\mathbb{R}} g_{F_\theta}^{1,\beta}(x) \ln r_\theta(x) dx.$$

Using Lemma 1(b) of Zografos and Balakrishnan [19] and Eq. (7), we obtain

$$\int_{\mathbb{R}} g_{F_\theta}^{1,\beta}(x) \ln(1 - F_\theta(x)) dx = \Psi(\beta) - \Psi(\beta+1) = -\frac{1}{\beta}.$$

Then, the above formula for $\mathcal{H}_{Sh}\left(g_{F_\theta}^{1,\beta}\right)$ is simplified as follows,

$$
\begin{aligned}
\mathcal{H}_{Sh}\left(g_{F_\theta}^{1,\beta}\right) &= -\ln \beta + 1 - \int_{\mathbb{R}} g_{F_\theta}^{1,\beta}(x) \ln r_\theta(x) dx \\
&= -\ln \beta + 1 - \int_{\mathbb{R}} \frac{1}{B(1,\beta)} f_\theta(x)(1 - F_\theta(x))^{\beta-1} \ln r_\theta(x) dx \\
&= -\ln \beta + 1 - \beta \int_{\mathbb{R}} f_\theta(x)(1 - F_\theta(x))^{\beta-1} \ln r_\theta(x) dx.
\end{aligned}
$$

Because, by hypothesis, $\mathcal{H}_{Sh}(f_\theta) - \mathcal{H}_{Sh}(g_{F_\theta}^{1,\beta}) = \ln \beta$, using the previous expression for $\mathcal{H}_{Sh}\left(g_{F_\theta}^{1,\beta}\right)$, it is obtained

$$\mathcal{H}_{Sh}(f_\theta) - 1 + \beta \int_{\mathbb{R}} f_\theta(x)\,(1 - F_\theta(x))^{\beta-1} \ln r_\theta(x)\,dx = 0. \tag{11}$$

Applying the transformation $u = 1 - F_\theta(x)$ to the last integral

$$\int_{\mathbb{R}} f_\theta(x)\,(1 - F_\theta(x))^{\beta-1} \ln r_\theta(x)\,dx = \int_0^1 u^{\beta-1} \ln r_\theta \left(F_\theta^{-1}(1 - u)\right) du. \tag{12}$$

Equations (11) and (12) lead to

$$\mathcal{H}_{Sh}(f_\theta) - 1 + \beta \int_0^1 u^{\beta-1} \ln r_\theta \left(F_\theta^{-1}(1 - u)\right) du = 0$$

and taking into account that $\mathcal{H}_{Sh}(f_\theta)$ does not depend on x or u,

$$\beta \int_0^1 u^{\beta-1} \left\{ \mathcal{H}_{Sh}(f_\theta) - 1 + \ln r_\theta \left(F_\theta^{-1}(1 - u)\right) \right\} du = 0.$$

In view of Exercise 4 of Aliprantis and Burkinshaw (see [4, p. 90]) for $n = \beta - 1$, $n \geq 0$, it is obtained that

$$\ln r_\theta \left(F_\theta^{-1}(1 - u)\right) + \mathcal{H}_{Sh}(f_\theta) - 1 = 0, \quad \text{for } \beta \geq 1.$$

The result now follows by using a similar argument as that of the last two lines in the proof of Theorem 2.1 of Barapour et al. [7]. $\qquad\square$

References

1. Akinsete A, Famoye F, Lee C (2008) The beta-Pareto distribution. Statistics 42:547–563
2. Alexander C, Sarabia JM (2010) Generalized beta generated distributions. ICMA centre discussion papers in finance DP2010-09
3. Alexander C, Cordeiro GM, Ortega EMM, Sarabia JM (2012) Generalized beta-generated distributions. Comput Stat Data Anal 56:1880–1897
4. Aliprantis CD, Burkinshaw O (1998) Principles of real analysis. Academic Press, San Diego
5. Alzaatreh A, Lee C, Famoye F (2014) The gamma-normal distribution: properties and applications. Comput Stat Data Anal 69:67–80
6. Asadi M, Ebrahimi N, Hamedani GG, Soofi E (2006) Information measures for Pareto distributions and order statistics. In: Balakrishnan N, Castillo E, Sarabia JM (eds) Advances in distribution theory, order statistics, and inference. Birkhäuser, Boston
7. Baratpour S, Ahmadi J, Arghami NR (2007) Some characterizations based on entropy of order statistics and record values. Commun Stat-Theor Meth 36:47–57

8. Barreto-Souza W, Santos AHS, Cordeiro GM (2010) The beta generalized exponential distribution. J Stat Comput Simul 80:159–172
9. Cordeiro GM, de Castro M (2011) A new family of generalized distributions. J Stat Comput Simul 81:883–893
10. Eugene N, Lee C, Famoye F (2002) Beta-normal distribution and its applications. Commun Stat Theor Meth 31:497–512
11. Gradshteyn IS, Ryzhik IM (1965) Table of integrals series and products. Academic Press, New York
12. Jones MC (2004) Families of distributions arising from distributions of order statistics. Test 13:1–43
13. Kullback S (1959) Information theory and statistics. Wiley, New York
14. Lee C, Famoye F, Alzaatreh A (2013) Methods for generating families of continuous distribution in the recent decades. WIREs Comput Stat 5:219–238
15. Nadarajah S, Kotz S (2006) The beta exponential distribution. Reliab Eng Syst Saf 91:689–697
16. Paranaiba PF, Ortega EMM, Cordeiro GM, Pescim RR (2010) The beta burr XII distribution with application to lifetime data. Comput Stat Data Anal 55:1118–1136
17. Zografos K (2008) On some beta generated distributions and their maximum entropy characterization: the beta-Weibull distribution. In: Barnett NS, Dragomir SS (eds) Advances in inequalities from probability theory & statistics. Nova Science Publishers, New Jersey
18. Zografos K (2011) Generalized beta generated-II distributions. In: Pardo L, Balakrishnan N, Gil MA (eds) Modern mathematical tools and techniques in capturing complexity. Springer, Berlin
19. Zografos K, Balakrishnan N (2009) On families of beta- and generalized gamma- generated distributions and associated inference. Stat Meth 6:344–362

Part IV
The Mathematics of Imprecision: Fuzzy Sets and other Theories and Approaches

Fuzzy Multi-criteria Support for Sustainable and Social Responsible Investments: The Case of Investors with Loss Aversion

Amelia Bilbao-Terol, Mariano Jiménez-López, Mar Arenas-Parra
and M. Victoria Rodríguez-Uría

Abstract The aim of this paper is to construct a support decision-making system for portfolio selection combining financial and sustainable objectives. The model we propose allows us to obtain a socially responsible portfolio which tracks the portfolio that an investor could have chosen if she did not take into account social, ethical and ecological (SEE) considerations in her investment decisions. For this purpose, we propose a fuzzy multi-criteria model that runs on two levels of decision-making: in the first stage, the preferred portfolio considering only financial objectives is obtained. In the second stage, we will use the preferred portfolio as a reference point with respect the financial behavior and then a socially responsible portfolio is obtained. In this paper linguistic labels have been used to model the parameters of the value function proposed by Kahneman and Tversky (Econometrica XVL(II):263–291, [10]). These linguistic labels determine in a "soft" way the loss and risk aversion. The developed methodology has been applied to eight Spanish companies, which have been selected for their relevance in the Spanish stock market.

1 Introduction

The concept of Corporate Social Responsibility (CSR) may be defined as the commitment by firms to contribute to sustainable economic development while improving the quality of the life of the workforce as well as the local community and society

A. Bilbao-Terol (✉) · M. Arenas-Parra · M. V. Rodríguez-Uría
Departamento de Economía Cuantitativa, Universidad de Oviedo, Oviedo, Spain
e-mail: ameliab@uniovi.es

M. Arenas-Parra
e-mail: mariamar@uniovi.es

M. V. Rodríguez-Uría
e-mail: vrodri@uniovi.es

M. Jiménez-López
Departamento de Economía Aplicada I, Universidad del País Vasco UPV/EHU,
Donostia-San Sebastián, Spain
e-mail: mariano.jimenez@ehu.es

© Springer International Publishing AG 2018
E. Gil et al. (eds.), *The Mathematics of the Uncertain*, Studies in Systems,
Decision and Control 142, https://doi.org/10.1007/978-3-319-73848-2_51

at large. Socially Responsible Investing (SRI) is an investment way that integrates environmental, social and corporate governance (ESG) considerations to generate long-term competitive financial returns and positive societal impact. It is a process of looking for companies with good CSR performance.

The aim of this paper is to design a model for selecting SR portfolios considering financial and sustainability objectives. The measuring of the SR degree of individual investments is complicated, given that a welfare function, which includes all social aspects, is not available (Hallerbach et al. [8]). On the other hand, the UK Social Investment Forum website (UKSIF) indicates that though the majority of people agree with what are the ethical issues, each individual must decide if a specific investment meets her criteria:"different people have different views as to what is acceptable and how important a set issue is for them". However, the UKSIF identifies what characterizes good SRI: "What they have in common is that they clearly state views on SEE issues. The key is to deliver and use information on the investments in such a way that those who invest may decide whether such investments are suitable for their clients and their pension funds" (UK Social Investment Forum, http://www.uksif.org/).

Our modelling is based on to find a SR portfolio that is as close as possible to a conventional reference portfolio that the investor would have chosen if she had not taken into account social issues in her investment decisions. For this purpose, we propose a fuzzy multi-criteria model that runs on two levels of decision-making. So, in the first stage, a conventional portfolio is obtained by the Kahneman–Tversky's prospect theory [10] with net profits as the financial objective. A key element of prospect theory is an S-shaped value function that is concave (risk aversion) in the domain of gains and convex (risk seeking) in the domain of losses, both measured relative to a reference point. In the second stage, we are faced with a multi-objective problem which is solved by using an extended Goal Programming methodology and the financial characteristics of the reference portfolio as targets.

The financial information has been obtained from the financial agency Morningstar Ltd. and the one related to the CSR from the analysis of the companies sustainability reports (Obama-Eyang [13], Bilbao-Terol et al. [3]). A sustainability report is a report published by a company or organization about the economic, environmental and social impacts caused by its business activities. In this paper, we have been analysed the CSR companies following the Global Reporting Initiative (GRI) guidelines that appear to be the most popular and comprehensive CSR reporting framework in use today (Brown et al. [4], GRI [6], Searcy and Elkhawas [16]). The GRI guidelines aim is to create a common social and sustainability reporting for the companies similar to financial reporting, in order to increase corporate transparency. It uses sustainability reporting in three Dimensions: Economic, Environmental and Social Sustainability. These Dimensions are broken down in 3 (Economic Performance, Market Presence and Indirect Economic Impacts), 8 (Materials, Energy, Water, Biodiversity, Emissions, Effluents and Waste, Compliance, Transport and Overall) and 4 (Labour Practices and Decent Work, Human Rights, Society and Product Responsibility) Categories, respectively.

In this paper, we have applied the Bilbao-Terol et al. [3] algorithm to construct the SR objective. The authors present a system based on the GRI Categories to obtain a normalised performance measure of each GRI Category that shows how far/close a company is to a sustainable behaviour. To do this, they consider the aggregation of the company performance on all aspects inside the Categories.

The paper is organised as follows. In Sect. 2, a sustainable portfolio selection model based on Goal Programming (GP) and fuzzy technology is shown. Section 3 presents an empirical application to the Spanish market. The paper ends with the main conclusions.

2 A Model for Selecting Socially Responsible Portfolios

We propose a fuzzy multi-criteria model that runs on two levels of decision-making: in the first stage, the reference portfolio considering only financial objectives is obtained. In the second stage, we will use the reference portfolio as an ideal point and a socially responsible portfolio is designed.

2.1 Obtaining the Reference Efficient Portfolio with Fuzzy Prospect Theory

For the construction of the financial preferred portfolio, i.e., the reference portfolio, the Prospect Theory has been used (net profits as the financial objective and error function as the utility function). The net profits random variable is represented on the real number line by means of the expected value at the end (EVE) and the Conditional Value at Risk (CVaR) associated with net profits at a certain confidence level. The minimization of the CVaR is carried out following the model proposed by Rockafellar and Uryasev [14]. With regard to the constraints, the budget constraint and short sales are not allowed.

In order to select the optimum portfolio under the EVE and the CVaR criteria we generate an approximation of the financial efficient frontier CVaR-EVE by applying the ε-constraint method, proposed by Haimes et al. [7].

On the efficient frontier (Fig. 1), we calculate the financial preferred portfolio with maximum certainty-equivalent. In order to do this an error function is used with an S-shaped according to that prescribed in Prospect Theory (Kahneman and Tversky [10]). We have using the following utility functions:

$$u(f(x)) = \begin{cases} f(x)^\alpha & \text{if } f(x) \geq 0 \\ -\lambda(-f(x))^\alpha & \text{if } f(x) < 0 \end{cases}$$

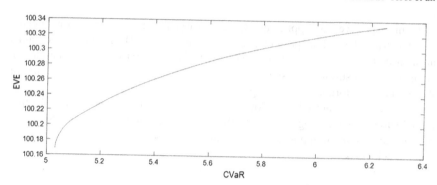

Fig. 1 The financial efficient frontier

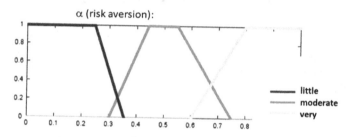

Fig. 2 Linguistic labels to model the parameter α

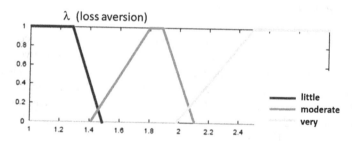

Fig. 3 Linguistic labels to model the parameter λ

with $0 < \alpha \leq 1$ and $\lambda \geq 1$ are the parameters of the utility functions that fit the investor profile (risk aversion, α, and loss aversion, λ).

Determining these values is a difficult task. Linguistic labels of the type 'little/moderate/very' averse to risk (see Fig. 2) and to losses (see Fig. 3), represented by fuzzy numbers, has been used in order to model the investor's profile.

From this information, we assign discrete values (within its support) to α and λ. For each (α, λ) pair, we obtain the corresponding certainty-equivalent maximum portfolio on the efficient frontier (see Fig. 4). The certainty-equivalent CE of the portfolio is calculated by the following expression:

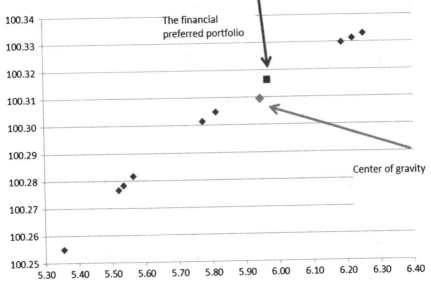

Fig. 4 The center of gravity of the (CVaR-EVE) pair and the preferred portfolio

$$
CE = \begin{cases} \overline{U}^{1/\alpha} & \text{if } \overline{U} \geq 0 \\[2ex] -\left(-\dfrac{1}{\lambda}\,\overline{U}\right)^{1/\alpha} & \text{if } \overline{U} < 0 \end{cases}
$$

where $\overline{U}$ is the expected utility (the average of utilities).

With these portfolios the center of gravity of the CVaR-EVE pair is obtained (see Fig. 4). The CVaR (EVE) of the center of gravity is obtained using as weights the membership degree of each portfolio (the t-norm product of the membership degrees of the parameters α and λ). The financial preferred portfolio will be the one closest to the center of gravity (see Fig. 4). We denote by CVaR* and EVE*, the CVaR and the EVE of this portfolio, respectively.

2.2 Extended GP Model for SR-Portfolio Selection

In order to use a GP approach for a multi-objective problem it is necessary that the DM establishes aspiration levels for each goal (Ignizio [9], Tamiz et al. [17]). For the financial objectives, these levels are the corresponding values of the reference portfolio: (CVaR*, EVE*). For the third goal, the socially responsible (SR) target, the investor determines its aspiration level.

By considering an investment universe of N companies, a portfolio is represented by the N-dimensional vector $x = (x_1, \ldots, x_N)$ where x_i denotes the number of shares invested in the company i. Once defined the goals and the constraints we propose the following Extended GP model (Romero [15], Bilbao-Terol et al. [3]) in order to obtain the SR portfolio tracking the reference one

$$\min \gamma \left[\omega_1 p_1 + \omega_2 n_2 + \omega_3 n_3 \right] + (1 - \gamma) D$$

s.t.

$$\psi + (1 - a)^{-1} \sum_{j=1}^{J} \pi_j z_j + n_1 - p_1 = \text{CVaR}^*, \quad z_j \geq \sum_{i=1}^{N} (-P_{ij} x_i) + C_0 - \psi, \quad z_j \geq 0$$

$$\sum_{i=1}^{N} E[P_i] x_i + n_2 - p_2 = \text{EVE}^*, \quad \sum_{i=1}^{N} \text{SR}_i x_i + n_3 - p_3 = \text{SR}^*,$$

where C_0 is the initial available capital, $E[P_i]$ is the expected price of the share of the generic ith company, P_{ij} denotes the share price of the ith company in the scenario j, π_j is the probability of the scenario j, z_j are dummy variables to solve the optimization problem and $\gamma \in [0, 1]$ is the tradeoff between the weighted goal programming model and minmax goal programming one.

3 Application to the Spanish Market

In order to apply the previously exposed methodology we have constructed a database containing 4,667 observations corresponding to daily closing prices for eight large Spanish companies included in the IBEX-35 (see Table 1).

We have set an estimation interval equal to one week and the investment date is 27/02/2014, therefore, 662 weekly observations are available (i.e. T is equal to 662). The investment horizon has been set at one month (4 weeks). We have worked with non-overlapping weekly-compounded returns.

To get the conventional portfolio, which will serve as a reference portfolio, the investor profile is setting by a pair of linguistic labels: medium for risk (support (α) = [0.3, 0.75]) and lower for losses (support (λ) = [1, 1.5]). From this information, we assign discrete values (within its support) to α and λ, obtaining the corresponding certainty-equivalent maximum portfolios on the efficient frontier (Table 2). The financial preferred portfolio will be the portfolio 90 on the efficient frontier (see Fig. 4).

Table 1 SR-index of the companies (Source: Bilbao-Terol et al. [3])

Company	Sector	Subsector	Market capitalization (in Meuros)(5/2015)	SR-Index
Acciona SA	Basic materials, industry and construction	Construction	4,002.44	0.6290
BBVA SA	Financials and real estate	Banks and saving banks	57,201.12	0.3526
Iberdrola SA	Oil and energy	Electricity and gas	38,624.77	0.6574
Inditex SA	Consumer goods	Textile, clothing and footwear	89,930.99	0.1713
NH Hotels SA	Consumer services	Leisure, Tourism and hospitality	1,831.921	0.3021
B. Santander SA	Financials and real estate	Banks and saving banks	94,993.32	0.3944
Sacyr SA	Basic materials, Industry and Construction	Construction	2,024.42	0.2860
Telefónica SA	Technology and Telecommunications	Telecommunications and others	67,952.62	0.2027

Table 2 The certainty-equivalent maximum portfolios on the efficient frontier

Portfolio	53	66	67	69	81	83	**90**	98	99	100
CVaR	5.35	5.52	5.53	5.56	5.77	5.81	**5.97**	6.19	6.22	6.26
EVE		100.25	100.28	100.28	100.28	100.30	100.30	**100.32**	100.33	100.33	100.33
C. Gravity	CVaR = 5.95					EVE = 100.31				

3.1 The SR-Criterion for the Portfolio

There are different ways in the literature to obtain a measure of the Corporate Social Responsible (CSR) performance of the companies (see, Bilbao-Terol et al. [1, 2], Cabello et al. [5], Liern et al. [12], Lamata et al. [11]).

In general, an SR criterion for a portfolio x is defined as a weighted average of the SR-indexes of the individual companies (see Table 1) contained in this portfolio, that is,

$$SR(x) = \sum_{i=1}^{N} SR_i P_{iT} x_i,$$

where SR_i represents a measure of the ith company's compliance.

In this paper, we have applied the Bilbao-Terol et al. [3] algorithm to construct the SR objective. The authors obtain a normalized performance measure of the GRI

Labor Practices and Decent Work Category that shows how far/close a company is to a sustainable behaviour. To do this, they consider the aggregation of the company performance on all aspects inside the Labor Practices and Decent Work Category.

3.2 Results

For the financial objectives, the aspiration levels are the corresponding values of the portfolio 90: $(CVaR_{90}, EVE_{90}) = (5.95, 100.32)$. Therefore, a very financial risky profile is used. For the third goal, the socially responsible target, the midpoint of the range of the sustainability criterion is chosen. Once defined the goals and the constraints included in the set X we solve our model in order to obtain the SR portfolio tracking the reference one.

Table 3 Optimal portfolio $(\gamma = 0.5)$

COMPANY	SR portfolio	Reference portfolio
Acciona	43.037	7.34
Iberdrola	9.48	2.22
Inditex	47.49	90.44
CRITERION		
CVaR	5.9683	5.97
EVE	100.2358	100.32
SR	41.4338	21.57

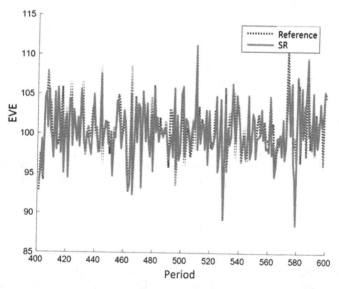

Fig. 5 Historical series of the portfolios

Table 3 and Fig. 5 show the SR and reference portfolios when parameter $\gamma = 0.5$. As we can see, the two portfolios invest in the same companies but with different amounts. The compositions displayed in Table 3 show that the introduction of the SR objective increases the investment in Acciona and Iberdrola that present the best results for the SR-index. Financial satisfaction in the SR portfolio is lower than a reference portfolio as a result of the lower SR score of the company with better financial performance (Inditex). On the other hand, the Inditex investment in the SR portfolio has been considerably reduced as a cause of that fact. The Acciona investment in the SR portfolio has increased due to its good socially responsible behaviour.

4 Conclusions

A new methodology is presented for those SR investors that prefer to make their own decisions about the securities in which to invest. Our proposal allows to obtain a portfolio that satisfies the financial and sustainability concerns of these investors. The modeling by means of the theory of the fuzzy subsets allows a flexible treatment of the parameters that define the financial utility (or value) of the investor with a behavior that follows the Kahneman–Tversky Theory. With our methodology the investor knows the possible financial sacrifice that involves his desire to invest in companies concerned to social responsibility. Our design of investment portfolios uses the Corporate Social Responsibility performance of the companies that have been obtained analysing its sustainability reporting.

Acknowledgements The authors wish to gratefully acknowledge the financial support from the Spanish Ministry of Economy, Industry and Competitiveness, Project ECO2015-66521-P.

References

1. Bilbao-Terol A, Arenas-Parra M, Cañal-Fernández V (2012) Selection of socially responsible portfolios using goal programming and fuzzy technology. Inf Sci 189:110–125
2. Bilbao-Terol A, Arenas-Parra M, Cañal-Fernández V, Bilbao-Terol C (2013) Selection of socially responsible portfolios using hedonic prices. J Bus Ethics 115(3):515–529
3. Bilbao-Terol A, Arenas-Parra M, Cañal-Fernández V, Obama-Eyang, PN (2017) Multi-criteria analysis of the GRI sustainability reports: An application to socially responsible investment. J Oper Res Soc (in press). https://doi.org/10.1057/s41274-017-0229-0
4. Brown HS, de Jong M, Levy DL (2009) Building institutions based on information disclosure: lessons from GRIs sustainability reporting. J Clean Prod 17(6):571–580
5. Cabello JM, Ruiz F, Pérez-Gladish B (2014) Synthetic indicators of mutual funds' environmental responsibility: an application of the reference point method. Eur J Oper Res 236(1):313–325
6. Global Reporting Initiative (GRI) (2011): Sustainability reporting guidelines. Available at https://www.globalreporting.org/resourcelibrary/G3.1-Guidelines-Incl-Technical-Protocol.pdf. Accessed 25 March 2017

7. Haimes YY, Lasdon LS, Wismer DA (1971) On a bicriterion formulation of the problems of integrated systems identication and system optimization. IEEE Trans Syst Man Cybern 1(3):296–297
8. Hallerbach W, Ning H, Soppe A, Spronk J (2004) A framework for managing a portfolio of socially responsible investments. Eur J Oper Res 153:517–529
9. Ignizio JP (1983) Generalized goal programming: an overview. Comput Oper Res 10(4):277–289
10. Kahneman D, Tversky A (1979) Prospect theory: An analysis of decision under risk. Econometrica XVL(II):263–291
11. Lamata MT, Liern V, Pérez-Gladish B (2016) Doing good by doing well: a MCDM framework for evaluating corporate social responsibility attractiveness. Ann Oper Res (in press). https://doi.org/10.1007/s10479-016-2271-8
12. Liern V, Méndez-Rodríguez P, Pérez-Gladish B (2015) A soft computing approach for ranking firms based on different corporate social responsibility valuations. Appl Math Inf Sci 9(3):1113–1122
13. Obama-Eyang PN (2014) Modelos multicriterio para inversiones socialmente responsables. Trabajo Fin de Máster en Modelización e Investigación Matemática Estadística y Computación, MSc Thesis. Universidad de Oviedo
14. Rockafellar RT, Uryasev S (2000) Optimization of conditional value-at-risk. J. Risk 2:21–41
15. Romero C (2001) Extended lexicographic goal programming: a unifying approach. Omega-Int J Manag Sci 29:63–71
16. Searcy C, Elkhawas D (2012) Corporate sustainability ratings: an investigation into how corporations use the Dow Jones sustainability index. J Clean Prod 35:79–92
17. Tamiz M, Jones DF, Romero C (1998) Goal Programming for decision-making: an overview of the current state-of-the-art. Eur J Oper Res 111:569–581
18. UK Social Investment Forum. Available at http://www.uksif.org/. Accessed 20 March 2017

Graphical Exploratory Analysis of Fuzzy Data as a Teaching Tool

Inés Couso, Luis Junco, José Otero and Luciano Sánchez

Abstract Graphical exploratory analysis for fuzzy data allows us to represent sets of individuals whose attributes are perceived with imprecision on a map so that the degree of dissimilarity between two objects is somehow compatible with the distances between their respective representations. This study will discuss the use of this tool to jointly analyze the evolution of a group of students during a course, and to select the most suitable personnel of a company to receive a training course, according to a catalog of competencies and considering the reliability of information sources.

1 Introduction

Graphical exploratory analysis consists of the projection of a set of individuals on a plane, so that the similarities between pairs of individuals are compatible with the distances between their corresponding representations [9]. There are different techniques to perform this projection, depending on the model used to link the similarities between objects to the distances between their representations on the map. The most frequent method is applied to objects described by a vector of numerical properties, and uses the Euclidean distance both to calculate the distances between objects and between their projections.

Different generalizations of graphical exploratory analysis to the case of imprecise data have been considered in the literature. More concretely, the case where every instance is characterized by means of a vector of fuzzy numbers has been considered

I. Couso (✉) · L. Junco · J. Otero · L. Sánchez
University of Oviedo, Oviedo, Spain
e-mail: couso@uniovi.es

L. Junco
e-mail: lajunco@uniovi.es

J. Otero
e-mail: jotero@uniovi.es

L. Sánchez
e-mail: luciano@uniovi.es

© Springer International Publishing AG 2018
E. Gil et al. (eds.), *The Mathematics of the Uncertain*, Studies in Systems,
Decision and Control 142, https://doi.org/10.1007/978-3-319-73848-2_52

by different authors [5–7, 13]. Just as in the exploratory analysis of crisp data each individual is associated with a point on the map, the projection of a fuzzy vector is a geometric figure that depends on the transformations between the spatial distances and the distances on the map.

Exploratory data analysis techniques are part of the general knowledge and routinely used in a multitude of knowledge discovery problems. However, the use of exploratory analysis of fuzzy data is not very widespread. In relation to information mining in a teaching context, the algorithm defined in [11] has been applied to the analysis of tests solved by groups of children with learning difficulties (early diagnosis of dyslexia [12]) and to the analysis of questionnaires of follow-up by the students of different undergraduate and master's degrees. In both cases, in addition to obtaining visual information about how many different types of students are in each group, it is possible to measure the variations between the learning success of the different subgroups, and the evolution of their relative returns over the course of the course. This study reviews these applications and introduces a new one, selecting the right people to attend in-company training courses, taking into account different sources of information about each employee's competencies (statement internal examinations, questionnaires, previous work, etc.) which are, in turn, affected by the credibility of each source.

The rest of the paper is organized as follows: Sect. 2 describes the usefulness of graphical data analysis techniques in a teaching context. Section 3 discusses the need for a representation based on fuzzy sets, and describes some technical aspects of the algorithm. Section 4 discusses several case studies. The paper ends with some concluding remarks and future work.

2 Usefulness of Graphical Exploratory Analysis in Teaching Problems

When organizing training courses for employees of a company, it is important to study what training skills need to be filled, so that you can choose the most appropriate content for the courses. A similar study also serves to select the best attendees for a course with a limited number of places, or to compare the results of the study before and after the end of the course, in order to check whether this effort is paying off and translates into an improvement of the global capacity of the company.

As mentioned in the Introduction, the different graphical exploratory analysis algorithms (Sammon maps, PCA, Multidimensional Scaling (MDS), Self Organized Maps (SOM), etc. [5]) are statistical techniques that project objects as points on a plane, so that the proximity of the projections of two objects (instances) on the map reflects the similarity between their respective properties (seen as functions of the corresponding vectors of attributes). If each of these objects is associated with an employee, and the properties of the employee are assumed to consist of numerical measures of their proficiency level in a competency catalog, the graphic exploratory analysis is one of the most adequate techniques to evidence groups of employees with different skills. By adding fictional individuals (hypothetical employees with perfect knowledge of one technology and no other technologies), the positions of actual employees over those of fictitious employees make it possible to detect training gaps. Finally, comparing several maps of the same individuals on different dates, it is possible to evaluate the impact of the courses received.

Notwithstanding, these techniques are not directly applicable to the case where there is some uncertainty about the values of some of the attributes of an individual (e.g. missing data) and also do not consider the reliability of the different sources of information used to characterize individuals. For example, one of the most accessible sources for checking the level of formation of a group is the follow-up questionnaire [10]. Unlike the exam or interview, it is the student or employee who declares their knowledge, so this information may be inaccurate: an individual can either declare an "advanced" knowledge of English in the curriculum or having passed an examination of that level; in the first case, the uncertainty about his/her language proficiency is greater.

A simple way to quantify the uncertainty associated with a questionnaire is to associate different questions with the same item (a value of a single attribute for a single individual); the dispersion of the corresponding responses is an indication of the reliability of the test. For example, if a student is approached about his or her knowledge of probability theory, he/she may declare a "high" knowledge of the concept of the "density function" and "null" knowledge about the notion of "Radon-Nikodym derivative", while another one can answer "medium" to both questions: the dispersion of the answers associated with the same item is an indication of their reliability. Clearly, if just a central location measure is selected in order to summarize all the responses associated with the same property, valuable information is lost.

Another frequent problem with the data collection phase is the problem with missing data [8]. The most frequent solutions are either to remove the individual from the sample or to follow some imputation technique. The latter is the preferred solution when the sample size is not sufficiently high, and generally consists in finding the closest individuals to calculate their average values. Again, the variability of these values is being ignored, which may mean that the distribution of the completed data is probably far from reality, so the analysis would be distorted. Other imputation methods do not affect the variance but only work well under some assumptions about the coarsening process [2, 3].

3   Use of Fuzzy Sets in Competency Analysis

The use of fuzzy sets allows a homogenous representation of all previous types of uncertainty in the data. By means of the possibilistic interpretation of the membership function of a fuzzy set [1, 4], each value of the attribute for a specific object can be associated with a fuzzy number $\tilde{X}$ whose α-cuts are interpreted as nested confidence intervals in the sense that:

$$P([\tilde{X}]_\alpha \ni x) \geq 1 - \alpha, \ \forall \alpha \in (0, 1).$$

Thus, for example, a missing value can be replaced by a fuzzy set that models the distribution of the corresponding attribute in other similar objects (even though these in turn are perceived imprecisely), and the statements of an "advanced" knowledge of English or "average" knowledge of Probability Theory will be associated with two fuzzy numbers, whose specificities will be linked to the reliabilities of information sources. Therefore, in this study it will be considered that the knowledge about an individual can be quantified by means of a vector of fuzzy numbers. This generalization has two fundamental consequences in order to perform a graphical analysis of the population:

(a) The spatial coordinates of each individual are unknown, except for a nested family of (multivariate) confidence intervals.
(b) The Euclidean distance between two individuals whose coordinates are uncertain, is uncertain in turn.

From the first statement it can be concluded that the projections of each individual on the map will not be points, but families of nested sets, whose form will depend on the distortion of the spatial geometry, proper to each technique, in the flat projection. From the second, it follows that it is not reasonable to use a distance between fuzzy sets and calculate a numerical array of distances between individuals, nor between their projections. In general, a distance like that will not induce a total order among projections that is consistent with the ordering between the actual values of the attributes, since such an order between the actual values is just partially known.

In previous works, different simplifications have been made to achieve an approximate projection. For example, in the method described in [5, 6] multidimensional scaling (MDS) is extended to allow distance matrices to contain ranks or fuzzy numbers. The standard version of MDS consists of finding the scatter plot that minimizes a stress function, defined by the quadratic difference between the matrix of distances between the data and the matrix of distances of the points included in the scatter plot. In the generalized version, a fuzzy-valued stress function is defined, which measures the fit between the set of distances compatible with the map figures and the set of distances between the fuzzy descriptions of the individuals. In this method it is assumed that the projections are circles on the map, which is not always correct, since the attributes are not allowed to have different levels of uncertainty. Subsequent extensions [13] removed this restriction, as explained below.

3.1 How to Determine the Shape of Projections

Let $\tilde{X}_i = (\tilde{X}_{i1} \times \ldots \times \tilde{X}_{if})$ and $\tilde{X}_j = (\tilde{X}_{j1} \times \ldots \times \tilde{X}_{jf})$ be two tuples of fuzzy sets representing our incomplete knowledge about the f attributes of individuals

Fig. 1 The α-cuts of the projected data are polygons defined by the distances R_{ij} in the directions that pairwise join the examples

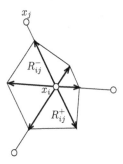

number i and j. Let $[\tilde{X}_i]_\alpha = [x_{i1}^-, x_{i1}^+] \times \ldots \times [x_{if}^-, x_{if}^+]$ and $[\tilde{X}_j]_\alpha = [x_{j1}^-, x_{j1}^+] \times \ldots \times [x_{jf}^-, x_{jf}^+]$ be in turn two cuts at the same level α of $\tilde{X}_i$ and $\tilde{X}_j$.

The set of possible values for the distances at a $1 - \alpha$-confidence level is:

$$D_{ij}^\alpha = \left\{ \sqrt{\textstyle\sum_{k=1}^f (x_{ik} - x_{jk})^2} \,\Big|\, x_{ik} \in [x_{ik}^-, x_{ik}^+], x_{jk} \in [x_{jk}^-, x_{jk}^+], 1 \le k \le f \right\}. \quad (1)$$

Some authors have used a distance similar to this one before [6], and further assumed that the shape of the projection of an imprecise case was a circle. We have found that, in our problem, this last is a too restrictive hypothesis. Instead, and according to [13], we propose to approximate the shape of the projections by a polygon (see Fig. 1) whose radii R_{ij}^+ and R_{ij}^- are not free variables, but depend on the distances between the cases.

Let us now consider a multivariate tuple of imprecise data $(\tilde{X}_1, \ldots, \tilde{X}_N)$, where $\overline{x}_i$ is the mode of $\tilde{X}_i$ and let $\{(z_{11}, \ldots, z_{1r}), \ldots, (z_{N1}, \ldots, z_{Nr})\}$ be the projection on a map of dimension r of that N-dimensional vector.

We propose that the radii R_{ij}^+ and R_{ij}^- depend on the distance between $[\tilde{X}_i]_\alpha$ and $\overline{x}_j$ (see Fig. 2 for a graphical explanation) as follows:

$$R_{ij}^+ = d_{ij} \left(\frac{\delta_{ij}^+}{\delta_{ij}} - 1 \right) \quad R_{ij}^- = d_{ij} \left(\frac{\delta_{ij}}{\delta_{ij}^-} - 1 \right) \quad (2)$$

where $d_{ij} = \sqrt{\sum_{k=1}^r (z_{ik} - z_{jk})^2}$, $\delta_{ij} = \{d(\overline{x}_i, \overline{x}_j)\}$, $\delta_{ij}^+ = \max\{d(x, \overline{x}_j) \mid x \in [\tilde{X}_i]_\alpha\}$, and $\delta_{ij}^- = \min\{d(x, \overline{x}_j) \mid x \in [\tilde{X}_i]_\alpha\}$.

3.2 Stress Function

According to the above, the available knowledge about the value of the effort function associated with the projection of the data is given by the following fuzzy-valued function, defined by its cuts:

Fig. 2 The distance between the respective projections of $[\tilde{X}_i]_\alpha$ and $[\tilde{X}_j]_\alpha$ is between the values $d_{ij} - R_{ij}^- - R_{ji}^-$ and $d_{ij} + R_{ij}^+ + R_{ji}^+$

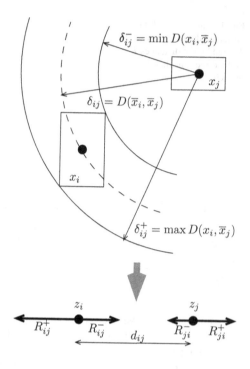

$$S_\alpha = \left\{ \sum_{i=1}^{N} \sum_{j=i+1}^{N} ||d(t, u) - \beta|| \mid \right.$$
$$t \in [\tilde{X}_i]_\alpha, u \in [\tilde{X}_j]_\alpha, \tag{3}$$
$$\left. \beta \in [d_{ij} - R_{ij}^- - R_{ji}^-, d_{ij} + R_{ij}^+ + R_{ji}^+] \right\}.$$

As an alternative to minimizing the previous fuzzy-valued function, it is possible to define a measure that quantifies how different the collection of spatial distances and the collection of the distances between their projections are, in terms of their corresponding rankings.

Let $D(X_i, X_j)$ be the fuzzy set whose α-cuts are the sets of distances between individuals, and let $D'(Z_i, Z_j)$ be the set of distances between their respective projections, Z_i and Z_j.

If the map correctly reflects the distances between individuals of the population, it must be observed that the rank of the distance between the i-th and j-th objects are the same as the rank of the distance between their projections, in the corresponding matrix. Therefore, the number of pairs of objects for which it is not true that it defines an alternative cost function.

In our case, the rank of a fuzzy-valued distance within its own matrix is not completely defined, since there are non-comparable pairs of distances. However, for each level α a relation between the α-distances (which are intervals) can be defined; given two interval-valued distances $[d^-, d^+]$ and $[e^-, e^+]$, they are not comparable when

$$[d^-, d^+] \parallel [e^-, e^+] \iff (d^+ > e^-) \wedge (e^+ \geq d^-). \tag{4}$$

Otherwise, we can say that one of them precedes the other (i.e., either $[d^-, d^+] \prec [e^-, e^+]$ or $([e^-, e^+] \prec [d^-, d^+])$. The rank of a distance will be defined, for the α level, according to the following iterative procedure: We take all the distances and we select those that are not preceded by any other in the collection. All of them are assigned rank equal to 1. We remove those distances from the initial collection. We take the remaining ones and iterate the process, by assigning a rank equal to 2 to those that are not dominated by any other one. We continue with the process until we get the empty set.

The purpose of the numerical algorithm (which will not be made explicit, due to extension limitation) is to obtain a map for which the ranks of each of the terms of the matrices of distances between individuals and between projections coincide for every α level. The value of the stress function is the infimum of those α levels for which both collections of ranks do coincide.

4   Numerical and Graphical Results

In this section we will illustrate, with the help of three real-world datasets, how to identify groups of students and how to stack two maps from the same individuals at different times, for showing the temporal evolution of the learning.

4.1   Variation of Individual Capacities in the Same Group and Between Groups

In the left part of Fig. 3 a diagram for 30 students of subject "Statistics" in *Ingeniería Telemática* at Oviedo University, taken at the beginning of the 2009–2010 course is shown. This survey is related to students' previous knowledge in other subjects.

In particular, this survey evaluates previous knowledge in Algebra (A), Logic (B), Electronics (C), Numerical Analysis (D), Probability (E) and Physics (F). The positions of the characteristic points have been marked with labels. Those points are of the type "A" (all the questions about the subject "A" are correct, the others are erroneous) "NO A" (all the questions except "A" ones are correct, the opposite situation), etc.

In the right part of Fig. 3 we have plotted together the results of three different groups, attending lectures by the same teacher. Each intensification has been coded with a distinctive class. This teacher has evaluated, as before, the initial knowledge

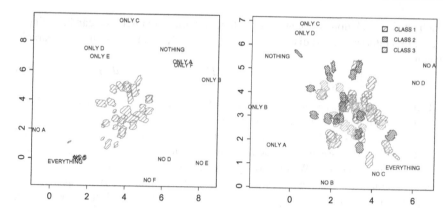

Fig. 3 Left part: Differences in knowledge of statistics for students in Ingeniería Telemática. Right part: Differences in knowledge about computer science between the students of Ingeniería Técnica Industrial specialized in chemistry, electricity and mechanics

of the students in subjects that are a prerequisite. From the graphic in that figure the most relevant fact is that the students of the intensification coded as CLASS 1 (*Ingeniería Industrial*) consider themselves better prepared than those coded as CLASS 2 (*Ingeniería Técnica Industrial Eléctrica*), with the CLASS 3 group in an intermediate position, closer to CLASS 1 (*Ingeniería Técnica Industrial Química*).

All the students of all the groups have a neutral orientation to math subjects, and some students in the CLASS 2 group think that their background is adequate only in subjects C (Operating Systems) and D (Internet).

4.2 Evaluation of Learning Results

Ten pre-doctoral students in Computer Science, Physics and Mathematics attending a research master were analyzed. The background of these students is heterogeneous. In the survey the students were asked about 36 subjects classified in "Control Algorithms" (A), "Statistical Data Analysis" (B), "Numerical Algorithms" (C) and "Linear Models" (D). On the left of Fig. 4 we can see that there is a large dispersion between the initial knowledges. Since the subject had strong theoretic foundations, students from technical degrees like Computer Science evaluated themselves with the lowest scores (shapes in the right part of each figure).

The same survey, at the end of the course, shows that all the students moved to the left, closer to characteristic point "EVERYTHING". Additionally, the displacement has been larger for the students in the group at the right. This displacement can be seen clearly in the right part of the same figure, where the shapes obtained from the final survey were replaced by arrows that begin in the initial position and end in the

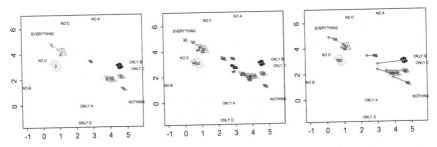

Fig. 4 Evolution of the learning of pre-doctoral students. Left part: Initial survey. Center: superposition of initial and final maps. Right part: The displacement has been shown by arrows

final center. The length of the arrows is related with the progress of the student during the course.

5 Conclusions

We have proposed the use of graphical exploratory maps to analyze the characteristics of groups of students, when those attributes are observed with some uncertainty either due to inconsistencies in the collection of data or missing data. The map of a group consists of several figures and a list of characteristic points. The proximity of an individual to one of these points means that the balance of such an individual with respect to different areas of knowledge resembles the value represented by this indicator. This technique can be used to corroborate the improvement of the abilities after receiving a training course: combining in the same graph the results of two tests, separated in time, it is possible to determine the displacement of each individual towards other characteristic points, and thus to detect the individuals who have best taken advantage of the course.

Acknowledgements We thank the editors of this volume for their kind invitation to participate. The research in this work has been supported by projects TIN2014-56967-R, TIN2017-84804-R and FC-15-GRUPIN14-073.

References

1. Couso I, Dubois D (2014) Statistical reasoning with set-valued information: ontic vs epistemic views. Int J Approx Reason 55:1502–1518
2. Couso I, Dubois D (2017) A general framework for maximizing likelihood under incomplete data. Int J Approx Reas 93:238–260
3. Couso I, Dubois D, Hüllermeier E (2017) Maximum likelihood estimation and coarse data. In: Moral S, Pivert O, Sánchez D, Marín N (eds) Scalable uncertainty management (Lecture notes in artificial intelligence), vol 10564. Springer, Cham

4. Couso I, Sánchez L (2008) Higher order models for fuzzy random variables. Fuzzy Sets Syst 159:237–258

5. Denœux T, Masson MH (2000) Multidimensional scaling of interval-valued dissimilarity data. Pattern Recognit Lett 21:83–92

6. Hebert PA, Masson MH, Denœux T (2006) Fuzzy multidimensional scaling. Comput Stat Data Anal 51:335–359

7. Honda K, Ichihashi H (2006) Fuzzy local independent component analysis with external criteria and its application to knowledge discovery in databases. Int J Approx Reason 42:159–173

8. Kim W, Choi B, Hong E-K, Kim S-K (2003) A taxonomy of dirty data. Data Min Knowl Discov 7:81–99

9. Kruskal JB (1964) Nonmetric multidimensional scaling: a numerical method. Psychometrika 29:115–129

10. Nuhfer E, Knipp D (2006) The use of a knowledge survey as an indicator of student learning in an introductory biology course. CBE Life Sci Educ 5:313–316

11. Mazza R, Milani C (2005) Exploring usage analysis in learning systems: Gaining insights from visualisations. In: proceedings of workshop usage analysis in learning systems. AEID, Amsterdam

12. Palacios A, Sánchez L, Couso I (2010) Diagnosis of dyslexia with low quality data with genetic fuzzy systems. Int J Approx Reason 51:993–1009

13. Sánchez L, Couso I, Otero J, Palacios A (2010) Assessing the evolution of learning capabilities and disorders with a graphical exploratory analysis of surveys containing missing and conflicting answers. Neural Netw World 20:825–838

Some Results and Applications Using Fuzzy Logic in Artificial Intelligence

Irene Díaz, Anca L. Ralescu, Dan A. Ralescu and
Luis J. Rodríguez-Muñiz

> *Algunas veces encuentras en la vida*
> *una amistad especial:*
> *ese alguien que al entrar en tu vida*
> *la cambia por completo.*
>
> Mario Benedetti, *"Algunas amistades"*

Abstract In this chapter, several results and applications in Artificial Intelligence based on Fuzzy Logic are presented. We aim to highlight some recent works that are connected with the topic of this book, both in the theoretical and in the applied fields.

1 Introduction

Pedro Gil was one of the frontrunners of Information Theory in Spain [13], many of his works were devoted to this topic. Apart from his pioneering work in information theory, in the 80s he established links with uncertainty and imprecision by exploring research topics related to fuzzy sets and fuzzy logic. One of the first connections he established focused on the need of aggregating information; t-norms and t-conorms

I. Díaz
Dpto. Informática, Universidad de Oviedo, Oviedo, Spain
e-mail: sirene@uniovi.es

A. L. Ralescu
School of Computing Sciences & Infomatics, University of Cincinnati, Ohio, USA
e-mail: anca.ralescu@uc.edu

D. A. Ralescu
Dept Mathematical Sciences, University of Cincinnati, Ohio, USA
e-mail: dan.ralescu@uc.edu

L. J. Rodríguez-Muñiz (✉)
Dpto. Estadística e I.O. y Didáctica de la Matemática,
Universidad de Oviedo, Oviedo, Spain
e-mail: luisj@uniovi.es

become a useful tool for some situations related to fuzzy measures and fuzzy integrals (e.g. [12, 15, 17, 38]). These are the topics we choose to cover in this memorial article dedicated to our beloved teacher, colleague, and friend. In particular, for two of the authors, Irene Díaz and Luis J. Rodríguez-Muñiz, this is a very special work, since they were both students in Pedro Gil's course on Information Theory at the Faculty of Sciences in the University of Oviedo. But the other two authors, Dan Ralescu and Anca Ralescu, had also the opportunity to meet Pedro Gil in their first contacts with the University of Oviedo, more than twenty years ago. We all cherished Pedro as a friend, a colleague, and a master.

The structure of this chapter is divided into six sections. Section 1 corresponds to the introduction. Sections 2–4 are mainly focused on theoretical results while Sect. 5 draws some applications. Finally, some concluding remarks are outlined.

2 Aggregation Operators

Aggregation operators are very useful tools in many fields when the need of combining different values or information inputs arises [2]. Mainstream research approaches are based on the study of theoretical properties of aggregation operators, as well as the analysis of different values when the operator is depending on a parameter.

In [40] a new approach to the study of aggregations functions, based not only on the theoretical properties, but on the statistical behavior of the outputs was initiated. A procedure to carry out statistical comparisons between outputs of different aggregation functions was adopted, which was used to compare several t-norms (minimum, product, and Łukasiewicz), as well as some means (geometric, arithmetic, and quadratic). Based on a large number of simulations, the conditions under which outputs of aggregations are distinguishable (or not) were studied. Hence, the focus for practitioners moves from the theoretical issues to the practical behavior of results.

The procedure introduced in [40] was further pursued in [42], where comparisons were made for parametric t-norms. The goal in this work was to provide simple rules that can be used for selecting the parameter value. A strong graphic support was introduced to help visualize the similarities and differences between t-norm outputs. Also, the effects of the t-norm arity, the sample size, the input distribution were analyzed. The main result of this work is the identification of different combinations of conditions for which, from a practical point of view, the choice of the parameter value is irrelevant.

In addition, in [39] properties related to Dujmovic's Iterative OWA operator were studied. Iterative OWA (ItOWA) operators as proposed by Dujmovic, is a two-stage procedure for computing the weighting vector by a double nested iteration: (i) weights at step h are computed as limit to infinity of a matrix power, (ii) the result is used to start the computation at step $h + 1$, until the OWA operator arity n is reached. Troiano and Díaz [39] propose an analytical solution to this procedure. This theoretical result enables a faster computation of the weighting vector and characterization in terms of weight values, attitudinal character and entropy.

Ralescu et al. in [31] study the concept of optimal aggregation as it plays an important role towards developing a theory of aggregation of fuzzy concepts, based on non-additive set-functions and nonlinear (Choquet or fuzzy) integrals. A framework supporting such aggregations, that are useful to decision making based on distributed sources of evidence is developed. In [45] arithmetic operations for LR mixed fuzzy random variables commonly used in practice for modeling fuzzy stochastic phenomena are studied. The operations are proposed based on mean chance measure, which as a natural extension of both the probability of a random event and the credibility of a fuzzy event, measures the mean or expected (in the sense of probability) credibility that the fuzzy random event occurs. Ralescu and Ralescu [29] studies the equivalence between aggregation of fuzzy sets and integration with respect to non additive set functions. The concept of co-monotonic maxitivity is a more desirable requirement than co-monotonic additivity.

The aggregation of rankings is a recurrent task in several applications, especially in the context of social choice theory. In this framework, in [22] we study some ranking aggregation strategies to represent group's opinion. In particular the strength with which $a \geq b \geq c$ is supported, should not be less than both the strength with which $a \geq b$ and $b \geq c$ are supported. A first approach to this ranking rule considering totally specified monotone reciprocal relations on a bipolar qualitative scale has already been taken. In this paper, a more general setting is considered: each voter is allowed to provide a partially specified reciprocal relation (that may not be monotone) on the unit interval. In this work we have also explored new ways of measuring the cost of imposing monotonicity. In [22] a more general setting is considered: each voter is allowed to provide a partially specified reciprocal relation (that may not be monotone) on the unit interval. In [17] different representations of votes (the votrix and the votex) are introduced. The former is a formalization of the well-known reciprocal matrix of pairwise comparisons between candidates already introduced by Condorcet. The latter is an extension of this reciprocal matrix considering hitherto unexploited information. These two representations lead to two monotonicity-based ranking rules.

In [21] we study an aspect of monotonicity that complements previous studies. In particular, it is studied if there is a true ranking on the set of candidates and every voter expresses a ranking on the set of candidates, then the number of times that each ranking is expressed should decrease when we move away from this true ranking in terms of pairwise discordances. In addition, we propose a probabilistic model that allows to formulate the choice of the best ranking as a maximum likelihood estimation problem.

3 Extension of Fuzzy Sets

Some extensions of the standard fuzzy sets have been extensively studied during the last 20 years. Interval, Atanassov's intuitionistic or hesitant fuzzy sets are representative examples of these extensions. In [18] an ordering framework required to work with interval-valued hesitant fuzzy sets is presented. In fact, ordering sets is a long-standing open problem due to its remarkable importance in many areas such as decision making, image processing or human reliability. Methods for ordering finitely generated sets as a generalization of those methods previously defined for ordering intervals were introduced in [18]. In addition, these orderings between finitely generated sets are also improved to present ordering between finite interval-valued hesitant fuzzy sets.

Finally, finite interval-valued hesitant fuzzy preference relations are introduced and used to define a new order between finite interval-valued hesitant fuzzy sets. Along the same lines, in [26] some concepts related to partitioning for interval-valued fuzzy sets are studied. Partitioning is a long-standing open problem due to its remarkable importance in many areas such as clustering. The definition of this partitioning method involves a definition of an ordering relation for finite interval-valued fuzzy sets membership degrees, i.e., finitely generated sets, as well as the definitions of t-norm and t-conorm for these kinds of sets.

Still focusing on interval-valued fuzzy sets, in [25] a definition of entropy for an interval-valued hesitant fuzzy environment is provided. As the properties of this kind of sets are more complex, the entropy is built by three different functions, where each one represents a different measure: fuzziness, lack of knowledge, and hesitance. Using all, an entropy measure for interval-valued hesitant fuzzy sets is obtained, quantifying various types of uncertainty.

Yang et al. [46] introduce the definition of type-2 uncertain variables within the framework of uncertainty theory through introduction of generalized uncertain measures and focuses on more complex twofold uncertainties. Some uncertainty reduction methods associated with type-2 uncertain variables are also proposed for convenience of applicability, including reduction of optimistic value, pessimistic value and expected value. Moreover, four classes of type-2 uncertain variables are reduced to type-1 uncertain variables with specific uncertainty distributions.

4 Classification

Going from aggregation operators to a broader knowledge domain, as Artificial Intelligence is, we have also used fuzzy logic to extend crisp classification algorithms. In particular, in [33] the fuzzy extension of a previous crisp measure is introduced. The implementation of this measure into an algorithm proved to improve results obtained with similar algorithms, as well as it introduced an efficient tool to deal with imprecision in human judgments.

Extracting rules from databases is one of the main tasks in Artificial Intelligence. Different algorithms for learning rules both from crisp and fuzzy databases have been proposed in the literature. This way, learning information from databases is opposed to general policies about data protection. Thus, statistical disclosure control is a paradigm consisting of finding the optimal balance between releasing statistical data and protecting confidentiality of sensitive information.

Fuzzy logic is a useful tool in finding this balance, since fuzzy association rules can provide interpretable rules while protecting sensitive records in the database. In [41] this problem was investigated with the aim of establishing the existence of a set of rules able to break protection, assessing the interpretability of the model and identifying the ability of attributes for revealing sensitive information. The results showed that fuzzy rules are, in general, simpler and easier to interpret than others.

Based on the results obtained in [41] a new approach for dealing with fuzzy sets in data protection was developed in [11]. In particular, the notion of fuzzy cardinality becomes a key idea in [41]: counting the number of elements in a given class is the basis to determine if data of individuals within that class could be disclosed. The type of fuzzy cardinality used in this work is the simplest σ-count, being an open problem how to compute the cardinality with more interpretable definitions of fuzzy counting as those in [27, 28].

Following this path, in [23, 24] fuzzy notions were introduced in well-known measures in the privacy framework as k-anonymity, l-diversity and t-closeness. These works propose the extension of these three measures when the data are protected using fuzzy sets instead of intervals or representative elements. This methodology was tested using different fuzzy partition methods, obtaining an improvement in protecting data encoded using fuzzy sets. In addition, [10] provides a brief overview of the emerging research in privacy issues in social networks.

With regard to classification, [32] proposes a classification algorithm which considers explicitly geometric and statistical characteristics of the data and combines them into a class representation. The obtained method shows that the proposed algorithm is less sensitive to the training data set than other classifiers, which is an important property for a classification algorithm.

5 Applications

This section describes some of our recent results concerning applications of fuzzy logic to different knowledge domains, further from mathematics, statistics or theoretical artificial intelligence. The main novelty of these applications consists in introducing fuzzy logic, in many cases for the first time, in areas usually considered the domain of crisp techniques, demonstrating how fuzzy logic based procedures perform better than their crisp counterparts.

One important application of fuzzy sets focused on Artificial Vision. It has been initially developed for binary and gray scale images. Nevertheless, the color is an important source of information. For this reason, during the last years these techniques have been developed for color images. However, nowadays, the representation and the treatment of color images are still open problems [1].

Mathematical Morphology is the natural area for a rigorous formulation of many problems in image analysis, as well as a powerful non-linear technique which includes operators for the filtering, texture analysis, shape analysis, edge detection or segmentation. In the eighties, Matheron [14] and Serra [37] proposed the latest mathematical formulation of morphology within the algebraic framework of the lattices. This means that the definition of morphological operators needs a totally ordered complete lattice structure. In that context, before defining the basic morphological operators (erosion and dilation) it is necessary to define an order on the space used for processing the images.

Fuzzy Mathematical Morphology [5, 44] is an extension of the Mathematical Morphology's binary operators to gray level images, by redefining the set operations as fuzzy set operations. In [7–9] we define the operators of the Fuzzy Mathematical Morphology for color images through the use of a fuzzy order. Other important works related to image processing are highlighted: in [16] information about spatial organization in an image is considered to improve object recognition and scene analysis tasks. Also, in [6] directional relative position relations are considered since they provide an important information about the spatial arrangement of objects in the scene. Such concepts are rather ambiguous, they defy precise definitions, but human beings have a rather intuitive and common way of understanding and interpreting them. Therefore in this context, fuzzy methods are appropriate to provide consistent definitions that integrate both quantitative and qualitative knowledge, thus providing a computational representation and interpretation of imprecise spatial relations, expressed in a linguistic way, and including quantitative knowledge. Bloch and Ralescu [6] review and compares different fuzzy approaches according to their properties and according to the types of questions they seek to answer.

Risk assessment in human reliability is another application area where fuzzy sets can provide solutions. Risk is usually mathematically formalized as a matrix that allows the classification of different kinds of errors according to their importance. This classification can help in decision making about the most important or urgent one. Usually the risk matrix takes into account only one criterion (most of the cases: economic impact). However, decision making in a company often considers more than one criterion. Therefore, it is interesting to consider at the same time more than one different risk matrix, each one associated with different criteria for consequences (for example, effects on people, environment, assets or reputation). In [19] we have developed a method to combine this information in order to classify the errors according to more than one criterion.

We have developed a general method to combine the information about different alternatives given by several experts or taking into account several criteria and the choice of the set of the best ones. This method can be applied in any environment where there exists interaction among the different alternatives or some experts are more reliable than others. This will be done by the definition of fuzzy preference relations and the use of different aggregation functions, in particular the ordered weighted averaging operator.

Research in Recommendation Systems has been exponentially growing since the development of the Internet, more precisely, with the development of e-commerce. A Recommender System (RS) usually provides a rating or a preference for each user. To provide this recommendation an RS requires information about the preferences of the user in relation to the website (movies, books, songs, hotels, etc.). This information can be acquired explicitly by asking the users to rate items or implicitly by monitoring users' behavior (booked hotels or heard songs). RS can also use other kinds of information as demographic features (e.g., age, gender) or social information.

The research related to RS has been focused on movies, music and books recommendations, being music recommendation the most studied topic although, recently, it has been applied in other e-commerce domains. In [43] we face the problem of filtering from an e-shop catalog a set of products which might be interesting for the customer on the basis of preferences expressed by a group of users within a market segment. This problem is studied from a theoretical point of view by means of Dempster-Shafer Theory of Evidence (D-S Theory). The purpose of this work is to show how the D-S Theory can be used in the context of RSs. In this work we propose to move from items to features in order to (i) reduce problem dimensionality and (ii) to infer user preferences even when they are not made explicit. Preferences induced by each feature are considered as an independent source of information, then combined by a rule. We also studied how to explore the subset inclusion lattice, once Belief and Plausibility are mapped over it. In addition, we outline some efficient algorithms to perform such an exploration.

Profile similarity is a key point in recommendation systems. In [34] it is studied how to assess the similarity between node profiles in a social network. Several approaches exist to study profile similarity, including semantic approaches and natural language processing. However, in this work we combine these aspects into a unified measure of profile similarity. Traditionally, semantic similarity is assessed using keywords, that is, formatted text information, with no natural language processing component. In this study we propose an alternative approach, whereby the similarity assessment based on keywords is applied to the output of natural language processing of profiles. A unified similarity measure results from this approach.

Ralescu and Ralescu in [30] study the representation of an optimization with inexact constraints, as a fuzzy integral with respect to a capacity (or to an outer measure) rather than to a fuzzy measure. It is proved a mean-value theorem for the fuzzy integral, which has the reduction of optimization with inexact constraints to classical optimization as a consequence. Sufficient conditions for this reduction to hold are also provided in this work.

Another recent field of application in which we have started to work is the use of artificial intelligence techniques for detecting and predicting dropout in higher education. This is a key problem for higher education institutions, and in the vast educational literature about it only classical statistical methods were present. In [35] we started a new line of research for considering other type of approaches, by applying our previous work on extracting rules from databases.

Finally, we also wanted to underline other recent research line because it connects two of Pedro's main beloved occupations: fuzzy logic and teaching. We have started an analysis of curricular guidelines in primary and secondary education (mainly focused on the Spanish framework) in order to determine how imprecision belongs to the so called "hidden curriculum". While uncertainty has been introduced in the analysis of the curricular standards in secondary school from the 80s, and in primary school from the beginning of 21st century, imprecision is used in many procedures: approximate and mental calculus, classification of events as sure, likely or unlikely, ranking events depending on their possibility, etc. We have started to study this type of situations from the educational point of view both for primary [4] and secondary school [36], and jointly [3] and, in a forthcoming work, we plan to develop experimental studies at schools.

6 Concluding Remarks

We have collected different results recently achieved in our research lines with the aim of presenting connections of works developed by Pedro Gil in his multiple research interests. We are sure that Pedro Gil would like these works, and particularly he would appreciate applications in education, as the last one quoted, since he was always very concerned with teaching and learning processes not only at university level but also at previous educational stages. Most of all, our main goal was to pay a tribute to the researcher, the professor and, especially, the person of Pedro Gil. Recalling the poet, our beloved Pedro, we hope that in these pages we have attained singing the praises of *"la madurez insigne de tu conocimiento"*.[1]

References

1. Aptoula E, Lefevre S (2008) On lexicographical ordering in multivariate mathematical morphology. Pattern Recognit Lett 29(2):109–118
2. Beliakov G, Pradera A, Calvo T (2007) Aggregation functions: a guide for practitioners. Springer, Heidelberg
3. Blanco-Fernández A, Díaz-Díaz P, García-Honrado I, Ramos-Guajardo AB, Rodríguez-Muñiz LJ (2016) A proposal for assessing imprecise concepts in spanish primary and secondary schools. Int J Uncertain Fuzziness Knowl-Based Syst 24(Suppl. 2):71–91
4. Blanco-Fernández A, García-Honrado I, Ramos-Guajardo AB, Rodríguez-Muñiz LJ (2014) Reflexiones sobre el tratamiento de lo incierto en Educación Primaria en España. In: Bobillo F, Bustince H, Fernández FJ, Herrera-Viedma E (eds) Actas del XVII congreso español sobre tecnologías y Lógica fuzzy. Universidad de Zaragoza
5. Bloch I, Maitre H (1993) Fuzzy mathematical morphology. Ann Math Artif Intell 10(1–2):55–84
6. Bloch I, Ralescu AL (2003) Directional relative position between objects in image processing: a comparison between fuzzy approaches. Pattern Recognit 36(7):1563–1582
7. Bouchet A, Quirós P, Alonso P, Ballarin V, Díaz I, Montes S (2015) Gray scale edge detection using interval-valued fuzzy relations. Int J Comput Intell Syst 8(2):16–27

---

[1]From *"Llanto por Ignacio Sánchez-Mejías"*, Federico García Lorca, 1935.

8. Bouchet A, Alonso P, Pastore JI, Díaz I, Montes S (2016) Fuzzy mathematical morphology for color images defined by fuzzy preference relations. Pattern Recognit 60:720–733
9. Bouchet A, Alonso P, Díaz I, Montes S (2017) On the performance of some edge detectors for gray scale images. Submitted
10. Díaz I, Ralescu AL (2012) Privacy issues in social networks a brief survey. Commun Comput Inf Sci 300:509–518
11. Díaz I, Rodríguez-Muñiz LJ, Troiano L (2012) Fuzzy sets in data protection: strategies and cardinalities. Logic J IGPL 20:657–666
12. Fernández MJ, Suárez F, Gil P (1993) T-eigen fuzzy sets. Inf Sci 75(1–2):63–80
13. Gil P (1981) Teoría matemática de la información. ICE, Madrid
14. Matheron G (1975) Random sets and integral geometry. Wiley, New York
15. Miranda P, Grabisch M, Gil P (2000) Divergence measures and aggregation operators. Int J Uncertain Fuzziness Knowl-Based Syst 8(6):677–690
16. Miyajima K, Ralescu AL (1994) Spatial organization in 2D segmented images: representation and recognition of primitive spatial relations. Fuzzy Sets Syst 65(2–3):225–236
17. Montes S, Couso I, Gil P, Bertoluzza C (2002) Divergence measure between fuzzy sets. Int J Approx Reason 30(2):91–105
18. Pérez-Fernández R, Alonso P, Bustince H, Díaz I, Jurío A, Montes S (2015) Ordering finitely generated sets and finite interval-valued hesitant fuzzy sets. Inf Sci 325:375–392
19. Pérez-Fernández R, Alonso P, Díaz I, Montes S (2015) Multi-factorial risk assessment: an approach based on fuzzy preference relations. Fuzzy Sets Syst 278:67–80
20. Pérez-Fernández R, Rademaker M, Alonso P, Díaz I, Montes S, De Baets B (2016) Representations of votes facilitating monotonicity-based ranking rules: from votrix to votex. Int J Approx Reason 73:87–107
21. Pérez-Fernández R, Rademaker M, Alonso P, Díaz I, Montes S, De Baets B (2017) Monotonicity as a tool for differentiating between truth and optimality in the aggregation of rankings. J Math Psychol 77:1–9
22. Pérez-Fernández R, Rademaker M, Alonso P, Díaz I, Montes S, De Baets B (2017) Monotonicity-based ranking on the basis of multiple partially specified reciprocal relations. Fuzzy Sets Syst 325:69–96
23. Quirós P, Alonso P, Díaz I, Montes S (2014) On the use of fuzzy partitions to protect data. Integr Comput-Aided Eng 21:355–366
24. Quirós P, Alonso P, Díaz I, Montes S (2015) Protecting data: a fuzzy approach. Int J Comput Math 92(9):1989–2000
25. Quirós P, Alonso P, Bustince H, Díaz I, Montes S (2015) An entropy measure definition for finite interval-valued hesitant fuzzy sets. Knowl-Based Syst 84:121–133
26. Quirós P, Alonso P, Díaz I, Montes S (2016) On delta-epsilon partitions for finite interval-valued hesitant fuzzy sets. Int J Uncertain Fuzziness Knowl-Based Syst 24(2):145–163
27. Ralescu AL (1986) A note on rule representation in expert systems. Inf Sci 38(2):193–203
28. Ralescu DA (1995) Cardinality, quantifiers, and the aggregation of fuzzy criteria. Fuzzy Sets Syst 69:355–365
29. Ralescu AL, Ralescu DA (1997) Extensions of fuzzy aggregations. Fuzzy Sets Syst 86(3):321–330
30. Ralescu DA, Ralescu AL (2015) Optimization in a fuzzy environment. Libertas Math (new series) 35(2):51–59
31. Ralescu AL, Ralescu DA, Yamakata Y (2007) Inference by aggregation of evidence with applications to fuzzy probabilities. Inf Sci 177(2):378–387
32. Ralescu AL, Díaz I, Rodríguez-Muñiz LJ (2015) A classification algorithm based on geometric and statistical information. J Comput Appl Math 275:335–344
33. Ranilla J, Rodríguez-Muñiz LJ (2007) A heuristic approach to learning rules from fuzzy databases. IEEE Intell Syst 22(2):62–68
34. Rawashdeh A, Rawashdeh M, Díaz I, Ralescu AL (2012) Semantic similarity between nodes in a social network. Commun Comput Inf Sci 443:76–85

35. Rodríguez-Muñiz LJ, Bernardo A, Esteban M, Díaz I (2017) University dropout: Discovering rules with machine learning methods. Submitted
36. Rodríguez-Muñiz LJ, Díaz-Díaz P (2013) Imprecision, uncertainty and probability in spanish secondary school: a working proposal. In: De Baets B, Fodor J, Montes S (eds) Proceedings EUROFUSE 2013 Works. Universidad de Oviedo
37. Serra J (1982) Image analysis and mathematical morphology, vol I. Academic Press, London
38. Suárez F, Gil P (1986) Fuzzy expected value with semiconormed integrals. Trab Estadíst 1(1):127–139
39. Troiano L, Díaz I (2016) An analytical solution to Dujmovic's Iterative OWA. Int J Uncertain Fuzziness Knowl-Based Syst 24(2):165–179
40. Troiano L, Rodríguez-Muñiz LJ (2011) A statistical study of differences and similarities among aggregation functions. Logic J IGPL 19(2):415–424
41. Troiano L, Rodríguez-Muñiz LJ, Ranilla J, Díaz I (2012) Interpretability of fuzzy association rules as means of discovering threats to privacy. Int J Comput Math 89(3):325–333
42. Troiano L, Rodríguez-Muñiz LJ, Marinaro P, Díaz I (2014) Statistical analysis of parametric t-norms. Inf Sci 257:138–162
43. Troiano L, Rodríguez-Muñiz LJ, Díaz I (2015) Discovering user preferences using Dempster-Shafer theory. Fuzzy Sets Syst 278:98–117
44. Vanegas MC, Bloch I, Inglada J (2016) Fuzzy constraint satisfaction problem for model-based image interpretation. Fuzzy Sets Syst 286:1–29
45. Wang K, Zhou J, Ralescu DA (2017) Arithmetic operations for LR mixed fuzzy random variables via mean chance measure with applications. J Intell Fuzzy Syst 32(1):451–466
46. Yang L, Liu P, Li S, Gao Y, Ralescu DA (2015) Reduction methods of type-2 uncertain variables and their applications to solid transportation problem. Inf Sci 291:204–237

On Extending Fuzzy Preorders to Sets and Their Corresponding Strict Orders

Francesc Esteva and Lluís Godo

Abstract In this paper we first consider the problem of extending a fuzzy (weak) preorder on a set W to a fuzzy relation (preorder) on subsets of W, and consider different possibilities using different forms of quantification. For each of them we propose possible definitions of corresponding indistinguishability and strict preorder relations associated to the initial preorder, both on W and on its power set $\mathscr{P}(W)$. We compare them and we study conditions under which the strict relation is transitive.

1 Introduction

In the classical setting, from a preorder $\leq$ on a universe W we can define:

- an equivalence relation $\equiv$, where $x \equiv y$ if $x \leq y$ and $y \leq x$,
- a strict order $<$, where $x < y$ if $x \leq y$ and $y \not\leq x$.

Observe that, so defined, these relations satisfy the condition $x \leq y$ iff $x \equiv y$ or $x < y$, so roughly speaking we can say that $\leq$ is the union of $\equiv$ and $<$.

An interesting topic is how we can obtain relations on the power set $\mathscr{P}(W)$ from a preorder on the universe W. With a logic-based approach (using quantifiers), there are six ways of doing such an extension, see for example [10, 11].

Definition 1.1 Given a set W together with a preorder $\leq$, one can define the following six relations on $\mathscr{P}(W)$:

- $A \leq_{\exists\exists} B$ iff there exist $u \in A$ and $v \in B$ such that $u \leq v$,
- $A \leq_{\exists\forall} B$ iff there exists $u \in A$, such that for all $v \in B$, $u \leq v$,
- $A \leq_{\forall\exists} B$ iff for all $u \in A$, there exists $v \in B$ such that $u \leq v$,
- $A \leq_{\forall\forall} B$ iff for all $u \in A$ and $v \in B$, then $u \leq v$,
- $A \leq_{\exists\forall 2} B$ iff there exists $v \in B$ such that, for all $u \in A$, $u \leq v$,

F. Esteva · L. Godo (✉)
IIIA - CSIC, 08193 Barcelona, Bellaterra, Spain
e-mail: godo@iiia.csic.es

F. Esteva
e-mail: esteva@iiia.csic.es

© Springer International Publishing AG 2018
E. Gil et al. (eds.), *The Mathematics of the Uncertain*, Studies in Systems, Decision and Control 142, https://doi.org/10.1007/978-3-319-73848-2_54

- $A \leq_{\forall \exists 2} B$ iff for all $v \in B$, there exists $u \in A$ such that $u \leq v$.

Notice that additional different preorders over subsets can be obtained as combination of the previously defined relations. As an example take a totally pre-ordered set $(W, \leq)$ and suppose we want to extend the preorder in W to an ordering on the set of intervals of W. Two very usual preorders on intervals are the following ones:

(i) $[a, b] \leq_1 [c, d]$ when $a \leq c$ and $b \leq d$,
(ii) $[a, b] \leq_2 [c, d]$ when $b \leq c$.

The relation $\leq_1$ coincides with the intersection of the relations $\leq_{\forall \exists}$ and $\leq_{\forall \exists 2}$, while the second, $\leq_2$, directly coincides with the relation $\leq_{\forall \forall}$. Observe that, strictly speaking, $\leq_{\forall \forall}$ is not a preorder because it is only reflexive for singletons.

The above six relations can be compared with respect to set inclusion.

Proposition 1.1 [8, 11] *The following inclusions hold:*

$$\leq_{\forall \forall} \subseteq \leq_{\forall \exists} \subseteq \leq_{\exists \exists}, \quad \leq_{\forall \forall} \subseteq \leq_{\exists \forall} \subseteq \leq_{\exists \exists}, \quad \leq_{\forall \forall} \subseteq \leq_{\forall \exists 2} \subseteq \leq_{\exists \exists}, \quad \leq_{\forall \forall} \subseteq \leq_{\exists \forall 2} \subseteq \leq_{\exists \exists}$$

Moreover, the four intermediate relations are not comparable, except for the following inclusions:

$$\leq_{\exists \forall 2} \subseteq \leq_{\forall \exists}, \quad \leq_{\exists \forall} \subseteq \leq_{\forall \exists 2}.$$

In this paper we cope with the case where the initial preorder is fuzzy, as a follow-up of our previous papers [8, 9]. After this brief introduction, in Sect. 2 we recall different forms of extending a fuzzy preorder on a set W to fuzzy relations on the set $\mathscr{P}(W)$ of subsets of W, in a similar way to classical preorders. In Sect. 3 we consider the problem of defining an indistinguishability relation and a strict fuzzy order in a set from a given fuzzy preorder, while in Sect. 4 we deal with the problem of how to lift the strict fuzzy order to subsets. In this sense we have used and recovered some results in [1, 2] and in [4–7] and focus on the transitivity property of the strict fuzzy orders in both settings. The paper ends with some conclusions and comments on further research.

2 Extending a Fuzzy Preorder on a Set W to a Fuzzy Relation on $\mathscr{P}(W)$

Let $\odot$ be a t-norm. In this section we study the extension of a fuzzy $\odot$-preorder on a set W to a relation on $\mathscr{P}(W)$. Remember that a fuzzy $\odot$-preorder is a relation $\leq : W \times W \longrightarrow [0, 1]$ satisfying reflexivity, i.e., $[u \leq u] = 1$ for all $u \in W$, and $\odot$-transitivity, i.e., for all $u, v, w \in W$, $[u \leq v] \odot [v \leq w] \leq [u \leq w]$, where $[u \leq v]$ denotes the value in $[0, 1]$ of the fuzzy relation $\leq$ applied to the ordered pair of elements $u, v \in W$. Moreover we will assume that W is a finite set, and we will denote by δ_u the singleton $\{u\}$.

Generalizing the classical case, in [8] we have introduced the following fuzzy relations on $\mathscr{P}(W)$ (using inf and sup to interpret the universal and existential quantifiers).

Definition 2.1 Given a fuzzy relation $\leq$ on W, we can define the following six fuzzy relations on $\mathscr{P}(W)$ by letting, for any $A, B \in \mathscr{P}(W)$:

- $[A \leq_{\exists\exists} B] = \sup_{u\in A} \sup_{v\in B} [u \leq v]$,
- $[A \leq_{\exists\forall} B] = \sup_{u\in A} \inf_{v\in B} [u \leq v]$,
- $[A \leq_{\forall\exists} B] = \inf_{u\in A} \sup_{v\in B} [u \leq v]$,
- $[A \leq_{\forall\forall} B] = \inf_{u\in A} \inf_{v\in B} [u \leq v]$,
- $[A \leq_{\forall\exists 2} B] = \inf_{v\in B} \sup_{u\in A} [u \leq v]$,
- $[A \leq_{\exists\forall 2} B] = \sup_{v\in B} \inf_{u\in A} [u \leq v]$.

In the same paper, we have proved similar comparisons to the classical case for these six relations.

Proposition 2.1 *For any sets $A, B \in \mathscr{P}(W)$, we have:*

- $[A \leq_{\forall\forall} B] \leq [A \leq_{\forall\exists} B] \leq [A \leq_{\exists\exists} B]$,
- $[A \leq_{\forall\forall} B] \leq [A \leq_{\forall\exists 2} B] \leq [A \leq_{\exists\exists} B]$,
- $[A \leq_{\forall\forall} B] \leq [A \leq_{\exists\forall} B] \leq [A \leq_{\exists\exists} B]$, *and*
- $[A \leq_{\forall\forall} B] \leq [A \leq_{\exists\forall 2} B] \leq [A \leq_{\exists\exists} B]$.

Moreover the four intermediate relations are not comparable, except for the same two cases of Proposition 1.1 changing inclusions by inequalities.

Moreover, in [8] we have also given characteristic properties for each one of these relations. All of them are reflexive (at least for singletons) and transitive, i.e., they are very close to be fuzzy preorders. As a matter of example, we give next the characterization results for the relation $\forall\exists$. The other relations can be characterized in a similar way.

Proposition 2.2 *The relation $\leq_{\forall\exists}$ satisfies the following properties, for all $A, B, C \in \mathscr{P}(W)$:*

1. *Inclusion:* $[A \leq_{\forall\exists} B] = 1$, *if $A \subseteq B$,*
2. *$\odot$-Transitivity:* $[A \leq_{\forall\exists} B] \odot [B \leq_{\forall\exists} C] \leq [A \leq_{\forall\exists} C]$,
3. *Left-OR:* $[(A \cup B) \leq_{\forall\exists} C] = \min([A \leq_{\forall\exists} C], [B \leq_{\forall\exists} C])$,
4. *Restricted Right-OR:* $[A \leq_{\forall\exists} (B \cup C)] \geq \max([A \leq_{\forall\exists} B], [A \leq_{\forall\exists} C])$. *The inequality becomes an equality if A is a singleton.*

Theorem 2.1 *Let $\preceq_{AE}$ be a relation between sets of $\mathscr{P}(W)$ satisfying Properties 1, 2, 3 and 4 of Proposition 2.2. Then there exists a fuzzy $\odot$-preorder $\leq$ on the set W such that $\preceq_{AE}$ coincides with $\leq_{\forall\exists}$ as defined in Definition 2.1.*

3 About the Decomposition of a Fuzzy Preorder and Its Associated Strict Fuzzy Order

In this section we recall from [8] a possible generalization of the decomposition of a crisp preorder given in the introduction to the case that the preorder be fuzzy, and we prove new results about the $\odot$-transitivity of the strict associated order.

In the fuzzy setting (see for example [1, 3, 4]), from a fuzzy $\odot$-preorder $\leq$: $W \times W \rightarrow [0, 1]$ we can define:

- the maximal indistinguishability relation $\equiv$ contained in the fuzzy preorder, defined by $[x \equiv y] = [x \leq y] \wedge [y \leq x]$;
- the minimal strict fuzzy $\odot$-order $<$ that satisfies the following equation

$$[x \leq y] = [x < y] \oplus [x \equiv y] \tag{1}$$

where $\oplus$ is a t-conorm (for example the maximum or the bounded sum).

So defined, the relation $\equiv$ is reflexive, symmetric and $\odot$-transitive, and thus it is a $\odot$-indistinguishability relation.

On the other hand, regarding (1), the minimal solution for b of the equation $a \leq b \oplus c$ in $[0, 1]$, is the so-called *dual residuated implication*, or implication associated to the t-conorm $\oplus$, which is defined as,

$$a \rightarrow^{\oplus} c = \inf\{b \mid a \oplus b \geq c\}.$$

Therefore, one can define the strict fuzzy order relation $<_{\oplus}$ associated to $\leq$ and to the t-conorm $\oplus$ as the fuzzy relation defined as

$$[x <_{\oplus} y] = [x \equiv y] \rightarrow^{\oplus} [x \leq y] = [y \leq x] \rightarrow^{\oplus} [x \leq y].$$

The following are the particular expressions of $[x <_{\oplus} y]$ for the three most prominent examples of $\oplus$.

(i) An easy computation shows that the strict fuzzy order relation for $\oplus = \max$ is defined as

$$[x <_{\max} y] = \begin{cases} [x \leq y], & \text{if } [x \leq y] > [y \leq x], \\ 0, & \text{otherwise.} \end{cases} \tag{2}$$

(ii) For $\oplus$ being the bounded sum (i.e. Łukasiewicz t-conorm), the corresponding strict fuzzy order is[1]:

$$[x <_{\oplus} y] = \left\{ \begin{array}{l} [x \leq y] - [y \leq x], \text{ if } [x \leq y] > [y \leq x] \\ 0, \qquad\qquad\qquad \text{otherwise} \end{array} \right\}$$
$$= \max([x \leq y] - [y \leq x], 0).$$

---

[1]This is the strict order companion defined and studied in [4].

(iii) And for $\oplus$ being the probabilistic sum (i.e. the dual of product t-norm by the negation $N(x) = 1 - x$), we have:

$$[x <_\oplus y] = \begin{cases} \frac{[x \leq y] - [y \leq x]}{1 - [y \leq x]}, & \text{if } [x \leq y] > [y \leq x], \\ 0, & \text{otherwise.} \end{cases}$$

It is well known (see, for example [5]) that the strict relation $<_\oplus$ obtained by the dual residuated implication satisfies the following form $\oplus$-transitivity:

$$[x <_\oplus y] \oplus [y <_\oplus z] \geq [x <_\oplus z].$$

But in general it is not $\odot$-transitive, even in cases where $\odot$ is a continuous t-norm and $\oplus = \max$, as the following examples show.

Example 3.1 Take a set $A = \{u, v, w\}$ and let $\odot$ be either the Łukassiewicz or product t-norm. Let $a, b, c, d \in (0, 1]$, with $a > c, b > d$ and such that $a \odot b = c \odot d > 0$. Suppose now $\leq$ is a fuzzy preorder on A defined by reflexivity plus $[u \leq v] = a > b = [v \leq u], [v \leq w] = c > d = [w \leq v]$ and $[u \leq w] = a \odot b = [w \leq u]$. This relation is transitive if $a \odot a \odot c \leq d$ and $a \odot c \odot c \leq b$ (for example if $a = c = 0.9$ and $b = d = 0.8$). Then it is obvious that the strict relation w.r.t. $\oplus = \max$ is defined as $[u <_{\max} v] = a, [v <_{\max} w] = b$ and $[u <_{\max} w] = 0 < a \odot b = [u <_{\max} v] \odot [v <_{\max} w]$.

Example 3.2 Take a set $A = \{u, v, w\}$ and let $\odot$ be the t-norm which is the ordinal sum of a copy of Łukasiewicz t-norm plus a copy of an arbitrary continuous t-norm, with e being the idempotent element separating the two components. Let $a, b, c, d \in (e, 1]$, with $a > c, b > d$ and such that $a \odot b = c \odot d = e$. Suppose now $\leq$ is a fuzzy preorder on A defined by reflexivity plus the conditions $[u \leq v] = a > c = [v \leq u], [v \leq w] = b > d = [w \leq v]$ and $[u \leq w] = a \odot b = e = c \odot d = [w \leq u]$. Then it is obvious that the strict relation w.r.t. $\oplus = \max$ is defined as $[u <_{\max} v] = a, [v <_{\max} w] = b$ and $[u <_{\max} w] = 0 < e = a \odot b = [u <_{\max} v] \odot [v <_{\max} w]$.

Nevertheless, as we show in the next proposition, we have the following positive results for the cases: (i) $\odot = \min$ and $\oplus = \max$ and (ii) $\odot$ and $\oplus$ being Łukasiewicz t-norm and t-conorm respectively. The case $\odot = \min$ and $\oplus = $ Łukasiewicz t-conorm is already proven in [4, Theorem 15].

Proposition 3.1 *Let $\leq$ be a $\odot$-preorder on a universe W and let $<_\oplus$ be the associated strict relation w.r.t. $\oplus$. Then*

(i) *$<_{\max}$ is min-transitive.*
(ii) *If $\odot$ and $\oplus$ are Łukasiewicz t-norm and t-conorm, then $<_\oplus$ is $\odot$-transitive.*

Proof To show (i), i.e. to show that $\min([u <_{\max} v], [v <_{\max} w]) \leq [u <_{\max} w])$, it is enough to check:

(1) $\min([u <_{\max} v], [v <_{\max} w]) \le \min([u \le v], [v \le w]) \le [u \le w]$,

(2) if $[u <_{\max} v] = [u \le v]$ and $[v <_{\max} w] = [v \le w]$ then $[u <_{\max} w] = [u \le w]$.

On the one hand, (1) holds since we assume $\le$ is min-transitive. We will prove (2) by contradiction. Suppose there exist elements $u, v, w \in W$ such that $[u <_{\max} v] > 0$, $[v <_{\max} w] > 0$ and $[u <_{\max} w] = 0$ which is equivalent that $[u \le v] = a > b = [v \le u]$, $[v \le w] = c > d = [w \le v]$ and $[u \le w] = [w \le u] = f$. Thus we have five values a, b, c, d, f and we know that

$$a > b \quad \text{and} \quad c > d. \tag{$*$}$$

We can now reason by cases:

(1) Suppose $a \ge c$ and $b \ge d$. Combining this assumption with $(*)$ we have that $a \ge c > d$. By transitivity, $f \ge \min(a, c) = c$ and $f \ge \min(d, b) = d$ by hypothesis. Moreover $\min([w \le u], [u \le v]) = \min(f, a) \le d = [w \le v]$. This implies that $a \le d$, in contradiction with the fact that $d < a$.

(2) Suppose $a \ge c$ and $b < d$. Combining this assumption with $(*)$ we have that $d < c \le a$. By transitivity, $f \ge \min(a, c) = c$ and $f \ge \min(d, b) = b$ by hypothesis. Moreover $\min([w \le u], [u \le v]) = \min(f, a) \le d = [w \le v]$. This implies that $f \le d$, and by hypothesis $f \le d < c$, in contradiction with $f \ge c$ previously proved.

(3) Suppose $a \le c$ and $b \ge d$. Combining this assumption with $(*)$ we have that $b < a \le c$. By transitivity, $f \ge \min(a, c) = a$ and $f \ge \min(d, b) = d$ by hypothesis. Moreover $\min([v \le w], [w \le u]) = \min(c, f) \le b = [v \le u]$. This implies that $f \le b$ and by hypothesis $f \le b < a$, in contradiction with $f \ge a$ previously proved.

(4) Suppose $a \le c$ and $b \le d$. Combining this assumption with $(*)$ we have that $b < a \le c$. By transitivity, $f \ge \min(a, c) = a$ and $f \ge \min(d, b) = b$ by hypothesis. Moreover $\min([v \le w], [w \le u]) = \min(c, f) \le b = [v \le u]$. This implies that $f \le b$, and by hypothesis $f \le b < a$, in contradiction with $f \ge a$ previously proved.

Now we will prove (ii). For all $u, v, w \in W$, suppose that $[u \le v] = a$, $[v \le w] = b$, $[v \le u] = c$, $[w \le v] = d$, $[u \le w] = e$, $[w \le u] = f$. We have to prove transitivity of $<_\oplus$ in case that $a > c, b > d$ and $e > f$. The other cases are obviously transitive. Then the associated strict relation $<_\oplus$ contains only the pairs, $[u <_\oplus v] = a - c$, $[v <_\oplus w] = b - d$, $[u <_\oplus w] = e - f$. Then $<_\oplus$ is Łukasiewicz transitive if $(a - c) \odot (b - d) \le (e - f)$. We have two cases:

- If $a - c \le 1 - (b - d)$ then $(a - c) \odot (b - d) = 0$, and thus the inequality holds.
- Otherwise $(a - c) \odot (b - d) = ((a - c) + (b - d)) - 1 = (a + b) - 1 - (c + d)$. Since $\le$ is Łukasiewicz transitive, then $a \odot b \le e$ and $d \odot c \le f$. Therefore $(a - c) \odot (b - d) \le e - (c + d) \le e - (c + d - 1) \le e - f$. $\qquad\square$

Related results about the transitivity of the strict relation associated to a fuzzy preorder can be found in [1, 2, 4–7].

4 Extending the Decomposition to Fuzzy Relations on the Power Set of the Universe

In this section we are interested in how to define a strict fuzzy order relation on sets of $\mathscr{P}(W)$ induced by a fuzzy preorder in W. Halpern notices in [10] that there are two different methods to define (in the crisp case) a strict relation on $\mathscr{P}(W)$ from a preorder on W. The extensions to the fuzzy case are straightforward and give us the following definitions (where $\leq_{\circ}$ denotes one of the six relations $\leq_{\exists\exists}$, $\leq_{\exists\forall}$, $\leq_{\exists\forall 2}$, $\leq_{\forall\exists}$, $\leq_{\forall\exists 2}$ or $\leq_{\forall\forall}$):

- The *standard method*, that amounts to define

$$[A <_{\circ}^{st} B] = \begin{cases} [A \leq_{\circ}^{st} B], & \text{if } [A \leq_{\circ} B] > [B \leq_{\circ} A] \\ 0, & \text{otherwise} \end{cases}.$$

This means in fact to define $[A <_{\circ}^{st} B]$ as the value of the strict order associated to the preorder $\leq_{\circ}$,

- The *alternative method*, that first considers the strict order $<$ on W, and then defines $<_{\circ}^{alt}$ on $\mathscr{P}(W)$ according to Definition 2.1, but replacing $\leq$ by $<$.

In general, these two methods give rise to *two different* irreflexive and (restricted) antisymmetric strict relations. Nevertheless the following inequality always holds: $[A <_{\circ}^{alt} B] \leq [A \leq_{\circ} B]$. Therefore, by definition, if $[A \leq_{\circ} B] > [B \leq_{\circ} A]$ (when $[A <_{\circ}^{st} B] \neq 0$), then $[A <_{\circ}^{alt} B] \leq [A \leq_{\circ}^{st} B]$ holds as well. But different possibilities arise depending on the quantified extension of the original preorder on W as studied below.

Proposition 4.1 *Given a fuzzy preorder $\leq$ on a universe W, let $\leq_{\circ}$ be the induced relation on $\mathscr{P}(W)$ (where $\circ \in \{\exists\exists, \exists\forall, \forall\exists, \exists\forall 2, \forall\exists 2, \forall\forall\}$). Then the strict relations $<_{\circ}^{st}$ and $<_{\circ}^{alt}$ obtained by using standard and alternative method respectively are not comparable in general, but the following relationships hold:*

(i) *Let $\circ = \exists\exists$. Then, for all $A, B \in \mathscr{P}(W)$, $[A <_{\exists\exists}^{st} B] \leq [A <_{\exists\exists}^{alt} B]$. In fact we have:*

$$\begin{cases} [A <_{\exists\exists}^{alt} B] = [A <_{\exists\exists}^{st} B], & \text{if } [A <_{\exists\exists}^{st} B] \neq 0 \\ [A <_{\exists\exists}^{alt} B] \geq [A <_{\exists\exists}^{st} B], & \text{otherwise} \end{cases}.$$

Moreover there are examples where $[A <_{\exists\exists}^{st} B] = 0$ and $[A <_{\exists\exists}^{alt} B] > 0$.

(ii) *Let $\circ = \forall\forall$. Then, for all $A, B \in \mathscr{P}(W)$, $[A <_{\forall\forall}^{st} B] \geq [A <_{\forall\forall}^{alt} B]$. In fact we have:*

$$\begin{cases} [A <_{\forall\forall}^{alt} B] \leq [A <_{\forall\forall}^{st} B], & \text{if } [A <_{\forall\forall}^{st} B] \neq 0 \\ [A <_{\forall\forall}^{alt} B] = [A <_{\forall\forall}^{st} B] = 0, & \text{otherwise} \end{cases}.$$

Moreover there are examples where $[A <_{\forall\forall}^{alt} B] < [A <_{\forall\forall}^{st} B]$.

(iii) For all intermediate cases, i.e. when $\circ \in \{\exists\forall, \forall\exists, \exists\forall 2, \forall\exists 2l\}$, the values of $[A <_{\circ}^{alt} B]$ and $[A <_{\circ}^{st} B]$ are incomparable in general.

Proof We prove (i) by cases, where for simplicity we will write $\leq$ instead of $\leq_{\exists\exists}$:

- Suppose that $[A <^{st} B] = [A \leq B] \neq 0$, equivalent to $[A \leq B] > [B \leq A]$. Therefore there exists $u_1 \in A, v_1 \in B$ such that $[u_1 \leq v_1] = \inf_{u \in A, v \in B}[u \leq v] > \inf_{v \in B, u \in A}[v \leq u]$, which implies that $[u_1 \leq v_1] > [v_1 \leq u_1]$. Therefore $[u_1 < v_1] = [u_1 \leq v_1]$, and so an easy computation proves that $[A <^{alt} B] = [u_1 < v_1] = [A \leq B] = [A <^{st} B]$.

- Suppose $[A <^{st} B] = 0$ due to the fact that $[A \leq B] \leq [B \leq A]$. In such a case it is possible that $[A <^{alt} B] > 0$, as the following example shows:
  Take $A = \{u_1, u_2\}, B = \{v_1, v_2\}$ and let $\leq$ be the relation that is reflexive and contains the following pairs, $[u_1 \leq v_1] = a \leq b = [v_2 \leq u_2]$ with $a \neq 0$. Then $[A \leq B] = a \leq b = [B \leq A]$ and so $[A <^{st} B] = 0$, but one can check that $[A <^{alt} B] = a > 0$.

We also prove item (ii) by cases, and as before, we will write $\leq$ instead of $\leq_{\forall\forall}$ for simplicity:

- Suppose that $[A <^{st} B] = 0$ due to the fact that $[A \leq B] \leq [B \leq A]$. In such a case there exists $u_1 \in A, v_1 \in B$ such that $[u_1 \leq v_1] = \inf_{u \in A, v \in B}[u \leq v] \leq \inf_{v \in B, u \in A}[v \leq u]$, which implies that $[u_1 \leq v_1] \leq [v_1 \leq u_1]$. Therefore $[u_1 < v_1] = 0$ and thus $[A <^{alt} B] = 0 = [A <^{st} B]$.

- Suppose that $[A <^{st} B] = [A \leq B]$. In such a case the following example shows that there exist cases where $[A <^{alt} B] < [A <^{st} B]$:
  Take $W = \{w_1, w_2\}$ with the preorder defined by reflexivity plus $[w_1 \leq w_2] = 1$. Further, take $A = \{w_1\}$ and $B = W$. Then it is obvious that $[A \leq B] = 1 > 0 = [B \leq A]$. Therefore we have $[A <^{st} B] = 1$, while $[A <^{alt} B] = \inf_{u \in A} \sup_{v \in B}[u < v] = 0$, since $[u < u] = 0$.

As for item (iii), and unlike the previous cases where we have shown that $<_{\circ}^{st}$ and $<_{\circ}^{alt}$ are comparable for $\circ \in \{\exists\exists, \forall\forall\}$, we will show the incomparability of the relations for the intermediate cases. Actually, we will prove it for the case $\circ = \forall\exists$, but similar results can be obtained for the other intermediate cases. As above, in the rest of the proof we will write $\leq$ instead of $\leq_{\forall\exists}$ for simplicity.

(a) First we give an example where $[A <^{alt} B] < [A <^{st} B]$. Take $A = \{u_1, u_2\}$, $B = \{v_1, v_2\}$ with the following fuzzy preorder: reflexivity ($[x \leq x] = 1$) plus $[u_1 \leq v_1] = [v_1 \leq u_1] = a$ and $[u_2 \leq v_2] = b$, with $a, b \neq 0$. The associated strict relation on W is the one having only $[u_2 < v_2] = b$. Let $A = \{u_1, u_2\}$ and let $B = \{u_3, u_4\}$. Then it is clear that $[A \leq B] = a \wedge b > 0 = [B \leq A]$ and thus, by definition, $[A <^{st} B] = a \wedge b \neq 0$. Finally $[A <^{alt} B] = \inf_{u \in A} \sup_{v \in B}[u < v] = 0$. Thus $[A <^{alt} B] < [A <^{st} B]$.

(b) Finally we give an example where $[A <^{alt} B] > [A <^{st} B]$.
Take the same sets as in the previous example and the relation defined by reflexivity plus $[u_1 \leq v_1] = [u_2 \leq v_1] = [u_1 \leq u_2] = a$, $[v_1 \leq u_2] = [u_2 \leq v_2] = b$ and $[v_2 \leq v_1] = c$ where $0 < a < b < c$. Then it is easy to compute that $[A \leq_{\forall\exists} B] = a < b =$

$[B \leq_{\forall\exists} A]$ and thus $[A <^{st}_{\forall\exists} B] = 0$ while $[A \leq^{alt}_{\forall\exists} B] = \min([u_1 < v_1], [u_2 < v_1]) = a$. $\qquad\square$

Notice that if the strict order on W is $\odot$-transitive, so are the strict relations obtained by the alternative method (they are strict orders), but this is not clear for strict relations obtained by the standard method. In fact we have the following open problems:

- Let $\leq$ be a fuzzy preorder on W and let $\leq_\circ$ be one of the fuzzy preorders defined on $\mathscr{P}(W)$ considered in the previous sections. Is the strict relation obtained by the standard method $\odot$-transitive?
- It is obvious that the strict order $<$ on W and the strict order on $\mathscr{P}(W)$ obtained from the preorder by the standard method satisfies the following anti-symmetry property: for all $A, B \in \mathscr{P}(W)$, $\min([A <_\circ B], [B <_\circ A]) = 0$. It is clear that for singletons the strict order obtained by the alternative method satisfies the same anti-symmetry property but, is this true for the strict order obtained by the alternative method in general? Otherwise, what type of anti-symmetry property does it satisfy?

Therefore, as far as we are interested in obtaining strict fuzzy orders (irreflexive and $\odot$-transitive relations), it seems reasonable to consider the strict relations obtained by the *alternative method* from a strict order over W and its characteristics properties. Next theorem provides a characterization result for these strict orders. Like for the relations associated to a fuzzy preorder, we give the characterization for the case of $\forall\exists$. The other cases can be characterized in a similar way.

Theorem 4.1 *Let $\prec_{AE}$ be a relation on $\mathscr{P}(W)$ satisfying Properties 2, 3 and 4 of Proposition 2.2 plus irreflexivity ($[A \prec_{AE} A] = 0$) and restricted antisymmetry ($\min([A \prec_{AE} B], [B \prec_{AE} A]) = 0$ for all singletons $A, B \in \mathscr{P}(W)$). Then there exists a strict fuzzy order $<$ on the set W such that $\prec_{AE}$ coincides with the strict fuzzy order associated to $\leq_{\forall\exists}$ obtained by the alternative method.*

5 Conclusions and Further Research

In this paper we have explored (crisply quantified) extensions of fuzzy preorders on a universe W to relations on $\mathscr{P}(W)$. Moreover, extending [8, 9], we have further studied the decomposition of a fuzzy preorder in an indistinguishability relation and a strict (order) relation as a generalization of the well known decomposition in the crisp case.

As for future work, we are interested in applications to preference modelling and reasoning, see [8] for some initial ideas in this direction. Moreover we plan the use of fuzzy quantifications like "nearly all" or "someone", etc. to obtain new extensions of the initial preorder to relations on $\mathscr{P}(W)$. This seems to be a challenging topic, specially related to applications.

Acknowledgements The authors acknowledges partial support by the Spanish FEDER/MINECO project TIN2015-71799-C2-1-P.

Dedication

This paper is our humble homage to the memory of Pedro Gil. Excellent researcher and better person, he has been one of the pioneers of fuzzy sets in Spain and founder and driving force of the research group on fuzzy sets at the University of Oviedo. Our contribution is devoted to fuzzy preorders and their decomposition, a subject that was very close to the research interests of Pedro Gil. Along many years, we jointly participated in many events and we have enjoyed his friendship and shared many unforgettable moments.

References

1. Bodenhofer U (2008) Orderings of fuzzy sets based on fuzzy orderings. Part I: the basic approach. Mathw Soft Comput 15(2):201–218
2. Bodenhofer U (2008) Orderings of fuzzy sets based on fuzzy orderings. Part II: generalizations. Mathw Soft Comput 15(3):219–249
3. Bodenhofer U, Demirci M (2008) Strict fuzzy orderings with a given context of similarity. Int J Uncertain Fuzziness Knowl-Based Syst 16(2):147–178
4. Díaz S, De Baets B, Montes S (2007) Additive decomposition of fuzzy pre-orders. Fuzzy Sets Syst 158(8):830–842
5. Díaz S, De Baets B, Montes S (2010) General results on the decomposition of transitive fuzzy relations. Fuzzy Optim Decis Making 9(1):1–29
6. Díaz S, Montes S, De Baets B (2004) Transitive decomposition of fuzzy preference relations: the case of nilpotent minimum. Kybernetika 40(1):71–88
7. Díaz S, De Baets B, Montes S (2008) On the compositional characterization of complete fuzzy pre-orders. Fuzzy Sets Syst 159(17):2221–2239
8. Esteva F, Godo L, Vidal A (2017) A modal account of preference in a fuzzy setting. In: Pelta D, Cruz Corona C (eds) Soft computing based optimization and decision models, studies in fuzziness and soft computing, vol 360. Springer, Heidelberg
9. Esteva F, Godo L, Vidal A (2017) On a graded modal logic approach to reason with fuzzy preferences. In: Aguiló I, Alquézar R, Angulo C, Ortiz A, Torrens J (eds) Recent advances in artificial intelligence research and development. Series frontiers in artificial intelligence and applications, vol 300. IOS Press, Amsterdam
10. Halpern JY (1997) Defining relative likelihood in partially-ordered preference structures. J Artif Intel Res 7:1–24
11. van Benthem J, Girard P, Roy O (2009) Everything else being equal: a modal logic for ceteris paribus preferences. J Philos Log 38:83–125

Resolution of Fuzzy Relation Equations When the Fuzzy Relation is Defined on Fuzzy Subsets

Manuel José Fernández and Fermín Suárez

Abstract In this paper some results of the theory of fuzzy relation equations are generalized, when fuzzy relations defined on fuzzy sets instead of crisp sets are considered. We find the biggest (or the smallest) solution of a fuzzy relation equation with different types of compositions.

1 Introduction

The theory of fuzzy relation equations was introduced in Sanchez [19], where the biggest solution of the equation $Q \circ R = S$ (for the sup-min composition) is obtained when either Q and S or R and S are assumed to be known, and the "biggest" is viewed as the maximum (in the set-theoretic inclusion sense) of the set of solutions of the equation.

This paper deals with the study of fuzzy relation equations, when the fuzzy relation is defined on fuzzy subsets instead of classical sets. The work is organized as follows: Basic definitions and notations are given in Sect. 2, where the properties which are required along next sections are introduced. Section 3 introduces the notions of sup-T and inf-S compositions of two fuzzy relations defined on fuzzy sets as a generalization of the definitions given when the fuzzy relations are defined on crisp sets (see [15]). In Sect. 4 the fuzzy equations of the form $Q_T R = W$, where $Q_T R$ is the

It has been a deep honour having Pedro Gil as a master and mentor. And, besides it, we have been certainly lucky because his support, mentorship and advice having been truly influential for us. He taught us not only how to research and to publish an article but also how to pursue ideas that eventually could make an impact. He taught us so many things,... the most important ones: the worth of honesty and friendship. Thank you Pedro! we feel proud having learned from you!

M. J. Fernández (✉)
Departamento de Matemáticas, Universidad de Oviedo, Oviedo, Spain
e-mail: mjfg@uniovi.es

F. Suárez
Departamento de Estadística, I.O. y D.M., Universidad de Oviedo, Oviedo, Spain
e-mail: fasuarez@uniovi.es

© Springer International Publishing AG 2018
E. Gil et al. (eds.), *The Mathematics of the Uncertain*, Studies in Systems, Decision and Control 142, https://doi.org/10.1007/978-3-319-73848-2_55

595

sup-T composition defined in Sect. 3, are proposed and solved. In Sect. 5 the fuzzy equations of the form $Q_\varphi R = W$ are proposed and solved.

2 Preliminaries

In this section we present the basic concepts upon which our work is based.

Definition 2.1 ([12, 21]) Let $I = [0, 1]$ and let $T : I \times I \to I$. Consider the conditions

(i) $T(x, 1) = T(1, x) = x$,

(i') $T(x, 0) = T(0, x) = x$,

(ii) If $x_1 \le x_2$ and $y_1 \le y_2$, then $T(x_1, y_1) \le T(x_2, y_2)$,

(iii) $T(x, y) = T(y, x)$,

(iv) $T[T(x, y), z] = T[x, T(y, z)]$.

If T fulfills properties i, ii, iii and iv, then it is said to be a *triangular norm* (or *t-norm*). If T fulfills properties i', ii, iii and iv, then it is said to be a *triangular conorm* (or *t-conorm*).

Henceforth, we will use T to denote a t-norm and S to denote a t-conorm. Given a t-norm T, the function $S(x, y) = 1 - T(1 - x, 1 - y)$ for all $x, y \in I$ is a t-conorm which is called the *dual* of T.

In the study of equations on fuzzy relations a crucial role is played by operator φ associated with a t-norm, whose definition is given as follows:

Definition 2.2 ([16]) An *operator* $\varphi : I \times I \to I$ is *associated with a t-norm* T if for all $a, b, c \in I$

(i) $\varphi(a, \max\{b, c\}) = \max\{\varphi(a, b), \varphi(a, c)\}$,

(ii) $T(a, \varphi(a, b)) \le b$,

(iii) $\varphi(a, T(a, b)) \ge b$.

Theorem 2.1 ([8]) *An operator φ is associated with a t-norm T if, and only if, T is lower semicontinuous.*

Theorem 2.2 ([2, 4, 13, 14]) *Let T be a lower semicontinuous t-norm and φ the operator associated with T. It is verified that*

$$\varphi(a, b_1) \le \varphi(a, b_2) \text{ if } b_1 \le b_2, \quad \varphi(a_1, b) \ge \varphi(a_2, b) \text{ if } a_1 \le a_2,$$

$$\varphi(\varphi(a, b), b) \ge a, \quad \varphi(a, b) = \sup\{c \in I : T(a, c) \le b\}.$$

Let us denote by $\mathscr{F}(\mathbb{X})$ the family of all the fuzzy subsets of $\mathbb{X}$.

Definition 2.3 ([3, 17]) The sup-T (resp. inf-S) *composition of* $Q \in \mathscr{F}(\mathbb{X} \times \mathbb{Y})$ *and* $R \in \mathscr{F}(\mathbb{Y} \times \mathbb{Z})$, denoted by $Q_T R$ (resp. $Q_S R$), is defined as a fuzzy relation on $\mathbb{X} \times \mathbb{Z}$ whose membership function is given for all $x \in \mathbb{X}, z \in \mathbb{Z}$ by

$$(Q_T R)(x, z) = \sup_{y \in \mathbb{Y}}[T(Q(x, y), R(y, z))] \ (\text{resp.} \ (Q_S R)(x, z) = \inf_{y \in \mathbb{Y}}[S(Q(x, y), R(y, z))]).$$

Definition 2.4 ([14]) Let φ be the operator associated with a t-norm T. The inf-φ *composition of* $Q \in \mathscr{F}(\mathbb{X} \times \mathbb{Y})$ *and* $R \in \mathscr{F}(\mathbb{Y} \times \mathbb{Z})$, denoted by $Q_\varphi R$, is defined as a fuzzy relation on $\mathbb{X} \times \mathbb{Z}$ whose membership function is given for all $x \in \mathbb{X}, z \in \mathbb{Z}$ by

$$(Q_\varphi R)(x, z) = \inf_{y \in \mathbb{Y}}[\varphi(Q(x.y), R(y, z))].$$

Let $\mathbb{X}$ be the reference set and $A, B \in \mathscr{F}(\mathbb{X})$. We will use the definition of *cartesian product* proposed by Zadeh [23], so that $A \times B$ is the fuzzy subset of $\mathbb{X} \times \mathbb{X}$ given for all $x, y \in \mathbb{X}$ by $(A \times B)(x, y) = \min\{A(x), B(y)\}$.

Definition 2.5 ([1, 23]) Let $R \subset A \times B$, that is, $R(x, y) \le \min\{A(x), B(y)\}$ for all $x, y \in \mathbb{X}$. Then, R is a fuzzy relation from A in B.

3 Composition of Fuzzy Relations

The max-min (sup-min) composition of two fuzzy relations is closely linked with the extension principle [24] as stated by Sanchez [20] and Pedrycz [18]. In [6, 7] we propose a generalization of this type of composition, using t-norms, as in Dubois and Prade [5] to define fuzzy operations between fuzzy numbers.

Due to the associative property of the t-conorms in [9], it is defined for $n \ge 2$

$$S\{x_1, \ldots, x_n\} = S\{S\{x_1, \ldots, x_{n-1}\}, x_n\}.$$

As by the monotony property of t-conorms and by their definition on the interval I is closed and limited, we can define [10] on a countable set

$$S_{i \in \mathbb{N}}\{x_i\} = \lim_{n \to \infty} S\{x_1, \ldots, x_n\}.$$

Let A, B and C be fuzzy subsets of $\mathbb{X}$, T a t-norm and S a t-conorm which are not necessarily duals. If the referential $\mathbb{X}$ is either finite or countable, we can define the composition of fuzzy relations as follows:

Definition 3.1 Let $Q \subset A \times B$ and $R \subset B \times C$. $Q_{ST} R$ is defined to be the fuzzy subset whose membership function is given for all $x, y \in \mathbb{X}$ by

$$(Q_{ST} R)(x, y) = S_{z \in \mathbb{X}}[T(Q(x, z), R(z, y))].$$

In the case of $S = \max$ and $T = \min$, we are in the classical definition of max-min composition of fuzzy relations.

Theorem 3.1 ([6]) *For all t-norm* T, $Q_{ST}R$ *will be a fuzzy relation on fuzzy subsets, for all pairs of fuzzy relations* $Q \subset A \times B$ *and* $R \subset B \times C$ *if, and only if,* $S = \max$.

We know, from the previous theorem, that given three fuzzy subsets $A, B, C \in \mathscr{F}(\mathbb{X})$ and two fuzzy relations $Q \subset A \times B$ and $R \subset B \times C$, the fuzzy relation $Q_T R$ given for all $x, y \in \mathbb{X}$ by $(Q_T R)(x, y) = \sup_{z \in \mathbb{X}}[T(Q(x, z), R(z, y))]$ where T is a t-norm, satisfies that $Q_T R \subset A \times C$.

When the fuzzy relations are defined on crisp sets, the sup-T composition is associated with the inf-S composition (denoted by $Q_S R$, where S is the dual t-conorm of T) according to the relationship $(Q_S R)^c = Q_T^c R^c$.

In fact, this was one of the reasons for attempting the extension of the inf-S definition to the case of fuzzy relations defined on fuzzy subsets.

Unfortunately, the extension cannot be made so straightforwardly because, in general, the inf-S composition of two fuzzy relations defined on fuzzy sets is not a fuzzy relation on fuzzy subsets, as we can see in the next counterexample (where $*$ denotes the min-max composition).

Let $\mathbb{X} = \{x, y\}$ and let A, B and C be defined as

$$A(x) = 0.3, \quad A(y) = 0.5, \quad B(x) = 0.1, \quad B(y) = 0.7, \quad C(x) = 0, \quad C(y) = 0.1.$$

Then,

$$A \times B = \begin{pmatrix} 0.1 & 0.3 \\ 0.1 & 0.5 \end{pmatrix}, \quad B \times C = \begin{pmatrix} 0 & 0.1 \\ 0 & 0.1 \end{pmatrix} = A \times C.$$

Assuming that

$$Q = \begin{pmatrix} 0.1 & 0.1 \\ 0 & 0.5 \end{pmatrix}, \quad R = \begin{pmatrix} 0 & 0.1 \\ 0 & 0.1 \end{pmatrix}$$

we have that

$$Q * R = \begin{pmatrix} 0.1 & 0.1 \\ 0 & 0.1 \end{pmatrix}.$$

And we see that $Q \subset A \times B$ and $R \subset B \times C$, whereas $Q * R \not\subset A \times C$.

Then, we need to modify the definition of inf-S composition in a manner such that the composition continues to be a fuzzy relation on fuzzy subsets and coincides with the classical definition when the subsets are crisp.

Definition 3.2 ([6]) Let $A, B, C \in \mathscr{F}(\mathbb{X})$. The inf-$S$ *composition* of $Q \subset A \times B$ and $R \subset B \times C$ is the fuzzy relation $Q_S R \subset A \times C$ given for all $x, y \in \mathbb{X}$ by the membership function

$$(Q_S R)(x, y) = \min \left\{ \inf_{z \in \mathbb{X}}[S(Q(x, z), R(z, y))], (A \times C)(x, y) \right\}.$$

The following theorem gives the main properties of the inf-S composition.

Theorem 3.2 ([6])

(i) $Q_S(R \cap W) = (Q_S R) \cap (Q_S W)$ *where* $Q \subset A \times B$ *and* $R, W \subset B \times C$.
(ii) *If* $Q \subset R$, *then* $Q_S W \subset R_S W$ *where* $Q, R \subset A \times B$ *and* $W \subset B \times C$.
(iii) $(Q_S R)^{-1} = R^{-1}{}_S Q^{-1}$ *where* $Q \subset A \times B$ *and* $R \subset B \times C$.
(iv) $Q * (R * W) = (Q * R) * W$ *where* $Q, R, W \subset A \times A$.

4 Solving Equations on Fuzzy Relations on Fuzzy Subsets (sup-T Composition)

Let $\mathbb{X}$ be a set, A, B and C be three fuzzy subsets of $\mathbb{X}$, and let T be a lower semicontinuous t-norm. Consider the fuzzy relation equations $Q_T R = W$, where $Q \subset A \times B$, $R \subset B \times C$ and $W \subset A \times C$, and point out the following typical problems:

(p1) Determine R when Q and W are given.
(p2) Determine Q when R and W are given.

Let φ be the operator associated with the t-norm T. To broaden these questions we extend the definition of inf-φ composition [14], to fuzzy relations defined on fuzzy subsets, in the following way:

Definition 4.1 ([6]) Let $Q \subset A \times B$ and $R \subset B \times C$. The fuzzy relation $Q_\varphi R \subset A \times C$ is given for all $x, y \in \mathbb{X}$ by

$$(Q_\varphi R)(x, y) = \min \left\{ \inf_{z \in \mathbb{X}} [\varphi(Q(x, z), R(z, y))], (A \times C)(x, y) \right\}.$$

Note that we have modified the definition in [14] in order to get $Q_\varphi R$ becoming a fuzzy relation on the fuzzy set $A \times C$.

The previous definition satisfies the following properties:

Theorem 4.1 *(i)* *Let* $Q \subset A \times B$ *and* $R, W \subset B \times C$ *such that* $R \subset W$. *Then,* $Q_\varphi R \subset Q_\varphi W$.
(ii) *Let* $Q, R \subset A \times B$ *such that* $Q \subset R$ *and* $W \subset B \times C$. *Then,* $Q_\varphi W \supset R_\varphi W$.

The proof is a straightforward verification.
In the next theorem we characterize the solutions of Problem $p1$. It is based on

Lemma 4.1 *Let* $Q \subset A \times B$, $R \subset B \times C$ *and* $W \subset A \times C$. *Then,*

(i) $R \subset Q^{-1}{}_\varphi(Q_T R)$.
(ii) $W \supset Q_T(Q^{-1}{}_\varphi W)$.

Proof (*i*) For all $x, y \in \mathbb{X}$

$$[Q^{-1}{}_\varphi (Q_T R)](x, y) = \min \left\{ \inf_{z \in \mathbb{X}} \left[\varphi(Q^{-1}(x, z), (Q_T R)(z, y)) \right], (B \times C)(x, y) \right\}$$

$$= \min \left\{ \inf_{z \in \mathbb{X}} \left[\varphi \left(Q(z, x), \sup_{t \in \mathbb{X}} \{ T(Q(z, t), R(t, y)) \} \right) \right], (B \times C)(x, y) \right\}$$

$$= \min \left\{ \inf_{z \in \mathbb{X}} \left[\varphi \left(Q(z, x), \max \left\{ T(Q(z, x), R(x, y)), \right. \right. \right. \right.$$

$$\left. \left. \left. \sup_{t \neq x} \{ T(Q(z, t), R(t, y)) \} \right\} \right) \right], (B \times C)(x, y) \right\}$$

$$\geq \min \left\{ \inf_{z \in \mathbb{X}} \left[\varphi(Q(z, x), T(Q(z, x), R(x, y))) \right], (B \times C)(x, y) \right\}$$

$$\geq \min \{ R(x, y), (B \times C)(x, y) \} = R(x, y).$$

(*ii*) For all $x, y \in \mathbb{X}$

$$[Q_T(Q^{-1}{}_\varphi W)](x, y) = \sup_{z \in \mathbb{X}} \left[T \left(Q(x, z), (Q^{-1}{}_\varphi W)(z, y) \right) \right]$$

$$= \sup_{z \in \mathbb{X}} \left[T \left(Q(x, z), \min \left\{ \inf_{t \in \mathbb{X}} [\varphi(Q(t, z), W(t, y))], (B \times C)(x, y) \right\} \right) \right]$$

$$\leq \sup_{z \in \mathbb{X}} \left[T \left(Q(x, z), \inf_{t \in \mathbb{X}} [\varphi(Q(t, z), W(t, y))] \right) \right]$$

$$\leq \sup_{z \in \mathbb{X}} [T (Q(x, z), \varphi(Q(x, z), W(x, y)))] \leq W(x, y). \qquad \square$$

Theorem 4.2 *Let* $\mathfrak{R} = \{ R \subset B \times C : Q_T R = W \}$. *If* $\mathfrak{R} \neq \emptyset$, *then the fuzzy relation* $Q^{-1}{}_\varphi W \subset B \times C$ *is the biggest element of* $\mathfrak{R}$.

Proof Let $R \in \mathfrak{R}$. On the basis of Lemma 4.1*i*, $R \subset Q^{-1}{}_\varphi W$. Then, $W = Q_T R \subset Q_T(Q^{-1}{}_\varphi W) \subset W$, from *ii* in Lemma 4.1. $\qquad \square$

In an analogous way, by means of the following lemma we will prove results that allow us to obtain the biggest solution of Problem *p2*.

Lemma 4.2 *Let* $Q \subset A \times B$, $R \subset B \times C$ *and* $W \subset A \times C$. *Then,*

(*i*) $Q \subset \left(R_\varphi (Q_T R)^{-1} \right)^{-1}$.

(*ii*) $W \supset (R_\varphi W^{-1})^{-1}{}_T R$.

The proof is left to the interested reader.

Theorem 4.3 *Let* $\Omega = \{Q \subset A \times B : Q_T R = W\}$. *If* $\Omega \neq \emptyset$, *then the fuzzy relation* $(R_\varphi W^{-1})^{-1} \subset A \times B$ *is the biggest element of* Ω.

Proof Let $Q \in \Omega$. On the basis of Lemma 4.2i, $Q \subset (R_\varphi W^{-1})^{-1}$. Then, $W = Q_T R \subset (R_\varphi W^{-1})^{-1}{}_T R \subset W$, from ii in Lemma 4.2. $\qquad\square$

5 Solving Equations on Fuzzy Relations on Fuzzy Subsets (inf-φ Composition)

Let $\mathbb{X}$ be a set, A, B and C be three fuzzy subsets of $\mathbb{X}$, and let φ be the operator associated with a lower semicontinuous t-norm T. Consider the fuzzy relation equations $Q_\varphi R = W$, where

$$W(x, y) = \min \left\{ \inf_{z \in \mathbb{X}} [\varphi(Q(x, z), R(z, y))], (A \times C)(x, y) \right\},$$

$Q \subset A \times B, R \subset B \times C$ and $W \subset A \times C$, and point out the following typical problems:

$(p'1)$ Determine R when Q and W are given.
$(p'2)$ Determine Q when R and W are given.

Problems $p'1$ and $p'2$ are solved by using the following lemmas:

Lemma 5.1 *Let* $Q \subset A \times B$, $R \subset B \times C$ *and* $W \subset A \times C$. *Then,*

(i) $R \supset Q^{-1}{}_T (Q_\varphi R)$.
(ii) $W \subset Q_\varphi (Q^{-1}{}_T W)$.

Proof (i) For all $x, y \in \mathbb{X}$

$$[Q^{-1}{}_T(Q_\varphi R)](x, y) = \sup_{z \in \mathbb{X}} \left[T \left(Q^{-1}(x, z), (Q_\varphi R)(z, y) \right) \right]$$

$$= \sup_{z \in \mathbb{X}} \left[T \left(Q(z, x), \min \left\{ \inf_{t \in \mathbb{X}} [\varphi(Q(z, t), R(t, y))], (A \times C)(z, y) \right\} \right) \right]$$

$$\leq \sup_{z \in \mathbb{X}} \left[T \left(Q(z, x), \inf_{t \in \mathbb{X}} [\varphi(Q(z, t), R(t, y))] \right) \right]$$

$$\leq \sup_{z \in \mathbb{X}} \left[T \left(Q(z, x), \varphi(Q(z, x), R(x, y)) \right) \right] \leq R(x, y).$$

(ii) For all $x, y \in \mathbb{X}$

$$[Q_\varphi(Q^{-1}{}_T W)](x, y) = \min\left\{\inf_{z \in X}\left[\varphi\big(Q(x, z), (Q^{-1}{}_T W)(z, y)\big)\right], (A \times C)(x, y)\right\}$$

$$= \min\left\{\inf_{z \in X}\left[\varphi\left(Q(x, z), \sup_{t \in X}[T(Q(t, z), W(t, y))]\right)\right], (A \times C)(x, y)\right\}$$

$$= \min\left\{\inf_{z \in X}\left[\varphi\left(Q(x, z), \max\left\{T(Q(x, z), W(x, y)),\right.\right.\right.\right.$$

$$\left.\left.\left.\left.\sup_{t \neq x}\{T(Q(t, z), W(t, y))\}\right\}\right)\right], (A \times C)(x, y)\right\}$$

$$\geq \min\left\{\inf_{z \in X}\left[\varphi\big(Q(x, z), T(Q(x, z), W(x, y))\big)\right], (A \times C)(x, y)\right\}$$

$$\geq \min\{W(x, y), (A \times C)(x, y)\} = W(x, y). \qquad \square$$

Lemma 5.2 *Let* $Q \subset A \times B$, $R \subset B \times C$ *and* $W \subset A \times C$. *Then,*

(i) $Q \subset (Q_\varphi R)_\varphi R^{-1}$.
(ii) $W \subset (W_\varphi R^{-1})_\varphi R$.

The proof is left to the interested reader.

With these lemmas we prove the following theorems that give us the smaller solution of the Problem $p'1$ and the biggest solution of the Problem $p'2$.

Theorem 5.1 *Let* $Q \subset A \times B$, $W \subset A \times C$ *and* $\mathfrak{R} = \{R \subset B \times C : Q_\varphi R = W\}$. *If* $\mathfrak{R} \neq \emptyset$, *then the fuzzy relation* $Q^{-1}{}_T W \subset B \times C$ *is the smallest element of* $\mathfrak{R}$.

Proof Let $R \in \mathfrak{R}$. On the basis of Lemma 5.1i, $R \supset Q^{-1}{}_T W$. Then, $W = Q_\varphi R \supset Q_\varphi(Q^{-1}{}_T W) \supset W$, from ii in Lemma 5.1. $\qquad \square$

Theorem 5.2 *Let* $R \subset B \times C$, $W \subset A \times C$ *and* $\Omega = \{Q \subset A \times B : Q_\varphi R = W\}$. *If* $\Omega \neq \emptyset$, *then the fuzzy relation* $W_\varphi R^{-1} \subset A \times B$ *is the biggest element of* Ω.

Proof Let $Q \in \Omega$. On the basis of Lemma 5.2i, $Q \subset W_\varphi R^{-1}$. Then, $W = Q_\varphi R \supset (W_\varphi R^{-1})_\varphi R \supset W$, from ii in Lemma 5.2. $\qquad \square$

6 Concluding Remarks

The solution of fuzzy equations when the fuzzy relation is defined on fuzzy subsets has been investigated. The procedures for solving different types of fuzzy relation equations have been established. We trust that their diversity allows us to find a suitable model for several real-world problems. In this way, these results may be applicable to fuzzy inference under compositional rules of inference. In this sense, if two propositions, P and Q, whose predicates are defined on universes of discourse A and B, are fuzzy sets, then an implication $P \rightarrow Q$ is defined, in general, as a fuzzy relation on $A \times B$ (see [2, 11, 14]).

References

1. Chakraborty MK, Das M (1983) Studies in fuzzy relations over fuzzy subsets. Fuzzy Sets Syst 9:79–89
2. Di Nola A, Sessa S, Pedrycz W, Sanchez E (1989) Fuzzy relation equations and their applications to knowledge engineering. Kluwer Ac Pub, Dordrecht
3. Di Nola A, Pedrycz W, Sessa S (1995) Fuzzy relational structures: the state-of-art. Fuzzy Sets Syst 75:241–262
4. Drewniak J (1982) Note on fuzzy relation equations. Busefal 12:50–51
5. Dubois D, Prade H (1981) Additions of interactive fuzzy numbers. IEEE Trans Auto Contr AC 26:926–936
6. Fernández MJ, Suárez F, Gil P (1992) Equations of fuzzy relations defined on fuzzy subsets. Fuzzy Sets Syst 52:319–336
7. Fernández MJ, Gil P (2004) Some specific types of fuzzy relation equations. Inform Sci 164:189–195
8. Gottwald S (1986) Fuzzy set theory with T-norms and φ-operators. In: Di Nola A, Ventre A (eds) Topics in the mathematics of fuzzy systems. Verlag TÜV Rheinland, Köln
9. Kimberling C (1973) On a class of associative functions. Public Math Debeces 20:21–39
10. Klement EP (1982) Construction of fuzzy σ-algebras using triangular norms. J Math Anal Appl 85:543–565
11. Klir GJ, Yuan B (1995) Fuzzy sets and fuzzy logic theory and applications. Prentice Hall, Upper Saddle River
12. Menger K (1942) Statistical metric spaces. Proc Natl Acad Sci USA 28:535–537
13. Miyakoshi M, Shimbo M (1984) Composite fuzzy relations with T-norms. Trans IECE Jpn J 67-D(4):391–398
14. Miyakoshi M, Shimbo M (1985) Solutions of composite fuzzy relational equations with triangular norms. Fuzzy Sets Syst 16:53–63
15. Pedrycz W (1982) Fuzzy relational equations with triangular norms and their resolutions. Busefal 11:24–32
16. Pedrycz W (1982) Some aspects of fuzzy decision-making. Kybernetes 11:297–301
17. Pedrycz W (1983) Fuzzy relational equations with generalized connectives and their applications. Fuzzy Sets Syst 10:185–201
18. Pedrycz W (1989) Fuzzy control and fuzzy systems. Wiley/Research Studies Press, London
19. Sanchez E (1976) Resolution of composite fuzzy relation equations. Inform Contr 30:38–48
20. Sanchez E (1979) Compositions of fuzzy relations. In: Gupta MM, Ragade RK, Yager RR (eds) Advances in fuzzy set theory and applications. North-Holland, Amsterdam
21. Schweizer B, Sklar A (1983) Probabilistic metric spaces. North-Holland, Amsterdam
22. Suárez F, Gil P (1986) Two families of fuzzy integrals. Fuzzy Sets Syst 18:67–81
23. Zadeh LA (1971) Similarity relations and fuzzy orderings. Inform Sci 3:177–200
24. Zadeh LA (1975) The concept of a linguistic variable and its application to approximate reasoning. Inform Sci Part I, 8:199–249, Part II, 8:301–357; Part III, 9:43–80

Measuring Dispersion in the Context of Ordered Qualitative Scales

José Luis García-Lapresta and Luis Borge

Abstract In this contribution, we introduce two families of dispersion measures in the context of ordered qualitative scales. They are based on the notions of ordinal proximity measure and the median.

1 Introduction

The concept of dispersion has been used in history since the first manifestations of the use of Statistics. Harter [10] attributes its use to the second century BC in a quote of Ptolemy, who mentions the estimation of the maximum variation of some observations to the half of the range by the Greek astronomer Hipparchus.

But we must wait until the 18 and 19th centuries in which, with the growth of statistical methods and their relationships with probability theory, there is a great interest in the study of dispersion. A large number of measures are presented, including their statistical properties, such as the standard deviation, the mean deviation, the mean absolute deviation, the interquartile range, and so on. Stigler [16] attributes this proliferation of measures to the efforts made by astronomers and geodists for agreeing on the observations obtained, because the measurement errors of their observations produced different estimates (see David [4]).

In the beginning of the 20th century a great movement in statistics began. The what nowadays is known as Classical Statistics: hypothesis testing, estimation, experiment design, and sample surveys, begins (see Lehman [12]). It can be attributed to Fisher [5] as the starting point of the current Statistics. This statistical revolution represented a great growth of this area due to the extension of the use of its methods to other fields of research.

J. L. García-Lapresta (✉)
PRESAD Research Group, BORDA Research Unit, IMUVA, Departamento de Economía Aplicada, Universidad de Valladolid, Valladolid, Spain
e-mail: lapresta@eco.uva.es

L. Borge
Departamento de Economía Aplicada, Universidad de Valladolid, Valladolid, Spain
e-mail: borge@eco.uva.es

© Springer International Publishing AG 2018
E. Gil et al. (eds.), *The Mathematics of the Uncertain*, Studies in Systems, Decision and Control 142, https://doi.org/10.1007/978-3-319-73848-2_56

These new statistical methods were based on parametric models and generally assumed a normal distribution. When these methods are used in new problems in which the distribution differs from normality or when applied to data that presented asymmetries, the outliers appear and the optimal properties of the estimators are no longer optimal. In the middle of the last century there was an important revival of robust statistics. For Hubert [11],"robustness signifies insensitivity to small deviations of the assumption". There are different ways of measuring the robustness as the breaking point, the influence curve or gross sensitivity error. The robustness gradually grew to become one of the fields of Statistics.

This classical theory of statistical inference assumes that observations are normally distributed or come from a parametric distribution, and the most widely used estimator for dispersion has been the standard deviation. This estimator, together with the sample mean, has provided important results in this field, both from the theoretical point of view and the data adjustments. However, when the data comes from models that deviate from normality, small proportions of atypical data, called outliers, appear. The statistics used no longer have optimal properties, they are no longer efficient.

The most used statistic and with best robust properties as dispersion estimator is the median of the absolute deviations from the median of the data, also known as MAD. It gives us the best possible breaking point (twice the interquartile range), and its influence function is bounded.

Rousseeuw and Croux [14] presented two alternative robust estimators to the MAD. One of the reasons why they consider alternatives to MAD is because this estimator has a symmetric point of view with respect to a central measure, in this case the median, which does not appear to be an adequate estimator for asymmetric distributions. The first one is S_n, and its construction consists in calculating for each element the median of all absolute deviations to the rest of the sample values, and to obtain the median of all the medians calculated in the last step. This estimator presents the difficulty of obtaining the value of the median twice, and in addition to having the inconvenience that the influence function has discontinuities.

The second estimator, Q_n, is analogous to the scale estimator obtained by Bickel and Lehman [3], where the approach to arrive at the estimator is to consider the dispersion in the distribution of the variable. Q_n is based only on the differences between sample observations. If $\{x_{(1)}, \ldots, x_{(n)}\}$ is the ordered sample, in a decreasing fashion, then the statistic is defined from the set of the differences of the form $x_{(i)} - x_{(j)}, \; j < i$.

The main purpose of this contribution is to measure the dispersion associated with the different results of a variable when they form a set of linguistic terms obtained after examining different qualities of people, services, etc. Typical examples of linguistic terms are: very bad, bad, regular, good, very good and excellent. We can order the different levels, but we assume that the proximities among them are not necessarily uniform. These proximities are expressed by a set of ordinal degrees that can not be quantified through real numbers (in this framework, distances between linguistic terms are not defined). Therefore, it is not possible to use classical dispersion measures (see Franceschini et al. [6]).

Our proposal for measuring the dispersion in a list of linguistic terms of an ordered qualitative scale is based on the notion of ordinal proximity measure, introduced by García-Lapresta and Pérez-Román [7]. It is a mapping that assigns an ordinal degree of proximity to each pair of linguistic terms of the scale. In this setting we introduce two families of dispersion measures, one based on the notion of range, and the second one related to the Gini index [9].

Given a vector of linguistic terms, range-based dispersion measures assign the ordinal degree of proximity between the minimum and the maximum linguistic terms of that vector as the dispersion degree of the vector. In turn, Gini-based dispersion measures assign the pair of medians of the ordinal degrees of proximity among all the pairs of linguistic terms of that vector as the dispersion degree of the vector. In the last case, a linear order on the set of feasible medians is needed for comparing the dispersion between pairs of vectors. Additionally, a tie-breaking process is provided for rank order the dispersion of vectors of linguistic terms.

We have also established some properties of the proposed dispersion measures.

The rest of the contribution is organized as follows. Section 2 is devoted to introduce the notation and some basic notions that are necessary for defining the two families of dispersion measures we propose in the setting of ordered qualitative scales. Section 3 presents these two families of dispersion measures and includes some illustrative examples. Section 4 contains some properties. Finally, Sect. 5 concludes with some remarks.

2 Notation and Basic Notions

In this section we present the two notions from which we construct our proposal for measuring the dispersion in the setting of ordered qualitative scales: ordinal proximity measures and the median operator.

2.1 Ordinal Proximity Measures

Let $\mathscr{L} = \{l_1, \ldots, l_g\}$ be an ordered qualitative scale, with $g \geq 3$, arranged from the lowest to the highest linguistic terms: $l_1 < l_2 < \cdots < l_g$.

We now recall the notion of ordinal proximity measure on $\mathscr{L}$, introduced by García-Lapresta and Pérez-Román [7]. We shall use a linear order $\Delta = \{\delta_1, \ldots, \delta_h\}$, with $\delta_1 \succ \cdots \succ \delta_h$, for representing different ordinal degrees of proximity among the terms of $\mathscr{L}$, being δ_1 and δ_h the maximum and minimum degrees, respectively.

It is important emphasizing that the elements of Δ are not numbers, but abstract elements of a linear order that represent different ordinal degrees of proximity among the linguistic terms of an ordered qualitative scale.

As usual in the setting of linear orders, $\delta_r \succeq \delta_s$ means $\delta_r \succ \delta_s$ or $\delta_r = \delta_s$; and $\delta_r \prec \delta_s$ means $\delta_s \succ \delta_r$. Given a weak order $\trianglelefteq$ on $\mathscr{L}^n$, with $\triangleleft$ we denote the asymmetric part of $\trianglelefteq$, i.e., $\boldsymbol{x} \triangleleft \boldsymbol{y} \Leftrightarrow$ not $\boldsymbol{y} \trianglelefteq \boldsymbol{x}$.

Definition 2.1 ([7]) An *ordinal proximity measure on $\mathscr{L}$ with values in Δ* is a mapping $\pi : \mathscr{L}^2 \longrightarrow \Delta$, where $\pi(l_r, l_s) = \pi_{rs}$ means the degree of proximity between l_r and l_s, satisfying the following conditions:

(a) *Exhaustiveness*: For every $\delta \in \Delta$, there exist $l_r, l_s \in \mathscr{L}$ such that $\delta = \pi_{rs}$.
(b) *Symmetry*: $\pi_{sr} = \pi_{rs}$, for all $r, s \in \{1, \ldots, g\}$.
(c) *Maximum proximity*: $\pi_{rs} = \delta_1 \Leftrightarrow r = s$, for all $r, s \in \{1, \ldots, g\}$.
(d) *Monotonicity*: $\pi_{rs} \succ \pi_{rt}$ and $\pi_{st} \succ \pi_{rt}$, for all $r, s, t \in \{1, \ldots, g\}$ such that $r < s < t$.

We note that the previous conditions are independent (see García-Lapresta and Pérez-Román [7, Proposition 1]).

Every ordinal proximity measure $\pi : \mathscr{L}^2 \longrightarrow \Delta$ can be represented by a $g \times g$ symmetric matrix with coefficients in Δ,

$$\begin{pmatrix} \pi_{11} & \cdots & \pi_{1s} & \cdots & \pi_{1g} \\ \cdots & \cdots & \cdots & \cdots & \cdots \\ \pi_{r1} & \cdots & \pi_{rs} & \cdots & \pi_{rg} \\ \cdots & \cdots & \cdots & \cdots & \cdots \\ \pi_{g1} & \cdots & \pi_{gs} & \cdots & \pi_{gg} \end{pmatrix},$$

where the elements in the main diagonal are $\pi_{rr} = \delta_1$, for every $r = 1, \ldots, g$. This matrix is called the *proximity matrix associated with π*.

Taking into account the conditions appearing in Definition 2.1, it is only necessary to show the proximity upper half matrix

$$\begin{pmatrix} \delta_1 & \pi_{12} & \pi_{13} & \cdots & \pi_{1(g-1)} & \pi_{1g} \\ & \delta_1 & \pi_{23} & \cdots & \pi_{2(g-1)} & \pi_{2g} \\ & & \cdots & \cdots & \cdots \\ & & & & \delta_1 & \pi_{(g-1)g} \\ & & & & & \delta_1 \end{pmatrix}.$$

As shown in García-Lapresta and Pérez-Román [7, Proposition 2], the minimum proximity between linguistic terms is only reached when comparing the extreme linguistic terms: $\pi_{rs} = \delta_h \Leftrightarrow (r, s) \in \{(1, g), (g, 1)\}$.

The cardinality of Δ is located between the cardinality of $\mathscr{L}$ and a polynomial of degree 2 of that cardinality (see García-Lapresta and Pérez-Román [7, Proposition 4]):

$$g \leq h \leq \frac{g \cdot (g - 1)}{2} + 1.$$

2.2 The Median Operator

Following García-Lapresta and Pérez-Román [8], we now introduce the median operator in the setting of ordinal degrees of proximity.

Given a vector of ordinal degrees of proximity $\delta = (\delta_1, \ldots, \delta_p) \in \Delta^p$, we arrange its components in a decreasing fashion, from the highest to the lowest degrees. If p is odd, then the median of δ is unique, say $\delta_r \in \Delta$. However, if p is even, then δ has two medians, say $\delta_s, \delta_t \in \Delta$ such that $s \leq t$, i.e., $\delta_s \succeq \delta_t$. In order to unify the assignment of medians, we consider the pair of medians (δ_r, δ_r) and (δ_s, δ_t) whenever p is odd and even, respectively.

More formally, given the *set of feasible medians* $\Delta_2 = \{(\delta_r, \delta_s) \in \Delta^2 \mid r \leq s\}$, the *median operator* is the mapping

$$M : \bigcup_{p=1}^{\infty} \Delta^p \longrightarrow \Delta_2$$

that assigns the corresponding pair of medians to each vector of ordinal degrees of proximity.

For ordering the pairs of medians of ordinal degrees of proximity, consider the linear order $\succeq_2$ on Δ_2 defined as

$$(\delta_r, \delta_s) \succeq_2 (\delta_t, \delta_u) \Leftrightarrow \begin{cases} r + s < t + u \\ \text{or} \\ r + s = t + u \text{ and } s - r \leq u - t, \end{cases} \tag{1}$$

for all $(\delta_r, \delta_s), (\delta_t, \delta_u) \in \Delta_2$.

It is easy to see that if $r + s = t + u$, then $s - r \leq u - t \Leftrightarrow r \geq t \Leftrightarrow s \leq u$.

3 Dispersion Measures

In this section we introduce two families of dispersion measures in the setting of ordered qualitative scales equipped with ordinal proximity measures. The first family generalizes the most basic dispersion measure, the range.

In what follows, vectors in $\mathscr{L}^n$ are denoted $x = (x_1, \ldots, x_n)$.

3.1 Range-Based Dispersion Measures

Given $n \geq 2$, let $D_R : \mathscr{L}^n \longrightarrow \Delta$ be the mapping defined as

$$D_R(x) = \pi(\min x, \max x),\tag{2}$$

for every $x \in \mathcal{L}^n$.

Based on the linear order $\succeq$ on Δ, we introduce the weak order $\trianglelefteq_R$ on $\mathcal{L}^n$ defined as

$$x \trianglelefteq_R y \Leftrightarrow D_R(x) \succeq D_R(y),$$

for all $x, y \in \mathcal{L}^n$, with the meaning of the dispersion in x is lower than or equal to in y (with respect to D_R).

Example 3.1 Consider the ordered qualitative scale $\mathcal{L} = \{l_1, l_2, l_3, l_4\}$ and the vectors $x = (l_1, l_2, l_2, l_3)$, $y = (l_3, l_3, l_4, l_4) \in \mathcal{L}^4$. We want to compare the dispersion in these vectors with respect to three different ordinal proximity measures.

(a) If $\mathcal{L}$ is equipped with the ordinal proximity measure

$$\pi : \mathcal{L}^2 \longrightarrow \Delta = \{\delta_1, \ldots, \delta_7\}$$

with associated proximity matrix

$$\begin{pmatrix} \delta_1 & \delta_2 & \delta_4 & \delta_7 \\ & \delta_1 & \delta_3 & \delta_6 \\ & & \delta_1 & \delta_5 \\ & & & \delta_1 \end{pmatrix},$$

we have $D_R(x) = \pi_{13} = \delta_4 \succ \delta_5 = \pi_{34} = D_R(y)$. Thus, $x \lhd_R y$.

(b) If $\mathcal{L}$ is equipped with the ordinal proximity measure

$$\pi : \mathcal{L}^2 \longrightarrow \Delta = \{\delta_1, \ldots, \delta_7\}$$

with associated proximity matrix

$$\begin{pmatrix} \delta_1 & \delta_4 & \delta_6 & \delta_7 \\ & \delta_1 & \delta_3 & \delta_5 \\ & & \delta_1 & \delta_2 \\ & & & \delta_1 \end{pmatrix},$$

we have $D_R(x) = \pi_{13} = \delta_6 \prec \delta_2 = \pi_{34} = D_R(y)$. Thus, $y \lhd_R x$.

(c) If $\mathcal{L}$ is equipped with the ordinal proximity measure

$$\pi : \mathcal{L}^2 \longrightarrow \Delta = \{\delta_1, \ldots, \delta_4\}$$

with associated proximity matrix

$$\begin{pmatrix} \delta_1 & \delta_2 & \delta_3 & \delta_4 \\ & \delta_1 & \delta_2 & \delta_3 \\ & & \delta_1 & \delta_2 \\ & & & \delta_1 \end{pmatrix},$$

we have $D_R(x) = \pi_{13} = \delta_3 \prec \delta_2 = \pi_{34} = D_R(y)$. Thus, $y \lhd_R x$.

3.2 Gini-Based Dispersion Measures

We now introduce a new family of dispersion measures in the mentioned framework. It is based on the Gini index [9] and it is closely related to the scale estimator appearing in Shamos [15, p. 260] and Bickel and Lehmann [3, p. 38] in the setting of real numbers (see Rousseeuw and Croux [14, p. 1277]).

Given $n \geq 2$, let $D_G : \mathscr{L}^n \longrightarrow \Delta_2$ be the mapping defined as

$$D_G(x) = M(\pi(x_i, x_j)_{i<j}), \tag{3}$$

for every $x \in \mathscr{L}^n$.

Based on the linear order $\succeq_2$ on Δ_2 defined in (1), we introduce the weak order $\unlhd_G$ on $\mathscr{L}^n$ defined as

$$x \unlhd_G y \iff D_G(x) \succeq_2 D_G(y),$$

for all $x, y \in \mathscr{L}^n$, with the meaning of the dispersion in x is lower than or equal to in y (with respect to D_G).

Since some vectors can share the same pair of medians, it is necessary to devise a tie-breaking process for ordering the vectors. We propose to use a sequential procedure based on Balinski and Laraki [1] (see Balinski and Laraki [2] for practical examples). It consists of withdrawing the pair of medians of the vectors that are in a tie, and then selecting the new pairs of medians of the remaining ordinal degrees of proximity for the corresponding vectors. The process continues until the ties are broken. It is important to note that different vectors never are in a final tie.

Example 3.2 Consider Example 3.1 and the same three ordinal proximity measures.

(a) We have

$$D_G(x) = M(\pi_{12}, \pi_{12}, \pi_{13}, \pi_{22}, \pi_{23}, \pi_{23}) = M(\delta_2, \delta_2, \delta_4, \delta_1, \delta_3, \delta_3) = (\delta_2, \delta_3)$$

and

$$D_G(y) = M(\pi_{33}, \pi_{34}, \pi_{34}, \pi_{34}, \pi_{34}, \pi_{44}) = M(\delta_1, \delta_5, \delta_5, \delta_5, \delta_5, \delta_1) = (\delta_5, \delta_5).$$

Since $(\delta_2, \delta_3) \succ_2 (\delta_5, \delta_5)$, we have $x \lhd_G y$.

(b) We have

$$D_G(x) = M(\pi_{12}, \pi_{12}, \pi_{13}, \pi_{22}, \pi_{23}, \pi_{23}) = M(\delta_4, \delta_4, \delta_6, \delta_1, \delta_3, \delta_3) = (\delta_3, \delta_4)$$

and

$$D_G(y) = M(\pi_{33}, \pi_{34}, \pi_{34}, \pi_{34}, \pi_{34}, \pi_{44}) = M(\delta_1, \delta_2, \delta_2, \delta_2, \delta_2, \delta_1) = (\delta_2, \delta_2).$$

Since $(\delta_3, \delta_4) \prec_2 (\delta_2, \delta_2)$, we have $y \lhd_G x$.

(c) We have

$$D_G(x) = M(\pi_{12}, \pi_{12}, \pi_{13}, \pi_{22}, \pi_{23}, \pi_{23}) = M(\delta_2, \delta_2, \delta_3, \delta_1, \delta_2, \delta_2) = (\delta_2, \delta_2)$$

and

$$D_G(y) = M(\pi_{33}, \pi_{34}, \pi_{34}, \pi_{34}, \pi_{34}, \pi_{44}) = M(\delta_1, \delta_2, \delta_2, \delta_2, \delta_2, \delta_1) = (\delta_2, \delta_2).$$

Consequently, in x and y the dispersion is the same. If we apply the tie-breaking procedure, then we have $D_G(x) = M(\delta_1, \delta_2, \delta_2, \delta_3) = (\delta_2, \delta_2)$ and $D_G(y) = M(\delta_1, \delta_1, \delta_2, \delta_2) = (\delta_1, \delta_2)$. Since $(\delta_2, \delta_2) \prec_2 (\delta_1, \delta_2)$, we finally have that the dispersion in y is lower than in x.

4 Properties

Let $\mathscr{L}$ be an ordered qualitative scale equipped with an ordinal proximity measure $\pi : \mathscr{L}^2 \longrightarrow \Delta$. We say that π is *totally uniform* if $\pi_{r\,(r+t)} = \pi_{s\,(s+t)}$ for all $r, s, t \in \{1, \ldots, g-1\}$ such that $r + t \le g$ and $s + t \le g$.

Let $N : \mathscr{L} \longrightarrow \mathscr{L}$ be the *negation operator* defined as $N(l_r) = l_{g+1-r}$, for every $r \in \{1, \ldots, g\}$.

Given $k \in \{1 - g, \ldots, g - 1\}$, let $T_k : \mathscr{L} \longrightarrow \mathscr{L}$ be the *translation operator* defined as $T_k(l_r) = l_{r+k}$, for every $r \in \{1, \ldots, g\}$ such that $r + k \le g$.

In the following proposition we establish some properties of the mappings introduced in (2) and (3). They are related to the ones considered in Martínez–Panero et al. [13] in a quantitative context.

Proposition 4.1 Let $D_R : \mathscr{L}^n \longrightarrow \Delta$ and $D_G : \mathscr{L}^n \longrightarrow \Delta_2$ be the mappings defined in (2) and (3), respectively, and their extensions $\widetilde{D}_R : \bigcup_{n=2}^{\infty} \mathscr{L}^n \longrightarrow \Delta$ and $\widetilde{D}_G : \bigcup_{n=2}^{\infty} \mathscr{L}^n \longrightarrow \Delta_2$. The following properties hold:

(a) Symmetry: $D_R(x_{\sigma(1)}, \ldots, x_{\sigma(n)}) = D_R(x)$ and $D_G(x_{\sigma(1)}, \ldots, x_{\sigma(n)}) = D_G(x)$, for every permutation $\sigma : \{1, \ldots, n\} \longrightarrow \{1, \ldots, n\}$ and every $x \in \mathscr{L}^n$.

(b) Invariance for replications: $\widetilde{D_R}(\overbrace{x, \ldots, x}^{m}) = D_R(x)$ and $\widetilde{D_G}(\overbrace{x, \ldots, x}^{m}) = D_G(x)$, for every $x \in \mathscr{L}^n$ and any number $m \in \mathbb{N}$ of replications of x.

(c) Minimum dispersion: $D_R(x) = \delta_1 \Leftrightarrow x_1 = \cdots = x_n$, for every $x \in \mathscr{L}^n$; and $D_G(l_r, \ldots, l_r) = (\delta_1, \delta_1)$, for every $l_r \in \mathscr{L}$.

(d) Anti-self-duality: if π is totally uniform, then $D_R(N(x_1), \ldots, N(x_n)) = D_R(x)$ and $D_G(N(x_1), \ldots, N(x_n)) = D_G(x)$, for every $x \in \mathscr{L}^n$.

(e) Invariance for translations: if π is totally uniform, then $D_R(T_k(x_1), \ldots, T_k(x_n)) = D_R(x)$ and $D_G(T_k(x_1), \ldots, T_k(x_n)) = D_G(x)$, for every $x \in \mathscr{L}^n$ and every $k \in \{1, \ldots, g-1\}$ such that $(T_k(x_1), \ldots, T_k(x_n)) \in \mathscr{L}^n$.

5 Concluding Remarks

There exists in the literature a number of dispersion measures in the context of real numbers. In this contribution, we have provided a proposal for measuring the dispersion of data belonging to ordered qualitative scales, non-necessarily uniform. We have considered ordinal proximity measures that assign different ordinal degrees of proximity to the pairs of linguistic terms of an ordered qualitative scale. In this framework, we have proposed and analyzed two families of dispersion measures, one based on the range and another one based on the Gini index, where the Euclidean distance between pairs of numbers has been changed to ordinal degrees of proximity between linguistic terms, and the average to the median.

Acknowledgements This contribution is dedicated to the memory of our dear friend Pedro Gil. We extend our homage to his great family. The first author gratefully acknowledges the funding support of the Spanish *Ministerio de Economía y Competitividad* (Project ECO2016-77900-P) and ERDF.

References

1. Balinski M, Laraki R (2007) A theory of measuring, electing and ranking. Proc Natl Acad Sci USA 104:8720–8725
2. Balinski M, Laraki R (2013) How best to rank wines: Majority Judgment. In: Güvenen O, Serbat H, Giraud-Héraud E, Pichery M (eds) Gvenen O. Wine economics. Quantitative studies and empirical observations. Palgrave-MacMillan, London, pp 149–172
3. Bickel PJ, Lehmann EL (1979) Descriptive statistics for nonparametric models IV: Spread. In: Jurečková J (ed) Contributions to statistics-Jaroslav Hájek memorial volume. Academia, Prague, pp 33–40
4. David HA (1998) Early sample measures of variability. Stat Sci 13:368–377
5. Fisher RA (1922) On the mathematical foundations of theoretical statistics. Philos Trans R Soc Lond Ser A 222:309–368

6. Franceschini F, Galletto M, Varetto M (2004) Qualitative ordinal scales: the concept of ordinal range. Qual Eng 16:515–524
7. García-Lapresta JL, Pérez-Román D (2015) Ordinal proximity measures in the context of unbalanced qualitative scales and some applications to consensus and clustering. Appl Soft Comput 35:864–872
8. García-Lapresta JL, Pérez-Román D (2017) A consensus reaching process in the context of non-uniform ordered qualitative scales. Fuzzy Optim Decis Mak 16(4):449–461
9. Gini C (1912) Variabilità e Mutabilità. Tipografia di Paolo Cuppini, Bologna
10. Harter HL (1978) A chronological annotated bibliography of order statistics, vol 1. Pre-1950. U.S. Government Printing Office, Washington
11. Hubert PJ (1981) Robust statistics. Wiley, New York
12. Lehmann EL (2011) Fisher, Neyman, and the creation of classical statistics. Springer, New York
13. Martínez-Panero M, García-Lapresta JL, Meneses LC (2016) Multidistances and dispersion measures. In: Calvo-Sánchez T, Torrens-Sastre J (eds) Fuzzy logic and information fusion, series studies in fuzziness and soft computing. Springer, Heidelberg, pp 123–134
14. Rousseeuw PJ, Croux C (1993) Alternatives to the median absolute deviation. J Am Stat Assoc 88:1273–1283
15. Shamos MI (1976) Geometry and statistics: problems at the interface. In: Traub JF (ed) New directions and recent results in algorithms and complexity. New York Academic Press, New York
16. Stigler SM (1986) The history of statistics: the measurement of uncertainty before 1900. Harvard University Press, Cambridge

The Kolmogorov–Smirnov Goodness-of-Fit Test for Interval-Valued Data

Przemysław Grzegorzewski

Abstract The Kolmogorov–Smirnov goodness-of-fit test for equality of two distributions is considered. Two generalizations of this test for interval-valued data are proposed. Each version correspond to a different view on the interval outcomes of the experiment – either the epistemic or the ontic one. Each view yield its own approaches to data analysis and statistical inference.

1 Introduction

Both practitioners and data analysts sometimes do not realize that interval-valued data may deliver two different types of information: the imprecise description of a point-valued quantity or the precise description of a set-valued entity. Firstly, let us realize that quite often the experimental results cannot be observed precisely. It may also happen that the results are too uncertain to be recorded as real numbers. In such situations one can utilize intervals containing the precise outcomes as the experiment results. Moreover, sometimes even having precise data one may be interested in hiding the exact value of some variables deliberately because of confidentiality reasons (see [8]). All these cases illustrate the so-called *epistemic view* on intervals considered as disjunctive sets representing incomplete information. More formally, an *epistemic set* A contains an ill-known actual value of a point-valued quantity x, so we can write $x \in A$. Of course, it represents only the epistemic state of an agent but it does not exist per se (see [2]).

However, we can also distinguish situations when the experiment outcomes are just intervals describing a precise information. As a typical example we may consider a range of fluctuations of some physical measurements or time interval spanned

P. Grzegorzewski (✉)
Systems Research Institute, Polish Academy of Sciences, Newelska 6,
01-447 Warsaw, Poland
e-mail: pgrzeg@ibspan.waw.pl

P. Grzegorzewski
Faculty of Mathematics and Information Science, Warsaw University of Technology,
Koszykowa 75, 00-662 Warsaw, Poland

© Springer International Publishing AG 2018
E. Gil et al. (eds.), *The Mathematics of the Uncertain*, Studies in Systems,
Decision and Control 142, https://doi.org/10.1007/978-3-319-73848-2_57

615

by some activity. Here we meet another perception of intervals, i.e. the *ontic view* on intervals. The *ontic set* (or conjunctive set) A is a precise representation of an objective entity, which means that A is an actual value of a set-valued variable X, so we can write $X = A$ (see [2]).

Now suppose our data set consist of two samples. A typical problem we face in the two-sample problem is to conclude whether the distributions of these samples differ significantly or not. In other words, we want to know if our samples come from the same distribution or from two different distributions. To solve the problem one should utilize an appropriate two-sample goodness of fit test. One of the most popular test for comparing real-valued data is the famous Kolmogorov–Smirnov test. The main goal of this paper is to generalize the Kolmogorov–Smirnov test for interval-valued data keeping in mind the two possible views on such type of the data, as we have discussed it above. We shown that each perspective requires its own approach to data analysis that implies different restrictions and problems starting just from the way of hypotheses stating, through the test construction, right up to the decision making and its interpretation.

The paper is organized as follows: In Sect. 2 we recall the classical Kolmogorov–Smirnov goodness-of-fit test. In Sect. 3 we introduce basic notations and concepts related to interval-valued data. Next, we propose two generalizations of the goodness-of-fit tests adequate to each type of interval data: for epistemic sets in Sect. 4 and for ontic sets in Sect. 5. The suggested methods are illustrated by numerical examples.

2　The Kolmogorov–Smirnov Test

Suppose, we observe independently two random samples $X_1, \ldots, X_n$ and $Y_1, \ldots, Y_m$ drawn from populations with unknown cumulative distributions function (c.d.f.) F and G, respectively. We want to verify the null hypothesis that both samples come from the same distribution, i.e.

$$H_0 : F(t) = G(t) \quad \text{for all} \quad t \in \mathbb{R}, \tag{1}$$

against the alternative hypothesis that the population distributions differ, i.e. $H_1 : F(t) \neq G(t)$ *for some* $t \in \mathbb{R}$.

Several goodness-of-fit tests can be used to solve this problem, including the Wald–Wolfowitz runs test, the Brown-Mood median test and various tests based on ranks, like the well-known Mann–Whitney–Wilcoxon tests (see [3]). But it seems that that most popular and famous test dedicated to the considered problem is the *Kolmogorov–Smirnov* test based on the empirical distribution function. Let us recall that the *empirical distribution function* (e.d.f.) for a given sample $X_1, \ldots, X_n$ is defined by

$$\widehat{F}_n(t) = \frac{1}{n} \sum_{i=1}^{n} \mathbb{I}(X_i \leqslant t), \tag{2}$$

where $\mathbb{I}$ denotes the indicator function. Thus $\widehat{F}_n(t)$ simply reflects the proportion of sample values less than or equal to t.

Since $F(t)$ is the probability of an observation less than or equal to t thus one may expect $\widehat{F}_n(t)$ to estimate $F(t)$. Indeed, for any fixed real value t, the e.d.f. $\widehat{F}_n(t)$ is a consistent estimator of $F(t)$, i.e. $\widehat{F}_n(t)$ converges to $F(t)$ in probability. Moreover, $\widehat{F}_n$ converges to F with probability 1 by the following theorem (called sometimes the Glivenko–Cantelli lemma).

Theorem 2.1 *Let $X_1, \ldots, X_n$ denote independent and identically distributed random variables from the distribution F. Then*

$$\mathbb{P}\left(\lim_{n \to \infty} \sup_{t \in \mathbb{R}} |\widehat{F}_n(t) - F(t)| = 0 \right) = 1.$$

Therefore, the divergence between e.d.f. $\widehat{F}_n$ and c.d.f. F might be used for testing whether a sample fits to some hypothetical distribution F. However, the similar reasoning applied for two samples may lead to the conclusion that the discrepancy between two e.d.f. $\widehat{F}_n$ and $\widehat{G}_m$ based on $X_1, \ldots, X_n$ and $Y_1, \ldots, Y_m$, respectively, might be used for testing whether the true population distributions of these two samples, although unknown, differ significantly, i.e. whether the null hypothesis (1) holds.

Thus, it seems natural to consider the following test statistic

$$D_{n,m} = \sup_{t \in \mathbb{R}} |\widehat{F}_n(t) - \widehat{G}_m(t)|, \tag{3}$$

where $\widehat{F}_n$ and $\widehat{G}_m$ are the e.d.f. based on the first and the second sample, respectively. The usefulness of statistic (3) is motivated by fact that the null distribution of $D_{n,m}$ does not depend on the true sample distribution as long as it is continuous. Therefore, we may utilize $D_{n,m}$ effectively in our goodness-of-fit testing problem. Indeed, the differences between $\widehat{F}_n(t)$ and $\widehat{G}_m(t)$ should be small for all t if the null hypothesis (1) holds. Conversely, large differences between $\widehat{F}_n(t)$ and $\widehat{G}_m(t)$ would discredit hypothesis H_0. It means, that assuming some significance level α we reject H_0 if $D_{n,m} > D_{n,m}(\alpha)$, where $D_{n,m}(\alpha)$ denotes the critical value such that $\mathbb{P}(D_{n,m} > D_{n,m}(\alpha)) = \alpha$. In practice, for small or moderate sample sizes n and m the critical values $D_{n,m}(\alpha)$ can be found in statistical tables, while for large samples one may obtain them using the following approximation of $D_{n,m}$ distribution (see [10])

$$\lim_{n,m \to \infty} \mathbb{P}\left(\sqrt{\frac{nm}{n+m}} D_{n,m} \leqslant d \right) = 1 - 2 \sum_{i=1}^{\infty} (-1)^{i-1} e^{-2i^2 d^2} \tag{4}$$

for every $d \geqslant 0$. Alternatively, a final decision whether reject or accept given null hypothesis may be taken using the p-value given by

$$p = \mathbb{P}_{H_0}(D_{n,m} \geqslant d), \tag{5}$$

where d stands for the actual value of the test statistic $D_{n,m}$. Then we reject H_0 if p is small enough, say $p < \alpha$, where α is the assumed significance level (typically $\alpha = 0.05$) or do not reject H_0 (accept H_0) otherwise.

Some generalizations of the Kolmogorov–Smirnov test for fuzzy data were suggested in [6, 7]. Further on we generalize the Kolmogorov–Smirnov test for the interval-valued data. However, firstly let us introduce basic notation and operations on such data.

3 Interval-Valued Data

Let $\mathscr{K}_c(\mathbb{R}) = \{[u, v] : u, v \in \mathbb{R},\ u \leqslant v\}$ denote the family of all non-empty closed and bounded intervals in the real line $\mathbb{R}$. Each compact interval $A \in \mathscr{K}_c(\mathbb{R})$ can be expressed by its endpoints, i.e. $A = [\underline{a}, \overline{a}]$. Alternatively, the notation $A = [\mathrm{mid}\,A \pm \mathrm{spr}\,A]$, with $\mathrm{spr}\,A \geqslant 0$, can be considered, where $\mathrm{mid}\,A = \frac{1}{2}(\underline{a} + \overline{a})$ is the mid-point (center) of the interval A and $\mathrm{spr}\,A = \frac{1}{2}(\overline{a} - \underline{a})$ is the spread (radius) of A.

To handle intervals a natural arithmetic on $\mathscr{K}_c(\mathbb{R})$ is defined by means of the Minkowski addition and the product by scalars, given by

$$A + B = \{a + b : a \in A, b \in B\}, \quad \lambda A = \{\lambda a : a \in A\},$$

for any $A, B \in \mathscr{K}_c(\mathbb{R})$ and $\lambda \in \mathbb{R}$. These two operations can be jointly expressed in terms of the mid/spr representation of the intervals as

$$A + \lambda B = [(\mathrm{mid}\,A + \lambda \mathrm{mid}\,B) \pm (\mathrm{spr}\,A + |\lambda|\mathrm{spr}\,B)],$$

while using the endpoints of the intervals we obtain $A + B = [\underline{a} + \underline{b}, \overline{a} + \overline{b}]$, $A - B = [\underline{a} - \overline{b}, \overline{a} - \underline{b}]$ and $\lambda A = [\min\{\lambda\underline{a}, \lambda\overline{a}\}, \max\{\lambda\underline{a}, \lambda\overline{a}\}]$.

It should be noted that the space $(\mathscr{K}_c(\mathbb{R}), +, \cdot)$ is not linear but semi linear, due to the lack of the opposite element with respect to the Minkowski addition: in general, $A + (-1)A \neq \{0\}$, unless $A = \{a\}$ is a singleton.

Both for the epistemic and ontic approach we use the same notation and basic operations on intervals. However, it does not mean that the type of interval data is of no significance for the data analysis. Actually there are significant differences in statistics of interval-valued data perceived from these two perspectives. In the epistemic approach we deal with usual random variables which attribute to each random event a real value. The only problem is that its perception is not known precisely but exact to interval. On the other hand, the ontic view on intervals require no longer usual real-valued random variables but random intervals. This is the reason that we have to consider two generalizations of the Kolmogorov–Smirnov test for interval-valued data – suitable for each vie on the data.

4 The Kolmogorov–Smirnov Test for the Epistemic Data

Let us consider a sequence of interval observations $[\underline{x}_1, \overline{x}_1], \ldots, [\underline{x}_n, \overline{x}_n]$, which are perceptions of the unknown true outcomes $x_1, \ldots, x_n$ of the experiment, where $x_i \in [\underline{x}_i, \overline{x}_i]$. We also observe the second sample $[\underline{y}_1, \overline{y}_1], \ldots, [\underline{y}_m, \overline{y}_m]$, which are perceptions of the unknown true outcomes $y_1, \ldots, y_m$, where $y_j \in [\underline{y}_j, \overline{y}_j]$. We assume that the samples and all observations are independent. As in the classical case we assume that our samples come from the unknown continuous distributions F and G, respectively, and our goal is to verify the null hypothesis H_0 vs H_1. Since the Kolmogorov–Smirnov test is based on e.d.f. we have to generalize this concept for interval-valued data. Let $\mathcal{Q}_n$ denote a family of all finite subsets of rational numbers of the form $\{\frac{i}{n} : 0 \leqslant i \leqslant n\}$. The following definition was given in [5].

Definition 4.1 The *interval-valued empirical distribution function* based on a sample of interval-valued observations $[\underline{x}_1, \overline{x}_1], \ldots, [\underline{x}_n, \overline{x}_n]$ is a multifunction $\widehat{\mathscr{F}}_n : \mathbb{R} \to \mathcal{Q}_n$ defined for each t as follows

$$\widehat{\mathscr{F}}_n(t) = \left\{ \frac{1}{n} \sum_{i=1}^{n} \mathbb{I}(x_i \leqslant t) : x_i \in [\underline{x}_i, \overline{x}_i] \right\}. \tag{6}$$

The proof of the following lemma is immediate (see [5]).

Lemma 4.1 *For each* $\widehat{F} \in \widehat{\mathscr{F}}_n$ *and for any* $t \in \mathbb{R}$

$$\widehat{F}_n^U(t) \leqslant \widehat{F}(t) \leqslant \widehat{F}_n^L(t), \tag{7}$$

where

$$\widehat{F}_n^L(t) = \frac{1}{n} \sum_{i=1}^{n} \mathbb{I}(\underline{x}_i \leqslant t), \quad \widehat{F}_n^U(t) = \frac{1}{n} \sum_{i=1}^{n} \mathbb{I}(\overline{x}_i \leqslant t). \tag{8}$$

By Lemma 4.1 the interval-valued empirical distribution function is a set of e.d.f. bounded by $\widehat{F}_n^L$ and $\widehat{F}_n^U$. Indices L and U applied in the aforementioned notation correspond to the stochastic order between two borderline random variables.

Example 4.1 Consider the following interval-valued sample: $[-2, 0.5], [1, 4], [2, 6], [5, 7]$ and $[8, 10]$. The upper and lower bounds of the interval-valued e.d.f. for this sample, i.e. $\widehat{F}_n^U$ and $\widehat{F}_n^L$, are given in Fig. 1 (the dashed rectangles between $\widehat{F}_n^U$ and $\widehat{F}_n^L$ have no meaning here but they are drawn just to visualize better the distance between these two bounds).

Obviously, $[\underline{y}_1, \overline{y}_1], \ldots, [\underline{y}_m, \overline{y}_m]$ generates its own interval-valued e.d.f. given by

$$\widehat{\mathscr{G}}_m(t) = \left\{ \frac{1}{m} \sum_{j=1}^{m} \mathbb{I}(y_j \leqslant t) : x_i \in [\underline{y}_j, \overline{y}_j] \right\}. \tag{9}$$

Fig. 1 The bounds of the interval-valued e.d.f., where $\widehat{F}_n^L$ is depicted by a solid line while $\widehat{F}_n^U$ is marked by a dashed one

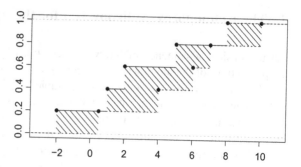

Here, of course, for each $\widehat{G} \in \mathscr{G}_m$ and for any $t \in \mathbb{R}$ we have $\widehat{G}_m^U(t) \leqslant \widehat{G}(t) \leqslant \widehat{G}_m^L(t)$, where $\widehat{G}_m^L(t) = \frac{1}{m} \sum_{j=1}^{m} \mathbb{I}(\underline{y}_j \leqslant t)$ and $\widehat{G}_m^U(t) = \frac{1}{m} \sum_{j=1}^{m} \mathbb{I}(\overline{y}_j \leqslant t)$.

Now we can return to our testing problem. In the classical Kolmogorov–Smirnov test we accept H_0 if the distance between the e.d.f. obtained for each sample is not too big and we reject H_0 otherwise. However, for interval-valued observations we have two multifunctions $\mathscr{F}_n$ and $\mathscr{G}_m$ which can be treated as two families of empirical distribution functions that might be obtained for different possible choices of $x_i \in [\underline{x}_i, \overline{x}_i]$ and $y_j \in [\underline{y}_j, \overline{y}_j]$, respectively. But since we know neither which x_i nor which y_j are the true outcomes of the experiment, we have to consider all possible choices of outcomes.

We can distinguish three general situations.

(a) It may happen that $D_{n,m}(\widehat{F}_n, \widehat{G}_m)$ is small for all outcomes of the experiment, i.e.

$$D_{n,m}(\widehat{F}_n, \widehat{G}_m) \leqslant D_{n,m}(\alpha) \quad \forall \widehat{F}_n \in \mathscr{F}_n \text{ and } \forall \widehat{G}_m \in \mathscr{G}_m. \tag{10}$$

If so, one may decide there are no reasons to reject H_0 for any possible realizations of the experiment which suggests the acceptance of the null hypothesis. Of course, checking (10) seems to be troublesome so it would be desirable to substitute it by some more concise conditions. In particular, by (7) we find that the most distant functions $\widehat{F}_n$ and $\widehat{G}_m$ such that $\widehat{F}_n \in \mathscr{F}_n$ and $\widehat{G}_m \in \mathscr{G}_m$, are either $\widehat{F}_n^L$ and $\widehat{G}_m^U$ or $\widehat{F}_n^U$ and $\widehat{G}_m^L$. Therefore, one may easily conclude that (10) is equivalent to the following condition

$$\max \left\{ D_{n,m}(\widehat{F}_n^L, \widehat{G}_m^U), D_{n,m}(\widehat{F}_n^U, \widehat{G}_m^L) \right\} \leqslant D_{n,m}(\alpha). \tag{11}$$

(b) On the contrary, it may happen that $D_{n,m}(\widehat{F}_n, \widehat{G}_m)$ is significantly large for any possible outcome of the experiment, i.e.

$$D_{n,m}(\widehat{F}_n, \widehat{G}_m) > D_{n,m}(\alpha) \quad \forall \widehat{F}_n \in \mathscr{F}_n \text{ and } \forall \widehat{G}_m \in \mathscr{G}_m. \tag{12}$$

Since it suggests to reject the null hypothesis whatever is the true realization of the experiment, it simply means that we should reject H_0. Keeping in mind the aforementioned considerations one may notice that (12) is equivalent to

$$\min \left\{ D_{n,m}(\widehat{F}_n^L, \widehat{G}_m^U), D_{n,m}(\widehat{F}_n^U, \widehat{G}_m^L) \right\} > D_{n,m}(\alpha). \tag{13}$$

(c) Finally, it may happen that $D_{n,m}(\widehat{F}_n, \widehat{G}_m)$ is neither to big nor to small in general, but it is small for some $\widehat{F}_n \in \mathscr{F}_n$ and $\widehat{G}_m \in \mathscr{G}_m$ while it is big for other functions in $\widehat{F}_n \in \mathscr{F}_n$ and $\widehat{G}_m \in \mathscr{G}_m$. Hence, we cannot decide definitely whether to reject or accept H_0 since it depend on the actual outcomes which are not available. In other words, if

$$\min \left\{ D_{n,m}(\widehat{F}_n^L, \widehat{G}_m^U), D_{n,m}(\widehat{F}_n^U, \widehat{G}_m^L) \right\} \leqslant D_{n,m}(\alpha) \quad \text{or} \tag{14}$$

$$\max \left\{ D_{n,m}(\widehat{F}_n^L, \widehat{G}_m^U), D_{n,m}(\widehat{F}_n^U, \widehat{G}_m^L) \right\} > D_{n,m}(\alpha) \tag{15}$$

then we abstain from making the final decision. It might be annotated that we suspend the decision and demand either more numerous or more accurate observations to make a well-based conclusion.

As in classical statistics, we may take the final decision using the p-value. However, in the case of epistemic interval-valued data we have no longer a single real value of the test statistic to calculate the p-value according to Eq. (5). Indeed, according to the aforementioned discussion we have to distinguish different situations related to possible outcomes of the experiment represented by interval data. Firstly, let us introduce the following notation

$$d_{\min} = \min \left\{ D_{n,m}(\widehat{F}_n^L, \widehat{G}_m^U), D_{n,m}(\widehat{F}_n^U, \widehat{G}_m^L) \right\} \tag{16}$$

$$= \min \left\{ \sup_{t \in \mathbb{R}} \left| \frac{1}{n} \sum_{i=1}^{n} \mathbb{I}(\underline{x}_i \leqslant t) - \frac{1}{m} \sum_{j=1}^{m} \mathbb{I}(\overline{y}_j \leqslant t) \right|, \right.$$

$$\left. \sup_{t \in \mathbb{R}} \left| \frac{1}{n} \sum_{i=1}^{n} \mathbb{I}(\overline{x}_i \leqslant t) - \frac{1}{m} \sum_{j=1}^{m} \mathbb{I}(\underline{y}_j \leqslant t) \right| \right\},$$

$$d_{\max} = \max \left\{ D_{n,m}(\widehat{F}_n^L, \widehat{G}_m^U), D_{n,m}(\widehat{F}_n^U, \widehat{G}_m^L) \right\} \tag{17}$$

$$= \max \left\{ \sup_{t \in \mathbb{R}} \left| \frac{1}{n} \sum_{i=1}^{n} \mathbb{I}(\underline{x}_i \leqslant t) - \frac{1}{m} \sum_{j=1}^{m} \mathbb{I}(\overline{y}_j \leqslant t) \right|, \right.$$

$$\left. \sup_{t \in \mathbb{R}} \left| \frac{1}{n} \sum_{i=1}^{n} \mathbb{I}(\overline{x}_i \leqslant t) - \frac{1}{m} \sum_{j=1}^{m} \mathbb{I}(\underline{y}_j \leqslant t) \right| \right\}.$$

Now, we can compute the following two probabilities

$$\underline{p} = \min\{\mathbb{P}_{H_0}(D_{n,m} \geqslant d_{\max})\}, \quad \overline{p} = \max\{\mathbb{P}_{H_0}(D_{n,m} \geqslant d_{\min})\}, \tag{18}$$

where the distribution of a random variable $D_{n,m}$ is given in (4).

Finally, keeping in mind Formulas (11)–(15) we obtain the following p-value based decision criteria for the generalized Kolmogorov–Smirnov test:

- if $\overline{p} < \alpha$ then reject H_0,
- if $\alpha < \underline{p}$ then accept H_0,
- otherwise (i.e. if $\underline{p} \leqslant \alpha \leqslant \overline{p}$) we abstain.

Although this type of the decision algorithm is well-grounded and may be recommended to practitioners, if one requires just a binary decisions (to reject or accept H_0) we suggest an appropriate randomization [4] or a method applied for testing hypotheses in fuzzy environment [6].

At a first glance it seems that calculations required in (16)–(17) are time consuming because we have to consider the distance between the e.d.f. and the hypothetical c.d.f. for each point of the real line. Actually, it is quite simple since $D_{n,m}$ assumes its maximum in one of the e.d.f. jumps. Therefore, it is enough to compute the differences between both e.d.f. in all points where they are not constant. Since we have to consider two e.d.f. $\widehat{F}_n^L$ and $\widehat{F}_n^U$ with jumps in $\underline{x}_i$ and $\overline{x}_i$, respectively, where $i = 1, \ldots, n$, and two e.d.f. $\widehat{G}_m^L$ and $\widehat{G}_m^U$ with jumps in $\underline{y}_j$ and $\overline{y}_j$, respectively, where $j = 1, \ldots, m$, thus altogether we have to determine the distances in $N = 2n + 2m$ points only.

Example 4.2 Consider two independent interval-valued samples $X_1, \ldots, X_{20}$ and $Y_1, \ldots, Y_{18}$ coming from the distributions F and G, respectively. All 38 observations are given in Table 1.

We verify a null hypothesis that these two samples come from the same distribution, i.e. $H : F(t) = G(t)$ for all $t \in \mathbb{R}$ versus $H_1 : F(t) \neq G(t)$ for some $t \in \mathbb{R}$. The interval-valued e.d.f. for samples $X_1, \ldots, X_{20}$ and $Y_1, \ldots, Y_{18}$ are given in Figs. 2 and 3, respectively.

After some calculations we obtain $D_{n,m}(\widehat{F}_n^L, \widehat{G}_m^U) = 0.4333$ and $D_{n,m}(\widehat{F}_n^U, \widehat{G}_m^L) = 0.4722$, which means that $d_{\min} = 0.4333$ and $d_{\max} = 0.4722$. Thus, by (18), we obtain $\underline{p} = 0.0194$ and $\overline{p} = 0.0381$. Therefore, since $\overline{p} < 0.05$, we reject H_0 at significance level 0.05.

Table 1 Interval-valued observations considered in Example 4.2

X	[32.13, 36.04], [28.81, 30.27], [24.52, 27.92], [29.57, 32.30], [31.89, 32.89]
	[29.11, 29.93], [27.30, 29.43], [31.17, 34.65], [28.70, 30.62], [33.01, 35.08]
	[29.23, 29.58], [32.42, 35.87], [26.65, 28.03], [29.39, 30.30], [28.73, 30.63]
	[32.60, 34.85], [29.70, 31.46], [28.28, 31.41], [30.49, 32.50], [31.73, 32.22]
Y	[29.51, 29.99], [29.54, 29.90], [30.25, 31.56], [32.75, 33.45], [28.75, 30.75]
	[26.60, 27.92], [29.66, 31.30], [29.20, 30.91], [29.56, 30.72], [32.74, 33.46]
	[32.83, 34.55], [29.03, 30.46], [30.50, 31.63], [28.16, 28.87], [29.17, 30.57]
	[33.47, 34.62], [28.57, 29.94], [28.07, 29.23]

Fig. 2 The bounds of the interval-valued e.d.f. based on $X_1, \ldots, X_{20}$

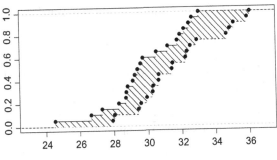

Fig. 3 The bounds of the interval-valued e.d.f. based on $Y_1, \ldots, Y_{18}$

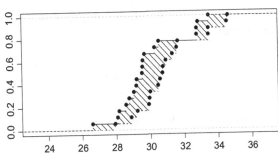

5   The Kolmogorov–Smirnov for the Ontic Data

In the epistemic approach we deal with usual random variables which attribute a real value to each random event. However, in the ontic approach we deal with random intervals defined as follows.

Definition 5.1 Given a probability space $(\Omega, \mathscr{A}, P)$, a mapping $X : \Omega \to \mathscr{K}_c(\mathbb{R})$ is said to be a *random interval* (interval-valued random set) if it is Borel-measurable with the Borel σ-field generated by the topology associated with by the Hausdorff metric on $\mathscr{K}_c(\mathbb{R})$.

Equivalently, a mapping $X : \Omega \to \mathscr{K}_c(\mathbb{R})$ is a random interval if mid $X : \Omega \to \mathbb{R}$ and spr $X : \Omega \to \mathbb{R}_+ \cup \{0\}$ are random variables defined as the mid-point and the spread of the interval $X(\omega)$, respectively, for each $\omega \in \Omega$. Consequently, to describe the distribution of a random interval we need the distribution of its mid-point and the distribution of its spread. Therefore, we will utilize these two distributions in formulating hypotheses for random intervals.

Let us consider two independent samples of independent random intervals: $X_1, \ldots, X_n$ and $Y_1, \ldots, Y_n$. Suppose, these samples come from populations characterized by the distributions: $F_{\mathrm{mid}\,X}$ and $F_{\mathrm{spr}\,X}$ denoting the c.d.f. of mid X and spr X, respectively, and by $G_{\mathrm{mid}\,Y}$ and $G_{\mathrm{spr}\,G}$, which are the c.d.f. of mid Y and spr Y, respectively. Now we have to formulate the null hypothesis being the counterpart of (1) and stating that the distributions of random intervals coming from both samples do not differ. It can be done as follows

$$H_0 : F_{\text{mid } X} = G_{\text{mid } Y} \text{ and } F_{\text{spr } X} = G_{\text{spr } Y}. \qquad (19)$$

As the alternative hypothesis we consider $H_1 : \neg H_0$ that at least one of the equalities in (19) fails.

If H_0 holds then the e.d.f. $\widehat{F}_{n,\text{mid}}$ and $\widehat{F}_{n,\text{spr}}$ based on mid-points and spreads of X's, where

$$\widehat{F}_{n,\text{mid}}(t) = \frac{1}{n} \sum_{i=1}^{n} \mathbb{I}(\text{mid } X_i \leqslant t), \quad \widehat{F}_{n,\text{spr}}(t) = \frac{1}{n} \sum_{i=1}^{n} \mathbb{I}(\text{spr } X_i \leqslant t),$$

should be "close" to the corresponding e.d.f. $\widehat{G}_{m,\text{mid}}$ and $\widehat{G}_{m,\text{spr}}$ based on mid-points and spreads of Y's, given by

$$\widehat{G}_{m,\text{mid}}(t) = \frac{1}{m} \sum_{j=1}^{n} \mathbb{I}(\text{mid } Y_j \leqslant t), \quad \widehat{G}_{m,\text{spr}}(t) = \frac{1}{m} \sum_{j=1}^{n} \mathbb{I}(\text{spr } Y_j \leqslant t).$$

Let us define the following two test statistics:

$$T_1 = D_n(\widehat{F}_{n,\text{mid}}, \widehat{G}_{m,\text{mid}}) = \sup_{t \in \mathbb{R}} |\widehat{F}_{n,\text{mid}}(t) - \widehat{G}_{m,\text{mid}}(t)|, \qquad (20)$$

$$T_2 = D_n(\widehat{F}_{n,\text{spr}}, \widehat{G}_{m,\text{spr}}) = \sup_{t \in \mathbb{R}} |\widehat{F}_{n,\text{spr}}(t) - \widehat{G}_{m,\text{spr}}(t)|. \qquad (21)$$

This way our test for the random intervals consists of two usual Kolmogorov–Smirnov tests. Suppose t_1 and t_2 denote a value of the test statistic T_1 and T_2, respectively. However, two test statistics imply two p-values

$$p_1 = \mathbb{P}_{H_0}(T_1 \geqslant t_1), \quad p_2 = \mathbb{P}_{H_0}(T_2 \geqslant t_2), \qquad (22)$$

related to T_1 and T_2, respectively. Hence the following question arises immediately: *How to combine p_1 and p_2 to find the overall p-value of the generalized Kolmogorov–Smirnov test for random intervals?*

The answer to this question in not straightforward. Several approaches for combining p-values were suggested in the literature (see, e.g. [11]). Unfortunately, most of them assume that the combined tests are independent which is usually not our case. It seems that a natural and efficient way for combining p-values can be reached as the effect of the following reasoning. Since p-value always belong to the unit interval, let $\xi : [0, 1]^2 \to [0, 1]$ denote a function combining two p-values obtained in our testing problem. Then $p = \xi(p_1, p_2)$ will denote the overall p-value. Assuming that $\xi(0, 0) = 0$, $\xi(1, 1) = 1$ and that $\xi(p_1, p_2) \leqslant \xi(p'_1, p'_2)$ for $p_i \leqslant p'_i$ ($i = 1, 2$), we conclude that ξ is an *aggregation function* (see [1]). If we additionally assume that ξ is symmetric (i.e. $\xi(p_1, p_2) = \xi(p_2, p_1)$), associative and 1 is its neutral element 1 (i.e. $\xi(1, p) = p$), it appears that ξ is a *triangular norm* (t-norm). Moreover, if we also assume the idempotency (i.e. $\xi(p, p) = p$) then we obtain $\xi(p_1, p_2) = \min\{p_1, p_2\}$, since minimum is the unique idempotent t-norm (see [1]). Therefore, we suggest to make a final decision on the rejection or acceptance of the null hypothesis (19) using the overall p-value obtained by combining p-values p_1 and p_2 given by (22) as follows

$$p = \min\{p_1, p_2\}. \tag{23}$$

It is worth noting that by combining p-value we obtain a single p-value, so our generalized Kolmogorov–Smirnov test for random intervals always leads to the binary decision: reject/accept H_0, as it is in the classical situation, contrary to what happens in the epistemic approach.

Example 5.1 Let us consider once more observations given in Table 1 and discussed in Example 4.2. However, now we will consider our two interval-valued samples $X_1, \ldots, X_{20}$ and $Y_1, \ldots, Y_{18}$ as realizations of random intervals characterized by the distributions $F_{\text{mid}}, F_{\text{spr}}$ and $G_{\text{mid}}, G_{\text{spr}}$, respectively.

We verify a null hypothesis that these two samples come from the same distribution, i.e. $H_0 : F_{\text{mid}\,X} = G_{\text{mid}\,Y}$ *and* $F_{\text{spr}\,X} = G_{\text{spr}\,Y}$ versus $H_1 : \neg H_0$ that at least one of these two equalities fails. E.d.f. $\widehat{F}_{n,\text{mid}}$ and $\widehat{G}_{m,\text{mid}}$ are given in Fig. 4, while e.d.f. $\widehat{F}_{n,\text{spr}}$ and $\widehat{G}_{m,\text{spr}}$ are given in Fig. 5.

By (20)–(21) we obtain $T_1 = 0.1778$ and $T_2 = 0.5944$ and the corresponding p-values $p_1 = 0.9256$ and $p_2 = 0.00096$. Using (23) we obtain the overall p-value $p = \min\{0.9256, 0.00096\} = 0.00096$ indicating the rejection of H_0, i.e. we conclude that our two samples do not come from the same distribution. However, if

Fig. 4 $\widehat{F}_{n,\text{mid}}$ and $\widehat{G}_{m,\text{mid}}$ based on interval-valued data

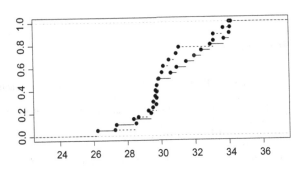

Fig. 5 $\widehat{F}_{n,\text{spr}}$ and $\widehat{G}_{m,\text{spr}}$ based on interval-valued data

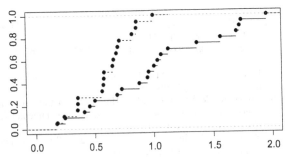

we consider the same null hypothesis against the alternative that their distribution differ in location, i.e. $H'_1 : F_{\text{mid}\,X} \neq G_{\text{mid}\,Y}$, we would obtain a quite high p-value $p = 0.9256$ indicating no reason for the rejection.

6 Conclusions

We have proposed two generalizations of the Kolmogorov–Smirnov test designed for two different views on interval-valued data. One should be aware of the distinction between ontic and epistemic sets because otherwise there is a risk of misusing even basic notions and tools. Both ontic and epistemic view yield different approach to data analysis. In particular, for epistemic interval-valued data we obtain a set of possible test statistic values which generate a set of p-values. As a consequence the test may lead either to decision on rejection or acceptance or to situation when no definite decision can be made, especially if the intervals are too broad. This problem does not concern the ontic interval-valued data modeled by random intervals. However, in this case we have to aggregate two subtests which is always subjective.

Another important aspect of the suggested tests is their computational simplicity. As it is known, epistemic data may cause computational problems if a sample is large enough. For example, the sample variance computation for the epistemic intervals is NP-hard (see [9]). However, fortunately, interval uncertainty in the considered Kolmogorov–Smirnov test does not increase the computational complexity when comparing it with real data analysis.

References

1. Beliakov G, Pradera A, Calvo T (2007) Aggregation functions: a guide for practitioners. Springer, Heidelberg
2. Couso I, Dubois D (2014) Statistical reasoning with set-valued information: ontic vs epistemic views. Int J Approx Reas 55:1502–1518
3. Gibbons JD, Chakraborti S (2003) Nonparametric Statistical Inference, 4th edn. Marcel Dekker, New York
4. Grzegorzewski P (2001) Fuzzy tests - defuzzification and randomization. Fuzzy Sets Syst 118:437–446
5. Grzegorzewski P (2017) The Kolmogorov goodness-of-fit test for interval-valued data. In: 2017 IEEE international conference on fuzzy system, (FUZZ-IEEE), pp 1–6, https://doi.org/10.1109/FUZZ-IEEE.2017.801557
6. Grzegorzewski P, Szymanowski H (2014) Goodness-of-fit tests for fuzzy data. Inf Sci 288:374–386
7. Hesamian G, Taheri SM (2013) Fuzzy empirical distribution function: properties and application. Kybernetika 49:962–982
8. Kreinovich V, Servin C (2015) How to test hypotheses when exact values are replaced by intervals to protect privacy: Case of t-tests. Departamental Technical Reports (CS), Paper 892, University of Texas at El Paso
9. Nguyen HT, Kreinovich V, Wu B, Xiang G (2012) Computing statistics under interval and fuzzy uncertainty. Springer, Berlin

10. Smirnov NV (1939) Estimate of deviation between empirical distribution functions in two independent samples (in Russian). Bull Moscow Univ 2:3–16
11. Westfall PH (2005) Combining p-values. In: Armitage P, Colton T (eds) Encyclopedia of biostatistics, 2nd edn. Wiley, Chichester

On New Frameworks for Decision Making and Optimization

María Teresa Lamata and José Luis Verdegay

To Pedro, Professor Pedro Gil Álvarez,
a fine example of a scientist, an academic and a scholar,
a very human person and a friend from whom we both learned
and who we miss so much

Abstract Problems of decision and optimization are of an importance beyond all doubt. However, the new life plans that the current Digital Society entails compel us to clarify some aspects of the essential elements of a decision problem, and therefore also of optimization. This paper presents a formulation of a decision problem that takes into account these new parameters and, based on them, proposes new models to optimize decision making.

1 Introduction

Decision Theory has traditionally been associated with the fields of Economics and Statistics and Operational Research. In any of these, the main problem that has been addressed, rather than being how to make a decision, has been to make the best decision, so that implicitly, the main problem underlying a decision problem is an optimization problem, i.e. to select the best alternative to a given situation.

Decision making is in the essence of human persons, that is, with people. We spend all day long doing nothing but make decisions, trying to make them the best possible,

M. T. Lamata · J. L. Verdegay (✉)
Departamento de Ciencias de la Computación e Inteligencia Artificial,
E.T.S. de Ingenierías Informática y de Telecomunicación, Universidad de Granada,
18014 Granada, Spain
e-mail: verdegay@decsai.ugr.es

M. T. Lamata
e-mail: mtl@decsai.ugr.es

© Springer International Publishing AG 2018
E. Gil et al. (eds.), *The Mathematics of the Uncertain*, Studies in Systems,
Decision and Control 142, https://doi.org/10.1007/978-3-319-73848-2_58

in any area of human activity. There is no professional field that escapes necessary and continuous decision-making. Whether it be in the family, legal, medical, or artistic contexts, we have to make decisions. But we also have to make the best decisions, so we have to consider a double aspect: normative and descriptive, or put another way, we have to consider how to decide, the normative approach of the decision process, and what the decision to be taken is, i.e. the descriptive version of the process.

For a long time, progress in Decision Theory has run along parallel paths depending on whether the study was focused from one point of view (normative) or another (descriptive), so it seemed that the "theorists" of Decision Theory worked with their backs to the "practitioners" of the same and, as if they were different worlds, both to the margin of what happened in the economic world. In fact, being identical basic models, and therefore results, what was translated in each case as a solution of the problem could have different versions, and therefore seem different solutions.

The successive social transformations of the last years, where we have passed through the Information Society, the Knowledge Society, the current Digital Society and the immediate Smart Society, has meant the omnipresence of computers, robots, mobile phones and, ultimately, all types of intelligent devices in our daily life, and with this has arisen the need to teach them to decide, and to do so optimally. Therefore, knowing as specifically as possible all the elements that take part in a decision process in detail is as important as having a guarantee that the results of the actions of a Smart System will be as correct as possible.

In this context, Decision Theory and Optimization Methods emerge as essential elements that play a major role in this new Smart Society, in which Artificial Intelligence will change our habits of life.

However, combining decision and optimization in the area of Artificial Intelligence requires knowing as far as possible how to develop a decision-making process; Herbert Simon identified this in three stages [9]:

- Intelligence stage, in which we try to determine how we can decide, that is, to specify as much as possible what data are available, the type of information that we have to manage, or the mechanisms of logical reasoning that can be used by decision makers.
- Design stage, which deals with everything related to modeling the problem so as to be able to clearly define the options, their consequences or the mechanisms of comparison to be used. It is a markedly theoretical stage in which the knowledge of other similar situations is fundamental in order to distinguish similarities and differences.
- Choice stage, which includes the application of the necessary methods to optimize our decision, but considering possible revisions of our choice in the light of what the model proposes us as the first option. The choice, although guided by an optimization process, can be subject to negotiations and modifications, which may be suggested or imposed by external factors before finally being adopted.

As is evident, the overcoming of these three stages supposes carrying them out sequentially and in a circular manner since at the end of the third stage, it is mandatory to again reconsider from the first stage of the model that we have to solve.

Therefore, concepts such as decision maker, consequences, order, context of optimization, etc. arise which, although known, in the new Smart Society may adopt new versions, and therefore may lead to decision mechanisms that deserve to be analyzed in the light of this new situation, so the mechanisms that people employ to make decisions are exportable to Intelligent Systems with the utmost reliability, thus ensuring safety, effectiveness and efficiency.

Thus, the main aim of this paper is to formalize a framework that serves to identify the elements necessary to address a decision-making problem with the subsequent optimization of the interests of the decision maker. To do this, the following section establishes the components that describe a decision problem and introduces some new concepts. The third section deals with how to represent these new concepts and, finally, Sect. 4 deals with the optimization of the decisions to be made depending on the type of information available.

2 Elements of a General Decision Problem

Classically the approach of a decision problem requires knowing the following essential elements [8]:

- One decision-maker, which can be an individual or a group,
- a set of actions on which the decision maker can choose,
- a set, called the Environment, which is constituted by the situations (states) that the decision maker can encounter when choosing, and cannot control,
- a set of consequences, associated with each action and each state,
- a criterion that orders the consequences,
- the nature of the information available,
- the duration of the process, and
- the (social) framework in which the decision-making process takes place.

Thus, if we assume in all that follows: (a) an individual decision maker, that is, we do not consider Group Decision Making (GDM) problems, (b) a single management criterion, thus avoiding the Multi-criteria Decision Making (MCDM) problems, and (c) the duration of the process is limited to one stage, so as not to enter into problems typical from Control Theory, a decision problem is described by a sextet $(X, I, E, C, \leq, K)$ that includes the set X of possible actions for the decision maker; the available information I; the environment E; the set C of the consequences of actions; the criterion $\leq$ that orders the consequences; and the framework K, which is the context in which the decision maker decides.

On two of these elements, the available information and the framework, we have to specify certain aspects that will help us to better model the problem to be solved in each case.

Starting with the information available, although it is traditionally supposed to be of a probabilistic nature because it is uncertain in nature, so that uncertainty about the information to which it is truly related is to "not knowing exactly what would happen if a particular course of action were to be adopted" [5]. But, of course, this Incompleteness may have different characteristics from the probabilistic ones. In this sense, the Smithson Taxonomy [10], according to which the situation of Incompleteness can be due to the problem being established in one of the following two situations, can be particularly useful and revealing:

- there is Uncertainty about the data, understood as the information being gradual and therefore not complete,
- a true Absence of information is given (a case we shall not consider here).

In turn, we can consider three different types of Uncertainty, depending on whether its nature is probabilistic, ambiguous or vague, that is to say, according to:

- chance intervention (probabilistic): the duration of the trip depends on the means of transport that we take,
- there is a finite number of options for each value (ambiguity): the hotels are in the metropolitan area, or
- data are given by value ranges (vagueness): We will travel by day, that is, between 7 a.m. and 5 p.m.

In this work, in all that follows, we will focus on decision problems for which the available information is vague in nature, and therefore can be managed with methodologies and techniques of Fuzzy Sets and Systems [1, 12]. In that case, parallel to when the information is random in nature, we can identify three Environments:

- Certainty Environment: It is characterized because the state of nature is known to be given, but that state is defined by a fuzzy set or a linguistic variable. For instance, the states of nature for the weather that there will be on one particular day, can be: cool, pleasant or hot. In the certainty environment we can know that it will be hot tomorrow, but that does not mean that we know exactly (i.e., with certitude) what the temperature will be.
- Possibility Environment: It is given when a perfectly known distribution of possibility (somewhat parallel to a probability distribution, but more associated with the concept of feasibility than that of randomness and therefore without any axiomatic verification) exists on the states of nature.
- Indeterminacy Environment: It arises when we completely ignore any information about which of the states of nature is presented, that is, we know that there is a distribution of possibility over the states of nature, but we do not know the rest of the information.

Secondly, regarding the Frameworks, as we have mentioned above in defining the Choice Stage described by Simon, the action that the decision maker ultimately chooses as optimal can be conditioned by the Framework, F, in which the problem is developed.

A Framework, regardless of the nature of the information available, is defined as a set of rules, often established in the form of logical predicates, that establish the qualitative characteristics that the decisions we choose for solving our problem must have.

The most known Frameworks are the classics of Ethics and Concurrence [8]. Concretely:

- Ethical Framework, which can appear in decision processes that are developed in very specific and professional contexts, such as legal, military, medical or, in general, any other where the final decision is subject to compliance with a certain "code". In this Framework, it is not only a matter of making decisions that fit certain moral behaviors, but rather that decisions are based on interests that conform to ethical codes. This is the typical case when a particular course of action is decided upon in a legal department because it is the best, but it must be abandoned for reasons of professional ethics. The Ethical Framework is usually defined by a set of "good practices" to which the decision makers must conform.

- Concurrence or Competitiveness Framework, suitable for decision processes in which several decision makers compete to achieve a result that is the best possible for each one at the cost of the damage that their decision may cause to others. This occurs in situations mainly associated with games in which what a player wins is what his opponent loses. It is important to note that the fact that there is more than one decision maker does not mean that it is a GDM problem, since there is no intention to reach a decision for the group, but each one acts on its own. Along with these frames, it is also important to consider what we call.

- Neutral Framework, which is considered when the context of the problem is free of peculiarities that can influence the decisions to be made. In general, this framework, in which the decision maker chooses his courses of action with rational criteria, without external incidents and for variables that take positive real values, will be the one that we assume for the purposes of theoretical study of solutions to the problem.

In addition to these frames, there are others that have emerged which are associated with the development of the Information Society and which deserve a mention. Specifically, they are new frameworks that can condition our decisions the following ones:

- Decision Making in the Presence of Adversaries Framework, which occurs when the decisions we make are known by our adversaries. Decision-making in the presence of adversaries poses the difficulties of a situation that sometimes requires recourse to sub-optimal decisions merely in order to confuse adversaries. Such situations arise clearly in military contexts, but also in areas such as perimeter surveillance, computer game development, intelligent systems design for personnel training, cyber-crime, etc. In general terms, an adversary is an entity whose benefits (in some sense) are inversely proportional to ours. This adversary is able to alter our benefits by taking certain actions and in addition, he can observe our actions/decisions thus having the opportunity to learn our pattern of behavior. This learning will lead you to be more effective in your attempt to maximize your profits and minimize ours. The Decision Making in the Presence of Adversaries Framework is different from the Concurrence Framework, since in the first one we do not know if there is or is not an adversary and, if it indeed exists, we do not know anything about it, whilst in the second it is known that there is an opponent competing against us to diminish our profits.

- Crisis Framework, which occurs when there are exceptional circumstances (e.g. catastrophes, accidents, ...) and in which you have to make the best possible decision among those available, which are usually not all those possible. Sometimes, that best decision among the available ones may coincide with the solution of the problem. However, in most cases this will not be the case, due to various factors such as the lack of resources to explore the entire space of alternatives, the possible disappearance of alternatives, the sudden infeasibility of some others, etc. In these cases, a good solution strategy, inspired by the design of preconditioning algorithms [2], can be to protocol the problem, so that when the emergency arises it is possible to consult a protocol of action that minimizes the risks of a bad performance as far as possible. This can thus increase the possibility of matching the solution to the problem in cases of emergency and the optimal solution of the problem without emergencies.

- Sustainability Framework, associated to what is understood by sustainable decisions in a specific ecosystem. Parallel to what is defined as "sustainable development" [7], in order for a decision to be sustainable, it must meet the expectations of the moment when it is taken, that is to say, be optimal in a certain sense established by the decision maker, and at the same time not compromise the choices that may be made about the problem in the future. It therefore makes perfect sense that we consider the frames to make decisions that, fitting the needs of the problem in question, enable us to solve the same problem once again when it presents itself again, without being conditioned by the previous decisions. This frame, although generally associated with environmental issues, is not limited to that context. The "occasional" acquisition of equipment without a minimal analysis of its sustainability, even if it is the result of a perfectly developed decision-making process, more often than one would desire, produces undesired results which, ultimately, demonstrate that the decision-making has been performed poorly. On the other hand, and although sustainable behavior is always ethically plausible, this frame does not resemble the previous Ethical Framework, since the latter is more oriented to questions of conflicts of interest and moral issues.

- Dynamic Framework, where the conditions that have led to the best decision being made at the moment it was made have changed, and therefore may cause that first optimal decision to no longer be so. A simple example explains these situations in which the best solution can change as the decision-making process develops. We wish to buy a certain complement over the internet. We find a model that satisfies us and that is our best option (it may even be a temporary special offer). We carry out all the procedures requested and we pay the amount due. However, more often than might be expected, we are informed shortly after from the e-supplier that the requested add-on cannot be sent to us because the stock has "run out" (the result of a massive avalanche of buyers who, independently but simultaneously, opted for the complement in question). This frame is typical of Transportation, Management or Investment Problems [11], and increasingly occurs in social networks.

- Corporate Social Responsibility (CSR) Framework, understood as a way of directing companies based on the management of the impacts that their activity generates on their customers, employees, shareholders, local communities, the environment and on society in general [6]. In short, CSR is a concept through which the company voluntarily integrates social and environmental dimensions into its business operations and its relations with stake-holders [4], which is a very current trend and which can - and must - modify the courses of action of decision processes. The frame that defines the CSR is not the same as the Ethical one, although on occasions they may share some similarities, since the latter focuses more on the moral consequences of the decisions. Decision making in a CSR framework, especially in the case of public corporations, is conditioned by what is called "accountability", which in all cases can change the selection of the best action that would be chosen if one did not contemplate this RSC Framework.

Obviously, each of these frames will depend on each specific situation and thus it is not easy to specify much more regarding each. But in all cases, and whether or not we have complete information, these frames shall be defined by rules, by logical predicates, that describe each specific situation. The next section is dedicated to describing these aspects in greater detail.

3 Particular Decision-Making Problems

Whenever we determine the characteristics of the elements involved in the definition of a decision problem, particularly the type of information available and the framework in which it will be developed, an optimization problem arises.

If the available information is represented by I and the framework is K, that optimization problem can be represented by a quartet $(X_K^I, E_K^I, R_K^I, \leq)$, where each element has a clear meaning, and where the consequences are measured by means of a function that gives the reward associated with each alternative, for each state of nature that is considered

$$R_K^I : X_K^I \times E_K^I \to C_K^I,$$

where C_K^I is the set (numeric, linguistic, visual, etc.) in which the consequences of our actions are valued: usually an interval of the real line.

Then the problem we have is to find an alternative $x^* \in X$ such that

$$R_K^I(x^*, e) = \max_K^I \{R_K^I(x, e) : x \in X_K^I, e \in E_K^I; R_K^I : X_K^I \times E_K^I \to C_K^I\}.$$

Particularly, if the available information is incomplete and fuzzy in nature, the set of alternatives is fuzzy, as well as the results, which can be valued by fuzzy numbers, and we have an Uncertainty Environment; in the sense defined above, the problem that remains is to find the alternative x^* such that,

$$R_K^f(x^*) = \max_K^f \{R_K^f(x) : x \in X_K^f; R_K^f : X_K^f \times E_K^f \to C_K^f\}$$

as the true state of the nature is perfectly known.

Thus, a general fuzzy optimization problem is formulated, with clear meaning, in the following terms

$$\begin{aligned}
&\text{Maximize} \quad R(x)^f \\
&\text{Subject to:} \quad g_i(x)^f \leq^f 0, i \in M = \{1, \ldots, m\} \quad\quad (1) \\
&\quad\quad\quad\quad\quad x \in K
\end{aligned}$$

where the (super)symbol "f" refers to the fact that the objective function, the set of constraints or the coefficients taking part in the problem, one by one, in part, or all at the same time are fuzzy in nature, and where $R(x)^f$ and $g_i(x)^f, i \in M = \{1, \ldots, m\}$, are functions that take values in the real line and in the set of fuzzy numbers, and K is a set of rules that define the framework in which the problem must be developed.

However, as mentioned before, frames are defined by sets of rules, which are usually expressed as logical predicates, which define the concrete context to which each one refers.

Consider, for example, the case of a Sustainable Framework. Then we could find rules of the following type,

- R1: IF (CO_2 emitted/unit of fuel consumed $\geq x$) AND (kg of waste generated/Mt of finished product $\leq y$) THEN (Subsidy $\geq z$),
- R2: IF (vehicle is from 1990 or earlier) THEN (Hydrocarbon emission ≤ 150 parts per million) AND (Carbon Monoxide Emission $\leq 1.5\%$),
- R3: IF (vehicle is from 1991 or later) THEN (Hydrocarbon Emission ≤ 100 parts per million) AND (Carbon Monoxide Emission $\leq 1.0\%$),

which, of course, we can express with a clear meaning as

- R1: IF X_1 AND X_2 THEN S,
- R2: IF Y_1 THEN Z_1 AND Z_2,
- R3: IF NOT Y_1 THEN Z_3 AND Z_4.

In short, the concrete framework that the decision process we were considering would be developed in would be given by a knowledge base K that would gather all the rules that defined it.

But these rules do not have to be strictly Boolean, but have much more meaning if we admit them as fuzzy, since it is difficult to imagine that an important decision, that a framework, favors or prevents an action because a certain marker fails to reach a preset value by remaining above it by one thousandth. Therefore, in what follows we will assume that the rules that define each of our frames are of a fuzzy type.

However, as stated above, our interest is in formulating decision problems as optimization problems. In this latter context, the frames should be properly formulated to them, that is, as compatible and coherent constraints and referring to the same terms as the set of possible actions for the problem in question and, in any case,

Fig. 1 Membership function for $x_1 - x_2 \leq^f b$

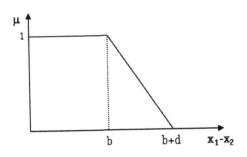

fuzzy formulation. Therefore, and not for the sake of completeness, we shall now describe how these rules can be formulated as linear constraints.

As is well known, the rule "IF X_1 THEN X_2" assumes that if X_1 is 1, that is, if X_1 is true, then X_2 must also be equal to 1, that is to say, it must also be true. Likewise, if X_1 is 0, that is to say, it is false, then X_2 can be 1 or 0, in other words, it can be true or false. This can be expressed as a linear constraint without further writing that rule as

$$X_1 - X_2 \leq 0.$$

Now, generalizing a little further, if X_1 and X_2 are expressed as fuzzy predicates (for example "the hydrocarbon emission is not significantly greater than 150 parts per million" or "the vehicle is old", etc.) a constraint like the previous one can be smoothed to allow violations of the same, that is, the decision maker admits minor breaches of the rule, up to a certain value of intolerance, with his degree of satisfaction on said compliance being inversely proportional to the magnitude of the violation, which is measured by means of a membership function as in Fig. 1 (where the symbol $\leq^f$ represents that the accomplishment of the constraint is fuzzy).

As it is clear, if $b + d \in [0, 1]$, is the value that is taken as the maximum limit to allow violation of the rule (constraint), that rule is represented as

$$X_1 - X_2 \leq^f b$$

which, taking into account the Representation Theorem, if we consider its corresponding α-cuts, $\alpha \in [0, 1]$, can also be represented as

$$X_1 - X_2 \leq b + d(1 - \alpha), \quad \text{for all } \alpha \in [0, 1].$$

Thus, the basic rule $X_1 - X_2 \leq 0$, when it is formulated as a fuzzy rule, is expressed as,

$$X_1 - X_2 \leq^f 0$$

and, therefore, as

$$X_1 - X_2 \leq d(1 - \alpha), \quad \text{for all } \alpha \in [0, 1]$$

with $b + d$ being the value from which the decision maker permits no further violations [3].

Now, from the two last inequations, which is the most elemental representation of a rule as a linear constraint, we can consider its generalization. Thus,

(a) IF X_1 AND X_2 AND ...AND X_n, THEN Z is formulated as

$$X_1 + X_2 + \ldots + X_n - Z \leq^f n - 1$$

or, equivalently,

$$X_1 + X_2 + \ldots + X_n - Z \leq (n - 1) + d(1 - \alpha), \quad \text{for all } \alpha \in [0, 1].$$

(b) IF X_1 OR X_2, THEN Z can be expressed by means of two constraints:

$$X_1 - Z \leq^f 0 \text{ AND } X_2 - Z \leq^f 0$$

or, using its corresponding α-cuts, $\alpha \in [0, 1]$, as

$$X_1 - Z \leq d(1 - \alpha) \text{ AND } X_2 - Z \leq d(1 - \alpha), \quad \text{for all } \alpha \in [0, 1].$$

(c) IF X_1 THEN W OR Z, can be translated to

$$X_1 - (W + Z) \leq^f 0$$

that is to say,

$$X_1 - (W + Z) \leq d(1 - \alpha), \quad \text{for all } \alpha \in [0, 1].$$

(d) In general, a rule such as IF X_1 AND X_2 AND ...AND X_n, THEN Y_1 OR Y_2 OR ...OR Y_m can be expressed by the constraint

$$X_1 + X_2 + \ldots + X_n - (Y_1 + Y_2 + \ldots + Y_m) \leq^f n - 1.$$

Or, by means of its α-cuts, $\alpha \in [0, 1]$, as

$$X_1 + X_2 + \ldots + X_n - (Y_1 + Y_2 + \ldots + Y_m) \leq (n - 1) + d(1 - \alpha), \quad \text{for all } \alpha \in [0, 1].$$

(e) The rule IF X_1 THEN Y_1 AND Y_2 can be expressed by means of two constraints, as $X_1 - Y_1 \leq^f 0$ AND $X_1 - Y_2 \leq^f 0$, which is equivalent to

$$2X_1 - (Y_1 + Y_2) \leq^f 0$$

and, therefore, to

$$2X_1 - (Y_1 + Y_2) \leq d(1 - \alpha), \quad \text{for all } \alpha \in [0, 1].$$

(f) Thus, in general, a rule such as X_1 AND X_2 AND ...AND X_n, THEN Y_1 AND Y_2 AND ...AND Y_m becomes

$$m(X_1 + X_2 + \ldots + X_n) - (Y_1 + Y_2 + \ldots + Y_m) \leq^f m(n - 1)$$

and therefore also becomes,

$$m(X_1 + X_2 + \ldots + X_n) - (Y_1 + Y_2 + \ldots + Y_m) \leq m(n-1) + d(1-\alpha) \text{ for all } \alpha \in [0, 1].$$

(g) IF X_1 OR X_2 THEN Y_1 can be expressed by means of two constraints, $X_1 - Y_1 \leq^f 0$ AND $X_2 - Y_1 \leq^f 0$, that is to say, as

$$X_1 + X_2 - 2Y_1 \leq^f 0$$

or, equivalently,

$$X_1 + X_2 - 2Y_1 \leq d(1 - \alpha), \quad \text{for all } \alpha \in [0, 1].$$

(h) Thus, in general, a rule such as IF X_1 OR X_2 OR ...OR X_n THEN Y_1 can be written as

$$X_1 + X_2 + \ldots + X_n - nY_1 \leq^f 0$$

which, by means of its α-cuts, $\alpha \in [0, 1]$, becomes

$$X_1 + X_2 + \ldots + X_n - nY_1 \leq d(1 - \alpha), \quad \text{for all } \alpha \in [0, 1].$$

Obviously, we could continue to put examples that illustrate the transformation of rules into linear constraints, but in view of the objectives of this paper, the previous cases given are sufficient.

Let us take the previous model (1) and suppose the simple case, which we develop for the sake of illustration, in which (a) both $R(x)^f$ and $g_i(x)^f, i \in M = \{1, \ldots, m\}$ are linear; (b) the coefficients of the objective function are given by fuzzy numbers; and (c) neither the constraints, nor the coefficients that define them, are fuzzy. Then (1) becomes

$$\begin{aligned}
\text{Maximize} \quad & c_1^f x_1 + c_2^f x_2 + \ldots + c_n^f x_n \\
\text{Subject to:} \quad & a_{1i} x_1 + a_{2i} x_2 + \ldots a_{ni} x_n \leq^f b_i, \, i \in M = \{1, \ldots, m\} \\
& x \in K
\end{aligned}$$

where the accomplishment of each fuzzy restriction is measured by a membership function of the form

$$\mu_i(a_i x, b_i) = \begin{cases} 1 & \text{if } a_i x \leq b_i \\ 1 - (a_i x - b_i)/d_i & \text{if } d_i \leq a_i x \leq b_i + d_i \\ 0 & \text{if } a_i x > b_i + d_i \end{cases} .$$

Thus, if for example we assume that our Framework is what we previously called Neutral, then $K = \{x \in \mathbb{R}^n : x \geq 0\}$, and the last problem is nothing but a classical Linear Programming problem with fuzzy costs and fuzzy constraints.

But K could also be defined, for example because the evolution of a market were conical, that is, because K were a cone $x \in K$ ensures that $\lambda x \in K$ for all $\lambda > 0$, and so on for any set of rules that would define each particular Framework that we considered.

4 Conclusions

In this paper we have generalized the concept of Framework in the definition of a decision problem, presenting and introducing some new types of frames. We have also proposed that the nature of the information available on the problem in question is also a basic element of the definition of the decision process that we are analyzing. As a consequence of the frame in which the problem develops and the information available, the choice of the best alternative has been shown as the solution of an optimization problem that has been illustrated for the particular case of the information being fuzzy in nature.

Acknowledgements Research supported by the project TIN2014-55024-P from the Spanish Government as well as by the project TIC-8001from the Andalusian Government (both financed with FEDER funds).

References

1. Bellman RE, Zadeh LA (1970) Decision-making in a fuzzy environment. Man Sci 1:B-141–B-164
2. Brassard G, Bratley P (1988) Algorithmics. Theory and practice. Prentice Hall, Englewood Cliffs
3. Delgado M, Verdegay JL, Vila MA (1990) Optimization models in fuzzy-logic-based decision support systems. In: Kulikowski R, Rudnicki J (eds) Proceedings of 2nd Polish-Spanish Conference. Omnitech Press, Warsaw
4. Libro verde: Fomentar un marco europeo para la responsabilidad social de las empresas (2001) Comisión de la Comunidad Europea, Bruselas, 18.7.2001. http://eur-lex.europa.eu/legal-content/ES/TXT/PDF/?uri=CELEX:52001DC0366&from=ES
5. Lindley DV (1971) Making decisions. Wiley, London
6. Observatorio de Responsabilidad Social Corporativa. http://observatoriorsc.org/

7. Report Brundtland (1987) Our Common Future (20 March 1987). World Commission on Environment and Development, United Nations
8. Ríos S (1976) Análisis de Decisiones. ICE Ediciones, Madrid
9. Simon H (1960) The new science of management decision. Harper & Brothers, New York
10. Smithson M (1989) Ignorance and uncertainty: emerging paradigms. Springer, New York
11. Yankelevich D (2017) Prediciendo un futuro predicho. https://www.linkedin.com/pulse/prediciendo-un-futuro-predicho-daniel-yankelevich?trk=hp-feed-article-title-like
12. Zadeh LA (1965) Fuzzy sets. Inf Control 8:338–353

The Additively Decomposable Fuzzy Shannon Type Inequality Index for Positive Random Fuzzy Numbers

Hortensia López-García and M. Amparo Gil-Martín

Abstract In comparing the extent of inequality for two social situations involving human perceptions, it is often unclear how to accomplish such a comparison. It has been highlighted in the literature that the concept of variable inequality is itself imprecise. In case one deals with intrinsically imprecise-valued variables, this last assertion is even more radical. This paper aims to define a fuzzy-valued inequality index for fuzzy-valued random elements. Relevant properties of the index are also given and a real-life example illustrates its potential applicability.

1 A Story Around the Research Topic

Before introducing the core of the paper, we would like to tell a story of how we came to the research in it. In 1975, on the occasion oh his Ph.D. Thesis [9] that combines Information and Decision Theories, Pedro Gil started researching on the idea of defining an entropy-like measure for the (to some extent paradoxically) so-called "useful uncertainty". Actually, this notion tried to jointly measure the uncertainty associated with random experiments, commonly quantified in terms of Shannon's entropy, with the 'uncertainty' that can be due to the variability of the 'utilities' (or simply values) associated with different experimental outcomes. As a first attempt, a measure was suggested in [8], and properties of this measure were discussed.

In 1977, along the supervision of a Ph.D. Thesis, he suggested to think about measuring separately the two above-mentioned uncertainties. This leads to the measure introduced in [3] (see also [4]), that concerned a so-called measure of uncertainty associated with the utilities.

In the eighties of the last century, Pedro Gil and collaborators considered, on one hand, the application of this measure to quantify income inequality [7]. On the other hand, they have tackled the problem of developing statistics and decision-making for

H. López-García (✉)
Departamento de Estadística, I.O. y D.M., Universidad de Oviedo, Oviedo, Spain
e-mail: hortensia@uniovi.es

M. A. Gil-Martín
Notaría Vicente Sánchez Segura, 06700 Villanueva de la Serena, Badajoz, Spain

© Springer International Publishing AG 2018
E. Gil et al. (eds.), *The Mathematics of the Uncertain*, Studies in Systems, Decision and Control 142, https://doi.org/10.1007/978-3-319-73848-2_59

643

fuzzy-valued experimental data. While this research was in progress, we both met in the University of Valladolid along our BSc in Math. A few years later López-García joined the University of Oviedo and Gil-Martín started working on an investment company.

This book means a nice opportunity for both of us to meet again and paying a tribute to our beloved and admired scientific (López-García) and biological (Gil-Martín) ancestor, Pedro Gil.

2 Introduction and Preliminaries

Imprecise data can be found in many real-life situations. Social Sciences often deal with this kind of data (for instance, the customer valuation about a product, the quality rating of a service, and so on), especially in dealing with human perceptions. Fuzzy numbers are usually considered to be an appropriate model to express this type of data (see [2, 10, 11, 13]).

In the literature on the inequality of real-valued random variables, it is pointed out that the concept of inequality is ill-defined (see, for instance, [1, 12]). This assertion becomes clearer when we deal with fuzzy-valued data.

In this paper the *Shannon type inequality index* in [4, 7], which is given by

$$I_{Sh}(X) = E\left[\log \frac{E(X)}{X}\right]$$

for positive random variables X, is extended to treat fuzzy data, and properties are analyzed, some of them like the additive decomposability not being fulfilled for most of the already published indices [5]. Income indices (when they are invariant by scale) measure the relative dispersion/variability of a positive random variable.

We first recall the preliminaries to formally present the extended index.

Definition 2.1 A (bounded) *fuzzy number* is a function $\widetilde{U} : \mathbb{R} \to [0, 1]$ such that it is upper semi-continuous, quasi-concave, normal and its support is a bounded interval, that is, for any $\alpha \in [0, 1]$ the α-level set defined as

$$\widetilde{U}_\alpha = \begin{cases} \{x \in \mathbb{R} : \widetilde{U}(x) \geq \alpha\} & \text{if } \alpha \in (0, 1] \\ \mathrm{cl}\{x \in \mathbb{R} : \widetilde{U}(x) > 0\} & \text{if } \alpha = 0 \end{cases}$$

with 'cl' denoting the closure of the set, is a nonempty compact interval. The space of (bounded) fuzzy numbers will be denoted by $\mathscr{F}_c^*(\mathbb{R})$. In case $\widetilde{U}_0 \subset (0, \infty)$, $\widetilde{U}$ will be said to be a *positive fuzzy number*; the space of positive fuzzy numbers will be denoted by $\mathscr{F}_c^*(0, \infty)$.

Definition 2.2 (*Puri and Ralescu* [13]) Let $(\Omega, \mathscr{A}, P)$ be a probability space modeling a random experiment. A mapping $\mathscr{X} : \Omega \to \mathscr{F}_c^*(\mathbb{R})$ is said to be an associated *random fuzzy number* (for short RFN) if and only if for all $\alpha \in [0, 1]$

the interval-valued mapping $\mathscr{X}_\alpha$, such that $\mathscr{X}_\alpha(\omega) = \big(\mathscr{X}(\omega)\big)_\alpha$ for all $\omega \in \Omega$, is a compact random interval (i.e., a Borel-measurable mapping w.r.t. the topology induced by Hausdorff metric in the space of the nonempty compact intervals). In case $\mathscr{X}(\Omega) \subset \mathscr{F}_c^*(0, \infty)$, $\mathscr{X}$ is said to be a *positive random fuzzy number* (PRFN).

Equivalently, $\mathscr{X}$ is an RFN if and only if for each $\alpha \in [0, 1]$ the real-valued mappings inf $\mathscr{X}_\alpha$ and sup $\mathscr{X}_\alpha$ are real-valued random variables.

In summarizing the location of a random fuzzy number the best known summary measure is the Aumann-type mean (Puri and Ralescu [13]), which is formalized as follows:

Definition 2.3 Let $\mathscr{X}$ be a random fuzzy number associated with the probability space $(\Omega, \mathscr{A}, P)$. The *Aumann-type mean* of $\mathscr{X}$ is the fuzzy number $\widetilde{E}(\mathscr{X}) \in \mathscr{F}_c^*(\mathbb{R})$, if it exists, such that for each $\alpha \in [0, 1]$

$$\big(\widetilde{E}(\mathscr{X})\big)_\alpha = \big[E(\inf \mathscr{X}_\alpha), E(\sup \mathscr{X}_\alpha)\big]$$

with E denoting the expected value of a real-valued random variable.

Definition 2.4 Given $\widetilde{U}, \widetilde{V} \in \mathscr{F}_c^*(0, \infty)$, the *sum* of $\widetilde{U}$ and $\widetilde{V}$ is the fuzzy number $\widetilde{U} + \widetilde{V} \in \mathscr{F}_c^*(0, \infty)$ such that for each $\alpha \in [0, 1]$

$$(\widetilde{U} + \widetilde{V})_\alpha = \big[\inf \widetilde{U}_\alpha + \inf \widetilde{V}_\alpha, \sup \widetilde{U}_\alpha + \sup \widetilde{V}_\alpha\big],$$

the *product* of $\widetilde{U}$ *by a positive scalar* γ is the fuzzy number $\gamma \cdot \widetilde{U} \in \mathscr{F}_c^*(0, \infty)$ such that for each $\alpha \in [0, 1]$

$$(\gamma \cdot \widetilde{U})_\alpha = \big[\gamma \cdot \inf \widetilde{U}_\alpha, \gamma \cdot \sup \widetilde{U}_\alpha\big],$$

and the *quotient* of $\widetilde{U}$ and $\widetilde{V}$ is the fuzzy number $\widetilde{U} \oslash \widetilde{V} \in \mathscr{F}_c^*(0, \infty)$ such that for each $\alpha \in [0, 1]$

$$(\widetilde{U} \oslash \widetilde{V})_\alpha = \left[\frac{\inf U_\alpha}{\sup \widetilde{V}_\alpha}, \frac{\sup \widetilde{U}_\alpha}{\inf \widetilde{V}_\alpha}\right].$$

3 The Shannon Type Fuzzy-Valued Inequality Index for PRFNs. Definition and Properties

Consider a random experiment formalized by the probability space $(\Omega, \mathscr{A}, P)$ and an associated PRFN $\mathscr{X} : \Omega \to \mathscr{F}_c^*(0, +\infty)$ for which $\widetilde{E}(\mathscr{X})$ exists.

Definition 3.1 The *Shannon type fuzzy-valued inequality index* associated with $\mathscr{X}$ is the fuzzy number, if it exists, given by

$$\widetilde{I}_{Sh}(\mathscr{X}) = \widetilde{E}\big[\log\big(\widetilde{E}(\mathscr{X}) \oslash \mathscr{X}\big)\big],$$

that is, for each $\alpha \in [0, 1]$

$$\left(\widetilde{I}_{Sh}(\mathscr{X})\right)_\alpha = \left[\inf\left(\widetilde{I}_{Sh}(\mathscr{X})\right)_\alpha, \sup\left(\widetilde{I}_{Sh}(\mathscr{X})\right)_\alpha\right],$$

where

$$\inf\left(\widetilde{I}_{Sh}(\mathscr{X})\right)_\alpha = E\left(\log\frac{E(\inf \mathscr{X}_\alpha)}{\sup \mathscr{X}_\alpha}\right),$$

$$\sup\left(\widetilde{I}_{Sh}(\mathscr{X})\right)_\alpha = E\left(\log\frac{E(\sup \mathscr{X}_\alpha)}{\inf \mathscr{X}_\alpha}\right).$$

The following properties, extending valuable ones from the real-valued case are fulfilled by the Shannon type fuzzy-valued inequality index.

Theorem 3.1 ("Average" non negativity) *Let* $\mathscr{X} : \Omega \to \mathscr{F}_c^*(0, \infty)$ *be a PRFN. Then, if* $\widetilde{I}_{Sh}(\mathscr{X})$ *exists, for each* $\alpha \in [0, 1]$ *we have that*

$$\inf\left(\widetilde{I}_{Sh}(\mathscr{X})\right)_\alpha + \sup\left(\widetilde{I}_{Sh}(\mathscr{X})\right)_\alpha \geq 0.$$

Theorem 3.2 (Minimalities) *Let* $\mathscr{X} : \Omega \to \mathscr{F}_c^*(0, \infty)$ *be a PRFN for which* $\widetilde{I}_{Sh}(\mathscr{X})$ *exists.* $\widetilde{I}_{Sh}(\mathscr{X})$ *is 'additively equivalent' to* $\mathbb{1}_{\{0\}}$ *(i.e., it is a fuzzy number which is symmetric w.r.t. 0) if, and only if,* $\mathscr{X}$ *is a degenerate RFN. Furthermore,* $\widetilde{I}_{Sh}(\mathscr{X}) = \mathbb{1}_{\{0\}}$ *if, and only if,* $\mathscr{X}$ *is an RFN which is degenerate at a positive real value.*

Theorem 3.3 (Mean independence) *Let* $\mathscr{X} : \Omega \to \mathscr{F}_c^*(0, \infty)$ *be a PRFN for which* $\widetilde{I}_{Sh}(\mathscr{X})$ *exists. Then, for all* $k \in (0, +\infty)$ *we have that*

$$\widetilde{I}_{Sh}(k \cdot \mathscr{X}) = \widetilde{I}_{Sh}(\mathscr{X}).$$

Theorem 3.4 (Population homogeneity) *Let* $\mathscr{X} : \Omega \to \mathscr{F}_c^*(0, \infty)$ *be a PRFN, and consider a population/sample of n individuals for which* $\mathscr{X}$ *takes on the fuzzy data* $\widetilde{x}_1, \ldots, \widetilde{x}_n \in \mathscr{F}_c^*(0, +\infty)$. *Consider the population/sample of* $n \times r$ *individuals in which each individual in the original population/sample is replicated r times, and let* $\mathscr{X}^{*r}$ *denote the PRFN corresponding to the r-replication of* $\mathscr{X}$ *(i.e., each of the original fuzzy observations* $\widetilde{x}_i$ *arises r times in the new population/sample). Then,*

$$\widetilde{I}_{Sh}(\mathscr{X}^{*r}) = \widetilde{I}_{Sh}(\mathscr{X}).$$

Along the next three properties an ordering should be considered on $\mathscr{F}_c^*(0, \infty)$ to extend the principles of transfers. Since there is no total ordering on $\mathscr{F}_c^*(0, \infty)$ which can be considered as universally accepted we will make use of a partial ordering, like the *strong dominance*. In accordance with such a partial ordering, given two positive fuzzy numbers $\widetilde{U}, \widetilde{V} \in \mathscr{F}_c^*(0, \infty)$, $\widetilde{U}$ is said to strongly dominate $\widetilde{V}$ if for each $\alpha \in [0, 1]$ one has that $\inf \widetilde{U}_\alpha \geq \inf \widetilde{V}_\alpha$ and $\sup \widetilde{U}_\alpha \geq \sup \widetilde{V}_\alpha$.

Theorem 3.5 (Progressive principle of transfers under "strong dominance") *Let* $\mathscr{X} : \Omega \to \mathscr{F}_c^*(0, \infty)$ *be a PRFN, and consider a population/sample of n individuals for which* $\mathscr{X}$ *takes on the fuzzy data* $\widetilde{x}_1, \ldots, \widetilde{x}_n \in \mathscr{F}_c^*(0, +\infty)$. *Assume that there exist* $h, l \in \{1, 2, \ldots, N\}$, *such that* $\widetilde{x}_h$ *strongly dominates* $\widetilde{x}_l$. *Let* $\varepsilon \in [0, +\infty)$ *such that for all* $\alpha \in [0, 1]$

$$(\inf \widetilde{x}_h)_\alpha \geq (\inf \widetilde{x}_h)_\alpha - \varepsilon \geq (\inf \widetilde{x}_l)_\alpha + \varepsilon \geq (\inf \widetilde{x}_l)_\alpha,$$

$$(\sup \widetilde{x}_h)_\alpha \geq (\sup \widetilde{x}_h)_\alpha - \varepsilon \geq (\sup \widetilde{x}_l)_\alpha + \varepsilon \geq (\sup \widetilde{x}_l)_\alpha.$$

If $\mathscr{Y}$ *is the PRFN defined from* $\mathscr{X}$ *so that for each* $\alpha \in [0, 1]$

$$\inf(\widetilde{y}_h)_\alpha = \inf(\widetilde{x}_h)_\alpha - \varepsilon, \ \inf(\widetilde{y}_l)_\alpha = \inf(\widetilde{x}_l)_\alpha + \varepsilon,$$

$$\sup(\widetilde{y}_h)_\alpha = \sup(\widetilde{x}_h)_\alpha - \varepsilon, \ \sup(\widetilde{y}_l)_\alpha = \sup(\widetilde{x}_l)_\alpha + \varepsilon,$$

and

$$\widetilde{y}_j = \widetilde{x}_j \ \text{for all} \ j \in \{1, 2, \ldots, N\} \setminus \{h, l\},$$

then $\widetilde{I}_{Sh}(\mathscr{X})$ *strongly dominates* $\widetilde{I}_{Sh}(\mathscr{Y})$ *(i.e., if data are 'approached' the inequality decreases).*

Theorem 3.6 (Regressive principle of transfers under "strong dominance") *Let* $\mathscr{X} : \Omega \to \mathscr{F}_c^*(0, \infty)$ *be a PRFN, and consider a population/sample of n individuals for which* $\mathscr{X}$ *takes on the fuzzy data* $\widetilde{x}_1, \ldots, \widetilde{x}_n \in \mathscr{F}_c^*(0, +\infty)$. *Assume that there exist* $h, l \in \{1, 2, \ldots, N\}$, *such that* $\widetilde{x}_h$ *strongly dominates* $\widetilde{x}_l$. *Let* $\varepsilon \in [0, +\infty)$ *such that* $(\inf \widetilde{x}_l)_0 - \varepsilon > 0$.

If $\mathscr{Y}$ *is the PRFN defined from* $\mathscr{X}$ *so that for each* $\alpha \in [0, 1]$

$$\inf(\widetilde{y}_h)_\alpha = \inf(\widetilde{x}_h)_\alpha + \varepsilon, \ \inf(\widetilde{y}_l)_\alpha = \inf(\widetilde{x}_l)_\alpha - \varepsilon,$$

$$\sup(\widetilde{y}_h)_\alpha = \sup(\widetilde{x}_h)_\alpha + \varepsilon, \ \sup(\widetilde{y}_l)_\alpha = \sup(\widetilde{x}_l)_\alpha - \varepsilon,$$

and

$$\widetilde{y}_j = \widetilde{x}_j \ \text{for all} \ j \in \{1, 2, \ldots, N\} \setminus \{h, l\},$$

then $\widetilde{I}_{Sh}(\mathscr{Y})$ *strongly dominates* $\widetilde{I}_{Sh}(\mathscr{X})$ *(i.e., if data are 'moved away' the inequality increases).*

Theorem 3.7 (Schur-convexity under "strong dominance") *Let* $\mathscr{X} : \Omega \to \mathscr{F}_c^*(0, \infty)$ *be a PRFN, and consider a population/sample of n individuals for which* $\mathscr{X}$ *takes on the fuzzy data* $\widetilde{x}_1, \ldots, \widetilde{x}_n \in \mathscr{F}_c^*(0, +\infty)$.

Let (μ_{jl}) *be a doubly stochastic* $n \times n$ *matrix. If* $\mathscr{Y}$ *is the PRFN such that it takes on data* $\widetilde{y}_1, \ldots, \widetilde{y}_n \in \mathscr{F}_c^*(0, +\infty)$ *such that*

$$\begin{pmatrix} \widetilde{y}_1 \\ \vdots \\ \widetilde{y}_n \end{pmatrix} = \begin{pmatrix} \mu_{11} \cdots \mu_{1n} \\ \vdots \ddots \vdots \\ \mu_{n1} \cdots \mu_{nn} \end{pmatrix} \odot \begin{pmatrix} \widetilde{x}_1 \\ \vdots \\ \widetilde{x}_n \end{pmatrix},$$

then, $\widetilde{I}_{Sh}(\mathscr{X})$ strongly dominates $\widetilde{I}_{Sh}(\mathscr{Y})$ (i.e., if data are 'approached', now through a double process of convex linear combination, the inequality decreases).

Theorem 3.8 (Anonimity) Let $\mathscr{X} : \Omega \to \mathscr{F}_c^*(0, \infty)$ be a PRFN, and consider a population/sample of n individuals for which $\mathscr{X}$ takes on the fuzzy data $\widetilde{x}_1, \ldots, \widetilde{x}_n \in \mathscr{F}_c^*(0, +\infty)$. Let $\mathscr{X}_\sigma$ be a PRFN taking on the same population/sample data $\widetilde{x}_{\sigma(1)}, \ldots, \widetilde{x}_{\sigma(n)} \in \mathscr{F}_c^*(0, +\infty)$, where σ is a permutation of $\{1, \ldots, n\}$. Then,

$$\widetilde{I}_{Sh}(\mathscr{X}_\sigma) = \widetilde{I}_{Sh}(\mathscr{X})$$

(i.e., the inequality index does not depend on the order according with which individuals are chosen).

To verify whether the index is continuous, and since it is a fuzzy-valued measure, a distance on the space of fuzzy numbers should be considered. For this purpose, we are going to make use of the supremum metric, since it is an upper bound for all the most usual metrics on $\mathscr{F}_c^*(\mathbb{R})$. This metric corresponds in the one-dimensional fuzzy-valued case to the one associating with $\widetilde{U}, \widetilde{V} \in \mathscr{F}_c^*(\mathbb{R})$ the value

$$d_\infty(\widetilde{U}, \widetilde{V}) = \sup_{\alpha \in [0,1]} \max\{|\inf \widetilde{U}_\alpha - \inf \widetilde{V}_\alpha|, |\sup \widetilde{U}_\alpha - \sup \widetilde{V}_\alpha|\}.$$

Theorem 3.9 (d_∞-continuity) Let $\mathscr{X} : \Omega \to \mathscr{F}_c^*(0, \infty)$ be a PRFN, and consider a population/sample of n individuals for which $\mathscr{X}$ takes on the fuzzy data $\widetilde{x}_1, \ldots, \widetilde{x}_n \in \mathscr{F}_c^*(0, +\infty)$. Let $\varepsilon \in \mathbb{R}$ such that $\widetilde{x}_h + \varepsilon \in \mathscr{F}_c^*(0, +\infty)$ for some $h \in \{1, 2, \ldots, n\}$, and assume that $\mathscr{X}'$ is a PRFN defined on the same population/sample so that it takes on fuzzy data $\widetilde{x}'_1, \ldots, \widetilde{x}'_n \in \mathscr{F}_c^*(0, +\infty)$ with $\widetilde{x}'_h = \widetilde{x}_h + \varepsilon$ and $\widetilde{x}'_j = \widetilde{x}_j$ for all $j \in \{1, 2, \ldots, N\} \setminus \{h\}$. Then,

$$\lim_{\varepsilon \to 0} d_\infty\left(\widetilde{I}_{Sh}(\mathscr{X}'), \widetilde{I}_{Sh}(\mathscr{X})\right) = 0.$$

Finally, the following result states that Shannon type fuzzy-valued income index satisfies a convenient property, in accordance with which it is additively decomposable but for an additive equivalence $\sim_\oplus$ (i.e., but for a translation consisting of adding a symmetric w.r.t. 0 fuzzy number). More concretely,

Theorem 3.10 (Additive decomposability) Let $\mathscr{X} : \Omega \to \mathscr{F}_c^*(0, \infty)$ be a PRFN, and consider a population/sample of n individuals for which $\mathscr{X}$ takes on the fuzzy data $\widetilde{x}_1, \ldots, \widetilde{x}_n \in \mathscr{F}_c^*(0, +\infty)$. Assume that the population/sample is classified according to a partition $\mathscr{P}$ grouping individuals in the population/sample into M groups of $n_1, \ldots, n_M$ individuals (with $n = n_1 + \ldots + n_M$). If for individuals in the m-th group $\mathscr{X}$ takes on data $\widetilde{x}_{m1}, \ldots, \widetilde{x}_{mn_m} \in \mathscr{F}_c^*(0, +\infty)$, for $m = 1, \ldots, M$, then

$$\widetilde{I}_{Sh}(\mathscr{X}) \sim_{\oplus} \widetilde{I}_{Sh}^{bg}(\mathscr{X}; \mathscr{P}) + \widetilde{E}\left(\widetilde{I}_{Sh}^{wg}(\mathscr{X}; \mathscr{P})\right),$$

where $\widetilde{I}_{Sh}^{bg}(\mathscr{X}; \mathscr{P})$ *can be viewed as a kind of between groups inequality (in fact, it corresponds to the Shannon type fuzzy-valued inequality of a PRFN defined so that it taken on values* $\widetilde{E}_m(\mathscr{X}) = (1/n_m) \cdot [\widetilde{x}_{m1} + \ldots + \widetilde{x}_{mn_m}]$ *with probabilities* n_m/n *for* $m = 1, \ldots, M$, *respectively), and* $\widetilde{E}\left(\widetilde{I}_{Sh}^{wg}(\mathscr{X}; \mathscr{P})\right)$ *is the Aumann type mean over the m groups of the Shannon type inequality within each group, which is given by*

$$\widetilde{E}\left(\widetilde{I}_{Sh}^{wg}(\mathscr{X}; \mathscr{P})\right) = \frac{n_1}{n} \cdot \widetilde{I}_{Sh}(\widetilde{x}_{11}, \ldots, \widetilde{x}_{1n_1}) + \ldots + \frac{n_M}{n} \cdot \widetilde{I}_{Sh}(\widetilde{x}_{M1}, \ldots, \widetilde{x}_{Mn_M}).$$

The additive equivalence $\sim_{\oplus}$ *reduces to equality of fuzzy numbers if, and only if,* $\mathscr{X}$ *is a positive real-valued random variable. Furthermore,*

$$\widetilde{I}_{Sh}(\mathscr{X}) \sim_{\oplus} \widetilde{I}_{Sh}^{bg}(\mathscr{X}; \mathscr{P})$$

if, and only if, in each of the M groups $\mathscr{X}$ *is a degenerate PRFN.*

4 Illustrative Example

The following real-life example illustrates the computation of the Shannon type fuzzy-valued inequality index.

Example 4.1 By using an online (computerized) application, an experiment has been conducted online in which people have been asked for "their perception of the relative length of different line segments with respect to a pattern longer one". This perception can be formalized as a PRFN. People contacted for this purpose have participated by providing with their perception of relative length for each of several line segments.

More concretely, on the center top of the screen the longest line segment has been drawn in black. This segment is fixed for all the trials, so that there is always the same reference for the maximum length. At each trial, a grey shorter line segment is generated and placed below the longest one, parallel and without considering a concrete location.

For each respondent, line segments are generated at random although, to avoid that the variation in the perception of different respondents can mainly be due to the variation in length of different generated segments, the (27 first) trials for each of the respondents concern the same segments that usually appear in different positions.

Each of the perceptions have been expressed by using the (trapezoidally shaped) fuzzy rating scale by Hesketh et al. [10] (see also [2, 11]), that is, a fuzzy number-valued free assessment with reference interval [0, 100] (see Fig. 1 for a screenshot).

Several respondents have been contacted, all of them with a university scientific background and only needing a minor training mostly consisiting of reading the instructions added to the online appilication.

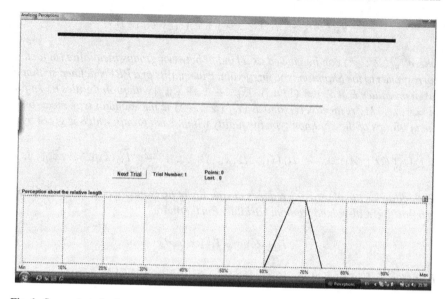

Fig. 1 Screenshot of a fuzzy rating scale-based response from the online application

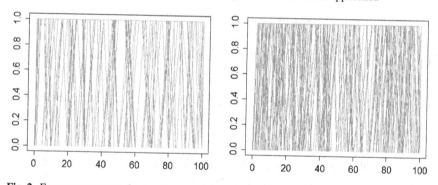

Fig. 2 Fuzzy responses to the question posed from women (on the left) and men (on the right)

The fuzzy responses have been displayed in Fig. 2 in which the 125 men responses and the 71 ones from women have been separately represented.

By examining the Shannon type inequality per sex, and taking into account that if $\mathscr{X} = \mathrm{Tra}(a_{\mathscr{X}}, b_{\mathscr{X}}, c_{\mathscr{X}}, d_{\mathscr{X}})$, then for all $\alpha \in [0, 1]$

$$\inf\left(\widetilde{I}_{Sh}(\mathscr{X})\right)_{\alpha} = \log\left[E(a_{\mathscr{X}}) + \alpha(E(b_{\mathscr{X}}) - E(a_{\mathscr{X}}))\right] - E\left[\log(d_{\mathscr{X}} + \alpha(c_{\mathscr{X}} - d_{\mathscr{X}}))\right],$$

we obtain the fuzzy values in Fig. 3.

Fig. 3 Shannon type fuzzy-valued inequalities for men responses (thick line) and women responses (dotted line)

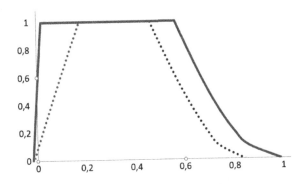

The fuzzy-valued inequalities in this example cannot be immediately compared. Actually, the inequality of responses for women is narrower (more precise to some extent) than the one for men, but a ranking comparison of the relative variability of responses through the Shannon type fuzzy-valued index is also imprecise.

References

1. Basu K (1987) Axioms for a fuzzy measure of inequality. Math Soc Sci 14(3):275–288
2. De la Rosa de Sáa S, Gil MA, González-Rodríguez G, López MT, Lubiano MA (2015) Fuzzy rating scale-based questionnaires and their statistical analysis. IEEE Trans Fuzzy Syst 23:111–126
3. Gil MA (1979) Incertidumbre y Utilidad. PhD thesis, Universidad de Oviedo
4. Gil MA (1981) Estudio de una medida para la incertidumbre correspondiente a las utilidades. Trab Est Inv Op 32:45–66
5. Gil MA, López-Díaz M, López-García H (1998) The fuzzy hyperbolic inequality index associated with fuzzy random variables. Eur J Oper Res 110:377–391
6. Gil MA, López-Díaz M, López-García H, Lubiano MA (1997) Extension of an inequality index for fuzzy-valued attributes. In: Proc 6th IEEE Int Conf Fuzzy Syst, FUZZ-IEEE'1997, Vol. 1. IEEE, Piscataway
7. Gil MA, Pérez R, Gil P (1989) A family of measures of uncertainty involving utilities: definition, properties, applications and statistical inferences. Metrika 36:129–147
8. Gil P (1975) Información útil suministrada por una variable aleatoria sobre un campo de probabilidad y utilidad. Trab Est Inv Op 26:187–204
9. Gil P (1975) Medidas de incertidumbre e información en problemas de decisión estadística. RACSAM Rev R Acad LXIX(1):549–610
10. Hesketh T, Pryor R, Hesketh B (1988) An application of a computerized fuzzy graphic rating scale to the psychological measurement of individual differences. Int J Man-Mach Stud 29:21–35
11. Lubiano MA, Montenegro M, Sinova B, de la Rosa de Sáa S, Gil MA (2016) Hypothesis testing for means in connection with fuzzy rating scale-based data: algorithms and applications. Eur J Oper Res 251:918–929
12. Ok EA (1996) Fuzzy measurement of income inequality: some possibility results on the fuzzification of the Lorenz ordering. Econ Theory 7:513–530
13. Puri ML, Ralescu DA (1986) Fuzzy random variables. J Math Anal Appl 114:409–422

An Incipient Fuzzy Logic-Based Analysis of the Medical Specialty Influence on the Perception About Mental Patients

María Asunción Lubiano, Pilar González-Gil, Helena Sánchez-Pastor, Carmen Pradas and Henar Arnillas

Abstract Analyses of the stigma associated with mental patients have been exhaustively developed. Some of these analyses refer to the general population in different countries, some others compare conclusions from these countries, and some others discuss the attitutes of either current or future (psychiatric and non-psychiatric) health professionals with respect to mental illness. Most of the studies are based on well-known questionnaires (usually on their country-adapted versions), each of them corresponding to a multi-item scale evaluated using either a 5-, 6- or 7-point Likert scale and focussing on different attitudinal factors or constructs. This paper introduces a quite preliminary study in this setting, aiming to examine the influence of the medical specialty on the perception about mental patients and involving a more expressive and flexible scale to rate attitudes: the fuzzy rating scale (allowing a free fuzzy set-valued response assessment to items).

The representatives of the Three Wise Men from Agones brought illusion to my home every Christmas. Over the years, oneself perceives and understands that the legacy of their Majesties cannot be compared even with gold, incense or myrrh. The representative embodied the humanity and captaincy of genius, the contagiousness of dreams, and Gil's excellence. Thank you tío Pedro! (Pilar).

M. A. Lubiano (✉)
Departamento de Estadística, I.O. y D.M., Universidad de Oviedo, Oviedo, Spain
e-mail: lubiano@uniovi.es

P. González-Gil · H. Sánchez-Pastor · C. Pradas · H. Arnillas
Unidad de Psiquiatría, Hospital Son Llàtzer, Palma de Mallorca, Spain
e-mail: pilar.gonzalez@hsll.es

H. Sánchez-Pastor
e-mail: helena.sanchez-pastor@hsll.es

C. Pradas
e-mail: mcpradas@hsll.es

H. Arnillas
e-mail: marnilla@hsll.es

© Springer International Publishing AG 2018
E. Gil et al. (eds.), *The Mathematics of the Uncertain*, Studies in Systems, Decision and Control 142, https://doi.org/10.1007/978-3-319-73848-2_60

653

1 Introduction and Background

The stigmatization associated with mental illness is a topic receiving an increasing attention along the last decades. People with mental disorders are often facing prejudices and discrimination, that could be removed to a great extent if nonexperts become more sensitive and they get achieve a deeper knowledge and health care about this illness.

To evaluate this stigma many studies have been developed. Some of them have been focussed in rating attitudes towards mental illness at the social-level on general populations/communities by simply analyzing the effect of sex, age, country, and so on (see, for instance, [10, 20, 21, 28, 30, 38]). Other studies have concerned attitudes of medical/health care students ('tomorrow's' doctors) and non-psychiatric physicians (see, for instance, [1–3, 8, 9, 12, 13, 15, 16, 21, 22, 24, 34–36, 41, 43, 47]) and some few are devoted to compare attitudes of psychiatrists, psychiatric nurses and relatives of mental patients in contrast to the general population (see, for instance, [40]). Also some others, like [39], deal with a comparative analysis of beliefs and attitudes among different countries.

The most common instruments to evaluate stigmatization and discrimination of mental illness in either general populations or health care professionals are questionnaires. The *Opinions about Mental Illness Scale* (*OMI*) has shown satisfactory psychometric properties and a long history of usage in different populations. It was originally developed by Cohen and Struening [6] (see, among others, [23, 32, 44] for some comments about).

Most of items in this questionnaire are based on a 6-point Likert scale ranging from STRONGLY DISAGREE to STRONGLY AGREE. The items have been conceived to evaluate five main dimensions, namely, *interpersonal etiology* (the belief that mental illness is due to problematic interpersonal relations and experiences), *authoritarianism* (the belief that obedience to authority is critical and mentally ill persons require coercive handling), *social restrictiveness* (the idea that mental patients should be restricted in some social domains such as voting, jobs, parenting, etc.), *negativism* (that can be viewed as opposite to the so-called mental hygiene ideology, this one supporting the idea that mental illness is an illness like any other, it should be treated by specialists and most of mental patients are not dangerous), and *prejudice* (that can be indented to be contrary to the so-called benevolence, the belief that they are not different from others).

This scale (as well as others like the well-known Community Attitudes Toward the Mentally Ill scale, CAMI, by Taylor and Dear [42], a briefer revised updated version of OMI which additionally involves the community mental health ideology) has been translated, adapted and validated in various languages (see, for instance, [32, 33, 46, 48–50]). In particular, the translation, adaptation and validation of the OMI scale to Spanish has been carried out by Yllá [48] and Ozámiz [33] by including a few new items, and leading to the so-denoted OMI-R.

This paper aims to perform a comparison of attitudes towards mental illness of three groups of nonpsychiatric physicians (more concretely, primary care doctors, neurologists and internists). Although a few attempts to such a comparative analysis can be found in the literature (see [4, 16, 31, 37]), and the need for eliminating the stigmatization and discrimination within the medical profession has been often claimed (see [29]), the topic has not yet received a deep research attention.

The comparison is to be based on an innovative tool in this setting, namely, the psychometric *fuzzy rating scale (FRS)* by Hesketh et al. [17–19]. This tool can be immediately applied to deal with classical questionnaires like OMI (or OMI-R) and it simply affects the way responses to items are given. Thus, instead of choosing a point in a Likert scale, the fuzzy rating scale allows respondent to draw a fuzzy value with a total freedom, whence the variability, diversity, subjectivity and the intrinsic imprecision corresponding to attitudes can be much better captured and expressed.

FRS-based data can be statistically analyzed by using some already developed methods (see [5, 7, 11, 26, 27]) and their implementation in R through the statistical packages for fuzzy data SAFD [45] and FuzzyStatTra [25].

2 Methods

To develop the comparative analysis a 14-item excerpt from the OMI-R questionnaire has been considered. The excerpt has been conducted on a sample of 22 physicians from the Hospital Son Llàtzer in Palma de Mallorca. The composition of the sample has been as follows: 7 primary care doctors, 4 neurologists and 11 internists. The excerpt, the fuzzy scale and the method to get statistical conclusions are to be briefly commented in this section.

2.1 Excerpt from the OMI-R Questionnaire

The (sub)questionnaire includes 14 items selected from the OMI-R. These 14 items have been viewed as very informative for the considered target, and they are the following:

I.1. Mental illness is an illness like any other
I.2. Most of the patients who are hospitalized either in the psychiatric units of general hospitals or in mental hospitals are not dangerous
I.3. Mental illness is a way to react to social demands and pressures
I.4. People who are mentally ill let their emotions control them; normal people think things out
I.5. People who are mentally ill are so worried by their own problems that do not care about what others can think about them
I.6. A heart patient has just one thing wrong with him, while a mentally ill person is completely different from other patients

I.7. People with mental illness should never be treated in the same hospital as people with physical illness

I.8. People who have been patients in a mental hospital will never be their own selves again

I.9. Mental illness is usually caused by some disease of the nervous system

I.10. Regardless of how you look at it, patients with severe mental illness are no longer really human

I.11. One of the main causes of mental illness is a lack of moral strength or will power

I.12. Most of us feel a bit uncomfortable or restless in the presence of mentally ill people

I.13. People who suicide are always mentally ill

I.14. What do you think about the convenience of the fact that along the last decades psychiatric units are being opened in general hospitals?

Items *I*.3, *I*.4, *I*.5 and *I*.6 relate to interpersonal etiology, *I*.9 concerns authoritarianism, *I*.2 affects social restrictiveness, *I*.8, *I*.10, *I*.13 and *I*.14 refer to negativism, and *I*.12 regards prejudice.

2.2 The Fuzzy Rating Scale and the Adapted Form

Respondents to this type of OMI-based questionnaires are asked to rate their level of agreement with each of the statements in the items. The level of agreement is usually assessed by considering the 6-point Likert scale consisting of STRONGLY DISAGREE, DISAGREE, SOMEWHAT DISAGREE, SOMEWHAT AGREE, AGREE and STRONGLY AGREE.

Since the number of possible 'values' to choose among is small, variability, adjustment, diversity, subjectivity of the natural level of agreement is lost. Moreover, the choice of the 'value' that best represents respondent level of agreement is not easy.

To avoid such a restrictive scale, Hesketh et al. [17–19] have suggested to consider a *fuzzy rating scale* (*FRS*) allowing respondents to draw the fuzzy number that best represents their score. In case the rating concerns the level of agreement with a given statement, the FRS-based level is to be stated as follows: firstly, a reference bounded interval (for instance [0, 100]) is considered, with 0 = STRONGLY DISAGREE and 100 = STRONGLY AGREE; the interval of real numbers which are considered to be 'fully compatible' with the level of agreement of the respondent is drawn with height 1 (this corresponds to the so-called core of the fuzzy number); the interval of real numbers which are considered to be 'compatible to some extent' with the level of agreement of the respondent is drawn with height 0 (this corresponds to the so-called support of the fuzzy number); these two intervals are linked to get a trapezium (see Fig. 1).

A FRS can cope to a full extent with the intrinsic imprecision associated with the level of agreement with a statement, it means a double continuum (w.r.t. both location and imprecision), its flexibility allows raters to properly capture individual

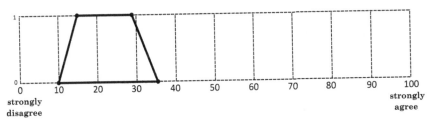

Fig. 1 Example of a FRS-based level of agreement with the statement of a given item

differences, whence the intrinsic variability, diversity and subjectivity are not lost, and it is much richer and more expressive than any one based on a (unavoidably finite) natural language (or its numerical encoding).

2.3 The Statistical Methodology

FRS-based responses can be mathematically and computationally handled in a suitable way, since one can state arithmetic and distances preserving their meaning and allowing us to extend/adapt/develop many concepts, results and procedures from the real-valued data analysis (see, for instance, [5, 7, 11, 26, 27] for more details).

To analyze Likert-type data a posterior numerical encoding of Likert's 'values' is usually considered. This makes all differences between consecutive 'values' to coincide, which is often inappropriate, and the transition from a value to another within the scale is rather abrupt. Moreover, only a few statistical techniques are rigorously applicable (they being mainly based on the frequencies of different values or their position in accordance with a certain ranking) and, as a consequence, relevant statistical information is often lost.

In this paper, to compare the influence of the medical specialty on the attitude towards mental illness an ANOVA test for fuzzy data introduced in [14] and implemented in [45] is to be applied. It should be pointed out that the mean value of a trapezoidal fuzzy dataset is a trapezoidal fuzzy number in which each of the four characterizing vertices is given by the mean of the real-valued dataset corresponding to these vertices.

3   Results and Discussion

Figure 2 displays the FRS-based datasets and means for each of the considered 14 items.

On the basis of the outputs in Fig. 2, one can empirically conclude that non-psychiatric physicians in the considered sample have shown a unequivocally high

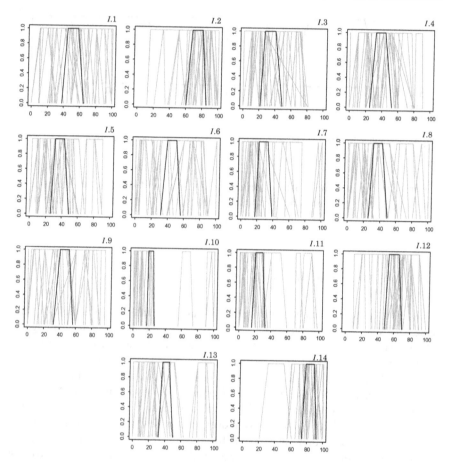

Fig. 2 Datasets (in gray) and fuzzy-valued sample means (in black) of the FRS-based responses to the considered 14 items

average agreement with the assertion in Item I.14 and a rather low average agreement with the assertion in Item I.7 (which seems quite coherent and indicate these physicians are quite in favour of avoiding discriminate/separate mental patients by hospitalizing them in non-general hospitals). The physicians have also shown on the average a quite low level of agreement with the statements in Items I.10 and I.11. Actually, the average behavior cane be associated with a rather high sensitivity w.r.t. mental illness.

In connection with the analysis of the influence of the medical specialty on the attitude towards the mental illness, Fig. 3 displays the three FRS-based means for each of the considered 14 items.

At the first glance, we do not expect differences in attitude to be significant. Actually, p-values of the ANOVA test for FRS-based data have been collected in Table 1.

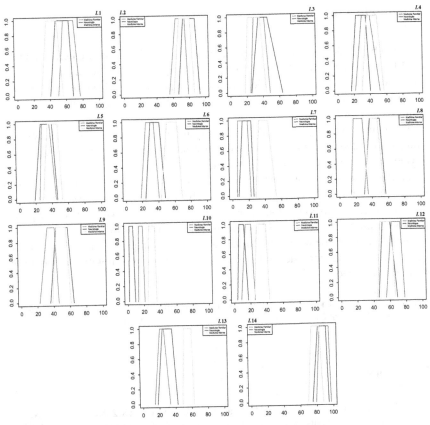

Fig. 3 Fuzzy-valued sample means for the three specialties of the FRS-based responses to the considered 14 items

Table 1 ANOVA p-values for the influence of the medical specialty on the attitude towards mental illness (with FRS-based responses)

Item	$I.1$	$I.2$	$I.3$	$I.4$	$I.5$	$I.6$	$I.7$	$I.8$	$I.9$	$I.10$	$I.11$	$I.12$	$I.13$	$I.14$
p-value	0.331	0.501	0.590	0.427	0.656	0.162	**0.052**	0.324	0.596	0.299	0.106	0.490	**0.044**	0.261

For most of the items, but $I.7$ and $I.13$, these p-values are greater than 0.1, so that we cannot consider the medical specialty as influential for these items.

However, if we look at p-values of the ANOVA test for FRS-based data concerning items $I.7$ and $I.13$, we get them to be either close to or lower than 0.05, whence we can consider the medical specialty affects the responses to these items.

In fact, for both items the lowest average level of agreement is the one associated with primary care doctors, whereas the highest (which is much greater than the other two) is the one associated with internists.

4 Conclusions

It should be pointed out that, because of the novelty of the considered scale, and the need for a certain training before filling questionnaire forms, the sample has been small, so conclusions are not as clear as one can expect for larger samples. In fact, the claimed advantages of the use of a (continuous) FRS are not as visible when small samples are considered.

In addition to having a larger sample, it would be also interesting to analyze the influence of other factors, like medical expertise or age, sex, as well as comparing psychiatricians' opinions with non-psychiatric physicians' ones.

Furthermore, since the same items can be responded simultaneously with both scales, a Likert- and a FRS-type one, it would be valuable to compare statistical conclusions (e.g., p-values) for both scales. Thus, in accordance with previous analyses with other problems, [11, 26, 27], conclusions sometimes differ.

Finally, it would be worthy to validate the 14-item subquestionnaire, as well as to quantify its internal consistency for the FRS, and to compare it with the one for the 6-point Likert one.

Acknowledgements The research in this paper has been partially supported by the Principality of Asturias/FEDER Grant GRUPIN14-101. We acknowledge this support.

References

1. Al-Awadhi A, Atawneh F, Alalyan MY, Shahid AA, Al-Alkhadhari S, Zahid MA (2017) Nurses' attitude towards patients with mental illness in a general hospital in Kuwait. Saudi J Med Med Sci 5:31–37
2. Arrillaga-Arizaga M, Sarasqueta-Eizaguirre C, Ruiz-Feliu M, Sánchez-Etxeberria A (2004) Actitudes del personal sanitario de atención primaria hacia el enfermo mental, la psiquiatría y el equipo de salud mental. Aten Primaria 33(9):491–495
3. Aruna G, Mittal S, Yadihal MB, Acharya C, Acharya S, Uppulan C (2016) Perception, knowledge, and attitude toward mental disorders and psychiatry among medical undergraduates in Karnataka: a cross-sectional study. Ind Psychiatry J 58(1):70–76
4. Bianchi EF, Bhattacharyya MR, Meakin R (2016) Exploring senior doctors' beliefs and attitudes regarding mental illness within the medical profession: a qualitative study. BMJ Open 6:e012598. https://doi.org/10.1136/mjopen-2016-012598
5. Blanco-Fernández A, Casals MR, Colubi A, Corral N, García-Bárzana M, Gil MA, González-Rodríguez G, López MT, Lubiano MA, Montenegro M, Ramos-Guajardo AB, de la Rosa de Sáa S, Sinova B (2014) A distance-based statistical analysis of fuzzy number-valued data. Int J Approx Reason 55(7):1487–1501
6. Cohen J, Struening E (1962) Opinions about mental illness in the personnel of two mental hospitals. J Abnorm Soc Psychol 64(5):349–360
7. De la Rosa de Sáa S, Gil MA, González-Rodríguez G, López MT, Lubiano MA (2015) Fuzzy rating scale-based questionnaires and their statistical analysis. IEEE Trans Fuzzy Syst 23:111–126
8. Dmour HH, Jumai'an AA, Al-Said HM (2012) Comparing attitudes of medical practitioners and allied medical professionals towards mental illness and patients with mental health disorders in Jordan. J Roy Med Serv 19(4):60–65

9. Failde I, Salazar A, Elorza J, Casais L, Pérez V, Caballero-Martínez L, Gilaberte I (2014) Spanish medical students attitudes and views towards mental health and Psychiatry: a multicentric cross-sectional study. Acad Psychiatry 38(3):332–338

10. Gibbons RJ, Thorsteinsson EB, Loi NM (2015) Beliefs and attitudes towards mental illness: an examination of the sex differences in mental health literacy in a community sample. Peer J 3:e1004t. https://doi.org/10.7717/peerj.1004

11. Gil MA, Lubiano MA, de la Rosa de Sáa S, Sinova B (2015) Analyzing data from a fuzzy rating scale-based questionnaire. A case study. Psicothema 27:182–191

12. Gil-Santiago H, Winter-Navarro M, León-Pérez P, Navarrete-Betancort E (2016) Stigma towards people with mental illness in general hospital health professionals. Norte Salud Ment XIV(55):103–111

13. Goldberg ID, Babigian HM, Locke BZ, Rosen BM (1978) Role of nonpsychiatrist physicians in the delivery of mental health services: implications from three studies. Public Health Rep 93(3):240–245

14. González-Rodríguez G, Colubi A, Gil MA (2012) Fuzzy data treated as functional data. A one-way ANOVA test approach. Comput Stat Data Anal 56(4):943–955

15. Haddad M, Waqas A, Qayyum W, Shams M, Malik S (2016) The attitudes and beliefs of Pakistani medical practitioners about depression: a cross-sectional study in Lahore using the Revised Depression Attitude Questionnaire (R-DAQ). BMC Psychiatry 16:349

16. Hassan TM, Asmer MS, Mazhar N, Munshi T, Tran T, Groll DL (2016) Canadian physicians' attitudes towards accessing mental health resources. Psychiatr J, Article ID 9850473. https://doi.org/10.1155/2016/9850473

17. Hesketh T, Pryor R, Hesketh B (1988) An application of a computerized fuzzy graphic rating scale to the psychological measurement of individual differences. Int J Man Mach Stud 29: 21–35

18. Hesketh B, Hesketh T, Hansen J-I, Goranson D (1995) Use of fuzzy variables in developing new scales from the strong interest inventory. J Couns Psychol 42:85–99

19. Hesketh B, Griffin B, Loh V (2011) A future-oriented retirement transition adjustment framework. J Vocat Behav 79:303–314

20. Högberg T, Magnusson A, Lützén K, Ewalds-Kvist B (2012) Swedish attitudes towards persons with mental illness. Nord J Psychiatry 66:86–96

21. Hori H, Sartorius N (2014) Cross-cultural comparisons of attitudes toward schizophrenia amongst the general population and physicians: a series of web-based surveys in Japan and the United States. Psychistry Res 215:300–307

22. Ighodaro A, Stefanovics E, Makanjuola V, Rosenheck R (2015) An assessment of attitudes towards people with mental illness among medical students and physicians in Ibadan, Nigeria. Acad Psychiatry 39:280–285

23. Link BQ, Yang LH, Phelan JC, Collins PY (2004) Measuring mental illness stigma. Schizophr Bull 30(3):511–541

24. Liu S-I, Lu R-B, Lee M-B (2008) Non-psychiatric physicians' knowledge, attitudes and behavior toward depression. J Formos Med Assoc 107(12):921–931

25. Lubiano MA, de la Rosa de Sáa S (2017) FuzzyStatTra: statistical methods for trapezoidal fuzzy numbers. https://CRAN.R-project.org/package=FuzzyStatTra

26. Lubiano MA, de la Rosa de Sáa S, Montenegro M, Sinova B, Gil MA (2016) Descriptive analysis of responses to items in questionnaires. Why not using a fuzzy rating scale? Inf Sci 360:131–148

27. Lubiano MA, Montenegro M, Sinova B, de la Rosa de Sáa S, Gil MA (2016) Hypothesis testing for means in connection with fuzzy rating scale-based data: algorithms and applications. Eur J Oper Res 251:918–929

28. Martínez-Zambrano F, García-Morales E, García-Franco M, Miguel J, Villellas R, Pascual G, Arenas O, Ochoa S (2013) Intervention for reducing stigma: assessing the influence of gender and knowledge. World J Psychiatry 3(2):18–24

29. Mental illness: Stigmatisation and discrimination within the medical profession (2001) Council report CR91. Royal College of Psychiatrists, Royal College of Physicians of London, British Medical Association

30. Mora-Ríos J, Bautista-Aguilar N, Natera G, Pedersen D (2013) Cultural adaptation of instruments to measure stigma and mental illness in Mexico city. Salud Ment 36:9–18

31. Mudarra-Duarte NC (2014) Actitudes de los Médicos Residentes hacia el Paciente con Trastorno Mental (Medical Residents Attitudes Toward The Patient With Mental Disorder). BSc in Mental Clinics, Caracas. http://saber.ucv.ve/handle/123456789/15491

32. Ochoa S, Martínez-Zambrano F, Vila-Badia R, Arenas O, Casas-Anguera E, García-Morales E, Villellas R, Martín JR, Pérez-Franco MB, Valduciel T, García-Franco M, Miguel J, Balsera J, Pascual G, Julia E, Casellas D, Haro JM (2016) Spanish validation of the social stigma scale: community attitudes towards mental illness. Rev Psiquiatr Salud Ment (Barc.) 9:150–157

33. Ozámiz JA (1980) Actitudes hacia las enfermedades mentales en el País Vasco. PhD thesis, Universidad del País Vasco

34. Pande V, Saini R, Chaudhury S (2011) Attitude toward mental illness amongst urban nonpsychiatric health professionals. Ind Psychiatry J 20(1):17–20

35. Pascucci M, Ventriglio A, Stella E, Di Sabatino D, La Montagna M, Nicastro R, Parente P, De Angelis A, Pozzi G, Janiri L, Bellomo A (2017) Empathy and attitudes towards mental illness among Italian medical students. Int J Cult Ment Health 10(2):174–184

36. Poreddi V, Thimmaiah R, Math SB (2015) Attitudes toward people with mental illness among medical students. Neurosci Rural Pract 6(3):349–354

37. Smith AL, Cashwell CS (2011) Stigma and mental illness: investigating attitudes of mental health and non mental health professionals and trainees. J Humanist Couns Educ Dev 49:189–202

38. Sørensen T, Sørensen A (2013) Dimensions of attitudes towards the mentally ill in the general population stability and change over time at urban and rural sites. Psychiatr J 2013:1–11, Article ID 319429

39. Stefanovics E, He H, Ofori-Atta A, Tavares-Cavalcanti M, Rocha-Neto H, Makanjuola V, Ighodaro A, Leddy M, Rosenheck R (2016) Cross-national analysis of beliefs and attitude toward mental illness among medical professionals from five countries. Psychiatr Q 87:63–73

40. Sun B, Fan N, Nie S, Zhang M, Huang X, He H, Rosenheck RA (2014) Attitudes towards people with mental illness among psychiatrists, psychiatric nurses, involved family members and the general population in a large city in Guangzhou, China. Int J Ment Health Syst 8:26

41. Svensson B, Brunt D, Bejerholm U, Eklund M, Gyllensten AL, Leufstadius C, Markström U, Sandlund M, Östman M, Hansson L (2014) Health care students' attitudes towards people with schizophrenia- A survey of eight university training programs. Open J Psychiatr 4:309–316

42. Taylor SM, Dear MJ (1981) Scaling community attitudes toward the mentally ill. Schizophr Bull 7:225–240

43. Tharyan P, Tharyan A (2001) Attitudes of 'tomorrow's doctors' towards psychiatry and mental illness. Nat Med J India 14(6):355–359

44. Todor I (2013) Opinions about mental illness. Procedia Soc Behav Sci 82:209–214

45. Trutschnig W, Lubiano MA (2015) SAFD: statistical analysis of fuzzy data. https://CRAN.R-project.org/package=SAFD

46. Tzouvara V, Papadopoulos C, Randhawa G (2016) Systematic review of the prevalence of mental illness stigma within the Greek culture. Int J Soc Psychiatr 62(3):292–305

47. Wallace JE (2010) Mental health and stigma in the medical profession. Health 16:3–18

48. Yllá L (1979) Estudio psicosocial de la actitud hacia el enfermo mental de la población general de Vizcaya y de la típica de Bermeo. PhD thesis, Universidad del País Vasco

49. Yllá L, González-Pinto A, Ballesteros J, Guillén V (2007) Evolución de las actitudes de la población frente al enfermo mental. Actas Esp Psiquiatr 35(5):323–335

50. Yllá L, Ozámiz A, Guimón J (1982) Sociedad, cultura y actitudes hacia la enfermedad mental. Psiquis 3:30–44

Some Results About the Vertices of k-Additive Dominating Measures

Pedro Miranda and Michel Grabisch

Abstract In this chapter we deal with the problem of deriving the vertices of the polyhedron of k-additive dominating measures. This set generalizes the polyhedron of dominating probabilities and has many potential applications in many different fields, as Decision Making or Game Theory. The results in this chapter are based on a theorem that obtains a superset of the set of dominating k-additive measures; this superset is built in a two-step process, and in each step a polyhedron of set functions is obtained; the first one does not cope with all dominating measures and the second includes any dominating measure but in general it also contains other functions. In this chapter we show that each of these polyhedra has a more intuitive structure, in the sense of their mathematical definition, than the polyhedron of dominating k-additive measures and thus, it is possible to obtain their corresponding set of vertices. From these sets it is possible to derive some results about the set of vertices of the polyhedron of dominating k-additive measures.

Keywords Non-additive measures · Dominance · k-monotone core · Polyhedron

1 Introduction

Let X be a set of n players, and consider its power set $\mathscr{P}(X)$, i.e., the set of all subsets of X, called *coalitions*. A *TU-game* on X is a set function $\mu : \mathscr{P}(X) \to \mathbb{R}$ satisfying $\mu(\emptyset) = 0$; for each coalition of players $A \subseteq X$, the value $\mu(A)$ represents the payoff that A can guarantee for itself, no matter what other players outside the coalition might do; in this chapter we focus on monotone games, i.e., games satisfying $\mu(A) \leq \mu(B)$ whenever $A \subseteq B$. A game is *normalized if* $\mu(X) = 1$; monotone games are also

P. Miranda (✉)
Universidad Complutense de Madrid, Plaza de Ciencias, 3, 28040 Madrid, Spain
e-mail: pmiranda@mat.ucm.es

M. Grabisch
Paris School of Economics, Université Paris I-Panthéon-Sorbonne,
106-112 Bd. de l'Hôpital, 75013 Paris, France
e-mail: michel.grabisch@univ-paris1.fr

© Springer International Publishing AG 2018
E. Gil et al. (eds.), *The Mathematics of the Uncertain*, Studies in Systems, Decision and Control 142, https://doi.org/10.1007/978-3-319-73848-2_61

known as fuzzy measures, non-additive measures or capacities. We will denote by $\mathcal{FM}(X)$ the set of all normalized monotone games on X.

One of the most important problems in Game Theory is to derive from μ a *solution* for the game, i.e. assuming that all players agree to form the grand coalition X, we look for a sharing vector $(x_1, \ldots, x_n)$ such that $x_1 + \cdots + x_n = \mu(X) = 1$ (if normalized), where x_i represents the payoff player i receives, $i = 1, \ldots, n$. Any $(x_1, \ldots, x_n)$ is a possible solution of the game, and we aim for a solution being rational and equitable in some sense. There are many solution concepts that have been proposed in the literature (see, e.g., Driessen [4]); among them, one of the most popular is the *core* of the game [19], which is defined as the set of additive games x dominating μ, i.e. such that

$$x(A) := \sum_{i \in A} x_i \geq \mu(A), \forall A \subseteq X, x(X) = \mu(X). \tag{1}$$

However, the core of a game could be empty [1]; when the core is not empty, it is a bounded polyhedron, and much research has been devoted to its study (see a survey in [8]).

Then, it seems interesting to look for extensions of the core that could be applied when it is empty. In this sense, a natural extension of probability is the concept of k-additive measure [6, 7]. These measures are defined in terms of the *Möbius transform* [17] (also called *dividends* [11] in Game Theory, or *Möbius inverse*); from a mathematical point of view, given a fuzzy measure μ, its Möbius transform m_μ is defined by

$$m_\mu(A) := \sum_{B \subseteq A} (-1)^{|A \setminus B|} \mu(B), \forall A \subseteq X. \tag{2}$$

The Möbius transform is an equivalent representation of the measure, in the sense that given m_μ it is possible to recover μ through

$$\mu(A) = \sum_{B \subseteq A} m_\mu(B). \tag{3}$$

The Möbius transform can therefore be seen as a linear and invertible operator on the set of set functions (hence its name), which we denote by m. With this notation, we have $\mathsf{m}(\mu) = m_\mu$, $\mathsf{m}^{-1}(m_\mu) = \mu$ for every game μ.

Table 1 Lower and upper bounds for the Möbius transform of a normalized fuzzy measure

$|A|$	1	2	3	4	5	6	7	8	9	10	11	12
u.b. of $m(A)$	1	1	1	3	6	10	15	35	70	126	210	462
l.b. of $m(A)$	1(0)	−1	−2	−3	−4	−10	−20	−35	−56	−126	−252	−462

Notice that m_μ can attain negative values; upper and lower bounds of the Möbius transform of a normalized fuzzy measure have been obtained in [10] and are presented in Table 1.

When μ is normalized and m_μ is nonnegative, we say that μ is a *belief function* (see Dempster [3] and Shafer [18]); belief functions play an important role in Evidence Theory; we will denote by $\mathcal{BEL}(X)$ the set of all belief functions on X.

It can be easily seen that probability measures have their Möbius transform vanishing for any coalition except singletons. Based on this property, a fuzzy measure μ is said to be k-*additive* if its Möbius transform m_μ vanishes for any $A \subseteq X$ such that $|A| > k$ and there exists at least one subset A with exactly k elements such that $m_\mu(A) \neq 0$. Thus defined, k-additive measures generalize additive measures (that are 1-additive fuzzy measures) and constitute a gradation between additive measures (e.g., probability measures) and general fuzzy measures. We will denote by $\mathcal{FM}^k(X)$ (resp. $\mathcal{BEL}^k(X)$), the set of normalized fuzzy measures (resp. belief functions) μ whose corresponding Möbius transform m_μ satisfies $m_\mu(A) = 0$ if $|A| > k$, i.e., the set of normalized fuzzy measures on X being k'-additive, with $k' \leq k$.

Now, focusing on normalized fuzzy measures μ, when the core of μ is empty, i.e., there is no probability measure dominating μ, it makes sense to look for 2-additive measures dominating the game; if there are no such 2-additive measures, then search for 3-additive measures and so on. This leads us to the concept of k-*additive monotone core* [13] for monotone normalized games, that is defined as the set of measures in $\mathcal{FM}^k(X)$ dominating μ. The k-additive monotone core of μ will be denoted by $\mathcal{FM}^k_\geq(\mu)$ (or $\mathcal{BEL}^k_\geq(\mu)$ if we restrict to belief functions); in particular, the core of the game is $\mathcal{FM}^1_\geq(\mu)$. It can be proved that for any monotone game μ, there exists a value of k such that $\mathcal{FM}^k_\geq(\mu) \neq \emptyset$. As for the classical core, when $\mathcal{FM}^k_\geq(\mu)$ is nonempty, it is a polytope (bounded polyhedron); more results about the $\mathcal{FM}^k_\geq(\mu)$ can be found in [13].

As $\mathcal{FM}^k_\geq(\mu)$ is a polytope, it is determined by its vertices. Therefore, an interesting problem for the k-additive monotone core of a monotone game is to characterize the set of vertices of $\mathcal{FM}^k_\geq(\mu)$. Some results for this polyhedron can be found in [9, 14]. In this paper, we will tackle the problem following the line developed by Chateauneuf and Jaffray [2] and that is related to the well-known Shapley–Ichiishi theorem [12, 19].

2 $\mathcal{FM}^k_\geq(\mu)$ as a Flow Problem

Let us denote by $\mathcal{P}_*(X)$ the power set of X without the empty set; similarly, we denote by $\mathcal{P}^k_*(X)$ the set of subsets of X with cardinality up to k without the empty set. Chateauneuf and Jaffray have studied in [2] the problem of dominating a fuzzy measure by probabilities. They obtain the following result:

Theorem 2.1 *Let μ be a normalized fuzzy measure on X, m its Möbius transform, and suppose $P \in \mathscr{FM}^1_{\geq}(\mu)$. Then, P can be put under the following form:*

$$P(\{i\}) = \sum_{B \ni i} \lambda(B, i) m(B), \ \forall i \in X, \tag{4}$$

where function $\lambda : \mathscr{P}_*(X) \times X \to [0, 1]$ is a weight function satisfying:

$$\sum_{i \in B} \lambda(B, i) = 1, \ \forall B \subseteq X, \tag{5}$$

$$\lambda(B, i) = 0 \ whenever \ i \notin B. \tag{6}$$

This result was also obtained by Dempster in [3] and also by Shapley in [19], but both of them only for belief functions. Let us introduce

$$\Lambda^k_{\subseteq,+} := \left\{ \lambda : \mathscr{P}_*(X) \times \mathscr{P}^k_*(X) \to [0,1] \mid \forall B \in \mathscr{P}(X), \ \sum_{A \subseteq B} \lambda(B, A) = 1, \right.$$
$$\left. \lambda(B, A) = 0 \ if \ A \not\subseteq B \right\},$$

$$\mathscr{M}_{\Lambda^k_{\subseteq,+}}(\mu) := \left\{ \mu_\lambda \mid m_\lambda(A) = \sum_{B \in \mathscr{P}(X)} \lambda(B, A) m(B), \lambda \in \Lambda^k_{\subseteq,+} \right\}, \tag{7}$$

for any fuzzy measure μ (not necessarily normalized) and where $m := \mathsf{m}(\mu)$ and $m_\lambda := \mathsf{m}(\mu_\lambda)$. Then, Theorem 2.1 establishes that $\mathscr{FM}^1_{\geq}(\mu) \subseteq \mathscr{M}_{\Lambda^1_{\subseteq,+}}(\mu)$.

Similarly, let us consider

$$\Lambda^k_\cap := \left\{ \lambda : \mathscr{P}_*(X) \times \mathscr{P}^k_*(X) \to \mathbb{R} \mid \forall B \in \mathscr{P}(X), \ \sum_{A \cap B \neq \emptyset} \lambda(B, A) = 1, \right.$$
$$\left. \lambda(B, A) = 0 \ if \ A \cap B = \emptyset \right\},$$

$$\mathscr{M}_{\Lambda^k_\cap}(\mu) := \left\{ \mu_\lambda \mid m_\lambda(A) = \sum_{B \in \mathscr{P}(X)} \lambda(B, A) m(B), \lambda \in \Lambda^k_\cap \right\}, \tag{8}$$

$$\Lambda^k_{\supseteq,+} := \left\{ \lambda : \mathscr{P}_*(X) \times \mathscr{P}^k_*(X) \to [0,1] \mid \forall B \in \mathscr{P}(X), \ \sum_{A \supseteq B} \lambda(B, A) = 1, \right.$$
$$\left. \lambda(B, A) = 0 \ if \ A \not\supseteq B \right\},$$

Table 2 Example of a dominating measure outside $\mathcal{M}_{\Lambda_\cap^k}(\mu)$

Subset	{1}	{2}	{3}	{1, 2}	{1, 3}	{2, 3}	{1, 2, 3}
μ	0	0	0	1	0	0	1
m_μ	0	0	0	1	0	0	0
μ^*	1	0	1	1	1	1	1
m_{μ^*}	1	0	1	0	−1	0	0

$$\mathcal{M}_{\Lambda_{\geq,+}^k}(\mu) := \left\{ \mu_\lambda \mid m_\lambda(A) = \sum_{B \in \mathscr{P}(X)} \lambda(B, A) m(B), \lambda \in \Lambda_{\geq,+}^k \right\}. \quad (9)$$

When attempting to extend Theorem 2.1 for the general k-additive case, natural extensions of the condition $i \in B$ are $A \subseteq B$ or more general $A \cap B \neq \emptyset$; however, even allowing λ to attain negative values and considering a non-empty intersection condition, i.e., $\lambda \in \Lambda_\cap^k$, it does not suffice to obtain a similar result to Theorem 2.1, as next example shows.

Example 2.1 Consider $|X| = 3$ and the capacities given in Table 2.

Then, $\mu^* \geq \mu$ but $\mu^* \notin \mathcal{M}_{\Lambda_\cap^2}(\mu)$ as the only subset in μ with non-null Möbius value is $\{1, 2\}$, and $\{1, 2\} \cap \{3\} = \emptyset$.

On the other hand, when dealing with belief functions, the following can be shown:

Theorem 2.2 ([16]) *Let* $\mu, \mu^* : \mathscr{P}(X) \to \mathbb{R}$, *where* μ *is a fuzzy measure, and* $\mu^* \in \mathscr{BEL}_\geq^k(\mu)$. *Then, necessarily* $\mu^* \in \mathcal{M}_{\Lambda_{\cap,+}^k}(\mu)$.

The proof of the result is based on Gale's Theorem for a transshipment network [5]. Moreover, if $\mu \in \mathscr{BEL}(X)$, the following can be proved:

Theorem 2.3 ([16]) *Let* $\mu, \mu^* : \mathscr{P}(X) \to \mathbb{R}$, *where* $\mu \in \mathscr{BEL}(X)$ *and* $\mu^* \in \mathscr{BEL}_\geq^k(\mu)$. *Then, there exists* $\mu' \in \mathscr{BEL}_\geq^k(\mu)$ *such that* $\mu' \in \mathcal{M}_{\Lambda_{\subseteq,+}^k}(\mu)$ *and* $\mu^* \in \mathcal{M}_{\Lambda_{\geq,+}^k}(\mu')$.

From this result, $\mathcal{M}_{\Lambda_{\subseteq,+}^k}(\mu) \subseteq \mathscr{BEL}_\geq^k(\mu) \subseteq \bigcup_{\mu' \in \mathcal{M}_{\Lambda_{\subseteq,+}^k}(\mu)} \mathcal{M}_{\Lambda_{\geq,+}^k}(\mu')$ for every

$\mu \in \mathscr{BEL}(X)$.

These results cannot be extended to the general case because the Möbius transform can attain negative values and this complicates the network so that Gale's theorem does not apply. Therefore, in [15], the so-called *shifted Möbius transform* is introduced. Specifically, for any game μ with Möbius transform m, its shifted Möbius transform is defined for any $A \subseteq X$ by $m_{sh}(A) = m(A) - l_{|A|}$, where l_i denotes the lower bound for the Möbius transform of subsets of cardinality i, with $i = 1, \ldots, n$. Observe that this also defines a linear and invertible operator on the set of set functions, which we denote by m_{sh}. Introducing the set function l defined by $l(A) = l_{|A|}$ for any $A \in \mathscr{P}_*(X)$ and $l(\emptyset) = 0$, we may write $\mathsf{m}_{sh} = \mathsf{m} - l$.

Although it is possible to consider any $\mu \in \mathscr{FM}(X)$ and $\mu^* \in \mathscr{FM}^k_{\geq}(\mu)$, it is convenient to suppose that $\mu \in \mathscr{FM}^k(X)$ for the sake of simplicity; note that this is not any constraint, as $\mathscr{FM}^k(X) \subseteq \mathscr{FM}^{k'}(X)$ if $k < k'$ and there exists k' such that $\mu \in \mathscr{FM}^{k'}(X)$. Let us define

$$\Lambda'^k_{\subseteq,+} := \Big\{ \lambda : \mathscr{P}^k_*(X) \times \mathscr{P}^k_*(X) \to [0,1] \mid \forall B \in \mathscr{P}^k_*(X), \quad \sum_{A \subseteq B} \lambda(B,A) = 1,$$

$$\lambda(B,A) = 0 \text{ if } A \nsubseteq B \Big\},$$

by analogy with $\Lambda^k_{\subseteq,+}$; we define similarly $\Lambda'^k_{\supseteq,+}$ and $\Lambda'^k_{\cap,+}$ and consider the sets $\mathscr{M}_{\Lambda^k_{\subseteq,+}}(\mu_{sh})$, $\mathscr{M}_{\Lambda^k_{\cap,+}}(\mu_{sh})$ and $\mathscr{M}_{\Lambda^k_{\subseteq,+}}(\mu_{sh})$, where $\mu_{sh} = \mathsf{m}^{-1}(\mathsf{m}_{sh}(\mu))$, $\mu \in \mathscr{FM}^k(X)$. Observe that

$$\mu_{sh} = \mathsf{m}^{-1}(\mathsf{m}(\mu) - l) = \mu + \mathsf{m}^{-1}(-l).$$

Remarking that $\mathsf{m}(\mu) - l$ is nonnegative so that $\mathsf{m}^{-1}(\mathsf{m}(\mu) - l)$ is a belief function (but not normalized!), it follows that μ_{sh} is a (nonnormalized) fuzzy measure. Now, the following can be shown:

Theorem 2.4 ([15]) *Let* $\mu, \mu^* : \mathscr{P}(X) \to \mathbb{R}$, *where* $\mu \in \mathscr{FM}^k(X)$ *and* $\mu^* \in \mathscr{FM}^k_{\geq}(\mu)$, *for* $k = 1, \ldots, n$, *and let us denote by* m_{sh}, m^*_{sh} *the corresponding shifted Möbius transforms. Then, necessarily* $\mu^*_{sh} \in \mathscr{M}_{\Lambda^k_{\cap,+}}(\mu_{sh})$.

Briefly speaking, the essence of the result is to take advantage of the fact that the shifted Möbius transform is a nonnegative function and thus it is possible to apply Gale's theorem. Finally,

Theorem 2.5 ([15]) *Let* $\mu, \mu^* : \mathscr{P}(X) \to \mathbb{R}$, *where* $\mu \in \mathscr{FM}^k(X)$, $\mu^* \in \mathscr{FM}^k_{\geq}(\mu)$ *and let us denote by* m_{sh}, m^*_{sh} *the corresponding shifted Möbius transforms. Then, there exists* $\mu' \in \mathscr{M}_{\Lambda'^k_{\subseteq,+}}(\mu_{sh})$ *such that* $\mu^*_{sh} \in \mathscr{M}_{\Lambda'^k_{\supseteq,+}}(\mu')$.

3　On the Set of Vertices

Consider a normalized fuzzy measure $\mu \in \mathscr{FM}^k(X)$ whose corresponding Möbius transform is m, and let us consider m_{sh} and μ_{sh} as defined before. In this section we tackle the problem of obtaining the set of vertices of $\mathscr{M}_{\Lambda^k_{\subseteq,+}}(\mu_{sh})$, taking advantage of its mathematical form.

Consider a total order $\prec$ over $\mathscr{P}^k_*(X)$, with $B_1 \prec B_2 \prec \cdots \prec B_r$ and $r := \sum_{i=1}^k \binom{n}{i}$. For each B_j, the *achievable family associated to* B_j by the order $\prec$ is the collection of sets $A \in \mathscr{P}^k_*(X)$ satisfying $A \supseteq B_j$ and $\forall B \subseteq A$, $B \in \mathscr{P}^k_*(X)$, it holds that $B \in \{B_1, \ldots, B_j\}$. We denote this collection by $\mathscr{ACH}_{j,\subseteq}$.

Note that $\mathscr{ACH}_{j,\subseteq}$ may be empty and that the collections $\mathscr{ACH}_{j,\subseteq}$ that are nonempty determine a partition on $\mathscr{P}^k_*(X)$. Remark also that for any $\mu^* \in \mathscr{FM}^k(X)$, and $A \in \mathscr{ACH}_{j,\subseteq}$, the value $\mu^*_{sh}(A)$ can be derived from $m^*_{sh}(B_1), \ldots, m^*_{sh}(B_j)$.

Now, for any $\mu \in \mathscr{FM}^k(X)$, we define $m_{\prec,sh} : \mathscr{P}(X) \to \mathbb{R}$ by

$$m_{\prec,sh}(B_j) := \sum_{A \in \mathscr{ACH}_{j,\subseteq}} m_{sh}(A), \forall B_j \in \mathscr{P}^k_*(X), \tag{10}$$

and 0 otherwise, where m_{sh} is the shifted Möbius transform of μ.

Proposition 3.1 *For any $\mu \in \mathscr{FM}^k(X)$ and any order $\prec$, the set function $m_{\prec,sh}$ defined by Eq. (10) is such that $\mu_{\prec,sh} \in \mathscr{M}_{\Lambda^{'k}_{\subseteq,+}}(\mu_{sh})$.*

Proof We just need to find a sharing function $\lambda \in \Lambda^{'k}_{\subseteq,+}$ such that $m_{\prec,sh}$ can be put under the form

$$m_{\prec,sh}(B_j) = \sum_{A | B_j \subseteq A} \lambda(A, B_j) m_{sh}(A). \tag{11}$$

And to show this, it suffices to consider

$$\lambda(A, B_j) = \begin{cases} 1 & \text{if } A \in \mathscr{ACH}_{j,\subseteq} \\ 0 & \text{otherwise} \end{cases}. \tag{12}$$

Then, the result holds.   □

Let us now show a technical lemma that will be needed later.

Lemma 3.1 *Let $\mu \in \mathscr{FM}^k(X)$ and $\prec$ be an order over $\mathscr{P}^k_*(X)$ with $B_1 \prec B_2 \prec \cdots \prec B_r$. Then, using the previous notation, for any $l = 1, \ldots, r$,*

$$\sum_{i=1}^{l} m_{\prec,sh}(B_i) \le \sum_{i=1}^{l} m'(B_i), \forall \mu' \in \mathscr{M}_{\Lambda^{'k}_{\subseteq,+}}(\mu_{sh}), \tag{13}$$

where $m' = \mathsf{m}(\mu')$.

Proof Suppose that the result does not hold. Then, $\exists \mu' \in \mathscr{M}_{\Lambda^{'k}_{\subseteq,+}}(\mu_{sh})$ and $l \in \{1, \ldots, r\}$ such that

$$\sum_{i=1}^{l} m_{\prec,sh}(B_i) > \sum_{i=1}^{l} m'(B_i). \tag{14}$$

This implies that in the definition of m', a quantity from, say $m_{sh}(A)$, such that $A \in \bigcup_{i=1}^{l} \mathscr{ACH}_{i,\subseteq}$ has not been assigned to any B_i, $i = 1, \ldots, l$, and this quantity has been assigned to a B_p with $p > l$.

However, as $\mu' \in \mathscr{M}_{\Lambda^{'k}_{\subseteq,+}}(\mu_{sh})$, this quantity must be assigned to a B_p such that $B_p \subseteq A$ and this contradicts $A \in \bigcup_{i=1}^{l} \mathscr{ACH}_{i,\subseteq}$.   □

For any $\mu \in \mathcal{FM}^k(X)$, consider the family

$$\mathcal{V}_\subseteq(\mu) := \{ m_{\prec,sh} | \prec \text{ total order over } \mathcal{P}_*^k(X) \}. \tag{15}$$

Proposition 3.2 *Let* $\mu \in \mathcal{FM}^k(X)$. *For any* $m_{\prec,sh} \in \mathcal{V}_\subseteq(\mu)$, $\mu_{\prec,sh}$ *is a vertex of* $\mathcal{M}_{\Lambda_{\subseteq,+}^k}(\mu_{sh})$, *i.e.,* $\mathsf{m}^{-1}\big(\mathcal{V}_\subseteq(\mu)\big) \subseteq \mathsf{ext}\big(\mathcal{M}_{\Lambda_{\subseteq,+}^k}(\mu_{sh})\big)$.

Proof Suppose there are $\mu_{sh}^1, \mu_{sh}^2 \in \mathcal{M}_{\Lambda_{\subseteq,+}^k}(\mu_{sh})$ such that $m_{sh}^1 = \mathsf{m}(\mu_{sh}^1)$ and $m_{sh}^2 = \mathsf{m}(\mu_{sh}^2)$ satisfy

$$m_{\prec,sh} = \alpha m_{sh}^1 + (1 - \alpha) m_{sh}^2, \ \alpha \in (0, 1). \tag{16}$$

By Lemma 3.1, considering order $\prec$,

$$\sum_{i=1}^l m_{\prec,sh}(B_i) \le \sum_{i=1}^l m_{sh}^1(B_i), \ \sum_{i=1}^l m_{\prec,sh}(B_i) \le \sum_{i=1}^l m_{sh}^2(B_i), \ \forall l = 1, \dots, r. \tag{17}$$

Then, all inequalities turn into equalities and $m_{\prec,sh} = m_{sh}^1 = m_{sh}^2$. $\qquad\square$

Proposition 3.3 *Let* $\mu \in \mathcal{FM}^k(X)$. *If* μ^* *is a vertex of* $\mathcal{M}_{\Lambda_{\subseteq,+}^k}(\mu_{sh})$, *then there exists an order* $\prec$ *such that* $m_{\prec,sh} = \mathsf{m}(\mu^*)$, *i.e.,* $\mathsf{ext}\big(\mathcal{M}_{\Lambda_{\subseteq,+}^k}(\mu_{sh})\big) \subseteq \mathsf{m}^{-1}\big(\mathcal{V}_\subseteq(\mu)\big)$.

Proof It suffices to prove that the convex closure of $\mathcal{V}_\subseteq(\mu)$ is the set of the shifted Möbius transforms of $\mathcal{M}_{\Lambda_{\subseteq,+}^{\prime k}}(\mu_{sh})$. If this is not the case, then there exists $\mu^* \in \mathcal{M}_{\Lambda_{\subseteq,+}^k}(\mu_{sh})$ such that its Möbius transform m^* is outside the convex closure of $\mathcal{V}_\subseteq(\mu)$. Then, there exists $\mathbf{v} = (v_1, \dots, v_r)$ with $r = \sum_{i=1}^k \binom{n}{i}$ such that for $\mathcal{P}_*^k(X) = \{ B_1, \dots, B_r \}$ we have

$$\sum_{i=1}^r m_{\prec,sh}(B_i) v_i < \sum_{i=1}^r m^*(B_i) v_i, \ \forall m_{\prec,sh} \in \mathcal{V}_\subseteq(\mu). \tag{18}$$

Let us suppose without loss of generality that $v_1 \le v_2 \cdots \le v_r$. Remark that we can also assume that $v_i \ge 0$, $\forall i$ because

$$\sum_{A \in \mathcal{P}_*^k(X)} m_{\prec,sh}(A) = \sum_{A \in \mathcal{P}_*^k(X)} m^*(A) = 1 + \sum_{j=0}^k \binom{n}{j} l_j, \ \forall m_{\prec,sh} \in \mathcal{V}_\subseteq(\mu). \tag{19}$$

Let us denote $c = 1 + \sum_{j=0}^k \binom{n}{j} l_j$. To prove that it is not possible to find a vector $\mathbf{v}$ in these conditions, we will prove by induction on $i \in \{1, \dots, r\}$ that for any $\lambda \ge 0$,

$$m_{\prec,sh}(B_1) v_1 + \cdots + m_{\prec,sh}(B_i) v_i + \Big(c - \sum_{j=1}^i m_{\prec,sh}(B_j) \Big)(v_i + \lambda)$$

$$\ge m^*(B_1) v_1 + \cdots + m^*(B_i) v_i + \Big(c - \sum_{j=1}^i m^*(B_j) \Big)(v_i + \lambda), \tag{20}$$

where $\prec$ is the order that coincides with the order in the coordinates of $\mathbf{v}$.

For $i = 1$, we have

$$m_{\prec,sh}(B_1)v_1 + \left(c - m_{\prec,sh}(B_1)\right)(v_1+\lambda) \geq m^*(B_1)v_1 + \left(c - m^*(B_1)\right)(v_1+\lambda)$$
$$\Leftrightarrow \left(c - m_{\prec,sh}(B_1)\right)\lambda \geq (c - m^*(B_1))\lambda \Leftrightarrow m_{\prec,sh}(B_1) \leq m^*(B_1),$$

and this holds by Lemma 3.1.

Let $i > 1$ and suppose the result holds till $i - 1$. Then,

$$m_{\prec,sh}(B_1)v_1 + \cdots + m_{\prec,sh}(B_i)v_i + \left(c - \sum_{j=1}^{i} m_{\prec,sh}(B_j)\right)(v_i + \lambda)$$

$$= m_{\prec,sh}(B_1)v_1 + \cdots + m_{\prec,sh}(B_{i-1})v_{i-1}$$
$$+ \left(c - \sum_{j=1}^{i-1} m_{\prec,sh}(B_j)\right)(v_{i-1} + v_i - v_{i-1}) + \left(c - \sum_{j=1}^{i} m_{\prec,sh}(B_j)\right)\lambda. \quad (21)$$

Now, applying the induction hypothesis,

$$m_{\prec,sh}(B_1)v_1 + \cdots + m_{\prec,sh}(B_{i-1})v_{i-1} + \left(c - \sum_{j=1}^{i-1} m_{\prec,sh}(B_j)\right)(v_{i-1} + v_i - v_{i-1})$$

$$\geq m^*(B_1)v_1 + \cdots + m^*(B_{i-1})v_{i-1} + \left(c - \sum_{j=1}^{i-1} m^*(B_j)\right)(v_{i-1} + v_i - v_{i-1}),$$
$$(22)$$

whence we obtain that (21) is greater or equal than

$$m^*(B_1)v_1 + \cdots + m^*(B_{i-1})v_{i-1} +$$
$$\left(c - \sum_{j=1}^{i-1} m^*(B_j)\right)(v_{i-1} + v_i - v_{i-1}) + \left(c - \sum_{j=1}^{i} m_{\prec,sh}(B_j)\right)\lambda.$$

Then, it suffices to show that

$$\left(c - \sum_{j=1}^{i} m_{\prec,sh}(B_j)\right)\lambda \geq \left(c - \sum_{j=1}^{i} m^*(B_j)\right)\lambda,$$

but this is true by Lemma 3.1 as $\lambda \geq 0$. Then, the result holds. Now, taking the assertion with $i = r$, the proposition is proved. $\qquad\square$

Remark that in these proofs we just need that the shifted Möbius transform is nonnegative; thus, we can apply this result for the corresponding polytope $\mathcal{M}_{\Lambda^k_{\subseteq,+}}(\mu)$ appearing in Theorem 2.3. Moreover, given μ'_{sh}, the same can be applied to obtain the vertices $\mathcal{M}_{\Lambda'^k_{\supseteq,+}}(\mu'_{sh})$ (and $\mathcal{M}_{\Lambda^k_{\supseteq,+}}(\mu')$) just changing the definition of achievable family: Given a total order $\prec$ over $\mathscr{P}^k_*(X)$, so that $B_1 \prec B_2 \prec \cdots \prec B_r$, we say that $A \in \mathscr{ACH}_{j,\supseteq}$ if and only if $A \subseteq B_j$ and $\forall B \supseteq A$, $B \in \mathscr{P}^k_*(X)$, it holds that $B \in \{B_1, \ldots, B_j\}$. Now, for any $\mu \in \mathscr{FM}^k(X)$, we define $m_{\prec,sh}$ by

$$m_{\prec,sh}(B_j) := \sum_{A \in \mathscr{ACH}_{j,\supseteq}} m_{sh}(A), \forall B_j \in \mathscr{P}^k_*(X) \tag{23}$$

and $\mathscr{V}_{\supseteq}(\mu) := \{m_{\prec,sh} \mid \prec \text{ total order over } \mathscr{P}^k_*(X)\}$. Then,

Theorem 3.1 For any $\mu \in \mathscr{FM}^k(X)$, $\mathsf{ext}\left(\mathcal{M}_{\Lambda'^k_{\supseteq,+}}(\mu_{sh})\right) = \mathsf{m}^{-1}\left(\mathscr{V}_{\supseteq}(\mu)\right)$.

Finally, let us study some relationships between the vertices of $\mathcal{M}_{\Lambda'^k_{\subseteq,+}}(\mu_{sh})$ and $\mathcal{M}_{\Lambda'^k_{\supseteq,+}}(\mu'_{sh})$ with the vertices of $\mathscr{FM}^k_{\supseteq}(\mu)$. The following can be shown:

Lemma 3.2 Let us consider $\mu \in \mathscr{FM}^k(X)$ and μ' such that $\mu'_{sh} \in \mathcal{M}_{\Lambda'^k_{\subseteq,+}}(\mu_{sh})$. Then, $\mu' \geq \mu$.

Proof For any $B \subseteq X$, $\mu'(B)$ is given by

$$\sum_{D \subseteq B} m'(D) = \sum_{D \subseteq B} m'_{sh}(D) + \sum_{i=1}^{|B|} \binom{|B|}{i} l_i = \sum_{D \subseteq B} \sum_{A \mid D \subseteq A} \lambda'(A, D) m_{sh}(A) + \sum_{i=1}^{|B|} \binom{|B|}{i} l_i.$$

If $A \subseteq B$, we have that whenever $D \subseteq A$, then $D \subseteq B$, and hence, $m_{sh}(A)$ is multiplied by

$$\sum_{D \mid D \subseteq A} \lambda'(A, D) = 1. \tag{24}$$

Consequently,

$$\sum_{D \subseteq B} m'_{sh}(D) = \sum_{A \subseteq B} m_{sh}(A) + \sum_{A \not\subseteq B} \sum_{D \subseteq A \cap B} \lambda'(A, D) m_{sh}(A). \tag{25}$$

Finally, as $m_{sh}(A) \geq 0$, $\forall A \in \mathscr{P}^k_*(X)$, $\lambda'(A, B) \geq 0$, $\forall A, B \in \mathscr{P}^k_*(X)$,

$$\mu'(B) \geq \sum_{A \subseteq B} m_{sh}(A) + \sum_{i=1}^{|B|} \binom{|B|}{k} l_k = \sum_{A \subseteq B} m(A) = \mu(B). \tag{26}$$

This finishes the proof. □

Similarly, the following holds.

Lemma 3.3 *Let us consider* $\mu \in \mathscr{FM}^k(X)$ *and* μ' *such that* $\mu'_{sh} \in \mathscr{M}_{\Lambda^{'k}_{\geq,+}}(\mu_{sh})$. *Then,* $\mu' \leq \mu$.

However, not all μ' such that $\mu'_{sh} \in \mathscr{M}_{\Lambda^{'k}_{\subseteq,+}}(\mu_{sh})$ are monotone measures. Indeed, the following can be shown:

Lemma 3.4 *For* $|X| \geq 2$, *there exists* $\mu'_{sh} \in \mathscr{M}_{\Lambda^{'k}_{\geq,+}}(\mu_{sh})$ *s. t.* μ' *is not monotone except for* $|X| = 2$ *and* μ *given by* $\mu(1) = \mu(2) = \mu(1,2) = 1$.

Finally, let us show an application of the previous results, that allows us to derive some of the vertices of $\mathscr{FM}^k_{\geq}(\mu)$ from the vertices of $\mathscr{M}_{\Lambda^{'k}_{\subseteq,+}}(\mu_{sh})$.

Proposition 3.4 *Consider* $\mu \in \mathscr{FM}^k(X)$ *and* $\mu' \in \mathscr{FM}^k(X)$ *such that* $m'_{sh} \in \mathscr{V}_{\subseteq}(\mu)$ *and* $\mu'' \not\geq \mu'$ *for any* $\mu''_{sh} \in \mathscr{M}_{\Lambda^{'k}_{\subseteq,+}}(\mu_{sh})$. *Then,* μ' *is a vertex of* $\mathscr{FM}^k_{\geq}(\mu)$.

Proof From Lemma 3.2, we know that $\mu' \geq \mu$ and by hypothesis $\mu' \in \mathscr{FM}^k(X)$, whence $\mu' \in \mathscr{FM}^k_{\geq}(\mu)$. If μ' is not a vertex, then there exists $\mu_1, \mu_2 \in \mathscr{FM}^k_{\geq}(\mu)$ such that $\mu' = \lambda\mu_1 + (1-\lambda)\mu_2$.

As $\mu_1 \in \mathscr{FM}^k_{\geq}(\mu)$, then $\mu_{1,sh} \in \mathscr{M}_{\Lambda^{'k}_{\geq,+}}(\mu'_{1,sh})$ with $\mu'_{1,sh} \in \mathscr{M}_{\Lambda^{'k}_{\subseteq,+}}(\mu_{sh})$; and $\mu_{2,sh} \in \mathscr{M}_{\Lambda^{'k}_{\geq,+}}(\mu'_{2,sh})$ with $\mu'_{2,sh} \in \mathscr{M}_{\Lambda^{'k}_{\subseteq,+}}(\mu_{sh})$. By Lemma 3.3, $\mu_{1,sh} \leq \mu'_{1,sh}$, $\mu_{2,sh} \leq \mu'_{2,sh}$, whence

$$\lambda\mu'_{1,sh} + (1 - \lambda\mu'_{2,sh}) \geq \lambda\mu_{1,sh} + (1 - \lambda\mu_{2,sh}) = \mu'_{sh} \tag{27}$$

But $\lambda\mu'_{1,sh} + (1 - \lambda\mu'_{2,sh}) \in \mathscr{M}_{\Lambda^{'k}_{\subseteq,+}}(\mu_{sh})$ because it is a polyhedron, a contradiction. $\square$

Acknowledgements This research has been partially supported by Spanish Grant MTM2012-33740.

This chapter has been done to pay homage to Prof. Pedro Gil. During many years, Prof. Pedro Gil has collaborated with us in our research and this chapter follows the line of one of our joint papers. We will miss his kindness and understanding, and his care of any aspect of the academic life.

References

1. Bondareva O (1963) Some applications of linear programming to the theory of cooperative games. Probl Kibernet 10:119–139
2. Chateauneuf A, Jaffray JY (1989) Some characterizations of lower probabilities and other monotone capacities through the use of Möbius inversion. Math Soc Sci 17:263–283
3. Dempster AP (1967) Upper and lower probabilities induced by a multivalued mapping. Ann Math Stat 38:325–339
4. Driessen T (1988) Cooperative games, solutions and applications. Kluwer Academic Publisher, Dordrecht
5. Gale D (1960) The theory of linear economic models. McGraw Hill, New York

6. Grabisch M (1996) k-order additive discrete fuzzy measures. In: Proceedings of the 6th international conference information processing and management of uncertainity in knowledge-based systems, IPMU'1996. Universidad de Granada

7. Grabisch M (1997) k-order additive discrete fuzzy measures and their representation. Fuzzy Sets Syst 92:167–189

8. Grabisch M (2013) The core of games on ordered structures and graphs. Ann Oper Res 204:33–64

9. Grabisch M, Miranda P (2008) On the vertices of the k-additive core. Discr Math 308:5204–5217

10. Grabisch M, Miranda P (2015) Exact bounds of the Möbius inverse of monotone set functions. Discr Appl Math 186:7–12

11. Harsanyi JC (1963) A simplified bargaining model for the n-person cooperative game. Int Econ Rev 4:194–220

12. Ichiishi T (1981) Super-modularity: applications to convex games and to the Greedy algorithm for LP. J Econ Theory 25:283–286

13. Miranda P, Grabisch M (2010) k-balanced games and capacities. Eur J Oper Res 200:465–472

14. Miranda P, Grabisch M (2012) An algorithm for finding the vertices of the k-additive monotone core. Discr Appl Math 160(4–5):628–639

15. Miranda P, Grabisch M (2016) Finding the set of k-additive dominating measures viewed as a flow problem. In: Carvalho JP, Lesot MJ, Kaymak U, Vieira S, Bouchon-Meunier B, Yager RR (eds) Information processing and management of uncertainty in knowledge-based systems. Springer, Cham

16. Miranda P, Grabisch M, Gil P (2006) Dominance of capacities by k-additive belief functions. Eur J Oper Res 175:912–930

17. Rota GC (1964) On the foundations of combinatorial theory I. Theory of Möbius functions. Zeitsch Wahrsch Verw Geb 2:340–368

18. Shafer G (1976) A mathematical theory of evidence. Princeton University Press, Princeton

19. Shapley LS (1971) Cores of convex games. Int J Game Theory 1:11–26

Divergence Measures: From Uncertainty to Imprecision

Susana Montes, Susana Díaz and Davide Martinetti

Abstract The link between information theory and fuzzy logic has been proven in several previous papers. From this starting point, we propose here a review about the concept of divergence measures, which was proposed as a tool for comparing two fuzzy sets. The initial definition comes from the ideas behind the classical concept of divergence between two probability distributions. Following a path similar to the one considered to obtain fuzziness measures from uncertainty measures, we are able to define fuzzy divergences. Apart from that, some possible generalizations are considered.

1 Introduction

Dealing with lack of information is a usual problem in many areas. This lack of information can be given in two different ways: uncertainty or imprecision. In the first case, we deal with experiments where we can have more than one possible outcome, each possible outcome can be specified in advance, but the outcome of the experiment depends on chance. For instance, in a coin toss, we know the two possible outcomes, head or tail, but we do not know the final result. In the second case, we have no uncertainty about the result of the experiment but imprecision. Thus, for instance, if we consider again the experiment of the coin toss, the coin could be already thrown but maybe it is too old and we are not sure that the face it shows is clearly a head.

S. Montes (✉) · S. Díaz
Department of Statistics and Operational Research and TM, University of Oviedo, Oviedo, Spain
e-mail: montes@uniovi.es

S. Díaz
e-mail: diazsusana@uniovi.es

D. Martinetti
INRA-PACA, UR Ecodeveloppement, Avignon Cedex 9, France
e-mail: davide.martinetti@inra.fr

© Springer International Publishing AG 2018
E. Gil et al. (eds.), *The Mathematics of the Uncertain*, Studies in Systems, Decision and Control 142, https://doi.org/10.1007/978-3-319-73848-2_62

Information theory studies the quantification and communication of the information and, in particular, it measures the amount of uncertainty involved in the value of a random experiment. It was originally proposed by Shannon [31] in 1948 as a tool in signal processing. Thus, this theory combines a lot of different fields such as mathematics, statistics, computer science, physics and electrical engineering. From the beginning, this theory was revealed as an interesting tool in many other areas and therefore a lot of researchers started to work on it (Rényi [30], Oniçescu [27], Sharma and Mittal [32], Havrda and Charvát [11], etc.). Later, an important step was given by Kampé de Fèriet and Forte [12] with an axiomatic definition of the information with or without a probability measure. From the theoretical aspects of this theory, Kullback [17] found a lot of interesting applications in statistical inference. From this initial application a lot of papers have been developed in this area. In particular, some very important achievements have been obtained by Pardo (see, among others, [28, 29]). An important review about all these theories can be found in Gil [7], since he was one of the most important researchers in this area in Spain. Divergence measures between probability distributions were an important topic on this monograph and it is the starting point of this chapter, as we will see later.

On the other hand, Zadeh [34] introduced in 1965 the concept of fuzzy set, as a way to model vague or poorly defined properties for situations in which it is not possible to fully discriminate between having and not having the said properties. From that, a whole mathematical and applied theory to deal with imprecision was developed. It is known as Fuzzy Logic Theory. Two interesting monographs about this theory were written by Dubois and Prade [6] and Klir and Folger [13].

As we can see from the title of this last book, the concepts fuzzy sets, uncertainty and information are mixed. This is not by chance, since these topics are very related, as we can see in [8–10]. In particular, we have studied [24] the relationship between uncertainty measures defined in Information Theory [12] and the fuzziness measures introduced by De Luca and Termini [5] and later analyzed in a deeper way by Knopfmacher [14]. The link between measures of uncertainty and imprecision in fuzzy environments will lie in what we will refer as divergence measure, because of the analogy with the classical meaning of the term used in comparing two probability distributions (see, for instance, [29]). The main purpose of this chapter is to use these measures to compare two fuzzy sets.

As introductory notions, we present two axiomatic definitions to measure the entropy–uncertainty measures and fuzziness measures in Sect. 2. A study on the relationship between them, in the most general context, is given there. The definition of divergence measure between fuzzy sets is given in Sect. 3 following the ideas considered previously. The most important results are contained in that section where we also comment some extensions. Finally, we conclude the work with some comments in Sect. 4.

2 Preliminary

Necessary concepts to understand the remaining parts of this work are given in this section. In particular, we will focus on the definitions and notations for uncertainty measures and fuzziness measures.

2.1 Uncertainty Measures

The first probabilistic uncertainty measure (also called entropy) was given by Shannon [31] in the context of Communication Theory. That initial definition considered that the uncertainty for a random experiment can be measured by means of the quantity

$$H(P) = -\sum_{i=1}^{n} p_i log_2(p_i)$$

where values p_i represent the probabilities of the possible results of the experiment.

From that initial definition, a lot of generalizations have been proposed in the literature.

Thus, Menéndez et al. [19] proved that all these measures of entropy are part of a wider family, which are named h-ϕ-entropies.

This family is slightly more general than Ben Bassat's family of f-entropies that were defined as those functions that can be expressed like

$$H(P) = \sum_{i=1}^{n} f(p_i)$$

where f is a concave function.

Later, the quasi-ϕ-entropies were introduced and characterized in the case of discrete distributions [3]. Thus, it is a family more general than Ben Bassat's one but different from the family of h-ϕ-entropies. More precisely, they are defined by

$$H(P) = \sum_{i=1}^{n} \phi(p_i)$$

where ϕ is a function such that $\phi(\lambda x + (1 - \lambda)y) \geq \lambda\phi(x) + (1 - \lambda)\phi(y), \forall x, y \in [0, 1], x + y \leq 1$.

An important property of uncertainty measures is the Principle of Transfer or Pigou–Dalton's condition. An uncertainty measure H fulfils this property if given two probability distributions P and P' with parameters $(p_1, p_2, \ldots, p_n)$ and $(p'_1, p'_2, \ldots, p'_n)$ respectively, then $H(P) \leq H(P')$, where, $p_k = p'_k, \forall k \notin \{i, j\}$ and $p'_i = p_i + \delta$, $p'_j = p_j - \delta$ for some $\delta \leq (p_i - p_j)/2$.

It is a very logical property, since it means that the more similar the probabilities of two outcomes of an experiment are, the higher the uncertainty is.

2.2 Fuzziness Measures

After having commented some results about uncertainty measures or probabilistic entropies, we now introduce fuzzy sets and the measures of their fuzziness, i.e., the non-probabilistic entropies.

They are well-known and can be found in a wide range of sources (see, for instance, the classical books [6, 13]).

The universal set is denoted by X. A fuzzy subset of X is a mapping from X into the unit interval [0, 1].

In this framework, we use the following notations:

- $\mathscr{P}(X)$ is the set of all subsets of X,
- $\mathscr{F}(X)$ is the set of all fuzzy subsets of X,
- $A \in \mathscr{P}(X)$ will denote any crisp set,
- $\widetilde{A} \in \mathscr{F}(X)$ will denote any fuzzy set.

We identify a fuzzy set and its membership function. Thus we have that $X(x) = 1$ for all $x \in X$ and for the empty set we have $\emptyset(x) = 0$ for all $x \in X$.

Two further important concepts are the containment relation and the complement set. We consider the standard Zadeh's negation for the complement (see [34]).

Definition 2.1 Let $\widetilde{A}, \widetilde{B} \in \mathscr{F}(X)$. The complement of $\widetilde{A}$ is the fuzzy set $\widetilde{A}^c(x) = 1 - \widetilde{A}(x)$, $x \in X$. $\widetilde{A}$ is contained in $\widetilde{B}$, denoted by $\widetilde{A} \subseteq \widetilde{B}$ if $\widetilde{A}(x) \leq \widetilde{B}(x)$ for all $x \in X$.

Apart from the previous relation of containment, we consider the concepts of intersection and union of fuzzy sets. The initial definitions were also given in [34] by means of the minimum and the maximum operators.

However, they are not the only way to generalize the classical set operations, since there exists a broader class of functions to represent them. In fact, for the intersection, this class is referred as t-norm and for the union as t-conorm.

A triangular norm (t-norm) is a function $T : [0, 1] \times [0, 1] \rightarrow [0, 1]$ satisfying the following properties:

(T1) $T(a, b) = T(b, a)$, for all $a, b \in [0, 1]$,
(T2) $T(T(a, b), c) = T(a, T(b, c))$, for all $a, b, c \in [0, 1]$,
(T3) $b \leq c \Rightarrow T(a, b) \leq T(a, c)$, for all $a, b, c \in [0, 1]$,
(T4) $T(a, 1) = a$, for all $a \in [0, 1]$.

Some important examples of t-norms are:

- Minimum: $T_M(a, b) = \min(a, b)$, for all $a, b \in [0, 1]$,
- Product: $T_P(a, b) = a \cdot b$, for all $a, b \in [0, 1]$,

- Łukasiewicz t-norm: $T_L(a, b) = \max(a + b - 1, 0)$, for all $a, b \in [0, 1]$,
- Drastic t-norm:

$$T_D(a, b) = \begin{cases} \min(a, b), & \text{if } \max(a, b) = 1 \\ 0, & \text{otherwise} \end{cases}.$$

For these basic t-norms, it holds that $T_D \leq T_L \leq T_P \leq T_M$. In fact, for any t-norm T, it is true that $T_D \leq T \leq T_M$. By changing the neutral element from 1 to 0, we obtain the triangular conorms (t-conorm).

A t-norm T and a t-conorm S are dual iff for each $a, b \in [0, 1]$ it holds that $T(a, b) = 1 - S(1 - a, 1 - b)$.

The dual conorms of the t-norms presented earlier are the following:

- Maximum: $S_M(a, b) = \max(a, b)$, for all $a, b \in [0, 1]$,
- Probabilistic sum: $S_P(a, b) = a + b - a \cdot b$, for all $a, b \in [0, 1]$,
- Łukasiewicz t-conorm: $S_L(a, b) = \min(a + b, 1)$, for all $a, b \in [0, 1]$,
- Drastic t-conorm:

$$S_D(a, b) = \begin{cases} \max(a, b), & \text{if } \min(a, b) = 0 \\ 1, & \text{otherwise} \end{cases}.$$

Using t-norms and t-conorms, we can define the intersection and union of two fuzzy sets as follows.

Definition 2.2 Let $\widetilde{A}, \widetilde{B} \in \mathscr{F}(X)$. Given a t-norm T and a t-conorm S,

- $\widetilde{A} \cap \widetilde{B}(x) = T(\widetilde{A}(x), \widetilde{B}(x)), \forall x \in X$;
- $\widetilde{A} \cup \widetilde{B}(x) = S(\widetilde{A}(x), B(x)), \forall x \in X$.

Thus, we can denote by (X, T, S) the triple formed by the universe with the t-norm and the t-conorm defining the intersection and the union, respectively.

The entropy for a fuzzy set is quantified by means of the non-probabilistic entropies or fuzziness measures (see, for instance, [33]), which are defined as follows.

Definition 2.3 A fuzziness measure is a real function f defined on $\mathscr{F}(X)$, fulfilling the following requirements:

(a) $f(\widetilde{A}) = 0 \iff \widetilde{A}$ is a crisp set.
(b) If $\widetilde{A}, \widetilde{B} \in \mathscr{F}(X)$ and $\widetilde{A}$ is "sharper" than $\widetilde{B}$, then $f(\widetilde{A}) \leq f(\widetilde{B})$.
(c) $f(\widetilde{A})$ takes maximum value if and only if $\widetilde{A}$ is "maximally fuzzy".

This last definition is based on the concepts "sharper than" and "maximally fuzzy", although the second one follows from the former. Thus, the most usual criteria to define the relation "to be sharper than" are the following:

- $\widetilde{A}$ is sharper than $\widetilde{B}$ iff either $\widetilde{A}(x) \leq \widetilde{B}(x) \leq 1/2$ or $\widetilde{A}(x) \geq \widetilde{B}(x) \geq 1/2$ for any x in X (see [13]) or
- $\widetilde{A}$ is sharper than $\widetilde{B}$ iff $|\widetilde{A}(x) - 1/2| \leq |\widetilde{B}(x) - 1/2|$ for any x in X (see [6]).

It is clear that the first one is a particular case of the second one and therefore we are going to consider the most general definition.

Knopfmacher introduced in 1975 a very important family of fuzziness measures, the Knopfmacher class [14], which is given by the functions f such that

$$f(\widetilde{A}) = F\left(\sum_{x \in X} c_x \cdot g_x(\widetilde{A}(x))\right)$$

for any $\widetilde{A}$ in $\mathscr{F}(X)$ where $c_x \in \mathbb{R}^+$; g_x is a real-valued function such that $g_x(0) = g_x(1) = 0$, $g_x(t) = g_x(1 - t)$, $\forall t \in [0, 1]$ and g_x is strictly increasing on $[0, 1/2]$; F is a positive strictly increasing function with $F(0) = 0$.

Later, we consider a particular class of Knopfmacher fuzziness measure (see [20, 22]) when F is the identity, g_x is the same for all $x \in X$ (we denoted g_x by u_f or simply u) and u is concave. Any function in this family was named local fuzziness measure.

2.3 From Uncertainty to Fuzziness

Proposition 2.1 ([24]) *Let $(X, \mathscr{A}, \mu)$ be a measurable space and let H be an uncertainty measure fulfilling the Pigou-Dalton's condition and such that $H(P) = 0 \iff P$ is degenerate. The map f defined as follows:*

$$f : \mathscr{A}^* \longrightarrow \mathbb{R}^+$$
$$\widetilde{A} \longrightarrow \int_X H(\widetilde{A}(x), \widetilde{A}^c(x)) d\mu(x)$$

is a fuzziness measure and it belongs to the Knopfmacher's class.

If we work on some particular spaces, we are also able to establish a one-to-one correspondence between fuzziness measures and uncertainty measures.

Thus, if we consider the subset of uncertainty measures given by

$$\mathscr{H}_2 = \{H | H \text{ is a quasi-}\phi\text{-entropy with } \phi \text{ continue}, \phi(x) = \phi(1 - x),$$

$$\forall x \in [0, \tfrac{1}{2}] \text{ and } \phi(x) = 0 \Leftrightarrow x = 0\}$$

we have the injective property, as we can see in the following proposition.

Proposition 2.2 ([24]) *If F_1 is a map from $\mathscr{H}_2$ in $\mathscr{F}$ such that*

$$F_1(H)(\widetilde{A}) = \int_X H(\widetilde{A}(x), \widetilde{A}^c(x)) d\mu(x),$$

where $\mathcal{F}$ denotes the Knopfmacher's class of fuzziness measures, then we have that F_1 is injective.

If we restrict our study to the family of ϕ-entropies given by $\mathcal{H}_\phi = \{H \in \mathcal{H}_2 | \phi \text{ is concave}\}$ and the family of fuzziness measures given by $\mathcal{F}_1 = \{f \in \mathcal{F} \text{ with } g \text{ continue}\}$ we have the bijection.

Theorem 2.1 ([24]) *There exists a one-to-one correspondence between the family of uncertainty measures $\mathcal{H}_\phi$ and the family of fuzziness measures $\mathcal{F}_1$.*

3 Divergence Measures

From the previous section, we could notice that the imprecision about the membership of any element $x \in X$ in a fuzzy set $\widetilde{A}$ could be represented by a probability distribution $\{\widetilde{A}(x), \widetilde{A}^c(x)\}$. Then, we looked at the classical divergence measures between probability distributions (see, for instance, [7, 29]) to try to compare two fuzzy sets.

Thus, from this starting point, we proposed a new way to compare two fuzzy sets [20], the divergence, with the following properties:

- It becomes zero when the two sets coincide.
- It is a nonnegative and symmetric function.
- It decreases when the two sets become "more similar" in some sense.

While it is easy to formulate the first and the second conditions analytically, the third one depends on the formalization of the concept "more similar". We base our approach on the fact that if we add a set $\widetilde{C}$ to both fuzzy sets $\widetilde{A}, \widetilde{B}$, we obtain two subsets which are closer to each other; the same with the intersection.

Definition 3.1 Let (X, T, S) be a triple with X a universe and T and S any t-norm and t-conorm, respectively. A map $D : \mathcal{F}(X) \times \mathcal{F}(X) \to \mathbb{R}$ is a divergence measure with respect to (X, T, S) iff for all $\widetilde{A}, \widetilde{B} \in \mathcal{F}(X)$, D satisfies the following conditions:

(a) $D(\widetilde{A}, \widetilde{A}) = 0$;
(b) $D(\widetilde{A}, \widetilde{B}) = D(\widetilde{B}, \widetilde{A})$;
(c) $\max\{D(\widetilde{A} \cup \widetilde{C}, \widetilde{B} \cup \widetilde{C}), D(\widetilde{A} \cap \widetilde{C}, \widetilde{B} \cap \widetilde{C})\} \leq D(\widetilde{A}, \widetilde{B})$, for all $\widetilde{C} \in \mathcal{F}(X)$,
where the union and intersection are defined by means of S and T, respectively.

It is clear that a divergence measure is associated to a triple (X, T, S) and a map D can be a divergence measure with respect to a t-norm and it cannot be a divergence measure with respect to a different t-norm.

However, when there is not ambiguity, we will call just divergence measure without specifying the used t-norm and t-conorm.

After different studies of this concept [2, 20, 22–24], we presented the most general study in [15], where we can also find the following examples.

Example 3.1 ([15]) The map

$$D(\widetilde{A}, \widetilde{B}) = \begin{cases} 0, & \text{if } \widetilde{A} = \widetilde{B} \\ 1, & \text{if } \widetilde{A} \neq \widetilde{B} \end{cases}.$$

is a divergence for any triple (X, T, S).

On the other hand, if we consider the map

$$D(\widetilde{A}, \widetilde{B}) = \sum_{x \in X} \alpha_x \cdot |\widetilde{A}(x) - \widetilde{B}(x)|$$

where $\alpha_x \geq 0$ for any $x \in X$, $\sum_{x \in X} \alpha_x = 1$ and X is a finite space, D is a divergence for the minimum t-norm, the product t-norm or the Łukasiewicz t-norm, but it is not for the drastic t-norm.

A divergence measure can be seen as a particular case of dissimilarity when the minimum t-norm is considered, which is the most usual way to compare two fuzzy sets [18].

Moreover, it avoids some counterintuitive examples for dissimilarities, while both divergence and dissimilarity measures can be seen as a particular case of the general measures of comparison given by Bouchon–Meunier et al. [1] in 1996. An interesting study about different ways to compare fuzzy sets can be found in [4].

From this starting point, we have been able to generalize this concept to define the divergence measure for comparing two intuitionistic fuzzy sets [25].

The particular case of local divergences for intuitionistic fuzzy sets was studied in [26]. There, we presented interesting applications of this concept in Pattern Recognition and Decision Theory.

A similar generalization has been done for hesitant fuzzy sets in [16].

Moreover, we have been able to use the divergences to measure the fuzziness of a fuzzy set by comparing it with the closest crisp set and conversely, we have used fuzziness measures to define a divergence measure [21].

All these definitions and results can be considered as a heritage of the classical divergence measures, and more precisely, of the knowledge about them conveyed by Prof. Gil to the authors of this work.

4 Conclusion

In this paper we have studied some relationships among different ways to compare two elements, under uncertainty and imprecision.

Thus, we have used the classical divergence measures between two probability distributions to obtain a new way to compare two fuzzy sets. This is a particularly interesting case of dissimilarity in some cases and it has very interesting and specific properties.

The link between randomness and fuzziness is proven one more time, as we did previously for probabilistic and non-probabilistic entropies.

Acknowledgements We would like to acknowledge the help and support of Prof. Gil to initiate us to the wonderful world of research. He used to say he was our scientific father and we are honored he really was.

From the economical point of view, this work was partially supported by the research project TIN2014-59543-P (Spain).

References

1. Bouchon-Meunier B, Rifqi M, Bothorel S (1996) Towards general measures of comparison of objects. Fuzzy Sets Syst 84:143–153
2. Couso I, Montes S (2008) An axiomatic definition of fuzzy divergence measures. Int J Uncertain Fuzziness Knowl Based Syst 16(1):1–17
3. Couso I, Gil P (1998) Characterization of a family of entropy measures. In: Proceedings of IPMU'98. Editions EDK, Paris
4. Couso I, Garrido L, Sánchez L (2013) Similarity and dissimilarity measures between fuzzy sets: a formal relational study. Inform Sci 229:122–141
5. De Luca A, Termini S (1972) A definition of a nonprobabilistic entropy in the setting of fuzzy sets theory. Inf Control 20:301–312
6. Dubois D, Prade H (1980) Fuzzy sets and systems: theory and applications. Academic Press, New York
7. Gil P (1981) Teoría Matemática de la Información. ICE, Madrid
8. Gil MA, Gil P (2015) Randomness and fuzziness: combined better than unified. In: Magdalena L, Verdegay JL, Esteva F (eds) Enric trillas: a passion for fuzzy sets, studies in fuzziness and soft computing, vol 322. Springer, Cham
9. Gil MA, López MT, Gil P (1985) Quantity of information; comparison between information systems: 1. Non-fuzzy states. Fuzzy Sets Syst 15:65–78
10. Gil MA, López MT, Gil P (1985) Quantity of information; comparison between information systems: 2. Fuzzy states. Fuzzy Sets Syst 15:129–145
11. Havrda J, Charvát F (1967) Quantification method of classification processes. Concept of structural α-entropy. Kybernetika 3(1):30–35
12. Kampé de Fériet J, Forte B (1967) Information et probabilité. CR Acad Sci Paris Ser A 265:110–114, 142–146, 350–353
13. Klir GJ, Folger TA (1988) Fuzzy sets, uncertainty, and information. Prentice Hall, Upper Saddle River
14. Knopfmacher J (1975) On measures of fuzziness. J Math Anal Appl 49:529–534
15. Kobza V, Janis V, Montes S (2017) Generalizated local divergence measures. J Intel Fuzzy Syst 33:337–350
16. Kobza V, Janis V, Montes S (2017) Divergence measures on hesitant fuzzy sets. J Intel Fuzzy Syst 33:1589–1601
17. Kullback S (1959) Information theory and statistics. Wiley, New York
18. Lui X (1992) Entropy, distance measure and similarity measure of fuzzy sets and their relations. Fuzzy Sets Syst 52:305–318
19. Menéndez ML, Morales D, Pardo L, Salicrú M (1993) Asymptotic distribution of (h, ϕ)-entropies. Commun Stat Theory Meth 22(7):2015–2031
20. Montes S, Gil P (1998) Some classes of divergence measures between fuzzy subsets and between fuzzy partitions. Mathw Soft Comput 5:253–265
21. Montes S, Couso I, Bertoluzza C (1998) Some classes of fuzziness measures from local divergences. Belg J Oper Res Stat Comput Sci 38:37–49

22. Montes S, Gil P, Bertoluzza C (1998) Divergence between fuzzy sets and fuzziness. In: Proceedings of IPMU'98. Editions EDK, Paris
23. Montes S, Couso I, Gil P, Bertoluzza C (2002) Divergence measures between fuzzy sets. Int J Approx Reason 30(2):91–105
24. Montes S, Couso I, Jimenez J, Gil P (2005) Las medidas de incertidumbre probabilística y no probabilística como herramienta en la comparación de conjuntos. In: Volumen homenaje al Profesor Ildefonso Yáñez de Diego. UNED, Madrid
25. Montes I, Pal N, Janis V, Montes S (2015) Divergence measures for intuitionistic fuzzy sets. IEEE Trans Fuzzy Syst 23(2):444–456
26. Montes I, Pal N, Janis V, Montes S (2016) Local divergences for Atanassov intuitionistic fuzzy sets. IEEE Trans Fuzzy Syst 24(2):360–373
27. Oniçescu O (1966) Theorie de l'information. Energie informationelle. CR Acad Sci Paris Ser A 263:841–842
28. Pardo L (1997) Teoría de la Información Estadística. Hespérides, Salamanca
29. Pardo L (2006) Statistical inference based on divergence measures. Chapman & Hall, Boca Raton
30. Rényi A (1966) Calcul des Probabilitiés. Dunod, Paris
31. Shannon CE (1948) A mathematical theory of communication. Bell Syst Tech J 27(379–423):623–656
32. Sharma BD, Mittal DP (1975) New nonadditive measures of entropy for discrete probability distributions. J Math Sci 10:28–40
33. Trillas E, Riera T (1978) Entropies in finite fuzzy sets. Inf Sci 15:159–168
34. Zadeh L (1965) Fuzzy sets. Inf Contr 8:338–353

Discounting Imprecise Probabilities

Serafín Moral

Abstract In this paper it is considered the problem of discounting a credal set of probability distributions by a factor α representing a degree of unreliability of the information source providing the imprecise probabilistic information. An axiomatic approach is followed by giving a set of properties that this operator should satisfy. It is shown that discounting can be defined taking a divergence measure between probabilities as basis. Several examples are given starting from different divergence measures, as the Kullback-Leibler divergence or the total variance divergence. Finally, a characterization of the associated discounting is given in terms of sets of almost desirable gambles for two of these measures, providing a behavioral interpretation on them. The usual discounting of belief functions based on a convex combination with the ignorance is associated with the use of what it is called the minimum ratio divergence.

1 Introduction

In belief functions the discounting of a belief function by an unreliability degree α is a basic operator introduced by Shafer [11]. This operator consists in computing the convex combination of the original belief (with weight $1 - \alpha$) and the vacuous belief function (with weight α), obtaining in this way a belief function which is less informative than the original one (with wider belief-plausibility intervals) and which is equal to the vacuous belief when $\alpha = 1$. This discounting can be immediately extended to credal sets as the convex combination is also defined in this setting and there is also a vacuous credal set. In fact, it has been proposed in [5, 9] as the basis for discounting credal sets and in [6] it has been generalized to the case of contextual discounting. But other rules have been also proposed as for example in [7] where a convex combination with the uniform distribution is considered.

S. Moral (✉)
Dpto. Ciencias de la Computación e Inteligencia Artificial,
Universidad de Granada, 18071 Granada, Spain
e-mail: smc@decsai.ugr.es

© Springer International Publishing AG 2018
E. Gil et al. (eds.), *The Mathematics of the Uncertain*, Studies in Systems, Decision and Control 142, https://doi.org/10.1007/978-3-319-73848-2_63

In this paper it is considered the study of discounting operators for credal sets, following a general framework in which general properties are assumed and then different examples of methods satisfying these properties are given. The discounting of a credal set is related with a divergence measure: given a divergence measure, then the discounting of a credal set can be defined as the set of probabilities such that there is a probability in the original credal set with a divergence less or equal than a given threshold. We then give the properties associated to a divergence measure between probabilities. These properties are not the ones associated to metric distances as in [1] and are closely related to those of Csiszár f-divergence measures [4].

Credal sets have a behavioral interpretation in terms of gambles desirability [2, 12]. The interpretation of a discounting rule in terms of desirability can be very useful for determining which method should be applied in a particular situation, as this determines how the agent behavior changes when discounting is applied. In this paper we give two results associated to two important methods: the generalization of the belief functions discounting and the one associated width the total variance divergence [10].

The paper is organized as follows: Sect. 2 is devoted to the problem definition and notation and gives the required basic properties for the discounting of a credal set; Sect. 3 considers the properties of divergence measures between probabilities and relates divergence and discounting; Sect. 4 studies specific procedures for discounting credal sets giving a behavioral interpretation for two of them. Finally Sect. 5 is devoted to conclusions and future work.

2 Problem Definition and Notation

Assume that we have a finite referential $U = \{u_1, \ldots, u_n\}$. A *probability distribution* on this set is a mapping $p : U \to [0, 1]$ satisfying $\sum_{i=1}^{n} p(u_i) = 1$. If $a \subseteq U$, its probability is equal to $p(a) = \sum_{u_i \in a} p(u_i)$.

A *credal set* $\mathcal{K}$ is a convex and closed set of probability distributions. If $\mathcal{K} \neq \emptyset$ it will be said to be *coherent*. The credal set of all the probability distributions on U will be denoted by $\mathcal{K}_0$.

We will consider two rules to combine credal sets. The *conjunctive combination* of two credal sets $AND(\mathcal{K}_1, \mathcal{K}_2)$ is the intersection of them ($\mathcal{K}_1 \cap \mathcal{K}_2$). Two credal sets are *mutually inconsistent* when $AND(\mathcal{K}_1, \mathcal{K}_2) = \emptyset$. The *disjunctive combination* of two credal sets $OR(\mathcal{K}_1, \mathcal{K}_2)$ is the convex hull of the union $CH(\mathcal{K}_1 \cup \mathcal{K}_2)$.

A *gamble* on U is a mapping $f : U \to \mathbb{R}$. The expected value of f with respect to a probability distribution p will be denoted as $p(f)$. If $\mathcal{K}$ is a credal set, then the lower probability of a gamble will be denoted by $\underline{p}(f)$ and it is defined as $\min_{p \in \mathcal{K}} p(f)$. The lower probability $\underline{p}(f)$ can interpreted as the upper buying price for gamble f.

Given a coherent credal set $\mathcal{K}$, we can associated with it a *coherent set of almost desirable gambles* $\mathscr{D}_{\mathcal{K}}$:

$$\mathscr{D}_{\mathcal{K}} = \{f : p(f) \geq 0, \quad \forall p \in \mathcal{K}\}.$$

A coherent set of almost desirable gambles $\mathscr{D}$ satisfies a set of properties [12]:

C1.   If $f < 0$, then $f \notin \mathscr{D}$.
C2.   If $f \geq 0$, then $f \in \mathscr{D}$.
C3.   If $f_1, f_2 \in \mathscr{D}$, then $f_1 + f_2 \in \mathscr{D}$.
C4.   If $f \in \mathscr{D}, \alpha \geq 0$, then $\alpha f \in \mathscr{D}$.
C5.   If $f + \epsilon \in \mathscr{D}, \forall \epsilon > 0$, then $f \in \mathscr{D}$.

Given a generic set of gambles $\mathscr{D}$ we can associated with it a credal set $\mathcal{K}$:

$$\mathcal{K}_{\mathscr{D}} = \{p : p(f) \geq 0, \quad \forall f \in \mathscr{D}\}.$$

A generic set of gambles $\mathscr{D}$ *avoids sure loss* when $\mathcal{K}_{\mathscr{D}}$ is coherent. In that case, the natural extension of $\mathscr{D}$ is the set of gambles $\overline{\mathscr{D}} = \mathscr{D}_{\mathcal{K}_{\mathscr{D}}}$. It is immediate to prove that $\overline{\mathscr{D}}$ is the intersection of all the coherent set of gambles containing $\mathscr{D}$, and therefore satisfies properties C1–C5.

If $\mathscr{A} = \{a_1, \ldots, a_m\}$ is a partition of U and p is a probability distribution on U, then the *marginalization* of p to $\mathscr{A}$, is the probability distribution on $\mathscr{A}$ given by $p^{\downarrow \mathscr{A}}(a_i) = p(a_i) = \sum_{u_j \in a_i} p(u_j)$.

If $\mathcal{K}$ is a credal set, then the marginalization of $\mathcal{K}$ to $\mathscr{A}$ is the credal set $\mathcal{K}^{\downarrow \mathscr{A}} = \{p^{\downarrow \mathscr{A}} : p \in \mathcal{K}\}$.

If p is a probability on U and σ a permutation on U, then p^{σ} is the probability on U given by $p^{\sigma}(u_i) = p(u_{\sigma(i)})$. For a credal set, $\mathcal{K}^{\sigma} = \{p^{\sigma} : p \in \mathcal{K}\}$.

We will assume that the information about an unknown value of U is represented by means of a credal set $\mathcal{K}$ or by a single probability distribution p. Our problem is the following: given a credal set $\mathcal{K}$ and a discounting value $\alpha \in [0, 1]$ to compute the discounting $D(\mathcal{K}, \alpha)$. The discounting is an extension of the set of probability distributions by a degree α. The value $1 - \alpha$ can be considered as the *degree of reliability* of our information: we make our set of probability distributions less specific if we are unsure about them to a degree α. More concrete meanings will be given to this degree for specific discounting functions. If $\mathcal{K}$ contains a single probability p, then the discounting $D(\{p\}, \alpha)$ will be denoted as $D(p, \alpha)$.

We shall assume the following basic properties for discounting:

D1. $D(\mathcal{K}, \alpha)$ is a credal set containing $\mathcal{K}$.
D2.   If $\alpha_1 \leq \alpha_2$, then $D(\mathcal{K}, \alpha_1) \subseteq D(\mathcal{K}, \alpha_2)$.
D3. $D(\mathcal{K}, 0) = \mathcal{K}$.
D4. $D(\emptyset, \alpha) = \emptyset, \quad \forall \alpha \in [0, 1]$.
D5.   If $\mathcal{K}$ is coherent, then $D(\mathcal{K}, 1) = \mathcal{K}_0$.
D6. $D(OR(\mathcal{K}_1, \mathcal{K}_2), \alpha) = OR(D(\mathcal{K}_1, \alpha), D(\mathcal{K}_2, \alpha))$.
D7. $D(\mathcal{K}, \alpha)^{\downarrow \mathscr{A}} = D(\mathcal{K}^{\downarrow \mathscr{A}}, \alpha)$, for any partition $\mathscr{A}$ of U.

D8.   If σ is a permutation on U, then $D(\mathscr{K}^\sigma, \alpha) = D(\mathscr{K}, \alpha)^\sigma$.

D9.   If $q \in D(p, \alpha)$, and $a \subset U$, then $q_a = qI_a + pI_{a^c}(q(a^c)/p(a^c)) \in D(p, \alpha)$,
where a^c is the complementary of a, and I_a is the indicator function of a.

Properties D1–D5 are intuitive. We have chosen a scale for discounting in $[0, 1]$. We could have chosen $\mathbb{R}_0^*$ (the set of non-negative reals) or including an infinite value, $\mathbb{R}_0^* \cup \{+\infty\}$, but we have preferred this option as some discounting methods will have a very concrete meaning when $\alpha \in [0, 1]$ and for other discounting procedures we can always change the scale $\mathbb{R}_0^* \cup \{+\infty\}$ to $[0, 1]$, with some increasing function as $g(x) = 1 - e^{-x}$.

Property D6 is based on the following idea: $OR(\mathscr{K}_1, \mathscr{K}_2)$ is the credal set resulting of a doubt between $\mathscr{K}_1$ and $\mathscr{K}_2$: any of them can be possible. If we have an unreliability of α on these credal sets, then this is an additional doubt about the credal sets. We have to discount them by a degree α. We have a double source of imprecision (disjunction and unreliability) and D6 says that the order in which the operators associated with these sources are applied is irrelevant: it is the same to make the disjunction and then to discount, than to discount and then apply the disjunction. This property implies the monotonicity property, i.e. if $\mathscr{K}_1 \subseteq \mathscr{K}_2$, then we have that $D(\mathscr{K}_1, \alpha) \subseteq D(\mathscr{K}_2, \alpha)$, as $\mathscr{K}_1 \subseteq \mathscr{K}_2$ is equivalent to $OR(\mathscr{K}_1, \mathscr{K}_2) = \mathscr{K}_2$. However, D6 is stronger than the monotonicity property.

We do not have a similar property for conjunction: let us observe that we can have two mutually inconsistent credal sets which can be coherent after discounting, so $D(AND(\mathscr{K}_1, \mathscr{K}_2), \alpha)$ can be the empty set, and $AND(D(\mathscr{K}_1, \alpha), D(\mathscr{K}_2, \alpha))$ can be coherent. In fact, one of the applications of the discounting operator will be to define a method to combine mutually inconsistent credal sets, by carrying out a previous discounting operator.

A consequence of Property D6 is that to define a discounting operator in credal sets with a finite set of extreme points ($\mathscr{K} = \{p_1, \ldots, p_k\}$), we only have to define it in probability distributions, as we have that $D(\mathscr{K}, \alpha) = OR(D(p_1, \alpha), \ldots, D(p_k, \alpha))$. For this reason, we shall concentrate in defining the discounting for single probability distributions.

Property D7, says that discounting commutes with marginalization. The marginalization of a probability distribution does not assume a change of knowledge, but only a focusing of this knowledge in some events (those represented by subsets of $\mathscr{A}$) and then if we are interested only in these events, the result is the same if we discount in the more fine-grained setting, U, or in the more coarse-grained framework $\mathscr{A}$.

Property D9, says that if we are discounting a single probability, then if $q \in D(p, \alpha)$, and we transform q by making it more similar to p in the complementary of a, while keeping it equal to q in a, then the result should also be in $D(p, \alpha)$.

3 Discounting a Probability

As a consequence of Property D6, for finitely generated credal sets it is enough to determine the discounting of all the extreme probabilities of $\mathcal{K}$. So, in this section we shall concentrate in defining the discounting for a single probability, $D(p, \alpha)$. In general, this discounting will be a credal set and not a single probability. We find a direct correspondence between discounting of probabilities and divergence measures between probabilities [3, 10]. A *divergence measure* $Di(p, q)$ is a non-negative real number quantifying how different is q from p. We will consider the following properties for a divergence measure,

I1. $Di(p, q) = 0$ if and only if $p = q$.
I2. $Di(p, \alpha q_1 + (1 - \alpha)q_2) \leq \max\{Di(p, q_1), Di(p, q_2)\}$.
I3. If $\mathcal{A} = \{a_1, \ldots, a_m\}$ is a finite partition of U, $Di(p^{\downarrow \mathcal{A}}, q^{\downarrow \mathcal{A}}) \leq Di(p, q)$ with equality if p and q are such that $\frac{p(u_i)}{p^{\downarrow \mathcal{A}}(a_j)} = \frac{q(u_i)}{q^{\downarrow \mathcal{A}}(a_j)}$, when $u_i \in a_j$.
I4. If σ is a permutation, then $Di(p, q) = Di(p^\sigma, q^\sigma)$.

Divergence measures have been studied in probability [4]. These properties are inspired on the properties given in [10], however I2 is weaker than the convexity property given in that reference. It is important to remark that divergences are not necessarily symmetrical as here it is assumed the existence of a reference *true* distribution, p, and an *approximate* one q. The use of metric distances, such as Euclidean distance is not always a good idea, as shown in the following example.

Example 3.1 Assume $U = \{u_1, u_2, u_3\}$ and the probabilities $p = (0.4, 0.3, 0.3)$, $q = (0.2, 0.4, 0.4)$. The Euclidean distance between them is $\sqrt{0.6}$. Consider now the partition $\mathcal{A} = \{a_1, a_2\}$, given by $a_1 = \{u_1\}, a_2 = \{u_2, u_3\}$, i.e. u_2 and u_3 have been joined in one value. The induced probabilities in $\mathcal{A}$ are $p^{\downarrow \mathcal{A}} = (0.4. 0.6), q^{\downarrow \mathcal{A}} = (0.2. 0.8)$. The Euclidean distance is now $\sqrt{0.8}$. We can notice that the distance increases when the information have been marginalized: we have lost some information in the marginalization process, but the distance has increased.

A general class of divergence measures are given by the Csiszár f-divergences [4, 10] given by

$$Di_f(p, q) = \sum_{u_i \in U} p(u_i) f\left(\frac{q(u_i)}{p(u_i)}\right) \tag{1}$$

where f is a strictly convex function defined on the non-negative reals with $f(1) = 0$, and $0/0$ is assumed to be 0. Examples of f-divergences are:

- *Kullback-Leibler divergence* [8], when $f(x) = -\log(x)$. The final formula is $KL(p, q) = \sum_{u_i \in U} p(u_i) \log\left(\frac{p(u_i)}{q(u_i)}\right)$.
- *Likelihood disparity*, when $f(x) = x \log(x)$. The formula is $LD(p, q) = \sum_{u_i \in U} q(u_i) \log\left(\frac{q(u_i)}{p(u_i)}\right)$.

- *Total variance divergence*, when $f(x) = (1/2)|1 - x|$. It is immediate to show that, in this case we obtain $TV(p, q) = (1/2) \sum_{u_i \in U} |p(u_i) - q(u_i)| = \max_{a \subseteq U} |p(a) - q(a)|$.
- *Hellinger divergence*. When $f(x) = (1/2)(1 - \sqrt{(x)})^2$, we obtain the square of the Hellinger distance $H^2(p, q)$.

There are also divergence measures that are not f-divergences. Among then, we shall consider what we call the min-ratio divergence:

$$MR(p, q) = 1 - \min_{u_i \in U} \frac{q(u_i)}{p(u_i)}. \tag{2}$$

If $Di(p, q)$ is a divergence measure and g is a strictly increasing function with $g(0) = 0$, then $g(Di(p, q))$ is also a divergence measure. So, we can transform any divergence so that its range of values is in $[0, 1]$. In that case, the divergence will be said to be *normalized*. We shall assume that we have done this transformation and that, for example, the normalized Kullback-Leibler divergence, $nKL(p, q)$ will take values in $[0, 1]$, transforming the original Kullback-Leibler divergence taking values on $[0, +\infty]$ by means of $g(x) = 1 - e^{-x}$, i.e. $nKL(p, q) = 1 - e^{-KL(p,q)}$, where $g(+\infty) = 1$.

Now, we can define the discounting of a probability distribution by using a (normalized) divergence measure. If Di is a normalized divergence measure, p is a probability measure and $\alpha \in [0, 1]$, then the discounting of p is given by

$$D(p, \alpha) = \overline{\{q \mid Di(p, q) \leq \alpha\}}, \tag{3}$$

where the overline stands for the topological closure.

Proposition 3.1 *If $D(p, \alpha)$ is the discounting of a probability associated to a normalized divergence measure Di, then*

(a) $D(p, \alpha)$ is a credal set.
(b) If $\alpha_1 \leq \alpha_2$, then $D(p, \alpha_1) \subseteq D(p, \alpha_2)$.
(c) $D(p, 0) = \{p\}$.
(d) $D(p, 1) = \mathcal{K}_0$.
(e) $D(p, \alpha)^{\downarrow \mathscr{A}} = D(p^{\downarrow \mathscr{A}}, \alpha)$, for any partition $\mathscr{A}$ of U.
(f) If σ is a permutation on U, then $D(p^\sigma, \alpha) = D(p, \alpha)^\sigma$.
(g) If $q \in D(p, \alpha)$, and $a \subset U$, then $q_a = qI_a + pI_{a^c}(q(a^c)/p(a^c)) \in D(p, \alpha)$.

Proof For the property in (a), $D(p, \alpha)$ is convex from I2. It is closed, as we are taking closure (also the closure of a convex set is always convex).

Property in (b) is a consequence of Eq. (3): if $\alpha_1 \leq \alpha_2$ and $Di(p, q) \leq \alpha_1$, then $Di(p, q) \leq \alpha_2$.

Property in (c) is a consequence of I1.

Property in (d) is a consequence of the fact that the discounting is normalized and $Di(p, q) \leq 1$ is always satisfied.

For the property (e), $D(p, \alpha)^{\downarrow \mathscr{A}} \subseteq D(p^{\downarrow \mathscr{A}}, \alpha)$ because $Di(p^{\downarrow \mathscr{A}}, q^{\downarrow \mathscr{A}}) \leq Di(p, q)$.

On the other hand, if q defined on $\mathscr{A}$ is such that $Di(p^{\downarrow \mathscr{A}}, q) \leq \alpha$, let us consider q' in U given by $q'(u_i) = q^{\downarrow \mathscr{A}}(a_j) . p(u_i)/p^{\downarrow \mathscr{A}}(a_j)$, when $u_i \in a_j$. By I3, we have that $Di(p, q') = Di(p^{\downarrow \mathscr{A}}, q) \leq \alpha$, and $q' \in \{r : Di(p, r) \leq \alpha\}$ and $q = q'^{\downarrow \mathscr{A}} \in \{r : Di(p, r) \leq \alpha\}^{\downarrow \mathscr{A}}$.

Then, we have that $\{r : Di(p^{\downarrow \mathscr{A}}, r) \leq \alpha\} \subseteq \{r : Di(p, r) \leq \alpha\}^{\downarrow \mathscr{A}}$, and the inclusion is kept if we take the closure operator.

Property (f) is an immediate consequence of I4.

For property (g), consider the partition of U given by $\mathscr{A} = \{\{u_i\} : u_i \in a\} \cup \{a^c\}$. In this case, $Di(p, q) \geq Di(p, q^{\downarrow \mathscr{A}}) = Di(p, q_a)$. The last equality is a consequence of the fact that $q^{\downarrow \mathscr{A}} = q_a^{\downarrow \mathscr{A}}$ and that $\frac{p(u_i)}{p(a_j)} = \frac{q_a(u_i)}{q_a(a_j)}$ for any $u_i \in a_j$. Then $Di(p, q) \geq Di(p, q_a)$ and the property is obtained. $\qquad \square$

On the other hand, a divergence measure can be obtained from a discounting function for probabilities.

Proposition 3.2 *If $D(p, \alpha)$ is a discounting of probabilities satisfying the following properties:*

(a) $D(p, \alpha)$ *is a credal set.*
(b) *If $\alpha_1 \leq \alpha_2$, then $D(p, \alpha_1) \subseteq D(p, \alpha_2)$.*
(c) $D(p, 0) = \{p\}$.
(d) $D(p, 1) = \mathscr{K}_0$.
(e) $D(p, \alpha)^{\downarrow \mathscr{A}} = D(p^{\downarrow \mathscr{A}}, \alpha)$, *for any partition $\mathscr{A}$ of U.*
(f) *If σ is a permutation on U, then $D(p^\sigma, \alpha) = D(p, \alpha)^\sigma$.*
(g) *If $q \in D(p, \alpha)$, and $a \subset U$, then $q_a = q I_a + p I_{a^c}(q(a^c)/p(a^c)) \in D(p, \alpha)$.*

Then, the function

$$Di(p, q) = \inf\{\alpha \in [0, 1] : q \in D(p, \alpha)\} \qquad (4)$$

is a divergence measure.

Proof Property I1 is an immediate consequence of $D(p, 0) = \{p\}$.

For I2, consider $\alpha < Di(p, \beta q_1 + (1 - \beta)q_2)$, then $(\beta q_1 + (1 - \beta)q_2) \notin D(p, \alpha)$. As $D(p, \alpha)$ is a convex set, then either $q_1 \notin D(p, \alpha)$ or $q_2 \notin D(p, \alpha)$. Therefore, $\max\{Di(p, q_1), Di(p, q_2)\} \geq \alpha$. And $\max\{Di(p, q_1), Di(p, q_2)\} \geq Di(p, \beta q_1 + (1 - \beta)q_2)$.

For I3, consider a partition $\mathscr{A}$, if $q \in D(p, \alpha)$, then $q^{\downarrow \mathscr{A}} \in D(p^{\downarrow \mathscr{A}}, \alpha)$ by Assumption $e)$ in this proposition, and $Di(p^{\downarrow \mathscr{A}}, q^{\downarrow \mathscr{A}}) \leq Di(p, q)$.

Consider now, p and q in such a way that $\frac{p(u_i)}{p^{\downarrow \mathscr{A}}(a_j)} = \frac{q(u_i)}{q^{\downarrow \mathscr{A}}(a_j)}$, when $u_i \in a_j$.

If $q^{\downarrow \mathscr{A}} \in D(p^{\downarrow \mathscr{A}}, \alpha)$, then as $D(p, \alpha)^{\downarrow \mathscr{A}} = D(p^{\downarrow \mathscr{A}}, \alpha)$, we have that there is a probability $q' \in D(p, \alpha)$ such that $q'^{\downarrow \mathscr{A}} = q^{\downarrow \mathscr{A}}$. Consider now a repeated transformation of q' by applying Assumption $g)$ in this proposition for events $a = a_j^c$ where $a_j \in \mathscr{A}$. Each probability that is obtained in this transformation is in $D(p, \alpha)$ and the final result is $q'' = \sum_{a_j \in \mathscr{A}} \frac{q(a_j)}{p(a_j)} I_{a_j} p$. But this probability is equal to q, as if $u_i \in a_j$, we have that $q(u_i) = p(u_i) \frac{q^{\downarrow \mathscr{A}}(a_j)}{p^{\downarrow \mathscr{A}}(a_j)} = p(u_i) \frac{q(a_j)}{p(a_j)} = q''(u_i)$. So $q \in D(p, \alpha)$, and therefore $Di(p, q) \leq Di(p^{\downarrow \mathscr{A}}, q^{\downarrow \mathscr{A}})$ and the equality holds as we had already proved the other inequality.

Finally, I4 is an immediate consequence of Assumption f) in this proposition. $\square$

4 Discounting Credal Sets

Consider that Di is a divergence measure, then the discounting of a credal set is defined as

$$D(\mathscr{K}, \alpha) = \overline{\{q \; : \; \exists p \in \mathscr{K} \text{ with } Di(p, q) \leq \alpha\}}. \tag{5}$$

When, $Di(p, q)$ is a continuous function, then $\{q \; : \; \exists p \in \mathscr{K} \text{ with } Di(p, q) \leq \alpha\}$ will be closed and then it is not necessary to take the closure. All the examples we have given are continuous. So, it is not necessary to apply the topological closure in these cases.

One important question when selecting a divergence measure to define a discounting procedure, is what its meaning is. In this section, we give an answer for the total variance divergence and the minimum-ratio divergence, by studying their implications in the associated set of almost desirable gambles. We start with the total variance divergence, $TV(p, q)$. The discounting of $\mathscr{K}$ by means of this divergence will be denoted by $D_{TV}(\mathscr{K}, \alpha)$.

If $\mathscr{D}$ is a coherent set of almost desirable gambles, then the discounting associated to $TV(p, q)$ is defined by

$$D_{TV}(\mathscr{D}, \alpha) = \overline{\{f + \alpha(\sup(f) - \inf(f)) \; : \; f \in \mathscr{D}_{\mathscr{K}}\}},$$

where overline stands for the natural extension operator. This makes sense, as according to the following theorem the credal set associated to $D_{TV}(\mathscr{D}, \alpha)$ is $D_{TV}(\mathscr{K}, \alpha)$.

Theorem 4.1 *Consider a credal set $\mathscr{K}$, then*

$$D_{TV}(\mathscr{K}, \alpha) = \mathscr{K}_{D_{TV}(\mathscr{D}_{\mathscr{K}}, \alpha)}. \tag{6}$$

Proof If $q \in D_{TV}(\mathscr{K}, \alpha)$, then there is $p \in \mathscr{K}$ with $TV(p, q) \leq \alpha$. This implies that $\sum_{p(u_i) > q(u_i)} (p(u_i) - q(u_i)) \, \alpha' \leq \alpha$ and $\sum_{p(u_i) < q(u_i)} (q(u_i) - p(u_i)) \, \alpha' \leq \alpha$.
If $f \in \mathscr{D}_{\mathscr{K}}$, then $p(f) \geq 0$.
On the other hand,

$$p(f) - q(f) = \sum_{u_i \in U} (p(u_i) - q(u_i)) f(u_i)$$

$$\leq \sum_{p(u_i) > q(u_i)} (p(u_i) - q(u_i)) f(u_i) - \sum_{p(u_i) < q(u_i)} (q(u_i) - p(u_i)) f(u_i)$$

$$\leq \sup(f) \sum_{p(u_i) > q(u_i)} (p(u_i) - q(u_i)) - \inf(f) \sum_{p(u_i) < q(u_i)} (q(u_i) - p(u_i))$$

$$= \sup(f) \alpha' - \inf(f) \alpha' \leq \alpha(\sup(f) - \inf(f)). \tag{7}$$

So,

$$q(f + \alpha(\sup(f) - \inf(f))) = q(f) + \alpha(\sup(f) - \inf(f))$$

$$\geq p(f) - \alpha(\sup(f) - \inf(f)) + \alpha(\sup(f) - \inf(f)) = p(f) \geq 0,$$

and $q \in \mathcal{K}_{D_{TV}(\mathcal{D}_{\mathcal{K}}, \alpha)}$.

Assume now that $p \notin D_{TV}(\mathcal{K}, \alpha)$. As $TV(p, q)$ is symmetrical, we have that $D_{TV}(p, \alpha) \cap \mathcal{K} = \emptyset$. As these sets are closed and convex, there is an hyperplane strictly separating them, i.e. a gamble f, such that $q(f) < 0$ for any $q \in D_{TV}(p, \alpha)$ and $q(f) > 0$ for any $q \in \mathcal{K}$.

As $q(f) > 0$ for any $q \in \mathcal{K}$, $f \in \mathcal{D}_{\mathcal{K}}$.

Let $M_1 = \{u_i \mid f(u_i) = \sup(f)\}$ and $M_2 = \{u_i \mid f(u_i) = \inf(f)\}$ and q' the point in $D_{TV}(\mathcal{K}, \alpha)$ where f is maximized. We have that $q'(M_1) < 1.0$, as $\sup(f) > 0$ and $q'(f) < 0$.

Also, if $q'(M_2) = 0$, we can modify f to f' which is equal to f in points not in M_2 and equal to $\inf_{u_i \in U \setminus M_2}$ in M_2. We have that $f' \geq f$, and therefore $f' \in \mathcal{D}_{\mathcal{K}}$. On the other hand, the maximum of $q(f')$ in $q \in D_{TV}(p, \alpha)$ is the same than the maximum of $q(f)$ in the same set: as given the nature of $D_{TV}(p, \alpha)$ this maximum is achieved at the same point q' and $q'(M_2) = 0$.

The fact that the maximum values of $q(f)$ and $q(f')$ are obtained at the same point $q' \in D_{TV}(p, \alpha)$, is a consequence of the fact that the point in which the maximum of $q(f)$ is achieved in $D_{TV}(p, \alpha)$ can be computed by considering a permutation σ in $\{1, \ldots, n\}$ such that $f(u_{\sigma(1)}) \geq f(u_{\sigma(2)}) \geq \cdots \geq f(u_{\sigma(n)})$, considering the sets $a_i = \{u_{\sigma(1)}, \ldots, u_{\sigma(i)}\}$ and considering $q'(a_i) = \min\{p(a_i) + \alpha, 1.0\}$. These values define an only probability measure q' at which $q(f)$ is maximized: it tries to assign larger probabilities to the points in which f is larger. As we can associate the same σ to f and f' the functional can be maximized at the same point q'.

As $q(f) < 0$ for any $q \in D_{TV}(p, \alpha)$, we also have that $q(f') < 0$ for any $q \in D_{TV}(p, \alpha)$ (the maximum negative value is the same in both cases).

We have proved that for any separating hyperplane f for which the probability maximizing it $q' \in D_{TV}(p, \alpha)$ is such that $q'(M_2) = 0$, there is another separating hyperplane, f' in which the set of points M_2 is enlarged. As in a separating hyperplane, we can not have $M_1 = M_2 = U$, we can conclude that there is a separating hyperplane f of $\mathcal{K}$ and $D_{TV}(p, \alpha)$, such that the probability maximizing it in $D_{TV}(p, \alpha)$ is such that $q'(M_1) < 1.0$, and $q'(M_2) > 0$.

We have that $q'(M_1) - p(M_1) = \alpha$, as the maximum of probability of $q(f)$ where $q \in D_{TV}(p, \alpha)$ is achieved in a point q' with $q'(M_1) = 1.0$ or $q'(M_1) = p(M_1) + \alpha$ (we have to move the maximum of probability to the point in which the function is maximized) and $q'(M_1) < 1.0$. Also, we can conclude that $p(M_2) - q'(M_2) = \alpha$ and $q'(x) = p(x)$, $\forall u_i \notin M_1 \cup M_2$.

Therefore

$$p(f) = q(f') - \alpha(\sup(f) - \inf(f)) < 0, \quad p(f + \alpha(\sup(f) - \inf(f))) < 0.$$

As $f \in \mathscr{D}_\mathscr{K}$, we have that $f + \alpha(\sup(f) - \inf(f)) \in D_{TV}(\mathscr{D}_\mathscr{K}, \alpha)$, and therefore $p \notin \mathscr{K}_{D_{TV}(\mathscr{D}_\mathscr{K}, \alpha)}$. $\qquad\square$

According to this result, the discounting of degree α taking as basis the total variation divergence, implies that old desirable gambles f must be increased by a value $\alpha(\sup(f) - \inf(f))$ in order to be desirable after discounting.

It is not simple to obtain an expression for discounting in terms of lower probabilities associated to generic gambles (the difficulty comes from the presence of the natural extension operator in the associated set of almost desirable gambles). However for events, it is easy to obtain that if $\underline{p}$ is the lower probability associated with $\mathscr{K}$ and $\underline{p}_\alpha$ is the lower probability associated with $D_{TV}(\mathscr{K}, \alpha)$, the for any $a \subseteq U$, then $\underline{p}_\alpha(a) = \max\{0, \underline{p}(a) - \alpha\}$.

Now, we give a similar characterization for the discounting based on the minimum ratio divergence. We call $D_{MR}(\mathscr{K}, \alpha)$ to the set $\{q \mid \exists p \in \mathscr{K} \text{ with } MR(p, q) \le \alpha\}$ and now the equivalent discounting for coherent sets of almost desirable gambles is $D_{MR}(\mathscr{D}, \alpha) = \{f - \alpha \inf(f) \mid f \in \mathscr{D}\}$. We do not have to take the natural extension as this set is always coherent if $\mathscr{D}$ is coherent. The following theorem, shows that these definitions are consistent in the sense that the credal set associated to the discounting of a coherent set of desirable gambles is the discounting of the credal set. Before, we give a characterization of the credal set $D_{MR}(\mathscr{K}, \alpha)$.

Proposition 4.1 *If $\mathscr{K}$ is a credal set, then*

$$D_{MR}(\mathscr{K}, \alpha) = \{(1 - \alpha)p_1 + \alpha p_2 \mid p_1 \in \mathscr{K}, \ p_2 \in \mathscr{K}_0\}.$$

Proof It is easy to prove that $p \in \{(1 - \alpha)p_1 + \alpha p_2 \mid p_1 \in \mathscr{K}, \ p_2 \in \mathscr{K}_0\}$ if and only if there is a probability $p_1 \in \mathscr{K}$ with $p \ge (1 - \alpha)p_1$ (as p_2 is an arbitrary probability). And this is equivalent, to $\frac{p}{p_1} \ge 1 - \alpha$, i.e. $1 - \frac{p}{p_1} \le \alpha$, that is $MR(p_1, p) \le \alpha$. $\qquad\square$

Theorem 4.2 *Consider a credal set $\mathscr{K}$, then*

$$D_{MR}(\mathscr{K}, \alpha) = \mathscr{K}_{D_{MR}(\mathscr{D}_\mathscr{K}, \alpha)}. \tag{8}$$

Proof If $\alpha = 1$, the result is trivial as both sets are $\mathscr{K}_0$. Assume now $\alpha < 1$.

Assume $p \in D_{MR}(\mathscr{K}, \alpha)$. then $p = (1 - \alpha)p_1 + \alpha p_2$ where $p_1 \in \mathscr{K}$ and p_2 is an arbitrary probability measure.

If $g \in D_{MR}(\mathscr{D}_{\mathscr{K}}, \alpha)$, then $g = f - \alpha(\inf(f))$, where $f \in \mathscr{D}_{\mathscr{K}}$.
As $f \in \mathscr{D}_{\mathscr{K}}$ and $p_1 \in \mathscr{K}$, we have that $p_1(f) \geq 0$.
In these conditions,

$$p(g) = p(f) - \alpha(\inf(f)) = (1 - \alpha)p_1(f) + \alpha p_2(f) - \alpha(\inf(f)).$$

As $p_2(f) \geq \inf(f)$, we have

$$p(g) \geq (1 - \alpha)p_1(f) + \alpha \inf(f) - \alpha(\inf(f)) = (1 - \alpha)p_1(f) \geq 0.$$

As $p(g) \geq 0$ for any $g \in D_{TR}(\mathscr{D}_{\mathscr{K}}, \alpha)$, we have that $p \in \mathscr{K}_{D_{TR}(\mathscr{D}_{\mathscr{K}}, \alpha)}$.

Assume now $p \notin D_{TR}(\mathscr{K}, \alpha)$. As $D_{TR}(\mathscr{K}, \alpha)$ is a closed and convex set, there is an hyperplane strictly separating p and $D_{TR}(\mathscr{K}, \alpha)$. Assume that f is the gamble associated with this hyperplane and that $p(f) < 0$ and $p'(f) > 0$ for any $p' \in D_{TR}(\mathscr{K}, \alpha)$.

This implies that,

$$(1 - \alpha)p_1(f) + \alpha p_2(f) > 0, \quad \forall p_1 \in \mathscr{K}, \ p_2 \in \mathscr{K}_0.$$

As $\mathscr{K}_0$ contains all the probability measures, consider that p_2 is the probability that assigns probability 1.0 to the point in which f takes its minimum. We obtain that

$$(1 - \alpha)p_1(f) + \alpha \inf(f) > 0, \quad \forall p_1 \in \mathscr{K},$$

and as a consequence

$$p_1((1 - \alpha)f + \alpha \inf(f)) > 0, \quad \forall p_1 \in \mathscr{K}.$$

The restriction associated to gamble $(1 - \alpha)f + \alpha \inf(f)$ is strictly satisfied by any $p_1 \in \mathscr{K}$, and then $g = (1 - \alpha)f + \alpha \inf(f) \in \mathscr{D}$.

We have that $\inf(g) = (1 - \alpha)\inf(f) + \alpha \inf(f) = \inf(f)$. Therefore, $g - \alpha \inf(g) = (1 - \alpha)f + \alpha \inf(f) - \alpha \inf(f) = (1 - \alpha)f \in D_{TR}(\mathscr{D}_{\mathscr{K}}, \alpha)$.

As $p(g - \alpha \inf(g)) = (1 - \alpha)p(f) < 0$ and $g - \alpha \inf(g) \in D_{TR}(\mathscr{D}_{\mathscr{K}}, \alpha)$, we have that $p \notin \mathscr{K}_{D_{TR}(\mathscr{D}_{\mathscr{K}}, \alpha)}$.

We have proved that $p \notin D_{TR}(\mathscr{K}, \alpha)$ implies that $p \notin \mathscr{K}_{D_{TR}(\mathscr{D}_{\mathscr{K}}, \alpha)}$, which completes the proof. □

Given Proposition 4.1, it is easy to prove that if $\underline{p}$ is the lower probability associated with $\mathscr{K}$ and $\underline{p}_\alpha$ the lower probability associated with $D_{MR}(\mathscr{K}, \alpha)$, then we have that $\underline{p}_\alpha(f) = \underline{p}(f) - \alpha(\inf(f))$, for any gamble f. In this case discounting by α implies that the upper buying price for any gamble f should be decreased by $\alpha(\inf(f))$.

5 Conclusions and Future Work

In this paper we have studied the problem of discounting imprecise probabilities. We have given a set of properties that this operator should satisfy and we have given several examples. We have shown that this problem has a strong connection with the computation of divergence between probability measures. Finally, we have shown that some of the discounting operators can have a behavioral interpretation.

In the future, we plan to provide interpretations in terms of desirability for other discounting operators. In particular, we are specially interested in the one associated with the use of Kullback-Leibler divergence, which is the most usual in probability theory. We also want to apply discounting to the problem of combination of information [9] when the information provided by different sources is mutually inconsistent. In this case the reliability of the sources decreases and it is natural to discount them till consistency is attained.

Acknowledgements This paper is devoted to the memory of Prof. Pedro Gil. In his book, I learned my first concepts of entropy and information, which have been one of the most important topics of my research. He was in the committee of my Ph.D. defense and since them he has been a master not only in the scientific side, but also he has been a guide in other aspects of academic and general life, giving very valuable opinions grounded in his vast experience and wisdom. For all these reasons it has been an honour for me to be able of participating in this homage book.

This research was supported by the Spanish Ministry of Economy and Competitiveness under projects TIN2013-46638-C3-2-P and TIN2016-77902-C3-2-P, and the European Regional Development Fund (FEDER).

References

1. Abellán J, Gómez-Olmedo M (2006) Measures of divergence on credal sets. Fuzzy Sets Syst 157:1514–1531
2. Couso I, Moral S (2011) Sets of desirable gambles: conditioning, representation, and precise probabilities. Int J Approx Reas 52:1034–1055
3. Csiszár I (1967) Information-type measures of difference of probability distributions and indirect observation. Studia Scient Math Hungarica 2:229–318
4. Csiszár I (2008) Axiomatic characterizations of information measures. Entropy 10:261–273
5. Del Sagrado J (2000) Fusión Topológica y Cuantitativa de Redes Causales. PhD thesis. University of Granada
6. Destercke S (2010) A new contextual discounting rule for lower probabilities. In: Hüllermeier E, Kruse R, Hoffmann F (eds) Information processing and management of uncertainty in knowledge-based systems. applications, IPMU'2010. Series Communications in Computer and Information Science, vol 81. Springer, Berlin
7. Karlsson A, Johansson R, Andler SF (2009) On the behavior of the robust Bayesian combination operator and the significance of discounting. In: Augustin T, Coolen FPA, Moral S, Troffaes MCM (eds) SIPTA'09: Proceedings of the 6th international symposium on imprecise probability: theories and applications. SIPTA, Durham
8. Kullback S (1997) Information theory and statistics. Dover Pub, New York
9. Moral S, Del Sagrado J (1998) Aggregation of imprecise probabilities. In: Bouchon-Meunier B (ed) Aggregation and fusion of imperfect information. Physica-Verlag, Heidelberg

10. Österreicher F (2002) Csiszárs f-Divergences—basic properties. Talk presented at workshop of the Research group in mathematical inequalities and applications at the Victoria University, Melbourne, Australia, October. https://www.uni-salzburg.at/fileadmin/oracle_file_imports/246178.PDF
11. Shafer G (1976) A mathematical theory of evidence. Princeton University Press, Princeton
12. Walley P (1991) Statistical reasoning with imprecise probabilities. Chapman & Hall, New York

On Some Concepts Related to Star-Shaped Sets

Ana Belén Ramos-Guajardo, Gil González-Rodríguez, Ana Colubi,
Maria Brigida Ferraro and Ángela Blanco-Fernández

Abstract The convenient theoretical properties of the support function and the Minkowski addition-based arithmetic have been shown to be useful when dealing with compact and convex sets on $\mathbb{R}^p$. However, both concepts present several drawbacks in certain contexts. The use of the radial function instead of the support function is suggested as an alternative to characterize a wider class of sets—the so-called star-shaped sets—which contains the class of compact and convex sets as a particular case. The concept of random star-shaped set is considered, and some statistics for this kind of variable are shown. Finally, some measures for comparing star-shaped sets are introduced.

1 Introduction

Random sets, also called set-valued random variables and denoted by RSs for short, have been used in different fields. For instance, they have been shown to be useful in spatial data analysis [18], in Econometrics [3] and in Structural Engineering [25], to name but a few. RSs can also be viewed as imprecise random variables, as Pedro Gil and his colleagues have pointed out in [17]. Several results for RSs have been accomplished, such as limit theorems [1, 19], confidence sets for the (Aumann) expected value [4], hypothesis testing for the expected value or the (Fréchet) variance [11–13, 15, 20, 21] and inference on regression models [2, 8, 10].

A. B. Ramos-Guajardo (✉) · G. González-Rodríguez · A. Colubi · Á. Blanco-Fernández
INDUROT/Dept of Statistics, OR and MD, University of Oviedo, Oviedo, Spain
e-mail: ramosana@uniovi.es

G. González-Rodríguez
e-mail: gil@uniovi.es

A. Colubi
e-mail: colubi@uniovi.es

M. B. Ferraro
Dipartimento di Scienze Statistiche, Sapienza Università di Roma, Rome, Italy
e-mail: mariabrigida.ferraro@uniroma1.it

© Springer International Publishing AG 2018
E. Gil et al. (eds.), *The Mathematics of the Uncertain*, Studies in Systems,
Decision and Control 142, https://doi.org/10.1007/978-3-319-73848-2_64

In the one-dimensional case, the compact intervals A of $\mathbb{R}$ can be characterized by either the infima and suprema of A, (inf A, sup A) so that inf A < sup A, or by the mid-point and radius of A, (mid A, spr A) $\in \mathbb{R} \times \mathbb{R}^+$. The usual interval arithmetic is based on the Minkowski addition [16] and the product by a scalar, and it preserves the length of the resulting intervals. Many statistical results concerning interval data are based on the Minkowski arithmetic (see, for instance, [2, 6, 8, 10, 15, 21]).

The statistical studies developed until now for the p-dimensional situation (with $p > 1$) frequently take advantage of some convenient theoretical properties of the support function, but they have several drawbacks in certain situations. To overcome these drawbacks, an alternative to the support function for characterizing star-shaped sets by means of the so-called *radial/polar function* has been investigated in [5, 14].

Basic concepts related to this new representation and some statistical results are addressed. More concretely, the concept of random star-shaped set—i.e. a random variable taking star-shaped sets as outcomes -, and those of expected value and variance are considered. The corresponding sample moments are defined, and the consistency with respect to their population counterparts is highlighted. In addition, the concept of mean directional length is introduced and some comparative measures of centered star-shaped sets are suggested.

The rest of the paper is organized as follows. Section 2 is devoted to the introduction of some preliminaries regarding compact and convex sets and star-shaped sets. The concept of random star-shaped sets and their moments are recalled in Sect. 3. A basic example illustrating these sample moments is provided. The notions related to the mean directional length are discussed in Sect. 4. Finally, some conclusions and open problems are provided in Sect. 5.

2 Preliminaries

Let the space $\mathbb{R}^p$ be endowed with the Euclidean norm $\| \cdot \|$ and the corresponding inner product $\langle \cdot, \cdot \rangle$. Let $\mathbb{S}^{p-1} = \{u \in \mathbb{R}^p : \|u\| = 1\}$ be the hypersphere with radius 1. The space of all non-empty compact and convex subsets of $\mathbb{R}^p$ is denoted by $\mathcal{K}_c(\mathbb{R}^p)$. If $A \in \mathcal{K}_c(\mathbb{R}^p)$, then the *support function* of U is defined such that $s_A(u) = \sup_{a \in A} \langle u, a \rangle$ for $u \in \mathbb{S}^{p-1}$ [9, 19].

The location and the imprecision of a set $A \in \mathcal{K}_c(\mathbb{R}^p)$ can be determined in terms of the support function by the so-called mid-spread representation in such a way that $s_A = \text{mid}_A + \text{spr}_A$, where $\text{mid}_A(u) = (s_A(u) - s_A(-u))/2$ and $\text{spr}_A(u) = (s_A(u) + s_A(-u))/2$ for all $u \in \mathbb{S}^{p-1}$.

As shown in Fig. 1, the support function identifies the boundary of the corresponding set, but the obtained result is not easy to relate with the original shape of the set. Actually, it is very difficult to identify which is the original set associated with a function verifying the properties of the support function (if any). This could be a drawback in some applied problems in which it is necessary to clearly identify the shape of the sets (as, for instance, in image processing).

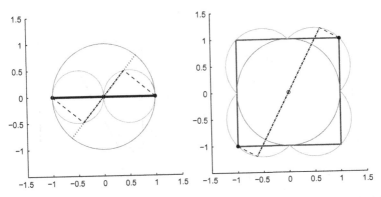

Fig. 1 Support function (distance from (0,0) to the contour of the gray line, marked in dashed-dotted black) of a line and a square in $\mathbb{R}^2$ (in black)

To overcome this disadvantage, other characterizations of sets can be taken into account. For instance, a useful tool in this framework is the so-called *radial function* [24]. It is defined on the class of *star-shaped sets* of $\mathbb{R}^p$, denoted by $\mathscr{K}_s(\mathbb{R}^p)$, which is an extension of $\mathscr{K}_c(\mathbb{R}^p)$—i.e. $\mathscr{K}_c(\mathbb{R}^p) \subset \mathscr{K}_s(\mathbb{R}^p)$.

A star-shaped set $A \in \mathscr{K}_s(\mathbb{R}^p)$ with respect to k_A, where k_A is *a center* of A, is a nonempty compact subset of $\mathbb{R}^p$ such that for all $a \in A$, $\lambda k_A + (1 - \lambda)a \in A$ for all $\lambda \in [0, 1]$. The radial function of a star-shaped set A is defined as $\rho_A : \mathbb{S}^{p-1} \to \mathbb{R}^+$ so that $\rho_A(u) = \sup\{\lambda \geq 0 : k_A + \lambda u \in A\}$. In this context, k_A can be viewed as a location point of the star-shaped set A whereas ρ_A is related to the imprecision of the set. The formal definition of k_A and ρ_A to be used in statistical problems is not trivial. This problem has been recently addressed in [14]. From now on, this representation of sets will be called *center-radial characterization*.

In contrast to the support function, the radial function identifies the shape of the sets in an intuitive way, as it is shown in Fig. 2, because it is simply based on the well-known polar coordinates over the unit sphere. In the case of the line (left side image in Fig. 2), the radial function is equal to 0 for all $u \in \mathbb{S}$ except for $u_1 = (1, 0)$ and $u_2 = (-1, 0)$, with $\rho_A(u_1) = \rho_A(u_2) = 1$. Further advantages of the radial function with respect to the support function are pointed out in [14].

The space $\mathscr{K}_s(\mathbb{R}^p)$ can be embedded into a cone on the Hilbert space $\mathscr{H}_r = \mathbb{R}^p \times \mathscr{L}^2(\mathbb{S}^{p-1})$ through the center-radial characterization. For the theoretical developments, from now on, star-shaped sets in $\mathscr{K}_s^*(\mathbb{R}^p)$ will be considered, where

$$\mathscr{K}_s^*(\mathbb{R}^p) = \left\{ A \in \mathscr{K}_s(\mathbb{R}^p) \mid \rho_A \in \mathscr{L}^2(\mathbb{S}^{p-1}) \right\}. \tag{1}$$

Regarding the arithmetic, we could consider the Minkowski addition between two star-shaped sets A and B, $A + B = \{a + b \mid a \in A, b \in B\}$. However, it has been shown that the Minkowski addition is not always meaningful (see [5, 18]), and it does not agree with the natural arithmetic induced by the center-radial characterization from the Hilbert space. That is, $A +_r \lambda B$ should be the element in $\mathscr{K}_s^*(\mathbb{R}^p)$ satisfying

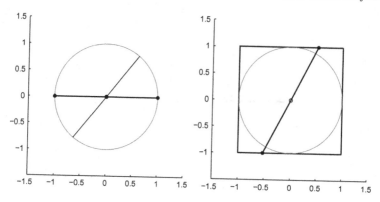

Fig. 2 Radial function of a line (black dots) and a square (which corresponds exactly to the square) in $\mathbb{R}^2$

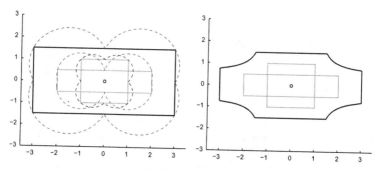

Fig. 3 Minkowski (left) and radial (right) sums (in black) of the gray quadrilaterals

that $k_{A+_r\lambda B} = k_A + \lambda k_B$ and $\rho_{A+_r\lambda B} = \rho_A + \lambda\rho_B$, where $+$ denotes the usual sum of two points in $\mathbb{R}^p$ and the usual sum of two functions in $\mathcal{L}^2(\mathbb{S}^{p-1})$ and $+_r$ denotes the sum in $\mathcal{H}_r$.

An example of the differences between the Minkowski sum and the center-radial sum of two star-shaped sets in $\mathcal{K}_s^*(\mathbb{R}^p)$ is provided in Fig. 3. We observe that the center-radial sum preserves, *directionally*, the lengths, whereas the Minkowski sum *dilates* them.

Regarding the metric structure in $\mathcal{K}_s^*(\mathbb{R}^p)$, the center-radial characterization induces a natural family of distances from the corresponding one in the associated Hilbert space. Thus, for any two star-shaped sets $A, B \in \mathcal{K}_s^*(\mathbb{R}^p)$, the τ-metric is defined as

$$d_\tau(A, B) = \sqrt{\tau\|k_A - k_B\|^2 + (1 - \tau)\|\rho_A - \rho_B\|_p^2}, \tag{2}$$

where $\tau \in (0, 1)$ determines the importance given to the location in contrast to the imprecision, $\|\cdot\|$ denotes the usual norm in $\mathbb{R}^p$ and $\|\cdot\|_p$ is the usual L^2-type norm in $\mathcal{L}^2(\mathbb{S}^{p-1})$ [14].

3 Random Star-Shaped Sets

Given a probability space $(\Omega, \mathscr{A}, P)$, a mapping $X : \Omega \to \mathbb{R}^p \times \mathscr{K}_s^*(\mathbb{R}^p)$ is a *random star-shaped set* if it is a Borel measurable mapping with respect to $\mathscr{A}$ and the Borel σ-field generated by the topology induced by the metric d_τ on $\mathbb{R}^p \times \mathscr{K}_s^*(\mathbb{R}^p)$. Equivalently, X can be decomposed in terms of its center-radial characterization, that is, $X = (k_X, \rho_X)$, and X can be defined to be a random star-shaped set iff k_X and ρ_X are random elements in the real and functional framework respectively [14].

Now we are in a position to define some summarizing measures for random star-shaped sets. On one hand, if $E(\|k_X\|) < \infty$ and $E(\|\rho_X\|_p) < \infty$, then the *expected value* of X is defined as the element $E(X) \in \mathbb{R}^p \times \mathscr{K}_s^*(\mathbb{R}^p)$ so that $k_{E(X)} = E(k_X)$ and $\rho_{E(X)} = E(\rho_X)$—this last expectation being considered in terms of the Bochner integral in $\mathscr{L}^2(\mathbb{S}^{p-1})$.

From an empirical point of view, given X a random star-shaped set and $\{X_i\}_{i=1}^n$ an i.i.d. sequence of random star-shaped sets drawn from X, the sample expectation of X can be defined in terms of the arithmetic in $\mathbb{R}^p \times \mathscr{K}_s^*(\mathbb{R}^p)$ as follows:

$$\overline{X} = \frac{1}{n} \sum_{i=1}^n X_i . \tag{3}$$

It is easy to show that $(k_{\overline{X}}, \rho_{\overline{X}}) = (\overline{k_X}, \overline{\rho_X})$.

If $E(\|k_X\|^2) < \infty$ and $E(\|\rho_X\|_p^2) < \infty$, then $E(X)$ is the unique element in $\mathbb{R}^p \times \mathscr{K}_s^*(\mathbb{R}^p)$ satisfying that

$$E(d_\tau^2(X, E(X))) = \min_{(k, A) \in \mathbb{R}^p \times \mathscr{K}_s^*(\mathbb{R}^p)} E(d_\tau^2(X, (k, A))). \tag{4}$$

Thus, by following the Fréchet approach, the *(scalar) variance* of a random star-shaped set X, denoted by σ_X^2, is defined as

$$\sigma_X^2 = E(d_\tau^2(X, E(X))). \tag{5}$$

The sample variance is also defined in terms of the distance d_τ, or equivalently, in terms of the corresponding variances in $\mathbb{R}^p$ and $\mathscr{L}^2(\mathbb{S}^{p-1})$, as follows:

$$\widehat{\sigma}_X^2 = \frac{1}{n} \sum_{i=1}^n d_\tau^2(X_i, \overline{X}) = \tau \widehat{\sigma}_{k_X}^2 + (1 - \tau) \widehat{\sigma}_{\rho_X}^2 . \tag{6}$$

The consistency of the estimators (3) and (5) for the mean and the variance of random star-shaped sets, respectively, is provided in the following result. It is an immediate consequence of the Strong Law for Large Numbers in Banach spaces.

Theorem 3.1 *[14] Let X be a random star-shaped set and $\{X_i\}_{i=1}^n$ be an i.i.d. sequence of random star-shaped sets drawn from X. Then,*

Fig. 4 Sample mean (in black) of a sample of 10 (gray-coloured) rectangles

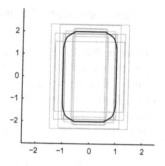

(a) If $E(\|k_X\|) < \infty$ and $E(\|\rho_X\|_p) < \infty$, then
$\overline{k_X} \xrightarrow{a.s.-P} E(k_X)$ and $\overline{\rho_X} \xrightarrow{a.s.-P} E(\rho_X)$. Therefore, $\overline{X} \xrightarrow{a.s.-P} E(X)$.

(b) If $E(\|k_X\|^2) < \infty$ and $E(\|\rho_X\|_p^2) < \infty$, then
$\widehat{\sigma}^2_{k_X} \xrightarrow{a.s.-P} \sigma^2_{k_X}$ and $\widehat{\sigma}^2_{\rho_X} \xrightarrow{a.s.-P} \sigma^2_{\rho_X}$. Therefore, $\widehat{\sigma}^2_{\overline{X}} \xrightarrow{a.s.-P} \sigma^2_X$.

Example 3.1 Let X be a random rectangle-shaped set (a particular case of a random star-shaped set) so that the upper right vertex is generated by following real normal distributions of means 2 and 3, respectively, and variance equal to 1; the longest side is distributed as an $U(1, 3)$ and the shortest one as an $U(3, 5)$. A sample $\{X_i\}_{i=1}^{10}$ of rectangle-shaped sets i.i.d. as X is generated. The rectangles are centered on their center of gravity. The centered sample and the corresponding sample mean are represented in Fig. 4.

It should be noticed that the sample mean is not a rectangle, as the corners are rounded due to the directional averaging.

The sample variance is computed for $\tau = 1$ (the sets are centered so that the importance is given to the imprecision) providing $\widehat{\sigma}^2_X = \widehat{\sigma}^2_{\rho_X} = 0.1192$.

4 Comparison of Centered Star-Shaped Sets

Let $A, B \in \mathcal{K}_s^*(\mathbb{R}^p)$ be two *centered* star-shaped sets (i.e., two star-shaped sets with common center which, without lack of generality, can be assumed to be 0) and let $\widetilde{\mathcal{K}_s^*}(\mathbb{R}^p)$ be the space of, either non-empty or empty, centered star-shaped sets of $\mathbb{R}^p$. In the same way that the so-called *length* of intervals and (fuzzy) sets has been used previously to develop statistics to compare convex and compact sets (see [21, 22]), the analogous concept can be considered for star-shaped sets. Thus, the *mean directional length* of A is defined in terms of the radial function by

$$S(A) = 2 \int_{\mathbb{S}^{p-1}} \rho_A(u) \, d\lambda_{\mathbb{S}^{p-1}}(u), \tag{7}$$

where $\lambda_{\mathbb{S}^{p-1}}$ denotes the normalized Lebesgue measure on the sphere. It should be noted that $S(A)$ is not the area of A, but an average of the magnitude of ρ_A over the unit sphere. It generalizes, in this way, the length of the intervals directionally, as it is always the case for the radial function.

Regarding the intersection, it is clear that $A \cap B \in \widetilde{\mathcal{K}_s^*}(\mathbb{R}^p)$. The *mean directional length of* $A \cap B$ can be expressed as follows:

$$S(A \cap B) = 2 \int_{\mathbb{S}^{p-1}} \min\left(\rho_A(u), \rho_B(u)\right) \, d\lambda_{\mathbb{S}^{p-1}}(u). \tag{8}$$

Based on the ideas in [23], the *degree of inclusion of A in B*, denoted by $Inc(A, B)$, is a value in $[0, 1]$ which can be defined by considering the quotient between the mean directional length of the intersection of A and B and the mean directional length of the reference set A, i.e.

$$Inc(A, B) = \frac{S(A \cap B)}{S(A)}. \tag{9}$$

If A is included in B, then it is clear that $S(A \cap B) = S(A)$ and $Inc(A, B) = 1$; otherwise, $S(A \cap B) < S(A)$ and $Inc(A, B) < 1$.

It is also possible to define the *degree of similarity of A and B*, denoted by $Sim(A, B)$, by following the ideas in [7], as the quotient between the shape of the intersection of A and B and the shape of the union of A and B, i.e.

$$Sim(A, B) = \frac{S(A \cap B)}{S(A \cup B)}, \tag{10}$$

where

$$S(A \cup B) = 2 \int_{\mathbb{S}^{p-1}} \max\left(\rho_A(u), \rho_B(u)\right) \, d\lambda_{\mathbb{S}^{p-1}}(u).$$

In this case, if A is equal to B, then $S(A \cap B) = S(A \cup B)$ and $Sim(A, B) = 1$; otherwise, $S(A \cap B) < S(A \cup B)$ and $Sim(A, B) < 1$. Moreover, $Sim(A, B) < Inc(A, B)$ in all the situations.

Two illustrative examples concerning rectangle-shaped sets are shown in Fig. 5. On the left part of the graphic, two partially overlapping centered rectangles A (in gray) and B (in black) are depicted. If we compute both the inclusion degree of A in B and the similarity degree between A and B, we obtain that $Inc(A, B) = 0.6822$ whereas $Sim(A, B) = 0.459$. On the right part of the graphic, the rectangle A (in gray) is completely contained in the rectangle B (in black). The computation of both indexes in this case leads us to the following results: $Inc(A, B) = 1$ and $Sim(A, B) = 0.5509$.

The measures presented in this section might be greatly useful in the context of image processing. Therefore, it would be interesting to develop a deep statistical analysis about these measures in the near future.

A. B. Ramos-Guajardo et al.

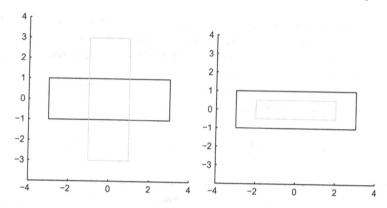

Fig. 5 Comparison of two rectangle-shaped sets in two different situations

5   Conclusions

An alternative representation for the class of star-shaped sets, called *center-radial characterization*, has been described. It has been shown to be useful for identifying intuitively the original shape of the sets. On the basis of this representation, some descriptive statistics for random star-shaped sets have been provided. Additionally, comparison measures based on the concept of mean directional length have been proposed. These measures are expected to be the starting point of an interesting research line in the area of image analysis. Furthermore, all the concepts provided in this work will be extended to the case of fuzzy subsets of $\mathbb{R}^p$ in a near future.

Acknowledgements The research in this paper has been partially supported by MTM2013-44212-P, GRUPIN14-005 and the COST Action IC1408.

Dedication

To Pedro, our research father and grandfather, a good man who encouraged us to follow this road. Thanks for giving us so much. Always in our hearts.

References

1. Artstein Z, Vitale R (1975) A strong law of large numbers for random compact sets. Ann Probab 3:879–882
2. Blanco-Fernández A, Corral N, González-Rodríguez G (2011) Estimation of a flexible simple linear model for interval data based on set arithmetic. Comput Stat Data Anal 55(9):2568–2578
3. Beresteanu A, Molchanov I, Molinari F (2011) Sharp identification regions in models with convex moment predictions. Econometrica 79:1785–1821
4. Choirat C, Seri R (2013) Bootstrap confidence sets for the Aumann mean of a random closed set. Comput Stat Data Anal 71:803–817
5. Colubi A, González-Rodríguez G (2017) On some functional characterizations of (Fuzzy) set-valued random elements. In: Ferraro MB, Giordani P, Vantaggi B, Gagolewski M, Gil MA, Grzegorzewski P, Hryniewicz O (eds) Soft methods for data science. series advances in intelligent systems and computing, vol 456. Springer, Cham
6. Corral N, Gil MA, Gil P (2011) Interval and fuzzy-valued approaches to the statistical management of imprecise data. In: Balakrishnan N, Gil MA, Pardo L (eds) Modern mathematical tools and techniques in capturing complexity. Series Understanding Complex Systems. Springer, Heidelberg
7. Dubois D, Prade H (1980) Fuzzy sets and systems: theory and applications. Academic Press, New York
8. Ferraro MB, Coppi R, González-Rodríguez G, Colubi A (2010) A linear regression model for imprecise response. Int J Approx Reas 51(7):759–770
9. Ghosh PK, Kumar KV (1998) Support function representation of convex bodies, its application in geometric computing, and some related representations. Comp Vis Image Unders 72(3):379–403
10. Gil MA, González-Rodríguez G, Colubi A, Montenegro M (2007) Testing linear independence in linear models with interval-valued data. Comp Statist Data Anal 51(6):3002–3015
11. Gil MA, Montenegro M, González-Rodríguez G, Colubi A, Casals MR (2006) Bootstrap approach to the multi-sample test of means with imprecise data. Comp Statist Data Anal 51:148–162
12. González-Rodríguez G, Colubi A, Gil MA (2011) Fuzzy data treated as functional data: a one-way ANOVA test approach. Comp Statist Data Anal 56:943–955
13. González-Rodríguez G, Montenegro M, Colubi A, Gil MA (2006) Bootstrap techniques and fuzzy random variables: synergy in hypothesis testing with fuzzy data. Fuzzy Sets Syst 157:2608–2613
14. González-Rodríguez G, Ramos-Guajardo AB, Colubi A, Blanco-Fernández A (2017) A new framework for the statistical analysis of set-valued random elements. Int J Approx Reas 92:279–294
15. Körner R (2000) An asymptotic α-test for the expectation of random fuzzy variables. J Statist Plan Infer 83:331–346
16. Minkowski H (1896) Geometrie der Zahlen. Teubner, Leipzig
17. Miranda E, Couso I, Gil P (2005) Random sets as imprecise random variables. J Math Annal Appl 307(1):32–47
18. Molchanov I (2005) Theory of random sets. Springer, London
19. Puri ML, Ralescu DA (1985) Limit theorems for random compact sets in Banach space. Math Proc Cambridge Philos 97:151–158
20. Ramos-Guajardo AB, Lubiano MA (2012) K-sample tests for equality of variances of random fuzzy sets. Comp Statist Data Anal 56(4):956–966
21. Ramos-Guajardo AB, Colubi A, González-Rodríguez G (2014) Inclusion degree tests for the Aumann expectation of a random interval. Inf Sci 288:412–422
22. Ramos-Guajardo AB, Colubi A, González-Rodríguez G (2014) Inclusion and exclusion hypothesis tests for the fuzzy mean. Fuzzy Sets Syst 243:70–83
23. Sanchez E (1979) Inverses of fuzzy relations. Application to possibility distributions and medical diagnosis. Fuzzy Sets Syst 2:75–86

24. Schneider R (1993) Convex bodies: the Brunn-Minkowski Theory. Cambridge University Press, Cambridge
25. Tonon F, Bernardini A (1998) A random set approach to the optimization of uncertain structures. Comp Struct 68:583–600

A Case Study-Based Analysis of the Influence of the Fuzzy Data Shape in Quantifying Their Fréchet's Variance

Sara de la Rosa de Sáa, Carlos Carleos, María Teresa López and Manuel Montenegro

Abstract In previous studies it has been shown that assumming a trapezoidal shape to model fuzzy number-valued data is not statistically restrictive in case we focus on the (Aumann-type) means of these data. The assertion has been supported by both case and simulation studies. This paper aims to analyze by means of a case study whether the same assertion applies in dealing with the Fréchet-type variance. More concretely, the p-values of tests have been compared for trapezoidal assessment versus other frequently used ones, like some LR assessments. The analysis is illustrated and corroborated with a real-life example. This analysis indicates that the shape of the fuzzy assessment scarcely affects statistical conclusions.

1 Introduction

Trapezoidal fuzzy numbers have been shown to be an easy-to-use/draw choice to model data from intrinsically imprecise-valued magnitudes. This is issued to the ease to handle for most of the computations, the ease to understand their meaning, and the ease of their elicitation, especially when people assessing fuzzy numbers have a low expertise/background about.

Thus, trapezoidal fuzzy numbers are characterized by simply giving their core (interval of the real values which are considered to be 'fully compatible' with the valuation to be elicited) and their support (interval of the real values which are

S. de la Rosa de Sáa
Oficina de Evaluación de Tecnologías Sanitarias, Servicio de Salud
del Principado de Asturias, Asturias, Spain
e-mail: sara.delarosa@sespa.es

C. Carleos · M. T. López · M. Montenegro (✉)
Departamento de Estadística, I.O. y D.M., Universidad de Oviedo, Oviedo, Spain
e-mail: mmontenegro@uniovi.es

C. Carleos
e-mail: carleos@uniovi.es

M. T. López
e-mail: mtlopez@uniovi.es

© Springer International Publishing AG 2018
E. Gil et al. (eds.), *The Mathematics of the Uncertain*, Studies in Systems,
Decision and Control 142, https://doi.org/10.1007/978-3-319-73848-2_65

709

considered to be 'compatible to some extent' with the valuation to be elicited). The remaining values can be directly obtained by a kind of 'linear interpolation' of these two intervals.

In Lubiano et al. [8] it has been empirically shown that, when fuzzy datasets are summarized by their fuzzy-valued means, the considered data shape is mostly not statistically relevant.

In this chapter we are going to check that the same conclusion can be drawn when fuzzy datasets are summarized by their real-valued variances. For this purpose, the real-life example analyzed in detail in Gil et al. [4] and Lubiano et al. [7, 8], and later recalled in Sect. 3, is considered for the comparative discussion. Some two-and k-samples hypothesis test about means p-values in Ramos-Guajardo and Lubiano [10] have been computed by considering all data as either being trapezoidal (as assumed in [4, 7, 8]) or belonging to any of the LR classes recalled in the next section.

2 Preliminaries

A (bounded) *fuzzy number* (also referred to by some authors as a fuzzy interval) is an imprecise-valued amount that is formalized as a mapping $\widetilde{U} : \mathbb{R} \to [0, 1]$ such that for all $\alpha \in [0, 1]$, the α-level set, defined as

$$\widetilde{U}_\alpha = \begin{cases} \{x \in \mathbb{R} : \widetilde{U}(x) \geq \alpha\} & \text{if } \alpha \in (0, 1] \\ \text{cl}\{x \in \mathbb{R} : \widetilde{U}(x) > 0\} & \text{if } \alpha = 0 \end{cases}$$

with 'cl' denoting the closure of the set, is a nonempty compact interval. $\widetilde{U}(x)$ is intuitively interpreted as the 'degree of compatibility' of the real number x with $\widetilde{U}$.

The space of (bounded) fuzzy numbers will be denoted by $\mathscr{F}_c^*(\mathbb{R})$.

A well-known and frequently used family of fuzzy numbers is that of *trapezoidal fuzzy numbers*. If $a, b, c, d \in \mathbb{R}$ with $a \leq b \leq c \leq d$, the trapezoidal fuzzy number $\text{Tra}(a, b, c, d)$ is such that for each $\alpha \in [0, 1]$ the α-level set equals

$$(\text{Tra}(a, b, c, d))_\alpha = [a + \alpha(b - a), d + \alpha(c - d)].$$

A wider interesting family of fuzzy numbers, including the one of trapezoidal fuzzy numbers, is that of the *LR-fuzzy numbers* (see Dubois and Prade [3]) with L and R invertible functions. If $a, b, c, d \in \mathbb{R}$ with $a \leq b \leq c \leq d$, the LR-fuzzy number $LR(a, b, c, d)$ is such that for each $\alpha \in [0, 1]$ the α-level set is such that

$$\inf \left(LR(a, b, c, d)\right)_\alpha = a + (b - a)\, L^{-1}(\alpha), \sup \left(LR(a, b, c, d)\right)_\alpha = d - (d - c)\, R^{-1}(\alpha).$$

Along this work we are going to consider quadratic functions (the so-called Π-curves) and functions with parametric monotonic Hermite-type interpolation, either using (2,2)-rational splines (LU_{1A} and LU_{1B}) or mixed exponential splines

$(LU_{2A}$ and $LU_{2B})$ (see Fig. 1) (see, for more details about Stefanini et al. [11]). More specifically, if $\tilde{U} \equiv LR(a, b, c, d)$, and $LR \in \{\text{Tra}, \Pi, LU_{1A}, LU_{1B}, LU_{2A}, LU_{2B}\}$, then for each $\alpha \in [0, 1]$

$$\tilde{U}_\alpha = [a + l_{LR}(\alpha)(b - a), c + r_{LR}(\alpha)(d - c)],$$

where the functions involved in the left and right arms can be seen in detail in Table 1.

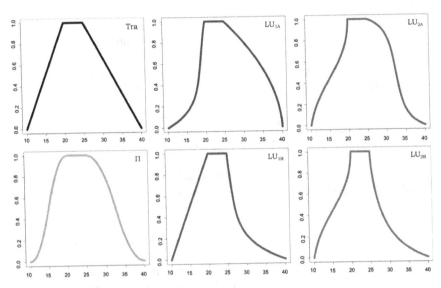

Fig. 1 Six types of fuzzy numbers sharing core [20, 25] and support (10, 40) and differing in shape. On the left, trapezoidal (top) and Π-curve (bottom), along with four different LR-fuzzy numbers on the middle and the right

Table 1 Expressions for functions l_{LR} and r_{LR} in the horizontal view of LR-fuzzy numbers with LR ranging on $\{\text{Tra}, \Pi, LU_{1A}, LU_{1B}, LU_{2A}, LU_{2B}\}$

LU	$l_{LU}(\alpha)$	$r_{LU}(\alpha)$
Tra	α	$1 - \alpha$
Π	$\begin{cases} \sqrt{\alpha/2} & \text{if } \alpha < 1/2 \\ 1 - \sqrt{(1-\alpha)/2} & \text{otherwise} \end{cases}$	$\begin{cases} 1 - \sqrt{\alpha/2} & \text{if } \alpha < 1/2 \\ \sqrt{(1-\alpha)/2} & \text{otherwise} \end{cases}$
LU_{1A}	$\dfrac{\alpha^2 + 5\alpha(1-\alpha)}{1 + 3.5\alpha(1-\alpha)}$	$(1-\alpha)(1 + 0.9\alpha)$
LU_{1B}	α	$1 - \dfrac{\alpha^2 + 5\alpha(1-\alpha)}{1 + 3.2\alpha(1-\alpha)}$
LU_{2A}	$\dfrac{\alpha^2(3 - 2\alpha) - 0.5(1-\alpha)^{1.55} + 0.5 + 0.05\alpha^{1.55}}{1.55}$	$1 - \dfrac{\alpha^2(3 - 2\alpha) - 5(1-\alpha)^{11} + 5 + 5\alpha^{11}}{11}$
LU_{2B}	$\dfrac{\alpha^2(3 - 2\alpha) - 0.5(1-\alpha)^{1.55} + 0.5 + 0.05\alpha^{1.55}}{1.55}$	$1 - \dfrac{\alpha^2(3 - 2\alpha) - 5(1-\alpha)^{6.05} + 5 + 0.05\alpha^{6.05}}{6.05}$

Key tools for the statistical analysis of fuzzy data are the following:

- the arithmetic with fuzzy numbers;
- the metric between fuzzy numbers;
- the model for the random mechanism generating fuzzy data.

Regarding the arithmetic, we will make use of the one based on Zadeh's extension principle [13].

Given $\widetilde{U}, \widetilde{V} \in \mathscr{F}_c^*(\mathbb{R})$, the *sum* of $\widetilde{U}$ and $\widetilde{V}$ is the fuzzy number $\widetilde{U} + \widetilde{V} \in \mathscr{F}_c^*(\mathbb{R})$ such that for each $\alpha \in [0, 1]$

$$(\widetilde{U} + \widetilde{V})_\alpha = \left[\inf \widetilde{U}_\alpha + \inf \widetilde{V}_\alpha, \sup \widetilde{U}_\alpha + \sup \widetilde{V}_\alpha \right].$$

Given $\widetilde{U} \in \mathscr{F}_c^*(\mathbb{R})$ and a scalar $\gamma \in \mathbb{R}$, the *product* of $\widetilde{U}$ *by the scalar* γ is the fuzzy number $\gamma \cdot \widetilde{U} \in \mathscr{F}_c^*(\mathbb{R})$ such that for each $\alpha \in [0, 1]$

$$(\gamma \cdot \widetilde{U})_\alpha = \begin{cases} \left[\gamma \cdot \inf \widetilde{U}_\alpha, \gamma \cdot \sup \widetilde{U}_\alpha \right] & \text{if } \gamma \geq 0 \\ \left[\gamma \cdot \sup \widetilde{U}_\alpha, \gamma \cdot \inf \widetilde{U}_\alpha \right] & \text{otherwise.} \end{cases}$$

It can be easily proved that for fixed invertible functions L and R, the family of LR-fuzzy numbers is closed under the sum and the product by scalars. More concretely,

$$LR(a, b, c, d) + LR(a', b', c', d') = LR(a + a', b + b', c + c', d + d'),$$

$$\gamma \cdot LR(a, b, c, d) = \begin{cases} LR(\gamma a, \gamma b, \gamma c, \gamma d) & \text{if } \gamma \geq 0 \\ LR(\gamma d, \gamma c, \gamma b, \gamma a) & \text{otherwise.} \end{cases}$$

These two operations do not endow $\mathscr{F}_c^*(\mathbb{R})$ with a linear, but with a conical structure, so special care should be taken in attempting to extend and deal with difference between fuzzy numbers. Actually, some of the inconveniencies associated with the nonlinearity have been substantially overcome in developing statistics with fuzzy data by incorporating suitable distances between them. In this respect, the metric given below has been introduced by Bertoluzza et al. [1], and it is a quite convenient choice for many statistical developments.

Given $\widetilde{U}, \widetilde{V} \in \mathscr{F}_c^*(\mathbb{R})$ and $\theta \in (0, 1]$, *Bertoluzza et al.'s θ-distance between* $\widetilde{U}$ *and* $\widetilde{V}$ is the real number

$$D_\theta(\widetilde{U}, \widetilde{V}) = \sqrt{\int_{[0,1]} \left([\text{mid } \widetilde{U}_\alpha - \text{mid } \widetilde{V}_\alpha]^2 + \theta [\text{spr } \widetilde{U}_\alpha - \text{spr } \widetilde{V}_\alpha]^2 \right) d\alpha},$$

with mid = mid-point/center, spr = spread/radius.

The most common particular choices of the parameter θ are $\theta = 1$ and $1/3$, since the first one corresponds to only taking into account and uniformly the squared distances between the extremes of the level sets, and the second one corresponds to taking into account and uniformly the squared distances between all the convex linear combinations of the extremes of the level sets.

Fuzzy number-valued data set in the case study to be considered, come from a so-called *fuzzy rating scale* (FRS), as introduced by Hesketh et al. [5], that allows a rater to draw the fuzzy number that best represents his/her score. The guideline for the mechanism to draw such a fuzzy number is as follows:

Step 1.   A reference bounded interval/segment is first considered. This is often chosen to be [0, 10] or [0, 100], but the choice of the interval is not at all a constraint. The end-points are often labeled in accordance with their meaning referring to the degree of agreement, satisfaction, quality, and so on.

Step 2.   The *core*, or 1-level set, associated with the response is determined. It corresponds to the interval consisting of the real values within the reference one which are considered to be as 'fully compatible' with the response.

Step 3.   The *support*, or its closure or 0-level set, associated with the response is determined. It corresponds to the interval consisting of the real values within the referential that are considered to be as 'compatible to some extent' with the response, and it should be always included in the reference interval.

Step 4.   The two intervals are 'interpolated' to get a fuzzy number. For instance, if a linear interpolation is considered a trapezoidal fuzzy number is obtained.

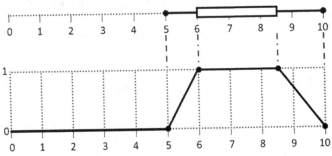

In developing statistics with fuzzy data coming from intrinsically imprecise-valued attributes, random fuzzy numbers constitute a well-formalized model within the probabilistic setting for the random mechanisms generating such data. Random fuzzy numbers, as defined by Puri and Ralescu [9] (in a more general context and dimension), integrate randomness (associated with the data generation) and fuzziness (associated with data nature).

Given a probability space $(\Omega, \mathscr{A}, P)$, an associated *random fuzzy number* (for short RFN) is a mapping $\mathscr{X} : \Omega \rightarrow \mathscr{F}_c^*(\mathbb{R})$ such that for all $\alpha \in [0, 1]$ the interval-valued mapping $\mathscr{X}_\alpha$ is a compact random interval (i.e., the real-valued mappings $\inf \mathscr{X}_\alpha$ and $\sup \mathscr{X}_\alpha$ are real-valued random variables). Equivalently, a mapping $\mathscr{X} : \Omega \rightarrow \mathscr{F}_c^*(\mathbb{R})$ is said to be an RFN if and only if it is a Borel-measurable mapping w.r.t. the Borel σ-field generated on $\mathscr{F}_c^*(\mathbb{R})$ by the topology induced by D_θ^φ; this Borel-measurability ensures that one can properly refer to the distribution induced by an RFN, the stochastic independence of RFNs, and so on, without needing to state expressly these notions.

In summarizing the induced distribution of a random fuzzy number, two measures/parameters are the most commonly used, namely, the Aumann-type mean (see Puri and Ralescu [9]) and the Fréchet-type variance (see, for instance, Lubiano et al. [6]).

Given an RFN $\mathscr{X}$ associated with the probability space $(\Omega, \mathscr{A}, P)$, the *(population) Aumann-type mean* of $\mathscr{X}$ is the fuzzy number $\widetilde{E}(\mathscr{X}) \in \mathscr{F}_c^*(\mathbb{R})$, if it exists, such that for each $\alpha \in [0, 1]$

$$\left(\widetilde{E}(\mathscr{X})\right)_\alpha = \left[E(\inf \mathscr{X}_\alpha), E(\sup \mathscr{X}_\alpha)\right]$$

with E denoting the expected value of a real-valued random variable.

If $\mathscr{X}$ is an LR-valued random fuzzy number for fixed invertible functions L and R, then $\widetilde{E}(\mathscr{X}) = LR\big(E(\inf \mathscr{X}_0), E(\inf \mathscr{X}_1), E(\sup \mathscr{X}_1), E(\sup \mathscr{X}_0)\big)$.

The Aumann-type mean preserves the main valuable properties from the real-valued case (i.e., additivity, equivariance under affine transformations, coherence with the above-described fuzzy arithmetic, and support by SLLN's).

In extending the variance of real-valued random variables to RFNs, Fréchet's approach has been considered, so that it can be interpreted as a measure of the 'least squares error/distance' in approximating the values of the RFN by a (non-random) fuzzy number.

The *(population) Fréchet-type variance* is the real number $\sigma_\mathscr{X}^2$, if it exists, given by

$$\sigma_\mathscr{X}^2 = E\left(\left[D_\theta\big(\mathscr{X}, \widetilde{E}(\mathscr{X})\big)\right]^2\right) = \int_{[0,1]} \left[\mathrm{Var}(\mathrm{mid}\ \mathscr{X}_\alpha)\, d\alpha + \theta \cdot \mathrm{Var}(\mathrm{spr}\ \mathscr{X}_\alpha)\right] d\alpha.$$

The Fréchet variance of an RFN satisfies the usual properties for this concept (i.e., nonnegativity, and vanishment for degenerate RFNs, invariance under translation, and additivity under independence of the involved RFNs).

3   Case Study to Be Analyzed

The case study to be analyzed is related to the well-known questionnaire TIMSS-PIRLS 2011 which is conducted on the population of Grade 4 students (i.e., nine to ten years old) and concerns their opinion and feeling on aspects regarding reading, math, and science. This questionnaire is rather standard and most of the involved questions have to be answered according to a 4-point Likert scale, responses being DISAGREE A LOT, DISAGREE A LITTLE, AGREE A LITTLE, and AGREE A LOT.

To get more expressive responses and informative conclusions, the original questionnaire form has been adapted to allow a double-type response: the original Likert and a fuzzy rating scale-based one with reference interval [0, 10] (see Fig. 2 for one of the items in the questionnaire).

The questionnaire involving these double response questions has been conducted in 2014 on a sample of 69 fourth grade students from Colegio San Ignacio (Oviedo-Asturias, Spain). These students have been distributed in accordance with (their usual) three groups, so that the teachers have decided that the 24 students in one of the three classrooms have to fill out the paper-and-pencil format and the 45 students from the other two groups have to complete the computerized version. To 'ease' the relationship between the two scales for these very young respondents, each numerically encoded Likert response has been superimposed upon the reference interval of the fuzzy rating scale part, as we can see in Fig. 2.

The training of the students to let them know about the meaning and purpose of the case study, as well as the aim of the double response, has been carried out in up to 15 min, and three researchers from our Department have been in charge of the explanation and conduction of the survey. At this point, it should be remarked that the students had no idea on the concept of real-valued functions and they have just learned that of a trapezium. The students have not had understanding problems, they

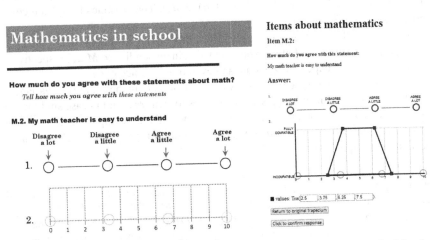

Fig. 2 Example of the double response paper-and-pencil (on the left) and computerized (on the right) form to an item in the case study

have catched the philosophy behind and they have been able to provide us with quite coherent responses in most of the cases. Actually, for all the questions, the number of 'no response''s has been very small and smaller for the fuzzy rating than for the Likert scale. In summary, the training has been surprisingly much easier and more effective than we had expected.

The complete questionnaire and dataset can be found in http://bellman.ciencias.uniovi.es/SMIRE/FuzzyRatingScaleQuestionnaire-SanIgnacio.html.

4 Comparative Analysis

The analysis of the influence of fuzzy data shape on the dataset variance is to be based on the test about the equality of variances with fuzzy data developed by Ramos-Guajardo and Lubiano [10] (see also Blanco-Fernández et al. [2]), which is a bootstrapped homoscedasticity test of k independent RFNs, which can be algorithmically summarized.

The analysis is carried out aiming to test the influence of the shape of fuzzy data on the Fréchet variance. By means of some of the data in the considered case study, this section follows two different comparative approaches. More concretely, it is first devoted to compare the p-values of two-sample and k-sample test about the equality of variances for different choices of the shape.

Table 2 gathers the p-values of the two-sample test about the equality of variances on the basis of the fuzzy rating scale responses to Item $M.2$ in the adapted questionnaire (that is, "My math teacher is easy to understand") when the two considered populations are 'boys' and 'girls' and the 4-tuples are associated not only with trapezoidal fuzzy numbers (as it has actually been made) but also with other LRs (those in Fig. 1 along with $\mathrm{Tri}(a, b, c, d) = \mathrm{Tra}(a, (b + c)/2, (b + c)/2, d)$ and $\mathrm{TriS}(a, b, c, d) = \mathrm{Tra}(a, (a + d)/2, (a + d)/2, d)$). The p-values have been computed for $\theta = 1/3$ and 1.

For the usually selected significance levels (those being lower than 0.25), there are no significant differences between boys and girls in responding to $M.2$, irrespectively of the considered shape of fuzzy data and even of the choice of θ.

Table 3 gathers the p-values of the two-sample test about the equality of variances on the basis of the fuzzy rating scale responses to Item $M.2$ in the case study when the two considered populations are 'paper-and-pencil' and 'computerized' form and the 4-tuples are associated with several LR-valued fuzzy numbers. The p-values have been computed for $\theta = 1/3$ and 1.

Table 2 p-Values for the equality of population Fréchet's variances ($\theta = 1/3$, $\theta = 1$) of boys' and girls' LRs responses to Item $M.2$ in the case study, depending on the considered shape

	Tra	Π	LU_{1A}	LU_{1B}	LU_{2A}	LU_{2B}	Tri	TriS
$\theta = 1/3$	0.416	0.478	0.539	0.466	0.473	0.456	0.466	0.397
$\theta = 1$	0.414	0.443	0.512	0.452	0.467	0.456	0.450	0.376

Table 3 p-Values for the equality of population Fréchet's variances ($\theta = 1/3, \theta = 1$) of 'paper-and-pencil' and 'computerized' form's LRs responses to Item $M.2$ in the case study, depending on the considered shape

	Tra	Π	LU_{1A}	LU_{1B}	LU_{2A}	LU_{2B}	Tri	TriS
$\theta = 1/3$	0.215	0.220	0.239	0.217	0.233	0.198	0.234	0.190
$\theta = 1$	0.149	0.166	0.186	0.165	0.161	0.165	0.184	0.161

Table 4 p-Values for the equality of population Fréchet's variances ($\theta = 1/3, \theta = 1$) of the four groups, $G1$ to $G4$, LRs responses to Item $M.2$ in the case study, depending on the considered shape

	Tra	Π	LU_{1A}	LU_{1B}	LU_{2A}	LU_{2B}	Tri	TriS
$\theta = 1/3$	0.270	0.255	0.255	0.282	0.263	0.275	0.247	0.218
$\theta = 1$	0.258	0.260	0.274	0.239	0.247	0.268	0.251	0.241

Table 5 p-Values for the equality of population Fréchet's variances ($\theta = 1/3$) of trapezoidal vs other LRs responses

Groups \ LR	Π	LU_{1A}	LU_{1B}	LU_{2A}	LU_{2B}	Tri	TriS
Boys	0.998	0.909	0.916	0.994	0.902	0.970	0.897
Girls	1.000	0.974	0.951	0.998	0.940	0.972	0.843
Paper-and-pencil	0.997	0.998	0.964	0.997	0.938	0.994	0.890
Computerized form	0.995	0.902	0.914	0.992	0.901	0.961	0.852
$G1$	0.991	0.866	0.923	0.980	0.904	0.832	0.820
$G2$	0.997	0.922	0.927	0.995	0.920	0.978	0.923
$G3$	1.000	0.979	0.949	0.998	0.949	0.986	0.927
$G4$	0.996	0.999	0.975	0.995	0.968	0.986	0.917

In this second situation, the effect of the choice of $\theta \in (0, 1]$ is not very relevant. Statistical conclusions scarcely depend on the considered shape of fuzzy data.

Table 4 gathers the p-values of the four-sample test about the equality of variances on the basis of the fuzzy rating scale responses to Item $M.2$ in the case study when the four considered populations are four groups of students based on their 'mark taken in the last examination of math' given by $G1 = [0, 6]$, $G2 = (6, 8]$, $G3 = (8, 9]$ and $G4 = (9, 10]$, according to the usual range $[0, 10]$ which is considered in Spain. The p-values have been computed for $\theta = 1/3$ and 1 and the 4-tuples are associated with several LR-valued fuzzy numbers.

Once more, in this third situation statistical conclusions scarcely depend on the considered shape of fuzzy data.

A second way to analyze the influence of the shape of fuzzy data by means of the case study, is to compare by means of the two-sample test about the equality of variances trapezoidal data vs other LR data in the responses to Item $M.2$ for different populations involved in the preceding tables in this section. Table 5 collects the corresponding p-values for $\theta = 1/3$.

Consequently, there are no significant differences between population Fréchet's variances for almost all the significance levels one can consider and all the seven developed comparisons.

Acknowledgements The research in this paper has been partially supported by the Principality of Asturias/FEDER Grant GRUPIN14-101, and the Ministerio de Economía y Competitividad Grants MTM2013-44212-P and MTM2015-63971-P. We acknowledge this financial support.

Dedication

Dear Pedro,

How difficult is to write a dedication to you!

How difficult assuming there is more than a year since you left us!

How difficult to find the right and accurate words to remember you!; to remember how much you meant (and still mean) for all of us. Words that will be little compared to what you deserve.

How difficult to write a dedication to a very honest man, a good man!

You have been a unique and unrepeatable man, with whom we feel deeply honored to have shared part of our lives.

You have been a generous man, always supporting those who have shared with you the love for Mathematics and, in particular, for Statistics and Operations Research. Fleing from imposition and taxation, you have taught your many disciples the need to devote a significant part of their working time to conduct forefront research. This vision and policy, which can seem to be obvious nowadays, was especially laudable at the time you created the embryo of our current university department.

Beyond your professional stature, your stature as a human being has been even greater. You have been permanently available to all your colleagues and students, willing to help and advice with a warm attitude. In each scenario, you have been always there to provide them with your calm, thoughtful and insightful opinion.

Pedro, you have been a good man, even in the discrepancy. In a rather hierarchical world, where disagreement can often entail a certain degree of punishment or discrimination of the weakest members, you have never made use of your heading position in such an unfair way.

So, in remembering you, we necessarily have to say that you have been the good friend, the magnificent researcher and teacher, and the wonderful colleague we all would like to be. The type of person all of us are delighted and honored to have found along our lives. We only regret that our joint life has not lasted longer.

Oviedo, August 2017

References

1. Bertoluzza C, Corral N, Salas A (1995) On a new class of distances between fuzzy numbers. Mathw & Soft Comput 2:71–84
2. Blanco-Fernández A, Casals MR, Colubi A, Corral N, García-Bárzana M, Gil MA, González-Rodríguez G, López MT, Lubiano MA, Montenegro M, Ramos-Guajardo AB, de la Rosa de Sáa S, Sinova B, (2014) A distance-based statistical analysis of fuzzy number-valued data. Int J Approx Reas 55(7):1487–1501
3. Dubois D, Prade H (1981) Additions of interactive fuzzy numbers. IEEE Trans Automat Contr 26:926–936
4. Gil MA, Lubiano MA, de la Rosa de Sáa S, Sinova B, (2015) Analyzing data from a fuzzy rating scale-based questionnaire. A case study. Psicothema 27:182–191
5. Hesketh T, Pryor R, Hesketh B (1988) An application of a computerized fuzzy graphic rating scale to the psychological measurement of individual differences. Int J Man-Mach Stud 29:21–35
6. Lubiano MA, Gil MA, López-Díaz M, López MT (2000) The λ-mean squared dispersion associated with a fuzzy random variable. Fuzzy Sets Syst 111:307–317
7. Lubiano MA, Montenegro M, Sinova B, de la Rosa de Sáa S, Gil MA, (2016) Hypothesis testing for means in connection with fuzzy rating scale-based data: algorithms and applications. Eur J Oper Res 251:918–929
8. Lubiano MA, Salas A, Gil MA (2017) A hypothesis testing-based discussion on the sensitivity of means of fuzzy data with respect to data shape. Fuzzy Sets Syst 328:54–69
9. Puri ML, Ralescu DA (1986) Fuzzy random variables. J Math Anal Appl 114:409–422
10. Ramos-Guajardo AB, Lubiano MA (2012) K-sample tests for equality of variances of random fuzzy sets. Comput Stat Data Anal 56(4):956–966
11. Stefanini L, Sorini L, Guerra ML (2006) Parametric representation of fuzzy numbers and applications to fuzzy calculus. Fuzzy Sets Syst 157:2423–2455
12. Trutschnig W, Lubiano MA (2015) SAFD: statistical analysis of fuzzy data. http://CRAN.R-project.org/web/packages/SAFD/index.html
13. Zadeh LA (1975) The concept of a linguistic variable and its application to approximate reasoning. Part 1. Inf Sci 8:199–249; Part 2. Inf Sci 8:301–353; Part 3. Inf Sci 9:43–80

A Note on Generalized Convexity for Fuzzy Mappings Through a Linear Ordering

Antonio Rufián-Lizana, Yurilev Chalco-Cano, Gabriel Ruiz-Garzón and M. Dolores Jiménez-Gamero

Abstract In De Campos Ibáñez and González-Muñoz (Fuzzy Sets Syst 29:145–154, 1989, [6]), Goetschel and Voxman (Fuzzy Sets Syst 18:31–43, 1986, [7]) the authors considered a linear ordering on the space of fuzzy intervals. For each fuzzy mapping (fuzzy interval-valued mapping) F, based on the aforementioned linear ordering, they introduced a real-valued function T_F on the domain of the fuzzy mapping F. Recently, Chalco-Cano et al. (Fuzzy Sets Syst 231:70–83, 2013, [4]) have studiedbreak the relationship between the generalized Hukuhara differentiability of a fuzzybreak mapping F (G-differentiability, for short) and the differentiability of T_F, and some properties of local-global minima. This paper studies such properties for fuzzy mappings, using new concepts which generalize the existing ones.

1 Notation and Basic Definitions

A fuzzy set on $\mathbb{R}^n$ is a mapping $u : \mathbb{R}^n \to [0, 1]$. For each fuzzy set u and for any $\alpha \in (0, 1]$, we denote $[u]^\alpha = \{x \in \mathbb{R}^n | u(x) \geq \alpha\}$ its α-level set. By *supp u* we denote the support of u, i.e. $\{x \in \mathbb{R}^n | u(x) > 0\}$. By $[u]^0$ we denote the closure of *supp u*. Let $\mathscr{F}_C$ denote the family of all fuzzy intervals, that is, $\mathscr{F}_C$ denotes the family of all compact and convex fuzzy sets on $\mathbb{R}$. Obviously, $[u]^\alpha$ is a nonempty compact and convex subset of $\mathbb{R}$ (denoted $\left[\underline{u}_\alpha, \overline{u}_\alpha\right]$) for any $u \in \mathscr{F}_C$ and $\alpha \in [0, 1]$.

A. Rufián-Lizana (✉) · M. D. Jiménez-Gamero
Departamento de Estadística e I.O., Universidad de Sevilla, Sevilla, Spain
e-mail: rufian@us.es

M. D. Jiménez-Gamero
e-mail: dolores@us.es

Y. Chalco-Cano
Instituto de Alta Investigación, Universidad de Tarapacá, Arica, Chile
e-mail: yurichalco@gmail.com

G. Ruiz-Garzón
Departamento de Estadística e I.O., Universidad de Cádiz, Cádiz, Spain
e-mail: gabriel.ruiz@uca.es

© Springer International Publishing AG 2018
E. Gil et al. (eds.), *The Mathematics of the Uncertain*, Studies in Systems, Decision and Control 142, https://doi.org/10.1007/978-3-319-73848-2_66

721

For fuzzy intervals $u, v \in \mathscr{F}_C$, represented by $[\underline{u}_\alpha, \overline{u}_\alpha]$ and $[\underline{v}_\alpha, \overline{v}_\alpha]$, respectively, and for any real number λ, we define the addition $u + v$ and scalar multiplication λu as follows:

$$(u + v)(x) = \sup_{y+z=x} \min\{u(y), v(z)\} \; , \quad (\lambda u)(x) = \begin{cases} u(\lambda^{-1}x), & \text{if } \lambda \neq 0, \\ 0, & \text{if } \lambda = 0. \end{cases}$$

It is well known that

$$[u + v]^\alpha = \left[(\underline{u+v})_\alpha, (\overline{u+v})_\alpha\right] = \left[\underline{u}_\alpha + \underline{v}_\alpha \, , \, \overline{u}_\alpha + \overline{v}_\alpha\right],$$

$$[\lambda u]^\alpha = \left[(\underline{\lambda u})_\alpha, (\overline{\lambda u})_\alpha\right] = \left[\min\{\lambda \underline{u}_\alpha, \lambda \overline{u}_\alpha\}, \max\{\lambda \underline{u}_\alpha, \lambda \overline{u}_\alpha\}\right],$$

for every $\alpha \in [0, 1]$.

We denote by $\mathscr{F}_C^C$ the family of all level-continuous fuzzy intervals: $u \in \mathscr{F}_C^C$ if the application $\alpha \mapsto [u]^\alpha$ is continuous, i.e. given $\varepsilon > 0$ and $\alpha_0 \in (0, 1)$ there exists $\delta > 0$ such that $|\alpha - \alpha_0| < \delta$ implies $H([u]^\alpha, [u]^{\alpha_0}) < \varepsilon$, where H stands for the Hausdorff distance.

It is well known that $(\mathscr{F}_C^C, H)$ is a separable and complete metric space (see [9]). Moreover, $\mathscr{F}_C^C$ is a closed subspace of $\mathscr{F}_C$.

Many authors have studied different methods for ranking fuzzy intervals. Most of these authors suggest mapping each fuzzy interval into the real line to define a ranking function. In [6] a ranking function called the average index (AI) was introduced. It is based on a function that selects a point of each level set, serving as a representative of the level set, and then integrates them weighted by some nonnegative function, which represents the subjective importance of each level set for the decision-maker.

Definition 1.1 ([6]) Let $\lambda \in [0, 1]$, D a distribution function with support contained in [0,1] and $u \in \mathscr{F}_C$. The average index of u is defined as

$$V_{\lambda,D}(u) = \int \left\{\lambda \underline{u}_\alpha + (1 - \lambda)\overline{u}_\alpha\right\} dD(\alpha) \in \mathbb{R}.$$

Although the definition in [6] is a bit more general, in the sense that instead of $\lambda \underline{u}_\alpha(x) + (1 - \lambda)\overline{u}_\alpha(x)$ they integrate a general function $f : [0, 1] \to \mathbb{R}$, in their paper they propose to choose f as in the definition given above. The parameter λ allows to vary the α-cut representative and, following [6], it may be interpreted as an optimism-pessimism degree, which must be selected by the decision-maker, depending on the context. The other choice is the distribution function D, which has the following interpretation: a D assigning a high probability to high values of α gives more weight to very precise points, where by precise points we mean points with a high degree of belonging; by contrast, a D assigning a high probability to low values of α gives more weight to rather imprecise points.

From now on, we will assume that λ and D are arbitrary but fixed. Now, by means of $V_{\lambda,D}(\cdot)$ an order relation on $\mathscr{F}_C$ is built as follows:

Definition 1.2 Let $u, v \in \mathscr{F}_C$. Then, u precedes v ($u \preceq v$) if $V_{\lambda,D}(u) \leq V_{\lambda,D}(v)$; u strictly precedes v ($u \prec v$) if $V_{\lambda,D}(u) < V_{\lambda,D}(v)$; u is indifferent to v ($u \simeq v$) if $V_{\lambda,D}(u) = V_{\lambda,D}(v)$.

Note that the order relation $\preceq$ is reflexive and transitive. Moreover, any two elements of $\mathscr{F}_C$ are comparable under the ordering $\preceq$. At this point it is interesting to note that the ordering considered in [4, 7] is just a particular case of the one in Definition 1.2 for $\lambda = 0.5$ and D the distribution function of a beta distribution with parameters 2 and 1, where a beta distribution with parameters a and b, for some $a, b > 0$, has probability density function

$$d_{\beta(a,b)}(x) = \frac{\Gamma(a+b)}{\Gamma(a)\Gamma(b)} x^{a-1}(1-x)^{b-1}, \quad x \in [0, 1],$$

$\Gamma(z)$ denoting the gamma function. This choice for F assigns more weight to very precise points than to imprecise points because

$$\int_{[0,\varepsilon]} dD_{\beta(a,b)} = \varepsilon^2 < 1 - \varepsilon^2 = \int_{[1-\varepsilon,1]} dD_{\beta(a,b)}, \quad \forall \varepsilon \in [0, 1/\sqrt{2}).$$

Example 1.1 Let us consider a trapezoidal fuzzy number u with membership function

$$u(x) = \begin{cases} (x-a)/(b-a) & \text{if } a \leq x < b, \\ 1 & \text{if } b \leq x \leq c, \\ (x-d)/(c-d) & \text{if } c < x \leq d. \end{cases}$$

The α-level sets are $[\underline{u}_\alpha, \overline{u}_\alpha] = [a + \alpha(b-a), d + \alpha(c-d)]$. Suppose that $D(\alpha) = \alpha^\theta$, $0 < \alpha < 1$, for some $\theta > 0$, which contains as a particular case the distribution function considered in [4, 7, 11] for $\theta = 2$. We have that

$$V_{\lambda,D}(u) = \int [\lambda a + (1-\lambda)d + \alpha \{\lambda(b-a) + (1-\lambda)(c-d)\}] \theta \alpha^{\theta-1} d\alpha$$

$$= \lambda a + (1-\lambda)d + \frac{\theta}{\theta+1} \{\lambda(b-a) + (1-\lambda)(c-d)\}.$$

Therefore, $V_{\lambda,D}(u)$ is just the representative of the α_D-level for $\alpha_D = \theta/(\theta+1)$. This is also true for any distribution function F, for $\alpha_D = \int \alpha dD(\alpha) \in [0, 1]$.

2 Differentiable Fuzzy Mappings

Henceforth, K denotes an open subset of $\mathbb{R}^n$ and T denotes an open interval in $\mathbb{R}$. A mapping $F : K \to \mathscr{F}_C$ is said to be a fuzzy mapping. For each $\alpha \in [0, 1]$, associated with F, we define the family of interval-valued functions $F_\alpha : K \to \mathscr{K}_C$, where $\mathscr{K}_C$ denotes the space of all compact intervals, given by $F_\alpha(x) = [F(x)]^\alpha$. For any $\alpha \in [0, 1]$, we denote

$$F_\alpha(x) = \left[\underline{f}_\alpha(x), \overline{f}_\alpha(x) \right].$$

Here, for each $\alpha \in [0, 1]$, the endpoint functions $\underline{f}_\alpha, \overline{f}_\alpha : K \to \mathbb{R}$ are called upper and lower functions of F, respectively.

Bede and Stefanani [1] introduced the following definition of derivative for fuzzy mappings.

Definition 2.1 ([1]) Let $F : T \to \mathscr{F}_C$ be a fuzzy mapping, we say that F is generalized Hukuhara differentiable (gH-differentiable) at $t_0 \in T$ if there exists an element $F'(t_0) \in \mathscr{F}_C$ such that (using the usual Hausdorff metric):

$$\lim_{h \to 0} \frac{F(t_0 + h) -_{gH} F(t_0)}{h} = F'(t_0).$$

We say that F is gH-differentiable on T if F is gH-differentiable at each point $t_0 \in T$.

Next example shows that a fuzzy mapping F can be differentiable, but their upper and lower functions are not.

Example 2.1 Let us consider the fuzzy mapping $F : \mathbb{R} \to \mathscr{F}_C$ defined by $F(t) = C \cdot t$, where C is a fuzzy interval defined via its α-levels by $[C]^\alpha = [1 + \alpha, 2(3 - \alpha)]$. Then

$$F_\alpha(t) = \begin{cases} [(1 + \alpha)t, 2(3 - \alpha)t] & \text{if } t \geq 0 \\ [2(3 - \alpha)t, (1 + \alpha)t] & \text{if } t < 0 \end{cases}.$$

We can see that the endpoint functions $\underline{f}_\alpha$ and $\overline{f}_\alpha$ are not differentiable at $t = 0$. However F is gH-differentiable on $\mathbb{R}$ and $F'(t) = C$.

We have the following result on the connection between the differentiability of F and its endpoint functions $\underline{f}_\alpha$ and $\overline{f}_\alpha$.

Theorem 2.1 ([5]) *Let $F : T \to \mathscr{F}_C$ be a fuzzy mapping. F is gH-differentiable at $t_0 \in T$, if and only if, we have the following cases:*

(a) $\underline{f}_\alpha$ *and* $\overline{f}_\alpha$ *are differentiable at x_0 and*

$$[F'(x_0)]^\alpha = \left[\min\left\{(\underline{f}_\alpha)'(x_0), (\overline{f}_\alpha)'(x_0)\right\}, \max\left\{(\underline{f}_\alpha)'(x_0), (\overline{f}_\alpha)'(x_0)\right\}\right];$$

(b) $(\underline{f}_\alpha)'_-(x_0), (\underline{f}_\alpha)'_+(x_0), (\overline{f}_\alpha)'_-(x_0)$ and $(\overline{f}_\alpha)'_+(x_0)$ exist and satisfy $(\underline{f}_\alpha)'_-(x_0) = (\overline{f}_\alpha)'_+(x_0)$ and $(\underline{f}_\alpha)'_+(x_0) = (\overline{f}_\alpha)'_-(x_0)$. Moreover

$$[F'(t_0)]^\alpha = \left[\min\left\{(\underline{f}_\alpha)'_-(x_0), (\overline{f}_\alpha)'_-(x_0)\right\}, \max\left\{(\underline{f}_\alpha)'_-(x_0), (\overline{f}_\alpha)'_-(x_0)\right\}\right]$$

$$= \left[\min\left\{(\underline{f}_\alpha)'_+(x_0), (\overline{f}_\alpha)'_+(x_0)\right\}, \max\left\{(\underline{f}_\alpha)'_+(x_0), (\overline{f}_\alpha)'_+(x_0)\right\}\right].$$

The above theorem leads us to define a stronger concept of derivate based on the differentiability of the endpoint functions.

Definition 2.2 Let $F : T \to \mathscr{F}_C$ be a fuzzy mapping. We say that F is level-wise gH-differentiable if the extreme functions $\underline{f}_\alpha$ and $\overline{f}_\alpha$ are differentiable for all $\alpha \in [0, 1]$.

Next we give an example where F is gH-differentiable at t_0, but $\lambda \underline{f}_\alpha + (1 - \lambda)\overline{f}_\alpha$ is not a differentiable function at t_0. This fact happens when F is $g\overline{H}$-differentiable at t_0 and part (b) from Theorem 2.1 holds.

Example 2.2 Let us consider the fuzzy mapping $F : \mathbb{R} \to \mathscr{F}_C$ in Example 2.1 The endpoint functions $\underline{f}_\alpha$ and $\overline{f}_\alpha$ are not differentiable at $t = 0$. However F is gH-differentiable on $\mathbb{R}$ and $F'(t) = C$. In this case, F is gH-differentiable and part (b) from Theorem 2.1 holds. Notice that

$$\lambda \underline{f}_\alpha(t) + (1 - \lambda)\overline{f}_\alpha(t) = \begin{cases} \{\lambda(1 + \alpha) + (1 - \lambda)2(3 - \alpha)\}t & \text{if } t \geq 0 \\ \{\lambda 2(3 - \alpha) + (1 - \lambda)(1 + \alpha)\}t & \text{if } t \geq 0 \end{cases}.$$

It is clear that $\lambda \underline{f}_\alpha(t) + (1 - \lambda)\overline{f}_\alpha(t)$ is not differentiable at $t = 0 \; \forall \lambda \neq 0.5$.

Next we define the partial derivative for a fuzzy mapping F defined on $K \subset \mathbb{R}^n$, i.e., $F(x) = F(x_1, \ldots, x_n) \in \mathscr{F}_C$ for each $x = (x_1, \ldots, x_n) \in K$. With this aim, given a fuzzy mapping $F : K \to \mathscr{F}_C$, we denote the fuzzy interval $F(x)$ by $F(x) = \left[\underline{f}(x), \overline{f}(x)\right]$ and, for each $\alpha \in [0, 1]$,

$$F_\alpha(x) = \left[\underline{f}_\alpha(x), \overline{f}_\alpha(x)\right] = \left[\underline{f}(\alpha, x), \overline{f}(\alpha, x)\right].$$

Definition 2.3 Let F be a fuzzy mapping defined on K and let $x_0 = (x_1^{(0)}, \ldots, x_n^{(0)}) \in K$ be fixed. Consider the fuzzy mapping $h_i(x_i) = F(x_1^{(0)}, \ldots, x_{i-1}^{(0)}, x_i, x_{i+1}^{(0)}, \ldots, x_n^{(0)})$. If h_i is gH-differentiable at $x_i^{(0)}$, then we say that F has ith *partial* gH-derivative at x_0 (denoted by $(\partial F/\partial x_i)(x_0)$) and $(\partial F/\partial x_i)(x_0) = (h_i)'(x_i^{(0)})$.

Definition 2.4 Let F be a fuzzy mapping defined on K and let $x_0 = (x_1^{(0)}, \ldots, x_n^{(0)}) \in K$ be fixed. We say that F is gH-differentiable at x_0 if all the partial gH-derivatives $(\partial F / \partial x_1)(x_0), \ldots, (\partial F / \partial x_n)(x_0)$ exist in some neighborhood of x_0 and they are continuous at x_0.

Some authors considered the differentiability in the sense of Seikkala [10] which has the same property of monotonicity of length of F as the H-differentiability. Nevertheless, this concept of differentiability is very restrictive (for more details see [2–4]).

3 Ranking Functions Associated with Fuzzy Mappings

Motivated by Goetschel and Voxman [7], we give the following definition of an average index function associated with a fuzzy mapping, which extends Definition 3.6 in [7].

Definition 3.1 Let $\lambda \in [0, 1]$ and let D be a distribution function with support contained in $[0, 1]$ fixed. For each fuzzy mapping $F : K \to \mathscr{F}_C$, we define the ranking function $T_F^{\lambda, D} : K \to \mathbb{R}$ associated with F by

$$T_F^{\lambda, D}(x) = V_{\lambda, D}(F(x)).$$

The next theorem shows the connection between the gH-differentiability of F and the differentiability of the ranking function $T_F^{\lambda, D}$.

With this aim, we introduce the following notation: given a fuzzy mapping $F : K \to \mathscr{F}_C$ and $\lambda \in [0, 1]$ we define the function $f^\lambda : [0, 1] \times K \to \mathbb{R}$ by

$$f^\lambda(\alpha, x) = \lambda \underline{f}(\alpha, x) + (1 - \lambda)\overline{f}(\alpha, x).$$

Theorem 3.1 Let $K \subset \mathbb{R}^n$ be an open set. If $F : K \to \mathscr{F}_C^C$ is gH-differentiable at $x \in K$, $\alpha \mapsto \frac{\partial}{\partial x_i} f^\lambda(\alpha, x)$ exists, for almost all (D) $\alpha \in [0, 1]$, there exists $\theta_i : [0, 1] \to \mathbb{R}$ such that $\left| \frac{\partial}{\partial x_i} f^\lambda(\alpha, x) \right| \leq \theta_i(\alpha)$, for almost all (D) $\alpha \in [0, 1]$ and $\int \theta_i(\alpha) d D(\alpha) < \infty$, $i = 1, \ldots, n$, then $T_F^{\lambda, D} : K \to \mathbb{R}$ is differentiable at x and

$$\frac{\partial}{\partial x_i} T_F^{\lambda, D(x)} = \int \frac{\partial}{\partial x_i} f^\lambda(\alpha, x) d D(\alpha),$$

$i = 1, \ldots, n$.

Example 3.1 Let $F : \mathbb{R}^2_+ \to \mathscr{F}^C_C$ be a fuzzy mapping defined by

$$F(x_1, x_2) = \begin{cases} \frac{2(x_1^2 + x_2^2) - z}{x_1^2 + x_2^2} & \text{if } x_1^2 + x_2^2 \leq z \leq 2\left(x_1^2 + x_2^2\right) \\ 1 & \text{if } -2x_1 x_2 \leq z \leq x_1^2 + x_2^2 \\ 0 & \text{if } z \notin \left[-2x_1 x_2, 2\left(x_1^2 + x_2^2\right)\right] \end{cases}$$

or equivalently

$$F(x_1, x_2) = (-2, -2, 0, 0)x_1 x_2 + (0, 0, 2)x_1^2 + (0, 0, 2)x_2^2.$$

Then, for each $\alpha \in [0, 1]$ we have that

$$F_\alpha(x_1, x_2) = \left[-2x_1 x_2, (2 - \alpha)x_1^2 + (2 - \alpha)x_2^2\right].$$

Since, for each $\alpha \in [0, 1]$, the endpoint functions $\underline{f}_\alpha$ and $\overline{f}_\alpha$ are differentiable functions then F is level-wise gH-differentiable. Moreover,

$$\frac{\partial}{\partial x_1}\left(\lambda \underline{f}_\alpha + (1 - \lambda)\overline{f}_\alpha\right)(x_1, x_2) = -2\lambda x_2 + 2(1 - \lambda)(2 - \alpha)x_1,$$

$$\frac{\partial}{\partial x_2}\left(\lambda \underline{f}_\alpha + (1 - \lambda)\overline{f}_\alpha\right)(x_1, x_2) = -2\lambda x_1 + 2(1 - \lambda)(2 - \alpha)x_2.$$

Note that $0 \leq \mu_D = \int \alpha d D(\alpha) \leq \int d D(\alpha) = 1 < \infty$. Thus, from Theorem 3.1, $T_F^{\lambda, D}$ is differentiable and

$$\frac{\partial T_F^{\lambda, D}}{\partial x_1}(x_1, x_2) = -2\lambda x_2 + 2(1 - \lambda)(2 - \mu_D)x_1,$$

$$\frac{\partial T_F^{\lambda, D}}{\partial x_1}(x_1, x_2) = -2\lambda x_1 + 2(1 - \lambda)(2 - \mu_D)x_2.$$

Theorem 3.2 *Let $F : K \subset \mathbb{R}^n \to \mathscr{F}_C$ be a fuzzy mapping defined by*

$$F(x) = g(x) \cdot C,$$

where $g : K \to \mathbb{R}$ is a function and $C = [\underline{c}, \overline{c}]$ is a fuzzy interval. Let $\mu_1 = \int \{\lambda \underline{c}(\alpha) + (1 - \lambda)\overline{c}(\alpha)\} d D(\alpha)$ and $\mu_2 = \int \{\overline{c}(\alpha) - \underline{c}(\alpha)\} d D(\alpha)$. We have the following cases:

(a) If $\lambda = 0.5$, g is differentiable at x iff $T_F^{\lambda, D}$ is differentiable at x and

$$\frac{\partial}{\partial x_i} T_F^{\lambda, D}(x) = \frac{\partial}{\partial x_i} g(x)\mu_1.$$

(b) If either $\lambda \neq 0.5$ and $g(x) \neq 0$ or $\lambda \neq 0.5$, $g(x) = 0$ and $\mu_2 = 0$, g is differentiable at x iff $T_F^{\lambda, D}$ is differentiable at x and

$$\frac{\partial}{\partial x_i} T_F^{\lambda, D}(x) = \frac{\partial}{\partial x_i} g(x)\{\theta_1 + I(g(x) < 0)(1 - 2\lambda)\theta_2\}$$

where $I(g(x) < 0) = 1$ if $g(x) < 0$ and $I(g(x) < 0) = 0$ otherwise.

Proof Note that, for each $\alpha \in [0, 1]$ we have

$$\lambda \underline{f}_\alpha(x) + (1 - \lambda)\overline{f}_\alpha(x)$$

$$= \lambda \min \{g(x)\underline{c}(\alpha), \overline{c}(\alpha)g(x)\} + (1 - \lambda) \max \{g(x)\underline{c}(\alpha), \overline{c}(\alpha)g(x)\}$$

$$= g(x) \begin{cases} \lambda \underline{c}(\alpha) + (1 - \lambda)\overline{c}(\alpha), & \text{if } g(x) \geq 0 \\ \lambda \overline{c}(\alpha) + (1 - \lambda)\underline{c}(\alpha), & \text{if } g(x) < 0 \end{cases}.$$

Thus, $T_F^{\lambda, D}(x) = g(x)\{\theta_1 + I(g(x) < 0)(1 - 2\lambda)\theta_2\}$. The result clearly follows from this expression. □

4 Generalized Convex Fuzzy Mappings and Fuzzy Optimization

Following the concept of solutions of fuzzy optimization problems via ranking function (see [11]), we present the following definition of fuzzy optimization.

Definition 4.1 Let $K \subset \mathbb{R}^n$, $\lambda \in [0, 1]$ and let D be a distribution function with support contained in $[0, 1]$. An element $x^* \in K$ is called minimum point of a fuzzy mapping F if and only if $T_F^{\lambda, D}(x^*) \leq T_F^{\lambda, D}(x)$ for all $x \in K$.

In contrast to [11], which define a stationary point for F by means of the gradient of F, next we give a stationary point concept for F through the ranking function $T_F^{\lambda, D}(x)$.

Definition 4.2 Let $K \subset \mathbb{R}^n$, $\lambda \in [0, 1]$ and D be a distribution function with support contained in $[0, 1]$. We say that $x^* \in K$ is a stationary point for a gH-differentiable fuzzy mapping $F : K \to \mathscr{F}_C$ if and only if the gradient $\nabla T_F^{\lambda, D}(x^*) = 0$.

Concepts of convexity and generalized convexity for fuzzy mappings based on ranking valued function were introduced in [11]. Those concepts are equivalent to convexity of $T_F^{1/2, D}$ with $D = D_{\beta(2,1)}$ (for more details see [4]). More precisely.

Theorem 4.1 ([4]) *Let $F : K \to \mathscr{F}_C$ be a fuzzy mapping. Then, F is convex (preinvex, prequasiinvex) if and only if $T_F^{1/2, D}$ is convex (preinvex, prequasiinvex, respectively).*

Inspired by Theorem 4.1 we give the following definition.

Definition 4.3 Let $K \subset \mathbb{R}^n$, $\lambda \in [0, 1]$ and let D be a distribution function with support contained in $[0, 1]$. Let $F : K \rightarrow \mathscr{F}_C$ be a fuzzy mapping and let $\lambda \in [0, 1]$ be fixed. If $T_F^{\lambda, D}$ is differentiable, then F is said to be invex if $T_F^{\lambda, D}$ is an invex function, i.e. there exists a function $\eta : K \times K \rightarrow K$ such that for $x, y \in K$

$$T_F^{\lambda, D}(x) - T_F^{\lambda, D}(y) \geq \nabla T_F^{\lambda, D}(y)\eta(x, y).$$

The following result characterizes the invexity of a fuzzy mapping F by using the stationary point concept in Definition 4.2.

Theorem 4.2 Let $F : K \rightarrow \mathscr{F}_C$ be a fuzzy mapping. Then, F is invex if and only if every stationary point is a minimum point of F.

Proof Since $T_F^{\lambda, D} : K \rightarrow \mathbb{R}$ is a real-valued function and it is invex, then from Theorem 2.1 in [8] we have the result. $\qquad\square$

Next we study the convexity properties of the ranking function $T_F^{\lambda, D}$ by means of the convexity of the extreme functions of the fuzzy mapping F.

Theorem 4.3 Suppose that assumptions in Theorem 3.1 hold. If $x^* \in K$ is a stationary point for all $\alpha \in (0, 1)$, then x^* is a stationary point for F.

Proof Suppose that $x^* \in K$ is a stationary point for $x \mapsto f^\lambda(\alpha, x)$. Taking into account Theorem 3.1 we have

$$\nabla T_F^{\lambda, D}(x^*) = \nabla \left(\int f^\lambda(\alpha, x^*)dD(\alpha) \right) = \int \nabla f^\lambda(\alpha, x^*)dD(\alpha) = 0.$$

Therefore x^* is a stationary point for F. $\qquad\square$

Theorem 4.4 Let $F : K \rightarrow \mathscr{F}_C$ be a fuzzy mapping such that $x \mapsto f^\lambda(\alpha, x)$ is a convex function for all $\alpha \in [0, 1]$. Then $T_F^{\lambda, D}$ is a convex mapping.

Proof Since $x \mapsto f^\lambda(\alpha, x)$ is a convex function for all $\alpha \in [0, 1]$, then, for all $\gamma \in [0, 1]$, we have

$$T_F^{\lambda, D}(\alpha, \gamma x + (1 - \gamma)y)$$

$$= \int f^\lambda(\alpha, \gamma x + (1 - \gamma)y)dD(\alpha)$$

$$\leq \gamma \int f^\lambda(\alpha, x)dD(\alpha) + (1 - \gamma) \int f^\lambda(\alpha, y)dD(\alpha)$$

$$= \gamma T_F^{\lambda, D}(x) + (1 - \gamma)T_F^{\lambda, D}(y).$$

Therefore $T_F^{\lambda, D}$ is convex. $\qquad\square$

As a consequence of Theorem 4.4 we have the following.

Corollary 4.1 *Let $F : K \to \mathscr{F}_C$ be a fuzzy mapping such that the endpoint functions $\underline{f}(\alpha, x)$ and $\overline{f}(\alpha, x)$ are convex functions for all $\alpha \in [0, 1]$. Then $T_F^{\lambda, D}$ is a convex fuzzy mapping.*

For the case of invexity we obtain the following.

Corollary 4.2 *Let $F : K \to \mathscr{F}_C^C$ be a gH-differentiable fuzzy mapping such that:*

(a) $\alpha \mapsto \frac{\partial}{\partial x_i} f^{\lambda}(\alpha, x)$ is continuous for all $x \in K$ and $i = 1, \ldots, n$.

(b) There exists a vector function η such that $x \mapsto \underline{f}(\alpha, x)$ and $x \mapsto \overline{f}(\alpha, x)$ are invex respect to same η for all $\alpha \in [0, 1]$.

Then, $T_F^{\lambda, D}$ is an invex fuzzy mapping respect to η.

5 Conclusions

This paper presents new and more general alternatives to study the optimization for fuzzy mappings, using its ranking function. We achieve the characterization of the optimum through the concept of stationary point.

This paper is full of theorems, definitions and corollaries, but it is also full of affection and fondness for Pedro.

Acknowledgements The research in this paper has been partially supported by Fondecyt-Chile 1151154, and by the Ministerio de Economía y Competitividad, Spain, through Grant MTM2015-66185-P.

References

1. Bede B, Stefanini L (2013) Generalized differentibility of fuzzy-valued function. Fuzzy Sets Syst 230:119–141
2. Chalco-Cano Y, Román-Flores H (2008) On new solutions of fuzzy differential equations. Chaos Solitons Fractals 38:112–119
3. Chalco-Cano Y, Román-Flores H, Jiménez-Gamero MD (2011) Generalized derivative and π-derivative for set-valued functions. Inf Sci 181:2177–2188
4. Chalco-Cano Y, Rufián-Lizana A, Román-Flores H, Osuna-Gómez R (2013) A note on generalixed convexity for fuzzy mappings through a linear ordering. Fuzzy Sets Syst 231:70–83
5. Chalco-Cano Y, Rodríguez-López R, Jiménez-Gamero MD (2016) Characterization of generalized differentiable fuzzy functions. Fuzzy Sets Syst 295:37–56
6. De Campos Ibáñez L, González-Muñoz M (1989) A subjective approach for ranking fuzzy numbers. Fuzzy Sets Syst 29:145–154
7. Goestschel R, Voxman W (1986) Elementary fuzzy calculus. Fuzzy Sets Syst 18:31–43
8. Martin DH (1985) The essence of invexity. J Optim Theory Appl 47:65–76

9. Román-Flores H, Rojas-Medar M (2002) Embedding of level-continuous fuzzy sets on Banach spaces. Inf. Sci 144:227–247
10. Seikkala S (1987) On the fuzzy initial value problem. Fuzzy Sets Syst 24:319–330
11. Syau YR, Stanley Lee E (2006) Fuzzy Weirstrass theorem and convex fuzzy mappings. Comput Math Appl 51:1741–1750

Scale Equivariant Alternative for Fuzzy M-Estimators of Location

Beatriz Sinova

Abstract The Aumann-type mean fulfills very convenient properties as a location measure of a random fuzzy number, but its high sensitivity to outliers makes other alternatives, such as fuzzy M-estimators of location, more suitable to describe contaminated data sets. Under some conditions, fuzzy M-estimators fulfill properties such as the strong consistency and the translation equivariance. However, the scale equivariance does not hold in general and the choice of the measurement units may have too much influence on the results. A first solution to solve this was the selection of the tuning parameters involved in the most used loss functions (Huber's, Tukey's and Hampel's) in terms of the distribution of distances of the observed data to the considered initial location estimate. Now a second solution is proposed including a robust estimate of the unknown dispersion in the definition of fuzzy M-estimators of location. The empirical comparison of both proposals shows that the latter solution may be more suitable for dealing with extreme data, and therefore it could better identify which observations should be considered outliers indeed.

1 Introduction

Fuzzy numbers can model experiments characterized by an underlying imprecision, such as ratings, opinions or perceptions (see e.g. De la Rosa de Sáa et al. [3]). Due to their interest, statistical methodology is being adapted to analyze this kind of data. With respect to central tendency measures, the best-known one is the Aumann-type mean [8], which generalizes the notion of mean of a real-valued random variable. Even when the Aumann-type mean fulfills very good statistical and

This paper is dedicated to the memory of Prof. Pedro Gil, who not only taught my mother and I Statistics in an interesting and calm way, but also left me bright memories in relation to our condition of neighbours and the conferences we both attended, as well as the *Champanadas* he cheered up with his accordion. I am deeply grateful for such moments and lessons.

B. Sinova (✉)
Departamento de Estadística e I.O. y D.M., Universidad de Oviedo, Oviedo, Spain
e-mail: sinovabeatriz@uniovi.es

© Springer International Publishing AG 2018
E. Gil et al. (eds.), *The Mathematics of the Uncertain*, Studies in Systems, Decision and Control 142, https://doi.org/10.1007/978-3-319-73848-2_67

733

probabilistic properties, outliers have too much impact on its estimate. For that reason, M-estimators of location have been recently defined in the fuzzy number-valued case by Sinova et al. [10] and their robustness has been shown. This is not the only robust location measure for fuzzy numbers proposed in the literature (we could think, for example, about some extensions of the concept of median like the ones introduced in [9, 11]), but their performance seems to be the best in general. Although the empirical study addressed in [10] concludes that there is no uniformly best location estimator, it highlights the good behavior of fuzzy M-estimators of location.

Scale equivariance, on the contrary, does not hold for fuzzy M-estimators of location unless the loss function involved in their definition is a power function. This is an important drawback, since it means that measurement units could have a lot of impact on the results. A first solution was provided in [10] and consists in choosing the tuning parameters in the used loss functions (Huber's and Hampel's, and it could be also applied to other functions such as Tukey's) taking into account the distribution of distances from the observed data to the initial estimate considered for the computation of the corresponding M-estimator. Therefore, the loss function is adapted to the magnitude of the data we are working with.

The aim of this paper is to present an alternative to solve the lack of scale equivariance. In the classical settings, where the same problem has had to be dealt with, a robust estimate of the dispersion is introduced in the definition of the M-estimator of location to make it scale equivariant. Recently, a robust estimate of the dispersion of a random fuzzy number, the median distance deviation about the median, has been analyzed (see [4]). The idea is, in consequence, to use a similar median distance deviation about the median to extend M-estimators of location with unknown dispersion to the fuzzy number-valued settings.

The rest of the paper is structured as follows. The preliminaries on the space of fuzzy numbers and fuzzy M-estimators of location are recalled in Sect. 2. Section 3 presents the concept of fuzzy M-estimators of location with unknown dispersion and the study of their scale equivariance, whereas their empirical comparison with the previously defined fuzzy M-estimators of location is presented in Sect. 4. Finally, some concluding remarks are provided in Sect. 5.

2 Preliminaries on the Space of Fuzzy Numbers and Fuzzy M-Estimation of Location

In this section, the most important characteristics of the space of fuzzy numbers will be recalled, as well as the adaptation of M-estimators of location to the fuzzy-valued settings.

$\mathscr{F}_c(\mathbb{R})$ will denote the class of (bounded) fuzzy numbers, which are mappings $\widetilde{U} : \mathbb{R} \to [0, 1]$ such that their α-levels

$$\widetilde{U}_\alpha = \begin{cases} \{x \in \mathbb{R} \ : \ \widetilde{U}(x) \geq \alpha\} & \text{if } \alpha \in (0, 1] \\ \text{cl}\{x \in \mathbb{R} \ : \ \widetilde{U}(x) > 0\} & \text{if } \alpha = 0, \end{cases}$$

are nonempty compact intervals. It is possible to interpret $\widetilde{U}(x)$ as the 'degree of compatibility' of x with $\widetilde{U}$ (or 'degree of truth' of the assertion "x is $\widetilde{U}$").

Fuzzy data are very useful to model those phenomena such as human perceptions or valuations which present an underlying imprecision. Indeed, their α-levels incorporate a certain gradualness that does not appear when dealing with interval-valued data.

Concerning the mathematical operations among these kinds of data, the sum and the product are defined by means of Zadeh's extension principle, which extends level-wise the usual interval arithmetic.

Definition 2.1 Let $\widetilde{U}, \widetilde{V} \in \mathscr{F}_c(\mathbb{R})$. The **sum** of $\widetilde{U}$ and $\widetilde{V}$ is defined as the fuzzy number $\widetilde{U} + \widetilde{V} \in \mathscr{F}_c(\mathbb{R})$ given for each $\alpha \in [0, 1]$ by

$$(\widetilde{U} + \widetilde{V})_\alpha = \text{Minkowski sum of } \widetilde{U}_\alpha \text{ and } \widetilde{V}_\alpha = \left[\inf \widetilde{U}_\alpha + \inf \widetilde{V}_\alpha, \sup \widetilde{U}_\alpha + \sup \widetilde{V}_\alpha\right].$$

Let $\widetilde{U} \in \mathscr{F}_c(\mathbb{R})$ and $\gamma \in \mathbb{R}$. The **product** of $\widetilde{U}$ **by the scalar** γ is defined as the fuzzy number $\gamma \cdot \widetilde{U} \in \mathscr{F}_c(\mathbb{R})$ given for each $\alpha \in [0, 1]$ by

$$(\gamma \cdot \widetilde{U})_\alpha = \gamma \cdot \widetilde{U}_\alpha = \begin{cases} \left[\gamma \cdot \inf \widetilde{U}_\alpha, \gamma \cdot \sup \widetilde{U}_\alpha\right] \text{ if } \gamma \geq 0, \\ \left[\gamma \cdot \sup \widetilde{U}_\alpha, \gamma \cdot \inf \widetilde{U}_\alpha\right] \text{ otherwise.} \end{cases}$$

Now the family of distances between fuzzy numbers introduced by Montenegro et al. [7], which extends the one proposed by Bertoluzza et al. [1], will be recalled. Note that its use is very convenient due to the lack of linearity in the space $(\mathscr{F}_c(\mathbb{R}), +, \cdot)$ as explained in e.g. [2].

Definition 2.2 Let $\theta \in (0, +\infty)$ and let φ be an absolutely continuous probability measure on $([0, 1], \mathscr{B}_{[0,1]})$ with the mass function being positive on $(0, 1)$. The **mid/spr-based L^2 distance** between any two fuzzy numbers $\widetilde{U}, \widetilde{V} \in \mathscr{F}_c(\mathbb{R})$ is defined as

$$D_\theta^\varphi(\widetilde{U}, \widetilde{V}) = \left[\int_{[0,1]} \left\{\left(\text{mid } \widetilde{U}_\alpha - \text{mid } \widetilde{V}_\alpha\right)^2 + \theta \left(\text{spr } \widetilde{U}_\alpha - \text{spr } \widetilde{V}_\alpha\right)^2\right\} d\varphi(\alpha)\right]^{1/2},$$

where mid $\widetilde{U}_\alpha = (\inf \widetilde{U}_\alpha + \sup \widetilde{U}_\alpha)/2$ and spr $\widetilde{U}_\alpha = (\sup \widetilde{U}_\alpha - \inf \widetilde{U}_\alpha)/2$.

The role of θ and φ is not stochastic, but to weigh the importance of the deviation 'in shape' in contrast to the deviation 'in center', and the relevance of the different α-levels, respectively. It can be proven that the usual choice $1/3$ makes all the points in the intervals (once fixed any α) equally important.

M-estimators of location will be defined in terms of the mid/spr-based L^2 distance since the space $(\mathscr{F}_c(\mathbb{R}), D_\theta^\varphi)$ can be isometrically embedded into a convex cone of a certain Hilbert space by means of the so-called support function (in Puri and Ralescu's sense [8]).

The notion of random fuzzy number in Puri and Ralescu's sense [8] mathematically formalizes the random mechanism generating fuzzy data.

Definition 2.3 Let $(\Omega, \mathscr{A}, P)$ be a probability space modeling a random experiment. A mapping $\mathscr{X} : \Omega \to \mathscr{F}_c(\mathbb{R})$ is said to be a **random fuzzy number** associated with the random experiment if, and only if, for each $\alpha \in [0, 1]$ the interval-valued mapping $\mathscr{X}_\alpha$ (where $\mathscr{X}_\alpha(\omega) = \big(\mathscr{X}(\omega)\big)_\alpha$ for all $\omega \in \Omega$) is a random compact interval or, equivalently, the real-valued functions $\inf \mathscr{X}_\alpha$ and $\sup \mathscr{X}_\alpha$ are random variables.

A random fuzzy number is Borel-measurable with respect to the Borel σ-field associated with the D_θ^φ distance, so the induced distribution can be trivially induced. In order to summarize the central tendency of the distribution of a random fuzzy number, one of the best-known measures is the Aumann-type mean, an extension of the concept of mean for real-valued random variables, which inherits very convenient statistical and probabilistic properties, but also the high sensitivity to outliers. For this reason, other location measures for fuzzy-valued data with a more robust behavior have already been proposed in the literature. In particular, the concept of median has been extended to the fuzzy number-valued settings as follows.

Definition 2.4 Let $\mathscr{X}$ be a random fuzzy number and $(\mathscr{X}_1, \dots, \mathscr{X}_n)$ be a simple random sample from $\mathscr{X}$. The (sample) **1-norm median** is the fuzzy number $\widehat{\mathrm{Me}}(\mathscr{X}_1, \dots, \mathscr{X}_n)$, for short $\widehat{\mathrm{Me}}$, such that for each $\alpha \in [0, 1]$ it coincides with

$$[\mathrm{Me}\{\inf (\mathscr{X}_1)_\alpha, \dots, \inf (\mathscr{X}_n)_\alpha\}, \mathrm{Me}\{\sup (\mathscr{X}_1)_\alpha, \dots, \sup (\mathscr{X}_n)_\alpha\}],$$

with Me denoting the median of a real-valued random variable. In case any of the medians is non-unique, the convention of considering the midpoint of the interval of possible medians is used.

Among the robust location measures for fuzzy-valued data, the performance of fuzzy M-estimators of location is certainly remarkable, achieving the best results in many of the situations studied in [10].

Definition 2.5 Let $(\Omega, \mathscr{A}, P)$ be a probability space and $\mathscr{X} : \Omega \to \mathscr{F}_c(\mathbb{R})$ be an associated random fuzzy number. Moreover, let $(\mathscr{X}_1, \dots, \mathscr{X}_n)$ be a simple random sample from $\mathscr{X}$. Then, the (sample) **fuzzy M-estimator of location** is the fuzzy number-valued statistic $\widehat{\widetilde{g}^M}(\mathscr{X}_1, \dots, \mathscr{X}_n)$, given by

$$\widehat{\widetilde{g}^M}(\mathscr{X}_1, \dots, \mathscr{X}_n) = \arg \min_{\widetilde{U} \in \mathscr{F}_c(\mathbb{R})} \frac{1}{n} \sum_{i=1}^n \rho(D_\theta^\varphi(\mathscr{X}_i, \widetilde{U})),$$

if it exists, where the loss function $\rho : \mathbb{R}^+ \to \mathbb{R}$ is assumed to be continuous and non-decreasing and to vanish at 0.

In [10, 12] it has been proven that, even when they may fulfill very good properties, fuzzy M-estimators of location are not scale equivariant unless ρ is a power function, which is not a possible choice if we are looking for robustness. In [10] the tuning parameters involved in some well-known loss functions (Huber's and Hampel's) are selected depending on the distribution of distances from the observed data to the initial estimate considered for the computation of the M-estimator to avoid the bad influence of the measurement units. A second alternative is introduced in Sect. 3 of this paper by extending the classical M-estimators of location with unknown dispersion, which are based on a robust estimate of the dispersion. The median distance deviation about the median of a random fuzzy number has been defined in [4] using the ρ_1 distance, which is an L^1 metric based on the infimum/supremum characterization of fuzzy numbers. However, a new alternative is now considered, by replacing the ρ_1 distance by the D_θ^φ metric, since fuzzy M-estimators of location are defined in terms of the latter (due to the isometrical embedding mentioned above).

Definition 2.6 Let $\mathscr{X}$ be a random fuzzy number and $(\mathscr{X}_1, \ldots, \mathscr{X}_n)$ be a simple random sample from $\mathscr{X}$. The (sample) **median** D_θ^φ**–distance deviation about the 1-norm median** (MDD) is the following real number

$$\widehat{\sigma}_\theta^\varphi(\mathscr{X}_1, \ldots, \mathscr{X}_n) = \mathrm{Me}\left\{ D_\theta^\varphi(\mathscr{X}_1, \widehat{\mathrm{Me}}), \ldots, D_\theta^\varphi(\mathscr{X}_n, \widehat{\mathrm{Me}}) \right\},$$

applying the same convention as in Definition 2.4.

It can be proven that the median D_θ^φ-distance deviation about the 1-norm median satisfies the scale equivariance property.

Proposition 2.1 *Let* $(\Omega, \mathscr{A}, P)$ *be a probability space and* $\mathscr{X} : \Omega \to \mathscr{F}_c(\mathbb{R})$ *be an associated random fuzzy number. Moreover, let* $(\mathscr{X}_1, \ldots, \mathscr{X}_n)$ *be a simple random sample from* $\mathscr{X}$. *The median* D_θ^φ-*distance deviation about the 1-norm median is scale equivariant, that is, given any* $\gamma \in \mathbb{R}$,

$$\widehat{\sigma}_\theta^\varphi(\gamma \cdot \mathscr{X}_1, \ldots, \gamma \cdot \mathscr{X}_n) = |\gamma| \cdot \widehat{\sigma}_\theta^\varphi(\mathscr{X}_1, \ldots, \mathscr{X}_n).$$

Proof First, due to the properties of the 1-norm median (see [9]),

$$\widehat{\mathrm{Me}}(\gamma \cdot \mathscr{X}_1, \ldots, \gamma \cdot \mathscr{X}_n) = \gamma \cdot \widehat{\mathrm{Me}}(\mathscr{X}_1, \ldots, \mathscr{X}_n).$$

Therefore, for all $i \in \{1, \ldots, n\}$,

$$D_\theta^\varphi(\gamma \cdot \mathscr{X}_i, \widehat{\mathrm{Me}}(\gamma \cdot \mathscr{X}_1, \ldots, \gamma \cdot \mathscr{X}_n)) = |\gamma| \cdot D_\theta^\varphi(\mathscr{X}_i, \widehat{\mathrm{Me}}(\mathscr{X}_1, \ldots, \mathscr{X}_n))$$

since $\mathrm{mid}\,(\gamma \cdot \widetilde{U})_\alpha = \gamma \cdot \mathrm{mid}\,\widetilde{U}_\alpha$ and $\mathrm{spr}\,(\gamma \cdot \widetilde{U})_\alpha = |\gamma| \cdot \mathrm{spr}\,\widetilde{U}_\alpha$ for all $\widetilde{U} \in \mathscr{F}_c(\mathbb{R})$ and all $\alpha \in [0, 1]$. Finally,

$$\widehat{\sigma}_\theta^\varphi (\gamma \cdot \mathscr{X}_1, \ldots, \gamma \cdot \mathscr{X}_n)$$

$$= \mathrm{Me} \left\{ D_\theta^\varphi (\gamma \cdot \mathscr{X}_1, \widehat{\mathrm{Me}}(\gamma \cdot (\mathscr{X}_1, \ldots, \mathscr{X}_n))), \ldots, D_\theta^\varphi (\gamma \cdot \mathscr{X}_n, \widehat{\mathrm{Me}}(\gamma \cdot (\mathscr{X}_1, \ldots, \mathscr{X}_n))) \right\}$$

$$= \mathrm{Me} \left\{ |\gamma| \cdot D_\theta^\varphi (\mathscr{X}_1, \widehat{\mathrm{Me}}(\mathscr{X}_1, \ldots, \mathscr{X}_n)), \ldots, |\gamma| \cdot D_\theta^\varphi (\mathscr{X}_n, \widehat{\mathrm{Me}}(\mathscr{X}_1, \ldots, \mathscr{X}_n)) \right\}$$

$$= |\gamma| \cdot \widehat{\sigma}_\theta^\varphi (\mathscr{X}_1, \ldots, \mathscr{X}_n).$$

$\square$

3 Location M-Estimators with Unknown Dispersion for Random Fuzzy Numbers

Due to the lack of scale equivariance of fuzzy M-estimators of location, the classical M-estimators of location with unknown dispersion will be now extended to the fuzzy number-valued case. The alternative of simultaneously estimating both the location and scale will be not considered in this paper, since this procedure is already not satisfactory in the classical settings from the robustness point of view and also due to the numerical inconvenience of solving the system of two non-linear equations.

Definition 3.1 Let $(\Omega, \mathscr{A}, P)$ be a probability space and $\mathscr{X} : \Omega \to \mathscr{F}_c(\mathbb{R})$ be an associated random fuzzy number. Moreover, let $(\mathscr{X}_1, \ldots, \mathscr{X}_n)$ be a simple random sample from $\mathscr{X}$. Then, the (sample) **fuzzy MDD-based M-estimator of location** is the fuzzy number-valued statistic $\widetilde{g}_{MDD}^M (\mathscr{X}_1, \ldots, \mathscr{X}_n)$, given by

$$\widetilde{g}_{MDD}^M (\mathscr{X}_1, \ldots, \mathscr{X}_n) = \arg \min_{\widetilde{U} \in \mathscr{F}_c(\mathbb{R})} \frac{1}{n} \sum_{i=1}^n \rho \left(\frac{D_\theta^\varphi (\mathscr{X}_i, \widetilde{U})}{\widehat{\sigma}_\theta^\varphi (\mathscr{X}_1, \ldots, \mathscr{X}_n)} \right),$$

if it exists, where the loss function $\rho : \mathbb{R}^+ \to \mathbb{R}$ is assumed to be continuous and non-decreasing and to vanish at 0.

It can be shown that the fuzzy MDD-based M-estimator of location is indeed scale equivariant as it happens in the real-valued settings.

Proposition 3.1 *Let $(\Omega, \mathscr{A}, P)$ be a probability space and $\mathscr{X} : \Omega \to \mathscr{F}_c(\mathbb{R})$ be an associated random fuzzy number. Moreover, let $(\mathscr{X}_1, \ldots, \mathscr{X}_n)$ be a simple random sample from $\mathscr{X}$. The fuzzy MDD-based M-estimator of location is scale equivariant, that is, given any $\gamma \in \mathbb{R}$,*

$$\widetilde{g}_{MDD}^M (\gamma \cdot \mathscr{X}_1, \ldots, \gamma \cdot \mathscr{X}_n) = |\gamma| \cdot \widetilde{g}_{MDD}^M (\mathscr{X}_1, \ldots, \mathscr{X}_n).$$

4   Simulation Study

This section aims to empirically compare the two alternatives proposed to extend M-estimators of location to the fuzzy-valued case and avoid any problem with the measurement units. First, the tuning parameters involved in the most used loss functions, such as Huber's, Tukey's or Hampel's could be selected in terms of the distribution of distances of the observed data to the considered initial location estimate as in [10]. Secondly, fuzzy MDD-based M-estimators of location are a scale equivariant measure that has been introduced in Sect. 3.

Among the usual loss functions, we will consider the Hampel loss function since its suitability was shown in [10] for many of the studied cases. The *Hampel loss function* [5] corresponds to

$$
\rho_{a,b,c}(x) = \begin{cases} x^2/2 & \text{if } 0 \le x < a, \\ a(x - a/2) & \text{if } a \le x < b, \\ \dfrac{a(x - c)^2}{2(b - c)} + \dfrac{a(b + c - a)}{2} & \text{if } b \le x < c, \\ \dfrac{a(b + c - a)}{2} & \text{if } c \le x, \end{cases}
$$

where the nonnegative parameters $a < b < c$ allow us to control the degree of suppression of large errors. The smaller their values, the higher this degree. Note that apart from not being convex, this function can cope with extreme outliers, since observations far from the center ($x \ge c$) all contribute equally to the loss.

Recall that the first alternative consists of fuzzy M-estimators of location carefully choosing the values of the tuning parameters. Following Kim and Scott [6], we will take a, b and c as the median, 75th and 85th percentiles of the distances between the observations and an initial estimate, which throughout this paper will be the 1-norm median.

100 trapezoidal fuzzy data are generated according to four real-valued random variables: $\mathscr{X} = \mathrm{Tra}(X_1 - X_2 - X_3, X_1 - X_2, X_1 + X_2, X_1 + X_2 + X_4)$, so $\inf \mathscr{X}_0 = X_1 - X_2 - X_3$, $\inf \mathscr{X}_1 = X_1 - X_2$, $\sup \mathscr{X}_1 = X_1 + X_2$ and $\sup \mathscr{X}_0 = X_1 + X_2 + X_4$.

A contamination proportion equal to $c_p \in \{0, 0.1, 0.2, 0.4\}$ is introduced in each sample. Any kind of outlier is allowed in these simulation studies: all the random variables detailed above (X_1, X_2, X_3 and X_4) can follow the corresponding distributions for the contaminated observations or just some (at least one) of them. This means that we deal with outliers in location, outliers in shape and/or outliers in both location and shape. A second parameter, $C_D \in \{0, 1, 5, 10, 100\}$, determines the distance between the distribution of the regular and contaminated observations.

In CASE 1 the variables X_i are independent. In particular,

- $X_1 \sim \mathscr{N}(0, 1)$ and $X_2, X_3, X_4 \sim \chi_1^2$ for the regular observations.

- $X_1 \sim \mathcal{N}(0, 3) + C_D$ and $X_2, X_3, X_4 \sim \chi_4^2 + C_D$ for the contaminated observations.

In CASE 2 dependence between the variables X_i is introduced as follows:

- $X_1 \sim \mathcal{N}(0, 1)$ and $X_2, X_3, X_4 \sim 1/(X_1^2 + 1)^2 + \sqrt{\chi_1^2}$ for the non-contaminated subsample (with χ_1^2 independent of X_1),
- $X_1 \sim \mathcal{N}(0, 3) + C_D$ and $X_2, X_3, X_4 \sim 1/(X_1^2 + 1)^2 + \sqrt{\chi_1^2} + C_D$ for the contaminated subsample (with χ_1^2 independent of X_1).

Both the fuzzy M-estimate of location and the fuzzy MDD-based M-estimate of location using the Hampel loss function are computed for each contaminated sample using an algorithm as in [10]. Their population values are approximated by Monte Carlo with 10,000 iterations and the performance of both proposals is compared in terms of the corresponding mean square error also approximated by Monte Carlo with 1000 iterations.

4.1 Results

Table 1 contains the results of the comparative analysis. In order to understand these results better, Fig. 4 shows some samples generated using the procedure explained

Table 1 Empirical comparison of the fuzzy M-estimate of location (Hampel) and the fuzzy MDD-based M-estimate of location (MDD-H) using the Hampel loss function

c_p	C_D	Case 1		Case 2	
		Hampel	MDD-H	Hampel	MDD-H
0	0	0.03157	**0.02077**	0.39544	**0.11508**
0.1	0	0.03242	**0.02790**	0.38452	**0.11492**
0.1	1	**0.03272**	0.03460	0.40867	**0.13536**
0.1	5	**0.03079**	0.04271	0.36526	**0.18030**
0.1	10	**0.02903**	0.02995	0.35188	**0.14413**
0.1	100	0.02977	**0.02258**	0.35691	**0.11714**
0.2	0	**0.04257**	0.05940	0.37045	**0.11383**
0.2	1	**0.04595**	0.08613	0.43351	**0.15930**
0.2	5	**0.04685**	0.16308	0.36094	**0.33573**
0.2	10	**0.04444**	0.08832	0.32239	**0.23062**
0.2	100	0.04649	**0.02491**	0.31370	**0.11416**
0.4	0	**0.10742**	0.20288	0.34975	**0.11612**
0.4	1	**0.15262**	0.37964	0.50878	**0.23767**
0.4	5	**0.35651**	1.82037	**0.69800**	1.55067
0.4	10	**0.54289**	3.21184	**0.87081**	2.37917
0.4	100	0.70982	**0.03470**	0.97474	**0.11623**

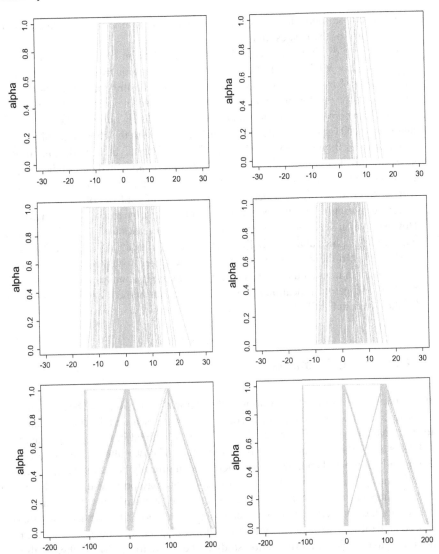

Fig. 1 A sample of generated fuzzy numbers from CASE 1 (left column) and CASE 2 (right column) is compared when (c_p, C_D) is chosen to be $(0.1, 5)$ –top–, $(0.4, 5)$ –middle– and $(0.4, 100)$ –bottom–

in this section with some illustrative choices of the contamination parameters c_p and C_D.

It can be concluded that

- there is no uniformly best estimator, since both the fuzzy M-estimator of location and the fuzzy MDD-based M-estimator of location can provide us with the most convenient estimate depending on the analyzed situation.

- The distribution seems to have more influence than the amount of contamination on the choice of the best estimator.
- The larger the difference between the outlier and the non-contaminated data, the more the fuzzy MDD-based M-estimate of location seems to improve the results of the fuzzy M-estimate of location, since even after scaling the observations, the difference between the outlier and the rest of data is still clear and the Hampel loss function can conveniently cope with it. This can be clearly noticed when $c_p = 0.4$ and $C_D = 100$ (graphics at the bottom in Fig. 4). Due to the chosen distributions, the opposite is shown when $c_p = 0.4$ and $C_D = 5$ (graphics at the middle in Fig. 4), since outliers do not lie so far away from the non-contaminated part of the sample in both CASE 1 and CASE 2 and adjusting the tuning parameters in the Hampel loss function seems to be a better option than scaling all the observations (and therefore shortening the distances between outliers and the rest of data even more). However, when the contamination proportion c_p decreases to 0.1 and C_D remains equal to 5 (graphics at the top in Fig. 4), it can be seen that CASES 1 and 2 are not so similar as in the previous situations, and the larger distance between outliers and non-contaminated data in CASE 2 than in CASE 1 makes the fuzzy MDD-based M-estimator of location improve the behaviour of the fuzzy M-estimator of location only in CASE 2.

5   Concluding Remarks

In this paper, M-estimators of location with unknown dispersion have been extended to the fuzzy number-valued settings in order to provide a scale equivariant alternative to fuzzy M-estimators of location. They have been defined in terms of the median (D_θ^φ)-distance deviation about the 1-norm median. The two alternatives have been empirically compared and, even when there is no uniformly best estimator, it seems that the fuzzy MDD-based M-estimator of location may provide us with the best results when the distance between the outliers and the 'standard' data is large enough. Therefore, it could better identify which observations should be considered outliers indeed since it uses the information of the global dispersion. However, it would be advisable to complete this simulation study in the future to deal with other kinds of distributions and check whether these preliminary conclusions would remain.

Acknowledgements This research has been partially supported by the Spanish Ministry of Economy and Competitiveness through the Grant MTM2013-44212-P and the Principality of Asturias/FEDER Grant GRUPIN14-101. Their support is gratefully acknowledged.

References

1. Bertoluzza C, Corral N, Salas A (1995) On a new class of distances between fuzzy numbers. Math Soft Comput 2:71–84
2. Blanco-Fernández A, Casals MR, Colubi A, Corral N, García-Bárzana M, Gil MA, González-Rodríguez G, López MT, Lubiano MA, Montenegro M, Ramos-Guajardo AB, de la Rosa de Sáa S, Sinova B (2014) A distance-based statistical analysis of fuzzy number-valued data. Int J Approx Reason 55:1487–1501
3. De la Rosa de Sáa S, Gil MA, González-Rodríguez G, López MT, Lubiano MA (2015) Fuzzy rating scale-based questionnaires and their statistical analysis. IEEE Trans Fuzzy Syst 23(1): 111–126
4. De la Rosa de Sáa S, Lubiano MA, Sinova B, Filzmoser P (2017) Robust scale estimators for fuzzy data. Adv Data Anal Classif 11(4):731–758
5. Hampel FR (1974) The influence curve and its role in robust estimation. J Am Stat Assoc 69:383–393
6. Kim JS, Scott CD (2012) Robust kernel density estimation. J Mach Learn Res 13:2529–2565
7. Montenegro M, Casals MR, Lubiano MA, Gil MA (2001) Two-sample hypothesis tests of means of a fuzzy random variable. Inf Sci 133:89–100
8. Puri ML, Ralescu DA (1986) Fuzzy random variables. J Math Anal Appl 114:409–422
9. Sinova B, Gil MA, Colubi A, Van Aelst S (2012) The median of a random fuzzy number. The 1-norm distance approach. Fuzzy Sets Syst 200:99–115
10. Sinova B, Gil MA, Van Aelst S (2016) M-estimates of location for the robust central tendency of fuzzy data. IEEE Trans Fuzzy Syst 24(4):945–956
11. Sinova B, Pérez-Fernández S, Montenegro M (2015) The Wabl/Ldev/Rdev median of a random fuzzy number and statistical properties. In: Grzegorzewski P, Gagolewski M, Hryniewicz O, Gil MA (eds) Strengthening links between data analysis and soft computing, vol 315. Advances in intelligent systems and computing. Springer, Heidelberg
12. Sinova B, Van Aelst S (2017) Tukey's biweight loss function for fuzzy set-valued M-estimators of location. In: Ferraro MB, Giordani P, Vantaggi B, Gagolewski M, Gil MA, Grzegorzewski P, Hryniewicz O (eds) Soft methods for data science, vol 456. Advances in Intelligent systems and computing. Springer, Cham

Measures and Vagueness

Alejandro Sobrino

Abstract In this paper we will discuss some aspects concerning the application of the concept of measure to quantitative and qualitative or vague properties, highlighting the limitations caused by the high order vagueness and pointing to tuning and propagation as factors to take into account when the measure of a vague predicate is involved.

1 Introduction

Many objects of our interest are so because, in some way, we can access their properties. The properties can be quantitative or qualitative and while the quantitative aspects admit measurement, the qualitative ones are usually refractory to their individuation.

Physical concepts, such as mass, speed, energy or volume, are frequently substantiated in numbers or vectors, but others properties, as resilience or intelligence, are more difficult to quantify. In spite of this, proposals have been made for their quantification, including some type of numerical assignment.

Qualitative properties are described with words, avoiding the use of numbers. The use of non-numerical language allows us refer to objects or situations in an appropriate but not always accurate way. Note the contrast between saying that that a metal bar measures 2 m or weighs 10 kg and that an individual is very strong or speaks such language fluently, denoting 'very strong' and 'fluently' vague qualifications. In cases that do not admit as much precision as their numbering presupposes, the adverbs or adjectives are linguistic resources to quantify in what extent a property hold.

There are properties that admit a numerical specification but not a crisp verbalization. They are both quantitative and qualitative, depending on we refer them. The predicate 'tall' is one of them: Perhaps a meter accurately determines the height of a person (1.76 cm), but even so serious doubts emerge if we decide to name him as

A. Sobrino (✉)
Faculty of Philosophy, Universidade de Santiago de Compostela,
15782 Santiago de Compostela, Spain
e-mail: alejandro.sobrino@usc.es

E. Gil et al. (eds.), *The Mathematics of the Uncertain*, Studies in Systems, Decision and Control 142, https://doi.org/10.1007/978-3-319-73848-2_68

'tall'. While measuring in cm is context-independent, the predicate 'tall' is context-sensitive and goal-directed. So, a person measuring 1.76 cm is tall to be a horse rider but not to be a basketball player; but in the context of a basketball team, 1.76 cm perhaps is a normal height for a point guard player but not at all for a forward one.

When there is more than one criterion for evaluating the quantity of an object, linguistic predicates come to our aid, making possible to mitigate the expression by supposing that there may be some disagreement about the use of it.

Generality, imprecision, indeterminacy and vagueness are different linguistic categories. Generality arises from the non-specification of a fact or situation due to the lack of data or motivated by a conversation does not demand more detail. For example, when we say that 'a lot of people came to the basketball game' the sentence is general because 'a lot of' perhaps could be replaced by a number, but that unspecific quantity is considered sufficient to maintain the thread of the conversation.

Imprecision occurs when we do not have all the information at hand to determine a situation but the date exist: it is inaccurate for me how long does it take for the sunlight to impact my retina, (I think it spend around 7 min), but a physicist could determine the lapse accurately.

Indeterminacy arises when, even having all the information at hand, we are not able to guess the future: a die is thrown and it is indeterminate whether the result is and odd or an even number. While imprecision and generality have to do with vagueness, indeterminacy is close to probability.

Finally, a term is vague if it has irreducible borderline cases so that we are not able say if an object definitely lies within the positive or negative extension of the predicate or in its indeterminate area. An example of vague predicate is 'tall', because we have not a rule to mark if a person is tall or not.

2 Measure and Science

Usually, precise quantifiable predicates are associated to scientific activity, as the Physics one. Physical quantities are susceptible to be measured and the measurements are carried out assigning numbers to these magnitudes [4]. Thus, we say that a person weighs 70 kg or that a bottle houses 50 ml of liquid as they are possible values of the length of a person or the capacity of a bottle. 70 kg and 50 ml are quantities. Once we have quantities, it is possible to make operations or establish relationships between them.

A typical operation is addition. Thus, if the volume of a 25 ml bottle is increased with 10 ml more, it can accommodate up to 35 ml. In this case it is said that the measure of volumes is additive.

A quantity resulting from the repeated sums of the same number is denoted by its ratio. Thus, if m is the measure of a magnitude M and the value of $m(a)$ is 10,000 and that of $m(b)$ is 1000 or the value of $m(a)$ is 1000 when the value of $m(b)$ is 100, the ratio is 10:1.

The relationships between physical quantities or between instances of them can be metric and submetric. A metric relationship happens when the link is numerical -the Greek beauty canon demands that the body must measure 7 times the measure of the head- and it is sub-metric when the connection is made using comparative terms -the human body measures more than head-.

So, if M is a magnitude and m its quantity, $m(a) < m(b)$ denotes that the quantity of a is less than the quantity of b. Then, objects representing quantities can be ordered regarding its quantity.

Numbers are a key skill of human cognitive system. With greater or lesser sophistication, all known people use numbering systems. So, while western culture has a remarkable numerical calculus, the Brazilian tribe of Pirahas only use one, two and many to quantify [2].

Numbering systems emerge at the early age of development -around 4 years-, although experiments with evoked potentials pointed to that we already make comparisons the first months of our life. Several experiments show that the numerical cognitive substrate is located, in both human and nonhuman primates, in the parietal cortex of both hemispheres and, more specifically, in the interparietal sulcus [1].

3 Measures, Instruments and Its Varieties

Measurements are often performed using measuring instruments. Thus, the time passed is measured checking the material changes of an object, as the position of the Sun or the sand movement from one vessel to another. Using instruments we determine possible mutually incompatible values for a quantity -for instance, different positions of the dishes on a scale or different levels of liquid in a container-. If an order among measured entities is established, the measure is successful; i.e., regarding weight, if the plate a is a higher, lower or in a balance position with the plate b.

Measuring instruments are harmless when applied to large objects (celestial mechanics), but very influential if used to quantify small objects (quantum mechanics). In quantum mechanics the instruments of observation irremediably modify the observed things conditioning in an essential way our epistemic access to them.

Different physical magnitudes require different ways of approaching their measurements. This becomes clear if instead of describing a measure we try to explain it. In this case, magnitudes such as mass or length show different characteristics. While the measure of mass can not contravene the physical laws of nature -laws of motion and gravitation-, the determination of length is a result of convention. In sum, while the mass must address the quantitative properties of the object, the instruments of measurement and the laws of physics, the length only attend the first two. In the case of the length, since there are no laws, a pattern is needed to measure and any object is quantified in relation to it.

The measure determine the relation between the object and the pattern: if it is greater or less, and in what proportion. Thus, the measure of length is based on conventions and observational dispositions that allow the comparison of an object with another one.

These dispositions are:

1. If you match the beginning of two objects and the end of one coincides with the end of the other, they have the same length.
2. An object must be larger than any of its own parts.

Both precepts enable to determine whether an object is more or less than another and so, to compare them (for instance, more or less tall). But this outcome is based on three fundamental assumptions:

1. That it is possible always to match, at the beginning or at the end, two objects.
2. That any part of an object is entirely contained in it.
3. That any object is divisible into any number of its own parts.

These three postulates are objectionable if the measure to be considered is one of a vague predicate.

4   Measures and Vague Properties

Prima facie In cursive it seemed that crisp predicates are linked to scientific activity and vague predicates to life tasks. This conviction held for a long time until the quantum mechanics emerged, introducing indeterminacy into the root of the knowledge and the explanation of the smallest or elementary physical entities.

But even in the realm of celestial mechanics, the use of vague predicates is more frequent than might at first glance seem. An inspection of the short works included in Hawcking's *Physics Colloquium* shows that.

In Puente et al. [6] a semi-automatic process recovering causal sentences in texts was approached and the result of applying it to the Hawking's texts is shown below:

- If the field is **nearly** constant in a region, the gradient terms will be **small**, and the energy momentum tensor, will be minus half V, times the metric.
- However, Lindeh pointed out that if the potential is not **too** steep, the expansion of the universe will slow down the rate at which the field rolls down the potential, to the minimum.
- The observations do not yet indicate that the universe is **definitely** open, or that lambda is non zero, but it is beginning to look like one or the other, if not both
- A **very** small perturbation if you are with one of the big banks.
- If the dilation had a **low** value, the effective coupling would be **weak**, and string theory would be a **good** quantum theory.

- If one can determine that there is **enough** matter in the universe, to focus our past light cone, one can then apply the singularity theorems, to show that time must have a beginning.
- It then follows that if there is **enough** matter to make the universe opaque, there is also **enough** matter to focus our past light cone.
- If the sphere were *very large*, space would be **nearly** flat, and Euclidean geometry would be a **very good** approximation over **small** distances.

As noted, mined sentences included a profuse vague lexicon, as approximate quantifiers or linguistic hedges. That inquiry shows that even if vagueness is frequent in ordinary language, it is not absolutely foreign in scientific language.

Science and measurement have historically gone hand in hand. As imprecision is present in scientific discourse, a goal of scientists has been and still is to provide a certain estimation of it.

Inductive logic, probability theory and statistics are mathematical tools for giving an accurate estimate of imprecise sentences and arguments. But it is still a problem in progress to offer measures for qualitative properties where language in general, and vague language in particular, plays a relevant role.

Next, we will discuss the three assumptions made in the previous section:

- The postulate that it is possible to match two objects is related to the principle of identity, which is problematic if we consider the magnitude itself instead of some of its instantiations. It is possible to align a ruler and a table in order to determine how much it measures, but it is doubtful how to compare the length of two news by matching their beginning or their end. Where information begins or ends is often a blurred task. And the same happens with some physical objects, like a mountain: there is no doubt that Everest is part of the Himalayas, but it is doubtful where this mountain begins; i.e., where is the definite line separating the Everest from the Shisha Pangma.

The equality of two entities has traditionally been linked to the principle of the identity of the indiscernible, which says that two objects that have the same properties are the same thing. But objects have essential or defining, and accidental, or circumstantial properties.

This distinction call for a weakening of the principle: if two objects have the same essential properties and differ in the accidental ones they will be much more indistinguishable than if they share accidental but non-essential properties.

- The second postulate refers to the notion of subset. In a traditional sense, a set B is a subset of a set A if any element of B is also an element of A. Thus, the set $\{1, 2, 3\}$ is contained in the set $\{1, 2, \ldots, 10\}$ and an object that measures 1, 2 or 3 cm can be placed in correspondence with a rule that measures 10 cm and so calculate its length.

But there are sets contained in another sets in a way that is not precise, but rather blurred. Some grammatical categories and some biological taxonomy are a examples of this, because they have border elements refractory to be clearly separated and thus included in the taxon or its complement.

Ross showed that among the grammatical categories of verb and name there is a continuum: *Verb > Present participle > Adjective > Preposition > Adjective name > Name*, so it is doubtful whether the word 'amusing' should be classified as 'Adjective name' or as 'Name' [7]. In Biology is usually also controversial where to locate the missing links between primates and humans.

- Finally, to guarantee that any object is divisible into any number of its own parts assumes that all objects are separable and such divisions are recognizable as such. Elementary particles are an objection to this thesis, as they represent the smallest elements of matter, lacking of any internal structure. But leaving aside this case and focusing now on the complex particles, even by reducing their parts to precise divisions in an imaginary rule, problems arise.

Observationally, we can measure an object aligning it with the beginning of a rule and determining its measurement in terms of the coincidence of its end and one of the separations of the rule. But each division of the rule occupies a space. Using a more powerful observation instrument, this separation, which at first glance seemed extremely fine, now appears grossly thick, so that it can be subdivided into more units in a process that, theoretically, is recursively unlimited.

Observational predicates are at a good extent vague predicates and the moral of the previous story is that it is not possible to measure them definitively. The open texture affecting vagueness demands new ways of precisification [3]. But precisification without limits leads to higher order vagueness. That one discloses how unnatural it is to try to be absolutely precise dealing with imprecision.

Higher order vagueness uncovers the problems emerging when it is intended to give a precise measure of an imprecise property. A property can be inaccurate because there is still no instrument to determine it precisely or because it is convenient use a vague term in order to maintain the dialogue.

Human being is social and communicates in turns of conversation with his peers. In most human dialogues, the main goal is to support the flow communication, not to mean precisely what is said. Far from being bad or pernicious, vagueness lubricates our dialogues, facilitating interaction and community among humans, making possible a polite coexistence and self-protection, inhibiting possible strong refutations to what is said. Vague language, rather than quantifying something precisely, is used to qualify approximately many of the actions and objects around us.

Vagueness should be differentiated whether it happens in ordinary language or in scientific language. In ordinary language any measure should show tolerance and, for that purpose, the use of vague predicates is inevitable. The margin of tolerance implies that the dialogue continues and is not interrupted by the use of too specific vocabulary.

Scientific language, on the other hand, demands precision, convenient to guarantee secure or at least reliable knowledge. In that case, there are two main challenges regarding vagueness: tuning and propagation. Tuning of vague values must meet the Popper's requirement that it should not be more precise than what the situation demands [5]; i.e., what the measuring instrument allows and what the problem-solving requires. Propagation involves controlling the dissemination of vagueness in argumentation, so that it does not increase until being non-informative or unmanageable. Measuring imprecision requires managing imprecise measures but not to the point of rendering them useless. As Popper said the important thing in that scenario is clarity, not precision [7]. We should to be clear measuring the vagueness, but not at the cost of converting something inherently vague in precise. The measure of vagueness requires being vague.

Dedication

Kindness seems to be a qualitative property that some people have. Pedro Gil possessed it in the highest degree. Kindness perhaps could not be measured, but experienced by who needed it. And there I met Pedro. I will always remember him. This work is dedicated to his memory.

Santiago de Compostela, September 2017

References

1. Cantlon JF, Platt ML, Brannon EM (2009) Beyond the number domain. Trends Cogn Sci 13(2):83–91
2. Everett D (2009) Don't sleep. There are snakes. Life and language in the Amazonian jungle. Profile Books Ltd, London
3. Fine K (1975) Vagueness, truth and logic. Synthese 30:265–300
4. Hempel CG (1966) Philosophy of natural science. Prentice Hall, Upper Saddle River
5. Popper K (1963) Conjectures and refutations: the growth of scientific knowledge. Routledge Classics, London
6. Puente C, Sobrino A, Olivas JA, Merlo R (2010) Extraction, analysis and representation of imperfect conditional and causal sentences by means of a semi-automatic process. In: Proceedings of FUZZ-IEEE 2010. IEEE, Piscataway
7. Ross JR (1972) The Category Squish: Endstation Hauptwort. In: Peranteau PM, Levi JN, Phares GC (eds) Proceedings of 8th Region Meeting Chicago Ling Soc. Chicago Linguistic Society, Chicago

A Short Reflection on the Word 'Probable' in Language

Enric Trillas and Itziar García-Honrado

Abstract This short paper is devoted to reflect about the meaning of the predicate 'probable' when it is used in plain language. When a predicate is said in a universe of discourse, an order is introduced. Then, we propose to capture this order through a measure and we show that this measure do not always verify the additive law of Kolmogorov's axioms of probability.

1 Introduction

The important theoretical and practical successes of Kolmogorov's Probability Theory [5] can easily conduct to believe that it constitutes a model for the use, in plain language, of the word 'probable'. Nevertheless, in plain language the statements 'p is probable' does not only refer to precise statements p but often to imprecise ones like, for instance, 'It is probable that forest is a huge one', in which $p = $ 'Huge forest' cannot be represented by a subset in a big enough universe of forests, as it is with the typically probabilistic statement 'It is probable to obtain an even number of points in throwing a single die', in which $p = $ 'Even number' is perfectly represented by the subset $P = \{2, 4, 6\}$ in the universe $X = \{1, 2, \ldots, 6\}$ of elemental events representing the universe of discourse.

It should be recalled that only precise statements on X are allowed to be represented by subsets in X, as it is stated by the Specification Axiom of Set Theory [4], but, as Sorites's type arguments show [10], the imprecise ones cannot; nevertheless they can be represented by membership functions of the fuzzy sets generated by the linguistic label p (see [9]) in the corresponding universe of discourse. For instance, this is the case with 'It is probable that 3 is a small number in the interval [0, 10] of

E. Trillas · I. García-Honrado (✉)
Department of Statistics, OR and Teaching Mathematics, University of Oviedo,
Oviedo, Spain
e-mail: garciaitziar@uniovi.es

E. Trillas
e-mail: trillasenric@uniovi.es

the real line', where 'small' admits to be represented by the membership function $1 - \frac{x}{10}$, among others, in [0, 10].

This paper is but an unended reflection on the subject of measuring the meaning of the word 'probable' in plain language, with the goal of not always confusing it with a probability measure.

2 On Meaning

To understand a statement 'p is probable' it is necessary to capture the meanings of the statement p, and that of the predicate 'probable', both in the context in which such statement is uttered. Meaning is context-dependent and purpose-driven; not only depends on the universe of discourse at which it refers to, but also on the context surrounding the particular situation in which the statement is placed, and on the purpose hold for its use. Since very often people learn meaning by polarity, that is, by jointly using a word and one of its antonyms, for capturing the meaning of 'probable' it also could be suitable to jointly consider the meaning of 'improbable', and not only the negation 'not probable'.

2.1 Qualitative Meaning

The meaning of a predicate P in a universe X is captured once known the relationship 'x is less P than y' (shortened, in X, by $x \leq_P y$); such relationship, obviously reflexive when it exists, reduces to 'x is equally P than y' (shortened, in X, by $x =_P y$) whenever its use is precise [8]. It is so because in the precise case, elements in X just are P, or they are not P. For instance, in the universe of positive integers, the numbers 5 and 2017 are equally odd and without the possibility of stating that one of them is (strictly) less odd than the other. In principle, it can be said that the precise use of words is definable through 'if and only if' conditions, but that it is not for the imprecise uses; in this precise case, $=_P$ is an equivalence relation partitioning X in two classes, the subset specified by P and its complement.

Notice that $x \leq_P y$, just designates that x shows the property p named P less than y shows it; hence, in plain language, such relation is often recognized empirically, and P is said to be 'meaningless' when $\leq_P = \emptyset$. The graph $(X, \leq_P)$ is called the *qualitative meaning* of P in X (see [1, 2]); if existing, its *maximal elements* are the 'prototypes' of P in X, and its *minimal elements* its 'anti-prototypes'.

Notice that $=_P$ is nothing else than the intersection of the subsets $\leq_P$ and $\leq_P^{-1}$ (translating the inverse relationship by 'more P than') in the Cartesian product $X \times X$, and whose coincidence makes $=_P$ coincidental with $\leq_P$.

For example, taking $P = small$ in the universe of the real numbers in the interval [0, 10], endowed with the usual linear order ($\leq$) of the real line; it is $x \leq_{small} y \Leftrightarrow y \leq x$, that is, $\leq_{small} = \leq^{-1}$. The graph $([0, 10], <_{small})$, the qualitative meaning of 'small' in $[0, 10]$ coincides with the reversed linearly ordered set $([0, 10], \leq^{-1})$; hence, under $\leq_{small}$ there is only a maximal or maximum 0, and a unique minimal or minimum 10. The relation $=_{small}$ is just the equality of number, since $x =_{small} x \Leftrightarrow x \leq y$ and $y \leq x$, that is, $x = y$.

2.2 Meaning's Measure

Once captured how P is primarily used in X by the graph $(X, \leq_P)$ it is the right moment to introduce the concept of a meaning's measure, a mapping $m_P : X \to [0, 1]$, verifying the three basic axioms [1]:

(a) If $x \leq_P y$, then $m(x) \leq m(y)$,
(b) If z is a maximal under $\leq_P$, then $m_P(z) = 1$,
(c) If z is a minimal under $\leq_P$, then $m_P(z) = 0$.

It should be noticed that, in general, these three axioms are not sufficient for specifying a single measure. For instance, in the case of 'small' in $[0, 10]$, the measures m_{small} are those functions from $[0, 10]$ into $[0, 1]$ that are decreasing, and satisfy the two border conditions $m_{small}(0) = 1$, and $m_{small}(10) = 0$. That is, all the decreasing functions joining the points $(0, 1)$ and $(10, 0)$, of which there is an enormous amount. If it is known that the measure is linear, $m_{small}(x) = ax + b$, then from the above axioms it follows $a = -\frac{1}{10}$ and $b = 1$: there is only the linear measure $m_{small}(x) = 1 - \frac{x}{10}$. Nevertheless, if it was known that the measure is quadratic, $ax^2 + bx + c$, the axioms facilitate the conditions $100a + 10b + c = 0$, and $c = 1$, or $100a + 10b + 1 = 0$ that, with the derivative $2ax + b < 0$ (because of its decreasing character), show that there are many quadratic measures. One of them is $(1 - \frac{x}{10})^2 = \frac{x^2}{100} - \frac{2x}{10} + 1$ that, decreasingly, joins the point $(0, 1)$ with the point $(10, 0)$, and whose derivative $\frac{2x}{100} - \frac{2}{10} = \frac{2}{10}(\frac{x}{10} - 1)$ is obviously negative for all $x \in [0, 1]$. Another possibility for specifying a quadratic measure is by knowing that it passes for some point; for instance, if this is the point $(0.5, 0.7)$ the new condition $0.7 = 0.25a + 0.5b + 1$ is obtained, and that, with the former one, allows finding a and b.

Hence, some additional information, and/or some reasonable hypotheses (like is the former linear character of m_{small}), is necessary for specifying a meaning's measure, and, for what concerns words in plain language, such information should be induced from the context surrounding the corresponding use's words.

That is, meaning's measures should be designed accordingly with their current linguistic use, and no universal measure exists unless such use is precise, since measures do preserve the relation $=_P$. In fact, if $x =_P y \Leftrightarrow x \leq_P y$ & $y \leq_P x \Rightarrow m_P(x) \leq m_P(y)$ & $m_P(y) \leq m_P(x) \Leftrightarrow m_P(x) = m_P(y)$. With it, and if the use of P is precise in X, with $\mathscr{P}$ the subset of X specified by P and consisting in its prototypes, it follows $m_P(x) = 1$ if $x \in \mathscr{P}$, and $m_P(x) = 0$ if $x \in \mathscr{P}^c$ the subset containing its anti-prototypes. In conclusion, if P is precisely used in X, the characteristic function of $\mathscr{P}$ is the unique measure for its meaning in X, but if P is imprecisely used then the meaning's measure is not unique [3, 8].

A full, or qualitative-quantitative, meaning of P in X is thus given each time by an specific quantity $(X, \leq_P, m_P)$ [3, 8]; the meaning of imprecise words is not a single and universal concept, and it is scientifically domesticated by means of such quantities that, in addition, should be carefully designed. Notice that a wrong design of the quantity can mean a bad specification of P and, hence, can conduct to erroneous results by presumably referring to a different use from the current one.

Nowadays, only considering the meaning of a word P through a measure by forgetting the relation $\leq_P$, can conduct to badly appreciate its qualitative meaning. In fact, given m_P (now shortened by just m) a new relation defined by $x \leq_m y \Leftrightarrow m(x) \leq m(y)$, appears in X, and, contrarily to $\leq_P$, is a linear or total relation that, in addition, is a partial order. Notice that there can perfectly exist elements x and y for which neither $x \leq_P y$, nor $y \leq_P x$ hold, that is, x and y are not comparable under $\leq_P$, something that cannot happen under $\leq_m$ since one of the real numbers $m(x)$, $m(y)$ is necessarily the greatest of both. It is $\leq_P \subseteq \leq_m$, and if both relations were coincidental the first should be linear, without not comparable elements, something that in general does not happen. Once a measure is designed, confusing $\leq_P$ with $\leq_m$ implies the risk of considering a pretended qualitative meaning larger than the original; it can be said that the act of measuring can, if only the measure is considered, alter the qualitative meaning, and a measure m is said to 'perfectly describe the use of P' [8] whenever $\leq_P$ coincides with $\leq_m$, as it happens with 'small' in [0, 10].

2.3 Remarks

(a) Precise words, defined by 'if and only if' conditions, have a single measure valued in $\{0, 1\}$ but provided those conditions do not contain imprecise words. For instance, the Hardy-Ramanujan's concept of a 'round number' [6] is defined by an *if and only if* condition that, affecting the number's unique decomposition in prime factors, includes the imprecise words 'considerable' and 'comparatively small'; consequently, neither its use is precise in the set of positive integers, nor it has a single measure.

(b) The antonyms P^a of P are represented by quantities $(X, \leq_{P^a}, m_{P^a})$ such that $\leq_{P^a} = \leq_P^{-1}$, and $m_{P^a} = m_P \circ s_P$, with s_P a symmetry $X \to X$ that reverses $\leq_P$ (see [8, 9]). For instance, provided 'small' is represented in [0, 10] by $\leq_{small} = \leq^{-1}$, and $m_{small}(x) = 1 - \frac{x}{10}$, then 'big' can be represented by $\leq_{big} = \leq_{small}^{-1} = \leq$, and $m_{big}(x) = m_{small}(10 - x) = \frac{x}{10}$.

3 On the Meaning of Probable

In what follows, some hints on the plain language's use of the word 'probable', affecting statements p uttered (or written, or gestured) on the elements of a universe of discourse X, will be presented.

3.1 First Approach to the Meaning of Probable

Since the corresponding elemental statements are of the form 'p is probable', the corresponding universe for 'probable' is not X, but that $S(X)$ of the statements on X. Hence, to capture the meaning of 'p is probable' it will be presumed that a full meaning of p in X is known by a quantity $(X, \leq_p, m_p)$, and then it will lack to capture a meaning of 'probable' $(S(X), \leq_{prob}, m_{prob})$ that, thanks to that of p, can allow to reach the meaning of 'p is probable'.

Questioning if the composition $m_{prob}(m_P(x)) = (m_{prob} \circ m_P)(x)$ can measure the meaning of 'p is probable', forces to define m_{prob} in the range of values of m_P in $[0, 1]$ instead of in $S(X)$; something very odd. Indeed, when the use of p is precise such range is $\{0, 1\}$, the unique subset of $[0, 1]$ that is a Boolean algebra where m_{prob} is only definable by $m_{prob}(0) = 0$, and $m_{prob}(1) = 1$, the only possible probability. Hence, 'p is probable' will be just with meaning's measure 0 or 1; for instance, 'It is probable to get five points' cannot be $\frac{1}{6}$. It seems to be a wrong way for focusing the problem since, actually, room for probabilities will never exist when p is precisely used and m_P necessarily takes the values 0 and 1.

3.2 Relation with Kolmogorov's Probability

Let's turn around the concept of Kolmogorov's probability. It requires that, for counting with a measure of probability, the 'events' do belong to a Boolean algebra (namely, a σ-algebra in the case of requiring the denumerable, or σ-additive law) [5]. Such Boolean algebra, and thanks to the Marshall Stone's characterization theorem [7], is isomorphic to a Boolean algebra of subsets of some universe; let $\Omega = \{A, B, C\}$ be such algebra. For simplicity, and since it will not be against our arguments, we will just consider the so-called 'finite probability' case, that of probabilities with just the finite additive law. A mapping $prob : \Omega \to [0, 1]$ is a probability, if:

(a) $prob(\Omega) = 1$, and,
(b) If A and B are in Ω, and their intersection $A \cap B$ is empty, then $prob(A \cup B) = prob(A) + prob(B)$.

From it, and since $\Omega^C = \emptyset$, it follows $1 = prob(\Omega) = prob(\Omega \cup \emptyset) = prob(\Omega) + prob(\emptyset)$, and $prob(\emptyset) = 0$; an analogous argument with A and A^C, immediately conducts to the law $prob(A^C) = 1 - prob(A)$, for all A in Ω.

Notice that Ω is the maximum and $\emptyset$ is the minimum of the Boolean algebra under its 'natural' lattice's order $A \subseteq B \Leftrightarrow A \cup B = B$. Hence, and for $prob$ being a measure in the graph $(\Omega, \subseteq)$, it only lacks to prove that it is non-decreasing: $A \subseteq B \Leftrightarrow B = A \cup (B \cap A^C)$, and since it is $A \cap (B \cap A^C) = \emptyset$, it follows $prob(B) = prob(A) + prob(B \cap A^C) \geq prob(A)$.

That is, thanks to the existence of the relative complement $B - A = B \cap A^C$, $prob$ verifies the three laws of a measure in the graph $(\Omega, \subseteq)$, and it is a measure of the meaning of the word 'probable' provided it could be presumed $\leq_{prob} = \subseteq$; this hypothesis, close to the former $\leq_{small} = \leq^{-1}$ in [0, 10], has with it the difference that this last is accepted since there cannot be doubt whatsoever on the equivalence $x \leq_{small} y \Leftrightarrow y \leq x$. But, what with the set's inclusion ($\subseteq$), coming from considering that 'less elements imply less probability', by, perhaps, presuming that $A \subseteq B \Leftrightarrow Card(A) \leq Card(B)$, when it is only clear that the first formula implies the second, and that it only can be accepted that if $A \subseteq B$, then $A \leq_{prob} B$, but not reciprocally.

What has been just said implies that, if m_{prob} was a measure for the graph $(\Omega, \leq_{prob})$ it is a measure for the graph $(\Omega, \subseteq)$ but not reciprocally, and that provided $prob$ is a probability in the algebra Ω, it is not necessarily a measure in the graph $(\Omega, \leq_{prob})$. The identification of probabilities with meaning measures of the word 'probable', when it is used in a Boolean algebra of events, is actually risky (see Example 4.1 in Sect. 4.6).

In plain language, 'probable' is not only predicated of precise statements, but also of imprecise ones that neither can be represented by subsets of the universe of discourse, nor constitute a Boolean algebra ([8, 9]).

These facts introduce a serious difficulty for holding the hypothesis that the Kolmogorov's probabilities can always measure the meaning of the word 'probable'. What should be known for this goal is a quantity $(X, \leq_{prob}, m_{prob})$, and a similar trouble to that in Sect. 3.2 is that the relation given by $m_{prob} : x \leq_{m_{prob}} y \Leftrightarrow m_{prob}(x) \leq m_{prob}(y)$ is just the linear order of the interval [0, 1] that, only when 'probable' is precise can coincide with the set's inclusion. It is necessary to know the relation $\leq_{prob}$, and distinguishing it from the larger order of the unit interval. It does not seem that such relation can be a universal one, like $\subseteq$ and $\leq$ are; it is something that should be recognized from the contextual use of the word 'probable' in the universe of discourse X, something that can only be done at each particular situation.

4 On the Relation Between Probabilities and Measures Of 'probable'

4.1 In Boolean Algebras

Provided it is known a full meaning $(X, \leq_{prob}, m_{prob})$ of the word 'probable' in a universe of discourse X, is it possible that m_{prob} be a Kolmogorov's probability? Firstly, it is necessary that X is endowed with a Boolean algebra's structure, and second that m_{prob} verifies the laws of a probability.

Concerning the first condition it should be remarked that plain language is, in general, free from any known algebraic structure, and, in particular, from that of Boolean algebra. For instance, to presume that conjunction is commutative is abusive in language where time almost always intervenes, and to suppose that conjunction and disjunction are associative in language, often means that commas can be avoided, something that does not always happen. A lattice structure can hold in plain language, but not universally and only in some part of it that can be formalized [8]; consequently, the actually open problem is to search for the meaning of 'probable' in unstructured parts of plain language.

4.2 The Case in Some Strongly Structured Parts of Language

If, for instance, a part of language can be structured as an Ortho-modular lattice (like it seems to happen with the language of Quantum Physics [1]), there is a slight modification of the theory of probability to be used for modelling 'probable', and if it is a De Morgan algebra of fuzzy sets, there exists the Zadeh's theory for the probability of imprecise events [11]; both verify Kolmogorov's axioms, even if in the second case empty intersection (incompatibility), seems to be too restrictive and, in both cases, incompatibility and contradiction are not equivalent like it happens in Boolean algebras. A mathematical model of meaning should be applicable to all parts of plain language; without meaning words cannot be understood. In any case, for identifying a measure of 'probable' with a probability, the coincidence of the qualitative meaning's relation $\leq_{prob}$ and the corresponding partial order $\subseteq$, or $\leq$, of the algebraic structure, should be accepted; as it is, for instance, to assert that a suitable probability of getting five points when throwing a die, is a meaning's measure of the statement 'Getting five points is probable'. Such probability should be chosen accordingly with the physical characteristics of the current die and surface on which it will land.

4.3 Measuring the Meaning of 'Probable' Based on Probabilities

Suppose $(S(X), \leq_{prob})$ is a Boolean algebra, and a quantity $(S(X), \leq_{prob}, m_{prob})$ modelling the meaning of 'probable' is known. Provided $S(X)$ were finite, and m_{prob} a probability, the addition of its values for the atoms in $S(X)$ should equal one. Hence, not all meaning's measure is a probability, unless the total sum of its atoms values is one. Notwithstanding, it does not mean that each one of the measure's values can be obtained by means of a statistical procedure. For instance, provided $S(X)$ contains four atomic statements $p_1, \ldots, p_4$, whose contextually obtained measure is:

$$m(p_1) = 0.5, m(p_2) = 0.7, m(p_3) = 1, m(p_4) = 0,$$

(that is not a probability, since the sum of its values is not one), it is easy to find many quadruplets of probabilities $prob_1, \ldots, prob_4$, each one giving the corresponding value of m; that is, verifying $m(p_k) = prob_k(p_k)$, for $1 \leq k \leq 4$. For instance,

- $prob_1$, with respective values 0.5, 0.3, 0.2, 0 (all in $[0, 1]$, and with total sum one), gives $m(p_1) = prob_1(p_1) = 0.5$;
- $prob_2$, with respective values 0.2, 0.7, 0, 0.1, gives $m(p_2) = prob_2(p_2) = 0.7$;
- $prob_3$, with respective values 0, 0, 1, 0, gives $m(p_3) = prob_3(p_3) = 1$;
- $prob_4$, with respective values 0.5, 0.2, 0.3, 0, gives $m(p_4) = prob_4(p_4) = 0$,

thus, $m(p_k) = prob_k(p_k)$, for all k. The measure is not a probability, but each one of its values comes from a probability. What seems possible is the design of four random experiments (four random variables) able to obtain the four values of m. Of course, it does not avoid the possibility that just one random variable can identify the measure with a probability.

In general, for each statement p in $S(X)$, and a meaning's measure m of it, random experiments on p can be designed in such a way that, from the corresponding random variable, a probabilistic value of $m(p)$ is obtained. Would this open problem be solved, it will be known which measures can be either a probability, or be equivalent to a family of probabilities.

To pose the problem in finer mathematical terms, for each p in $S(X)$, and each measure m of its meaning, it should be found:

(a) A σ-algebra $\Omega(p, m)$ in some universe associated to the context surrounding the meaning of p,
(b) A representation p^* of p in $\Omega(p, m)$, and,
(c) A probability $prob_{p,m}$ on $\Omega(p, m)$, such that, $m(p) = prob_{p,m}(p^*)$.

4.4 Some Differences of Dealing with a Measure of Probable and a Probability

Provided $S(X)$ is finite and consists in n statements $p_1, \ldots, p_n$, the equality to one of the sum of all the values $m(p_1), \ldots, m(p_n)$, provided m is a probability, comes from both the additive law of probability and that the maximum element of the Boolean algebra is with probability 1, that it is a normalized measure. Since, under the hypotheses that $S(X)$ is a Boolean algebra, and that $\leq_{prob}$ coincides with its partial order $\subseteq$, the maximum is the only maximal, and the normalized character of measures of meaning is obvious, the problem just lies in the additive law and, consequently, in the possibility of decomposing elements in two 'separate' pieces. Although such law can be easily accepted in cases with a precise use of the statements p, when this use is imprecise it is not so clear [8]. For instance, if the statements were represented by membership functions μ in $[0, 1]^X$, then decompositions $\mu = \mu_1 + \mu_2$, with $\mu_1 \cdot \mu_2 = \mu_0$ (the membership function of the empty set $\emptyset$), should be computed (in a Standard Algebra of Fuzzy Sets [9]) through solving, for each x in X, the functional equation corresponding to $T(\mu_1(x), \mu_2(x)) = 0$ (that is, $T(a, b) = 0 \Leftrightarrow T$ is in the Łukasiewicz family) in the unknown t-norm T, with the conditioning $S(\mu_1(x), \mu_2(x)) = \mu(x)$, for a suitable t-conorm S for each μ. Alternatively, and instead of considering the incompatibility $\mu_1 \cdot \mu_2 = \mu_0$, it can be considered the contradiction $\mu_1 \leq \mu_2' \Leftrightarrow \mu_1(x) \leq N(\mu_2(x))$, with the same restriction with a t-conorm S for each μ. In all cases, it requires designing a suitable algebra of fuzzy sets in which decomposing the membership functions in either 'disjoint', or contradictory pieces, can be done. The additive law depends, at least, on fixing such algebra.

4.5 Remark

Let's recall that a t-norm T is in the Łukasiewicz family [9], that of t-norms with zero divisors, if and only if it is $T = W_f$, that is, $T(a, b) = f^{-1}(\max(0, f(a) + f(b) - 1))$, with $f : [0, 1] \rightarrow [0, 1]$ a strictly non-decreasing function such that $f(0) = 0$, and $f(1) = 1$. Then $T(a, b) = 0$ means $\max(0, f(a) + f(b) - 1) = 0 \Leftrightarrow f(a) \leq 1 - f((b) \Leftrightarrow a \leq f^{-1}(1 - f(b)) \Leftrightarrow a \leq N_f(b)$, with N_f a strong negation function. Hence, $T = W_f$ and $N = N_f$, constitute the only t-norms and strong negation functions under which incompatibility and contradiction are equivalent, like they are in Boolean algebras. In general, with fuzzy sets such properties are one independent from the other, as it is in language and when using imprecise words [8]. With other t-norms, these equivalence is not preserved; for instance, with $T = \min$, it is $\min(a, b) = 0 \Leftrightarrow a = 0$ or $b = 0$, and the contradiction between a and b clearly follows. But $a \leq N_{id}(b) = 1 - b \Leftrightarrow a + b \leq 1$, does not imply $\min(a, b) = 0$.

4.6 Examples

In this section we will show two examples related with the additive law for two different measures of probable. The first one dealing with the qualitative meaning represented by a Boolean algebra and the last one dealing with fuzzy sets.

Example 4.1 Let us consider the universe of discourse $S(X)$ collecting the precise sentences that show possible places where a family can spend the afternoon of the following day. Three possibilities are considered as the atoms of $S(X)$: The family goes to the mountain (p_1), the family goes to the beach (p_2) or the family remains at home (p_3). Then, a Boolean algebra structure (see Fig. 1) that represents the family possible decisions is considered, taking into account the unions and the null intersections of the atoms, and supposing that the qualitative meaning, $\leq_{prob}$, is $\subseteq$. In this framework, different measures, m_P, of the predicate $P = probable$ could be defined in $S(X)$. It is enough to verify that each m_P is a measure that is: (a) $m(p_0) = 0$, (b) $m_P(p_7) = 1$, and (c) if $p_i \leq p_j$, then $m_P(p_i) \leq m_P(p_j)$, for all $i, j = 0, 1, \ldots, 7$.

In the case of defining m_P by $m_P(p_0) = 0$, $m_P(p_1) = 0.2$, $m_P(p_2) = 0.3$, $m_P(p_3) = 0.5, m_P(p_4) = 0.3, m_P(p_5) = 0.5, m_P(p_6) = 0.5$, and $m_P(p_7) = 1, m_P$ is a measure, but it is not a probability since additive law is not verified. It is enough

Fig. 1 The representation of
$(S(x), \leq_{prob})$ as a Boolean
algebra

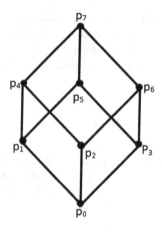

Fig. 2 The representation of
fuzzy sets translating the
meaning of the sentences in
$S(X)$

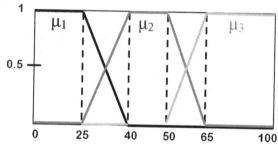

to take p_1 and p_2: it is $p_1 \cdot p_2 = p_0$ but $m_P(p_1 + p_2) = m_P(p_4) = 0.3$ is not equal
to $m_P(p_1) + m_P(p_2) = 0.2 + 0.3 = 0.5$.

Example 4.2 Let us consider the set of sentences $S(X)$ with three imprecise sentences p_1 = he is young, p_2 = he is medium aged, p_3 = he is old. Each sentence p_i
can be represented in the universe of discourse $X = [0, 100]$ by a fuzzy set μ_i for
all $i = 1, 2, 3$. They are shown in Fig. 2.

The predicate probable introduces an order in the universe of discourse of these
sentences, the measure of how probable each sentence is which is represented by a
fuzzy set μ_i, is computed through $m_P(\mu_i) = \frac{1}{100} \int_0^{100} \mu_i(x)dx$, for all $i = 1, 2, 3$.

In the example: $m_P(\mu_1) = 0.325$, $m_P(\mu_2) = 0.25$, and $m_P(\mu_3) = 0.425$. It is
clear that m_P is a measure since $m_P(\mu_0) = 0$, $m_P(\mu_1) = 1$, and it verifies the
monotonic condition for the order $\leq_{prob}$ (in this case, $p_2 \leq_{prob} p_1 \leq_{prob} p_3$, then
$m_P(\mu_2) \leq m_P(\mu_1) \leq m_P(\mu_3)$). In the same way as the previous example, this measure is not a probability since additive law does not hold. It is enough to take μ_1
and μ_2: it is $W_f(\mu_1, \mu_2) = 0$ with $f(x) = x^2$, but $m_P(W_f^*(\mu_1, \mu_2)) = 0.5467$ is not
equal to $m_P(\mu_1) + m_P(\mu_2) = 0.325 + 0.25 = 0.575$.

5 Conclusion

This short paper cannot be conclusive; it just tries to present some reflections that, eventually, can allow to mathematically modelling the use, in plain language, of the word probable, whose meaning cannot be fully captured by the Kolmogorov's theory of probability.

Basic reasons for it are the (necessary) Boolean, or Ortho-modular, lattice's ground such theory requires, and that the relation of subsets inclusion (that orders Boolean algebras) seems to be, in general, just a part of the qualitative meaning of the word 'probable'. In addition, the basic additive law of probability requires that what is submitted to compute its probability, should be decomposable in either disjoint, or contradictory pieces, something not actually clear with the imprecise words that permeate language, as it is shown by the algebras of the fuzzy sets' [9] membership functions with which imprecise words are represented.

Since the meaning of words in plain language is both context-dependent, and purpose-driven, more knowledge on specific uses of the word 'probable' is still necessary for obtaining quantities reflecting its meaning. Some instances of its use should be previously studied before counting with actual possibilities for going ahead with a theory on the meaning of the word 'probable' as a possible scientific domestication of it; it is something similar to what happened with probabilities, whose complete theory came from many practical experiences in, for instance, random games. Eventually, it could be done by also and jointly considering the uses of the antonym word 'improbable' that, once known a quantity $(X, <_{prob}, m_{prob})$, will be given by another quantity $(X, \leq_{prob}^{-1}, m_{prob} \circ s_{prob})$, with $\leq_{improb} = \leq_{prob}^{-1}$, and s_{prob} a symmetry in X reversing $\leq_{prob}$. Those instances should be studied by experimenting, in plain language, with the goal of testing them with the presented model.

A possible alternative way for conducting such study could be done, perhaps, by following the 'Subjective Probability' approach started by Bruno de Finetti [2]. Based on some 'a priori' knowledge of who tries to assign a probability, and on the Bayes formula, it seems closer to how people reasons; consequently, it can be closer to plain language than Kolmogorov's axiomatic theories.

References

1. Bodiou G (1964) Théorie dialectique des probabilités (englobant leurs calculs classique et quantique). Gauthier-Villars, Paris
2. de Finetti B (1931) Sul significato soggetivo della probabilità. Fund Mathematica XVII:298–329
3. García-Honrado I, Trillas E (2011) An essay on the linguistic roots of fuzzy sets. Inform Sci 181:4061–4074
4. Halmos PR (1960) Naïve set theory. Van Nostrand, Princeton
5. Kolmogorov AN (1956) Foundations of the theory of probability. Chelsea, New York
6. Renedo E, Sobrino A (2007) Round numbers revisited. A new fuzzy approach. Fuzzy Sets Syst 158:1618–1629

7. Stone M (1936) The theory of representation of Boolean algebras. Trans AMS 40:37–111
8. Trillas E (2017) On the logos: a naïve view on ordinary reasoning and fuzzy logic. Springer, Cham
9. Trillas E, Eciolaza L (2015) Fuzzy logic: an introductory course for engineering students. Springer, Heidelberg
10. Trillas E, García-Honrado I (2013) A layperson reflection on sorites. In: Seising R, Tabacchi ME (eds) Fuzziness and medicine: philosophical reflections and application systems in health care, studies in fuzziness and soft computing, vol 32. Springer, Berlin
11. Zadeh LA (1968) Probability measures of fuzzy events. J Math Anal Appls 23:421–427

Measuring Uncertainty in the Portfolio Selection Problem

Enriqueta Vercher and José D. Bermúdez

Abstract In this paper, we propose a new index for ranking portfolios based on the credibility expected return and loss on their investment. We assume that the return on a given portfolio is modeled as a trapezoidal fuzzy variable, whose credibility distribution is built using the data set of its historical returns. The credibilistic loss on the investment for a given portfolio is measured by means of a suitable loss function. In order to take risk-adverse investor attitudes into account, we analyze the performance of some credibility measures related to loss and risk on the investment for a given portfolio and their relationship with similar possibility measures. A numerical example is presented showing the performance of different fuzzy ranking indices for real portfolios in the Spanish stock market.

1 Introduction

From the seminal work of Markowitz [20] many researchers have developed optimization models and procedures based on random variables for selecting optimal portfolios, taking into consideration the fact that the future return on assets can be approximated by using random variables. The mean-variance (MV) portfolio selection model is considered the beginning of modern portfolio theory, where it is assumed that the expected means and covariance matrix of the return on assets suitably represent the risk-return tradeoff. Concerning the uncertainty of returns, alternative portfolio selection models have been developed either by modifying the risk measure or by adding statistical higher moments [15, 16]. An alternative approach to this probabilistic framework, making it possible to incorporate incomplete information about the returns on assets, knowledge of financial experts and vagueness leads to many fuzzy portfolio selection models which have been developed mainly based

E. Vercher (✉) · J. D. Bermúdez
Department of Statistics and Operational Research, Universitat de València,
C/ Dr. Moliner 50, 46100 Burjassot, Spain
e-mail: enriqueta.vercher@uv.es

J. D. Bermúdez
e-mail: bermudez@uv.es

© Springer International Publishing AG 2018
E. Gil et al. (eds.), *The Mathematics of the Uncertain*, Studies in Systems, Decision and Control 142, https://doi.org/10.1007/978-3-319-73848-2_70

on possibility measures [5, 13, 26, 27, 30]. Recently, Liu [18] has introduced the credibility theory for fuzzy variables and defined their moments and values making it possible to obtain new fuzzy portfolio selection models, where the returns on assets are approximated by means of fuzzy variables [9, 11, 31]. For dealing simultaneously with randomness and fuzziness, returns on assets in the portfolio selection problem have also been modeled by using random fuzzy variables [10, 12].

On the other hand, linear and quadratic programming problems have been applied for selecting optimal portfolios based on MV or mean-absolute deviation (MAD) models; however, in order to manage more realistic constraints for the portfolio selection problem several multi-objective evolutionary procedures have been explicitly introduced in a Soft Computing framework [1, 24, 34]. It is important to note that the goals and constraints of all these portfolio selection models are defined by means of expected values, either for probability or possibility and credibility distributions.

In our research, we propose dealing with the historical returns on a given portfolio, instead of considering the returns on the individual assets as a data set. This change of paradigm allows us to measure the uncertainty of the risk and the benefit of a given investment incorporating the contemporary relationship of the returns on the individual assets that make up a portfolio. In a first approach, we use the possibility moments of trapezoidal and LR fuzzy numbers to approximate the expected return and fuzzy downside risk of every portfolio, which are the goals of various multi-objective optimization problems with cardinality constraints [3, 24, 28]. More recently, for approximating the future return on a given portfolio we use fuzzy variables, because of the credibility distributions which allow us to measure the uncertainty of the return, the risk of the investment and the Value-at-Risk (which is a measure of the worst expected loss over a given horizon) by means of credibility moments, fuzzy quantities and crisp values [14, 29, 31] respectively.

The paper is planned as follows: Sect. 2 reviews some useful notation and results of credibility measures. Section 3 presents several relationships between possibilistic and credibilistic moments, which are the goals of the fuzzy portfolio selection problem. A new ranking procedure for scoring portfolios is presented in Sect. 4. There we also deal with measures of loss and risk for a portfolio whose returns have been approximated using a trapezoidal credibility distribution. Finally, we illustrate the performance of our proposal by comparing different fuzzy ranking indices for a given set of portfolios from the Spanish stock market in Sect. 5.

2 Preliminaries

A fuzzy number Q is a convex normalized fuzzy set of the real line, whose membership function μ_Q is piecewise continuous. For $y \in \mathbb{R}$, $\mu_Q(y)$ is the degree of membership of y in Q; the closer the value of $\mu_Q(y)$ is to 1, the more it belongs to Q.

The possibility and necessity of every fuzzy event (for instance, $\{Q \geq y\}$ where is y a real number) can be evaluated based on the possibility distribution associated with the corresponding membership function [33]. In particular,

$$\text{Pos}\{Q \geq y\} = \sup_{t \geq y} \mu_Q(t), \text{ and Nec}\{Q \geq y\} = 1 - \sup_{t < y} \mu_Q(t). \qquad (1)$$

Let us recall some definitions of fuzzy variables and credibility measures [18]. The credibility measure of a fuzzy event as an alternative form of measuring its uncertainty was introduced by Liu and Liu [19]:

$$\text{Cr}\{Q \geq y\} = \frac{1}{2}(\text{Pos}\{Q \geq y\} + \text{Nec}\{Q \geq y\}). \qquad (2)$$

Definition 2.1 Let ξ be a fuzzy variable described on the set of real numbers $\mathbb{R}$ by its membership function, which is given by:

$$\mu_\xi(y) = \min(1, 2\,\text{Cr}\{\xi = y\}), \text{ for } y \in \mathbb{R}. \qquad (3)$$

Definition 2.2 The credibility distribution $\Phi_\xi : \mathbb{R} \to [0, 1]$ of a fuzzy variable ξ is defined as:

$$\Phi_\xi(y) = \text{Cr}\{\xi \leq y\} = \frac{1}{2}(\sup_{t \leq y} \mu_\xi(t) + 1 - \sup_{t > y} \mu_\xi(t)). \qquad (4)$$

Definition 2.3 The expected value of a fuzzy variable ξ is defined as:

$$E(\xi) = \int_0^{+\infty} (1 - \Phi_\xi(y))\,\mathrm{d}y - \int_{-\infty}^0 \Phi_\xi(y)\,\mathrm{d}y \qquad (5)$$

provided that at least one of those integrals is finite.

In a similar way, for $k = 2, 3, \ldots$, the kth central moment of a fuzzy variable ξ is defined as: $E^{(k)}(\xi) = E((\xi - E(\xi))^k)$ [17].

Recently we have proposed credibility measures of return and risk for generating efficient portfolios within a fuzzy mean-absolute deviation framework [29] for LR-type fuzzy variables. Jalota et al. [14] have in turn also provided a complete review of uncertain portfolio parameters for these families of fuzzy variables.

3 Fuzzy Expected Values for Portfolio Selection

In the portfolio selection problem it is assumed that competing investment portfolios are ranked based on the expected return and risk of those portfolios. There is widespread agreement on the formula for calculating the expected return on an investment; however, many discrepancies arise with regard to the representation of the risk, also taking into account the decision maker's attitude. Throughout this paper we alternately use possibility and credibility distributions in order to evaluate expected values of fuzzy numbers and fuzzy variables respectively.

Possibilistic approaches to portfolio selection use possibilistic moments based on the α-level sets of a fuzzy number Q: $Q^\alpha = \{y : \mu_Q(y) \geq \alpha\} = [Q_-^\alpha, Q_+^\alpha]$, for $\alpha \in [0, 1]$. The crisp possibilistic mean value of Q has been introduced in [7], as follows:

$$E(Q) = \frac{1}{2} \int_0^1 [Q_-^\alpha + Q_+^\alpha] d\alpha \tag{6}$$

provided that at least one of those integrals is finite. Note that the integrals involved in Eq. (6) are the lower and upper probability mean values of the fuzzy number Q, and they also define the endpoints of the interval-valued mean of Q, allowing an alternative approach for the portfolio selection based on interval optimization [32]. Additionally, the use of lower and upper probabilities of a random set for modeling the information about the values of the probability distribution of a random variable has been extensively analyzed in [21].

In a different context, Delgado et al. [6] introduce the value and ambiguity of a fuzzy number, taking into account the influence of the α-cuts, using a reducing function $s(\alpha)$. The ambiguity of Q, denoted by $\tilde{A}(Q)$, is defined as:

$$\tilde{A}(Q) = \int_0^1 s(\alpha)[Q_+^\alpha - Q_-^\alpha] d\alpha. \tag{7}$$

Alternatively, Carlsson and Fullér [4] introduce the possibilistic mean value and variance of Q, respectively, as:

$$m(Q) = \int_0^1 \alpha[Q_-^\alpha + Q_+^\alpha] d\alpha, \text{ and } \sigma^2(Q) = \frac{1}{2} \int_0^1 \alpha[Q_+^\alpha - Q_-^\alpha]^2 d\alpha \tag{8}$$

making it possible to represent the decision makers' attitude in the face of uncertainty given by a weighting function which affects the endpoints of the α-level sets. An extensive analysis on possibilistic moments can be found in [25].

Recently it has been proved that credibility expected value and central moments of any LR-type fuzzy variable coincide with the crisp possibilistic moments for the LR-type fuzzy number with the same membership function, when the possibilistic expected values are calculated applying Eq. (6) [29]. Therefore fuzzy MV portfolio selection models using one of the two approaches (interval-valued possibility expectations and credibility expected values) for measuring uncertainty of the future return on a portfolio provide the same model, while the fuzzy quantity is represented using membership functions of any fuzzy LR-type. Additionally, when symmetric behaviour of the portfolio returns is not assumed, higher moments can be included in the modeling of the optimization problem, by means of credibilistic skewness considered as a goal or as a constraint [24, 28].

It is commonly assumed that investors are rationally risk-adverse and that the variance of the return on assets can suitably represent the risk of the investment, although the variance is a measure of the variability (spread) around the mean, which does not take into account whether the returns were below or above the expected value. Other measures of the risk have been suggested in the literature under a credibilistic framework, for example: semi-variance of a fuzzy variable [17], fuzzy Value-at-Risk (FVaR) [31] and below-mean absolute semi-deviation [29]. In particular, for a fuzzy variable ξ with finite expected value $E(\xi)$, its below-mean absolute semi-deviation is defined by $\text{MASd}(\xi) = E((\xi - E(\xi))^-)$, where:

$$(\xi - E(\xi))^- = \begin{cases} E(\xi) - \xi & \text{if } \xi \leq E(\xi) \\ 0 & \text{if } \xi > E(\xi) \end{cases}. \tag{9}$$

The main difference between the above downside risk measure and FVaR, under a credibility distribution, is that the investor does not need to establish previous credibility levels for a suitable loss function.

4   A New Fuzzy Ranking Index for Portfolio Selection

Some authors consider that the risk of the investment is better perceived by investors by means of measures that represent losses [22, 23] than with risk measures. In any case, from the investor's point of view it is important to clarify the differences between these alternative measures of the uncertainty of the future return in the meaning and information they provide.

Let us consider a portfolio X, where $X = (x_1, ..., x_N)$ is the vector of sharing proportions, and N is the number of assets. Let us suppose that the uncertainty of the return on a given portfolio X is represented by means of a fuzzy variable ξ_X. In order to select a suitable portfolio we propose a new credibility rank-index that uses a credibilistic loss as an alternative measure of risk aversion.

Definition 4.1 Let X be a given portfolio, whose uncertain return is approximated by a fuzzy variable ξ_X. Its credibility rank-index is:

$$CRI(\xi_X) = E(\xi_X) - \gamma L_0(\xi_X) \tag{10}$$

where $L_0(\xi_X) = \text{Cr}\{\xi_X \leq 0\}$ measures the credibility of achieving a non-positive return, and $\gamma > 0$ is a representative weight of the investor's loss aversion.

Note that if the investor requires other levels of loss, an alternative credibilistic loss function can be used: $L_r(\xi_X) = \text{Cr}\{\xi_X \leq r\}$, which measures the credibility of achieving a return below a given value r.

Let us suppose that the uncertainty of the return on a given portfolio X is represented by means of a trapezoidal membership function μ_X, for which the relationships between credibility and possibility expected values are well established. Then, comparing the above credibility rank-index with other fuzzy indices (previously introduced for solving the portfolio selection problem under a possibilistic framework [2, 5]) allows us to analyze its explicit performance. The observed differences between the indices should properly lie in the measures of risk and loss.

For a trapezoidal fuzzy variable $\xi_X = (a, A, B, b)$, with support $\{y : \mu_X(y) \geq 0\} = [a, b]$, $a < b$ and core $[A, B] = \{y : \mu_X(y) = 1\}$, where $\mu_X(y)$ is a straight line on $[a, A]$ and $[B, b]$, its credibility distribution is defined by:

$$\Phi_{\xi_X}(x) = \mathrm{Cr}\{\xi_X \leq x\} = \begin{cases} 0 & \text{if } x \leq a \\ \frac{1}{2}\left[1 - \left(\frac{A-x}{A-a}\right)\right] & \text{if } a \leq x \leq A \\ \frac{1}{2} & \text{if } A \leq x \leq B \\ \frac{1}{2}\left[1 + \left(\frac{x-B}{b-B}\right)\right] & \text{if } B \leq x \leq b \\ 1 & \text{if } x \geq b \end{cases} . \tag{11}$$

It is easy to see that the expected value of ξ_X is:

$$E(\xi_X) = \frac{A+B}{4} + \frac{a+b}{4}. \tag{12}$$

The expected below-mean absolute semi-deviation is given by [29]:

$$\mathrm{MASd}(\xi_X) = \begin{cases} \frac{1}{2}\left(E(\xi_X) - a\right) - \frac{1}{4}(A-a)\left(1 - \left(\frac{A-E(\xi)}{A-a}\right)^2\right) & \text{if } a \leq E(\xi_X) \leq A \\ \frac{1}{8}\left((B-A) + (b-a)\right) & \text{if } A \leq E(\xi_X) \leq B \\ \frac{1}{2}\left(b - E(\xi_X)\right) - \frac{1}{4}(b-B)\left(1 - \left(\frac{E(\xi)-B}{b-B}\right)^2\right) & \text{if } B \leq E(\xi_X) \leq b \end{cases} .$$

In order to select an optimal portfolio using a fuzzy ranking strategy, Carlsson et al. [5] proposed rating the investment portfolios using the following utility function, based on the possibilistic expected return and risk of the portfolio return P_X:

$$U(P_X) = m(P_X) - 0.005 \, a \, \sigma^2(P_X) \tag{13}$$

where a is the Arrow-Pratt index of the risk aversion ($a \approx 2.46$). There, P_X was built as a convex linear combination of the returns on individual assets that make up the portfolio X, whose possibility distributions are approximated by means of trapezoidal fuzzy numbers. Finally, it provides the return on the portfolio, P_X, which is also a trapezoidal fuzzy number. Recently, Georgescu [8] has proposed other definitions of possibilistic risk aversion, related to the Arrow-Pratt index a.

On the other hand, a linear alternative to the above quadratic utility function is introduced in [2] for ranking portfolios, where the uncertainty of the return on every portfolio X is approximated using a trapezoidal membership function for the fuzzy number P_X:

$$RI(P_X) = E(P_X) - \beta w(P_X) \tag{14}$$

where $\beta > 0$ is a given constant, and $w(P_X)$ is a fuzzy downside risk [30], which coincides with $w(P_X) = 2\tilde{A}(P_X)$.

Additionally, if the return on a given portfolio X is represented by a trapezoidal fuzzy number $P_X = (a, A, B, b)$, it is easy to see that:

$$E(P_X) = \frac{A + B}{4} + \frac{a + b}{4}, \quad m(P_X) = \frac{A + B}{3} + \frac{a + b}{6} \tag{15}$$

and

$$\sigma^2(P_X) = \left[\frac{B - A}{3} + \frac{b - a}{6}\right] + \frac{[(A - B) + (b - a)]^2}{72}. \tag{16}$$

Note that for a trapezoidal fuzzy number P_X, $E(P_X)$ corresponds to the midpoint of the 0.5-level set, while $m(P_X)$ is the middle point of its $\frac{2}{3}$-level set; and that for a constant reducing function $s(\alpha) = \frac{1}{2}$, the value of P_X coincides with the crisp possibilistic mean value $E(P_X)$, the ambiguity being the semi-amplitude of the interval-valued mean [2]:

$$\tilde{A}(P_X) = \frac{B - A}{4} + \frac{b - a}{4}. \tag{17}$$

5 Numerical Comparisons

In this section we report the results obtained on 1000 randomly generated portfolios, $X = (x_1, x_2, ..., x_N)$, whose returns have been evaluated using a historical data set from the Spanish stock market in Madrid, between January 2013 and March 2016. The number of positive proportions in each portfolio is set to $k = 9$, and for the simulation on the 9-dimensional simplex we apply a routine introduced in [3].

We consider the weekly returns $r_{ti} = \frac{p_{(t+1)i} - p_{ti}}{p_{ti}}$ on $N = 33$ assets from the Spanish IBEX35, where p_{ti} is the weekly closing price of asset i of week t, $t = 1, ..., 165$. For every portfolio X, its historical returns are computed as:

$$r_t(X) = \sum_{i=1}^{33} r_{ti} x_i, \tag{18}$$

and we assume a trapezoidal fuzzy representation μ_X of the uncertainty associated with its future return. The sample percentiles q_h of the set $\{r_t(X)\}_{t=1}^{165}$ are used to determine the core, $[q_{48}, q_{52}]$, and the support of the trapezoidal membership function of P_X and ξ_X, that is $[a, b] = [q_3, q_{97}]$.

Table 1 Correlation Coefficients between several fuzzy measures of risk and loss

Measures	$\sigma^2(P_X)$	$w(P_X)$	MASd(ξ_X)	$L_0(\xi_X)$
$\sigma^2(P_X)$	1	0.9926	0.9926	0.2863
$w(P_X)$		1	0.9999	0.2987
MASd(ξ_X)			1	0.2933

To show the relationships between the fuzzy expected values used in the afore-mentioned rank indices (Eqs. 10, 13 and 14), we have calculated the correlation coefficient between those measures of uncertainty for the 1000 simulated portfolios. First of all, it must be noted that there is a high correlation between the two fuzzy measures of expected return: $\rho(E(\xi_X), m(P_X)) = 0.9421$.

Secondly, concerning the correlation between the fuzzy measures of risk of the investment: variance of P_X, fuzzy downside risk ($w(P_X)$), and the expected below-mean absolute semi-deviation (MASd(ξ_X)), they also achieve high correlation coefficient values. The first three columns of Table 1 show these correlation coefficients, while the fourth presents the correlation between the credibilistic loss and the above risk measures. Note the low correlation achieved for the loss function ($L_0(\xi_X)$) with respect to all risk measures.

From the above results it seems that the portfolios provided by the possibilistic utility function ($U(P_X)$) should be close to those determined by the fuzzy ranking index ($RI(P_X)$). However, they could lead to different sets of efficient portfolios. Additionally, it must be noted that, by definition, $U(P_X)$ provides the best portfolio in some sense. On the other hand, the ranking index $RI(P_X)$ has been introduced for detecting non-dominated portfolios, using different values for β, which gives the investor a set of good solutions for different risk-aversion levels. Both indices have been compared with the simulated portfolios, and we see that the best portfolio according to the utility function, $X1$, has also been detected as a non-dominated portfolio using the other ranking indices, $RI(P_X)$ and $CRI(\xi_X)$.

Figure 1 shows the expected return, $E(P_X)$, and downside risk, $w(P_X)$, values of the 1000 portfolios randomly generated. The non-dominated portfolios are also shown; they are connected by the lines representing different β values, allowing us to build the approximated Pareto frontier associated with these return and risk measures.

In order to facilitate the comparison of the performance of the new credibilistic index, Fig. 1 also shows the non-dominated portfolios provided by using Eq. (10). Their polygonal is represented in the $w(P_X) - E(P_X)$ coordinate axis, although the dominance rules have been established using the measure of loss, $L_0(\xi_X)$.

The comparison between these fuzzy ranking indices shows that different sets of non-dominated portfolios have been obtained, given rise to different approximated Pareto frontiers, although the non-dominated portfolios produced by both indices present higher expected returns and higher levels of risk-aversion simultaneously. The

Fig. 1 Downside risk and expected return of the 1000 simulated portfolios. The biggest dot represents portfolio $X1$. The gray squares correspond to the non-dominated portfolios using $RI(P_X)$, and the black dots represent the non-dominated ones generated by the $CRI(\xi_X)$ index

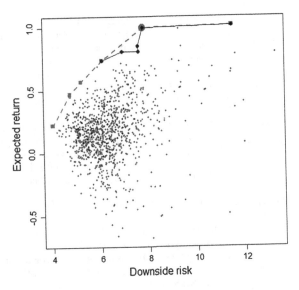

difference between the two indices is probably due to the fact that each index utilizes different measures of risk $(w(P_X))$ and loss $(L_0(\xi_X))$ respectively, and consequently relies on a different set drawn from the sample of simulated portfolios.

6 Conclusions

We introduce a new credibility rank-index for ranking portfolios based on the expected mean and a loss function for fuzzy variables. We analyze the relationship between possibilistic rank-indices based on measures of risk and the new credibilistic index for a set of simulated portfolios, whose returns are approximated by means of trapezoidal membership functions.

Acknowledgements This research was partially supported by the Project MTM2017-83850-P (Ministerio de Economía y Competitividad, Spain), co-financed by FEDER funds.

References

1. Anagnostopoulos KP, Mamanis G (2011) The mean-variance cardinality constrained portfolio optimization problem: an experimental evaluation of five multiobjective evolutionary algorithms. Expert Syst Appl 38:14208–14217
2. Bermúdez JD, Segura JV, Vercher E (2007) A fuzzy ranking strategy for portfolio selection applied to the Spanish stock market. In: Proceedings Fuzz-IEEE 2007, 2007 IEEE International Fuzzy Systems Conference IEEE, Piscataway

3. Bermúdez JD, Segura JV, Vercher E (2012) A multi-objective genetic algorithm for cardinality constrained fuzzy portfolio selection. Fuzzy Set Syst 188:16–26
4. Carlsson C, Fullér R (2001) On possibilistic mean value and variance of fuzzy numbers. Fuzzy Set Syst 122:315–326
5. Carlsson C, Fullér R, Majlender P (2002) A possibilistic approach to selecting portfolios with highest utility score. Fuzzy Set Syst 131:13–21
6. Delgado M, Vila MA, Voxman W (1998) On a canonical representation of fuzzy numbers. Fuzzy Set Syst 93:125–135
7. Dubois D, Prade H (1987) The mean value of a fuzzy number. Fuzzy Set Syst 24:279–300
8. Georgescu I (2009) Possibilistic risk aversion. Fuzzy Set Syst 160:2608–2619
9. Gupta P, Inuiguchi M, Mehlavat MH, Mittal G (2013) Multiobjective credibilistic portfolio selection model with fuzzy chance-constraints. Inf Sci 229:1–17
10. Huang X (2007) A new perspective for optimal selection with random fuzzy returns. Inf Sci 177:5404–5414
11. Huang X (2009) A review of credibilistic portfolio selection. Fuzzy Optim Decis Making 8:263–281
12. Hasuike T, Katagiri H, Ishii H (2009) Portfolio selection problems with random fuzzy variable returns. Fuzzy Set Syst 160:2579–2596
13. Inuiguchi M, Ramík J (2000) Possibilistic linear programming: a brief review of fuzzy mathematical programming and a comparison with stochastic programming in portfolio selection problem. Fuzzy Set Syst 111:3–28
14. Jalota H, Thakur B, Mittal G (2017) Modelling and constructing membership function for uncertain portfolio parameters: a credibilistic framework. Expert Syst Appl 71:40–56
15. Konno H, Yamazaki H (1991) Mean-absolute deviation portfolio optimization model and its application to Tokyo stock market. Manag Sci 37:519–531
16. Lai T (1991) Portfolio selection with skewness: a multiple-objective approach. Rev Quant Financ Acc 1:293–305
17. Liu B (2003) Fuzzy random variables: a scalar expected value operator. Fuzzy Optim Decis Making 2:87–100
18. Liu B (2006) A survey of credibility theory. Fuzzy Optim Decis Making 5:387–408
19. Liu B, Liu Y-K (2002) Expected value of fuzzy variable and fuzzy expected value models. IEEE Trans Fuzzy Syst 10:445–450
20. Markowitz HM (1952) Portfolio selection. J Financ 7:77–91
21. Miranda E, Couso I, Gil P (2010) Approximations of upper and lower probabilities by measurable selections. Inf Sci 180:1407–1417
22. Ormos M, Timotity D (2016) Generalized asset pricing: expected downside risk-based equilibrium modeling. Econ Model 52:967–980
23. Rockafellar RT, Uryasev S (2000) Optimization of conditional value at risk. J Risk 2:21–41
24. Saborido R, Ruiz AB, Bermúdez JD, Vercher E, Luque M (2016) Evolutionary multi-objective optimization algorithms for fuzzy portfolio selection. Appl Soft Comput 39:48–63
25. Saedifar A, Pasha E (2009) The possibilistic moments of fuzzy numbers and their applications. J Comput Appl Math 223:1028–1042
26. Tanaka H, Guo P (1999) Possibilistic data analysis for operations research. Physica, Heidelberg
27. Vercher E, Bermúdez JD (2012) Fuzzy portfolio selection models: a numerical study. In: Doumpos M, Zopounidis C, Pardalos PM (eds) Financial decision making using computational intelligence. Springer, New York
28. Vercher E, Bermúdez JD (2013) A possibilistic mean-downside risk-skewness model for efficient portfolio selection. IEEE Trans Fuzzy Syst 21(3):585–595
29. Vercher E, Bermúdez JD (2015) Portfolio optimization using a credibility mean-absolute semideviation model. Expert Syst Appl 42:7121–7131
30. Vercher E, Bermúdez JD, Segura JV (2007) Fuzzy portfolio optimization under downside risk measures. Fuzzy Set Syst 158:769–782
31. Wang B, Wang S, Watada J (2011) Fuzzy-portfolio-selection models with value-at-risk. IEEE Trans Fuzzy Syst 19(4):758–769

32. Wu M, Kong D-W, Xu J-P, Huang N-J (2013) On interval portfolio selection problem. Fuzzy Optim Decis Making 12:289–304
33. Zadeh LA (1978) Fuzzy sets as a basis for a theory of possibility. Fuzzy Set Syst 1:3–28
34. Zhang WG, Wang YL, Chen ZP, Nie ZK (2007) Possibilistic mean-variance models and efficient frontiers for portfolio selection problem. Inf Sci 177:2787–2801

Part V
Miscellanea: Operations Research and Other (More or Less Uncertain) Mathematics and Applications

Quantities Characterizing Lower Semi-Fredholm Linear Relations

Teresa Álvarez, Antonio Martínez-Abejón and Javier Pello

Abstract Certain operational quantities derived from the norm and from the injection modulus in the context of multivalued linear operators are considered in order to obtain characterizations of lower semi-Fredholm multivalued linear operators.

This paper shows a little part of the research interests of the team of Mathematical Analysis of the Department of Mathematics when Pedro Gil was his vice-head. Clearly, there is a large scientific distance between his research subjects and ours, but this gap makes the work of Pedro in constructing the School of Mathematics of the University of Oviedo even more remarkable, and the three of us feel greatly indebted to him for that. Indeed, nowadays it is not easy to keep a healthy balance between the different branches of mathematics when very few mathematicians have the right perspective on general mathematics, and Pedro was very successful at this task.

Each of the three authors of this paper met Pedro Gil at different phases of the construction of the School of Mathematics of Oviedo: Teresa did it when this school was a mere project; Antonio was hired when its construction began; and Javier was one of the first students of this School.

Unfortunately, Pedro passed away when his presence was more necessary than ever.

T. Álvarez · A. Martínez-Abejón (✉)
Department of Mathematics, University of Oviedo, Oviedo, Spain
e-mail: ama@uniovi.es

T. Álvarez
e-mail: seco@uniovi.es

J. Pello
Escuela Superior de Ciencias Experimentales y Tecnología,
Universidad Rey Juan Carlos, Móstoles, Spain
e-mail: javier.pello@urjc.es

© Springer International Publishing AG 2018
E. Gil et al. (eds.), *The Mathematics of the Uncertain*, Studies in Systems, Decision and Control 142, https://doi.org/10.1007/978-3-319-73848-2_71

1 Introduction

Many authors have considered operational quantities in order to study bounded semi-Fredholm operators on Banach spaces and other semigroups of operators on Banach spaces (for operational quantities, [4, 11]; for semigroups [14, 16] developed at the University of Oviedo, and [1]). The generalization of this study to the class of unbounded operators in arbitrary normed spaces was initiated in R.W. Cross [7] and continued by several authors (see [2, 3]).

On the other hand, we note that the linear relations (sometimes called multivalued linear operators) were introduced in Functional Analysis by J. von Neumann [15], motivated by the need to consider adjoints of non-densely defined operators used in applications to the theory of generalized equations [6] and also by the need to consider inverses of certain operators, as used, for example, in the study of some Cauchy problems associated to parabolic type equations [10].

At present, the investigation of linear relations in normed spaces or Hilbert spaces is of significance since it has applications in many problems in Physics and in other areas of Applied Mathematics. We cite some of them.

(a) Applications of perturbation results for linear relations to the study of degenerate elliptic-parabolic evolution equations [9].
(b) Applications of the fixed point theory of linear relations in Game Theory and Mathematical Economics. Discontinuous differential equations occurring in Biological Sciences (for instance, population in dynamics and epidemiology), Optimal Control and Digital Imaging (a systematic collection of references can be found in [13]).
(c) Applications of the semi-Fredholm theory of linear relations to the study of many problems of Operator Theory as, for example, the theory of linear bundles and also the theory of pseudoresolvents [5].
(d) Applications to the problem of the invariant subspace [12].

In regard of the subjects mentioned above, an attempt to generalize the existing results for operators to the general situation of linear relations seems to be worthy and necessary in view of scientific progress in this field.

The purpose of this note is to extend some of the main results of [3, 7] to the case of multivalued linear operators in arbitrary normed spaces. In Sect. 2 we characterize the lower semi-Fredholm linear relations in terms of operational quantities derived from the norm.

Throughout this paper an operational quantity or simply a quantity will be a function defined on the class of all multivalued linear operators in normed spaces, taking values in the interval $[0, \infty]$. Furthermore, X, Y and Z will denote infinite dimensional normed spaces and T will always be a linear relation from X to Y. We adhere to the notation and terminology of [8], so $LR(X, Y)$ denotes the class of all linear relations from X to Y.

2 A Characterization of F_--Relations

Let $T \in LR(X, Y)$ and let F be an arbitrary finite dimensional subspace of Y. Then

$$\dim D(T')/(D(T') \cap F^\perp) = \dim(D(T') + F^\perp)/F^\perp$$
$$\leq \dim Y'/F^\perp = \dim F' < \infty$$

and hence $\dim D(T') \cap F^\perp = \infty$ as $\dim D(T') = \infty$. In consequence, the set $D(T') \cap S_{F^\perp}$ is non-empty whenever $\dim D(T') = \infty$, where $S_{F^\perp} := \{ y' \in F^\perp : \|y'\| = 1 \}$. This property suggests the following notion.

Definition 2.1 For a given linear relation $T \in LR(X, Y)$ with $\dim D(T') = \infty$ we define

$$\tau_0'(T) := \sup\{ j(T'|_{F^\perp}) : F \in \mathscr{F}(Y) \}$$

where $\mathscr{F}(Y)$ denotes the class of all finite dimensional subspaces of Y.

In the sequel, $\mathscr{E}(X)$ denotes the class of all closed infinite codimensional subspaces of X. We write J_X for the injection of X into its completion and if M is a closed subspace of X then Q_M denotes the quotient map from X onto X/M.

We shall use the following quantities:

Definition 2.2 [8, IV.1.5] Let $T \in LR(X, Y)$. We define

$$\Gamma_0'(T) = \inf\{\|Q_F T\| : F \in \mathscr{F}(Y)\},$$
$$\Gamma'(T) = \inf\{\|Q_M J_Y T\| : M \in \mathscr{E}(\widetilde{Y})\}.$$

Now we can state the following result:

Proposition 2.1 Let $T \in LR(X, Y)$ and let $K \in \mathscr{F}(Y)$. Then

(i) $\Gamma_0'(Q_K T) = \inf\{\|Q_{K+F} T\| : F \in \mathscr{F}(Y)\} = \Gamma_0'(T)$.
(ii) $\tau_0'(Q_K T) = \sup\{j(T'|_{(K+F)^\perp}) : F \in \mathscr{F}(Y)\} = \tau_0'(T)$.

Proof (i) This statement is proved in [8, IV.5.3] and [8, IV.5.6].

(ii) Let $K \in \mathscr{F}(Y)$. Then $(Q_K T)' = T' J_{K^\perp}$ by [8, III.1.6] and $\dim(D(T') \cap K^\perp) = \infty$, so that we deduce from the fact that if $M \in \mathscr{E}(X)$ then $\mathscr{F}(X/M) = \{Q_M(M + F) : F \in \mathscr{F}(X)\}$ that

$$\tau_0'(Q_K T) = \sup\{j(T' J_{K^\perp}|_{((K+F)/K)^\perp}) : F \in \mathscr{F}(Y)\}.$$

But we have that, up to isometry, $((K + F)/K)^\perp$ is just $(K + F)^\perp$ considered as a subspace of $K^\perp$. Hence the first equality of (ii) is true. Consequently, $\tau_0'(Q_K T) = \tau_0'(T)$ since trivially $K + F \in \mathscr{F}(Y)$ and $(K + F)^\perp \subset F^\perp \cap K^\perp$. Therefore (ii) holds. □

Proposition 2.2 *Let T be closable. Then for $f \in \{\Gamma_0', \Gamma', \tau_0'\}$ we have $f(T) = f(\overline{T})$.*

Proof The result for the cases $f \in \{\Gamma_0', \Gamma'\}$ is proved in [8, IV.5.7]. For $f = \tau_0'$, the desired equality follows from the definitions observing that $T' = \overline{T}'$ by [8, III.1.3].
□

Definition 2.3 [8, V.1.1] We say that $T \in LR(X, Y)$ is lower semi-Fredholm, denoted $T \in F_-(X, Y)$, if its conjugate is upper semi-Fredholm, that is, there exists a finite codimensional subspace M of Y' for which $T'|_M$ is injective and open.

Proposition 2.3 [8, V.5.2 and V.5.5] *We have*

(i) $T \in F_-(X, Y)$ *if and only if* $\dim Y/\overline{R(T)} < \infty$ *and* $\gamma(T') > 0$.
(ii) $T \notin F_-(X, Y)$ *if and only if there is no finite dimensional subspace F of Y for which $T'|_{F^\perp}$ is bounded below if and only if for each $\epsilon > 0$ there exists a compact operator $K \in LR(X, \widetilde{Y})$ with norm not exceeding ϵ such that $D(T) \subset D(K)$ and $\dim \widetilde{Y}/\overline{R(J_Y T + K)} = \infty$.*

Proposition 2.4 [8, V.5.14 and V.5.17] *We have*

(i) *There is $F \in \mathscr{F}(Y)$ such that $Q_F T$ is precompact if and only if $\Gamma_0'(T) = 0$.*
(ii) *If $\dim T(0) < \infty$, then $T \in F_-(X, Y)$ if and only if $\Gamma'(T) > 0$.*

The following result concerning the behaviour of the quantities Γ_0', Γ' and τ_0' with respect to the product of the linear relations is fundamental to obtain the main result of this section (Theorem 2.1 below).

Proposition 2.5 *Let $T \in LR(X, Y)$ be closed and let $S \in LR(Y, Z)$ be continuous such that $R(T) \subset D(S)$ and $S(0)$ is finite dimensional. If $f \in \{\Gamma', \Gamma_0', \tau_0'\}$, then $f(ST) \geq f(S)\tau_0'(T)$ $(0 \cdot \infty$ excluded$)$.*

Proof We may clearly assume that $f(ST) < \infty$ and that $f(S)$ and $\tau_0'(T)$ are both positive numbers.

We first note that

$$f(S) < \infty, \quad \dim D(S') = \infty \text{ and } SK \in \mathscr{F}(Z) \text{ whenever } K \in \mathscr{F}(Y). \quad (1)$$

Indeed, that $f(S) < \infty$ follows immediately from the definitions observing that S and S' are both continuous [8, III.1.9]. Moreover, by virtue of [8, III.2.3], $\dim D(S') = \dim S(0)^\perp = \dim(Z/S(0))' = \infty$.

Finally, the property $SK \in \mathscr{F}(Z)$ if $F \in \mathscr{F}(Y)$ follows from [8, I.2.14] combined with the fact that $S(0)$ is finite dimensional.

Let us consider two cases for S:

Case 1: *S is bounded.*

We first see that

$$S' \text{ is single valued and } (ST)' = T'S'. \quad (2)$$

Indeed, since $D(S) = Y$ we have that $S'(0) = D(S)^\perp$ ([8, III.1.4]) = $\{0\}$ so that by virtue of [8, I.2.9], S' is single valued. Furthermore the hypothesis $D(S) = Y$ implies that $D(ST)$ is just $D(T)$ and thus the assumption $T = \overline{T}$ combined with [3, 3.2] leads to $(ST)' = T'S'$, as desired.

Now, select $K \in \mathscr{F}(Y)$ arbitrarily. Since $D(S) = Y$, by virtue of [8, III.1.13] and [8, I.2.11] we get

$$
\begin{aligned}
S'(SK)^\perp &= R(S') \cap (S^{-1}SK)^\perp \\
&= R(S') \cap [\{K \cap D(S)\} + N(S)]^\perp = R(S') \cap K^\perp \cap N(S)^\perp.
\end{aligned}
$$

But, by [8, II.3.14], S is closed and so we deduce from [8, III.1.13] that $N(S) = R(S')^\top$ and hence $R(S') \subset N(S)^\perp$. Consequently

$$
S'(SK)^\perp = R(S') \cap K^\perp. \tag{3}
$$

If $f = \tau_0'$: Let ϵ be arbitrarily chosen in the interval $0 < \epsilon < \tau_0'(S)$. By Proposition 2.1(ii) we may choose $F \in \mathscr{F}(Z)$ such that

$$
SK \subset F, 0 < \tau_0'(S) - \epsilon = \tau_0'(Q_{SK}S) - \varepsilon < j(S'|_{F\perp}). \tag{4}
$$

In consequence, as $F^\perp \subset (SK)^\perp$, $S'|_F$ is injective by (4) and recalling that S' is single valued by (2), it follows from (1), (2), (3) and (4) that

$$
\begin{aligned}
\tau_0'(ST) &= \sup\{j((ST)'|_{A\perp}) : A \in \mathscr{F}(Z)\} \\
&= \sup\{j(T'S'|_{A\perp}) : A \in \mathscr{F}(Z)\} \geq j(T'S'|_{F\perp}) \\
&= \inf\{\|T'S'u'\|/\|u'\| : 0 \neq u' \in (D(T'S') \cap F^\perp)\} \\
&= \inf\{(\|T'(S'u')\|/\|S'u'\|)(\|S'u'\|/\|u'\|) : 0 \neq u' \in (D(T'S') \cap F^\perp)\} \\
&\geq \inf\{\|T'(S'u')\|/\|S'u'\| : 0 \neq u' \in (D(T'S') \cap F^\perp)\} \\
&\quad \cdot \inf\{\|S'u'\|/\|u'\| : 0 \neq u' \in (D(S') \cap F^\perp)\} \\
&\geq \inf\{\|T'(S'u')\|/\|S'u'\| : 0 \neq u' \in (SK)^\perp \setminus N(S')\}(\tau_0'(S) - \varepsilon) \\
&\geq \inf\{\|T'w'\|/\|w'\| : 0 \neq w' \in (K^\perp \cap D(T'))\}(\tau_0'(S) - \varepsilon).
\end{aligned}
$$

Taking the supremum over all $K \in \mathscr{F}(Y)$ we get

$$
\tau_0'(ST) \geq (\tau_0'(S) - \varepsilon)\tau_0'(T)
$$

whence $\tau_0'(ST) \geq \tau_0'(S)\tau_0'(T)$.

If $f = \Gamma_0'$: Let $\varepsilon > 0$. By Proposition 2.1(i) we may choose $F \in \mathscr{F}(Z)$ such that

$$
F \supset K, \ 0 < \|Q_F ST\| < \Gamma_0'(Q_{SK}ST) + \varepsilon = \Gamma_0'(ST) + \varepsilon. \tag{5}
$$

Therefore $Q_F ST$ is continuous and thus, by means of suitable applications of [8, III.1.6] and [8, II.3.11],

$$\infty > \|Q_F ST\| = \|(Q_F ST)'\|$$
$$= \|(ST)' J_{F^\perp}\| = \|(ST)' J_{F^\perp}\| \|J_{F^\perp}^{-1}\|$$
$$\geq \|(ST)' J_{F^\perp} J_{F^\perp}^{-1}\| = \|(ST)'|_{F^\perp}\|.$$

Therefore

$$\|Q_F ST\| \geq \|(ST)'|_{F^\perp}\|. \tag{6}$$

Observe that $\|Q_F S\| > 0$ and that S' is single valued by (2). Hence we deduce that

$$\|Q_F S\| = \|S' J_{F^\perp}\| = \|S'|_{F^\perp}\|. \tag{7}$$

It now follows from (1), (2), (3), (5), (6) and (7) that

$$\Gamma_0'(ST) + \varepsilon \geq \|(ST)'|_{F^\perp}\| = \|T'S'|_{F^\perp}\|$$
$$= \sup\{\|T'(S'u')\|/\|u'\| : 0 \neq u' \in (D(T'S') \cap F^\perp)\}$$
$$= \sup\{(\|T'(S'u')\|/\|S'u'\|)(\|S'u'\|/\|u'\|) : u' \in F^\perp \setminus N(S')\} \cdot$$
$$\quad \cdot \inf\{\|T'(S'u')\|/\|S'u'\| : u' \in F^\perp \setminus N(S')\}$$
$$\geq \|S'|_{F^\perp}\| \inf\{\|T'(S'u')\|/\|S'u'\| : u' \in (SK)^\perp \setminus N(S')\}$$
$$= \|Q_F S\| \inf\{\|T'v'\|/\|v'\| : 0 \neq v' \in S'(SK)^\perp = K^\perp \cap R(S')\}$$
$$\geq \Gamma_0'(S) \inf\{\|T'w'\| : w' \in K^\perp \cap D(T'), \|w'\| = 1\}.$$

Taking the supremum over all $K \in \mathscr{F}(Y)$ we obtain that

$$\Gamma_0'(ST) + \varepsilon \geq \Gamma_0'(S)\tau_0'(T)$$

whence $\Gamma_0'(ST) \geq \Gamma_0'(S)\tau_0'(T)$ as $\varepsilon > 0$ was arbitrary.

Case 2: *S is continuous with $R(T) \subset D(S)$ and* $\dim S(0) < \infty$.
Denote $T_1 := T|_{D(ST)}$. Then we have that

$$\overline{S}T_1 = ST. \tag{8}$$

Indeed, $D(\overline{S}T_1) = D(ST)$ since $D(\overline{S}T_1) := \{x \in D(T_1) = D(ST) : Tx \cap D(\overline{S}) \neq \emptyset\}$, $D(ST) = \{x \in D(T) : Tx \cap D(S) \neq \emptyset\}$ and $D(\overline{S}) = \overline{D(S)}$.

Clearly $ST \subset \overline{S}T_1$ so that $\overline{S}T_1(0) \supset ST(0)$. On the other hand, let $z \in \overline{S}T_1(0)$, that is, $(0, z) \in G(\overline{S}T_1)$, so that $(0, y) \in G(T_1)$ and $(y, z) \in G(\overline{S})$ for some $y \in D(\overline{S}) = D(S)$ and since $T(0) \subset R(T) \subset D(S)$ by hypothesis and S is closable by virtue of [8, III.5.1] we deduce that $(0, y) \in G(T)$ and $(y, z) \in G(S)$ and hence $z \in ST(0)$. Now, these properties together with [8, I.2.10] lead to (8).

If $f = \tau_0'$: Since $\tau_0'(T) > 0$ is $T \in F_-(X, Y)$ by Proposition 2.3(ii) and thus we infer from Proposition 2.3(i) that $\overline{R(T)}$ is a closed finite codimensional subspace of Y and since $R(T)$ is contained in $D(S)$ by assumption we have that $D(\overline{S})$ is also a closed finite codimensional subspace of Y. Let P denote the bounded projection of Y onto $D(\overline{S})$ considered as an element of $LR(Y, D(\overline{S}))$. It then follows by the Case 1 for $f = \tau_0'$ and Proposition 2.2 that

$$\tau_0'(ST) = \tau_0'(\overline{S}T_1) \geq \tau_0'(\overline{S}J_{D(\overline{S})})\tau_0'(PT_1) \geq \tau_0'(\overline{S})\tau_0'(P)\tau_0'(T_1) \geq \tau_0'(S)\tau_0'(T)$$

as desired.

Note that trivially $T_1 \subset T$ so that $T^{-1} \subset T_1^{-1}$ and hence we conclude from the definition of the adjoint of a linear relation that $T' \subset T_1'$ and consequently $\tau_0'(T_1) \geq \tau_0'(T)$. Moreover as $N(P') = R(P)^\perp$ (see [8, III.1.4]) is $\tau_0'(P) \geq 1$.

If $f = \Gamma_0'$: Arguing as when $f = \tau_0'$, it follows from the Case 1 for $f = \Gamma_0'$ that

$$\Gamma_0'(ST) \geq \Gamma_0'(S)\tau_0'(T)$$

as desired.

We finally consider the case $f = \Gamma'$, which includes all remaining cases: Fix $M \in \mathscr{E}(\widetilde{Z})$. Since the result has been proved for $f = \Gamma_0'$, we have that

$$\Gamma_0'(Q_M J_Z ST) \geq \Gamma_0'(Q_M J_Z S)\tau_0'(T).$$

Taking the infimum over $M \in \mathscr{E}(\widetilde{Z})$ and observing that

$$\Gamma'(T) = \inf\{\Gamma_0'(Q_M J_Y T) : M \in \mathscr{E}(\widetilde{Y})\}$$

(see [7, IV.5.5]) we conclude that $\Gamma'(ST) \geq \Gamma'(S)\tau_0'(T)$. $\qquad\square$

We are now in a position to prove the main result of this paper.

Theorem 2.1 *Let $T \in LR(X, Y)$ with $\dim D(T') = \infty$. The following statements are equivalent:*

(i) $T \in F_-(X, Y)$.
(ii) *There exists a positive constant c such that for each normed space Z, each $f \in \{\Gamma', \Gamma_0'\}$ and each continuous linear relation $S \in LR(Y, Z)$ with $\overline{R(T)} \subset D(S)$ and $S(0)$ finite dimensional, we have $f(S\overline{T}) \geq cf(S)$.*

Proof We first note that from the definition of lower semi-Fredholm linear relation and the fact that $T' = \overline{T}'$ [8, III.1.3] we have that T is F_- if and only if so is $\overline{T}$. Furthermore we note that $\overline{T}(0) \in \mathscr{E}(Y)$. Indeed, by means of [8, III.1.4], we have that

$$D(T') \subset (D(T')^\top)^\perp = (\overline{T}(0))^\perp = (Y/\overline{T}(0))'$$

and hence $\overline{T}(0)$ is a closed infinite codimensional subspace of Y whenever $\dim D(T') = \infty$.

(i) $\Rightarrow$ (ii) Assume that $T \in F_-(X, Y)$. Then the closure of T is lower semi-Fredholm and thus $\tau_0'(\overline{T}) > 0$ by Proposition 2.3(ii).

Let us consider two cases for $\tau_0'(\overline{T})$:

Case 1: $\tau_0'(\overline{T}) < \infty$. Let $c := \tau_0'(\overline{T})$. Then the validity of (ii) follows immediately from the above Proposition 2.5 by noting that $R(\overline{T}) \subset \overline{R(T)}$.

Case 2: $\tau_0'(\overline{T}) = \infty$. Let c be any positive real number. Without loss of generality we may suppose that $f(S) > 0$. As S is continuous, $f(S) < \infty$. Now, since $f(S\overline{T}) \geq f(S)\tau_0'(\overline{T}) = \infty$ by Proposition 2.5, assertion (ii) holds for c.

$(ii) \Rightarrow (i)$ Suppose that $T \notin F_-(X, Y)$. Then dim $D(T') = \infty$ and it follows from Proposition 2.3(ii) that there exists a compact operator $K \in LR(X, \widetilde{Y})$ such that $\overline{R(J_Y\overline{T} - K)}$ is a closed infinite codimensional subspace of $\widetilde{Y}$. Put $M := \overline{R(J_Y\overline{T} - K)}$.

Then it is clear that $Q_M J_Y \overline{T} = Q_M K$ where $Q_M K$ is compact and single valued. Hence we have the chain of implications $Q_M J_Y \overline{T}$ compact $\Rightarrow \Gamma_0'(Q_M J_Y \overline{T}) = 0$ (Proposition 2.4(i)) $\Rightarrow \Gamma'(Q_M J_Y \overline{T}) = 0$. Therefore, there exists $M \in \mathscr{E}(\widetilde{Y})$ such that $f(Q_M J_Y \overline{T}) = 0$. On the other hand, since trivially $Q_M J_Y$ is a lower semi-Fredholm operator we have that $\tau_0'(Q_M J_Y) > 0$ by Proposition 2.3(ii) and thus Proposition 2.4(ii) ensures that $\Gamma'(Q_M J_Y) > 0$ and hence $\Gamma_0'(Q_M J_Y) > 0$. Consequently (ii) fails. $\qquad\square$

References

1. Aiena P, González M, Martínez-Abejón A (2001) Operator semigroups in Banach spaces theory. Boll Unione Mat Ital Ser B 4:157–205
2. Álvarez T (1998) Perturbation and coperturbation functions characterising semi-Fredholm type operators. Math Proc Royal Irish Acad 98 A:41–46
3. Álvarez T, Cross RW (1999) Operator quantities characterizing upper and lower semi-Fredholm type operators. Math J Toyama Univ 22:205–224
4. Astala K, Tylli HO (1990) Seminorms related to weak compactness and to Tauberian operators. Math Proc Camb Philos Soc 107:367–375
5. Baskakov AG, Chernyshov KI (2002) Spectral analysis of linear relations and degenerate semigroups of operators. Sbornik Math 193:1573–1610
6. Coddington EA (1971) Multivalued Operators of Boundary Value Problems, vol 183. Lecture notes in mathematics. Springer, Berlin
7. Cross RW (1998) Properties of some norm related functions of unbounded linear operators. Math Z 159:285–302
8. Cross RW (1998) Multivalued linear operators. Marcel Dekker, New York
9. Cross RW, Favini A, Yakukov Y (2011) Perturbation results for multivalued linear operators. Prog Nonlinear Differ Equ Their Appl 80:111–130
10. Favini A, Yagi A (1993) Multivalued operators and degenerate evolution equations. Annali Matem Pura Appl 163:353–384
11. González M, Martinón A (2012) Quantities characterizing semi-fredholm operators and perturbation radii. J Math Anal Appl 390:362–367
12. Grixti-Cheng D (2008) The invariant subspace problem for linear relations in Hilbert spaces. J Austr Math Anal Appl 5:4–7
13. Kaczynski T (2008) Multivalued maps as a tool in modeling and rigorous numerics. J Fixed Point Theory Appl 4:151–176
14. Martínez-Abejón A (1994) Semigrupos de Operadores y Ultrapotencias. Ph.D. thesis. Universidad de Cantabria

15. von Neumann J (1951) Functional operators, vol. 2: the geometry of orthogonal spaces. Annals of mathematics studies. Princeton University Press, Princeton
16. Pello J (2005) Semigrupos de operadores asociados a la propiedad de Radon-Nikodym. Ph.D. thesis. Universidad de Oviedo

Some Examples of First Exit Times

Jesús Antonio Álvarez López and Alberto Candel

To Pedro Gil, in memoriam.

Abstract The purpose of this article is to compute the expected first exit times of Brownian motion from a variety of domains in the Euclidean plane and in the hyperbolic plane.

1 Introduction

The theory of Brownian motion on Riemannian manifolds allows probabilistic interpretations of solutions to second order differential equations on them via the so called Dynkin formula. One example of such equation and solution is the following: if D is a regular domain on a Riemannian manifold M with attending Laplace operator $\triangle$, then the expected value of the "first exit time" from D for Brownian paths in the Wiener space of M, if finite, is the minimal solution to the differential equation

$$\triangle f \equiv -1$$

on D, with $f > 0$ on D and $f \equiv 0$ on ∂D. The number $\rho(D) = 4 \int_D f$ is called the torsional rigidity of D, a sort of isoperimetric constant whose study originated with Saint-Venant memoir [3].

Ghys [5] gave a spectacular application of first exit times to the topology and dynamics of foliated spaces. The examples and calculations presented here are the

J. A. Álvarez López
Facultade de Matemáticas, Departamento de Xeometría e Topoloxía,
Universidade de Santiago de Compostela, 15782 Santiago de Compostela, Spain
e-mail: jesus.alvarez@usc.es

A. Candel (✉)
Department of Mathematics, California State University, Northridge, CA 91330, USA
e-mail: alberto.candel@csun.edu

© Springer International Publishing AG 2018
E. Gil et al. (eds.), *The Mathematics of the Uncertain*, Studies in Systems,
Decision and Control 142, https://doi.org/10.1007/978-3-319-73848-2_72

base of some exercises in [2, Exercises 2.7.14, C.9.4 and C.9.5], and were motivated by a discussion on Ghys theorem.

2 Generalities on First Exit Times and on Harmonic Functions

More details about first exit times and probabilistic solutions to differential equations can be found in Dynkin [4]; the theory needed for what follows is detailed in [2, Appendix C].

Consider a Riemannian manifold, M, with attending Laplacian Δ. These data permit to construct Brownian motion on M, a continuous-time stochastic process taking place in the space of continuous paths $\omega : [0, \infty) \to M$ that is regulated by a set of probability measures $\{P_x \mid x \in M\}$ (with P_x supported on paths $\{\omega(0) = x\}$) which are constructed via the heat kernel density of the Laplacian Δ.

Solutions to a variety of differential equations on M admit probabilistic interpretations via Brownian motion. One such example is the following. Let $D \subset M$ be a regular domain (a connected open set with piecewise smooth boundary), and consider the first exit time T_D from D, the function on paths given by $T_D(\omega) = \inf\{t > 0 \mid \omega(t) \notin D\}$ (with the standard convention that the infimum of the empty set is ∞).

The expected first exit time with respect to the Brownian measures $\{P_x\}$ defines a function $E_\bullet[T_D] : x \mapsto E_x[T_D] = \int T_D(\omega) \cdot P_x(\omega)$ which is 0 for all $x \notin D$, and which is either $E_x[T_D] < \infty$ for all $x \in D$, or $E_x[T_D] \equiv \infty$ for all $x \in D$.

If D is a relatively compact domain, then $E_x[T_D] < \infty$ for all $x \in D$. In this case, Dynkin's formula shows that this function is a solution to the differential equation problem (Saint-Venant problem)

$$\begin{cases} \Delta f \equiv -1 & \text{on } D, \\ f > 0 & \text{on } D, \text{ and} \\ f \equiv 0 & \text{on } \partial D. \end{cases} \tag{1}$$

Because of the maximum modulus principle for harmonic functions (to the effect that a function that is harmonic on a relatively compact domain and continuous on its closure must attain its extreme values on the boundary of the domain), the solution to the differential equation (1) is unique: the difference of two solutions is harmonic and equal to 0 on ∂D, so it must be 0 on all of D. Therefore, if D is relatively compact, the expected first exit time from D is the unique solution to the differential equation (1).

If D is not relatively compact, then $E_\bullet[T_D]$ may or may not be a finite function. At any rate, there is an increasing sequence of relatively compact domains $D_1 \subset D_2 \subset \ldots \subset D$ that exhaust D. The first exit time functions, T_n, from D_n increase pointwise to the first exit time function T_D and thus the monotone convergence theorem implies that the sequence of expected first exit times $E_\bullet[T_n]$ increases to the expected first exit time $E_\bullet[T_D]$.

Furthermore, if $E_x[T_D] < \infty$ for one $x \in D$, then $E_\bullet[T_D] < \infty$ everywhere and is a solution to Eq. (1). In fact, if the expected first exit time $f = E_\bullet[T_D] < \infty$, then f is the minimal solution to that equation on D. Indeed, from the above paragraph, you infer that $f = E_\bullet[T_D]$ is given by

$$f = \sup_{B \subset D} f_B$$

where $\{B \subset D\}$ is the set of relatively compact regular domains contained in D, and where $f_B = E_\bullet[T_B]$ is the expected first exit time from B. If g is any positive function on D such that $\triangle g = -1$ and $g \equiv 0$ on ∂D, then $g \geq f$ because, if that was not the case, then $g < f$ on an open subset of D, and so it follows from the definition $f = \sup_B f_B$ that there exists a relatively compact domain $B \subset D$ where $g < f_B$. Then $f_B - g$ is harmonic and > 0 on B but ≤ 0 on ∂B, in contradiction to the maximum modulus principle.

Moving on to a brief review of harmonic functions, besides the already mentioned maximum modulus principle, two other well-known facts will be repeatedly used below. Both concern harmonic functions on domains in $\mathbb{R}^2$ endowed with Riemannian metrics conformal to the standard metric, that is, of the form $\varphi(dx \otimes dx + dy \otimes dy)$. The second fact is then that, since the Laplacian for this metric is $\triangle_\varphi u = (1/\varphi)(u_{xx} + u_{yy})$, a function u is harmonic on a domain of this type if and only if u is harmonic in the classical sense that $u_{xx} + u_{yy} = 0$ all throughout the domain. Because of this, if ϕ is a holomorphic function with range in the domain of the harmonic function u, then the composite $u \circ \phi$ is harmonic in the domain of ϕ, as is easily verified via the chain rule, utilizing the harmonicity of u and the Cauchy–Riemann equations for ϕ.

The third fact about harmonic functions is deeper and concerns their integral representations. To each function $u \geq 0$ that is harmonic on the right half plane $\{x > 0\}$ there corresponds a measure μ on the line $\Im = \{(0, t) \mid -\infty < t < \infty\}$ and a constant $C \geq 0$ so that

$$u(x, y) = Cx + \int_\Im \frac{x}{x^2 + (y - t)^2} \cdot \mu(t),$$

for all (x, y) with $x > 0$.

In particular, if $u \geq 0$ is harmonic and extends continuously by 0 to all but finitely many points $t_1, t_2, \ldots, t_n$ on the line $x = 0$, then the measure μ is supported on the set $\{t_k\}$, and u may be expressed as

$$u(x, y) = C_\infty x + \sum_{k=1}^n \frac{C_k x}{x^2 + (y - t_k)^2}, \tag{2}$$

for some constants $C_\infty, C_1, C_2, \ldots, C_n \geq 0$.

3 Domains in the Euclidean Plane

In Cartesian coordinates $(x, y) \in \mathbb{R}^2$, the Euclidean metric is $dx \otimes dx + dy \otimes dy$, and the Laplacian is given by

$$\Delta_e f = f_{xx} + f_{yy}.$$

In polar coordinates (r, θ), the Laplacian is given by

$$\Delta_e f = f_{rr} + \frac{1}{r} f_r + \frac{1}{r^2} f_{\theta\theta}. \tag{3}$$

3.1 Domain Bounded by an Ellipse

Let D be the domain enclosed by an ellipse in the Euclidean plane with axes of lengths $a, b > 0$ and center (h, k). Up to isometry (a rotation), D consists of all points (x, y) such that $\dfrac{(x - h)^2}{a^2} + \dfrac{(y - k)^2}{b^2} < 1$.

The expected first exit time from D is given by the function

$$f(x, y) = \frac{a^2 b^2}{2a^2 + 2b^2} \left(1 - \frac{(x - h)^2}{a^2} - \frac{(y - k)^2}{b^2} \right). \tag{4}$$

Indeed, f is positive on D, identically 0 on the ellipse ∂D, and satisfies the differential equation $\Delta f = -1$ on D. Therefore $E_{(x,y)}[T_D] = f(x, y)$ because, D being relatively compact, Eq. (1) has exactly one solution.

3.2 Domain Bounded by a Parabola

Up to isometry, a parabola has an equation of the form $y^2 = 4px$ (focus at $(p, 0)$ and focal distance p), and a convex domain bounded by a parabola is isometric to the domain, D, consisting of all $(x, y) \in \mathbb{R}^2$ such that $4px > y^2$.

The expected first exit time from D is

$$E_{(x,y)}[T_D] = 2px - \frac{y^2}{2}, \tag{5}$$

for all $(x, y) \in D$.

It is plain that $f(x, y) = 2px - y^2/2$ is a solution to Eq. (1) on D, but to prove that the expected first exit time $E_{(x,y)}[T_D] = f(x, y)$ on D requires some extra work because D is not relatively compact. Let D_n, $n > [p]$, denote the domain enclosed by the ellipse with foci at $(p, 0)$ and $(2n - p, 0)$ and eccentricity $e = 1 - p/n$. An equation for this ellipse is

$$\frac{(x - n)^2}{n^2} + \frac{y^2}{2p(n - p)} = 1.$$

For $n > [p]$, the domains $D_n \subset D_{n+1} \subset \cdots \subset D$ increase to the domain D, and so the first exit times functions T_n of D_n increase pointwise to the first exit time function T_D. By the dominated convergence theorem, $E_{(x,y)}[T_n]$ converges to $E_{(x,y)}[T_D]$. The expected first exit time from D_n was shown to be (4)

$$
\begin{aligned}
E_{(x,y)}[T_n] &= \frac{n^2 p(n - p)}{2p(n - p) + n^2}\left(1 - \frac{(x - n)^2}{n^2} - \frac{y^2}{2p(n - p)}\right) \\
&= -\frac{p(n - p)}{2p(n - p) + n^2}x^2 + \frac{2np(n - p)}{2p(n - p) + n^2}x - \frac{n^2}{4p(n - p) + 2n^2}y^2.
\end{aligned}
$$

It follows immediately that $E_{(x,y)}[T_n] \to 2px - y^2/2$, as $n \to \infty$, which is the expression for the solution to Eq. (1) shown at (5), uniformly on compact subsets of D.

Any solution to Eq. (1) is of the form $E_\bullet[T_D] + u$, where u is harmonic and ≥ 0 on D and identically 0 on ∂D. The function $\phi : z \mapsto \cosh\frac{\pi}{2}\sqrt{\frac{z}{p} - 1}$ is a conformal representation of D onto the right half plane $\Re z > 0$. If u is harmonic and ≥ 0 on D and identically 0 on ∂D, then $u \circ \phi^{-1}$ is harmonic and ≥ 0 on $\Re z > 0$ and identically 0 on $\Re z = 0$, so, by (2), $u \circ \phi^{-1}(w) = C\Re w$, for some $C \geq 0$, or $u(z) = C\Re\phi(z)$ after the switch $w = \phi(z)$.

That is, any solution to Eq. (1) on D is given by (using complex coordinates $z = x + yi$):

$$z \mapsto 2p\Re(z) - \frac{\Im z^2}{2} + C\Re\left(\cosh\frac{\pi}{2}\sqrt{\frac{z}{p} - 1}\right), \tag{6}$$

for some constant $C \geq 0$.

3.3 Domain Between Two Concentric Circles

Let D be a domain bounded by two concentric circles of radii $a < b$. Rotations about the common center of the circles are isometries that leave D invariant. Therefore the expected first exit time, $f = E_\bullet[T_D]$, from D, in polar coordinates (r, θ) about its center is a function of r only, and so, in those coordinates, the equation $\Delta_e f = -1$ becomes, by (3), $f''(r) + (1/r)f'(r) = -1$ in (a, b). The general solution is $f(r) = -r^2/4 + A \log r + B$, and the boundary conditions $f(a) = f(b) = 0$ make

$$A = \frac{b^2 - a^2}{4(\log b - \log a)} \text{ and } B = \frac{a^2 \log b - b^2 \log a}{4(\log b - \log a)}.$$

3.4 Angular Domain

Let $V \subset \mathbb{R}^2$ be an angular domain of angle α. In polar coordinates, V is, up to isometry, the set of (r, θ) with $r > 0$ and $\theta \in (-\alpha/2, \alpha/2)$.

The expected first exit time from an angular domain V of angle α is infinite if $\alpha \geq \pi/2$, and is finite if $\alpha < \pi/2$ and given by

$$E_{(r,\theta)}[T_V] = \frac{r^2}{4}\left(\frac{\cos 2\theta}{\cos \alpha} - 1\right).$$

Because an angular domain is not a relatively compact domain, there is no guarantee of existence or of uniqueness of solutions to Eq. (1). The expected first exit time, $f = E_\bullet[T_V]$, from V has two other properties that, if finite, will characterize it uniquely among the solutions to that equation.

(a) The expected first exit time function is homogeneous of order 2. Indeed, for $\lambda > 0$, the mapping $(x, y) \mapsto (\lambda x, \lambda y)$ is a dilation of the Euclidean metric that leaves V invariant. The Euclidean heat kernel density at $\lambda p = (\lambda x, \lambda y)$ at time t deposited in $\lambda p' = (\lambda x', \lambda y')$ is the Euclidean heat kernel density at $p = (x, y)$ at time t/λ^2 deposited in $p' = (x', y')$:

$$\frac{1}{4\pi t}e^{-|\lambda p - \lambda p'|^2/4t}d(\lambda x')d(\lambda y') = \frac{1}{4\pi(t/\lambda^2)}e^{-|p-p'|^2/4(t/\lambda^2)}dx'dy'.$$

The effect of such dilation is to rescale Brownian motion times by a factor of λ^2, and so the expected first exit time f must satisfy

$$f(\lambda x, \lambda y) = \lambda^2 f(x, y);$$

in polar coordinates

$$f(\lambda r, \theta) = \lambda^2 f(r, \theta).$$

(b) In polar coordinates, the expected first exit time satisfies $f(r, \theta) = f(r, -\theta)$ because reflection about the axis of V is an isometry that leaves V invariant.

Consequently, by (a), the expected first exit time f is completely determined by the values $f(1, \theta)$, and thus it can be written as $f(r, \theta) = r^2 h(\theta)$, where h is positive and symmetric on $(-\alpha/2, \alpha/2)$. Writing out the differential equation $\Delta_e f = -1$ for $f(r, \theta) = r^2 h(\theta)$ in polar coordinates (3) results in the following differential equation for h on $(-\alpha/2, \alpha/2)$:

$$h''(\theta) + 4h(\theta) = -1. \tag{7}$$

The boundary condition $f \equiv 0$ on ∂C results in the boundary condition $h(\pm\alpha/2) = 0$. Note also that $h > 0$ and that, by (b), h is symmetric about $0 \in [-\alpha/2, \alpha/2]$.

The general solution to Eq. (7) is of the form

$$h(\theta) = A \cos 2\theta + B \sin 2\theta - 1/4.$$

The symmetry of h about 0 in $[-\alpha/2, \alpha/2]$ implies that $B = 0$, and the initial conditions impose that $A = 1/(4 \cos \alpha)$. Therefore,

$$h(\theta) = [\cos 2\theta - \cos \alpha]/4 \cos \alpha.$$

Because this holds for all θ between $-\alpha/2$ and $\alpha/2$, and $h > 0$, you must have $\cos \alpha > 0$ and with the same sign as $\cos 2\theta$, and so $\alpha < \pi/2$. Writing $f(r, \theta) = r^2 h(\theta)$ confirms the statement at the beginning of this section.

In rectangular coordinates, an angular domain of angle $\alpha < \pi/2$ is isometric to the domain, V, consisting of all (x, y) such that $x > m|y|$, where $0 < m = \tan \alpha/2 < 1$.

The expected first exit time from $V = \{(x, y) \mid x > m|y|\}$, with $0 < m < 1$, is

$$E_{(x,y)}[T_V] = \frac{1}{2 - 2m^2} \left(m^2 x^2 - y^2\right). \tag{8}$$

Any solution, g, to Eq. (1) on V is given by

$$g(x, y) = \frac{1}{2 - 2m^2} \left(m^2 x^2 - y^2\right) + C(x^2 + y^2)^{\pi/\alpha} \cos \frac{\pi \arctan(y/x)}{\alpha},$$

for some constant $C \geq 0$.

The second summand in the expression for g is justified in a manner similar to that of the case of the parabola at the end of Sect. 3.2. In this case, you consider the conformal representation, ϕ, of the angular domain V above onto the right half plane given by $\phi(z) = z^{\pi/\alpha}$, and then you use (2) to show that any non-negative harmonic function on V that is identically 0 on ∂V is of the form $z \mapsto C\Re(z^{\pi/\alpha})$, which is as stated above.

3.5 Domain Bounded by a Hyperbola

3.5.1 Convex Domain

Let D be a convex domain in $\mathbb{R}^2$ bounded by a hyperbola. Up to isometry, this hyperbola is given by an equation of the form $(x/a)^2 - (y/b)^2 = 1$, with $a, b > 0$, and D consists of all (x, y) such that $(x/a)^2 - (y/b)^2 > 1$ and $x > 0$.

If $b \geq a$, then the expected first exit time from D is infinite.

Indeed, D contains the angular domain V bounded by the lines $y = \pm m(x - a)$ with $m = b/a$ and $x > a$, and so the expected first exit times $E_\bullet[T_D] \geq E_\bullet[T_V]$. If $b \geq a$, then V has angle $2 \arctan m \geq \pi/2$, and so, as established in Sect. 3.4, the expected first exit time from V is infinite.

If $b < a$, then the expected first exit time from D is given by the function

$$g(x, y) = \frac{1}{2 - 2m^2} \left(m^2 x^2 - y^2 - b^2 \right),$$

for all $(x, y) \in D$.

It is plain that g is a solution to Eq. (1). If g is not the expected first exit time from D, then it is not the minimal solution to (1), and so the expected first exit time from D is of the form $E_\bullet[T_D] = g - u$, where u is a positive, harmonic function on D satisfying $u \equiv 0$ on ∂D.

If $m = b/a < 1$, the expected first exit time $E_\bullet[T_V]$ is finite and given by (8) (after a horizontal shift), and is a minorant for $E_\bullet[T_D] = g - u$. Thus

$$\frac{1}{2 - 2m^2} \left(m^2 (x - a)^2 - y^2 \right) \leq \frac{1}{2 - 2m^2} \left(m^2 x^2 - y^2 - b^2 \right) - u(x, y),$$

or

$$u(x, y) < \frac{b^2}{1 - m^2} \left(\frac{x}{a} - 1 \right), \tag{9}$$

for all (x, y) in V.

You will now reach a contradiction as follows. The function ϕ given by (appropriate branches taken)

$$\phi(z) = c \cosh\left(\frac{2\mu}{\pi} \operatorname{arcosh} z \right) = \frac{c}{2} \left((z + \sqrt{z^2 - 1})^{2\mu/\pi} + (z + \sqrt{z^2 - 1})^{-2\mu/\pi} \right),$$

where $\mu = \arctan m$ and $c = \sqrt{a^2 + b^2}$ the linear eccentricity, is a conformal representation of the right half plane $\{x = \Re z > 0\}$ onto D which takes the boundary $\{x = \Re z = 0\}$ onto ∂D and the ray $\{x \geq 0\}$ onto the ray $\{x \geq a\}$. (As an aid in visualizing this mapping, you recall that cosh takes the horizontal line through βi $(0 < \beta < \pi/2)$ onto the right branch of the hyperbola of equation $x^2/\cos^2\beta - y^2/\sin^2\beta = 1$, cf. [1, 3.4.2] or [6] for more background.)

Then, on the one hand, you deduce from Inequality (9) that

$$u(\phi(x)) < \frac{b^2}{1 - m^2}\left(\frac{\phi(x)}{a} - 1\right),$$

because $\phi(z)$ is real for z real, and from this inequality that $\displaystyle\lim_{x\to\infty} \frac{u(\phi(x))}{x} = 0$, because, for real $x > 1$, $\phi(x) = \frac{c}{2}\left((x + \sqrt{x^2 - 1})^{2\mu/\pi} + (x + \sqrt{x^2 - 1})^{-2\mu/\pi}\right)$, hence $\phi(x) = O(x^{2\mu/\pi})$, and so $\displaystyle\lim_{x\to\infty}\frac{\phi(x)}{x} = 0$ because $\mu = \arctan m < \pi/2$.

On the other hand, the composite function $u \circ \phi$ is positive and harmonic on $\Re z > 0$ and is identically 0 on the boundary $\Re z = 0$. Therefore, by (2), $u \circ \phi(x, y) = Cx$, for some constant $C > 0$, and so $\displaystyle\lim_{x\to\infty}\frac{u(\phi(x))}{x} = C > 0$.

3.5.2 Concave Domain

A (concave) domain bounded by the two branches of a hyperbola is isometric to the domain $D = \{(x, y) \in \mathbb{R}^2 \mid x^2/a^2 - y^2/b^2 > -1\}$, for some $a, b > 0$. Let $m = b/a$ and $\mu = \arctan m$. If $m \geq 1$, then the expected first exit time from D is infinite because D contains an angular domain of angle $2\mu \geq \pi/2$.

If $b < a$, the expected first exit time from D is given by the function

$$g(x, y) = \frac{1}{2 - 2m^2}\left(m^2 x^2 - y^2 + b^2\right), \tag{10}$$

for all $(x, y) \in D$.

Certainly, g is a solution to Eq. (1) on D, so $g = E_{(x,y)}[T_D] + v$, where v is ≥ 0 and harmonic on D and extends continuously to 0 on ∂D.

The domain D contains the angular domains $V_- = \{x < 0 \ \& \ b^2 x^2 > a^2 y^2\}$ and $V_+ = \{x > 0 \ \& \ b^2 x^2 > a^2 y^2\}$. These domains have angle $2\mu < \pi/2$, so their expected first exit time is finite and given by $f_\pm(x, y) = \frac{a^2 b^2}{2a^2 - 2b^2}\left(b^2 x^2 - a^2 y^2\right)$ (same expression, different domain). By comparison,

$$\frac{a^2 b^2}{2a^2 - 2b^2} \left(b^2 x^2 - a^2 y^2 \right) \leq \frac{a^2 b^2}{2a^2 - 2b^2} \left(b^2 x^2 - a^2 y^2 + 1 \right) - v(x, y),$$

i.e.,

$$v(x, y) \leq \frac{a^2 b^2}{2a^2 - 2b^2}, \tag{11}$$

for all $(x, y) \in V_- \cup V_+$. In particular, v is bounded on the real axis $\{y = 0\}$.
The mapping ψ given by

$$\psi(z) = \frac{c}{2} \left(z^{2\mu/\pi} - \frac{1}{z^{2\mu/\pi}} \right) = c \sinh \left(\log z^{2\mu/\pi} \right),$$

where $c = \sqrt{a^2 + b^2}$, is a conformal representation of the right half plane $\Re z > 0$
onto D that takes the positive imaginary axis $\Im z > 0$ to the upper branch of the
hyperbola bounding D, takes the negative imaginary axis to the lower branch of that
hyperbola, and takes the positive real axis onto the real axis (cf. [1, 3.4.2] or [6]
for help on constructing this mapping). Therefore, the composite function $v \circ \psi$ is a
positive, harmonic function on $\Re z > 0$ and that extends continuously by 0 to $\Im z \neq 0$
on the right half plane. Therefore, by (2), $v \circ \psi$ is of the form

$$v \circ \psi(z) = A\Re z + B\Re(1/z)$$

for some constants $A, B \geq 0$. Because of (11), the composite $v \circ \psi$ must be bounded
on the positive real axis $\Im z = 0$, and therefore both $A = 0$ and $B = 0$, that is, $v \equiv 0$,
which shows that the expected first exit time $E_\bullet[T_D] = g$ as was stated at (10).

4 Domains in the Hyperbolic Plane

The unit disk model for the hyperbolic plane is realized by the unit disk $\{x^2 + y^2 < 1\}$
in $\mathbb{R}^2$, endowed with the metric $4(1 - x^2 - y^2)^{-2} (dx \otimes dx + dy \otimes dy)$.
    In geodesic polar coordinates (r, θ) about a point, the hyperbolic metric is $dr \otimes dr + \sinh^2 r d\theta \otimes d\theta$, and the corresponding Laplacian is

$$\Delta_h f = f_{rr} + 2 \cotanh r \, f_r + f_{\theta\theta}. \tag{12}$$

The right half plane model for the hyperbolic plane is realized on the domain
$\{x > 0\}$ with the metric $x^{-2}(dx \otimes dx + dy \otimes dy)$, with Laplacian

$$\Delta_h f = x^2 (f_{xx} + f_{yy}). \tag{13}$$

In Euclidean polar coordinates (r, θ), with $r > 0$ and $-\pi/2 < \theta < \pi/2$, the Laplacian is

$$\Delta_h f = (\cos^2 \theta) \left(r^2 f_{rr} + r f_r + f_{\theta\theta} \right). \tag{14}$$

4.1 Hyperbolic Disks

Let D be a disk of radius R in the hyperbolic plane. In geodesic coordinates (r, θ) based at the center of D, the Laplacian is $\triangle_h f = f_{rr} + 2 \coth r \, f_r + f_{\theta\theta}$. The expected first exit time from D is invariant by (hyperbolic) rotations about the center of D, therefore $f = f(r)$ is a function of the distance r only (the hyperbolic distance to the center of the disk), and so the equation $\triangle_h f = -1$ on D becomes, by (12), $f''(r) + 2 \coth(r) \, f'(r) = -1$ on $(0, R)$, with the boundary conditions that $\lim_{r \to 0} f(r)$ exists and that $f(R) = 0$. The general solution is $f(r) = -(r/2) \coth r + A \coth r + B$, and the boundary conditions imply that $A = 0$ and that $B = (R/2) \coth R$.

The expected first exit time from a hyperbolic disk of radius R is

$$f(r, \theta) = -\frac{r}{2} \coth r + \frac{R}{2} \coth R. \tag{15}$$

4.2 Horodisks

A horodisk is a domain in the hyperbolic plane that may be visualized as a hyperbolic disk with center at a point on the ideal boundary of the hyperbolic plane. Any horodisk is isometric to the domain, D, in the right half plane model consisting of the points (x, y) with $x > R$, for some constant $R > 0$.

The expected first exit time from the horodisk $D = \{x > R\}$ is given by

$$E_{(x,y)}[T_D] = \log \frac{x}{R}. \tag{16}$$

The expected first exit time, $f = E_\bullet[T_D]$, from D must be invariant under vertical translations because these are hyperbolic isometries that leave D invariant; that is, $f(x, y + t) = f(x, y)$ for all t, or $f(x, y) = f(x)$ for all y. Given this, the equation $\triangle_h f = -1$, with $f \equiv 0$ on $x = R$, reduces to $f''(x) = -x^{-2}$ on (R, ∞), with $f(R) = 0$, that is $f(x, y) = \log \frac{x}{R}$, as advertised at (16).

4.3 Neighborhoods of Geodesics

A neighborhood, D, of a geodesic in the hyperbolic plane is determined, up to isometry, by its radius: $D = D(R)$. In the right half plane model and in Euclidean polar coordinates (r, θ) as in (14), the geodesic is $\theta = 0$, and $D(R)$ is the set of all (r, θ) with $r > 0$ and $-\alpha < \theta < \alpha$, where $\log \dfrac{\cos \alpha}{1 - \sin \alpha} = R$.

Because the mapping $(r, \theta) \mapsto (\lambda r, \theta)$ is an isometry of the hyperbolic plane that preserves D (it is a hyperbolic translation along the geodesic $\theta = 0$), the expected first exit time, $f = E_{\bullet}[T_D]$, must satisfy $f(r, \theta) = f(\lambda r, \theta)$ for all $\lambda > 0$. That is, $f(r, \theta) = f(\theta)$ is a function of θ only. So writing out the equation $\Delta_h f = -1$ in coordinates (r, θ), results in, cf. (14), $f''(\theta) = -1/\cos^2 \theta$. The general solution is $f(\theta) = \log \cos \theta + A\theta + B$. The expected first exit time is invariant under reflection about $\theta = 0$, so the constant $A = 0$, and the boundary condition $f(\alpha) = 0$ forces $B = -\log \cos \alpha$.

The expected first exit time from $D = D(\alpha)$ is given by

$$f(r, \theta) = \log \frac{\cos \theta}{\cos \alpha}.$$

Any solution to Eq. (1) on D is of the form

$$f(r, \theta) + \left(Ar^{\pi/2\theta} + \frac{B}{r^{\pi/2\theta}} \right) \cos \frac{\pi \theta}{2\alpha},$$

for some constants $A, B \geq 0$.

Indeed, the composite of a non-negative harmonic function, u, on D that is 0 on the boundary ∂D and the conformal representation of the right half plane $\{x > 0\}$ onto D given by $\phi : (r, \theta) \mapsto (r^{2\alpha/\pi}, 2\alpha/\pi\theta)$ is a non-negative harmonic function, $u \circ \phi$, on the right half plane that is identically 0 on the boundary $\theta = \pm\pi/2$, except perhaps at the origin. By (2), such function is of the form $(r, \theta) \mapsto (Ar + B/r) \cos \theta$, for some constants $A, B \geq 0$.

4.4 Neighborhood of Ideal Point

You may adjust the calculation in the previous section to cover the case of a one-sided neighborhood of a geodesic. In Euclidean polar coordinates as above, the one-sided neighborhood of radius R is isometric to the domain $D = D(R)$ of all (r, θ) with $r > 0$ and $0 < \theta < \alpha(R)$, with $\alpha(R)$ as in the first paragraph of Sect. 4.3. The expected first exit time from $D(R)$ is

$$E_{(r,\theta)}[T_{D(R)}] = \log \cos \theta - \frac{\theta}{\alpha} \log \cos \alpha.$$

A neighborhood of an ideal point is isometric to the domain $D(R)$ when $R = \infty$, or $\alpha = \pi/2$. Hence, the expected first exit time from such domain is infinite because there is no positive solution to the differential equation $f''(\theta) = -1/\cos^2\theta$ on $(0, \pi/2)$ with boundary conditions $f(0) = f(\pi/2) = 0$.

References

1. Ahlfors L (1979) Complex analysis. International series in pure and applied mathematics, 3rd edn. McGraw-Hill, New York
2. Candel A, Conlon L (2003) Foliations II. Graduate studies in mathematics, vol. 60. AMS, Providence
3. De Saint-Venant AJ-CB (1856) Mémoire sur la torsion des prismes. Mémoires présentés par divers savants à l'Académie des Sciences (S.I.) 14:233–560. Available at https://babel.hathitrust.org/cgi/pt?id=hvd.32044091959866;view=1up;seq=286
4. Dynkin E (1965) Markov Processes. Series Die Grundlehren der Mathematischen Wissenschaftteumen, vol. 121–122. Springer, Berlin
5. Ghys É (1995) Topologie des feuilles génériques. Ann Math 141:387–422
6. Kober H (1962) Dictionary of conformal representations. Dover Pub, New York

Jordan Systems: A Search for the Exceptional

José A. Anquela and Teresa Cortés

Abstract From the problem posed in 1934 about the possibility of generalizing the mathematical framework of Quantum Mechanics, we give an overview of the algebraic theory of Jordan systems.

1 The Origin

In 1934, Jordan, von Neumann and Wigner introduced in [8] the notion of Jordan algebra in an attempt to generalize the usual approach to Quantum Mechanics.

In Quantum Mechanics, observables are represented by self adjoint linear operators in a Hilbert space. To simplify, we can go to a finite dimensional setting so that observables are given by symmetric matrices. Given two symmetric matrices $A, B \in \mathcal{M}_n(\mathbf{R})$, the usual matrix product AB need not be a symmetric matrix. The way to obtain again a symmetric matrix is given by the *symmetrized product*

$$A \circ B = \frac{1}{2}(AB + BA).$$

The idea of [8] was finding the relevant properties of the symmetrized product to define a new algebraic structure which would provide new examples, different from that of symmetric matrices (or self adjoint liner operators), which could be used to obtain a generalization or an alternative to Quantum Mechanics.

They decided that there were two relevant properties of the product $\circ$: one was the obvious *commutativity* $A \circ B = B \circ A$, and the other one was the less obvious identity

$$\big((A \circ A) \circ B\big) \circ A = (A \circ A) \circ (A \circ B),$$

J. A. Anquela (✉) · T. Cortés
Departamento de Matemáticas, Universidad de Oviedo,
C/ Federico García Lorca 18, 33007 Oviedo, Spain
e-mail: anque@uniovi.es

T. Cortés
e-mail: cortes@uniovi.es

© Springer International Publishing AG 2018
E. Gil et al. (eds.), *The Mathematics of the Uncertain*, Studies in Systems, Decision and Control 142, https://doi.org/10.1007/978-3-319-73848-2_73

later named the *Jordan identity*.

Definition 1.1 A *Jordan algebra* over a field K of characteristic not two is a K-algebra $(J, \cdot)$ satisfying:

(J1) $x \cdot y = y \cdot x$,   (J2) $(x^2 \cdot y) \cdot x = x^2 \cdot (y \cdot x)$,

for any $x, y \in J$, where x^2 stands for $x \cdot x$.

In [8], the following classification was obtained:

Theorem 1.1 *If J is a simple, finite dimensional, formally real Jordan $\mathbf{R}$-algebra, then J is isomorphic to one of the following:*

(i) $J(Q) = \mathbf{R} + V$ the Jordan algebra of a quadratic form on a finite dimensional real vector space V,

(ii) $H_n(\mathbf{R})$, $H_n(\mathbf{C})$, or $H_n(\mathbf{Q})$, for a positive integer n,

(iii) $H_3(\mathbf{O})$,

where $\mathbf{C}$ stands for the algebra of complex numbers, $\mathbf{Q}$ is the Hamilton's quaternion algebra, $\mathbf{O}$ is the Cayley octonion algebra, and $H_n(D) = \{A \in \mathcal{M}_n(D) \mid \bar{A}' = A\}$ is the set of "hermitian" or symmetric elements in $\mathcal{M}_n(D)$ with respect to the conjugate transpose involution built out of the natural conjugation in D.

Definition 1.2 The *symmetrization* $A^{(+)}$ of an associative algebra A is an algebra over the same linear structure equipped the symmetrized product $\circ$ defined by $x \circ y = \frac{1}{2}(xy + yx)$, where the product in A is denoted by juxtaposition, as in the example of symmetric matrices above.

Looking at the list of algebras of Theorem 1.1, it turns out that, with the exception of (iii), all of them can be imbedded as subalgebras of $A^{(+)}$ for an associative $\mathbf{R}$-algebra A. In [1], Albert showed that an algebra of the form $H_3(\mathbf{O})$ cannot be imbedded in the symmetrization of an associative algebra. Later on, algebras directly related to $H_3(\mathbf{O})$ would be called *Albert algebras*.

Definition 1.3 A Jordan K-algebra J is said to be *special* if there exists an associative K-algebra A such that J is a subalgebra of $A^{(+)}$. Otherwise J is called *exceptional*.

Theorem 1.1 only shows an exceptional example of dimension 27, which offers almost no option to built an infinite dimension exceptional Jordan algebra, which would provide the desired generalization of Quantum Mechanics.

2   From Finite Linear to Infinite Quadratic

After 1934, very relevant algebraists joined the search for exceptional Jordan algebras. The first step was allowing more general fields of characteristic not two, then studying finiteness conditions less restrictive than finite-dimensionality. As a result artinian linear Jordan algebras were studied, obtaining nice theorems like the following description (see the Second Structure Theorem, on [7, p. 179]):

Theorem 2.1 *If J is a unital nondegenerate Jordan algebra with minimum conditions over a field K of characteristic not two, then J satisfies one of the following assertions:*

(i) J is a division algebra

(ii) J is isomorphic to $J(Q) = F + V$, the Jordan algebra of a quadratic form on vector space V over an extension F of K,

(iii) J is isomorphic to the Jordan subalgebra of hermitian elements $H(A, *)$ of $A^{(+)}$, where A is an associative artinian algebra with involution,

(iv) J is isomorphic to $H_3(\mathbf{O}, \gamma)$, the albert algebra of three by three hermitian matrices over the octonions, on an extension of K.

The theorem and its proof were as beautiful as disappointing. Again the only exceptional object was just a mere scalar extension of the Albert algebra of Theorem 1.1.

Still there was the hope that minimum conditions were not far enough from finite-dimensionality. Maybe studying purely infinite dimensional objects would bring new exceptional examples to light.

The knowledge of purely infinite dimensional objects came from the work of very brilliant Russian mathematicians in what was called "The Russian Revolution in Jordan Algebras" (see [10]). In particular, a young mathematician called Efim Zelmanov introduced in the late 1970s and early 1980s two new tools: *absorbers* and *hermitian polynomials*. Both were used to link arbitrary linear Jordan algebras to special Jordan algebras.

Hermitian polynomials are associative polynomials that produce Jordan products out of associative symmetric products. This may not sound very impressive, but its level of complexity is far beyond its degree which is, on the other hand, very big.

The use of these tools gives rise to a description of strongly prime quadratic Jordan algebras over an arbitrary field of characteristic not two with no restriction on the dimension, and no minimum condition at all (see [14]):

Theorem 2.2 *Let J be a strongly prime Jordan algebra over a field K of characteristic not two. Then either*

(i) J is a special Jordan algebra of a quadratic form with central closure of the form $J(Q)$ as in Theorem 2.1 (ii), or

(ii) J is a special Jordan algebra containing an ideal of the from $H(A, *)$, where A is a prime associative algebra with involution and J is a subalgebra of $H(Q(A), *)$, where $Q(A)$ is the Martindale algebra of quotients of A, or

(iii) J is an Albert algebra in the sense that its central closure is the 27-dimensional algebra of three by three symmetric matrices over the octonions.

Once more, no new exceptional objects appeared.

The only remaining possibility of generalization was allowing fields of characteristic two or even arbitrary associative commutative rings of scalars.

In that sense, Kevin McCrimmon proposed the notion of *quadratic Jordan algebra* in which dividing by two plays no role. The idea is focusing in *U*-operators rather than in the bilinear product. In that sense, a Jordan algebra J over an arbitrary ring of scalars Φ, is a Φ-module equipped with two quadratic maps

$$(\ )^2 : J \longrightarrow J \quad \text{and} \quad U : J \longrightarrow \text{End}(J)$$

satisfying certain identities.

Given an associative algebra A over Φ, one can consider the quadratic symmetrization of A, also written $A^{(+)}$, over the same linear structure of A with quadratic products given by $x^2 = xx$ and $U_x y = xyx$. It turns out that $A^{(+)}$ is an example of quadratic Jordan algebra. Moreover, any linear Jordan algebra is a quadratic Jordan algebra with usual squares, and U operators as defined in [7].

Zelmanov and McCrimmon worked together to extend [14] to a general quadratic setting, and succeeded in [12].

By that time, nobody expected new examples of exceptional algebras, and they did not come out: every exceptional strongly prime quadratic Jordan algebra is a form of an Albert algebra.

Afterwards, using suitable hermitian polynomials, a structure theorem of primitive quadratic Jordan algebras was given in [4].

3 Pairs and Triple Systems

In the mean time, applications to Lie algebras, Differential Geometry, Riemannian symmetric spaces, bounded symmetric domains, Functional Analysis, and Projective Geometry came up. These applications gave rise to new Jordan objects.

Definition 3.1 *(i)* A *Jordan pair* consists of a pair of Φ-modules $V = (V^+, V^-)$, with products $Q_x y \in V^\varepsilon$, for any $x \in V^\varepsilon$, $y \in V^{-\varepsilon}$, $\varepsilon = \pm$.

(ii) A *Jordan triple system* J is a Φ-module with products $P_x y \in J$, for any $x, y \in J$.

In both (i) and (ii) the products are quadratic in x and linear in y and satisfy certain identities (see [9]).

A Jordan algebra gives rise to a Jordan triple system by simply forgetting the squaring and letting $P = U$. By doubling any Jordan triple system T one obtains the *double Jordan pair* $V(T) = (T, T)$ with products $Q_x y = P_x y$, for any $x, y \in T$. From a Jordan pair $V = (V^+, V^-)$ one can get a (*polarized*) Jordan triple system $T(V) = V^+ \oplus V^-$ by defining $P_{x^+ \oplus x^-}(y^+ \oplus y^-) = Q_{x^+} y^- \oplus Q_{x^-} y^+$ (see [9, pages 1.13 and 1.14]).

An associative system (pair or triple system) R gives rise to a Jordan system $R^{(+)}$ by *symmetrization*: over the same Φ-module or pair of Φ-modules, we define $P_x y = xyx$, for any $x, y \in R$, in the case of triple systems, and $Q_{x^\sigma} y^{-\sigma} = x^\sigma y^{-\sigma} x^\sigma$, $\sigma = \pm$, in the pair case. This constructions lead us to suitable notions of speciality and exceptionality for Jordan pairs and triple systems as in Definition 1.3.

Not every Jordan triple systems comes from a Jordan algebra by forgetting the squaring. However, there is an additional important connection between algebras and pairs and triple systems through the following definitions based on the work of Kurt Meyberg [13].

Definition 3.2 *(i)* Given a Jordan triple system J, the *homotope* $J^{(a)}$ of J at $a \in J$ is the Jordan algebra over the same Φ-module as J with products $x^{(2,a)} = x^2 = P_x a$, $U_x^{(a)} y = U_x y = P_x P_a y$, for any $x, y \in J$.

The subset $\mathrm{Ker}\, a = \mathrm{Ker}_J a = \{x \in J \mid P_a x = P_a P_x a = 0\}$ is an ideal of $J^{(a)}$ and the quotient $J_a = J^{(a)}/\mathrm{Ker}\, a$ is called *the local algebra of J at a*. When J is nondegenerate, $\mathrm{Ker}\, a = \{x \in J \mid P_a x = 0\}$.

(ii) Given a Jordan pair V, the *homotope* $V^{\sigma(a)}$ of V at $a \in V^{-\sigma}$ $(\sigma = \pm)$ is the Jordan algebra over the same Φ-module as V^{σ} with products $x^{(2,a)} = x^2 = Q_x a$, $U_x^{(a)} y = U_x y = Q_x Q_a y$, for any $x, y \in J$.

The subset $\mathrm{Ker}\, a = \mathrm{Ker}_V a = \{x \in V^{\sigma} \mid Q_a x = Q_a Q_x a = 0\}$ is an ideal of $V^{\sigma(a)}$ and the quotient $V_a^{\sigma} = V^{\sigma(a)}/\mathrm{Ker}\, a$ is called *the local algebra of V at a*. When V is nondegenerate, $\mathrm{Ker}\, a = \{x \in V^{-\sigma} \mid Q_a x = 0\}$.

The above notion of homotope appears naturally when studying primitivity of Jordan algebras (cf. [4]), and is extensively studied by D'Amour and McCrimmon in [5]. The idea consists of relating regularity conditions in Jordan pairs and triple systems to their analogues in Jordan algebras, and then using the structure theory of Jordan algebras to obtain the corresponding descriptions of pairs and triple systems. To do that a new tool was developed: homotope hermitian polynomials. These creatures are associative algebra polynomials living in the free special Jordan algebra, that when evaluated in the homotope of a Jordan pair o triple system, behave as hermitian polynomials, i.e., they produce Jordan products out of symmetric associative ones. The proof of their existence, given in [2, 6], is rather involved (following the constructive arguments of [2], their degree should be around 2×10^{10}).

Among other results, we can look at the following similar descriptions of primitive Jordan pairs (Corollary 2.7 of [3]) and primitive Jordan triples systems (Corollary 2.8 of [3]).

Theorem 3.1 *A Jordan pair V is primitive if and only if one of the following holds:*

(i) V is a simple Jordan pair equaling its socle,

(ii) V consists of hermitian elements: V has an ideal $0 \neq H_0(R, *)$, which is an ample subpair of a $*$-primitive associative pair R with polarized involution $*$ and V is a subpair of $H(Q(R), *)$, where $Q(R)$ is the Martindale pair of symmetric quotients of R.

Theorem 3.2 *A Jordan triple system T is primitive if and only if one of the following holds:*

(i) T is a simple Jordan triple system equaling its socle,

(ii) T consists of hermitian elements: T has an ideal $0 \neq H_0(R, *)$, which is an ample subspace of a $*$-primitive associative triple system R with involution $*$ and T is a subsystem of $H(Q(R), *)$, where $Q(R)$ is the Martindale triple system of symmetric quotients of R.

4   A Failure?

Regarding the search for exceptional Jordan algebras in the spirit of [8], taking into account Theorem 2.2, one could think that the study of Jordan systems has been just a very long waste of time. However, the structure theory of Jordan systems is far from being a mathematical failure.

Indeed, the applications to different areas of Mathematics are numerous an relevant. Apart from those mentioned at the beginning of Sect. 3, the structure theory of strongly prime quadratic Jordan algebras [12] plays a fundamental role in group theory: Zelmanov solved the Restricted Burnside Problem and got a Fields Medal in 1994.

On the other hand, the study of Jordan systems has produced the development of beautiful tools and techniques, which have brought joy and fun to those who have had the opportunity to work in this field

As it frequently happens in Mathematics, and, even in the usual life, reaching the destination turns less important than enjoying the trip.

5   Disclaimer

This paper is far too simple and undetailed to be considered a survey on the Structure Theory of Jordan systems. For that, the reader should have a look at [10] or to the historical survey included in [11].

Those readers eager for details can get their curiosity satisfied with several nice monographs which can be found in the mathematical literature. Among others, we recommend [7, 9, 11].

Acknowledgements Partially supported by the Spanish Ministerio de Economía y Competitividad and Fondos FEDER, MTM2014-52470-P and MTM2017-84194-P.

References

1. Albert AA (1934) On an algebraic generalization of the quantum mechanical formalism. Ann Math 36:29–64
2. Anquela JA, Cortés T (1996) Primitive Jordan pairs and triple systems. J Algebr 184:632–678
3. Anquela JA, Cortés T (1998) Primitivity in Jordan systems is ubiquitous. J Algebr 202:295–314
4. Anquela JA, Cortés T, Montaner F (1995) The structure of primitive quadratic Jordan algebras. J Algebr 172:530–553
5. D'Amour A, McCrimmon K (1995) The local algebras of Jordan systems. J Algebr 177:199–239
6. D'Amour A, McCrimmon K (2000) The structure of quadratic Jordan systems of Clifford type. J Algebr 234:31–89
7. Jacobson N (1968) Structure and representations of Jordan algebras, vol XXXIX. AMS Colloquium Publications, Providence

8. Jordan P, von Neumann J, Wigner E (1934) On a certain algebra of quantum mechanics. Ann Math 36:65–73
9. Loos O (1975) Jordan pairs, vol 460. Lecture notes in mathematics. Springer, Berlin
10. McCrimmon K (1984) The Russian revolution in Jordan algebras. Algebr Groups Geom 1:1–61
11. McCrimmon K (2004) A taste of Jordan algebras. Springer, Berlin
12. McCrimmon K, Zelmanov E (1988) The structure of strongly prime quadratic Jordan algebras. Adv Math 69:133–222
13. Meyberg K (1972) Lectures on algebras and triple systems. Lecture notes University of Virginia, Charlottesville
14. Zelmanov E (1983) On prime Jordan algebras II. Sib Math J 24:89–104

The Cyclic Universe: From the Cyclical Night to the Latest Cosmological Data

Francisco Argüeso, Luigi Toffolatti and Concha Masa

Sol chi non lascia eredità d'affetti, poca gioia ha dell'urna
Ugo Foscolo *"Dei Sepolcri, a Ippolito Pindemonte"*, 1807

Abstract The idea of a cyclic universe has stimulated the imagination of philosophers and scientists throughout history. The notion can be traced back to ancient myths and religions and in the twentieth century it was put forward in a scientific frame, especially as part of the relativistic models of the universe. Although it is no longer fashionable and current cosmological data seems to rule it out, it has recently been proposed in a new guise. The authors present a brief history of the concept of the cyclic universe in these pages and pay special attention to the latest measurements of the cosmic microwave background (CMB) that seem essential for discriminating between a cyclic cosmos and inflationary universes.

1 Cyclic Universe in Myths, Religion and Philosophy

They knew it, the fervent pupils of Pythagoras:
That stars and men revolve in a cycle

This is the beginning of Jorge Luis Borges' poem 'The cyclical night' [3], that refers, beautifully and evocatively, to this old idea: the universe repeats itself, the same events happen again and again in a framework of infinite time.

F. Argüeso (✉) · C. Masa
Departamento de Matemáticas Universidad de Oviedo,
C/ Federico García Lorca 18, 33007 Oviedo, Spain
e-mail: argueso@uniovi.es

C. Masa
e-mail: masa@uniovi.es

L. Toffolatti
Departamento de Física, Universidad de Oviedo,
C/ Federico García Lorca 18, 33007 Oviedo, Spain
e-mail: ltoffolatti@uniovi.es

© Springer International Publishing AG 2018
E. Gil et al. (eds.), *The Mathematics of the Uncertain*, Studies in Systems,
Decision and Control 142, https://doi.org/10.1007/978-3-319-73848-2_74

This notion is fascinating and has often appeared in history from ancient myths to the most recent theories of scientific cosmology. When did the idea of a cyclic universe begin? Ironically, if the universe is really recurrent, the idea would not have had a beginning. To be specific, one of the great advantages of these models is that they avoid the problem of the origin, the problem of thinking how something appeared from nothing or from some original state of the cosmos, such as quantum vacuum in current physics.

But ironies apart, the idea of a cyclic universe dates back to ancient Hindu cosmologies. In Hinduism the universe goes through extremely long stages which are repeated again and again, as in Brahma's day, the 'kalpa' which lasts 4.32 billion years and is the duration of a cycle of the universe [27]. Then, the universe is destroyed by fire or water and Brahma rests for a night as long as Brahma's day. These numbers correspond, in order of magnitude, with the 13.8 billion years of our universe, according to the Big Bang theory.

The universe recreates itself continuously during Brahma's life, a life that lasts 100 of Brahma's years, each of them equivalent to 3.11 trillion of our years, a figure whose order of magnitude is the length of a cycle in the model developed by Steinhardt and Turok [26], the most recent cyclical model in scientific cosmology, which we will discuss at the end of this article.

According to Hindu religion, there have been many Brahmas and there will be more in the future. In these stages, the universe creates itself, destroys itself and is created again, though it is unclear whether the cycles are similar or there are variations.

The reader, like ourselves, might be surprised by the coincidence, that we presume is by chance, between the Hindu figures and those of modern Cosmology. At any rate, it is amazing that ancient Hindus could express such large numbers and they had the imagination to think about such extraordinary lengths of time. It reminds us, by comparison, of the creation of the universe about five thousand years ago according to Christian genealogies.

Among ancient mythologies, we also find worlds that are created and destroyed in Aztec religion, where human sacrifices were held in order to placate the gods and allow the world to continue.

The idea of cycles comes, in a natural way, from the succession of days and nights, the phases of the moon, the seasons, etc. Some civilizations have extrapolated the observations of continuous repetitions to the world as a whole.

We find our next cyclical model in Greek philosophy and although we expected to cite Pythagoras, given the reference in Borges' poem, after some investigation we must mention Empedocles [14] as an advocate of a recurrent cosmos. It is true that Pythagoras, at least in a reference found in "History of Western Philosophy" by Bertrand Russell [25], argued that: "all that is born will be born again in the revolutions of a given cycle, since nothing is completely new"; hence perhaps the first line in the poem of Borges.

Empedocles, born around 490 BC in Acragas (Sicily), was the Greek philosopher who came up with the theory that matter was formed from four elements: earth, air, water and fire. These elements were ruled by two forces: Love and Strife, which mixed and separated them in a cyclic pattern. When Love ruled the elements combined and they separated completely when Strife was the dominant force. In this way, creations and destructions of the cosmos occurred. Cycles were symmetric, the events of a phase repeated themselves in the next, according to an inverse temporal order.

The theory of the four elements was taken up and adapted by Aristotle who, argued, however, that the universe was eternal, spatially finite and did not go through phases of creation and destruction. In a certain way, the world of Aristotle is similar to the static model developed by Einstein in 1917.

Stoic philosophers such as Zeno of Citium (4th century BC) or later Chrysippus and Posidonius imagined a cosmos that oscillated slowly, going through phases of condensation and rarefaction, a world that contracts and expands in an external vacuum, destroyed by fire and reborn again from its ashes (ekpyrosis). The similarities with the cyclical models that appear in the relativistic cosmologies of the 20th century are evident.

Besides, these philosophers argued that the Earth could not be eternal, because in that case mountains would have disappeared due to erosion phenomena. In this way, they opposed the model of the eternal, unchanging cosmos advocated by Aristotle.

Cicero (1st century BC) describes the processes of destruction and creation of the world in his treatise "On the Nature of the Gods" [5]: "I mean the final conflagration of the whole universe;...; nothing, therefore, they say, is left except fire as the agency, vivifying and divine, by which the universe should be renewed again, and the same external order called into being". This is not Cicero's view but that of one of the characters in the book, Quintus Lucilius Balbus, who defends the theories of the Stoics.

The cyclic model seems to disappear from philosophy for centuries, since the predominant Christian view is that of a universe created by God out of nothing. So, a long time goes by until we find another cyclical universe and this appears surprisingly in the work of a well-known writer: Edgar Allan Poe (1809–1849), who was indeed very interested in cosmological matters. In his prose poem 'Eureka' [24], Poe defends the idea of a universe that begins with the explosion of a uniform state of matter in a sudden burst, this matter expands and then contracts due to attractive forces that lead it to the original state to make it explode again.

Obviously, the poem is not a scientific work, but a solution of Olbers' paradox (Why is the sky dark at night if there are an infinite number of stars?) can be found in its pages. This solution is similar to the one accepted today: we receive the light of a finite number of stars since there has not been enough time for the light of the remotest stars to reach us. Although Poe is often cited as the first supporter of this solution, he rejects it in his poem as very unlikely, defending a finite universe as a more reasonable alternative [16].

Eureka is a work of art and as in philosophical or mythological models, we cannot speak of science inasmuch as cyclical models are not contrasted specifically by observations that would enable the scientific community to support them or reject them.

Also in the 19th century the German philosopher Friedrich Nietzsche proposed what he called 'The Eternal Return' [19], that is, a cyclical universe. Nietzsche's vision is based on the idea that there is a finite number of possible combinations of events and if time is infinite, those events will repeat themselves as is necessary, giving rise to an eternal recurrence. Nietzsche called this idea the 'heaviest burden' because we would be forced to repeat our lives again and again. Whereas in other cyclical models the repetition of the cycles is not exact, in Nietzsche's view we are doomed to live the same lives eternally. Similar ideas can be found in the works of other 19th century thinkers and writers such as the French socialist Louis Auguste Blanqui or the German poet Heinrich Heine.

Nietzsche's idea leads us to Poincaré's Recurrence Theorem. This theorem states that a dynamic system, under certain mathematical conditions, for instance if it moves in a finite phase space volume (positions and velocities), will end up in a neighbourhood as close as we wish to the initial conditions. The time of recurrence, that it will take to return to the initial state, will be extremely long, but not infinite. This theorem seems to be a precise, mathematical form of the Eternal Return. However, we can avoid the conclusion if we have an infinite phase space -the possibilities of events are infinitely many- or also if time is finite.

2 Cycles and Cosmology in the 20th Century

At the beginning of the 20th century Cosmology started to be scientific, especially with the publication of Einstein's General Theory of Relativity (1916) and the discovery of the expansion of the universe by Hubble.

General Relativity is a theory of the gravitational field and enables us to treat the universe as a whole from the physical and mathematical point of view, thus already in 1917 Einstein was exploring the cosmological implications of his theory. He searched for a static solution for his equations, a non-evolving universe; we must remember that the Milky Way was considered to be the whole cosmos at that time and the cosmological expansion had yet to be observed. But Einstein was forced to introduce a new constant into his equations, the famous cosmological constant, as the source of a repulsive force that prevents the otherwise inevitable collapse of a stationary universe [8].

In 1922 the Russian physicist Alexander Friedmann solved Einstein's equations assuming the possibility of an evolving universe. In this way he discovered three possible solutions: one with negative curvature, another flat and the third with positive curvature. In the simplest cases (those without a cosmological constant) the geometry determines the evolution of the universe in a unique way: the universe expands forever in the first two solutions, but in the third solution it expands to a maximum radius and then contracts. This was the birth of the expanding universe model and also of a new cyclical universe [10] as a fascinating possibility linked to a physical theory which had been confirmed in 1919 by measuring the deflection of the light from several stars by the Sun.

In his article Friedmann presented as an illustration the example of a cyclical world with a period of ten billion years. In addition, the models depend on the density of the universe, if this is greater than a critical value we live in a cyclical cosmos.

At first Friedmann's models were viewed by Einstein as wrong mathematically. When he checked that the solutions were correct, he dismissed them as physically unacceptable. The idea of an expanding universe was still shocking even for a physicist with revolutionary ideas.

In 1927 the Belgian priest and physicist Georges Lemaître published a model similar to that of Friedmann in a little known scientific journal [15]. Lemaître was unaware of Friedmann's paper and connected the expansion of the universe in his model with the recent measurements, made in the 1920s, of redshifts in the spectra of galaxies that indicated a recession velocity. Lemaître found a linear relationship between this velocity and the distance to the galaxies.

We must remember that in 1925 the American astronomer Edwin Hubble had found that nebulae are galaxies like the Milky Way, thus increasing the size of the known universe. In 1929 Hubble proved that there was a linear relationship between the recession velocity of galaxies and their distance, the proportionality constant is called the Hubble constant. This drew increasing attention to Friedmann and Lemaître's models; unfortunately, Friedmann had died in 1925. Lemaître prefigures the Big Bang theory with his hypothesis of a primordial atom: that the universe had evolved from a very hot and dense phase.

These theories were still very speculative at that time. The discovery of the cosmic microwave background by Penzias and Wilson in 1964 [20] was the essential breakthrough for the confirmation of the Big Bang theory: the universe had been extremely dense and hot in the past and would have expanded from a state of high density, emitting a radiation decoupled from matter that we receive now from every direction. The cyclic universe, one of the possibilities in this theory, has a series of problems that make it a marginal model [26]:

(1) The measurements of the Hubble constant in 1929 indicated that the age of the universe was less than the age of the Earth. This was a problem for the Big Bang theory which cosmologists tried to alleviate by using the cosmological constant [7] as a free parameter. The cyclical closed models (without a cosmological constant) produced the shortest times in each cycle and hence were the most rejected. Even when Baade corrected the value of the Hubble constant in the 1950s and the age of the universe became more compatible with the Big Bang theory, the closed model still had problems. The discovery of the acceleration of the universe, based on the brightness data of distant supernovae obtained in the 1990s, seemed to have eliminated the cyclic models, since the accelerating expansion, probably produced by a cosmological constant, would give rise to an endless expansion.

(2) Richard Tolman, a cosmologist at Caltech, studied the cyclic model in 1934 [28] and concluded that the entropy increase in each phase generates longer cycles in the future and, therefore, shorter in the past, leading inevitably to a beginning of the universe.

(3) In the 1960s several Russian cosmologists (Belinskii, Khalatnikov, Lifshitz) discovered that in a contraction phase any small anisotropy would be amplified, bringing about chaotic behaviour.

(4) In the 1960s and 1970s Hawking and Penrose proved their singularity theorems in General Relativity [13], making the idea of a beginning of time apparently inevitable. If quantum effects, present at the incredibly high energies of the early universe, are taken into account, this conclusion would not be sure.

3 Cyclic Universe and Current Cosmology

Nowadays, observational data seems to support a flat universe dominated by dark energy -probably due to a cosmological constant- that contributes about 70% of the energy content of the universe. Another 25% consists of dark matter -a weird form of matter only detected by its gravitational effect- and the final 5% is made up of ordinary atoms. Dark energy causes the acceleration of the universe, and prevents its future contraction if it is the energy associated with a cosmological constant. However, if dark energy is due to a field that changes with time (called sometimes quintessence) the future of the universe would be unpredictable now, since we still do not know the characteristics of that field.

The standard cosmological model also establishes an ultrafast expansion at the first moments of the universe (about 10^{-35} s after the hypothetical $t = 0$!!), during the stage called inflation [12]. The inflationary model, which began as a generic method to explain the homogeneity and isotropy of our universe and the fact that in the first moments it was flat with high precision, has had remarkable success: its prediction of a flat universe has been confirmed as has its prediction of the generic distribution of the matter-energy fluctuations that gave rise to the current structures: clusters and galaxies.

The measurements of the differences of intensity in different directions (anisotropies) of the cosmic microwave background (CMB) carried out by the COBE, WMAP and Planck satellites, fit the inflationary model well. We must remember that, according to the inflationary theory, these anisotropies are due to quantum fluctuations of a field, the inflation, which produced the exponential expansion of a tiny patch of the universe from subatomic to macroscopic size. This patch finally gave rise to our observable universe -a sphere with a diameter of nearly 100 billion light years-. The CMB anisotropies we observe today are connected with the fluctuations that generated the galaxies and clusters of galaxies.

However, the inflationary model is not free of serious problems: the physical justification of the theory is not clear and fine tuning in the properties of the inflaton is required in order to produce a universe such as ours. Many researchers have invoked the anthropic principle and the multiverse: that there are many universes, maybe an infinite number, with diverse properties and we observe this because it has an inflation field with the suitable properties to originate a universe with life.

This explanation seems impossible to test [18]. Besides which, many different potentials have been proposed for the inflaton field, making the whole inflationary scenario difficult to falsify [11].

Due to his dissatisfaction with the inflationary model, Paul Steinhardt, one of the creators of the theory of inflation, has, along with Neil Turok, another distinguished theoretical physicist, developed the most recent model of a cyclic universe [26].

This model uses the so-called M theory, a generalization of string theory in which the fundamental objects are branes. These branes are multidimensional objects moving in extra dimensions. In the cyclic model at hand, our world is a tetradimensional brane (space-time) filled with matter, radiation and dark energy and there is another brane, a sort of parallel world, which collides with our universe every trillion years.

These two branes recede from each other along another dimension and after reaching a maximum distance, carried by a force similar to that of a spring, they attract, approach each other and collide again. Each of these collisions is a big bang in which new matter and radiation are created. Our world expands from the big bang, continues its expansion due to dark energy and does not contract, since the contraction occurs in another dimension. In this way, problems 1–4 that plagued conventional cyclic models are avoided. Thus, there are an infinite number of cycles with a big bang, a recession of the branes, an approach and a collision, and a new big bang that starts the whole process again and again.

Although this model is supported by a minority of scientists, it is philosophically more comforting than the inflationary model, since the inflationary universe is doomed to be dominated by dark energy in the long term, a cold and empty universe, that will not be born again from its ashes like the new cyclic universe, which was given the Greek name ekpyrotic (from the fire) by its creators.

Trying to be objective, we see the physics of both models as highly speculative, because the energies involved are exceedingly high and will not be reached even in the most powerful particle accelerators.

From the point of view of agreement with observational data, both models fit the data of current Cosmology well, although the agreement with inflation was not built a posteriori as was the case with the ekpyrotic model. The fact of the universe being flat and the generic fluctuation spectrum were both observed after the theory was developed.

Which model, if any, is true? There is an observational test that can be decisive: the detection of the B mode, the magnetic mode, in the CMB polarization. This mode, associated with gravitational waves, appears in the inflationary models and not in the ekpyrotic. If it is discovered it will represent a success for inflation, if not, the cyclic model will gain influence.

There are diverse experiments under way that are trying to find the B mode. One of them, BICEP2, the second of a series, operated from Antarctica and was a collaboration between several American and European universities and research institutions. In March 2014, the BICEP2 principal investigators announced, with great fanfare, the detection of the cosmological B mode with a level that, according to them, confirmed the inflationary model [1]. The tensor mode contribution (gravitational waves) with respect to the scalar fluctuations was given by the number $r = 0.2$. There was a detection of $r > 0$ with a 7 sigma level, 5.9 sigma after foreground subtraction (suffice it to say that a 5 sigma level is considered enough to confirm a detection).

However, there were two problematic issues: (1) although the simplest inflationary models predict a value of r similar to the one obtained, it is also possible to obtain it in different non-inflationary models [4] and especially, (2) the Planck satellite had put an upper bound of $r < 0.11$ with 95% confidence [22].

From that moment on there have been many attempts to reconcile BICEP2 and Planck data by using more complicated inflationary models or with other hypotheses.

In May 2014, only two months after the celebrated detection, and making things even more complicated, two papers appeared, [9, 17], in which the method of elimination of the Galactic dust contribution to the detected signal was questioned and it was claimed that all or part of the signal detected was due to the dust in our Galaxy.

In the article published by the BICEP2 team in Physical Review Letters in June 2014 [2], it was recognized that a Galactic origin of the signal could not be ruled out.

This was confirmed by the Planck data that was released in September [21]: the detection could be due completely to Galactic dust and more careful analyses, conducted by the BICEP2 and Planck teams, were required to conclude whether at least a part of the detection had a cosmological origin.

The story finished (for now) in February 2015, with the publication of a joint analysis of the BICEP2 and Planck data by both teams [23]. The conclusion is this: there was no cosmological detection and the bound obtained on the parameter r was similar to that previously put by Planck, $r < 0.12$.

For the moment, the detection heralded as one of the most important discoveries of 2014 has ended up as an example of an excessive haste to win a Nobel Prize. Therefore, the battle between the inflationary and cyclic models continues.

At any rate, although future observations can make a cyclic universe very unlikely, we doubt that this kind of universe will disappear from cosmologists' imaginations. It will always be ready, to be reborn, phoenix-like from the ashes.

We finish with the last lines of Borges' poem:

In my human flesh, eternity keeps recurring
And the memory, or plan, of an endless poem beginning:
"They knew it, the fervent pupils of Pythagoras . . ."

4   Addendum

This book is devoted to mathematical aspects of the uncertainty mainly in connection with complex systems and the most complex one is, by far, the whole universe, albeit limited to a spherical region of the same universe comprising all matter that can be observed from Earth at the present time: i.e., from inside our currently "observable" volume of the universe. The late Pedro Gil, a great companion and a wise scholar, was also interested in this topic -the complex and largely unknown large-scale structure of the cosmos- and we discussed it with him sometimes, in the corridors and in the entrance hall of the Faculty of Sciences. He used to stop us, in his unfailingly friendly manner -with a touching smile- and then with his sharp wit, he used to question what we claimed about the standard Hot Big Bang model and other cosmological topics. However, in spite of our somewhat different views on how to statistically describe the observed large-scale structure of the universe, on one specific topic we agreed with Pedro: the possibility of describing the whole 'observable' universe by some -obviously partially (or mostly...) undiscovered- physical theory, based on a fully statistical description, grounded in robust mathematical methods. Sometimes, we also discussed with him the possible existence of other living beings on faraway planets.[1]

At the most general level, he believed in an underlying common knowledge, a common intellectual base connecting all scientific disciplines as well as all the other branches of our globalized culture. Moreover, he also believed in a sort of eternal reincarnation until a time in the distant future when everybody treats other living beings correctly. This way of thinking, more characteristic of eastern cultures, is directly linked to the cyclical universe and with the Hindu idea of kalpas, the eternal cycles of creation and destruction. Most importantly, we consider this way of thinking is a clear indication of Pedro Gil's great humanity and modesty. What is doubtlessly true is that you did not have to look for Pedro on other worlds or in faraway galaxies because he was always very approachable and helpful towards his friends and colleagues, as well as being well grounded in everyday reality. It was enough to simply call him any time you needed his valuable help. Pedro, what do you say if we...? Pedro, what do you think about...? And you always got a kind look, close attention to your question and a quick, sensible answer, full of wisdom and affection.

We were all colleagues in the Faculty of Sciences and we worked on the same floor. Pedro was that kind of calm, quiet man loved by everybody but never criticized by anybody. He was the wise patriarch par excellence, in the best sense. When fate led me (Concha Masa) to the position of Dean of the Faculty, we started to have a much closer relationship. Pedro immediately demonstrated his confidence in my capabilities: this simple fact proved to be a fundamental support to my work, given his acknowledged good sense and caution. I can still remember asking his advice about my possible colleagues for the Faculty governing body. He promptly gave an excellent suggestion, guiding me to the best possible choice. From that precise moment I realized how lucky I had been to find such a great companion in the deanship -given that Rosa proved to be an invaluable asset- and at the same time,

---

[1]The existence of other living beings should be currently acceptable by everybody by only considering the most recent data collected, e.g., by the NASA Kepler satellite and by the HARPS spectrograph mounted on the 3.6 m Telescope operating at La Silla, in the Atacama desert (Chile). This data, together with very simple statistical considerations about the nature and distribution of stars inside our galaxy, suggests that there are potentially millions -but, more probably, a few billions- of planets with dimensions and equilibrium temperatures closely matching those of Earth, in the Milky Way alone. Moreover, taking into account that the most recent estimates [6] of the total number of galaxies currently in the observable universe gives the figure of $(2.0 \pm 0.6) \times 10^{12}$ (!), it is, obviously, very difficult to think we are alone in the universe.

such a fine advisor as Pedro, a man I could rely on whenever I needed him, then, and for many years afterwards. A deep friendship was cemented that is still alive and that will never disappear.

Acknowledgements We thank Liz and John Wyke for their careful revision of the first English manuscript. They have helped a lot to improve this contribution.

References

1. BICEP2 Collaboration: Ade PAR et al. (2014) BICEP2: Detection of B-mode polarization at degree angular scales. arXiv:1403.3985v1
2. BICEP2 Collaboration: Ade PAR, et al (2014) BICEP2 I: Detection of B-mode polarization at degree angular scales by BICEP2. Phys Rev Lett 112:241101
3. Borges JL (2000) Selected Poems. Trans Reid A, Penguin Modern Classics, London
4. Brandenberger RH (2011) Is the spectrum of gravitational waves the Holy Grail of inflation? arXiv:1104.3581v1
5. Cicero MT (1896) De natura deorum (On the nature of the gods). Trans Brooks F, Methuen, London
6. Conselice CJ, Wilkinson A, Duncan K, Mortlock A (2016) The evolution of galaxy number density at $z<8$ and its implications. Astroph J 830(2):83
7. Earman J (2001) Lambda. The constant that refuses to die. Arch Hist Ex Sci 55(3):189–220
8. Einstein A (1917) Kosmologische Betrachtungen zur allgemeinen Relativitätstheorie. Sitzungsberichte Königlich Preus Akad Wissensch:142–152
9. Flauger R, Hill JC, Spergel DN (2014) Toward an understanding of Foreground Emission in the BICEP2 Region. J Cosmol Astropart Phys 8:039
10. Friedmann A (1922) Über die Krümmung des Raumes. Zeitscr Phys 10:377–386
11. Gubitosi G, Lagos M, Magueijo J, Allison R (2016) Bayesian evidence and predictability of the inflationary paradigm. J Cosmol Astropart Phys 6:002
12. Guth AH (1997) The inflationary universe. Perseus Books, New York
13. Hawking SW, Penrose R (1996) Nature of space and time. Princeton University Press, Princeton
14. Kragh H (2007) Conceptions of Cosmos. From myths to the accelerating Universe: a history of Cosmology. Oxford University Press, Oxford
15. Lemaître G (1927) Un univers homogène de masse constant et de rayon croissant rendant compte de la vitesse des nébuleuses extra-galactiques. Ann Société Sci Bruxel A 47:49–59
16. Molaro P, Cappi A (2012) Edgar Allan Poe: the first man to conceive a Newtonian evolving universe. In: Champion N, Sinclair R (eds) Culture and Cosmos 16(1–2):225–239
17. Mortonson MJ, Seljak U (2014) A joint analysis of Planck and BICEP2 B modes including dust polarization uncertainty. J Cosmol Astropart Phys 10:035
18. Mosterín J (2000) The anthropic principle in cosmology. A critical review. Acta Instit Philos Aesthet 18:111–139
19. Nietzsche F (1883–1891) Also sprach Zaratustra. Ein Buch für Alle und Keinen. Ernst Schmeitzner, Chemnitz
20. Penzias AA, Wilson RW (1965) A measurement of excess antenna temperature at 4080 Mc/s. Astrophys J Lett 142:419–421
21. Planck Collaboration: Adam R et al. (2016) Planck Intermediate results. XXX. The angular power spectrum of polarized dust emission at intermediate and high galactic latitudes. Astron Astrophys 586:A133
22. Planck Collaboration: Ade PAR (2014) Planck, et al (2013) results. XVI. Cosmological parameters. Astron Astrophys 571:A16
23. Planck Collaborations: Ade PAR, et al (2015) A joint analysis of BICEP2/Keck Array and Planck data. Phys Rev Lett 114:101301

24. Poe EA (1848) Eureka, a prose poem. GP Putnam, New York
25. Russell B (1994) History of western philosophy (1st edn. 1946). Routledge, London
26. Steinhardt PJ, Turok N (2007) Endless universe. Beyond the big bang, Doubleday, New York
27. Tardan-Masquelier Y (2011) L'Origine du Monde. Le Nouvel Observateur, Paris
28. Tolman R (1934) Relativity, thermodynamics and cosmology. Oxford University Press, Oxford

Phasors, Always in the Real World

Nilo Bobillo-Ares and M. Luisa Garzón

Abstract Complex phasors are reinterpreted as real operators in a real vector space of functions. This structure is linked with dilative rotations of the Euclidean plane. Finally we conclude with some methodological ideas related to the teaching of the complex numbers.

1 Introduction

The phasor technique, introduced by Steinmetz [7] in 1893, is frequently used in engineering (see [2, 3] and [8]), physics and applied mathematics [4], mainly to facilitate the algebra of circular functions and also to obtain a particular solution of linear differential equations whose excitation is a circular function.

The application of phasors in the standard method requires to alternate between two vector spaces, a real space and a complex space. The use of complex numbers seems inevitable if circular functions are expected to share certain property, similar to that of the exponential functions:

$$f(t)f(s) = f(t+s); \tag{1}$$

this property, to convert products into sums, is the *leit motiv* of this work; we call it *logarithmic property*.[1] Here we suggest a possible way out of this small inconvenient by reinterpreting phasors as elements of a certain subspace of linear operators.

---

[1]The methods to replace the operation of multiplication by a sum, called *prosthaphaeresis*, have a long history, even before the discovery of logarithms by Napier. See, for example, the article by Clavius in [5, p. 459].

N. Bobillo-Ares (✉) · M. L. Garzón
Departamento de Matemáticas, Universidad de Oviedo, Oviedo, Spain
e-mail: nilo@uniovi.es

M. L. Garzón
e-mail: maria@uniovi.es

© Springer International Publishing AG 2018
E. Gil et al. (eds.), *The Mathematics of the Uncertain*, Studies in Systems, Decision and Control 142, https://doi.org/10.1007/978-3-319-73848-2_75

823

The formal character of this paper requires a clearly specified starting point; thus, we must devote enough space to motivate an analytical definition of circular functions.[2]

The phasor space is constructed from the derivation operator, defined on the two-dimensional space of the circular functions. The new technique is get to work in a couple of typical examples.

The algebraic structure of phasors in the space of circular functions can be translated to the Euclidean vector plane; in this way the usual geometric interpretation of circular functions and phasors is obtained.

From this simple algebraic game, some methodological conclusions (perhaps also some psychological ones!) about how we understand the algebra of complex numbers can be drawn.

2   The Beloved Logarithmic Property

Hyperbolic and circular functions have similar properties; so, for example, for the sum of arguments we have the formulas:

$$\cosh(t \pm s) = \cosh t \cosh s \pm \sinh t \sinh s, \tag{2}$$
$$\cos(t \pm s) = \cos t \cos s \mp \sin t \sin s, \tag{3}$$

and similar ones, only with little differences, for $\sinh(t \pm s)$ and $\sin(t \pm s)$. These formulas are quite complicated if we compare them with the logarithmic property (1); however, for the hyperbolic functions certain linear combinations can be obtained:

$$e^{\pm t} = \cosh t \pm \sinh t, \tag{4}$$

which, as we know, have the desired logarithmic property (1). We wonder if there will be a similar trick for circular functions; specifically, we would like to find a function, let us call it cis(t), as a linear combination of the base circular functions

$$\operatorname{cis}(t) = \lambda \cos t + \mu \sin t, \tag{5}$$

where λ and μ are *certain numbers*, which verify the logarithmic property:

$$\operatorname{cis}(t + s) = \operatorname{cis}(t)\operatorname{cis}(s). \tag{6}$$

A simple calculation, substituting (5) into (6), leads to the conditions:

$$\lambda = 1, \qquad \mu^2 = -1; \tag{7}$$

---

[2]We have learned in the critical analysis by Spivak [6] that we need an analytical definition of circular functions.

the second one shows that the coefficients of the proposed problem can not be real numbers. On the other hand, if we allow complex coefficients the solution of the stated problem leads to the venerable Euler formulas:

$$\operatorname{cis}(\pm t) = \cos t \pm i \sin t =: e^{\pm it}. \tag{8}$$

Thus, to enjoy the logarithmic property within the circular functions, we have to replace the real vector space, whose elements are easily interpretable functions but with complicated algebra, by a complex space with simple algebra but lacking any direct interpretation. This is the price to pay when adopting phasors as a calculation method. In order to illustrate this fact, let us add two circular functions using phasors, following the traditional method:

$$A \cos(t + \alpha) + B \cos(t + \beta) = \Re\left[Ae^{i(t+\alpha)} + Be^{i(t+\beta)}\right]$$
$$= \Re\left[(Ae^{i\alpha} + Be^{i\beta})e^{it}\right] = \Re\left[Ce^{i\gamma}e^{it}\right] = C \cos(t + \gamma), \tag{9}$$

where we have set $Ae^{i\alpha} + Be^{i\beta} = Ce^{i\gamma}$. Note the persistent "back and forth" between the real and the complex spaces.

A reinterpretation of phasors will be proposed so that we keep the logarithmic property without abandoning the real vector space.

3 Circular Functions

Circular functions arise in a geometric context: let $\mathbf{E}$ be the vector Euclidean plane and $(\mathbf{e}_1, \mathbf{e}_2)$ a direct orthonormal basis.

The *square operator* $\mathbf{J}$ rotates the vectors at a right angle in a positive sense; as $\mathbf{J}\mathbf{e}_1 = \mathbf{e}_2$ and $\mathbf{J}\mathbf{e}_2 = -\mathbf{e}_1$, this operator is represented by the matrix

$$(j) = \begin{pmatrix} 0 & -1 \\ 1 & 0 \end{pmatrix}. \tag{10}$$

The set $S_1 = \{\mathbf{r} \in \mathbf{E} | \mathbf{r}^2 = 1\}$ is called *trigonometric circle*. On S_1 consider the differentiable curve $\mathbf{r}(t)$ that verifies

$$\mathbf{r}(t) \cdot \mathbf{r}(t) = 1, \quad \mathbf{r}(t) \cdot \mathbf{r}'(t) = 0, \quad \mathbf{r}'(t) \cdot \mathbf{r}'(t) = 1; \tag{11}$$

here the second equation is obtained by derivation of the first one and the third selects the parameter t as the arc length. Now, from these three relations, adding the initial condition, we obtain the linear problem:

$$\mathbf{r}'(t) = \pm\mathbf{J}\mathbf{r}(t), \quad \mathbf{r}(0) = \mathbf{e}_1. \tag{12}$$

If we assume to move along the curve S_1 in the positive sense, we have to choose the sign "+". Now we substitute $\mathbf{r}(t) = x(t)\mathbf{e}_1 + y(t)\mathbf{e}_2$ into Eq. 12 and we get the system:

$$x'(t) = -y(t), \quad y'(t) = x(t), \quad x(0) = 1, \quad y(0) = 0; \tag{13}$$

by means of its solution we define $\cos(t) := x(t)$ and $\sin(t) := y(t)$. From this definition the formulas for the sum of arguments (3), their periodic character, parity and, even, the definition of the real number π can be derived [6].

Now, in terms of the functions cos and sin, we define the space $\mathscr{E}$ of *circular functions* as the solution set of the differential equation:

$$f'' + f = 0; \tag{14}$$

any of these functions can be expressed in the form, see [6]:

$$f(t) = R\cos(t + \alpha) = \lambda \cos t + \mu \sin t, \quad R, \alpha, \lambda, \mu \in \mathbb{R}. \tag{15}$$

4 Phasors

We define the *derivative operator* $j : \mathscr{E} \to \mathscr{E}$, as $jf = f'$. The Eq. 14 can be written in the form $j^2 f = -f$ which leads us to the fundamental relation:

$$j^2 = -1, \tag{16}$$

where $1 : \mathscr{E} \to \mathscr{E}$ is the *identity operator* and $j^2 := j \circ j$. Despite d/dt is not invertible in other spaces, j is indeed invertible, $j^{-1} = -j$. The set[3]

$$\mathscr{C} = \{\lambda 1 + \mu j | \lambda, \mu \in \mathbb{R}\} \tag{17}$$

is isomorphic to the set $\mathbb{C}$ of complex numbers; here j plays the role of the imaginary unit i. The elements of $\mathscr{C}$ are the (new!) phasors.

Let us consider now a circular function $R\cos(t + \alpha)$, we have:

$$R\cos(t + \alpha) = R(\cos\alpha\cos t - \sin\alpha\sin t) = R(\cos\alpha 1 + \sin\alpha j)\cos t, \tag{18}$$

since $-\sin = j\cos$. It is convenient to use the notation:

$$e^{j\alpha} := \cos\alpha 1 + \sin\alpha j. \tag{19}$$

---

[3]As customary, the linear combination of two linear operators $u, v : \mathscr{E} \to \mathscr{E}$ is defined as $(\lambda u + \mu v)f := \lambda(uf) + \mu(vf)$, with $\lambda, \mu \in \mathbb{R}$.

Taking into account the sum of arguments relations of the circular functions, it is straightforward to demonstrate the following useful formulas:

$$e^{j\alpha}e^{j\beta} = e^{j(\alpha+\beta)} \quad \text{and from here} \quad e^{-j\alpha} = (e^{j\alpha})^{-1}. \tag{20}$$

Thus, we have the relations

$$R\cos(t+\alpha) = Re^{j\alpha}\cos t, \quad R\sin(t+\alpha) = Re^{j\alpha}\sin t. \tag{21}$$

4.1 Phasors for Circular Functions of Frequency ω

These results can be easily generalized to the set $\mathscr{E}_\omega$ of circular functions of frequency ω, the solution set of the differential equation $f'' + \omega^2 f = 0$; in this space we define the operator $j : \mathscr{E}_\omega \to \mathscr{E}_\omega$ as[4]

$$jf(t) := \frac{1}{\omega}f'(t). \tag{22}$$

We have the relations:

$$j\cos\omega t = -\sin\omega t, \quad j\sin\omega t = \cos\omega t. \tag{23}$$

From here it follows, for circular functions of frequency ω, equations which are analogous to (21):

$$R\cos(\omega t+\alpha) = Re^{j\alpha}\cos\omega t, \quad R\sin(\omega t+\alpha) = Re^{j\alpha}\sin\omega t. \tag{24}$$

5 Main Examples

To see phasors "in action", let's detail a couple of examples. First, we aim to calculate the sum of two circular functions:

$$\begin{aligned}
R_1\cos(\omega t+\alpha_1) + R_2\sin(\omega t+\alpha_2) &= R_1 e^{j\alpha_1}\cos\omega t + R_2 e^{j\alpha_2}\sin\omega t\\
&= R_1 e^{j\alpha_1}\cos\omega t - R_2 e^{j\alpha_2}j\cos\omega t = (R_1 e^{j\alpha_1} - R_2 e^{j\alpha_2}j)\cos\omega t\\
&= Re^{j\alpha}\cos\omega t = R\cos(\omega t+\alpha),
\end{aligned}$$

---

[4]There is certain abuse of notation here; we would distinguish the operator $j : \mathscr{E} \to \mathscr{E}$ from the operator $j : \mathscr{E}_\omega \to \mathscr{E}_\omega$. This complication, that for each frequency we need a different operator j, makes the phasor technique useless when dealing with circular functions of several frequencies.

where we have set $Re^{j\alpha} = R_1 e^{j\alpha_1} - R_2 e^{j\alpha_2} j$, which is a simple calculation with complex numbers (although, actually, we are dealing with linear operators!); you should compare this process with (9).

As a second example, let us consider the equation of motion of a damped oscillator with a forcing term $f \in \mathscr{E}_\omega$:

$$a\frac{d^2x}{dt^2} + b\frac{dx}{dt} + cx = f, \quad a, b, c \in \mathbb{R}; \tag{25}$$

as usual we have to obtain a particular solution $x \in \mathscr{E}_\omega$. Writing the differential equation derivatives in terms of the ωj operator:

$$a(\omega j)^2 x + b\omega j x + cx = \left((c - a\omega^2)1 + b\omega j\right) x = f; \tag{26}$$

and setting $(c - a\omega^2)1 + b\omega j = Re^{j\alpha}$, we can easily solve for x multiplying by the inverse operator:

$$x(t) = \frac{1}{R}e^{-j\alpha} f(t). \tag{27}$$

Now, suppose that $f(t) = A_0 \cos(\omega t + \beta_0)$, then we have the (unique) particular solution in $\mathscr{E}_\omega$:

$$x(t) = \frac{1}{R}e^{-j\alpha} A_0 \cos(\omega t + \beta_0) = \frac{A_0}{R} \cos(\omega t + \beta_0 - \alpha). \tag{28}$$

6   Back to Geometry

Let's go back to the Euclidean plane **E**. The operators identity I and square J verify the relation $J^2 = -I$; consequently, the set

$$\mathbf{C} = \{\lambda I + \mu J | \lambda, \mu \in \mathbb{R}\} \tag{29}$$

is isomorphic to the set $\mathbb{C}$ of the complex numbers. It is easy to be convinced that elements of **C** of the form $\cos \alpha I + \sin \alpha J =: e^{\alpha J}$ constitute the rotation group $\mathcal{O}(2)$. The set **C** also includes the homothecies, since any element can be expressed as

$$\lambda I + \mu J = R(\cos \alpha I + \sin \alpha J) = Re^{\alpha J}. \tag{30}$$

We have, therefore, three canonically isomorphic sets:

$$\mathscr{E}_\omega \simeq \mathbf{C} \simeq \mathbb{C}, \tag{31}$$

respectively, phasors, homothetic rotations and abstract complex numbers.

It is an easy task to establish an isomorphism $\Phi : \mathcal{E}_\omega \to \mathbf{E}$ that, in addition, would be compatible with the operators isomorphism (31); to this end, we choose a circular function, i.e. $\cos \omega t$, and we can affirm that

$$\Phi(\cos \omega t) = \mathbf{e}_1, \quad \Phi(j \cos \omega t) = \mathsf{J}\mathbf{e}_1 = \mathbf{e}_2. \tag{32}$$

This isomorphism allows the geometric interpretation, by means of vector operations, of any calculation with circular functions of same frequency ω (see for example the classic text for electronic engineers [1]). It should be noted that the isomorphism Φ is not canonical since it depends on the discretionary choice of $\cos \omega t$ as the antecedent of the vector $\mathbf{e}_1$.

7 Why Complex Numbers Seem Alien?

Complex numbers appeared in the attempt to solve certain algebraic equations in terms of the square root of -1, named the imaginary unit, as it did not match any previously known number. The daring of certain mathematicians like Cardano, who continued dealing with this new-born monster, gave rise to the creation and development of the complex numbers algebra. Complex numbers were born in the "backyard" of the real numbers and remained questionable until Gauss constructed an abstract theory in which the formal manipulation of these numbers was clearly specified.

As shocking as it can be, these two introductory methods for the complex numbers are still in use, confronting the student either to consider these numbers as monstrous objects within the real numbers, or as completely abstract objects, with perfectly ruled use, but no connexion to any real known object. Trying to avoid this problem and expecting to make complex numbers "more real", the geometric interpretation of these numbers is emphasized in such a manner that it could become even suspicious. Considering that the origin of the complex numbers has nothing to do with geometry, the above mentioned interpretation acquires a metaphorical nature, leaving always doubts about its "real character".

One of the clues to explain the complex numbers weirdness is that, in general, no particular realization is elaborated from the first principles. To illustrate this fact, let us imagine that our first approach to vectors would be the statement of the vector space structure, without any previous knowledge of vectors in physics or geometry. Needless to say that under this scenario such structure would appear awfully artificial or even gratuitous. That is the same situation when the complex numbers are introduced as Gauss pairs.

Phasors and homothetic rotations are specific models of the complex structure and can be of enormous help to acquire familiarity with such structure; as an example let us consider the enigmatic relation $i^2 = -1$ in $\mathbb{C}$:

- On the plane we have $J^2 = -I$ and it means that two right angle rotations are a flat angle rotation;
- In the set $\mathscr{E}_\omega$ of the circular functions of frequency ω, the relation $j^2 = -I$ means that $\cos(\omega t + \alpha + \frac{\pi}{2} + \frac{\pi}{2}) = -\cos(\omega t + \alpha)$.

Nothing mysterious!

The meaning of an abstract mathematical theory is revealed when a specific realization is known; to understand its *abstract nature* requires the knowledge of two or more realizations, without them, possibly, we will understand nothing.

References

1. Angot A (1961) Compléments de Mathématiques à l'usage des Ingénieurs de l'Électrotechnique et des Télécomunications (Sect. 1.2). Éditions de la Revue d'Optique, Paris
2. Expósito A, Bachiller-Soler A, Rosendo-Macías JA (2006) Application of generalized phasors to eigenvector and natural response computation of LTI circuits. IEEE Trans Circuit Syst 53(7):1533–1543
3. Forrest E (1985) A structured approach to phasors. Int J Math Educ Sci Technol 16(1):11–14
4. Pontryagin LS (1962) Ordinary differential equations (Sect. 2.7). Pergamon Press, London
5. Smith DE (1959) A source book in mathematics. Dover Publications, New York
6. Spivak M (1994) Calculus (Chap. 15). Publish or Perish, Houston
7. Steinmetz CP (1893) Complex quantities and their use in Electrical Engineering. In: Proceedings of International Electrical Congress of the American Institute of Electrical Engineers, New York
8. Wolf DA (2010) Complex numbers and phasors without vectors or $\sqrt{-1}$. IEEE Antennas Propag Mag 52(1):218–220

Indexation Strategies and Calmness Constants for Uncertain Linear Inequality Systems

M. Josefa Cánovas, René Henrion, Marco A. López and Juan Parra

Dedicated to the memory of Pedro Gil

Abstract The present paper deals with uncertain linear inequality systems viewed as nonempty closed coefficient sets in the $(n + 1)$-dimensional Euclidean space. The perturbation size of these uncertainty sets is measured by the (extended) Hausdorff distance. We focus on calmness constants—and their associated neighborhoods—for the feasible set mapping at a given point of its graph. To this aim, the paper introduces an appropriate indexation function which allows us to provide our aimed calmness constants through their counterparts in the setting of linear inequality systems with a fixed index set, where a wide background exists in the literature.

This research has been partially supported by Grant MTM2014-59179-C2-(1-2)-P from MINECO, Spain, and FEDER "Una manera de hacer Europa", European Union, and by the Australian Research Council, Project DP160100854.

M. J. Cánovas · J. Parra
Center of Operations Research, Miguel Hernández University of Elche,
03202 Elche, Alicante, Spain
e-mail: canovas@umh.es

J. Parra
e-mail: parra@umh.es

R. Henrion
Weierstrass Institute for Applied Analysis and Stochastics, Mohrenstr. 39,
10117 Berlin, Germany
e-mail: rene.henrion@wias-berlin.de

M. A. López (✉)
Department of Mathematics, University of Alicante, 03071 Alicante, Spain
e-mail: marco.antonio@ua.es

© Springer International Publishing AG 2018
E. Gil et al. (eds.), *The Mathematics of the Uncertain*, Studies in Systems, Decision and Control 142, https://doi.org/10.1007/978-3-319-73848-2_76

831

1 Introduction

We consider the *uncertain linear inequality system*

$$\left\{ a'x \le b, \ \begin{pmatrix} a \\ b \end{pmatrix} \in U \right\},$$ (1)

where $x \in \mathbb{R}^n$ is the vector of variables and the *uncertainty set* U is assumed to be a nonempty closed subset of $\mathbb{R}^{n+1}$. All elements in $\mathbb{R}^n$ are regarded as column-vectors and y' denotes the transpose of $y \in \mathbb{R}^n$. Accordingly, elements in $\mathbb{R}^{n+1}$ will be written in the form $\binom{u}{v}$, where $u \in \mathbb{R}^n$ and $v \in \mathbb{R}$. Observe that (1) is in general a linear *semi-infinite* inequality system (i.e., with finitely many decision variables but possibly infinitely many constraints). Linear semi-infinite inequality systems have been extensively studied in [11].

The uncertainty set U is considered as the parameter to be perturbed. So, formally, we are considering the *parameter space* $CL\left(\mathbb{R}^{n+1}\right)$ of all nonempty closed subsets in $\mathbb{R}^{n+1}$. From the topological side, the space of variables $\mathbb{R}^n$ is endowed with an arbitrary norm $\|\cdot\|$, and the parameter space is equipped with the (extended) Hausdorff distance, d_H, specified in Sect. 2.1 (see e.g. [1] for a comprehensive analysis of the Hausdorff metric).

Associated with the parametrized system (1), roughly speaking referred to as system U, we consider the *feasible set mapping*, $\mathscr{F} : CL\left(\mathbb{R}^{n+1}\right) \rightrightarrows \mathbb{R}^n$, given by

$$\mathscr{F}(U) := \left\{ x \in \mathbb{R}^n \mid a'x \le b \text{ for all } \begin{pmatrix} a \\ b \end{pmatrix} \in U \right\}.$$

Observe that the closedness assumption on U is not restrictive since the feasible set mapping has the same values if general sets are replaced with their closures, and the same happens with the definition of excess—see (3)—and hence with d_H.

Our main goal consists of providing calmness constants (cf. Sect. 2.2) for $\mathscr{F}$ at a nominal (fixed) element of its graph (U_0, x_0). We can find in the literature different contributions to the calmness of the feasible set mapping in the context of linear systems with a fixed index set T, say $\mathscr{F}^T : \left(\mathbb{R}^{n+1}\right)^T \rightrightarrows \mathbb{R}^n$, which is given by

$$\mathscr{F}^T(\sigma) := \left\{ x \in \mathbb{R}^n \mid a_t'x \le b_t, \ t \in T \right\},$$ (2)

where

$$\sigma(t) = \begin{pmatrix} a_t \\ b_t \end{pmatrix} \in \mathbb{R}^{n+1}, t \in T.$$

In this framework, the parameter space $\left(\mathbb{R}^{n+1}\right)^T$ is assumed to be endowed with the uniform converge topology; see Sect. 2.1 for details.

With the aim of taking advantage of the vaste literature about calmness for mappings in the format $\mathscr{F}^T$ to derive calmness constants for $\mathscr{F}$, we introduce in Sect. 3.1

a specific *indexation function*, $\mathscr{I} : CL\left(\mathbb{R}^{n+1}\right) \to \left(\mathbb{R}^{n+1}\right)^{\mathbb{R}^{n+1}}$, which assigns to each set $U \in CL\left(\mathbb{R}^{n+1}\right)$ a certain function $\mathscr{I}_U \in \left(\mathbb{R}^{n+1}\right)^{\mathbb{R}^{n+1}}$ with rge $\mathscr{I}_U = U$, where rge stands for range (image). In this way, if U is the set of coefficient vectors of system (1), $\sigma = \mathscr{I}_U$ (whose index set is the whole $\mathbb{R}^{n+1}$) can be interpreted as enssentialy the same system but with the addition of repeated constraints.

The definition of our indexation function $\mathscr{I}$ is inspired, but sensibly different (see Sect. 3 for details), by the one introduced in [6] with the aim of analyzing the stability of the *optimal value function* of linear optimization problems with uncertain constraints. In the present paper the properties of $\mathscr{I}$ will enable us to derive calmness constants (and associated neighborhoods) for $\mathscr{F}$ from those for $\mathscr{F}^{\mathbb{R}^{n+1}}$. In a second step we wonder whether $\mathbb{R}^{n+1}$ may be replaced with a smaller index set $T \subset \mathbb{R}^{n+1}$.

Paper [4] provides the calmness modulus of $\mathscr{F}^T$ in the particular case when T is finite (see [2] for an extension to the nonlinear case), whereas [5] proves that this calmness modulus is in fact a calmness constant for a certain neighborhood (specified therein) when we restrict ourslevs to right-hand-side perturbations. In Sect. 5 of the present paper we show how to extend this result to perturbations of all coefficients. Coming back to our framework of uncertain linear systems, the reader is addressed to [7, 8] for the study of robust local and global error bounds, respectively. Recall that, the local error bound property is closely related to calmness of feasible solutions when only right-hand-side perturbations are allowed. See also [10] for the development of dualy theory in robust linear optimization with infinitely many uncertain constraints.

Now we summarize the main original contributions of the paper. Section 3 motivates (see Example 3.1) and introduces the announced indexation function $\mathscr{I}$ which allows us to derive calmness constants for $\mathscr{F}$ at $(U_0, x_0) \in$ gph$\mathscr{F}$ (the graph of $\mathscr{F}$) via calmness constants for $\mathscr{F}^{\mathbb{R}^{n+1}}$ at (σ_0, x_0), with $\sigma_0 := \mathscr{I}_{U_0}$. After that, Sect. 4 solves the question of whether or not $\mathbb{R}^{n+1}$ may be replaced with a smaller subset T. Specifically, for $U_0 \subset T \subset \mathbb{R}^{n+1}$, we prove that the calmness of $\mathscr{F}^T$ is equivalent to the calmness of $\mathscr{F}^{\mathbb{R}^{n+1}}$, with the same calmness constants and closely related neighborhoods. We also analyze the particular case when U_0 is the convex hull of some subset in $\mathbb{R}^{n+1}$. Finally, Sect. 5 allows to derive from [5] operative point-based expressions (in terms of the nominal data) for a tight calmness constant for $\mathscr{F}$ and a neighborhood where it works.

2 Preliminaries

Given $X \subset \mathbb{R}^k$, $k \in \mathbb{N}$, we denote by conv$X$ and cone$X$ the *convex hull* and the *conical convex hull* of X, respectively. It is assumed that cone$X$ always contains the zero-vector 0_k, in particular cone$(\emptyset) = \{0_k\}$. If X is a subset of any topological space, int$X$, cl$X$ and bd$X$ stand, respectively, for the *interior*, the *closure* and the *boundary* of X.

2.1 Hausdorff and Chebyshev Distances

The space $CL\left(\mathbb{R}^{n+1}\right)$ will be endowed with the (extended) Hausdorff distance d_H : $CL\left(\mathbb{R}^{n+1}\right) \times CL\left(\mathbb{R}^{n+1}\right) \to [0, +\infty]$ given by

$$d_H\left(U_1, U_2\right) := \max\{e\left(U_1, U_2\right), e\left(U_2, U_1\right)\},$$

where $e\left(U_i, U_j\right)$, $i, j = 1, 2$, represents the *excess of* U_i *over* U_j. Recall that (see [1, Lemma 1.5.1] for the last equality)

$$\begin{aligned} e\left(U_i, U_j\right) &:= \inf\left\{\varepsilon > 0 \mid U_i \subset U_j + \varepsilon\mathbb{B}\right\} \\ &= \sup\left\{d\left(x, U_j\right) \mid x \in U_i\right\} \\ &= \sup\left\{d\left(x, U_j\right) - d\left(x, U_i\right) \mid x \in \mathbb{R}^{n+1}\right\}. \end{aligned} \tag{3}$$

Here $\mathbb{B}$ represents the unit open ball in $\mathbb{R}^{n+1}$ endowed with the norm

$$\left\| \begin{pmatrix} u \\ v \end{pmatrix} \right\| = \max\left\{\|u\|_*, |v|\right\}, \quad \begin{pmatrix} u \\ v \end{pmatrix} \in \mathbb{R}^{n+1}, \tag{4}$$

where $\|\cdot\|_*$ represents the dual norm in $\mathbb{R}^n$ given by $\|u\|_* = \sup_{\|x\| \le 1} u'x$.

For any set T, the space of functions $\left(\mathbb{R}^{n+1}\right)^T$ is endowed with the uniform convergence topology, through the (extended) Chebyshev (supremum) distance $d_\infty : \left(\mathbb{R}^{n+1}\right)^T \times \left(\mathbb{R}^{n+1}\right)^T \to [0, +\infty]$, given by

$$d_\infty\left(\sigma_1, \sigma_2\right) := \sup_{t \in T} \|\sigma_1(t) - \sigma_2(t)\|.$$

From now on, $\mathbb{B}_H\left(U; \varepsilon\right)$ and $\mathbb{B}_\infty\left(\sigma; \varepsilon\right)$ represent the open balls of radius $\varepsilon > 0$ centered at $U \in CL\left(\mathbb{R}^{n+1}\right)$ and $\sigma \in \left(\mathbb{R}^{n+1}\right)^T$, respectively, with respect to the Hausdorff and Chebyshev distances (for the sake of simplicity, $\mathbb{B}_\infty\left(\sigma; \varepsilon\right)$ represents a ball in all spaces $\left(\mathbb{R}^{n+1}\right)^T$, for any T, which will be distinghished by the context).

2.2 Calmness of Multifunctions

Consider a generic multifunction between metric spaces Y and X (with distances denoted indistinctly by d), $\mathscr{M} : Y \rightrightarrows X$. The multifunction $\mathscr{M}$ is said to be *calm* at $(\bar{y}, \bar{x}) \in \text{gph}\mathscr{M}$ if there exist a constant $\kappa \ge 0$ and neighborhoods W of $\bar{x}$ and V of $\bar{y}$ such that

$$d\left(x, \mathscr{M}\left(\bar{y}\right)\right) \le \kappa d\left(y, \bar{y}\right), \quad \text{whenever } x \in \mathscr{M}(y) \cap W \text{ and } y \in V. \tag{5}$$

Sometimes we will be interested in finding some specific neighborhoods and calmness constants; in order to make explicit reference to these elements, we say that $\mathcal{M}$ is *calm at* $(\bar{y}, \bar{x})$ *with constant* κ *on* $V \times W$ when (5) holds.

The calmness property is known to be equivalent to the *metric subregularity* of the inverse multifunction $\mathcal{M}^{-1} : X \rightrightarrows Y$, given by $\mathcal{M}^{-1}(x) := \{y \in Y \mid x \in \mathcal{M}(y)\}$; the metric subregularity of $\mathcal{M}^{-1}$ at $(\bar{x}, \bar{y}) \in \mathrm{gph}\,\mathcal{M}^{-1}$ is stated in terms of the existence of a (possibly smaller) neighborhood W of $\bar{x}$, as well as a constant $\kappa \geq 0$, such that

$$d(x, \mathcal{M}(\bar{y})) \leq \kappa d\left(\bar{y}, \mathcal{M}^{-1}(x)\right), \text{ for all } x \in W. \tag{6}$$

In other words, (6) can be read as: $\mathcal{M}$ is calm at $(\bar{y}, \bar{x})$ with constant κ on $Y \times W$. The reader is addressed to the monographs [9, 12–14] for a comprehensive analysis of these notions among others variational concepts.

The infimum of all possible constants κ in (5) (for some associated W and V) is equal to the infimum of constants κ in (6) and is called the *calmness modulus* of $\mathcal{M}$ at $(\bar{y}, \bar{x})$, denoted as $\mathrm{clm}\,\mathcal{M}\,(\bar{y}, \bar{x})$, defined as ∞ if $\mathcal{M}$ is not calm at $(\bar{y}, \bar{x})$.

3 Calmness via an Indexation Strategy

In this section we discuss three indexation strategies. The first one, the projection strategy $\mathcal{J}$, at a first glance seems to be the most natural, but it turns out not to be adequate as far as

$$d_\infty\left(\mathcal{J}_U, \mathcal{J}_{U_0}\right) \gg d_H(U, U_0) \tag{7}$$

may occur in any neighborhood of a given $U_0 \in CL\left(\mathbb{R}^{n+1}\right)$; where the notation $\gg$ means $\limsup_{U \to U_0}\left(d_\infty\left(\mathcal{J}_U, \mathcal{J}_{U_0}\right)/d_H(U, U_0)\right) = \infty$; see Example 3.1 below. The second strategy, traced out from [6], acts on pairs of closed subsets, say $(U_1, U_2) \mapsto \mathcal{J}_{U_1;U_2}$, and satisfies

$$d_\infty\left(\mathcal{J}_{U_1;U_2}, \mathcal{J}_{U_2;U_1}\right) = d_H(U_1, U_2). \tag{8}$$

The main drawback of this strategy is that, for a given $U_0 \in CL\left(\mathbb{R}^{n+1}\right)$, the indexation of the nominal system U_0 depends on the system U we are comparing with. The third strategy, giving rise to the aimed indexation mapping $\mathcal{I}$, gathers the good features of the other two, as far as it provides an indexation of any system U exclusively in terms of U and the nominal system U_0 and satisfies

$$d_\infty\left(\mathcal{I}_U, \mathcal{I}_{U_0}\right) = d_H(U, U_0) \tag{9}$$

(see Theorem 3.1), which turns out to be enough for the study of the calmness of $\mathcal{F}$ at (U_0, x_0) for any given $x_0 \in \mathcal{F}(U_0)$.

Hereafter in the paper we consider a given nominal set (or system) $U_0 \in CL\left(\mathbb{R}^{n+1}\right)$ and an arbitrarily chosen selection, P, of the *metric projection* multifunction, Π : $\mathbb{R}^{n+1} \times CL\left(\mathbb{R}^{n+1}\right) \rightrightarrows \mathbb{R}^{n+1}$, which is given by

$$\Pi\left(t, U\right) := \{z \in U \mid \|t - z\| = d\left(t, U\right)\}, \quad \left(t, U\right) \in \mathbb{R}^{n+1} \times CL\left(\mathbb{R}^{n+1}\right).$$

Observe that $\Pi\left(t, U\right)$ is always non-empty by the closedness of U. For simplicity we will write $P_U\left(t\right)$ instead of $P\left(t, U\right)$.

3.1   The Projection Strategy

We define $\mathscr{J} : CL\left(\mathbb{R}^{n+1}\right) \to \left(\mathbb{R}^{n+1}\right)^{\mathbb{R}^{n+1}}$ as $\mathscr{J}_U := P_U$ for all $U \in CL\left(\mathbb{R}^{n+1}\right)$. Now we are going to show an example where (7) happens even for compact convex sets.

Example 3.1 Consider $\mathbb{R}^2$ endowed with the Euclidean norm and let $\mathbb{R}^3$ be equipped with the norm (4).

$$U_0 := \{(x_1, x_2, 0)' \in \mathbb{R}^3 \mid 0 \leq x_1 \leq 1, \ 0 \leq x_2 \leq x_1^{2/3}\}$$

and pick any $\varepsilon > 0$. If we move from $\left(\varepsilon, \varepsilon^{2/3}, 0\right)'$ orthogonally to the surface $x_2 = x_1^{2/3}$ until we meet the plane $x_1 = 0$, then we reach $u_\varepsilon := \left(0, \varepsilon^{2/3} + \frac{3}{2}\varepsilon^{4/3}, 0\right)'$. If our orthogonal movement starts at $\left(\frac{8}{27}\varepsilon, \frac{4}{9}\varepsilon^{2/3}, 0\right)'$ and ends at the plane $x_1 = -\varepsilon^{1/3}$, then we reach $z_\varepsilon := \left(-\varepsilon^{1/3}, \frac{13}{9}\varepsilon^{2/3} + \frac{8}{27}\varepsilon^{4/3}, 0\right)'$. For each $\varepsilon > 0$ let

$$U_\varepsilon = \text{conv}\left(U_0 \cup \{u_\varepsilon\}\right).$$

In this case, for $0 < \varepsilon \leq (24/65)^{3/2}$ in order to guarantee $\frac{13}{9}\varepsilon^{2/3} + \frac{8}{27}\varepsilon^{4/3} \geq \varepsilon^{2/3} + \frac{3}{2}\varepsilon^{4/3}$, we have

$$P_{U_0}\left(z_\varepsilon\right) = \left(\frac{8}{27}\varepsilon, \frac{4}{9}\varepsilon^{2/3}, 0\right)' \quad \text{and} \quad P_{U_\varepsilon}\left(z_\varepsilon\right) = u_\varepsilon.$$

Accordingly, as $\varepsilon \downarrow 0$ we have

$$d_\infty\left(\mathscr{J}_{U_\varepsilon}, \mathscr{J}_{U_0}\right) \geq \left\|P_{U_\varepsilon}\left(z_\varepsilon\right) - P_{U_0}\left(z_\varepsilon\right)\right\| \approx \frac{5}{9}\varepsilon^{2/3},$$

$$d_H\left(U_\varepsilon, U_0\right) = \left\|\left(0, \varepsilon^{2/3} + \frac{3}{2}\varepsilon^{4/3}, 0\right)' - \left(\varepsilon, \varepsilon^{2/3}, 0\right)'\right\| \approx \varepsilon,$$

$$d\left(z_\varepsilon, U_0\right) \approx \varepsilon^{1/3},$$

where $\approx$ means (as usual) that the quotient between left-hand and right-hand sides tends to 1 as $\varepsilon \downarrow 0$. This clearly entails $d_\infty \left(\mathscr{J}_{U_\varepsilon}, \mathscr{J}_{U_0} \right) \gg d_H \left(U_\varepsilon, U_0 \right)$, even if $\mathbb{R}^{n+1}$ is replaced with any neighborhood of U_0 (i.e., a set of the form $U_0 + \delta \mathbb{B}$, for any $\delta > 0$).

3.2 The Pairwise Strategy

The following indexation strategy is inspired in [6, Theorem 4.2]. For each $U_1, U_2 \in CL\left(\mathbb{R}^{n+1} \right)$ let us define $\mathscr{J}_{U_1;U_2} \in \left(\mathbb{R}^{n+1} \right)^{\mathbb{R}^{n+1}}$ given by

$$\mathscr{J}_{U_1;U_2}(t) := \begin{cases} P_{U_1}(t) & \text{if } t \in U_1 \cup U_2, \\ \binom{0_n}{1} & \text{if } t \notin U_1 \cup U_2. \end{cases}$$

Observe that $\binom{0_n}{1}$ is associated with the trivial inequality $0'_n x \leq 1$.

The proof of (8) for this pairwise indexation mapping is essentially given in [6, Theorem 4.2], although in that theorem the uncertainty is confined to the left-hand-side coefficients. Example 3.1 shows that points $t \notin U_1 \cup U_2$ may 'spoil' $d_\infty \left(\mathscr{J}_{U_1}, \mathscr{J}_{U_2} \right)$ in relation to the projection strategy. As said at the beginning of this section, the main drawback of the current pairwise strategy is that the indexation of the nominal system U_0 depends on the system U we are comparing with. In other words, when U varies around a fixed U_0, the indexations of U_0 vary with U, so that we cannot apply the literature background to a fixed $\sigma_0 \in \left(\mathbb{R}^{n+1} \right)^T$. Recalling the indexation mapping $\mathscr{J}$ providing the projection strategy in Sect. 3.1, we immediately observe that $\mathscr{J}_U$ coincides with our current $\mathscr{J}_{U;\mathbb{R}^{n+1}}$.

3.3 The U_0-Based Strategy

Now we are going to define the indexation function $\mathscr{I}$ announced at the beginning of this section. Recall that we are considering a given nominal set $U_0 \in CL\left(\mathbb{R}^{n+1} \right)$, although, for the sake of simplicity, the notation does not reflect the dependence on U_0. We define $\mathscr{I} : CL\left(\mathbb{R}^{n+1} \right) \to \left(\mathbb{R}^{n+1} \right)^{\mathbb{R}^{n+1}}$ as follows: For each $U \in CL\left(\mathbb{R}^{n+1} \right)$, let $\sigma := \mathscr{I}_U \in \left(\mathbb{R}^{n+1} \right)^{\mathbb{R}^{n+1}}$ be given by

$$\sigma(t) := \begin{cases} t, & \text{if } t \in U, \\ (P_U \circ P_{U_0})(t), & \text{if } t \notin U. \end{cases} \tag{10}$$

In this way, one easily checks that $\mathrm{rge}\, \mathscr{I}_U = U$, for all $U \in CL\left(\mathbb{R}^{n+1} \right)$, Obviously, $\mathscr{I}_{U_0} = P_{U_0}$. Next we establish (9).

Theorem 3.1 *Let $\mathscr{I} : CL\left(\mathbb{R}^{n+1}\right) \to \left(\mathbb{R}^{n+1}\right)^{\mathbb{R}^{n+1}}$ be the indexation function defined in (10) and let $\sigma_0 := \mathscr{I}_{U_0}$. Then,*

$$d_\infty\left(\sigma, \sigma_0\right) = d_H\left(U, U_0\right), \quad \text{whenever } \sigma = \mathscr{I}_U, \ U \in CL\left(\mathbb{R}^{n+1}\right). \tag{11}$$

Proof Consider any $U \in CL\left(\mathbb{R}^{n+1}\right)$. In order to establish the inequality '$\leq$' in (11), take any $t \in \mathbb{R}^{n+1}$ and distinguish two cases: If $t \in U$, then

$$\left\|\sigma\left(t\right) - \sigma_0\left(t\right)\right\| = \left\|t - P_{U_0}\left(t\right)\right\| = d\left(t, U_0\right) \leq e\left(U, U_0\right),$$

where we have taken (3) into account. Otherwise, if $t \notin U$, then

$$\left\|\sigma\left(t\right) - \sigma_0\left(t\right)\right\| = \left\|(P_U \circ P_{U_0})(t) - P_{U_0}\left(t\right)\right\|$$
$$= d\left(P_{U_0}\left(t\right), U\right) \leq e\left(U_0, U\right).$$

So, $\left\|\sigma\left(t\right) - \sigma_0\left(t\right)\right\| \leq d_H\left(U, U_0\right)$, for all $t \in \mathbb{R}^{n+1}$, and then $d_\infty\left(\sigma, \sigma_0\right) \leq d_H\left(U, U_0\right)$. Let us see the opposite inequality. We have that

$$e\left(U, U_0\right) = \sup_{t \in U} d\left(t, U_0\right) = \sup_{t \in U} d\left(t, P_{U_0}\left(t\right)\right)$$
$$= \sup_{t \in U} d\left(\sigma\left(t\right), \sigma_0\left(t\right)\right) \leq d_\infty\left(\sigma, \sigma_0\right).$$

$$e\left(U_0, U\right) = \sup_{t \in U_0} d\left(t, U\right) = \sup_{t \in U_0} d\left(t, P_U\left(t\right)\right)$$
$$= \sup_{t \in U_0} d\left(P_{U_0}\left(t\right), P_U\left(P_{U_0}\left(t\right)\right)\right)$$
$$= \sup_{t \in U_0} d\left(\sigma_0\left(t\right), \sigma\left(t\right)\right) \leq d_\infty\left(\sigma, \sigma_0\right).$$

Consequently, $d_\infty\left(\sigma, \sigma_0\right) \geq d_H\left(U, U_0\right)$. $\qquad\square$

Finally, the following result formalizes the fact that the calmness of $\mathscr{F}^{\mathbb{R}^{n+1}}$ turns out to be equivalent to the calmness of $\mathscr{F}$, with the same constants and closely related neighborhoods.

Theorem 3.2 *Let $x_0 \in \mathscr{F}\left(U_0\right)$, $W \subset \mathbb{R}^n$ be a neighborhood of x_0, and $\sigma_0 = \mathscr{I}_{U_0}$. Then $\mathscr{F}^{\mathbb{R}^{n+1}}$ is calm at (σ_0, x_0) with constant $\kappa \geq 0$ on $\mathbb{B}_\infty\left(\sigma_0; \varepsilon\right) \times W$ if and only if $\mathscr{F}$ is calm at (U_0, x_0) with the same constant κ on $\mathbb{B}_H\left(U_0; \varepsilon\right) \times W$.*

Proof First assume that $\mathscr{F}^{\mathbb{R}^{n+1}}$ is calm at (σ_0, x_0) with constant κ on $\mathbb{B}_\infty\left(\sigma_0; \varepsilon\right) \times W$. From Theorem 3.1 we get $\mathscr{I}^{-1}\left(\mathbb{B}_\infty\left(\sigma_0; \varepsilon\right)\right) = \mathbb{B}_H\left(U_0; \varepsilon\right)$. Take any $(U, x) \in \mathbb{B}_H\left(U_0; \varepsilon\right) \times W$, such that $x \in \mathscr{F}\left(U\right)$ and let $\sigma = \mathscr{I}_U \in \mathbb{B}_\infty\left(\sigma_0; \varepsilon\right)$.
Then, applying Theorem 3.1 we have

$$d\left(x, \mathscr{F}\left(U_0\right)\right) = d\left(x, \mathscr{F}^{\mathbb{R}^{n+1}}\left(\sigma_0\right)\right) \leq \kappa d_\infty\left(\sigma, \sigma_0\right) = \kappa d_H\left(U, U_0\right).$$

On the other hand, assume that $\mathscr{F}$ is calm at (U_0, x_0) with constant κ on $\mathbb{B}_H(U_0; \varepsilon) \times W$. Picking any $\sigma \in \mathbb{B}_\infty(\sigma_0; \varepsilon)$ and defining $U := \mathrm{cl}\sigma\left(\mathbb{R}^{n+1}\right)$, i.e., $U = \mathrm{cl}\{\sigma(t), t \in \mathbb{R}^{n+1}\}$, it is clear from the definitions that $d_\infty(\sigma, \sigma_0) \geq d_H(U, U_0)$. More in detail, for each $t \in \mathbb{R}^{n+1}$ we have $\|\sigma(t) - \sigma_0(t)\| \geq d(\sigma(t), U_0)$ and, accordingly,

$$d_\infty(\sigma, \sigma_0) = \sup_{t \in \mathbb{R}^{n+1}} \|\sigma(t) - \sigma_0(t)\| \geq \sup_{t \in \mathbb{R}^{n+1}} d(\sigma(t), U_0) = e(U, U_0).$$

In a completely analogous way se obtain $d_\infty(\sigma, \sigma_0) \geq e(U_0, U)$. Consequently $d_H(U, U_0) \leq \varepsilon$ and

$$d\left(x, \mathscr{F}^{\mathbb{R}^{n+1}}(\sigma_0)\right) = d(x, \mathscr{F}(U_0)) \leq \kappa d_H(U, U_0) \leq \kappa d_\infty(\sigma, \sigma_0).$$

$$\square$$

4 Calmness and Minimal Indexations

This section tackles the question of replacing $\mathbb{R}^{n+1}$ with a smaller index set. In fact, keeping the notation of the previous sections, if we consider $U_0 \in CL\left(\mathbb{R}^{n+1}\right)$, and the corresponding indexed system $\sigma_0 = \mathscr{I}_{U_0} \in \left(\mathbb{R}^{n+1}\right)^{\mathbb{R}^{n+1}}$, we wonder if U_0 itself could play the role of the index set, yielding to a certain minimal indexation (where repetitions of constraints are eliminated).

Let us consider $U_0 \subset T \subset \mathbb{R}^{n+1}$, and the corresponding feasible set mapping, $\mathscr{F}^T : \left(\mathbb{R}^{n+1}\right)^T \rightrightarrows \mathbb{R}^n$ defined in (2). From now on $\sigma_0|_T : T \to \mathbb{R}^{n+1}$ represents the usual rectriction of function σ_0 to the domain T. Obviously

$$\sigma_0|_{U_0}(t) = \sigma_0|_T(t) = \sigma_0(t) = P_{U_0}(t) = t, \text{ for all } t \in U_0.$$

Accordingly, rge $\sigma_0|_{U_0} = $ rge $\sigma_0|_T = $ rge$\sigma_0 = U_0$, which entails

$$\mathscr{F}^{U_0}\left(\sigma_0|_{U_0}\right) = \mathscr{F}^T(\sigma_0|_T) = \mathscr{F}^{\mathbb{R}^{n+1}}(\sigma_0) = \mathscr{F}(U_0).$$

Roughly speaking, $\sigma_0|_{U_0}$, $\sigma_0|_T$ and σ_0 correspond to three systems with different index sets but having the same coefficient vector set, U_0. So, $\sigma_0|_T$ and σ_0 are formed by the same inequalities as $\sigma_0|_{U_0}$ but with different amount of repetitions. In order to identify the repetitions of constraints in σ_0, we define the following sets of indices:

$$R_{t_0} := \left\{t \in \mathbb{R}^{n+1} \mid \sigma_0(t) = t_0\right\}, \ t_0 \in U_0;$$

so, $t \in R_{t_0}$ is indexing an inequality which is a repetition of the one associated with $t_0 \in U_0$. Clearly $\{R_{t_0}\}_{t_0 \in U_0}$ constitutes a partition of $\mathbb{R}^{n+1}$.

Theorem 4.1 *Let $U_0, T \in CL\left(\mathbb{R}^{n+1}\right)$ with $U_0 \subset T$. Let $x_0 \in \mathscr{F}(U_0)$ and $W \subset \mathbb{R}^n$ be a neighborhood of x_0. Let $\sigma_0 = \mathscr{I}_{U_0}$. Then, the following conditions are equivalent:*

(i) $\mathscr{F}^{\mathbb{R}^{n+1}}$ is calm at (σ_0, x_0) with constant κ on $\mathbb{B}_\infty(\sigma_0; \varepsilon) \times W$;
(ii) $\mathscr{F}^T$ is calm at $(\sigma_0|_T, x_0)$ with constant κ on $\mathbb{B}_\infty(\sigma_0|_T; \varepsilon) \times W$;
(iii) $\mathscr{F}^{U_0}$ is calm at $\left(\sigma_0|_{U_0}, x_0\right)$ with constant κ on $\mathbb{B}_\infty\left(\sigma_0|_{U_0}; \varepsilon\right) \times W$.

Moreover, in the case when $U_0 = \text{conv}(T_0)$ with $T_0 \in CL\left(\mathbb{R}^{n+1}\right)$, the following condition is also equivalent to the previous ones:

(iv) $\mathscr{F}^{T_0}$ is calm at $\left(\sigma_0|_{T_0}, x_0\right)$ with constant κ on $\mathbb{B}_\infty\left(\sigma_0|_{T_0}; \varepsilon\right) \times W$.

Proof $(i) \Rightarrow (ii)$. Consider any $(\sigma, x) \in \text{gph} \mathscr{F}^T \cap (\mathbb{B}_\infty(\sigma_0|_T; \varepsilon) \times W)$ and let us see that $d\left(x, \mathscr{F}^T(\sigma_0|_T)\right) \leq \kappa d_\infty(\sigma, \sigma_0|_T)$. Define $\widetilde{\sigma} \in \left(\mathbb{R}^{n+1}\right)^{\mathbb{R}^{n+1}}$ as an extension of σ in the following natural way:

$$\widetilde{\sigma}(t) := \sigma(t_0), \text{ whenever } t \in R_{t_0} \backslash T, \ t_0 \in U_0.$$

Note that, $\widetilde{\sigma}(t)$ is well defined since for each $t \in \mathbb{R}^{n+1}$ there exists a unique $t_0 \in U_0$ such that $t \in R_{t_0}$ (because of the definition of R_{t_0}).

In this way, one easily checks that $\mathscr{F}^T(\sigma) = \mathscr{F}^{\mathbb{R}^{n+1}}(\widetilde{\sigma})$ and $d_\infty(\widetilde{\sigma}, \sigma_0) = d_\infty(\sigma, \sigma_0|_T) < \varepsilon$. In fact, for each $t_0 \in U_0$ and each $\widetilde{t} \in R_{t_0} \backslash T$ we have

$$\left\|\widetilde{\sigma}\left(\widetilde{t}\right) - \sigma_0\left(\widetilde{t}\right)\right\| = \|\sigma(t_0) - \sigma_0(t_0)\| \leq \sup_{t \in U_0} \|\sigma(t) - \sigma_0(t)\|$$

$$\leq \sup_{t \in T} \|\sigma(t) - \sigma_0(t)\| = d_\infty(\sigma, \sigma_0|_T).$$

Accordingly, $d_\infty(\widetilde{\sigma}, \sigma_0) = d_\infty(\sigma, \sigma_0|_T)$. Then, applying (i) we have our aimed inequality

$$d\left(x, \mathscr{F}^T(\sigma_0|_T)\right) = d\left(x, \mathscr{F}^{\mathbb{R}^{n+1}}(\sigma_0)\right) \leq \kappa d_\infty(\widetilde{\sigma}, \sigma_0) = k d_\infty(\sigma, \sigma_0|_T).$$

$(ii) \Rightarrow (iii)$. It is completely analogous to $(i) \Rightarrow (ii)$.

$(iii) \Rightarrow (i)$. Assume (iii), take any $(\sigma, x) \in \text{gph} \mathscr{F}^{\mathbb{R}^{n+1}} \cap (\mathbb{B}_\infty(\sigma_0; \varepsilon) \times W)$, and let us show that

$$d\left(x, \mathscr{F}^{\mathbb{R}^{n+1}}(\sigma_0)\right) \leq \kappa d_\infty(\sigma, \sigma_0).$$

Since $\sigma|_{U_0}$ may be seen as a subsystem of system σ, we immediately have that

$$\mathscr{F}^{\mathbb{R}^{n+1}}(\sigma) \subset \mathscr{F}^{U_0}\left(\sigma|_{U_0}\right) \text{ and } d_\infty\left(\sigma|_{U_0}, \sigma_0|_{U_0}\right) \leq d_\infty(\sigma, \sigma_0) < \varepsilon.$$

So, we have $\left(\sigma|_{U_0}, x\right) \in \text{gph} \mathscr{F}^{U_0} \cap \left(\mathbb{B}_\infty\left(\sigma_0|_{U_0}; \varepsilon\right) \times W\right)$ and

$$d\left(x, \mathscr{F}^{\mathbb{R}^{n+1}}(\sigma_0)\right) = d\left(x, \mathscr{F}^{U_0}\left(\sigma_0|_{U_0}\right)\right) \leq k d_\infty\left(\sigma|_{U_0}, \sigma_0|_{U_0}\right) \leq k d_\infty(\sigma, \sigma_0).$$

From now on we assume that $U_0 = \mathrm{conv}\,(T_0)$ for some $T_0 \in CL\left(\mathbb{R}^{n+1}\right)$. In this case, we are going to establish $(iii) \Leftrightarrow (iv)$.

$(iii) \Rightarrow (iv)$. Assume that $\mathscr{F}^{U_0}$ is calm at $\left(\sigma_0|_{U_0}, x_0\right)$ with constant κ on $\mathbb{B}_\infty\left(\sigma_0|_{U_0}; \varepsilon\right) \times W$.

Take any $(\sigma, x) \in \mathrm{gph}\mathscr{F}^{T_0} \cap \left(\mathbb{B}_\infty\left(\sigma_0|_{T_0}; \varepsilon\right) \times W\right)$ and let us see that

$$d\left(x, \mathscr{F}^{T_0}\left(\sigma_0|_{T_0}\right)\right) \le \kappa d_\infty\left(\sigma, \sigma_0|_{T_0}\right).$$

To do that, we are going to define an appropiate extension of $\sigma \in \left(\mathbb{R}^{n+1}\right)^{T_0}$, say $\widetilde\sigma$, to the domain U_0. Let $\mathbb{R}_+^{(T_0)}$ denote the set of functions from T_0 to $\mathbb{R}_+$ which are zero except at finitely many elements of T_0. For each $t \in T_0$ we define $\lambda^t = \left(\lambda_{t_0}^t\right)_{t_0 \in T_0}$ by $\lambda_{t_0}^t = 1$ if $t_0 = t$ and $\lambda_{t_0}^t = 0$ otherwise. For each $t \in U_0 \backslash T_0$, recalling $U_0 = \mathrm{conv}\,(T_0)$, choose arbitrarily $\lambda^t \in \mathbb{R}_+^{(T_0)}$ satisfying $t = \sum_{t_0 \in T_0} \lambda_{t_0}^t t_0$, and $\sum_{t_0 \in T_0} \lambda_{t_0}^t = 1$. Then define

$$\widetilde\sigma\,(t) := \sum_{t_0 \in T_0} \lambda_{t_0}^t \sigma\,(t_0), \quad \text{for all } t \in U_0.$$

In this way, any inequality in $\widetilde\sigma$ is a consequence of σ, and σ is a subsystem of $\widetilde\sigma$. Therefore, $\mathscr{F}^{U_0}\left(\widetilde\sigma\right) = \mathscr{F}^{T_0}\left(\sigma\right)$. Moreover

$$d_\infty\left(\widetilde\sigma, \sigma_0|_{U_0}\right) = \sup_{t \in U_0} \left\|\widetilde\sigma\,(t) - \sigma_0|_{U_0}\,(t)\right\| = \sup_{t \in U_0} \left\|\sum_{t_0 \in T_0} \lambda_{t_0}^t \sigma\,(t_0) - t\right\|$$

$$= \sup_{t \in U_0} \left\|\sum_{t_0 \in T_0} \lambda_{t_0}^t(\sigma\,(t_0) - t_0)\right\|$$

$$= \sup_{t \in U_0} \left\|\sum_{t_0 \in T_0} \lambda_{t_0}^t(\sigma\,(t_0) - \sigma_0|_{T_0}\,(t_0))\right\| = d_\infty\left(\sigma, \sigma_0|_{T_0}\right).$$

The last equality comes from the triangular inequality together with the definiton of λ^t for $t \in T_0$. Consequently, $(\widetilde\sigma, x) \in \mathrm{gph}\mathscr{F}^{U_0} \cap \left(\mathbb{B}_\infty\left(\sigma_0|_{U_0}; \varepsilon\right) \times W\right)$ and, then,

$$d\left(x, \mathscr{F}^{T_0}\left(\sigma_0|_{T_0}\right)\right) = d\left(x, \mathscr{F}^{U_0}\left(\sigma_0|_{U_0}\right)\right) \le k d_\infty\left(\widetilde\sigma, \sigma_0|_{U_0}\right) = k d_\infty\left(\sigma, \sigma_0|_{T_0}\right).$$

Finally, the proof of $(iv) \Rightarrow (iii)$ follows exactly the same argument as $(iii) \Rightarrow (i)$ just by replacing $\mathbb{R}^{n+1}$ and U_0 with U_0 and T_0, respectively. $\qquad\square$

5 Calmness Constants for Polyhedral Uncertainty Sets

Throughout this section we assume that $U_0 := \text{conv}(T_0)$, where $\emptyset \neq T_0 \subset \mathbb{R}^{n+1}$ is a finite set, say

$$T_0 := \left\{ \begin{pmatrix} \overline{a}_i \\ \overline{b}_i \end{pmatrix} : i = 1, \ldots, m \right\},$$

with m standing for the cardinality of T_0 (i.e., there are no repetitions). Obviously, as an index set T_0 can be identified with $\{1, \ldots, m\}$. Let us denote by $\mathscr{F}_{\overline{a}}^{T_0} : \mathbb{R}^m \rightrightarrows \mathbb{R}^n$ the feasible set mapping associated with the system

$$\left\{ \overline{a}_i' x \leq b_i, i = 1, \ldots, m \right\},$$

with $b = (b_i)_{i=1,\ldots,m}$ being the parameter to be perturbed around $\overline{b}$. Theorem 4 in [4] provides a point-based formula (depending exclusively on the nominal data) for $\text{clm}\mathscr{F}_{\overline{a}}^{T_0}\left(\overline{b}, x_0\right)$, with $\left(\overline{b}, x_0\right) \in \text{gph}\mathscr{F}_{\overline{a}}^{T_0}$. Further, [5, Theorem 3] provides a point-based neighborhood $U_{\overline{b}}(x_0)$ such that $\text{clm}\mathscr{F}_{\overline{a}}^{T_0}\left(\overline{b}, x_0\right)$ is indeed a calmness constant for $\mathscr{F}_{\overline{a}}^{T_0}$ at $\left(\overline{b}, x_0\right)$ on $\mathbb{R}^m \times U_{\overline{b}}(x_0)$; see also the comment just after (6). Denoting $\overline{\sigma} = \begin{pmatrix} \overline{a}_i \\ \overline{b}_i \end{pmatrix}_{i=1,\ldots,m} \in \left(\mathbb{R}^{n+1}\right)^m \equiv \left(\mathbb{R}^{n+1}\right)^{T_0}$, Theorem 5.1 below provides a way to construct, from [5, Theorem 3], a calmness constant for $\mathscr{F}^{T_0}$ at $(\overline{\sigma}, x_0)$ on a certain neighborhood of $(\overline{\sigma}, x_0)$, which is also provided by Theorem 5.1.

Theorem 5.1 *Assume that $\kappa \geq 0$ is a calmness constant for $\mathscr{F}_{\overline{a}}^{T_0}$ at $\left(\overline{b}, x_0\right) \in \text{gph}\mathscr{F}_{\overline{a}}^{T_0}$ on $\mathbb{R}^{T_0} \times W$, where W is a neighborhood of x_0. Then, for any given $\varepsilon > 0$ and $\overline{\sigma}$ being defined as above, $\kappa\left(\|x_0\| + 1 + \varepsilon\right)$ is a calmness constant for $\mathscr{F}^{T_0}$ at $(\overline{\sigma}, x_0)$ on $\left(\mathbb{R}^{n+1}\right)^{T_0} \times (W \cap \mathbb{B}(x_0, \varepsilon))$.*

Proof Lemma 10 in [3] establishes, for our norm choice (4),

$$d\left(\overline{\sigma}, \left(\mathscr{F}^{T_0}\right)^{-1}(x)\right) = \frac{\max_{i \in \{1,\ldots,m\}}\left[\overline{a}_i' x - \overline{b}_i\right]_+}{\|x\| + 1} \quad \text{for all } x \in \mathbb{R}^n,$$

where $[\alpha]_+ := \max\{\alpha, 0\}$ stands for the positive part of $\alpha \in \mathbb{R}$. Also observe that $\max_{i \in \{1,\ldots,m\}}\left[\overline{a}_i' x - \overline{b}_i\right]_+$ may be written as $d_\infty\left(\overline{b}, \left(\mathscr{F}_{\overline{a}}^{T_0}\right)^{-1}(x)\right)$.

Accordingly, for all $x \in W \cap \mathbb{B}(x_0, \varepsilon)$ we have

$$d\left(x, \mathscr{F}^{T_0}(\overline{\sigma})\right) = d\left(x, \mathscr{F}_{\overline{a}}^{T_0}(\overline{b})\right) \leq \kappa d_\infty\left(\overline{b}, \left(\mathscr{F}_{\overline{a}}^{T_0}\right)^{-1}(x)\right)$$

$$= \left(\|x\| + 1\right)\kappa \frac{\max_{i \in \{1,\dots,m\}}\left[\overline{a}_t' x - \overline{b}_i\right]_+}{\|x\| + 1}$$

$$\leq \kappa\left(\|x_0\| + 1 + \varepsilon\right) d_\infty\left(\overline{\sigma}, \left(\mathscr{F}^{T_0}\right)^{-1}(x)\right). \qquad \square$$

Remark 5.1 (*i*) A straightforward combination of Theorems 3.2, 4.1 and 5.1 provides a calmness constant and an asociated neighborhood for $\mathscr{F}$ at (U_0, x_0). For comparative purposes see also [4, Theorem 5] in relation to $\mathrm{clm}\mathscr{F}^{T_0}(\overline{\sigma}, x_0)$.

(*ii*) The previous theorem may be applied in the context when T_0 is our nominal (finite) system and the uncertainty on these coefficient vectors leads to a robust counterpart where each coefficient $\binom{\overline{a}_i}{\overline{b}_i}$, $i = 1, \dots, m$, may move in a closed box centered at such a point. In this way, in the robust counterpart of T_0 we may replace the union of such boxes with its convex hull, which is a polyhedral set. Also observe that the perturbed uncertainty sets need not to be polyhedral.

References

1. Beer G (1993) Topologies on closed and closed convex sets. Kluwer Acad Pub, Dordrecht
2. Cánovas MJ, Henrion R, López MA, Parra J (2016) Outer limit of subdifferentials and calmness moduli in linear and nonlinear programming. J Optim Theor Appl 169:925–952
3. Cánovas MJ, López MA, Parra J, Toledo FJ (2005) Distance to ill-posedness and the consistency value of linear semi-infinite inequality systems. Math Program Ser A 103:95–126
4. Cánovas MJ, López MA, Parra J, Toledo FJ (2014) Calmness of the feasible set mapping for linear inequality systems. Set-Valued Var Anal 22:375–389
5. Cánovas MA, Parra J, Rückmann JJ, Toledo FJ (2017) Point-based neighborhoods for sharp calmness constants in linear programming. Set-Valued Var Anal 25:757–772. https://doi.org/10.1007/s11228-017-0427-6
6. Chan TCY, Mar PA (2017) Stability and continuity in robust optimization. SIAM J Optim 27:817–841
7. Chuong TD, Jeyakumar V (2016) Characterizing robust local error bounds for linear inequality systems under data uncertainty. Linear Algebra Appl 489:199–216
8. Chuong TD, Jeyakumar V (2016) Robust global error bounds for uncertain linear inequality systems with applications. Linear Algebra Appl 493:183–205
9. Dontchev AL, Rockafellar RT (2009) Implicit functions and solution mappings: a view from variational analysis. Springer, New York
10. Goberna MA, Jeyakumar V, Li G, López MA (2013) Robust linear semi-infinite programming under uncertainty. Math Program B 139:185–203
11. Goberna MA, López MA (1998) Linear semi-infinite optimization. Wiley, Chichester
12. Klatte D, Kummer B (2002) Nonsmooth equations in optimization: regularity, calculus, methods and applications. Series nonconvex optimization and applications, vol 60. Kluwer Acad Pub, Dordrecht
13. Mordukhovich BS (2006) Variational analysis and generalized differentiation, I: basic theory. Springer, Berlin
14. Rockafellar RT, Wets RJ-B (1998) Variational analysis. Springer, Berlin

The LFM Data Qualification in Convex Multiobjective Semi-infinite Programming

Miguel Ángel Goberna, Margarita M. L. Rodríguez
and Virginia N. Vera de Serio

Abstract Given a semi-infinite multiobjective convex problem we introduce a data qualification that enables to characterize optimality in terms of Lagrange multipliers. We show that this condition characterizes the weak efficient solutions through the weak Karush-Kuhn-Tucker (KKT) condition, and identifies the proper efficient solutions through the strong KKT condition. We also address the question in relation to a gap function.

1 Introduction

We consider convex multiobjective semi-infinite programming (MOSIP) problems in $\mathbb{R}^n$ of the type:

$$(P) \text{ minimize } f(x) = \left(f_1(x), ..., f_p(x)\right) \text{ subject to } g_t(x) \leq 0, \ t \in T,$$

where the set of indices T is possibly infinite, the criteria (or objective) functions $f_i : \mathbb{R}^n \to \mathbb{R}, i = 1, ..., p$, are finite-valued and convex, and the constraint functions $g_t : \mathbb{R}^n \to \mathbb{R} \cup \{+\infty\}$, $t \in T$, are extended real-valued, lower semi-continuous, proper and convex. The space $\mathbb{R}^n$ is the decision space, while $\mathbb{R}^p$ is the objective (or criterion) space. We only consider feasible problems (P), and S will denote its non-empty feasible set; clearly S is a closed convex set.

This type of problems arise in a natural way in decision making situations under uncertainty. For instance, the classical portfolio problem can be seen as a particular instance of (P) where $p = 2$, the vector x denotes a portfolio composed by n different assets, $f_1(x)$ and $f_2(x)$ represent the expected return of x (a linear function)

M. Á. Goberna (✉) · M. M. L. Rodríguez
Dept of Mathematics, Universidad de Alicante, Alicante, Spain
e-mail: mgoberna@ua.es

M. M. L. Rodríguez
e-mail: marga.rodriguez@ua.es

V. N. Vera de Serio
Faculty of Economics, Universidad Nacional de Cuyo, Mendoza, Argentina
e-mail: virginia.vera@fce.uncu.edu.ar

© Springer International Publishing AG 2018
E. Gil et al. (eds.), *The Mathematics of the Uncertain*, Studies in Systems, Decision and Control 142, https://doi.org/10.1007/978-3-319-73848-2_77

845

and its variance (a quadratic convex function), respectively, while the constraint functions g_j, $j \in J$, are usually affine. These problems cannot be replaced by a unique scalarization due to the uncertain risk-aversion of the decision makers (typically, a group of managers of an investment fund). In practice, some of the involved functions, f_1, f_2, g_j, are uncertain. Adopting the robust approach, one assumes that $f_i : \mathbb{R}^n \times U_i \to \mathbb{R}$ and $g_j : \mathbb{R}^n \times V_j \to \mathbb{R}$ are given functions, where U_i, $i = 1, 2$, and V_j, $j \in J$, are the (possibly infinite or singleton) uncertainty sets. The pessimistic (or worst-case) counterpart of the portfolio problem is then formulated as the MOSIP problem

$$(P) \ \text{minimize} \ \left(\sup_{u_1 \in U_1} f_1(x, u_1), \sup_{u_2 \in U_2} f_2(x, u_2) \right)$$
$$\text{subject to} \ g_j(x, v_j) \leq 0, \ v_j \in V_j, j \in J,$$

where the objective functions are real-valued and convex whenever $f_i(x, \cdot)$ is continuous on U_i for all x and U_i is a compact topological space, $i = 1, 2$, while the linear constraints are now indexed by the infinite set $\bigcup_{j \in J} V_j$.

We consider different types of optimal solutions for (P) : efficient (or Pareto) solutions, weak efficient solutions, and proper efficient solutions. For the corresponding definitions, we need to fix the following notation: given $y, z \in \mathbb{R}^p$, we write $y \leq z$ when $y_i \leq z_i$ for all $i = 1, ..., p$, $y \leq z$ when $y \leq z$ and $y \neq z$, and $y < z$ when $y_i < z_i$ for all $i = 1, ..., p$.

Consider a point $\widehat{x} \in S$. Then $\widehat{x}$ is an *efficient (weak efficient) solution* for (P) if there is no $x \in S$ such that $f(x) \leq f(\widehat{x})$ $(f(x) < f(\widehat{x}))$. Furthermore, following Geoffrion's definition, $\widehat{x}$ is a *proper efficient solution* if there exists $M > 0$ such that for any $i \in \{1, ..., p\}$ and any $x \in S$ with $f_i(x) < f_i(\widehat{x})$, there is some $j \in \{1, ..., p\}$ such that $f_j(x) > f_j(\widehat{x})$ and

$$\frac{f_i(\widehat{x}) - f_i(x)}{f_j(x) - f_j(\widehat{x})} < M.$$

By the convexity assumptions on (P), all concepts of proper efficient solution spread in the literature are equivalent to this one (see e.g. [2]). Clearly any proper efficient solution is an efficient (or Pareto) solution, and every efficient solution is weak efficient.

As in scalar optimization, the characterization of the optimal solutions of (P) in terms of Lagrange multipliers requires the fulfillment of some conditions by the data (the functions f_i and g_t) which are generically called *data qualifications* (DQs in short) or, more specifically, *constraint qualifications* (CQs) when they only involve the constraint functions. Recent reviews can be found in [9] (on CQs in scalar convex optimization), [3] (on CQs in *continuous* convex MOSIP, meaning that the index set T is a compact topological space and the function $(x, t) \mapsto g_t(x)$ is continuous) and [4] (on DQs in general convex MOSIP).

This note is focused on a new data qualification called Local Farkas Minkowski (LFMDQ) which is weaker than the CQ with the same name introduced in [3], and extends, to convex MOSIP, the homonymous CQ introduced by the third author [7] for scalar linear semi-infinite programs (actually, the weakest CQ allowing to characterize optimality in terms of Lagrange multipliers). Indeed, we show that the LFMDQ allows to characterize the weak efficient solutions through the (weak) Karush-Kuhn-Tucker (KKT) condition, and to identify proper efficient solutions through the strong KKT condition. (See the next section for the definitions.)

This paper is organized as follows. Section 2 presents the necessary notation and preliminaries, including the Locally Farkas Minkowski Constraint Qualification. Section 3 introduces the Locally Farkas Minkowski Data Qualification. Section 4 provides a characterization of the weak-efficient solutions through the (weak) KKT condition. The connection with a gap function is also discussed. Finally, Sect. 5 analyzes relationships between proper efficient solutions and the strong KKT condition in a similar fashion.

2 The Local Farkas Minkowski Constraint Qualification and Other Preliminaries

Throughout the paper we use the following notation. The zero vector in $\mathbb{R}^n$ is denoted by $\mathbf{0}_n$. We denote $\Delta_+^n = \left\{ x \in \mathbb{R}_+^n : \sum_{i=1}^n x_i = 1 \right\}$ and $\Delta_{++}^n = \left\{ x \in \Delta_+^n : x > \mathbf{0}_n \right\}$. The subgradient (convex subdifferential) of any convex function h at some point x of its domain is written as $\partial h(x)$. Given $X \subset \mathbb{R}^n$, we denote by conv X, cone $X :=$ $\mathbb{R}_+ \operatorname{conv}(X \cup \{\mathbf{0}_n\})$, and cl X the convex hull of X, the convex conical hull of X, and the closure of X, respectively. If X is non empty, then its *negative polar cone* *(strictly negative polar cone)* is $X^0 := \left\{ y \in \mathbb{R}^n : y'x \le 0 \text{ for all } x \in X \right\} (X^- :=$ $\left\{ y \in \mathbb{R}^n : y'x < 0 \text{ for all } x \in X \right\}$, respectively), where $y'x$ stands for the Euclidean product $\langle y, x \rangle$. For $\widehat{x} \in \operatorname{cl} X$, the *cone of feasible directions of X at $\widehat{x}$* is

$$D(X; \widehat{x}) := \left\{ d \in \mathbb{R}^n : \exists \theta > 0 \text{ such that } \widehat{x} + \varepsilon d \in X \text{ for all } \varepsilon \in (0, \theta) \right\},$$

and the *contingent cone* (or *Bouligand tangent cone*) at $\widehat{x}$, $T(X; \widehat{x})$, is the cone formed by all $v \in \mathbb{R}^n$ such that there exist sequences $\left\{ v^k \right\}_{k \in \mathbb{N}} \subset \mathbb{R}^n$ and $\left\{ t^k \right\}_{k \in \mathbb{N}} \subset \mathbb{R}_+$ with $v^k \to v$, $t^k \to 0$, and $\widehat{x} + t^k v^k \in X$ for all $k \in \mathbb{N}$. If X is a convex set, then it is easy to see that

$$T(X; \widehat{x}) = \operatorname{cl} D(X; \widehat{x}).$$

In relation with the constraints of (P), another important cone will be used. Suppose that $\widehat{x} \in S$ and let

$$T(\widehat{x}) := \{t \in T : g_t(\widehat{x}) = 0\}$$

be the set of active indices at $\widehat{x}$. The *active cone at $\widehat{x}$* is the convex conical hull of the set of subgradients at $\widehat{x}$ of the active constraints at $\widehat{x}$:

$$G(\widehat{x}) := \text{cone}\left(\bigcup_{t \in T(\widehat{x})} \partial g_t(\widehat{x})\right).$$

It is well known that

$$G(\widehat{x}) \subset D^0(S; \widehat{x}),$$

which gives the inclusion

$$\text{cl } D(S; \widehat{x}) = D^{00}(S; \widehat{x}) \subset G^0(\widehat{x}). \tag{1}$$

From the objectives of (P), we will make use of the convex hull of the set of subgradients at $\widehat{x}$ of the objective functions:

$$F(\widehat{x}) := \text{conv}\left(\bigcup_{i=1}^{p} \partial f_i(\widehat{x})\right).$$

We also associate with (P) the so-called *gap function*:

$$\vartheta : \bigcup_{x \in \mathbb{R}^n}\left(\{x\} \times \prod_{i=1}^{p} \partial f_i(x)\right) \times \Delta_+^p \to \overline{\mathbb{R}} = [-\infty, +\infty]$$
$$\vartheta(x, \xi, \lambda) := \sup_{y \in S} \sum_{i=1}^{p} \lambda_i \xi_i'(x - y),$$

where $\xi = (\xi_1, ..., \xi_p)$. This gap function is quite useful in solving the problem (P) because it is well known (see e.g. [2, Propositions 3.9 and 3.10, and Theorem 3.15]) that a weak (or proper) efficient solution can be found by solving a weighted sum scalarization of (P),

$$\text{minimize} \sum_{i=1}^{p} \lambda_i f_i(x), \text{ subject to } g_t(x) \le 0, \ t \in T,$$

for some adequate $\lambda = (\lambda_1, ..., \lambda_p) \ge \mathbf{0}_p$ (or $\lambda > 0_p$).

The aim of this paper is to give conditions for the optimality of a given $\widehat{x} \in S$ in terms of the existence of multipliers satisfying any of the following conditions related to $F(\widehat{x})$, $G(\widehat{x})$ and $\vartheta(\widehat{x}, \cdot, \cdot)$:

- The *weak Karush-Kuhn-Tucker* (KKT) condition holds at $\widehat{x} \in S$ if

$$\mathbf{0}_n \in F(\widehat{x}) + G(\widehat{x}), \tag{2}$$

i.e., there exist $\alpha_i \ge 0$, $i = 1, ..., p$, $\sum_{i=1}^{p} \alpha_i = 1$, and $\beta_t \ge 0$, with $\beta_t > 0$ for finitely many indexes in $T(\widehat{x})$, such that

$$\mathbf{0}_n \in \sum_{i=1}^{p} \alpha_i \partial f_i(\widehat{x}) + \sum_{t \in T(\widehat{x})} \beta_t \partial g_t(\widehat{x}). \tag{3}$$

- The *strong* KKT condition holds at $\widehat{x}$ when all the coefficients α_i in (3) are positive.
- $\vartheta\,(\widehat{x}, \cdot, \cdot)$ *attains zero value* on Δ_+^p (Δ_{++}^p) if there exist $\xi \in \prod_{i=1}^p \partial f_i\,(\widehat{x})$ and $\lambda \in \Delta_+^p$ $(\lambda \in \Delta_{++}^p$, respectively) such that $\vartheta\,(\widehat{x}, \xi, \lambda) = 0$.

Optimality theorems are given in [3] (where (P) is assumed to be continuous) and [4] for different types of solutions and under different constraint qualifications. We now recall some of them.

The *Abadie constraint qualification* at $\widehat{x}$, ACQ, requires the validity of the reverse inclusion in (1), i.e., $G^0\,(\widehat{x}) \subset \mathrm{cl}\ D\,(S; \widehat{x})$. Indeed ACQ is equivalent to the equality

$$G^0\,(\widehat{x}) = \mathrm{cl}\ D\,(S; \widehat{x}).\qquad(4)$$

In our convex setting, the *Guignard constraint qualification* at $\widehat{x}$, GCQ, defined by the equation $\mathrm{cl}\,G\,(\widehat{x}) = D^0(S; \widehat{x})$, is nothing else than ACQ written by taking polars in both sides of (4). The *Zangwill* constraint qualification, ZCQ, holds when $\mathrm{cl}\,G^-\,(\widehat{x}) \subset D(S; \widehat{x})$, which implies ACQ. Also related to ACQ, in [3] it is introduced the so called *Local Farkas Minkowski constraint qualification* at $\widehat{x}$, LFMCQ, defined by the equation

$$G\,(\widehat{x}) = D^0(S; \widehat{x}),$$

which subsumes the LFMCQ for linear SIP [7] and the *Basic constraint qualification* for ordinary convex optimization [5]. In the case that $G\,(\widehat{x})$ is closed, ACQ and LFMCQ are equivalent; otherwise LFMCQ $\Rightarrow$ ACQ.

3 The Local Farkas Minkowski Data Qualification

Recall that LFMCQ at $\widehat{x} \in S$ requires the equality $G\,(\widehat{x}) = D^0(S; \widehat{x})$. Since it is always true that $G\,(\widehat{x}) \subset \mathrm{cl}\,G\,(\widehat{x}) \subset D^0(S; \widehat{x})$, we only need to consider the condition

$$D^0(S; \widehat{x}) \subset G\,(\widehat{x}).\qquad(5)$$

Here we introduce a weaker condition which will allow us to characterize the weak efficient solutions of (P).

Definition 3.1 The Local Farkas-Minkowski data qualification (LFMDQ) holds at $\widehat{x} \in S$ when

$$[-F\,(\widehat{x})] \cap D^0(S; \widehat{x}) \subset G\,(\widehat{x}).\qquad(6)$$

This LFMDQ property is clearly weaker than LFMCQ. To the best of our knowledge, the closest data qualification is the so-called weak Abadie data qualification, WADQ, consisting in $F^-\,(\widehat{x}) \cap G^0\,(\widehat{x}) \subset \mathrm{cl}\ D(S; \widehat{x})$. Once again, the implication ACQ$\Longrightarrow$WADQ is trivially true; the problem in [4, Example 3], where $G\,(\widehat{x})$ is not closed, satisfies ACQ but not LFMDQ, so that ACQ$\nRightarrow$LFMDQ and WADQ$\nRightarrow$LFMDQ.

The following example shows a problem with an efficient solution such that LFMDQ holds at this solution, while LFMCQ and ACQ fail.

Example 3.1 Consider the linear MOSIP problem in $\mathbb{R}^2$ given by

$$(P) \quad \text{minimize } f(x_1, x_2) = (x_2, x_1 + x_2)$$
$$\text{s.t.} \quad -x_1 - t^2 x_2 + 2t \leq 0, t \geq 0,$$
$$-x_1 + 1 + t \leq 0, t < 0.$$

The feasible set of (P) is

$$S = \left\{(x_1, x_2) \in \mathbb{R}^2 : x_1 \geq 1, \; x_1 x_2 \geq 1\right\}, \tag{7}$$

and the image of S by f, is

$$f(S) = \left\{(z_1, z_2) \in \mathbb{R}^2 : z_1 > 0, \; z_2 \geq 1 + z_1, \; z_2 \geq z_1 + 1/z_1\right\}.$$

It is easy to see that any point $(x_1, 1/x_1)$ with $x_1 \geq 1$ is an efficient solution for (P). In particular we will consider $\widehat{x} = (1, 1)$ and $\widetilde{x} = (2, 1/2)$. We have

$$G(\widehat{x}) = \text{cone}\{(-1, -1)\}, \quad F(\widehat{x}) = \text{conv}\{(0, 1), (1, 1)\},$$

$$\text{cl } D(S; \widehat{x}) = \text{cone}\{(0, 1), (1, -1)\} \quad \text{and} \quad D^0(S; \widehat{x}) = \text{cone}\{(-1, 0), (-1, -1)\}.$$

Thus $[-F(\widehat{x})] \cap D^0(S; \widehat{x}) = \{(-1, -1)\} \subset G(\widehat{x})$, but $G^0(\widehat{x}) \neq \text{cl } D(S; \widehat{x})$. Hence (P) satisfies the LFMDQ condition at $\widehat{x}$, but it does not satisfy ACQ at $\widehat{x}$, neither LFMCQ. Observe that $G(\widehat{x})$ is closed, so ACQ and LFMCQ are equivalent. For $\widetilde{x}$, we have $G(\widetilde{x}) = \text{cone}\{(-1, -4)\} = D^0(S; \widetilde{x})$ and $G(\widetilde{x})$ is closed, so LFMCQ, ACQ and LFMDQ hold at $\widetilde{x}$.

In the scalar linear case, i.e., when the objective and the constraint functions of (P) have the form $f(x) = c'x$ and $g_t(x) = a_t'x - b_t$, with $c, a_t \in \mathbb{R}^n$ and $b_t \in \mathbb{R}$, respectively, in the same fashion as in [6] it is easy to prove that LFMDQ at $\widehat{x} \in S$ can be expressed in terms of the data as follows:

$$c'\widehat{x} = d \text{ and } c'x \geq d \text{ for all } x \in S \Rightarrow -(c, d) \in \text{cone}\{(a_t, b_t) : t \in T\}.$$

4 Optimality Conditions for Weak Efficiency

We now characterize the weak efficient solution to (P) under the LFMDQ property by means of the weak KKT condition and the gap function. To this aim we need a simple property of the directional derivatives of the objective functions at a weak efficient solution. We include the proof for completeness purposes.

Lemma 4.1 *If $\widehat{x}$ is a weak efficient solution to (P), then for all $d \in$ cl $D(S; \widehat{x})$ it holds that*

$$\max_{i \in \{1, \ldots, p\}} f_i'(\widehat{x}; d) \geq 0.$$

Proof If there is some $d \in D(S; \widehat{x})$ such that $\max_{i \in \{1, \ldots, p\}} f_i'(\widehat{x}; d) < 0$, then for all $i = 1, \ldots, p$, and all $\varepsilon > 0$ sufficiently small, we have

$$\frac{f_i(\widehat{x} + \varepsilon d) - f_i(\widehat{x})}{\varepsilon} < 0$$

contradicting the fact that $\widehat{x}$ is a weak efficient solution to (P) because $\widehat{x} + \varepsilon d \in S$ for some small $\varepsilon > 0$. So $\max_{i \in \{1, \ldots, p\}} f_i'(\widehat{x}; d) \geq 0$ for all $d \in D(S; \widehat{x})$. Now the proof concludes by observing that each $f_i'(\widehat{x}; \cdot)$ is a continuous finite convex function since f_i is a finite convex function (see [5, Remark VI.1.1.3]). $\qquad \square$

Theorem 4.1 *Assume that (P) satisfies the LFMDQ condition at $\widehat{x} \in S$. Then, $\widehat{x}$ is a weak efficient solution for (P) if and only if the weak KKT condition holds at $\widehat{x}$.*

Proof The converse is known and does not require LFMDQ (see, e.g., [3, Theorem 27(i)]). So it is enough to assume that $\widehat{x}$ is a weak efficient solution to (P), and show that (2) holds.

Claim: $[-F(\widehat{x})] \cap D^0(S; \widehat{x}) \neq \emptyset$. Indeed, if $[-F(\widehat{x})] \cap D^0(S; \widehat{x}) = \emptyset$, by observing that $F(\widehat{x})$ is a compact convex subset of $\mathbb{R}^n$ and $D^0(S; \widehat{x})$ is a closed convex cone in $\mathbb{R}^n$, a separation argument yields some $w \in \mathbb{R}^n$ such that

$$w'z > 0 \text{ for all } z \in -F(\widehat{x}) \text{ and } w'z \leq 0 \text{ for all } z \in D^0(S; \widehat{x}).$$

Thus, $w \in D^{00}(S; \widehat{x}) = \text{cl } D(S; \widehat{x})$ while $w'y < 0$ for all $y \in F(\widehat{x})$. Then, from [8, Theorem 23.4] and the assumptions on each $f_i, i = 1, \ldots, p$,

$$f_i'(\widehat{x}; w) = \max \left\{ w'y_i : y_i \in \partial f_i(\widehat{x}) \right\} < 0$$

contradicting Lemma 4.1. Hence, $[-F(\widehat{x})] \cap D^0(S; \widehat{x}) \neq \emptyset$.
Therefore, by taking any $z \in [-F(\widehat{x})] \cap D^0(S; \widehat{x}) \subset G(\widehat{x})$, we obtain that the problem (P) satisfies the weak KKT condition at $\widehat{x}$:

$$0_n = (-z) + z \in F(\widehat{x}) + G(\widehat{x}).$$

$\qquad \square$

For the point $\widehat{x}$ in Example 3.1, we have

$$0_2 = (1, 1) + (-1, -1) \in F(\widehat{x}) + G(\widehat{x}),$$

satisfying the weak KKT condition.

Theorem 4.2 *Assume that* (P) *satisfies LFMDQ at* $\widehat{x} \in S$. *Then,* $\widehat{x}$ *is a weak efficient solution to* (P) *if and only if* $\vartheta\,(\widehat{x}, \cdot, \cdot)$ *attains zero value on* Δ_+^p.

Proof The "if" part does not require LFMDQ. Suppose $\vartheta\,(\widehat{x}, \xi, \lambda) = 0$ for some $\xi \in \prod_{i=1}^p \partial f_i\,(\widehat{x})$ and $\lambda \in \Delta_+^p$. Since $0 = \vartheta\,(\widehat{x}, \xi, \lambda) = \sup_{x \in S} \sum_{i=1}^p \lambda_i \xi_i'\,(\widehat{x} - x)$, we get that $\sum_{i=1}^p \lambda_i \xi_i'\,(x - \widehat{x}) \geq 0$ for all $x \in S$. Then $\widehat{x}$ is an optimal solution of the weighted sum scalarization of (P),

$$\text{minimize } \sum_{i=1}^p \lambda_i f_i\,(x), \text{ subject to } g_t\,(x) \leq 0,\ t \in T,$$

because

$$\sum_{i=1}^p \lambda_i f_i\,(x) \geq \sum_{i=1}^p \lambda_i f_i\,(\widehat{x}) + \sum_{i=1}^p \lambda_i \xi_i'\,(x - \widehat{x}) \geq \sum_{i=1}^p \lambda_i f_i\,(\widehat{x}),$$

for any $x \in S$. From [2, Proposition 3.9] we obtain that $\widehat{x}$ is a weak efficient solution to (P).

Next, the proof follows the lines in [4, Theorem 4]. In order to show the necessity of $\vartheta\,(\widehat{x}, \xi, \lambda) = 0$ for some $\xi \in \prod_{i=1}^p \partial f_i\,(\widehat{x})$ and $\lambda \in \Delta_+^p$ when $\widehat{x}$ is a weak efficient solution to (P), we need to appeal to Theorem 4.1 above to get some multipliers $\lambda_i \geq 0$, $i = 1, ..., p$, $\sum_{i=1}^p \lambda_i = 1$, and $\beta_t \geq 0$, with $\beta_t > 0$ only for finitely many indexes in a finite set $T' \subset T\,(\widehat{x})$, such that

$$\mathbf{0}_n \in \sum_{i=1}^p \lambda_i \partial f_i\,(\widehat{x}) + \sum_{t \in T'} \beta_t \partial g_t\,(\widehat{x}).$$

Then, there exist $\xi_i \in \partial f_i\,(\widehat{x})$, $i = 1, ..., p$, and $u_t \in \partial g_t\,(\widehat{x})$, $t \in T'$, with

$$\mathbf{0}_n = \sum_{i=1}^p \lambda_i \xi_i + \sum_{t \in T'} \beta_t u_t. \tag{8}$$

Now, take any $y \in S$ and observe that the convexity of the function g_t gives, for all $t \in T' \subset T\,(\widehat{x})$,

$$g_t\,(y) \geq g_t\,(\widehat{x}) + u_t'\,(y - \widehat{x}) = u_t'\,(y - \widehat{x}).$$

Recalling that $g_t\,(y) \leq 0$ for all $t \in T$, we obtain

$$u_t'\,(y - \widehat{x}) \leq 0$$

for $t \in T'$. Taking into account (8) and multiplying by $(y - \widehat{x})$ we get

$$-\sum_{i=1}^{p} \lambda_i \xi_i' (y - \widehat{x}) = \sum_{t \in T'} \beta_t u_t' (y - \widehat{x}) \leq 0. \tag{9}$$

Then,

$$\vartheta (\widehat{x}, \xi, \lambda) = \sup_{y \in S} \sum_{i=1}^{p} \lambda_i \xi_i' (\widehat{x} - y) = 0$$

by virtue of (9) and the fact that $\widehat{x} \in S$.

$\square$

Theorem 4.1 extends, to convex MOSIP, the KKT characterization of optimal solutions in linear SIP in [6, Theorem 3.6] as well as the weak KKT characterization of weak efficient solutions to (P) in [4, Corollary 1] (under LFMCQ) and [1, Theorem 6] (under ZQC together the closedness of $G(\widehat{x})$, with the same proof). Similarly, Theorem 4.2 generalizes [4, Theorem 4 under LFMCQ].

5 Optimality Conditions for Proper Efficiency

In [3, Theorem 27] it is shown that the strong KKT condition at any feasible point $\widehat{x}$ is a sufficient condition for assuring that $\widehat{x}$ is a proper efficient solution in any convex MOSIP. Moreover it is also shown that, under LFMCQ, the strong KKT is a necessary condition for $\widehat{x}$ being a proper efficient solution for (P). Here we will prove a similar result on proper efficiency under the weaker LFMDC condition.

Theorem 5.1 *Assume that (P) satisfies the LFMDQ condition at $\widehat{x} \in S$. Then, $\widehat{x}$ is a proper efficient solution to (P) if and only if the strong KKT condition holds at $\widehat{x}$.*

Proof The "if" part is known (see, e.g., [3, Theorem 27(i)]) and does not require LFMDQ. So, we only need to show the necessary condition. Assume that $\widehat{x}$ is a proper efficient solution to (P). According to [2, Theorem 3.15] $\widehat{x}$ is an optimal solution to a weighted sum scalarization of (P),

$$\text{minimize} \sum_{i=1}^{p} \lambda_i f_i (x), \text{ subject to} g_t (x) \leq 0, \ t \in T,$$

for some adequate $\lambda \in \Delta_{++}^{p}$. From [8, Theorem 27.4], there exists a vector $u \in \partial \left(\sum_{i=1}^{p} \lambda_i f_i (\widehat{x}) \right)$ such that $-u \in D^0(S; \widehat{x})$ (the normal cone to S at $\widehat{x}$). Moreover, [8, Theorem 23.8] gives $\partial \left(\sum_{i=1}^{p} \lambda_i f_i (\widehat{x}) \right) = \sum_{i=1}^{p} \lambda_i \partial f_i (\widehat{x})$, which is a subset of $F(\widehat{x})$. Therefore, taking into account the LFMDQ condition at $\widehat{x}$, we obtain

$$-u \in [-F(\widehat{x})] \cap D^0(S; \widehat{x}) \subset G(\widehat{x}),$$

thus

$$0_n = u + (-u) \in F(\widehat{x}) + G(\widehat{x}),$$

which shows that the strong KKT condition holds by the positivity of λ.

□

According to Theorem 5.1, we can conclude that the efficient solution $\widehat{x} = (1, 1)$ to problem (P) in Example 3.1 is not proper efficient because it does not satisfy the strong KKT condition. However, for the point $\widetilde{x} = (2, 1/2)$, we have

$$0_2 = \tfrac{1}{4}(1, 1) + \tfrac{3}{4}(0, 1) + \tfrac{1}{4}(-1, -4) \in F(\widetilde{x}) + G(\widetilde{x}),$$

so the strong KKT condition holds at $\widetilde{x}$. Thus $\widetilde{x}$ is a proper efficient solution to (P).

Theorem 5.2 *Assume that (P) satisfies the LFMDQ condition at $\widehat{x} \in S$. Then, $\widehat{x}$ is a proper efficient solution to (P) if and only if $\vartheta(\widehat{x}, \cdot, \cdot)$ attains zero value on Δ_{++}^p.*

Proof It is the same as the proof of Theorem 4.2, by appealing to [2, Theorem 3.15] and to Theorem 5.1.

□

Finally, let us observe that the proofs in this work rely on the possibility of identifying weak and proper efficient solution via scalarization. Obtaining similar results for other types of solutions is a challenging problem requiring the use of a totally different methodology.

Acknowledgements This work has been supported by MINECO of Spain and ERDF of EU, Grant MTM2014-59179-C2-1-P, and SECTyP-UNCuyo, Argentina, Res. 3853/2016-R.

References

1. Barilla D, Caristi G, Puglisi A (2016) Optimality conditions for nondifferentiable multiobjective semi-infinite programming problems. Abstr Appl Anal 2016:Art.ID5367190
2. Ehrgott M (2005) Multicriteria optimization, 2nd end. Springer, Berlin
3. Goberna MA, Guerra-Vázquez F, Todorov MI (2016) Constraint qualifications in convex vector semi-infinite optimization. Eur J Oper Res 249:32–40
4. Goberna MA, Kanzi N (2017) Optimality conditions in convex multiobjective SIP. Math Program Ser A 164:167–191
5. Hiriart-Urruty JB, Lemaréchal C (1991) Convex analysis and minimization algorithms. I. Springer, Berlin
6. Liu Y (2016) New constraint qualification and optimality for linear semi-infinite programing. Pac J Optim 12:223–232
7. Puente R, Vera de Serio VN (1999) Locally Farkas-Minkowski linear inequality systems. Top 7:103–121
8. Rockafellar RT (1970) Convex analysis. Princeton University Press, Princeton
9. Yamamoto S, Kuroiwa D (2016) Constraint qualifications for KKT optimality condition in convex optimization with locally Lipschitz inequality constraints. Linear Nonlinear Anal 2:101–111

Complexity and Dynamical Uncertainty

Santiago Ibáñez, Antonio Pumariño and José Ángel Rodríguez

Abstract Uncertainty is usually linked to non-deterministic evolutions. Nevertheless, along the second half of the past century deterministic phenomena with unpredictable behaviour were discover and the notion of strange attractor emerged as the new paradigm to describe chaotic behaviours. The goal of this paper is to review all this story and to provide a perspective of the state of the art regarding this subject.

1 Introduction

This paper is dedicated to the memory of Professor Pedro Gil. In 1996 he gave the inaugural lecture to open the academic year at the University of Oviedo. It was entitled "Las matemáticas de lo incierto" [17] and his first reflection, "uncertainty is inherent in nature", serves as a starting point for the topics that we develop below.

Uncertainty is understood as the impossibility to predict the evolution of a process. Therefore, uncertainty has to be apparent in any mathematical model chosen for the study of this process. Initially, models are stated by taking deterministic variables over spaces M (often differentiable manifolds), endowed with good properties. Then, on M, the laws governing the evolution of the process are introduced in terms of well-defined maps or operators $f : M \to M$. In this way, differential equations, evolutionary partial differential equations or delayed functional equations arise, to cite relevant examples which can be included in the more general context of dynamical systems. When variables can be established only in terms of probabilities or when the evolution laws cannot be stated in precise terms, appealing to random phenomena, deterministic models give rise to stochastic processes. In this way, we

S. Ibáñez (✉) · A. Pumariño · J. Á. Rodríguez
Departamento de Matemáticas, Universidad de Oviedo,
Avda. Federico García Lorca, 18, 33007 Oviedo, Spain
e-mail: mesa@uniovi.es

A. Pumariño
e-mail: apv@uniovi.es

J. Á. Rodríguez
e-mail: jarodriguez@uniovi.es

© Springer International Publishing AG 2018
E. Gil et al. (eds.), *The Mathematics of the Uncertain*, Studies in Systems,
Decision and Control 142, https://doi.org/10.1007/978-3-319-73848-2_78

855

give up on predicting with certainty the dynamics of the process and in fact, *we accept the uncertainty of the process.*

Thus, a big gap appears between dynamical (deterministic) systems and stochastic processes. Dynamical uncertainty seems to be, exclusively, a consequence of chance and, if chance is not present in a given process, dynamics should be simple and predictable. However, this is not the case. Our goal is to explain very briefly how to cover the gap. We will understand how simple and fully deterministic models present dynamical uncertainty, giving rise to complex structures. These structures are the so called strange attractors, which since the second half of the last century are the new paradigm to describe the notions of chaos and turbulence (see [34] and references therein).

Roughly speaking, strange attractors are invariant sets with fractal structure (fractional dimension) and with inner dynamics which is exponentially sensitive to changes of the initial conditions. These dynamics imply uncertainty. This uncertainty, together with the fractal structure, justifies the complexity mentioned above. Once we have proved the abundance of dynamical systems exhibiting strange attractors, we will conclude, according to Pedro Gil, that complexity and uncertainty are inherent in nature.

For the sake of simplicity, we will define in Sect. 2 the notion of strange attractor for the iteration of a diffeomorphism (or a map in general) defined on a differentiable manifold M. The close relationship between the flow of a vector field and the iteration of a diffeomorphism through the concept of Poincaré maps, allow us to easily extend the notion of strange attractor to vector fields. Strange attractors, like any other dynamical property, reach their true physical meaning if they persist for small perturbations of the system (structural stability) or they occur for generic families over a set of parameters with positive Lebesgue measure (observability with positive probability). By finding this persistence throughout this second section, we will conclude the abundance of strange attractors and, as a conclusion, the abundance of dynamical uncertainty.

In Sect. 3 we will consider vector fields to understand how simple geometric configurations lead to the existence of persistent strange attractors. Since these configurations will be present in generic unfoldings of low codimension singularities and since singularities are the simplest elements which can be determined in the study of vector fields, this third section provides a criterion of existence of strange attractors. This criterion is easy to apply and makes unnecessary many of the numerical simulations suggesting the existence of possible strange attractors.

All the above mentioned strange attractors are one-dimensional. However, the progress in establishing a hierarchy in dynamical complexity involves studying the existence of strange attractors of increasing dimension. The proof of the existence of two-dimensional strange attractors for diffeomorphisms or vector fields is nowadays a very interesting challenge. Section 4 includes results pointing in that direction. With the same motivation as Sect. 3, we conclude Sect. 4 proposing the search of a criterion for the existence of two-dimensional strange attractors in generic unfoldings of low codimension singularities.

2 Strange Attractors

We will consider throughout this section a map or a diffeomorphism $f : M \to M$ defined on a differentiable manifold M. A set $A \subset M$ is said to be invariant by f if $f(A) \subseteq A$. The inner dynamics within an invariant set A will be observable if it has a stable set

$$W^s(A) = \{z \in M : d(f^n(z), A) \to 0 \text{ as } n \to \infty\}$$

with non-empty interior or, at least, with positive Lebesgue measure. The invariant set A will be minimal if it is transitive, that is, if it contains a dense orbit.

Definition 2.1 An attractor for a transformation f defined on a manifold M is a compact, f-invariant and transitive set $A \subset M$ whose stable set $W^s(A)$ has nonempty interior. An attractor is said to be strange if it contains a dense orbit $\{f^n(x) : n \geq 0\}$ displaying exponential growth of the derivative, that is, if there exists a constant $c > 0$ such that, for every $n \geq 0$,

$$\|Df^n(x)\| \geq e^{cn}.$$

In particular, the above definition implies that strange attractors display, in a dense orbit, at least one positive Lyapunov exponent.

Attracting fixed points or attracting periodic orbits provide the simplest examples of attractors. Attracting tori with a dense orbit also provide examples of recurrent attractors on which there is no expansivity. The inner dynamics of a strange attractor is quite different. On a strange attractor, the exponential growth of the derivative of f along a dense orbit implies an exponential growth of the deviations between any pair of different orbits, this fact not depending on the distance between the respective initial conditions. This is the reason why f is said to have high sensitivity to the initial conditions and the evolution of the dynamics inside the attractor becomes unpredictable. This uncertainty also extends to every point in the stable manifold of the attractor.

From a physical point of view, a certain degree of persistence is as relevant as the unpredictability of the dynamics resulting from the aforementioned sensitivity to initial conditions. So, if a family f_μ of diffeomorphisms exhibits a strange attractor when $\mu = \mu_0$, the dynamics of the attractor should be only considered if, for every $\delta > 0$, strange attractors still exist for values of the parameter in a positive Lebesgue measure set $E \subset B(\mu_0, \delta)$. In this case, the attractor is said to be **persistent** for the family f_μ, and it is said to be **fully persistent** if $E = B(\mu_0, \delta)$ for some $\delta > 0$.

The possible existence of persistent strange attractors for parameter families of diffeomorphism takes an exceptional interest. A first strange attractor, the solenoid, was built by Smale, see [36], inspired by its well-known horseshoe map. For this map, Smale found a compact set Λ which is invariant, transitive and hyperbolic, but whose stable set $W^s(\Lambda)$ has empty interior. However, the solenoid is a hyperbolic attractor whose stable set has non-empty interior. Every hyperbolic attractor is fully persistent and even structurally stable: arbitrarily small perturbations of f have attractors which are topologically equivalent. Unfortunately, there are no natural mechanisms providing abundance of hyperbolic strange attractors.

The first strange attractors appearing in the literature can be detected numerically in very simple scenarios: the Lorenz attractor [23], for a family of three-dimensional quadratic fields, and the Hénon attractor [18], for the family of diffeomorphisms

$$H_{a,b}(x, y) = (1 - ax^2 + y, bx). \tag{1}$$

From numerical analysis, Lorenz attractor seems to be strange, fully persistent, but not structurally stable. On the other hand, the attractors of (1) seem to be strange, persistent, but not fully persistent. Therefore, none of these attractors seem to be hyperbolic. But, does there exist non hyperbolic strange attractors?

At early 1990s, in a historical and very involved paper [9], M. Benedicks and L. Carleson proved that the Hénon family $H_{a,b}$ given in (1) has strange attractors for parameter values close to $a = 2$ and $b = 0$. In fact, these attractors coincide with the closure of the unstable manifold of a saddle point of the map. This one-dimensional manifold folds time after time arbitrarily close (if b is small enough) to the line $y = 0$. When $b = 0$ this line is invariant by $H_{a,0}$ and the dynamics of the limit family $H_{a,0}$ on this invariant set coincides with the dynamics of the one-dimensional quadratic family $f_a : [-1, 1] \to [-1, 1]$, $a \in [1, 2]$, defined by

$$f_a(x) = 1 - ax^2. \tag{2}$$

In [8] it was proved that f_a displays persistent strange attractors for parameter values close enough to $a = 2$: There exists a positive Lebesgue measure set of parameters E such that for every $a \in E$ the map f_a exhibits an invariant set where the orbit of the critical value is not only dense but also displays a positive Lyapunov exponent. The proof given in [9] is a laborious extension of the previous results given in [8] in order to get, in a positive Lebesgue measure set of parameters (a, b) close to $(2, 0)$, a dense orbit in the aforementioned unstable manifold displaying a positive Lyapunov exponent.

The idea of taking advantage of the dynamics of limit families was used in [24] to prove the persistence of strange attractors (like the ones obtained for the Hénon family) when a generic two-dimensional homoclinic tangency is unfolded. To be precise, we must recall that the main result in [24] is strongly based on the existence of limit families of return maps associated to the unfolding of homoclinic tangencies. Under an appropriate change of coordinates, these return maps are defined in a neighbourhood of the homoclinic points and they are very similar to the ones defined in (1) and hence, the proof developed in [9] can be adapted. In short, since unfoldings of homoclinic tangencies are present in generic families of difeomorphisms, the abundance of Hénon-like strange attractors follows.

In order to prove the persistence of strange attractors in families X_μ of three-dimensional vector fields one might consider the Poincaré map f_μ associate to a saddle type periodic orbit Γ_μ of X_μ. Then, for $\mu = \mu_0$ one may assume also that f_{μ_0} has a generic homoclinic tangency which is generically unfolded by f_μ. In this way, the previous results can be applied to conclude the existence of Hénon-like strange attractors. However, there exists a simpler and more natural route to reach strange attractors in families of vector fields in $\mathbb{R}^3$. Doing so, we may even prove the coexistence of any arbitrarily large number of strange attractors (see [25, 26]). These are the strange attractors emerging in a neighbourhood of a Shilnikov homoclinic orbit.

Let $0 \in \mathbb{R}^3$ be an equilibrium point of the vector field X_{μ_0} whose eigenvalues are $\lambda > 0$ and $-\rho \pm i\omega$ with $\lambda > \rho > 0$. Let us suppose that the unstable manifold $W^u(0)$ and the stable manifold $W^s(0)$ of the equilibrium point intersect in a homoclinic orbit Γ. This orbit is named Shilnikov homoclinic orbit and in every neighbourhood of Γ there exist a countable set of periodic orbits [35]. In fact, Tresser [38] proved that in every neighbourhood of such a homoclinic orbit, an infinity of linked horseshoes can be defined in such a way that the dynamics is conjugated to a subshift of finite type on an infinite number of symbols. When $\mu \in \mathbb{R}^s$ with $s \geq 2$, the Shilnikov homoclinic orbit for $\mu = \mu_0$ remains, generically, for values of μ on a manifold $\mathcal{H} \subset \mathbb{R}^s$ of codimension one. Just off this manifold $\mathcal{H}$ the homoclinic orbit disappears and an infinite number of horseshoes given in [38] are destroyed. Then, homoclinic tangencies take place and, as a consequence of [24], Hénon like strange attractors arise.

Let us now suppose that μ varies on the manifold $\mathcal{H}$ and assume that $\lambda = \rho$. In this non generic context there exists a one-parameter family X_a of piecewise regular vector fields such that for every neighbourhood V of the homoclinic orbit Γ, for each $k \in \mathbb{N}$ and for every value of the parameter a in a set of positive Lebesgue measure depending on k, at least k strange attractors coexist in V (see Theorem A in [25]). In order to prove this result, it is necessary to choose a suitable section Π_0 transversal to the flow of X_a in V and to define the corresponding transformation $T : \Pi_0 \to \Pi_0$ associated to the flow. After splitting Π_0 into a countable union of rectangles R_m and carrying out adequate changes of variable, we get the following sequence of families of diffeomorphisms

$$T_{\lambda,a,b}(x, y) = (f_{\lambda,a}(x) + \frac{1}{\lambda} \log(1 + \sqrt{b}y), \sqrt{b}(1 + \sqrt{b}y)e^{\lambda x} \sin x)$$

defined on certain rectangles $U_m \subset R_m$, with $b = e^{-2\pi\lambda m}$ and $m \in \mathbb{N}$. For a large enough m, each $T_{\lambda,a,b}$ is a small perturbation of the limit family $\Psi_{\lambda,a}(x, y) = (f_{\lambda,a}(x), 0)$ where

$$f_{\lambda,a}(x) = \lambda^{-1} \log a + x + \lambda^{-1} \log \cos x. \tag{3}$$

In spite of $f_{\lambda,a}$ is not the quadratic family (2), the proof of the existence of strange attractors for the diffeomorphisms $T_{\lambda,a,b}$ can be developed by means of a cautious adaptation of the ideas and the arguments in [9, 24]. Then, these attractors correspond to suspended strange attractors for the family of vector fields X_a.

We conclude this section by stressing that, indeed, strange attractors are abundant for diffeomorphisms and vector fields as well.

3 Germs of Strange Attractors

Let $\mathscr{X}$ be a set of dynamical systems endowed with a topology and an equivalence relation (usually the topological equivalence). A system is said to be structurally stable if it belongs to the interior of its equivalence class. The set $\mathscr{B}$ of non-structurally stable systems is called **bifurcation set**.

The most ambitious goal in studying $\mathscr{X}$ is to know the structure of $\mathscr{B}$. However, even when $\mathscr{X}$ is the set of regular vector fields in a so low dimension as $n = 3$, the structure of $\mathscr{B}$ can become very intricate and only a few of its elements, namely vector fields having non-hyperbolic singularities, can be easily found. If no specific restriction is imposed to $\mathscr{X}$, vector fields with a homoclinic orbit are also elements of $\mathscr{B}$ which often explain transitions between different and sometimes intrigued dynamics as, for example, persistence of strange attractors, as seen in the previous section. Unfortunately, the existence of a homoclinic orbit for a given vector field (or even for a given family of vector fields) is not easy to prove. So, the set $\mathscr{B}_1 \subset \mathscr{B}$ of vector fields with non-hyperbolic singularities seems to be the only part of $\mathscr{B}$ which can be analytically determined. In order to obtain more information about the structure and the layout of $\mathscr{B}$, we can look at $\mathscr{B}_1$ as a set of rivers and landmarks in the middle of the jungle. Then we wonder if each equivalence class $\mathscr{C}$, corresponding to a relevant dynamics, is adjacent to $\mathscr{B}_1$ and, in such a case, which are those elements of $\mathscr{B}_1$ that are adjacent to a given class.

As we have just said in the previous section, the simplest homoclinic orbits that yield infinitely many transitions and complicated dynamics (existence of persistent strange attractors) are the Shilnikov homoclinic orbits. Therefore, we aim now to find the most general elements of B_1 which are adjacent to the set of vector fields having such orbits; that is, to seek for the lowest codimension singularity from which these vector fields can be unfolded generically.

Consider a C^∞ vector field X defined in a neighbourhood of $0 \in \mathbb{R}^3$ such that X vanishes at 0 and the linear part of X is linearly conjugate to

$$y\frac{\partial}{\partial x} + z\frac{\partial}{\partial y}.$$

As proved in [13], X can be written in the following normal form

$$y\frac{\partial}{\partial x} + z\frac{\partial}{\partial y} + \left(ax^2 + bxy + cxz + dy^2 + O(\|(x, y, z)\|^3)\right)\frac{\partial}{\partial z}. \tag{4}$$

The condition $a \neq 0$ defines a stratum of codimension three in the space of germs of C^∞ vector fields in $\mathbb{R}^3$ having a singularity at the origen, where, as proved in [13], there is a unique topological type. Then, it makes sense to refer to (4) as the **nilpotent singularity of codimension three** or **the nilpotent singularity** for short. The condition $a = 0$ characterizes a stratum of codimension four with, as proved in [13], five different topological types.

Let $\eta \in \mathbb{R}^s$, where s is the codimension of X in (4). A family X_η of vector fields is said to be an unfolding of X if for some value $\eta = \eta_0$ it holds that $X = X_{\eta_0}$. Without loss of generalidad we suppose $\eta_0 = 0$. Again from [13] we know that any generic 3-parameter unfolding of the nilpotent singularity of codimension three can be written in the following normal form:

$$y\frac{\partial}{\partial x} + z\frac{\partial}{\partial y} + (\lambda + \mu x + \nu z + x^2 + bxy + cxz + dy^2 + eyz + \alpha(x, y, z, \eta))\frac{\partial}{\partial z}$$

where

$$\alpha(x, y, z, \eta) = O(\|(x, y, z, \lambda, \mu, \nu)\|^3) = O(\|(y, z)\|)$$

and λ, μ and ν represent the exact coefficients in the Taylor expansion with respect to (x, y, z), more precisely

$$\alpha(0, \eta) = \frac{\partial \alpha}{\partial y}(0, \eta) = \frac{\partial \alpha}{\partial z}(0, \eta) = 0.$$

By means of the rescaling

$$\lambda = u^6\widehat{\lambda}, \mu = u^2\widehat{\mu}, \nu = u\widehat{\nu}$$
$$x = u^3\widehat{x}, y = u^4\widehat{y}, z = u^5\widehat{z}$$

where $\widehat{\lambda}^2 + \widehat{\mu}^2 + \widehat{\nu}^2 = 1$ and $(\widehat{x}, \widehat{y}, \widehat{z})$ belong to an arbitrarily big ball $A \subset \mathbb{R}^3$ centred at $0 \in \mathbb{R}^3$, the previous family reduces to

$$\widehat{y}\frac{\partial}{\partial \widehat{x}} + \widehat{z}\frac{\partial}{\partial \widehat{y}} + (\widehat{\lambda} + \widehat{\mu}\widehat{x} + \widehat{\nu}\widehat{z} + \widehat{x}^2 + O(u))\frac{\partial}{\partial \widehat{z}}. \tag{5}$$

This family has been studied in [13–15]. By using the so-called family blowing-up technique, it was proven that all dynamics in the unfolding of the nilpotent singularity are detectable in (5). Let $Y_{\widehat{\eta}}$ be the family given by (5) when $u = 0$. Since all the structurally stable behaviours of $Y_{\widehat{\eta}}$ are persistent for u small enough, it is clear that $Y_{\widehat{\eta}}$ plays an essential role in the study of the unfolding. Extending the advances achieved in the cited references [13–15], and after a lengthy and careful study of the limit family $Y_{\widehat{\eta}}$, it was proven in [21] (see also [5]) that Shilnikov homoclinic orbits appear in any generic unfolding of the three dimensional nilpotent singularity of codimension three. Earlier, in [20], it had been proven that Shilnikov homoclinic orbits appear in any generic unfolding of any three dimensional nilpotent singularity of codimension four.

With the goal of determining the singularity of lowest codimension which generically unfolds Shilnikov homoclinic orbits, the paper [16] dealt with Hopf-Zero singularities of codimension two. We say that a singularity is of Hopf-Zero type if the linear part has spectrum $\{0, \pm\omega i\}$. Sufficient conditions for the existence of Shilnikov homoclinic orbits were obtained. Later, in [2–4], such sufficient conditions were translated to expressions which depend on the full jet of the singularity. Nowadays, it can be said that any generic unfolding of the Hopf-Zero singularity unfolds Shilnikov homoclinic orbits. Nevertheless, generic conditions to be checked are algebraic only in the case of the nilpotent singularity.

These singularities of low codimension that generically unfold persistent strange attractors are understood as germs of these attractors or seeds of chaos. These germs are never present in families of vector fields of dimension $n < 3$. However, vector fields of low dimension can be coupled to generate dynamics in greater dimension (coupling cells to generate tissues, for example). This is a natural route towards dynamic complexity in which the analysis of the singularities of greater and greater codimension can be as useful as it is essential. It was proven in [10] that the coupling by a simple linear diffusion of a two-dimensional model of a simple chemical reaction (the brusselator) leads to chaotic dynamics (see also [11]). The proof is essentially reduced to verifying that the coupling defines in dimension four a family which is a generic unfolding of the nilpotent singularity. Alternative organizing centers (Hopf-pitchfork bifurcations) of chaotic dynamics in coupled systems are discussed in [12].

The role of singularities as germs of complexity still needs to be extensively exploited in applications. Neuronal models are the next challenge. Reference [7] shows an outline of the extremely rich catalogue of homoclinic phenomena arising in the 3-dimensional neuronal model of Hindmarsh-Rose and it is a first step to tackle coupled neurons where singularities are expected to emerge as organizing centers of a great variety of complex dynamics.

4 Two Dimensional Strange Attractors: An Open question

The strange attractors mentioned in Sect. 2 are one-dimensional. Actually, they coincide with the closure of a one-dimensional unstable manifold folding over and over again to define a fractal set of dimension d, with $1 < d < 2$. The inner dynamics of the attractor expands in the direction of the tangent space of the manifold. It has, therefore, a unique positive Lyapunov exponent. In order to get a strange attractor with two positive Lyapunov exponents one has to consider homoclinic tangencies where the involved unstable manifold is two dimensional. With this premise in mind, we consider in [31] a generic two-parameter family $f_{a,b} : M \to M$ of three-dimensional diffeomorphisms unfolding a generalized homoclinic tangency, as it was defined in [37]. Then, the unstable manifold involved in the homoclinic tangency has dimension two and the limit family is conjugate to the family of quadratic endomorphisms defined on $\mathbb{R}^2$ by

$$T_{a,b}(x, y) = (a + y^2, x + by). \tag{6}$$

Nowadays, in the two dimensional setting, it is usual to define two-dimensional strange attractors as those ones for which the sum of the two Lyapunov exponents is positive. Hence, if one tries to show the existence of two-dimensional strange attractors when a three-dimensional homoclinic tangency is unfolded, the first step should be to prove, as in the one-dimensional setting, the existence of such attractors for the limit family $T_{a,b}$. Only after this, it makes sense to lift the dynamics to the closure of the unstable manifold, which is the candidate to be the two-dimensional strange attractor arising in the unfolding of the tangency.

The dynamical behavior of the family $T_{a,b}$ is rather complicated as was numerically pointed out in [28] and, in particular, the attractors exhibited by $T_{a,b}$ for a large set of parameters seem to be two-dimensional strange attractors. Moreover, in [27], a curve of parameters

$$\gamma = \left\{ (a(s), b(s)) = -\frac{1}{4}s^3(s^3 - 2s^2 + 2s - 2, -s^2 + s) : s \in \mathbb{R} \right\} \quad (7)$$

has been constructed in such a way that the respective transformation $T_{a(s),b(s)}$ has an invariant region in $\mathbb{R}^2$ homeomorphic to a triangle. This curve contains the point $(-4, -2)$ (by taking $s = 2$) and the map $T_{-4,-2}$ is conjugate to the non-invertible piecewise affine map

$$\Lambda(x, y) = \begin{cases} (x + y, x - y) & \text{if } (x, y) \in \mathcal{T}_0 \\ (2 - x + y, 2 - x - y) & \text{if } (x, y) \in \mathcal{T}_1 \end{cases} \quad (8)$$

defined on the triangle $\mathcal{T} = \mathcal{T}_0 \cup \mathcal{T}_1$, where

$$\mathcal{T}_0 = \{(x, y) : 0 \le x \le 1, 0 \le y \le x\},$$
$$\mathcal{T}_1 = \{(x, y) : 1 \le x \le 2, 0 \le y \le 2 - x\}.$$

As a first approach to the study of the dynamics of $T_{a(s),b(s)}$ for s close to 2, a family of piecewise linear maps was introduced in [32]. These maps are defined on the triangle $\mathcal{T}$ by means of

$$\Lambda_t(x, y) = \begin{cases} (t(x + y), t(x - y)) & \text{if } (x, y) \in \mathcal{T}_0 \\ (t(2 - x + y), t(2 - x - y)) & \text{if } (x, y) \in \mathcal{T}_1 \end{cases}. \quad (9)$$

These maps can be seen as the composition of linear maps defined by the matrices

$$A_t = \begin{pmatrix} t & t \\ t & -t \end{pmatrix}$$

with the fold of the whole plane along the line $\{(x, y) \in \mathbb{R}^2 : x = 1\}$ defined by

$$S(x, y) = \begin{cases} (x, y) & \text{if } x < 1 \\ (2 - x, y) & \text{if } x \ge 1 \end{cases}.$$

Notice that Λ in (8) coincides with Λ_t for $t = 1$.

The triangle $\mathcal{T}$ is invariant for the map Λ_t whenever $0 \le t \le 1$. If $t > \frac{1}{\sqrt{2}}$, then the eigenvalues of the matrix A_t have modulus bigger than one. In this case, the map Λ_t is called, according to [29], an Expanding Baker Map (EBM for short). The study of the dynamics of the family Λ_t is mainly justified when one compares its attractors (numerically obtained in [31]) with the attractors (numerically obtained in [28]) for the family $T_{a(s),b(s)}$ with $(a(s), b(s)) \in \gamma$. Both families of maps display convex strange attractors, connected (but non simply-connected) strange attractors and non-connected strange attractors (these last ones formed by numerous connected pieces).

A rigorous analytical proof of the existence of a strange attractor $\mathcal{R}_t \subset \mathcal{T}$ of the map Λ_t for every $t \in (t_0, 1]$, with $t_0 = \frac{1}{\sqrt{2}}(1 + \sqrt{2})^{\frac{1}{4}}$, was given in [33], see Theorem 1.1. Moreover, it was proven that the map Λ_t is strongly topologically mixing in $\mathcal{R}_t$, the periodic orbits are dense in $\mathcal{R}_t$, and $\mathcal{R}_t$ supports a unique absolutely continuous invariant and ergodic measure μ_t. The continuity of the map

$$t \in (t_0, 1] \rightarrow \frac{d\mu_t}{dm} \in L^1(\mathcal{T})$$

was proved in [1], where $d\mu_t/dm$ stands for the density function associated to the absolutely continuous measure μ_t.

The appearance of attractors with several pieces motivates in [29], see Sect. 2, the definition of renormalizable EBM. Then, from a renormalization process, the Main Theorem in [29] proves the existence of three values of the parameter t, $\frac{1}{\sqrt{2}} < t_3 < t_2 < t_1 = \frac{1}{\sqrt[5]{4}}$, for which:

(a) Λ_t is a n times renormalizable EBM for every $t \in (\frac{1}{\sqrt{2}}, t_n)$, $n = 1, 2, 3$.
(b) For every $n = 1, 2, 3$ there exists an interval of parameters $I_n \subset (\frac{1}{\sqrt{2}}, t_n)$ such that Λ_t displays, at least, 2^{n-1} different strange attractors.

Once Λ_t is renormalized, the new map is no longer an EBM with a single fold, but an EBM with two folds, each one of them determined by the same parameter. So, the set $\mathbb{F}$ of these EBM's with two folds can be identified with a subset $\mathcal{P} \subset \mathbb{R}^2$ of pairs of parameters (a, b). Therefore, in order to perform successive renormalizations, it is necessary to extend the process of renormalization to the set $\mathbb{F}$. Fortunately, there is a region $\mathcal{P}_3 \subset \mathcal{P}$ such that, for every $(a, b) \in \mathcal{P}_3$, the corresponding $\Psi_{a,b} \in \mathbb{F}$ can be renormalized on two disjoint restrictive domains $\Delta_{a,b}$ and $\Pi_{a,b}$ at the same time. In this way, we can define two different renormalization operators H_Δ and H_Π from $\mathcal{P}_3$ to $\mathcal{P}$ and prove Theorem A in [30]. This result state that for every $(a, b) \in \mathcal{P}_3$ the map $\Psi_{a,b}$ is simultaneously renormalizable in $\mathbb{F}$. More precisely, it holds that:

(i) The restriction of $\Psi_{a,b}^4$ to $\Delta_{a,b}$ is conjugate by means of an affine change in coordinates to $\Psi_{H_\Delta(a,b)}$ restricted to $\Psi_{H_\Delta(a,b)}(\mathcal{T})$.

(ii) The restriction of $\Psi_{a,b}^4$ to $\Pi_{a,b}$ is conjugate by means of an affine change in coordinates to $\Psi_{H_\Pi(a,b)}$.

This renormalization was an useful tool to prove, also in [30], the coexistence of arbitrarily large number of strange attractors of Λ_t (see Theorem B). Namely, it was proven that for each natural number n there exists an interval I_n such that Λ_t displays, at least, 2^n different strange attractors whenever $t \in I_n$. Notice that both, these attractors and those ones found in [29, 33], are fully persistent.

The study of the attractors of the families of EBM's Λ_t and $\Psi_{a,b}$ is a first step to obtain information about the possible global structures of the attractors of the quadratic family $T_{a,b}$ given in (6). However, one has to expect further complications to prove the existence of strange attractors in this nonlinear case because the maps $T_{a,b}$ are not expansive near the fold line $y = 0$. As a first consequence, as happens in the one-dimensional case, the strange attractors, if they exist, will not be completely persistent. Then, as was also done in the one-dimensional case, see [22], one may first have to investigate the existence of absolutely continuous invariant measures.

We would like to conclude this section by taking up again the ideas of the previous ones to propose the search for germs of strange attractors of dimension two. Once it has become clear that these attractors can only appear in the dynamics of diffeomorphisms in dimension $n \geq 3$, we have to consider families of vector fields in $\mathbb{R}^4$ with some kind of homoclinic cycle involving an unstable manifold of dimension two. One of these cycles could be a bifocal homoclinic orbit of an equilibrium point with eigenvalues $-\rho \pm i\omega_1$ and $\lambda \pm i\omega_2$ where ρ and λ are positive.

In [5] it was proven that bifocal homoclinic orbits are present in generic unfoldings of the nilpotent singularity of codimension four in $\mathbb{R}^4$. In [6] it was proven that suspended robust heterodimensional cycles and suspended robust homoclinic tangencies can be found arbitrarily close to any non-degenerate bifocal homoclinic orbit of a Hamiltonian vector field. Since these vector fields are found in the limit family of any generic unfolding of the nilpotent singularity of codimension four in $\mathbb{R}^4$, it was conjectured that suspended robust cycles can be generically unfolded from such singularity. Perhaps with some extra effort it can be shown in the future that the generalized homoclinic tangency, as was defined in [37], is also found in generic unfolding of the nilpotent singularity of codimension four. Then the last and most ambitious task is to prove that in any generic unfolding of this homoclinic tangency there exist persistent strange attractors. Right here the results obtained for the limit family $T_{a,b}$ given in (6) will be of great help. With this regard, the discussion about the existence of three dimensional horseshoes around a bifocal homoclinic orbit included in [19] should be also helpful.

Acknowledgements Authors have been supported by the Spanish Research project MTM2014-56953-P.

References

1. Alves J, Pumariño A, Vigil E (2017) Statistical stability for multidimensional piecewise expanding maps. Proc Am Math Soc 145(7):3057–3068
2. Baldomá I, Castejón O, Seara TM (2013) Exponentially small heteroclinic breakdown in the generic Hopf-zero singularity. J Dyn Differ Equ 25(2):335–392

3. Baldomá I, Castejón O, Seara TM (2016) Breakdown of a 2D heteroclinic connection in the Hopf-zero singularity (I). arXiv:1608.01115
4. Baldomá I, Castejón O, Seara TM (2016) Breakdown of a 2D heteroclinic connection in the Hopf-zero singularity (II). The generic case. arXiv:1608.01116
5. Barrientos PG, Ibáñez S, Rodríguez JA (2011) Heteroclinic cycles arising in generic unfoldings of nilpotent singularities. J Dyn Differ Equ 23(4):999–1028
6. Barrientos PG, Ibáñez S, Rodríguez JA (2016) Robust cycles unfolding from conservative bifocal homoclinic orbits. Dyn Syst 31(4):546–579
7. Barrio R, Ibáñez S, Pérez L (2017) Hindmarsh-Rose model: close and far to the singular limit. Phys Lett A 381(6):597–603
8. Benedicks M, Carleson L (1985) On iterations of $1 - ax^2$ on $(-1, 1)$. Ann Math 122:1–25
9. Benedicks M, Carleson L (1991) The dynamics of the Hénon map. Ann Math 133:73–169
10. Drubi F, Ibáñez S, Rodríguez JA (2007) Coupling leads to chaos. J Differ Equ 239(2):371–385
11. Drubi F, Ibáñez S, Rodríguez JA (2008) Singularities in coupled systems. Bull Belg Math Soc Simon Stevin 15:797–808
12. Drubi F, Ibáñez S, Rodríguez JA (2011) Hopf-pitchfork singularities in coupled systems. Phys D 240(9–10):825–840
13. Dumortier F, Ibáñez S (1996) Nilpotent singularities in generic 4-parameter families of 3-dimensional vector fields. J Differ Equ 127(2):590–647
14. Dumortier F, Ibáñez S, Kokubu H (2001) New aspects in the unfolding of the nilpotent singularity of codimension three. Dyn Syst 16(1):63–95
15. Dumortier F, Ibáñez S, Kokubu H (2006) Cocoon bifurcation in three-dimensional reversible vector fields. Nonlinearity 19(2):305–328
16. Dumortier F, Ibáñez S, Kokubu H, Simó C (2013) About the unfolding of a Hopf-zero singularity. Discrete Contin Dyn Syst Ser A 33(10):4435–4471
17. Gil Álvarez P (1996) Las matemáticas de lo incierto. Servicio de publicaciones. Universidad de Oviedo. http://digibuo.uniovi.es/dspace/bitstream/10651/28625/1/matematicasincierto.pdf
18. Hénon M (1976) A two-dimensional mapping with a strange attractor. Commun Math Phys 50:69–77
19. Ibáñez S, Rodrígues A (2015) On the dynamics near a homoclinic network to a bifocus: switching and horseshoes. Int J Bif Chaos 25(11):1530030
20. Ibáñez S, Rodríguez JA (1995) Shil'nikov bifurcations in generic 4-unfoldings of a codimension-4 singularity. J Differ Equ 120(2):411–428
21. Ibáñez S, Rodríguez JA (2005) Shil'nikov configurations in any generic unfolding of the nilpotent singularity of codimension three on $\mathbb{R}^3$. J Differ Equ 208(1):147–175
22. Jacobson MV (1981) Absolutely continuous invariant measures for one-parameter families of one-dimensional maps. Commun Math Phys 81:39–88
23. Lorenz EN (1963) Deterministic non-periodic flow. J Atmos Sci 20:130–141
24. Mora L, Viana M (1993) Abundance of strange attractors. Acta Math 171:1–71
25. Pumariño A, Rodríguez JA (1997) Coexistence and persistence of strange attractors, vol 1658. Lecture notes in mathematics. Springer, Berlin
26. Pumariño A, Rodríguez JA (2001) Coexistence and persistence of infinitely many strange attractors. Ergod Theory Dyn Syst 21(5):1511–1523
27. Pumariño A, Tatjer JC (2006) Dynamics near homoclinic bifurcations of three-dimensional dissipative diffeomorphisms. Nonlinearity 19:2833–2852
28. Pumariño A, Tatjer JC (2007) Attractors for return maps near homoclinic tangencies of three-dimensional dissipative diffeomorphisms. Discrete Contin Dyn Syst Ser B 8:971–1005
29. Pumariño A, Rodríguez JA, Vigil E (2017) Renormalizable expanding baker maps: coexistence of strange attractors. Discrete Contin Dyn Syst Ser A 37(3):1651–1678
30. Pumariño A, Rodríguez JA, Vigil E (2018) Renormalization of two-dimensional piecewise linear maps: abundance of 2-D strange attractors. Discrete Contin Dyn Syst Ser A 38(2):941–966

31. Pumariño A, Rodríguez JA, Tatjer JC, Vigil E (2013) Piecewise linear bidimensional maps as models of return maps for 3D diffeomorphisms. In: Ibáñez S, Pérez del Río J, Pumariño A, Rodríguez J (eds) Progress and challenges in dynamical systems. Springer proceedings in mathematics & statistics, vol 54. Springer, Berlin
32. Pumariño A, Rodríguez JA, Tatjer JC, Vigil E (2014) Expanding baker maps as models for the dynamics emerging from 3D-homoclinic bifurcations. Discrete Contin Dyn Syst Series B 19(2):523–541
33. Pumariño A, Rodríguez JA, Tatjer JC, Vigil E (2015) Chaotic dynamics for two-dimensional tent maps. Nonlinearity 28(2):407–434
34. Ruelle D (1989) Chaotic evolution and strange attractors. Cambridge University Press, Cambridge
35. Shilnikov LP (1965) A case of the existence of a denumerable set of periodic motions. Sov Math Dokl 6:163–166
36. Smale S (1967) Differentiable dynamical systems. Bull Am Math Soc 73:747–817
37. Tatjer JC (2001) Three-dimensional dissipative diffeomorphisms with homoclinic tangencies. Ergod Theory Dyn Syst 21:249–302
38. Tresser C (1981) Modéles simples de transitions vers la turbulence. Universitè de Nice, Thèse d'Etat

Finite Orthogonal Laurent Polynomials

Francisco Marcellán and Anbhu Swaminathan

Abstract In this work, orthogonal Laurent polynomials on the real line associated with a finite family of classical orthogonal polynomials (the so called Romanovski–Hermite polynomial sequence) are discussed. Their explicit representation, a second order linear differential equation they satisfy as well as their orthogonality relations are obtained. The connection with a strong Stieltjes moment problem is discussed. The strong Gaussian quadrature formulae are also given. For such a family of orthogonal Laurent polynomials, a comparison with the classical Gaussian quadrature formulae is illustrated with some examples.

1 Introduction

The analysis of the "Strong Stieltjes moment problem" started in 1980 in the seminal paper by Jones et al. [12], where orthogonal Laurent polynomial sequences [11] were used as a main tool. For further development and basic results on orthogonal Laurent polynomial sequences, see [2, 3, 8–10] and references therein. The strong moment problem related to the theory of orthogonal Laurent polynomial sequences is similar to the "classical moment problem" related to the theory of standard orthogonal polynomials. Note that many results of classical orthogonal polynomials can be extended to orthogonal Laurent polynomial sequences. Nevertheless, some results that are true for orthogonal Laurent polynomial sequences do not hold for standard orthogonal polynomials [2].

F. Marcellán (✉)
Instituto de Ciencias Matemáticas (ICMAT) and Departamento de Matemáticas, Universidad Carlos III de Madrid, Avenida de la Universidad 30, 28911 Leganés, Spain
e-mail: pacomarc@ing.uc3m.es

A. Swaminathan
Department of Mathematics, Indian Institute of Technology, Roorkee 247 667, Uttarkhand, India
e-mail: swamifma@iitr.ernet.in; mathswami@gmail.com

© Springer International Publishing AG 2018
E. Gil et al. (eds.), *The Mathematics of the Uncertain*, Studies in Systems, Decision and Control 142, https://doi.org/10.1007/978-3-319-73848-2_79

869

In [6, 7] Hagler and others have shown the connection between orthogonal polynomials and orthogonal Laurent polynomials by computing orthogonal Laurent polynomial sequence on the real line $\mathbb{R}$ from the well known classical orthogonal polynomials (COPS, in short) associated with the Normal, Gamma and Beta distributions, i.e. Hermite, Laguerre and Jacobi, polynomials.

Notice that the above three COPS constitute an infinite sequence of classical orthogonal polynomials, since the recent contributions of Masjed-Jamei [15] show three other classes of orthogonal polynomials, due to Romanovski [19], which are finite in the sense that their parameters yield a finite family of polynomials satisfying orthogonality conditions. These finite polynomials were initially analyzed by Routh [20] and by Romanovski [19] and deeply studied by Masjed–Jamei in [15]. For details on these polynomials see [15–18] and references therein. For a more updated information, we recommend to read Chap. 4 of the monograph [13].

Due to the restriction on the finiteness of these families of orthogonal polynomials and the fact that they can be reduced to Jacobi polynomials [17], these polynomials are not studied in detail as the other three well-known COPS. However it is understood (see [17]) that these finite orthogonal polynomials are analyzed deeply in the framework of spectral analysis of second order differential operators. In this work, we focus our attention on the orthogonal Laurent polynomials associated with the Romanovski–Hermite finite classical orthogonal polynomials.

We further remark that another type of transformation was considered in [1] to obtain Laurent orthogonal polynomials which are much useful while considering the corresponding Szegő polynomials and related results. In this work we consider the transformation given in [6] which is very useful for obtaining Gaussian quadrature rules.

Definition 1.1 [2] A **Laurent polynomial** (L-polynomial, in short) is a function $R(x) = \Sigma_{j=m}^{n} r_j x^j$ where x is a real variable and $m, n \in \mathbb{Z}$ with $m \leq n$, $r_j \in \mathbb{R}$. We will denote $\mathscr{R}_{m,n} = span\{x^j\}_{j=m}^{n}$. Two important classes of L-polynomials are

$$\mathscr{R}_{2n} = \{R \in \mathscr{R}_{-n,n} : \text{the coefficient of } x^n \text{ is non zero}\}$$

$$\mathscr{R}_{2n+1} = \{R \in \mathscr{R}_{-n-1,n} : \text{the coefficient of } x^{-n-1} \text{is non zero}\}$$

for all $n \in \mathbb{Z}_0^+$.

We will denote by Λ the linear space of Laurent polynomials with real coefficients.

Definition 1.2 [2] An L-polynomial R is said to be of **L-degree** m if $R \in \mathscr{R}_m$. The coefficients of x^n and x^{-n-1} are called the L-leading coefficients and coefficient of x^{-n} and x^n are called the L-trailing coefficients for $\mathscr{R}_{2n}$ and $\mathscr{R}_{2n+1}$, respectively. An L-polynomial is said to be **regular** if the respective trailing coefficients are non zero and **monic** if the respective leading coefficients are one.

Let $\mathbb{M}$ be a linear functional defined on Λ. The corresponding moments are given by $\mu_n = \mathbb{M}(x^n)$, $n \in \mathbb{Z}$. In the sequel we will call $\mathbb{M}$ a moment functional. For each moment functional, we can define a bilinear form as $\langle P, Q \rangle_{\mathbb{M}} = \mathbb{M}(P(x)Q(x))$.

Definition 1.3 [2] A sequence of polynomials $\{R_n\}_{n=0}^{\infty}$ is said to be an orthogonal Laurent polynomial sequence with respect to a moment functional $\mathbb{M}$ if the following three properties hold for every $m, n \in \mathbb{Z}_0^+$

- $R_n \in \Lambda$, R_n has L-degree n,
- $\langle R_n, R_m \rangle_{\mathbb{M}} = 0$ if $m \neq n$,
- $\langle R_n, R_n \rangle_{\mathbb{M}} = \|R_n\|_{\mathbb{M}}^2 > 0$.

The orthogonality of an orthogonal Laurent polynomial sequence can be characterized as

$$\langle R_{2m}(x), x^k \rangle_{\mathbb{M}} = 0 \quad \text{for} \quad k = -m, -m+1, \cdots, m-1,$$

$$\langle R_{2m+1}(x), x^k \rangle_{\mathbb{M}} = 0 \quad \text{for} \quad k = -m, -m+1, \cdots, m,$$

$$\|R_{2m}\|_{\mathbb{M}}^2 = \langle R_{2m}(x), x^m \rangle_{\mathbb{M}} = \frac{H_{2m+1}^{(-2m)}}{H_{2m}^{(-2m)}} > 0,$$

$$\|R_{2m+1}\|_{\mathbb{M}}^2 = \langle R_{2m+1}(x), x^{-m-1} \rangle_{\mathbb{M}} = \frac{H_{2m+2}^{(-2m-2)}}{H_{2m+1}^{(-2m)}} > 0.$$

Here $H_k^{(n)} = det[\mu_{k+j+l}]_{j,l=0}^{n-1}$ denotes the Hankel determinant associated with the moment sequence $\{\mu_n\}_{n=-\infty}^{\infty}$. By a "Strong Moment Problem" we mean the following: Given a bilateral sequence of moments $\{\mu_n\}_{n=-\infty}^{\infty}$ of real numbers to find a distribution function ψ such that

$$\mu_n = \int_a^b t^n d\psi(t), \quad n \in \mathbb{Z}, \quad -\infty \leq a < b \leq \infty. \tag{1}$$

Notice that for the classical moment problem, given a sequence of real numbers $\{\mu_n\}_{n=0}^{\infty}$ one deals with the existence of a distribution function ψ such that (1) holds only for non negative integers.

In the sequel, we use the notation ψ as the moment distribution function for the monic orthogonal polynomials (MOPS) $\{\hat{P}_n(x)\}_{n=0}^{\infty}$ with spectrum $\sigma(\psi)$ and $\tilde{\psi}$ as the strong moment distribution function for the monic orthogonal Laurent polynomials $\{\tilde{P}_n(x)\}_{n=0}^{\infty}$ with spectrum $\sigma(\tilde{\psi})$.

The structure of this contribution is as follows. In Sect. 2, the corresponding orthogonal Laurent polynomial sequence for the Romanovski–Hermite polynomials is obtained. We focus our attention on the second order linear differential equation they satisfy as well as their orthogonality properties. Gaussian quadrature rules with respect to the strong weight function are given in Sect. 3. Some numerical tests are presented.

2   Finite Laurent Orthogonal Polynomials

In this section, we consider the Romanovski–Hermite polynomials introduced in [15, 16] as well as we use the transformation formulae given in [6]. Our aim is to obtain analytic properties of corresponding family of finite orthogonal Laurent polynomials. Furthermore, for these polynomials $J_n^{(p,q)}(x; a, b, c, d)$, we will deal with the monic polynomials, which will be denoted by $\hat{J}_n^{(p,q)}(x; a, b, c, d)$, as well as the corresponding monic Laurent polynomials which will be denoted by $\widetilde{J}_n^{(p,q)}(x; a, b, c, d)$.

2.1   Generalized Romanovski–Hermite Type Polynomials

These polynomials are closely related to Jacobi polynomials and hence they are a particular case of Romanovski pseudo-Jacobi polynomials. This family was introduced in [16] and extensively studied in [17]. It is said to be a generalized Romanovski–Hermite class because it reduces to the Romanovski class given in [15] for the choice of parameters $\hat{J}_n^{(p-\frac{1}{2},0)}(x; 1, 0, 0, 1)$, which is related to Hermite polynomials.

The Generalized Romanovski–Hermite moment distribution function $\psi_I^{(p)}$ (see [16]) is given by

$$\frac{d\psi_J^{(p,q)}}{dx} = ((ax + b)^2 + (cx + d)^2)^{-p} \exp\left(q \arctan \frac{ax + b}{cx + d}\right), \quad x \in \mathbb{R}. \quad (2)$$

Applying the transformation $\upsilon(x) = \frac{1}{\lambda}\left(x - \frac{\gamma}{x}\right)$ [6, Theorem 2.3.1] to the moment distribution function (2) we get a strong moment distribution function for each choice of $\lambda > 0$ and $\gamma > 0$. The spectrum $\sigma(\widetilde{\psi})$ is

$$\sigma(\widetilde{\psi}) = v_+^{-1}(\sigma(\psi)) \cup v_-^{-1}(\sigma(\psi)) = \mathbb{R}\backslash\{0\}.$$

To find the generalized Romanovski–Hermite strong moment distribution function $\widetilde{\psi}_J^{(p,q)}$, we use the fact that $\frac{d\widetilde{\psi}}{dx} = w(\upsilon(x))$ with $w(x) = \frac{d\psi}{dx}$, so that

$$\frac{d\widetilde{\psi}_J^{(p,q)}}{dx} = \left[\left(\frac{a(x^2 - \gamma)}{\lambda x} + b\right)^2 + \left(\frac{c(x^2 - \gamma)}{\lambda x} + d\right)^2\right]^{-p}$$

$$\times \exp\left(q \arctan \frac{a(x^2 - \gamma) + b\lambda x}{c(x^2 - \gamma) + d\lambda x}\right), \quad x \in \mathbb{R}\backslash\{0\},$$

is a strong moment distribution function for each choice of the parameters λ and γ.

From the expression of $J_n^{(p,q)}(x; a, b, c, d)$ given in [16, p. 139, (6)], for $\hat{J}_n^{(p,q)}(x; a, b, c, d)$ with parameters $p > N + \frac{1}{2}$ and $ad - bc \neq 0$, where $N = \max\{m, n\}$, we get

$$\hat{J}_n^{(p,q)}(x; a, b, c, d)$$

$$= (-1)^n (a^2 + c^2)^n (n + 1 - 2p)_n \sum_{k=0}^{n} \binom{n}{k} \left(\frac{a^2 + c^2}{(ab + cd) + i(ad - bc)} \right)^k$$

$$\times {}_2F_1 \left(k - n, p - n - iq/2; 2p - 2n; \frac{2(ad-bc)}{(ab+cd)+i(ad-bc)} \right) x^k, \qquad (3)$$

where $i = \sqrt{-1}$ and ${}_2F_1(-, -; -)$ is the Gaussian hypergeometric function.

The explicit representation for $\tilde{J}_n^{(p,q)}(x; a, b, c, d)$ can be obtained by using the fact that $\tilde{P}_{2n}(x) = \lambda^n P_n(\upsilon(x))$ and $\tilde{P}_{2n+1}(x) = (\frac{\lambda}{\gamma})^n \frac{1}{x} P_n(\upsilon(x))$ together with (3). Indeed,

$$\tilde{J}_{2n}^{(p,q)}(x; a, b, c, d)$$

$$= (-1)^n (a^2 + c^2)^n (n + 1 - 2p)_n \sum_{k=0}^{n} \binom{n}{k} \left(\frac{a^2 + c^2}{(ab + cd) + i(ad - bc)} \right)^k$$

$$\times {}_2F_1 \left(k - n, p - n - iq/2; 2p - 2n; \frac{2(ad-bc)}{(ab+cd)+i(ad-bc)} \right) \lambda^{n-k} \times (x - \frac{\gamma}{x})^k,$$

and

$$\tilde{J}_{2n+1}^{(p,q)}(x; a, b, c, d)$$

$$= \frac{(-1)^n}{\gamma^n x} (a^2 + c^2)^n (n + 1 - 2p)_n \sum_{k=0}^{n} \binom{n}{k} \left(\frac{a^2 + c^2}{(ab + cd) + i(ad - bc)} \right)^k$$

$$\times {}_2F_1 \left(k - n, p - n - iq/2; 2p - 2n; \frac{2(ad-bc)}{(ab+cd)+i(ad-bc)} \right) \lambda^{n-k} (x - \frac{\gamma}{x})^k,$$

for $n = 0, 1, 2, 3, \cdots$.

In a self-adjoint form, the second order linear differential equation for $\hat{J}_n^{(p,q)}$ $(x; a, b, c, d)$ is

$$\frac{d}{dx} \left[\left((ax + b)^2 + (cx + d)^2 \right)^{-p+1} \exp \left(q \arctan \frac{ax+b}{cx+d} \right) y_n'(x) \right]$$

$$- n(n + 1 - 2p)(a^2 + c^2) \left((ax + b)^2 \right.$$

$$\left. + (cx + d)^2 \right)^{-p} \exp \left(q \arctan \frac{ax+b}{cx+d} \right) y_n(x) = 0,$$

which is the analog of the self-adjoint form for $J_n^{(p,q)}(x; a, b, c, d)$ [16, p. 140, (12)]. Applying [6, Theorem 3.8.1] with the above self-adjoint form, we get that the function $Y = \tilde{J}_n^{(p,q)}(x; a, b, c, d)$ satisfies the following differential equation

$$\left[\left(\left(\frac{a}{\lambda}\left(x - \frac{\gamma}{x}\right) + b\right)^2\right.\right.$$

$$\left.\left.+ \left(\frac{c}{\lambda}\left(x - \frac{\gamma}{x}\right) + d\right)^2\right)^{-p+1} Y'(x) \exp\left(q \arctan \frac{a(x^2-\gamma)+b\lambda x}{c(x^2-\gamma)+d\lambda x}\right)\right]'$$

$$-n(n+1-2p)(a^2+c^2)\left(\left(\frac{a}{\lambda}\left(x - \frac{\gamma}{x}\right) + b\right)^2\right.$$

$$\left.+ \left(\frac{c}{\lambda}\left(x - \frac{\gamma}{x}\right) + d\right)^2\right)^{-p} Y(x) \exp\left(q \arctan \frac{a(x^2-\gamma)+b\lambda x}{c(x^2-\gamma)+d\lambda x}\right) = 0.$$

The corresponding orthogonality relation for $\hat{J}_n^{(p,q)}(x; a, b, c, d)$ becomes (see [16, p. 142, (23)])

$$\langle \hat{J}_n^{(p,q)}(x; a, b, c, d), \hat{J}_m^{(p,q)}(x; a, b, c, d)\rangle_{\psi_j^{(p,q)}}$$

$$= \int_{-\infty}^{\infty} \hat{J}_n^{(p,q)}(x; a, b, c, d)\hat{J}_m^{(p,q)}(x; a, b, c, d)((ax + b)^2 + (cx + d)^2)^{-p}$$

$$\times \exp\left(q \arctan \frac{ax+b}{cx+d}\right) dx = \frac{2^{2n+1-2p}(ad - bc)^{2n-2p+1} \exp(-q \arctan(c/a))}{(2p - 2n - 1)(a^2 + c^2)^{-p+1}}$$

$$\times \frac{n!\Gamma(2p - n)}{\Gamma(p - n + iq/2)\Gamma(p - n - iq/2)} \times \left(\frac{(ab+cd)+i(ad-bc)}{(a^2+c^2)}\right)^{2n} \delta_{mn}. \qquad (4)$$

Applying [6, Theorem 2.2.8] to (4), the orthogonality relation for $\widetilde{J}_n^{(p,q)}(x; a, b, c, d)$ is

$$\langle \widetilde{J}_n^{(p,q)}(x; a, b, c, d), \widetilde{J}_m^{(p,q)}(x; a, b, c, d)\rangle_{\widetilde{\psi}_j^{(p,q)}} = K_n\delta_{mn},$$

where

$$K_n = \begin{cases} \lambda^{2n+1}\,\overline{K}_n & \text{if } n \text{ is even,} \\ (\frac{\lambda}{\gamma})^{2n+1}\overline{K}_n & \text{if } n \text{ is odd,} \end{cases}$$

with

$$\overline{K}_n = \frac{2^{2n+1-2p}(ad - bc)^{2n-2p+1}\exp(-q\arctan(c/a))}{(2p - 2n - 1)(a^2 + c^2)^{-p+1}}$$

$$\times \frac{n!\Gamma(2p - n)}{\Gamma(p - n + iq/2)\Gamma(p - n - iq/2)}\left(\frac{(ab + cd) + i(ad - bc)}{(a^2 + c^2)}\right)^{2n}\delta_{nm}.$$

3 Strong Gaussian Quadrature Rules

In a general form, a weighted quadrature formula is given by

$$\int_a^b f(x)w(x)dx = \sum_{j=1}^n A_j f(x_j) + \sum_{k=1}^m v_k f(z_k) + R_{n,m}(f),   (5)$$

where $w(x)$ is positive function on $[a, b]$, $\{A_j\}_{j=1}^n$, $\{v_k\}_{k=1}^m$ are unknown coefficients, $\{x_j\}_{j=1}^n$ are unknown nodes, and $\{z_k\}_{k=1}^m$ are pre-determined nodes. This quadrature rule (5) is known in the literature as Gauss–Kronrod formula.

The error function $R_{n,m}(f)$ for (5) is given by

$$R_{n,m}(f) = \frac{f^{(2n+m)}(\xi)}{(2n+m)!} \int_a^b w(x) \prod_{k=1}^m (x - z_k) \prod_{i=1}^n (x - x_i)^2 dx,$$

where $f^{(m)}$ denotes the mth derivative of f and $a < \xi < b$.

In general, the above quadrature rule has $2n + m - 1$ as highest degree of precision and it reduces to the Gaussian quadrature formula for $m = 0$ if and only if $\{x_j\}_{j=1}^n$ are the zeros of the polynomials of degree n orthogonal with respect to the weight function $w(x)$ (see [14]).

Moreover, to obtain the coefficients $\{A_j\}_{j=1}^n$, it is necessary to solve the following linear system

$$\sum_{j=1}^n A_j x_j^k = \int_a^b x^k w(x)dx,   k = 0, 1, 2, \cdots, 2n - 1.$$

Instead of this approach, one can use this very well known expression (see [4])

$$\frac{1}{A_j} = \sum_{l=0}^{n-1} P_l^{*2}(x_j),   j = 1, 2, \cdots, n,$$

where $P_l^*(x)$ denotes the orthonormal polynomial of degree l with respect to w, i.e.,

$$P_l^*(x) = \frac{P_l(x)}{\langle P_l(x), P_l(x) \rangle^{\frac{1}{2}}}.$$

We refer to [5] for exhibiting the relation between the quadrature nodes and quadrature weights for the orthogonal polynomial sequence and orthogonal Laurent polynomial sequence, respectively. We state two important theorems that will be used in the sequel.

Theorem 3.1 [5] (Zeros) *Let n be a positive integer number and assume $\{x_{n,k}\}_{k=1}^n$ are the zeros of $P_n(x)$ in increasing order, i.e. $x_{n,1} < x_{n,2} < x_{n,3} < \cdots < x_{n,n}$. Then* *the zeros of $\widetilde{P}_{2n}(x)$ and $\widetilde{P}_{2n+1}(x)$ are $x^*_{n,j} = v^{-1}_*(x_{n,j})$, for $* = +, -$ and have the ordering $x^-_{n,1} < x^-_{n,2} < \cdots < x^-_{n,n} < 0 < x^+_{n,1} < x^+_{n,2} < \cdots < x^+_{n,n}$.*

Theorem 3.2 [5] (Weights) *Let $\{x_{n,j}\}^n_{j=1}$ and $\{x^*_{n,j}\}^n_{j=1}$ be the quadrature nodes given in Theorem 3.1, and let $\{A_{n,k}\}^n_{k=1}$ and $\{A^*_{n,k}\}^n_{k=1}$, denote the corresponding Gauss quadrature weights, respectively. Then $A_{n,k} = v'(x^*_{n,k})A^*_{n,k}$, $k = 1, 2, 3, \cdots, n$.*

3.1 L-Quadrature Formulae for Generalized Romanovski–Hermite Polynomials $\hat{J}^{(p,q)}_n(x; a, b, c, d)$

Consider the two term quadrature formula for $\hat{J}^{(p,q)}_n(x; a, b, c, d)$ with parameters $p = 4, q = 1, a = d = 1, b = c = 0$. given in [16] as

$$\int_{-\infty}^{\infty} \frac{\exp(\arctan x)}{(1+x^2)^4} f(x)dx$$

$$\cong 0.6220884910 f(-0.2109772229) + 0.4316063958 f(0.7109772229),$$

where the nodes are the zeros of $J^{(4,1)}_2(x; 1, 0, 0, 1)$ and the weights are given by $\frac{1}{A_j} = \sum_{k=0}^{1} \left(J^{*(4,1)}_k(x; 1, 0, 0, 1) \right)^2 (x_j)$. The respective L-quadrature formulae for $\lambda = \gamma = 1$ is

$$\int_{-\infty}^{\infty} \frac{\exp(\arctan x)}{(1+x^2)^4} f(x)dx$$

$$\cong 0.3436748193 f(-1.1110371419) + 0.1435191249 f(-0.7058182022)$$
$$+0.2784136716 f(0.9000599190) + 0.2880872708 f(1.4167954251),$$

where the nodes and the weights are calculated by using Theorems 3.1 and 3.2, respectively (Table 1).

Table 1 Two point Gaussian quadrature approximation with respect to generalized strong Romanovski–Hermite distribution for $p = 4, q = 1, a = d = 1, b = c = 0$

$f(x)$	$\int_{\sigma(\tilde{\psi})} f(x)d\tilde{\psi}(x)$	L-quadrature	Rel. error
$\frac{1}{1+e^x}$	0.4910992188	0.4913335249	0.0004771052
$\frac{1}{\sqrt{1+2x^2}}$	0.5884790363	0.58674796350	0.0029416049
$\cos x$	0.48410953800	0.47897149960	0.0106133798

Table 2 Three point Gaussian quadrature approximation with respect to generalized strong Romanovski–Hermite distribution for $p = 4$, $q = 1$, $a = d = 1$, $b = c = 0$

$f(x)$	$\int_{\sigma(\tilde{\psi})} f(x) d\tilde{\psi}(x)$	L-quadrature	Rel.error
$\frac{1}{1+e^x}$	0.4910992188	0.4910014528	0.0001990759
$\frac{1}{\sqrt{1+2x^2}}$	0.5884790363	0.5915109163	0.0051520611
$\cos x$	0.48410953800	0.4845999436	0.0010130055

The three term quadrature formula for $\hat{J}_n^{(p,q)}(x; a, b, c, d)$ with parameters $p = 4$, $q = 1$, $a = d = 1$, $b = c = 0$ is given by

$$\int_{-\infty}^{\infty} \frac{\exp(\arctan x)}{(1 + x^2)^4} f(x) dx$$

$$\cong 0.25322550273 f(-0.5229034027) + 0.7679319861 f(0.3293582536)$$
$$+ 0.0325373985 f(1.693545149),$$

where the nodes are the zeros of $J_3^{(4,1)}(x; 1, 0, 0, 1)$ and the weights are given by $\dfrac{1}{A_j} = \displaystyle\sum_{k=0}^{1} \left(J_k^{*(4,1)}(x; 1, 0, 0, 1) \right)^2 (x_j)$. The respective L-quadrature formula for $\lambda = \gamma = 1$ is

$$\int_{-\infty}^{\infty} \frac{\exp(\arctan x)}{(1 + x^2)^4} f(x) dx$$

$$\cong 0.158639342846 f(-1.2950652616) + 0.3215751447 f(-0.8487897749)$$
$$+ 0.0057555832 f(-0.4635799707) + 0.0945861597 f(0.7721618589)$$
$$+ 0.4463568385 f(1.1781480285) + 0.0267818152 f(2.1571251197),$$

where the nodes and the weights are calculated by using Theorems 3.1 and 3.2, respectively (Table 2).

Acknowledgements The work of the first author (FM) has been supported by Dirección General de Investigación Científica y Técnica, Ministerio de Economía, Industria y Competitividad of Spain, grant MTM2015-65888-C4-2-P.

References

1. De Andrade EXL, Bracciali CF, Sri Ranga A (2007) Another connection between orthogonal polynomials and L-orthogonal polynomials. J Math Anal Appl 330(1):114–132
2. Cochran L, Cooper SC (1992) Orthogonal Laurent polynomials on the real line. In: Cooper SC, Thron WJ (eds) Continued fractions and orthogonal functions, vol 154. Lecture notes in pure and applied mathematics. Marcel Dekker, New York, pp 47–100
3. Costa MS, Godoy E, Lamblém RL, Sri Ranga A (2012) Basic hypergeometric functions and orthogonal Laurent polynomials. Proc Amer Math Soc 140(6):2075–2089
4. Chihara TS (1978) An introduction to orthogonal polynomials. Gordon and Breach, New York
5. Gustafson PE, Hagler BA (1999) Gaussian quadrature rules and numerical examples for strong extensions of mass distribution functions. J Comput Appl Math 105(1–2):317–326
6. Hagler BA (1997) A transformation of orthogonal polynomial sequences into orthogonal Laurent polynomial sequences. PhD Thesis, University of Colorado, Colorado
7. Hagler BA, Jones WB, Thron WJ (1996) Orthogonal Laurent polynomials of Jacobi, Hermite, and Laguerre types. In: Jones WB, Sri Ranga A (eds) Orthogonal functions, moment theory, and continued fractions, vol 199. Lecture notes in pure and applied mathematics. Marcel Dekker, New York, pp 179–186
8. Hendriksen E, van Rossum H (1986) Orthogonal Laurent polynomials. Nederl Akad Wetensch Indag Math 48(1):17–36
9. Jones WB, Njåstad O (1999) Orthogonal Laurent polynomials and strong moment theory: a survey. J Comput Appl Math 105(1–2):51–91
10. Jones WB, Thron WJ (1981) Survey of continued fraction methods of solving moment problems and related topics. In: Jones WB, Thron WJ, Waadeland H (eds) Analytic theory of continued fractions, vol 932. Lecture notes in mathematics. Springer, Berlin, pp 4–37
11. Jones WB, Thron WJ, Njåstad O (1984) Orthogonal Laurent polynomials and the strong Hamburger moment problem. J Math Anal Appl 98(2):528–554
12. Jones WB, Thron WJ, Waadeland H (1980) A strong Stieltjes moment problem. Trans Am Math Soc 261(2):503–528
13. Koekoek R, Lesky PA, Swarttouw RF (2010) Hypergeometric orthogonal polynomials and their q-analogues. Springer monographs in mathematics. Springer, Berlin
14. Krylov VI (1962) Approximate calculation of integrals (translated by Stroud AH). Macmillan, New York
15. Masjed-Jamei M (2002) Three finite classes of hypergeometric orthogonal polynomials and their application in functions approximation. Integ Transf Spec Funct 13(2):169–191
16. Masjed-Jamei M (2004) Classical orthogonal polynomials with weight function $((ax + b)^2 + (cx + d)^2)^{-p} \exp(q \arctg((ax + b)/(cx + d)))$, $x \in (-\infty, \infty)$ and a generalization of T and F distributions. Integ Transf Spec Funct 15(2):137–153
17. Masjed-Jamei M, Marcellán F, Huertas EJ (2014) A finite class of orthogonal functions generated by Routh–Romanovski polynomials. Compl Var Ellip Equ 59(2):162–171
18. Malik P, Swaminathan A (2012) Derivatives of a finite class of orthogonal polynomials related to inverse gamma distribution. Appl Math Comput 218(11):6251–6262
19. Romanovski VI (1929) Sur quelques classes nouvelles de polynômes orthogonaux. C R Acad Sci Paris 188:1023–1025
20. Routh EJ (1884) On some properties of certain solutions of a differential equation of the second order. Proc Lond Math Soc 16:245–261

From Semi-infinite Programming to Efficiency Analysis: Forty Years of Research Experiences

Jesús T. Pastor

Abstract Life is mostly surrounded by uncertainties. Mathematics, as part of life, deals also with them. As an applied mathematician I have decided to relate my own research experiences, a relevant part of my life, so that people from younger generations can appreciate how research topics may appear just by chance. The second message I would like to transmit is that if someone really likes doing research he has to fully commit to it. The reason for it is easy: it delivers plenty of moments of happiness and intellectual satisfaction.

In Memoriam

I have written this paper in honour of Prof. Pedro Gil, a well-known Spanish mathematician who passed away in 2016 leaving behind his wife Pilar and three children, Eva, Juan and Eduardo. He was a good man, but also a very good teacher and a clever university professor. He was the founder of the Statistics, Operations Research and Didacticism group at the University of Oviedo. He achieved his academic background at the Universidad Complutense, being one of the distinguished scholars of Prof. Sixto Ríos. He was just two years older than I was and we met regularly, for a long period of time every one and a half years, at the national meetings organised by our scientific society (SEIO). I was the seventh president of SEIO, and he was the ninth. Although he was basically a statistician and I have basically been an operations researcher, we have had the opportunity to share and contrast our points of view in relation to our society and our common profession. He inspired confidence and was always ready to help his students and colleagues. His legacy will always be part of us.

1 Introduction

This paper is intended to be an informational paper, based on my research and related vital experiences over the last forty years.

I finished my five year Mathematics degree at the University of Valencia in June 1972. The first phase of my university teaching career lasted for only three years, from

J. T. Pastor (✉)
Universidad Miguel Hernández, UMH Elche (Alicante), Spain
e-mail: jtpastor@umh.es

© Springer International Publishing AG 2018
E. Gil et al. (eds.), *The Mathematics of the Uncertain*, Studies in Systems, Decision and Control 142, https://doi.org/10.1007/978-3-319-73848-2_80

879

October 1972 to September 1975. One of my colleagues was Miguel A. Goberna. We decided to apply for a Professorship in secondary education in 1974. At that time, the Spanish Education System was still centralised, and we had to compete with people coming from all over Spain. Since both of us succeeded, we left our teaching position at Valencia University and moved to a different state secondary school. In 1975 we met Marco A. López, only one year older than us, who became professor at the University of Valencia quite early in his career. In 1975, the three of us decided to share a seminar on a new topic for all of us, Semi-infinite Programming. Marco was our leader and we were able to present our PhD dissertation in 1979. Since we wanted to move again to the University and were not admitted to Marco's department, we decided to apply for a more stable position at a Spanish University. We succeeded in 1983 and in 1984 we moved to the University of Alicante as associate professors. Marco also moved with us but as a professor. There, we continued to research Semi-infinite Programming, publishing papers from 1984 to 1987, year in which we started doing some consulting work for several saving banks. One of the topics we used for locating branches was Location Theory and as I found it so appealing, I decided to move from Semi-infinite Programming to Location.

I started doing some work on my own and soon two younger colleagues in my department wanted to start working with me in Location Analysis. I ended up being the Director of their PhD dissertations. I started publishing papers on Location Theory in 1990 and ended in 2000.

Meanwhile, in 1991, I was invited to deliver a couple of talks in the Summer Seminars organised by the Spanish Open University (UNED) at its centre in Denia, a small coastal Mediterranean town located 100 Km north of Alicante. Marco, Miguel and I were invited almost every summer to participate in this enjoyable seminar. The year before, I had already given two talks on Location Analysis, and I had got some additional material for delivering a third talk in 1991. Therefore, at Easter, I went to our science library and started reading recent articles appearing in top Operations Research journals. After a couple of hours I found an interesting paper comparing pharmaceutical companies and evaluating their efficiency using a method called DEA (Data Envelopment Analysis). I found it highly appealing and decided to give my second talk on DEA. Curiously enough, the same year I was going to attend, as usual, the EURO meeting, to be held in Aachen, Germany; when I got the programme, some months later, I realised that a stream of six sessions were devoted to DEA. I was sure that it was the first appearance of this topic in a EURO meeting. During the EURO meeting in Aachen, I had the opportunity to talk to distinguished researchers in the area of Efficiency and Productivity Analysis, and I realised that there was a real disembarkation of U.S. researchers in Europe. I became more and more interested in DEA, ending up devoting my research efforts to it. I published my first paper in 1995 and have not stopped since then. Looking back, I am happy to have moved to this applied area with a strong economic flavour that has produced, since its start in 1978, over 7.000 papers. Moreover, I have had diverse and interesting research activities and opportunities all over the world, as I will explain later on.

The paper is organised as follows. Sections 2, 3 and 4 are devoted to my three research areas, in chronological order. Section 5 concludes.

2 Semi-infinite Programming

As mentioned before, we started the second phase of our University career at the University of Alicante, where, as said before, Miguel A. and I arrived in 1984. It took some time until Miguel and I became full-time professors at the University of Alicante, in 1990. At that moment, Marco, Miguel and I were the only professors in our department. Since we started working together in 1975, we devoted the first 12 years of our joint research to investigating several topics in Semi-infinite Programming. We presented our first international communication at the meeting organised by the Mathematical Programming Society in Budapest, during the summer of 1975. At that meeting there were 4 official languages: English, French, German and Russian. I prepared the presentation in German since, at that time, I was more fluent in German than in English as I had been educated at the German School in Valencia. When we arrived at the meeting we were told that our presentation would be the very last day, with a predictably small audience. Since we had not been informed in advance, they offered to move us to a slot on the first day with the compulsory condition that we had to deliver our talk in English. We decided to do it, and as a result only had a few hours sleep the night before our presentation, as we had to rewrite all our slides. There, we understood directly that the scientific language was English, even behind the "iron curtain". Besides that incident, we had a nice stay and enjoyed the charming restaurants in Buda, where we had very pleasant dinners thanks to the presence of some local violinists and several Hungarian dancers.

I published my first Semi-infinite paper in [Applied Mathematics and Optimization 7:295–308] together with my two colleagues in 1980, and the last one, alone, in 1987 in [Trabajos de Investigación Operativa 2:69–80]. The last mentioned journal, devoted to Operations Research and owned by SEIO, was transferred to Springer, changed its name to TOP, and, after several years, met the standards to be considered as a JCR journal. Previously, the same process was applied to TEST, the other journal of SEIO devoted to Statistics, and as president of our society, I was witness to it. TEST was the first one to appear in the JCR list.

I am indebted to Prof. Marco A. López for encouraging me when I started doing research, for being the director of my PhD dissertation, and for helping me in my academic evolution and promotion. We have experimented together unforgettable moments travelling around the globe to attend scientific meetings, as well as visiting Madrid on a regular basis when I was a member of the Executive Committee of the SEIO, and he was the editor of one of our journals (TOP). We enjoyed the night life of Madrid, including visits to the best restaurants and to cinemas and theatres. We ended up being clients of "Casa Botín", close to the Plaza Mayor, the oldest restaurant in the world founded in 1725, according to the Guinness World Records. Its owner was a good friend of Marco's father in law.

3   Location Analysis

My first location paper was published, jointly with Marco and Miguel, in 1990, in the series [Research Papers in Banking and Finance of the Institute of European Finance in Bangor, Wales 90/15:1–53]. It was devoted to the performance and location of bank branches. Two younger members of my department, Marc Almiñana and Fernando Borrás, started cooperating with me on Location Analysis. Marc and I published our first research paper in 1993, in [Brazilian Journal of Operations Research 3(1):25–39], revising the problem of maximal coverage. One year later, in 1994, I designed a bi-criterion programme for locating bank branches, and published it in [JORS 45(12):1351–1362]. The same year, Marc Almiñana got his PhD which dealt with the total covering problem, obtaining the highest qualification, and we published another paper in [Top 2(2):315–328]. Two years later I published two papers, co-authored by Marc and Fernando, in "Studies in Locational Analysis", studying the pq-median problem in the first one and the centralised probabilistic location set covering problem in the second. In 1997, I published two papers with Marc, one in [Top 5(2):213–219], the other one in [EJOR 100:586–593]. In 2000, Fernando Borras brilliantly presented his PhD on a completely new family of models for locating emergency services, which had the singularity of being probabilistic. Finally, we published my last location paper two years later in [Annals of Operations Research 111:51–74]. The twelve years of research devoted to locational analysis had been very fruitful and interesting, not only for solving real problems and promoting two PhD students, but also for having the opportunity to visit new places around the world. These places included Edmonton, the capital city of the Canadian province of Alberta, where the first part of the congress was held at the University. The second part was held in Jasper where all the participants joined a tour visiting its National Park and one of its glaciers. The fact that we used multiple criteria optimisation in our research gave me the opportunity to get in touch with Ralph E. Steuer, the author of the, at that time, most famous book on this topic. I met him at the University of Georgia, in 1994, and we had time to talk about our research interests. He invited me for lunch and we enjoyed eating chicken wings, one of his preferred meals.

4   Efficiency and Productivity Analysis

In Sect. 1, I have already explained how I got in touch with DEA. It was clearly by chance. Nonetheless, I started my studies in this area in 1991 and am still active in 2017. Such a long period – 27 years – can easily be assumed and explained if different positive factors have appeared at the right moment, contributing to deepen and broaden my interest in this research subject.

Let me start with the 1991 EURO meeting in Aachen. I attended all the sessions and was impressed with the scope of Efficiency and Productivity Analysis. In fact, it started in the late seventies and I was being introduced to all the staff generated after thirteen years of research. There were – and still are – mainly two streams: the nonparametric approach, based on programming models, where the best practice frontier is identified by running the model as many times as the number of units being rated; and the parametric approach, also called SFA (stochastic frontier analysis), which needs to assume beforehand the shape of the frontier and adjust the corresponding econometric model to the data. The first approach does not account for random or statistical noise, while the second does. I have mainly done my research using nonparametric techniques within a typical DEA framework, and have only signed a few papers where statistical tools are used or developed.

Going back to the Aachen meeting, I was able to identify two outstanding researchers. The first one was Prof. William Cooper from the University of Texas in Austin, one of the founders of linear, goal and fractional programming, in tandem with Prof. Abraham Charnes, and the other was Prof. Knox Lovell from the University of Georgia in Athens, one of the founders of SFA. I proposed that each of them do future research with me, and both gave me the same answer: "when, in your opinion, you have got some worthy idea, please write it down and send it to me". It was the first time I had met Prof. Lovell, and the second time I had met Prof. Cooper. In fact, Charnes and Cooper had organised an international workshop on Semi-infinite Programming at their University in summer 1981, where I went together with Marco, Miguel and some other colleagues from Valencia University. In 1993, I sent to each of them my first proposal. With Bill, I decided to work on the definition of efficiency measures for the additive model; with Knox, I wanted to revise the three basic DEA models and study their ability for dealing with negative data. Both problems were unsolved at that time. We kept in touch by email. With the support of my department I proposed Prof. Cooper as PhD Honoris Causa at the University of Alicante, and he was invested in 1995. Prof. Bill Cooper was a charming and very positive person. He was close to his retirement age and had had an exciting scientific life. He was very proud, as Dean at Carnegie-Mellon University, of the negotiations with the "black panthers", a violent civil rights movement which defended the black community and lasted from 1966 to 1982. He always wore a ring which was a gift by the black panthers after reaching a satisfactory agreement with them. In 1993, I managed to convince Prof. Knox Lovell to present a paper which was published in 1994 in [Top 2(2):175–248], with its associated comments. The same happened with Prof. Bill Cooper. His paper and corresponding comments was published in the same Top issue [Top 2(2):249–314]. At Easter 1994, I visited Prof. Knox Lovell at the University of Georgia. As a visiting researcher I was given an office and we started working together. I had already solved the "translation invariant" property and the corresponding paper was already sent out to a journal. Therefore, Knox and I started working on a basic economic property known as "units invariant". The result was a short paper, published in 1995 in [Operations Research Letters 18(3):147–151]. My former paper was published one year later in [Annals of Operations Research 66:93–102]. Furthermore, we worked on a second paper devoted to measuring the Macroeconomic Performance of OECD countries.

It was a pioneer paper because we considered not only economic variables, but also undesirable environmental variables. The paper was published in 1995 in [EJOR 87(3):507–518] and was co-authored by Judy Turner. I enjoyed my stay very much, including the access to the library facilities and to the papers gathered and neatly ordered by Prof. Lovell, as well as the possibility of using the golf course owned by the University. Knox and I frequently went to a sports bar in Athens to watch the NBA final games or the NASCAR races.

We met again some years later, in 1998, and spent a couple of weeks together at the University of Lovaine le Neuve, Belgium, where the prestigious Belgian Professor Henry Tulkens used to organise European efficiency meetings. Prof. Tulkens is known for having created a non-convex technique for measuring efficiency, known as FDH. During our stay, besides drinking all kinds of Belgian beer, we wrote two papers. The first one studied, for the first time, radial DEA models without inputs or outputs, and was published in [EJOR 118(1):46–51] in 1999. The second one dealt with the evaluation of the financial performance of bank branches and compared DEA and FDH. It was co-authored by Tulkens and was published much later in 2006, in [Annals of Operations Research 145(1):321–337]. In 1994, I was invited by Dr. Teodoro Ravelo, from the University of La Laguna, to give a talk on DEA and to help them evaluate the managerial efficiency of the Canary Islands municipalities. We published a first working paper in 1995, and I continued visiting them for the next two years. The result was a paper which was published in 1997 in the Spanish Journal [Gestión y Análisis de Políticas Públicas 10:87–98].

I kept meeting Bill Cooper at the Efficiency International Meetings and in 1997 we finished our first paper defining, for the first time, an efficiency measure – called RAM – associated to the additive model. It was a long paper, co-authored by one of Cooper's collaborators, and published two years later in [JPA 11:5–42]. Based on this, we delivered a second paper evaluating the water supply services in the Kanto region of Japan. This paper was co-authored by two Japanese people, Dr. Aida and Prof. Sueyoshi. It was published in 1998 in [OMEGA 26(2):207–232] and received an award from an English evaluation agency as "the best applied paper of the year devoted to the Asian-Pacific countries".

In 1997 I moved from the University of Alicante to the newly created "University Miguel Hernández", with four campuses, the central one situated in Elche, the second biggest town and the biggest industrial centre of the Alicante province. I was elected as Vice-Rector for Research and, in parallel, started working with two young colleagues from my new department who wanted to obtain their PhD degree. I wrote several papers with José Luis Ruiz and with Inmaculada Sirvent. They delivered their respective PhD dissertations in 2000 and in 2001. In 1999, we published two DEA papers in the same journal. The first one, [EJOR 115(3):187–198], presented a new efficiency measure defined through a linear fractional programme, the ERG measure, which was expressed by K. Tone as a "slacks-based measure", SBM, and published two years later in the same journal. We had met Prof. Tone one year before at an international meeting in Lausanne and discovered that we were pursuing the same efficiency measure. I was rather disappointed when I saw that his 2001 paper did not include any reference to our 1999 paper. The second paper, published in [EJOR 115(3):132–144], tackled the problem of detecting influential observations in DEA and was based on a new statistical test which we developed in a third paper dealing with the deletion of non-relevant inputs or outputs. The third paper was published three years later in [Operations Research 50(4):728–735], after suffering a long and strange refereeing process. A fourth paper, where F. Borras appeared as the fourth author, presented a Montecarlo evaluation of the process of deleting variables within a DEA model, and was published in 2005 in [Information Technology and Decision Making 4(3):325–343].

I was able to organise a Banking Efficiency meeting in the year 2000 in Alicante, with the help of Ana Lozano, at that time associate professor at the University of Málaga, and Ifthekar Hasan, professor at the Rensselaer Polytechnic Institute in Troy, New York. Prof. Gary Becker, awarded the Nobel Prize in Economics in 1992, "for having extended the domain of microeconomic analysis to a wide range of human behaviours and interaction, including nonmarket behaviour", was incorporated onto our list of participants. It was an interesting meeting, sponsored by the CAM savings bank, and gathered around 20 significant scientists. This was the beginning of a fruitful collaboration with Ana and, from time to time, with Ifthekar, who is always ready to join us when we visit New York City. My first paper in banking was a joint piece of work with Ana and José Manuel Pastor, an economic researcher from IVIE and Valencia University, and it took some time before we published it. It was eventually published in 2002 in [JPA 18:59–77] and it analysed and compared the European Banking systems by taking into account not only the typical banking variables but also the macroeconomic, regulatory and accessibility conditions of each country. A second paper was published by Ana, Ifthekar and I in 2001 on the subject [European Finance Review 5(1/2):141–165], studying the functioning of banks across borders. In 2006, Ana and I published two more papers on banking, in [JPA 25:67–78] and in [The Manchester School 74(4):469–482]. Our last published paper came out in 2010 in [OMEGA 38(5):275–282], and for the last two years we have been working together with Juan Aparicio and Miguel Angel Durán on a new paper that evaluates banking risks using segmentation analysis. I have often visited Ana at the University of Málaga, giving seminars, working together and enjoying the Andalusian way of life.

After my 6 years as Vice-Rector, I was able to visit Knox again in Athens in 2004. We started working on productivity, trying to define a new Malmquist Index with better properties. Previously, in 1998, we had published a paper in [JPA 10(1):7–20] with Emily Grifell, where the additive model was considered. Very recently, in 2016, Fernando Vidal, Juan Aparicio and I published an improved related paper formulating any weighted additive model as a distance function in [EJOR 254:338–346]. As a result of my visit to Knox, we were able to define a new circular Malmquist Index, baptised the Global Malmquist Index, and published it in 2005 in [Economics Letters 88(2):266–271]. We then published a related paper in 2011, in [Socio-Economic Planning Sciences 45:10—15] co-authored by Prof. M. Asmild which introduced the Biennial Malmquist Index that has the nice property of avoiding infeasibilities.

The same year Knox and I went together to Toronto, where the North American Productivity Workshop took place. After the workshop, I stayed in Toronto by way of invitation of Prof. Joseph Paradi. It was an interesting stay because I gave some seminars to the PhD students, I had plenty of time for doing research and I participated in an applied project, required by the Toronto police, with Joe and Mette Asmild. Both Joe, a former Hungarian immigrant, and Mette, a young Danish lady hired by Joe, were the organisers of the aforementioned Workshop. It was fun to work on the last mentioned applied project and we were able to increase significantly the efficiency of the police stations just by reallocating their personnel. The methodology and results were published in 2012 in [OR Spectrum 34(4):921–941]. We wrote another paper which studied the BCC models with centralised resource allocation, which was published in 2009 in [OMEGA 37:40–49]. I was very pleased that Joe was also an avid golfer like me. We were able to play together several times and from that moment on, we organise a round of golf each time we attend an international meeting. Prof. Robin Sickles from Rice University and Prof. Hal Fried from Union College are usually our partners.

In 2008, after having served for 3 years as General Director for Research and Technological Transference in the Regional Government of the Valencia Region, I was able to visit Queensland University in Brisbane, Australia. I stayed there with my former PhD student and member of my department, Juan Aparicio, for one and a half months. Knox Lovell had already retired, was working at the University as an honorary professor, had married a charming Australian lady, July, and had his own jazz music programme on a local radio station. We worked with Knox and, in 2012, we published a paper in [JPA 38:109–120] that provided a novel formulation of all the DEA models based on Debreu's loss function. Juan and I developed the "multiplicative directional distance functions" that was first published in 2010 in [Indian Economic Review 45:193–231], and then in 2015, as a chapter of a Springer book edited by S. Ray, S. Kumbhakar and P.Dua entitled "Benchmarking for Performance Evaluation: A Production Frontier Approach". We had an interesting stay in Brisbane, drinking beer almost every day and doing some sightseeing. We visited the Golden Coast and Moreton Island, the second biggest sand island in the world. We stayed there for a weekend and had the opportunity to feed the dolphins inside the water, do water sports and drive round the island on a 4 × 4 quad.

The same year, 2008, Mette Asmild had moved to Warwick University which is situated in The Midlands in England, and I visited her during the summer. We worked together and published a new paper defining new slack free efficiency measures – MEA and RDM – in 2010 in [OMEGA 38(6):475–483].

Years 2009 and 2010 were extremely productive years, having published 9 papers. Even more so were 2011 and 2012, with 15 published papers. In 2011 I was elected as Rector of my university, and in 2015 I was re-elected for a second – and final – period of four years. Fortunately, I have had many collaborators during the past 6 years, which explains why we have been able to publish 48 papers or book chapters. Despite the duties of my new position, I still attend science meetings for two main reasons: to communicate my latest findings and to keep up to date.

5 Conclusions

In this paper, I have tried to transmit the idea that if you like doing research, you can enjoy life. You need, as usual, a bit of luck to work with colleagues who are reliable and bright enough. I have met some very nice people from all around the world and I have always tried to collaborate with people who belong to this group. I have also taken the opportunity to meet people who have different skills, and I have learnt a lot from engineers and economists. I have benefited from being an academic, a profession which I am still in love with. I also have the feeling that I became a world citizen, without renouncing my strong Spanish roots. Taking a nap whenever possible is definitely part of my life, as well as sitting outdoors talking to my friends and sharing a drink and, of course, doing research! I hope that anyone reading this paper is able to conclude, like I do, that if you love your subject, you can work hard and enjoy your work, which is one of the basic requirements for enjoying life.

Coupling of a Mixed-FEM and a Spectral BEM for the Approximation of an Exterior Elliptic Problem

Javier Valdés, Omar Menéndez, Salim Meddahi and Pablo Pérez

Abstract We present a Galerkin method based on the combination of a mixed finite element method and a spectral boundary element method to solve an exterior second order elliptic boundary value problem. we prove that the scheme is stable with respect to the discretization parameters and derive optimal a priori error estimates.

1 Introduction

Our aim is to solve a second order elliptic partial differential equation posed in an unbounded planar domain. We assume that we need to approximate simultaneously two unknowns: the scalar potential and the vector field representing the flux variable. The combination of the mixed finite element method (mixed-FEM) and the boundary element method (BEM) introduced in [8] is well suited for such a purpose. Nevertheless, the nearly singular boundary integrals appearing in FEM/BEM formulations are difficult to approximate via simple quadratures, especially for high order methods. In the two-dimensional case, a remedy to this drawback has been addressed in [6] by reformulating the classical Johnson-Nédélec [4] and Costabel [3] FEM/BEM methods in terms of a smooth coupling boundary. This permits one to change all terms on this interface to periodic functions by means of a parameterization and employ trigonometric polynomials to approximate the periodic representation of the boundary unknown. It is shown that the resulting scheme, combining the (standard) finite element method and a spectral method, is unconditionally stable and convergent. It

J. Valdés · O. Menéndez · S. Meddahi (✉) · P. Pérez
Facultad de Ciencias, Universidad de Oviedo, C/ Federico García Lorca, 18, 33007 Oviedo, Spain
e-mail: salim@uniovi.es

J. Valdés
e-mail: javier.valdes@uniovi.es

O. Menéndez
e-mail: omar@uniovi.es

P. Pérez
e-mail: riera@uniovi.es

© Springer International Publishing AG 2018
E. Gil et al. (eds.), *The Mathematics of the Uncertain*, Studies in Systems, Decision and Control 142, https://doi.org/10.1007/978-3-319-73848-2_81

889

turns out that this numerical method is amenable to the application of elementary quadrature formulas for the approximation of the weakly singular boundary integrals. Moreover, as the approximation is spectral on the boundary, only few BEM degrees of freedom are needed in order to attain the order of convergence imposed by the FEM method, which reduces the complexity of the linear systems of equations arising from the discrete formulation, as shown in [6]. We also refer to [7] where a similar strategy is used to deal with an acoustic scattering problem.

Our purpose here is to show that the mixed-FEM/BEM formulation presented in [8] with a parameterized coupling interface benefits from the same advantages obtained in [6] for the standard FEM/BEM formulations. Indeed, we prove that the scheme resulting from the combination of a mixed-FEM and a spectral BEM is stable with respect to the finite element mesh size h and the spectral parameter n. Moreover, we deduce optimal asymptotic error estimates.

The paper is organized as follows. We introduce the model problem in Sect. 2 and reformulate it in terms of a problem posed on a bounded domain with nonlocal boundary conditions imposed on a smooth artificial boundary Γ. In Sect. 3, we derive the mixed-FEM/BEM variational formulation and prove its well-posedness. Finally, in Sect. 4, we present our finite element discretization method and give its convergence analysis.

Notations. We end this section by recalling some standard notations related with Sobolev spaces. Given a Lipschitz domain $\Omega \subset \mathbb{R}^2$, we let $H^m(\Omega)$, $m \geq 1$, be the usual Sobolev space with norm $\|\cdot\|_{m,\Omega}$. The mixed formulation of the model problem relays on the space

$$H(\text{div}, \Omega) := \left\{ \mathbf{q} \in [L^2(\Omega)]^2; \quad \text{div } \mathbf{q} \in L^2(\Omega) \right\},$$

which is endowed with the usual Hilbertian norm

$$\|\mathbf{q}\|^2_{H(\text{div},\Omega)} := \|\mathbf{q}\|^2_{0,\Omega} + \|\text{div } \mathbf{q}\|^2_{0,\Omega}.$$

Finally, for the boundary unknown, we need to introduce periodic Sobolev spaces. Let $\mathscr{C}^\infty$ be the space of 2π–periodic infinitely often differentiable real valued functions of a single variable. Given $g \in \mathscr{C}^\infty$, we define its Fourier coefficients

$$\widehat{g}(k) := \frac{1}{2\pi} \int_0^{2\pi} g(s) e^{-kis} ds.$$

Then, for $r \in \mathbb{R}$, we define the Sobolev space H^r to be the completion of $\mathscr{C}^\infty$ with the norm

$$\|g\|_r := \left(\sum_{k \in \mathbb{Z}} (1 + |k|^2)^r |\widehat{g}(k)|^2 \right)^{1/2}.$$

Throughout this paper C (with or without subscript) will denote a generic positive constant that is independent of the discretization parameters. We use boldface small letters to denote vector-valued functions.

2 The Model Problem

We consider the following first order system of partial differential equations posed in the exterior of a Lipschitz bounded domain Ω_0:

$$
\begin{aligned}
\mathbf{p} &= a(x)\nabla u && \text{in } \Omega_0^c := \mathbb{R}^2 \setminus \overline{\Omega}_0 \\
\operatorname{div}\mathbf{p} + f &= 0 && \text{in } \Omega_0^c \\
u &= 0 && \text{on } \Gamma_0 = \partial\Omega_0,
\end{aligned}
\tag{1}
$$

and subject to the asymptotic behaviour

$$
u(x) = O(\tfrac{1}{|x|}) \quad \text{as} \quad |x| \to \infty.
\tag{2}
$$

Here, f is a given function with a bounded support and the coefficient $a(x)$ satisfies:

$$
0 < C_1 \le a(x) \le C_2 \quad \forall x \in \Omega_0^c \quad \text{and} \quad a(x) \equiv 1 \quad \text{if} \quad \operatorname{dist}(x, \Gamma_0) \ge C_3
$$

for some positive constants C_1, C_2 and C_3.

We introduce an artificial boundary Γ that contains in its interior the support of f and the set $\{x \in \Omega_0^c;\ \operatorname{dist}(x, \Gamma_0) \le C_3\}$ and consider the annular domain Ω delimited by Γ_0 and Γ. We denote by $\mathbf{n}$ the outward unit normal vector to Γ. Hereafter, we assume that the closed curve Γ is given by a 2π–periodic parametric representation $x : \mathbb{R} \to \mathbb{R}^2$ of class $\mathscr{C}^\infty$ such that

$$
|x'(s)| > 0, \quad \forall s \in \mathbb{R}, \quad \text{and} \quad x(t) \ne x(s), \quad 0 < |t - s| < 2\pi.
$$

We can define by means of $x(\cdot)$ the parameterized trace on Γ as the extension of

$$
\gamma : \mathscr{C}^\infty(\overline{\Omega}) \to L^2(0, 2\pi)
$$
$$
u \mapsto \gamma u(\cdot) := u|_\Gamma(x(\cdot))
$$

to $H^1(\Omega)$. The resulting linear application $\gamma : H^1(\Omega) \to H^{1/2}$ is bounded and onto; cf. Theorem 8.15 of [5]. Likewise, the parameterized normal trace on Γ is obtained by extending

$$
\gamma_n : \left(\mathscr{C}^\infty(\overline{\Omega})\right)^2 \to L^2(0, 2\pi)
$$
$$
\mathbf{q} \mapsto \gamma_n \mathbf{q}(\cdot) := \mathbf{q}(x(\cdot)) \cdot \mathbf{n}(x(\cdot))
$$

to H(div, Ω). The extended mapping (which is also denoted γ_n) is linear and bounded from H(div, Ω) onto $H^{-1/2}$.

A finite element formulation of Problem (1) requires the approximation of the asymptotic condition (2) by an homogeneous Dirichlet boundary condition satisfied by u on the artificial boundary Γ. In such a case, Γ should be located sufficiently far from Γ_0, which may originate a huge computational domain Ω. A more accurate and computationally efficient strategy consists in choosing Γ as described above (located as close as possible to Γ_0) and providing the problem in Ω with non-local boundary conditions on this artificial interface. These non-local boundary conditions consist in the following (parameterized) integral equations relating the Cauchy data $\lambda := \gamma_n \mathbf{p}$ and $\psi := \gamma u$ on Γ (see, e.g., [3]):

$$\psi = \left(\tfrac{1}{2}I + \mathscr{K}\right)\psi - \mathscr{V}\lambda \tag{3}$$

$$\lambda = -\mathscr{W}\psi + \left(\tfrac{1}{2}I - \mathscr{K}^{t}\right)\lambda. \tag{4}$$

Here, $\mathscr{V}$ and $\mathscr{K}$ are the parameterized versions of the boundary integral operators representing the single and double layer potentials,

$$\mathscr{V}g(\cdot) := \int_0^{2\pi} V(\cdot, t)g(t)\, dt, \qquad \mathscr{K}g(\cdot) := \int_0^{2\pi} K(\cdot, t)g(t)\, dt,$$

whose kernels are given by $V(s, t) := -\tfrac{1}{2\pi} \log |\mathbf{x}(s) - \mathbf{x}(t)|$ and

$$K(s, t) := \frac{1}{2\pi} \frac{(\mathbf{x}(s) - \mathbf{x}(t)) \cdot \mathbf{n}(\mathbf{x}(t))}{|\mathbf{x}(s) - \mathbf{x}(t)|^2} |\mathbf{x}'(t)|.$$

The operator K^{t} is the adjoint of $\mathscr{K}$ and the hypersingular operator $\mathscr{W}$ is related to $\mathscr{V}$ through the identity

$$\mathscr{W} = -\tfrac{d}{dt}\mathscr{V}\tfrac{d}{dt}.$$

The mixed-FEM/BEM formulation is obtained by combining the restriction of Problem (1) to Ω with (3)–(4) after merging the variables $\gamma_n\mathbf{p}$ and λ and introducing ψ as a further variable representing the (parameterized) restriction of the potential u to Γ. In this way, we obtain the following set of equations:

$$\begin{aligned}
\mathbf{p} &= a(x)\nabla u & \text{in } \Omega && (5)\\
\operatorname{div}\mathbf{p} + f &= 0 & \text{in } \Omega && (6)\\
u &= 0 & \text{in } \Gamma_0 && (7)\\
\gamma u &= \left(\tfrac{1}{2}I + K\right)\psi - V\gamma_n\mathbf{p} & \text{on } \Gamma && (8)\\
\gamma_n\mathbf{p} &= -W\psi + \left(\tfrac{1}{2}I - K^{t}\right)\gamma_n\mathbf{p} & \text{on } \Gamma && (9)
\end{aligned}$$

We end this section by recalling some well–known properties of the integral operators $\mathscr{V}$ and $\mathscr{K}$.

Lemma 2.1 *The operators* $\mathcal{K} : H^\theta \to H^\theta$ *and* $\mathcal{V} : H^\theta \to H^{\theta+1}$ *are bounded for all* $\theta \in \mathbb{R}$. *Moreover there exists* $\alpha > 0$ *such that*

$$\langle \mu, \mathcal{V}\mu \rangle \geq \alpha \|\mu\|^2_{-1/2}, \qquad \forall \mu \in H_0^{-1/2},$$

where $H_0^{-1/2} := \{\mu \in H^{-1/2}; \ \langle \mu, 1 \rangle = 0\}$.

Proof See for instance [5]. $\qquad \square$

3 The Mixed-FEM/BEM Variational Formulation

It is straightforward to deduce from (5)–(9) the following variational formulation of our problem (see [8] for the details):

$$\text{Find } (\mathbf{p}, \psi) \in H(\text{div}, \Omega) \times H_0^{1/2} \text{ and } u \in L^2(\Omega);$$
$$A\big((\mathbf{p}, \psi), (\mathbf{q}, \varphi)\big) + B\big((\mathbf{q}, \varphi), u\big) = 0, \quad \forall (\mathbf{q}, \varphi) \in H(\text{div}, \Omega) \times H_0^{1/2}$$

$$B\big((\mathbf{p}, \psi), v\big) = -\int_\Omega f v \, dx, \ \forall v \in L^2(\Omega), \tag{10}$$

where the bounded bilinear forms $A(\cdot, \cdot)$ and $B(\cdot, \cdot)$ are defined by

$$A((\mathbf{p}, \psi), (\mathbf{q}, \varphi)) := \int_\Omega a^{-1}\mathbf{p} \cdot \mathbf{q} \, dx + \langle \mathcal{V} \gamma_n \mathbf{p}, \gamma_n \mathbf{q} \rangle + \langle \mathcal{V} \tfrac{d}{dt}\psi, \tfrac{d}{dt}\varphi \rangle$$
$$+ \langle (\tfrac{1}{2}I + \mathcal{K})\varphi, \gamma_n \mathbf{p} \rangle - \langle (\tfrac{1}{2}I + \mathcal{K})\psi, \gamma_n \mathbf{q} \rangle$$

and $B((\mathbf{q}, \varphi), v) := \int_\Omega \text{div} \, \mathbf{q} \, v \, dx$.

Theorem 3.1 *Problem* (10) *is uniquely solvable.*

Proof We first notice the kernel of the bilinear form $B(\cdot, \cdot)$ is given by

$$\ker(B) := \left\{ (\mathbf{q}, \varphi) \in H(\text{div}, \Omega) \times H_0^{1/2}; \ \text{div} \, \mathbf{q} = 0 \text{ in } \Omega \right\}.$$

By definition, it holds

$$A\big((\mathbf{q}, \varphi), (\mathbf{q}, \varphi)\big) = \int_\Omega a^{-1}|\mathbf{q}|^2 \, dx + \langle \mathcal{V} \gamma_n \mathbf{q}, \gamma_n \mathbf{q} \rangle + \langle \mathcal{V} \tfrac{d}{dt}\psi, \tfrac{d}{dt}\varphi \rangle.$$

Using Lemma 2.1 and the fact that $\tfrac{d}{dt} : H_0^{1/2} \to H_0^{-1/2}$ is an isomorphism we deduce the existence of a constant $C_0 > 0$ such that

$$A\big((\mathbf{q}, \varphi), (\mathbf{q}, \varphi)\big) \geq C_0 \left(\|\mathbf{q}\|_{0,\Omega}^2 + \|\varphi\|_{1/2}^2\right) \quad \forall (\mathbf{q}, \varphi) \in H(\mathrm{div}, \Omega) \times H_0^{1/2}, \quad (11)$$

which proves that $A(\cdot, \cdot)$ is $H(\mathrm{div}, \Omega) \times H_0^{1/2}$-elliptic on $\ker(B)$.

On the other hand, the following inf-sup condition is proved in [8]: there exists a constant $\beta > 0$ such that

$$\sup_{(\mathbf{q}, \varphi) \in H(\mathrm{div}, \Omega) \times H_0^{1/2}} \frac{B\big((\mathbf{q}, \varphi), v\big)}{\|(\mathbf{q}, \varphi)\|_{H(\mathrm{div}, \Omega) \times H_0^{1/2}}} \geq \beta \|v\|_{0,\Omega} \quad \forall v \in L^2(\Omega). \quad (12)$$

We deduce from (11), (12) and the Babuška-Brezzi theory (cf. [1]) that the saddle point Problem (10) is well-posed and the result follows. $\qquad\square$

4   The Galerkin Method and Its Convergence Analysis

Let N be a given integer. We consider the equidistant subdivision $\{t_i := i\pi/N; \quad i = 0, \ldots, 2N - 1\}$ of the interval $[0, 2\pi]$ with $2N$ grid points. We denote by Ω_h the polygonal domain whose vertices lying on Γ are $\{\mathbf{x}(t_i) : i = 0, \ldots, 2N - 1\}$. Let $\mathscr{T}_h^0$ be a regular triangulation of $\overline{\Omega}_h$ by triangles T of diameter h_T not greater than $\max |\mathbf{x}'(s)| h$ with $h := \pi/N$ and whose vertices on Γ are given by $\{\mathbf{x}(t_i) : i = 0, \ldots, 2N - 1\}$. We obtain from $\mathscr{T}_h^0$ a triangulation $\mathscr{T}_h$ of $\overline{\Omega}$ by replacing each triangle of $\mathscr{T}_h^0$ with one side along the exterior part of $\partial \Omega_h$ by the corresponding curved triangle.

Let T be a curved triangle of $\mathscr{T}_h$. We denote its vertices by a_1^T, a_2^T, and a_3^T, numbered in such a way that a_2^T and a_3^T are endpoints of the curved side of T. Let $\hat{T}$ be the reference triangle with vertices $\hat{a}_1 := (0, 0)$, $\hat{a}_2 := (1, 0)$, and $\hat{a}_3 := (0, 1)$. Consider the affine map G_T defined by $G_T(\hat{a}_i) = a_i^T$ for $i \in 1, 2, 3$. For $h \in (0, h_0)$, with h_0 small enough, there exists a $\mathscr{C}^\infty(\hat{T})$ bijective mapping $F_T : \hat{T} \to \mathbb{R}^2$ given by $F_T := G_T + \Theta_T$ such that each side of $\hat{T}$ is mapped onto the corresponding side of T, i.e., $\Theta(0, t) = \Theta_T(t, 0) = (0, 0)$ and $F_T(t, 1 - t) = \varphi(t)$ for all $t \in [0, 1]$, (cf. [9]). We notice that if T is a straight (interior) triangle, then $\Theta_T \equiv 0$ and thus F_T is the usual affine map from the reference triangle. We denote by $B_T := DF_T$ the differential of the mapping F_T.

In the sequel, $P_m(\hat{T})$ stands for the space of piecewise polynomial functions of degree at most $m \geq 0$ on the reference triangle $\hat{T}$. For any $k \geq 1$ and $T \in \mathscr{T}_h$, we consider

$$\mathbf{W}(T) := \left\{ \mathbf{q} : T \to \mathbb{R}^2; \quad \mathbf{q} \circ F_T = \frac{1}{\det(B_T)} B_T \hat{\mathbf{q}}, \quad \hat{\mathbf{q}} \in P_k(\hat{T})^2 \right\},$$

and

$$U(T) := \left\{ v : T \to \mathbb{R}; \quad v \circ F_T = \hat{v}, \quad \hat{v} \in P_{k-1}(\hat{T}) \right\}$$

and introduce the global finite element spaces

$$\mathbf{W}_h := \{\mathbf{q} \in H(\mathrm{div}, \Omega); \quad \mathbf{q}|_T \in W(T); \quad \forall T \in \mathcal{T}_h\}$$

$$U_h := \{\mathbf{q} \in L^2(\Omega); \quad v|_T \in U(T); \quad \forall T \in \mathcal{T}_h\}.$$

Let us now recall some well-known properties of the Brezzi-Douglas-Marini (BDM) mixed finite element. The global BDM-interpolation operator $\Pi_h :$ $H(\mathrm{div}, \Omega) \cap (H^s(\Omega))^2 \to \mathbf{W}_h, s > 1/2$, satisfies the following classical error estimate, see [1, Proposition 2.5.4],

$$\|\mathbf{q} - \Pi_h \mathbf{q}\|_{0,\Omega} \leq Ch^{\min(s,k+1)} \|\mathbf{q}\|_{s,\Omega}, \qquad \forall \mathbf{q} \in H^s(\Omega)^2, \text{ with } s > 1/2. \quad (13)$$

Moreover, thanks to the commutativity property, if $\mathrm{div}\, \mathbf{q} \in H^s(\Omega)$, then

$$\|\mathrm{div}\, (\mathbf{q} - \Pi_h \mathbf{q})\|_{0,\Omega} = \|\mathrm{div}\, \mathbf{q} - \mathcal{U}_h \mathrm{div}\, \mathbf{q})\|_{0,\Omega} \leq Ch^{\min(s,k)} \|\mathrm{div}\mathbf{q}\|_{s,\Omega}, \quad (14)$$

where $\mathcal{U}_h$ is the $L^2(\Omega)$-orthogonal projection onto U_h.

Finally, let n be a given integer such that $n \leq N$. We consider the $2n$–dimensional space

$$T_n := \left\{ \sum_{j=1}^{n} a_j \cos jt + \sum_{j=1}^{n-1} b_j \sin jt; \quad a_j, b_j \in \mathbb{R} \right\}.$$

The following approximation property holds true, (cf. [2]):

$$\inf_{\mu \in T_n} \|\lambda - \mu\|_s \leq 2^{t-s} n^{s-t} \|\lambda\|_t \quad \forall \lambda \in H_0^t \quad \forall t \geq s. \quad (15)$$

We are now in a position to introduce the conforming Galerkin finite element approximation of Problem (10):

find $(\mathbf{p}_h, \psi_n) \in \mathbf{W}_h \times T_n$ and $u_h \in U_h$;

$$A\big((\mathbf{p}_h, \psi_h), (\mathbf{q}, \varphi)\big) + B\big((\mathbf{q}, \varphi), u_h\big) = 0, \qquad \forall (\mathbf{q}, \varphi) \in \mathbf{W}_h \times T_n \quad (16)$$

$$B\big((\mathbf{p}_h, \psi_h), v\big) = -\int_\Omega f v \, dx, \; \forall v \in U_h.$$

Theorem 4.1 *Problem* (16) *has a unique solution and there exists a constant* $C_1 > 0$, *independent of h and n such that*

$$\|\mathbf{p} - \mathbf{p}_h\|_{H(\mathrm{div},\Omega)} + \|\psi - \psi_n\|_{1/2} + \|u - u_h\|_{0,\Omega} \leq C_1$$

$$\left(\inf_{\mathbf{q} \in \mathbf{W}_h} \|\mathbf{p} - \mathbf{q}\|_{H(\mathrm{div},\Omega)} + \inf_{\varphi \in T_n} \|\psi - \varphi\|_{1/2} + \inf_{v \in U_h} \|u - v\|_{0,\Omega} \right).$$

Moreover, if the regularity assumptions $\mathbf{p} \in H^k(\Omega)^2$, div $\mathbf{p} \in H^k(\Omega)$, $u \in H^k(\Omega)$ and $\psi \in H^{\sigma+1/2}$ are satisfied for some $k \geq 1$ and $\sigma > 0$, then there exists a constant $C_2 > 0$ independent of h and n such that

$$\|\mathbf{p} - \mathbf{p}_h\|_{H(\text{div},\Omega)} + \|\psi - \psi_n\|_{1/2} + \|u - u_h\|_{0,\Omega} \leq C_2$$
$$\left(h^k (\|\mathbf{p}\|_{k,\Omega} + \|\text{div }\mathbf{p}\|_{k,\Omega} + \|u\|_{k,\Omega}) + n^{-\sigma} \|\psi\|_{1/2+\sigma} \right). \quad (17)$$

Proof We notice that the discrete kernel

$$\ker_h(B) = \{(\mathbf{q}, \varphi) \in \mathbf{W}_h \times T_n; \quad B((\mathbf{q}, \varphi), v) = 0 \quad \forall v \in U_h\}$$

of the bilinear form $B(\cdot, \cdot)$ is a subspace of its continuous counterpart $\ker(B)$. Hence, the ellipticity of $A(\cdot, \cdot)$ on $\ker_h(B)$ is directly inherited from (11). Moreover, the discrete analogue of the inf-sup condition (12) can be proved as in [8]. It follows again from the Babuška-Brezzi theory [1] that the discrete problem (16) is well-posed and we also have the Céa estimate (17). Moreover, using the interpolation error estimates (13), (14) and (15) we deduce the asymptotic behaviour of the numerical scheme when h goes to 0 and n tends to infinity, provided the analytical solution is smooth enough.

$\square$

References

1. Boffi D, Brezzi F, Fortin M (2013) Mixed finite element methods and applications. Springer series in computational mathematics. Springer, Heidelberg
2. Canuto C, Quarteroni A (1982) Approximation results for orthogonal polynomials in Sobolev spaces. Math Comput 38:67–86
3. Costabel M (1987) Symmetric methods for the coupling of finite elements and boundary elements. In: Brebbia CA, Wendland WL, Kuhn G (eds) Mathematical and computational aspects. Springer, Berlin
4. Johnson C, Nédélec JC (1980) On the coupling of boundary integral and finite element methods. Math Comput 35:1063–1079
5. Kress R (1999) Linear integral equations. Springer, New York
6. Meddahi S, Márquez A (2002) A combination of spectral and finite elements for an exterior problem in the plane. Appl Numer Math 43:275–295
7. Meddahi S, Márquez A, Selgas V (2003) Computing acoustic waves in an inhomogeneous medium of the plane by a coupling of spectral and finite elements. SIAM J Numer Anal 41:1729–1750
8. Meddahi S, Valdés J, Menéndez O, Pérez P (1996) On the coupling of boundary integral and mixed finite element methods. J Comput Appl Math 69:113–124
9. Ženíšek A (1990) Nonlinear elliptic and evolution problems and their finite element approximations. Academic Press, London

Part VI
Some Biographical Sketches and Outstanding Facts

Pedro Gil, Colleague and Friend: Experiences Shared in the Department of Statistics and Operations Research at the Complutense University of Madrid

Rafael Infante Macías

Abstract Pedro Gil was an undergrad student in Math and a colleague of mine in the Department of Probability and Statistics of the Complutense University of Madrid. This paper aims to describe what my memory remembers about those nice shared years.

Pedro arrived in Madrid to complete the course of studies for a Degree in Mathematics in the same year, 1965, in which my close friend Paco Cano and I joined, what was then, the Department of Probability Calculus and Mathematical Statistics, chaired by Professor Sixto Ríos García. Remember that before Faculties were organised in Departments, Faculties were made up of Branches and the Sections, Biology, Physics, Geology, Mathematics and Chemical Sciences, which formed part of the Faculty of Sciences, had not yet been converted into Faculties.

The permanent members of Professorial Staff were the Professor and the Caretaker, since Associate Professors and Assistants had a limited term within the University, therefore if they believed their promotion was long coming, in the best case scenario, a few years, they could choose to sit for an official examination to become a member of what was then the Body of Heads of Department, which explains the high level of this Faculty, by way of example, our colleagues, Ildefonso Yáñez and Miguel Martín, who obtained the Professorship of Mathematics in the 'Padre Luis Coloma' Secondary Institute in Jerez de la Frontera, although they requested unpaid leave in order to continue in the University.

When we completed our course of studies in June 1965, Paco Cano and I talked to Prof. Sixto to express our interest in him directing our Doctoral Dissertation. He asked us about our ambitions and told us to come back in September. We returned and he said he was going to propose our appointment as Supernumerary Assistant, i.e. unpaid, to the University Rector.

R. Infante Macías (✉)
Departamento de Estadística e Investigación Operativa, C/ Tarfia s/n, 41012 Sevilla, Spain
e-mail: infante@us.es

E. Gil et al. (eds.), *The Mathematics of the Uncertain*, Studies in Systems, Decision and Control 142, https://doi.org/10.1007/978-3-319-73848-2_82

Prof. Sixto, our Mentor and I am talking in present tense because, as we have commented on numerous occasions, Paco Cano, Ramiro Melendreras, Pedro and I, our Mentor-Disciple relationship, will never be broken, he taught us, not just scientifically, but more importantly, he taught us with compassion, a quality only attributable to genuine Mentor. With him we learnt the importance of exact data and precision; from him we learnt to have a critical and open attitude; to think without being subject to preconceived models, which are sometimes more limitative than explanatory. He taught us how to love a job well done and above all, he instilled in us a love for the University that he made the core and centre of his life.

The Associate Professor of the Branch was the beloved and admired Ildefonso Yáñez de Diego. He taught us, in addition to probability, that there is nothing more precious in this world than friendship. From him we learnt about the value of solidarity; he would tell us that we should give to others without expecting anything in return; we undoubtedly owe any qualities we may have to his fine example. The Assistants were Antonio López Damas and José Pérez Vilaplana, who replaced Miguel Martín Díaz, since he was in Paris that year doing his Doctoral Dissertation. The following year, López Damas enrolled in the School of Armament Engineering and Construction, a centre belonging to the Ministerial Department in charge of the Army and Pérez Vilaplana left the position of assistant professor a short time later. Therefore as two vacancies became available, Prof. Sixto proposed our candidature to the Rector of the University as Associate Professors, now with a salary. And during that year, 1966–67, we met Pedro; he was the student and we, as Assistants, taught the practical Calculus of Probability and Mathematical Statistics classes and soon we noticed how Pedro stood out from the rest of his classmates. I believe one of the best ways of identifying intelligence, speed of reflexes and rigour in a student is through problem solving and Pedro had all these attributes.

During those years, the classes for the Statistics speciality courses in the Bachelor's degree in Mathematics were taught in the building called Caserón San Bernardo, the Statistics School's main building where the classes for the Diploma in Statistics were given. It had two branches: Economics and Mathematics, the former directed by Prof. Arniz and the latter by Prof. Sixto. The subjects specific to the Faculty were taught together with those of the Diploma Course, therefore we hardly ever went to the Faculty, except for Prof. Sixto, who had an office in the Faculty, ours were in the Statistics and Operations Research Institute of the Spanish National Research Council (CSIC), the Director of which was also Prof. Sixto.

On 31 March 1966, Decree 1199/1966 on the Classification of Departments in Faculties of Sciences was published in the Official State Gazette and it included the creation of the Department of Mathematical Statistics integrating within this the Branch of Calculus of Probability and Mathematical Statistics. Once the Department had been created and was up and running, it had to be equipped with the infrastructure, both in terms of material resources and space, a task entrusted by Prof. Sixto to Cano and myself. Works were carried out in the Faculty building and the Department was granted half of the fourth floor left wing, distributed into three offices, a classroom, a seminary and a library and a place for the Department secretary's office was created in the entrance hall to Prof. Sixto's office. When Paco Cano moved to the University of Zaragoza and I went to the University of Granada, Ramiro Melendreras and Pedro were given this administrative management task and I believe it was an excellent source of learning for Pedro for carrying out the wonderful work he did in the University of Oviedo. Prof. Sixto decided to separate the classes for the Degree subjects from those of the Diploma program, so the former were taught in the Faculty and the latter in San Bernardo. Once the works were completed and we had a place in which to prepare our classes in the mornings, we spent the mornings in the Faculty and the afternoons in the Council, as Prof. Sixto had managed to get both Paco Cano and myself a position as Research Assistants in the Council.

Another positive effect for the Branch, now converted into a Department, was the increase in the number of teaching staff; positions were given to Teachers in Charge of the Course, Associates and Assistants. The position of Teacher in Charge of the Course was created in the Statistics School, which enabled new Teachers to join, including my dear friend, Ramiro Melendreras, who passed away in his prime, taking with him a part of me, Pilar Ibarrola, Vicente Quesada, Francisco Javier Girón and Pedro. Miguel Martín returned from Paris and his arrival had a decisive impact on Pedro, since Miguel introduced him to the study of Mathematical Theory of Information, which led to his Doctoral Dissertation, which he defended in 1974, on "Measures of uncertainty and information in statistical decision problems", published in 1981 by the ICE editorial, based on notes drawn up by Pedro and directed by Prof. Sixto Ríos. This book is an essential reference for students studying Mathematics and Computer Science.

As the number of Teachers was much higher than the number of offices, with the exception of Prof. Sixto's office, they were shared, but not only did we share offices, also tables and sometimes even chairs. The tables in the other office and those of the seminary were shared by Miguel Martín, Paco Cano, Ramiro, Girón and Pedro. Sharing such a small space had a very positive effect, as a great comradeship was formed and a profound friendship. I remember when Paco and I sat for the official exams for Assistantship in Oviedo and Granada, how the rest of our colleagues helped us to draw up the Report and prepare the fifteen subjects proposed by the tribunal for the fifth exam. Specifically, Pedro prepared the subject "Relationship between the Bayesian Statistics, and Information Theory" for me, a subject I still hold dear.

We were not only colleagues, we were great friends, we shared our joys and the reprimands aimed at some of us by Prof. Sixto. I look back with nostalgia at that golden age of our academic lives in the Department, since we used to help one another; there was no selfishness or personal aspirations and, above all, we felt care and respect for our Mentor, Prof. Sixto.

I remember Paco, Ramiro, Pedro and I going for breakfast in the Faculty bar located on the ground floor outside the building. Sometimes there was a moment of tension when one of us, normally Paco or myself, would get a telling off from Prof. Sixto and if he was unable to find us, Ramiro would get the telling off, so, to relieve this tension, Miguel would suggest going to the bar and we would relax with a glass of wine and a piece of tortilla.

In October 1971, I moved to the University of Granada as an Assistant Professor of Operations Research, but I continue to go to the Department in Madrid, where I was provided with bibliography and to the Council to check the Journals, therefore I remained in contact with my colleagues. Two years later Ramiro joined the University of Santiago de Compostela and as the Professorship vacancy had to be filled in the University of Granada as I had moved to Seville, Ramiro moved to this University to then join the Murcia University a short time later. Pedro joined the Oviedo University in 1976.

Although Pedro was in Oviedo and I in Granada first and then Seville, there was someone that acted as a linkage between us and that person was Ramiro Melendreras, who went out of his way to get us together in Madrid, which was quite easy back then because there were plenty of entrance examinations held during those years and we would either coincide on the Boards or, if this were not the case, we would make sure we met up in Madrid and I can remember us eating in Bar Manolo on Calle Princesa, where we discussed how the examinations being held were going and we did so because we shared our concerns and we wanted the best for the development of Statistics and Operations Research. We also met in the Statistics and Operations Research Conferences organised by the SEIO (Spanish Society of Statistics and Operations Research). In fact, Pedro replaced me in 2001 as Chairman of the Spanish Society of Statistics and Operations Research, therefore we held meetings at the beginning, him as Chairman elect and I as acting Chairman and then he, as acting Chairman and I as the outgoing Chairman.

An example of the sensitivity shared by Pedro and I, was the profound affection we both had for Prof. Sixto and the high esteem we had for his scientific work, to the point that in 2000 he proposed for him to be invested as Doctor *Honoris Causa* by the University of Oviedo and I proposed it in 2001 by the University of Seville. In both cases, Prof. Sixto was truly happy to be accompanied by his disciples in these solemn acts. I believe that when friendships are formed early on in life, they remain over time, which is what happened to us.

And to conclude these brief notes reflecting a very important period in our lives, I would like to remember that group of colleagues and friends that formed part of our Department of the Complutense University during the early years and who sadly are no longer with us. Prof. Sixto Ríos, Ildefonso Yáñez, Miguel Martín, Francisco Cano, Ramiro Melendreras and Pedro Gil, all of you are still present in my thoughts and I await the day in which we will be together again, because, in the words of Miguel Hernández in his *"Elegía a Ramón Sijé"*

> *A las aladas almas de las rosas*
> *del almendro de nata te requiero*
> *que tenemos que hablar de muchas cosas*
> *compañero del alma, compañero*

Pedro Gil at the Complutense University

Julián de la Horra

Abstract The stay of Pedro Gil in the Department of Statistics and O. R. at the Complutense University of Madrid was not very long, because he left that Department for moving to the University of Oviedo in 1976. But the mark of Pedro at the Complutense University was very important. I want to present here my personal view of this mark.

Pedro Gil did not stay for long in the Department of Statistics and O.R. at the Complutense University of Madrid. He left that Department for moving to the University of Oviedo in 1976. However, Pedro left a very deep mark on the Complutense University. At least, he left a very deep mark in me. I always say that Pedro is the 'guilty' of my professional dedication to the University.

But let me get back to the beginning: Pedro was my teacher on the subject 'Information Theory' in the fourth year of Mathematics. But the true influence of Pedro on my life was in the last year of my graduate studies, when I went to his office looking for his advice on career options and he suggested me the possibility of writing a PhD thesis and working at the University. At first, I completely discarded the idea, because that idea had never entered into my plans. But ... the seed was planted and it was some months later that I decided to try. And I enjoyed the experience ... very much.

And here I am, 42 years later, and I have to say that I do not regret at all this choice. In short, thanks to that conversation with Pedro, I found this wonderful profession. Well, before continuing, let me clarify the meaning of 'wonderful profession', because not everything is a rose garden:

J. de la Horra (✉)
Facultad de Ciencias, Departamento de Matemáticas,
Universidad Autónoma de Madrid, 28049 Madrid, Spain
e-mail: julian.delahorra@uam.es

© Springer International Publishing AG 2018
E. Gil et al. (eds.), *The Mathematics of the Uncertain*, Studies in Systems, Decision and Control 142, https://doi.org/10.1007/978-3-319-73848-2_83

- Research work is hard, really hard at times (results are difficult to obtain and certain referees' reports ...), but when, finally, you see your research accepted and published in a good journal ... you forget the problems.
- And the teaching work is also hard, really hard at times (very large classes, a lot of exams to read, ...), but when, years later, a student thanks you for your teachings, for your help ... you also forget the problems.

Of course, the mark of Pedro in the Complutense University is much deeper. I would especially like to mention his first doctoral student, Pilar García-Carrasco, who left us prematurely in 1987. Pilar began her research on some topics of Information Theory (with Pedro as supervisor), and the result of that research was her PhD thesis, "Criterios para la comparación de experimentos" [1], in July 1977.

Pilar became, in only a few years, one of the best Spanish researchers in Statistics. I was pleased to share with Pilar, office, research and, above all, friendship. The result of this research collaboration is one of my most satisfying papers, "Maximizing uncertainty functions under constraints on quantiles" posthumously published in Statistics & Decisions in 1988 [2]. The main theorem of this article is as follows:

"Let u be any permutation invariant, uncertainty function. The maximum of u (subject to constraints on several quantiles) is reached at a point which does not depend on u."

Moreover, the theorem gives an explicit expression for this point. This is the only article jointly published with Pilar, because there was no time for more. If life had been more generous with her, probably it would not have been the only one.

In short, I am very grateful to Pedro for many things but, especially, for two things:

- Pedro encouraged me to choose an exciting career that I have enjoyed for years and years.
- Pedro and this career give me the opportunity to know and to research with Pilar, one of the best friends I have had throughout my life.

Thanks for all, Pedro, and forever!

References

1. García-Carrasco MP (1977) Criterios para la comparación de experimentos, Ph.D. thesis. Complutense University of Madrid, Madrid (see also (1978) Trab Est IO 29(2):28–51)
2. García-Carrasco MP, De la Horra J (1988) Maximizing uncertainty functions under constrains on quantiles. Stat Decis 6:275–282

Pedro Gil: Teacher, Researcher, Mentor, Leader and Colleague at the University of Oviedo

Norberto Corral and Manuel Montenegro

Abstract Professor Pedro Gil launched the Department of Statistics, OR and Math Teaching in the University of Oviedo. He prompted the research Statistics and OR in Asturias and he did a lot for Mathematics in the region. This paper aims to show Gil's influence on all of us throughout the thirty-four years he was in the University of Oviedo, and how indebted we feel with him.

As it has been already mentioned by Infante Macías, Pedro Gil joined the Department of Statistics and OR in 1969 as a research and teaching assistant. Partially thanks to a fellowship from Fundación Juan March, he completed his Ph.D. Dissertation in 1974, under the supervision of Professor Sixto Ríos, at the Complutense University of Madrid (see [1] for a published version of this Ph.D. thesis). Professor Ríos was a doctoral student of Professor Julio Rey Pastor (see Part I of this book), and he is considered to be the main driver and promoter of Statistics and OR in Spain.

After getting the Ph.D. Degree in Mathematics, Pedro Gil continued working on researching and teaching with the ultimate goal of applying at any appropriate moment for a tenure-track position at the university. This opportunity arose possibly earlier than expected, and it concerned a position in between Associate and Full Professorship (coined as Profesor Agregado, a position that was extinguished many years ago). To directly apply to this high level permanent position was not certainly the typical and easiest path into an academic career in the field. Nevertheless, when Pedro was twenty-eight years old he became Profesor Agregado in Operations Research of the University of Oviedo.

In 1976, a twenty-eight years old Pedro Gil (see Fig. 1) moved along with his young family (at that moment composed of his wife, Pilar, his eldest daughter, Eva, and himself) from Madrid to Oviedo. This was the embryo of a family who has been the main engine of his life, and that was completing with two other children, Juan and Eduardo and, much later, with their grandchildren.

N. Corral · M. Montenegro (✉)
Departamento de Estadística, I.O. y D.M., Universidad de Oviedo, Oviedo, Spain
e-mail: mmontenegro@uniovi.es

N. Corral
e-mail: norbert@uniovi.es

905

Fig. 1 Pedro Gil at the time
of his arrival to Oviedo

Moving from Madrid meant, on one hand, abandoning the 'scientific nest' that had been created around Professor Ríos by many endearing colleagues and friends. Anyway, he had always kept in touch with them, and the periodic conferences of the Spanish Society of Statistics and OR (SEIO) were the perfect excuse not to lose that strong relationship.

On the other hand, moving from Madrid entailed too many responsibilities for a young professional. In the University of Oviedo he was affiliated to the Faculty of Sciences, where the presence of mathematicians had been very scarce and was very longed for. So, he was quite enthusiastically welcome. But, from the very beginning, he was aware about him having to assume a leadership position in all possible respects related to the Department of Mathematics of the Faculty of Sciences, namely:

- heading the Department, what means being involved in the management of many bureaucracies and commitments;
- recruiting people (a complex task in a university not having the B.Sc. Degree in Mathematics at that moment, see the paper by Dugnol and Valdés, just after this one in the book);
- guiding the teaching of the recruited people, since the first ones were young mathematicians (most of them with only a B.Sc. degree);
- supervising the research of the recruited people.

And Pedro wanted this to be made in the most rigorous possible way. And, in spite of his (starting) youth, he got it. And he became an excellent teacher, a good and generous researcher and mentor, a sympathetic boss, and deeply beloved by all of his colleagues. Let us now briefly detail each of these facets.

1   Pedro as Teacher

When Pedro arrived to Oviedo he had the expertise of having delivered courses on many different Statistics and OR disciplines. Actually, this expertise was crucial in both, guiding and helping the young collaborators he was recruiting in their teaching endeavors, and coordinating the launching and design of the B.Sc. in Mathematics for the University of Oviedo (as it will be later explained in more depth by Dugnol and Valdés).

In 1976, the Mathematics Department Pedro headed had to be in charge of General Mathematics (mainly involving Calculus and Algebra) and Statistics for Biology, Chemistry, Geology and also Statistics/OR for Economics. And he taught some of the courses, especially those concerning Statistics, being the first mathematician teaching some of them in Oviedo.

Although the students might initially be suspicious of the increase in rigor and proficiency level that surely occurred at that time, they soon realized that such an increase was accompanied by clear explanations, well-argued motivations and a strong empathy to students. Even those being highly averse to Mathematics admitted Pedro's teaching excellence.

Of course, this opinion was increasing in the nineties of the last century, as the B.Sc. in Math was launched in the University of Oviedo, and Pedro was teaching again to Math students (those for whom the aversion to Mathematics is assumed to be discarded). He had a strong background and knowledge, and he was especially skilled at transmitting them. In fact, most of students confess that because of the clarity of his explanations, they believed that the courses were simpler than they were.

Furthermore, Pedro was always ready to listen to the students, advice them, solve their doubts, answer their questions, and so on. Most of these questions were mathematical, but they also concerned their future working lives, and even some rather personal matters.

At the end of the first quarter, just before Christmas, students of the B.Sc. in Mathematics organized a special event in which they (often with some professors) sang, made gymkanas and other competitions, parodies, etc. And a permanent participant in all of these events was Pedro. With his accordion, he played a well-known Basque song 'El Alcalde de Arrigorriaga', after writing the lyrics on the blackboard and indicating students the movements of going up and down alternately in the chorus ("Pantaleón, Pantaleón, …"). If possible, this made him even closer to the students.

Pedro's warm attitude supported the frequent (informal albeit affectionate) nicknames students used in referring to him: *Father Pedro* or even *Saint Pedro* (as the Chancellor has already recalled in the foreword for this book).

2 Pedro as Researcher and Mentor

From the very beginning of his scientific career, Pedro was firmly convinced of the importance of research in academic life. And he tried to convey to his Ph.D. students and people at his Department this conviction. He also tried to show the benefits of the synergies between research and teaching.

He was the main example for us in this respect. Actually, he got the maximum possible number of six-year period research activity evaluation in Spain, the so-called 'sexenios de investigación'. And his expertise made much easier our first approach to research.

Research was probably the most visible and best test of Pedro's generosity. He tried to ensure the highest levels of rigour and quality in all research undertaken under his supervision, as well as to create a dynamic and ethic atmosphere around. Since he does not believe in hierarchical structures, he encouraged us to become gradually independent in research tasks (either in writing papers, or in applying for granted research projects, etc.). This based on gradual independence, evolution and generational change policy has certainly led to an enrichment of the research activities and directions in the Department.

Despite his many duties in heading first Mathematics Department of the former Faculty of Sciences and later the Statistics, OR and Teaching Mathematics Department of the University of Oviedo, Pedro was permanently involved in research tasks. He supervised 20 Ph.D. Thesis. His first supervision started at the Complutense University of Madrid, the Ph.D. Student being M. Pilar García-Carrasco. The Dissertation was completed in 1978 and dealt with the comparison of statistical experiments (see comments made about by de la Horra in the preceding paper in this book).

Pedro supervised other 19 Ph.D. theses in the University of Oviedo. He fervently defended that the Ph.D. Thesis should be a high level research work. But it should not be our highest level research. We should scientifically grow and do better jobs every time. As mentor, Pedro guaranteed support, guidance, encouragement and understanding at every stage.

3 Pedro as Leader and Colleague

Since Pedro arrived at Oviedo, he had to act as a leader and manager. Everyone noticed his high ability to hold management positions, and he realized that for a newly created Department at the University, it was advisable the head was involved in many management aspects. Moreover, he was also aware on his skills and fondness for the scientific research. So, with much effort and personal work he had to combine both leaderships. Before being thirty years old he was holding, among others, the Directions of the Mathematics Department at the Faculty of Sciences, the Academic Secretary for such a Faculty, and the supervision of four doctoral theses. We must admit that we have not yet been able to discover how he could do so in the right way in which he did it.

Some years later he was holding several other positions, supervising many other doctoral theses and research works, and heading the first research projects conducted in the Department. But, in view of the principles, this is not surprising at all. Among the positions he held, to be three-years President of the Spanish Society of Statistics and OR, SEIO, was quite grateful. Pedro and his wife were unconditional participants in all SEIO's Conferences.

After some educational changes in Spain, that are detailed in the next paper by Dugnol and Valdés in this book, in 1997 a new department was created in the University of Oviedo: the Department of Statistics, Operations Research and Teaching Mathematics. Since its creation, Pedro was the head (and indisputable soul) of this Department. And, much to his regret, he did not get anyone to relieve him until his retirement in 2010. We all felt safe and sound under his warm protection.

Pedro was also our colleague and friend, and we all knew he was there for any problem we could have. Nevertheless, we were not able to look at him exactly as one of us. In spite of some of us being only a few years younger than him, he had been taking care of us since the beginning, he was our boss, our supervisor, our scientific ancestor, and our personal support along the years in the University of Oviedo. So, for sure he was not one of us. He meant the creator, promoter, germ and heart of the project that began in 1976.

In this respect, the homage held in the Paraninfo of the University of Oviedo to tribute him on the occasion of his retirement in 2010 (see https://youtu.be/APP4j5azLwo for the whole ceremony, in Spanish), was a parade of affection, admiration and recognition. Even an *ad hoc* choir, called 'Pantaleón', was set up for the special occasion, the eighth singers being either alumni, or scientific descendants, or colleagues or relatives, and the songs sung were some of his favorites in different times of his life.

4 Pedro and the Mathematics in the University of Oviedo and in the Principality of Asturias

As we have already mentioned, Pedro arrived to Asturias in 1976, when he was twenty-eight years old. His tenure track position as 'Profesor Agregado' of the University of Oviedo was a necessary stage to get the Full Professorship. At that moment, after one to four years Agregados become Full Professors through an exposure of merits.

Maybe because of not having yet Mathematics B.Sc. studies, the University of Oviedo delayed sightly more than four years in convening the position of Full Professor in Statistics. Meanwhile, the University of Santiago de Compostela offered the same position, and Pedro could apply to it under quite favorable conditions. It would be a nice opportunity in many professional and personal respects, namely: the University of Santiago had studies in Mathematics, with a long reputed tradition; he knew several members in Statistics and OR of this university, thanks to his beloved friend Professor Ramiro Melendreras; his wife was born in La Puebla del Caramiñal (around 50 km far from Santiago), and joining Santiago University would mean approaching their family.

Table 1 Distribution of Pedro's 'scientific linear descent' in the Department of Statistics, OR and Math Teaching at the University of Oviedo

Line of 'scientific descent'	Pedro's child	Pedro's grandchild	Pedro's great-grandchild
Absolute freq.	11	12	1

In this contingency, the members of his Department in Oviedo feared that their very young academic lives would undergo a major change by losing Pedro's leadership and protection. As a result of Pedro and his wife's generosity along with their love to Asturias, they decided not to move. We never thanked them as they deserved.

And his influence in the University of Oviedo and the advancement of Mathematics in the Principality of Asturias is beyond question.

As a proof of this assertion we can mention two outstanding achievements related to the University of Oviedo:

- At present, in the Department of Statistics, OR and Teaching Mathematics at the University of Oviedo there are 24 Pedro's 'scientific linear descendants', the distribution of this descendants being summarized in Table 1. Furthermore, 6 Ph.D. students of Pedro hold tenured positions in other departments of the University of Oviedo.

- As it will be commented to a certain extent in the next paper by Dugnol and Valdés, Pedro's expertise, knowledge and his permanent relationship with experts from other universities, were definitely crucial in the launching of the B.Sc. in Mathematics in the University of Oviedo, as well as in the technical and academical elaboration and management of the starting B.Sc. Programmes, in which one of the two considered specialties (which were oriented to the Principality of Asturias main demands) was Statistics (and OR). Probably, this was an idea hovering his mind since he arrived in Asturias. And Professor Benjamín Dugnol motivated him to do it by arguing that this will make possible having Ph.D. in Mathematical Sciences by the University of Oviedo (something that prevented the laws in force at that time). And Professor Javier Valdés collaborated with him in this complex but exciting mission that, fortunately, succeeded.

Pedro did much for Math in Asturias, but not only for those at the university level. He always fostered cooperation between university education and previous education in Mathematics. As a proof of Pedro's influence on the Mathematics in the Principality of Asturias in connection with such a cooperation, let us refer just a few additional achievements:

- Since he arrived to Oviedo and for many years he coordinated Mathematics studies and examinations leading to university access. He visited many high schools, and called for many meetings with Math teachers, in order to explain what should be considered for students to know and handle as well as the basis for access exams. Readers should realize that Internet was not yet in our lives at that moment, so there is no chance for disseminating this key information but visits and meetings. Pedro believed in the need to make students and teachers more familiar with this information, and he added this new 'backpack' to the ones he already carried behind him.

- Also from the very beginning of his stay in Asturias till his retirement, Pedro was in charge of coordinating and organizing the Asturian Mathematical Olympiad for students of the one-two years before accessing the university. This was a national (as a second intermediate step between regional and international) competition seeking to improve Math education and creating greater interest in Mathematics careers among students. One more backpack on Pedro's shoulders.
- And, much more recently, and once he is retired, he has always participated in the scientific committees to select Asturian representatives for the Spanish Contest 'Incubator for Surveys and Experiments' sponsored by SEIO. A way to enhance the relevance of Statistics and OR in high school students, and making them more familiar with their techniques and tools by applying them to analyze data from different real-life problems.

When Pedro passed away in March 16, 2016, a headline of the newspaper 'La Nueva España' said that "Statistics become orphaned in Asturias". Well, we feel in this way to a great extent but, at the same time, he left us many scientific and personal reserves that will always remain with us.

Thank you, Pedro, for your care, your affection, your support, your modesty…in summary for you having been as you were (Fig. 2).

Fig. 2 Pedro Gil in front of the *Seminario Pedro Gil* (Pedro's office that became, after his retirement, the current Meeting Room of the Statistics, OR and Teaching Mathematics Department in the Faculty of Sciences)

Reference

1. Gil P (1975) Medidas de incertidumbre e información en problemas de decisión estadística. Rev Real Acad Cienc Exact Fis Natur Madrid LXIX:549–610. http://www.rac.es/ficheros/Revistas/REV_20091030_00137.pdf

BSc Mathematics at the University of Oviedo. On the Launching and Two First Bachellor's Programmes in Mathematics

Benjamín Dugnol and Javier Valdés

Abstract This paper aims to highlight the crucial role played by Professor Pedro Gil in launching, planning and designing undergraduate studies in Mathematics for the University of Oviedo (Spain).

1 Background

Up until the entry into force of the Organic Act on University Reform (LRU) in 1983, virtually each faculty/school within the University of Oviedo had its own Department of Mathematics, which was in charge of teaching all Mathematics subjects within the centre. Specifically, first in the Faculty of Sciences, where it was located, and later in the Biology building, there was a Department of Mathematics, led by Professor Pedro Gil Álvarez since 1976, who taught the subjects of Mathematics and Statistics in the sciences degrees (Geology, Biology and Chemistry) and Statistics in the Economics degree.

On September 1, 1983, the LRU was published in the Official Bulletin. It stipulated that: "the departments are the essential units responsible for organizing and developing research and also for teaching their respective areas of knowledge in one or more faculties and technical schools and, where appropriate, those other centres that may have been created under the provisions of this law". Later, in the same article it states that: "The departments will be constituted by areas of scientific, technical or artistic knowledge, and will bring together all teachers and researchers specialized in those areas."

This law meant a drastic change to the organization of the university: from a structure based around centres to one based around departments. In order to form a department it was necessary to have a minimum number of Lecturers and Professors

---

Dugnol and Valdés—Professors in Applied Matematics, both of them being currently retired.

B. Dugnol · J. Valdés (✉)
Departamento de Matemáticas, Facultad de Ciencias, Universidad de Oviedo,
C/ Federico García Lorca, 18, 33007 Oviedo, Spain
e-mail: javier.valdes@uniovi.es

© Springer International Publishing AG 2018
E. Gil et al. (eds.), *The Mathematics of the Uncertain*, Studies in Systems, Decision and Control 142, https://doi.org/10.1007/978-3-319-73848-2_85

913

and the only way to fulfill this requirement was to create a Department of Mathematics, grouping all the professors of the University of Oviedo who taught Mathematics and Computer Science in any centre. Thus a new department was established in October 1986 covering the following fields of knowledge: Computer Science and Artificial Intelligence, Teaching Mathematics, Statistics and Operations Research, Computer Languages and Systems, Applied Mathematics and Quantitative Methods for Economics.

At the beginning, there were only four Full Professors in the newly big created department (Pedro Gil in the Faculty of Sciences, Carlos Conde and Benjamín Dugnol in the Mining Engineering School, and Emilio Costa in the Faculty of Economics). This department had its headquarters in the Mining Engineering School and its only Director, while this structure was maintained, was Benjamín Dugnol, Pedro Gil being its Vice-head.

Years later, as new Professor and Lecturer positions were allocated to the department, independent departments were created on 'Quantitative Economics', 'Computer Science' and 'Statistics and OR and Teaching Mathematics'. For the last one, Pedro Gil became the Director.

At the time, Ph.D. students (who joined the Departments for either teaching or research assistance purposes) could attend doctoral courses and present their theses in Sciences (Math) leading to get their Ph.D. degrees in Mathematical Sciences by the University of Oviedo. There is no doubt that this was an unusual situation, namely that a university that did not offer a Mathematics degree could issue the Ph.D. in Mathematics. This was possible because there was no rule that prevented it…until the Official Bulletin published in 1985 a Royal Decree in accordance with which: "Universities may not offer the Ph.D. title corresponding to the official titles of Bachelor, Engineer or Architect whose studies can not be taken in the same university. However, they may enter into agreements with others which do so in accordance with the procedure laid down in this article". The publication of this Royal Decree marked the starting point of what later led to the implementation of the Mathematics' Bachelor Degree at the University of Oviedo.

A Committee was promptly created, composed by Professors José Ángel Huidobro (the first affected by the new regulations), Secundino López, Pedro Gil and Javier Valdés, in charge of establishing negotiations with the University of Cantabria. The first mission for this committee was mainly to solve the situation for several teaching assistants or predoctoral researchers who were at that time either developing their doctoral thesis or about to finish it. Since the University of Oviedo could not issue the Ph.D. title in Mathematics because of the new regulations, there was a need to overcome such an inconvenience. These negotiations concluded with a collaboration agreement that went beyond what was initially planned. In this respect, those who had already started their Ph.D. thesis and were now allowed to arrange both the bureaucracies and dissertation at the University of Oviedo, could arrange them in the University of Cantabria (so they get the Ph.D. Degree by Cantabria University). Additionally, as a consequence from the collaboration agreement, in the academic year 1987–1988, the first course of the Mathematics degree with the syllabus of the University of Cantabria was offered (for all purposes students were part of the University of Cantabria); in the 1987/1989 biennium, a third-cycle programme in Mathematics was offered by the University of Oviedo in collaboration with the Department of Mathematics, Statistics and Computing of the University of Cantabria.

At the beginning of the 1989–1990 academic year, the headquarters of the Department of Mathematics, along with the teachers in this Department from the fields of Statistics and OR, Computing Languages and Systems, and Applied Mathematics, moved to the Faculty of Sciences building. In this academic year, and in the new departmental headquarters, the first two years of the syllabus of the University of Cantabria were offered, alongside the third-cycle of the Mathematics programmes, in collaboration with the University of Cantabria, corresponding to the biennia of 1988/1990 and 1989/1991.

At this point, the next logical step was the introduction of a Bachelor in Mathematics degree at the University of Oviedo (see the related words by Pedro Gil in the Addendum to this paper). This would require new teaching staff in the areas of knowledge that did not exist at the University of Oviedo, namely, Algebra and Mathematical Analysis. In 1989 the Department of Mathematics was the most numerous of the University of Oviedo. It was composed by 137 Professors. These were the means of teaching that were available at that moment to face the task of developing the new syllabus.

2 1990 Bachelor's Programme in Mathematics of the University of Oviedo

From the very first moment the syllabus considered was well differentiated from the traditional Mathematics degree that was offered in almost all the Spanish Universities and, in particular from the ones closest to Asturias (Cantabria, Valladolid and Santiago de Compostela). The objective was to elaborate a syllabus with the specialties of "Applied Mathematics and Computing" and "Statistics" (actually, the last one involving Statistics, OR and Computing). Years later, in 1996, Alfredo Pérez Rubalcaba in a lecture at the Auditorium of the University of Oviedo, on the occasion of the delivery of Diplomas to the newly graduates, said that as Secretary of State for Education of the Ministry of Education and Science, he very much welcome the proposal of the University of Oviedo to implement a syllabus with these two new specialities. We guess that this was one of the reasons supporting and facilitating the implementation of Mathematics studies at the University of Oviedo.

In the 1989–1990 academic year, the department's management launched two working groups to develop a syllabus for the new Mathematics Degree, one coordinated by Pedro Gil for the Statistics and OR field (taking advantage of Pedro's previous teaching expertise in the BSc in Mathematics at the Complutense University of Madrid), and the other by Javier Valdés for the Applied Mathematics field. In these groups several professors of the department participated, among which it is worth mentioning for its special support and enthusiasm to Teófilo Brezmes, in the Statistics/OR side, and Omar Menéndez in the Applied Mathematics side. It also relied on the advice of professors from other Universities.

Let us recall some arguments that were used in the proposal:

"The need for mathematics teachers in Asturias is a burden that the region has suffered for many years. One just needs to see the problematic situations produced when covering for other teachers within the area, in the levels of Vocational Training, Baccalaureate or the University itself for the current academic year, in order to be convinced that the situation has not improved. To a large extent, this endemic disease has been fostered by the absence of these studies at the University of Oviedo."

"On the other hand, companies based in our region, the Principality of Asturias, and the public sector itself, need more and more experts in Mathematics, in any of the two specialties, which, at a higher level, could be trained in our classrooms. To fulfil this task, we can already count with a department able to teach the majority of the subjects, although it would be necessary to incorporate some new teachers, if possible with the title of Doctor, for certain subjects, as it will be necessary to progressively find substitutes for some professors in its current teaching, that is, in general mathematics subjects."

"The proposed syllabus includes two specialties: 'Applied Mathematics and Computing' and 'Statistics', both with a common first three-year cycle, and their respective compulsory and optional subjects from the fourth year onwards."

Recognitions of equivalences were also proposed to allow for the incorporation of students who had taken first or second year courses, following the syllabus of the University of Cantabria.

For the 1990–1991 academic year, the Ministry of Education and Science created the Faculty of Chemistry and Mathematics and authorized the Chemistry and Mathematics Degrees.

3 1991 Bachelor's Programme in Mathematics of the University of Oviedo

The 1990–1991 academic year had just barely begun when the Ministry of Education and Science published a new Royal Decree establishing the official University degree of Mathematics and the general guidelines of the syllabus leading to obtain such a degree.

The publication of this Royal Decree implied a change in the organization and structure of the syllabus: "The syllabus approved by the Universities should be articulated as first and second cycle education, with a total duration between four and five years, and a duration per cycle of at least two years."

A period of debate opened up throughout the Spanish Universities on the duration of these cycles: first cycle of three years and second of two, or two and three, or two and two. Although the aforementioned Royal Decree established a maximum period of three years to adapt to the new regulations, the University of Oviedo was in a great hurry to adapt its degrees to the new regulations and decided that all degrees last four years.

In these circumstances, the department of Mathematics commissioned the same persons who had drawn up the current plan, due to their 'recent experience', the definition of a new syllabus in line with the new regulations. A new syllabus was then proposed with the same specialties, with a common first cycle of two years, and a second cycle, also of two years, in which there would be a number of mathematics core subjects (mandatory) and subjects relating to each speciality (some mandatory and some optional).

This plan was approved by the University of Oviedo in 1991.

It should be noted that the contents of the 1991 plan are practically the same as the ones in 1990, but concentrated in four years instead of five. Both plans coexisted during four years, until the year 1994–1995, in which the first and only promotion of the 1990 plan and the first of that of 1991 plan graduated. The 1991 syllabus continued until the coming into force of the new Mathematics degree in the 2009–2010 academic year.

Addendum: Pedro Gil's Own Related Words on the Occassion of the Official Ceremony Held for His Retirement in November 2010

I wanted to thank my friend the Chancellor, Professor Vicente Gotor. Over the last many years, we have both been doing a job, a job with a purpose. We have been doing it without fanfare, but with results. And yes, looking back, you realize that you have achieved things. But I wanted to thank him today because, in my opinion, he was a key enabler behind the creation of the area of Mathematics.

Vicente was, at that time, area director (a kind of Vice-Chancellor assistant) within the Vice-chancellorship led by Professor Marita Aragón, the Chancellor of the University being Professor Juan S. López Arranz. Then, one fine day he called me and asked "would you prepare a syllabus for the diploma in Statistics?", and this is how it all started. Neither shy nor lazy, I sent him, as soon as I could, a complete syllabus for the diploma of Statistics, that never came to see the light.

Well, that degenerated somewhat because, Benjamín Dugnol was more ambitious than me and said, "and why not a Bachelor's degree in Mathematics?" So I made a plan for a Bachelor's degree in Mathematics based on what I know, which is the specialty of Statistics. And then, with help from Javier Valdés, who is also here today, we did, both of us, sitting at a table, hand in hand, a syllabus that was then published in the Official Bulletin. Any flaws it may have are exclusively our own, any strengths, too.

Printed in the United States
By Bookmasters